AF683683

Petrology and Plate Tectonics

An Earth Systems Approach

ALLEN F. GLAZNER
MATTHEW J. KOHN

OXFORD
UNIVERSITY PRESS

OXFORD
UNIVERSITY PRESS

Oxford University Press is a department of the University of Oxford. It furthers the University's objective of excellence in research, scholarship, and education by publishing worldwide. Oxford is a registered trade mark of Oxford University Press in the UK and in certain other countries.

Published in the United States of America by Oxford University Press
198 Madison Avenue, New York, NY 10016, United States of America.

For titles covered by Section 112 of the US Higher Education Opportunity Act, please visit www.oup.com/us/he for the latest information about pricing and alternate formats.

Library of Congress Cataloging-in-Publication Data
Names: Glazner, Allen F. author | Kohn, Matthew J. author
Title: Petrology and plate tectonics : an earth systems approach / Allen F. Glazner, Matthew J. Kohn.
Description: [New York] : [Oxford University Press], [2025] | Includes bibliographical references and index.
Identifiers: LCCN 2024055839 (print) | LCCN 2024055840 (ebook) | ISBN 9780190490645 paperback | ISBN 9780197788127 ebook | ISBN 9780190490676 epub
Subjects: LCSH: Petrology—Textbooks | Plate tectonics—Textbooks | Geochemical cycles—Textbooks
Classification: LCC QE431.2 .G53 2025 (print) | LCC QE431.2 (ebook) | DDC 552—dc23/eng/20250327
LC record available at https://lccn.loc.gov/2024055839
LC ebook record available at https://lccn.loc.gov/2024055840

Printed by Marquis, Canada

MIX
Paper | Supporting responsible forestry
FSC® C103567

Contents

Preface

Petrology and Plate Tectonics offers a new approach to teaching petrology. Rather than presenting igneous and metamorphic petrology as separate topics, coauthor Matt Kohn and I instead use plate tectonics as a unifying theme, interweaving igneous and metamorphic topics as we follow plate tectonic cycles. We use an Earth systems viewpoint, emphasizing the importance of petrology in global geochemical cycles and links between the lithosphere, hydrosphere, atmosphere, and biosphere.

The artificial division of hard-rock petrology into igneous and metamorphic subfields does not correspond with how the Earth works because igneous and metamorphic rocks are intimately associated in space and time. For example, basalts at oceanic spreading ridges are born by melting but immediately metamorphosed and hydrated; the water thus carried down in subduction zones participates in metamorphic and melting reactions that produce high-pressure rocks as well as the magmas of volcanic arcs. Some of these magmas erupt to form volcanoes, and others freeze in the crust to produce plutons that metamorphose their wall rocks. And on and on. The Earth is easier to understand when we study related igneous and metamorphic processes together rather than weeks apart.

Teaching petrology from an Earth systems viewpoint makes clearer the underappreciated role that petrology plays in many areas of inquiry, including causes of climate change on both long and short timescales; causes of, and recovery from, mass extinctions; sources of key materials critical to modern society; geologic hazards, such as volcanic eruptions and sector collapse; bedrock control of plant life; and the evolution and global dispersal of humans.

The idea for this new approach germinated at a 2003 workshop on teaching petrology that was sponsored by the National Science Foundation. I remember a strong feeling among the 75 attendees that something was wrong. Many of us had been students in the 1970s, give or take, when plate tectonics was a remarkable new field, astronauts were bringing rocks back from the Moon, ocean drilling was revealing secrets of the seafloor, and spacecraft were giving us close looks at other planets. Petrology was integral to all this, and we were excited and motivated to go out and solve problems. But by the 2000s, a lot of that energy had worn off, and our students just weren't as enthusiastic about petrology as we had been at that stage. What was wrong?

A month later, I hiked up Mount Whitney with John Bartley to collect dike samples, and the all-day affair gave us plenty of time to talk. I free-associated about events at the workshop much of the day, and by the time we returned to the trailhead, we had identified the main problem: Books and classes treat igneous and metamorphic petrology as separate fields, but the Earth doesn't. Thus, the longstanding and wholly artificial "Part 1: Igneous, Part 2: Metamorphic" division splits apart systems that belong together, such as plutons and contact metamorphism, and segregates petrologists into two domains that should be closely linked.

The basic outline of this book was born on that hike. Actual writing did not begin until 2016, after a solicitation from Oxford University Press. It has been a long process, as there was no template to follow. In 2022, with about a quarter of the book left to write, I asked Matt Kohn to help me with material that was far from my training in igneous petrology. He graciously agreed, and the collaboration has led to a far better book.

Although the structure of this book is a break from the past, I owe much to prior texts in petrology and other fields, particularly *Igneous and Metamorphic Petrology* by Myron Best. *Volcanoes, A Planetary Perspective* by Peter Francis and *Principles of Physical Sedimentology* by J. R. L. Allen are readable, concise, occasionally humorous, and erudite books about complex topics; I hope that some of that style shows up in this book.

—Allen Glazner, March 2025

Like Allen, I have long felt that something has been lacking in classes on igneous and metamorphic petrology. In 2017, I decided that I would try to integrate them into a plate tectonic context. I contacted numerous colleagues for advice, without much luck, until Allen sent me his syllabus outlining the petrologically integrated tectonic cycle that frames this book—just what I was looking for! Although Allen's syllabus helped, my attempts at teaching the newly integrated class were not completely successful. Why not? Because there was no corresponding unifying text. So, when Allen asked me in late 2022 whether I would consider contributing, I agreed, partly because I thought it was important, but also so that I would have something to use in my classes.

As someone whose scientific interests range widely, I especially wanted to write about petrologic connections to diverse fields outside of petrology. I am interested in paleontology and paleoclimate, and this book contains several connections sections on these topics. Why include this material? It's more than just a love of fossils, or curiosity about extinctions, or concerns about climate change. Like Allen, I wanted to make the point that petrology is about much more than just rocks or volcanoes or mountain ranges—it's an important contributor to scientific inquiry in diverse fields. My greatest hope for this book is that in addition to helping to educate future Earth scientists, it inspires new applications of petrology in ways we haven't yet conceived of for problems we haven't yet thought to wonder about.

—Matt Kohn, March 2025

Acknowledgments

Allen F. Glazner is grateful for the outstanding education he received at Pomona College and UCLA, with particular thanks to professors Donald McIntyre, Ian Carmichael, Gary Ernst, Art Montana, Gerald Schubert, Wayne Dollase, and Harry Mullikin. His long-time collaborations with John Bartley, Drew Coleman, Lang Farmer, and Doug Walker guided much of this book's content. Dan Kaveney at Oxford University Press got this book off the ground and into the air, and Tracey MacDonald and Suzanne Carter saw it through the cruise phase and to landing. Finally, he thanks the University of North Carolina for his appointment as the Mary Lily Kenan Flagler Bingham Distinguished Professor, providing him with the time and wherewithal to explore new ways to think about and teach petrology.

Matthew J. Kohn thanks all the faculty, students, postdocs, and friends who helped train his thinking in geology and science. As an undergraduate, John Southard, Frank Spear, Clark Burchfiel, Wiki Royden, and Peter Molnar helped inspire enthusiasm and dispel lazy thinking; as a PhD student, Frank Spear, Bruce Watson, and Mark Harrison provided robust foundations in metamorphic petrology, geochronology, and kinetics; as a postdoc, John Valley, Margaret Schoeninger, and Rick Ryerson guided his research in stable isotopes, especially in metamorphic fluid flow, ecology, and climate; as a professional, close collaborations with Huixia Ding, Xiaochi Liu, Sarah Penniston-Dorland, Sudip Paul, Sean Long, Delores Robinson, Nadine McQuarrie, and (of course) Frank Spear sustained his love of petrology, geochronology, and tectonics.

List of Reviewers

The publishers would like to acknowledge the following reviewers, and the anonymous reviewers, whose feedback was invaluable in the development of this text:

Zachary D. Atlas, *University of South Florida*
Matthew Brueseke, *Kansas State University*
Sarah Carmichael, *Appalachian State University*
Cameron Davidson, *Carleton College*
Michael DeAngelis, *University of Arkansas, Little Rock*
Mary Kate Fidler, *Clemson University*
Katherine F. Fornash, *Ohio University*
Lydia K. Fox, *University of the Pacific*
Aurélie Germa, *University of South Florida*
Judith L. Hannah, *Colorado State University*
Jo Laird, *University of New Hampshire*
Shawn Malone, *University of Wisconsin, Green Bay*
Harry Y. McSween, *University of Tennessee, Knoxville*
Carrie Menold, *Albion College*
Kaustubh Patwardhan, *State University of New York, New Paltz*
Jason Ricketts, *University of Texas, El Paso*
Michael Roden, *University of Georgia*
Joshua Schwartz, *California State University, Northridge*
Adrian Van Rythoven, *Bloomsburg University*
Barry Weaver, *University of Oklahoma*

About the Authors

Allen F. Glazner is Distinguished Professor of Geology Emeritus at the University of North Carolina at Chapel Hill. He earned a BA in Geology from Pomona College and a PhD in Geology from the University of California at Los Angeles. An award-winning teacher specializing in igneous petrology, tectonics, and geoinformatics, he is also author of five books in the popular *Geology Underfoot* series.

Matthew J. Kohn is Distinguished Professor in Geosciences at Boise State University. He earned a BS in Geology from the Massachusetts Institute of Technology and an MS and PhD in Geology from the Rensselaer Polytechnic Institute. He uses geochemistry of metamorphic rocks to investigate orogenesis and geochemistry of fossil teeth to study Cenozoic paleoclimate and paleoecology. His book *Tested* describes his adventures in China during the COVID-19 pandemic.

To our families, who made us who we are, and to our teachers, who taught us how to think.

CHAPTER 1

Plate Tectonics and Petrology

In 2015, the spacecraft *New Horizons* flew past Pluto, giving the first fly-by view of that most distant of the original nine planets. Pluto turns out to be unlike any other planetary body that has been studied, with nitrogen snow spawning nitrogen glaciers carving water-ice bedrock, and this was no surprise: the planets, minor planets, asteroids, and moons of our solar system are remarkably diverse, each their own geologic wonderland.

The wealth of extraterrestrial data sent back by spacecraft since the Soviet Luna 3 spacecraft first snapped pictures of the Moon's far side in 1959 makes it easy to forget that we live on the most remarkable planet of all. Even a distant telescopic view of the Earth would show striking blue color, abundant bright water clouds, oceans, continents, and surficial variety, with volcanic plumes, waxing and waning ice, seasonal color changes, and other signs of biologic life and geologic activity.

Let's imagine that we're Martians, observing Earth telescopically and with orbiters and landers. What would we see? Telescopes would show the highly uneven distribution of land, including odd features such as arcuate strings of islands, some of which are active volcanoes. Some of Earth's highest topography is located near the coast, not in the central parts of continents where one might expect it, and the curious fit of North and South America with Europe and Africa would be evident even from planetary distances. Spectral analysis would show abundant chlorophyll at the surface and abundant oxygen in the atmosphere, both clear signs of life.

Landers would find a huge variety of rocks and sediments at the surface and active depositional systems that move the sediment around. Spectrometers would show rock compositions including basalt, granite, limestone, shale, and sandstone—many of these rocks with no Martian names, because they are not found on Mars. Oceangoing craft would find more mysteries under the water's surface: chains of mountains tens of thousands of kilometers long, thousands of flat-topped seamounts that are almost islands, and most curious of all, deep ocean trenches, far deeper than the average ocean depth, right up against the land. There would be no mistaking that Earth is a geologically and biologically active planet, quite different from Mars.

View of Earth on January 22, 2016, centered on southern Africa. This image was taken by the Deep Space Climate Observatory (DSCOVR) satellite, located between Earth and the Sun, 1.5 million km away. Clouds, oceans, ice, and continents are all prominent, and spectroscopy would show the presence of oxygen and chlorophyll. The contrast between the Sahara Desert, equatorial Africa, and icy Antarctica (bottom) is especially clear in this image. Even from the distance of Mars, clouds, oceans, continents, and geologic and biologic activity are all easily seen. DSCOVR was placed at Lagrange point L1, a gravitationally favored spot in the Earth-Sun system where gravitational attractions are balanced, and the satellite therefore stays between the Earth and the Sun.

Source: NASA/DSCOVR: EPIC.

1.1 Distribution of Volcanoes, Earthquakes, and Topography on Earth

Learning Objective

By the end of this section you should be able to:

1.1.1 Explain how the physiography and seismicity of Earth reflect tectonic processes.

We Earthlings know that most of these features are a direct consequence of plate tectonics. Most volcanoes are located near plate boundaries, especially around the Pacific Rim and where eastern Africa is separating from the rest of the continent. Earthquakes similarly outline plate boundaries, delineating those underwater as well as those on land. Diffuse, dangerous zones of earthquakes occur throughout Europe and southern Asia, marking broad bands of deformation where Eurasia interacts with Africa and India. Plate boundaries are clearly defined by these geologic features (Figure 1.1).

The distribution of topography on Earth is a first-order consequence of plate tectonics and plays a fundamental role in how humans are distributed as well. The highest topography occurs along active and ancient subduction zones, including the Himalaya, **Cordillera** of North and South America, and Alps. These lofty ranges modify global weather patterns, trapping moisture as ice and snow and thereby supplying water to billions of people. Most of the Earth's living space is on continental crust at an elevation not much above sea level. Figure 1.2 compares elevation on the Earth with other planetary bodies. Unlike other terrestrial planets and the Moon, the Earth's topography is **bimodal**; average ocean depth is about 3800 m, whereas half of dry land is less than 400 m above sea level. The shoreline is located on the sharp continental peak in the distribution, so anything that changes the amount of water in the oceans (e.g., glaciation) or the volume of the ocean basins (e.g., changes in seafloor spreading rate) will have a large impact on terrestrial life by moving sea level up or down. One of the goals of petrology is to understand how the Earth generates two fundamentally different types of crust and how they evolve.

Cordillera
A system of generally parallel mountain ranges and basins; generally used for those along the western margins of the Americas, from Alaska to Patagonia.

Bimodal
Here, a statistical distribution with two maxima (modes).

1.2 The Range of Magmatism on Earth: The Basalt-Rhyolite Spectrum

Learning Objectives

By the end of this section you should be able to:

1.2.1 Apply bivariate diagrams (e.g., total alkalis-silica) to discriminate the most common volcanic rocks.

1.2.2 Calculate compositions from raw data and plot them on two-dimensional composition diagrams for discrimination purposes.

The Basalt-Andesite-Dacite-Rhyolite (BADR) Trend

One obvious consequence of plate tectonics is that the Earth has a much wider variety of igneous rock compositions than the other silicate planets and satellites. Igneous rocks on Mars mostly seem to be variants of basalt. The dark rocks on the Moon are basalt, and the light rocks are anorthosite, a rock formed from plagioclase that floated in a basalt magma ocean. Venus is still a mystery, but what little data we have from Soviet landers indicates that the rocks at the landing sites are basalt. Mercury seems to be basalt; the asteroid Vesta is largely basalt; igneous meteorites are mostly basalt. It seems that most of the non-ice, non-sulfur[1] igneous material in the solar system is basalt.

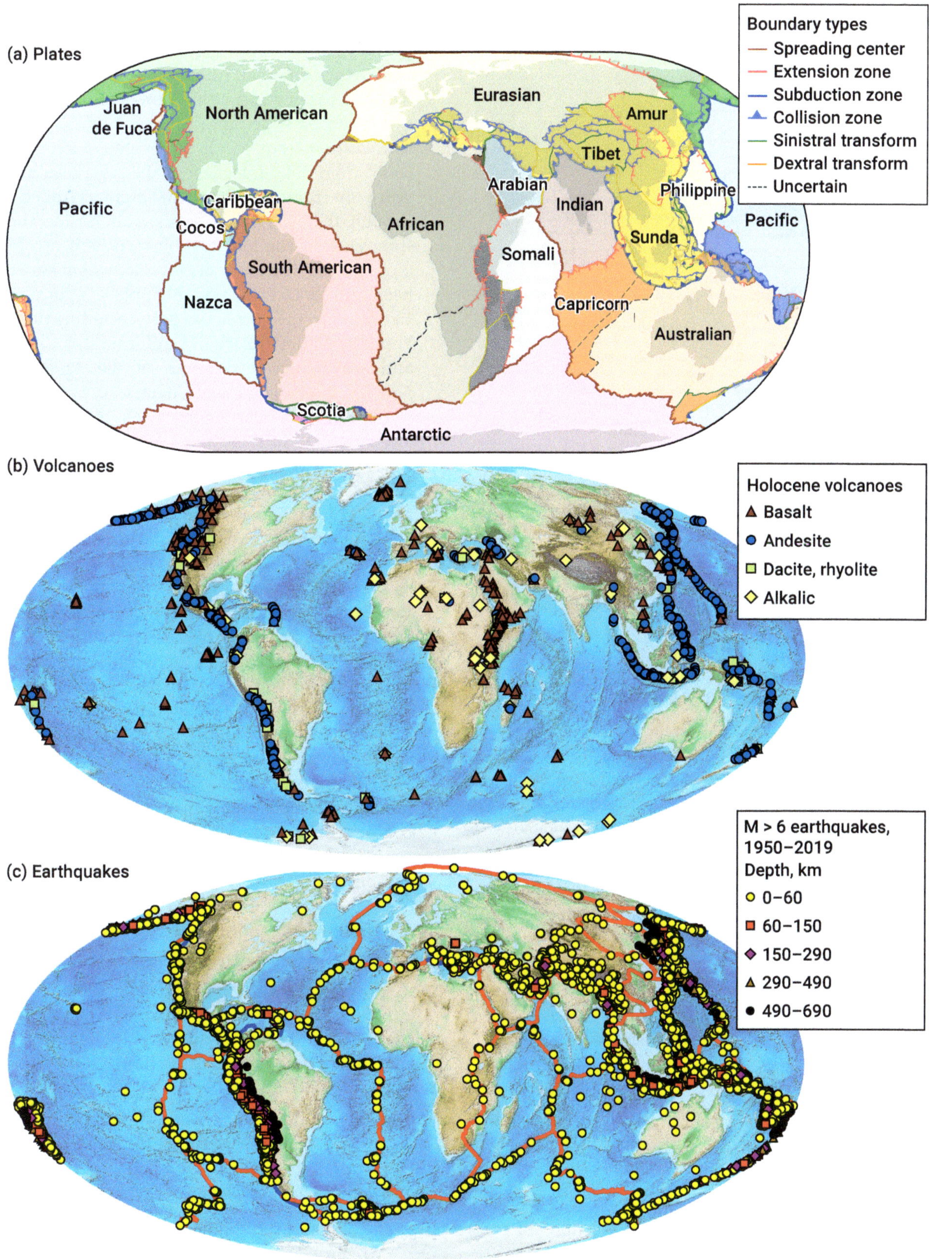

Figure 1.1 Maps of the configuration of the major plates and how volcanoes and earthquakes are focused along plate boundaries. Plate tectonics is a major driver of igneous and metamorphic processes, and most igneous and metamorphic rocks form along plate boundaries. (a) The major plates. Plates are typically viewed as rigid, but it is increasingly clear from geologic and geodetic studies that large regions are not effectively rigid. These are shown by darker shading and include much of western North America, much of Asia, and a zone leading west from the Himalaya, through Turkey, and across the Mediterranean. (b) Distribution of Holocene volcanoes, colored by composition. (c) Distribution of 70 years of earthquakes with moment magnitude >6, colored by depth.

Sources: (a) Map from Hasterok et al. (2022), © 2022 Elsevier B.V./Derrick Hasterok; (b) US Geological Survey.

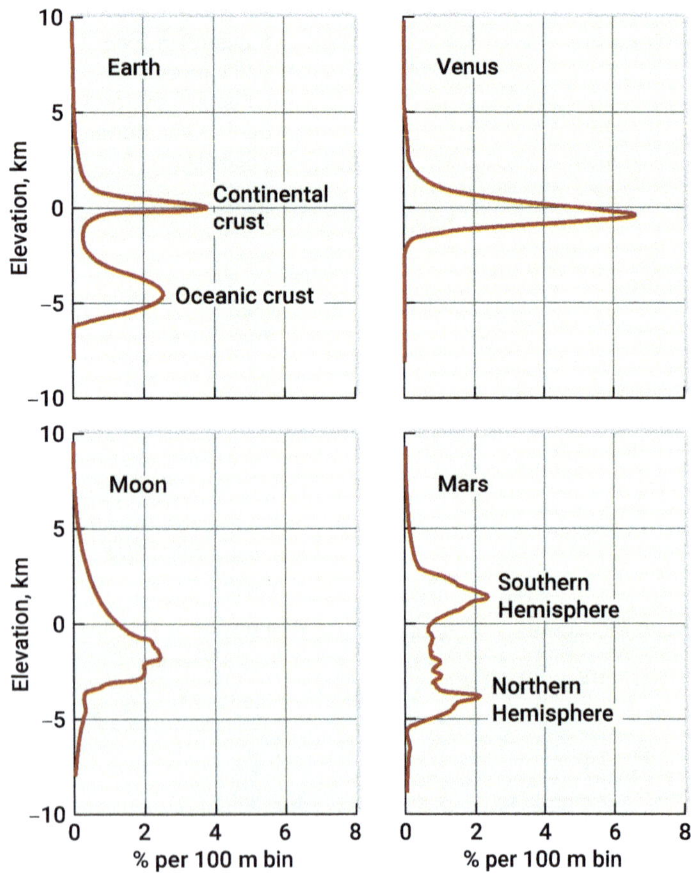

Figure 1.2 Histogram of elevations on Earth, Venus, Mars, and the Moon. Elevations on Earth are bimodal. Most of the ocean basins on Earth lie at 3–6 km depth, but most of the dry land on continents lies within 1 km of sea level. This fundamental distinction, so important to life on Earth and apparently absent on other planets, is a consequence of plate tectonics. Venus and the Moon are unimodal. Mars is bimodal, but this is largely an artifact of the profound differences between its northern and southern hemispheres (Aharonson et al., 2001). Zero points ("sea levels") are arbitrary for bodies other than Earth, as they lack seas.

Source: After Lorenz et al. (2011).

Variation diagram
A plot of one chemical parameter of rock composition against another; typically, SiO_2 is plotted on the abscissa.

Whole-rock analysis
A chemical analysis of a bulk rock sample rather than of one or more of its component parts (minerals or glass).

Earth is different. Although ocean floors are dominated by basalt under a thin film of sediment, the continents host a wide range of igneous rocks ranging in composition from basalt to rhyolite. Such rocks can result from quiet intrusion in the crust over a million years, or from lava flows that were extruded slowly enough that you could safely walk beside them, or from violent eruptions that blow volcanic ash into the stratosphere. The more silicic rocks, which have not yet been reliably identified on other planets, form some of Earth's most spectacular scenery, including the cliffs of Yosemite National Park in California (see Chapter 10) and soaring stratovolcanoes of the Andes, Pacific Northwest, and Japan (see Chapter 7).

The bulk chemical composition of a rock is generally described in terms of oxides such as SiO_2, otherwise known as silica, an archaic but useful term for the chemical component SiO_2. For the moment don't worry about why particular chemical components are used; just think of them as descriptors, like using weight and length to describe trucks (see Toolbox 1.1). Figure 1.3 shows a plot of silica against the sum of Na_2O and K_2O for 26,000 igneous rocks from western North America. This plot is an example of a **variation diagram**. These analyses were taken from the North American Volcanic and Intrusive Rock Database (NAVDAT) at earthchem.org. Na_2O and K_2O are alkalis, and this plot is called a total alkalis-silica, or TAS, plot and is used for chemically assigning rocks to the names on the figure (see Appendix 3).

Because there are so many analyses we've plotted contours of data density, with contours every 5 percent. Several interesting things are apparent from this compilation:

1. Total alkalis and silica are positively correlated.
2. Most of the analyses lie along an array that is shaped something like a narrow bowling pin and goes from the basalt field to the rhyolite field.
3. Most of the analyses (74 percent) lie in the fields of basalt, basaltic andesite, andesite, dacite, and rhyolite. We will refer to this basalt-andesite-dacite-rhyolite trend as the BADR ("badder") trend.
4. Nearly all the rest of the analyses lie just above those fields, in fields with the same names plus the qualifier "trachy" (from the Greek word for "rough," but connoting "alkali-rich").
5. Only a few percent of the analyses lie outside these fields.
6. Nearly half of the analyses lie in the basalt, basaltic andesite, and andesite fields.

If we plot MgO instead of total alkalis we get another strong correlation, but this one has a negative slope—using the truck analogy in Toolbox 1.1, these variables might be weight and fuel economy. There is no equivalent classification diagram here, but they are the same analyses and so most are basalt, andesite, etc. The bulls-eye concentration around 55 wt% silica remains.

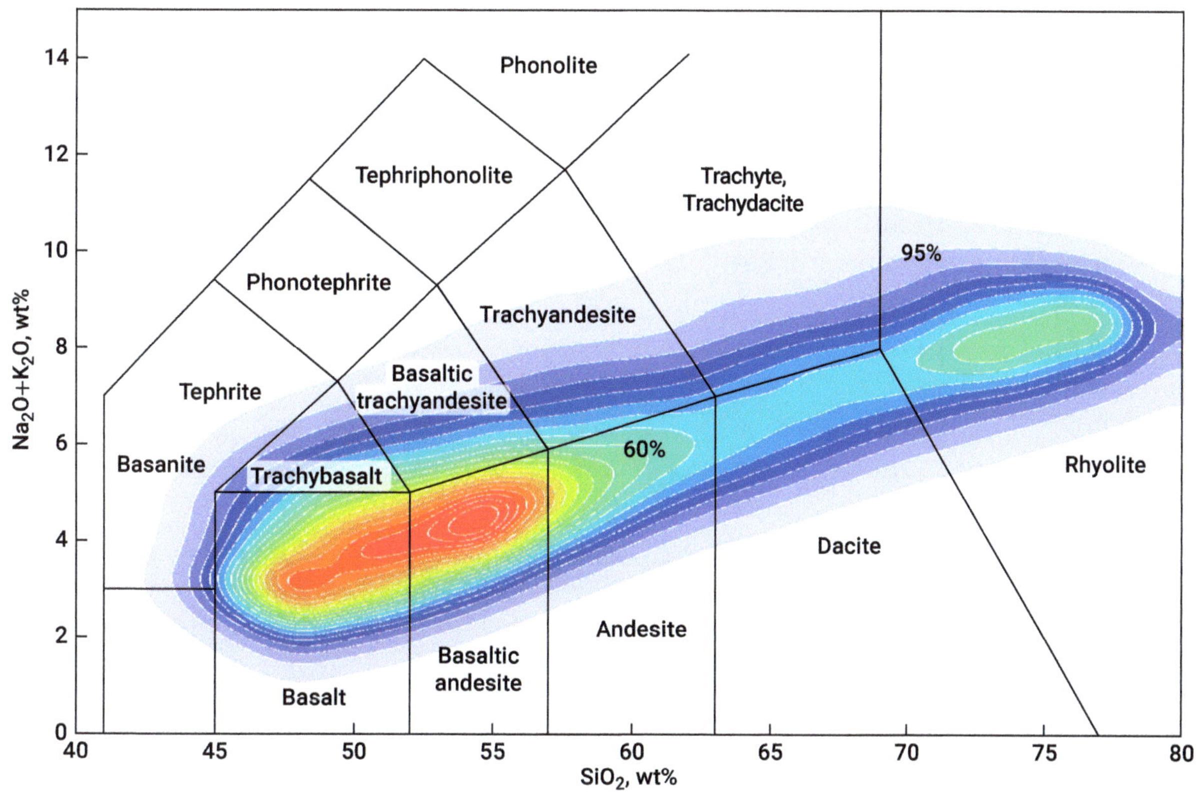

Figure 1.3 Total alkalis-silica (TAS) plot for 26,000 igneous rock analyses from western North America. Contours of data density are every 5% (e.g., 60% contours enclose 60% of the points). These variables are highly correlated, and three-quarters of them lie in the set of fields ranging from basalt to rhyolite.

Toolbox 1.1

How Chemical Analyses Are Depicted

Rocks are commonly described by their chemical composition, a listing of the elements present and their amounts. Chemical analyses can be determined by a wide range of methods, and geochemical databases contain analyses of hundreds of thousands of rocks, each of which took significant time and money to obtain.

A rather complete **whole-rock analysis**, so called because the entire rock sample was analyzed rather than a particular mineral, is shown in Table 1.1. This analysis is of US Geological Survey (USGS) rock standard BHVO-1, a basalt from the 1919 eruption of Kīlauea volcano in Hawai'i. Chemical components are generally broken into two groups: major elements and trace elements. Major elements are those present in more than trace amounts, and they are usually presented as weight percent (wt%) oxide (e.g., MgO) rather than elements. Oxygen, the most abundant element in a typical rock, is difficult to measure, and in most rocks it is the only abundant anion, so assuming that the cations are bound to it is usually valid. Trace elements are generally presented as the element in parts per million by weight (ppm, or μg/g) even though they, too, are generally bound to oxygen.

There are several things to keep in mind when reading a chemical analysis:

1. Both wt% and ppm have units that must balance in equations: wt% oxide can conveniently be interpreted as $\frac{g\ oxide}{100\ g\ rock}$, and ppm element as $\frac{\mu g\ element}{g\ rock}$.
2. You can convert an oxide to an element, or vice versa, by using a gravimetric factor, which is simply the ratio of an element's mass to the mass of the same element oxide. For example, the gravimetric factor for CaO is 0.7147, which is derived by 40.08/(40.08 + 16.00); for Na_2O it is $(2 \times 22.99)/(2 \times 22.99 + 16.00) = 0.7419$. A useful table of gravimetric factors is given in Appendix 9.

continued

Table 1.1 A Representative Chemical Analysis: Basalt BHVO-1, a USGS Rock Standard

Oxide	Original analysis wt%	All Fe as FeO^t wt%	All Fe as $Fe_2O_3{}^t$ wt%	All Fe as $Fe_2O_3{}^t$, volatile-free wt%	Element	Element ppm	Oxide wt%
SiO_2	49.60	49.60	49.60	49.87	Ba	753	0.0840
TiO_2	0.90	0.90	0.90	0.90	Co	46.4	0.0059
Al_2O_3	15.70	15.70	15.70	15.79	Cr	331	0.0484
Fe_2O_3	3.66		10.81	10.87	Rb	36	0.0039
FeO	6.50	9.79			Sr	685	0.0810
MgO	9.18	9.18	9.18	9.23	Zn	89	0.0111
CaO	9.09	9.09	9.09	9.14	Zr	160	0.0216
Na_2O	2.79	2.79	2.79	2.81			
K_2O	0.96	0.96	0.96	0.97			
P_2O_5	0.25	0.25	0.25	0.25			
MnO	0.17	0.17	0.17	0.17			
H_2O^+	0.64	0.64	0.64				
H_2O	0.32	0.32	0.32				
CO_2	0.01	0.01	0.01				
Total	99.77	99.40	100.42	100.00			

3. If everything in the rock is analyzed, the total should be 100 wt%, give or take a little analytical error. Quality analyses typically have totals within 1 wt% of 100, but there is no firm rule.
4. Fe is the only major element that can exist in multiple valence states: Fe^{2+} (FeO) and Fe^{3+} (Fe_2O_3). Analyzing for the two valence states of Fe is difficult and not commonly done. Fe can be reported as separate FeO and Fe_2O_3 (if both valence states were measured), FeO^t, or $Fe_2O_3{}^t$, these latter two indicating all the Fe in the rock expressed as the corresponding oxide.
5. Rocks contain minor amounts of volatile species such as F, Cl, CO_2, and H_2O, bound in minerals such as apatite, calcite, biotite, hornblende, and so on. These are not commonly analyzed. H_2O^+ is water that is structurally bound in minerals or dissolved in glass, whereas H_2O^- is adsorbed water stuck to the outside of grains of powder. This latter is generally driven off in a drying oven at 100 °C before weighing.
6. LOI stands for "loss on ignition," and is the weight change that occurs when a rock powder is roasted in a drying oven at a temperature high enough, typically 1000 °C, to drive off volatiles by breaking down the volatile-bearing minerals. This process also oxidizes Fe to Fe^{3+}, and so a rock with lots of Fe^{2+} and low volatiles can actually gain weight, giving a negative LOI. LOI is an imperfect measure of the amount of volatiles in the rock.
7. To compare analyses in which the issues mentioned here were treated differently, chemical analyses are commonly renormalized such that they total 100 wt% on a volatile-free basis with Fe as either $Fe_2O_3{}^t$ or FeO^t.

Examination of the other major oxides in igneous rocks shows that there is generally a significant correlation between any pair. Figure 1.4 shows a matrix of oxide-oxide plots for 615 analyses of volcanic rocks from the Medicine Lake volcanic field in northeastern California. This young volcanic field lies east of Mount Shasta and is typical of volcanic fields in the subduction-related Cascades chain. As with the entire NAVDAT dataset, these rocks follow the BADR trend from basalt to rhyolite and plot more or less down the axis of the bowling pin in Figure 1.3; 8 percent of the analyses fall just barely above the basalt-to-rhyolite field boundary, in the alkali-rich "trachy" fields. Silica shows excellent negative correlations with $Fe_2O_3{}^t$ (Fe expressed as entirely ferric Fe; see Toolbox 1.2) and CaO. TiO_2 and Al_2O_3 are also negatively correlated with silica, but

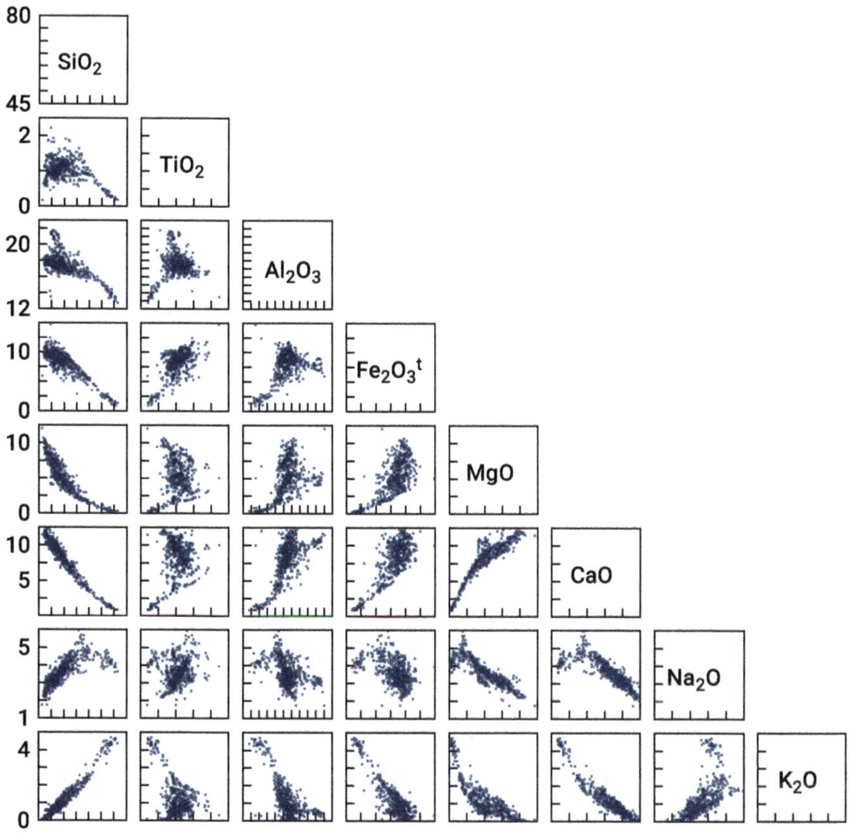

Figure 1.4 Matrix of oxide-oxide plots for the Medicine Lake volcanic field in northeastern California, showing that most of the elements correlate with one another. First column shows SiO_2 against the other oxides; second, TiO_2 against the others; and so on. Some pairs show positive correlations, some negative, and some are curved. Each plot is a 2D projection from *n*-dimensional composition space. All axes in wt%.

Toolbox 1.2

Converting from Oxides to Elements and Other Simple Calculations

There are several simple arithmetic procedures for converting values in a chemical analysis from one form to another:

1. *Elements to oxides*. For example, how much Ti is in a rock with 2.71 wt% TiO_2? This conversion is simple: wt% element = $G \times$ (wt% oxide), where G, the gravimetric factor, (see Appendix 9) is, in this case, 0.5995. To convert in the other direction, from oxide to element, divide by G.
2. *Gravimetric factors*. Gravimetric factors do not depend on stoichiometry—that is, on how an oxide formula is written. For example, G for Fe_2O_3 is 0.699, because this is the ratio of the mass of 2Fe to Fe_2O_3. If we used instead $FeO_{1.5}$ for ferric oxide, the ratio ($Fe/FeO_{1.5}$) is the same. A rock with 10 wt% Fe_2O_3 also has 10 wt% $FeO_{1.5}$.
3. *Wt% to ppm*. Wt% is g per 100 g, and ppm is µg per g (or g per megagram), so these quantities just differ by a factor of 10,000. Hence, 139 ppm is 0.0139 wt%.
4. *Iron*. Fe is the only major element that exists in multiple oxidation states: Fe^{2+} (FeO) and Fe^{3+} (Fe_2O_3). You might find Fe reported as separate FeO and Fe_2O_3 (if both were measured), FeO^t, or $Fe_2O_3{}^t$. These latter two express all Fe as either Fe^{2+} or Fe^{3+}. There is a separate gravimetric factor relating these quantities: $Fe_2O_3 = 1.111 \times FeO$. If you took a given mass of FeO and roasted it in air, oxidizing the Fe^{2+} to Fe^{3+}, this quantity expresses the 11 percent gain in mass as it took on O from the air. Note that in the analysis of BHVO-1, $Fe_2O_3{}^t = 1.111FeO + Fe_2O_3$.
5. *Trace elements*. In the analysis in Toolbox 1.1, several of the more abundant trace elements are listed as both ppm as metal and wt% as oxide. Note that their oxide total is about 0.25 wt%. This is typical for trace elements, and they are generally ignored when renormalizing.

funnel down from significant spreads at low silica to tight correlation at high silica. MgO shows a tight negative correlation but is distinctly bow-shaped, concave-up. K_2O and Na_2O are positively correlated, but Na_2O is bow-shaped, convex-up. Some oxides show a splatter pattern plotted against each other (e.g., TiO_2 and Na_2O), but for most pairs there are decent correlations.

In mathematical terms, these analyses plot along a scattered (locally very scattered) 1D curve in n-dimensional composition space. The curve, which connects basalt to rhyolite, is fairly straight but includes the curvatures noted earlier. That is, if you plot the Medicine Lake data in two dimensions, as, say, MgO vs. SiO_2, they are connected by a curve that connects the basalt end to the rhyolite end, as in Figure 1.3. If you plot them in three dimensions, as, say, K_2O vs. MgO vs. SiO_2, they are connected by a banana-shaped curve that connects basalt and rhyolites in three dimensions. If you could plot all eight oxides in Figure 1.4 on 8D graph paper, they would follow a hyperdimensional-banana-shaped curve (an 8-hyperbanana?) connecting basalt and rhyolite.

Just as you can draw a 1D line on a 2D piece of paper, you can stretch a piece of string out as a 1D line in the 3D space that we live in. This concept generalizes to any number of dimensions. Our brains are three-dimensional and cannot visualize this, but mathematically the concept is simple. We will jump up and down in dimensionality throughout the rest of this book, typically using 2D projections, as in Figures 1.3 and 1.4.

Much of igneous petrology is concerned with understanding how these variations come about, what they mean, and how to use them to understand how the Earth works. As we shall see, the bulk composition of a magma controls many of its physical properties, and these in turn control how magmas behave, how volcanoes erupt, how dikes intrude, timescales of magmatism, and many other features of the igneous world. In particular, the silica content of a magma (i.e., where it is on the basalt-rhyolite spectrum) is strongly correlated with its viscosity, and this in turn controls how explosive and destructive an eruption can be.

Spectrum of Properties

Figure 1.5 summarizes some of the important physical, chemical, and mineralogical changes that occur across the BADR spectrum. Many of these will not be covered until our discussion of volcanism in Chapter 7. Rocks at the high-silica and low-silica ends of the spectrum may be called by the mnemonics **felsic** and **mafic**, respectively.

Felsic
Denoting a rock with a high abundance of feldspars and quartz or, in alkalic rocks, feldspathoids such as leucite or nepheline.

Mafic
Denoting a rock with a high abundance of ferromagnesian minerals such as pyroxenes and amphiboles.

The Chemical Range of Sedimentary and Metamorphic Rocks

The differences in physical properties and behavior between magmas at the opposite ends of the compositional spectrum are vast, yet sedimentary rocks and the metamorphic rocks derived from them have a much greater range of compositional variability. Figure 1.6, showing CaO vs. SiO_2 from a database of sedimentary and metamorphic rock compositions, graphically shows just how broad the range of sedimentary rock compositions is compared to a typical igneous suite (the same Medicine Lake data as earlier). Sedimentary processes can, and commonly do, make rocks that are pure calcite, or pure dolomite, or pure quartz, and they can make rocks that have these minerals in any combination. Thus, the ranges of CaO and SiO_2 in sedimentary rocks are far greater than in igneous rocks, and this is true for most of the major oxides. The strong correlations seen in igneous data (see Figure 1.4) are less apparent.

It almost seems backward that the dramatic, high-temperature processes involving magmas can only drive igneous rock compositions through a small bit of compositional space, whereas sedimentary processes occurring at room temperature produce a huge spread of compositions. The range of igneous compositions is governed by the geochemistry of melting and crystallization, which drives compositions along the basalt-rhyolite trend, whereas sedimentary rock compositions are controlled by weathering, sorting in water and air, biologic processes, and other phenomena.

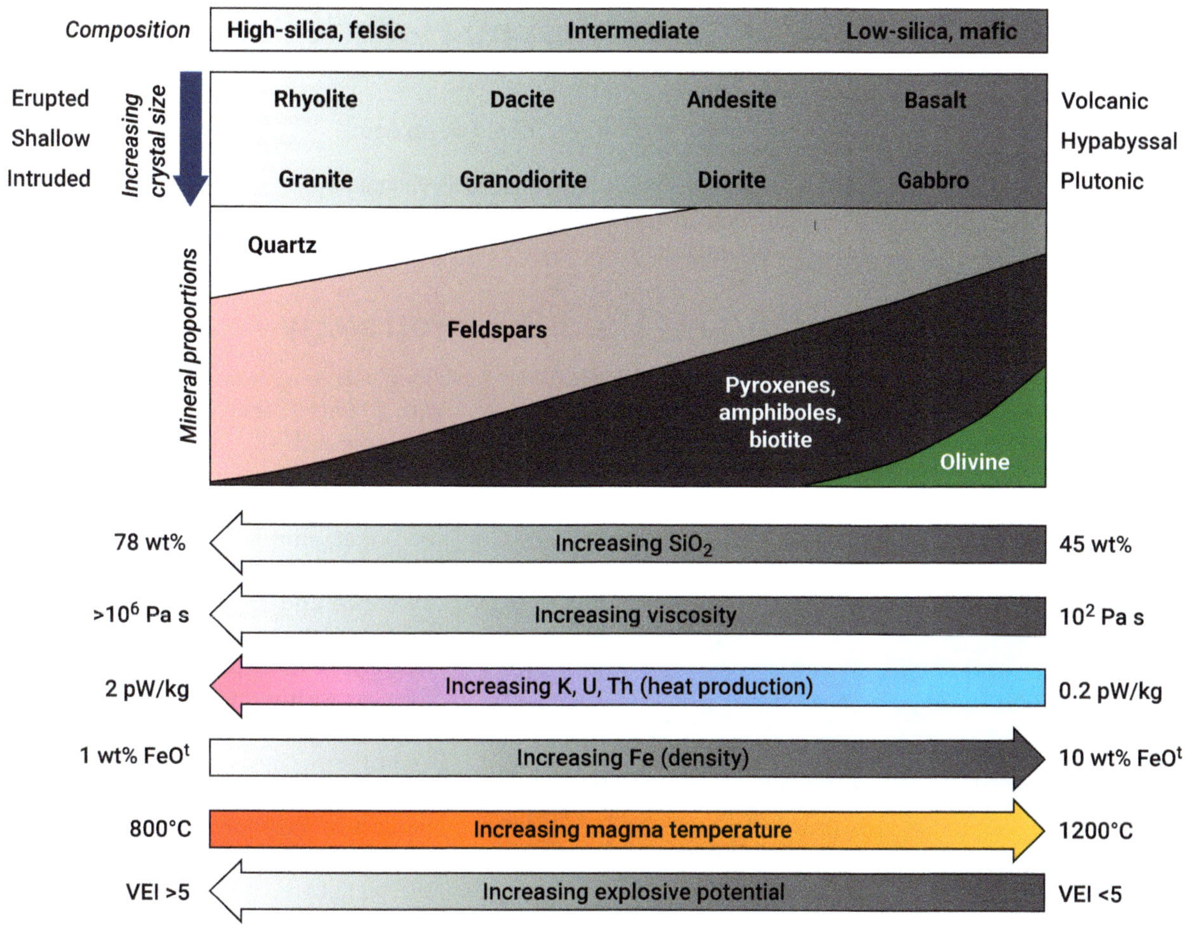

Figure 1.5 Simplified representation of the igneous rock compositional spectrum and the variations in physical properties that accompany compositional change. Most of the world's igneous rocks fall on the basalt-andesite-dacite-rhyolite (BADR) spectrum or its plutonic equivalent (gabbro-diorite-granodiorite-granite). VEI is volcanic explosivity index, a logarithmic measure of the intensity of an eruption (see Chapter 7).

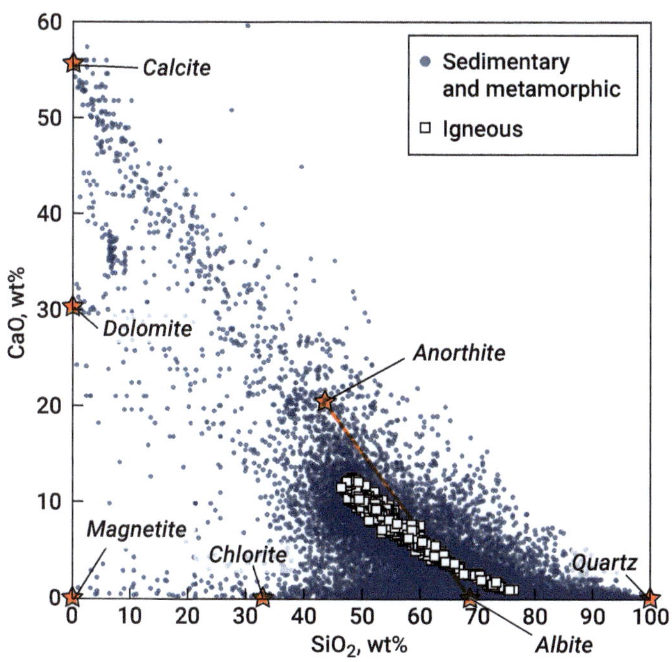

Figure 1.6 Typical igneous trend from basalt to rhyolite (Medicine Lake data as earlier) plotted with sedimentary and metamorphic rock analyses from the US Geological Survey's National Geochemical Database. Igneous compositions are tightly restricted to the basalt-rhyolite trend, whereas sedimentary rocks and the metamorphic rocks derived from them range off toward the compositions of pure minerals such as calcite and quartz. Note that metamorphic rocks include metamorphosed versions of all three rock classes.

1.3 Earth Systems and Holistic Petrology

Learning Objective

By the end of this section you should be able to:

1.3.1 Provide examples of how igneous and metamorphic processes affect diverse Earth systems ranging from Earth's interior to the atmosphere and including human activities and the biosphere.

The Nature of Hard-Rock Petrology

Igneous and metamorphic petrology have long been taught as separate subfields of hard-rock petrology, and books reinforce this with a "Part 1: Igneous, Part 2: Metamorphic" organization. In this book, we take a holistic approach that follows plate tectonics. Basalts are made at oceanic ridges, metamorphosed by reaction with seawater, and then metamorphosed again as they descend in a subduction zone. Plutons intrude the crust and metamorphose their surroundings. Igneous and metamorphic processes are deeply entwined and together make the continents; help regulate sea level; control long-term CO_2 and H_2O cycles; produce topography that influences weather, water, and ice; generate volcanic eruptions, megalandslides, tsunamis, and other hazards; produce economic deposits of metals and other substances necessary to modern civilization; and so on. Petrology thus interacts with other Earth systems to produce the planet we live on, and the grand system that ties petrology together is plate tectonics.

Figure 1.7 Some examples of petrology in action; see text for explanations. (a) Sarychev volcano in eruption, June 12, 2009. Sarychev lies on an island in the Kuril chain between Japan and Kamchatka. Photo taken by an astronaut aboard the International Space Station. (b) A portion of coastal California just north of San Francisco Bay, which empties into the Pacific Ocean through the Golden Gate (lower right); bright area is the urban part of the northwest corner of the San Francisco Peninsula. Landsat 8 natural-color image taken March 1, 2017. (c) The Matterhorn from Gornergrat, Swiss Alps. (d) Exceptionally well-preserved fossils of barrel-chested rhinoceroses encased in volcanic ash at Ashfall Fossil Beds State Historical Park, Nebraska. Skull in foreground is 33 cm long.

Sources: (a, b) NASA/JSC; (c) AFG photo; (d) Courtesy of University of Nebraska State Museum.

Connections

Petrology is closely tied to diverse fields in the natural world, such as ecology, anthropology, climatology, and environmental science, and to materials science and other aspects of the constructed world. Indeed, much of petrology is materials science of the solid Earth. In the following chapters, we highlight these connections to show the relevance of petrology to pursuits as diverse as tracing trade routes of ancient peoples, how mountain building affects global climate, cement manufacture and its environmental consequences, and rock climbing. So, although this book focuses on petrology and frames it in the context of plate tectonics—the fundamental system driving geologic processes in the solid Earth—each chapter contains at least one connection to another area of interest to show the diverse applications and implications of petrology.

Figure 1.7 illustrates the variety of Earth systems to which petrology can be applied. In Figure 1.7a, astronauts on the International Space Station got an oblique view, through a hole in the clouds, of an ongoing eruption of

Sarychev volcano on an island in the Kuril chain north of Japan. The eruption sent a column of ash 7 to 10 km skyward while a pyroclastic flow raced down the volcano's flank toward the bottom of the picture. This remote volcano has no ground monitoring stations but is under jetways from North America to east Asia. Volcanic ash clouds pose serious threats to jets that wander into them, so these volcanoes are constantly monitored by satellite.

Figure 1.7b shows how the San Andreas Fault has sliced a gash through coastal California north of San Francisco. Rocks on either side are dramatically different (granites of a magmatic arc on the west against a subduction assemblage on the east), and it was dramatic lithologic mismatches such as this one that led to the hypothesis that the fault has accumulated hundreds of kilometers of slip. The subduction assemblage contains abundant serpentinite (a rock composed mainly of serpentine), a substrate that hosts unusual flora including numerous rare plants. The sediment plume issuing from the mouth of San Francisco Bay in the lower right is derived primarily from erosion of the Sierra Nevada range, which owes its elevation to low-density granite bedrock and, perhaps, loss of a dense metamorphic root that detached and sank into the mantle. These sediments help define the chemistry of San Francisco Bay. Farther upstream, older deposits contain gold from the granites, and this spawned the California gold rush.

Figure 1.7c shows the Matterhorn, on the border between Italy and Switzerland. In the foreground are ocean-floor rocks that were metamorphosed under high-pressure, low-temperature conditions—a regime found only in subduction zones. These are remnants of an ocean basin that was crushed between converging Africa and Europe, a collision that produced the Alps. How did rocks that were deep in a subduction zone end up high in the Alps, and why are the Alps so tall? Good questions. In the background, the Matterhorn exposes a sheet of rock derived from the edge of Africa, thrust over the oceanic rocks in the foreground and the southern margin of Europe (behind the photographer) during the Alpine collision.

In Figure 1.7d, we see part of an exceptionally well-preserved and abundant fossil assemblage at Ashfall Fossil Beds State Historical Park in Nebraska. Animal groups at this ancient watering hole included rhinos, horses, camels, cranes, lizards, turtles, tortoises, and dogs. The animals were killed owing to a huge eruption in southern Idaho 12 million years ago. Ash from the eruption, whose source was nearly 1500 km away, buried the area under 30 cm of ash, which suffocated animals (in part through progressive lung damage caused by ash inhalation), killed plants, and ruined their water supply. Such gargantuan eruptions are common in the geologic record, but their effects on ecosystems are not well known.

1.4 Petrology and Tectonic Setting

Learning Objectives

By the end of this section you should be able to:

- **1.4.1** Predict the type of igneous rock expected in typical tectonic environments.
- **1.4.2** Discuss how the pressure-temperature paths that metamorphic rocks follow should vary with tectonic environment.
- **1.4.3** Explain the origin of the Wilson cycle using the concept of plate motions on a sphere (conservation of surface area).

Igneous Rocks

Correlations between tectonic setting and rock composition were noted long before anyone knew what those tectonic settings were. Over a century ago, Marshall (1911) defined an "andesite line" in the southwestern Pacific. On the ocean side of this line andesite was absent and nearly every igneous rock was basalt, but take a step in the other direction and you are in the land of tall, explosive andesite stratovolcanoes whose behavior is quite different from those basalt volcanoes. This was early recognition of a petrologic aspect to the circum-Pacific "Ring of Fire." Chayes (1964) demonstrated that basalts within the ocean basins have consistently higher TiO_2 than those just outside the oceans—across the andesite line, for example. These authors did not tie these boundaries to subduction zones, because that concept had not yet been invented, but they recognized the spatial patterns.

Chemical variability in igneous rocks sometimes correlates strongly with tectonic setting. Let's examine the chemical compositions of igneous rocks produced in four tectonic settings: mid-ocean ridges (or oceanic ridges; outside the Atlantic, most "mid-ocean" ridges are not in the middles of oceans), subduction zones, oceanic islands, and continental hotspot tracks. Representative plots of MgO vs. SiO_2 from representative areas are given in Figure 1.8. These two oxides were chosen because they show a large range of variability in igneous rocks, but most of the others would serve just as well.

These data, like most of the Earth's magmatism, span a range from about 40 to 80 wt% SiO_2, and MgO is negatively correlated with SiO_2. It is clear there is a great difference in spread among the tectonic settings. Subduction zone magmatism in the Aleutians spans and fills a range from about 45 to 75 wt% SiO_2; ocean island magmatism is concentrated down near the low-silica end of the spectrum; rifts are somewhat bimodal, with a concentration at low SiO_2 (basalt) and another at high SiO_2 (dacite, rhyolite); and oceanic ridges make basalt, and only basalt.

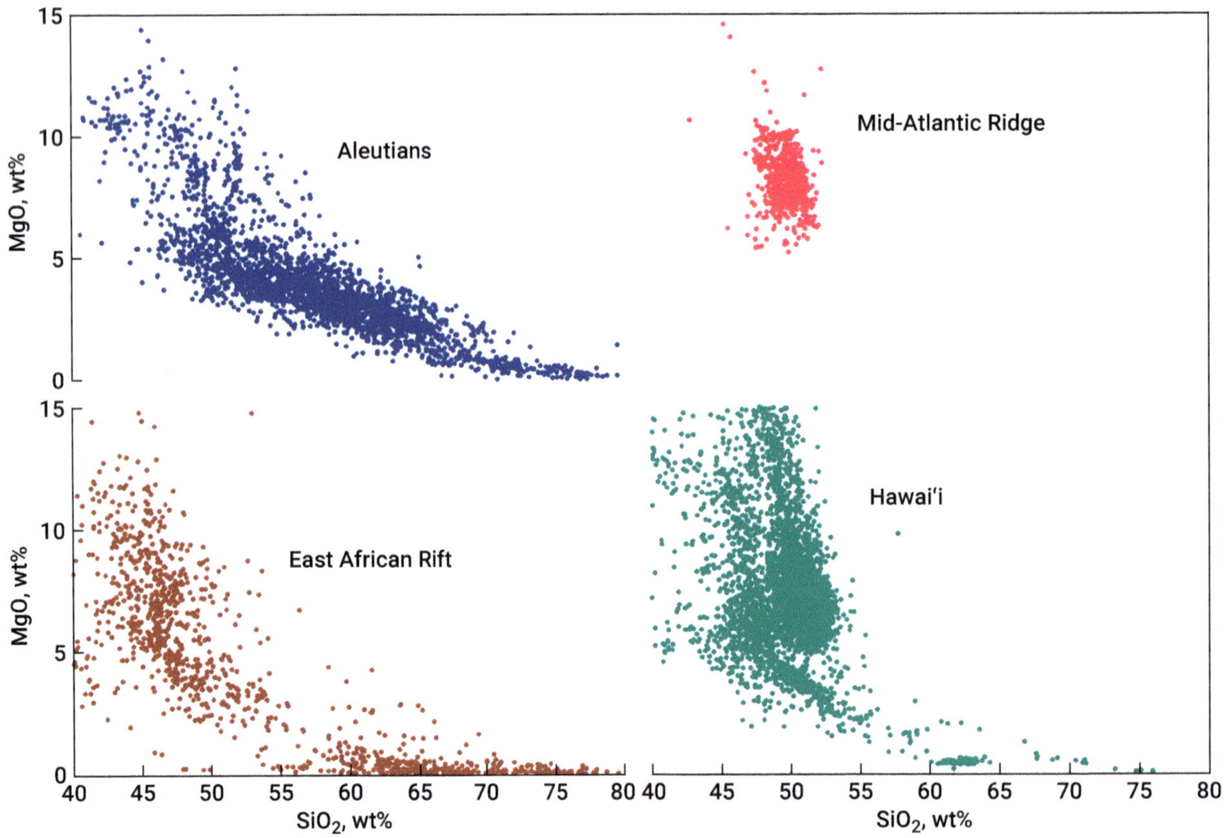

Figure 1.8 Plots of MgO vs. SiO_2 showing the significant correlation between tectonic setting and chemical behavior. Localities are a subduction zone (Aleutian Islands), an oceanic ridge (Mid-Atlantic Ridge), a continental rift (East African Rift), and an oceanic island hotspot complex (Hawai'i). Ridge basalts show a tight compositional range at the basalt end of the spectrum, whereas subduction zones show a continuous spectrum from basalt to rhyolite. Oceanic hotspots are typically basaltic but with significantly more range than mid-ocean ridge basalts, and continental rifts are typically bimodal, with abundant basalt and rhyolite but lesser amounts of andesite and dacite (or their more alkali-rich equivalents; see Chapter 14).

Sources: Data from Georoc and EarthChem.

Clearly there are some big differences in the compositional spread of magmatism between these different settings, and if we looked at other examples we would see the same patterns: subduction zones span the BADR trend, oceanic ridges and ocean islands tend to be concentrated at the basalt end, and continental hotspots tend to produce bimodal basalt-rhyolite assemblages that are conspicuously lacking in andesite and dacite. These differences are of more than just academic concern. For example, although the Hawaiian volcano Mauna Loa is vastly larger than a stratovolcano such as Mount Rainier, the latter is capable of far more violent and destructive eruptions, because it erupts much higher-silica, and therefore more viscous, magma (see Chapter 7). The continental hotspot volcanoes of Yellowstone have produced some of the largest eruptions in the geologic record.

Metamorphic Rocks

Metamorphic rock types also correlate with tectonic setting. In the late 1700s and early 1800s, James Hutton and Charles Lyell proposed that metamorphic rocks formed through the action of heat and pressure on preexisting igneous and sedimentary rocks—essentially the same explanation we use today. Other petrologists of the 1800s recognized differences between local contact metamorphism caused by proximity to plutons vs. regional metamorphism that developed over much larger areas. Regional metamorphism was also called "dynamo-thermal" to highlight the fact that these rocks were often pervasively deformed, and so must have experienced something other than just heating. Finally, right at the dawn of the plate tectonics era, Miyashiro (1961) showed that, around the Pacific Ocean, metamorphic rocks commonly occur in two parallel belts. One, on the ocean side, was metamorphosed at low *T* (temperature) and high *P* (pressure) and contained the unusual minerals glaucophane and jadeite, and another, on the landward side, was metamorphosed under relatively low-*P* conditions, with a more typical metamorphic mineral assemblage. None of these scientists had a plate tectonic framework within which to explain how these rocks formed, but within a decade Miyashiro's paired metamorphic belts were understood to have been generated at subduction zones. In particular, high-*P* jadeite and glaucophane are characteristic of rocks that have been down a subduction zone and brought back to the surface, whereas low-*P* contact metamorphic rocks form near intrusions in arcs. Regional metamorphic rocks commonly represent the product of continental collisions. Importantly, many systems transition from subduction to collision and can include contact metamorphic effects—the logical consequence of plate tectonics.

Figure 1.9 shows pressure-temperature (*P-T*) paths for metamorphic rock assemblages from several tectonic settings, including ocean-continent subduction (Franciscan, Monviso), continental collision (Himalaya, western Acadian), and

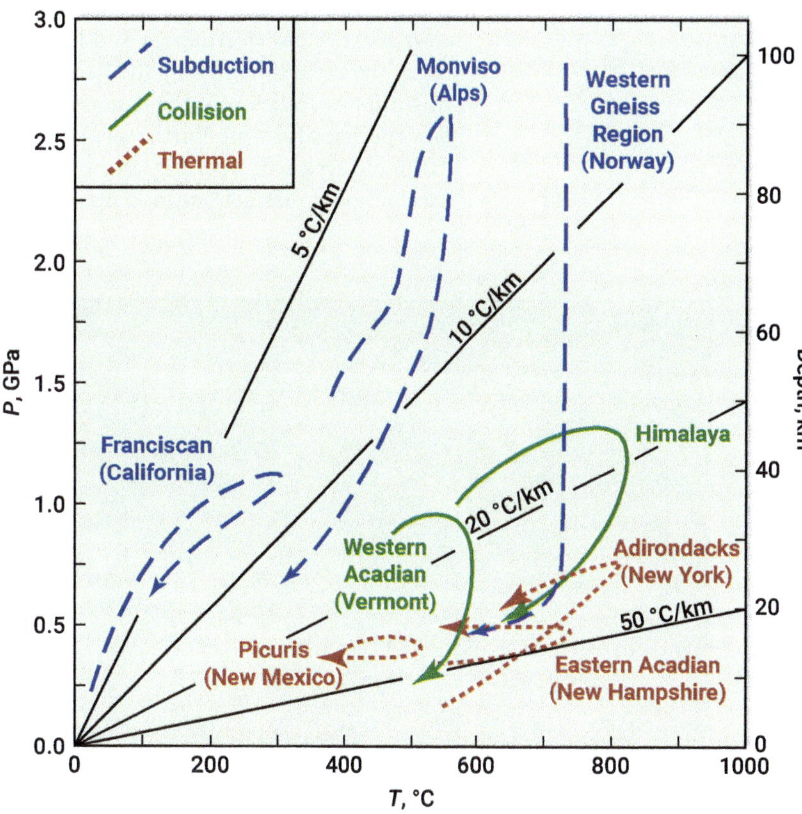

Figure 1.9 Pressure and depth vs. temperature diagram with reference temperature gradients. This diagram shows selected pressure-temperature (*P-T*) paths for rock assemblages from tectonic settings broadly representing subduction (both oceanic and thinned continental plates; ~5–10 °C/km), collisions (~20 °C/km), and thermal anomalies (up to 50 °C/km).

Sources: Data from Spear (1995), Williams & Karlstrom (1996), Spear & Markussen (1997), Hacker et al. (2010), Angiboust et al. (2012), and Kohn (2014).

areas that were anomalously hot (Picuris, eastern Acadian). These are depictions of *P-T* conditions experienced by the corresponding rocks and were determined using thermodynamic calculations based on mineral compositions, as described in Chapter 12. Clearly, different tectonic settings produce different *P-T* paths. Subduction produces characteristically high *P* at relatively low *T*, and in some regions (Monviso, Franciscan), but not in others (Western Gneiss Region), rocks return to the surface along a similar *P-T* path. Rocks from Vermont and the Himalaya reached much higher temperatures for their pressures than subduction zone rocks, whereas regions with thermal anomalies, either because of extension or heating from plutons, show very flat *P-T* paths dominated by heating and cooling at relatively restricted pressures.

The Wilson Cycle

Opening an ocean basin by seafloor spreading on one part of the globe requires closing ocean basins elsewhere, because the Earth's surface area is constant. Continental collision is an inevitable result of this process. The North Atlantic, which formed when the supercontinent Pangea split and separated North America from northern Africa and Europe, has been steadily widening for the past 200 million years; the South Atlantic, separating South America from southern Africa, has been widening for about 140 million years. The 5000 km of spreading represented by the Atlantic has been taken up by subduction around the Pacific. If you stand on the east coast of North America and look east across the Atlantic, you (theoretically) would see the Eurasian and African Plates moving away at 20 to 25 mm/yr. Going around the globe, though, this means that if you stand on the west coast of North America and look west, you would see the Eurasian Plate approaching at the same rate, closing the Pacific Ocean (Figure 1.10). Extrapolated forward in time, this implies collision of the Americas with island arcs and continents of the western Pacific, perhaps forming Earth's next supercontinent.

Such cycles of opening and closing of ocean basins rearrange plate boundaries—subduction zones get jammed up, new ones form, transform faults move things around,

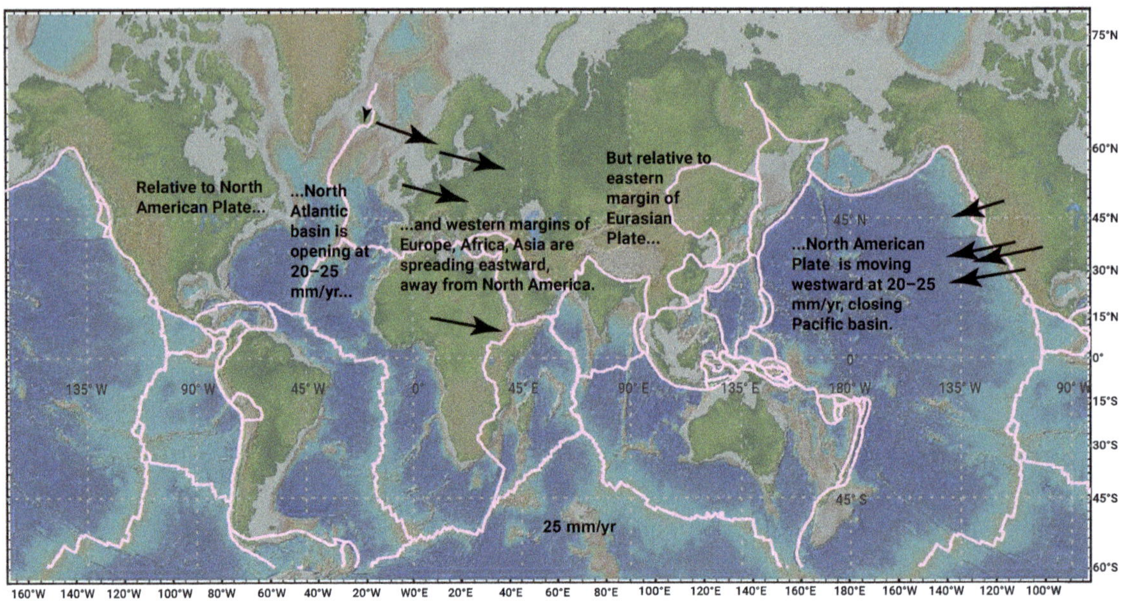

Figure 1.10 Motion on a globe means that plates can be moving apart and yet moving together at the same time. Relative to North America and looking eastward across the Atlantic Ocean, the Eurasian and African Plates are moving away at 20–25 mm/yr. Looking westward across the Pacific, the Eurasian Plate is approaching North America at about the same rate. Opening of the Atlantic is forcing closure of the Pacific and, in a simple forward extrapolation, will lead to formation of a future supercontinent. Exactly when and how will this happen? We just need to wait a few hundred million years to find out.

supercontinents form and disperse—and this leads to tectonic environments being superimposed on one another. This cycles of ocean basin opening and closing is called the **Wilson cycle** in honor of Canadian geophysicist J. Tuzo Wilson, who is credited with recognizing the phenomenon (Wilson, 1968). We return to this in Chapter 12.

Wilson cycle
Cycle of seafloor spreading, continental collision and amalgamation, and subsequent rifting to form new ocean basins.

Was There Ever *Not* Plate Tectonics on Earth?

Maybe. Although some geochemical evidence from Archean zircons does suggest subduction soon after Earth's birth (e.g., Harrison, 2009), there is at least a reasonable chance that Earth prior to about 2.5 billion years ago did not experience plate tectonics, or experienced a different style of plate tectonics than today. For example, metamorphic rocks today exhibit a profound dichotomy between cold and hot tectonic environments (subduction zones and arcs; see Figure 1.9). However, going back in time, this dichotomous metamorphic indicator of subduction virtually completely dies out by 2.5 billion years ago. How Earth might have worked before plate tectonics is largely the topic of modeling (e.g., Brown et al., 2020), and models are controversial. Nonetheless, it is worth keeping in mind that nearly all the discussion in this textbook is predicated on plate tectonics, which may have more applicability to the latter half of Earth's life than its first half.

1.5 The Terminology of Petrology

Learning Objectives

By the end of this section you should be able to:

- **1.5.1** Explain the value, but also the limitations, of petrologic names and classifications.
- **1.5.2** Apply time designations appropriately.

The Curse of Classification

One of the stumbling blocks that students encounter in studying petrology is the astonishing number of names used to describe rocks. Igneous petrology seems to be particularly stricken by excess names; if metamorphic petrology followed the same conventions there would be 100 varieties of schist, each named for the place where it was first described, with the classification based on quartz/plagioclase ratio or some other property that you might or might not be able to discern in the field and that might or might not be meaningful.

There are many reasons for this proliferation of names. One is simply that in the early days of petrology, the field grew as people examined new areas and found new rocks, and new rocks were given new names. Another is that a century ago, when petrology was a new field, the petrographic microscope opened up a new way to look at rocks and to quantify what you saw. Suddenly, "granite" and other rocks could be subdivided into many varieties based on abundances of major minerals and identities of minor ones. Yet another is that petrology tried to follow the lead of biologic taxonomy. The rank-based taxonomic system pioneered by Linnaeus and others proved to be highly useful for organizing and cataloging the world's life. If a new species was discovered it was generally clear that it was a new species, and it was added to the catalog. This procedure was successful because the branching classification ends at distinct branches.

Consider cats and dogs. A rank-based taxonomic classification for a cat is domain Eukarya, kingdom Animalia, phylum Chordata, class Mammalia, order Carnivora, family Felidae, genus *Felis*, species *catus* (and then, maybe, "Mittens"). A dog shares the same classification through order Carnivora, and then branches to family Canidae, genus *Canis*, species *familiaris* (and then, maybe, "Rex"). There are many kinds of carnivores, but the canid and felid families are distinct genetic lines that cannot mix, and it makes sense to put a taxonomic fence between cats and dogs.

Attempts to emulate this in petrology did not fare well. Rocks are different from flora and fauna; along the BADR spectrum, for example, there is every gradation from one end to the other. The TAS diagram puts boundaries between various names, but these are arbitrary and not respected by the rocks. A lava flow that is an andesite in one place might grade into dacite a few centimeters away. One could theoretically trace a surface basalt flow, which is clearly a volcanic rock, into the ground, down a feeder dike, and into a coarse-grained gabbro, which is clearly an intrusive rock, with no sharp break in texture that would tell when you had passed from the volcanic to the plutonic realm.

Over the history of petrology many schemes have been devised for classifying rocks, with most based on texture (e.g., fine-grained or coarse-grained) combined with chemical composition or mineral proportions. Some of these are covered in Appendix 3. There are hundreds of names in some of these schemes. Our advice for dealing with rock names is: DON'T PANIC. In this book, classification is deemphasized, and we have tried to use a minimal number of the most common rock names. There are many other resources for learning what less-common names such as "ankaramite" and "hawaiite" mean (hint: in this book, they're just basalts). Know the important ones, and look up the rest.

The Significance of Classification

Norman Bowen, who more than anyone else put petrology on a firm quantitative basis, distrusted quantitative classifications based on chemical composition or mineral proportions, and his comments are worth careful study:

> If any analogy is to be drawn between the animal and the mineral kingdom it should consist rather in a comparison of mineral species with animal species. And if it is to be carried further this can be done by comparing rock types with faunas. Just as a fauna is a collection of animal species resulting from a response to external conditions and to each other, so a rock type is a collection of mineral species resulting from a response to external conditions and to each other. With this analogy in mind we get a rational view of the importance of the quantitative element in rock classification. How artificial a classification of faunas would be which was based on the ratio of foxes to hares, of hares to moles and so on! To be sure it is by no means accidental that the ratio of hares to foxes is 10 in a certain area and only 2 in another, but as compared with the broad factors controlling life in the two areas it is relatively accidental. A classification of the two faunas on the basis of the ratios noted might serve a utilitarian purpose from the viewpoint of a trapper, but it would have little scientific value. So it is not accidental that a rock is nearly pure olivine here and only 75 per cent olivine a few feet distant, but it is relatively accidental and should not be made a fundamental factor in classification. (Bowen, 1928, 321–322)

What Bowen called a "fauna" we might now call an "ecosystem," and his "external conditions" are our "environment." Tell an ecologist that you have an assemblage of zebras, elephants, cheetahs, grassland, and acacia trees, and they will know a lot about the environment, where on Earth it might occur, and what other organisms to expect. Tell a geologist that you have an assemblage of eclogite, greenschist, metachert, and mélange, and they can tell you a lot about the tectonic history of the area and what else you might expect to find.

That is the reason it is worth describing and naming igneous and metamorphic rocks: they are petrology's equivalent of an ecosystem, and if you can read them you can infer much about the history and environment of formation of the Earth. Mineral and rock assemblages reflect these environments and histories, and in general it is the assemblages that are important, not the exact ratios of the minerals or chemical components.

A Definition of "Magma"

Magma is the essence of igneous petrology, but there is a curious lack of specificity in petrology about just what "magma" is. To some, a rock that is even a tiny bit molten is magma; to others, for partially molten material to be called magma it must be capable of flowing on some not-too-long timescale, such as the timescale of volcanic processes (e.g., seconds to years). We will use the latter interpretation with the following definition from Glazner et al. (2016):

> **Magma:** Naturally occurring, fully or partially molten rock material generated within a planetary body, consisting of melt with or without crystals and bubbles and containing a high enough proportion of melt to be capable of intrusion and extrusion.

A Note about Mineralogy

Basic Vocabulary of Petrology Rocks are composed almost entirely of minerals and glass. These materials are the fundamental building blocks of rocks, and rocks are the fundamental building blocks of the solid Earth. To understand petrology it is essential that you have or develop a solid grounding in the rock-forming minerals. Knowing the minerals is somewhat like knowing basic vocabulary of a foreign language; in a literature course you could look up every word you come to, but you won't make it very far if you have to do so. You must memorize the vocabulary. In this sense, minerals are like words, rocks are like paragraphs or chapters, and geologic regions are like biographies. Having a solid grounding in the vocabulary (mineralogy), which must be memorized at some level of detail, allows you to understand and appreciate what the rocks are telling you.

There are over 4000 distinct mineral species, but fewer than 1 percent are common. These should be memorized, or you should at least know what's in them (e.g., the important mineral glaucophane has a complicated formula, but knowing that it is a sodic amphibole in high-pressure rocks will get you far).

Abbreviations Throughout the book, for brevity, we use standard capitalized abbreviations for minerals or other phases (see Appendix 1; e.g., Ab for albite, Qz for quartz, Pl for plagioclase). We switch back and forth between spelling mineral names out and using abbreviations in order to help remind you what they mean.

A Note about Time

Many but not all organizations now use "a" to mean both years (a duration) and years ago (time since the present). However, similar to paleoceanographers, who distinguish meters (a unit of length) from meters below seafloor (a distance relative to a datum), in this book we will use "yr" for durations and "a" for years before present. Otherwise, we will use standard abbreviations for time, especially k (kilo-) for thousands, M (mega-) for millions and G (giga-) for billions. So, for example, 5 Myr means 5 million years in duration, regardless of absolute age, and 5 Ma means 5 million years ago. Both terms could be used in a single sentence, as in "Cooling from 800 to 600 °C took 10 Myr, from 250 to 240 Ma." OK, maybe that's not the world's greatest sentence, but you get the idea.

1.6 Lithosystems

Learning Objective

By the end of this section you should be able to:

1.6.1 Use the concept of a lithosystem to interrelate characteristic rock suites with geologic environments.

The concept of rock assemblages that indicate particular sets of pressure-temperature conditions, tectonic settings, source materials, geochemical cycles, and so forth is central to petrology but has not been well-systematized. In this book we will use the term **lithosystem** to refer to a particular collection of rocks, structures, and geochemical processes that are characteristic of a given geologic environment. The rock assemblages in

Lithosystem
A set of rock types and related structures, such as faults or fabrics, that are collectively diagnostic of a given environment of formation.

Table 1.2 Important Igneous and Metamorphic Lithosystems

Chapter	Lithosystem	Principal Rock Types and Structures	Interpretation
3	Ophiolite	Pillow lavas, sheeted dikes, gabbro, peridotite, thin-bedded chert	Oceanic crust; remnant of a closed ocean basin
3	Mantle	Olivine-rich ultramafic rocks such as peridotite and harzburgite; serpentinite	Upper mantle
5	Seafloor alteration	Basalt and basaltic products of hydrothermal alteration, greenschist, serpentinite, volcanogenic massive sulfides	Mid-ocean ridge with hydrothermal alteration
6	Arc volcanic	Extrusive rocks ranging from basalt to rhyolite with abundant andesite and dacite; pyroclastic flows; lahars	Magmatic arc above a subduction zone
6	High-pressure/ low-temperature (HP/LT)	Eclogite, blueschist, serpentinite, mélange	Material from subduction zone metamorphosed under high-pressure, low-temperature conditions
10	Arc batholith	Extensive bodies composed of numerous plutons dominated by granodiorite, granite, tonalite, or related rocks; typically I-type, calc-alkaline rocks in post-Archean terranes	Plutonic roots of a magmatic arc
11	Contact metamorphic	Hornfels, skarn, calc-silicate, marble; fabric (e.g., schistosity) weak to absent	Wall rocks of plutons
12	Regional metamorphic	Schist, gneiss, amphibolite, granulite, marble; fabric generally strong; thrust faults, folds, and shear zones indicative of shortening	Collision
12	Ultrahigh-pressure (UHP)	Eclogite and eclogite-facies crustal rocks with $P > \sim 2.7$ GPa, coesite-bearing, sometimes diamond-bearing	Subduction to >90 km
13	Continental rift	Typically bimodal (basalt-rhyolite) volcanic rocks, although andesite and dacite may be abundant; typically mildly to highly alkalic, with modal nepheline and leucite; rift-basin sedimentary rocks; normal faults	Continental extensional province
14	Oceanic hotspot	Basalt, commonly alkalic to highly alkalic	Hotspot in oceanic lithosphere

a given lithosystem are a product of the material available and the controlling variables, including pressure, temperature, heat flow, mass fluxes, and state of stress. The lithosystem concept is somewhat parallel to the biologic concept of an ecosystem, which is a set of organisms interacting with each other and their physical environment in a given area. Later in the book you will learn about diverse lithosystems, recognition of which tells you a lot about the geologic history, geochemical characteristics, and tectonic development of a given area. Lithosystems discussed in this book are listed in Table 1.2.

1.7 A Final Note

Learning Objective

By the end of this section you should be able to:

1.7.1 Understand that not everything you read in a textbook need be true.

Petrology is a young science; like all of geology, it was revolutionized by the understanding of Earth processes afforded by plate tectonics and new analytic methods. For example, when your authors were students, textbooks said that fractional crystallization of a parent, low-silica, basalt magma was how more silicic igneous rocks like andesite and rhyolite formed. In the 1970s, however, isotope studies blew up that idea and showed the importance of magma mixing and partial melting of the crust in producing silicic rocks. New data led to new models and interpretations. Similarly, for the past couple centuries, textbooks have depicted bodies of granite and other plutonic rocks as the frozen corpses of giant magma chambers. However, precise geochronology in the 2000s blew up that idea, too, and many large bodies of granite are now known to have accumulated in small increments over millions of years.

We have done our best to present up-to-date material in this book, and to say "we don't know" when we still don't know something. It will undoubtedly turn out, though, that some of the concepts or interpretations we present will someday be shown to be wrong. Question everything. Collect new data. Develop new ideas. Solve those unsolved problems. That's what makes science fun.

Summary

Key points of this chapter include the following:

- The Earth differs markedly from other planets, at least in part owing to plate tectonics and interactions of petrologic processes with other Earth systems.
- Bivariate (*x*-*y*) diagrams are commonly used to classify, name, and discriminate among common igneous rocks.
- Igneous rocks generally show strong correlations between compositional parameters, whereas sedimentary and metamorphic rocks are generally far more compositionally diverse.
- Igneous and metamorphic rocks are not randomly distributed either in geographic or compositional space but show distinct clusters or correlations between rock type and tectonic setting.
- The Wilson cycle (rifting of ocean basins, subduction of oceanic crust, and collision of continents, followed by rifting . . .) is the logical consequence of plate movements on a sphere.
- Plate tectonics might not have existed on early Earth.
- Rock and mineral names and mineral compositions are numerous and complex; it is important to learn the common names and compositions, and to not get lost in all the highly specific ones.
- Time in this book is expressed in multiples of "yr" (years in duration) and "a" (years ago).
- Lithosystems represent distinctive rock assemblages that are characteristic of important geologic environments. We investigate certain ancient lithosystems to understand their corresponding geologic environment.
- Although most material in textbooks is reliable, especially observations, equations, and data, interpretations should be viewed as working hypotheses, always subject to testing and replacement. Advances in science will eventually disprove some concepts.

Note

1. If we define "igneous rocks" as those crystallized from a melt, then cryovolcanoes that erupt nitrogen or methane, such as those on Neptune's moon Triton, and sulfur volcanoes, such as those on Jupiter's moon Io, qualify as igneous features.

CHAPTER 2

Raw Materials: Formation of the Earth

This book is largely concerned with how the materials that make up the Earth behave when processed through the plate tectonic system, and thus the materials that make up the planet are key. Earth's crust is particularly rich in O, Si, Al, Fe, Ca, Na, Mg, K, Ti, and H, and so the crust is composed largely of minerals that incorporate these elements. You can only make a planet out of the materials you had to start with, and this chapter is concerned with how Earth got its share of the elements, and with their gross distribution.

2.1 Formation of the Solar System and Earth

Learning Objectives

By the end of this section you should be able to:

2.1.1 Explain how gravitational contraction of a nebula can ignite fusion in the nebular cloud.

2.1.2 Sketch the giant-impact hypothesis for the origin of the Moon and explain how this accounts for the tilt of the Earth's axis, the relative core sizes of the Earth and Moon, and the similarity of their isotopic compositions.

How the solar system formed is one of the grand tales in science and is fundamental to petrology and geochemistry. What follows is a highly abbreviated version of current thinking. Five billion years ago (5 Ga) our galactic neighborhood (the "uncharted backwaters of the unfashionable end of the Western Spiral arm of the Galaxy" as Douglas Adams so memorably described it in *The Hitchhiker's Guide to the Galaxy*) was undeveloped. Matter in the area, the solar **nebula**, was unorganized gas and dust and consisted mostly of H and He, with a small proportion of heavier elements. Nebulae (from Latin, "mists") are abundant in the night sky, and a particularly beautiful one, visible with the naked eye, occurs in Orion's sword (Figure 2.1). If gravitational conditions are right, given the proper push, nebulae can condense into stars and planets.

Nebula
A cloud of gas and dust in space; gases are predominantly H and He.

Around 4.5 Ga some event caused the solar nebula, which was on the order of one light-year (about 10 trillion km) in diameter, to begin to contract, a process that accelerated because gravitational attraction of nebular matter for itself increases as the matter is packed into smaller and smaller volumes. Nebulae observed from Earth typically have a

The protostar within the dark cloud L1527, in this infrared image by NASA's James Webb Space Telescope, is embedded within a cloud of material feeding its growth. Ejections from the protostar have cleared out cavities above and below it, whose boundaries glow orange and blue. This young star-forming region has not yet begun nuclear fusion. The dark band crossing the bright center of the hourglass shape is an accretion disk—a planet nursery—that is about the size of our solar system. L1527 is in the constellation Taurus.

Sources: NASA, ESA, CSA, and STScI. Image processing: J. DePasquale, A. Pagan, and A. Koekemoer (STScI).

Figure 2.1 The Orion nebula, a nursery for stars, as seen from the Hubble Space Telescope. At 1500 light-years distance, this is the closest star-forming region to Earth. Bright areas contain massive stars that are shaping the nebula with intense ultraviolet light. Young stars that still have protoplanetary disks are present but too small to distinguish here. The image, a composite of 520, covers an area of the sky about the size of the full moon.

Sources: NASA, ESA, M. Robberto (Space Telescope Science Institute/ESA) and the Hubble Space Telescope Orion Treasury Project Team.

bit of spin, and the solar nebula did as well. As it contracted it spun faster and faster to conserve angular momentum. Presumably, the solar nebula was a semispherical blob early in the contraction process, but with further contraction it evolved into a disk, and its spin caused it to spread out perpendicular to the axis of spin into a protoplanetary nebula (Figure 2.2). Without that spin, the matter would have condensed down to a single focus—the Sun—instead of smearing out a little bit to form the planets—and us.

Gravitational forces within such disks—in particular, tidal forces from the central mass concentration—cause them to become unstable and break up into concentrations called **planetesimals**, where self-gravitation is stronger than the disrupting forces. These mass concentrations feed on the nebula, drawing material to themselves. Isotopic dating (discussed later) indicates that formation of planetesimals as big as Mars probably occurred within the first 5 Myr or so of solar system history.

Once planetesimals began to form, they started to affect one another gravitationally. A set of such planetesimals, if set in nice, coplanar, circular orbits, is unstable owing to this mutual gravitation, and the orbits become elliptical, and thus crossing. The orderly, concentric merry-go-round of the nebula shown in Figure 2.2 becomes more of a demolition derby as planetesimals are flung into crossing orbits that lead to collisions. As with colliding vehicles, collisions usually cause parts to go flying off, but some make them stick together, and the present solar system arose out of such a dynamic arrangement. Models of solar system evolution indicate that the giant planets Jupiter and Saturn play particularly important roles in governing the motions of everything orbiting the Sun, occasionally flinging asteroids into the inner solar system and into Earth-crossing orbits.

Planetesimals
Solid objects forming in a nebular disk owing to gravitational forces; essentially baby planets.

Figure 2.2 Remarkable image of a protoplanetary disk surrounding HL Tau, a Sun-like star ~450 light-years from Earth, also in the constellation Taurus. Gaps between dust rings are where protoplanets are sweeping their lanes of dust and gas, growing with every bit they gather. This image, from 2014, was made with the ALMA radio telescope array in Chile; the long separation of the individual radio telescopes involved, up to 15 km, gives this image resolution equivalent to seeing the width of a penny from over 100 km away.

Sources: ALMA (NRAO/ESO/NAOJ); C. Brogan, B. Saxton (NRAO/AUI/NSF).

The later stages of accretion involved collisions of massive objects because by that time most of the nebular material was concentrated into planetesimals whose orbits were continually distorted by the giant planets. It is likely that the Moon formed from the collision of a Mars-sized body with the proto-Earth (Figure 2.3; Canup & Asphaug, 2001; Stevenson, 2008). The Earth-Moon system has a number of characteristics that point to this origin: large angular momentum (a consequence of an off-center impact that set the system to spinning), a particularly large metallic core in the Earth and small core in the Moon, and most significantly, very similar isotopic compositions. In this hypothesis, the giant impactor left much of its core in the Earth, and the vaporized debris, depleted in metal, condensed to form the Moon. Both the Earth and Moon were probably melted

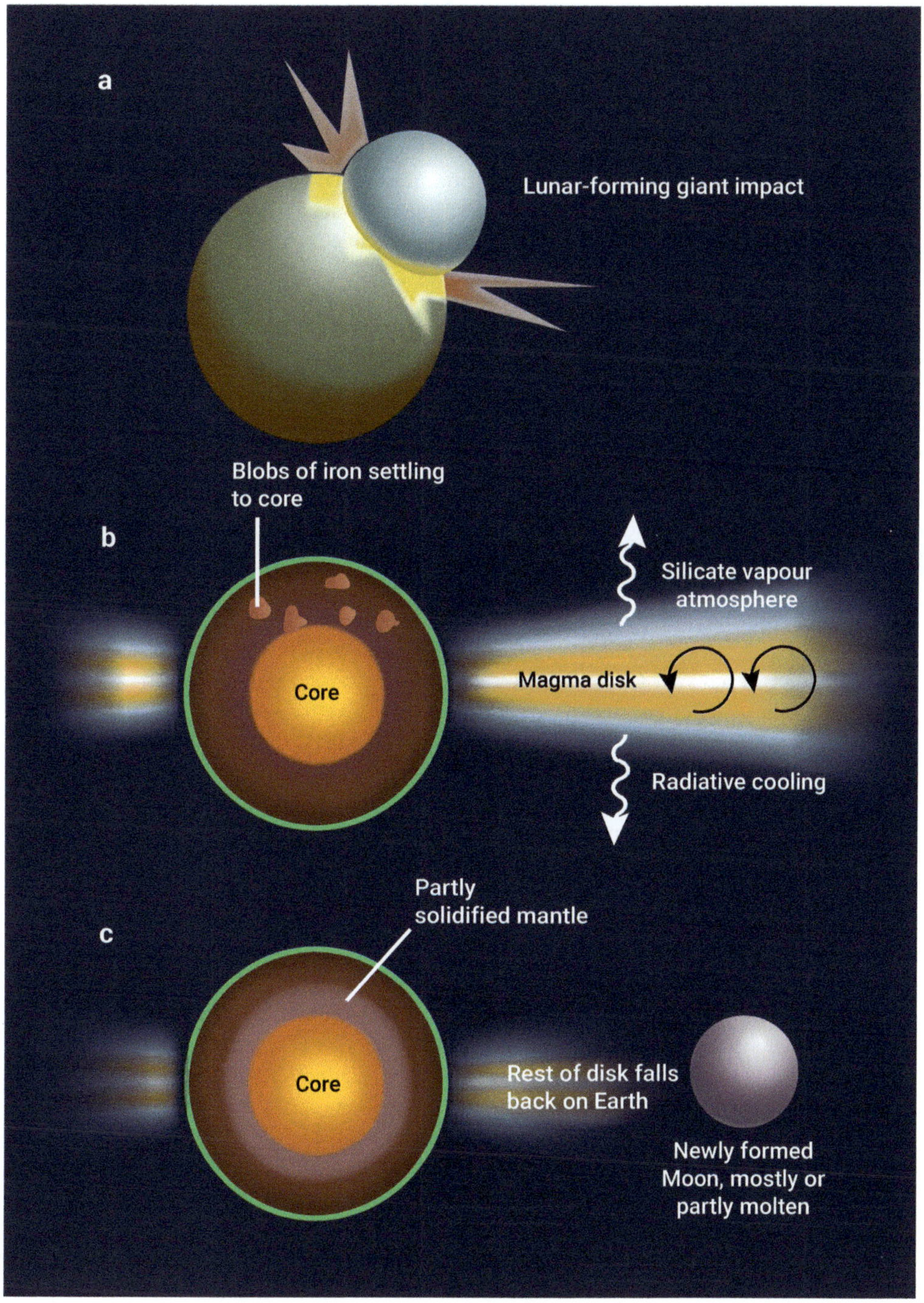

Figure 2.3 One version of the Moon-forming impact hypothesis. (a) A planetesimal with a mass comparable to that of Mars collides with the early Earth. (b) This collision sends vaporized silicates into orbit as a magma disk, with molten iron sinking in the shaken Earth to add to the core. (c) Some disk material condenses to form the Moon, and the rest falls back to Earth. Much of the impactor's core is left in the Earth, leaving a Moon depleted in Fe.

Source: From Stevenson (2008), © 2008, Springer Nature Limited.

completely and degassed during this event, and estimates of the post-impact temperature range up to 4000 °K (Stevenson, 2008). The Earth as we know it arose out of this impact, which knocked it askew, giving it a 23.4° axial tilt, and thus giving us seasons.

The Earth's core formed sometime during the first tens of millions of years. The geochemical balance of the Earth is reducing enough (i.e., is deficient in oxygen enough) that much of its Fe is in metallic form. Metallic, molten Fe is dense, fluid, and immiscible with molten silicates, so any metallic droplets that formed in a molten Earth quickly sank to the core—as if you made salad dressing with mercury and honey rather than oil and water. Thus, within some tens of millions of years after the solar system began, the Earth was a differentiated planet with a metallic core, silicate mantle, and some sort of silicate crust. The surface of this hellish, molten early Earth cooled quickly as it radiated heat into space; calculations indicate that it could have radiated away half its impact energy in as little as 1000 years. Such rapid cooling occurs because the rate of heat loss from a body undergoing black-body radiation is proportional to the fourth power of the Kelvin temperature:

$$Q \propto T^4 \qquad \textbf{(Eq. 2.1)}$$

This relationship, known as the Stefan-Boltzmann law, plays an important role in cooling of lava (see Chapter 7), and rapid radiative cooling of the super-hot outer part of the hypothetical magma disk (see Figure 2.3) led to its coalescence into the Moon.

2.2 Meteorites

Learning Objectives

By the end of this section you should be able to:

2.2.1 List the three main classes of meteorites and their cosmic significance.

2.2.2 Predict what one might look for in a meteorite collection to find the most primitive materials.

Planetary accretion continues today, as demonstrated by meteorite falls. Most of the big material has been swept up, but the Earth gains some tens of millions of kilograms of meteoritic material each year (Love & Brownlee, 1993). This is a non-trivial amount—about a gram per square meter per decade averaged over the Earth's surface. Fortunately for life on Earth, catastrophic large strikes, such as the 10-km object that struck Yucatan and terminated the contract of the dinosaurs, happen rarely. The object that exploded over Chelyabinsk, Russia, in February 2013, causing widespread damage and awe, was estimated to have been about 15 to 20 m in diameter, and the 1908 explosion over Tunguska, Siberia, which liberated energy equivalent to a large hydrogen bomb, was caused by an object 20 to 60 m in diameter.

Meteoroids, meteors, meteorites
Respectively, pieces of interplanetary matter, the incandescent streaks they leave in the sky, and the debris that hits a planet if the meteoroid survives passage through the atmosphere.

Meteoroids are objects that leave behind the glowing trails we see in the sky (**meteors**). They become **meteorites** if they survive passage through the atmosphere to strike the Earth. Collection and classification of thousands of meteorites demonstrate that most are quite unlike Earth rocks—they are literally out of this world, and our only tangible links to the earliest history of the solar system. Table 2.1 gives a simplified classification scheme for meteorites.

The common concept of a meteorite is of a dense, magnetic, lumpy hunk of iron. Only a small proportion of such objects brought to meteorite experts turn out to be meteorites—the rest are meteor-wrongs. However, iron-rich meteorites, although well-represented in the world's collections, make up less than 10 percent of observed meteorite falls. Most meteorites that were observed to fall are less-impressive, dark-brown stony meteorites.

Chondrites
Primitive meteorites composed of relatively unprocessed material from the solar nebula.

Chondrules
Formerly molten, spherical globules of olivine and pyroxene in chondritic meteorites.

These unassuming meteorites, **chondrites**, are the most primitive rocks known, samples of broken planetesimals that orbited the Sun and avoided being accreted again until they struck the Earth. They are defined by **chondrules**, spherical globules of once-molten

Table 2.1 Simplified Classification of Meteorites

Class	Composition	Distinguishing Characteristics	Density (kg/m^3)	Abundance among Falls (%)
Chondrite	Olivine, pyroxene, Fe-Ni alloy	Chondrules; some contain organic compounds	3400–3700; organic-rich much less	86
Achondrite	Basaltic to ultramafic	Stony but without chondrules	3000–3300	8
Iron	Fe-Ni alloy	Magnetic, dense	6900–7800	5
Stony-iron	Olivine and Fe-Ni alloy	Contain abundant Fe-Ni alloy as well as silicates	4200–5000	1

material that are now largely olivine and pyroxene. Chondrites are essentially cosmic sediment made up of a number of diverse components including chondrules, fine matrix materials (cosmic dust bunnies), and volatile compounds such as water (Figure 2.4). The texture of chondritic meteorites is quite unlike that of Earth rocks, and a quick look at a thin section is enough to tell you that you're looking at a chondrite. Chondrules do not stand out in **hand sample**, though, and most chondrites are ordinary dark-brown, smooth-surfaced rocks that could be mistaken for any number of Earth rocks.

Hand sample
Said of a rock specimen that fits conveniently in one's hand for close examination.

Aphelion, perihelion
Points in an orbit farthest and closest, respectively, to the body around which orbiting takes place; plurals end in "ia."

The matrix of many chondrites contains grains that are older than the solar system—literally dust from stars that seeded the solar nebula. These are typically minuscule (<1 mm) grains of refractory (difficult to melt) compounds such as SiC, diamond, graphite, and Al_2O_3, and are recognized as older than the Sun by their unusual isotopic compositions. Earth was homogenized by the melting that happened during giant impacts, but these tiny bits of material were not, and thus are tangible links to what was here before the solar system.

A number of chondritic meteorite falls have been caught on photographic networks with enough precision to calculate rough orbits. These orbits are all highly elliptical, with an **aphelion** (point farthest from the Sun; point closest to the Sun is **perihelion**) in the vicinity of the asteroid belt. This indicates that most are from broken planetesimals in the asteroid belt rather than just small bits that wandered around, never accreted into anything larger until they hit Earth.

Additional evidence that many meteorites were derived from larger planetesimals is that a significant number were heated enough to show thermal metamorphism or melting (Figure 2.5). Such thermal events would not happen in isolated little bits of material. Chondritic meteorites show a wide range of alteration and metamorphism, and this is interpreted to reflect exposure to water and heat on their parent planetesimal. The least-affected chondrites have pristine minerals (e.g., olivine and pyroxene) with a wide range of compositions that indicate assembly from various bits and pieces of the solar nebula. Many show effects of aqueous alteration in which primary minerals were converted to hydrous ones (e.g., olivine altered to serpentine). Others show effects of thermal metamorphism, which causes mineral compositions to homogenize and greatly affects the texture. Chondrules are most distinct in meteorites with the least thermal metamorphism and become increasingly difficult to find as the grade of metamorphism increases (Figures 2.4 and 2.5).

As with rocks on Earth, thermal metamorphism may culminate in melting. Meteorites that were produced by partial melting of the parent planetesimal are called achondrites

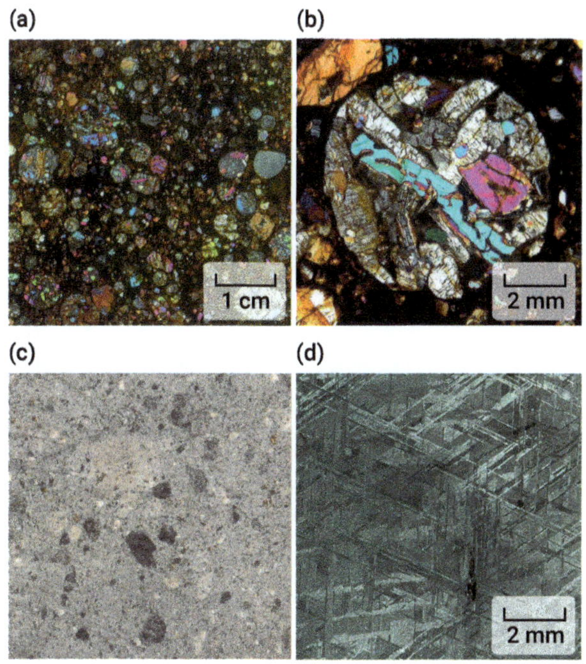

Figure 2.4 Meteorites. (a) Brownfield chondrite thin section in crossed polars. (b) Closer view of olivine-pyroxene chondrule from same. (c) Polished slab of achondrite PIA15137. This specimen is an impact breccia thought to be derived from asteroid Vesta. Width of field several centimeters. (d) Iron meteorite Gibeon showing exsolution of Fe-Ni alloys (Widmanstätten structure). Iron meteorites are pieces of planetary cores.

Sources: (a, b) jm-derochette.be; (c) NASA/JPL-Caltech/Hap McSween (University of Tennessee), and Andrew Beck and Tim McCoy (Smithsonian Institution); (d) Randy L. Korotev.

Figure 2.5 Thin section view (crossed polars) of a thermally metamorphosed chondrite found in the Sahara Desert (97095). Chondrules are significantly less apparent than those in the unmetamorphosed chondrite in Figure 2.4a. Metamorphism causes these textural changes and homogenizes mineral compositions. Width of field 8 cm.

Source: jm-derochette.be.

because they lack chondrules (see Figure 2.4c). Many are essentially basalts, and we now know that a large number of those in the world's collections probably came from the asteroid Vesta (McSween et al., 2014). Melting and formation of basalt is a fundamental process in the evolution of the terrestrial planets and is an example of differentiation of an originally homogeneous mass into multiple parts. How and why differentiation occurs is a theme we shall return to many times.

The most recognizably meteoritic of the meteorite classes are the irons (see Figure 2.4d). These are further evidence for planetesimal differentiation, as they are interpreted to be the Fe-rich cores of differentiated planetesimals—smaller versions of the Earth. A particularly beautiful class of meteorites is the pallasites, which consist of roughly equal parts Fe-Ni alloy and green olivine. These may be samples from near the planetesimal's core-mantle boundary, or mixtures of core and mantle formed by impact (Scott, 2007).

Metamorphism and differentiation of meteorites is so pervasive that any consistent theory of meteorite origins must account for these processes. What could heat up cold dust and rock enough to melt it? To understand the source of this thermal energy, we must delve into isotope geology.

2.3 Age of the Earth

Learning Objectives

By the end of this section you should be able to:

2.3.1 Critique the assumption that the age of the Earth is the same as the age of the solar system.

2.3.2 Demonstrate how to calculate an age using an isochron diagram.

The age of the Earth is generally taken as equal to the age of the solar system. The original determination of around 4.5 Ga was determined using the decay of U to Pb (Patterson et al., 1955). This is a complicated argument, so we will examine an alternative method that uses a different isotopic clock.

Radiogenic
Produced by radioactive decay.

Nuclide
A specific isotopic form of an element, defined by the numbers of protons and neutrons in the nucleus.

The rubidium-strontium (Rb-Sr) system illustrates how **radiogenic** isotopic systems work, and once you understand it you can understand any other radiogenic system. Sr, element 38, has four stable isotopes: ^{84}Sr, ^{86}Sr, ^{87}Sr, and ^{88}Sr (Figure 2.6). Each of these is a distinct **nuclide**—an atom defined by a given number of protons and neutrons; different forms of the same element (defined by the number of protons in the nucleus) are different isotopes of that element. Isotopes of Sr with other masses (e.g., ^{85}Sr) exist, but they decay so rapidly that any formed early in the solar system's history have long since disappeared. Rb, element 37, has only one stable isotope, ^{85}Rb, but ^{87}Rb decays so slowly that only a small fraction of the ^{87}Rb that was present at the solar system's birth has decayed to Sr by the reaction:

$$^{87}_{37}\mathrm{Rb} \rightarrow {}^{87}_{38}\mathrm{Sr} + \beta^- + \text{energy} \qquad \textbf{(Eq. 2.2)}$$

One can figure out the nuclear decay scheme from atomic numbers and weights. In this case, the atomic weights are the same, but the atomic number increases by one, so the process was beta decay, wherein a beta particle (β^-, a high-energy electron) is ejected from a neutron in the nucleus, turning it into a proton.

^{83}Sr	^{84}Sr	^{85}Sr	^{86}Sr	^{87}Sr	^{88}Sr	^{89}Sr	^{90}Sr	^{91}Sr	^{92}Sr
5 s	stable	68 m	stable	stable	stable	51 d	29 y	9.6 h	2.7 h
^{82}Rb	^{83}Rb	^{84}Rb	^{85}Rb	^{86}Rb	^{87}Rb	^{88}Rb	^{89}Rb	^{90}Rb	^{91}Rb
1.3 m	86 d	33 d	stable	19 d	50 Gyr	18 m	15 m	158 s	58 s

Figure 2.6 A piece of the chart of the nuclides, a compilation of data about isotopic abundances, radioactive decay schemes, and other information relevant to isotope geochemistry. Half-lives are given in seconds (s), minutes (m), hours (h), or days (d) except ^{87}Rb, with a half-life of 50 Gyr. This chart is arranged such that a given element makes up a row, and therefore has a constant atomic number *Z*; columns are constant neutron number *N*; and negative-sloping diagonals are constant atomic weight ($Z + N$). All the isotopes of Rb and Sr in this figure, and many more, were present at the birth of the solar system, but only six remain, the rest having long since decayed away. Of these six, all are stable except ^{87}Rb, which has such a long half-life that only a small percentage of the Earth's original stockpile has decayed away in the last 4.6 Gyr. ^{87}Rb decays to ^{87}Sr by β-decay (arrow), which decreases *N* by 1 and increases *Z* by 1.

Source: Brookhaven National Laboratory, National Nuclear Data Center; information extracted from the NuDat database (https://www.nndc.bnl.gov/nudat/).

Radioactive decay is a statistical process defined by the half-life, which is the time it takes half of a population of atoms of a given radioactive isotope to decay. The half-life of ^{87}Rb ($t_{1/2}$) is quite long, roughly 10 times the age of the Earth (50 Gyr). Radioactive atoms are unstable; you might think of them as vibrating blobs that can, randomly, vibrate strongly enough that a piece breaks off. The shorter the half-life the more violently a given nuclide "vibrates."

Because radioactive decay is a statistical process, we can't select a given atom and predict when that particular one is going to decay, but we can make confident predictions about large populations of such atoms. In such a process, the number of parent ^{87}Rb atoms decreases exponentially, asymptotically approaching zero as the half-lives pile up. The number of child ^{87}Sr atoms changes with time according to the decay equation:

$$^{87}\mathrm{Sr} = {}^{87}\mathrm{Sr}_0 + {}^{87}\mathrm{Rb}(e^{\lambda t} - 1) \qquad \textbf{(Eq. 2.3)}$$

where $^{87}\mathrm{Sr}_0$ is the number of initial ^{87}Sr atoms and λ, the decay constant, is

$$\lambda = \frac{\ln(2)}{t_{1/2}} \qquad \textbf{(Eq. 2.4)}$$

The reciprocal of the decay constant has a useful meaning: it is the average lifetime of an atom. The number of parent atoms at time *t* is given by

$$^{87}\mathrm{Rb}(t) = {}^{87}\mathrm{Rb}_0 e^{-\lambda t} \qquad \textbf{(Eq. 2.5)}$$

Thus, in the Earth's lifetime of 4.5 Ga, around 6 percent of its ^{87}Rb inventory has decayed, adding ^{87}Sr to its original stockpile of Sr.

It is difficult to determine the actual numbers of atoms of these isotopes in a given sample, so the decay equation is rendered into a more usable form by dividing by a non-radiogenic isotope whose amount is not changing. For this system, ^{86}Sr is the isotope of choice, although ^{84}Sr or ^{88}Sr would do:

$$\frac{^{87}\mathrm{Sr}}{^{86}\mathrm{Sr}} = \left(\frac{^{87}\mathrm{Sr}}{^{86}\mathrm{Sr}}\right)_0 + \frac{^{87}\mathrm{Rb}}{^{86}\mathrm{Sr}}\left(e^{\lambda t} - 1\right) \qquad \textbf{(Eq. 2.6)}$$

This is the form of the decay equation used in practice, and corresponding equations can be written for other radiogenic isotope systems (e.g., Sm-Nd or U-Pb). Thermal ionization

Table 2.2 Half-Lives and Decay Constants of Several Isotopic Systems of Use in Geochemistry

Parent	Child	Half-life $t_{1/2}$ (yr)	Decay Constant λ (1/yr)	Reference
^{87}Rb	^{87}Sr	4.962×10^{10}	1.397×10^{-11}	Rotenberg et al. (2012)
^{40}Ar	^{40}Ar, ^{40}Ca*	1.248×10^{9}	5.554×10^{-10}	NuDat 3.0, https://www.nndc.bnl.gov/nudat3/
^{238}U	^{206}Pb**	4.468×10^{9}	1.551×10^{-10}	NuDat 3.0
^{235}U	^{207}Pb**	7.040×10^{8}	9.846×10^{-10}	NuDat 3.0
^{232}Th	^{208}Pb**	1.400×10^{10}	4.951×10^{-11}	NuDat 3.0
^{147}Sm	^{143}Nd	1.060×10^{11}	6.539×10^{-12}	NuDat 3.0

* Branching decay
** Decay chains, of which Pb isotopes are the stable endpoints

Isochron
A linear array on an isotope diagram whose slope defines an age. The ordinate is the radiogenic isotope and the abscissa the parent isotope, both normalized to a stable isotope of the radiogenic element.

mass spectrometers can measure these isotope ratios with exceedingly high precision (five or six significant figures). This equation is fundamental to isotope geochemistry and should be committed to memory. Decay constants for several systems of geologic interest are given in Table 2.2.

Isotope decay can be used to estimate the timing of geologic events in a number of ways. The simplest is the **isochron** method, described in Toolbox 2.1. Minster and Allègre (1979) used the Rb-Sr isochron method (Figure 2.7) to determine the age of Tieschitz, a primitive chondrite that fell in eastern Europe in 1878. They determined precise ^{87}Sr/^{86}Sr and ^{87}Rb/^{86}Sr ratios for three different whole-meteorite fragments, two chondrules, and olivine and pyroxene crystals and found that they plot on a linear trend—an isochron—whose slope indicates an age of ~4.59 Ga (see Figure 2.7). Given the assumptions of the isochron method—in particular, that all the materials started out with the same ^{87}Sr/^{86}Sr and have been chemically unmodified since origin—this age is an estimate of the age of the solar system.

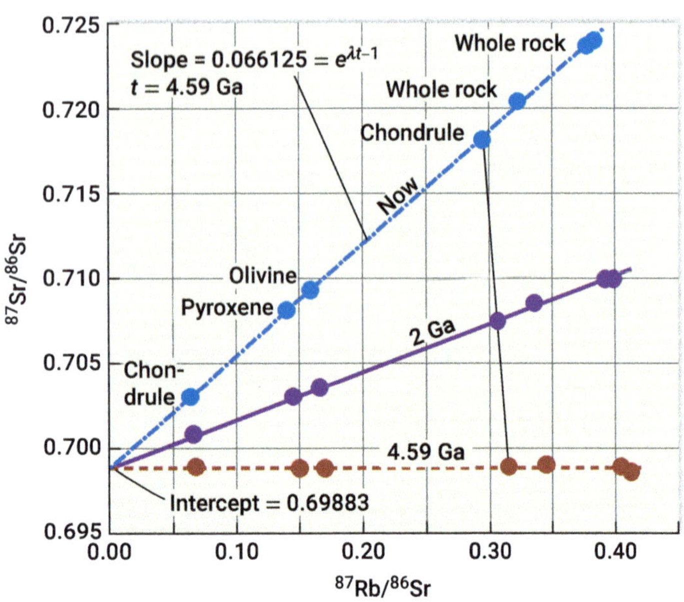

Figure 2.7 Isochron produced by analyzing components of the Tieschitz chondrite. The slope gives an estimate of the age of the solar system, and the intercept its initial ^{87}Sr/^{86}Sr ratio. Also shown are the predicted positions of the points 2 Gyr after formation of the solar system and at 4.59 Ga, the time of solar system formation predicted by this analysis. If points fall on an isochron, then when they are corrected back to the age of formation they will fall on a horizontal line (all the components have the same initial ^{87}Sr/^{86}Sr). As time progresses the points follow steep negative trajectories (solid arrow illustrates this for one chondrule) as ^{87}Sr/^{86}Sr increases and ^{87}Rb/^{86}Sr decreases.

Source: Data from Minster & Allègre (1979).

The initial ratio from this isochron is 0.699, a value that is taken as the original ^{87}Sr/^{86}Sr ratio of the solar system and the Earth. This value can only remain constant or increase with time (because the amount of ^{86}Sr is fixed, but ^{87}Sr is increasing owing to decay of ^{87}Rb), and therefore, if correct, nothing on Earth should have a ratio < 0.699. Isolated bits of pre-solar grains in undifferentiated, unmetamorphosed meteorites can, but the mixed Earth should not. This fact underlies the plot of an unproduced (well, okay, unwritten) *X-Files* script by the authors that starts out with the murder of an isotope geochemist by an alien after the geochemist finds material at a crash site with ^{87}Sr/^{86}Sr = 0.5 (cue eerie music). Not of this world!

The accepted age for initial accretion of the solar system is currently 4.567 Ga. A remarkably detailed account of its early history has been worked out with the aid of short-lived isotopes that are no longer found on Earth (Figure 2.8).

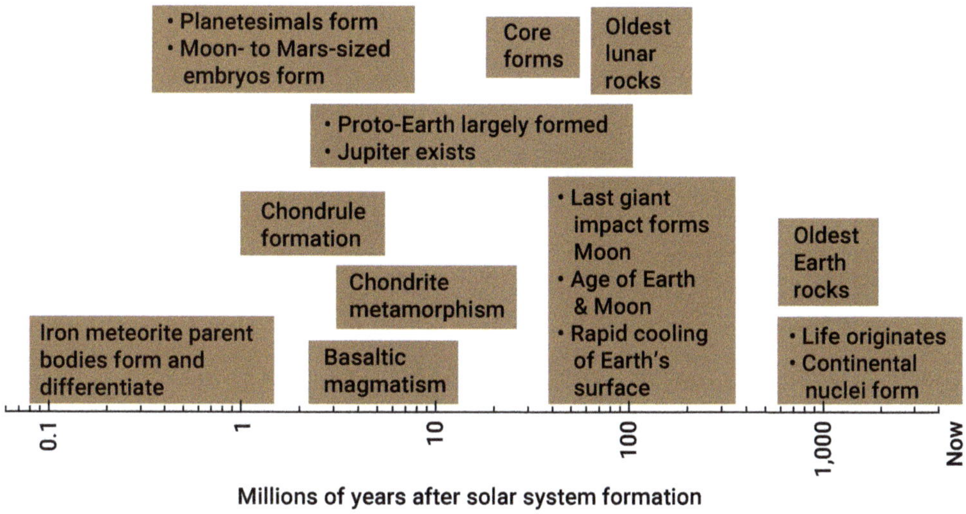

Figure 2.8 Events early in the solar system's history, plotted on a logarithmic timescale.

Sources: Data from Stevenson (2008) and Kleine & Rudge (2011).

Toolbox 2.1

Isochrons and Geologic Time

Given the right circumstances, the concentrations of parent and child isotopes can be used to determine an age that has geologic significance, and the Rb-Sr system has long been employed in this way. The decay equation for the Rb-Sr system is of the form

$$y = a + bx \quad \textbf{(Eq. 2.7)}$$

where a is the initial Sr-isotope ratio $(^{87}Sr/^{86}Sr)_0$, b is $e^{\lambda t} - 1$, and y and x are the measured ratios $^{87}Sr/^{86}Sr$ and $^{87}Rb/^{86}Sr$. This is one equation in two unknowns, and with measurements of at least two samples we can use it to determine time.

Consider a volcanic rock with crystals of plagioclase, sanidine, and hornblende in a glassy or fine-grained matrix. With several assumptions, we can use the Rb-Sr system to determine when the magma that formed this rock was erupted. Important assumptions are that all the phases in the rock had the same $^{87}Sr/^{86}Sr$ at "time zero" (in this case, when the rock was erupted), and that the system has been chemically undisturbed since that time. In particular, it is critical that the concentrations of Rb and Sr are the same now as they were then, and that there has been no exchange of Sr with an external source—a hydrothermal fluid, for example—that could modify $^{87}Sr/^{86}Sr$.

The first assumption relies on a characteristic of so-called heavy isotope systems such as Sr: at high temperatures all the separate parts of a magma will have the same $^{87}Sr/^{86}Sr$, because the isotopes are chemically so similar that minerals and melts can't tell them apart—a non-discrimination policy is in effect, and if a crystal wants an atom of Sr, either isotope will do equally well. Thus, before eruption, the crystals and melt all had the same $^{87}Sr/^{86}Sr$, but different values of Rb/Sr and thus $^{87}Rb/^{86}Sr$, because the various phases in the magma *do* discriminate between Rb and Sr. Plagioclase prefers Sr (which is geochemically like Ca) more than Rb (which is like K), whereas sanidine has the opposite preferences.

Table 2.3 Rb- and Sr-Isotope Data

Mineral	$^{87}Rb/^{86}Sr$	$^{87}Sr/^{86}Sr$
Plagioclase	0.05255	0.70407
Sanidine	1.8196	0.70638
Hornblende	2.8900	0.70778

Table 2.3 shows measured values from our rock. There is a wide range in $^{87}Sr/^{86}Sr$, which is helpful—the wider the spread, the more precise the age determination.

Plotting these values produces an isochron (Figure 2.9), a line that tells us two important things: time since the isotopic clock was started and the initial $^{87}Sr/^{86}Sr$ of the sample. Time can be determined because the slope is $e^{\lambda t} - 1$, a dimensionless number that can be converted to time using the relevant decay constant.

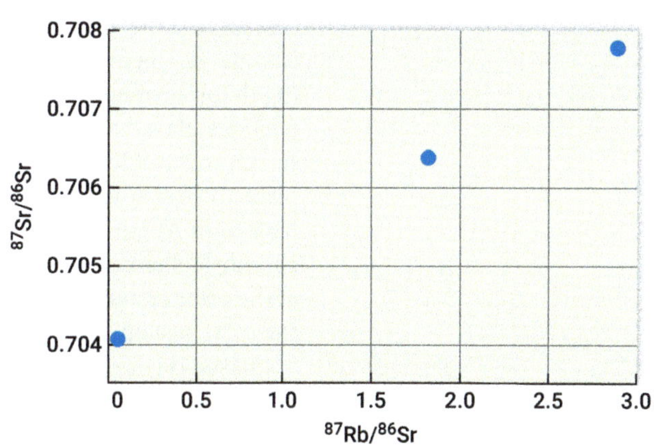

Figure 2.9 Sample data on an isochron plot. Using the ^{87}Rb decay constant in Table 2.2, what age and initial $^{87}Sr/^{86}Sr$ would you estimate for this sample?

2.4 Sources of Heat

Learning Objectives

By the end of this section you should be able to:

2.4.1 Explain why nuclides such as ^{26}Al were important sources of heat early in Earth history, but are no longer.

2.4.2 Deduce how the Earth would be different if an isotope of Fe were a significant source of radiogenic heat.

Heat flow
Flux of heat through a surface (e.g., surface of the Earth or wall of a dike), in units of energy per unit time per unit area (e.g., W/m^2).

Core infall
Postulated event during Earth formation in which immiscible globules of metallic Fe-Ni solution sank to form the core.

Radionuclides
Radioactive nuclides.

Internal heat drives plate tectonics, magmatism, and metamorphism. Average **heat flow** at the Earth's surface is around 50 mW/m^2. That's not very much. An area of 1 m^2 provides about enough energy to light up a small LED; if you could capture the energy coming up through an area the size of a football field, it would amount to a few hundred watts, or enough to power a box fan. But the Earth is big, and the total surface heat flow of the Earth is roughly equivalent to the output of 10,000 nuclear power plants. This heat arises from three main sources: heat generated by impacts, **core infall** (conversion of potential energy to heat that occurred as Fe sank to form the core), and radioactive decay. Radioactive heat today arises primarily from decay of ^{40}K, ^{238}U, ^{235}U, and ^{232}Th. Other radioisotopes, such as ^{87}Rb and ^{147}Sm, are so rare, so weak, or both, that they do not contribute a significant amount.

One might think that the primordial heat of accretion and core infall would have long since radiated away into space, because 4.5 Ga is a long time. However, rocks are terrible heat conductors—so terrible that if one were to perturb the temperature at the Earth's center it would take about a trillion years for that disturbance to propagate to the surface by conduction alone. Of course, there is more than conduction at work—in particular, convection in the liquid outer core and in the mantle—but various lines of evidence indicate that about half of the heat flowing out of the Earth, and thus driving plate tectonics and volcanism, is primordial heat from infall and accretion.

Early in the solar system's history, though, things were different. Primitive meteorites contain isotopic evidence for extinct radioactive nuclides (**radionuclides**)—isotopes that were present early in the solar system's history but have since decayed away. The most important of these is ^{26}Al, an isotope of aluminum (atomic number $Z = 13$) with 13 neutrons. All aluminum on Earth is ^{27}Al, which is stable; ^{26}Al is radioactive and decays to ^{26}Mg with a half-life of about 717 kyr, and so any ^{26}Al that was present in the solar system at its birth decayed away, with release of significant amounts of heat, during the first several million years.

^{26}Al and other exotic short-lived nuclides such as ^{129}I (which decays to ^{129}Xe) are synthesized in and around massive stars, and their former presence in chondrites means that they were somehow spread throughout the solar nebula during its formation. In the 1970s, isotope geochemists searched for evidence of ^{26}Al, the importance of which lay in testing theories of how the elements formed, how the solar system began, and what role such nuclides played as a heat source in the early solar system. ^{26}Al decays to ^{26}Mg, and so the search was not for the ^{26}Al itself, long since decayed, but for this decay product. A natural place to search was in feldspars and Ca-Al-rich inclusions in chondrites, both of which have high Al but low Mg concentrations. Analysis of Ca-Al-rich inclusions in the Allende chondrite showed marked enrichment in ^{26}Mg, verifying the presence of significant ^{26}Al at the solar system's birth (Lee et al., 1976). How it got there is a key to understanding how the solar system formed.

Based on current understanding of the nuclear physics of stars (Iliadis, 2015), ^{26}Al can only be produced in stellar interiors. The presence of this nuclide in the material from which the solar system formed indicates that a star must have spilled its guts into the solar nebula at this time. Given the short half-life of ^{26}Al, this event must have happened within a few million years of formation of the solar system. But how?

2.5 Supernovas, Kilonovas, and the Elements

Learning Objectives

By the end of this section you should be able to:

2.5.1 Understand the significance of decay products of ^{26}Al and other short-lived radionuclides in chondritic meteorites.

2.5.2 Summarize where heavy elements (those above Fe) in the periodic table were synthesized.

Stellar Evolution

One possible source of ^{26}Al was a **supernova** explosion. Although less than 1 percent of stars are massive enough to end as supernovas, they play a super role in forming and dispersing the heavier elements. The Sun is powered by fusion of H to form He in its core (fusion is commonly referred to as "burning," although it has nothing to do with combustion in the presence of oxygen). As it does so, an "ash" of denser He builds up in the core. Eventually, this buildup of inert He ash causes things to get ugly. Enough He will build up in the core to force H burning into a shell that surrounds the inert He core, and the Sun will expand into a **red giant** star, like Aldebaran, the bright orange star that forms the eye of Taurus the Bull. The Sun will enlarge enough to engulf Mercury, and the Earth will be burned down to a cinder or vaporized completely. Eventually, the temperature at the core becomes high enough to fuse He into C and O. The Sun is not large enough to maintain the temperatures needed to burn these elements, so it runs out of He fuel and contracts. At the end, it will fling off about half its mass into an expanding nebula, leaving behind a **white dwarf** star. Fortunately for us, this is several billion years in the future.

A fraction of a percent of stars, those at least about 10 times the mass of the Sun, end their lives as supernovas (specifically, Type II supernovas). As they evolve, they synthesize (as ash) and then burn (as fuel) successively heavier elements, building up shells from surface to core of H, He, C, Ne, O, Si, and Fe. At the incredibly high temperatures ($>10^9$ °C) and densities ($>10^{15}$ kg/m^3) in the star's core, the nuclear soup arranges itself into the most energetically favorable configuration, centered around Fe—the **iron peak elements**. These elements have the highest nuclear binding energy, are the most stable, and are the ultimate ash of nuclear fusion. A kind of nuclear slag, they pile up in the massive star's core. When the lighter elements have been consumed and there is not enough thermal energy to resist gravitational collapse, the final act begins. A critical point comes where the iron core collapses down to a tiny **neutron star** only tens of kilometers across. Matter contracts until it is so dense that the short-range, repulsive part of the strong nuclear force takes over and the core rebounds, blowing outward at tens of thousands of kilometers per second. For a few months this explosion can be as bright as a hundred billion Suns, outshining its entire galaxy. This is a supernova (Figure 2.10), and by this process elements such as oxygen were synthesized and then blown out into the universe to be incorporated into nebulae, where they could then be incorporated into new stars, planets, ecosystems, and you.

Many of the universe's heavy nuclides were made by the **r-process** (r for rapid), which involves machine-gunning relatively heavy seed nuclei such as ^{56}Fe with neutrons so fast that they pump up into much heavier, radioactive nuclides with short half-lives. The key to this process is adding neutrons faster than the unstable nuclides decay. Once the neutron flux stops, these unstable nuclides decay back to stable ones. Some heavy nuclides were made by a slower process, the **s-process**, in which nuclides heavier than Fe that were formed in preexisting stars were converted to even heavier ones.

The presence of decay products from nuclides such as ^{26}Al in chondrites is strong evidence that nuclides synthesized in stellar interiors were present at the birth of the solar

Supernova
A massive star that has undergone explosive death, either by the onset of carbon burning and runaway fusion in a white dwarf (Type I) or by catastrophic collapse and rebound of the depleted core of a massive star (Type II).

Red giant
A giant star with a relatively low-temperature outer shell that glows red.

White dwarf
Very dense stellar core remnant; may have the mass of the Sun in a volume comparable to the Earth.

Iron peak elements
The quintet of elements centered on Fe (Cr, Mn, Fe, Co, and Ni); these elements have the highest nuclear binding energy and thus the most stable nuclear configurations.

Neutron star
Dense ball of neutrons left after a supernova explosion; may have a mass comparable to the Sun in a volume 20 km across.

r-process, s-process
Processes that inject neutrons into nuclei; rapid (r) occurs during supernova explosions or neutron star collisions, whereas slow (s) occurs mainly late in the life of low-mass stars such as the Sun.

Figure 2.10 Supernova remnant Cassiopeia A, the remains of a Type II supernova 11,000 light-years from Earth. Light from this detonation probably first reached Earth around 1690, and the expanding shell is now 5–10 light-years across.

Source: NASA/JPL-Caltech/O. Krause (Steward Observatory).

system—outside of their parent stars rather than inside. For decades the preferred interpretation of these observations was that collapse of the solar nebula was precipitated by a nearby supernova detonation that (1) seeded the nebula with ^{26}Al and other short-lived nuclides and (2) triggered its condensation and organization into the solar system. Shock waves from such events travel more rapidly through thinly dispersed parts of a nebula than through denser parts, so they tend to wrap around and slam into the denser parts. This compression can be enough to trigger gravitational collapse.

Recent results from observational astronomy and theoretical physics, however, indicate that the simple supernova explanation is unlikely to be correct. Supernovas make a number of other important radioactive nuclides in addition to ^{26}Al, including ^{60}Fe, which decays to ^{60}Ni with a half-life of 2.6 Myr. Based on calculations of nuclide abundances during supernova evolution, the abundance of ^{60}Fe in chondrites (inferred from its decay product ^{60}Ni) should be far greater than that of ^{26}Al if both were derived from a supernova, but the opposite is observed. In addition, before they explode as supernovas, massive stars carve out large, ionized regions around themselves in which the gas density is too low for star formation.

A number of other scenarios have been proposed for seeding the solar nebula with ^{26}Al in excess of ^{60}Fe. A key process is ejection of ^{26}Al from the interior of a massive star by intense stellar wind (Gounelle & Meynet, 2012). These scenarios are consistent with observations of star-forming regions. If you're in the Northern Hemisphere some clear winter evening, take a look at the constellation Orion. The middle "star" of his sword looks a bit fuzzy, especially with binoculars, because it is a star nursery—a nebula of gas, dust, and thousands of stars in various stages of formation (see Figure 2.1). Some of the stars are large, 10 or more times as massive as the Sun, and some are much smaller. Like rock stars only more so, big stars live life hard and burn out fast. The Sun should last around 12 Gyr, whereas giant stars such as Rigel and Betelgeuse (the lower-right and upper-left corners of Orion, respectively) might last only 10 Myr. Thus, in a nebula populated with diverse stars, large ones will go through their life cycles quickly, turning into supernovas, blowing heavy elements into the nebula, and perhaps triggering collapse of parts of the nebula to form new star systems.

Recent remarkable observations confirm that supernovas may not go quietly into the night once they blow off much of their mass and settle down as a neutron star. The majority of stars in the universe are in binary systems, and when both stars are massive, they end up as a binary neutron star system once both have gone supernova. Astronomers had predicted that in such a system the neutron stars will spiral into one

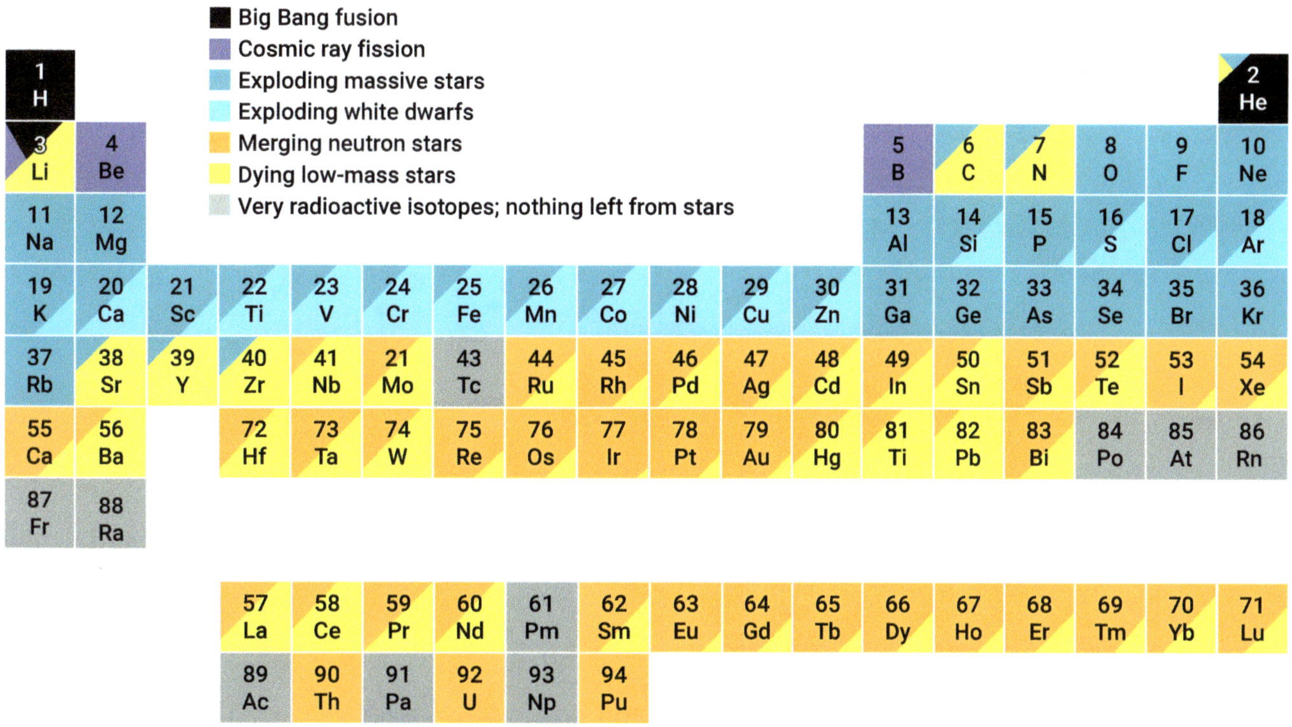

Figure 2.11 Periodic table of the elements showing processes by which they were synthesized. Note that most of the O in the universe was likely produced during supernova explosions.

Source: Jennifer A. Johnson.

another, merging and producing an event dubbed a **kilonova**, and in 2017 an event that matched theoretical predictions was observed. This remarkable occurrence was observed by the LIGO gravitational-wave observatory as well as at optical, X-ray, and gamma-ray wavelengths. Of particular geochemical interest is the observation that massive amounts of heavy elements were synthesized by the r-process in this event, and it is likely that such events are common and are responsible for synthesis of most of the heavier elements in the periodic table. Figure 2.11 shows current knowledge of the synthesis paths of the elements.

Kilonova
Explosive event produced by the merger of two neutron stars or a neutron star and black hole.

Element Abundances

The mixture of elements that makes up the Earth arose from what was available and condensable in the solar nebula at Earth's distance from the Sun. H and He are by far the most abundant elements in the universe and solar system, but these elements are concentrated in the Sun and gas giants. Aside from H and He, relative abundances of elements in the Earth are commonly assumed to follow those in chondrites (Figure 2.12). There is an obvious sawtooth pattern, where elements with even atomic numbers are anywhere from 2 to 100 times more abundant than adjacent odd-numbered elements. This observation is known, oddly enough, as the Oddo-Harkins rule and will come up later in discussions of trace-element geochemistry.

With a few exceptions, elements in Figure 2.12 up to Ti are ones that are abundant in the crust—the major elements of igneous petrology plus C and N, which are abundant in sedimentary and metamorphic rocks and the atmosphere. The exceptions are transition metals (Cr, Co, etc.) that are relatively depleted in the crust but concentrated in the Earth's mantle and core.

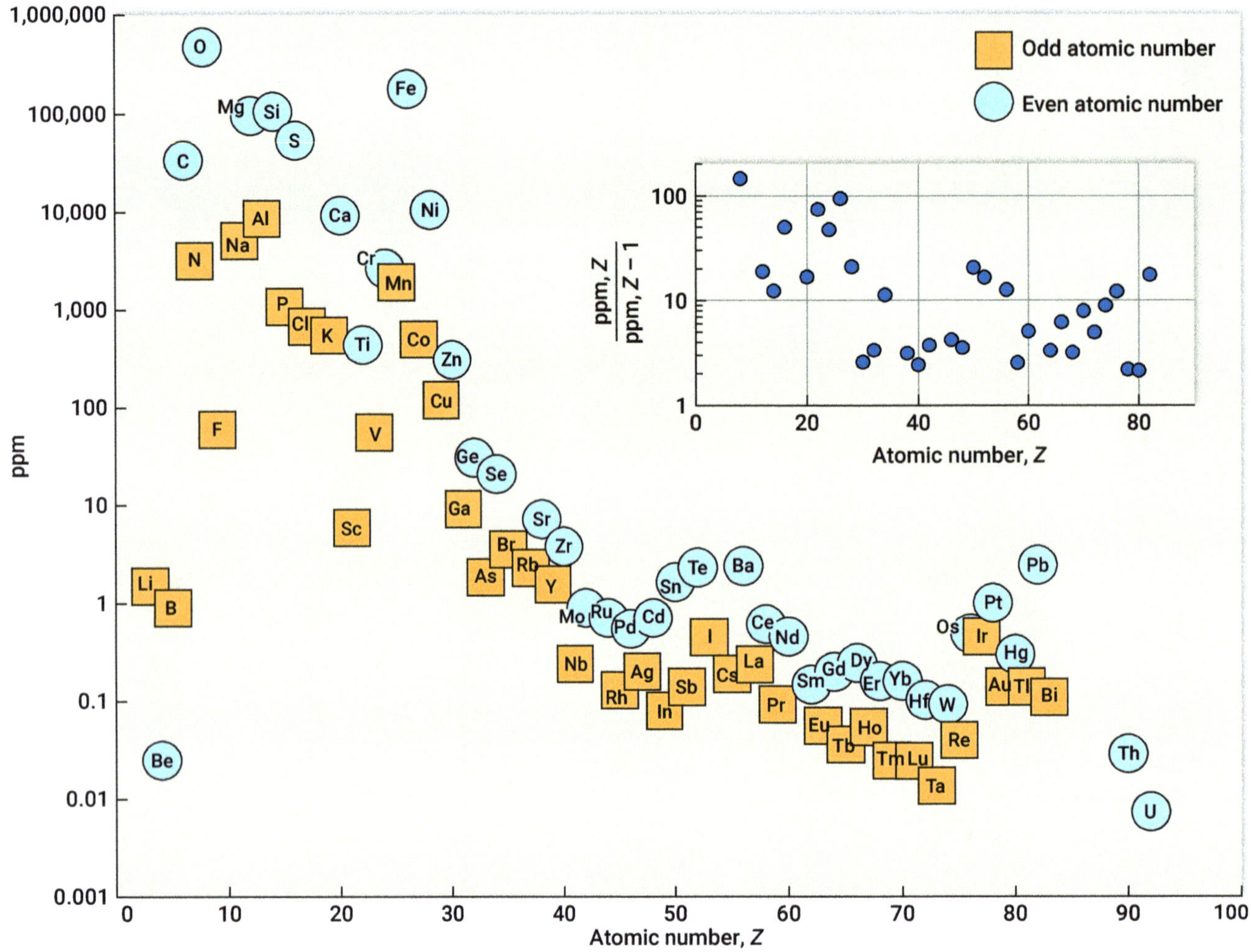

Figure 2.12 Estimated abundances of the elements from Li to U in chondritic meteorites. Note the general decrease over a factor of more than a million from O to U. Elements with even atomic numbers are more abundant than those with odd atomic numbers, leading to a sawtooth pattern. Inset shows the abundance ratios for even elements *Z* divided by the lower-adjacent odd element *Z* – 1 (e.g., Zn, *Z* = 30, over Cu, *Z* = 29). In general, the even elements are roughly five times more abundant than the nearby odd ones, with a more pronounced difference at lower atomic numbers.

Source: Data from McDonough & Sun (1995).

Making Connections 2.1

Where Does Lithium Come From?

Lithium, element 3, is critical to modern electronics; batteries based on Li provide power for consumer electronics, electric vehicles, robot vacuum cleaners, and other devices. Battery technology is changing rapidly, but Li-based batteries are currently dominant owing to their high-power density and long lifetimes even when recharged frequently. Li-ion batteries currently in use exchange Li^+ between an anode (typically graphite) and cathode (conventionally CoO_2). Li^+ is carried by an organic solvent and $LiPF_6$, and during discharge, Li^+ ions detach from graphite, migrate from anode to cathode through a porous plastic separator, and form $LiCoO_2$ (Figure 2.13). The electrons complementary to the Li^+ have done the work (moved the car, transmitted a text message, shown a cat video, etc.), and recharging the battery involves applying a reverse charge to pull the Li^+ back into the graphite matrix of the anode.

A glance at Figure 2.12 shows that Li, element 3, and the next two elements, B and Be, are far less abundant than other

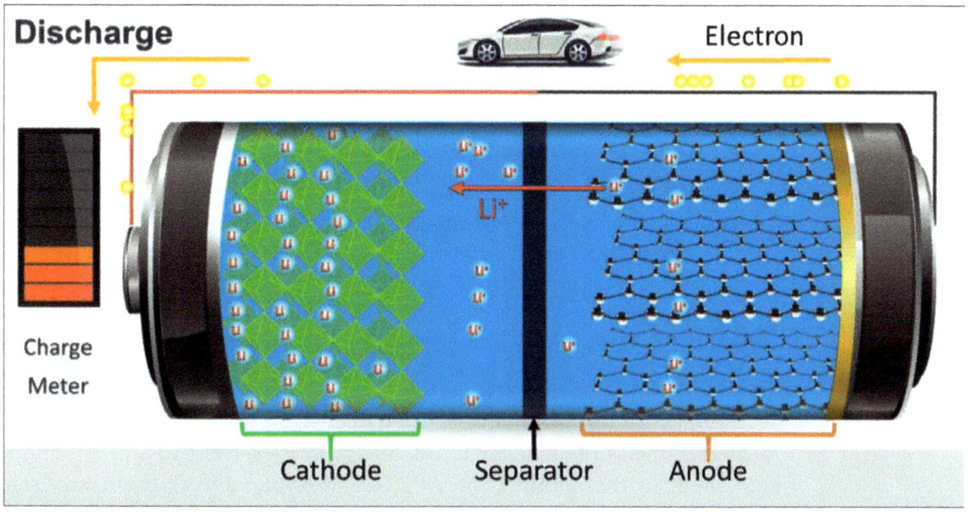

Figure 2.13 The basics of a Li-ion battery: in the graphite anode, electrons are drawn off and pulled through a device that does work (e.g., a motor). The resulting liberated Li^+ ions migrate through a plastic separator and react at the cathode to form $LiCoO_2$. Recharging involves applying a reverse voltage to drive Li^+ ions off the $LiCoO_2$. They then migrate back to the anode and reset the cycle.

Source: Drawing courtesy of roadmaps.mit.edu.

elements in that part of the periodic table. Li is ~10^9 times less abundant than H in the solar system, about as abundant as Pt and not much more abundant than Au. The elements after B, particularly C, N, and O, are far more abundant.

The low abundances of Li, B, and Be are a consequence of the energetics of fusion reactions in stars. For example, hydrogen burning in the Sun proceeds via this chain of reactions:

$$^1H + {}^1H = {}^2H + e^+ + \nu \quad \textbf{(Eq. 2.8)}$$

$$^2H + {}^1H = {}^3He + \gamma \quad \textbf{(Eq. 2.9)}$$

$$^3He + {}^3He = {}^4He + 2{}^1H \quad \textbf{(Eq. 2.10)}$$

where ν designates a neutrino, e^+ a positron (antimatter equivalent of an electron), and γ a gamma ray. Interesting fact: on a per-mass basis, the Sun produces energy via these reactions at a rate of about 0.0002 W/kg. A professional cyclist produces power at a rate of perhaps 3 W/kg, 15,000 times greater. However, whereas an elite cyclist might be able to do this for several hours a day over a 3-week race, the Sun can do it for 10 billion years.

The first of these reactions (Eq. 2.8) proceeds quite slowly (and is the reason that the Sun will stay in this phase for on the order of 10 Gyr); the next two proceed much more rapidly, with the result being an overall reaction of four ^1H nuclei combining to make one ^4He nucleus. One might think that the next logical step would be

$$^4He + {}^4He = {}^8Be \quad \textbf{(Eq. 2.11)}$$

but the ^8Be nucleus has higher energy than two ^4He nuclei, so any ^8Be made is ephemeral and is rapidly converted to more stable nuclides. This is true for the isotopes of Li, Be, and B; in effect, these elements are skipped in the reactions that occur in most stars.

Li stands apart among the low-abundance trio of elements, because it has stable nuclides left over from the Big Bang and made in exploding white dwarfs[1]. Be and B do not, and apparently Earth's inventory of these elements is entirely derived from spallation, in which a high-energy cosmic ray whacks a nucleus and cleaves it. If one of the cleaved pieces has four or five protons, it is a Be or B nucleus; if it has the right number of neutrons to make a stable nuclide (five for Be, five or six for B), then it sticks around. A piece with four protons and six neutrons cleaved off is ^{10}Be, an important isotope that has been used to trace sediment input into subduction zones (see Section 6.7).

Modern technology depends critically "under the hood" on elements scattered all over the periodic table, many of which are extremely rare in the universe and only practical to obtain owing to concentration by geologic process such as subduction magmatism, contact metamorphism, and hydrothermal processes (see Chapters 6 and 11). The next time you use your phone, think about how the Li and Co that power it were synthesized in stars, including in exploding white dwarfs and supernovas, and by cosmic rays of unearthly energy blasting atoms apart.

2.6 Gross Structure of the Earth

Learning Objectives

By the end of this section you should be able to:

2.6.1 Deduce how the gross structure of the Earth would differ if its oxygen content were lower.

2.6.2 Explain how measurements of the Earth's angular momentum support seismic evidence for an Fe-rich core.

Differentiation into Core, Mantle, and Crust

Differentiation of the Earth into a silicate part and a metallic part happened during accretion, with formation of the crust occurring somewhat later. How and when the crust formed is one of the key questions that petrology seeks to answer, and there is much disagreement in the details (Figure 2.14). Everyone agrees there was zero crust when the Earth formed, and that the amount of crust today is . . . well, the amount of crust today. But how we got from then to now is unclear. Much of the crust may have formed in the first few 100 Myr as the molten planet radiated heat away and formed a solid scum on the surface. Other studies infer that the crust forms episodically, or that the rate of crust production is increasing, or even that the proportion of crust has been decreasing for several billion years. We will return to this issue in later chapters.

How do we know the Earth consists of a core, mantle, and crust? Human observation of the Earth's interior is remarkably limited. The deepest mine in the world only reaches down 4 km, about 0.06 percent of the way to the center, and the deepest boreholes are only 12 km deep, about 0.2 percent of the way (relative to a standard 12-inch globe, that is 0.6 mm). Some information comes from geologic material brought up from depth (e.g., pieces of the mantle brought up in lava flows), but most of what we know about the bulk of the Earth is inferred from geophysical measurements.

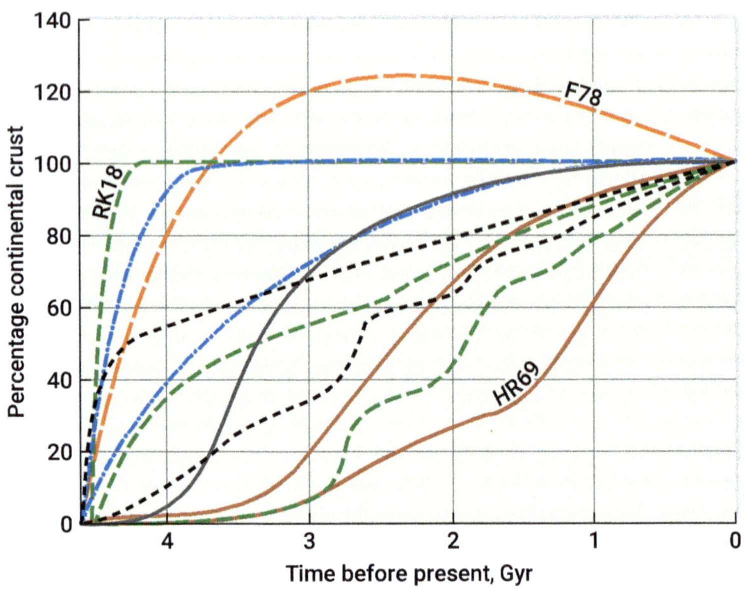

Figure 2.14 Various published models for growth of the continental crust over geologic time. Estimates of when the crust formed show significant disagreement. Hypotheses range from slow, incremental growth at an increasing rate (HR69; Hurley & Rand, 1969), to most crust formation in the earliest Archean with little change since then (RK18; Rosas & Korenaga, 2018), to a maximum at ~3 Ga that was ~20% greater than today's volume, following a steady decline as continental crust is recycled back into the mantle (F78; Fyfe, 1978).

Source: Data from Korenaga (2018).

Mass, Density, and Rotational Inertia

So what do we know about the Earth's interior? The radius has been known since Eratosthenes in the third century BCE. Once Henry Cavendish determined the gravitational constant G at the end of the 18th century, the mass and density were known, as the mass of the Earth is related to its radius and the acceleration of gravity, g (which can be determined with a pendulum) by

$$M = gR^2/G \qquad \textbf{(Eq. 2.12)}$$

where R is the Earth's radius. The mass can also be determined from the orbital periods of planetary satellites, a useful fact for determining the masses of planets that we have not yet visited and planted pendulums on.

This density (hereafter ρ) is surprisingly large: 5514 kg/m^3. Rocks at the Earth's surface are much less dense, typically in the range 2600 to 3000 kg/m^3. Part of this greater density arises from pressure within the Earth, which crushes rocks down to denser forms, but it has long been known that some denser material must be down there to account for the overall density of the planet, and Fe ($\rho = 7874$ kg/m^3 at standard T and P) is the main suspect owing to its abundance in meteorites.

The axial moment of inertia of a planet measures the radial distribution of mass: the more it is concentrated in the center, the lower the moment of inertia. This is a difficult quantity to estimate and is calculated indirectly from the rate of precession of the spin axis and from the shape of the gravity field. The Earth's moment of inertia is relatively low compared to a sphere of uniform density, confirming a concentration of mass toward the center.

Seismic Velocities

Much of our knowledge of the Earth's interior comes from studies of how seismic waves propagate through it. The presence of a liquid core in the Earth was inferred over a century ago (Oldham, 1906) from the P-wave shadow zone that encircles the Earth, and a solid inner core was discovered later. An iron-rich, dense core neatly explains the Earth's low moment of inertia. The concept of a crust was put on a firm basis by Andrija Mohorovičić in 1909, who used seismic refraction to show that many parts of the Earth are underlain at a depth of some tens of kilometers by a sharply defined transition at which P-wave velocity jumps from ~7 km/s to 8 km/s. This jump is now called the **Mohorovičić discontinuity** (**Moho** for short); by definition, the base of the crust is the Moho.

Mohorovičić discontinuity (Moho)
Downward jump in seismic velocity from 6–7 km/s to ~8 km/s; typically interpreted as the crust-mantle interface.

Studies of seismic wave travel times have led to a worldwide model of how P- and S-wave velocities vary with depth (Figure 2.15). A generalized model like this is commonly used as the starting point for more detailed seismic investigations. These measurements demonstrate that both P- and S-wave velocities generally increase with depth throughout the crust and mantle, drop precipitously at the core-mantle boundary, then rise gradually toward the center of the Earth. The Moho marks one sharp jump in P-wave velocity, and other sharp jumps occur at around 410 and 660 km depth (see Chapter 3).

Laboratory measurements of rock seismic velocities and densities allow us to model what rocks are present at depth. Figure 2.16 shows a set of such measurements made by Birch (1960) for various rocks at a pressure of 1 GPa (gigapascal), which is approximately

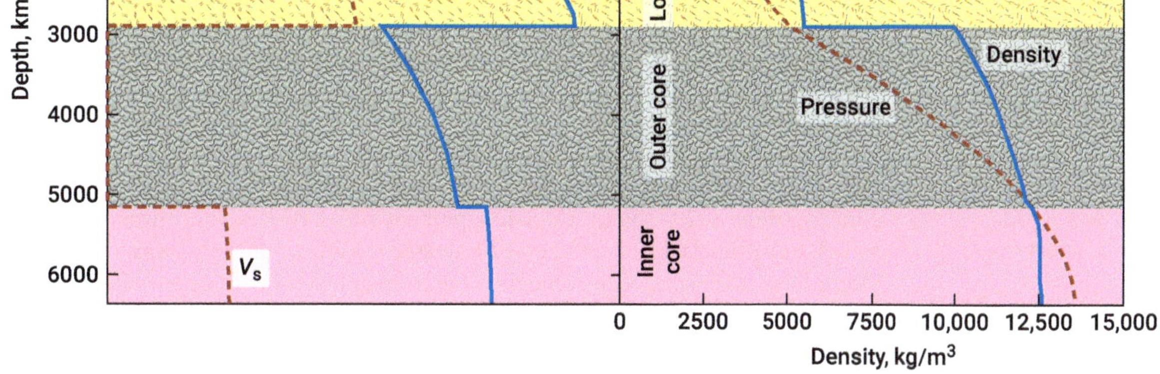

Figure 2.15 Velocities of P- and S-waves from the surface to the center of the Earth according to the IASPEI 1991 model, with corresponding estimates of density and pressure.

Sources: IASPEI data from Kennett (1991); density and pressure data from Anderson & Hart (1976).

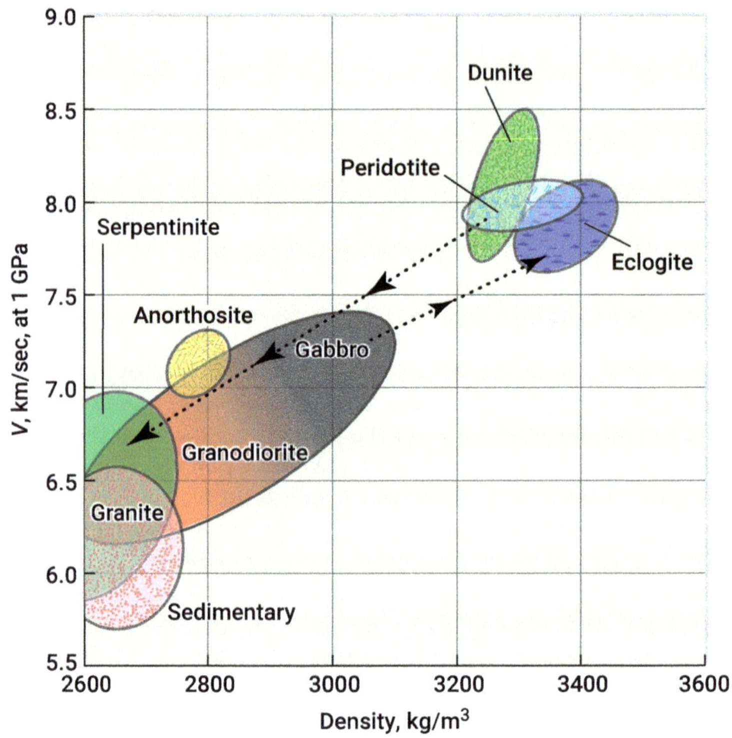

Figure 2.16 Fields showing P-wave velocity (V_P) versus density for various crustal and upper mantle rocks. There is a positive correlation despite the theoretical inverse dependence of V_P upon density. Gray field marks the range of plutonic rocks ranging from granite to gabbro. Dotted lines connect compositionally related rocks: eclogite, formed by high-*P*, low-*T* metamorphism of gabbro, and serpentinite, formed by hydration of peridotite. The gap between gabbro and the ultramafic rocks (peridotite, dunite) and eclogite corresponds to the Moho.

Source: Data from Birch (1960).

the pressure at the base of the crust. There is clearly a strong positive correlation between density and V_P, a relationship known as Birch's law.[2]

Rocks at the upper-right corner of the graph have V_P of about 8 km/s or greater, consistent with velocities below the Moho. There are two groups here: ultramafic rocks, composed mostly of olivine and pyroxenes, and eclogite, a high-pressure form of basalt made primarily of garnet and pyroxene. These groups are likely what the upper mantle is made of in much of the world. Granite, granodiorite, gabbro (the plutonic equivalent of the BADR trend), and various metamorphic and sedimentary rocks with V_P in the range of 6.0 to 7.5 km/s make up the lower-velocity portion of the diagram, with one oddball thrown in.

The oddball is serpentinite, a hydrated version of the ultramafic rocks in the upper-right corner of the diagram. Hydrating olivine to serpentine (discussed in Chapter 5) causes a dramatic decrease in both density and V_P, and one of the complications in interpreting regions of the crust with low velocities is determining whether serpentinite is present.

These crustal rocks are the familiar ones that we walk around on (and a lot of basalt under the oceans that we do not walk around on). Given the assumption that the core is predominantly Fe, then in the grossest sense the Earth comprises a metallic core (16% of its volume), a thin crust (~2%), and something in between that makes up the largest volume (~82%). This is the mantle, and it is composed largely of peridotite—arguably the world's most important rock.

Peridotite is a green rock (named for the gem variety of olivine, peridot; Figure 2.17), and consists mostly of four minerals: olivine (~70–80% by volume), orthopyroxene and clinopyroxene (~20–30%), and a few percent of an Al-bearing mineral—either garnet, spinel, or plagioclase. It was inferred long ago that the mantle is composed largely of peridotite, based on several arguments. One is cosmochemical: if you take a chondrite and remove the metallic material, the remainder is peridotite. Other reasons include (1) peridotite has the right seismic velocities, (2) melting peridotite yields basalt, and (3) many basaltic volcanoes bring up pieces of peridotite. Later, we will see how the *P* and *T* that these pieces were coughed up from can be determined and used to construct a *P-T* profile of the upper mantle.

Gross Earth structure based on these assumptions is shown in Figure 2.18. As we shall see, the mantle is only true peridotite in the upper part, because the minerals that make it up collapse to denser forms as pressure increases. The core-mantle-crust division is easy to comprehend, because it is largely based on composition, but these compositional differences were first found by their effects on seismic wave velocities.

Crustal thickness is highly variable, ranging from about 7 km for oceanic crust (Bown & White, 1994) to an average of 40 km for continental crust (Christensen & Mooney, 1995). Continental crustal thickness is strongly controlled by tectonics and ranges from as low as about 20 km in some rifts to more than 70 km in the Andes and the Himalaya. It is difficult to appreciate how thin this is when compared to the entire Earth; if you took a billiard ball and spray-painted it with one coat, the paint thickness would be relatively about twice as thick as oceanic crust. Two coats would give you an average continental crust.

Figure 2.17 A beautiful garnet peridotite from Norway, a piece of the mantle, consisting of red garnet, pale brown orthopyroxene, bright green clinopyroxene, and light green olivine. Olivine that was near the surface of the sample, as in upper right and upper left, is altered to orange-brown iddingsite. Peridotite is the source of most basalts via partial melting.

Source: AFG photo, sample provided by Stephen Ehrenberg.

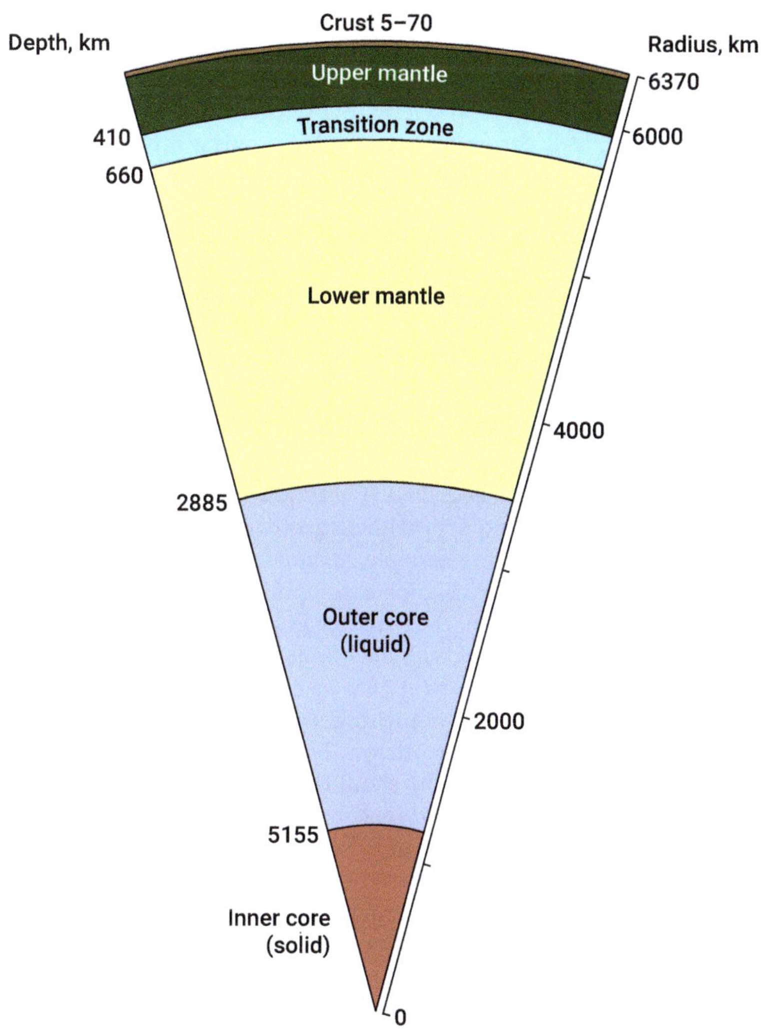

Figure 2.18 Gross Earth structure as inferred from seismic studies.

2.7 Temperature and Pressure in the Earth

Learning Objectives

By the end of this section you should be able to:

2.7.1 Make a sketch demonstrating why the geothermal gradient must decrease with depth from measured surface values.

2.7.2 Calculate the pressure at a given depth in a multilayer crust.

Temperature (T) and pressure (P) are fundamental environmental variables that control the minerals that make up a rock, how it melts, and how it deforms. They are examples of intensive variables. An **intensive variable** is a property that is the same in any part of a system that is at equilibrium (in contrast to an **extensive variable** such as volume). In a geologic system at equilibrium, P and T must be the same everywhere. What do we know about P and T in the Earth?

Intensive variable
One whose value is the same in all parts of a system at equilibrium and does not depend on the amount of the system considered (e.g., temperature).

Extensive variable
One whose value depends on the amount (extent) of the system or phase considered (e.g., mass).

Temperature

The rate at which temperature increases with depth is the **geothermal gradient**, and the resulting relationship is the **geotherm**. If z is depth (positive downward), then the geothermal gradient is the slope of a plot of T vs. z, or dT/dz. The Earth's mean surface temperature is around 14 °C, but the oceans cover about 70 percent of the surface and the deeper parts of the ocean are just a few °C above zero, so assuming that the mean surface temperature is 0 °C is not too far off. Deep mines are hot; at the bottom of a 4-km-deep mine in South Africa, the rocks are over 60 °C, giving a geothermal gradient of about 15 °C/km. The deep Kola (Russia) and KTB (Germany) scientific boreholes give gradients of 17 and 29 °C/km, respectively. So we know that at least some geotherms start off at about 0 °C at the surface and increase at a rate of 15 °C or more per km.

Geothermal gradient
Rate at which temperature increases with depth; typically given in °C/km.

Geotherm
Variation of temperature with depth.

Temperature measurements in boreholes are used to construct maps of heat flow (Figure 2.19). Heat flow across North America is highly variable and correlates well with surface geology. In particular, areas of active volcanism and rifting, such as the **Basin and Range province** of western North America, have much higher heat flow than geologically inactive areas, such as the eastern part of the country. Heat flow and geothermal gradient are related by Fourier's law:

Basin and Range province
A region of active crustal extension, volcanism, and high heat flow in western North America characterized by north-trending mountain ranges separated by valleys. The Basin and Range encompasses most of Nevada, eastern California, western Utah, western and southern Arizona, and part of northern Mexico.

$$q = -k\frac{dT}{dz} \quad \textbf{(Eq. 2.13)}$$

where k is **thermal conductivity** and q is heat flow. If heat flow is predominantly by conduction, then these can be converted into approximate geothermal gradients via Fourier's law (this is demonstrably not true if, for example, magma or hydrothermal fluids are carrying heat to the surface). Given the conduction assumption and assuming that $k = 3$ W/m/°C (a good middle-of-the-road value), then surface gradients range from approximately 10 to 50 °C across the map, consistent with data from mines and boreholes.

Thermal conductivity
A measure of how well a material conducts heat; units of W/m/°C.

If we extrapolate even a low surface gradient (say, 15 °C/km) to depth, we get into trouble fast. Figure 2.20 shows the beginning-of-melting curve (the **solidus**) for peridotite and granite, plotted against this extrapolated geotherm. The curves cross at about 100 km depth, indicating that the mantle below that depth should be molten. Partially molten rock has much lower seismic velocities than unmelted rock, but no such decrease in velocity is seen (see Figure 2.15). The extrapolated geotherm rapidly passes above the temperature at which peridotite is completely molten and into territory that is absurd. Clearly, the geothermal gradient must decrease to keep the geotherm below the melting curve of peridotite, if our assumption that the mantle is mostly peridotite is correct.

Solidus
For a given bulk composition at a given pressure, the temperature at which melting begins upon heating.

We also know that the outer core is molten but that there is a solid inner core,[3] so under the assumption that the core is predominantly Fe, the geotherm should be above the melting point of Fe at the top of the core but pass below it at the top of the inner core. We thus know the surface T, the shallow geothermal gradient, a curve that bounds the maximum T in the mantle (the peridotite solidus), a curve that bounds the minimum T in the outer core (the melting point of Fe), and a point on the curve (T at the inner-outer core boundary). Those constraints, coupled with the physics of heat flow, lead to the estimated crust-mantle geotherm shown in Figure 2.20. Keep in mind our earlier point that we can only measure T near the surface, and therefore everything below that is model-based.

The relatively sharp bend in the geotherm at 100 to 200 km depth is a function of two main effects. One is convection in the mantle, a topic that we will defer to the next chapter. The other is that the main heat-producing elements—K, U, and Th—are concentrated in the shallow crust, where they produce excess heat that raises crustal temperature and is eventually radiated away to space. This shallow concentration is a function of these elements' geochemical behavior as incompatible elements, another topic that we will defer. If, say, Fe were radioactive instead of K, then excess heat would be produced in the core, deep-Earth temperatures would be higher, mantle convection would be more vigorous, and the Earth would be a different planet.

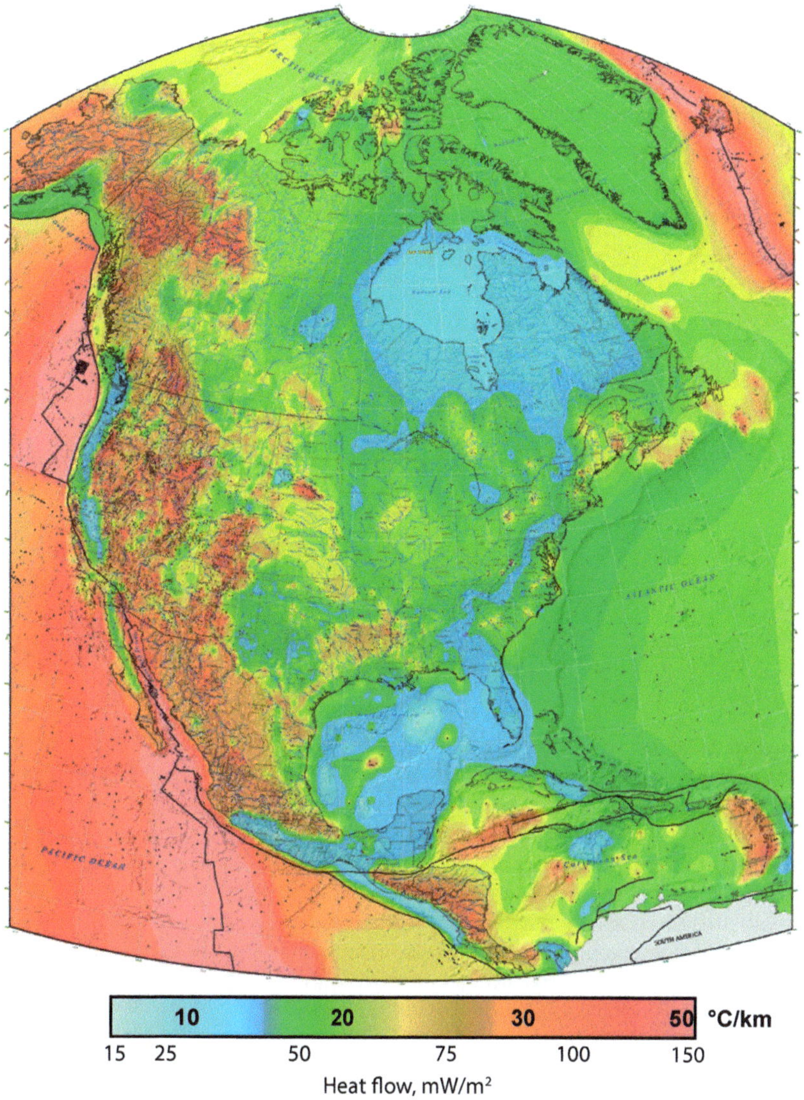

Figure 2.19 Map of heat flow in and around North America, showing a strong correlation with regional geology. In particular, the Basin and Range province, Rio Grande Rift, and Snake River Plain—areas of regional extension, active volcanism, or both—have much higher heat flow than less active parts of the continent. If heat flow is assumed to occur primarily by conduction (clearly not true in at least some areas), then these heat flow values can be converted into approximate geothermal gradients (see text).

Sources: Modified from Blackwell & Richards (2004). Geothermal map of North America, AAPG Map, scale 1:6,500,000.

Pressure

Pressure is force per unit area, and **lithostatic** pressure refers to the weight (force of gravity) of the overlying rocks at a given point in the Earth per unit area. For relatively shallow parts of the Earth, pressure is given by

$$P = \rho g z \qquad \textbf{(Eq. 2.14)}$$

where ρ is the average density of the rocks above that point, g is the acceleration due to gravity, and z is depth. The **geobaric gradient** is the slope of this relationship:

$$\frac{dP}{dz} = \rho g \qquad \textbf{(Eq. 2.15)}$$

Let's use this relationship to calculate the approximate pressure at the base of typical continental crust, using SI (Système International) units, as always, to keep things simple. If the crust is 35 km thick with an average density of 2900 kg/m^3, using g = 9.8 m/s^2 we get $P = 9.9 \times 10^8$ Pa, or about 1 GPa. This is a nice round number, easy to remember: the pressure at the base of typical continental crust is about 1 GPa. A pascal is a tiny unit, but that's what the SI gives us. Atmospheric pressure is about 10^5 Pa, or 0.1 MPa. Historical note: a bar is (among other things) a unit of pressure approximately equal to one atmosphere, and until the SI took over, geologists used the kilobar (kb) as a convenient unit of crustal pressure. And 1 kb = 0.1 GPa, so the pressure at the base of typical continental crust is about 10 kb.

Lithostatic
Confining pressure given by the weight of the overlying rocks per unit area.

Geobaric gradient
Rate at which pressure increases with depth.

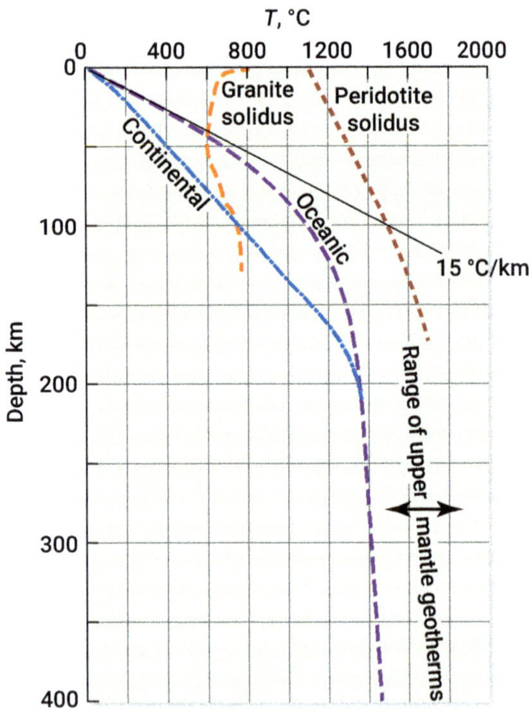

Figure 2.20 Estimated geotherms in the crust and upper mantle. Extrapolating even a low (15 °C/km) geothermal gradient downward implies temperatures at 100 km depth that are above the solidus of peridotite and well above the solidus of water-saturated granite. Continental and oceanic geotherms are schematic but demonstrate that the curves must turn steeply downward or else much of the upper mantle would be entirely molten, at odds with seismic data. Peridotite solidus, granite solidus, and a range of upper mantle geotherms are also shown.

Sources: Continental, oceanic data from Turcotte & Schubert (2002), peridotite from McKenzie & Bickle (1988); granite from Boettcher & Wyllie (1968); upper mantle from Hofmeister (1999).

Calculating P is easy when we're dealing with constant density. If the crust is layered, the calculation is not much harder: just add up the pressures contributed by each layer. For example, for a two-layer crust consisting of an upper layer, 10 km thick, with $\rho = 2750$ kg/m^3 and a lower layer, 25 km thick, with $\rho = 2900$ kg/m^3, $P = 980$ MPa. A general form of the pressure equation is

$$P = \int_0^h \rho(z) g dz \qquad \textbf{(Eq. 2.16)}$$

where h is total depth. Example: if crustal density increases linearly from 2700 kg/m^3 at the surface to 3000 kg/m^3 at a depth of 30 km, what is the pressure at 30 km? Here, $\rho(z) = 2700 + 0.01z$, where z is in meters, so $P = 798$ MPa.

In Figure 2.15, there is a discontinuity in density at the core-mantle boundary. This corresponds to a kink in the pressure profile, because the geobaric gradient is much larger within the core owing to its greater density. Pressure is calculated as before, but as you descend into the Earth, the value of g decreases, because there is significant mass above you. This effect is important at the planetary scale but can be ignored for processes occurring in the outer few hundred kilometers.

2.8 The Lithosphere

Learning Objectives

By the end of this section you should be able to:

2.8.1 Contrast different ways to define the base of the lithosphere.

2.8.2 Deduce how lithospheric thickness should depend upon (1) geothermal gradient, (2) rate of deformation, and (3) concentration of water in the upper mantle.

Lithosphere
Outer, rigid shell of the Earth, with a sharp upper boundary (the surface) and a gradational lower boundary that is typically equated to an isotherm such as 1200 °C; typical lithosphere is ~100 km thick.

The boundaries that separate crust, mantle, and core are well-defined and sharp. A less tangible yet equally important boundary in the Earth is the base of the **lithosphere**, the outer rigid shell of the Earth. The amount of differential stress that a rock can withstand ("strength" in the broad sense) before deforming by frictional sliding increases as confining pressure increases, but decreases exponentially as T increases and deformation by viscous creep mechanisms take over (Brace & Kohlstedt, 1980). As a result, for a given rock type along a given geotherm, there is a peak in rock strength whose depth depends on the rock type and geothermal gradient. For oceanic crust and upper mantle, this peak is estimated based on laboratory experiments (Kohlstedt et al., 1995) to be in peridotite in the upper mantle (Figure 2.21). The continents are more complicated, because quartz-rich crustal rocks such as granite, which are abundant in the upper crust, are much weaker than olivine-rich mantle rocks, and plagioclase-rich lower crustal rocks are in between. This leads to a double or triple peak in strength. For both oceanic and mantle settings, the strength drops dramatically with depth in the mantle owing to rising T.

The lithosphere is defined as the part of the Earth that behaves rigidly at plate tectonic strain rates (e.g., during deformation by mantle flow or plate-boundary sliding), and its base is generally equated with the depth at which mantle peridotite loses most of its strength. Thus, the base of a plate is defined by this vague, gradational definition, at odds with the typical depiction of a plate as having both a sharp upper boundary (the surface)

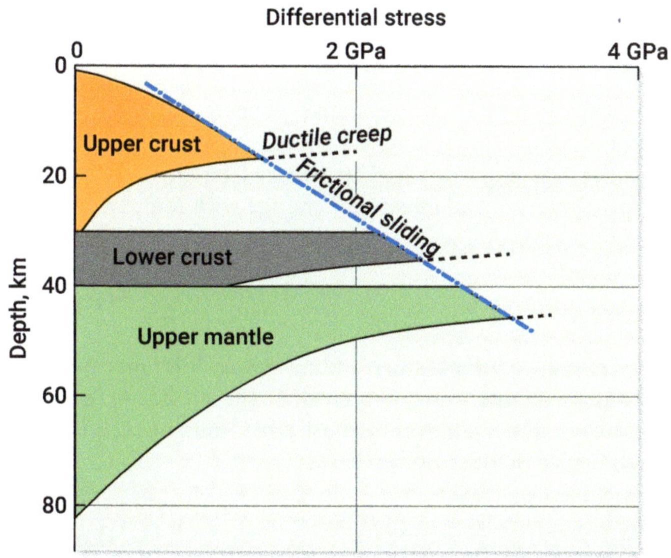

Figure 2.21 Schematic illustration of strength in the crust and upper mantle. Each envelope gives the maximum differential stress that the corresponding rock can withstand and is formed by the intersection of the frictional sliding strength (blue dash-dot line), which increases linearly with confining pressure (and thus depth), and a ductile creep curve (black dotted lines) in which the rock strength declines exponentially with temperature (and thus depth). The rock fails by faulting or flow depending on which failure curve is at lower stress (e.g., the strength of the upper crust increases more or less linearly to a depth of about 20 km, at which point the ambient temperatures are high enough to cause the rock to fail by ductile flow). The lower crust and mantle are made of minerals that are more resistant to thermally induced ductile failure. This particular model of continental crust comprises a granitic upper crust, gabbroic lower crust, and dry, olivine-rich mantle. A small amount of water in the upper mantle would decrease its strength to much lower values.

Source: Modified from Burov & Watts (2006).

and a sharp lower boundary (Figure 2.22). Defining the base of the lithosphere is like defining the top of the atmosphere or where space begins—all are fuzzy boundaries. Because strength is so strongly dependent upon temperature, the base of the lithosphere can be equated with a particular isotherm, such as 1200 °C or 1300 °C. Drawing the base of the lithosphere as a line in cross sections rather than as a gradational boundary is a historical problem that has more to do with the difficulty of drawing a curved, gradational boundary with a pen than with the mechanics of the system.

Given typical geothermal gradients, a very round number for lithospheric thickness away from mid-ocean ridges is around 100 km. This is about 1.5 percent of the way to the

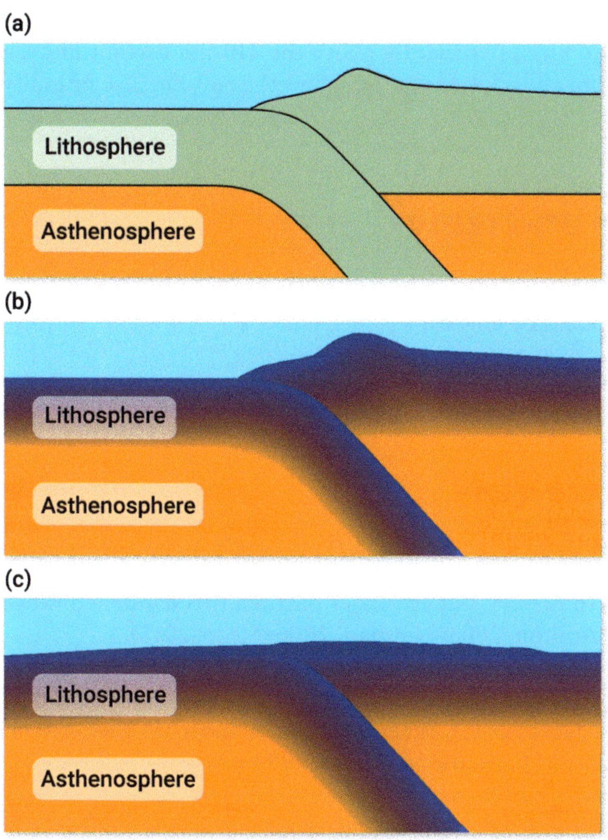

Figure 2.22 (a) Standard representation of the lithosphere as a layer with a well-defined base (e.g., introductory textbooks) and (b) more realistic depiction where the lower boundary is gradational. Note that in both of these cross sections, the vertical relief on the surface is extremely exaggerated, but slab dip is not. (c) As in (b) but with the Earth's surface drawn closer to scale and curvature and topography shown, modeled on the high Andes. Width of section 800 km.

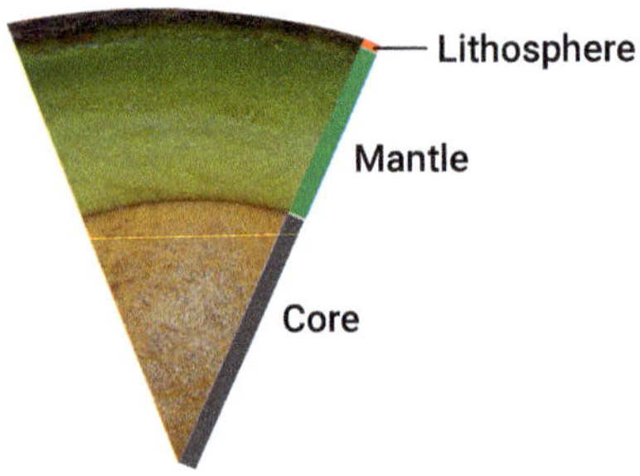

Figure 2.23 A round avocado with a medium-sized pit makes a decent Earth model. The skin is about 2% of the radius, similar to the scale of a lithospheric plate; the pit is roughly half the radius, similar to the radius of the core. The green of the interior echoes olivine and pyroxene, and the gradation in color from the skin toward the pit represents the increase in temperature that accompanies an increase in depth. A few coats of spray paint would mimic the relative thickness of the crust.

Source: AFG photo.

center of the Earth, so when we talk about plates on the Earth, we're talking about thin plates—more like the relative thickness of a chicken eggshell or avocado skin than an orange peel (Figure 2.23).

The lithosphere is generally denser than the mantle beneath it—we know this from the way it sinks into the mantle at subduction zones, and it makes sense because the outer part of the Earth is much cooler. The thermal expansion coefficient α describes the way that the dimensions of a material change with temperature; for volumetric changes, the coefficient is defined by

$$\frac{\Delta V}{V_0} = \alpha \Delta T \qquad \textbf{(Eq. 2.17)}$$

where V_0 is the volume at a reference temperature and ΔT is the difference between the reference temperature and the temperature of interest. As the left side of the equation is volume/volume and thus dimensionless, α must have units of reciprocal T. From this definition we can derive that density varies as

$$\rho = \rho_0 \left(\frac{1}{1 + \alpha \Delta T} \right) \qquad \textbf{(Eq. 2.18)}$$

Using a typical value of 3×10^{-5} K^{-1} for α (Wong & Brace, 1979), it takes about a 300 °C drop in T to increase the density of rock by 1 percent at constant pressure. This density increase helps drive plate tectonics.

Asthenosphere
Relatively weak zone in the upper mantle, softened by high temperatures and capable of flow at plate tectonic strain rates.

Upper mantle
Mantle above 410 km.

Transition zone
Mantle between 410 and 660 km.

Lower mantle
Mantle below 660 km.

The region of the upper mantle below the lithosphere is the **asthenosphere**. It logically follows from the definition of the lithosphere that the asthenosphere is that warmer part of the upper mantle that flows at plate tectonic strain rates. This mechanical definition has geochemical consequences, because the asthenosphere is better mixed than the lithosphere. Material stored in the lithosphere can sit by itself, removed chemically and physically from the rest of the mantle, for billions of years, while the asthenosphere undergoes homogenization.

The base of the asthenosphere is another poorly defined boundary. A more objective way to subdivide the mantle is into an **upper mantle** above the 410-km discontinuity, a **transition zone** from 410 to 660 km, and a **lower mantle** below the 660-km discontinuity (see Figures 2.15 and 2.18).

2.9 The Earth Then and Now

Learning Objective

By the end of this section you should be able to:

2.9.1 Explain how Earth's thermal state has varied since the birth of the solar system and how this could affect plate tectonics.

The Earth as we know it has evolved dramatically over its 4.6 Gyr lifetime, but many of its main features were established early, in Hadean and early Archean time. Separation of the core happened within the first 100 Myr (see Figure 2.8), and some sort of early silicic crust was present by 4 Ga or earlier (Wilde et al., 2001). The onset of plate tectonics is one of the key events in Earth history, but estimates of when this happened range from 4.4 Ga or earlier to 2 Ga (Hamilton, 2003; Nutman, 2006; Sleep, 2015).

Carbonate-silicate cycle
The weathering of silicates, transport to oceans, and precipitation of carbonates, to reduce CO_2 concentrations in the atmosphere, followed by subduction of carbonates and reaction with silicates to form CO_2, which escapes to the atmosphere.

The Earth was geophysically a much different place early in its history, a consequence of higher radioactive heat generation and more accretion and core-infall energy to radiate away (Stevenson, 2008; Sleep, 2015). Igneous and metamorphic rock assemblages reflect this. Archean rock assemblages are dominated by low-K granites (so-called tonalite-trondhjemite-granodiorite, or TTG, assemblages) and metamorphosed basalt. Classic plate tectonic assemblages produced in subduction zones in the more recent geologic past are generally not obvious.

Many unusual rock types are restricted, or nearly so, to the Archean and Proterozoic. These include komatiites, ultramafic lavas composed of olivine and pyroxene and indicative of high melting temperatures; anorthosites, rocks composed almost entirely of plagioclase, most of which were emplaced during the middle Proterozoic; and many of the world's layered mafic intrusions, gigantic intrusions that are basaltic in overall composition but locally show spectacular mineralogical layering.

In this book, we will focus on the petrology of Phanerozoic rocks, because such assemblages are much more widely represented in the rock record and the processes that produce them are active today, but it is always good to put these in the context of overall Earth evolution. There are many resources for exploring the pre-Phanerozoic history of the Earth (e.g., Condie, 1997; Best, 2003).

Negative feedback
A process whose effects counteract or reverse that process (e.g., weathering has a negative feedback on atmospheric CO_2 concentration).

Making Connections 2.2

Earth's Thermostat: The Carbonate-Silicate Cycle

Astrophysicists often remark on how unusual Earth is because it has liquid water (oceans) at its surface. But to geologists, what makes Earth's liquid water so remarkable is not that it exists, but that it has existed for *billions of years*, even back when the Sun was much dimmer than it is now. We know this from records of marine sediments and fossils, as well as from geochemical tracers. For example, 3.8-Ga gneisses in Greenland were metamorphosed from water-laid sedimentary rocks, so there must have been liquid water on Earth's surface at least that far back. This means that Earth's temperature has been regulated fairly closely—somewhere between 0 and 100 °C for 4 Gyr or more. How can this be? What regulates Earth's thermostat?

The main control appears to be atmospheric CO_2, which, on long timescales, is balanced through the **carbonate-silicate cycle** (Figure 2.24; Kasting, 2019). This cycle has two main components. First, a surface and near-surface component consumes atmospheric CO_2 through weathering of silicates, transport of dissolved chemical species to the oceans, and precipitation of carbonates. The overall reaction is (Urey, 1952):

$$CO_2(g) + CaSiO_3(s) = CaCO_3(s) + SiO_2(s) \quad \textbf{(Eq. 2.19)}$$

where the terms in parentheses represent gas (g) and solid (s). Here, $CaSiO_3$(s) doesn't really mean the mineral wollastonite, which is rare; rather we mean Ca in rocks, including Ca silicates such as pyroxenes, feldspars, and amphiboles, as well as volcanic glass. SiO_2(s) can mean silica (opal, quartz, etc.) or silicates in general. The wide variety of other elements in rocks and minerals (Al, Mg, Fe, etc.) go into making other minerals (clays, oxides, etc.).

The actual process is a little more complicated: CO_2 first dissolves into water to form carbonic acid, which "eats" (weathers) Ca-bearing silicate minerals and puts both calcium and silicon in solution. These ions are washed to the oceans, where organisms and hydrothermal systems precipitate calcium carbonate. More weathering means more conversion of atmospheric CO_2 to carbonate, and thus lower atmospheric CO_2.

If all the Earth ever did was weather silicates and precipitate carbonates, CO_2 concentrations would plummet to near zero. Something must return CO_2 to the atmosphere (Ebelman, 1845). That process is plate tectonics—the second component of the cycle. As carbonates subduct, they react with silicates to produce CO_2, which returns to Earth's atmosphere by diffuse degassing in forearcs and continental interiors and via direct volatile emissions from volcanic rocks at mid-ocean ridges, arcs, and rifts. Very generally, more (or faster) subduction increases atmospheric CO_2 (e.g., Berner et al., 1983).

There are numerous other components to Earth's C-cycle, including the organic carbon cycle (so-called "fast" C-cycle) and the sulfur cycle (Hilton & West, 2020). Nevertheless, the carbonate-silicate cycle has likely been the primary driver of Earth's CO_2 budget on timescales of millions to tens of millions of years.

The carbonate-silicate cycle regulates Earth's atmospheric thermostat, because weathering rates depend strongly on surface temperature, which in turn depends strongly on atmospheric CO_2. There is a **negative feedback** such that if atmospheric CO_2 increases, Earth warms and weathering rates increase, reducing CO_2 and (eventually) temperature. However, if CO_2 and temperature drop too low, weathering rates decrease, CO_2 increases, and Earth warms. Voila! Earth's thermostat.

The rise of trees and the Late Devonian mass extinction(s) illustrate how carbon cycles affect climate (Algeo et al., 1995). In this hypothesis, evolution of trees in the Late Devonian and their colonization of the continents, whose previous inhabitants were mostly just blue-green algae (the continents were probably not very hospitable places before trees came along), dramatically increased terrestrial biomass and intensified rock weathering, leading to drawdown of CO_2 nearly to modern levels. Release of nutrients from weathered rocks also fertilized the oceans, increasing their biomass, further reducing atmospheric CO_2, and producing abundant organic-rich black shales, now a major source of hydrocarbons in eastern North America. Decreased temperatures

continued

(a)

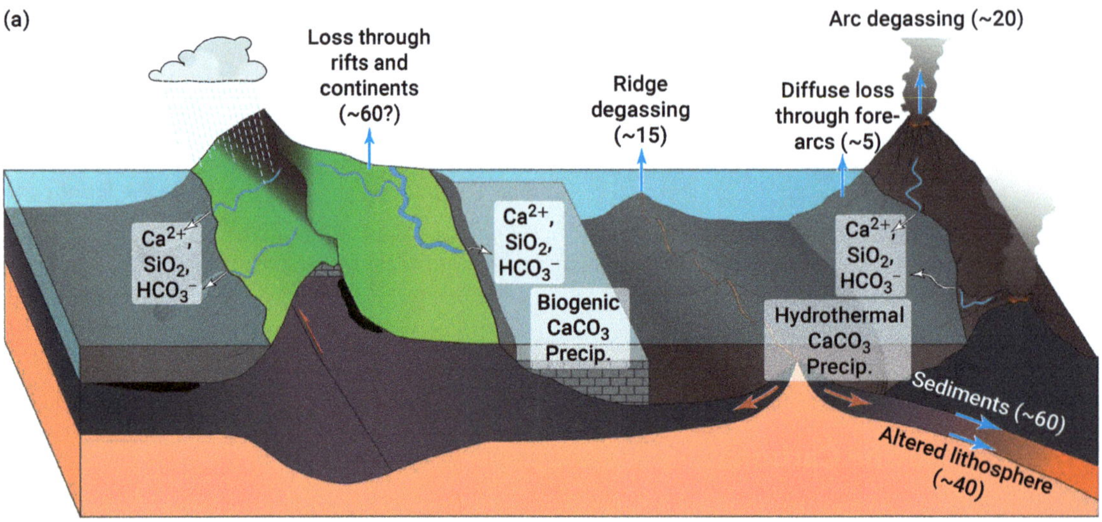

(b)

Figure 2.24 (a) Simplified version of the carbonate-silicate cycle. Weathering of Ca-silicates removes CO_2 from the atmosphere by delivering Ca^{2+}, SiO_2, and HCO_3^- to the oceans, where $CaCO_3$ precipitates in biogenic and hydrothermal systems. Subducted sediments and lithosphere react to form CO_2 that is emitted from arcs, ridges, and continents back into the atmosphere. Numbers in parentheses are net production or consumption of C in Mt/yr (megatons per year). (b) Artist's conception of a Devonian forest that could have catalyzed rapid weathering and CO_2 drawdown, leading to mass extinctions.

Sources: (a) Modified from Hilton & West (2020), © Springer Nature BV; (b) Richard Jones/Science Photo Library.

led to widespread glaciations, while changes to ocean chemistry led to widespread extinctions. Earth did eventually recover, but it took a few tens of millions of years, because the carbonate-silicate cycle is slow to act.

It is worth appreciating that the net fluxes involved in the two components of the carbonate-silicate cycle are only ~100 Mt C/yr, whereas anthropogenic fossil-fuel burning is 100 times higher, or 10,000 Mt C/yr. These anthropogenic emissions are driving Earth's C-cycle far out of equilibrium. Of course, Earth is responding with increased weathering, uptake of CO_2 into the oceans (acidifying them), and increased plant biomass. But these responses are relatively slow, so CO_2 relentlessly rises, warming Earth's surface and leading to changes in climate. Earth will eventually return to its normal CO_2 state, but absent human intervention, this will take thousands to hundreds of thousands of years (Archer, 2005).

Summary

Key points of this chapter include the following:

- Formation of the Earth involved collapse of a nebula to form the Sun, accretion of planetesimals to form a protoplanet, and collision with a Mars-sized impactor to form the Earth-Moon system. This impact essentially melted the Earth and Moon, coalesced nearly all of both cores into the core of the Earth, drove extensive degassing to form the atmosphere, and tilted Earth's spin axis so that we have seasons.
- Meteorites are mostly derived from the asteroid belt.
- Certain classes of meteorites are taken to represent the building blocks of Earth, and their ages equal the age of the Earth, assuming models of solar system and planetary evolution. The age of the Earth cannot be determined by the age of any terrestrial material.
- Geochronology relies on the predictable decay of radioactive isotopes ("parent," e.g., ^{87}Rb) to form stable isotopes ("child," e.g., ^{87}Sr). In most geochronologic systems, correction is necessary for the prior concentration of the daughter isotope.
- Heat flow through Earth's surface represents contributions from radioactive heat production and primordial heat loss from Earth's formation, including core infall.
- Supernovas are catalysts of nebular collapse and star formation; other than H and He, elements in the periodic table reflect contributions from dying stars.
- Elemental abundances in the solar system dramatically decrease with increasing atomic number, and elements with odd atomic numbers normally have lower concentrations than elements with even atomic numbers.
- Three light elements—Li, Be, and B—are less stable than ^{4}He, and so are in anomalously low abundance compared to overall abundance trends.
- Our understanding of Earth's interior is derived mostly from seismic observations, but also from general physical properties (mass, density, moment of inertia) in comparison with meteorite mineralogy and chemistry.
- The geotherm represents the temperature distribution with depth and reflects the thermal conductivity of rocks, concentrations of heat-producing elements, and rate of heat loss from deeper rocks. The geothermal gradient depends strongly on tectonic environment.
- Pressures are determined by integrating the density of rocks (which changes with depth) over the interval of interest. Pressure at the base of 35-km-thick crust is roughly 1 GPa (10 kilobars, kb, in older units).
- The lithosphere represents the outermost part of the solid Earth that behaves rigidly under plate tectonic strain rates. The upper boundary is sharp, but the lower boundary is gradational.
- Ancient Earth likely operated very differently from modern Earth, largely as a result of higher temperatures and magnitudes of heat loss.
- Weathering of silicates on Earth's surface is thought to control atmospheric CO_2 concentrations on timescales of millions to tens of millions of years; CO_2 concentrations are buffered through a negative feedback between temperature and weathering efficiency.

Notes

1. This would make a good name for a heavy metal band.
2. This positive correlation is odd, because V_p should decrease in denser material; it occurs because V_p correlates positively with the incompressibility and rigidity of a material, and in rocks these outweigh the density effect.
3. This is known from details of how P-waves pass through the core.

CHAPTER 3

Mid-Ocean Ridges and Production of Oceanic Crust

The main activity of plate tectonics is production of new oceanic lithosphere by seafloor spreading and its consumption at subduction zones. Nearly all seafloor spreading occurs under water, so direct observation generally requires expensive expeditions. However, narrow strips of newly forming ocean, such as the Red Sea and Gulf of California, allow direct observation of what newly rifted continental material looks like, and Iceland is unique in being the only place on Earth where a spreading ridge makes it to the surface (Figure 3.1). In this chapter, we examine the process of seafloor spreading, how the mantle melts, and the production of oceanic lithosphere.

3.1 The Mid-Ocean Ridge Network

Learning Objectives

By the end of this section you should be able to:

3.1.1 Calculate the average time (residence time) that a newly formed piece of oceanic lithosphere spends at the surface before being subducted.

3.1.2 Sketch cross sections of typical fast-spreading and slow-spreading ridges and give examples of each.

Production of Oceanic Lithosphere

Over 60 percent of the Earth's present lithosphere was formed relatively recently at **mid-ocean ridges (oceanic ridges)**. The plate tectonic system is continually generating new lithosphere at oceanic ridges, which is then consumed at subduction zones and recycled back into the mantle. The mid-ocean ridge (MOR) system stretches across the ocean floors for a distance of over 1.5 times the circumference of the Earth. Major branches split the Atlantic Ocean (the Mid-Atlantic Ridge), run down the eastern Pacific Ocean (the East Pacific Rise), and circle Antarctica (Figure 3.2).

Mid-ocean ridges (oceanic ridges)
Linear ridges in the ocean basins across which seafloor spreading takes place and new oceanic lithosphere is generated.

Looking northward from space at northwestern Mexico and the southwestern United States. A new ocean basin is forming as Baja California, the long peninsula, moves northwestward with the Pacific Plate, away from mainland Mexico and the rest of the North American Plate. The plate boundary consists of a series of short northeast-striking spreading ridges connected by short northwest-striking transform faults. Taken from the Moderate Resolution Imaging Spectroradiometer (MODIS) on NASA's Aqua satellite on November 27, 2011.

Source: NASA image by Norman Kuring, Ocean Color Team, (https://visibleearth.nasa.gov/images/76891/baja-california).

Figure 3.1 Looking across the North American-Eurasian plate boundary in Þingvellir (Thingvellir) National Park, Iceland. View to the northeast from the western rift shoulder across rifted lowland to the eastern rift shoulder—essentially, from North America to Europe. The low ground, which hosts a lake known as Þingvallavatn, is a series of north-striking grabens cut in basalt flows. These actively widening rifts accommodate spreading of the North Atlantic. Þingvellir is where the Alþing, an open-air assembly, was held annually, from ~930 to 1798, to make laws and mediate disputes. Divergence of the North American and European Plates has widened the rift valley fully 20 m since the first meeting of the Alþing.

Source: AFG photo.

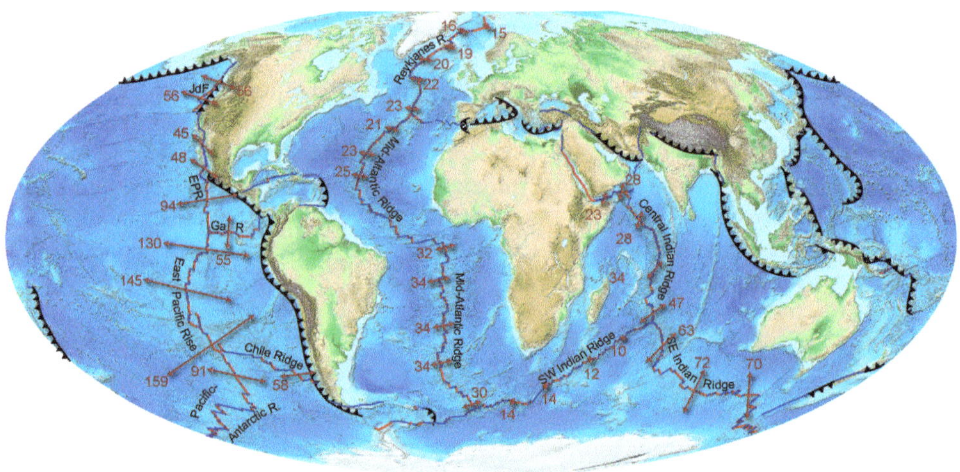

Figure 3.2 The oceanic ridge system with full spreading rates in mm/yr. Map projection distorts spreading directions at high latitudes. Fastest rates are on the East Pacific Rise west of South America. The ultraslow Gakkel Ridge is in the Arctic, off the top of the map. R = Ridge; JdF = Juan de Fuca; EPR = East Pacific Rise; Ga = Galápagos.

Source: Data from Bird (2003).

A few relevant facts about the MOR network (Bird, 2003; Müller et al., 2008; ETOPO Global Relief Model):

- Length: ~67,300 km
- Area of the ocean basins: 3.62×10^8 km^2
- Average age of the oceanic crust: 64 Myr
- Average thickness of oceanic crust: 7 km
- Average areal rate of production of oceanic lithosphere: 3.0 km^2/yr
- Average volumetric rate of production of oceanic crust: 21 km^3/yr

To put this in human perspective, the areal rate of production works out to the area of about 1.5 football fields, or a little over one soccer pitch, per day, and the volumetric rate would fill a pro football stadium in about half an hour. If we spread a year's basalt production over the ocean basins like cake frosting, it would be about 0.1 mm thick.

A little arithmetic with these numbers yields some useful results. The average areal production rate divided by the length of the MOR system gives an average spreading rate of about 45 mm/yr. This is an average full spreading rate (the rate at which points on either side of a ridge move apart); the rate at which a single plate moves away from its ridge is the half spreading rate. Spreading rates are quite variable across the globe, ranging from less than 10 mm/yr at the Gakkel Ridge in the Arctic Ocean to more than 150 mm/yr at the East Pacific Rise near Easter Island. In general, ridges in the eastern Pacific have much faster spreading rates than the Mid-Atlantic Ridge or those in the Indian Ocean. This is evident at a glance from a map of the age of the ocean floor (Figure 3.3), because the width of a given age band is proportional to the spreading rate (keeping in mind distortions caused by map projections).

Another useful number comes from dividing the area of the ocean basins by the areal production rate. The result, 121 Myr, has units of time and is an estimate of the mean **residence time**—the average time that a randomly chosen parcel of newly formed lithosphere spends traveling across the ocean basin before being subducted. The oldest oceanic crust formed about 170 Ma, ignoring a few older bits (see Figure 3.3),

Residence time
In a volume of something (e.g., water, air, glacial ice, oceanic lithosphere) subject to steady-state inflow and outflow, the average time that a unit of the substance spends in that volume. Residence time is given by volume divided by inflow rate (e.g., the residence time of water in an overflowing 200-liter bathtub being fed at 20 liters per minute is 10 minutes). The concept can be generalized to area.

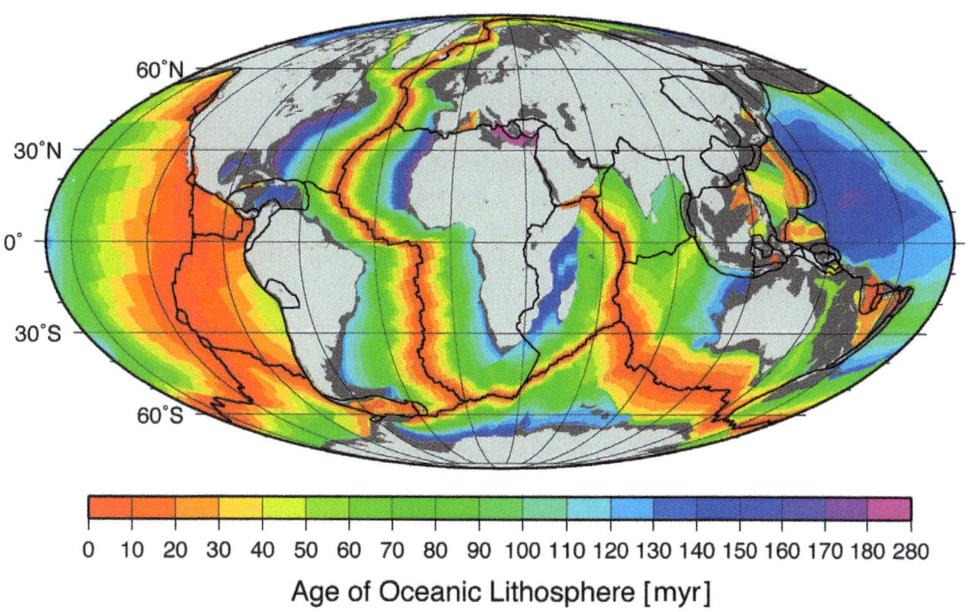

Figure 3.3 Age of oceanic lithosphere.

Source: NOAA.

and plate tectonics has been operating for at least a few billion years, so it is clear that oceanic lithosphere is, in the grand scheme, fleeting, transitory, and ephemeral, fated to be subducted or, for a few lucky pieces, thrust up on land and taken out of the oceanic conveyor system.

Under the Ocean Surface

Almost three-quarters of the Earth is covered by seawater, and under most of that seawater is lithosphere that was produced at mid-ocean ridges. Until the mid-1900s what was under all that water was mostly unknown. Lakes were better understood, and the general guess as to the topography and geology of the ocean basins was that they are like lakes, only bigger: deepest in the middle and carrying a thick, probably undisturbed sequence of sediments. It was also assumed that the ocean basins were old, primordial features. If true, then sediment was laid down over a long time, and drill cores through that sediment could pull up a record of much of Earth's history. The ocean basins were also convenient places to dispose of unwanted or lost pieces of continents, such as Atlantis, Lemuria, and various land bridges that were needed to explain the distribution of plants and animals before the discovery of plate tectonics made that unnecessary.

Until the invention of electronic echosounders, water depth was measured by dropping a line with a weight attached and trying to determine when it hit the bottom. Such measurements were time-consuming, labor-intensive, and not very useful in deep water. The existence of the Mid-Atlantic Ridge was known from a limited number of such soundings in the mid-19th century, but several technological developments in the first half of the 20th century, largely driven by military necessity, clarified the picture dramatically. In particular, the threat posed by submarines led to an urgent need to know the topography of the ocean floors (in *The Hunt for Red October* by Tom Clancy, submarines use the complex topography of the Mid-Atlantic Ridge south of Iceland to hide) and its magnetic signature (can we find a big hunk of metal down there with magnetometers?). Development of nuclear weapons led to an urgent need for precise monitoring of seismic events around the world. Soon ships were routinely carrying continuously recording depth sounders to map out the ocean floors, magnetic and gravity surveys were expanded, and a sensitive worldwide seismic network was established.

Data came flooding in, and it completely changed our understanding of the geologic history of the ocean basins. The oceans are not deepest in the middle, like lakes; the deepest parts, deeper than Mount Everest is tall, occur adjacent to continents and strings of islands. A worldwide chain of mountains runs through the oceans, including right down the middle of the Atlantic Ocean. These mountains stand up about 3 km above the surrounding **abyssal** plains. **Seamounts**, flat-topped mountains that reach almost, but not quite, to sea level, pepper the ocean floors. Drilling showed that sediment on the ocean floors is quite thin, typically less than a few hundred meters. And if you drill through it you almost always bring up basalt. That's what oceanic ridges make.

Abyssal
Deeper parts of the ocean, >2000–3000 m water depth.

Seamount
Isolated, commonly flat-topped submarine mountain; generally an extinct volcano.

Morphology of Mid-Ocean Ridges

Figure 3.2 shows that the rate of spreading at oceanic ridges varies by a factor of 10 between the fastest-spreading segments along the East Pacific Rise and the slowest-spreading segments along the Southwest Indian Ridge and north of Iceland. These spreading-rate differences are reflected in the topography of the ridges (Figure 3.4). Slow-spreading ridges have well-defined axial valleys and rugged topography produced by faulting, whereas fast-spreading ridges produce broader, smoother ridges without a clearly defined axial valley.

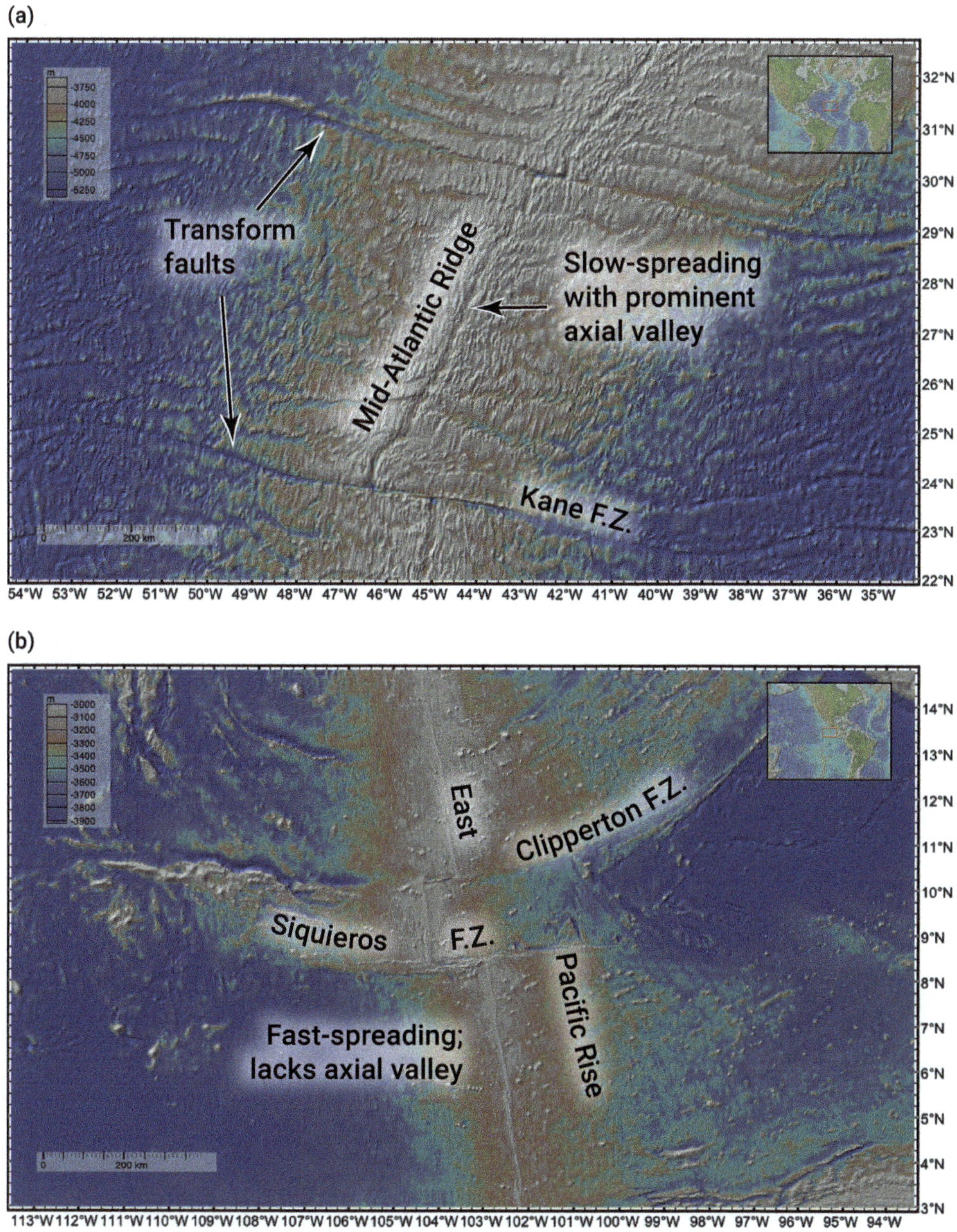

Figure 3.4 Morphology of two areas of oceanic ridge at the same scale. (a) Mid-Atlantic Ridge at 23–31°N, separating the North American (left) and African (right) Plates, showing well-developed axial valley characteristic of slow-spreading ridges. Spreading rate here is 24.5 mm/yr parallel to the transform faults, which are commonly referred to as fracture zones (F.Z.). (b) East Pacific Rise (109 mm/yr), separating the Pacific (left) and Cocos (right) Plates. The ridge steps across the Clipperton (north) and Siquieros (south) fracture zones. This fast-spreading ridge characteristically lacks an axial valley. Note different scales used for coding depth to color.

Source: Maps produced using GeoMapApp.

3.2 **Ophiolites and the Seismic Structure of Oceanic Crust**

Learning Objectives

By the end of this section you should be able to:

3.2.1 Interpret the components of a dismembered ophiolite in terms of the structure of typical oceanic crust and upper mantle.

3.2.2 Explain why the presence of an ophiolite sequence in a mountain belt requires closure of an ocean basin.

The general structure of oceanic crust is known primarily from three datasets: seismic studies, drilling, and geologic studies of oceanic assemblages that have been emplaced on land. The seismic structure of oceanic crust and upper mantle is known from seismic refraction studies in which compressed air explosions from an air gun are recorded by towed arrays of hydrophones and ocean-bottom seismometers. Seismic refraction is good at determining velocity layering, with a resolution on the order of 1 km. The velocity structure determined by these studies (Clague & Straley, 1977; Vera et al., 1990) shows very low velocity just below the ocean floor, with steady increases to V_P of about 7 km/s a few kilometers below the ocean floor, then a jump to about 8 km/s (the Moho) at a typical depth of about 7 km below the ocean floor (Figure 3.5). Drilling demonstrates that the uppermost layers are poorly consolidated sediments that are underlain by basaltic lavas and **dikes**. The deepest holes get into gabbro intrusions. This general structure nicely explains the observed seismic velocities down to the Moho.

Dike
A tabular, thin (generally meter-scale) intrusive body. Term is used for both active magma pathways and the frozen remains thereof.

Ophiolites

A more complete interpretation comes from observation of **ophiolites**. Over a century ago, the close association of serpentinite, basalt, and deep-sea chert composed of radiolaria in the Alps was recognized by Steinmann (1905), who proposed that they represent former ocean-floor rocks. Serpentinite is a rock composed largely of serpentine and produced by hydration of other ultramafic rocks; this will be discussed in detail in Chapter 5.

Ophiolite
Association of deep-sea sediments (typically chert), pillow lavas, basaltic dikes, gabbro, and peridotite; interpreted to be a piece of oceanic crust and upper mantle.

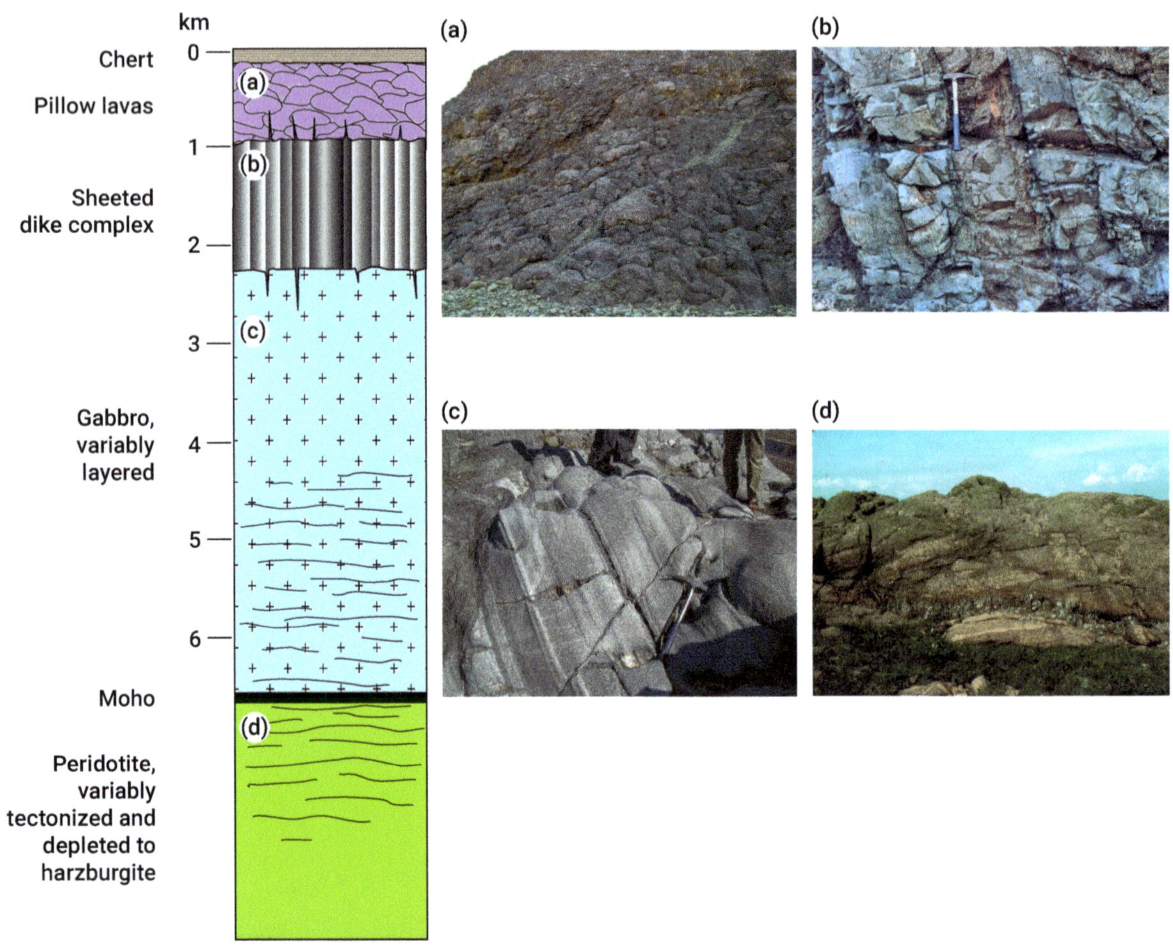

Figure 3.5 Interpretation of the structure of typical oceanic crust from seismic refraction, drilling, and studies of ophiolites. Field exposures of pillow lavas, sheeted dikes, and layered gabbro from the Oman ophiolite, and of peridotite from the Bay of Islands ophiolite in Newfoundland, are shown.

Source: Photos courtesy of Jeff Karson.

The term "ophiolite" had been used in the 19th century to refer to green rocks such as serpentine and metamorphosed basalt, and Steinmann repurposed it to refer to the ocean-floor serpentinite-basalt-chert assemblage.

Even before plate tectonic theory was developed, the oceans were known to be geologically different from the continents, so the presence of belts of ocean-floor rocks within continents and high in mountain ranges was recognized as requiring some out-of-the-ordinary process. Recognition of the Wilson cycle of ocean basin opening and closing (see Chapter 12) in the 1960s led to recognition that ophiolite belts mark places where former ocean basins have closed.

The ophiolite lithosystem was standardized in the early 1970s as an assemblage consisting of, from top to bottom (see Figure 3.5), chert, shale, and minor limestone; basalt lava flows; basalt dikes; gabbro; and ultramafic rocks in various stages of serpentinization. This lithosystem is interpreted as a section of oceanic crust and upper mantle.

The basaltic rocks in ophiolites have distinctive characteristics that link them to an undersea divergent plate boundary. The lavas are commonly pillow lavas, formed by eruption into water (Figure 3.6). Piles of pillow lavas in ophiolites can be spectacular. The dikes in ophiolites are commonly **sheeted dikes**, in which the wall rocks of a given dike are older dikes of the same set—dikes intruding dikes intruding dikes . . . This once-puzzling set of field relations is easily explained by injection of dikes into a dilating plate boundary, where space is made by divergence of the two plates.

The views of oceanic lithosphere obtained with seismic methods, drilling, and on-land studies of ophiolites are not independent but, taken together, indicate the general structure shown in Figure 3.5. Ophiolites are crime scenes where the victim was baked, squeezed, dismembered, and scattered across the landscape, and reconstructing them requires much forensic work. By their very nature, they are unusual (they're on land rather than under several kilometers of seawater), and various lines of evidence, particularly trace-element geochemistry, indicate that many formed near a subduction zone—for example, in a setting analogous to the Sea of Japan, which is underlain by new oceanic crust formed by spreading behind the subduction zone that dips west under Japan. Future closure of the Sea of Japan by continental collision could trap that oceanic lithosphere in the continent, forming an ophiolite.

(a)

(b)

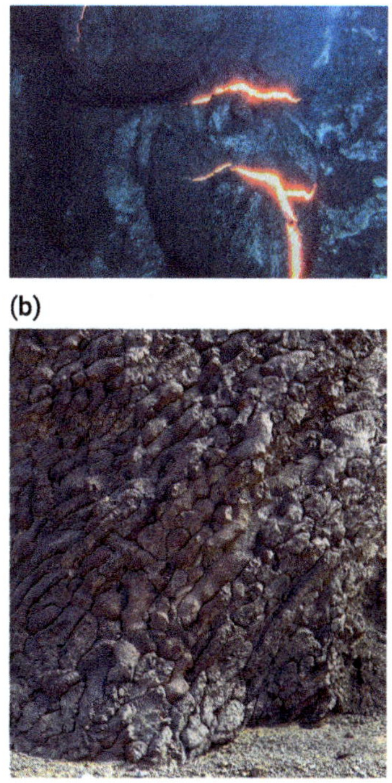

Figure 3.6 (a) Pillow lava forming in the Lau Basin, west of the Tonga arc, southwestern Pacific. A chilled pillow margin is in the process of splitting in several places; new pillows can form as lava squirts out of the crack and chills. Width of field ~1 m. (b) Pillow lavas in the Oman ophiolite, southeastern Arabian Peninsula. Individual pillows are 0.5–1 m across.

Sources: (a) NSF and NOAA; (b) Courtesy of Brad Hacker.

Sheeted dikes
Sets of parallel dikes in which the walls of each dike are older dikes.

Rocks of the Oceanic Ridge and Ophiolite Lithosystems

These systems are dominated by mafic volcanic and intrusive rocks, ultramafic rocks from both the upper mantle and accumulation of olivine and pyroxene in the crust, and deep-water sedimentary rocks.

- *Basalt*: Chiefly low-alkali, commonly referred to as MORB (mid-ocean ridge basalt)
- *Pillow basalt flows*: Characteristic lobes the size and shape of a pillow; indicate intrusion underwater
- *Sheeted basalt dikes*: Dikes whose walls are formed from earlier dikes, intruded at a spreading ridge
- *Gabbro*: Commonly with layering defined by variations in mineral abundances and sizes
- *Ultramafic rocks*: Include peridotite and harzburgite
- *Deep sea sediments*: Chert (siliceous ooze), carbonate muds, and clays, interlayered on a centimeter-scale
- *Serpentinite*: Hydrated peridotite
- *Greenstone*: Basalt metamorphosed in the greenschist facies
- *Massive sulfide deposits*: Abundant pyrite, sphalerite, galena, and chalcopyrite

Axial Magma Chambers

When the basic mechanism of oceanic ridges was worked out, it was assumed that ridges were underlain by tubular magma chambers, shaped in cross section something like a turnip, that ran along the length of the ridge system. In an influential paper, Cann (1974) drew such a vegetable-shaped magma chamber. Its top lay just below the sheeted dikes, perhaps a few kilometers below the seafloor, and went down to the Moho—a branching turnip 10 km wide, 5 km tall, and presumably over 60,000 km long when you add up all the branches. These magma chambers were thought to be the staging areas from which pillow lavas and dikes were erupted and gabbros were crystallized. Given that P-waves pass only slowly through magma and S-waves are blocked entirely, these postulated axial magma chambers made tempting targets for imaging by marine seismologists, and several expeditions went looking for them.

The seismologists came back empty-handed. At the fast-spreading East Pacific Rise, for example, multiple seismic studies using both towed arrays of seismometers and ocean-bottom seismometers failed to find a large magma chamber. Oceangoing seismology is a difficult endeavor, but these studies placed rather tight limits on the size of any active magma chambers, limiting them to no more than perhaps tens or a few hundred meters in height and, at most, a few kilometers in width (Sinton & Detrick, 1992). It appears that large quantities of magma are not stored at shallow depth under the ridges, and that accumulations of magma are likely ephemeral.

Serpentine soils
Soils developed on serpentinite and other ultramafic rocks (e.g., peridotites).

Serpentine endemic
A plant that grows exclusively on serpentine soils.

Making Connections 3.1

Serpentine Soils

Bedrock, when weathered, gives rise to soils, and it is the chemistry of rocks in concert with climate, topography, biota, and time that defines what that soil looks like and what plants live there. Of all bedrock sources, ultramafic rocks pose one of the greatest chemical challenges for plants. Soils developed on them are generally called **serpentine soils**, and they have especially low concentrations of the inorganic nutrients Ca, K, and P, and typically toxic concentrations of Mg, Ni, and Cr (Table 3.1). Consequently, the chemically harsh environment of serpentine soils suppresses growth of most plants (Figure 3.7a), except a small group of **serpentine endemic** plants (e.g., Kruckeberg, 2002; Figure 3.7b). These plants are highly diverse; for example, in California, serpentine soils account for only ~1% of total land surface area but represent ~10% of plant species diversity (Brady et al., 2005).

How do serpentine endemic plants survive? The most common coping mechanism is chemical exclusion. These plants have evolved to exclude Mg, Ni, and other toxic metals more effectively than other plants, while they still manage to take up enough Ca, K, etc. to survive. A small subset of serpentine endemics not only tolerates toxic metals, but actually accumulates them. These metals act as natural insecticides, because high metal concentrations are lethal to animals as well as plants. Plants that hyperaccumulate metals are important, because some can be used to clean areas that are contaminated (so-called "phytoremediation") or even "mine" metals by extracting them from soils ("phytomining"). Last, an even smaller subset of serpentine endemic plants obtains key nutrients from insects and even small reptiles and mammals. These carnivorous plants either wait for prey to walk over them,

Table 3.1 Soil Compositions

Material	FeO	MgO	CaO	Mg/Ca	K_2O	P_2O_5	Ni	Co	Cr
Average soil	3.5	0.76	0.65	1.6	2.0	0.13	20	9	52
Serpentine soil	8.8	13.4	0.17	57.8	0.04	0.03	2770	138	4330

Note: Oxides are in wt%; Mg/Ca is a molar ratio; Ni, Co, and Cr are in ppm (μg/g).
Sources: Data from Muhs et al. (2001), Oze et al. (2004), and Environmental Protection Agency (2005).

Figure 3.7 (a) View westward from sparsely vegetated serpentinite mélange on Ring Mountain, California, toward heavily vegetated sandstones of Mount Tamalpais, California. The unusual chemistry of serpentine soils inhibits plant growth. (b) An abrupt change in vegetation marks the boundary between soils developed on serpentine vs. sandstone at Jasper Ridge, San Francisco Bay Area, California. (c) California pitcher plant (*Darlingtonia californica*) is a carnivorous, serpentine endemic plant.

Sources: (a) AFG photo; (b) Photo by Bruce Bohm, from Brady et al. (2005); (c) Noah Elhardt/Wikimedia.

or they actively attract their prey with pleasing odors or plays of color. They then use either sticky sap with curling leaves (e.g., the sundew) or liquid-filled bowls with slippery sides and downward-pointing hairs (pitcher plants; Figure 3.7c) to trap, suffocate, and dissolve their hapless victims. The high nutrient content of small animals makes up for the nutrient-poor character of serpentine soils.

3.3 Melting and Generation of Magma

Learning Objectives

By the end of this section you should be able to:

3.3.1 Arrange a set of mantle xenolith samples, including peridotites with, variously, plagioclase, garnet, and spinel as the main Al-bearing mineral, in order of probable increasing depth of origin.

3.3.2 Design an experimental protocol to determine the minerals present with melt in a given basaltic composition at 1 atmosphere (atm) and 1150 °C.

How Do We Study Magma Generation?

How the Earth makes basalt at oceanic ridges is understood as well as anything in petrology, so we begin our study of magmatism here. As noted in the previous chapter, basalt is common stuff in the terrestrial planets, and the recipe for making it is well known—just partially melt peridotite. "Partially" is key; if you melt any rock 100 percent, the melt has the same composition as the starting material. Basalt is quite different from peridotite, so 100 percent melting won't do. To make basalt, you must only melt the peridotite part of the way (e.g., 5%, 10%, or 20%).

Rock melting experiments are the domain of experimental petrology, and such experiments are essential to testing hypotheses about the origin of particular magma types. As we will see results from experimental petrology throughout the book, a brief introduction to the topic is needed.

The simplest technique is to melt rocks in a furnace at surface pressures. The most popular 1-atm furnace (Figure 3.8a) consists of a vertical ceramic tube about 5 cm in diameter that is heated by U-shaped heating elements. A few tenths of a gram of rock powder

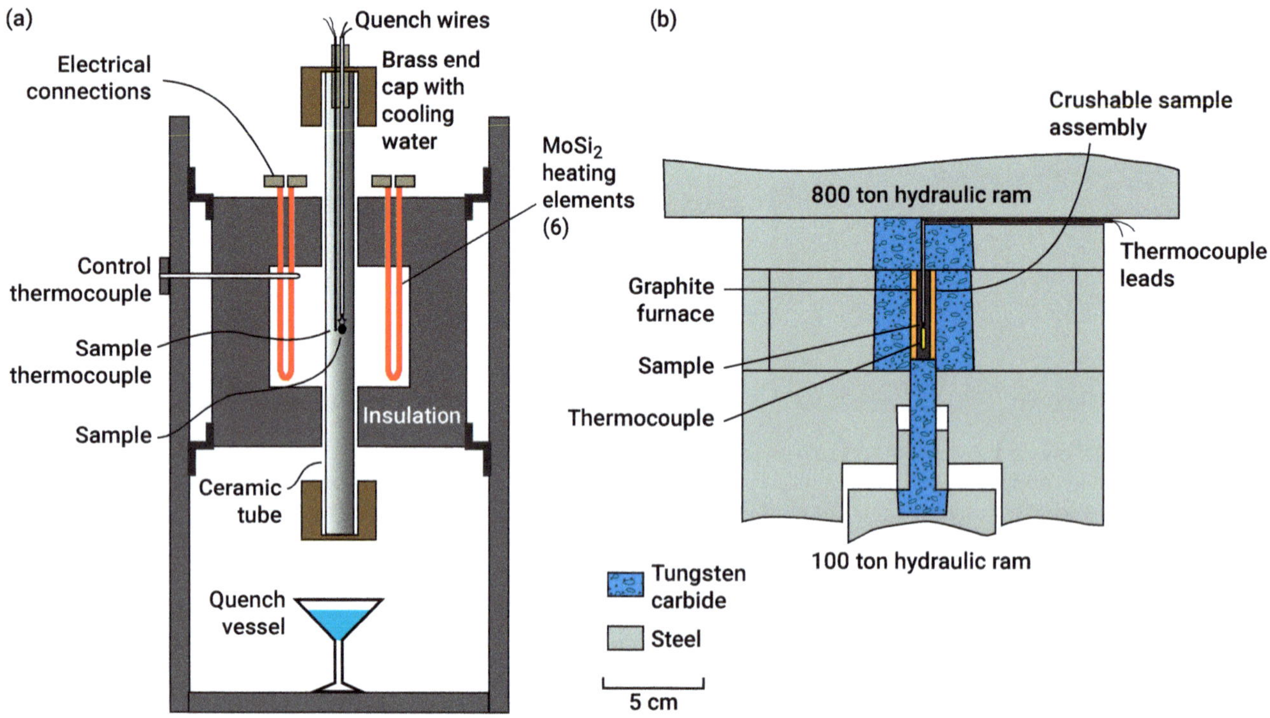

Figure 3.8 Workhorse instruments for experimental petrology. (a) Cross section of a 1-atm vertical-tube furnace. Sample is hung from fine platinum wires in the middle of a ceramic tube surrounded by six U-shaped heating elements. A thermocouple in the heating chamber gives feedback for controlling temperature, and a second thermocouple hangs next to the sample to monitor the temperature it experiences. A mixture of gases can be passed through the ceramic tube to control access of oxygen to the sample. The experiment is quenched by vaporizing the wire holding the charge by applying a large electric current to it. The magma drops into water in the conical quench vessel below, cooling at a rate of >200 °C/s. (b) Cross section of a piston-cylinder device; wavy lines indicate cut-off of schematic. This device is mounted in a steel frame; pressure is applied via hydraulic rams at either end and heat via a graphite furnace that surrounds the sample. Quenching is slower than in the 1-atm furnace and is accomplished by cutting power to the furnace.

Sources: (a) Drawn from plans by Deltech Furnaces; (b) from Boyd & England (1960).

starting material is glued onto a loop of fine platinum wire and then placed in the center of the furnace where it is subjected to a prescribed heating schedule, in a flow of gases that controls access of oxygen to the charge if necessary. After an appropriate amount of time, which can range from minutes to weeks, the experimental charge is quenched by dropping it into a water-filled vessel. (Historical note: a century ago, when such studies began, they quenched experimental charges in mercury!) The resulting mixture of crystals and glass is then analyzed under the microscope or by an electron beam technique. Such furnaces are typically operated at 1300 °C and below, but can reach temperatures over 2000 °C if heated by iridium windings.

The vertical-tube furnace described in Figure 3.8a above is fine for experiments at 1 atm (~10^5 Pa = 0.1 MPa), but a different apparatus is needed for doing experiments at higher pressures. High-pressure experiments relevant to basalt genesis are typically performed in a piston-cylinder apparatus (Figure 3.8b), which consists of a small noble-metal capsule (typically Au or Pt) embedded in a crushable medium and subjected to pressure from hydraulic rams at either end. Temperature is set via a small graphite furnace that jackets the sample. A piston-cylinder rig can reach *P-T* conditions appropriate to the lower crust and upper mantle—roughly 0.5 to 6 GPa and up to 1700 °C. Modifications of this basic design can reach pressures of over 20 GPa, corresponding to depths over 600 km.

To test the hypothesis that partial melting of peridotite yields basalt requires grinding up peridotite (or a chemical mixture of the same composition), melting it partway, and then analyzing the resulting liquid, which is present as glass in the quenched material. The glass in a typical experimental charge is present as tiny intergrain films that are perhaps only tens of micrometers across. Until the invention of the electron microprobe, analyzing this presented a difficult technical challenge. Reay and Harris (1964) circumvented this problem by melting large amounts of peridotite (30 g) and then painstakingly separating glass by hand for chemical analysis, but true advances were not made until the electron microprobe, which allows precise chemical analyses of spots as small as 1 μm or so. The microprobe revolutionized petrology, mineralogy, and geochemistry, starting in the 1960s.

What Melts? Peridotite

The cosmochemical, geophysical, and observational lines of evidence discussed earlier indicate that the upper mantle is composed largely of peridotite (see Figure 2.17). Table 3.2 shows a comparison of the compositions of a chondrite, Allende, with a typical peridotite sample from the mantle (in this case, from a piece brought up in a Hawaiian basalt). After metallic Fe and Ni and associated sulfides are removed from the meteorite, the match is excellent.

Allende, a primitive carbonaceous chondrite, has long been used for cosmochemical studies. A few tons of it fell on Pueblito de Allende, Mexico, on February 8, 1969. Because it is such a primitive meteorite and so plentiful, it was distributed to labs around the world and is one of the best-analyzed chondrites. Analysis of inclusions in Allende provided evidence for excess ^{26}Mg, the decay product of ^{26}Al (Lee et al., 1976; see Section 2.4).

There are many ways to describe the composition of a rock. The three most common are (1) via a chemical analysis such as those above; (2) by the volume percentages of the minerals present, a measure known as the **mode**; and (3) by recalculating the chemical composition into a set of hypothetical minerals that are geochemically compatible and, when added together, sum to the bulk chemical composition. Such a measure is known as a **norm** and is discussed in Toolbox 3.1.

Mode
Volume percentages of minerals in a rock.

Norm
A rock composition recalculated in terms of compatible end-member minerals, typically using the CIPW algorithm.

Peridotite is composed mostly of olivine, with about 20 percent pyroxenes and a few percent other minerals. End-member olivine and pyroxene molecules contain only Mg, Fe, Ca, Si, and O, but a quick look at the compositions in Table 3.2 shows significant concentrations of other elements—in particular, Al and Na. These must be present in the peridotite somewhere, and where they are accommodated turns out to be a strong function of pressure.

Table 3.2 Comparison of Compositions of a Chondritic Meteorite with a Peridotite

Oxide	Chondrite*	Peridotite
SiO_2	46.50	48.3
TiO_2	0.22	0.22
Al_2O_3	4.56	4.91
FeO^t	9.91	9.95
MnO	0.24	0.14
MgO	33.31	32.5
CaO	3.59	2.99
Na_2O	0.60	0.66
K_2O	0.04	0.07

* The silicate composition of the chondrite has been calculated by removing S and metallic Fe and Ni from the bulk analysis.
Sources: After Yoder (1976); data from Seitz & Kushiro (1974).

Phase Diagrams

Phase diagrams
Diagram that show stable phase assemblages for a given set of variables such as bulk composition, *P*, and *T*.

Phase
A particular form and composition of matter that is at least theoretically separable from other phases.

Component
One of a set of the smallest number of chemical constituents needed to describe the compositions of all the phases in a system.

Anhydrous
Water-free.

Mineral facies
A collection of minerals characteristic of a given set of *P* and *T* conditions for a given bulk composition.

Melting, mineral reactions, and other relationships between minerals and melts can be depicted graphically with **phase diagrams**. These are roadmaps to what one would find in a system under a given set of conditions. For example, a pressure-temperature (*P-T*) diagram for the system SiO_2 shows which **phase** (α-quartz, β-quartz, cristobalite, tridymite, etc.) is stable at a given pressure and temperature. This is a one-**component** system even though two different elements occur, because only one component, SiO_2, is needed to describe the compositions of all the phases.

Figure 3.9 is a pressure-temperature (*P-T*) phase diagram for peridotite under **anhydrous** conditions, showing **mineral facies** as a function of *P* and *T*. The phase diagram was determined by piston-cylinder experiments and is specific to a particular bulk composition: the diagram gives no information about compositions, and the positions of the boundaries will shift as the bulk composition changes. A phase diagram such as this one is used to predict the phase assemblage for a given bulk composition at a particular *P* and *T*. A phase in geochemistry is a particular form and composition of matter—for example, in a *P-T* diagram for the bulk composition SiO_2, α-quartz, β-quartz, cristobalite, tridymite, and molten SiO_2 are all possible phases.

In Figure 3.9, there are four main fields—plagioclase peridotite, spinel peridotite, garnet peridotite, and a field of crystals plus liquid (by 100–200 °C above the solidus, all crystals would be gone). All versions of peridotite are dominantly olivine with lesser amounts of orthopyroxene and clinopyroxene. Because these latter two names are lengthy and very common, hereafter we will refer to them as Opx and Cpx.

The peridotite fields reflect changes in mineralogy that accompany changes in *P* and *T*, and thus show different mineral facies. Consider a system at 1000 °C. At low pressures, it consists of plagioclase peridotite, but as *P* is increased, it shifts to spinel peridotite and then garnet peridotite. Plagioclase peridotite is only stable at pressures less than about 800 MPa, and spinel peridotite only at pressures less than about 2 GPa. All the peridotites have olivine plus two pyroxenes, so this shift in the Al-bearing mineral—from plagioclase to spinel and then garnet—reflects increasing pressure. Spinels and garnets are complicated mineral groups; in peridotites the spinel is typically a mixture of true spinel ($MgAl_2O_4$) and chromite ($FeCr_2O_4$), and garnets are predominantly pyrope (Mg-Al garnet) with significant almandine (Fe-Al), grossular (Ca-Al), and uvarovite (Ca-Cr) components. The olivine and pyroxenes do not change much over this pressure interval.

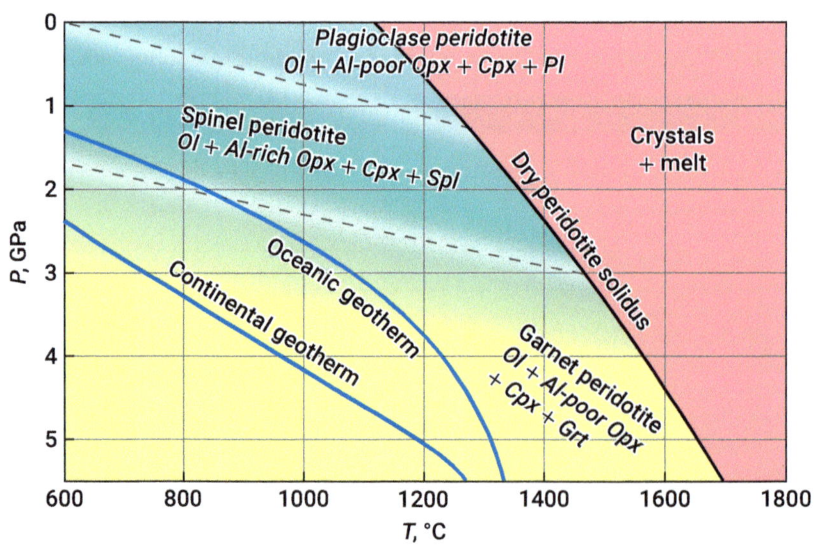

Figure 3.9 Mineral facies of the upper mantle, showing approximate solidus and stability fields of plagioclase, spinel, and garnet peridotite.

Source: Modified from Pearson & Wittig (2014).

Change in Aluminum Coordination with Increasing Pressure

The pressure-induced change in the aluminous phase in peridotite reflects a fundamental truth of petrology: Al can be in fourfold or sixfold coordination in typical crustal minerals, and the sixfold (octahedral) coordination is a denser packing than fourfold (tetrahedral). Thus, *Al shifts from tetrahedral to octahedral coordination with increasing pressure, and the transition occurs at crustal pressures.* This handy feature of Al geochemistry is the basis of most of the geobarometers used in both igneous and metamorphic petrology; you should store it in a convenient place in your head for easy access during your study of petrology. To that end, it is worth restating:

- Al shifts from tetrahedral coordination (as in feldspars) to octahedral coordination (as in garnets and many metamorphic minerals) with increasing pressure, and this transition occurs at crustal pressures.
- Most Al in the crust is in plagioclase, in tetrahedral coordination, and is eliminated by densification reactions at pressures less than about 1 GPa—conveniently (and not coincidentally) the approximate base of the crust.

If all is going according to plan you are thinking, *Wait a minute—you can't just turn plagioclase into spinel—they have different compositions!* (If not, then it's time to review Appendix 2, crack that mineralogy book open, and refresh your basic vocabulary.) The peridotite transitions are complex, but the relevant mineral assemblages with increasing pressure are as in Figure 3.10. Note that these assemblages acknowledge that real Opx can contain significant Al and ignore the abundant Cr in spinel. Life is complicated.

Seismic studies show there are jumps in seismic velocities worldwide at depths of about 410 and 660 km, and these reflect collapse of olivine to denser minerals. The 410-km transition is attributed to the olivine-wadsleyite transition, in which the rather loosely packed olivine structure shrinks down to a denser polymorph with significantly higher V_P and V_S. At depths of about 520 km, wadsleyite converts to another polymorph, ringwoodite, which has a spinel structure, and the overall transition is commonly referred to as the olivine-spinel transition. This transition takes place over a depth range of some tens of kilometers.

Ringwoodite is stable down to the next transition, where a more complicated reaction is responsible for the jump in seismic velocities at 660 km depth, the 660-km transition. Here, ringwoodite breaks down to two minerals, one with Opx composition but perovskite structure and the other (Mg,Fe)O, or ferropericlase. The perovskite minerals have Si in octahedral coordination—a higher pressure, upper mantle version of the Al^{IV}-Al^{VI} transition discussed earlier. To summarize:

$$\begin{aligned} P\,13-14\,\text{GPa; depth 410 km: } (\text{Mg,Fe})_2\text{SiO}_4 &= (\text{Mg,Fe})_2\text{SiO}_4 \\ \text{olivine} &= \text{wadsleyite} \end{aligned} \quad \textbf{(Eq. 3.1)}$$

$$\begin{aligned} P\,18\,\text{GPa; depth 520 km: } (\text{Mg,Fe})_2\text{SiO}_4 &= (\text{Mg,Fe})_2\text{SiO}_4 \\ \text{wadsleyite} &= \text{ringwoodite} \end{aligned} \quad \textbf{(Eq. 3.2)}$$

$$\begin{aligned} P\,23\,\text{GPa; depth 600 km: } (\text{Mg,Fe})_2\text{SiO}_4 &= (\text{Mg,Fe})\text{SiO}_3 + (\text{Mg,Fe})\text{O} \\ \text{ringwoodite} &= \text{perovskite} + \text{ferropericlase} \end{aligned} \quad \textbf{(Eq. 3.3)}$$

With these transitions, we can now complete the mineral facies map of the upper mantle (Figure 3.11).

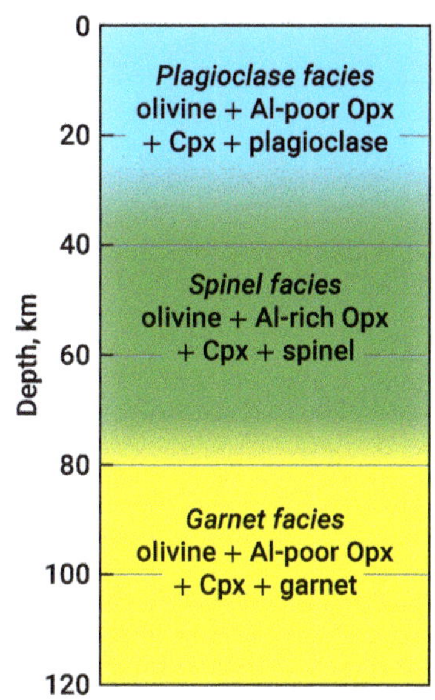

Figure 3.10 Column depicting the generally expected changes in mineralogy of the upper mantle with increasing pressure.

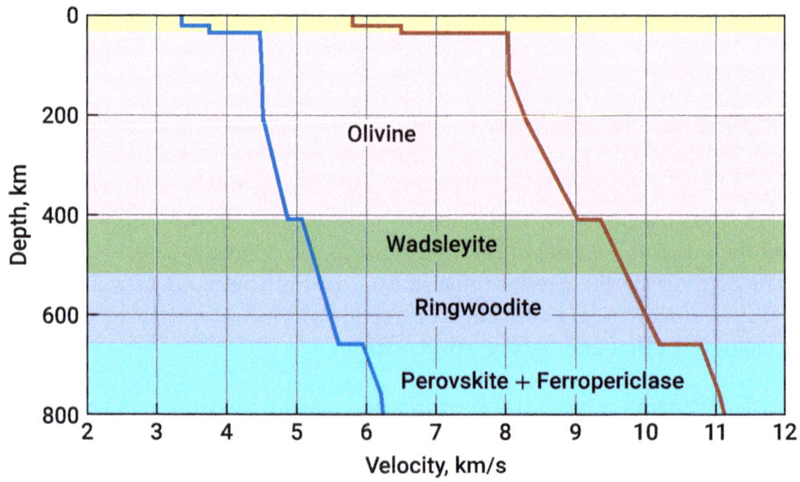

Figure 3.11 P- and S-wave velocities in the upper 800 km of the mantle according to the IASPEI 91 model (see Figure 2.15) and corresponding mineral transitions. Starting at the surface, the discontinuities are the Conrad (not seen everywhere), the Moho, the 410-km (olivine-spinel), and the 660-km (spinel-perovskite) discontinuities.

Source: Data from Kennett (1991).

Rocks of the Upper Mantle Lithosystem

Rocks of this lithosystem are ultramafic (essentially all mafic minerals) and are typically found in collision zones as part of ophiolite sequences, as xenoliths in basaltic lavas, or as xenoliths in more exotic explosive eruptions. They include the following combinations of olivine, Cpx, and Opx (see classification triangle in Appendix 3):

- *Peridotite*: Majority olivine with subequal amounts of Opx and Cpx
- *Harzburgite*: Majority olivine with Opx
- *Dunite*: >90% olivine
- *Clinopyroxenite*: >90% Cpx

Peridotites typically carry several percent of an aluminous phase whose identity changes depending on depth: plagioclase at less than ~30 km, spinel from ~30 to ~70 km, and garnet at greater than ~70 km.

3.4 Thermodynamics of Melting and Other Phase Transitions

Learning Objectives

By the end of this section you should be able to:

3.4.1 Use the Clapeyron equation to infer relative densities for assemblages on either side of a reaction on a *P-T* diagram.

3.4.2 Use a table of Gibbs free energies of formation at a given pressure to calculate the temperature of a reaction.

Predicting Melting and Other Phase Transitions

Phase transformations are the stuff of everyday life as well as petrology. Fill a tumbler with ice cubes and leave it out at typical room temperature, and the ice will melt. The ice, which comes out of the freezer at approximately −18 °C, melts at 0 °C and will eventually warm up to room temperature. Why does the ice melt, and what governs the melting temperature?

The answer lies in a fundamental precept of thermodynamics: that a system will evolve at constant T and P to minimize a measure of energy known as G, or the **Gibbs free energy**. A further condition is that the chemical system can exchange heat, but not matter, with its surroundings; a given volume of magma, or an experimental charge as in Figure 3.8, would qualify. This is a complicated topic that deserves a semester (really, a lifetime) of study, and here, we will just accept the existence of this quantity.

Gibbs free energy
The amount of chemical energy available to do work in a system at constant P and T.

G can only be defined relative to a reference state, just as gravitational potential energy can only be defined relative to a reference level (e.g., for a book on a table, you might define its gravitational potential energy relative to the floor, although you could use sea level, the center of the Earth, the top of Mount Everest, or any other relevant datum). All that matters in thermodynamics are changes in G.

Consider the polymorphic reaction Qz = Trd. At what temperature will this occur at 1-atm pressure? The numbers in Table 3.3 may look boring, but they contain the answer: they give free energy data for these two polymorphs of SiO_2 at different temperatures. These G values are relative to a standard state of the pure elements Si and O at 1-atm pressure and the stated temperatures (i.e., they are a measure of the change in G that occurs when Si and O are combined at the listed temperature to produce Qz or Trd). These are large negative numbers, because the reactions are highly favored. One cannot burn Si at these temperatures, and so the values of G are calculated from more easily measurable quantities, such as the **heat capacity**. These values, from Robie and Hemingway (1995), are known as the Gibbs free energy of formation from the elements.

Heat capacity
The amount of heat needed to produce a given increase in temperature for a given quantity of material; typical units are J °C^{-1} kg^{-1}.

Data for both Qz and Trd are relative to the same standard state, so we can predict which is stable at a given temperature by comparing them. We can define ΔG for the Qz = Trd reaction as $G_{Trd} - G_{Qz}$; a Δ term indicates products minus reactants. Remembering that lower G is favored, if $\Delta G > 0$, then Qz is favored, and if $\Delta G < 0$, then Trd is favored. Figure 3.12 shows ΔG vs. T; we would predict that at 1 atm, the reaction occurs at the temperature where the difference is zero, or about 880 °C. We have predicted our first phase transition.

At what temperature will this reaction occur at higher P? This requires knowing how G varies with P. An important result from thermodynamics is that

$$\frac{\partial G}{\partial P} = V \qquad \textbf{(Eq. 3.4)}$$

Table 3.3 Gibbs Free Energy of Formation from the Elements for Quartz and Tridymite at 10^5 Pa (1 atm) and 100 MPa

T (K)	T (°C)	Qz (1 atm)	Trd (1 atm)	ΔG (1 atm)	Qz (100 MPa)	Trd (100 MPa)	ΔG (100 MPa)
298	25	−856,300	−853,800	2500	−854,031	−851,147	2884
400	127	−837,700	−835,400	2300	−835,431	−832,747	2684
500	227	−819,400	−817,500	1900	−817,131	−814,847	2284
600	327	−801,200	−799,700	1500	−798,931	−797,047	1884
700	427	−783,200	−782,000	1200	−780,931	−779,347	1584
800	527	−765,300	−764,500	800	−763,031	−761,847	1184
900	627	−747,600	−747,000	600	−745,331	−744,347	984
1000	727	−730,000	−729,600	400	−727,731	−726,947	784
1100	827	−712,500	−712,300	200	−710,231	−709,647	584
1200	927	−695,000	−695,100	−100	−692,731	−692,447	284
1300	1027	−677,600	−677,900	−300	−675,331	−675,247	84
1400	1127	−660,300	−660,900	−600	−658,031	−658,247	−216
1500	1227	−643,100	−643,800	−700	−640,831	−641,147	−316
1600	1327	−625,900	−626,800	−900	−623,631	−624,147	−516

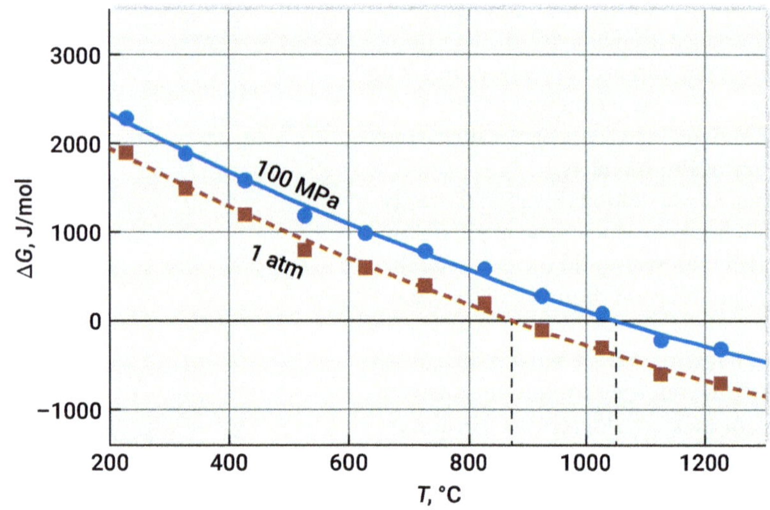

Figure 3.12 Plots of ΔG vs. *T* for the reaction Qz = Trd at two pressures. The predicted transition temperatures are those where the curve crosses the ΔG = 0 line.

where *V* is the molar volume of the phase and the partial derivative notation indicates that this relationship holds at a given, constant *T*. Robie and Hemingway (1995) give volumes of 22.688 and 26.540 cm^3/mole for Qz and Trd, respectively.

Assuming that *V* is constant is generally a good approximation for solid phases and makes integration trivial. Under this assumption, at constant *T* integration gives

$$G_{T,P} = G_{T,P_0} + V(P - P_0) \qquad \textbf{(Eq. 3.5)}$$

where P_0 is the reference pressure (1 atm, or 10^5 Pa). At first glance, this seems a bit problematic, because the units of the second term are in $cm^3{\cdot}Pa$, which seems mighty odd. However, since pressure times volume has units of energy (think of the ideal gas law, or the energy stored in a pressurized tank), $E = PV$, $V = E/P$, and thus volume can be expressed as energy divided by pressure. So, 1 cm^3 works out, handily, to 1 J/MPa.

We can apply this correction to each tabulated value of *G* once we choose a pressure. Values in Table 3.3 and Figure 3.12 are calculated for *P* = 100 MPa. At this pressure, the predicted reaction temperature is about 1050 °C.

Figure 3.13a is a *P-T* diagram for the system SiO_2, and Figure 3.13b shows detailed relationships around the Qz-Trd-Crs fields. Our calculated transition points at 1 atm and 100 MPa are shown, and they match the experimentally determined curve quite well.

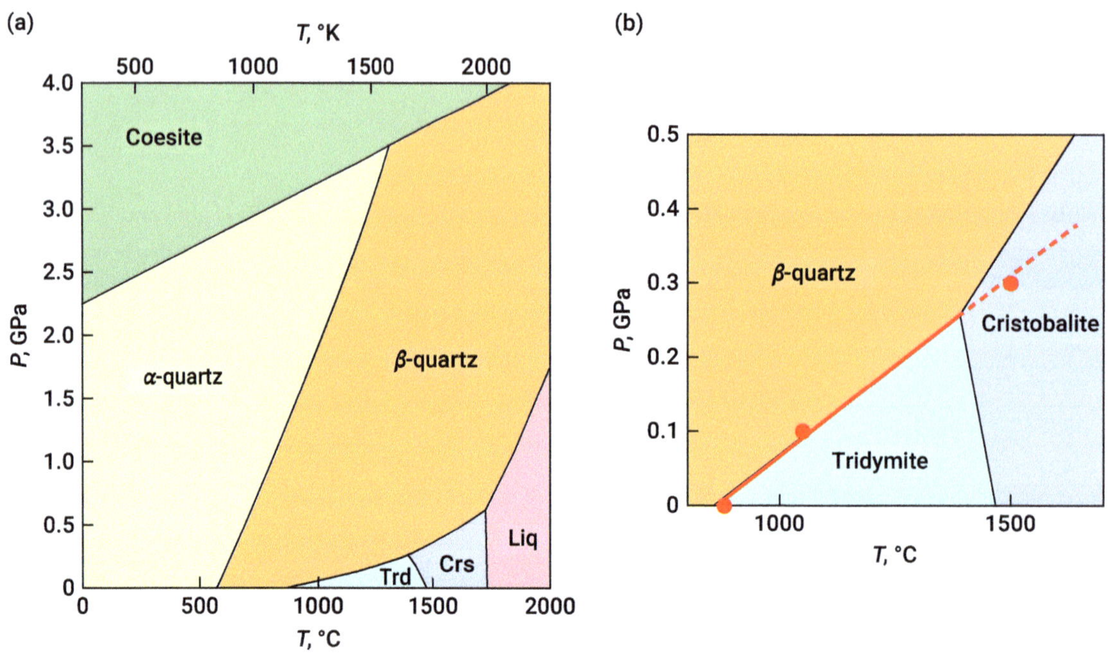

Figure 3.13 (a) Pressure-temperature phase diagram for the system SiO_2. Phase boundaries generally have positive slopes except for the reaction Trd = Crs. (b) Detail of the β-Qz = Trd reaction, with calculated points superimposed. There is good agreement between the calculated points and the published phase diagram. Our calculation at 1500 °C told us that the reaction should take place at ~300 MPa, but the phase diagram tells us that neither phase would be stable relative to Crs.

Source: From Swamy et al. (1994).

A calculated reaction pressure at 1500 °C is also shown—but it is in the Crs field. This illustrates an important consideration in thermodynamic calculations—we can only calculate *relative* stabilities. We predict that Trd is more stable than β-quartz at 1500 °C and 0.2 GPa, but Crs, ignored in our calculations, is more stable than either. The dashed red line in the Crs field is a metastable extension of the β-Qz = Trd reaction curve.

Metastability is a feature of the real world—for example, diamonds are metastable at surface pressure, but the high activation energy for the conversion of diamond to graphite indicates that spontaneous conversion takes on the order of a billion years at room temperature. In the 1963 novel *Cat's Cradle* by Kurt Vonnegut, a polymorph of ice known as ice-nine is found and kept secret for military use. Ice-nine has the disturbing property of being solid at room temperature (i.e., of having a lower G at 1 atm and 25 °C than liquid water). If any gets into the environment, it could act as a seed crystal and cause the world's surface water to freeze. Does that happen? Read the book and find out.

The system SiO_2 is a one-component system, because only one chemical component (SiO_2) is needed to describe the composition of any phase in the system. It is not particularly relevant to basalts, but it is an important system that illustrates many useful aspects of *P-T* diagrams.

Simple inspection reveals several features:

- Trd and Crs are only stable at low *P*, and we therefore should expect to see them in volcanic and shallow plutonic rocks, but not in rocks crystallized at higher *P*.
- Coesite is only present at high *P* (2 GPa and above); these are mantle pressures.
- If we heat up a sample of α-Qz (the familiar kind) at 1 atm, the diagram predicts that it will transform to β-Qz at 573 °C, Trd at 867 °C, Crs at 1470 °C, and Liq (molten SiO_2) at 1713 °C.
- The melting *T* increases somewhat with increasing *P*.
- Most of the phase boundaries (curves separating the stability fields of different phases) have positive slopes of varying steepness, but the Trd-Crs boundary has a negative slope.

Pressure-Temperature (*P-T*) Diagrams and the Clapeyron Equation

Melting of peridotite to generate basalt occurs over a range of *P-T* conditions and is commonly analyzed with the aid of *P-T* diagrams such as those in Figure 3.13. The slope of a phase boundary on a *P-T* diagram is given by the **Clapeyron equation** and is referred to as the Clapeyron slope:

Clapeyron equation
Used to predict the slope of a reaction boundary on a *P-T* diagram in terms of entropy and volume changes.

$$\frac{dP}{dT} = \frac{\Delta S}{\Delta V} \quad \textbf{(Eq. 3.6)}$$

The left side is the slope, and the right is the ratio of the entropy change to the volume change. "All well and good," you say, "but what is entropy?" Many have asked this question. Entropy is a thermodynamic measure of the disorder (lack of information) in a system; the less you know about something, such as the locations of atoms in a system, the greater the entropy (so you could say that the word "entropy" has high entropy). Thus, for any substance, $S_{solid} < S_{liquid} < S_{gas}$. For example, $S_{ice} < S_{water} < S_{steam}$. Values of S and V are tabulated for many solid phases, and so the slopes of solid-solid reactions can generally be calculated with some precision.

Consider the reaction curve for the transformation of α-quartz to β-quartz, which runs through the center of the diagram with a steep slope. For this reaction we can define:

$$\Delta S = S_{\beta-quartz} - S_{\alpha-quartz} \text{(entropy change)} \quad \textbf{(Eq. 3.7)}$$

$$\Delta V = V_{\beta-\text{quartz}} - V_{\alpha-\text{quartz}} \text{(volume change)} \quad \textbf{(Eq. 3.8)}$$

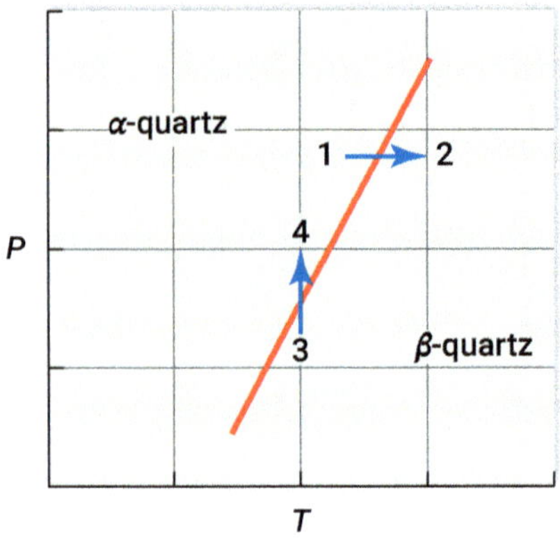

Figure 3.14 Analysis of Clapeyron slope.

There are two rules that help apply this to reactions on *P-T* diagrams (Figure 3.14):

- If, at constant *P*, you step across a reaction boundary in the direction of higher *T*, you are going in the direction of higher entropy.
- If, at constant *T*, you step across a reaction boundary in the direction of higher *P*, you are going in the direction of higher density.

Consider the boundary in Figure 3.14. Path 1-2 steps across the boundary isobarically (at constant *P*); path 3-4 steps across the boundary isothermally (at constant *T*). We can infer from this latter path that β-quartz is less dense than α-quartz. Figure 3.13 shows that all the solid-solid reactions have positive slopes except for the reaction tridymite = cristobalite; cristobalite must be denser, because it becomes stable as you step across the reaction in the direction of higher *P*.

A steep slope on a *P-T* diagram generally means that ΔV for the reaction is small. Such reactions can make good **geothermometers**, because they depend mostly on *T*, whereas flat slopes generally mean that ΔV is large, and such reactions can be used as **geobarometers** (see Section 12.4).

Geothermometer, geobarometer
A method, generally based on geochemical measurements of phase compositions, that can estimate *T* or *P* at a given point in a rock or magma's life cycle.

The Phase Rule

Phase diagrams are roadmaps of the phases that you will find for the given composition at a given set of conditions. If you were to throw a dart at the SiO_2 *P-T* diagram, it would most likely land in a single-phase field (e.g., α-quartz, cristobalite, etc.). You might happen to land within error of a phase boundary where two phases coexist, and you'd be very unlikely to hit a triple point where the fields of three phases come together. The number of phases to expect is given by the **phase rule**:

Phase rule
A thermodynamic formula relating the numbers of components, phases, and degrees of freedom in a system.

$$F + \pi = C + 2 \quad \textbf{(Eq. 3.9)}$$

where π is the number of phases, *C* is the number of components, and *F*, called the number of degrees of freedom, is the number of intensive variables that you can change a little and still maintain the same phase assemblage.

Assume that you throw your dart and it lands somewhere in the tridymite field. You can change either *P* or *T* or both a little and remain in the tridymite field. If you change either too far, you will cross into the field of another phase, but for small changes, the phase remains the same. You are in a 2D field and have two degrees of freedom; this is termed a **divariant assemblage**. Now, move to the tridymite-cristobalite boundary. Your phase assemblage has changed from tridymite ($\pi = 1$) to tridymite plus cristobalite ($\pi = 2$). By the phase rule, $F = 1$, and you can change one intensive variable, but not both arbitrarily, and maintain the phase assemblage. If you increase *T* a bit, you have to decrease *P* a bit to stay on the phase boundary; uncorrelated changes will push you into a one-phase field. Phase assemblages for which $F = 1$ are termed **univariant assemblages**, and on a *P-T* diagram they form 1D curves. Now, move to the tridymite-cristobalite-β-quartz triple point. Here, $F = 0$, and you are at an invariant point. If you change either *P* or *T* you destroy the assemblage.

Divariant assemblage
A phase assemblage in which you can make small independent changes in one or both of two intensive variables and maintain the same phase assemblage. On a *P-T* diagram, a divariant assemblage occupies an area between two reaction curves.

Univariant assemblages
Phase assemblages in which you can make small changes in only one intensive variable and maintain the same phase assemblage; other intensive variables must adjust to the change made. On a *P-T* diagram, a univariant assemblage lies along a reaction curve.

3.5 Melting Peridotite

Learning Objectives

By the end of this section you should be able to:

3.5.1 Contrast partial melting of peridotite with complete melting and explain the expected compositional differences between the melts so produced.

3.5.2 Interpret whether a given phase melts congruently or incongruently from a relevant phase diagram.

Solidus, Liquidus, and Melting Styles

Rocks and minerals have defined temperatures over which they melt. For a given composition at a given pressure (say, a certain peridotite at 1 GPa), the temperature at which first melt appears upon heating is called the solidus, and the temperature at which the last crystals disappear upon heating is the **liquidus**. Igneous petrology is concerned mostly with what happens between the solidus and the liquidus.

For some materials, such as water, Fe, or SiO_2, the solidus and liquidus are the same. At 1 atm, pure ice melts at 0 °C, and it melts entirely at 0 °C. For Fe and SiO_2, the corresponding temperatures are much higher, but the behavior is the same. Such behavior is called **congruent melting**; the solid and liquid have congruent compositions.

Rocks generally do not melt congruently. Rather, they undergo **incongruent melting**, over a range of temperatures, to a liquid that is different in composition from the rock composition. Experiments on peridotite under anhydrous conditions define a solidus that is roughly 1100 °C at 1 atm and increases by roughly 100 °C/GPa, with undulations or kinks where the peridotite mineralogy changes (Figure 3.15; also see Figure 3.9). Details depend upon the bulk composition of the starting peridotite, but regardless of the details, the melt formed at the solidus is basalt.

Figure 3.16 compares the starting material and partial melts from this study to a set of mid-ocean ridge basalts (MORBs) from the Mid-Atlantic Ridge. As we saw in Chapter 1, MORBs have a tight range of compositions compared to other tectonic environments, clustering in the basalt field around 50 wt% SiO_2 (see Figure 1.8). The starting peridotite is dramatically different from MORB in several oxides, especially MgO, Al_2O_3, Na_2O, and CaO. However, the partial melts are all basaltic, providing a good match.

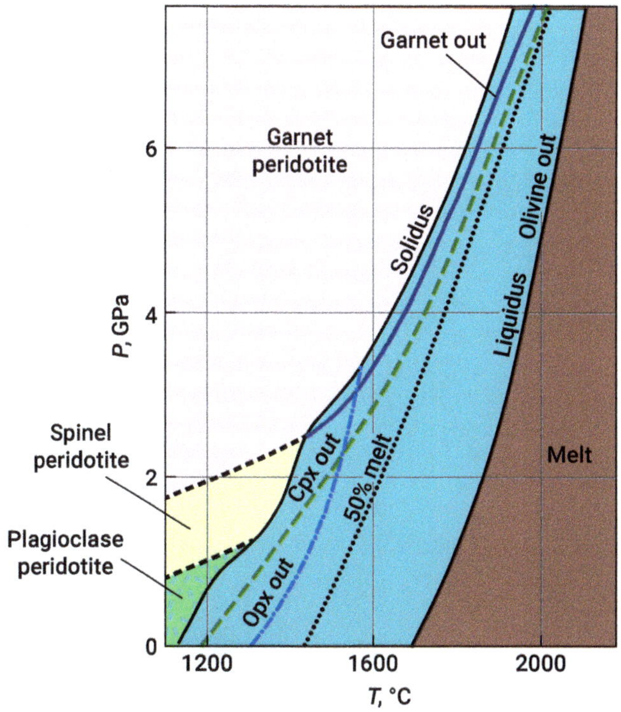

Figure 3.15 *P-T* diagram for melting of peridotite KLB-1. "Out" curves give the temperature at which the noted phase disappears upon heating; the last out curve, which is olivine-out at all pressures, is the liquidus. Plagioclase- and spinel-out curves are not shown but are near the solidus.

Source: Data from Takahashi et al. (1993).

Liquidus
For a given bulk composition at a given pressure, the temperature at which crystallization begins upon cooling.

Congruent melting
Melting of a phase in which the composition of the liquid formed is the same as the composition of the phase being melted.

Incongruent melting
Melting of a phase in which the composition of the melt differs from the composition of the phase being melted.

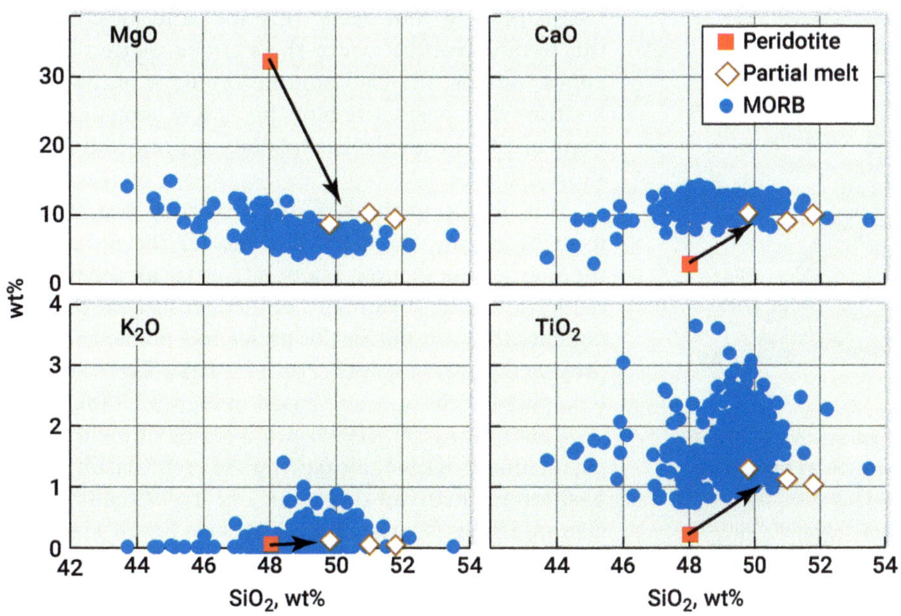

Figure 3.16 Basalts from the Mid-Atlantic Ridge compared with starting peridotite and partial melts. The experimental partial melts make a decent match for the basalts.

Source: Data from the experiments of Takahashi & Kushiro (1983).

The oxides Al_2O_3 and Na_2O are particularly interesting ones, because they are not found in end-member olivine, Cpx, or Opx. Both are found in plagioclase, the lowest-pressure form of peridotite, and Al_2O_3 is a major component of spinel and garnet, the Al-bearing phases in higher-pressure forms of peridotite. But what about Na_2O? When plagioclase breaks down under increasing pressure, the anorthite component can convert to grossular garnet (where Al is in sixfold coordination), but there is no Na-bearing garnet. Pressure-induced breakdown of albite occurs via the following reaction:

$$\text{albite} = \text{jadeite} + \text{quartz} \quad \textbf{(Eq. 3.10)}$$

Jadeite is a Na-bearing pyroxene, with Al in sixfold coordination. This reaction proceeds with a significant decrease in volume, and the former albite is stuffed into Cpx. In the spinel peridotite field, both Cpx and Opx contain significant Al_2O_3 (several wt%), and the Cpx contains significant Na_2O (a few wt%) as well. In this manner, Na is accommodated by minerals that nominally do not contain Na.

Residues of Melting, Classification of Ultramafic Rocks, and Several Skills

Consider a spinel peridotite such as the one melted by Takahashi et al. (1993; see Figure 3.15). Such peridotites typically consist of about 70 wt% olivine, 10 to 15 wt% each of Cpx and Opx, and a few wt% spinel. The particular peridotite used by Takahashi et al., named KLB-1 and collected from a xenolith in basalt from the Kilbourne Hole volcano in New Mexico, has approximately 60 wt% olivine, 23 wt% Opx, 16 wt% Cpx, and 2 wt% spinel. We can plot the three major constituents on a triangular diagram (Figure 3.17) used to classify ultramafic rocks and see that this rock plots in the peridotite field.

Figure 3.17 is our first triangular compositional diagram. These are used extensively in geochemistry, and basic skills for understanding them are in Toolbox 3.2.

What about spinel? Since it only makes up a few percent of the rock, we're ignoring it. Technically, we are projecting from all the mineral constituents that we're ignoring. Olivine, Opx, and Cpx make up the three components of the classification triangle, with all other mineral components at the fourth apex of the tetrahedron (see Figure 3.18). Projecting from these components means putting your eye at this apex and looking into the tetrahedron at where the various plotted points appear against the background (the floor of the tetrahedron). The peridotite plots just a few percent off the floor of the tetrahedron, and thus is projected only a short distance.

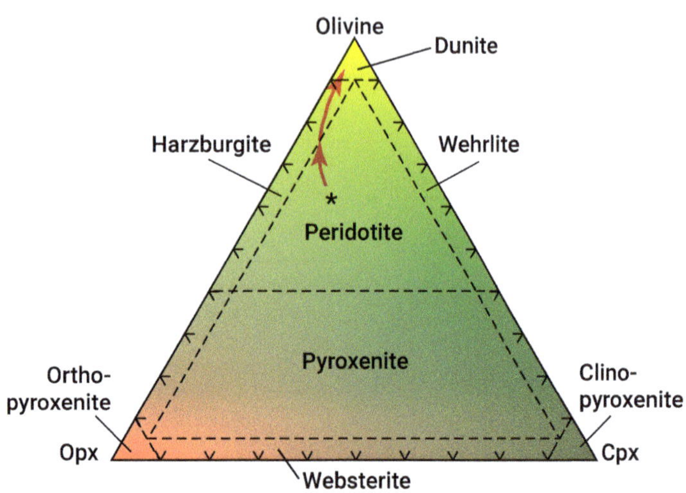

Figure 3.17 Simplified classification of ultramafic rocks, with a typical spinel peridotite plotted. Arrows give pathway of residue into harzburgite and dunite fields as basalt melt is extracted from it. Of the names on this diagram, peridotite, pyroxenite, harzburgite, and dunite are the most commonly used.

Source: Data from Streckeisen (1976).

It is helpful to think of this projection process as a form of casting shadows. The shorter the projection, the sharper the shadow. If a point lies far off the floor of the tetrahedron, meaning that it contains a lot of components that are not in the projection plane, then the projected position is less certain; it has a fuzzy shadow (compare harzburgite and basalt in Figure 3.18b).

Takahashi et al. (1993) ran a series of melting experiments on KLB-1, measuring the melt (basalt) composition for many of them. For one experiment, at 1.5 GPa and 1450 °C, they found 24 wt% basalt melt in a matrix consisting of Ol + Opx. All the Cpx and spinel had been melted away. What is the composition of the

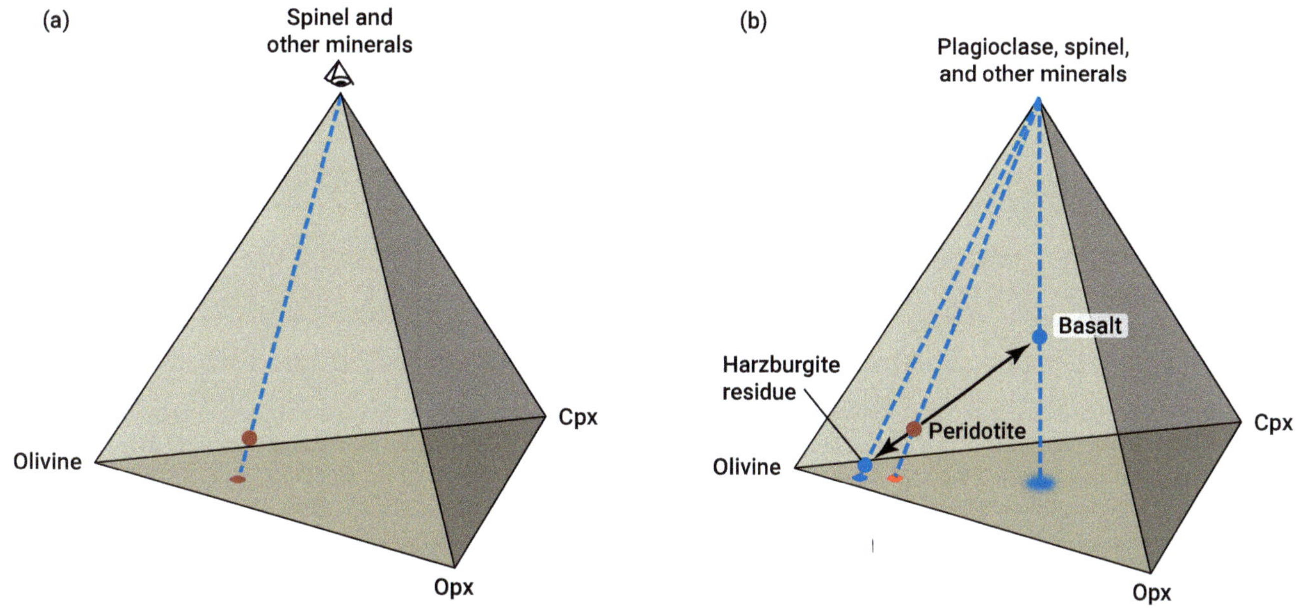

Figure 3.18 Projection from a tetrahedron to a triangle. (a) Just imagine putting your eye at the apex of the tetrahedron (ouch!) and looking into it; you'll see the bottom triangle, with the point in three dimensions projected onto it. (b) Here, a peridotite, a basalt extracted from it by melting, and the harzburgite residue are plotted within the tetrahedron and projected onto the Ol-Opx-Cpx base (dotted lines). Because the basalt is farther from the projection plane, its projection is fuzzy and less well-defined than the others.

Ol + Opx residue left behind? We can calculate it because we know the starting peridotite composition, the composition of the melt, and the proportions of each. This sort of calculation is demonstrated in Toolbox 3.3, and the resulting residue composition is in Table 3.4. It has a lot of MgO, low SiO_2, some FeO, and not much else. Sounds like it must have a lot of olivine.

We have the measured composition of the basalt and the calculated composition of the residue, but these are oxide compositions. How can we plot them on a mineralogical diagram such as Figure 3.17? One method is to calculate a **CIPW norm**, a set of standardized minerals that, added together, add up to the rock composition. The underpinnings of this tedious calculation[1] are given in Toolbox 3.1.

In Table 3.4, the mineralogical composition of the three components treated in the subtraction toolbox are presented as CIPW norms. The norm is one way to recalculate a rock composition into minerals, but not the only way. The norm commonly deviates significantly from the mode[2] of the rock, and this is true for KLB-1. The norm calculation involves a fixed set of minerals that may or may not be fully appropriate; in particular, there are no hydrous minerals such as biotite or hornblende in the norm.

These compositions are plotted in Figure 3.18 and projected onto the Ol-Opx-Cpx base of the tetrahedron. The calculated residue is harzburgite (Ol-Opx rock), and the

CIPW norm
An algorithm for converting the oxide composition of an igneous rock into a standard set of anhydrous minerals that describes a possible equilibrium assemblage for that bulk composition.

Table 3.4 Norm and Mode of a Peridotite and Norms of Its Partial Melting Products

Mineral	Peridotite KLB-1 Norm (wt%)	Peridotite KLB-1 Mode (vol%)	Basalt Norm	Harzburgite Residue Norm
Olivine	65.7	60	16.8	82.3
Opx	14.4	23	16.3	12.4
Cpx	6.7	16	18.1	3.4
Other	13.2		48.8	1.9

basalt plots well off the base toward the upper apex, because it contains nearly 50 percent plagioclase. Removing basalt from the peridotite drives the residue directly away from the basalt, toward harzburgite, and further extraction of basalt at higher temperature drives the residue into the dunite field as all the Cpx and most of the Opx is eliminated by melting.

Toolbox 3.1

Converting Chemical Analyses into Minerals: The CIPW Norm

Whole-rock compositions can be displayed in tables or graphs, but it is useful to think about compositions in terms of the minerals that could form if the rock fully crystallized. The hypothetical minerals determined in this process are called normative minerals. This norm calculation is particularly useful for volcanic rocks, where much of the rock is uncrystallized glass or the minerals are too small to identify. To get around this, the norm calculation uses the chemical composition to produce a hypothetical set of minerals for that composition.

The recipe for doing this was developed in the late 19th century and has been given the name CIPW norm in honor of its developers (Cross et al., 1902). They based their recipe upon what they could see in thin section—that is, which minerals tend to occur together, and which do not. The calculation can be tedious and prone to error but is easily done with a spreadsheet or other programming tool.

The following exercise illustrates how a norm is calculated, using a particularly simple rock. The analysis (Rock 1) is given in Table 3.5. You are able to identify plagioclase, diopside, and forsterite in a hand sample and/or thin section and have analyzed the rock for major elements. Given this analysis, what proportions of the observed minerals will add up to the rock composition?

The steps in the norm calculation consist of filling in Table 3.7. As we are only carrying three significant digits in this exercise, there will be minor rounding error.

1. Begin by writing out the formulas of the minerals in oxide form, splitting plagioclase into its end-members albite and anorthite. Oxide form is simply expanding the formula such that each cation is written as its oxide (note that this gives a doubled albite formula). Formula weights are included for convenience (Table 3.6).
2. Calculate the number of moles of each oxide in the rock and write your results in the second row of Table 3.7. Note that these numbers have units of moles oxide per 100 g rock.
3. Now, note that the only mineral that has Na in it is albite; thus, the number of moles of albite in the result will be the same as the number of moles of Na_2O in the rock (0.19). Write that in the third row under Na_2O and then fill in the rest of the row (e.g., if you take out *x* moles of Na_2O, then, by the formula from earlier, you need to take out 0 moles of CaO, *x* moles of Al_2O_3, 0 moles of MgO, and 6*x* moles of SiO_2). The column giving the moles of the mineral in the rock is simply the number of moles of Na_2O.
4. Continue the process for the other oxides. (Hint: note that, of the remaining minerals, only anorthite has Al in it, so that is the next one to deal with.) Using this logic, you can fill out the rest of the table, and if you've done the arithmetic correctly, you will end up with 0 moles of each oxide left (barring rounding error), meaning that they have all been allocated to minerals.
5. Finally, calculate the number of grams of each mineral in the norm by multiplying the number of moles (remember, this is now moles mineral per 100 g rock) by the formula weight of each mineral in step 1 above. The units are thus $\left(\frac{\textit{moles mineral}}{\textit{100 g rock}}\right)\left(\frac{\textit{g mineral}}{\textit{mole mineral}}\right)=\frac{\textit{g mineral}}{\textit{100 g rock}}$, or weight percent.

Table 3.5 Rock Composition for Norm Calculation

Oxide	Rock (wt%)	Oxide Molecular Weight
Na_2O	1.2	62.0
CaO	14.4	56.1
Al_2O_3	9.3	102.0
MgO	24.6	40.3
SiO_2	50.5	60.1
Total	100	

Table 3.6 Minerals in Oxide Form

Mineral	Mineral Formulas in Oxide Form	Oxide Molecular Weight
Albite	$Na_2O{\cdot}Al_2O_3{\cdot}6SiO_2$	524.4 g/mol
Anorthite	$CaO{\cdot}Al_2O_3{\cdot}2SiO_2$	278.2 g/mol
Diopside	$CaO{\cdot}MgO{\cdot}2SiO_2$	216.6 g/mol
Forsterite	$2MgO{\cdot}SiO_2$	140.7 g/mol

Table 3.7 Example of a Norm Calculation

	Na_2O	CaO	Al_2O_3	MgO	SiO_2	Moles mineral in rock	Grams mineral in rock	Notes
Moles oxide in rock	0.019	0.257	0.091	0.610	0.840			From step 2 above
In albite	0.019	0.000	0.019	0.000	0.116	0.019	10.15	Moles albite = moles Na_2O
Left after making albite	0.000	0.257	0.072	0.610	0.724			Na_2O zeroed out; Al_2O_3, SiO_2 decreased
In anorthite	0.000	0.072	0.072	0.000	0.144	0.072	19.98	Moles anorthite = moles Al_2O_3
Left after making anorthite	0.000	0.185	0.000	0.610	0.580			Al_2O_3 zeroed out; MgO still original quantity
In diopside	0.000	0.185	0.000	0.185	0.370	0.185	40.02	Moles diopside = moles CaO remaining
Left after making diopside	0.000	0.000	0.000	0.426	0.211			Only MgO and SiO_2 left
In forsterite	0.000	0.000	0.000	0.426	0.213	0.213	29.94	
Left after making forsterite	0.000	0.000	0.000	0.000	–0.002			Small negative SiO_2 is rounding error
Sum							100.1	Total >100 is rounding error

Toolbox 3.2

Plotting on Triangular Diagrams

There are many methods for plotting data on a triangular diagram. One is to get a computer to do it for you, but that's not always possible. Doing it by hand is easy, but takes a bit of thought. This is an important skill, because triangular compositional diagrams are ubiquitous in geochemistry, and practice plotting compositions is essential to understanding them.

Plotting on a triangular diagram (Figure 3.19) is different from a standard *x-y* plot, because the three components being plotted must add up to 100 percent. It is the 3D equivalent of plotting two components on a line segment, where the position of the point gives the proportions of the two ingredients. The first step in plotting a point on a triangular diagram is to normalize the three numbers to 100 percent (weight %, mole %, whatever units you're using). Let's use Ol, Opx, and Cpx from the peridotite norm in Table 3.4 as an example. These numbers are 65.7, 14.4, and 6.7, which, normalized to 100 and rounded, become 76, 16, 8. If you have gridded triangular graph paper handy and are plotting on an equilateral triangle, then you can plot these numbers directly. If not, read on.

Note that any line that is parallel to one side of the triangle is a line of constant percentage of the third component. Thus, line de in Figure 3.19b is 20 percent of A. Any line that radiates out of an apex is a line of constant ratio of the other two components. Thus, line fA has a constant ratio (7:3) of B to C.

An easy way to plot the olivine-Opx-Cpx data on such a triangle is to simply draw a light line 76 percent of the way from the Opx-Cpx side toward olivine (red line in Figure 3.19c). The rock must lie on this line. Then draw another light line 16 percent of the way toward Opx from the olivine-Cpx side (blue line). The rock must lie on this line as well, and thus at the intersection of the two lines. Done; it plots in the peridotite field.

Plotting on scalene triangles follows the same rules. For example, we can take the example just done and shear it (Figure 3.19d). The red and blue lines are the same ones plotted earlier. The angles are changed, but the relative distances have not.

continued

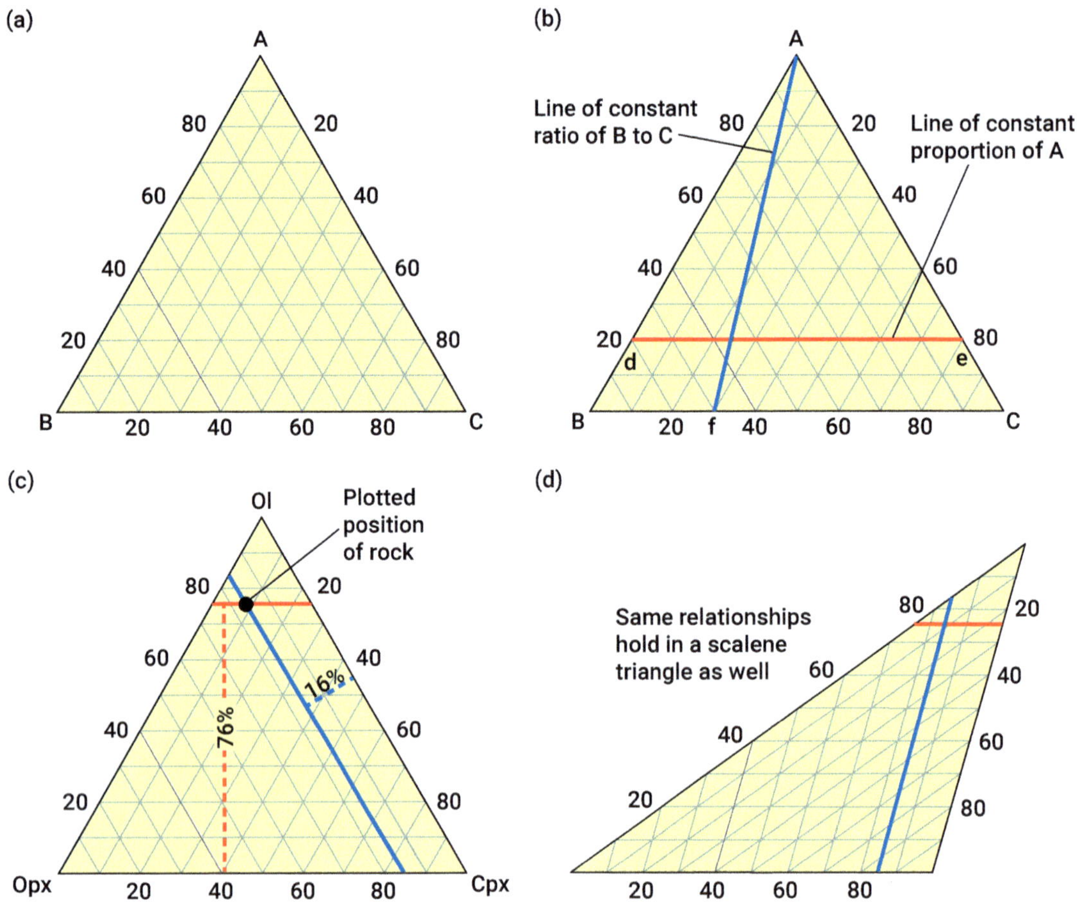

Figure 3.19 Plotting on triangular diagrams. (a) The raw triangle in a generic system ABC. (b) Lines radiating from one apex are compositions with a constant ratio of the other two components; lines parallel to one side of the triangle are compositions with a constant percentage of the third component. (c) Our peridotite composition plots where lines of 76 wt% Ol and 16 wt% Opx intersect; the percentage of Cpx is not needed but should agree with the plotted value. (d) The same geometric relationships hold in a scalene triangle.

Toolbox 3.3

Subtraction Calculations

We can write a mass-balance equation for the peridotite melting problem as

$$peridotite = basalt + residue \quad \textbf{(Eq. 3.11)}$$

or

$$P = B + R \quad \textbf{(Eq. 3.12)}$$

where *P*, *B*, and *R* are vectors representing the compositions of the peridotite, basalt, and residue, respectively (e.g., *P* is the first numeric column in Table 3.8). The elements of *P* are the number of grams of the corresponding oxide per 100 g of the rock. We are subtracting 24 wt% of *B* from *P*, so we multiply each element of *B* by 0.24 to get the number of grams of each oxide that is being subtracted from the starting 100 g of peridotite. This gives us the number of grams of each oxide in the residue *R*. However, these only add up to about 75.5 g. Chemical analyses are in wt%, so we need to normalize the resulting analysis to 100. The resulting vector is *R*.

Mixing calculations are handled the same way; the algebra is identical.

Table 3.8 Example of a Subtraction Calculation

	KLB-1	Melt at 1.5 GPa, 1450 °C	Subtract 24% Melt	Residue (previous column normalized to 100%)
SiO_2	44.48	48.8	32.77	43.40
TiO_2	0.16	0.5	0.04	0.05
Al_2O_3	3.59	14.7	0.06	0.08
FeO_t	8.1	7.6	6.28	8.31
MnO	0.12			
MgO	39.22	15	35.62	47.18
CaO	3.44	11.3	0.73	0.96
Na_2O	0.3	1.5	0.00	0.00
K_2O	0.02			
Total	99.43	99.4	75.49	100.00

3.6 Chemographic Diagrams and Mass Balance

Learning Objectives

By the end of this section you should be able to:

3.6.1 Demonstrate the principle of projection by projecting points within a triangle onto one side of the triangle.

3.6.2 Infer the form and stoichiometry of a chemical reaction from a chemographic diagram.

Mass Balance and Baldy's Law

Peridotite undergoes incongruent melting to form basalt and a solid residue of melting. Because the basalt and starting peridotite differ in composition, the residue must be different still, because of a fundamental principle of mass balance. This critically important principle, which we will call **Baldy's law** (the origin of this name is lost in the mists of time), is stated thus:

BALDY'S LAW

Some of it plus the rest of it equals all of it.

This is simply a statement of mass balance—everything must be accounted for. It may sound obvious, but it is a remarkably useful tool for analyzing geochemical problems. Conversely, it is easy to get into geochemical trouble if you do not honor it.

How Can We Visualize Chemical Reactions?

There is yet another way to examine these melting relationships: **chemographic diagrams**. These are diagrams used to graphically balance chemical reactions—essentially graphical expressions of Baldy's law. Chemographic diagrams are buried in many common representations of geochemical and phase equilibrium data, and it is worth the effort to learn how to recognize and exploit them.

Chemographic diagrams Graphical representations of a chemical reaction.

Consider the simple reaction of periclase plus silica to form forsterite: $2MgO + SiO_2 = Mg_2SiO_4$, or Per + Qz = Fo, where Qz is understood to mean whichever silica polymorph is stable at the relevant conditions. You have had significant experience in dealing with such reactions written as formulas, but we can represent this reaction with a line like this:

Forsterite is a mixture of the oxides Per and Qz, so it goes in between—but where? The answer depends on what kind of units we're using, weight or molar. In many geochemical calculations, weight units are more useful, in part because they are conserved: if there are *x* grams of reactants, then there are *x* grams of products. Molar units are not conserved—in our reaction there are three moles of reactant oxides, but only one mole of products.

One mole of Fo is made by mixing 2 moles of Per (40.3 g/mol) with 1 mole of Qz (60.1 g/mol). Fo thus has a molecular weight of 140.7 g/mol and is $(2 \times 40.3)/140.7 = 0.573$ Per and $60.1/140.7 = 0.427$ Qz by weight. We can plot Fo on the line by putting it 0.573 of the way (or 57.3% of the way) toward Per and 0.427 of the way toward Qz:

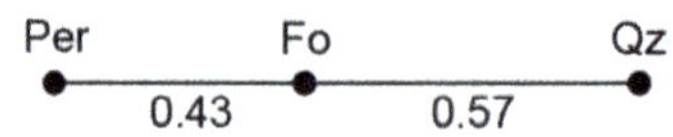

Where would enstatite plot on this diagram?

Note that you can see at a glance, without even knowing the formulas of the phases on the diagram, that the reaction is Per + Qz = Fo, because Fo lies between the other two. No other reaction is possible.

Lever rule
Graphical method for determining the proportions of phases in a system from a chemographic diagram.

If we were just given the diagram without the weight fractions, we could determine them using the **lever rule**. The distance from Fo to Qz (b) divided by the total line segment length, $b/(a + b)$, is the weight proportion of Per in Fo, and $a/(a + b)$ is the weight proportion of Qz.

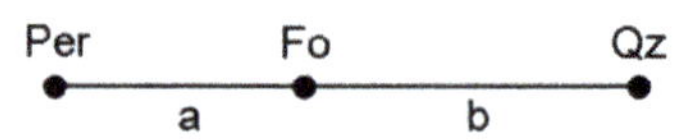

This is a simple 1D example of a system with two chemical components. With three components, you need a triangle, which is 2D; with four components, you need a tetrahedron, which is 3D; etc.

Figures 3.17 and 3.18 are chemographic diagrams, and so is Figure 3.20, a plot of MgO vs. SiO_2 for the peridotite, its minerals, the basalt, and the residue (we omit spinel for simplicity, because it is present in such a small amount). On this plot (or a plot of any other pair of oxides), the peridotite must be the sum of the minerals that make it up (Baldy's law), so the proportions of Ol, Opx, and Cpx in the scalene triangle bounded by those minerals must add up to the bulk composition. Using our method for plotting in scalene triangles (essentially, the lever rule in a ternary system), we find the proportions of Ol, Opx, and Cpx to be 63, 19, and 18—close to the proportions given in Table 3.4. Thus, the diagram is the chemographic expression of the reaction

$$0.63 olivine + 0.19 Opx + 0.18 Cpx = peridotite \qquad \textbf{(Eq. 3.13)}$$

The proportions are not identical to the proportions in Table 3.4, because we're only considering two oxides and not using all the data.

Interior reaction
A reaction where a single product is made by combining multiple reactants; on a chemographic diagram the product lies within the volume formed by the reactants.

The general form of the reaction is shown by extracting just the compositions of the relevant phases in their proper relative positions (Figure 3.20b). The peridotite lies within the triangle, consistent with the reaction described; we will call this sort of reaction, where the product lies within the volume formed by the reactants, an **interior reaction**. We use "volume" in a general sense here; the Per + Qz = Fo reaction is an interior reaction in which the relevant volume is simply the line segment.

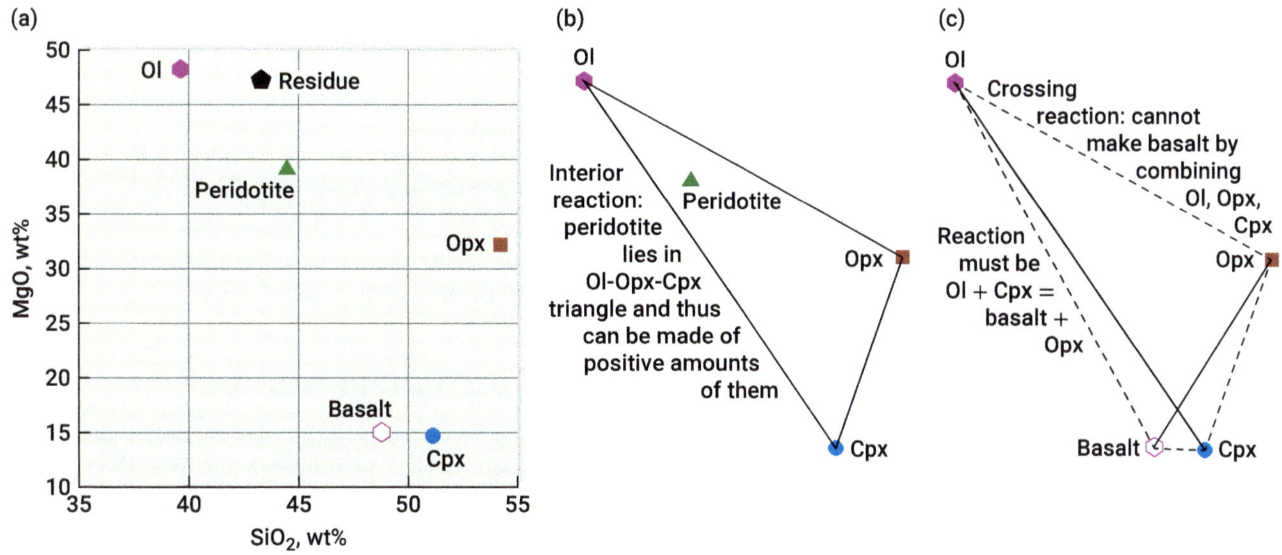

Figure 3.20 Chemography of basalt production from peridotite. (a) Plot of MgO vs. SiO_2 for peridotite KLB-1, its major constituent minerals, the basalt produced by melting, and the residue left after extraction of basalt (see Table 3.4). Many pairs of oxides would work for this sort of diagram as long as the different minerals have different amounts. K_2O, for example, would not be a good choice, because it is nearly absent in the minerals and whole rock. (b) Chemographic diagram extracted from (a) for the peridotite mineral composition. The plotted composition of the peridotite lies within the triangle defined by Ol, Opx, and Cpx, and thus the peridotite can be expressed as a positive combination of those minerals. The proportions of the minerals can be determined by using the ternary lever rule, because, by Baldy's law, the proportions of the minerals must add up to the whole rock (ignoring minor constituents such as spinel). If the peridotite does not fall within the triangle defined by the minerals, then something is wrong. (c) Chemographic diagram for production of basalt. The basalt does not lie within the Ol-Opx-Cpx triangle, and so it must be produced by reaction of Ol + Cpx to produce basalt + Opx. More about that in Section 4.2.

How about the basalt melting reaction? If we try the same exercise to find the proportions of Ol, Opx, and Cpx in the basalt, we run into a problem—the basalt plots outside the triangle defined by the minerals, and the four points form a quadrilateral with no point in the middle (Figure 3.20c). There is no reaction in which three phases can be combined to make a fourth, as in the reaction described.

The chemographic diagram shows there is only one reaction that can be made with the three minerals and basalt: basalt + Opx = Ol + Cpx. This is a **crossing reaction**. The intersection of the Cpx-olivine and basalt-Opx lines represents a point where those two assemblages have equivalent compositions, and the proportions of the line segments on either side of the intersection give the reaction coefficients:

Crossing reaction
A reaction where multiple products are made by combining multiple reactants; on a chemographic diagram, the geometric figure (line segment, triangle, etc.) crosses by the products intersects the figure formed by the reactants.

$$0.89cpx + 0.11olivine = 0.80basalt + 0.20opx \quad \textbf{(Eq. 3.14)}$$

Now, if we were to do this same exercise with a different pair of elements, we might get a different answer owing to experimental error, the fact that we're ignoring spinel and other phases, etc. In practice, these problems are done using least-squares analysis, which considers all the data.

The chemographic diagrams thus offer a graphical means for visualizing and solving the reactions. Note that in Figure 3.20a the line connecting the residue and basalt goes through peridotite and is itself a 1D chemographic diagram for the reaction

$$peridotite = basalt + residue \quad \textbf{(Eq. 3.15)}$$

What are the proportions of basalt and residue?

3.7 Upwelling and Melting

Learning Objectives

By the end of this section you should be able to:

3.7.1 On a *P-T* diagram, sketch the geotherm under a spreading ridge, the peridotite solidus, and the general path a rising parcel of mantle follows during adiabatic ascent.

3.7.2 Explain why the amount of melt produced during adiabatic ascent is positively correlated with the higher potential temperature of the mantle source.

Adiabatic Upwelling and the Péclet Number

There is little question[3] that MORB forms by decompression melting of the mantle. The key to this process lies in the positive Clapeyron slope (remember, this is the slope of a reaction on a *P-T* diagram) of the basalt solidus (see Figures 3.9 and 3.15). If a parcel of mantle that is near the solidus is lifted to shallower depth and thus lower pressure, it can cross the solidus and melt. The rising limbs of convection cells provide such an environment. Convection is (probably) driven by sinking of dense lithospheric plates at subduction zones; spreading occurs as a passive response to this sinking; mantle wells up to fill the space produced by spreading.

Rates of plate divergence at oceanic ridges are well-established (see Figure 3.3), but the velocity of upwelling under oceanic ridges depends upon the geometry of mantle flow, and this is not well known. One typical model is shown in Figure 3.21; in such

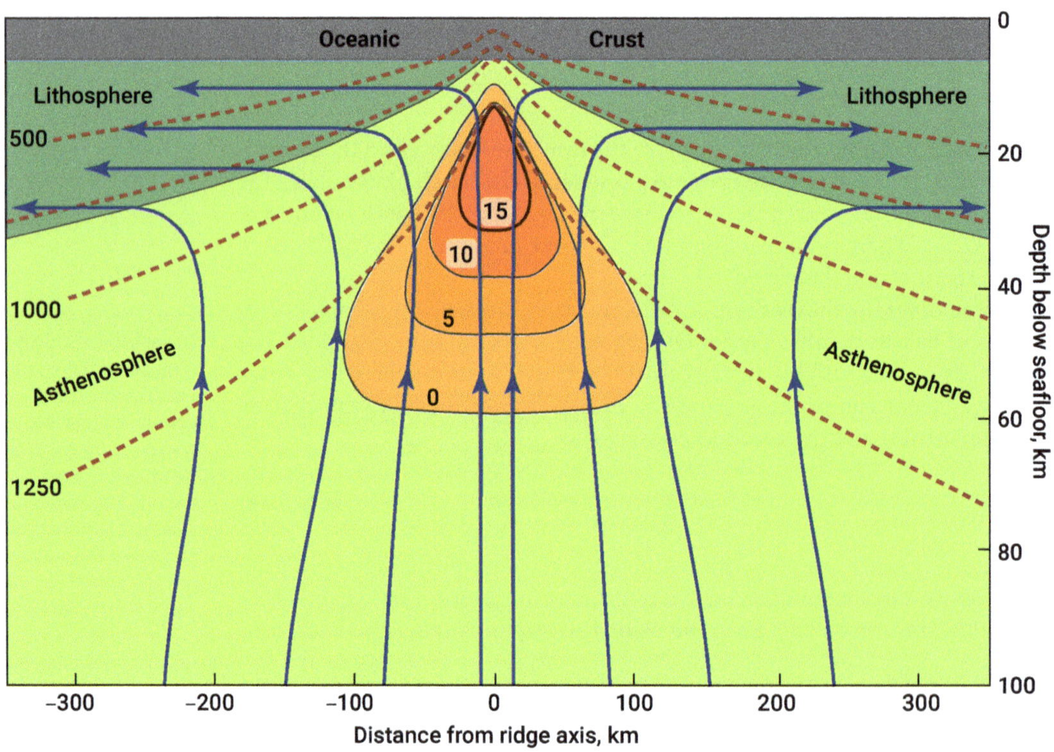

Figure 3.21 Schematic view of melting under an oceanic ridge showing calculated isotherms (dashed red lines) and schematic mantle flow lines (blue arrows). Orange-shaded regions enclose areas undergoing melting; contours are wt% melt. Oceanic crust is in gray, and growing lithospheric plate (defined by 800 °C isotherm) is in dark green. Isotherms are dragged up by mantle flow; their steepness is a reflection of the high Péclet number.

Source: Adapted from Ligi et al. (2008).

a geometry, the velocity of ascent under the ridge equals the surface velocity (the half spreading rate). This averages about 22 mm/yr worldwide, ranging up to about 75 mm/yr (see earlier).

This inch-per-year velocity is slow, but diffusion of heat in rock is even slower. The ratio of the rate at which material is transported by motion to that at which heat is transported by diffusion is a dimensionless number known as the **Péclet number**, given by

$$\text{Pe} = \frac{Lv}{\kappa} \quad \textbf{(Eq. 3.16)}$$

Péclet number
A dimensionless number giving the ratio of the rate at which material moves to the rate at which heat is transported by diffusion.

where v is velocity, κ is **thermal diffusivity**, and L is a length scale relevant to the problem at hand. If Pe is significantly greater than unity, then movement of heat via flow of the material dominates over thermal diffusion, and if Pe is significantly less than unity, then thermal diffusion swamps out material motion.

Thermal diffusivity
A coefficient that describes how rapidly a temperature perturbation moves through a material by conduction.

Dimensionless numbers are fuzzy things, because they depend upon parameters that are only loosely known and on vague concepts such as "length scale." Nevertheless, they are quite useful, and we will run into several of them in our study of petrology. For oceanic ridges, we can use the value of 10^{-6} mm^2/s for κ and $v = 20$ mm/yr. What about L? A lithospheric length scale seems appropriate, so let's use $L = 100$ km. Plugging into the definition for Pe (and not forgetting to change the velocity from mm/yr to m/s), we get Pe ~ 60. This is much greater than unity, high enough that we can consider the rate of uplift to outstrip the rate at which heat leaks out of a rising parcel of mantle. Note conversely that on the length scale of a dike (say, 1 m), our 20 mm/yr velocity is not fast enough to ignore heat flow, as Pe is on the order of 10^{-3}.

Such a process, in which the rate of heat flow can be ignored, is referred to by the thermodynamic term **adiabatic**, and upward flow of mantle under oceanic ridges is referred to as adiabatic upwelling. Ignoring heat flow does not mean that temperature is constant, any more than adding heat to a boiling pan of water means that the temperature of the water will rise. Temperature during adiabatic ascent of mantle, before it begins to melt, depends on a variety of parameters but should drop by about 1 °C for each kilometer of ascent; thus, a parcel rising from 100 km depth would cool by about 100 °C.

Adiabatic
An adiabatic process is one in which heat does not enter or leave the system.

The concept of an adiabatic process also applies to convecting air, where the drop in temperature that a rising parcel of air undergoes is referred to as the **lapse rate**, typically about 6 °C/km. The lapse rate is what causes mountain air to be cooler than valley air. It is strongly affected by humidity, and under conditions of higher humidity, the lapse rate can lead to atmospheric instability and thunderstorms.

Lapse rate
The rate at which temperature decreases with altitude.

Melting during adiabatic upwelling is shown schematically in Figure 3.22. The geotherm in the convecting part of the mantle follows the mantle lapse rate (a very low geothermal gradient) and is thus quite steep on a T-depth plot. Away from upwelling or downwelling limbs of convection cells, the geotherm is conductive and must end up near zero at the surface. The sharp bend in the geotherm between the convecting mantle and the lithosphere marks the base of the lithosphere (again, a gradational definition), and the lithosphere is sometimes referred to as a **thermal boundary layer**.

Thermal boundary layer
Region parallel to a boundary over which the variation in temperature is significant.

Under oceanic ridges, however, rapid ascent of mantle leads to much hotter mantle at shallow depths than occurs away from oceanic ridges. Rising mantle follows an adiabatic path (black dashed line) and is not constrained to follow the conductive geotherm through the lithosphere. If it crosses the peridotite solidus, melting occurs (see Figure 3.22). Extrapolating the adiabatic ascent path to the surface gives the **potential temperature**, T_p, the temperature that a mantle parcel would be at if brought rapidly to the surface in the absence of melting. The energy available for melting is proportional to the difference between T_p and the solidus at a given pressure. If T_p is below the 1-atm solidus, no melting occurs, and the higher T_p is, the more melting occurs. Thus, the total thickness of oceanic crust produced by

Potential temperature
The temperature that a rising parcel of mantle material would attain if brought to the surface adiabatically.

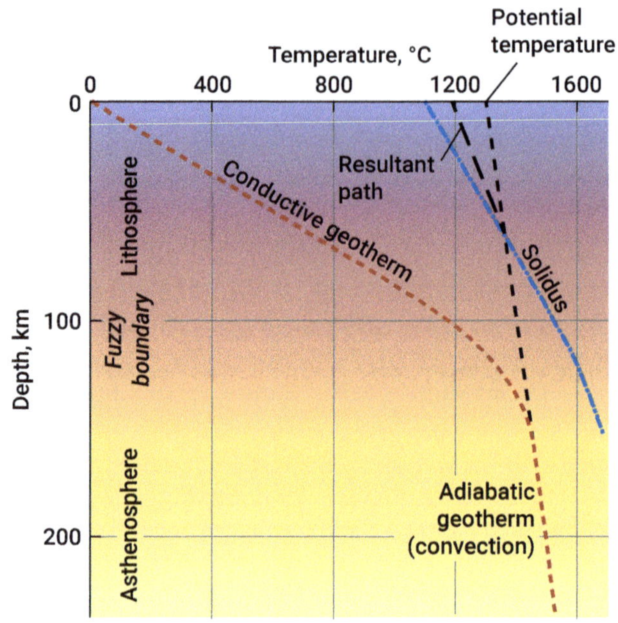

Figure 3.22 Idealization of the temperature profile in the asthenosphere and lithosphere. In the former, heat transfer by convection maintains a steep, near-adiabatic thermal gradient (note that a steep curve on this plot means a low geothermal gradient); in the latter, heat transfer by conduction causes a steeper geothermal gradient. Thus, most of the temperature variation occurs in the lithosphere, which is thus a thermal boundary layer. A rising parcel of magma encounters the solidus at about 60 km depth.

Source: Modified from McKenzie & Bickle (1988).

this mechanism depends on T_p. The higher T_p is, the deeper melting begins and the greater the thickness of basalt produced.

Consider a parcel rising along an adiabatic path like that in Figure 3.22. Melting begins when the parcel crosses the solidus (in this schematic example, at about 60 km depth). Melting is a phase change and consumes energy, so the rising parcel cools more rapidly than it would by adiabatic processes alone, following a path like the dashed resultant path in Figure 3.22. Flow paths taken by the rising and spreading mantle are generally idealized as in Figure 3.21.

Thickness of Oceanic Crust and Potential Temperature

A worldwide survey (Bown & White, 1994) shows that oceanic crustal thickness is fairly constant at about 7 ± 1 km, regardless of spreading rate, for full spreading rates greater than about 20 mm/yr (Figure 3.23). This result, perhaps surprising at first blush, is consistent with the notion that the total amount of melting in a rising column of mantle depends on T_p. The drop in thickness at slow spreading rates may result from failure of the adiabatic assumption; at slower ascent rates there is more heat loss to the wall rocks. Note that the calculated value of Pe for full spreading rates of 10 mm/yr is still much greater than unity, predicting that the adiabatic assumption should be good. As noted before, dimensionless number concepts are fuzzy, order-of-magnitude things.

Polybaric Melting and Focusing at Ridges

Polybaric melting
Mantle melting that occurs at successively shallower depths as a parcel of mantle rises; the resulting magma is a mixture of melts derived over a range of pressures.

Estimates of geotherms and solidus temperatures discussed earlier (see Figure 3.22) indicate that adiabatic melting typically commences at depths of around 60 km, and melting continues as the mantle parcel rises toward the surface. Thermal calculations indicate that rising mantle should melt a few tenths of a percent per kilometer of ascent (Sleep, 2015), and this adds up to around 7 km over the 60 km or so of the ascent path that is above the solidus. Melting over a range of depths (and thus pressures) is referred to as **polybaric melting**. How and where these bits of melt get to the surface, and how much

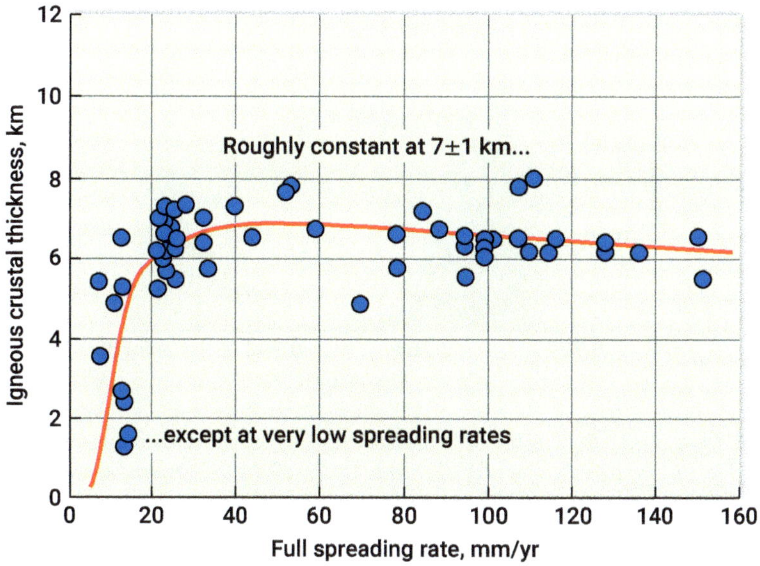

Figure 3.23 Oceanic crustal thickness vs. spreading rate, showing that thickness is fairly constant at 7 ± 1 km regardless of spreading rate, until the spreading rate falls below 20 mm/yr.

Source: Modified from Bown & White (1994), © 1994 Published by Elsevier B.V.

they mix upon ascent, is not well understood, but theoretical and observational data indicate that basaltic melt can escape its source even if the amount of melting is only on the order of 1 percent (Faul, 2001). As melt is generated and escapes to rise, it could have opportunities to mix with previously generated batches.

An important constraint on melt ascent is that the thickness of oceanic crust is relatively constant across oceanic ridges (see Figure 3.21). This implies that magma emplacement by lava flows, diking, and intrusion is focused at least somewhat at ridges—if not, then oceanic crust should thicken away from oceanic ridges. Consider the model in Figure 3.21. In this model, the pear-shaped melting zone extends over 100 km laterally from the ridge. A new bit of oceanic crust generated at the ridge will travel away from the ridge at the half-spreading rate, passing over the entire melting zone on that side of the ridge. If melt were to rise vertically from its point of generation and to the surface without freezing, then each bit would be added to the crust that was generated closer to the ridge, and the crust would thicken noticeably as it traveled away from the ridge. As any such thickening is minor, the majority of melt must reach the surface near the ridge.

3.8 Sr-Isotope Geochemistry of Oceanic Ridges

Learning Objective

By the end of this section you should be able to:

3.8.1 Explain why the estimated bulk-Earth $^{87}Sr/^{86}Sr$ ratio of approximately 0.7045 indicates that the Earth has been depleted in volatile elements relative to chondrites.

When we last left the $^{87}Sr/^{86}Sr$ ratio of the Earth, it was 4.6 Ga, and the ratio was 0.699. Assuming that the Earth has a chondritic composition overall and using chondrite values from McDonough and Sun (1995), the Rb/Sr ratio in weight units (e.g., ppm Rb over ppm Sr) is 0.317 (in keeping with the Oddo-Harkins rule, even-numbered Sr is several

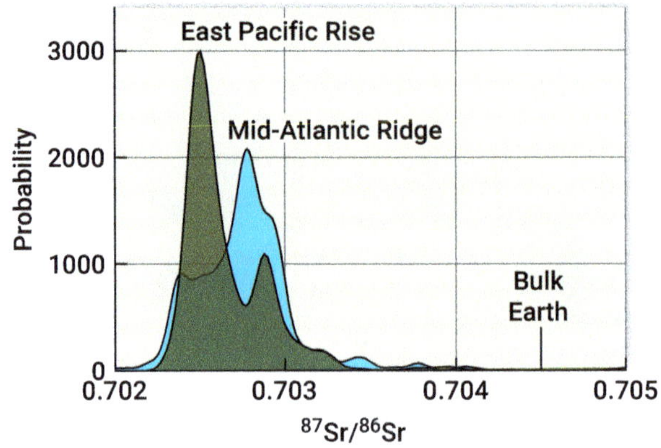

Figure 3.24 Measured $^{87}Sr/^{86}Sr$ for 1421 basalts from the East Pacific Rise and 928 from the Mid-Atlantic Ridge. The mean value is 0.7027, and nearly all the values are below 0.704. Given that the bulk Earth sits at about 0.7045, this indicates that MORBs are derived from a part of the mantle that has a Rb/Sr ratio lower than the bulk Earth.

Source: Data from EarthChem.

times more abundant than odd-numbered Rb). Converting this to atomic $^{87}Rb/^{86}Sr$ (a slightly complicated bit of arithmetic) gives 0.920, which we can use with the decay equation (Chapter 2) to estimate the present-day $^{87}Sr/^{86}Sr$ of the bulk Earth:

$$\frac{^{87}\mathrm{Sr}}{^{86}\mathrm{Sr}} = \left(\frac{^{87}\mathrm{Sr}}{^{86}\mathrm{Sr}}\right)_0 + \frac{^{87}\mathrm{Rb}}{^{86}\mathrm{Sr}}(e^{\lambda t} - 1) = 0.699 + 0.920(e^{(1.40\times10^{-11})(4.567\times10^{9})} - 1) = 0.760 \quad \textbf{(Eq. 3.17)}$$

This is a fine number, similar to present-day values for many chondritic meteorites (although significantly higher than ratios measured for the Tieschitz meteorite discussed in Chapter 2), but it is far higher than estimates of $^{87}Sr/^{86}Sr$ for the bulk Earth, and MORBs have $^{87}Sr/^{86}Sr$ values that are far lower still (Figure 3.24). Something is amiss.

The problem is that the chondrite model for the composition of the bulk Earth fails in many details—in particular, the Earth is depleted in many volatile elements relative to chondrites (Gast, 1960; McDonough & Sun, 1995). Rb, an alkali element like Na, is more volatile than Sr, an alkaline earth like Ca. Bulk-Earth Rb/Sr is considerably lower than chondrites and is estimated at 0.03 (McDonough & Sun, 1995). If we plug this value into the decay equation, we get a bulk-Earth $^{87}Sr/^{86}Sr$ of 0.7047. MORBs fall significantly below this value, so they must be derived from a source that has lower Rb/Sr than the bulk Earth, a mystery to which we shall return in Chapter 9.

Summary

Key points of this chapter include the following:

- Oceanic lithosphere is generated at oceanic ridges, which spread at an average rate of about 45 mm/yr.
- Oceanic crust comprises pillow basalt flows over basalt dikes over gabbro and averages about 7 km in thickness.
- The association of pillow basalt, radiolarian cherts, dikes, gabbro, peridotite, and serpentinite (altered peridotite) is known as an ophiolite sequence and is interpreted as a piece of oceanic lithosphere; finding this in the middle of a continent implies closure of an ocean basin.
- Partial melting of peridotite at surface to upper mantle pressures produces basalt.
- Phase diagrams depict the stable phase assemblage for a system for a given set of variables, such as *P*, *T*, and composition.

- The mineralogy of Al-bearing rocks is sensitive to *P* in the crust and upper mantle, because crustal minerals with Al in tetrahedral coordination, such as plagioclase, collapse to denser structures in which Al is in octahedral coordination, such as garnet. Many geobarometers depend on these reactions.
- Similar reactions in the mantle convert Si from its usual tetrahedral coordination to denser forms with Si in octahedral coordination.
- Phase transitions (e.g., quartz to tridymite) can be predicted if relevant and accurate thermodynamic data are available.
- Harzburgite and dunite found in ophiolites represent successively more extraction of basaltic partial melt from peridotite.
- Chemographic diagrams are graphical representations of chemical reactions.
- Melting under oceanic ridges is driven by decompression of rising peridotite.
- Adiabatic upwelling refers to ascent fast enough (as estimated by the Péclet number) that little heat is lost to the mantle surrounding the upwelling parcel.
- Oceanic crust is close to 7 km thick worldwide, regardless of spreading rate.
- MORBs have $^{87}Sr/^{86}Sr$ ratios of around 0.7045. This is far lower than the estimated bulk-Earth ratio and indicates that the mantle from which MORBs are derived has a lower Rb/Sr ratio than the bulk Earth.

Notes

1. It was long a rite of passage for petrology students to calculate by hand a norm for a particularly complicated rock. We would never do that to you.
2. This mode was calculated by least squares from the bulk composition and mineral compositions and is not a true mode measured at a microscope.
3. Note that statements such as "There is little question that . . ." or "We know that . . ." sometimes turn out to be wrong!

CHAPTER 4

How Rocks Melt and How Melts Crystallize

Igneous petrology is concerned with how magmatic liquids are generated by melting and how they are eliminated by crystallization and other processes. MORBs are generated by melting the mantle and turned into basalts or gabbros by cooling, and the details can be quite involved. Everyday experience is less helpful here than you might think. Humans have a lot of experience with melting and crystallization of ice, but ice melts congruently. Ice is H_2O; water is H_2O; if you have a mixture of ice and water in a beaker and remove some of the ice, the composition of what's in the beaker doesn't change: it's still H_2O. Nothing of interest happens to the composition.

Mix two such substances, however, and phenomena that are both interesting and counterintuitive can occur. If you put together two congruently melting compounds that do not dissolve in one another, such as ice and halite, the mixture melts at a much lower temperature than either of the two pure substances. This is the chemistry that underlies putting salt on an icy sidewalk. Ice melts congruently at 0 °C, and halite melts congruently at 801 °C, but an ice-salt mixture begins to melt at –21 °C, lower than either ice or halite. That is counterintuitive. Even more counterintuitive, however, is that the composition of the brine—its saltiness—does not depend on the ice/salt ratio. Why?

In this chapter, we explain why and use this process to account for the remarkable global uniformity in composition shown by MORBs. We then bring in other processes that allow magmas to reach the diverse compositions displayed by igneous rocks on the continents. This is the domain of **phase equilibria**, a topic that is fundamental to how the Earth works, and indeed to all materials science. The phase equilibrium concepts that apply to rocks also apply to ceramics, steel, semiconductors, superconductors, concrete, and many of the other products we interact with every day.

Phase equilibria
The nature of equilibrium phase assemblages that occur in a chemical system under specified conditions of *P*, *T*, and other intensive variables.

Thin section view of an olivine-rich vesicular basalt from Iceland, in cross-polarized light. Olivine crystals are several millimeters across and are the only phenocrysts; black regions are vesicles. Other phases (plagioclase, Cpx, magnetite) crystallized after olivine had been growing for a while and are only present in the groundmass. This illustrates a truism of magma crystallization: the crystallizing assemblage changes during crystallization as different phases become saturated in the melt. Width of field 20 mm.

Source: AFG photo.

4.1 Eutectic Melting: Forsterite-Diopside as a Model of the Mantle

Learning Objectives

By the end of this section you should be able to:

4.1.1 Sketch a generic binary eutectic phase diagram and explain the significance of solidus, liquidus, and eutectic.

4.1.2 Track the melting behavior of a given bulk composition on the diagram.

4.1.3 Explain how the worldwide uniformity of MORB composition indicates that the mantle melts in the manner of a eutectic.

A Brief Note about Notation

Hereafter, we will use capitalized abbreviations for minerals or other phases (see Appendix 1) and capital letters for bulk compositions. For example, Fo stands for the mineral forsterite, and X for a particular bulk composition. For simplicity or clarity, we will sometimes use the names of phases as stand-ins for formulas. For example, the chemical system Mg_2SiO_4-$CaMgSi_2O_6$ may be referred to by those formulas or by the corresponding mineral names or abbreviations (forsterite-diopside or Fo-Di). Because SiO_2 polymorphs are common and may be quartz (either α or β), cristobalite, tridymite, or another form, we will use Qtz as a generic symbol to represent whichever is appropriate (recall that the abbreviation for quartz itself is Qz).

Eutectic Behavior

Eutectic
In eutectic melting, two or more solid phases melt to a liquid whose composition is a mixture of the compositions of the phases, and this melting occurs at a fixed temperature and with fixed composition regardless of the proportions of the solid phases.

Ice-salt mixtures melt at a much lower temperature than either ice or salt owing to a phenomenon known as **eutectic** melting. MORBs are remarkably consistent in composition worldwide. This uniformity could result from melting the same material the same amount under the same conditions worldwide, but such global uniformity suggests that something more fundamental is at work. There is: the uniformity of MORB compositions is due largely to eutectic melting.

The same phenomenon occurs with many mineral mixtures. Consider the system Mg_2SiO_4-$CaMgSi_2O_6$. The corresponding minerals, Fo and Di, are the Mg end members of the olivine and Cpx series and form a reasonable two-component representation of the mantle. If we were going to simplify the mantle down to one component, that component would be Fo, but that would be a rather boring model—Fo melts congruently at a very high temperature. A good two-component model is Fo-Di, which shows much more interesting behavior.

Figure 4.1 shows a phase diagram for the system Fo-Di at 1-atm pressure (recall that this is an abbreviation for ambient lab conditions and is ~10^5 Pa or ~0.1 MPa). Such diagrams are referred to as *T-X* diagrams, where *X* stands for composition. This system, a binary (two-component) eutectic, was originally determined experimentally by Bowen (1914) using a furnace like that shown in Figure 3.8. The horizontal axis represents composition—pure Di on the left, pure Fo on the right—and the axis units are wt% so that "50" means a mixture that is half Di and half Fo by weight. The diagram in Figure 4.1 is simplified somewhat from the true diagram, which is more complicated than a simple binary eutectic. In the discussion that follows, % is understood to mean wt%, and temperatures are understood to mean °C.

The diagram is divided by lines and curves into four fields. The horizontal line at about 1375, the eutectic temperature, is the solidus for the binary system. Any mixture of Di and Fo begins to melt at this temperature. The curves above the Fo + Liq (Liq for liquid) and Di + Liq fields are the liquidus—the temperature at which crystals start to form

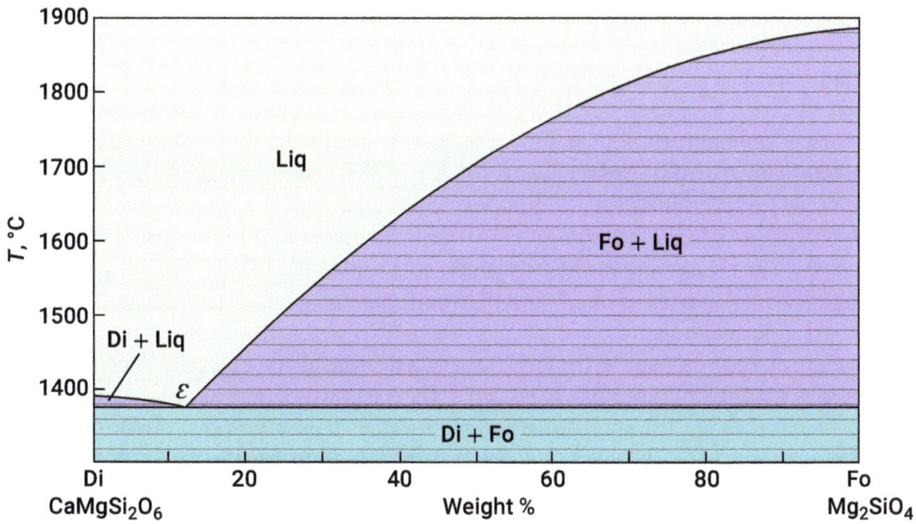

Figure 4.1 The system Di-Fo at 1 atm.

Source: From Bowen (1914).

upon cooling. The liquidus is a function of composition, whereas the solidus is not in this system. For compositions on either side of the eutectic, we say that "Di is on the liquidus" or "Fo is on the liquidus," referring to which phase is the first to crystallize. The field above the liquidus, labeled Liq, is a field of liquid (or melt; we'll use these terms interchangeably). The field below the solidus, Di + Fo, also known as the subsolidus field, consists of crystals of Di and Fo. The eutectic is designated by ε.

The Di + Fo, Di + Liq, and Fo + Liq fields themselves are **forbidden zones**—at equilibrium, no phase composition can plot within them. If you have a bulk composition-temperature pair that plots in one (say, 50% Fo at 1500), the stable assemblage will be found at either end of the horizontal (isothermal) line going through that point—in this case, Fo crystals and a liquid that is about 24 percent Fo. The horizontal rulings in these fields emphasize this point; the stable phase compositions lie at the ends of these **tie lines**, or **joins**. Any bulk composition below the eutectic temperature will consist of a mixture of Di and Fo crystals.

Forbidden zones
In this context, regions on a phase diagram in which no phase composition may plot at equilibrium.

Tie lines, joins
On a binary *T-X* phase diagram, isothermal lines connecting the phases in equilibrium in a two-phase field. This concept generalizes to higher dimensions (e.g., a tie plane in a three-component phase diagram), and to *P-X* and other diagrams.

The melting point of pure Di is taken as 1391.5 (read off the left-hand axis). The melting point of pure Fo is difficult to determine experimentally, because it is so high, but it is estimated as about 1890. The eutectic (solidus) temperature is about 1385, just a bit below the melting temperature of Di, but we have fudged a bit and moved it down to 1375 to make relationships a little clearer.

Eutectic Melting, Baldy's Law, and the Phase Rule

The eutectic point ε holds special honor. If a mixture of Fo and Di (our model mantle) is heated, melting begins at the eutectic temperature, and the liquid has the composition of the eutectic (ε) regardless of the proportions of Fo and Di in the bulk composition. Consider a peridotite with 70 percent olivine and 30 percent pyroxene (Opx and Cpx). In our simple binary system, we can model this with 70 percent Fo and 30 percent Di, and this bulk composition is marked as X in Figure 4.2. At temperatures below the eutectic, this bulk composition is subsolidus (and thus entirely solid). If we add enough heat to push it up to the solidus, it begins to melt, with liquid of composition ε forming. With further addition of heat, the temperature stays at the eutectic, and more melt of composition ε forms. The assemblage Di + Fo + L_ε (where L_ε is liquid of composition ε) is an invariant assemblage. L_ε is our model basalt—about 87 percent Di and 13 percent Fo.

Let's analyze this with the phase rule. The phase rule introduced earlier in the *P-T* diagram for SiO_2 is $F = C + 2 - \pi$. However, this system is isobaric, with pressure fixed at 1 atm. If an intensive variable is fixed, then the phase rule becomes $F = C + 1 - \pi$. This is a binary system, so $C = 2$; there are three phases, so $\pi = 3$; hence, $F = 0$. We cannot change temperature or the composition of the liquid and maintain the three-phase assemblage. If we move temperature up or down, even a little, the phase assemblage changes. For example, if we raise *T* a few degrees, bulk composition X goes into the Fo + Liq field; all the Di must dissolve. If we drop *T* a few degrees, we go subsolidus, and Liq disappears.

If we apply the phase rule to the two-phase subsolidus field (Di + Fo), we get $F = 1$; we can change one intensive variable, *T*, and maintain the two-phase assemblage. *T* is the only intensive variable available.

Let's continue to add heat to the invariant assemblage at the eutectic. Each little increment of Liq is made at the expense of the Di + Fo assemblage, and because Liq is so much richer in Di component than the bulk composition X, removing liquid with the composition of the eutectic drives the residual crystals to the right (see Figure 4.2). This brings us to important corollaries of Baldy's law (repeating the law to help you to remember it):

Baldy's law: Some of it plus the rest of it equals all of it.

Corollary 1A: The bulk composition of the residual crystals moves directly away from the liquid that is being removed from them.

Corollary 1B: The composition of the liquid moves directly away from the bulk composition of the crystals being removed from it.

At some point, the crystals have migrated from X to p (see Figure 4.2), but the liquid remains at the eutectic. The system consists of eutectic liquid plus Di crystals + Fo crystals, and these three phases add up to the bulk composition X. We can use the lever rule to figure out the proportions of the various items in this mix. It is helpful in visualizing lever rule problems to pull out the relevant items from the phase diagram. Here are relevant points for this problem:

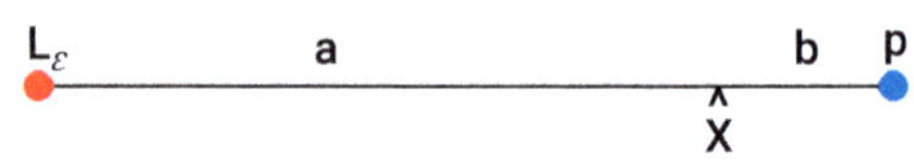

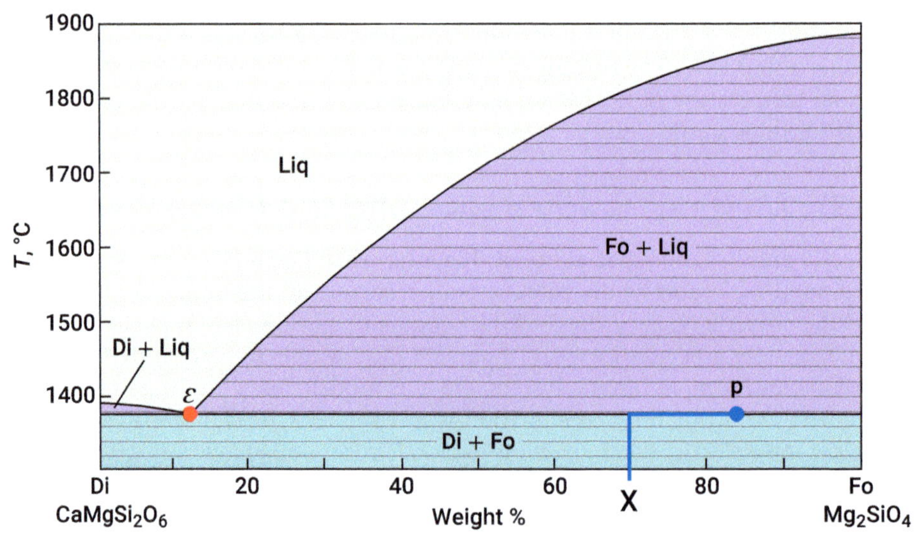

Figure 4.2 The system after some heat has been added. The temperature is stuck at the eutectic until all the diopside has been melted away.

Source: From Bowen (1914).

This is a chemographic diagram of the reaction X = L + p, and like all chemographic diagrams, it is a graphical statement of Baldy's law. The bulk composition divides the line segment into two subsegments, a and b. The lever rule tells us that the proportion of liquid is b/(a + b), or about 20 percent, and the proportion of crystals is a/(a + b), or about 80 percent. Further, the bulk crystals (point p) are about 84 percent Fo and 16 percent Di. Putting all this together, at this point in the magma's history the system is 20 percent eutectic liquid, 67 percent Fo, and 13 percent Di.

Note: It is very easy to get the wrong line segment in the numerator of lever rule problems, as countless students and your authors have learned repeatedly over the decades. This is because the relevant segment is away from the relevant phase—for example, the proportion of liquid has b in the numerator, even though b is adjacent to the crystal assemblage p. Always check!

If we continue to heat the magma, we continue to make eutectic liquid. If the liquid escapes its source region and erupts, the resulting lavas will have the same composition for eruption after eruption—the liquid is stuck at the eutectic. As heat is added and more and more eutectic liquid is made, the bulk composition of the residual crystals migrates more and more to the right (toward Fo), because we keep extracting Di-rich liquid. Eventually the crystals reach the Fo side of the diagram (point q in Figure 4.3), and all the Di component has been sucked out of the crystals—all the Di crystals have melted. We've lost a phase and now have one degree of freedom—either *T* or the liquid composition can vary, but they cannot vary independently. As *T* rises, the liquid must follow the liquidus curve that bounds the Fo + Liq field by dissolving Fo crystals into the liquid as *T* rises.

If we raise the temperature to 1500, the system looks like that in Figure 4.3. The system is liquid of composition s (~24% Fo) and Fo crystals. If we extract the relevant elements, we have this chemographic diagram:

The system consists of a/(a + b) Fo crystals and b/(a + b) liquid of composition s.

Continued heating shows that the last crystals are melted away at about 1810, because this is where the proportion of crystals goes to zero. We have eliminated another phase, and

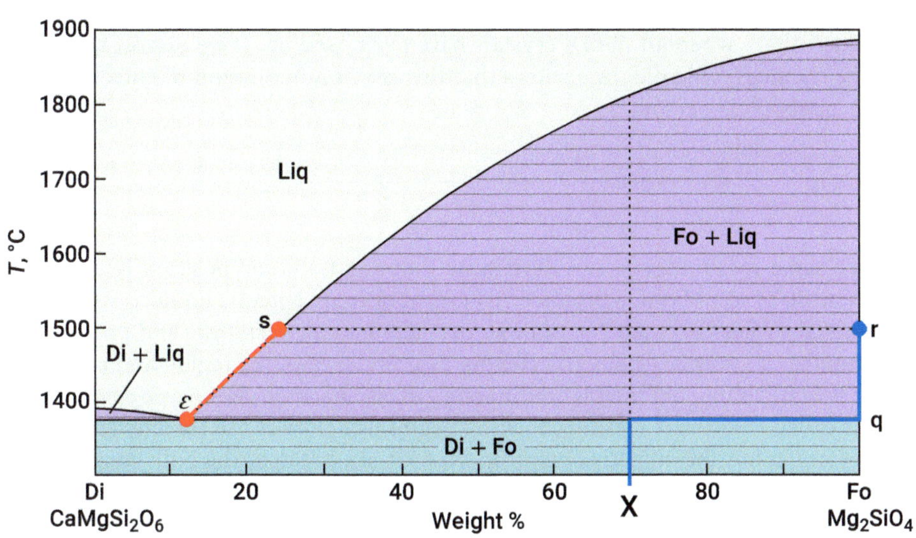

Figure 4.3 The system at 1500 °C. All the diopside has melted away.

Source: From Bowen (1914).

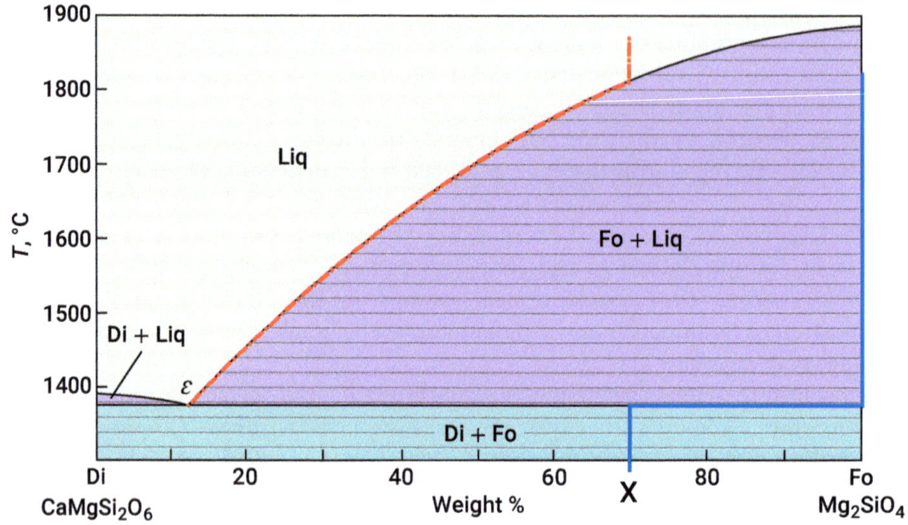

Figure 4.4 The system above the liquidus for bulk composition X.

Source: From Bowen (1914).

the system is now divariant: we can change T or liquid composition or both and maintain the phase assemblage. The full liquid and solid paths for heating are shown in Figure 4.4.

Application to MORB

One of the powerful things about phase diagrams is that principles that pertain in simple systems also apply in much more complicated real systems. That is why they are worth studying—principles from a two-component system that we can study on a flat piece of paper apply in 10-component systems that we can't envision. The real mantle is not a simple binary eutectic, but it exhibits behavior similar to such a eutectic. Note that any bulk composition in the binary system will yield a first melt of eutectic composition, at the eutectic temperature. The amount of such liquid that can be generated depends on the bulk composition, but the composition does not.

Determining how much of the eutectic liquid can be produced before the composition starts changing is a simple application of Baldy's law (in the guise of the lever rule). Note that the maximum happens when the residual crystal bulk composition just reaches the Fo sideline—that's when all the Di crystals have been used up. If we extract the relevant items, the chemographic diagram shows that this maximum amount of eutectic liquid is b/(a + b), or about 35 percent.

Using the same reasoning we can determine that any bulk composition lying between about 6 and 56 percent Fo will produce 50 percent or more eutectic liquid.

We noted earlier that melt is mobile in partially molten peridotite and probably does not accumulate in very large amounts before escaping the source region and rising. Consider what would happen if melt is continuously extracted from the source as it is formed, a process called **fractional melting**. Melting begins as before at the eutectic temperature, and as before, it has the eutectic composition. As it is extracted, the bulk composition of the residual system shifts to the right as in Figures 4.2 and 4.3, as if we had replaced the source after each little increment of melt production. We still produce about 35 percent eutectic melt, but after the last bit of Di is melted away, the system now consists of just Fo instead of Fo plus a lot of eutectic liquid. It has gone from two components to one. Forsterite melts at about 1890, and so the melt path, instead of crawling up the Fo liquidus as in

Fractional melting
A melting process in which melt is extracted as it is formed, rather than being allowed to continuously react with residual crystals.

Figure 4.4, jumps to the melting point of Fo, if sufficient heat (a lot!) is added. This is an invariant point in the one-component system. Fractional melting is characterized by jumps in temperature and composition from one invariant point to the next.

4.2 Olivine and Orthopyroxene: Peritectic Melting

Learning Objectives

By the end of this section you should be able to:

4.2.1 Deduce whether a reaction on a binary phase diagram is eutectic or peritectic.

4.2.2 Explain how partial melting of an olivine-Opx source rock can yield a magma that will crystallize quartz upon cooling.

The Fo-Di system is a reasonably well-behaved eutectic and explains a lot of the behavior we see in mantle melting. The other pyroxene phase in the mantle, Opx, introduces a new twist: **peritectic melting** involving an incongruently melting phase—in this case, Opx. This is illustrated by the system Mg_2SiO_4-SiO_2 (Fo-Qtz).

Peritectic melting
Melting in which a phase or phase assemblage melts to a liquid of a different composition plus a different phase or phase assemblage.

The 1-atm phase diagram for this system is shown in Figure 4.5. End-member components are Mg_2SiO_4 and SiO_2, and the phase En forms by the reaction

$$\begin{aligned} Mg_2SiO_4 + SiO_2 &= 2MgSiO_3 \\ \text{Fo} + \text{Qtz} &= 2\text{En} \end{aligned} \quad \textbf{(Eq. 4.1)}$$

Thermal divide
On a phase diagram, a ridge on the liquidus surface equivalent to a topographic divide; a cooling liquid will flow away from the divide along the liquidus surface.

Remember that for simplicity we use Qtz as a stand-in for whichever silica polymorph is stable. In these simple systems, temperatures are much higher than in real rocks, so cristobalite and tridymite are commonly the stable phases. Here, Crs is the relevant silica polymorph.

In this system, En lies between Fo and Crs; in the jargon of phase diagrams, we say it is on the join between them. That relationship is implicit in Equation 4.1. Now if En melted congruently, then the Fo-Qtz diagram would just be two binary eutectics joined at the intermediate compound, as in the system nepheline-silica (Nph-Si; Figure 4.6).

The presence of Ab along the Nph-Qtz join means that Nph and Qtz are not stable together. Albite melts congruently, so its composition marks a high point on the Ab liquidus, an arch that joins the two eutectics. Such a relationship is known as a **thermal divide**, because liquids on either side of the peak will roll into different eutectic valleys, like raindrops, as temperature falls, depending on where they land on the liquidus. A bulk composition to the left of Ab will end up at eutectic ε, whereas one to the right will end up at eutectic δ.

The Fo-Qtz diagram shows a few new twists. The first is the peritectic point π. This, and the eutectic ε, are invariant points in the binary system. The second is the field of two liquids on the SiO_2-rich side of the diagram. This is a field of liquid immiscibility, a topic that will be deferred for now. Our model mantle (70% olivine, 30% pyroxene) projects into this binary well to the left side of the diagram, around bulk composition X, far from the immiscibility field.

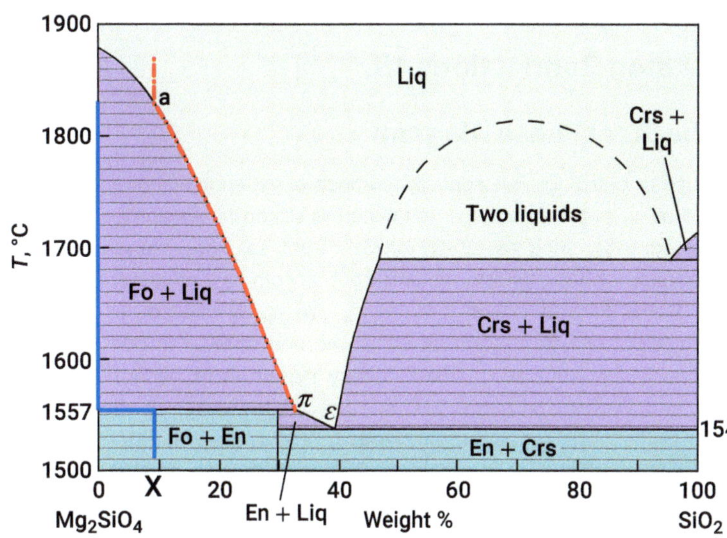

Figure 4.5 The system Mg_2SiO_4-SiO_2, summarized by Ernst (1976). Note that the scale on the abscissa refers to wt% SiO_2 relative to SiO_2 + Mg_2SiO_4, not wt% SiO_2 in the bulk composition. Fo has 42.7 wt% SiO_2, so a bulk composition of, say, 50 wt% has 71.4 wt% SiO_2. The two-liquid field of immiscibility is not well known; dashed line is speculative.

Source: From Bowen & Andersen (1914).

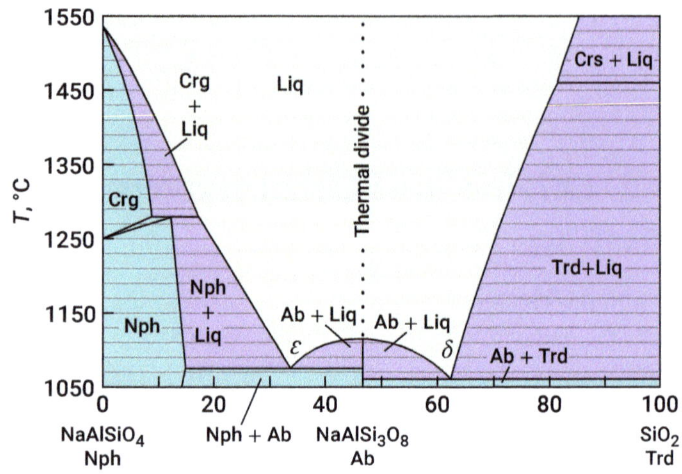

Figure 4.6 The system nepheline-silica at 1 atm. Nph-Qtz has an intermediate compound, albite (Ab), that melts congruently (a simplification), and thus this binary system consists of two binary eutectics joined at Ab. Diagram has been simplified, ignoring solution of Ab in Nph. Jadeite occurs on the Nph-Ab join but is not stable at 1 atm. Cng, carnegieite, is an artificial, high-*T* polymorph of Nph.

Source: From Greig & Barth (1938).

Consider heating of a rock made of Fo and En in the proportions given by X (see Figure 4.5). The solidus for this bulk composition is at 1557. Nothing happens until this temperature is reached, at which point melting begins. Pulling out the relevant chemographic diagram, we have

As with any reaction involving three phases, there is only one way to balance it: the two things on the outside react to form the thing in the middle. That means our reaction must be En = Fo + π, where the reaction goes from left to right as heat is added. This reaction proceeds at 1557 until all the En is used up by this incongruent melting; when En disappears, the system consists of Fo + π. At the invariant point, the system can be just Fo + En (when heating first brings it to 1557), Fo + En + π, or Fo + π (when En disappears and *T* can rise again). Where in this spectrum the system lies depends on the total heat content. Adding heat after En is used up causes the liquid to climb up the Fo liquidus until it reaches the composition of X at point a (Figure 4.5), where all the Fo is used up.

Cementite
Iron carbide, Fe_3C; has been found in a meteorite and given the name cohenite after its discoverer, so it is technically a mineral, but a vanishingly rare one.

Austenite
Also γ-Fe; medium-*T* Fe polymorph that can dissolve up to 2 wt% C.

Ferrite
Also α-Fe; low-*T* Fe polymorph with low C solubility.

Making Connections 4.1

The Steel Phase Diagram

Much of petrology is essentially a branch of materials science, and many of the processes that materials scientists are preoccupied with, such as crystal growth from solution, recrystallization, phase transformations that accompany changes in *T* and *P*, and crystal coarsening, are quite relevant to petrology. A system of great economic and societal importance is the iron-carbon (Fe-C) binary system, and particularly the low-carbon part with C < 7 wt% (Figure 4.7a). This is the domain of steel, and understanding phase relations in this system and the microstructures that develop during cooling is critical to controlling properties such as tensile strength, ductility, hardness, and toughness, and thus critical to keeping buildings and bridges from falling down and making rails that can stand repeated pounding by heavy trains, engine blocks that can run for years, and classic bicycle frames.

Figure 4.7a is not the true equilibrium diagram, which is essentially a eutectic between carbon and iron, but rather a metastable diagram, because a metastable phase, Fe_3C, rather than graphite, appears during cooling and precipitates directly from liquids that lie to the right of the eutectic at 4.3 wt% C. This compound, known as **cementite** (unrelated to the cement that binds concrete), anchors the right side of the diagram. The Fe side is anchored by the chemical component Fe, which comes in three forms at 1 atm: with decreasing *T*, δ-Fe above 1392 °C, γ-Fe (**austenite**) from 1392 to 911 °C, and α-Fe, or **ferrite**, below 911 °C. In spite of their names, none of these is a mineral, but they are analogous to the various *T*-dependent mineral forms that SiO_2 can take on at 1 atm (see Figure 3.13).

The amount of C in steel has a large effect on its mechanical properties. Cementite is quite hard, in the range of feldspar and quartz (~6.5 on the Mohs scale), whereas ferrite's hardness is closer to that of calcite (3). Ferrite is ductile, easily shaped, and tough, where toughness is the ability to absorb energy (as from, say, a hammer blow) without fracturing. Cementite is

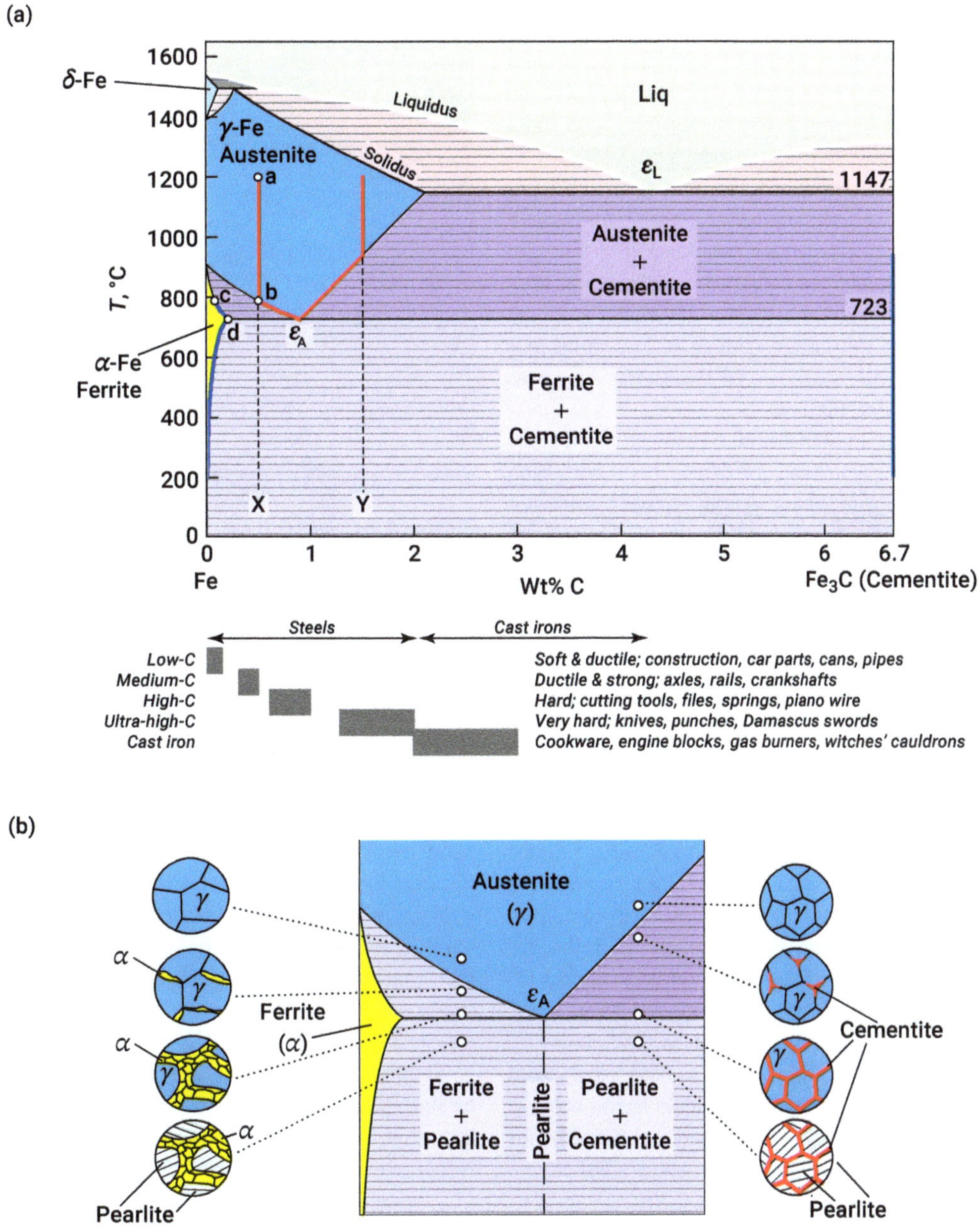

Figure 4.7 (a) Fe-Fe$_3$C (ferrite-cementite) portion of the Fe-C phase diagram at 1 atm. The system consists of three Fe phases; with decreasing T, these are δ-Fe above 1392 °C, γ-Fe (austenite) from 1392 to 911 °C, and α-Fe (ferrite) below 911 °C. Cooling liquid with C < ~2 wt% passes into the austenite field, and further cooling is discussed in the text. The chart below the phase diagram shows the C content of various types of steel and cast iron and the uses to which they are put. (b) Close-up of the austenite side of the diagram, with microstructures. See text for discussion.

Source: Adapted from Ohring (1995).

brittle and cannot be easily worked, and thus Fe with significant cementite is used for casting into shapes such as pipes, engine blocks, and frying pans.

Austenite, γ-Fe, has a face-centered cubic structure with interstices large enough to admit 2.06 wt% C at 1147 °C. The lowest-T polymorph, ferrite, has a body-centered cubic lattice with smaller interstices, and the maximum solubility of C is much lower—0.025 wt%. Producing the steel you want depends on manipulating its C content and crystal structure.

The Fe-Fe$_3$C phase diagram is essentially a eutectic with a little complication near pure Fe involving conversion of δ-Fe to γ-Fe, which we will ignore. Austenite and cementite coprecipitate from the liquid at the ε_L eutectic at 1147 °C, just like the

continued

Fo-Di diagram (see Figure 4.1). However, because austenite can absorb a significant amount of C, bulk compositions with less than 2 wt% C, such as X and Y in Figure 4.7a, never make it to the eutectic, but rather solidify completely to austenite, and that's where the fun begins. Consider bulk composition X at 1200 °C (point a), at which point it consists of austenite. It cools until it hits the curve at b, where ferrite of composition c begins to **exsolve** from the austenite, driving the latter's composition to the right until it reaches eutectic ε_A, where cementite coprecipitates from the austenite with ferrite of composition d until all the austenite has been used up, and the system consists of just ferrite plus cementite. This is just a eutectic like Fo-Di, except that it involves a solid (austenite) instead of a liquid.

Composition Y ends up at the eutectic as well, but it starts to precipitate cementite instead of ferrite first, and then ferrite plus cementite precipitate at the eutectic until the austenite is all used up.

Nothing too crazy about all of that, except that steels with less than 1 wt% C, such as X, have much different mechanical properties than those to the right of the eutectic, such as Y. Consider X at around 900 °C, where it consists of just austenite crystals (Figure 4.7b). When it cools to about 800 °C, it starts to precipitate ferrite, which grows along grain boundaries. At 725 °C, just above the eutectic, it consists of subequal amounts of ferrite and austenite, and at the eutectic, all the remaining austenite, whose composition has now migrated to the eutectic, exsolves to ferrite plus cementite in the eutectic proportions. This eutectic exsolution produces **pearlite**, a form of steel consisting of finely intergrown plates of cementite and ferrite (Figure 4.8). Using the lever rule and the full diagram (Figure 4.7a) shows that pearlite must be almost 90 wt% ferrite, with the rest cementite. Composition X thus ends up as colonies of pearlite among crystals of ferrite (Figure 4.8), and the resulting medium-carbon steel is both hard (from cementite in pearlite) and strong (from ductile ferrite holding the pearlite together). This kind of steel is suitable for demanding applications, such as rails, axles, and crankshafts.

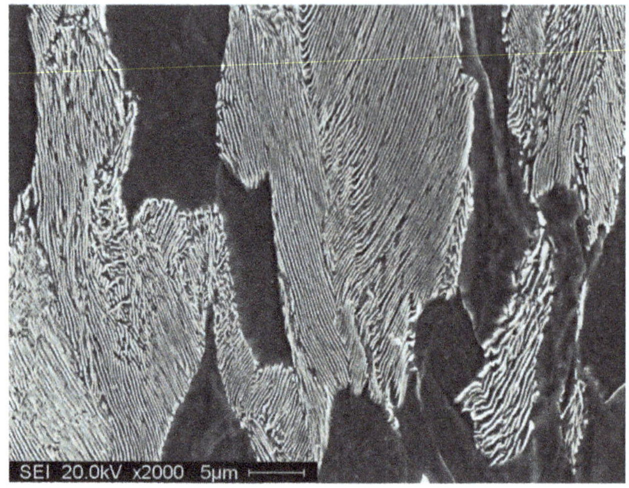

Figure 4.8 Scanning electron micrograph of a polished and etched steel surface showing pearlite (finely intergrown ferrite + cementite) in a dark matrix of ferrite crystals.

Source: McGill University/Wikimedia.

Composition Y, although not all that different from X, ends up with quite a different product. At about 900 °C, cementite starts to exsolve from austenite, again at grain boundaries, and just above the eutectic, it will consist of about 10 wt% cementite and 90 wt% austenite (be sure that you know where those numbers come from). Once processed through the eutectic, the end result will be about 90 wt% pearlite, with cementite holding the colonies together. This ultrahigh-carbon steel is significantly harder and more brittle than that produced by composition X, because the pearlite grains are held together with cementite, and it is suitable for files, knives, and cutting tools.

Pearlite is such a persistent feature of steels that it is treated as a single phase of eutectic composition even though it is clearly composed of two phases. Does an equivalent material exist in rocks?

Exsolve
Form by coming out of solution from another phase.

Pearlite
The product of eutectic exsolution of austenite to form intergrown plates of cementite and ferrite.

4.3 Crystal Fractionation

Learning Objectives

By the end of this section you should be able to:

4.3.1 Use the Fo-Qtz phase diagram to show how the liquid reaches a different end point under fractional crystallization than it does under equilibrium crystallization.

4.3.2 Create a process by which one could take a rock consisting of Fo + En and make quartz crystals.

In the system Fo-Di, the liquid starts or ends at the eutectic, depending on whether you're heating or cooling. Any bulk composition in the binary system starts to melt at the eutectic, and a liquid of any bulk composition ends up at the eutectic if you cool it enough. Peritectic melting of En gives this system much more compositional flexibility than a eutectic system has.

Consider again our bulk composition X (Figure 4.9). If you calculate a norm, you will get back the 70 percent Fo and 30 percent En that we put into our hypothetical mix. Such a composition is olivine-normative (it is also enstatite-normative, but that's not as important in the present instance). If you partially melt this composition to make eutectic melt of composition π and calculate a norm on it, you will get about 97 percent En and 3 percent Qz. The melt is thus quartz-normative. Peritectic melting has taken a bulk composition that is 70 percent Fo and made a slightly quartz-normative melt out of it. Note that in Fo-Di, any bulk composition is olivine-normative, and the eutectic melt is as well.

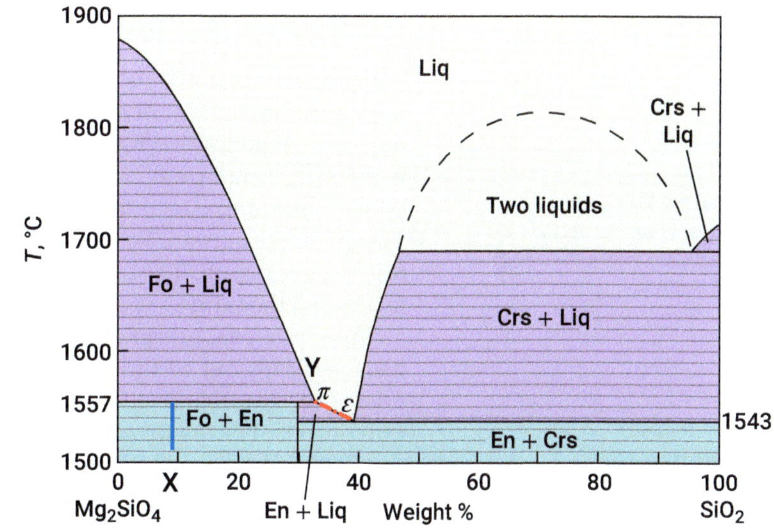

Figure 4.9 Crystal fractionation in the system Fo-Qtz. Melting bulk composition X (Fo + En) produces liquid Y. If this liquid is separated and then cooled, it will crystallize En and proceed down the En liquidus to eutectic ε, where it will coprecipitate En and Crs. In this way, a Fo-normative bulk composition can end up crystallizing as a quartz-normative rock.

Source: From Bowen & Andersen (1914).

If you partially melt bulk composition X, you generate liquid of composition π. If that liquid were to escape the source region (fractional melting) and then begin crystallizing on its own, it can be considered a new bulk composition (call it Y). When that melt cools, it has the peritectic melt composition but cannot undergo the reaction En = Fo + π, because there is no Fo present—we left it behind in the source region. Our liquid proceeds merrily down the En liquidus, oblivious to its Fo-rich parentage, to the eutectic ε, where Crs joins En in the crystallizing assemblage. That would be a weird rock in the real world, but we can equate it to an andesite, with a lot of Opx and a little Qz.

The same sort of evolution to eutectic ε can occur if the liquid is prevented from reacting with crystals during crystallization. If bulk composition X cools as in Figure 4.5 but crystals are removed (as by settling) such that the liquid is unaware of their presence, then the peritectic reaction again cannot occur—there are no Fo crystals for the melt to react with—and the liquid again proceeds to the eutectic ε.

Notice what happened here—we started with our mantle rock X, 70 percent Fo, and ended up crystallizing a rock with no Fo but some Crs (or Qz in a real rock at lower T). The rock composition evolved from Fo-rich X to a liquid all the way over at ε. This behavior was the basis of Bowen's scheme for producing the BADR trend from basalt, and gave its name to his groundbreaking 1928 book, *The Evolution of the Igneous Rocks*. The inferred process of separating liquid from crystals to produce a magma of a different composition is known as **fractional crystallization**, or **crystal fractionation**. Merely crystallizing minerals does not constitute crystal fractionation; they must be physically removed from a volume of melt (or melt must be physically removed) such that the bulk composition of that volume has changed.

Fractional crystallization, crystal fractionation
The process of separating magmatic liquid from crystals, or vice versa, resulting in liquid-enriched and liquid-depleted volumes of different composition.

4.4 Olivine-Orthopyroxene-Clinopyroxene: The Ternary System

Learning Objectives

By the end of this section you should be able to:

4.4.1 Examine a ternary phase diagram and determine the composition of the first melt for a variety of bulk compositions.

4.4.2 Determine if an invariant point is eutectic or peritectic by inspection.

4.4.3 Trace equilibrium crystallization and melting paths in a ternary system.

Anatomy of a Ternary Diagram

Ternary system
A system in which three components are needed to describe the compositions of all phases present.

We have seen that olivine and Cpx have a eutectic relationship and produce an olivine-normative liquid, but Ol and Opx have a peritectic relationship and produce a quartz-normative liquid. What happens if all three minerals are present, as in the mantle? What is the melt composition, and is it Ol- or quartz-normative? Such a system is a **ternary system**, meaning that it takes three components to describe all the phases that occur. These components are Fo, Di, and Qtz. We have examined two of the binary systems in this ternary system. The third, Di-Qtz, is a simple eutectic, so two of the binaries are eutectics and the third is a peritectic. How do they combine?

A ternary phase diagram is conceptually constructed by lifting the three binary systems vertically out of the page and linking them together in an equilateral triangular prism (Figure 4.10). Relationships in the interior of the prism need to be determined via experiments. The important Fo-Di-Qtz system was originally worked on by Bowen (1914), and further developed by Kushiro (1972). Kushiro took advantage of the newly developed electron microprobe to establish many of the details that Bowen could not.

Our isobaric binary *T-X* diagrams require two dimensions to portray (e.g., see Figure 4.5), one for the single compositional variable and one for temperature. A ternary system takes three dimensions, and to portray one on a flat sheet of paper or on a screen requires displaying liquidus temperature with contours, analogous to the way elevation is portrayed on topographic maps. Figure 4.10 shows a 3D model of a generic ternary eutectic diagram along with a flat portrayal of the same system contoured for temperature. The invariant eutectic points in the binary systems become univariant curves that descend into the middle of the diagram and meet in an invariant ternary eutectic point. These curves (known as cotectics) are cusps in the liquidus surface, analogous to stream valleys on a topographic surface. The ternary eutectic point is like a sinkhole that the streams pour into.

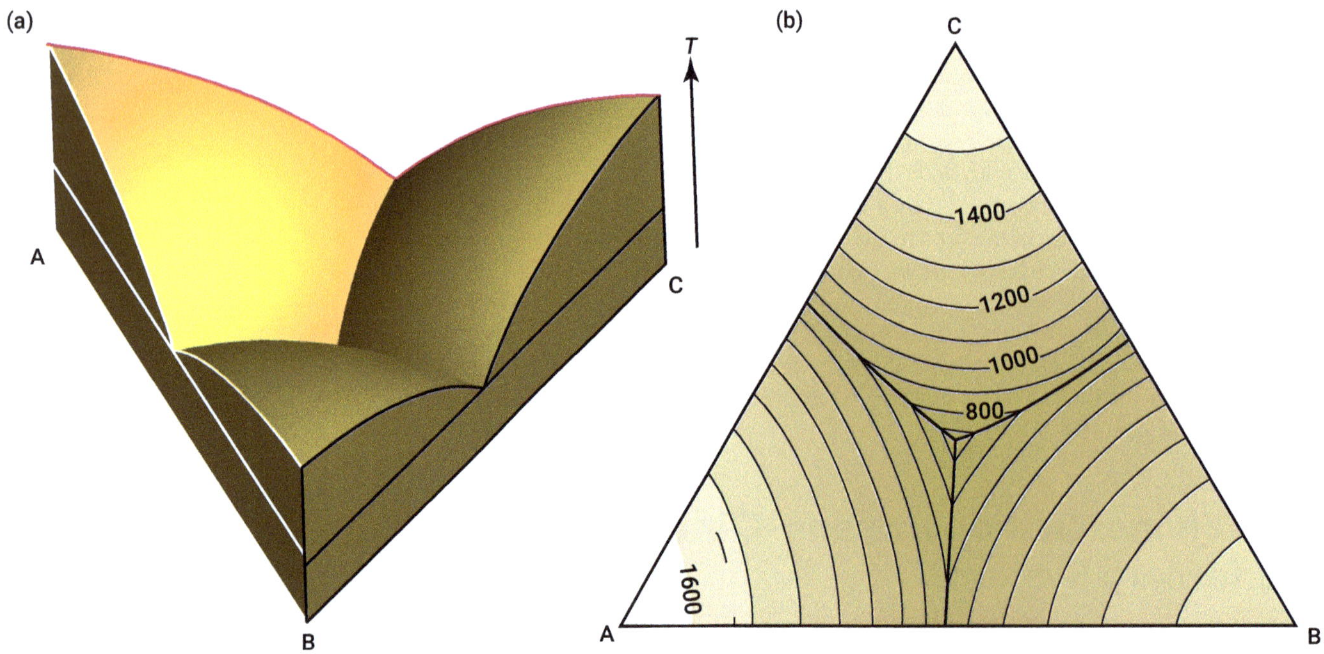

Figure 4.10 (a) 3D view of a ternary eutectic system, showing two of the binary eutectic systems that make up the walls, with temperature as the vertical axis. Eutectic valleys that run from the binary eutectic points are creases in the ternary liquidus surface. The ternary solidus surface is a horizontal plane within the volume, at the temperature of the ternary eutectic—the "drain" of the system. (b) Vertical view of the surface using contours to show values of temperature. Creases in the isotherms correspond to creases in the surface.

The System Forsterite-Diopside-Quartz

The system Fo-Di-Qtz is more complicated than a ternary eutectic owing to the Fo-Qtz peritectic. A simplified version of the ternary is shown in Figure 4.11. Complicating behavior such as incongruent melting of Di, the presence of pigeonite (a low-Ca Cpx), and most importantly, solid solution between Di and En are ignored, but the essential behavior is preserved (Figure 4.11a). This system has two ternary invariant points, π and ε. Invariant point ε occurs where the liquidus fields for Crs, Di, and En meet and is a simple ternary eutectic. As in the schematic system in Figure 4.10, at the eutectic there are three crystalline phases plus melt, giving $F = 0$. Invariant point π lies where the fields for Fo, Di, and En come together. Unlike ε, here two "streams" come in and one leaves; this type of invariant point is known as a **tributary reaction point**, with analogy to tributary streams that join to form a single stream.

Tributary reaction point
In a ternary system, a reaction point where, going down *T*, two curves that bound liquidus fields meet and a third emerges. At such a point, the reaction as heat is removed is always this: crystalline phase that lies between the tributaries + liquid = the other two phases.

At the ternary eutectic, the reaction is Di + En + Si = ε, an interior reaction, as heat is added; in the chemographic diagram (Figure 4.11b), the liquid lies within the triangle formed by the three solid phases. Invariant point π is different. Here, the four phases form a convex quadrilateral, and the reaction is the second kind possible with four phases: Di + En = Fo + π. This is a crossover reaction, the ternary equivalent of the binary peritectic discussed earlier (Figure 4.11c).

Let's test drive this system. Consider peridotite KLB-1 with 60 percent olivine, 23 percent Opx, and 16 percent Cpx. This plots at X in Figure 4.12 if we treat olivine as Fo, Opx as En, and Cpx as Di (remember that these percentages apply to the scalene Fo-En-Di triangle, not the entire triangle). What does this phase diagram predict if we heat this peridotite to melting?

The first order of business is to determine what minerals the rock consists of below the solidus. There are only two possible subsolidus assemblages: Fo-Di-En and Di-En-Crs. There is one more group of three minerals that you can make out of the four in this system, Fo-Di-Crs, but this is not a valid assemblage, because Fo and Crs react to form En (as shown by the phase diagram). X plots in the Fo-Di-En triangle, so the first melt must form at a place where these three minerals coexist. This is reaction point π, and the

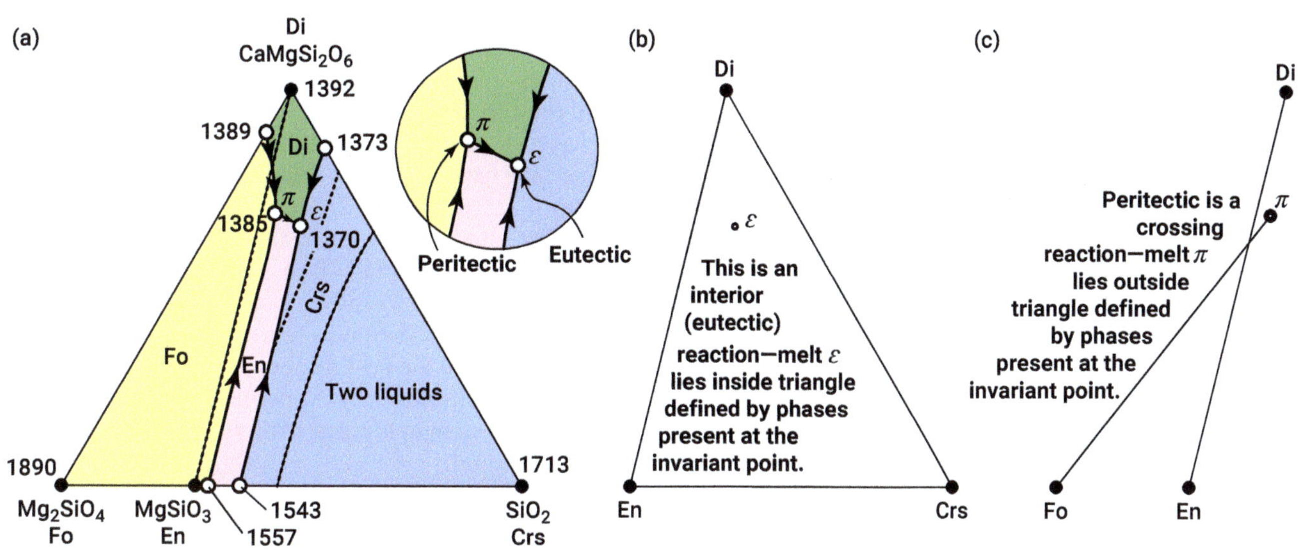

Figure 4.11 (a) Simplified version of the system Fo-Di-Qtz; temperature contours have been omitted for clarity. This diagram ignores solid solution and existence of pigeonite, and temperatures have been slightly adjusted for self-consistency. (b) Chemographic diagram for the ternary eutectic point involving Di, En, and Crs. (c) Chemographic diagram for the peritectic point involving Fo, Di, and En.

Source: From Morse (1980).

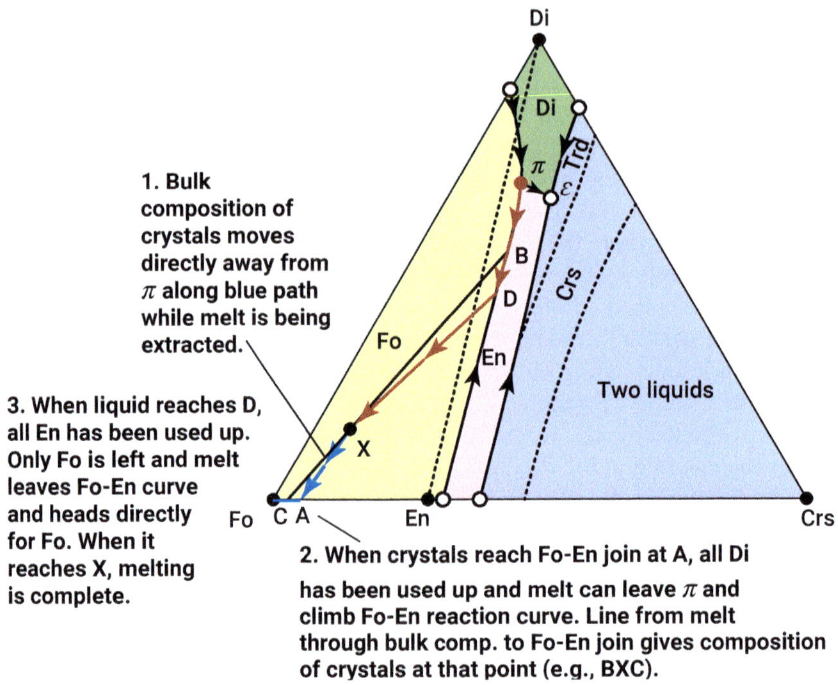

Figure 4.12 The system Fo-Di-Qtz showing the paths taken by the crystals (blue) and liquid (red) during equilibrium melting of bulk composition X, which represents peridotite KLB-1 projected into this ternary system. At each point during heating, the bulk crystals and liquid are at the same temperature and lie on the red and blue paths on a tie lie that goes through X. Melting begins at invariant point π, which is the only liquid composition that can coexist with the phases Fo, Di, and En.

Source: From Morse (1980).

resulting liquid is quartz-normative, because it lies to the right of the En-Di join. Adding heat causes more and more melt π to form, and the melt composition and temperature are stuck here as long as all four phases are present. By Corollary 1A of Baldy's law, the bulk crystal composition moves directly away from π along path X-A as more and more melt is formed. When the crystals reach A, they are on the Fo-En sideline, so all the Di has been eliminated, and the residual rock has transitioned from peridotite to harzburgite, following the progression seen in ophiolites. We have lost a phase, and so the system is now univariant and temperature can change.

With further addition of heat, the liquid climbs upstream ("south") in the Fo-En valley. As it does so, the bulk crystal composition migrates from point A toward Fo. Baldy's law requires that a line connecting the liquid (say, B) and bulk solid (C) must go through the bulk composition. When this line pivots to the point that the solid end is at pure Fo, the system becomes divariant and can leave the Fo-En valley to travel directly toward X. The residual crystals at this point are dunite, extending the peridotite-harzburgite trend and once again following the progression seen in ophiolites. Melting ends when the liquid reaches X, because the system is entirely liquid.

In the binary system Fo-Qtz (see Figure 4.5) the forsterite liquidus extends beyond En into quartz-normative compositions. Another way to state this is that melting of a Fo-bearing composition produces a liquid that is far from Fo in composition. In the ternary system, the same thing happens; melting a Fo-rich composition such as X produces a liquid that lies way across the diagram (π), is quartz-normative (to the right of the En-Di join), and remains quartz-normative for a significant portion of its life. But MORB source melting begins at pressures of around 2 GPa. Is that high-pressure liquid quartz-normative as well? Section 4.7 explores this question.

4.5 What Controls the Composition of the First Liquid Produced?

Learning Objectives

By the end of this section you should be able to:

4.5.1 Deduce from mineral structure the relative magnitude of a mineral's enthalpy of fusion.

4.5.2 Predict how the composition and temperature of a binary eutectic will shift if the melting temperature or enthalpy of fusion of one end member is changed.

Enthalpy of Fusion

Forsterite (or, more generally, olivine) has a massive liquidus surface in most phase diagrams, taking over much more real estate than it deserves at high temperature, then giving it back at lower temperature. In both Fo-Qtz and Fo-Di-Qtz, Fo is stable over a much larger part of the compositional space when liquid is present than when it is not. In the ternary, the Fo liquidus surface laps well over into compositions in which Fo is not stable at lower-temperature conditions (quartz-normative compositions, to the right of the Di-En join). It is this relationship that causes En to precipitate from the liquid at peritectics—it's forsterite's fault.

The locations of invariant points and univariant curves depend primarily on two thermodynamic properties: the temperature of fusion (or melting), T_f, and the enthalpy of fusion, ΔH_f. Enthalpy is a precisely defined thermodynamic energy term that is commonly equated with the more colloquial "heat content." The "heat" that we are familiar with in everyday life is commonly referred to as sensible heat: "sensible" because we can feel (sense) it, not because it is practical or wise. Enthalpy of fusion is referred to as latent heat, because it is an energy term that is hidden in a substance. ΔH_f is a delta term and refers to the enthalpy change that accompanies melting—in other words, the amount of thermal energy needed to melt a crystal. For example, Fo melts at 1890 °C, and so

$$\Delta H_f^{fo} = H_{1890}^{fo,liquid} - H_{1890}^{fo,crystal} \qquad \textbf{(Eq. 4.2)}$$

This is the enthalpy difference that accompanies melting—the energy needed to break bonds and turn a solid into a liquid.

Measuring ΔH_f is technically demanding, and data are sparse, but a knowledge of mineralogy allows us to predict general trends. Enthalpy of fusion is a measure of how much energy it takes to break some of the bonds in a crystal to turn it from a highly ordered crystalline substance to a disordered liquid. Tetrahedral Si-O and Al-O bonds are among the strongest in minerals, but when minerals rich in these elements melt, the melt retains a high proportion of such bonds in a polymerized network (Figure 4.13). The higher the polymerization, the lower the enthalpy of fusion. Minerals that have a lot of SiO_2 and tetrahedral Al produce highly polymerized melts, and thus have low ΔH_f, and minerals with low SiO_2 produce poorly polymerized melts. Thus, nesosilicates such as olivine have isolated SiO_4 tetrahedra in their structure, melt to poorly polymerized liquids, and have large ΔH_f, whereas tectosilicates such as quartz and feldspar have fully polymerized SiO_4 and AlO_4 tetrahedra, melt to highly polymerized liquids, and have low ΔH_f (see Figure 4.13). Enthalpies of fusion for some common igneous minerals are given in Table 4.1.

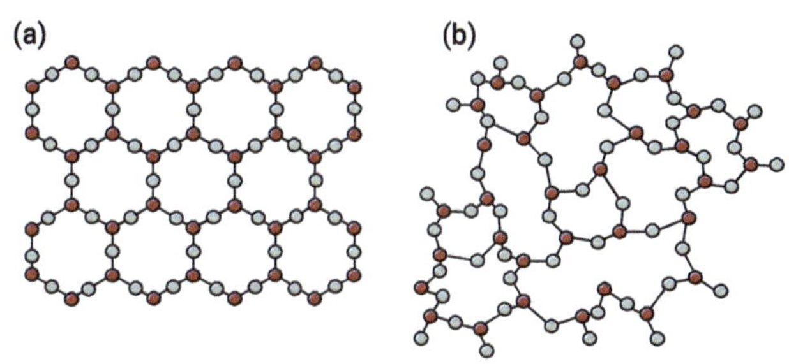

Figure 4.13 Schematic comparison of the structure of quartz (a) with SiO_2 liquid (b). The long-term crystalline order of quartz gives it low entropy. Molten SiO_2 lacks the long-range order of crystalline varieties but has significant short-range order owing to the remnants of a tetrahedral framework, so its entropy is relatively low for a liquid. The relatively small number of strong Si-O bonds that are broken when SiO_2 is melted gives a low enthalpy of fusion.

Liquidus curves are also solubility curves, giving the maximum amount of a given component that can be dissolved in

Table 4.1 Specific (per gram) Enthalpies of Fusion of Some Common Minerals*

Phase	Formula	T_f(°C)	ΔH_f(J/g)
Quartz**	SiO_2	1427	156
Orthoclase***	$KAlSi_3O_8$	1200	207
Albite	$NaAlSi_3O_8$	1118	246
Fayalite	Fe_2SiO_4	1217	438
Anorthite	$CaAl_2Si_2O_8$	1557	478
Titanite	$CaTiSiO_5$	1397	631
Diopside	$CaMgSi_2O_6$	1392	636
Enstatite***	$MgSiO_3$	1561	729
Phlogopite	$KMg_3AlSi_3O_{10}F_2$	1397	740
Forsterite	Mg_2SiO_4	1890	1210

* Ranked low to high. Values are at various temperatures and have not been corrected to a common temperature.
** Inferred metastable melting point.
*** Melts incongruently; estimated congruent melting *T* indicated.
Sources: Data from Lange & Carmichael (1990) and Nicholls & Stout (1982).

the liquid at a given temperature. Consider Figure 4.1. If we take a composition of pure Di at 1600 °C, it will be entirely liquid. If we drop in some Fo crystals and push the composition to 10 wt% Fo, the system is still in the liquid field, and the crystals will dissolve. We can keep adding Fo crystals up to about 35 wt%, at which point the system is saturated in Fo. Adding any more crystals will not change the composition of the liquid, which is stuck on the liquidus at about 35 wt% Fo.

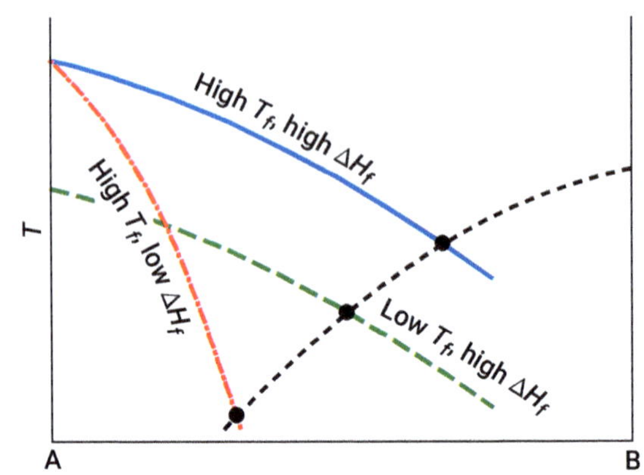

Figure 4.14 Schematic binary eutectic system A-B showing how differences in the melting temperature (T_f) and enthalpy of fusion (ΔH_f) affect the position of the eutectic (solid circle), located where the liquidus curves of each component intersect. An increase in either property pushes the eutectic away from the corresponding phase. Fo (and olivine in general) has high T_f and high ΔH_f, so it tends to have a large liquidus surface, pushing eutectics and peritectics away from itself. This in turn means that melting olivine-bearing assemblages commonly produces melts that are not very olivine-like, whereas feldspars and quartz behave the opposite way. This has a profound effect on the compositions of magmas.

Source: After Carmichael et al. (1974).

Size of a Mineral's Liquidus Share

Equilibrium between a chemical component (say, Fo) shared between a crystal and melt leads, with several simplifying assumptions, to a general equation for such equilibrium:

$$X_A^m = e^{\frac{-\Delta H_f}{kT}\left(1-\frac{T}{T_f}\right)} \qquad \textbf{(Eq. 4.3)}$$

where X_A^m is the mole fraction of component A in the melt, T is the temperature of interest, T_f is the melting temperature, and k is a constant (Carmichael et al., 1974). The parenthetical term measures how far below the melting temperature you are, and thus it is always positive and increasing as T goes down, and ΔH_f is always positive, so the exponent is a negative number. At a given temperature, the exponent is just $-\Delta H_f$ multiplied by a positive constant. In the words of Carmichael et al. (1974, p. 164), "It is profitable, despite the apprehension the reader may feel at looking at [Equation 4.3], to appreciate the important parameters." They were referring to a different equation,[1] but the sentiment is the same: this equation says that *at a given temperature, the smaller the value of* ΔH_f, *the higher the concentration of the corresponding component in the melt will be.*

These relationships lead to the general patterns in Figure 4.14, a schematic eutectic system. For a given liquidus curve for component B, raising T_f or ΔH_f for component A moves the eutectic away from A. Fo has high values for both components, and thus its liquidus surface is large. At the other end of the spectrum, albite and orthoclase have low

T_f and ΔH_f, their share of the liquidus surface is small, and invariant points crowd around them in compositional space. This has a profound effect on crustal melting, a topic we will return to in Chapter 10.

Quartz has the lowest ΔH_f of all the minerals in Table 4.1, but a relatively high melting temperature. However, in the presence of dissolved water the melting temperature of quartz drops dramatically (see Chapter 9), so quartz behaves like the alkali feldspars.

4.6 Types of Basalt: The Basalt Tetrahedron

Learning Objectives

By the end of this section you should be able to:

4.6.1 Classify a basalt in the basalt tetrahedron given its chemical composition.

4.6.2 Explain why the Ol-Pl-Cpx plane behaves as a thermal divide in basaltic systems.

Even though all basalts live at the low-silica end of the compositional spectrum, there are many varieties and hundreds of names (only a handful of which are important in most cases). The TAS diagram of Chapter 1 is one method for classifying rocks. Another enduring one is known as the basalt tetrahedron (Yoder & Tilley, 1962).

The basalt tetrahedron (Figure 4.15) is based on normative mineralogy and thus, like the TAS diagram, can only be used on rocks for which you have a chemical analysis. The tetrahedron is a generalization of experimental work in the quaternary (four-component) system Fo-Di-Nph-Qtz.

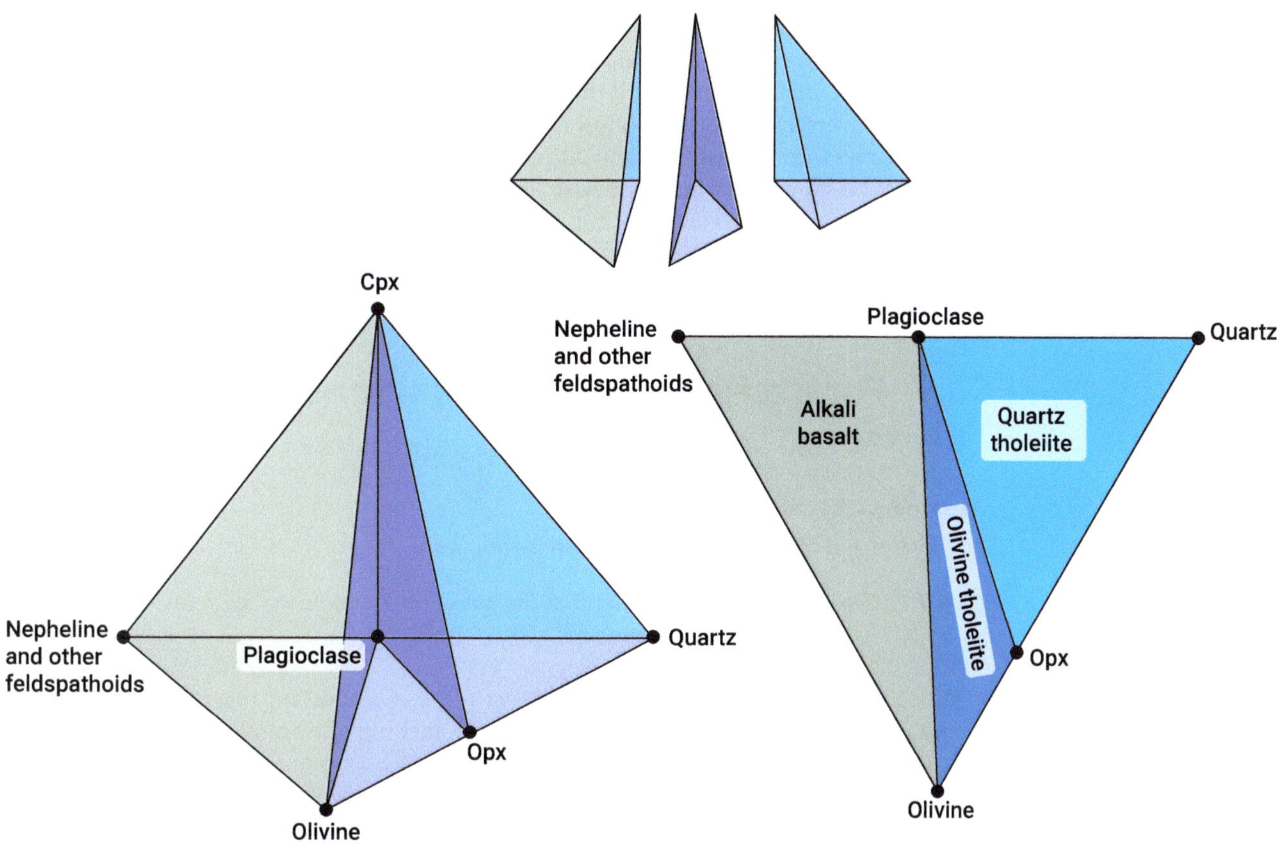

Figure 4.15 Basalt tetrahedron, commonly used to differentiate types of basalt. See text for discussion.

Source: After Yoder & Tilley (1962).

The components at the apices of the basalt tetrahedron are olivine, Cpx, quartz, and nepheline plus any other feldspathoids. Plagioclase lies on the nepheline-quartz join, and Opx lies on the olivine-quartz join. Note that Opx is a proper mixture of olivine and quartz, whereas plagioclase is not a mixture of just Nph and Qtz—the Ca component is ignored, a consequence of projecting from nine or so dimensions down to three. These intermediate compounds help define two internal planes (plagioclase-Cpx-Opx and plagioclase-Cpx-olivine) that split the main tetrahedron into three sub-tetrahedra, and a basalt is classified according to which of these three it falls in, as an alkali basalt, olivine tholeiite, or quartz tholeiite. Note that one can project from Cpx onto the base of the tetrahedron by normalizing the three remaining phases (e.g., olivine, Opx, plagioclase) to determine which tetrahedron a point falls in. This is a reasonable thing to do, because most basalts contain Cpx, and Cpx can coexist with the minerals at the apices of the tetrahedron. Projecting from, say, quartz would be less useful, because one would only see the Cpx-Opx-plagioclase triangle—everything behind it would be hidden, because quartz is not stable with olivine or nepheline.

The odd term "tholeiite" (tho′-lee-ite) is . . . well, it's not worth going into its history, because we're stuck with it. (OK, since you asked, the name comes from the town of Tholey in western Germany.) The term "tholeiitic" connotes various other chemical parameters as well as those in the basalt tetrahedron; suffice to say that most MORBs are tholeiites, and thus tholeiites are quite abundant.

The plane with vertices at plagioclase, Cpx, and olivine plays an important role in basalt evolution. Basalts in the sub-tetrahedra on either side of it are likely to crystallize Ol + Pl ± Cpx, and little else, throughout much of their history. The bulk composition of the crystallizing assemblage will lie within the plane, and thus the liquid composition will move away from the plane as temperature drops regardless of which side it starts on, just as a raindrop will flow downhill, away from a drainage divide, regardless of which side it starts on. It is a crest on the four-dimensional liquidus surface and difficult to envision. The plane behaves as a thermal divide, like the crest of the Ab liquidus in Figure 4.6. Crystallization of Ol, Cpx, and Pl will thus drive an alkali basalt deeper into alkali basalt territory. Conversely, crystallization of those phases from an olivine tholeiite can drive its liquid across the Pl-Cpx-Opx plane into quartz-normative compositions—just as crystallization of Fo from an olivine-normative (really, Fo-normative) liquid in the system Fo-Qtz (see Figure 4.5) can drive the liquid into quartz-normative compositions to the right of En. The Pl-Cpx-Opx plane is thus not a thermal divide; rather, it marks when an evolving liquid goes from olivine-normative to quartz-normative.

4.7 Forsterite-Diopside-Quartz at Higher Pressure

Learning Objective

By the end of this section you should be able to:

4.7.1 Use chemography to show that the peritectic point in the low-pressure version of Fo-Di-Qtz becomes a eutectic at 1 GPa.

The ternary peritectic point π in Figure 4.11 plays an important role in mantle melting, because it marks the first liquid produced upon melting of a Fo-Di-En source. At 1 atm, this point lies on the SiO_2 side of the Di-En join and is thus quartz-normative. However, at higher pressures (Figure 4.16) an interesting thing happens; the phase boundaries have shifted to the left such that by 1 GPa the invariant point where Fo, Di, and En coexist with Liq has moved into the Fo-Di-En subtriangle.

The reaction has gone from peritectic (Di + En = Fo + Liq) to eutectic (Di + Fo + En = Liq), and the liquid formed during first melting is now olivine-normative. At 1 GPa, the system

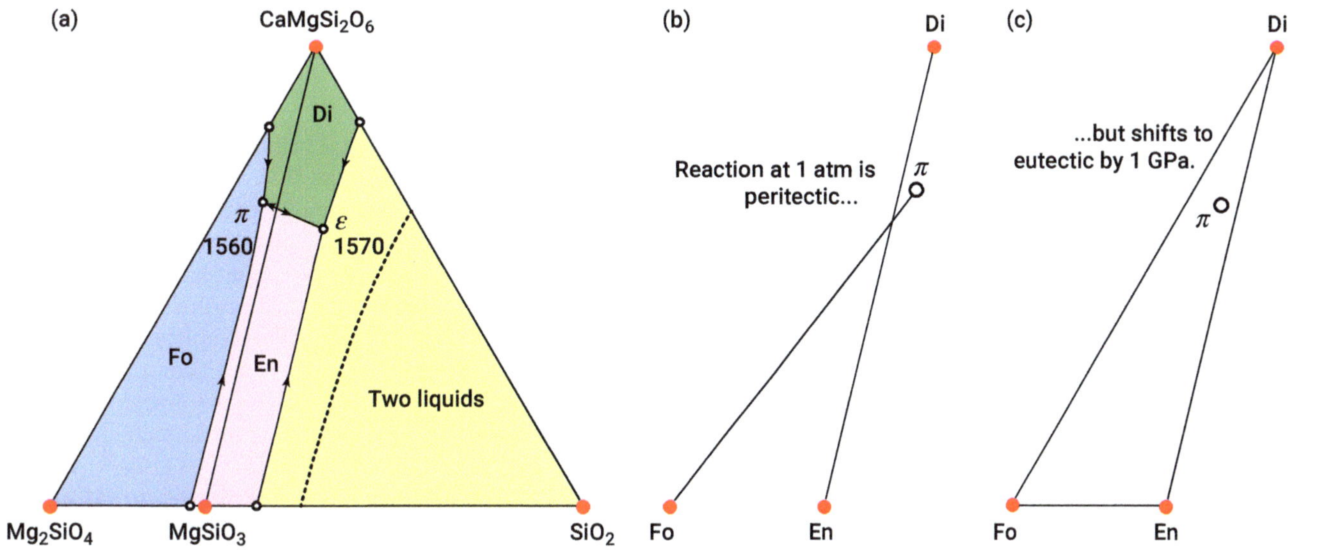

Figure 4.16 (a) Schematic of the Fo-Di-Qtz system at 1 GPa pressure, showing chemography of Fo-Di-En melting at 1 atm (b) and 1 GPa (c). With increasing *P*, the liquid migrates into the Fo-Di-En triangle, thus changing from a peritectic to a eutectic. The cotectic curve that ran downhill from π to ε in Figure 4.11 is now an arch that runs downhill to both π and ε from its crossing with the Di-En join.

Source: From Morse (1980).

Fo-Qtz is just two eutectics joined at En. The Di-En join is now a thermal divide; liquids on either side will migrate to the corresponding eutectic, one quartz-normative and one olivine-normative. At higher pressure, the liquids produced at the eutectic are alkali basalts; at low pressure they are tholeiites. This illustrates a recurring theme in igneous phase equilibria: increasing pressure drives liquids to more silica-undersaturated, and generally more alkalic, compositions. Remember that MORB melting probably begins at depths of around 60 km (see Figure 3.21), corresponding to pressures of about 2 GPa. Such liquids should be olivine-normative, but with ascent, the invariant point should shift to quartz-normative compositions. The compositions of individual packets of magma thus have depth information encoded within them.

4.8 Iron and Reality: The Olivine System

Learning Objectives

By the end of this section you should be able to:

- **4.8.1** Trace melting and crystallization paths in a binary solid solution phase diagram.
- **4.8.2** Deduce how an increase in oxygen fugacity will shift the compositions of Mg-Fe solid solutions such as olivine.
- **4.8.3** Use the Fo-Fa phase diagram to explain why peridotite that has partially melted and lost that melt is less dense than unmelted peridotite.

Solid Solution

The phase diagrams discussed earlier have been simplified by omitting solid solution, incongruent melting, and other things that occur in the real world. You are familiar with solid solution from mineralogy. One of the most important mixing pairs is Mg-Fe. Ferrous Fe (Fe^{2+}) and Mg have the same charge and nearly the same ionic radius, and both are abundant in the silicate Earth, so minerals that contain Mg

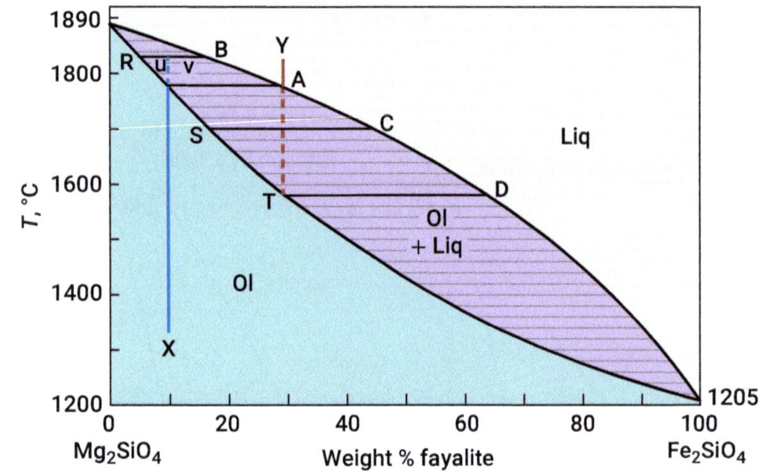

Figure 4.17 The olivine system Fo-Fa at 1 atm. This diagram is formed of a small number of experimental points in the Fe-rich (lower-temperature) part of the system extrapolated to the estimated melting point of Fo. The system was kept in equilibrium with metallic Fe and is not truly binary. [See Morse (1980) for a discussion of complications.]

Source: After Bowen & Schairer (1935).

typically show solid solution toward equivalent Fe^{2+} minerals. In basalts, Ol, Cpx, and Opx all show extensive solid solution between Mg and Fe end members.

The Ol system is a binary solid solution between Fo and fayalite (Fa). Such a phase diagram is quite distinct from a binary eutectic. A pure solid solution such as the Ol system (Figure 4.17) consists of a fish-shaped, two-phase area bounded by a solidus and liquidus that join the congruently melting end members. The key aspect of a eutectic system—that a mixture melts at a much lower temperature than either end member—is missing from such a system. Fa melts at a much lower temperature than Fo, and the solidus rises with each increment of Fo added. Adding Fa to Fo lowers the melting point of the system, but not below the melting point of Fa; adding Fo to Fa raises the melting point of the system.

This system exhibits many important behaviors that occur in (Mg, Fe)-bearing systems. Consider bulk composition X, which is 90 wt% Fo and 10 wt% Fa (a reasonable mantle olivine composition). The phase diagram shows that this composition will melt in this binary system at an exceedingly high temperature, over 1750 (temperatures are much lower in a real multicomponent basalt). The first melt must be at the same temperature, and its composition (A) is given by the liquidus; it is much richer in Fe, about 30 wt% Fa. This is how such systems behave; in equilibrium between an (Mg, Fe)-bearing mineral and liquid, the liquid is enriched in Fe compared to the mineral. For olivine, this relationship is reliable enough that the double ratio

$$K_D = \left(\frac{X_{\text{FeO}}}{X_{\text{MgO}}}\right)_{ol} \Big/ \left(\frac{X_{\text{FeO}}}{X_{\text{MgO}}}\right)_{liq} \quad \textbf{(Eq. 4.4)}$$

where X refers to atomic (or mole) fraction, is useful for predicting the composition of olivine that will precipitate from a liquid of a given composition (Roeder & Emslie, 1970). We say that olivine **fractionates** Mg from Fe, and Roeder and Emslie (1970) found that $K_D = 0.3$ for olivine-melt equilibrium. Using this relationship to follow olivine crystallization is covered in Toolbox 4.1.

Fractionate
Separate similar elements, chemicals, or isotopes from one another.

The Phase Rule and Solutions

Our first foray into the phase rule (see Section 3.4) dealt with phases of constant composition such as Qz, Trd, and silica melt. The Ol system is different—both Ol and coexisting melt are solutions. Let's try applying the phase rule. At constant pressure, $F = C + 1 - \pi$ (see Section 4.1). In the olivine system, $C = 2$, so for the two phases present, $F = 1$. We can vary either the composition of the liquid, or the composition of the crystals, or T, but not more than one independently, and keep a two-phase assemblage. As there is no three-phase assemblage in this system, there are no binary invariant points.

With addition of heat, the temperature rises, and the compositions of both crystals and liquid shift toward Fo (become more Mg-rich). The shift in crystal composition is a reaction that can be written schematically as

$$\text{Ol} = \text{slightly more Mg-rich Ol} + \text{Fe-rich Liq} \quad \textbf{(Eq. 4.5)}$$

The system is univariant ($F = C + 1 - P = 2 + 1 - 2 = 1$), and any one of the three intensive variables temperature, crystal composition, and liquid composition can be changed without destroying the phase assemblage Ol + Liq, as long as the other variables adapt to the new value of the one that was changed. Note that here the Ol composition is an intensive variable, whereas in a binary eutectic such as Fo-Di, the bulk composition (proportion of Fo to Di) is not (a thermodynamic point; the proportion of a chemical component in a phase is an intensive variable, but the proportions of phases are not). If we raise the temperature to 1820 °C, then the liquid is at B (~16% Fa), and the crystals are at R (5% Fa). By the lever rule, the proportion of crystals at this point is v/(u + v), a little over 50 percent.

With continued heating, the last bit of crystals melts away when the temperature is about 1850 °C, and the liquid composition reaches the bulk composition. The crystals are exceedingly Fo-rich by this point. If we were to remove liquid B when the system reaches 1820 °C, then we leave behind an Fe-depleted residue by taking away an Fe-enriched melt. Fe is significantly heavier than Mg (atomic weights of 55.8 and 25.3 g/mol, respectively), so mantle that has had melt extracted from it is less dense than undepleted mantle. This change in density can affect the buoyancy and long-term stability of the lithosphere.

The phase diagram shows that liquid in equilibrium with Ol will always be more Fe-rich (have a higher Fe/Mg ratio) than Ol crystals (this is shown by the K_D expression earlier as well). In a cooling system, crystallization of Ol rapidly depletes the liquid in Mg, leaving less for subsequently crystallized Ol. As a result, crystallizing Ol gets richer and richer in Fe as temperature drops.

Assume that our model mantle X melts a little, producing liquid A, and then this liquid escapes upward as basalt melt (remember, we're grossly simplifying here!), cools, and crystallizes. This separation of crystals from liquid is crystal fractionation. The volume of liquid A is now a new system; call this bulk composition Y. It starts to crystallize at the liquidus, which, if we're at the same pressure, is the same temperature as the solidus of X, and the first crystals to form are 10 percent Fa. As temperature drops, the equilibrium crystal composition slides down the solidus, and the liquid correspondingly slides down the liquidus. At 1700 °C, the crystals have evolved to S and the liquid to C, and when the last drop of liquid crystallizes, the crystals have evolved to T and the liquid to D—far more Fe-rich than the starting material X (remember, we made our basalt by melting the mantle). To stay at equilibrium, all the earlier-crystallized Ol must react with the liquid, exchanging Mg for Fe, as temperature drops. If the crystal cannot keep up with this process, then the cores of the Ol crystals will be more Mg-rich than predicted by equilibrium, and the crystals will be zoned from Mg-rich cores to Fe-rich rims.

As different as eutectic systems are from solid solutions, they are linked by the behavior exhibited by peritectic systems (e.g., see Figure 4.5). In a peritectic system, an intermediate compound (in this case, En) melts incongruently to a solid that is richer in one component (Fo) and a liquid that is richer in the other. A solid solution behaves the same way, except that the solid produced by melting is insignificantly different from the solid that is being melted; solid solution is an infinite series of such peritectics (Figure 4.18).

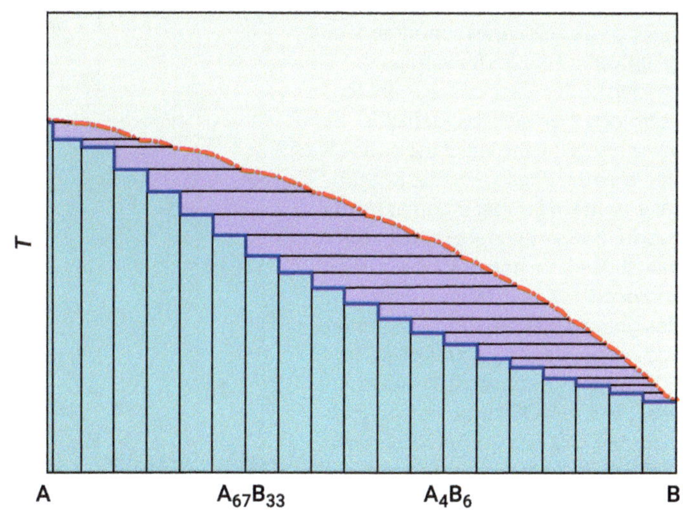

Figure 4.18 A solid solution can be viewed as an infinite series of peritectics. Here, in the hypothetical system A-B, there are 19 incongruently melting intermediate compounds, such as $A_{67}B_{33}$, A_4B_6, etc. Each melts incongruently to a more B-rich liquid and the next compound to the left, and the envelope described by the step-shaped solidus (blue) and cuspate liquidus (red) mimics the fish shape of a binary solid solution loop. As the number of intermediate compounds goes to infinity, this becomes a binary solid solution loop.

Oxidation State of Iron

Iron is the only major cation in igneous and metamorphic rocks that can take on different valence states. Imagine that we take the olivine system in Figure 4.17, which has only ferrous Fe, Fe^{2+}, and then oxidize all the Fe so that it is now ferric, Fe^{3+}. Olivine does not accept Fe^{3+}, so its composition must shift to pure Fo, booting the Fe^{3+} out. This occurs via the reaction

$$\begin{gathered} 3Fe_2SiO_4 + O_2 = 2Fe_3O_4 + 3SiO_2 \\ \text{fayalite} + \text{oxygen} = \text{magnetite} + \text{quartz} \end{gathered} \qquad \textbf{(Eq. 4.6)}$$

This famous reaction is known to petrologists as the FMQ buffer—a buffer because at constant *T* and *P*, if the three solid phases Fa, Mag (magnetite), and Qz (quartz) are present, then the partial pressure of oxygen is fixed by this equilibrium. Add more O_2 (increase the pressure), and more Fa is consumed to use up the excess; decrease it, and Mag and Qz react, releasing O_2.

Oxygen fugacity
The partial pressure of oxygen (O_2) in a system corrected for deviations from the ideal gas law; typically written as f_{O_2}.

The partial pressure of oxygen in rocks is known as the **oxygen fugacity**, which you can think of as its partial pressure corrected for non-ideality. Oxygen fugacity is known by the symbol f_{O_2} and is a critical parameter in Earth materials. The FMQ buffer approximates it in many igneous systems, and it is a surprisingly tiny value—about 10^{-8} Pa (10^{-13} atm) at 900 °C. That's about one-trillionth the pressure in the atmosphere. Thus, only a vanishingly small proportion of the oxygen in rocks is not bound to cations. That is why so many basaltic lavas are red; when they hit the atmosphere, which has an f_{O_2} of about 20 kPa, some of the Fe^{2+} in the magma or solid rock reacts to form red Fe^{3+} (see Chapter 7). Petrologists use a number of other buffer reactions to place themselves in T-f_{O_2} space (to use the typical jargon); these are shown in Figure 4.19. The Ni-NiO buffer is another important one, because it tends to approximate f_{O_2} in magmas produced in subduction zones.

Thermobarometer
Generic term for an estimate of temperature and pressure based on equilibrium thermodynamics of a chemical reaction; sometimes written as geothermobarometer.

Estimates of *T* and f_{O_2} for volcanic rocks, shown in Figure 4.19, were calculated using the Fe-Ti oxide **thermobarometer**. This tool takes advantage of two f_{O_2}-dependent reactions:

$$\begin{gathered} 3Fe_2TiO_4 + 0.5O_2 = Fe_3O_4 + 3FeTiO_3 \\ \text{ülvospinel} + \text{oxygen} = \text{magnetite} + \text{ilmenite} \end{gathered} \qquad \textbf{(Eq. 4.7)}$$

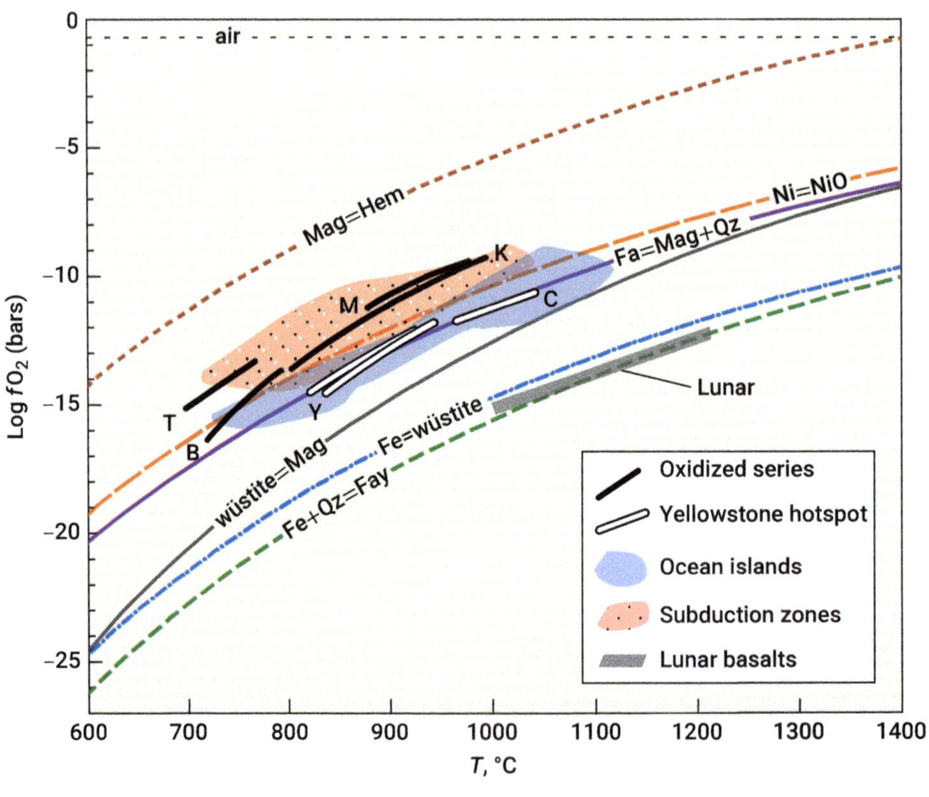

Figure 4.19 Variation in relative f_{O_2} among volcanic series superimposed on univariant curves for various oxygen buffer reactions relevant to petrology. For each, oxygen plus the phase(s) on the left of the equal sign give the phase(s) on the right. Take-home message: magmas from subduction settings are more oxidized than those from hotspot, ocean island, and oceanic ridge settings. Fields designate data from subduction zones and ocean islands (Cottrell et al., 2021) and from lunar basalts (Sato et al., 1973); MORB (not plotted) runs close to FMQ. Curves from Hildreth (1981) show variation among selected volcanic areas. [Letters key to names: oxidized series comprises the Bishop Tuff (B), Timber Mountain (T), Mazama (M), and Katmai (K); Yellowstone hotspot series comprises Yellowstone (Y; Lava Creek and Huckleberry Ridge tuffs) and Cougar Point (C).]

Sources: Data from Moretti & Neuville (2021); Cottrell et al. (2021); Sato et al. (1973); and Hildreth (1981).

$$Fe_3O_4 + 0.5O_2 = 3Fe_2O_3$$
$$\text{magnetite} + \text{oxygen} = \text{hematite} \quad \textbf{(Eq. 4.8)}$$

Each defines a univariant line on Figure 4.19, and for a given sample, these intersect in a point that gives a T-f_{O_2} estimate for that sample (Buddington & Lindsley, 1964). Compositions are determined by electron microprobe. Because the intersecting lines corresponding to reactions in Equations 4.7 and 4.8 are near-parallel, measurement error smears the estimates out along lines that parallel the buffer curves, and thus at least some of the elongation of the data arrays is a result of correlated error rather than geochemistry.

Regardless of data smearing, the arrays in Figure 4.19 make several important points. First, there are clearly distinct ranges in both T and f_{O_2}. Second, magmas from subduction zones, or areas where subduction had recently ended, are higher in f_{O_2} (near Ni-NiO) than those of hotspots, ocean islands, and MORB (near FMQ). And finally, lunar magmas are much lower in f_{O_2} than Earth magmas.

Adding Fe adds a dose of reality to the study of basalts. Because of the FMQ reaction and others like it, the actual mineral compositions of rocks depend on f_{O_2}, as does the CIPW norm calculation. The higher f_{O_2} is, the lower the quantities of olivine and other minerals that contain Fe^{2+}. Thus, in the basalt classification tetrahedron, changing the proportion of Fe^{2+} to total Fe in a rock analysis changes where the rock plots. A rock that is a little olivine-normative when the proportion is low can become quartz-normative if the proportion is raised.

What to do? Most methods of rock analysis do not distinguish the oxidation states of Fe (see Toolbox 1.1). Even if the proportions of Fe^{3+} and Fe^{2+} are determined, they probably do not reflect the magmatic state owing to late-stage oxidation. In most magmas, the atomic ratio $Fe^{3+}/(Fe^{3+} + Fe^{2+})$ is low, on the order of 0.1 or 0.2. There are formulas of varying sophistication for estimating this ratio from a chemical analysis (LeMaitre, 1976; Sack et al., 1980). In most cases, it is probably best to assume a low ratio (e.g., 0.1 or 0.2) and soldier on.

Toolbox 4.1

Modeling Olivine Fractionation

The partitioning of Mg and Fe between melt and olivine, characterized by the $K_D = 0.3$ relationship, allows us to model melt and crystal evolution during crystallization and removal of olivine from a melt. The arithmetic is easy to handle with a bit of programming.

Given the weight percents of FeO and MgO in a melt w_{FeO} and w_{MgO}, we can calculate the molar ratio of FeO to MgO in the melt (the denominator of K_D) by

$$r_m = \frac{X^m_{FeO}}{X^m_{MgO}} = \frac{w^m_{FeO}/71.85}{w^m_{MgO}/40.31} \quad \textbf{(Eq. 4.9)}$$

where the numeric denominators are the molecular weights of the two oxides. The corresponding ratio in the olivine is then

$$r_{ol} = \frac{X^{ol}_{FeO}}{X^{ol}_{MgO}} = 0.3r_m \quad \textbf{(Eq. 4.10)}$$

In the olivine, the number of moles of FeO and MgO add up to 2 for a four-oxygen formula:

$$n^{ol}_{FeO} + n^{ol}_{MgO} = 2 \quad \textbf{(Eq. 4.11)}$$

and thus

$$\frac{n^{ol}_{FeO}}{2 - n^{ol}_{FeO}} = r_{ol} \quad \textbf{(Eq. 4.12)}$$

With a little arithmetic, we can show that

$$n^{ol}_{FeO} = \frac{2r_{ol}}{1+r_{ol}} \text{ and } n^{ol}_{MgO} = \frac{2}{1+r_{ol}} \quad \textbf{(Eq. 4.13)}$$

These are the numbers of moles of FeO and MgO in one mole of olivine, and if we multiply them by their respective formula weights, add in one mole of SiO_2 (60.09 g), and normalize to 100, we have the composition of the olivine in wt% oxides.

4.9 The AFM Diagram and Calc-Alkaline Trend

Learning Objectives

By the end of this section you should be able to:

4.9.1 Classify a rock series as having a calc-alkaline or Fe-enrichment trend by its pattern on an AFM diagram.

4.9.2 Explain how increasing oxygen fugacity can shift a rock series from an Fe-enrichment trend to a calc-alkaline trend.

The term "tholeiitic" is defined in a number of ways, including by plotting an analysis in the basalt tetrahedron, but in practice it connotes two qualities: low alkalis and an Fe-enrichment trend. The latter refers to how Fe, or more specifically the Fe/Mg ratio, varies as you proceed from low silica to high silica along the BADR trend. It is time to introduce a useful ratio to our working vocabulary. The Mg-number is defined as

$$Mg\# = 100\frac{\text{Mg}}{\text{Mg} + \text{Fe}} \qquad \textbf{(Eq. 4.14)}$$

on an atomic basis; it is the atomic percentage of Mg relative to total Mg + Fe.

The AFM diagram (Figure 4.20) is a triangular plot of wt% total alkalis (Na_2O + K_2O), FeO, and MgO. It was seen long ago that rocks from what we now recognize as subduction zones are distinct from those of the oceans on such a plot. Figure 4.20 shows a typical subduction suite (Medicine Lake again) and rocks from Iceland, along with indications of where rocks plot as you go from low silica to high silica along the BADR trend. The two areas are quite distinct, and the dividing line separates them reasonably well. Rock suites that follow the Iceland pattern are tholeiitic, and those that follow the subduction trend are given the term "calc-alkaline"—another old term, from pre-plate tectonics days, that has been dragged into the present day. The term was introduced as "calc-alkalic" by Peacock (1931) along with the terms "alkalic," "alkali-calcic," and "calcic," but only one, usually written as "calc-alkaline," gets much use these days, because it means "probably subduction-related." As you can infer, these terms have to do with how Ca and alkalis covary, and were introduced as a purely descriptive way to separate different rock series. In a general way, igneous rocks from the oceans are tholeiitic, and those from the continents are calc-alkaline.

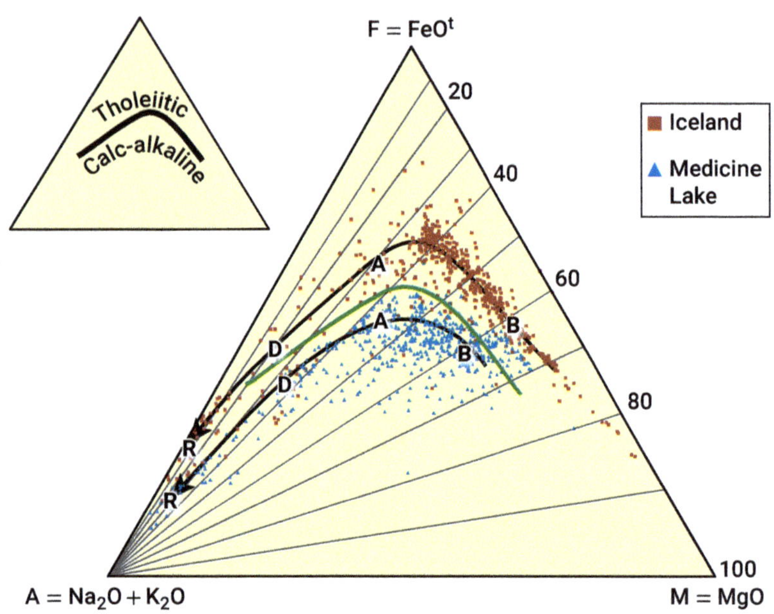

Figure 4.20 AFM diagram with ~600 analyses each from a subduction-related volcanic system (Medicine Lake) and an oceanic island (Iceland). B, A, D, and R refer to the approximate positions of basalt, andesite, dacite, and rhyolite (based on silica) for each system. Iceland follows the tholeiitic trend and Medicine Lake the calc-alkaline one. Gray lines radiating from A are Mg#.

The essence of the tholeiitic trend is Fe enrichment. Going from basalt to Iceland's version of andesite (called, not surprisingly, icelandite) takes you almost directly away from the M apex, perpendicular to the contours of Mg#, because of the strong fractionation of Mg over Fe into olivine (see Figure 4.17). Icelandite is andesite with a low Mg#. The Mg# drops dramatically from a typical value around 70 to about 35, and then a very scattered trend heads off toward the AF sideline. The calc-alkaline trend also starts off at an Mg# around 70 but heads toward the F apex more slowly, cutting across the contours.

This difference in behavior is most likely controlled by f_{O_2}. In a system at low f_{O_2}, most of the Fe is ferrous and goes into Ol and other ferromagnesian silicates. Increasing f_{O_2}, however, can cause magnetite (Mag), a mixed ferric-ferrous oxide, to join the crystallizing assemblage. As Mag is 100 percent Fe oxides, it rapidly sequesters Fe and prevents the liquid from following the tholeiitic trend, even though Ol, Cpx, and other crystallizing minerals are busy fractionating Mg from Fe. High oxygen fugacity probably plays a critical role in generating the range of magmas found in subduction zones, a topic we will return to in Chapter 6.

Summary

Key points of this chapter include the following:

- Compounds such as minerals can melt either congruently (to a liquid of the same composition; e.g., Qz = Liq) or incongruently (to a liquid of a different composition plus a different solid assemblage; e.g., Fo = En + Liq).
- Mixtures of two congruently melting compounds that do not exhibit solid solution toward one another melt at a lower temperature than either compound, and the composition of the liquid does not depend upon the proportions of minerals involved. This is eutectic melting. By Baldy's law, the composition of the eutectic melt lies between the compositions of the compounds melting. This generalizes to more than two components.
- The uniformity of MORB bulk compositions across Earth indicates that they are generated by pseudo-eutectic melting of the mantle. We say pseudo-eutectic because solid solution and other phases make this melting not strictly eutectic.
- An incongruently melting compound melts at a peritectic point. In a mixture of such a compound plus its breakdown product (e.g., En + Fo), the composition of the liquid lies outside of the solid-solid join. In this example, melt produced from a solid solution containing olivine (Fo) is quartz-normative and, if separated from the solids, will go on to crystallize Qz.
- If melt in a crystallizing system is mechanically separated from the crystals with which it is in equilibrium, then crystal fractionation has occurred, and the liquid can crystallize to a rock of a different composition, leaving behind a crystalline residue of yet a different composition.
- By Baldy's law, the separated liquid and residual crystals add up to the original composition.
- If melt is not separated, then regardless of how far the liquid composition evolves during crystallization, *no change in rock composition occurs.*
- For thermodynamic reasons, in a complex natural system undergoing melting, the first liquid tends to be rich in silica and feldspar components—essentially rhyolite.
- Pressure can shift the curves and invariant points on phase diagrams. An important example occurs in the system Fo-Di-Qtz, where the peritectic in the subsystem Fo-Di-En becomes a eutectic at higher temperature, trapping liquid in the subsystem.
- Solid solutions can be viewed as an infinite series of peritectics.
- Oxygen fugacity is important in Fe-bearing systems (e.g., higher values move Fe out of minerals such as olivine and into those with ferric Fe, such as magnetite).

Note

1. The equation they were referring to is more complicated and even includes the inscrutable Greek letter ζ.

CHAPTER 5

Metamorphism and Hydration of Oceanic Lithosphere

In Chapter 2, we learned that the Earth was likely melted completely early in its history—in particular, following the Moon-forming impact. Such events should have caused near-but-not-quite complete degassing and dehydration of the early Earth. If true, then where did Earth's water come from? Besides recondensation after the Moon-forming impact, a likely source is accretion of water-bearing planetesimals and other debris from outer parts of the solar system. Once the really large impacts ceased, impact of goopy, ice-rich material (in particular, comets) probably delivered water to the cooling early Earth (Morbidelli et al., 2000).

Water has a profound effect on magmatism, metamorphism, and plate tectonics. This chapter discusses the deep water cycle: hydration of oceanic lithosphere at oceanic ridges, metamorphism of that lithosphere in subduction zones, and eventual deep release of water, promoting magmatism at subduction zones.

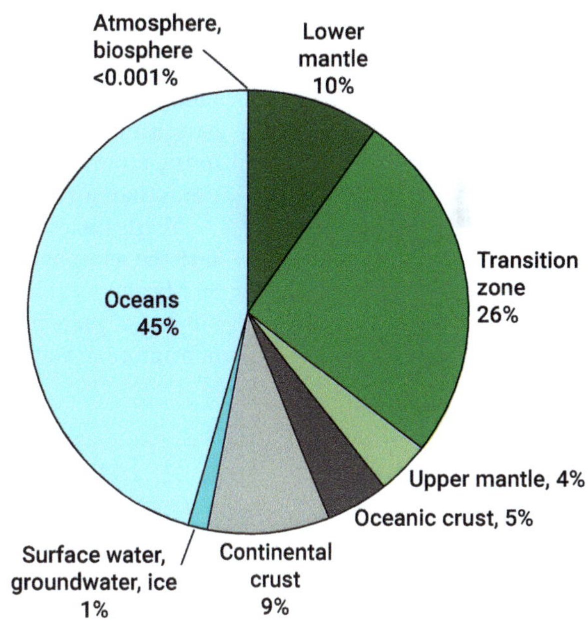

Figure 5.1 One estimate of the distribution of water among Earth's reservoirs.

Source: Data from Bodnar et al. (2013).

5.1 The Earth's Deep Water Cycle

Learning Objective

By the end of this section you should be able to:

5.1.1 Make a sketch showing the principal reservoirs of water on and in Earth and how they exchange with one another.

Water is as important to the lithosphere as it is to the biosphere, because it has dramatic effects on melting and mechanical behavior. The proportion of H_2O in the Earth is estimated at about 0.025 percent by weight (Bodnar et al., 2013). Figure 5.1 shows the distribution of H_2O among various reservoirs; surprisingly, the amount contained in the solid Earth (the geosphere) may be

Mineral-laden water emerging from a hydrothermal vent on the Niua underwater volcano in the Lau Basin, southwest Pacific Ocean. As the water cools, minerals precipitate to form tower-like chimneys and dark gray, sulfide-laden plumes. Crabs (lower right) and other life forms live near vents such as these.

Source: USGS.

Table 5.1 Distribution, Fluxes, and Residence Time of Water in Various Reservoirs

Reservoir	Amount of H_2O (10^{17} kg)	Amount of H_2O (km^3)	Percentage	Flux in/out (10^{12} kg/yr)	Residence Time (yr)
Atmosphere	0.13	1.3×10^4	0.00042	496,000	2.6×10^{-2}
Biosphere	0.047	4.7×10^3	0.00015	7,400	6.4×10^{-1}
Oceans	13,700	1.4×10^9	45	441,000	3.1×10^3
Surface water, groundwater, ice	440	4.4×10^7	1.43	154,500	2.8×10^2
Continental crust	2,800	2.8×10^8	9.2	0.4	7.0×10^8

Source: Modified from Bodnar et al. (2013).

comparable to the amount contained in the more conventional reservoirs of the hydrosphere, atmosphere, and biosphere. This is because the solid Earth has a much greater volume than the oceans and atmosphere, and because many nominally anhydrous minerals in the mantle, such as olivine, contain small but significant quantities of hydrogen. In particular, the transition-zone minerals wadsleyite and ringwoodite (see Chapter 3) can contain up to about 3 wt% H_2O (Smyth & Jacobsen, 2006)—a huge amount, given the volume of that Earth shell. If saturated, the transition zone could hold several oceans' worth of water.

Estimated fluxes in and out of these reservoirs are given in Table 5.1. The water cycle consists almost completely of exchanges between the atmosphere, oceans, and fresh water, with a little diverted in and out of the biosphere just to keep things interesting. Remember that dividing a mass or volume (e.g., 1.4×10^{21} kg H_2O in the oceans) by a flux (e.g., 4.4×10^{17} kg/yr in and out of the oceans) yields a residence time (see Chapter 3), an estimate of the average time that an aliquot of the substance spends in that reservoir. For the oceans, this works out to about 3000 years. Spit in the ocean, and your spit will hang around (on average) for a long time. Water's residence time is approximately a week in the atmosphere and half a year in the biosphere.[1] These are much shorter residence times than for the ocean.

Yearly fluxes in and out of geospheric reservoirs are minuscule compared to those outside the geosphere, and residence times in the geo-reservoirs are much longer: around 80 Myr in oceanic crust, 700 Myr in continental crust, and hundreds of millions or even billions of years in the mantle, meaning that the mantle could conceivably contain traces of primordial water and other volatile substances from when it first formed.

5.2 Hydrothermal Systems at Mid-Ocean Ridges

Learning Objectives

By the end of this section you should be able to:

5.2.1 Predict the physical behavior of water under different *P-T* conditions and trajectories.

5.2.2 Use paragenesis diagrams, petrogenetic grids, and chemographic diagrams to predict the assemblages observed during progressive metamorphism of hydrated mafic and ultramafic rocks.

Behavior of Hot Water

Subaerial
Literally below the air, so on the Earth's surface.

Basaltic magma reaches the surface approximately 1200 °C above ambient temperatures. **Subaerial** eruptions cause the air above the lava flow to convect dramatically, carrying heat away in columns of rising hot air. Air convection and radiation of infrared energy efficiently cool the surfaces of basalt flows, such that a rigid crust can form in minutes. The principal chemical reaction that accompanies cooling of subaerial eruptions is partial oxidation of Fe^{2+} to Fe^{3+}, turning FeO to Fe_2O_3. Depending on how slowly they cool, and thus how much of the magmatic liquid crystallizes instead of turning to glass, subaerial basalts end up composed of plagioclase, olivine, clinopyroxene, oxides (e.g., Cr-spinel, magnetite, ilmenite), and glass. Such basalts are nearly anhydrous, typically containing around 0.5 wt% H_2O^+ (water that is bound in minerals or held in glass; see Toolbox 1.1).

Nearly all MORB is erupted underwater, and such **submarine** eruptions lead to much more significant chemical reactions. Basaltic magma contains enough thermal energy to heat up and boil an approximately equivalent volume of liquid water, and that water, when boiled at 1-atm pressure, turns to steam with a volume about 1700 times greater than the liquid. Thus, basalt lava flows that reach the ocean are usually marked by huge plumes of white steam that rise from the point of entry, along with vigorously upwelling hot water (Figure 5.2). It is difficult to superheat the water (get it much above its boiling point), because boiling causes a huge volume expansion that drives low-density steam away from the heat source and into the air.

Figure 5.2 Explosive interaction of hot magma from Puʻuʻōʻō with seawater at the Kamokuna ocean entry, May 2017. Lava was entering the ocean at a flux of ~2 m^3/s, causing huge billows of steam and producing copious amounts of black sand from rapid quenching and fragmentation of the lava.

Source: AFG photo.

Submarine
Literally below the ocean, so in water.

Eruptions at MOR volcanoes are different, because oceanic ridges typically lie at depths of 2 to 4 km. At such depths, the high pressure prevents steam expansion that occurs at shallow depths, and the distinction between steam and water disappears. Figure 5.3 shows a *P-T* diagram for water at pressures relevant to the oceans. The oceans are not fresh water, but we will use this simpler system to illustrate some of the general aspects of the water system. Complications arise in the more realistic H_2O-NaCl system (Lowell et al., 1995), but the simpler system in Figure 5.3 captures relevant behavior.

This portion of the water phase diagram displays several important features. The first is that the ice-water boundary is nearly vertical, with a steep but negative slope (thus, the

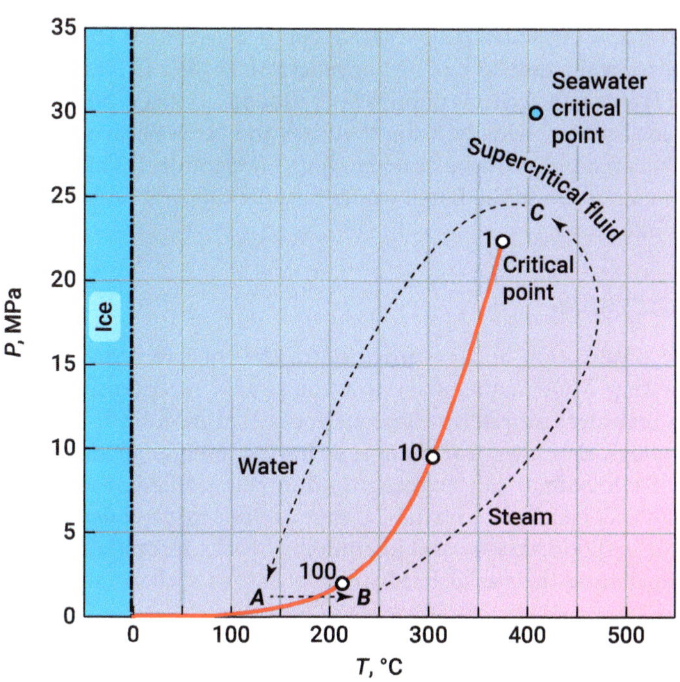

Figure 5.3 *P-T* diagram for water at conditions relevant to the oceans, down to a depth of ~3.5 km. The boiling curve separating liquid water from steam terminates at a critical point at 374 °C and 22.1 MPa; at greater pressures or temperatures, the distinction between water and steam breaks down, and we refer to this material as a supercritical fluid. The critical point for seawater is indicated and lies at somewhat higher *P* and *T*. Tics along the curve give the ratio of the volume of steam to water at that point. The high compressibility of steam means that explosive eruptions are restricted to shallow depths; consequently, it is easier to heat water to high temperatures at greater depths. Most oceanic ridges lie at 2–4 km depth, corresponding to pressures of roughly 20–40 MPa. Thus, much seawater heated by magmatic activity at oceanic ridges becomes a supercritical fluid.

melting point goes down as pressure goes up). Using the Clapeyron equation (see Chapter 3), we can infer that ice must be less dense than liquid water, at least near its melting point. This makes water one of a small class of substances in which the solid phase floats on the liquid phase,[2] and it is interesting to contemplate an alternative Earth in which ice sinks.[3]

Another characteristic of the H_2O phase diagram is that the water-steam phase boundary is strongly curved, being nearly flat (~1 Pa/°C) at low *P* and *T* but curving to a steep positive slope (> 200,000 Pa/°C) at higher *P* and *T*—a massive increase. This curvature occurs because steam is vastly more compressible than water. At low *P*, steam has a much larger molar volume than liquid water, leading to a large denominator ($\Delta V = V_{steam} - V_{water}$) in the Clapeyron equation and a low slope. As *P* increases, however, the molar volume of steam decreases dramatically, ΔV decreases, and the curve steepens. This shape generally characterizes dehydration reactions, too, and you will see that their slopes on *P-T* diagrams also steepen with increasing pressure.

The final important characteristic of the *P-T* diagram of water is that the water-steam curve terminates at a **critical point** at around 22 MPa and 374 °C. At pressures and temperatures above the critical point, the first-order distinction between water and steam disappears. Consider path *ABCA* in Figure 5.3. At a pressure of 1 MPa (corresponding to 100 m underwater), going from *A* to *B*, water boils at a temperature of about 177 °C. This is well above its boiling point in your kitchen, but otherwise the basic effects are the same—at that temperature, liquid water converts to much lower-density steam. Boiling and freezing are first-order phase transitions, where parameters such as density and entropy are discontinuous across the boundary. One could then take this parcel of steam along path $B \rightarrow C \rightarrow A$, turning it gradationally into what we would call a **supercritical fluid** at *C* and then gradationally back to water at *A*. We've turned steam back into water, but without it condensing or undergoing a first-order phase transition.

From Section 2.7, the geobaric gradient is $\frac{dP}{dz} = \rho g$. The density of water is approximately constant at 1000 kg/m^3, so the pressure gradient in the ocean works out to approximately 10^4 Pa/m, or 10^5 Pa/10 m (10^5 Pa is very close to 1 atm, so every 10 m of depth in water adds another atmosphere to the pressure). Thus, the pressure at MOR crests is in the range of 20 to 40 MPa, a range that encompasses the critical point. Magma erupting at a typical MOR depth of 3 km erupts into water that is at a pressure above the critical point.

Consider the phase diagram in Figure 5.3. At a pressure of 10 MPa, corresponding to about 1 km water depth, water boils at about 300 °C, with a volumetric expansion of about 10 times. This places a limit on how hot water can get at that depth, because any steam that is formed rapidly rises away from the heat source by buoyancy. At 35 MPa (3.5 km water depth), however, the system is above the critical point. Water expands as it gets hotter, but there is no dramatic expansion to lower-density steam that rises and carries heat away. So, water at this depth (pressure) can more easily be heated to high temperatures. This allows for significant water-rock interaction, and the Earth would be quite different without this mechanism for fixing water into rocks.

Water-Rock Interaction

Shipboard and submersible observation and examination of the rock record—in particular, ophiolites—indicate that MOR magmatism is accompanied by extensive reaction between hot rock and seawater; we refer to igneous rocks that have been affected by post-emplacement mineralogical or chemical changes as having undergone **alteration** in a **hydrothermal system**. Such systems can be recognized in many different ways, including seawater thermal and geochemical anomalies, areas of sulfide mineral deposition on the seafloor, alteration of seafloor basalts, and alteration of rocks in ophiolites. We will refer to such systems, including their associated ore deposits, as seafloor alteration lithosystems.

Cruises in the 1960s and 1970s located a number of areas with obvious characteristics of hydrothermal systems, but it was the discovery of dramatic "smoker" jets of hot

Critical point
The point on a liquid ⇢ gas transition curve on a *P-T* diagram at which the distinction between liquid and gas disappears.

Supercritical fluid
A fluid at *P-T* conditions above the critical point at which the gas-liquid distinction disappears.

Alteration
Chemical changes in a rock relative to a higher-temperature reference condition. Usually refers to the addition of water or other components, such as hydration of basalts or high-*T* anhydrous minerals (e.g., olivine) to produce hydrous amphiboles or sheet silicates (e.g., serpentine).

Hydrothermal system
Circulation of water through fractures or rocks, driven by temperature and buoyancy gradients. Hot, buoyant water rises above oceanic ridges and other heat sources and is replaced by cooler water, driving a flow system.

White smokers
Plumes of fluid in deep-sea hydrothermal systems, colored white by precipitating sulfates and silica.

Black smokers
Plumes of fluid in deep-sea hydrothermal systems, colored dark gray or black by precipitating sulfides.

Hydration
Addition of water as a structural component in minerals, either as H_2O or OH^- (or both). Hydration of olivine (Mg_2SiO_4) and enstatite ($MgSiO_3$) can form serpentine [$Mg_3Si_2O_5(OH)_4$]; hydration of anhydrite ($CaSO_4$) can produce gypsum ($CaSO_4 \cdot 2H_2O$). Glass can be hydrated simply by dissolving H_2O into it.

water (Figure 5.4) on the East Pacific Rise off the coast of Mexico that cemented the importance of hydrothermal systems on the ocean floor. These vents are of two types. **White smokers** send out cloudy white fluids at 100 to 350 °C, and **black smokers** shoot out dark gray jets at temperatures over 350 °C and at velocities over 1 m/s. White smokers are colored by suspended barite ($BaSO_4$) and silica, and black smokers by sulfides of Zn, Cu, and Fe. These vents are commonly surrounded by dense communities of deep-water organisms (see Making Connections 5.1).

Figure 5.4 Sully Vent in the Main Endeavour Vent Field, northeastern Pacific. A bed of tube worms (white stalks with red terminations) cover the base of the black smoker. Various instruments sit on and next to the tube worm colony, and several vents are releasing jets of black or gray sulfide-rich hot water.

Source: NOAA PMEL EOI.

Thermal and chemical studies indicate that oceanic-ridge hydrothermal systems process huge amounts of seawater—enough that a volume equivalent to the world's oceans passes through them every million years or so (Wolery & Sleep, 1976). Interaction of water and rock has significant effects on both. The main effect of hydrothermal circulation through oceanic lithosphere is **hydration**. MORBs and associated gabbros consist mostly of plagioclase, olivine, Cpx, and oxides, with variable amounts of glass that formed from liquid that was present during rapid cooling. Glass is abundant in the rims of pillow basalts owing to rapid quenching. The minerals in this list are anhydrous (water-free), and glass in fresh, unaltered basalts is relatively water-poor, such that a typical fresh MORB contains about 0.5 to 1.5 wt% H_2O^+ (remember from Chapter 1 that H_2O^+ is water that is held tightly enough that it is not driven off by drying a rock powder at 100 °C). Hydrothermally altered MORB contains 6 wt% H_2O^+ or more, much of it fixed in hydrous minerals. Such structurally bound water can be carried into the deeper Earth by subduction, leading to melting and other interesting behavior.

Hydrothermal alteration also affects ultramafic rocks such as peridotite and harzburgite, both by deep circulation and because ultramafic rocks are exposed at the surface over significant parts of the ocean floor, generally at slow-spreading ridges, by normal faults (Karson et al., 2006) or transform faults (Bonatti et al., 2001). In these environments, peridotite and other ultramafic rocks can be extensively converted to hydrous minerals. Exactly what minerals you see depends on *P-T* conditions (mostly temperature in this case) and bulk composition. This is illustrated in Figure 5.5 for hydrated metabasalts and metaperidotites. This type of diagram is called a **paragenesis diagram**, because it illustrates the mineralogy (paragenesis) you would expect across *P-T* conditions. For example, a fully hydrated metaperidotite at low temperature would contain brucite, serpentine, and chlorite. If we heat it, olivine, talc, and amphiboles will progressively develop as brucite, serpentine, and chlorite disappear. If a metaperidotite reaches high enough temperature, it will contain olivine, plagioclase, clinopyroxene, and orthopyroxene (and thus become a plagioclase peridotite). The bars are tapered to indicate that the appearance or loss of a mineral occurs over some range of *P-T* conditions; it is rarely abrupt for chemically complicated (that is to say, real) rocks.

Paragenesis diagram
A diagram that shows the appearance and disappearance of key minerals across a range of *P-T* conditions from low-grade to high-grade.

Metamorphism
Literally a change (meta) in shape or form (morph). Usually implies growth of minerals in response to a different pressure or temperature than the rock's original formation conditions (e.g., formation of a metamorphic schist from a sedimentary shale). Deformation can also form metamorphic rocks without changing mineral assemblage (e.g., shearing can transform a granite into a gneiss).

Metamorphism and Metasomatism

Hydrothermal alteration is a form of **metamorphism** that changes the mineralogical makeup and structure of the rock, transforming one mineral assemblage into another through the action of heat, pressure, and fluids. It is also a form of **metasomatism**. Metasomatism is a metamorphic process in which the bulk composition of a rock is changed by reaction with an external reservoir of chemical components—in this case, the ocean. Metamorphism and metasomatism are related and typically concurrent processes.

Metasomatism
Literally a change (meta) in body (soma). Usually refers to changes in whole-rock compositions because of fluid flow, which can either add or remove chemical constituents. Hydration is a form of metasomatism.

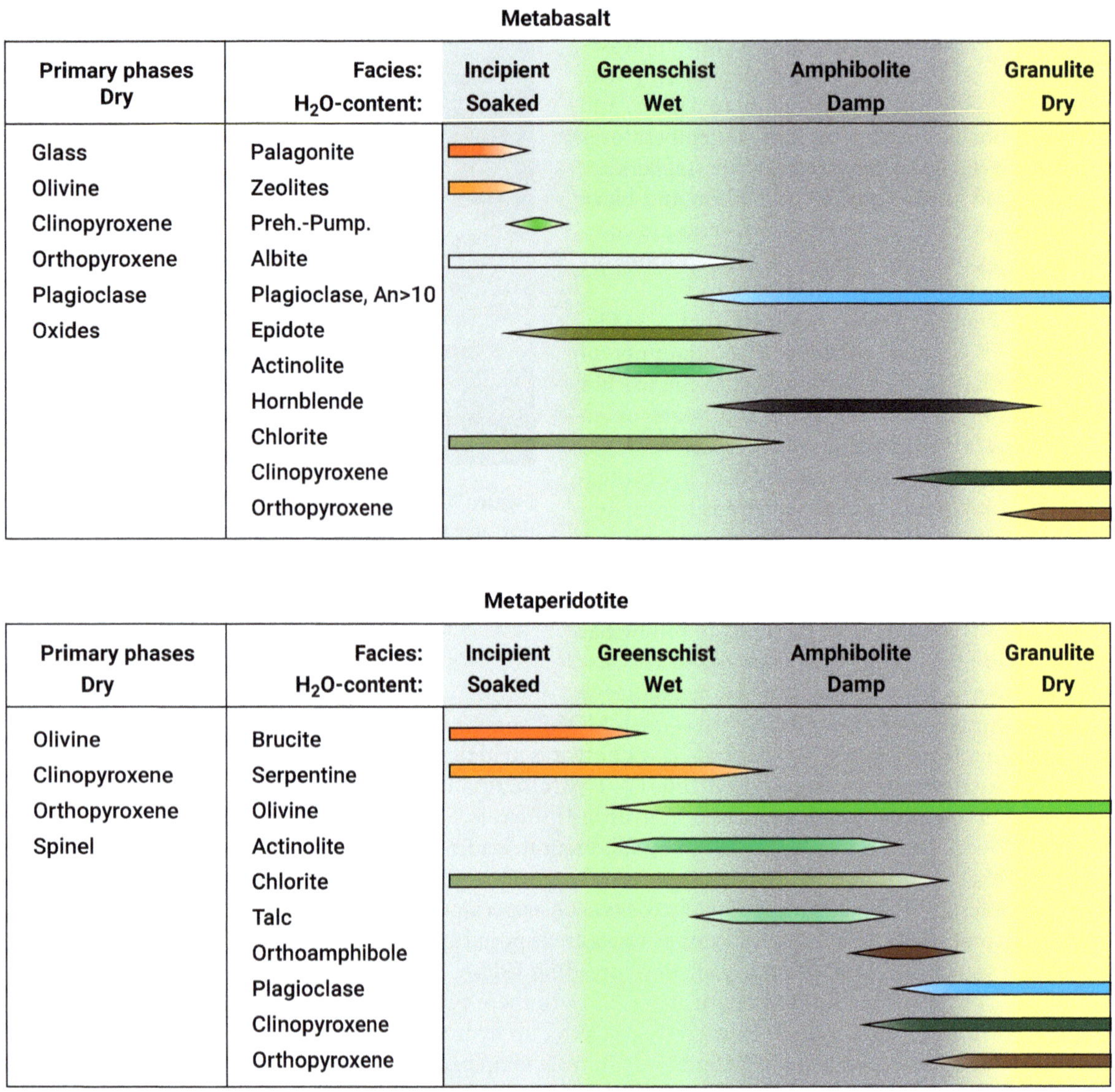

Figure 5.5 Paragenesis diagram showing approximate stabilities and qualitative water contents of mineral assemblages for hydrated basalt and peridotite as they are heated. Metamorphic facies and hydration reactions are described in the following sections. Tapered ends indicate that minerals usually do not appear or disappear abruptly, but rather over some range of temperature or pressure.

Rocks and Minerals of the Seafloor Alteration Lithosystem

This lithosystem forms at oceanic ridges by reaction of hot, newly erupted basalt with seawater and by hydrothermal circulation of seawater through oceanic lithosphere, both crust and mantle. It can also form off-ridge during serpentinization of peridotite.

The rocks and minerals produced by hydrothermal alteration at oceanic ridges are chemically and texturally complex, and most are fated to be recycled in subduction zones, where the hydration process is reversed—we'll learn more about that in the next chapter. Common minerals and their water contents are in Table 5.2. Determining the amount of H_2O (or any other component) from a mineral formula is a useful skill (see Toolbox 5.1).

In addition to producing hydrous minerals, hydration of basalt and ultramafic rocks produces the following rocks.

Table 5.2 Typical Hydrous Minerals in Altered Basalt and Ultramafic Rocks, and Their Water Contents

Group	Representative Mineral	Formula	Wt% H_2O
Chlorite	Clinochlore	$Mg_5Al(AlSi_3O_{10})(OH)_8$	13.0
Amphibole	Actinolite	$Ca_2(Mg,Fe)_5Si_8O_{22}(OH)_2$	2.1
Serpentine	Antigorite	$Mg_3Si_2O_5(OH)_4$	13.0
Hydrous Ca-Al silicates	Zoisite	$Ca_2Al_3Si_3O_{12}(OH)$	2.0
Hydrous Ca-Al silicates	Prehnite	$Ca_2Al_2Si_3O_{10}(OH)_2$	4.4
Hydrous Ca-Al silicates	Pumpellyite	$Ca_2MgAl_2Si_3O_{11}(OH)_2 \cdot H_2O$	7.6
Sheet silicates	Talc	$Mg_3Si_4O_{10}(OH)_2$	4.8
Zeolite	Stilbite	$NaCa_4(Si_{27}Al_9)O_{72} \cdot 28(H_2O)$	20.3
Zeolite	Laumontite	$CaAl_2Si_4O_{12} \cdot 4H_2O$	15.3

- *Palagonite*: A complex assemblage of orange to brown hydration and oxidation products of basalt glass (clays, zeolites, chlorite, oxyhydroxides) with up to 30 wt% H_2O
- *Spilite*: Hydrated, incipiently metamorphosed basalt with abundant zeolites, chlorite, and albite
- *Greenstone*: Hydrated, metamorphosed basalt (greenschist facies) composed primarily of chlorite, actinolite, and albite
- *Serpentinite*: Hydrated ultramafic rocks composed primarily of serpentine and magnetite (the state rock of California)
- *Massive sulfide deposits*: Seafloor hydrothermal deposits with abundant pyrite, pyrrhotite, sphalerite, galena, and chalcopyrite, as well as oxides and silicates

Toolbox 5.1

Calculating the Amount of Water in a Mineral

In this chapter, we focus on the importance of hydrous minerals in carrying water into the mantle. Some of these minerals contain surprisingly large weight fractions of water. Consider two of the minerals we've discussed:

- Stilbite, $NaCa_4(Si_{27}Al_9)O_{72} \cdot 28(H_2O)$
- Chlorite, $(Mg,Fe,Al)_6(Si,Al)_4O_{10}(OH)_8$

In one of these minerals, water is listed as discrete H_2O species, and in the other as OH. How do we calculate the amount of water in each?

For stilbite, the calculation is straightforward. A little arithmetic shows that this molecule, with 100 oxygens, has a very high molecular weight of 2840.83 g/mol. Of this mass, 504 g, the product of 28 times the molecular mass of H_2O (18), is H_2O, so H_2O makes up (504/2840.3) = 20.3 wt% of stilbite. That's a very wet mineral!

The calculation for chlorite is essentially the same even though the water is expressed as OH (hydroxyl) rather than H_2O. The molecular weight of the clinochlore end member, $Mg_5Al(AlSi_3O_{10})(OH)_8$, is 555.80 g/mol. Eight hydrogen atoms makes four H_2O molecules, so the water content is [(4 × 18)/555.80] = 13.0 wt%.

That's all there is to it. This reasoning works for any cation (e.g., MgO in clinochlore).

Hydration Reactions

A simple hydration reaction involves conversion of olivine to serpentine. Serpentine is a name for a group of minerals with the idealized formula $Mg_3Si_2O_5(OH)_4$, and we will use the mineral antigorite to represent this group (others are chrysotile, an asbestos-like fibrous phase, and lizardite, which is typically scaly). Figure 5.6 shows a chemographic diagram for relevant phases in the ternary system MgO-SiO_2-H_2O as an aid to working out hydration reactions involving Mg silicates. Several important hydrous minerals occur in this system.

Time to refresh our memories about the phase rule ($F = C + 2 - \pi$). This is a three-component system; how many phases can we expect in a given situation? For an arbitrary bulk composition (throwing a dart at the diagram), at arbitrary P and T (two degrees of freedom) we have $2 = 3 + 2 - \pi$, so we should have three phases. A little thought shows that at arbitrary P and T, $C = P$, and thus the number of phases expected equals the number of components; this is known as the mineralogical phase rule.[4] If you have good aim (or get lucky), your dart might land on a single phase, such as Qz, or on a tie line between two coexisting phases, but for most of us, our darts will land within a field where three phases would be stable. So, we can expect that our stable phase assemblages will be connected by triangles.[5]

What are those stable phase assemblages? Which dots should be connected by tie lines? For the dart in Figure 5.6, many are possible, such as Atg-Fo-Brc, Atg-Per-Brc, Tlc-Per-W, and many others. In fact, with eight points, there are 56 possible triangles,[6] counting degenerate ones where the compositions are collinear, such as Qz-En-Fo. Which are the true stable assemblages? The chemographic diagram by itself gives no information. They could be predicted from thermodynamic data or determined experimentally. Bowen and Tuttle (1949) worked out reactions in this system experimentally at relatively low pressure (<300 MPa, and thus relevant to shallow oceanic lithosphere) using a hydrothermal furnace. Such a device consists of a strong, cylindrical, steel pressure vessel within a tube furnace. The sample and thermocouple are placed in a thin cylindrical cavity along the axis of the steel pressure vessel, and pressure is applied by pumping a gas, typically Ar, into the sample chamber. Such furnaces are workhorses of metamorphic petrology but are limited to relatively low temperatures, because steel softens as the temperature is raised.

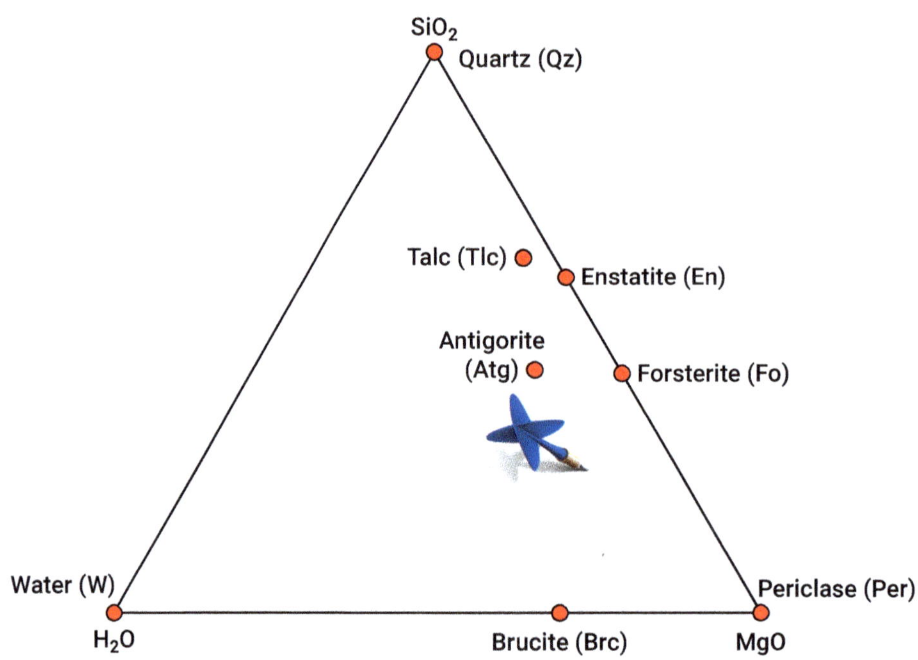

Figure 5.6 Chemographic diagram of phases in the system MgO-SiO_2-H_2O. Dart represents a random bulk composition.

Bowen and Tuttle used oxide starting materials (e.g., MgO) mixed in the proper mineralogical proportions to synthesize the relevant minerals. They found a set of assemblages bounded by reactions that are mainly *T*-dependent. These reactions and relevant tie lines, determined experimentally, are shown in Figure 5.7.

The lowest-temperature reaction (reaction #1), at about 400 °C, is the crossover reaction (also called a crossing tie-line reaction or a tie-line flip) Atg + Brc = Fo + W, where the Atg-Brc tie line is broken by the Fo-W tie line. This is a dehydration reaction going up-temperature, and conversely a hydration reaction going down-temperature, relevant to olivine in a cooling basalt. At temperatures above the reaction, Fo + W is stable, but below the reaction, they react to form Atg + Brc—a **serpentinization** reaction.

Serpentinization
Formation of serpentine by hydration of olivine, pyroxene, and/or glass.

The next reaction up-temperature (reaction #2) is Atg = Fo + Tlc + W. This is an interior reaction where Atg breaks down to form a triangle of three phases. This reaction, at about 500 °C, marks the high-temperature stability of antigorite at this pressure. Like the water-steam reaction in Figure 5.3, the temperature of this reaction increases at higher pressure. Continuing up-temperature we encounter the reactions Tlc + Fo = En + W (crossover reaction, #3), Tlc = En + Qz + W (interior reaction, #4), and Brc = Per + W (interior reaction, #5). These reactions represent progressive dehydration (up-temperature) of various hydrous assemblages. Serpentine (antigorite) is not stable above reaction #2, talc is not stable above reaction #4, and brucite is not stable above reaction #5.

It is important to understand that a rock will react only if its bulk composition overlaps the chemographic space in which that reaction occurs. For example, consider a rock containing a layer of pure Qz, another of Brc + Fo + Per, and a third of Brc + Fo + Atg (top-left diagram in Figure 5.7). If this rock is heated such that it crosses reaction #1, only the

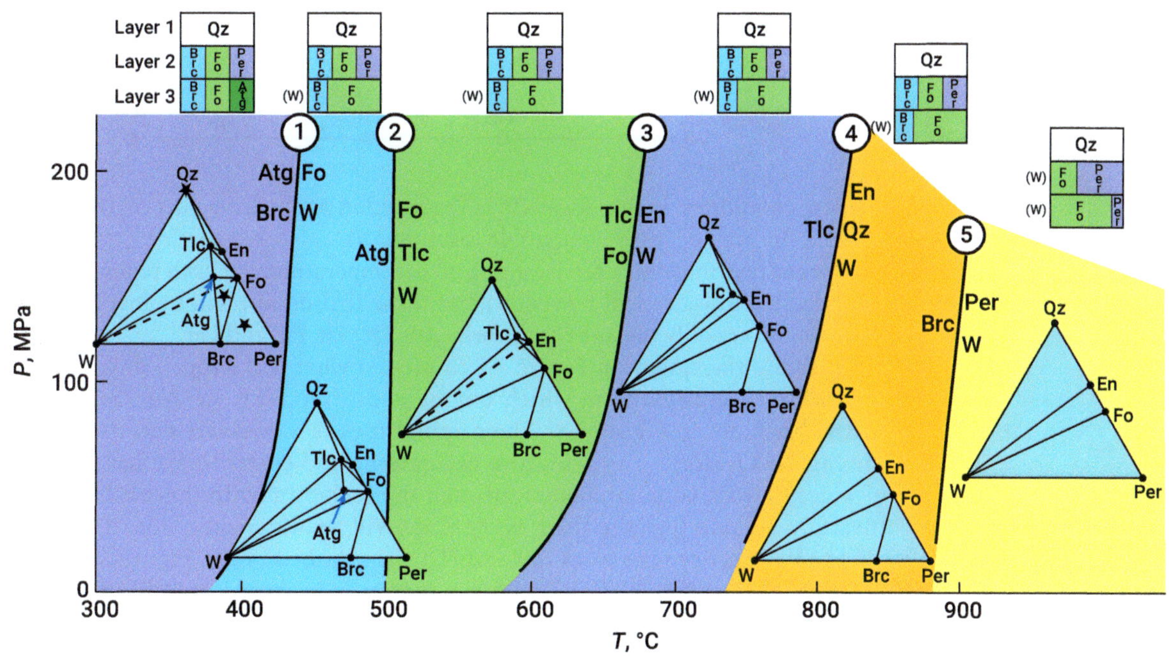

Figure 5.7 *P-T* diagram of reactions in the system $MgO-SiO_2-H_2O$. Labels along numbered reaction curves indicate phase compatibilities on either side. Inset triangles show phase compatibilities under different *P-T* conditions. Dashed black lines in lower temperature diagrams connect the next stable association at higher *T* to illustrate the geometry of the crossover reaction. Top diagrams show how mineral assemblages—indicated by small stars on first inset triangle on far left—could change in different bulk compositions with increasing temperature; "(W)" indicates that water would be stable, but very little would be present in the rock. Note progressive loss of hydrous minerals (dehydration) from left to right such that, at $T \geq \sim 900$ °C, only anhydrous minerals coexist with water. During cooling of MOR basalts in the presence of water, the opposite happens: high-*T* anhydrous assemblages become hydrated.

Source: After Bowen & Tuttle (1949). Geological Society of America.

third layer will show any reaction—the other layers are unaffected by reaction #1, because they lack the assemblage Atg + Brc. Quartz doesn't care that Atg and Brc react at 400 °C, nor does the Brc + Fo + Per layer, because there is no Atg for the Brc to react with; the contact between the second and third layers could, however, react wherever Brc and Atg are in contact. At around 900 °C, reaction #5 will eliminate the Brc, yielding Fo + Per + W (but the water will largely leave the rock). And even then the Qz layer will be unperturbed.

In Chapter 3, we explored projection of a chemographic diagram to lower dimensions. The diagram in Figure 5.7 has only three components, so it is already amenable to display on a flat piece of paper. But it can be further projected for practice as a skill we'll be needing later on. Since we're studying hydration, we will consider only those minerals and rock compositions that coexist with water (liquid, steam, or supercritical fluid, depending on *P* and *T*). The 2D chemographic diagram, with all phases in the system, looks like this:

Now, if we look at mineral compatibilities *in the presence of water* at temperatures below reaction #1, we put our eye at the water apex and peer into the triangle, seeing this:

We do not see W on the chemographic diagram, because we are projecting from it—it is present in every assemblage. Per is invisible, shielded from view by Brc; it is incompatible with water and would react with it to form Brc under these conditions. If we step up in temperature to the region between reactions #1 and #2, the mineral compatibilities now look like this:

At these conditions, Fo has formed via the reaction Atg + Brc = Fo + W (reaction #1), and Atg and Brc are no longer compatible together.

We can continue this exercise going up in temperature. For each region, any phase that has a tie line to water will be visible, and those that do not will be invisible. Eventually, water would see only the anhydrous minerals Qz, En, Fo, and Per.

Let's use the system in Figure 5.7 to predict what will happen during hydration of a harzburgite (peridotite in which olivine and orthopyroxene dominate), represented by the black dot in Figure 5.8. The original rock is anhydrous and therefore plots on the MgO-SiO_2 sideline as a mixture of forsterite and enstatite. We have a whole ocean of water to work with, circulated through the rock by a hydrothermal system, and can therefore assume that the final assemblage will be water-saturated. Thus, it must lie along a line between the black dot and H_2O (the dashed line in Figure 5.8), within a triangle that connects to H_2O; all triangles that include water as a stable phase are shaded in the figure. Stable assemblages at 550 °C, between reactions #2 and #3, are shown in Figure 5.8a. Under these conditions, Atg is not stable. Neither Fo-En (our original harzburgite) nor Tlc-Fo-En can be stable, because there is no free water in those assemblages; the stable assemblage must therefore be Tlc-Fo-W, along the portion of the dashed tie line that lies within the triangle that connects our original rock composition with water. Where exactly it lies is not important, because Tlc/Fo (roughly 0.25 for our example) is constant along the dashed line, and it is the solid rock (Tlc + Fo) that we observe, collect, and analyze.

At 350 °C, between reactions #1 and #2 (Figure 5.8b), mineral compatibilities are different. Again, our rock must fall along the dashed line between our original rock and water in a water-stable triangular field. In this case, our harzburgite now consists of Atg + Brc + W. The

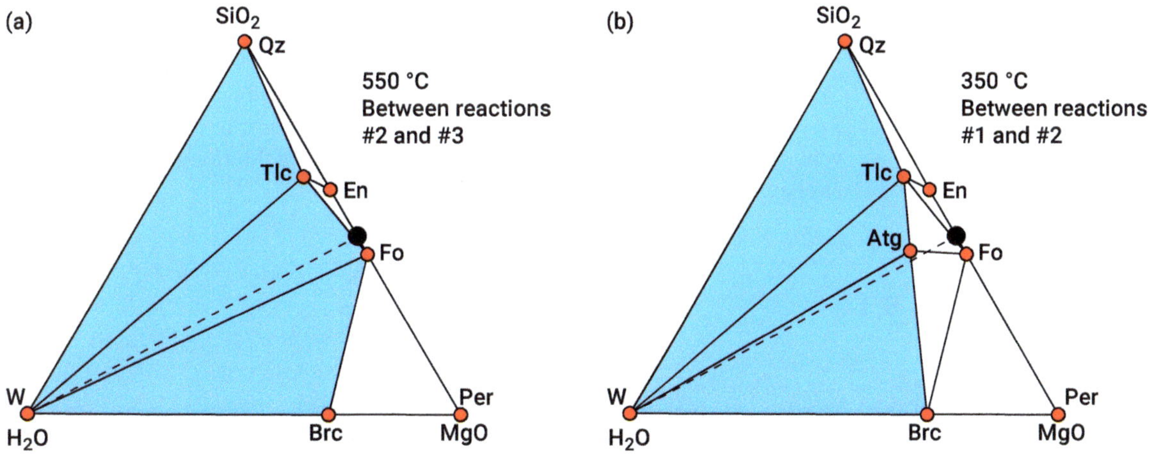

Figure 5.8 Phase compatibilities in the system MgO-SiO_2-H_2O at two different temperatures: (a) 550 °C and (b) 350 °C. The black dot between Fo and En represents a harzburgite.

dashed line cuts the Atg-Brc tie line almost at Atg, so the resulting rock will be almost entirely Atg. Note that if the starting rock were a bit richer in En (e.g., half Fo and half En), Tlc would be stable instead of Brc.

The reactions that take place during hydration of a real basalt or ultramafic rock are significantly more complicated than those just described, because most of the relevant minerals contain significant Fe in addition to Mg (recall from Chapter 4 that the Mg# of mantle minerals is around 90, and primitive basalts are around 70). One possible reaction allowed by chemography in the Fe-free system (see Figure 5.8) is Fo + Qz + W = Atg. Written out and balanced, this is

$$3Mg_2SiO_4 + SiO_2 + 4H_2O = 2Mg_3Si_2O_5(OH)_4 \quad \textbf{(Eq. 5.1)}$$

The Fe component of olivine, fayalite (Fa), can undergo the following oxidation reaction with water:

$$3Fe_2SiO_4 + 2H_2O = 2Fe_3O_4 + 3SiO_2 + 2H_2 \quad \textbf{(Eq. 5.2)}$$

Combining 3 moles forsterite and 1 mole fayalite and adding the reactions yields a total reaction that just balances SiO_2, such that silica consumed by forsterite equals that produced by oxidation of fayalite:

$$12Mg_{1.5}Fe_{0.5}SiO_4 + 14H_2O = 6Mg_3Si_2O_5(OH)_4 + 2Fe_3O_4 + 2H_2 \quad \textbf{(Eq. 5.3)}$$

The $Mg_{1.5}Fe_{0.5}$ olivine is Fo_{75}, not far from that found in primitive basalts and the mantle. The overall reaction says that hydrating olivine via this path produces serpentine, magnetite, and hydrogen, and it explains the common association of serpentine and magnetite in serpentinites.

Several aspects of these reactions make them exceedingly important in Earth science. These include:

- *Volume change.* Serpentinization reactions generally involve a significant increase in volume, because serpentine is much less dense than olivine (see Figure 2.16). For example, the reaction Fo + W = Atg + Brc involves a volume increase of the solid phases of over 50 percent (O'Hanley, 1992). In the early development of plate tectonics, expansion of mantle rocks via serpentinization was thought to be a significant driving force in vertical tectonics (Hess, 1955). Although this is now thought to be unlikely, pervasive serpentinization may locally affect tectonics and topography (e.g., Bonatti et al., 1996).

Asbestos
A commercial product made from fibrous (asbestiform) minerals.

Asbestiform
A textural term meaning fibrous or able to be split into fibers.

- *Heat generation.* Serpentine-forming reactions are exothermic (heat-producing) and contribute to the energy balance of seafloor hydrothermal systems (Lowell & Rona, 2002).
- *Production of hydrogen.* Re-oxidation of hydrogen produced via reaction of water with Fe-bearing minerals is an important energy source for microbes at some hydrothermal vent systems, such as the Lost City hydrothermal field along the Mid-Atlantic Ridge (Kelley et al., 2005).

Making Connections 5.1

Asbestos

Asbestos gets a bad rap from the public, mostly because of fears that inhalation causes a rare cancer called mesothelioma that affects the lining around your internal organs. Some of those fears are well-founded, others less so. But what is asbestos, and how does it relate to Earth processes? Why has it been used so much commercially, and what are its health concerns?

The word "asbestos" refers to a commercial product, made from fibrous minerals, with good flexibility and low heat transfer (e.g., Gunter et al., 2007; Strohmeier et al., 2010). Asbestos has been used most famously in insulation, but also in tiles, shingles, brakes, and various other heat-resistant materials, including (we shudder to think . . .) certain cigarette filters in the early 1950s. Asbestos is manufactured from **asbestiform** minerals that can be woven into cloth or made into fluffy, cotton-like insulation that does not burn. The fact that fibrous minerals occur naturally (they're mined to make asbestos, after all) has given rise to the non-scientific term "naturally occurring asbestos." Although mineralogists may wince at the term, some asbestiform minerals can cause diseases, and so it is important to understand where these minerals occur and how they might be inhaled. Here, we will use the term "asbestiform mineral" or "fibrous mineral" rather than "naturally occurring asbestos" when talking about natural materials.

Although virtually any mineral could theoretically form fibers, only a few classes of minerals commonly do. Many asbestiform minerals are chain silicates, especially amphiboles, but others include, less intuitively, the sheet silicate chrysotile and the framework silicate erionite. Chrysotile, a member of the serpentine group, is the most common source for asbestos. Structurally, chrysotile consists of two molecular sheets bonded tightly together, but to match atomic spacing in the different sheets, the sheets bend, curling into tiny tubes or "drinking straws." Each tube makes a tiny, hair-like (asbestiform) fiber. Serpentine is one of the primary products of hydrated oceanic crust and mantle, and so asbestos can be present in exposures of ophiolites and ultramafic rocks.

There are over 80 different varieties of amphiboles, but relatively few develop fibrous habits. The most notorious is crocidolite, a fibrous, Na-rich amphibole that was used to make so-called blue asbestos. It's the mineral that was used briefly in cigarette filters. Other fibrous amphiboles include tremolite and actinolite (non-aluminous, Ca-Fe-Mg amphiboles; Figure 5.9a,b), anthophyllite and grunerite (non-calcic, Fe-Mg amphiboles), and a few other sodic amphiboles. Most of these are not mined to make asbestos, but they can occur in other materials that are mined and thus present hazards.

Figure 5.9 (a, b) Scanning electron microscope images of fibrous tremolite collected near Sacramento, California. (c) Scanning electron microscope image of a tremolite fiber that white blood cells attacked and coated in Fe-rich material; (d) Early 1960s photo of the W. R. Grace vermiculite mine near Libby, Montana, showing widespread dust that was contaminated with asbestiform amphiboles.

Sources: (a, b) USGS/US Dept of Interior, from Meeker et al. (2006); (c) from Gunter et al. (2007), with kind permission of Elena Belluso & Annibale Mottana; (d) McGarvey Law.

We emphasize that few amphiboles are asbestiform, and even the amphiboles listed here are only rarely fibrous. Especially after the 1970s and 1980s, when governments removed asbestos-containing materials from schools and other public buildings and started imposing stricter regulations, the likelihood of encountering asbestiform minerals in dangerous quantities has dropped dramatically. Students today are much safer in this regard than when we (the authors) were kids.

So, what are the health risks? Keep in mind that, if you're exposed to enough of it, any material will kill you—beer, radiation, porcupines, meteors, etc. The same is true for minerals. Your body does a good job of eliminating dust, but the high surface area of asbestos is highly reactive, and fibers can puncture cells. Your body doesn't handle fibers very well (Figure 5.9c). Still, some are more dangerous than others, and this has led to the listing of specific asbestiform minerals for regulatory concern. Regulated minerals in their fibrous form include crocidolite, tremolite, actinolite, anthophyllite, grunerite, chrysotile, and erionite. Of these minerals, chrysotile is relatively benign, and some scientists argue it should not be listed. In contrast, crocidolite and erionite are clearly linked to disease and should be avoided. Back off if you see blue asbestos, and avoid cigarettes from the 1950s.[7] The other asbestiform amphiboles fall somewhere between these health extremes.

Nearly all asbestos today (yes, it is still used) is made from chrysotile, because it is apparently the safest form. But the other dangerous asbestiform minerals can be found in deposits of diverse strategic minerals. Vermiculite, a puffy, weathered biotite used for insulation and mined in Libby, Montana (Figure 5.9d), also contains asbestiform tremolite, actinolite, and other dangerous minerals. Mine workers and local residents were exposed to dust that contained asbestiform amphibole, because mine tailings were used in town for playgrounds, baseball fields, gardens, and roads. In 2009, the US Environmental Protection Agency declared Libby a Public Health Emergency, and since then, air quality has improved dramatically. However, the publicly disclosed costs of cleanup and litigation have run to hundreds of millions of dollars, and numerous cases have been settled out of court. Asbestiform minerals also occur in a few other commercial products (e.g., in the talc base of talcum powder). This has led to long-standing litigation and to the gradual replacement of mineral ingredients with substitutes; today, corn starch is commonly the main ingredient in "talcum" powder.

5.3 Metamorphic Facies

Learning Objectives

By the end of this section you should be able to:

5.3.1 Use a facies diagram to predict the mineral assemblages expected in metamorphosed mafic rocks.

5.3.2 Apply the Arrhenius relation to calculate rates of mineral reactions.

The Facies Concept

The systematic relationships between bulk composition, *P*, and *T* illustrated in Figures 5.7 and 5.8 give us a way to use mineral assemblages to infer the *P-T* conditions of rock genesis. For example, if we have a rock that is adequately represented by the system $MgO\text{-}SiO_2\text{-}H_2O$, then the presence of serpentine (antigorite) indicates that the mineral assemblage formed at temperatures below 500 °C, the upper temperature limit of stability of Atg. Atg + Brc gives an even lower upper limit (~425–450 °C, depending a bit on pressure). En + Qz in the presence of water is not stable below ~800 °C. Note that the reactions in Figure 5.7 all have steep Clapeyron slopes and so are relatively pressure-insensitive over this range of pressures.

These relationships were systematized by the concept of **metamorphic facies**, developed over a century ago by field-based petrologists such as Eskola (1915) in Finland, Goldschmidt (1911) in Norway, and Barrow (1893) in Scotland. Eskola proposed that differences in mineral assemblage in rocks of the same composition result from crystallization under different *P-T* conditions and used the term **facies** to characterize these different conditions.

Metamorphic grade normally refers to a qualitative measure of the maximum temperature that a rock has experienced. This is partly because of the exponential role that temperature plays in the rates of chemical reactions, but also because changes to mineral assemblages commonly depend more on temperature than pressure (e.g., Figure 5.7).

Metamorphic facies
A metamorphic mineral assemblage, developed in a rock of basaltic bulk composition, characteristic of a *P-T* region in the crust and upper mantle. Alternatively, the *P-T* region in which that mineral assemblage is found is stable in a metabasalt. The facies are named for metabasaltic rocks (e.g., greenschist, amphibolite), but relevant *P-T* conditions can be applied to other rock compositions (i.e., a metashale might be described as a schist metamorphosed under amphibolite-facies conditions).

Facies
The textural and/or mineralogical character of a rock, indicative of the conditions under which it formed.

Metamorphic grade
A qualitative characterization of metamorphic temperature, which often also correlates with pressure. A high-grade metamorphic rock forms at elevated temperature, whereas a low-grade metamorphic rock forms at lower temperature.

Pressure generally has a far smaller effect on reaction rates and mineral stabilities. We speak of low-grade assemblages that formed at relatively low temperatures (e.g., <300 °C) and high-grade assemblages that formed at higher temperatures (e.g., >700 °C). What temperatures and facies correspond with "high" vs. "low" depends a little on who you're talking to; the amphibolite facies may be "high" for some people and "low" for others.

A general guide to the *P-T* conditions corresponding to various metamorphic facies is given in Figure 5.10. Hydrothermal metamorphism at shallow depths under the seafloor takes place at low pressures—up to perhaps 100 MPa—and temperatures ranging up to 600 °C or so. A basalt erupted at 1200 °C and seafloor pressure (20–40 MPa) cools rapidly to ambient temperature, which is close to 0 °C. If it stays at that temperature, then nothing much more happens—reaction rates are too slow. It just sits there, wet and cold, maybe becoming slimy if microorganisms colonize it. If it is affected by hydrothermal circulation, though, it will be heated into more interesting conditions—greenschist or perhaps even amphibolite facies.

Humphris and Thompson (1978) studied the mineralogy and composition of altered pillow basalts from several locations along the Mid-Atlantic Ridge. They found that the rims of altered pillows are systematically richer in H_2O and MgO, and poorer in CaO, than their cores, and that the primary phases plagioclase, olivine, clinopyroxene, and glass were transformed by alteration into metamorphic assemblages consisting largely of albite, actinolite, chlorite, and epidote—in many cases retaining textural aspects of the original pillows.

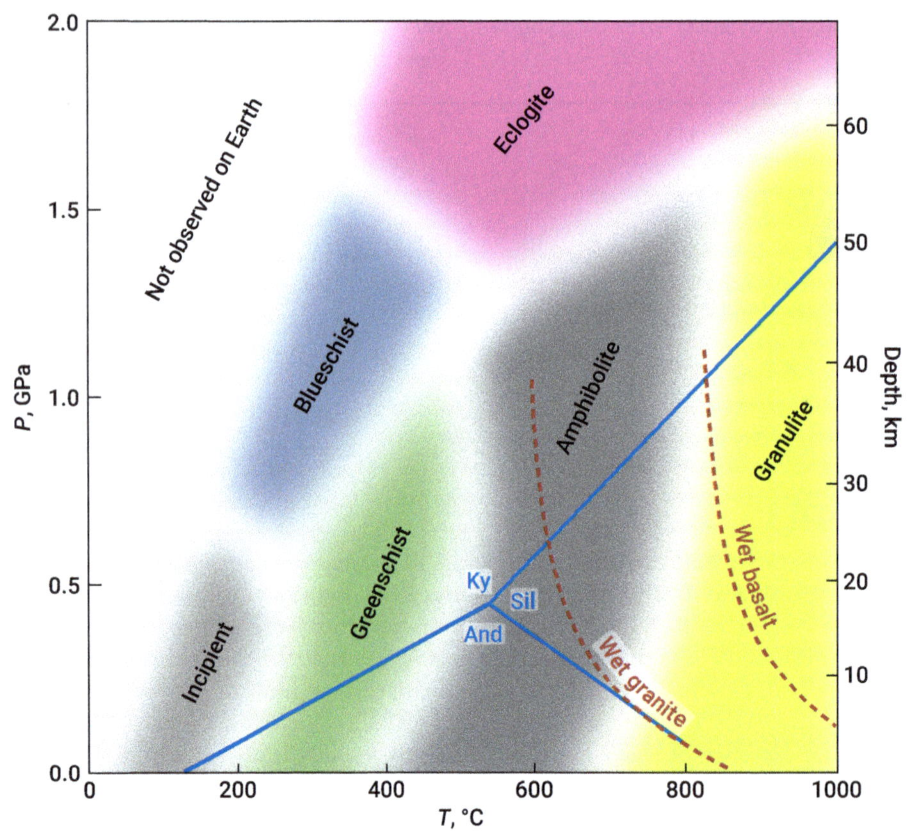

Figure 5.10 The five principal metamorphic facies. Use with caution; boundaries between facies are gradational and depend on composition. "Incipient" refers to rocks that straddle the boundary between sedimentary rocks and metamorphic rocks; these are sometimes divided into zeolite facies (~100 °C) and prehnite-pumpellyite facies (~200 °C). "Not observed on Earth" indicates that rocks that form at these *P-T* conditions are not generally exhumed for us to study, at least not on Earth. The blue, giraffe-shaped stick figure gives estimated boundaries between the Al_2SiO_5 polymorphs kyanite, andalusite, and sillimanite; red dashed curves are approximate solidi (beginning-of-melting curves) for water-present ("wet") granite and basalt. Pressure is converted to depth using a model crust, 35 km thick, with $\rho = 2700$ kg/m^3 overlying a mantle with $\rho = 3300$ kg/m^3.

Source: Adapted and simplified from Spear (1995).

Table 5.3 Mineral Assemblages in Altered Basalts from Active Mid-Ocean Ridges

Metamorphic facies	Minerals
Incipient	Zeolites, chlorite, smectite, calcite, prehnite, epidote
Greenschist	Albite, actinolite, chlorite, epidote, quartz, titanite, talc, magnetite
Amphibolite	Hornblende, plagioclase, quartz, chlorite, apatite, biotite, epidote, magnetite, titanite

Source: Simplified from Humphris & Thompson (1978).

Mineral assemblages in the metabasalts differ significantly from place to place even though the bulk compositions of the metabasalts are broadly similar. Typical assemblages for the facies most relevant to seafloor hydrothermal alteration are given in Table 5.3; key minerals include zeolites, chlorite, amphiboles (hornblende and tremolite-actinolite), and epidote. Given that the underlying rock compositions are similar (MORB with varying amounts of H_2O), these different mineral assemblages, divided into different facies, indicate different *P-T* conditions. Basalt that has been metamorphosed in the greenschist facies to an assemblage of albite, actinolite (green amphibole), chlorite (green sheet silicate), and epidote (green, too) is known as greenstone for some reason. If it is pushed into the amphibolite facies (higher *T*), a set of complex reactions turns it largely into black hornblende and compositionally intermediate plagioclase, a rock known as amphibolite.

Why does the facies diagram (Figure 5.10) have fuzzy boundaries? Why doesn't it look like Figure 5.7, with sharp boundaries between assemblages? There are three reasons. First, Figure 5.7 is for a very simple chemical system with end-member reactions. Basalts are chemically far more complicated, and chemical complexity smears out reactions (remember olivine melting and Figure 4.17). So, mineral assemblages don't suddenly transform from one facies to another, they gradually replace one another over a range of temperature or pressure. Second, different basalts may have *similar* compositions, but they're not *identical* (e.g., the rims of pillow basalts are not quite the same as their cores). So, different rocks—although still basaltic—may start and finish transforming at different temperatures or pressures. Last, some transformations depend on how much water was added to the original basalt. The granulite-eclogite boundary (especially) occurs at much lower pressures in wetter rocks than in drier rocks (roughly 1 GPa vs. 2 GPa at 800 °C). Depending on how much water a rock started with, the boundary can be higher or lower. We drew the boundary to more or less split the difference, and in other sources, you may see somewhat different *P-T* conditions for the facies. The bottom line is that rocks are complicated and don't like to be pinned down. Embrace the fuzziness.

Effect of Temperature on Reaction Rate: Arrhenius Relation

A given mineral assemblage (e.g., plagioclase + olivine + augite in a basalt) transforms to another (e.g., albite + actinolite + chlorite + epidote) via a series of chemical and mineralogical reactions. Such reactions take time, and therefore the kinetics (rates) of these reactions are important. The rate of a chemical reaction generally follows an equation of the form

$$k = Ae^{-E/RT} \qquad \textbf{(Eq. 5.4)}$$

where k is the rate, A is a constant, E is the activation energy, R is the universal gas constant, and T is the absolute temperature; this is known as the **Arrhenius equation**.[8] We will see the Arrhenius equation again when talking about geochronology. The activation energy is a measure of the temperature sensitivity of a reaction rate; as E goes to zero, the reaction rate becomes insensitive to temperature.

You might recall a rule of thumb from introductory chemistry that a 10 °C rise in temperature generally causes the rate of a chemical reaction to double, more or less. Figure 5.11a shows some experimental data that follow this relationship. Carlson and Rosenfeld (1981) measured the growth rate of calcite crystals forming from aragonite as a

Arrhenius equation, plot
Mathematical expression that describes the rate of a process (e.g., reaction or diffusion) as an exponential function of activation energy and absolute temperature; a graph that plots the logarithm of a rate vs. reciprocal temperature.

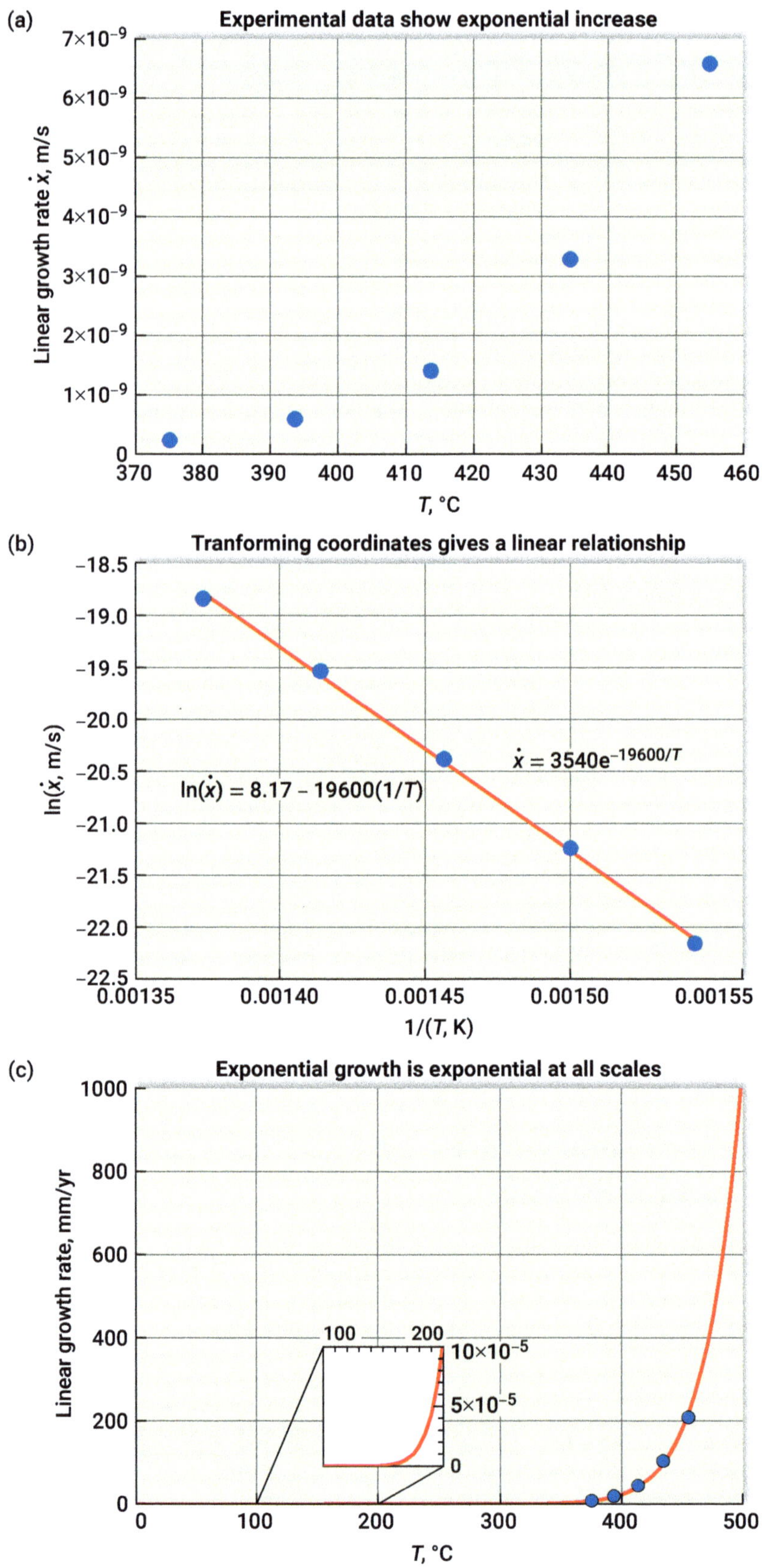

Figure 5.11 Rate of calcite crystal growth. (a) Experimental data on growth of calcite from aragonite, measured parallel to aragonite *a*-axis. (b) Transformed data produce a straight line (an Arrhenius plot) whose fit coefficients give the variables in the Arrhenius equation—the slope is $-E/R$, and the intercept is $\ln(A)$. (c) Calculated growth rate as a function of temperature, from data in (a). Note that exponential curves are self-similar; zooming in on the apparently flat part reveals that it too is an exponential curve.

Source: Data from Carlson & Rosenfeld (1981).

function of T (aragonite is a high-P form of calcite; although not stable at surface pressure, many organisms construct their shells from it, blissfully unaware of the relevant thermodynamics). The growth rate $\dot{x}$ (with the dot signifying a time derivative, and hence the rate of change of length x) increases exponentially with T, consistent with a thermally activated process. Taking the logarithm of both sides of Equation 5.4 transforms it into a linear equation, and replotting the data as $\ln(\dot{x})$ vs. $1/T$, where T is in K (absolute temperature; very important!) produces a straight line (Figure 5.11b) whose slope, from Equation 5.4, is $-E/R$. This gives the rate equation:

$$\dot{x} = 3540e^{-19600/T} \qquad \textbf{(Eq. 5.5)}$$

where the growth rate $\dot{x}$ is in m/s and T is in kelvins.

Figure 5.11c shows the linear crystal growth rate as a function of temperature, calculated from the equation above. The growth rate increases exponentially with T, and over much of the range in the figure, this amounts to roughly a doubling with every 10 °C increase in T, consistent with the Chem 101 rule of thumb. A doubling every 10 °C means 10 doublings every 100 °C, or $2^{10} = 1024$, a factor of about 1000. Another 100 °C increase means another factor of 1000, so a 200 °C increase in temperature means a rate increase of about a million.

Consider a piece of aragonite at room temperature. Based on the study above, the growth rate of any calcite nuclei (we will leave aside the problem of nucleation) is negligible, on the order of 10^{-16} mm/yr. No need to worry about your beautiful aragonite crystals turning into calcite. At 100 °C, the rate is still tiny, about 10^{-9} mm/yr; even if you boiled your aragonite for a year, nothing significant would happen. But at 300 °C, the rate is on the order of 0.15 mm/yr, and at 400 °C, it is about 25 mm/yr. Do not store your aragonite collection at 400 °C or let anyone stub out their cigarette (400–600 °C) on your prized seashells.

Now a thought experiment. Consider aragonite crystals with tiny calcite nuclei, sitting in the crust, maybe 5 km deep, at 100 °C. Intrusion of a dike nearby warms them to 300 °C in 1000 years, after which they cool back to ambient temperature at the same rate. Figure 5.12 shows this T-t (temperature-time) path, along with the total length of the growing calcite crystals plotted as a function of time. In this cooked-up example, the crystals grow to large size and, importantly, *nearly all the growth occurs during a short time window when the system was at and near the maximum temperature.* If we changed the maximum T reached, the curve would look the same, although the vertical scale would be different.

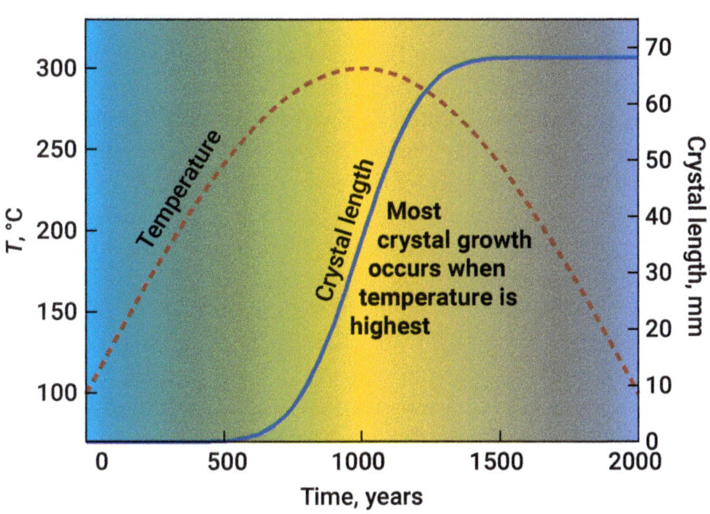

Figure 5.12 Growth of a calcite crystal (blue) over a heating-cooling cycle (red; see text for discussion). Note that nearly all the growth occurs in a small window of time when the system was at and near the maximum temperature.

This illustrates the fundamental principle that *metamorphic assemblages generally record the highest temperatures reached, if reaction kinetics are fast enough for reactions to proceed.*

5.4 Geochemical Evidence for the Source of Water in Hydrothermal Systems

Learning Objectives

By the end of this section you should be able to:

5.4.1 Identify the different types of water and how they interact with different Earth reservoirs.

5.4.2 Hypothesize how oxygen and strontium isotopes can be used to identify different hydrothermal processes in different scenarios.

Mantle, Meteoric, and Other Waters

Isotopic studies verify that the water circulated in seafloor hydrothermal systems is predominantly seawater, and not from some other source (e.g., water released as magmas crystallizes). To explain the logic behind this, we will discuss the various water reservoirs on the Earth and introduce the oxygen isotope system.

As noted earlier, the water that makes the Earth habitable was probably delivered to the planet early in its history, after it had cooled enough to accept it. This water now resides in several physical reservoirs, especially the oceans, mantle, and continental crust (see Figure 5.1 and Table 5.1). Another way to view these reservoirs is to divide them by type (Figure 5.13). Here, all the mantle reservoirs are lumped into mantle water (or **juvenile water**); water in the crust is magmatic, metamorphic, or **formation water** (buried with sediment), depending on its source and where it resides; and water in rivers, lakes, ice, the atmosphere, and the very near surface (e.g., soils and groundwater) is **meteoric water**, a term that comes from Greek ("of the atmosphere"), as in meteorology.[9] Given these divisions, several sources might be possible for the water circulated through oceanic-ridge hydrothermal systems, including magmatic, meteoric (if a section is subaerial), metamorphic, and seawater. How do we tell them apart? One way is to look at stable isotopes of oxygen.

Juvenile water
Water that is (theoretically) sourced from Earth's origin, so it hasn't participated in Earth's hydrologic cycle previously. More generally, water that comes from the mantle or from crystallization of magmas. Clearly, this is a problematic term, because Earth's mantle and its magmas do participate in the hydrologic cycle.

Formation water
Pore water in sediments or sedimentary rocks (i.e., in a sedimentary rock formation).

Meteoric water
Water that comes from the sky in any form—snow, sleet, rain, etc.

Oxygen and Its Isotopes

Oxygen is the most abundant atom in the Earth by number and a close second to Fe in weight proportion. There are three stable isotopes of O: ^{16}O (99.76 atomic %), ^{17}O (0.04%), and ^{18}O (0.2%). ^{18}O is a full 12.5 percent heavier than ^{16}O, so its chemical behavior is measurably different from its more abundant sibling. For example, when the Sun warms the ocean, water evaporates, and water built with ^{16}O ($H_2{}^{16}O$) is slightly more likely to jump into the sky than water with ^{18}O. Ocean water is well mixed and has essentially the same ratios of these oxygen isotopes everywhere at any given time. The slight mass differences mean that clouds have slightly lower $^{18}O/^{16}O$ than the ocean water they were evaporated from. When those clouds produce rain, the heavier isotopes condense a bit more easily than the lighter ones, so as the rain falls, the clouds become even "lighter"—with an even lower $^{18}O/^{16}O$.

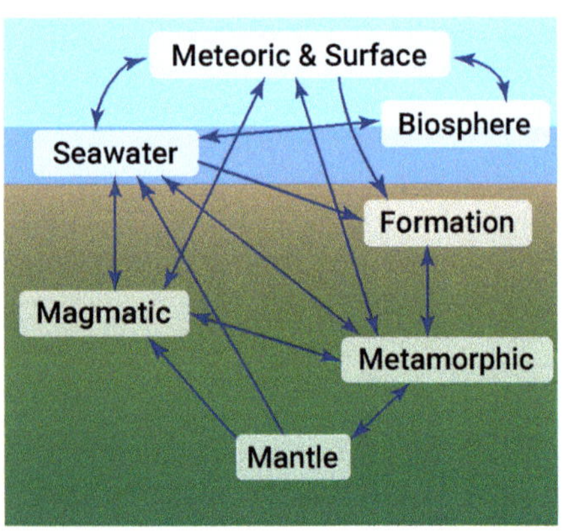

Figure 5.13 The main reservoirs of water on Earth; arrows indicate possible exchange paths.

By these processes rain and snow and various water bodies develop different $^{18}O/^{16}O$. When an animal uses oxygen from seawater to build its shell, it picks up a record of the isotopic composition of seawater at that time. During times of widespread glaciation (e.g., 20 ka), a lot of isotopically light (^{18}O-depleted) water is stored on land as ice. This leaves the oceans, and the animals that live in it, with a relative abundance of ^{18}O. Studies of the oxygen isotopic composition of ancient shells of a known age is one of the primary ways geologists have used to determine the timing of Earth's glacial cycles over the past several million years (Emiliani, 1957).

We could characterize the $^{18}O/^{16}O$ of a sample simply by that ratio (~0.002), but for several reasons, the oxygen isotopic composition of a sample is denoted by

$$\delta^{18}O = 1000\left(\frac{R_{\text{sample}}}{R_{\text{standard}}} - 1\right) \quad \textbf{(Eq. 5.6)}$$

where R_{sample} and R_{standard} are the ratios in those materials. This denotes the deviation of the ratio in a sample from that in a standard in parts per thousand, or per mil (‰, which is a percent symbol with an extra "o"; "mil" is from the Latin for 1000).

Does Equation 5.6 seem overly complicated? Maybe, but you probably encounter this kind of calculation every time you go to the store. If something normally costs $5.00, but today is 10 percent off, it must now be $4.50. How did the store calculate the percentage? One way is by taking the cost of the sale-price "sample" ($4.50) and dividing by the cost of the "standard" ($5.00; sample/standard = 0.90), subtracting 1.00 (0.90 – 1.00 = –0.10), and multiplying by 100 to get –10 percent. Equation 5.6 is the same thing, but isotopic differences are smaller, so we multiply by 1000 and call the units per mil (‰). The left side of Equation 5.6 is generally read "delta-oh-18"[10] even though "18" comes before "O" in the symbol. If a given sample has $\delta^{18}O$ = 10‰, then that sample has an $^{18}O/^{16}O$ that is 10 parts per thousand (10 per mil, 10‰), or 1 percent, greater than the standard. A negative value means that the sample has a lower ratio than the standard (the price is lower), whereas a positive value means it has a higher ratio (the price is higher). Unlike other isotopes, oxygen has two agreed-upon international standards. The one we will use here is VSMOW, or Vienna Standard Mean Ocean Water.[11] Instruments are mostly calibrated using secondary standards whose ratios are calibrated against VSMOW.

Isotopic signature
The characteristic isotope composition of a major reservoir (e.g., the isotopic signature of the mantle or ocean).

Different reservoirs have characteristically different $\delta^{18}O$ values (sometimes referred to as an **isotopic signature**[12]):

- *Seawater*: 0‰ today; probably between about –2 and +2‰ (it is customary to put + signs in front of positive values) through the Phanerozoic, but more negative before then (e.g., Galili et al., 2019).
- *Meteoric water*: Highly variable but virtually always negative, sometimes highly so; for example, surface water in the continental United States ranges from $\delta^{18}O$ values around –2 to –4‰ along the Gulf Coast to less than –16‰ in the Rockies (Figure 5.14). Meteoric water and ice from polar latitudes are lower still (e.g., –25 to –45‰ for Greenland and even –55‰ in Antarctica).
- *Mantle*: $\delta^{18}O$ in fresh mantle samples generally falls in a narrow range between +5 and +6‰.
- *Sediments* (which can be melted or metamorphosed to create other rocks): Highly variable, but coarser-grained siliciclastic sediments (sandstones, etc.) are typically +10 to +20‰, clay-rich sediments +15 to +25‰, carbonates +25 to +30‰, and Cenozoic cherts +30 to +35‰.
- *Magmatic and metamorphic*: Highly variable since they can be a mixture of the sources above, but generally positive and mostly about +7 to +20‰ for silicic rocks and up to +30‰ for very pure marbles.

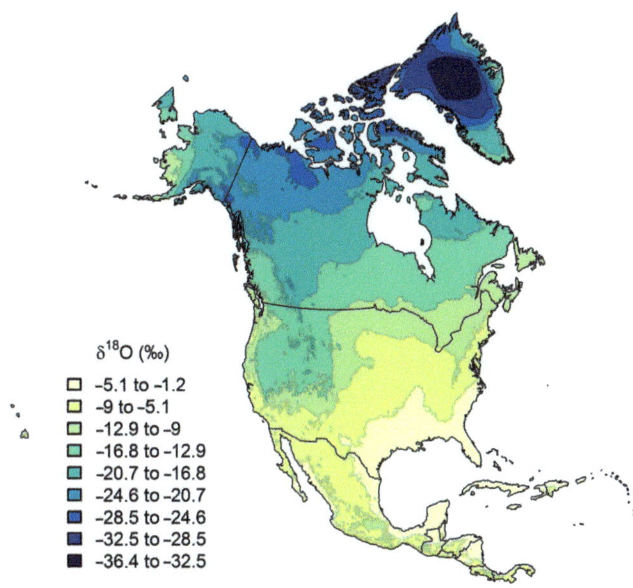

Figure 5.14 Shaded contour map of the interpolated oxygen isotope composition of meteoric water in Central and North America. Although there is a strong latitudinal pattern, with much lower values in Canada relative to points south, weather and topography leave their own imprints. The incursion of high-$\delta^{18}O$ water into the US Southeast reflects precipitation derived from the Gulf of Mexico and Caribbean. This precipitation travels northward easily over the flat Mississippi valley and East Coast, bending contours to the north. In between, the Appalachians impede this movement and, in combination with higher elevations (lower temperatures), bend contours to the south. The large area of low-$\delta^{18}O$ water in the Great Basin and Rocky Mountains arises from two effects: the rain shadow, caused by storms from the west dropping rain on the Sierra Nevada and other mountain ranges near the coast, and a dominance of winter precipitation.

Source: Gabriel Bowen, waterisotopes.org.

Fractionation between Phases and the Temperature Effect

To understand what happens isotopically when water reacts with a mineral, we first have to understand how oxygen isotopes partition themselves between minerals. ^{18}O and ^{16}O distribute themselves among phases according to statistical- and quantum-mechanical principles. For example, quartz and magnetite divide up oxygen isotopes in such a way that quartz gets more ^{18}O and magnetite more ^{16}O. We say that the phases fractionate the isotopes, because they take them up in different proportions, in a manner similar to the way that olivine and coexisting melt take up Mg and Fe in different proportions (see Section 4.8). The difference in their $\delta^{18}O$ values (e.g., $\delta^{18}O_{Qz} - \delta^{18}O_{Mt}$) is written using a capital delta as $\Delta_{Qz\text{-}Mt}$. Now, why do different minerals fractionate isotopes differently? Without getting into the pleasures of quantum mechanics, strong bonds prefer the heavier nuclide ^{18}O. The Si-O bond is one of the strongest in nature, whereas a metal-O bond (e.g., Fe-O, Mg-O, etc.) is relatively weak (bond strength; see Section 7.3). Consequently, tectosilicates such as quartz and alkali feldspar, with abundant Si-O bonds, should have the highest $\delta^{18}O$ values; non-silicates such as metal oxides should have the lowest $\delta^{18}O$ values; and intermediate-silica minerals (pyroxenes, micas) should have intermediate $\delta^{18}O$ values. The C-O bond is also strong, so carbonates tend to have high $\delta^{18}O$ values as well. There are many natural exceptions with interesting implications for Earth processes, but these basic equilibrium principles are consistent across the rock spectrum, especially at higher temperatures.

Like so many things in thermodynamics, a mineral-mineral fractionation decreases rapidly with increasing temperature, generally following a $1/T^2$ dependence (a result from statistical mechanics). As temperature goes up and atomic vibration increases, different phases are less able to distinguish among isotopes, and the magnitude of fractionation goes down—one isotope is as good as another. This is shown in Figure 5.15, which plots $\Delta_{\text{mineral-calcite}}$ for numerous minerals against $10^6/T^2$ (temperature is in kelvins,

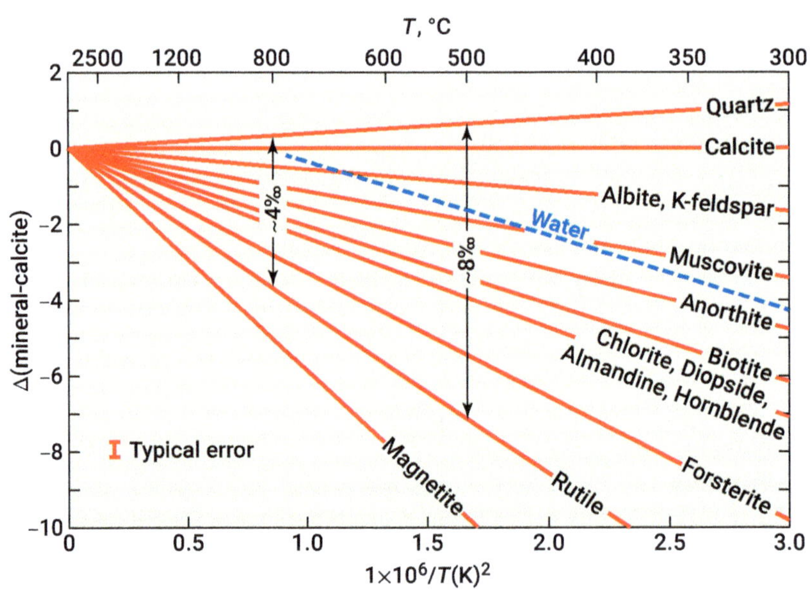

Figure 5.15 Oxygen isotope fractionation among minerals relative to calcite, showing generally systematic behavior with respect to chemistry and mineral structure. Tectosilicates and calcite with strong Si-O and C-O bonds prefer ^{18}O, whereas minerals with progressively less Si progressively prefer ^{16}O. Note that temperature decreases to the right.

Sources: Simplified from Chacko et al. (2001); data from Kohn & Valley (1998) and Hu & Clayton (2003).

so at T = 727 °C or 1000 K, $1.0 \times 10^6/T^2 = 1.0$). At lower temperatures, Δ can be large (e.g., $\Delta_{quartz\text{-}rutile}$ is ~8‰ at 500 °C), but with increasing temperature, Δ decreases (e.g., $\Delta_{quartz\text{-}rutile}$ is ~4‰ at 800 °C).

At one time, geochemists tried to estimate temperatures in rocks based on oxygen isotope compositions of coexisting minerals (e.g., Anderson et al., 1971). This approach may work well in quenched volcanic rocks but proves problematic in plutonic or metamorphic rocks, because minerals re-equilibrate during cooling by different amounts depending on oxygen diffusion rates (which are mineral-specific) and mineral abundances (e.g., Giletti, 1986). Other types of thermometers are less complicated and analytically easier. Nonetheless, oxygen isotopes are useful tracers of petrologic processes, including hydrothermal alteration.

Water has an unusual structure, so it doesn't fall into line with other minerals; rather, its fractionation cuts across the curves for other minerals (dashed line in Figure 5.15). Regardless, at low temperatures, many minerals prefer ^{18}O more than water (Figure 5.16), and thus, for them, $\delta^{18}O_{mineral} \gg \delta^{18}O_{water}$. This is shown in Figure 5.16, which plots $\Delta_{mineral\text{-}water}$ for quartz, serpentine, calcite, and talc against temperature. At 100 °C, Δ is large for the silicate minerals, +14‰ to +25‰, but it decreases to small or even negative numbers at 400 °C. From Figure 5.16, we can predict that if Qz, Tlc, and Srp were in equilibrium with seawater ($\delta^{18}O$ = 0‰) at 100 °C, all three would be significantly enriched in ^{18}O over seawater, with $\delta^{18}O_{Qz}$ = +23‰, $\delta^{18}O_{Tlc}$ = +28‰, and $\delta^{18}O_{Srp}$ = +16‰. At 400 °C, though, this enrichment would be much less, and serpentine would actually be depleted: $\delta^{18}O_{Qz}$ = +5‰, $\delta^{18}O_{Tlc}$ = +1‰, and $\delta^{18}O_{Srp}$ = −2‰. This dependence of mineral isotope composition on temperature and water composition gives us a method for testing the hypothesis that seawater circulates through seafloor hydrothermal systems and hydrates different rocks at different temperatures.

Figure 5.17 shows the variation of $\delta^{18}O$ with depth in the Samail ophiolite, a spectacularly exposed piece of oceanic lithosphere that overlooks the Gulf of Oman on the southeastern corner of the Arabian Peninsula. To understand patterns in the data, let's start with the mantle. Fresh peridotites and the MORBs derived by melting them have $\delta^{18}O$ ~ +5.5‰ (e.g., Eiler, 2001). MORBs are essentially the same as their source peridotite, because oxygen isotope fractionation is small at magmatic temperatures (see Figure 5.15). Now, let's jump up to the top of the section. Clays, carbonates, and opaline oozes (the raw material for chert) have very high $\delta^{18}O$, because they equilibrate with seawater at low temperatures, typically less than 20 °C. At such temperatures, fractionation is

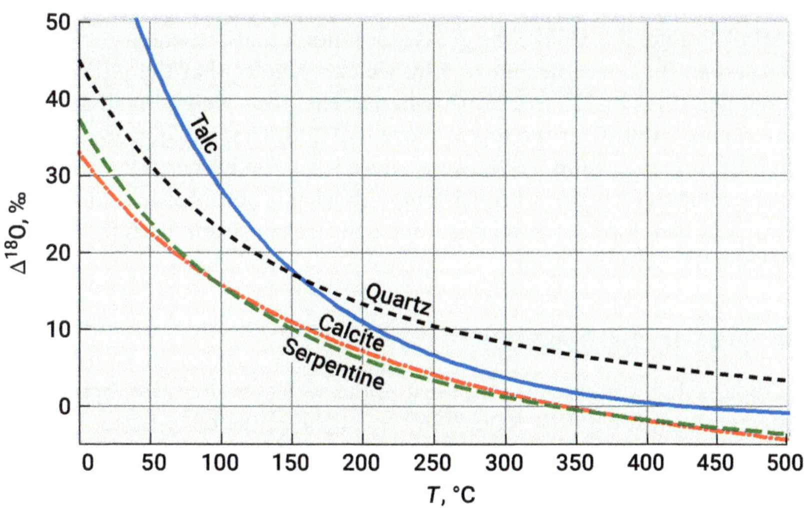

Figure 5.16 Oxygen isotope fractionation between water and quartz, talc, serpentine, and calcite.

Sources: Data from Saccocia et al. (2009), Chacko & Deines (2008), Cole et al. (2004).

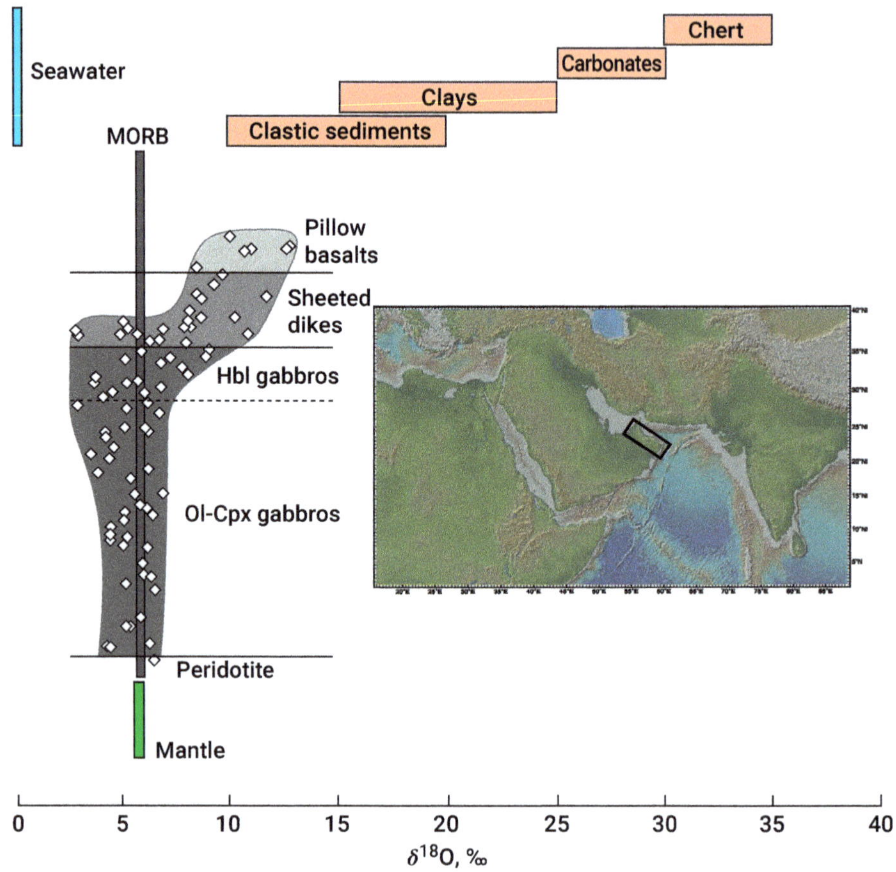

Figure 5.17 Representative isotope compositions of different Earth reservoirs in comparison with variation in $\delta^{18}O$ with depth in the Samail ophiolite, Oman. Diamonds are whole-rock or average of plagioclase and clinopyroxene compositions. Unaltered mafic rocks should be similar to MORB but have much higher $\delta^{18}O$ in the pillow basalt and upper sheeted dike sections, then ramp down to much lower values within the lower sheeted dike and gabbro sections. Inset map shows location of Oman.

Sources: Data from Gregory & Taylor (1981), McCulloch et al. (1981), Stakes & Taylor (1992).

huge (see Figure 5.16); for example, calcite precipitated from seawater at 10 °C should have $\delta^{18}O$ of around +33‰. Measured values from ocean-floor carbonates are typically about +30‰, broadly consistent with this prediction.

The igneous rocks under the sediments are more complicated. Pillow basalts have $\delta^{18}O$ around +8‰ to +13‰, whereas the dikes under them range from +12‰ down to +3‰, which is substantially lower than fresh basalt (+5.5‰). The gabbros mostly lie at or below the fresh MORB value. If we make the gross assumption that the whole-rock fractionation follows the serpentine curve in Figure 5.16, we can predict that the pillow basalts were altered at temperatures of 125 to 175 °C, whereas the dikes were altered at temperatures ranging from about 150 °C at the top to more like 200 °C at the bottom. Some gabbros, with minimum values around +3‰, were altered at even higher temperatures, at least 250 °C. These temperatures are all consistent with what is observed in seafloor hydrothermal systems, and therefore the oxygen isotope data are consistent with (but do not prove) widespread reaction with seawater.

Sr Isotopes in Seawater

The Sr-isotope system provides another test of the seawater alteration hypothesis. As we saw in Chapter 3, most MORBs have $^{87}Sr/^{86}Sr$ in the range of 0.702 to 0.704. Seawater is quite a bit higher—the present-day ratio is about 0.7092, and this ratio has bounced around in the range of 0.707 to 0.709 throughout the Phanerozoic (Figure 5.18). The residence time of Sr in the oceans is greater than 4 Myr, whereas the average mixing time of the oceans is about 3000 years, so the oceans are well-mixed with respect to Sr, and $^{87}Sr/^{86}Sr$ is thus essentially the same (to five decimal places) worldwide. The ratio is high because seawater receives detritus from the continents, whose age and high Rb/Sr generate high $^{87}Sr/^{86}Sr$ (see Section 9.4). Variations in the oceanic value depend on how much

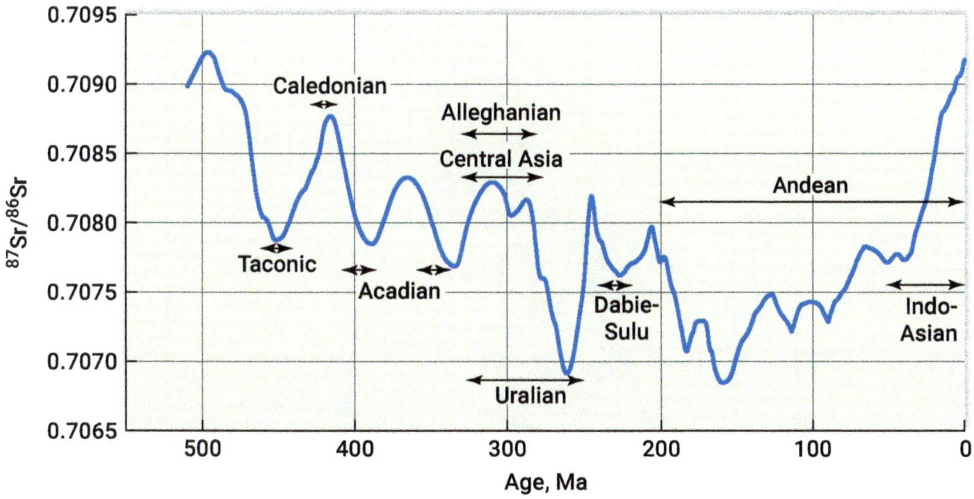

Figure 5.18 Variation in $^{87}Sr/^{86}Sr$ in the ocean during the Phanerozoic, annotated with some major orogenic events. Enhanced weathering from continental collisions between old cratons generally increases $^{87}Sr/^{86}Sr$, as in the Indo-Asian collision that produced the Himalaya, or the older Caledonian and Alleghanian orogenies. However, if the material subjected to weathering is largely oceanic and primitive island arc material, $^{87}Sr/^{86}Sr$ may decrease, as postulated for the Taconic orogeny (see Making Connections 12.1).

Source: Modified from McArthur et al. (2001).

continental detritus is being carried into the oceans by rivers; the rapid increase over the last about 50 Myr (see Figure 5.18) reflects the rapid rise and erosion of the Himalaya, which delivers a lot of material with high $^{87}Sr/^{86}Sr$ to the oceans. Enhanced continental erosion, arguably tied to continental collision (see Chapter 12), generally causes seawater $^{87}Sr/^{86}Sr$ to increase, but the opposite can happen if the material being eroded is dominated by island arcs and ophiolites trapped during collision (Ingram & DePaolo, 2022).

Typical MORB has around 100 ppm Sr, and seawater has about 8 ppm, so there is plenty of Sr in seawater to affect $^{87}Sr/^{86}Sr$ in basalts as seawater circulates through them. This is not true for meteoric water, which has much lower Sr concentrations; for example, typical rivers have less than 0.1 ppm Sr (Palmer & Edmond, 1989). It takes a much greater volume of fresh meteoric water to significantly affect $^{87}Sr/^{86}Sr$ in the rocks. Our hypothesis that seawater is the main source of water in altered oceanic lithosphere predicts that $^{87}Sr/^{86}Sr$ in those altered rocks should be shifted toward seawater, and that is just what is observed (Figure 5.19). In hydrothermally altered seafloor rocks, $^{87}Sr/^{86}Sr$ is shifted from mantle values toward the seawater with which it reacted (note from Figure 5.18 that $^{87}Sr/^{86}Sr$ in seawater reacting with the Cretaceous Samail ophiolite was much lower than that in present-day seawater reacting with the active system in Iceland).

To demonstrate the power of combined Sr and O isotope studies, consider Figure 5.20, which shows isotope data from a 3-km-deep drill hole on the Reykjanes Peninsula in Iceland, where abundant hydrothermal activity is tapped as a source of geothermal power. Iceland is an overthickened piece of oceanic lithosphere along the Mid-Atlantic Ridge, so thick that it sticks up above sea level. In the drill hole, $^{87}Sr/^{86}Sr$ is shifted significantly up toward present-day seawater, but $\delta^{18}O$ is shifted *down* several per mil, opposite the pattern seen in the Samail ophiolite. Recalling that pillow lavas in the Samail ophiolite have $\delta^{18}O$ of about +12‰, 12 units above seawater, the much lower values in Iceland (typically less than about +2‰) require either that (1) alteration happened at much higher temperature, such that $\Delta_{rock\text{-}water} \leq +2‰$, or (2) the reacting fluid was not seawater. The former would require temperatures on the order of 400 °C, which is well above the seawater boiling curve in the drill hole (Marks et al., 2015) and therefore unlikely. The second possibility is neatly explained if the reacting fluid is dominated by meteoric water (−8‰ in Iceland today; about −12‰ during the Last Glacial Maximum) rather than seawater. Input of meteoric

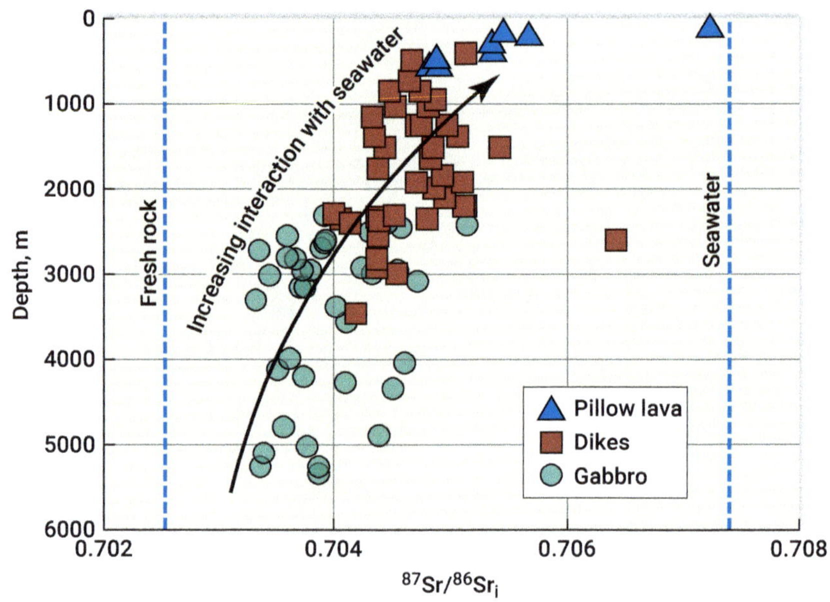

Figure 5.19 $^{87}Sr/^{86}Sr$ vs. approximate depth in the Samail ophiolite, Oman. Hydrothermal circulation has shifted values toward Cretaceous seawater from mantle values of ~0.703 (the authors inferred an original value of 0.70254). This shift is more pronounced toward the top of the column (pillow basalts), consistent with patterns from O isotopes. Even the deeper parts of the gabbro section seem to be shifted toward seawater.

Source: From Kawahata et al. (2001).

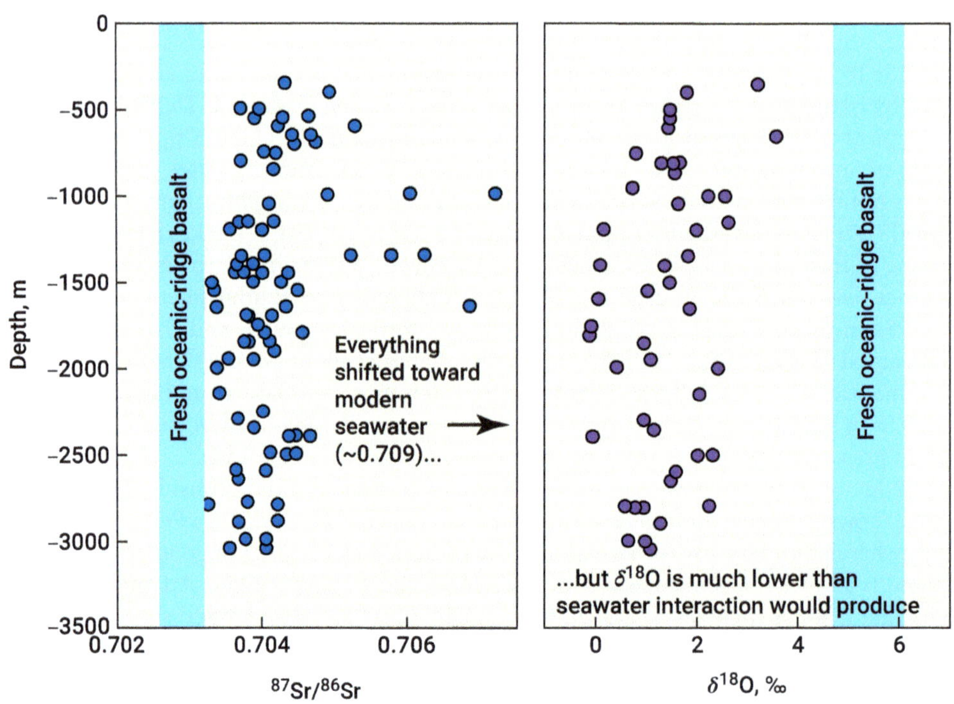

Figure 5.20 $^{87}Sr/^{86}Sr$ and $\delta^{18}O$ values of rocks and epidote veins in a 3-km drill hole from the Reykjanes Peninsula, southwestern Iceland. High $^{87}Sr/^{86}Sr$ indicates hydrothermal interaction with seawater, whereas low $\delta^{18}O$ indicates hydrothermal interaction with meteoric water (which has very low values in Iceland).

Source: From Marks et al. (2015).

water is not possible for most of the MOR system, but Iceland is one of the few places where the MOR system sticks up above sea level.

As a section of oceanic lithosphere is carried away from the spreading axis by plate motion, it moves farther from the magmatic heat source driving hydrothermal circulation, and activity shuts down. Lower-temperature hydrothermal systems driven by heat given off by serpentinization (e.g., Kelley et al., 2005) can exist anywhere peridotite can react with seawater, but most alteration happens near a spreading axis. After the lithosphere moves away from that, little happens until it meets a subduction zone.

5.5 Massive Sulfide Deposits

Learning Objective

By the end of this section you should be able to:

5.5.1 Make a sketch showing the geologic setting and processes that produce economic deposits of Cu, Zn, Pb, and other metals at spreading centers.

Volcanogenic massive sulfide (VMS) deposits
Deposits of massive sulfides, usually Fe-, Zn-, and Cu-rich, associated with volcanic rocks and interpreted to reflect concentration in localized hydrothermal systems.

One of the byproducts of hydrothermal activity at oceanic ridges is concentration of metals into **volcanogenic massive sulfide (VMS) deposits**. Such deposits are forming today at black smoker vents, and observation of this process in action (Francheteau et al., 1979) clarified the origin of such deposits, many of which have been mined for millennia. Massive sulfide deposits typically consist of a lens of sulfide-rich rock, tens to hundreds of meters thick, atop a pipe of altered breccia (Figure 5.21). Sulfide minerals present in the deposits include

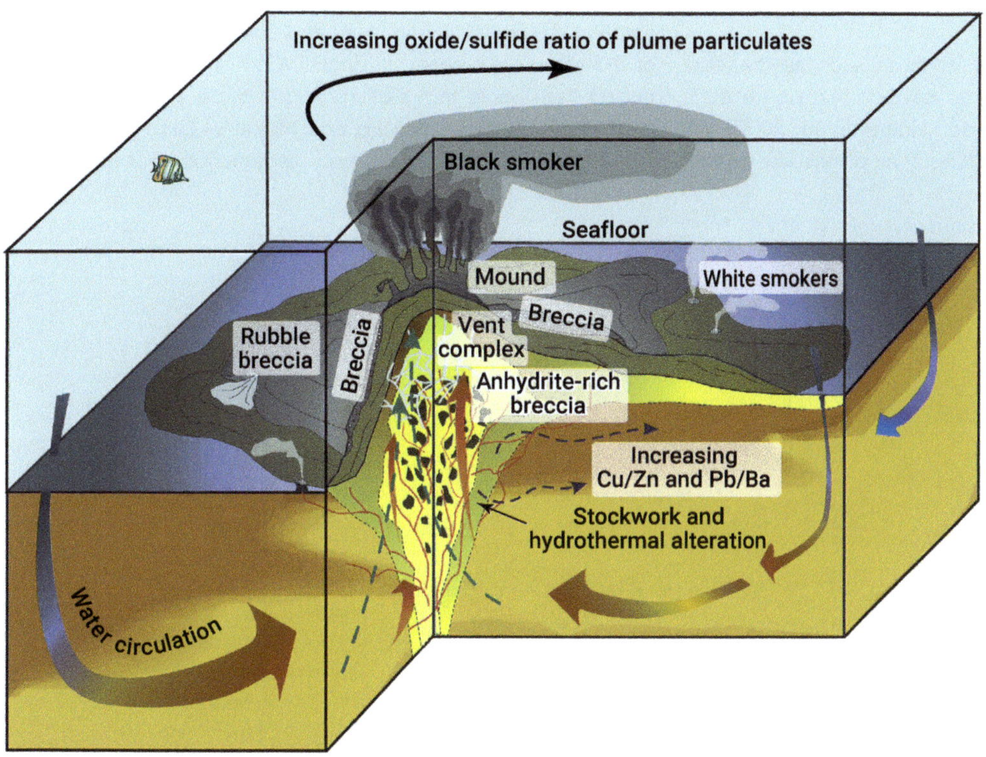

Figure 5.21 Schematic cross section of a massive sulfide deposit formed via hydrothermal circulation in oceanic crust. Deeper portions represent direct precipitation of sulfides from hydrothermal fluids, whereas shallower portions represent an apron of sulfide fallout from fluids that vent into the ocean.

Source: Modified from Tornos et al. (2015). Elsevier Science & Technology Journals.

pyrite (FeS_2), sphalerite (ZnS), galena (PbS), chalcopyrite ($CuFeS_2$), and pyrrhotite ($Fe_{1-x}S$). They are valuable sources of Cu, Zn, and Pb, and Au may be present in significant amounts.

The Troodos ophiolite on the island of Cyprus in the eastern Mediterranean contains abundant VMS deposits that have been mined since the Bronze Age, at least as early as the 18th century BCE (Knapp & Kassianidou, 2008). The (Cu, Zn)-rich deposits on Cyprus are hosted in pillow basalts and consist of pyrite, chalcopyrite, and sphalerite, with minor Au in overlying (Fe, Mn)-rich cherts and shales. They were generated by black smoker activity.

5.6 The Next 100 Million Years or So

Cold, wet, dark, and boring, until the plate gets subducted. Read on.

Making Connections 5.2

Hydrothermal Systems, Serpentine, and Deep Life

Early studies of hydrothermal vents with deep submersible craft such as *Alvin* (Corliss et al., 1979) showed that many such vents host ecosystems including clams, mussels, limpets, tube worms, and bacteria (see Figure 5.4). This was unexpected as vent environments were assumed to be devoid of life owing to high pressures, high temperatures, absence of light for photosynthesis, and toxic chemicals such as H_2S. How do these ecosystems work?

Deep-sea hydrothermal vent colonies such as the one in Figure 5.4 are largely based on bacteria that derive their energy from oxidation of sulfide. However, in 2000, an expedition to investigate deep rifting at an off-axis site on the Mid-Atlantic Ridge, far from the active spreading center, found an active hydrothermal field with extensive carbonate structures at depths of 700 to 800 m. This area, named Lost City, comprises carbonate structures up to 60 m tall built on extensively serpentinized and sheared mantle peridotite that had been brought to the seafloor by lithospheric extension (Figure 5.22; see Section 13.2). Organisms include small gastropods, amphipods, crabs, shrimp, eels, and corals, but the real stars of the show are archaeans (single-celled organisms that lack nuclei) that live in thin biofilms and provide energy for the food chain by processing H_2 and methane (CH_4) produced by serpentinization reactions such as Equations 5.2 and 5.3

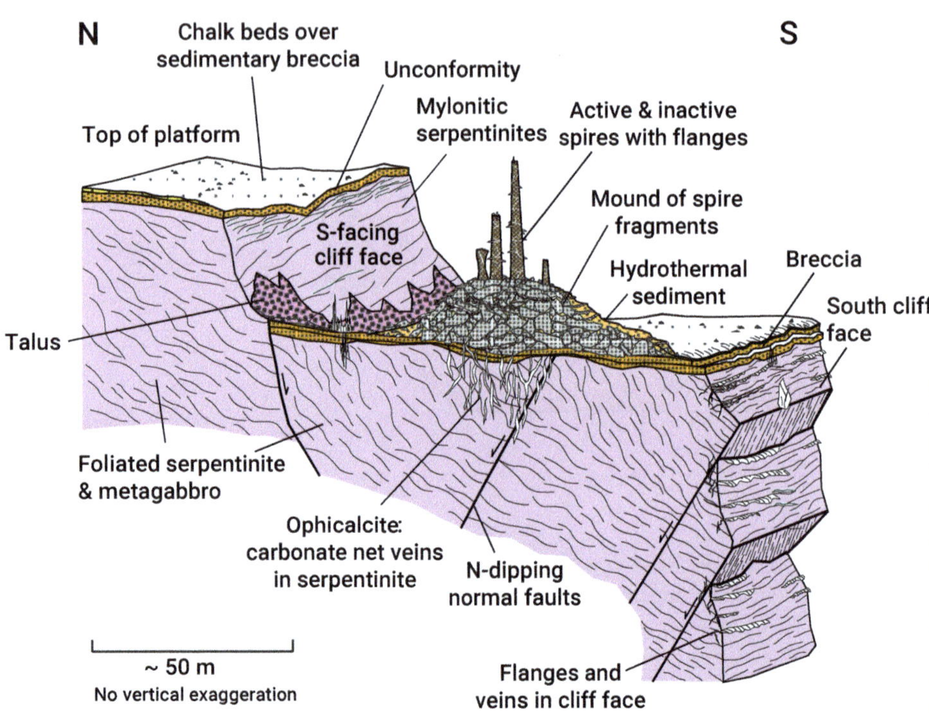

Figure 5.22 Schematic view of the geologic setting of the Lost City hydrothermal community. The site is developed on sheared peridotite. Hydration of the peridotite (serpentinization) creates H_2 and methane, which forms the basis for energy at the bottom of this complex food web.

Source: Modified from Kelley et al. (2005), © 2005 The American Association for the Advancement of Science.

(Kelley et al., 2005). Evidence of bacterial life has even been found in serpentine mud dredged from the Mariana Trench, over 10 km deep (Plümper et al., 2017).

Colonies of animals and bacteria in the deep ocean, or even in oceanic lithosphere beneath the ocean floor, may have provided pockets of survival during catastrophic events at the surface. For example, Sleep et al. (1989) argued that large impacts during the first several hundred million years of Earth history could have sterilized the surface and upper oceans repeatedly, knocking out any life that had gained a foothold there, but that deep-sea ecosystems may have survived. The deep oceans may have also provided a haven for life during proposed times of global glaciation in the Neoproterozoic (Hoffman & Schrag, 2000).

Summary

Key points of this chapter include the following:

- The crust and mantle contain quantities of water comparable to the world's oceans that cycle through the Earth on timescales of tens to hundreds of millions of years.
- Water at mid-ocean ridges is typically supercritical, allowing it to be heated to much higher temperatures than for shallower systems where it would boil.
- Hydrothermal circulation at mid-ocean ridges pervasively metamorphoses oceanic lithosphere.
- Chemographic diagrams illustrate mineral assemblage compatibilities that can be linked to reactions in *P-T* space; stable mineral assemblages can be used to estimate metamorphic temperatures.
- Metamorphic facies represents a mineral assemblage in a metabasaltic composition that is stable over a region of *P-T* space; "facies" may refer to both the mineral assemblage and the *P-T* region.
- Metamorphic grade generally represents the maximum temperature that a rock has reached.
- Reaction rates depend exponentially on temperature, according to the Arrhenius relation.
- Isotopes of O and Sr are used to track processes of hydrothermal alteration.
- Oxygen isotope fractionation between water and minerals is strongly *T*-dependent, driving dramatic differences in the isotopic behavior of low-*T* vs. high-*T* alteration.
- Hydrothermal systems at mid-ocean ridges form the foundation of volcanogenic massive sulfide deposits and, separately, deep-sea vent communities.

Notes

1. This sounds long for humans, and it is. Humans are around 50 percent water by mass and should drink about 2 to 3 liters of water per day. This works out to a residence time of about a week or two.
2. Another is plutonium. Word to the wise—don't pour liquid plutonium into a glass bottle, or the plutonium will expand and shatter the bottle when it freezes. Or, just don't pour liquid plutonium into anything. Actually, better to just avoid plutonium altogether.
3. Among other things, the oceans and many lakes would be frozen throughout most of their depths. Ponder how that would change how the Earth works.
4. Also "Goldschmidt's mineralogical phase rule" after Victor Goldschmidt (1888–1947), a Swiss-born geochemist considered the founder of modern geochemistry.
5. This is complicated by solid solution and other forms of compositional variability.
6. This is the number of combinations of three items chosen from a set of eight.
7. And every other decade, for that matter.
8. Named for the Swedish physical chemist Svante Arrhenius (1859–1927). Among other discoveries, he first proposed that increasing CO_2 contents in the atmosphere would drive global warming.
9. Meteors were originally thought to be weather phenomena.
10. *Not* "del-oh-18." To help you remember, Harmon Craig (a pioneering isotope geochemist) composed the following limerick: "There was a young man from Cornell, who pronounced every 'delta' as 'del,' [till] the spirit of Urey returned in a fury, and transferred that fellow to Hell." Harold Urey received the Nobel Prize for the discovery of deuterium (^2H), and he developed much of the theoretical foundation of stable isotope geochemistry.
11. Alert readers will ask how this can be, given that Austria is landlocked. The standard was developed with the International Atomic Energy Agency, which is based in Vienna. It replaced the original SMOW standard of the US Bureau of Standards. The actual VSMOW standard is made of distilled seawater from around the world, mixed in proportions so that VSMOW is as close to the original SMOW as possible. In case you're wondering, the other oxygen isotope standard is VPDB ("Vienna Peedee Belemnite"). Although Austria does have fossil belemnites, VPDB doesn't exist physically; rather, it represents a virtual standard—the best estimate of the composition of the original PDB standard from the 1950s. The original PDB was a fossil belemnite guard from the Peedee Formation in South Carolina. The Field Museum in Chicago gave one to Harold Urey (there's that guy again . . .) to use as an isotope standard for calcite. You can still collect "PDB" fossils, but we've never found one that had exactly the same composition as the original (close, but no cigar—look up an image of a belemnite, and you'll see why that's funny). VPDB is also the international standard for carbon isotopes.
12. Do not use this term (as, unfortunately, many do) to mean the isotopic composition of ordinary samples. For example, you might say, "The isotopic compositions of my samples reflect the isotopic signature of the mantle," but you should not say, "The isotopic signature of my samples . . ."

CHAPTER 6

Subduction and the Formation of Blueschists and Magmas

Not much happens to oceanic lithosphere during its long journey from its birth at a spreading ridge to the trench. Off-axis hydrothermal activity, such as that produced by serpentinization (see Chapter 5), can brighten up this dark journey, but most oceanic lithosphere is just slowly buried by a thin blanket of deep-sea sediment (siliceous or carbonate ooze, depending on whether the seafloor is below or above the **carbonate compensation depth**) as the cooling lithosphere moves away from the ridge, thickens, and subsides. Tens to hundreds of millions of years of quiet existence in the ocean depths comes to an end when the lithosphere reaches the hinge of the subduction zone. There, it fractures (slightly) and begins its descent into the mantle—subduction!

Carbonate compensation depth
Depth in the ocean at which the rate of calcite dissolution equals the rate at which calcite particles (e.g., foram shells) are supplied from above; typically, 4–5 km. Below this depth, calcite does not accumulate, and thus the abyssal plains are covered in siliceous mud rather than calcite mud.

Nearly all oceanic lithosphere is fated to be subducted—only the odd piece that gets stranded on a buoyant continent avoids being recycled into the mantle. As we saw in Chapter 3, the mean residence time of oceanic lithosphere is about 120 Myr. Some is subducted mere millions of years after being generated at a MOR (e.g., the Juan de Fuca Plate off the Pacific Northwest), whereas lithosphere in the western Pacific travels across half the globe before being subducted. When oceanic lithosphere goes down a subduction zone, its cold, wet, dark, boring life improves dramatically.

Subduction zones are the tectonic feature that sets the Earth apart from the other planets. Subduction of oceanic lithosphere brings water into the Earth's interior, where it plays a key role in making magma—especially silicic magma, which is rare or absent on other planets in the solar system and which, when cooled, makes up much of the continental crust that we live on. Partial subduction of continental crust buries shallow rocks to great depth, producing high-pressure metamorphic rocks and the Earth's highest topography.

Augustine Volcano, a classic subduction-zone stratovolcano in the Aleutian arc of Alaska, releasing a plume of gas in January 2006.

Source: Cyrus Read, Alaska Volcano Observatory, USGS.

6.1 Subduction Zones

Learning Objectives

By the end of this section you should be able to:

6.1.1 Draw a map view across a subduction zone, labeling all major geomorphic and geologic features.

6.1.2 Sketch a cross section of a subduction zone, to scale.

Accretionary wedges
The wedges of sediment and metamorphic rocks scraped off the subducting slab; the part of the slab that is not subducted.

Fore-arc
The region of a subduction zone between the arc and trench; includes the accretionary wedge and any fore-arc basins.

Island arc
A chain of volcanoes associated with subduction of oceanic lithosphere; may or may not be arcuate.

Back-arc
The part of a subduction zone that is behind the arc and farthest from the trench; the back-arc region can contain basins floored by actively spreading oceanic crust.

General Structure

Subduction zones are easily identifiable from space, because they produce characteristic landforms. On the one hand, subduction produces topographically obvious chains of volcanoes that can emit copious amounts of thermal energy and gases. The topography, temperature, and chemistry can all be observed remotely. On the other hand, subduction produces a deep-sea trench. This cannot (yet) be imaged directly from space, because we can't see through so much water. Subduction zones pile up vast stores of sedimentary rocks in **accretionary wedges**, which lie in the **fore-arc** region (between the trench and the arc). Most fore-arc rocks are only incipiently metamorphosed, but fore-arcs also expose rocks exhumed from deep in the subduction system and represent samples of the high-pressure/low-temperature (HP/LT) lithosystem.

Subduction of oceanic lithosphere beneath oceanic lithosphere typically results in an arcuate chain of volcanic islands known as an **island arc**. The Lesser Antilles, which bound the Caribbean Plate on its eastern side (Figure 6.1), are an excellent example. Other classic island arcs include the Aleutian (northern Pacific), Kurile and Mariana (western Pacific), and Scotia (south Atlantic) island chains (Figure 6.2). Some ocean-ocean subduction zones produce volcanic chains that are not particularly arcuate, such as many in the southwestern Pacific Ocean (e.g., Tonga-Kermadec). Geologists generally use the arc and trench as geographic references to describe rocks and processes as occurring in the trench, fore-arc, arc, and **back-arc**.

Figure 6.1 The Lesser Antilles, a classic island arc that bounds the eastern side of the Caribbean Plate. Most of the islands are volcanoes produced by westward subduction of a piece of the Atlantic Plate beneath the Caribbean Plate. Image was taken September 20, 2002, by NASA's MODIS Terra satellite, which travels in an orbit 705 km above the Earth's surface. The lush green islands are surrounded by water tinted gray by the sun's reflection; west of the islands, in the downwind direction, the interaction of wind and waves modifies this reflection to produce curious dark areas that look like cloud shadows and make the islands look as if they are floating above the water. The volcano on Montserrat was erupting, and the eruption plume is traveling to the northeast, at right angles to the winds shown by the clouds. All the islands except Barbados form a nice arc; Barbados is an uplifted part of the fore-arc.

Source: Jacques Descloitres, MODIS Rapid Response Team, NASA/GSFC.

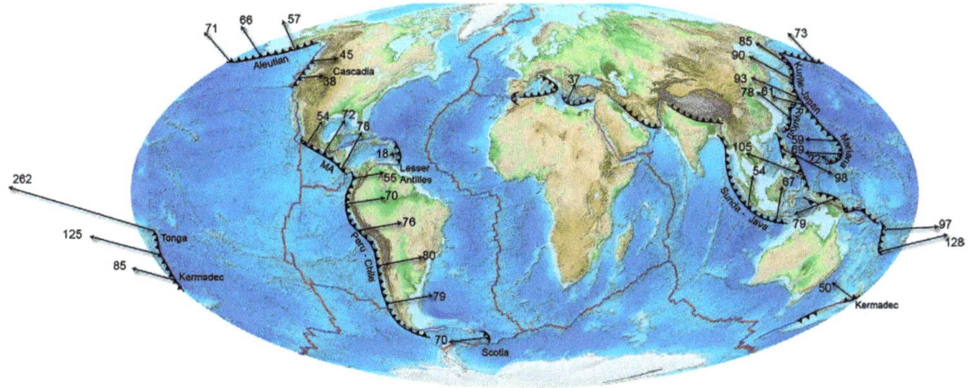

Figure 6.2 Map of the world showing plate boundaries and subduction zones, barbed lines with barbs on the hanging-wall plate. Arrows give azimuth and rate of convergence of the plate being subducted (map projection distorts azimuths at higher latitudes). Subduction zones without rates listed are mostly continental collision zones at which rates are far lower than when oceanic lithosphere is being subducted. MA = Middle America.

Subduction of oceanic lithosphere beneath continental lithosphere also produces chains of volcanoes, but these tend to be more linear than arcuate. Examples include the Cascades of Oregon/Washington/British Columbia, the Mexican volcanic belt, and the Andes. In spite of these variations in curvature, it has become common to refer to subduction-related volcanic systems as arc volcanism, and to sets of subduction-related volcanoes as an arc. We will use this shorthand, realizing that many arcs are not particularly arcuate.

One of the simpler ocean-ocean island arcs is the Mariana arc, in the western Pacific south of Japan. The strategically and historically important islands of Guam, Saipan, and Tinian are part of this arc, which was born about 45 Ma along a north-south transform fault that accommodated spreading between Asia and Australia. The Mariana arc is particularly tightly curved, and bathymetry (Figure 6.3) shows numerous interesting features. This simple arc is not so simple after all! There are actually three arcs (two of which are submarine—the West Mariana Ridge and the Palau-Kyushu Ridge) separated by two **back-arc basins** (split by an extinct and an active ridge) and bounded by the deep Mariana Trench on the east. This trench hosts the Challenger Deep, the deepest place in the world's oceans (nearly 11 km, or 2 km deeper than Mount Everest is tall). Most of the arc is underwater, with only the tops of some of the volcanoes poking out (Figure 6.4).

Back-arc basins
Basin in the back-arc; often a site of active spreading.

The "simple" Mariana system includes three arcs, because it has been rifted twice to form mini-ocean basins. One rifting event started about 30 Ma, splitting the first arc to form the Parece Vela Basin. The western part of the arc became inactive after this rifting event. A second rifting event, about 5 Ma, formed the Mariana trough, again stranding the western part as a dead arc. Such rifting is known as back-arc rifting, and formation of back-arc basins is common, especially in the western Pacific.

Figure 6.5 shows bathymetry and topography of the Cascade subduction system of the Pacific Northwest. Unlike the Mariana arc, this arc is entirely subaerial and is strikingly linear compared to typical island arcs. This arc is generated by subduction of young oceanic lithosphere produced at the nearby Juan de Fuca Ridge. The active arc terminates in the south at the Mendocino triple junction, which marks the southern end of subduction. Several of the taller volcanoes, such as Mount Shasta, Mount Rainier, and Mount Hood (Figure 6.6), are clearly visible as pimples on the arc, but along much of its length volcanoes are subdued, forming a long, linear ridge 150 to 200 km from the coastline and 200 to 300 km from the trench. Abundant sediment pours off the west coast of North America, nearly filling the Cascade arc trench so that it is shallow and barely noticeable, especially compared to the Mariana Trench. The off-scraped sediment from Cascade subduction piles up in the fore-arc accretionary wedge, which is exposed along the Oregon Coast

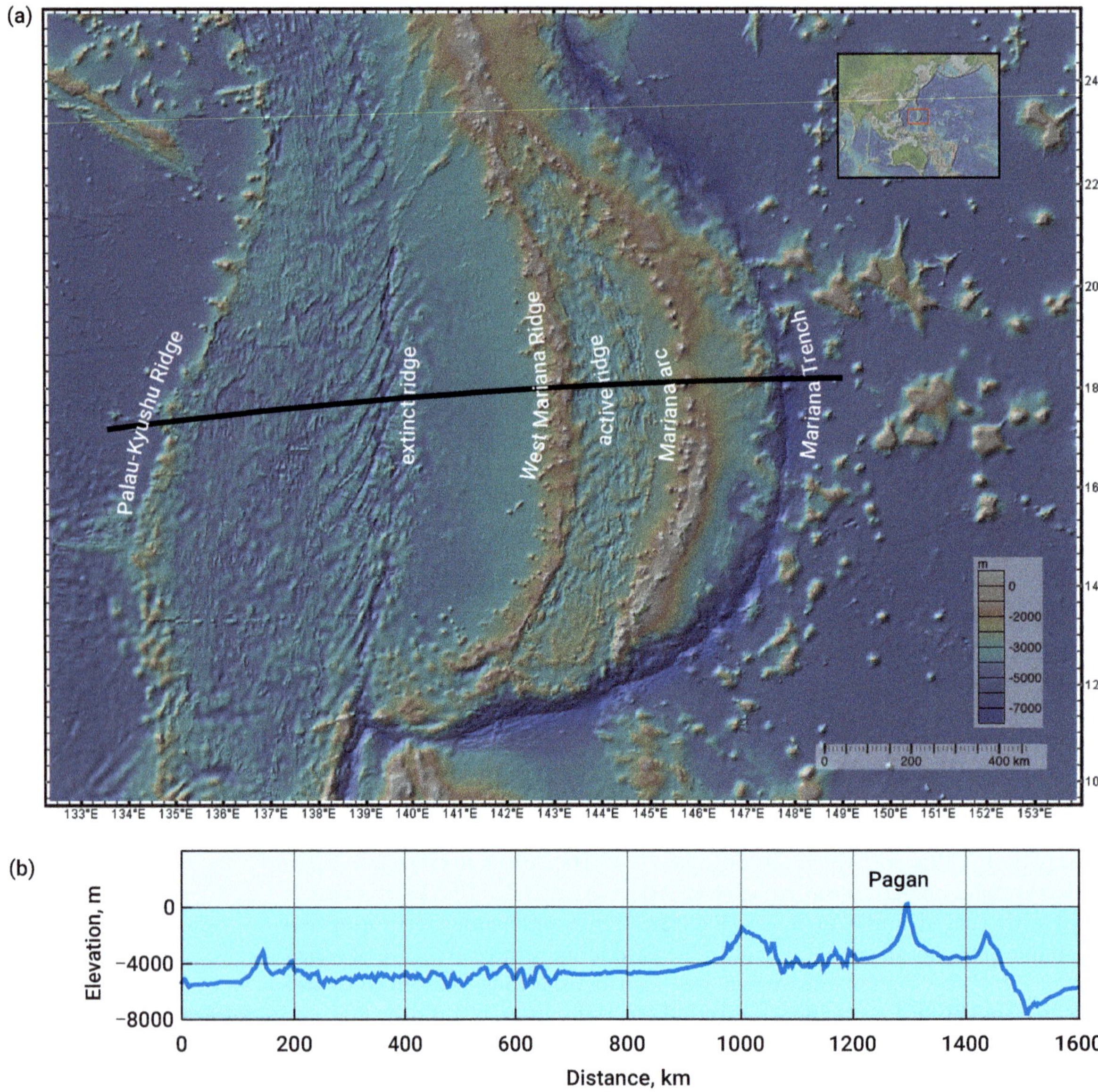

Figure 6.3 (a) Bathymetry of the Mariana arc in the western Pacific Ocean. Many of the seamounts just west of the trench are mud volcanoes. (b) Cross section through the Mariana fore-arc – arc – back-arc system at the latitude of Pagan, one of the northernmost volcanoes. Although Pagan barely breaks sea level, it rises over 4 km above the surrounding ocean floor.

Range, the Olympic Peninsula of Washington, and Vancouver Island. Clear HP/LT rocks (see Section 6.2) are relatively rare in modern subduction systems and are limited to a few blueschists in Washington that predate modern Cascade subduction.

Wadati-Benioff Zones

Wadati-Benioff zones
The curved planes of seismicity that dip away from a trench and delineate the subducting plate.

Most subduction zones are characterized by well-defined **Wadati-Benioff zones**, sheets of seismicity that dip away from, and are interpreted to mark the path of, the subducted plate (Figure 6.7). A common but incorrect assumption is that these earthquakes mark the plane of a megathrust fault along the top of the subducted slab, but earthquake focal mechanism solutions[1] contradict this; rather, they indicate that deformation occurs within the slab itself. In well-studied subduction zones, the Wadati-Benioff zone commonly consists

Figure 6.4 Farallon de Pajaros (Bird Island), northernmost island in the Mariana arc. The island is the tip of a large volcano that rises over 3 km above the surrounding ocean floor, but the island itself, a stratovolcano (see Chapter 7), only sticks up about 360 m above sea level. Many volcanoes in this arc are entirely submarine.

Source: Pedro Lebron, NOAA.

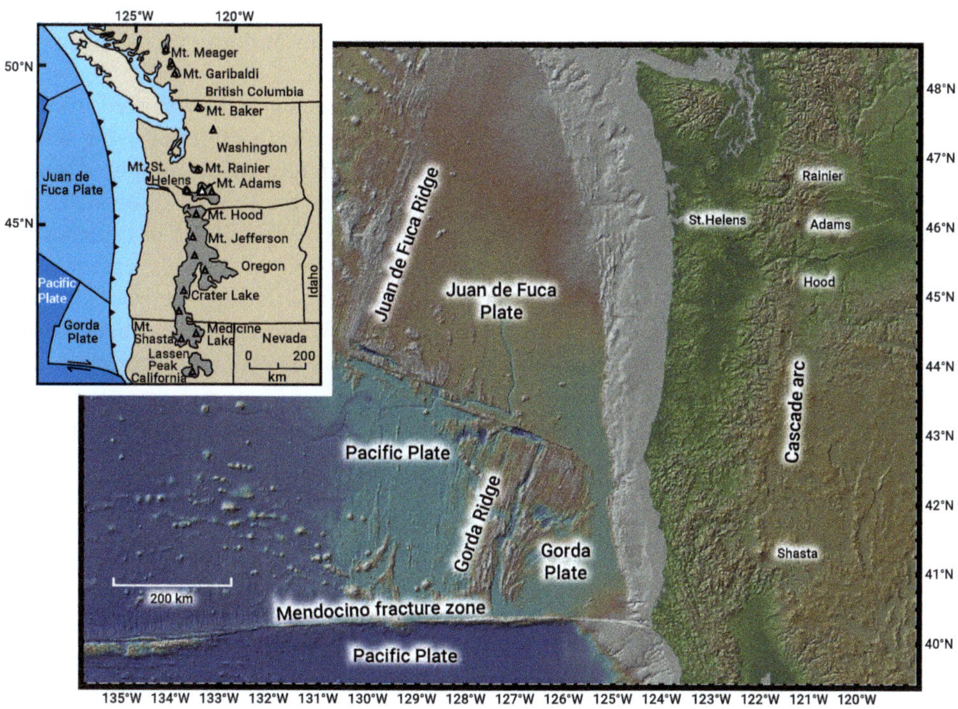

Figure 6.5 The Cascade arc of northern California, Oregon, and Washington. The arc lies inland of the small Juan de Fuca Plate; this plate and the Rivera Plate off western Mexico are remnants of the Farallon Plate, most of which was subducted under western North America in the Cenozoic. The arc, which continues to the north into British Columbia, forms a linear chain about 200–300 km from the barely recognizable trench. The Cascade arc terminates to the south at Lassen Peak, an active volcano in northern California that lies on the inland projection of the Mendocino fracture zone. This fracture zone is interpreted to bound the subducted slab on the south.

Source: Inset image © 2008 Springer-Verlag.

Figure 6.6 Mount Hood seen in a telephoto shot from Portland, Oregon. This stratovolcano lies 80 km east of the city.

Source: Ed Ruttledge, USGS.

of two parallel sheets 10 to 30 km apart, an upper one interpreted to be in the basaltic crust of the subducted slab and a lower one in its upper mantle.

The existence of deep (>70 km) earthquakes is puzzling, because the ability of a rock to deform by fracture diminishes as pressure and temperature go up. High confining pressure on a fault surface clamps it in place, and high temperatures cause rocks to flow rather than fracture. High *P-T* experiments designed to reproduce deep earthquakes have generally failed. Subducted lithosphere is cooler and therefore stronger than surrounding asthenosphere, but high confining pressure at depth should still inhibit generation of earthquakes. Several mechanisms for generating deep earthquakes have been proposed, including implosive phase transformations to denser phases, runaway shear heating, and rapid dehydration reactions that release water and lubricate fault planes.

The existence of deep earthquakes was first established by Wadati (1928), and Benioff (1949) proposed that inclined planes of deep earthquakes found by Wadati (1935) result from faulting between oceanic and continental material. The realization that these zones mark descent of relatively cool, dense lithosphere into the mantle came from studies of seismic waves generated by deep earthquakes. The subducted slab acts as a waveguide, transmitting seismic waves faster (because cooler rocks are stiffer) and with less attenuation than the hotter and weaker asthenosphere around it. Existence of lithospheric slabs under island arcs explained anomalously early arrival of seismic waves from deep earthquakes in stations that lie up-dip, and the greater intensity of such waves (Figure 6.8).

Seismic tomography
A mathematical technique that uses arrival times of a large number of earthquakes at a large number of stations to estimate velocity variations at depth.

Knowledge of the geometry of subduction zones has been greatly advanced in recent decades by **seismic tomography**, a method of imaging the Earth's interior using seismic waves (see Toolbox 6.1). Seismic tomography is closely akin to medical computed tomography (CT) scans; in each technique, the analyst measures the properties of waves that have passed through a volume (seismic waves and X-rays, respectively) to infer what is inside the volume. Figure 6.9 shows a particularly well-determined seismic cross section of the subduction zone under Honshu, the main island of Japan. This image was calculated ("inverted" in the jargon of tomography) using data from 315 seismometers that recorded over 200,000 ray paths generated by over 7000 earthquakes—a prodigious problem in data manipulation and equation solving.

It is important to keep in mind that tomographic models are non-unique; many different velocity structures could produce a good match to observed travel times. The velocity model in Figure 6.9 shows deviations from an assumed background seismic velocity model (generally increasing downward), with blue meaning faster than expected and red meaning slower. These velocity deviations are generally a result of either a change in rock type (see Figure 2.16) or a change in temperature, with cooler rocks having higher velocities. A change in mineralogy for a given rock type can also perturb seismic velocities, and this can be counterintuitive: a cold metabasalt with abundant hydrous minerals has a much lower velocity than a hot metabasalt with anhydrous minerals, even at the same pressure.

The subducted slab is clearly shown by the blue band traversing the diagram at a dip of about 30°, and the red blobs under the Honshu volcanoes are most likely a result of partially molten rock. The blue band is the waveguide that conducts deep seismic waves so efficiently to the surface. Earthquakes near the line of section are projected onto the figure and comprise two groups—one at shallow crustal depths and one within the upper half of the slab, extending to 400 km depth. This is the general geometry of subduction.

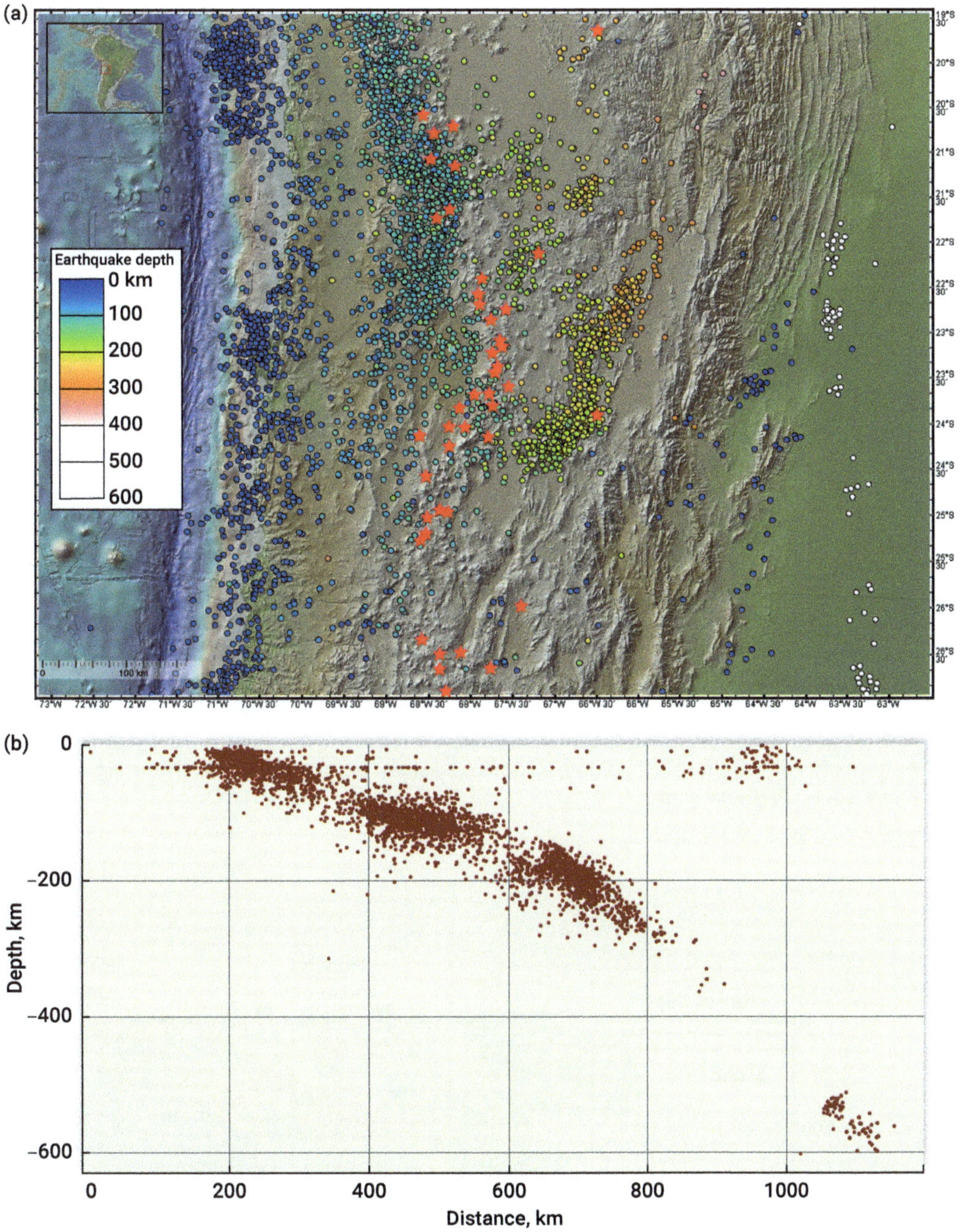

Figure 6.7 (a) Map of earthquakes in the central Andes, color-coded by depth. Red stars are volcanoes. (b) Cross section showing the corresponding Wadati-Benioff zone, with earthquakes projected onto an east-west plane. The junction of Chile, Argentina, and Bolivia is near the center of the map. In this area, earthquakes form three elongate clusters: a shallow one near the coast, a deeper one (~50–150 km) a few hundred kilometers inland, and a still deeper one (~150–300 km) yet further inland. A cluster of truly deep earthquakes, around 600 km below the surface, occurs deep under the western Argentine plains. Shallow earthquakes above the deep cluster are related to back-arc thrusting in the Andean foothills. The linear array of earthquakes at 33 km depth is an artifact; if the depth of an earthquake is not well-determined, it is commonly listed in catalogs as 33 km by default.

Source: (a) Map produced using GeoMapApp.

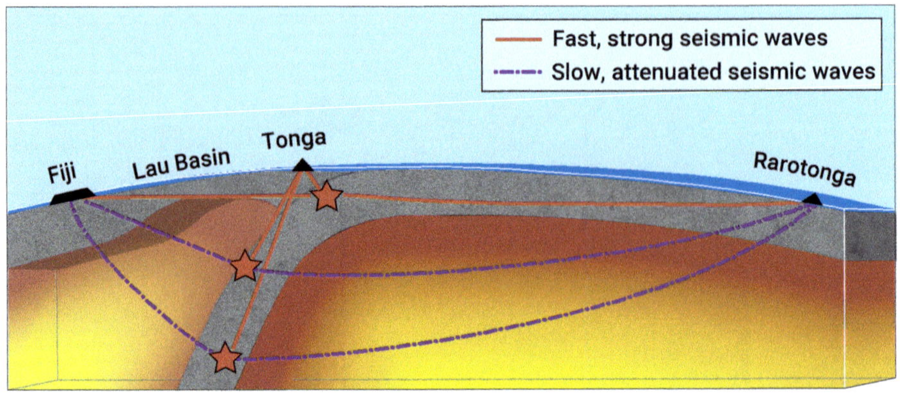

Figure 6.8 Schematic east-west cross section across the Tonga-Fiji region showing slow and fast paths for seismic waves and how they were used to infer the existence of the subducted slab. Earthquakes as deep as 700 km are common under the Tonga-Kermadec arc (see Figures 1.1 and 6.2). Before recognition of subduction, seismologists noted that seismic waves from these deep earthquakes arrive ahead of schedule in Tonga, behind schedule in Rarotonga (in the Cook Islands), and way behind schedule in Fiji, whereas shallow earthquakes arrive early at all the stations. Seismic energy from the deep earthquakes also arrives at Fiji and Rarotonga quite attenuated compared to that hitting the Tonga station. This led to the inference of a strong waveguide along the Wadati-Benioff zone under Tonga—what we now recognize as the subducted slab. Waves traversing the back-arc region to get to Fiji pass through hot upwelling mantle under the Lau Basin and are thus slowed and attenuated. This figure also illustrates an important constraint on seismologic studies of this area: most seismometers are on land, and land is scarce in the oceans, so observation points are limited to beautiful tropical islands. The bottom of the lithosphere in this figure is shown as a sharp boundary, but recall from Figure 2.22 that it is actually gradational.

Source: Modified from Molnar (2015).

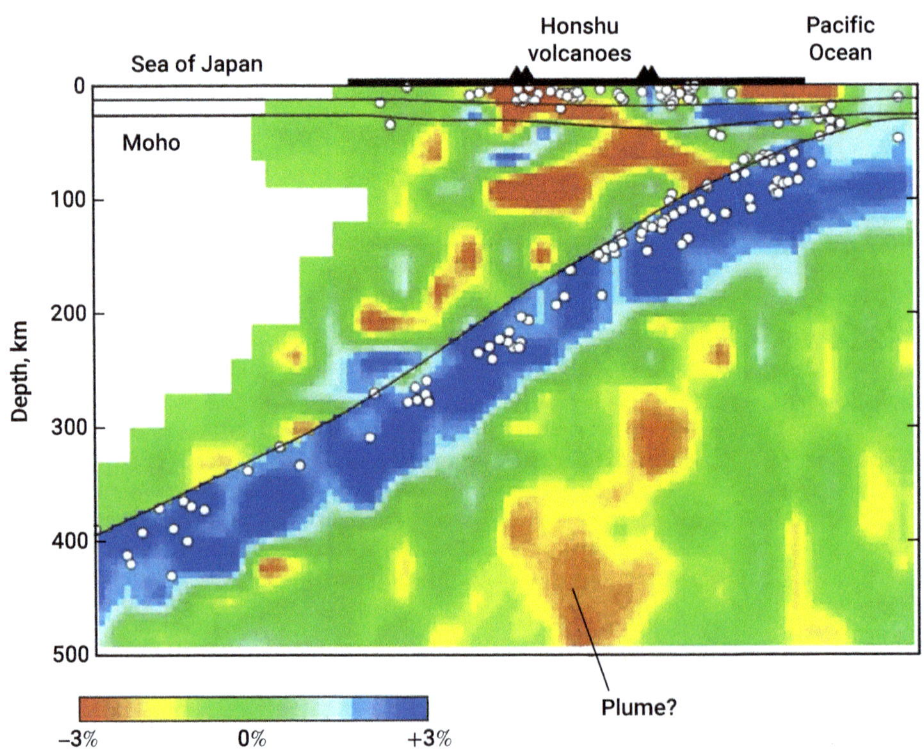

Figure 6.9 Tomographic cross section through central Japan, oriented WNW-ESE. Red and blue colors denote velocities that are slow and fast, respectively, compared to a reference model (scale at bottom). White dots denote earthquakes that occurred within 40 km of the profile; solid triangles are active volcanoes; black bar denotes land. The three curved lines show, with increasing depth, a midcrustal seismic discontinuity, the Moho discontinuity, and the upper boundary of the subducting Pacific Plate. Note that earthquakes occur within the slab rather than at its top surface.

Source: From Zhao (2004), © 2004 Elsevier B.V.

Key Questions

Although the basic anatomy of subduction zones is reasonably well understood from surface geology and geophysics, many aspects of subduction are poorly known. Key questions include:

- What is the thermal structure of subduction zones?
- How are rocks exhumed from deep in subduction zones? Are the ones brought to the surface representative or thermally anomalous?
- Where does subducted lithosphere release its water?
- How and where is magma generated?
- What are the sources of such magma—in other words, what melted to produce the magma? If you have a piece of andesite in your hand, where did the atoms in it come from? Possibilities include subducted oceanic crust, subducted sediment, the mantle above the subducted slab, and preexisting continental crust.
- How much crustal material, especially water, is carried into the deeper mantle?
- What are the driving forces for subduction—and indeed, all of plate tectonics?

Toolbox 6.1

Seismic Tomography

Seismic tomography is a procedure in which a large number of earthquake arrival times are used to infer the *P*-wave velocity structure under an area. The result of such a study is generally a 3D model in which the volume of interest is divided into a large number of blocks, each of which has a velocity calculated to best explain observed *P*-wave arrival times.

Consider an analogous situation. You are a traffic engineer in a city whose streets are gridded, with avenues (First, Second, etc.) running one way and streets (A, B, etc.) the other. You know that with no traffic, drivers can cover one block in one minute, and you monitor these speeds in real time so that you can direct drivers around areas of trouble. You come in to work one day, look at the Big Board, and find the situation illustrated in Figure 6.10.

Traffic is moving smoothly over much of the grid, but there is clearly something gumming up the works. A little inspection shows that the problem is likely on Fourth Avenue, involving the intersections with B and C Streets. With the data at hand, you cannot tell just how many cars are there, but you can tell that speeds are low on B and C Streets, and especially along Fourth Ave. This is not a unique solution by any means but is the simplest explanation.

Seismic tomography is similar. The analyst takes a starting model (perhaps a constant seismic velocity across the entire area, or a model in which velocity steadily increases downward), traces seismic rays from earthquakes through the model, computes the expected travel times (for the assumed velocity structure), and compares them to the observed travel times. The model is then changed to do a better job (e.g., if observed rays consistently come in earlier than the model in an area, the velocity there is increased), and after many iterations, the model ideally settles on a velocity structure that produces a good (but nonunique) match to observed travel times.

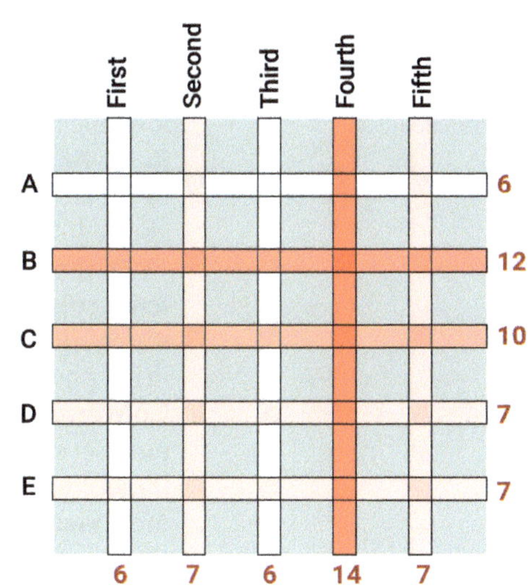

Figure 6.10 Traffic travel times in a hypothetical city. Numbers give time in minutes to traverse the corresponding six-block stretch. Roads are shaded according to travel time, with deeper shades indicating slower average speed over that stretch. Most likely, something is wrong in the vicinity of the intersections of Fourth Avenue with B and C Streets.

6.2 Lithosystems Associated with Subduction Zones

Learning Objectives

By the end of this section you should be able to:

6.2.1 Characterize rocks of the two main lithosystems associated with subduction zones (arc and HP/LT) and explain their origin.

6.2.2 Use characteristics of rocks of the subduction zone lithosystems to articulate geologic processes associated with subduction zones.

The Arc Lithosystem

Subduction zone magmatism produces characteristic rock associations that we may term the arc lithosystem. The arc lithosystem is characterized by calc-alkaline volcanic and plutonic rocks with a wide range of compositions along the BADR spectrum, along with associated sedimentary and volcaniclastic rocks such as immature sandstones and volcanic mudflows. The term calc-alkaline was discussed in Chapter 4 on the basis of AFM diagrams and Mg#; in particular, by how low the Mg# gets along the BADR path. Use of the term and its variants has a long and inconsistent history in petrology (Arculus, 2003), but most arc magmatism is, broadly speaking, calc-alkaline. We will get a closer look at the igneous components of the arc lithosystem in Chapters 7 and 9.

Terrane
A region, generally bounded by faults, with specific geologic characteristics that differ from those of adjacent regions; *terrain* refers to a parcel of land and its physical characteristics.

This lithosystem is commonly recognizable even when highly deformed. For example, Arizona contains a belt of rocks, the Yavapai **terrane**, that includes the metamorphosed equivalents of clastic sedimentary rocks and calc-alkaline volcanic rocks. This belt is interpreted as an island arc that was accreted to North America about 1.7 Ga. The less-deformed Carolina terrane of the eastern United States, which runs from Georgia to Virginia, contains abundant intermediate and silicic volcanic rocks, shallow marine sedimentary rocks, and abundant plutons ranging from gabbro to granite. All have been washed by a greenschist facies overprint, but they clearly appear to be an arc lithosystem that was accreted to the eastern margin of North America about 450 Ma.

Although oceanic ridges generate by far the largest area of igneous crust on Earth, few people live on oceanic crust or oceanic ridges, because they are mostly underwater—Iceland being an exception. Oceanic ridges are copious producers of oceanic lithosphere, but most of this is subducted within tens to a few hundred million years. Conversely, subduction zone magmatic systems are active on a relatively small proportion of the Earth's surface at any given time, but the material they produce is largely unsubductable and piles up at the surface, producing the continents. Plate movement shuffles this material around in a global shell game, but once a parcel of low-density continental crust is made, most of it stays at the Earth's surface for the duration.[2] Hundreds of millions of people live on and near active subduction zones, and most of the world's population lives on crust that was produced at a subduction zone at some time.

Lithospheric material made at subduction zones is able to survive because subduction zone volcanoes, unlike oceanic ridges, produce rocks from the entire BADR spectrum, including relatively felsic, buoyant rocks (andesites, dacites, and rhyolites), not just basalt. Indeed, it is this variety in magma composition that produces the extreme variety of eruptive styles seen in subduction volcanoes—everything from calm basaltic eruptions that one can view up close in relative safety to cataclysmic eruptions of rhyolite over 100 times more voluminous than the 1883 eruption of Krakatau.

Rocks of the Arc Volcanic Lithosystem

The arc volcanic lithosystem comprises the products of arc volcanoes. These range in composition across the entire calc-alkaline BADR suite and include lavas, pyroclastic rocks (anything blown out of a volcano), and mudflows. Important components of this lithosystem, not preserved in the rock record, are H_2O and CO_2.

The High-Pressure/Low-Temperature Lithosystem

Blueschist and Eclogite Much of our knowledge of what happens to subducted lithosphere comes from observation and analysis of the high-pressure/low-temperature (HP/LT) lithosystem, a highly recognizable set of rocks and structures commonly found in a belt running parallel to coeval igneous rocks of the arc lithosystem and their associated low-*P*, high-*T* metamorphic rocks (LP/HT; Figure 6.11). Some of the most intriguing rocks represent scattered occurrences where subducted material has been spit back up to the surface in a chaotic deposit known as a **mélange** (French for "mixture"; Figure 6.12). A mélange is a collection of hard blocks (up to kilometer-scale) of blueschist, eclogite, serpentinite, and other high-*P* rocks in a highly deformed, sheared, weak matrix of metasedimentary rocks, chiefly metashale and metagraywacke,[3] or a sheared mixture of metasedimentary rock and serpentine. Although this matrix looks in places like a normal sedimentary rock, attempts to map it any significant distance fail, because it is so highly sheared and deformed—stratigraphic units just disappear along strike. Because the matrix is so much weaker than the solid blocks it contains, mélange areas are characterized by

Mélange
A tectonic mixture of high-*P* rocks such as eclogite or blueschist in a sheared, weak matrix of metasedimentary rocks and serpentine.

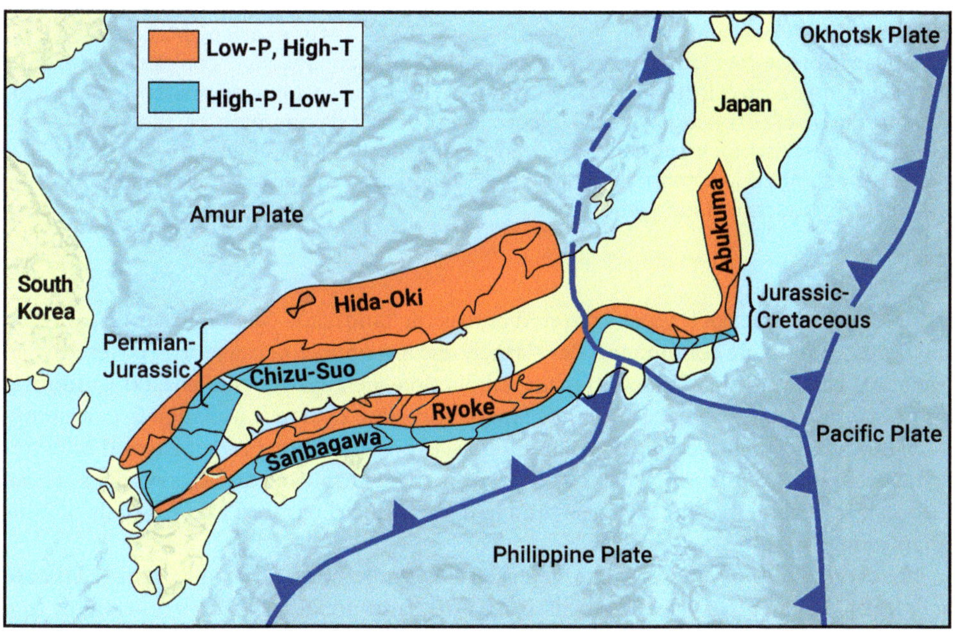

Figure 6.11 Map of the central islands of Japan showing two paired belts of high-*P*, low-*T* rocks (representing the subduction channel) juxtaposed against low-*P*, high-*T* rocks (representing the arc).

Source: Modified from Frisch (2016).

Figure 6.12 Classic mélange in the Franciscan Complex of coastal California. (a) Large blocks of eclogite and blueschist dot the beach at Jenner, north of San Francisco. These have weathered out of their weak, shaley, highly deformed matrix. (b) Mélange exposure at Ring Mountain, north of San Francisco. Resistant blocks of eclogite and blueschist stick up above the weak, shaley matrix, here covered in grass. The lack of trees at Ring Mountain is a consequence of soil developed on serpentine (see Making Connections 3.1 on serpentine soils). (c) Mélange in a beach cliff at San Simeon, central California coast. More competent (resistant to deformation) blocks of metagraywacke, blueschist, and chert sit in an intensely sheared, shaley matrix. (d) Blueschist block from Jenner showing characteristic blue amphibole (glaucophane) and red garnet. Any one of these outcrops is easily recognizable as an example of the HP/LT lithosystem, products of a subduction zone.

Source: AFG photos.

large blocks of exotic rock types scattered across the landscape (see Figure 6.12), and the matrix, the dominant component, can largely escape notice, because it is covered with vegetation and soil. Classic localities for subduction mélange include coastal California (Franciscan Complex), Newfoundland, Japan, and New Zealand. Mélange is commonly associated with dismembered parts of ophiolites, exposed as thin, blueschist- and eclogite-facies thrust slices of metamorphosed basaltic and oceanic metasedimentary rocks. These tracts can be several kilometers across and can sometimes be traced for hundreds of kilometers along strike.

The HP/LT lithosystem is characterized by sometimes strikingly blue schistose rocks called blueschists, as well as distinctive red-and-green rocks called eclogites (see Figure 6.12). The blue color is imparted by glaucophane, a Na-rich amphibole with the nominal formula $Na_2(Mg,Fe)_3Al_2(Si_8O_{22})(OH)_2$. Although glaucophane makes up only a small percentage of the Franciscan Complex and other blueschist terranes, its color, rare among minerals, makes it stand out. Red-and-green eclogites contain red garnet with a typical composition of $(Ca,Mg,Fe,Mn)_3Al_2Si_3O_{12}$ and a green, Na-rich pyroxene called omphacite, which is a solid solution between jadeite and calcium-rich clinopyroxene, with a nominal formula of $(Na,Ca)(Fe,Mg,Al)Si_2O_6$.

Several other important minerals characterize this lithosystem, including:

- Lawsonite, $CaAl_2(Si_2O_7)(OH)_2{\cdot}H_2O$
- Albite, $NaAlSi_3O_8$
- Jadeite, $NaAlSi_2O_6$ (sodic pyroxene)
- Aragonite, $CaCO_3$

Many of these minerals are relatively rare but are linked through progressive mineral parageneses (Figure 6.13).

These assemblages puzzled metamorphic petrologists in the middle of the 20th century. Excepting albite, all are dense minerals, favored by high pressure, an inference that was verified by experimental work (e.g., Ernst, 1964; Liou, 1971; Boettcher & Wyllie, 1968), and yet associated minerals (including albite, zeolites, and chlorite) define relatively

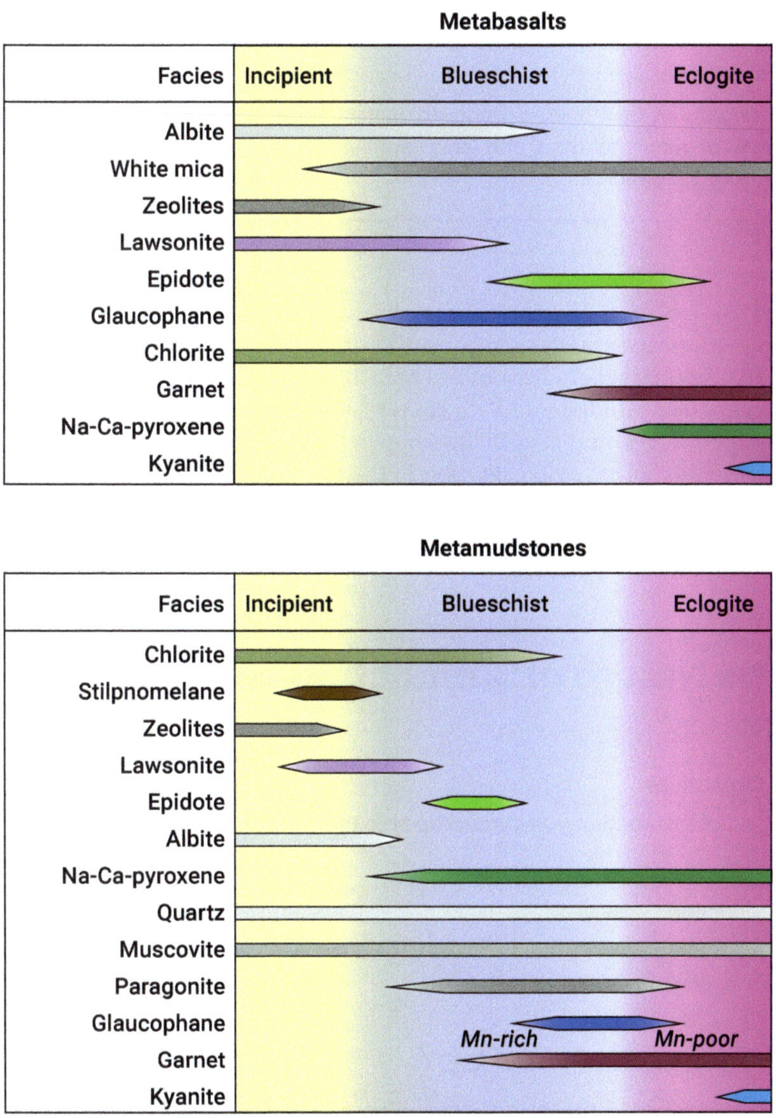

Figure 6.13 Paragenesis diagram for metabasalt and metamudstone compositions. Glaucophane, lawsonite, and Na-Ca-pyroxene generally indicate high-pressure, low-temperature (HP/LT) conditions. Other minerals can form over a wide range of *P-T* conditions, not necessarily only in subduction zones.

Rocks of the HP/LT Lithosystem

These systems are dominated by mafic volcanic and intrusive rocks, ultramafic rocks from the upper mantle and lower crust, and deep-water sedimentary rocks:

- *Blueschist*: A metamorphosed mafic rock that is blue because it contains abundant glaucophane. Blueschists are not necessarily schistose, but form at very low-*T* and high-*P* conditions, and also typically contain chlorite, albite, and either lawsonite or epidote.
- *Eclogite*: A high- to ultrahigh-pressure metamorphosed mafic rock that contains garnet and sodic clinopyroxene (omphacite).
- *Serpentinite*: A hydrated ultramafic rock mostly made of the mineral serpentine. Typically also contains magnetite.
- *Mélange*: A mixed rock, meaning a jumbled assemblage of resistant blocks ranging from centimeter- to kilometer-scale within a weak matrix of metasedimentary or serpentinite schist.
- *(Meta)chert*: A (metamorphosed) siliceous sedimentary rock, usually formed from accumulations of deep-water siliceous microfossils.
- *(Meta)graywacke*: A (metamorphosed) immature sandstone with broadly equal abundances of quartz, feldspar, and lithic fragments. Turbidity currents near trenches typically form graywackes.

low-grade[4] assemblages. Together, these observations imply a metamorphic path that reached high pressure (>1 GPa) at relatively low temperature (200–300 °C). In fact, 1 GPa at 300 °C is a geothermal gradient of less than 10 °C/km, below the low end of measured geothermal gradients. In the mid-20th century, no one knew how these rocks could have reached such high pressures at such low temperatures.

The puzzle was solved by recognition in the 1960s of subduction zones. Using a convergence rate of 20 mm/yr and a length scale of 100 km (a typical plate thickness) gives Pe ~60, indicating that subduction is fast enough to push isotherms down and refrigerate the subduction zone.

6.3 The Nature of Subduction and Plate-Driving Forces

Learning Objectives

By the end of this section you should be able to:

6.3.1 Use thermodynamics (Clapeyron equation) to predict how buoyancy drives or retards subduction.

6.3.2 Sketch simple cross sections of the Earth showing how plates interact and their velocities in different reference frames.

Top-Down or Bottom-Up?

In a broad sense, plate tectonics is driven by loss of heat from the Earth; as noted in Chapter 2, the Earth's internal heat is a mixture of primordial thermal energy from accretion and core infall and radioactive heat generated by decay of U, Th, and K. Plate tectonics is a way of dissipating this heat to space by bringing it to the surface, where it can be radiated away. However, details of this process are murky.

A subduction system is commonly equated with a one-way conveyor belt that carries oceanic lithosphere from a source (the mid-ocean ridge) to the subduction zone (Figure 6.14). This image can lead to the mistaken concept that the subducted plate has a well-defined bottom that passes over a hinge and then slides down a fixed slot in the mantle. Introductory geology books commonly equate convection in the mantle to a pot boiling on the stove, where heat supplied from below causes convection that drives hot liquid up and around the convection cycle. In this bottom-up view, plate motion is caused by drag of convecting mantle below the plates.

There is another way to envision this process, however. In this view, plate tectonics is a top-down process driven by cooling of the surface and consequent sinking of densified oceanic lithosphere. This is what happens at lava lakes (Figure 6.15), where hot magma at the surface loses heat rapidly by radiation and convection, forming a dense crust that sinks back into the molten material beneath when given a chance. Observation of active lava lakes shows that crustal plate foundering (sinking) causes a rapid increase in spreading rate at nearby spreading ridges, and that when foundering stops, spreading slows down or stops as well. In such systems, then, foundering of the subducting plate is the main driver of plate movement, and spreading is largely a passive response to what the foundering plates are doing.

Lava lakes are highly imperfect models of plate tectonics (Duffield, 1972), showing many features that do not have counterparts in lithospheric plates, but the concept that plate motion is largely driven by top-down cooling has merit.

Figure 6.14 Subduction is commonly presented as a conveyor-belt system in which oceanic lithosphere (the belt) is cranked over a fixed hinge and driven down a slot in the mantle. This is not how subduction works.

Figure 6.15 Lava lake in Kīlauea summit caldera, February 2017. Bright lines indicate upwelling of fresh magma between separating darker plates of cooler (but still smoking hot!) lava. Space for this fresh magma is made by sinking of surficial plates. This is happening at the bright area at right, where foundering of a plate is accompanied by significant upwelling.

Source: Michael Patrick, USGS.

How Far Do Slabs Descend?

A key question in geology concerns the depth to which subducted slabs descend. There are two main points of view: either they descend all the way to the core-mantle boundary, losing their identity on the way down as they heat up, or they bottom out at the 660-km discontinuity, the base of the transition zone (see Figure 2.18). The reasoning behind this second possibility makes use of the Clapeyron equation. Although the olivine → wadsleyite and ringwoodite → perovskite + ferropericlase reactions are complex, the former has a positive Clapeyron slope and the latter a negative one (Bina & Helffrich, 1994).

Remember that for a positive slope, reactions occur at shallower depth with decreasing *T*, whereas for a negative slope, they occur at greater depth (Figure 6.16). A descending slab carries relatively cool material (mostly mantle lithosphere, with a thin basaltic crust) to depth; at a given depth in the mantle, the coolest place to be is within a slab. As a result, material in the descending slab will undergo the olivine → wadsleyite transition, a densification reaction, above 410 km, giving the slab additional negative buoyancy that helps it to keep descending. Conversely, the ringwoodite → perovskite + ferropericlase reaction will occur deeper than 660 km in the slab. This causes decreased negative, and perhaps positive, buoyancy, slowing down or arresting the slab's descent.

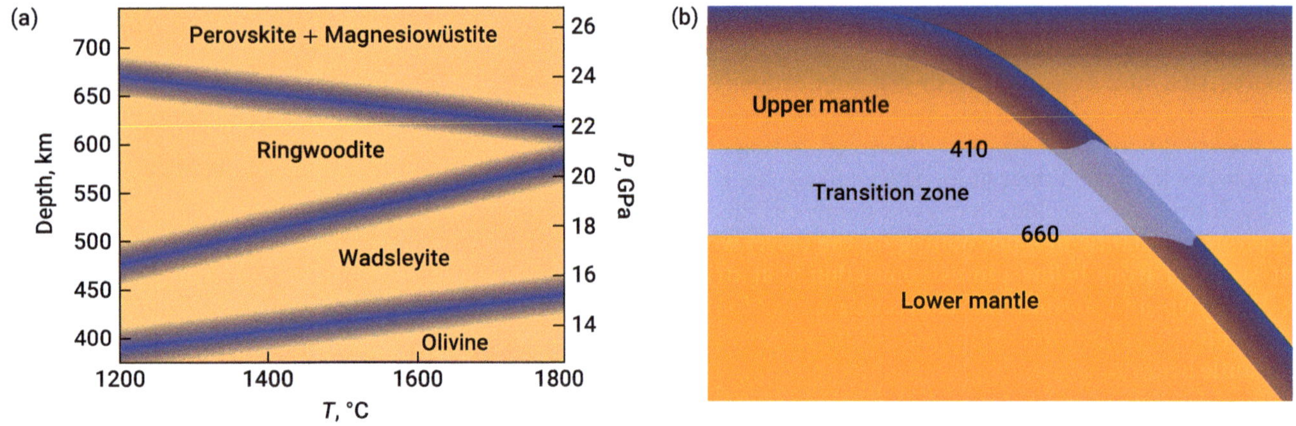

Figure 6.16 (a) Schematic *P-T* diagram of the reactions olivine → wadsleyite, wadsleyite → ringwoodite, and ringwoodite → perovskite + ferropericlase; the first has a positive slope, whereas the last has a negative slope. Along a typical mantle geotherm, these are encountered at 410 and 660 km, respectively. (b) Along the cooler geotherm inside a subducting slab, the olivine → wadsleyite reaction occurs at shallower depth. The subducting slab thus attains additional negative buoyancy, which aids its descent. Conversely, along the slab geotherm, the ringwoodite → perovskite + ferropericlase reaction occurs at greater depth, delaying its densification and opposing further descent.

Seismic tomography offers a method for testing the hypothesis that slabs bottom out at the 660-km discontinuity. Fukao and Obayashi (2013) made tomographic images of arcs using over 10 million observations of earthquake arrival times.[5] Under many arcs, the slab does indeed seem to bottom out at the 660-km discontinuity (Figure 6.17); under some, it seems to pile up between 660 and 1000 km; and in a few, it appears to penetrate even more deeply, into the lower mantle. These results indicate that the delayed density increase and negative Clapeyron slope at 660 km do constitute a barrier to slab sinking, but that in some circumstances, slabs can push through this barrier to greater depth.

A Simple Model of Subduction

Subduction zones are kinematically like thrust faults, and along them, surface area is consumed. If the surface area of the Earth is constant, then this loss of area must be made up by new surface area generated by rifting. As noted in Chapter 3, new oceanic lithosphere is made at oceanic ridges at a rate of about 3.0 km^2/yr; ignoring new area produced by rifting in places such as East Africa and the Basin and Range (see Chapter 13), then subduction must consume a like amount—again, about 1.5 football fields per day worldwide.

Constant surface area of the Earth was not always a given in geology. Both expanding and contracting Earth models were popular in the 19th and 20th centuries, the former to explain rifting and continental drift and the latter to explain fold belts. The expanding Earth hypothesis remained popular into the plate tectonic era, with its chief proponent, Australian geologist S. Warren Carey, writing a damning review of plate tectonic thought in 1975 (Carey, 1975). The expanding Earth hypothesis was put to rest by geodetic (in particular, GPS) data showing undeniable evidence for convergence at subduction zones.

Let's consider a simple 2D planet that consists of only two plates, one subduction zone, and one spreading ridge (Figure 6.18). Surface area (in our case, the length of the surface) is conserved, and the subducting plate sinks vertically downward to rest upon the top of the lower mantle.

Our sinking plate, which we will call the blue plate, because it's special, displaces the upper mantle below it as it sinks, pushing clockwise (to the right at the top of the model). Let's consider the horizontal velocities of the plates and mantle. Relative to the lower mantle (our reference frame), the blue plate has zero velocity. Call the average velocity of the displaced mantle *v*, positive clockwise. The hinge on the blue plate moves clockwise at *v*, and to maintain continuity in the mantle and constant surface area, the mantle and

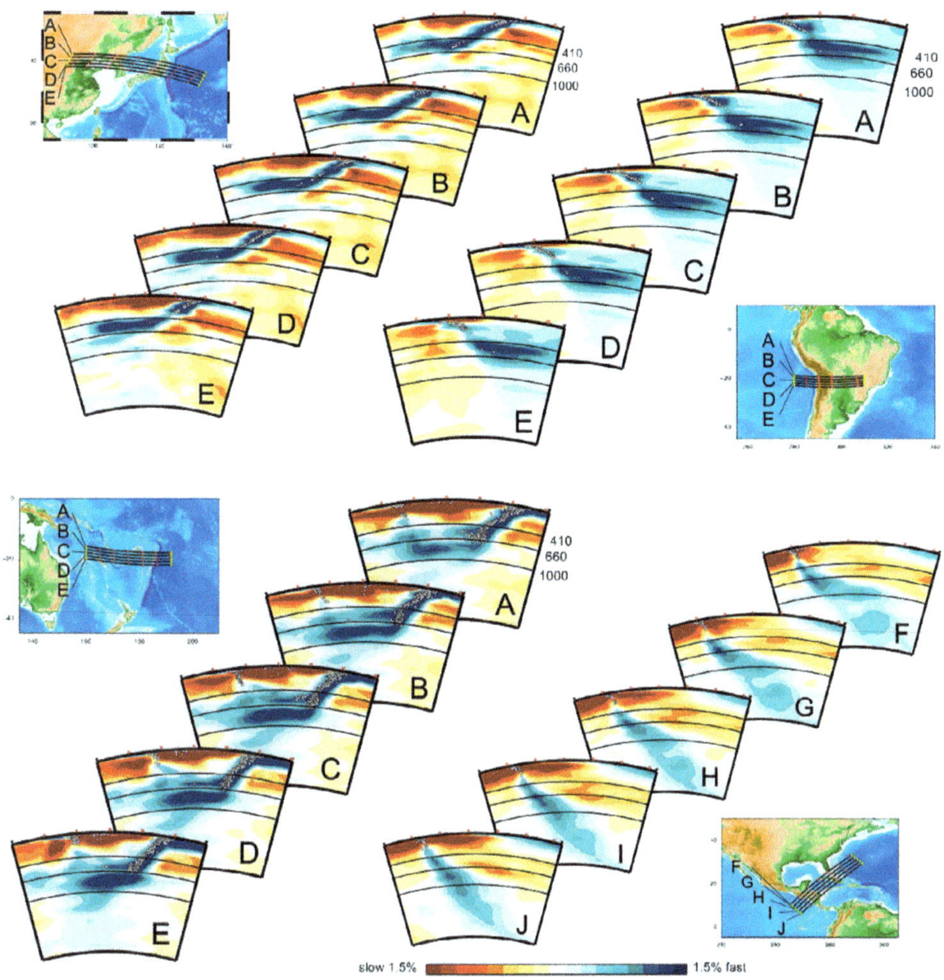

Figure 6.17 Seismic tomography cross sections of selected arcs, showing the path of the subducted slab. In each image, the most saturated red indicates 1.5% slower, and the most saturated blue 1.5% faster, relative to the reference velocity model. The subducted slabs are thus interpreted to be the blue bands. Curves indicate depths of 410, 660, and 1000 km. Under many arcs (e.g., Honshu and the Andes), the slab appears to bottom out at 660 km. Under others (Tonga), it penetrates to about 1000 km, and under some (Central America), it goes deeper, into the lower mantle.

Source: From Fukao & Obayashi (2013).

lithosphere in the hanging wall of the subduction zone must move clockwise at average velocity v as well. A second plate (purple) must follow the retreating hinge, making up the hanging wall of the subduction zone, and this requires a spreading center somewhere between the subduction hinge and the hanging wall plate. The full spreading rate at the ridge must be v (because spreading there balances shortening at the subduction zone), and so the half-spreading rate is $v/2$, and the ridge moves clockwise at $v/2$. Note that spreading at the ridge is driven by mantle displaced by the sinking plate and is a response to sinking.

Absolute Reference Frames

In the simple model of Figure 6.18, there is a clear choice for a reference frame—the lower mantle. Relative to this reference frame, the blue plate has a horizontal velocity of zero, and the ridge migrates with the subduction zone. Thus, the subducting plate in this simple model forms an equivalent reference frame.

On the real Earth, there is no clear choice for an absolute reference frame, given that the plates, upper mantle, outer core, and inner core are all translating and rotating on a spinning planet. There are several reference frames in use. The hotspot reference frame (Figure 6.19a) assumes that the lower mantle is strong, that hotspot plumes rise through the lower mantle from deep in the Earth, and that hotspots therefore are fixed relative to one another. These various assumptions will be examined in Chapter 14. One can always choose a given plate for reference and calculate relative velocities of the other plates. The Antarctic Plate is one such choice; it is bounded along nearly its entire circumference by

Figure 6.18 A simple 2D view of subduction in which plate movement is driven solely by sinking of the plate into the mantle (green) and onto the top of the lower mantle (gray). There are two plates (blue and purple). Sinking of the blue plate pushes mantle clockwise at average velocity *v*. Details of the flow pattern are not shown, but if the top and bottom boundaries are no-slip, then there must be some form of parabolic flow with highest velocities in the middle. Because the subduction hinge is moving clockwise at velocity *v*, there must be a second plate (purple) that is also moving clockwise at *v* and that is separated from the blue plate by a spreading ridge that moves clockwise at *v*/2.

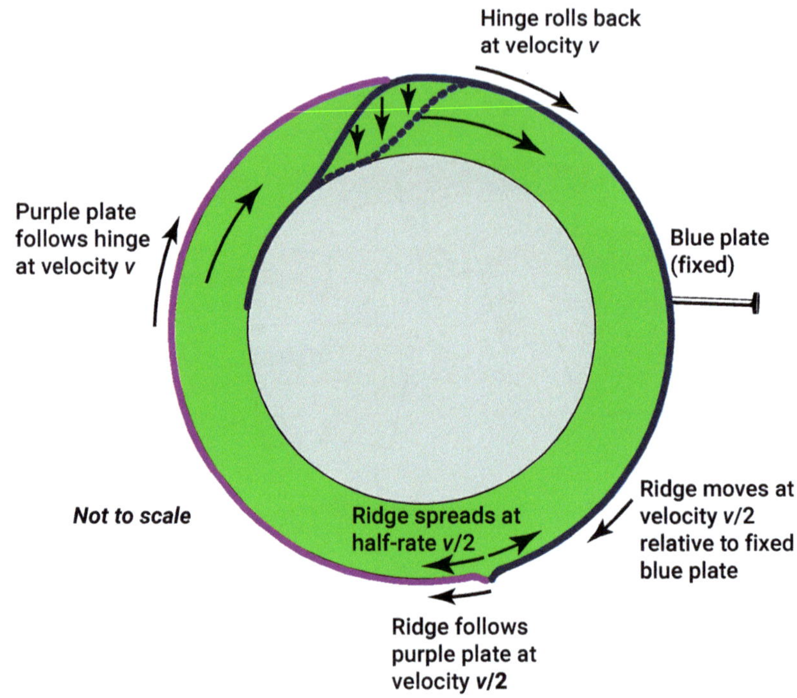

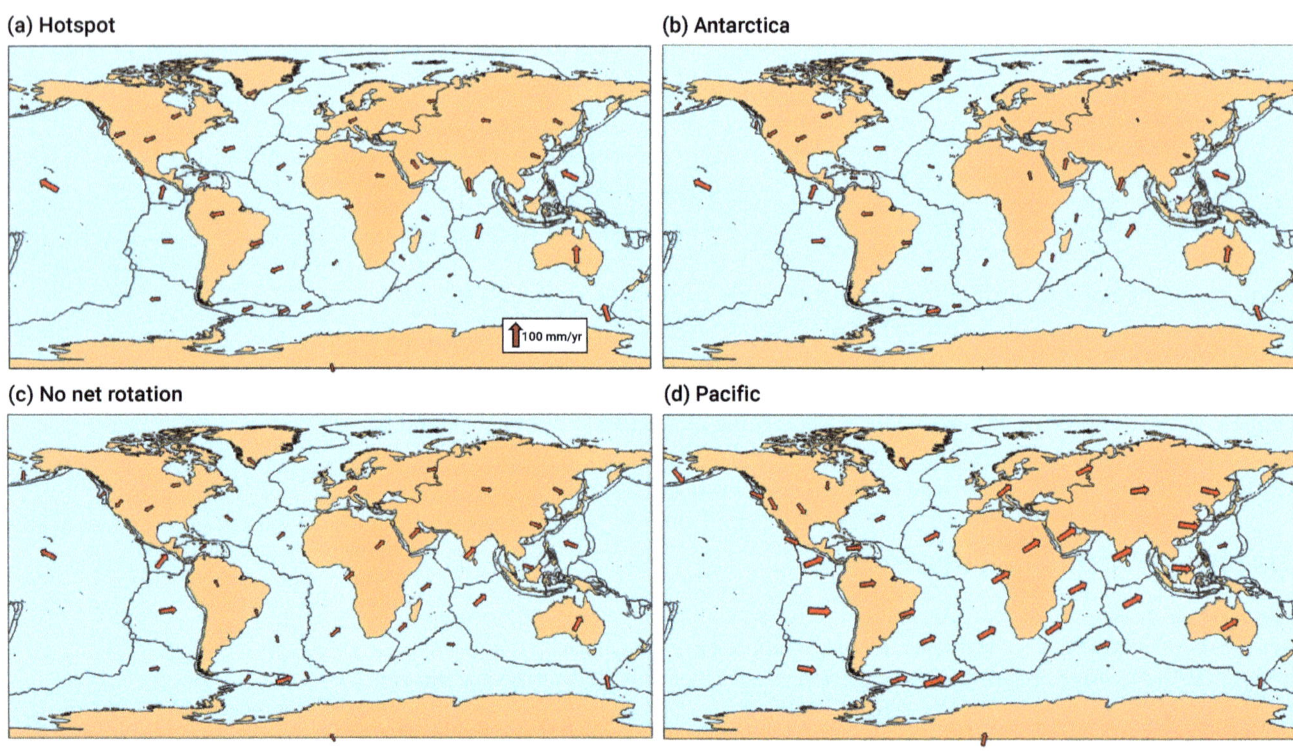

Figure 6.19 Plate motions relative to the (a) hotspot reference frame, (b) Antarctic Plate, (c) no-net-rotation reference frame, and (d) Pacific Plate. Vector length is proportional to velocity and is the same for all maps, which are shown in Mercator projection and thus greatly distorted at high latitudes. The relative motions in each figure are the same; only the reference frame is different. At the present state of knowledge, there is no clear choice among the various reference frames.

Sources: Vectors calculated from the UNAVCO plate motion calculator (MORVEL 2010 model); hotspot model is HS3, GAGE Plate Motion Calculator, EarthScope Consortium.

spreading ridges, and relative to it, other plate motions are relatively slow (Figure 6.19b). The no-net-rotation reference frame (Figure 6.19c) calculates plate motions such that the integral of plate velocities across the surface of the Earth is zero. Figure 6.19d shows that relative to the Pacific Plate, most other plates are moving rapidly to the east.

6.4 Anatomy of Subduction Zones

Learning Objective

By the end of this section you should be able to:

6.4.1 Sketch a cross section of a subduction zone, illustrating the main geomorphic, geologic, and tectonic elements.

Figure 6.20 shows a cross section of an idealized subduction zone where oceanic lithosphere is being subducted beneath continental lithosphere. Important components of this system include:

1. *Subducting oceanic lithosphere*: Lower plate.
2. *Continental lithosphere*: Upper plate.
3. *Magmatic arc*: The *volcanic front* is the trenchward limit of arc-related magmatism.
4. *Accretionary wedge*: Where sediments scraped off the subducting plate pile up.
5. *Trench*: The seaward edge of the accretionary wedge.
6. *Fore-arc basin*: Bounded by the shoreline and the *fore-arc ridge*, which is the outermost piece of the overriding plate, jacked up by accretionary wedge material that has been dragged under it.
7. *Subduction interface*: A region of variable thickness between the plates containing a mixture of accretionary wedge, upper-plate, and lower-plate material. Slip is concentrated here.
8. *Hinge*: The entire curving part of the oceanic plate where the plate rolls from a horizontal orientation to the dip that characterizes the subduction zone.

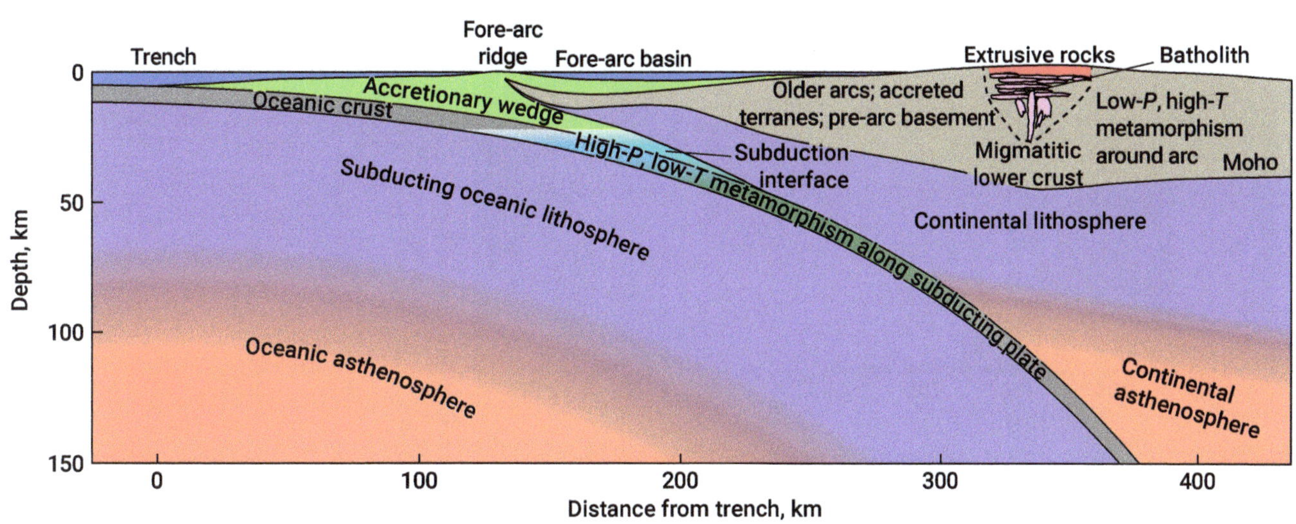

Figure 6.20 Cross section of an idealized ocean-continent subduction zone showing the main structural components.

Source: Modified from Hamilton (2007).

Sorry about all the new terminology, but this diagram encompasses a number of important lithosystems, discussed in this and succeeding chapters.

The Volcanic Front

Volcanic front
The abrupt appearance of volcanoes away from the trench in the direction of subduction.

In most arcs, the area between the trench and the arc is devoid of volcanoes (Gill, 1981). This changes dramatically at the **volcanic front**, where the greatest volumetric flux of magmatism and highest topography occur. In the Cascades, the volcanic front is marked by the chain of tall volcanoes; in the Mariana arc, the corresponding chain just barely peeks ("peaks"?) above the water. In a typical arc, most magmatism occurs in and near the arc, with only scattered volcanism behind the arc unless there is active back-arc spreading. Regions with significant back-arc volcanism include Japan (e.g., Oshima-Ōshima and Ichinomegata volcanoes, northern Honshu) and Argentina (e.g., Cerro Galán caldera).

A simple hypothesis about what leads to the volcanic front is that the relatively cold subducted slab descends until it gets hot enough to melt. This hypothesis makes several predictions about correlations between the position of the volcanic front and subduction parameters. In particular, (1) younger, warmer oceanic lithosphere should melt at shallower depth than old lithosphere, and (2) faster subduction should lead to deeper melting than slower subduction (although friction offsets this). These are reasonable predictions, but they are not met by observations, and in fact, slab melting seems to be rare and is not a general explanation for arc volcanism.

Active arc volcanoes are typically situated between about 70 and 200 km above the Wadati-Benioff zone, with a strong mode at around 100 km (Figure 6.21). This depth is not significantly correlated with the rate of plate convergence, angle of subduction, age of the subducted plate, or any other simple parameter that might be expected to affect heating of the subducted slab. Models of the temperature distribution in subducted slabs (discussed later) show that there should be wide variations in temperature at a given depth depending on convergence velocity, slab dip angle, and other parameters, and yet the depth to the seismic zone is relatively limited.

One solid link between a subduction parameter and position of the volcanic front is that if the subducted slab is shallow enough, there is no volcanic front—volcanism shuts off. A clear demonstration of this is in the Andes, where three arc segments with abundant

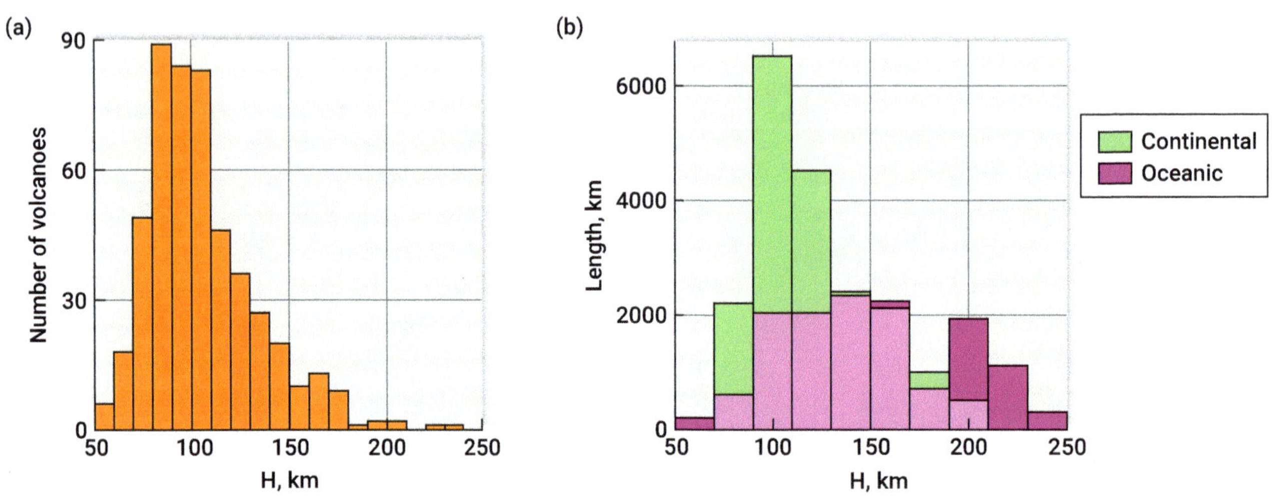

Figure 6.21 Global compilations of the depth to the Wadati-Benioff zone beneath arcs. (a) Histogram of number of volcanoes. There is a significant mode between 80 and 110 km. (b) Histogram of total arc length divided into continental and intraoceanic arcs. In this compilation, there is a significant mode at ~100 km depth for continental arcs, but no particular mode for intraoceanic arcs. The mode for continental arcs is largely defined by volcanoes in Sumatra and the southern Andes, which lie in each case along a major trench-parallel fault.

Sources: (a) From Syracuse & Abers (2006), © 2006 by the American Geophysical Union; (b) From Schmidt & Poli (2014).

active volcanoes are separated by two zones where active volcanoes are absent (Figure 6.22). Barazangi and Isacks (1976) correlated these gaps in the arc with anomalously shallow dips of the Wadati-Benioff zones that lie under them and reasoned that if the subducted slab does not reach the critical depth of about 100 km, then magma is not generated. This empirical observation lent new urgency to the question of what happens at that critical depth.

The Subduction Interface

The subduction interface is the slip region at relatively shallow levels (<~80 km) between the subducting and overriding plates. Exactly what the interface looks like is not entirely clear, as methods to investigate it are highly indirect. Although some rocks are expelled from subduction zones as thin thrust slices and blocks in mélange, they represent less than 1 percent of all subducted material. How representative are they? Geophysics tells us that the interface is highly anisotropic, meaning it is probably highly sheared (maybe not surprising, considering the tectonic setting). And thermal-mechanical models predict the interface consists of sliced off pieces of oceanic sediment, upper oceanic crust, bits of overriding mantle, and serpentinite, similar to mélange, but perhaps with larger slices of rock intermingled. In general, these converge on a consistent picture of a highly hydrated, sheared mass of blocks and sheets of rock that steadily grows in size as a subduction zone matures. The sheets and mélange can become stacked and stored at depth until some event, such as the attempted subduction of a larger mass like a seamount, arc, or continent, expels them. Like the mantle wedge, the subduction interface is broadly triangular in cross section, and material within it circulates. Circulation represents one mechanism for lifting dense blocks of rock from deeper in the subduction system closer to the surface (Cloos, 1982). It also implies that rocks could experience multiple cycles of increasing then decreasing P ("washing machine" behavior), which has been inferred from chemical zoning in garnets from some HP/LT rocks (e.g., Garcia-Casco et al., 2002).

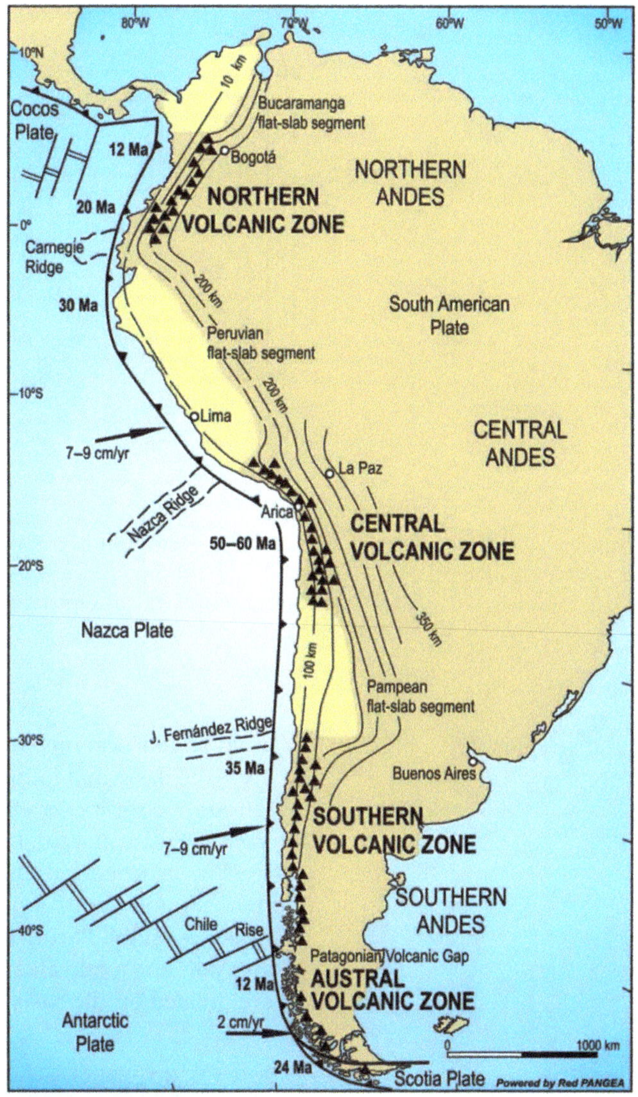

Figure 6.22 Active volcanoes in the Andes with isobaths (contours of constant depth) of the Wadati-Benioff zone (WBZ). The northern, central, and southern volcanic zones of the Andes are separated by two areas that lack active volcanism. Under these areas, the WBZ is gently dipping for several hundred kilometers inland of the trench before assuming a more typical dip. Note that there is no active arc far inland in these areas even though the WBZ, and by inference the slab, descends to several hundred kilometers depth there.

Source: From Ramos & Folguera (2009).

Cross-Arc Trends

Arcs are inherently asymmetrical. One plate overrides another, and the arc is arranged with the blueschist lithosystem and high-P/low-T metamorphism closest to the trench and the arc lithosystem and low-P/high-T metamorphism a few hundred kilometers inland (see Figures 6.11 and 6.20).

This asymmetry is manifested in several ways in arc magmatism, one obvious example being the distribution of volcanoes. In a given arc, there are no volcanoes between the volcanic front and the trench, a high concentration of volcanoes near the volcanic front, and a gradual die-off of volcanism into the back-arc (Figure 6.23). In the active Andes and much of the Cordilleran arc system, isolated volcanoes reach about 500 km inland from the trench. In the early and middle Cenozoic, however, arc-type volcanoes that erupted while subduction was occurring along the western margin of North America were found

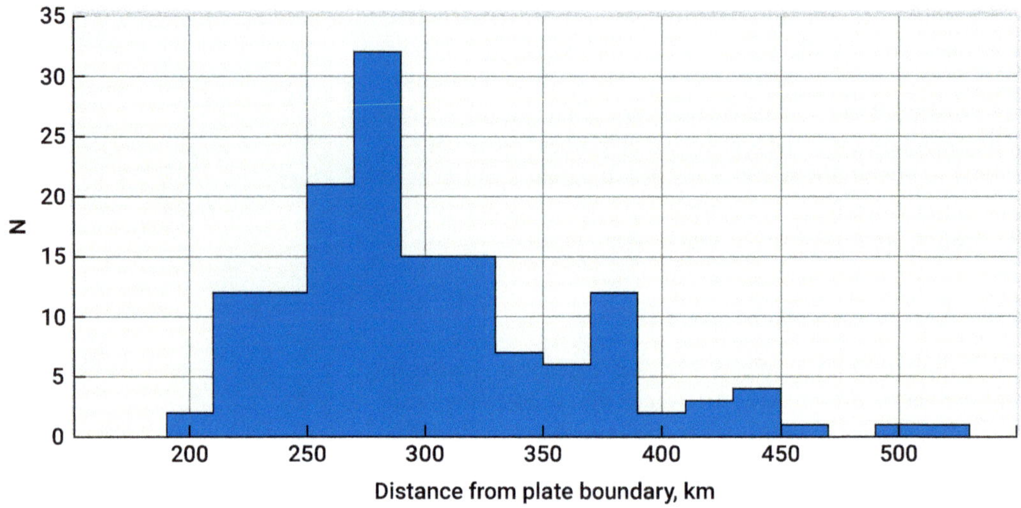

Figure 6.23 Histogram of distances between active volcanoes of the Andes and the plate boundary (trench). The distribution is right-skewed (positively skewed), with no volcanoes trenchward of the volcanic front in a given area, an abundance near the front, and a rapid dribbling off to isolated volcanism in the back-arc.

Source: Volcano list from the Smithsonian Global Volcanism Program.

as far as 1500 km inboard of the plate boundary (Lipman et al., 1972). A big chunk of this area (the Basin and Range) has been widened several hundred kilometers by Cenozoic extension, but restoring extension still leaves many of these volcanic centers over 1000 km from the plate margin, unlike any arcs in the modern world (Figure 6.24).

The geochemical character of arc volcanics also correlates with distance from the trench. Kuno (1959) divided Cenozoic volcanic rocks in the Japanese islands into a tholeiite series on the Pacific (eastern) side, an alkaline series on the Japan Sea (western, back-arc) side, and a calc-alkaline series that straddles the other two (Figure 6.25a). These series are divided on the basis of composition (e.g., via the basalt tetrahedron, see Figure 4.15;

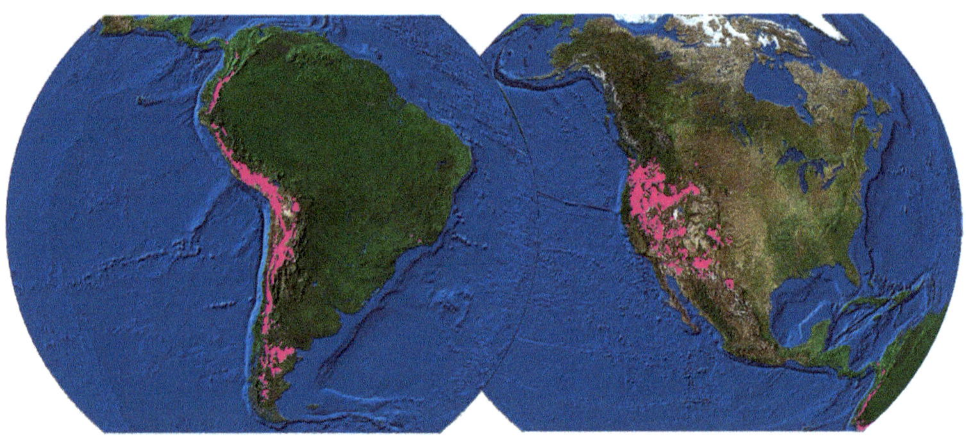

Figure 6.24 Distribution of Cenozoic igneous rocks in South America and the conterminous United States. Although such rocks reach inland from the plate boundary only about 500 km in the Andes, they extend nearly 1500 km inland in western North America—over 1000 km after Basin and Range extension is removed. Although some of the North American rocks clearly owe their origins to processes other than subduction (e.g., the Yellowstone-Snake River Plain hotspot track; see Chapter 14), most resemble subduction-related rocks petrologically and geochemically and were formed while subduction was occurring along the western margin of North America (Lipman et al., 1972). Western North America seems unique in how far inland subduction-related magmatism occurred.

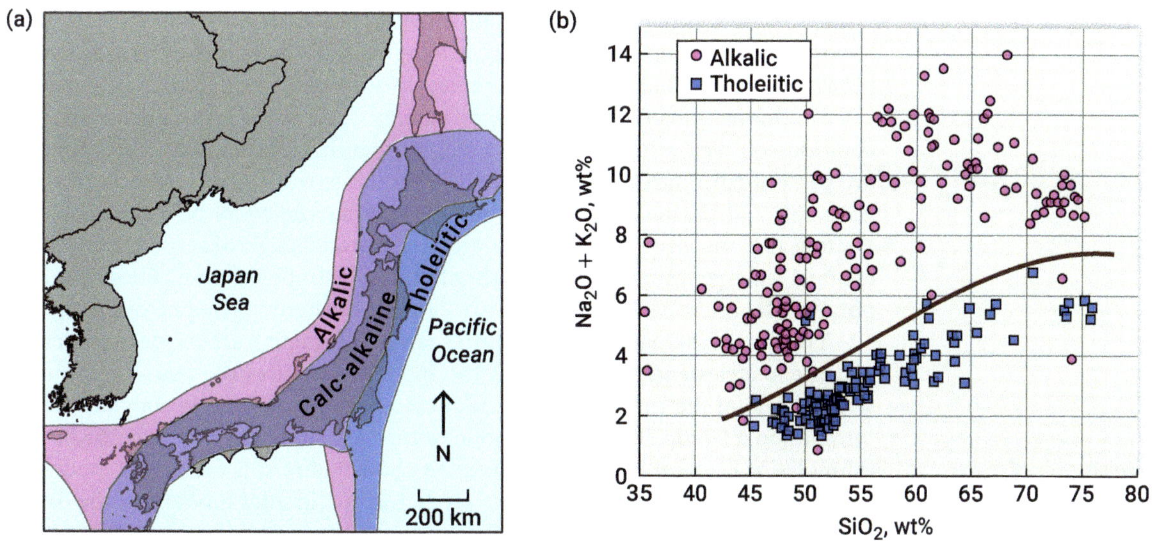

Figure 6.25 (a) Map of geochemically defined Cenozoic volcanic provinces in Japan. Kuno noted that tholeiitic (low-alkali) lavas are erupted on the Pacific Ocean side of Japan and alkalic lavas on the Japan Sea side. In an insightful leap, he correlated increasing alkalinity with increasing depth to the Wadati-Benioff zone, well before the concept of a subduction zone was conceived. (b) Plot showing that rocks on the inland (what we would now refer to as back-arc) side of the volcanic belt are significantly more alkalic than those on the ocean side. Rocks of the calc-alkaline series lie geographically and chemically between the tholeiitic and alkalic series.

Source: (a) From Kuno (1959), © 1959 Stabilimento Tipografico Francesco Giannini & Figli.

higher alkalis at a given value of silica; Figure 6.25b) and details of their mafic mineral assemblages. Kuno noted a consistent pattern of more alkalic rocks toward the Japan Sea and, in a bit of prescient thinking that anticipated plate tectonics, noted that the more alkalic rocks occur above the deeper parts of the Wadati-Benioff zone. He proposed that alkali basalts and other members of the alkalic suite were generated by deeper melting of the mantle than rocks of the tholeiitic series—an interpretation that holds true today.

6.5 Thermal Structure of Subduction Zones

Learning Objectives

By the end of this section you should be able to:

6.5.1 Sketch a cross section of a subduction zone, illustrating key changes in rock strength that affect its tectonic behavior.

6.5.2 Explain why the presence of coesite and diamond in collision zones requires subduction of crustal material to depths of 100 km or more.

Because we cannot observe temperatures in a subduction zone directly, they must be inferred from mineral assemblages and geophysical modeling. Mineral assemblages and compositions can provide relatively tight estimates of the *P-T* conditions in a subduction zone. However, remember that few rocks are ever expelled from a subduction zone. This raises questions about how representative these oddballs are of "typical" subduction *P-T* conditions. So, to get a fuller picture of the range of thermal structures that might characterize subduction zones, geophysicists and petrologists turn to numerical models of heat flow in a subduction system.

The driver of low temperatures in a subduction zone is the lower plate, which, as noted earlier, descends fast enough to chill the surrounding mantle. The essential geometry is that of a reverse fault, where the lower plate, whose upper surface starts out near 0 °C, is inserted into the mantle, where it refrigerates the hanging wall. Two types of models to describe this behavior are common: thermal-mechanical models and thermal-kinematic models.

In a thermal-mechanical model, parameters such as rock strengths, convergence rates, conditions of key mineral reactions (e.g., serpentine breakdown), and numerous others are prescribed in advance, and then the model is run until it evolves to a steady-state geometry and thermal structure. Features such as the coupling point between the subducting slab and circulating mantle, and the size and shape of serpentine-rich subduction channel, emerge from the model rather than being imposed on it. These models are computationally intensive, and a single run can require hours on supercomputers.

A thermal-kinematic model requires fewer parameters and is more prescriptive about plate geometries and motions (e.g., the modeler specifies the coupling point and shear strength along the interface). Such models are computationally simpler and can run in minutes on a laptop. Two typical model setups are shown in Figure 6.26a,b, with specific examples in Figure 6.26c–e. These models show that results are highly model-dependent, and the range in thermal conditions predicted by different models can be dramatic (Figures 6.26 and 6.27).

Despite numerous differences among models, Figures 6.26 and 6.27 illustrate many common features of such thermal models:

Coupling depth
The point along the top of the subducting plate where it binds with the overriding mantle, driving circulation of the mantle wedge. Dehydration of weak serpentine along the interface to form strong olivine is thought to control this depth.

Mantle wedge
A wedge of mantle toward the arc-side of a subduction zone with sufficiently high temperature in its core to flow readily.

- At shallow levels, the two plates slip past each other (often generating large earthquakes), but at greater depths, they bind and move together. The depth where the two plates mechanically couple (labeled the mechanical **coupling depth**) is thought to be about 80 km and to be controlled by dehydration of serpentine (Kerswell et al., 2021).
- Mechanical coupling drives flow in the **mantle wedge** in the hanging wall that brings a hot corner of convecting asthenospheric mantle under the volcanic front. Models that do not bring this hot material under the volcanic arc are generally viewed as failures.

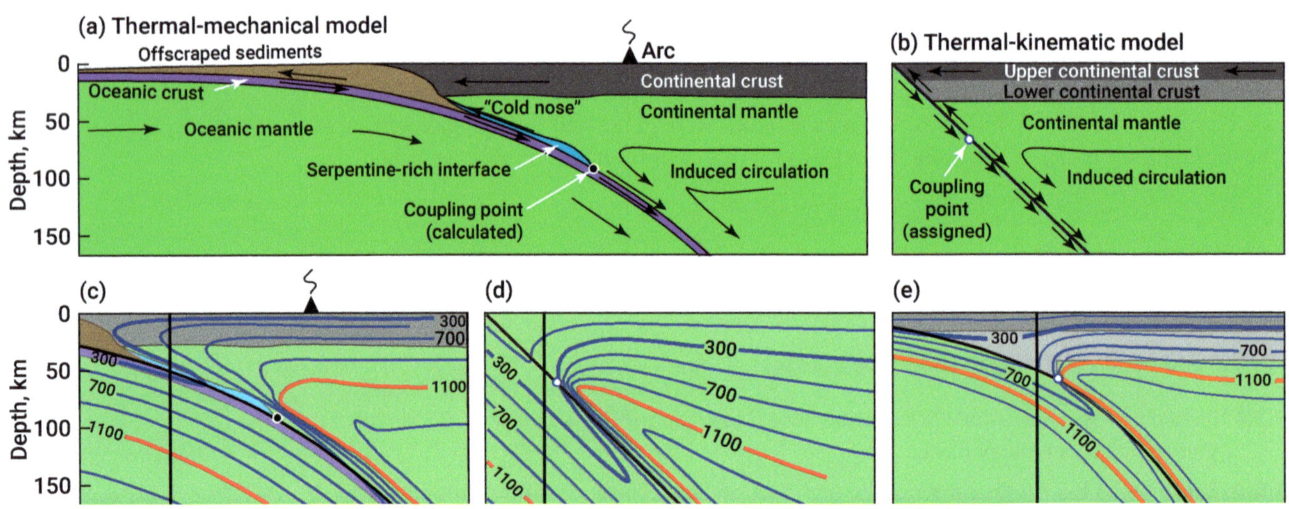

Figure 6.26 Examples of thermal models of the subduction of an oceanic plate beneath a continental plate to illustrate similarities and differences. (a) Thermal-mechanical model. Sediments and serpentine lubricate the subduction interface at shallow levels, allowing differential slip along the interface. The cool tip of the mantle wedge is sometimes called the "cold nose." Breakdown of serpentine to olivine, at inferred depths of ~80 km, mechanically couples the mantle to the subducting slab such that they move together, inducing circulation in the mantle wedge. The coupling point also creates a choke point, beyond which few rocks are recovered. (b) A simple thermal-kinematic model that specifies plate motion rates, interface geometry, and coupling point. (c–e) Isotherms (black lines) for three different models: (c) a thermal-mechanical model; (d) a simple thermal-kinematic model; (e) a thermal-kinematic model for the northern Cascades. Note stark differences in distributions of isotherms—for example, the geometry and location of the 300 °C and 1100 °C isotherms in the mantle wedge. Vertical lines intersect the subduction interface at 50 km depth and mark location of temperature profiles shown in Figure 6.27a.

Sources: (a, c) After Kerswell et al. (2021); (b, d) after England & Katz (2010); (e) after van Keken et al. (2002).

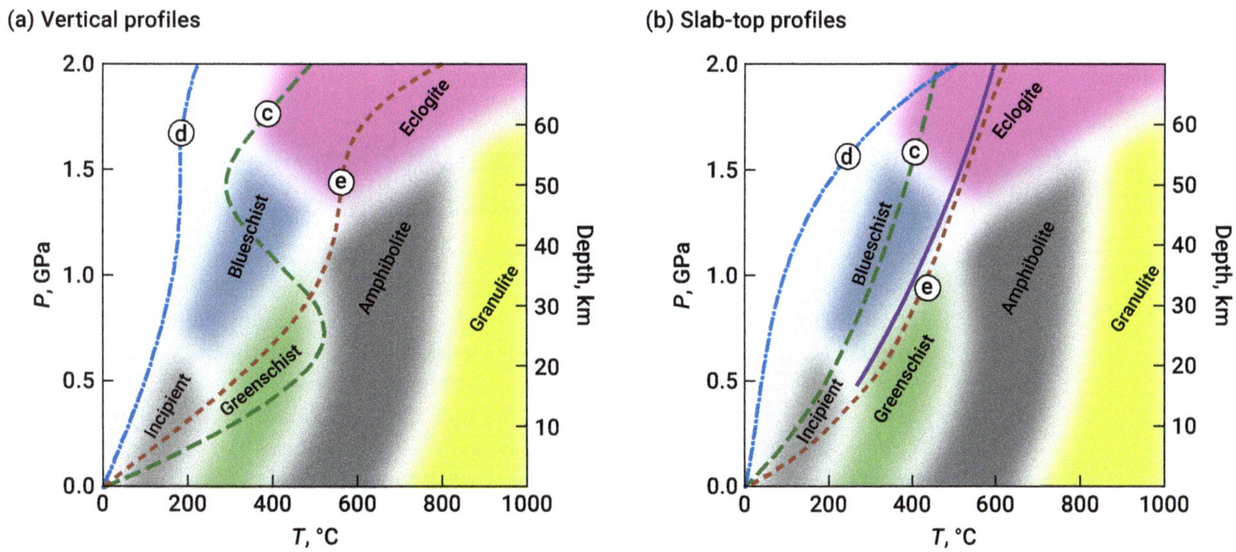

Figure 6.27 (a) Three different vertical profiles through thermal models of subduction zones (vertical lines in Figure 6.26c–e), showing the wide range of predicted temperatures. Temperature differences below ~5 km mainly reflect the assumed temperature of the subducting slab and the effects of friction. Key: Green curve with long dashes is Figure 6.26c, Kerswell et al. (2021); blue curve with dash-dot-dash is Figure 6.26d, England & Katz (2010); red curve with short dashes is Figure 6.26e, van Keken et al. (2002). (b) Slab-top geotherms (temperature along the subducting slab interface) corresponding with the thermal models of subduction zones illustrated in Figure 6.26c–e, showing wide range of predicted temperatures. The metamorphic rock record of *P-T* conditions in subduction zones (Penniston-Dorland et al., 2015) is also shown for reference (purple solid curve).

- The subducting slab carries relatively low temperatures with it owing to the high Péclet number (high velocity relative to heat diffusion). In cooler models, temperatures as low as 300 °C are encountered at depths exceeding 100 km; however, in warmer models, the 300 °C isotherm is dragged down only to about 25 km. Which model is closer to the truth? Could they both be right for different subduction zones?
- Near the hot mantle wedge, thermal gradients are extremely high—some models predict that mantle at 1300 °C in the wedge is only 10 km above subducting crust and mantle that are hundreds of degrees cooler. Subduction zones produce large thermal gradients.
- The subducted slab can develop a negative geothermal gradient such that temperatures first increase downward (in the overriding plate; this is normal), then decrease downward (into the top of the subducting plate; this is unusual), and then, finally, increase again (deeper in the subducting plate; see Figure 6.27). The magnitude of the thermal inversion varies widely among models and depends on the location of the profile.
- Temperatures along the top of the subducting plate are relatively low, but proximity to the hot mantle raises them to about 800 °C at depths below the coupling point (typically ~80 km, but model dependent).
- Temperatures at the Moho (base of the crust) in the subducted slab remain low to depths greater than 200 km.

Most thermal-mechanical models predict much lower temperatures along the slab top than what rocks record, more like 4 to 7 °C/km, rather than 7 to 10 °C/km (Penniston-Dorland et al., 2015). The "hot rocks, cold models" paradox could represent deficiencies in the models, especially how we model interface friction (Penniston-Dorland et al., 2015; Kohn et al., 2018), or it could be that the processes that cause rocks to detach and rise from the subducting slab-induced thermal anomalies. For example, the roughness of a seamount might increase its frictional resistance to shearing, increasing temperature (e.g., Gao & Wang, 2014), and simultaneously enhance its likelihood of detaching and being exhumed.

Ultrahigh-Pressure Metamorphism—How Deep Can Surface Material Be Subducted and Returned?

Up until the mid-1980s, geologists assumed that continental crust could not be subducted below about 70 km depth. That was about the maximum *P* that metamorphosed crustal rocks were known to record (roughly 2 GPa) and about the maximum thickness of continental crust known at the time (although thicker crust has since been found). That made sense, because felsic crust and sediments are far less dense than mantle, so buoyancy should prevent deeper subduction into the mantle, right? After all, that's why (felsic) Archean rocks persist on every continent. However, that paradigm began to fall apart in 1984 with the discovery of continental metamorphic rocks containing coesite in the Alps (Chopin, 1984) and Caledonides (Smith, 1984). Coesite is stable only at $P > 2.7$ GPa, or greater than 90 km depth, for temperatures of 600 °C or higher (see Figure 3.13). Such unusual rocks were labeled "UHP" for ultrahigh-pressure. Discovery of diamond-bearing crustal rocks in Kazakhstan (Sobolev & Shatsky, 1990) showed that some rocks reach depths exceeding 150 km, yet somehow return to the surface (Gilotti, 2013; Hermann & Rubatto, 2014). These remarkable rocks testify that although buoyancy matters in subduction, clearly it is not the only control on how deep something can be subducted.

What is so intriguing about UHP rocks is not simply that they reach such great depths, but that they manage to come back out. Only a tiny fraction of eclogites are exhumed from $P > 2.5$ GPa (Penniston-Dorland et al., 2015; Kerswell et al., 2021). The corresponding depth in thermal-mechanical models represents both the coupling depth and a choke point along the subduction interface. So, it is relatively easy to envision how shallower rocks return—the choke point creates a countercurrent that flows back up the interface. But if a choke point occurs, how do deeper UHP rocks squeeze past it? Tectonics probably plays an important role, because nearly all UHP metamorphism is associated with the early stages of continental collision. Still, how UHP rocks manage to return to the surface remains an area of active debate.

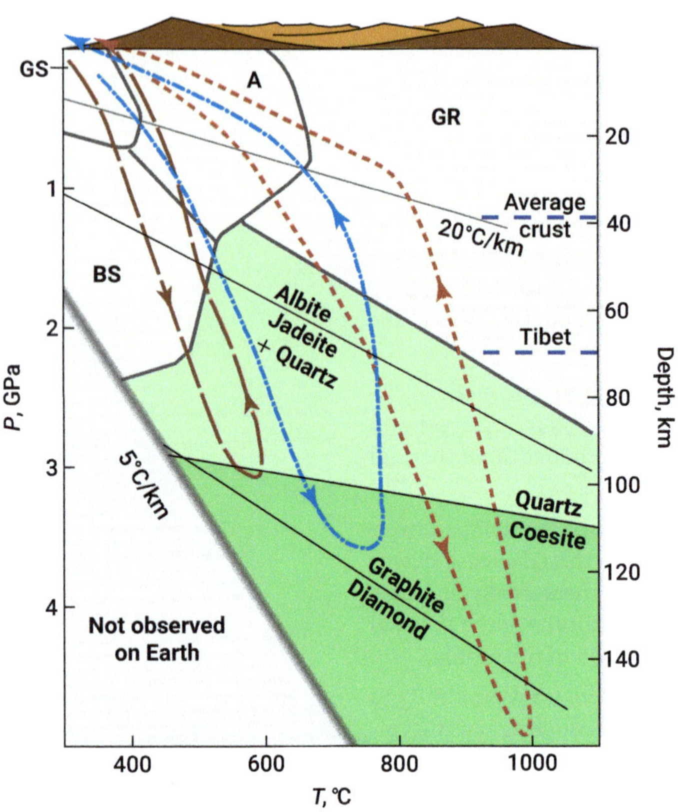

Figure 6.28 *P-T* paths of typical UHP rocks, based on observations on diverse UHP terranes worldwide. The eclogite facies is divided into high pressure (light shading) and ultrahigh-pressure (dark shading) regions. Facies: A = amphibolite; BS = blueschist; GR = granulite; GS = greenschist. "Not observed on Earth" represents *P-T* region from which metamorphic rocks are not recovered. Key reactions, geotherms, and crustal thicknesses are shown.

Source: From Gilotti (2013), with permission of Jane A. Gilotti.

UHP *P-T* Paths

P-T paths have been determined for numerous UHP terranes, and most are hairpin-shaped (Figure 6.28). *P-T* paths with this shape differ from those produced in thermal-mechanical models, except in cases where friction along the subduction interface is significant. Exhumation occurred either nearly isothermally or with cooling, forming hairpin *P-T* paths. Deep in subduction zones, *T* increases upward above the subduction interface. So, these kinds of *P-T* paths suggest that rocks can be exhumed along the subduction interface by mechanisms that we do not fully understand. Most UHP rocks represent slices of continental crust from thinned continental margins that were subducted and exhumed during the initial stages of collision. Although continents eventually jam subduction zones, UHP metamorphism may be a common precursor to collisional metamorphism, which is discussed in Chapter 12.

6.6 Melting in the Presence of Water

Learning Objectives

By the end of this section you should be able to:

6.6.1 Distinguish among dehydration, water-present melting, and dehydration-melting reactions.

6.6.2 Use the Clapeyron equation and an understanding of volumes and compressibilities of fluids to predict the change in slope of dehydration, water-present melting, and dehydration-melting reactions in *P-T* space.

Depression of Melting Temperatures

As we saw in Chapter 5, metamorphism at and near mid-ocean ridges fixes a lot of water into basalt, turning anhydrous minerals such as olivine, plagioclase, and clinopyroxene into serpentine, chlorite, amphibole, talc, and other hydrous minerals. The water in these phases is carried into the mantle by the subducted slab, where it promotes melting, even though subducted slabs are much cooler than ambient mantle at a given depth.

Dissolved water has a profound effect on melting in silicate systems, because it greatly depresses the temperature of melting. Figure 6.29a shows the magnitude of this effect. At 1 atm, in the absence of water, albite melts at about 1100 °C,[6] but at 1 GPa with water present, it melts at less than 700 °C. Any anhydrous silicate will show the same magnitude of melting-point depression. Recall that cristobalite melts at 1713 °C at 1 atm (see Figure 3.13); at 1 GPa in the presence of water, this is depressed to about 1100 °C, an even greater melting-point depression than that for albite.

It is worth the effort to tease out what these diagrams have to say, because water is of supreme importance in igneous petrology. Consider a platinum capsule with two components, water (W) and albite (Ab), in it. Our capsule is flexible such that *P* is controlled by whatever *P* we subject it to; in the lab, $P = 1$ atm. Ignoring any possible hydrous phases that could form in this system, we can assume that this is a binary eutectic. What might the phase diagram look like?

Let's put our capsule in a cold-seal furnace and examine what we get at 100 MPa, corresponding to a depth of 3 or 4 km in the crust. As we are well above the critical point of water, the vapor phase is supercritical. Our experiments will show that melting of Ab, which is the reaction V + Ab = M, where V is vapor and M is melt, occurs at much lower *T* than for melting of Ab at 1 atm. The phase diagram looks like the 100 MPa isobaric section in Figure 6.29c. Ab + V (remember that "vapor" here means supercritical H_2O with a little Ab dissolved in it) melts at about 880 °C, and the eutectic liquid has about 4.5 wt% W. If we continue to raise *T*, what we get depends on the water content. For W < 4.5 wt%, our capsule will contain Ab + M until *T* is raised to the liquidus *T* for that bulk composition (e.g., ~1000 °C for 2 wt% W). Above the liquidus, of course, Ab is no longer present, and the system consists of just melt M. For higher W contents, all the Ab melts out at the eutectic, and the system consists of M + V unless the system is extremely rich in W.

If we do another set of experiments at 500 MPa, melting occurs at even lower *T*, and the amount of W in the eutectic melt is higher (Figure 6.29b). The binary drops to ever-decreasing *T* as *P* is increased, although the effect becomes less and less pronounced, until by about 1 GPa there is little further decrease with increasing *P*.

The solidus sweeps out a curve (the water-present or "wet" solidus) on a *P-T* diagram (see Figure 6.29a) that defines the onset of melting when water is present. At what temperature the last crystals are melted (the liquidus) depends on the water content. At a given

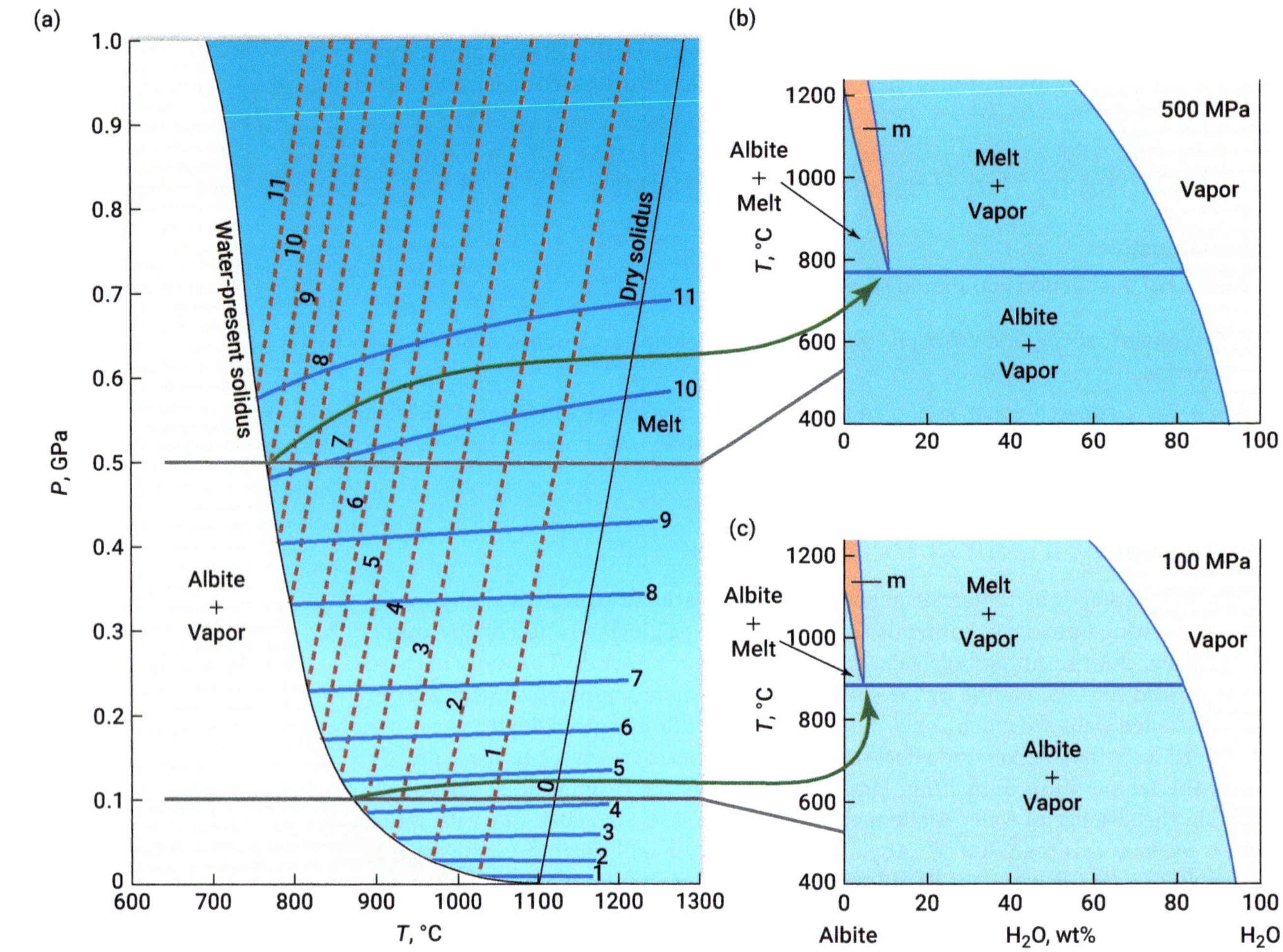

Figure 6.29 Phase equilibria in the system albite-H_2O. Panel (a) shows a *P-T* projection of the system and (b) and (c) show binary phase diagrams at two pressures (0.1 and 0.5 GPa). The *P-T* diagram is composed of an infinite number of binary *T-X* diagrams like (b) and (c), stacked so that you are looking down the *X* (composition) axis. The solidus in (a) traces out the temperature of the point at the bottom of the melt (m) field in the Ab-H_2O diagrams as a function of pressure. Steep red-dashed lines show the liquidus for different water contents (in wt%). Flat blue lines contour melt composition in weight percent water.

Source: Based on Burnham & Davis (1974).

P, for water contents greater than that of the eutectic, Ab is used up at the eutectic, and thus the solidus and liquidus are the same; this is water-saturated melting. For water contents less than that of the eutectic, the liquidus is somewhat higher. In Figure 6.29a, the red dashed contours on the melting surface give the liquidus for the corresponding water content.

To summarize, for albite-present bulk compositions, as long as the water content is greater than that at the eutectic, the solidus and liquidus are the same. For bulk compositions with less water than the eutectic (water-undersaturated melting), the situation is different. Melting still begins at the eutectic temperature, but V is used up before Ab. For such bulk compositions, the liquidus is at a higher temperature than the eutectic, rising higher and higher as the water content is decreased.

An important aspect of water solubility shown in Figure 6.29 is that the water content of albite melt (and other silicate melts) is primarily a function of *P*; **isopleths** of water content in the melt are nearly flat in *P-T* space.

Isopleths
Contours of constant value; most isopleths are for composition, but they can also be for modal abundance, seismic properties, etc.

Clapeyron Analysis

The rapid decrease in temperature of the water-present melting curve in Figure 6.29 happens because ΔV for the reaction Ab + V = M is strongly negative, giving the curve a flat negative slope. Water in the vapor phase has a large molar volume, but its volume when dissolved in the melt is significantly less. Figure 6.30 shows the rapid decrease in the specific volume (volume per unit mass) of water along the water-saturated solidus of Figure 6.29, along with its molar volume when dissolved in a silicate melt. With increasing *P*, the specific volume of the vapor phase approaches the specific molar volume of water in the melt, making the denominator of the Clapeyron equation small and the slope of the *P*-*T* curve steep.

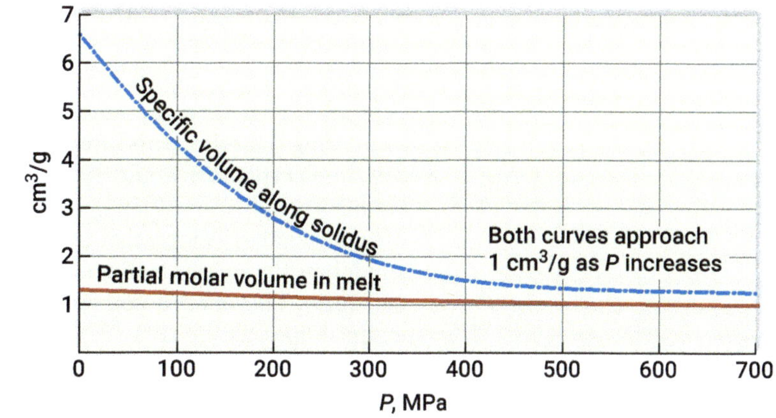

Figure 6.30 Specific volume of water at *P*-*T* conditions along the water-saturated solidus of Figure 6.29. Note the rapid decrease with *P* approaching values for the specific volume of water in silicate liquids, ~1.3 cm^3/g.

Sources: Data from Ochs & Lange (1997); specific volumes from Burnham et al. (1969).

Effect of Negative Slope on Magma Ascent

It was long assumed that the negative slope of the water-present solidus would prevent ascent of water-saturated magmas, because upon rising, they would pass into the subsolidus field (Cann, 1970; dotted path *A* in Figure 6.31). This inference was based on the

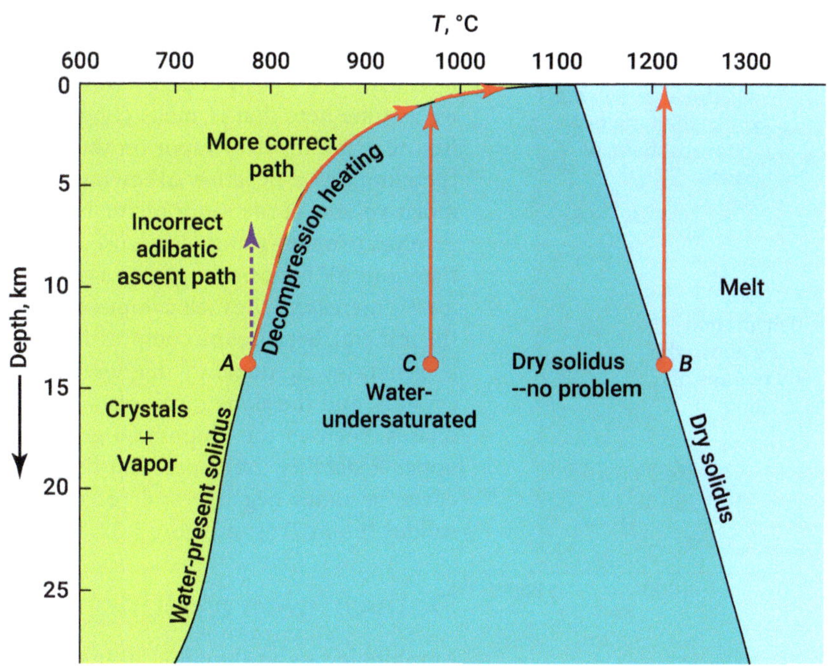

Figure 6.31 If water-saturated melt that was generated along the solidus (*A*) were to rise isothermally (dotted path), it would crystallize as it passed into the subsolidus field. It was once thought that this process would freeze the magmas in place. However, release of latent heat of crystallization heats the magma, which will follow the solidus as it ascends. Water-undersaturated melt at higher *T* (*C*) will start to crystallize upon reaching the solidus and also then heat up and follow the solidus to shallower depth. Dry melts simply ascend through the superliquidus field as do MORB magmas (*B*).

incorrect assumption that an adiabatic ascent path for such a magma parcel would be essentially isothermal, with analogy to magma rising beneath oceanic ridges (see Section 3.7; path *C* in Figure 6.31). By this logic, a water-undersaturated magma at a temperature higher than the water-present solidus could rise to shallower depth before encountering the solidus and freezing (path *C* in Figure 6.31). Dry magma rising from the dry solidus (path *B*) would encounter no barriers to ascent. These first two predictions are incorrect, because they ignore latent heat that is released by crystallization. Thermodynamic analysis shows that magma rising from along the solidus will slowly crystallize and follow the solidus (solid path *A*), heating dramatically as latent heat is released (Glazner, 2019). This effect accounts for the temperature increase experienced by many silicic magmas upon ascent (Blundy et al., 2006).

Melting of Hydrous Phases

Melting behavior of water-bearing minerals is different from that of anhydrous ones. A generic example is given in Figure 6.32. At low *P*, the hydrous phase (H) decomposes via a dehydration reaction such as those we studied in Figure 5.7 (e.g., talc = enstatite + quartz + water vapor). Dehydration produces an anhydrous assemblage (call it A) and liberates water as a vapor phase (V). No melting occurs at the dehydration curve, because the system is at a *T* below the wet solidus, but if the system is heated to the wet solidus, then melting occurs. However, at relatively low *P*, the decomposition curve of H, which has a positive slope, intersects the wet solidus at point 1. Now the breakdown (melting) *T* of H is higher than the wet solidus, so no melting occurs until the system is heated to the melting *T* of H, which can be significantly above the wet solidus.

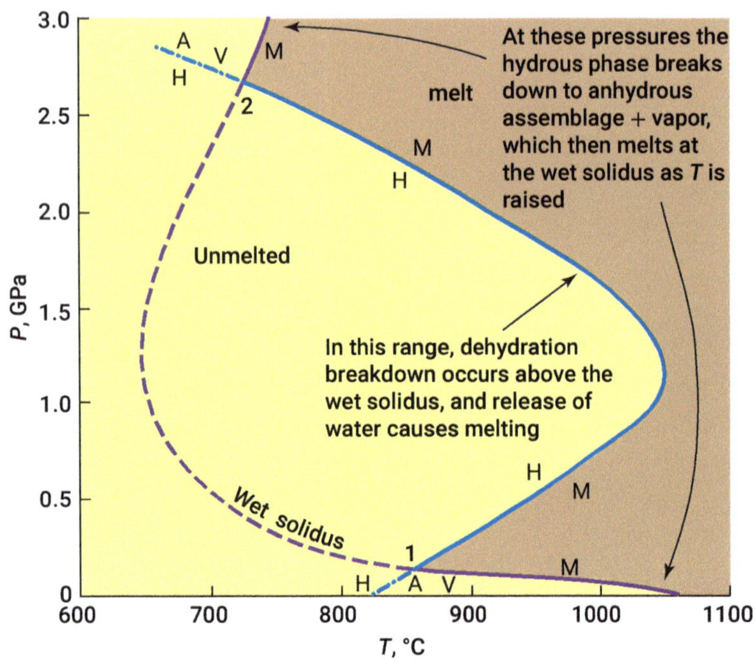

Figure 6.32 Schematic *P-T* diagram for melting of a rock containing a hydrous phase in the absence of free water. At low *P*, the hydrous phase breaks down at a *T* below that of the wet solidus, and melting commences when *T* is raised to that of the wet solidus. At higher *P*, breakdown of the hydrous phase occurs at temperatures above that of the wet solidus, so the melting *T* is higher than the wet solidus; melting commences when the hydrous phase dehydrates. The stability curve for the hydrous phase curves back to lower *T* at even higher *P*, so once again, melting occurs at the wet solidus.

Why does the water-present melting curve of H have a positive slope, whereas the water-present melting curve of an anhydrous phase has a negative slope? Consider the two reactions that form melt (M): A + V = M and H = M (for simplicity we assume here that H melts congruently). The first reaction involves vapor on the left side and therefore has a negative ΔV owing to the large molar volume of the vapor phase, but the second involves two non-vapor (condensed) phases, and therefore ΔV is positive, giving a positive slope. As *P* is increased, the melt compresses more than the mineral, because the compressibility of melts is generally significantly higher than that of crystals, and the slope increases. Many hydrous phases, such as the amphibole group, have an upper *P* stability limit such that the breakdown curve becomes negative and recrosses the wet solidus (point 2) as in Figure 6.32.

Dehydration Melting

Amphibole can be abundant in many rocks that are potential source regions for magma, such as subducted oceanic crust and lower-crustal gabbro. Consider melting such a rock under three different sets of conditions: abundant water present, little water present, and no water present. In the first case, the rock will melt extensively at the wet solidus of the corresponding rock type. In the second, the rock will begin melting at the wet solidus, but if only a small amount of free water is present, only a small amount of melt will be produced (see binaries in Figure 6.29). If there is no free water, the rock will not begin to melt until the hydrous phase breaks down, releasing water that will promote melting of the other phases. This is **dehydration melting**, and the amount of melt produced depends on how much water is in the hydrous phase. Amphibole, with only a few weight percent H_2O (see Table 5.2) is not particularly effective; chlorite and serpentine are far more water-rich. The ability of a parent source rock to produce a significant amount of "offsping" (melt) is called the rock's **fertility**. Rocks being melted under water-saturated conditions are generally much more fertile than those melting by dehydration.

Dehydration melting
The breakdown of a hydrous mineral to form a melt in the absence of free water.

Fertility
Here, the degree to which a rock is able to produce melt for a given thermal input.

6.7 Generation of Magma in Subduction Zones

Learning Objectives

By the end of this section you should be able to:

6.7.1 Identify different potential sources of arc magmas.

6.7.2 Propose different approaches for distinguishing potential sources of arc magmas.

The Problem of Subduction Zone Magmatism

Generation of andesite and other rocks more silicic than basalt has occupied petrologists for over a century. As we have seen, it's easy to make basalt; if alien geochemists[7] were to grind up the Earth and melt it a little, say 10 percent, they would make basalt. All of the terrestrial planets and larger asteroids seem to have made copious amounts of basalt, but evidence for magmas in the solar system that are more silicic than basalt is scarce.[8]

Marshall's description of the andesite line (Marshall, 1911; see Chapter 1) emphasized how regular the distribution of igneous rocks on the Earth's surface is. Inside the Pacific Basin, there is little that isn't basalt, but step across the line onto one of the stratovolcanoes that ring the Pacific, and you're into a whole different environment, with abundant andesite, dacite, and rhyolite—the arc lithosystem. Basalt is also abundant there, but the entire BADR spectrum is well-represented.

Possible Parent Materials for Arc Magmas

Oceanic ridges are relatively simple petrologically—as far as we can tell, they are underlain by rising peridotite that undergoes relatively dry decompression melting, and rising magma encounters only peridotite and related ultramafic rocks, gabbro, and basalt on the way to the surface. Subduction zones are far more complicated. Arcs sit atop a column (see Figure 6.20) that includes the following components—at least:

- Relatively thick and locally quite complex crust
- Lithospheric mantle of the upper plate
- Asthenospheric mantle of the mantle wedge
- Subducted oceanic crust, including sediment and pieces of the hanging wall carried down the subduction channel
- Oceanic lithospheric mantle of the subducting plate
- Asthenospheric mantle below that

A further complication is that many arcs are thrust for tens or hundreds of kilometers over the back-arc region (see Chapter 12), and this counter-subducted lithosphere must also get into the mix. Any and all of these reservoirs could supply material for arc magmas.

Primary Magma: Basalt or Andesite?

A key question in arc petrology is whether the primary magma generated in the subduction zone is basaltic or andesitic. Arcs are dominated topographically by andesites and dacites, but the lower topography around the volcanoes is commonly underlain by basalt.[9] There is clearly a basalt end member in arcs, because arc magmatism typically encompasses the entire BADR spectrum, ranging down to about 45 wt% SiO_2. We know how to make basalt, but what processes are responsible for generating andesite and more silicic rocks? In at least some arcs, there is geochemical evidence for parental magma of high Mg# (>60) andesite (Kelemen et al., 2014). If so, how is it made?

The system Fo-Di-Qtz (see Figure 4.11) provides an interesting lesson in the evolution of thought. At 1 atm, a model peridotite consisting of Fo, Di, and En melts to a quartz-normative liquid at the ternary peritectic (see Figure 4.11); as noted in Chapter 4, this occurs because the liquidus field for Fo extends into quartz-normative territory. This melt could charitably be called an andesite in this simple system (lacking plagioclase, this system makes a poor model of real rocks). In a real system, could this peritectic behavior generate andesite by melting an olivine-rich parent? Perhaps, but increasing *P* drives the peritectic to the left via shrinkage of the Fo field such that at 1 GPa, the invariant point has migrated into the Fo-Di-En triangle and is now a eutectic (see Figure 4.16). So much for generating a quartz-normative liquid by melting an olivine-normative parent. However, further high-*P* experiments in this system with water added (Kushiro, 1972) showed that the presence of water re-expands the Fo field, and the invariant point becomes, once again, a peritectic, generating quartz-normative melts. Based on this simple system, then, high-*P* melting of hydrous peridotite could generate andesitic melt.

Slab Melting

Basaltic crust of the subducted slab was one of the early suspects in the hunt for the source of andesites. Green and Ringwood (1966), in a pioneering set of piston-cylinder experiments, proposed that andesites and other calc-alkaline rocks (what we would call, today, arc magmatic rocks) are produced by dry melting of eclogite in orogenic belts. Their arguments were somewhat indirect; they studied several compositions along the BADR trend and found that andesite had the lowest liquidus *T*, from which they concluded that it was the low-melting fraction—the first thing to be produced upon melting.

Green and Ringwood's work just barely predated the plate tectonics revolution, and so they argued for generation of thick piles of basalt in orogenic belts, conversion of basalt to eclogite when the pile got thick enough, catastrophic sinking of these dense eclogitic roots into the mantle, and melting of eclogite to produce andesite. The idea of subduction was around at that time but was considered a wild idea, outside the mainstream of scientific thought. Coats (1962) drew a cross section showing a subduction zone that he called a megathrust going under the Aleutians. Coats's cross section looks remarkably like a modern rendering of a subduction zone, although he proposed that the megathrust tapped little pockets of basaltic magma in peridotite.

Early experimental work in which the synthesized melts were analyzed by electron microprobe cast doubt on melting of eclogite, either hydrous or anhydrous, as a general method for producing andesite (Stern et al., 1975). However, later experiments and calculations indicate that hydrous melting of eclogite can produce melts more silicic than basalts (e.g., Kay, 1978; Rapp et al., 1991)—a way to make andesite by melting the slab, if the slab gets hot enough.

Thermal models combined with experimental work and thermodynamic calculations indicate that oceanic crust in slabs should largely dehydrate before reaching the wet solidus. Thus, direct melting of basaltic crust is unlikely to generate arc magmas in most cases. Figure 6.33 shows one set of possible *P-T* paths for the surface of the subducted slab superimposed on calculated mineral assemblages in hydrous MORB. At relatively low *P* and *T*, hydrated MORB has 5 wt% or more H_2O, largely tied up in

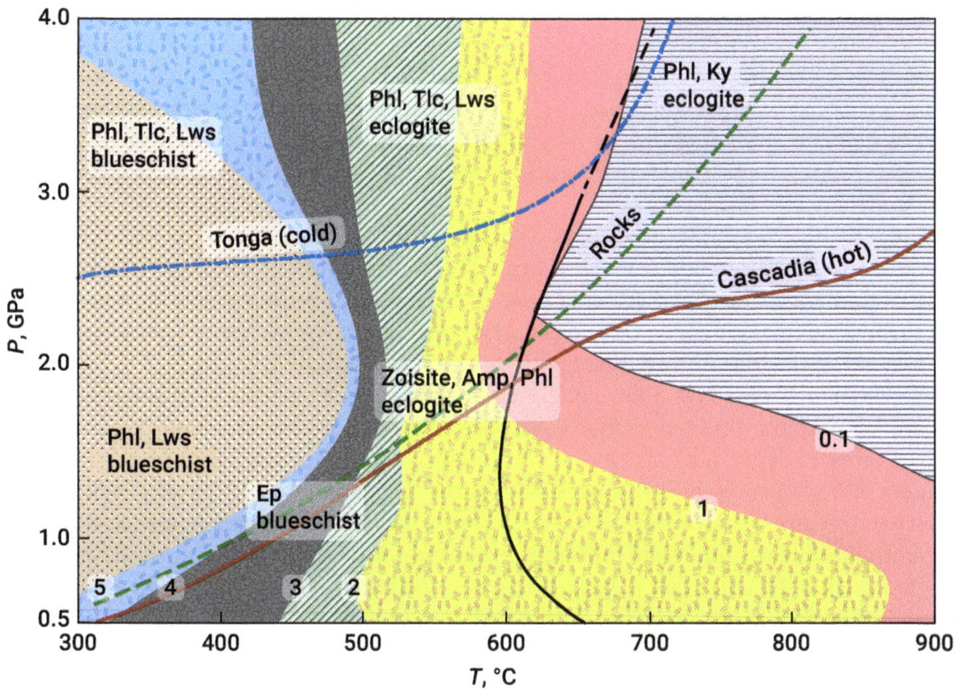

Figure 6.33 Calculated bulk-rock water content (color contours) and mineral assemblages for fresh MORB as a function of *P* and *T*; black curve is the water-present solidus for MORB. Tonga and Cascadia are top-of-slab temperature profiles along the subduction interface for the corresponding arcs and represent cold and hot subduction zones, respectively. According to these models and calculations, only subduction zones that involve young oceanic lithosphere (e.g., Cascades) reach temperatures for dehydration melting, and even in those cases, the slab is mostly dewatered (e.g., down to 0.1 wt% H_2O for Cascadia), so the source is relatively infertile. The curve for rocks is based on median blueschist and eclogite *P-T* conditions and is thought to represent the temperature profile along the subduction interface as represented by exhumed rocks.

Sources: Data from Hacker (2008), Syracuse et al. (2010), and Penniston-Dorland et al. (2015).

minerals such as talc, lawsonite, and **phengite**, but with increasing *T*, a series of dehydration reactions drops the water content such that at the wet solidus, about 600 °C, most of the water has been lost, and therefore the amount of melt that can be generated by dehydration melting is minimal. Melting of the slab is favored in cases where young (<5 Ma) oceanic lithosphere is being subducted, because such lithosphere is warmer and leads to higher temperatures in the subduction zone (Defant & Drummond, 1993; Peacock et al., 1994).

Phengite
Muscovitic mica, common in metamorphic rocks, with excess Si (Si > Al) that is charge-balanced by cations such as Mg.

Serpentine and the Mantle Wedge

Subducted oceanic lithosphere, made juicy by hydrothermal alteration at and near oceanic ridges and transform faults, as well as during faulting as the plate bends before descending into the mantle, releases water as it sinks into the mantle. These reactions in turn release water that travels upward owing to its low density and hydrates the mantle wedge.[10]

The following viewpoint on evolution of water from the subducted slab is summarized from Schmidt and Poli (2014). Free water in pore spaces is likely expelled during the first kilometer or two of burial as the subducted plate is overridden by the hanging wall. Bound water at this stage is largely in zeolites, prehnite, and pumpellyite, and amounts to 8 to 9 wt%. At around 15 km depth, subducted crust enters the blueschist facies; significant minerals are chlorite, glaucophane, phengite, and either lawsonite or zoisite—all of which are hydrous—but low-temperature reactions have already reduced the bulk H_2O content of the crust to about 6 wt%. Over the depth range of blueschist metamorphism, various reactions continue to reduce or eliminate water-rich minerals such as chlorite and lawsonite. At the amphibole-out reaction at pressures of 2.2 to 2.4 GPa (65–70 km), rocks retain less than 1 wt% water, possibly as little as 0.1 wt%, and are thus quite dry (see Figure 6.33).

Fully hydrated oceanic crust loses about two-thirds of its water while still under the cold fore-arc ($P \leq \sim 2$ GPa), so most of this water must either serpentinize the cool tip of the mantle wedge (the so-called "cold nose") or pass back up the subduction interface to the surface. Submarine mud volcanoes carrying bits of blueschist have been observed in the fore-arc of the Mariana arc (see Figure 6.3; Fryer et al., 1999).

Serpentine likely plays a key role in arc magmatism. In addition to serpentinizing the cool, immobile tip of the mantle wedge, fluids expelled from subducting oceanic crust at greater depth may hydrate the convecting part of the wedge or be carried even deeper. In addition, a significant portion of the upper mantle in the downgoing slab is serpentinized, and this, too, is carried to greater depth. Experimental work on the serpentine polymorph antigorite (Figure 6.34) demonstrates that antigorite is stable to $P > 7$ GPa at sufficiently low T. Because serpentine (1) is likely abundant in the mantle both above and within the slab; (2) contains significant H_2O, ~13 wt% (see Table 5.2); and (3) dehydrates at temperatures less than about 700 °C, dehydration of serpentinite in subducted lithosphere and the mantle wedge likely releases significant water under the arc. What these depths actually are depends critically on the P-T regime that the serpentinized slab and mantle wedge traverse, but broadly speaking, it should be in the range of about 70 to 200 km (see Figure 6.34).

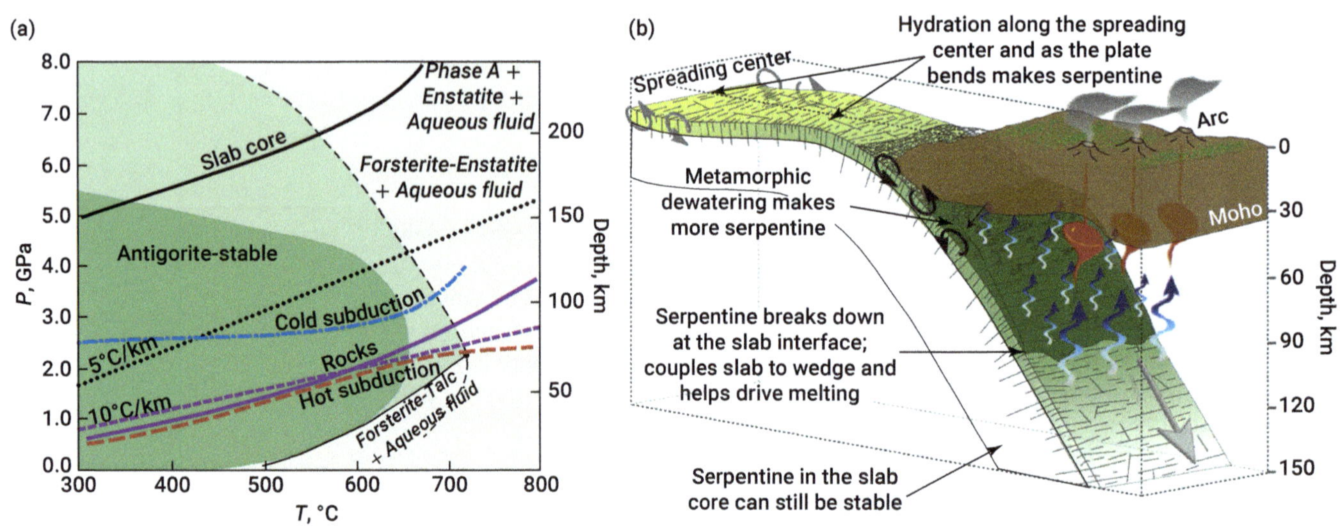

Figure 6.34 Importance of serpentine (antigorite) on water in subduction zones. (a) Different estimates of the P-T stability of serpentine. Below ~2 GPa (shallower than ~60 km), serpentine breaks down to forsterite, talc, and water, with talc dehydrating at higher T. From ~2 to ~6.6 GPa (~200 km), serpentine dehydrates to forsterite, enstatite, and water. Above ~6.6 GPa, serpentine breaks down to enstatite, water, and a dense, high-P phase known as phase A ($Mg_7Si_2O_8(OH)_6$). Curves for cold subduction, hot subduction, and rocks represent conditions along the slab interface; curve for slab core shows conditions along the central cold portion of the subducting slab. (b) Block model showing formation and demise of serpentine. Serpentine forms in multiple locations—at the spreading center, as the plate bends before entering the subduction zone, and above subducting sediments and hydrated crust. Serpentine breakdown along the subduction interface mechanically couples the slab to the mantle wedge and helps drive melting. In the cold core of the slab, serpentine can still be stable and carry water past the arc.

Sources: (a) Modified from Deschamps et al. (2013); dark green field from Bromiley & Pawley (2003); light green field from Ulmer & Trommsdorff (1995).

Equilibrium with Mantle Olivine

Ignoring melting of eclogite, the analysis in the foregoing section assumes that melting is driven by delivery of water to parent peridotite that is above its wet solidus but below the dry solidus. It is important to realize that adding water to rock that is above its wet solidus does not mean that the *T* of the rock suddenly drops to the wet solidus; *T* will drop somewhat in response to the melting reaction owing to the need to supply the required enthalpy of melting (latent heat; see Section 4.5), but the main effect will be to produce as much melt as that amount of water can sustain at that temperature.

There is another constraint that we can apply to test this hypothesis: magma produced from peridotite should be in equilibrium with the minerals from which it is produced. One simple test for such equilibrium uses Mg#. Mantle olivine is generally assumed to be about Fo_{89} in composition (Ringwood, 1975), and we can use this assumed reservoir composition and the consistent distribution of Mg and Fe between olivine and melt (see Section 4.8) to predict what the Mg# of a melt in equilibrium with mantle olivine should be. Using $K_D = 0.3$ and a bit of arithmetic detailed in Toolbox 4.1, this predicts a whole-rock Mg# (assuming that the rock is a cooled sample of the melt separated from peridotite) of about 60. Figure 4.20 shows that many basalts from both the Medicine Lake volcano (representing an arc volcano in the Cascades) and Iceland (representing a MOR) have Mg# $\geq$ 60; thus, this simple test is met.

Figure 6.35 Vesicular picrite (olivine-enriched basalt) from Hawai'i. This rock features abundant olivine phenocrysts in an aphanitic groundmass.

Source: AFG photo.

It is clear that many basalts in these datasets have Mg# > 60. How are such compositions produced, if mantle olivine is Fo_{89}? Such rocks likely contain an excess of olivine. The beautiful basalt variety known as **picrite** (Figure 6.35) is common around the world and likely forms by accumulation of olivine crystals that settled to the bottom of a lava lake or other magma chamber. We will explore these processes in Chapter 8.

Picrite
An unusually Mg-rich basalt with abundant phenocrysts of olivine.

Another test we can apply to see if a given magma might have been produced by melting of the mantle is to see if it would be in equilibrium with mantle phases at an appropriate *P* and *T*. Consider Figure 6.36, which shows the binary eutectic Fo-Di (see Figure 4.1) at 1 atm and

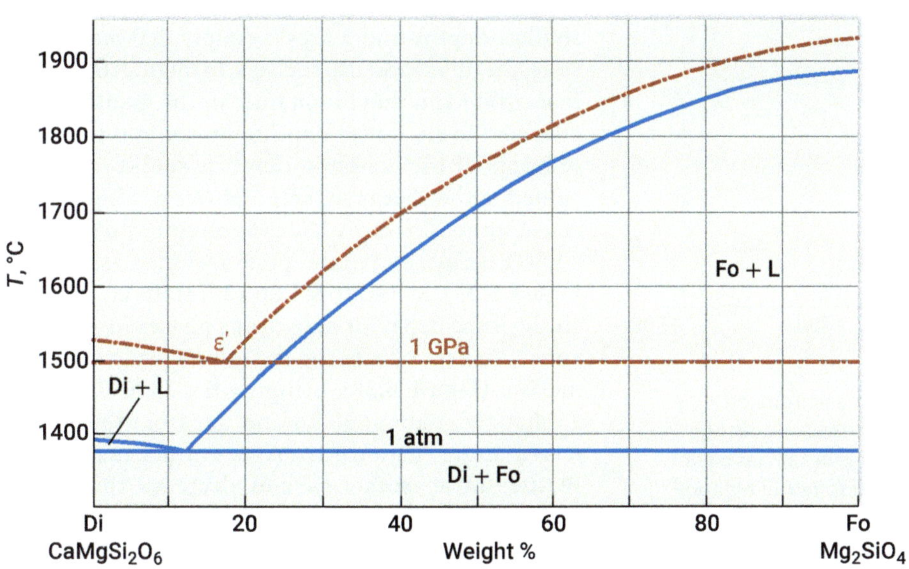

Figure 6.36 The system forsterite-diopside at 1 atm and 1 GPa.

1 GPa, both anhydrous. Raising *P* has the effect of raising the melting *T* of both end members and the eutectic, and of shifting the eutectic toward Fo. Assume that we make a basaltic magma by eutectic melting at 1 GPa, producing a melt with composition ε′. If we took the rock solidified from this melt out of this system and melted it at 1 atm, upon cooling it would start to crystallize Fo at about 1430 °C and would not crystallize Di until the eutectic at about 1380 °C. Clearly, it is not cosaturated in Fo and Di at 1 atm. If we do experiments at increasing *P*, we will find that at 1 GPa, the melt crystallizes Fo and Di simultaneously, and the solidus and liquidus are the same. Thus, we can infer that the rock could have been made by eutectic melting of Fo + Di at 1 GPa, and if we know the proper bulk composition (including H_2O, which is generally lost upon cooling), then we can use this technique to infer the *P* and *T* of melting.

Alas, such reasoning does not apply in the case of peritectic melting. Consider the system Fo-Si (see Figure 4.5). If we melt bulk composition X at 1 atm, we get peritectic melt π. If we separate this melt from its source and then remove heat from the system, it will crystallize En at the peritectic temperature 1557 °C and will not fully crystallize until reaching the eutectic. Fo is stable in the melt at the peritectic temperature, but no Fo will crystallize from it, and it will not be obvious from the phase equilibria that the magma was generated by melting of a Fo + En source.

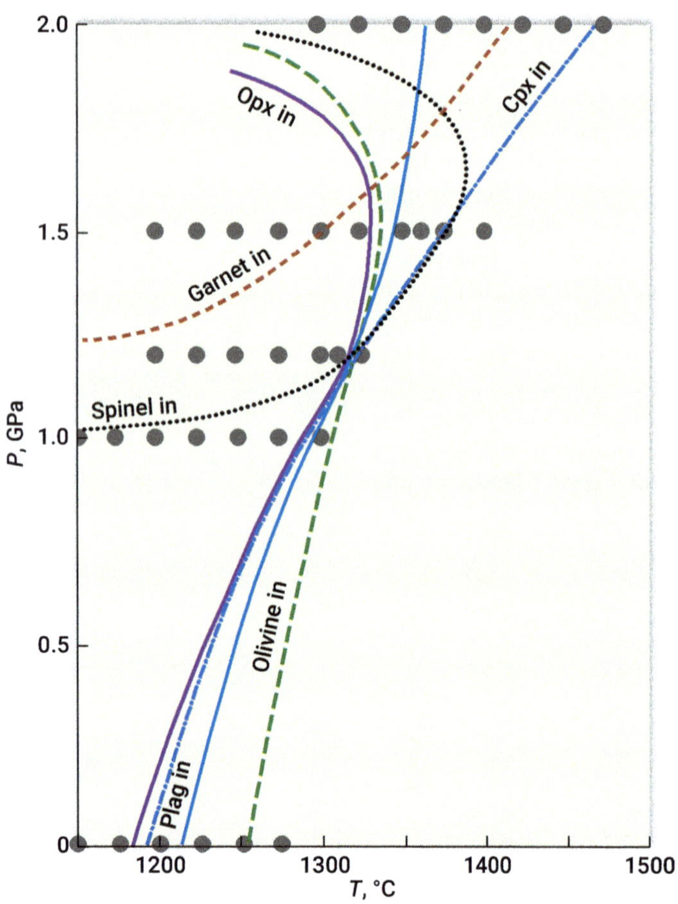

Figure 6.37 Phase equilibria of a primitive basalt from the Aleutian arc. Curves give the *T* at which a given phase starts to crystallize upon cooling. Like many basalts, this basalt has olivine on the liquidus at low-*P*, changing to Cpx (and spinel) at higher *P*. Convergence of the olivine, Opx, Cpx, plagioclase, and spinel curves at 1.2 GPa and 1300 °C indicates that a magma of this composition could have been in equilibrium with plagioclase-spinel peridotite at those conditions.

Source: From Johnston & Draper (1992); © 1992, Springer-Verlag.

P-T studies of multiple saturation indicate that some primitive magmas could have been generated by high-*T*, anhydrous melting of mantle peridotite. For example, Johnston and Draper (1992) found that a primitive basalt from Umnak Island in the Aleutian arc (49 wt% SiO_2, Mg# = 70) is multiply saturated in olivine, Cpx, Opx, plagioclase, and spinel—a spinel peridotite assemblage—at about 1300 °C and 1.2 GPa (Figure 6.37). They inferred that the basaltic magma separated from the mantle at these conditions. This pressure implies 1300 °C at a depth of about 35 km, far hotter than most thermal models (see Figure 6.26), and the authors invoked ascent of buoyant blobs of hot mantle to advect high temperatures to shallow depth. Thus, in their model, the *P-T* conditions inferred from experimental work represent the last equilibration of the basalt magma with the mantle, not the initial conditions of melting.

An alternative explanation for high temperatures at shallow depths under arcs is simply that many steady-state thermal models, such as in Figure 6.26c,d, are inaccurate and that hot mantle in the mantle wedge is advected to shallower depths. This might happen, for example, if the coupling depth is shallower than assumed, either because temperatures are higher or because serpentine dehydrates to olivine and enstatite at a lower temperature (and pressure) than assumed (see Figure 6.34). Schmidt & Poli (2014), in an exhaustive study of the behavior of hydrous phases in subduction zones, inferred such convection to explain arc magmatism (Figure 6.38), bringing the 1300 °C isotherm as shallow as about 50 km, not far from the depth inferred above. Upward transport of hot magma could further raise isotherms and decrease the depth of melting (England & Katz, 2010). If mantle wedge convection is arrayed as shown in Figure 6.38, then adiabatic upwelling—much as we infer under oceanic ridges—plus addition of water contributes to arc magmatism.

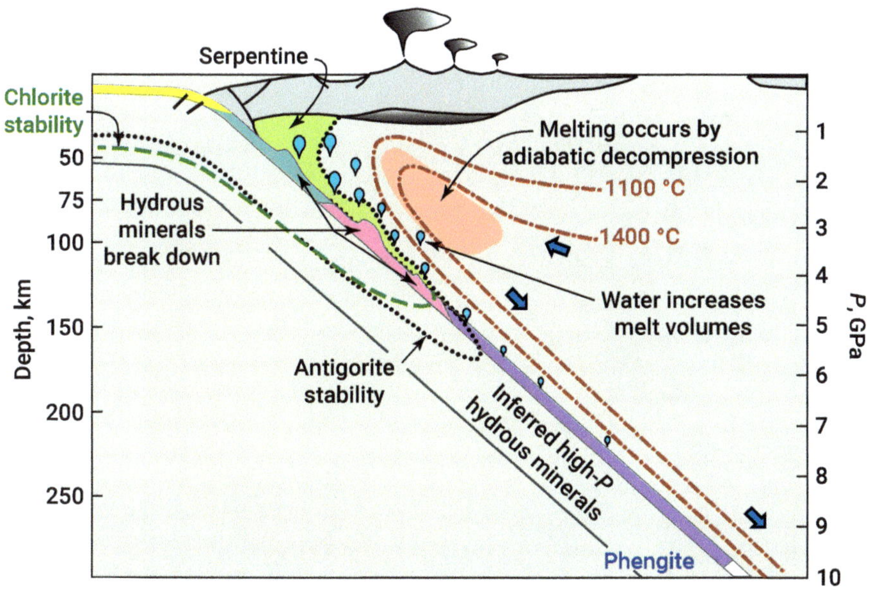

Figure 6.38 Cartoon of mineral stabilities and processes occurring in subduction zones. Breakdown of hydrous minerals generates abundant water in the shallow part of the subduction system to form abundant serpentine. Shallow upwelling in the mantle wedge drives melting, which is enhanced by addition of water from the subducting slab. Dotted and dashed curves show stabilities of chlorite and antigorite; dashed curves labeled with temperatures show representative isotherms.

Source: Modified from Schmidt & Poli (2014).

Role of Subducted Sediments in Arc Magmatism

There are several compelling lines of evidence demonstrating that subducted sediments are at least partially recycled into arc magmas. White and Dupré (1986) examined several radiogenic isotope systems in arc lavas for the Lesser Antilles arc (see Figure 6.1) and compared them to sediments delivered to the trench. The arc has a vaguely defined trench along its northern half, but the southern half is buried by submarine fan deposits from the Orinoco River (Figure 6.39). The Orinoco drains a portion of South America that contains Proterozoic and Archean bedrock, and the sediments of the fan thus have high present-day $^{87}Sr/^{86}Sr$ ratios. These sediments are delivered to the southern half of the subduction zone, where they have filled the trench, but the starved northern half contains just a thin section of sediments with lower $^{87}Sr/^{86}Sr$. These variations show up in the arc magmas; northern volcanoes erupt material with $^{87}Sr/^{86}Sr$ less than 0.704, whereas $^{87}Sr/^{86}Sr$ in southern volcanoes ranges up to almost 0.710. Similar spatial variations occur in other isotope systems such as $^{206}Pb/^{204}Pb$. Sediment must in some way contribute to arc magmas.

Trace elements can also be used to track subducted sediments. Plank and Langmuir (1993) compiled data from arcs around the globe and estimated the flux of sediments into the subduction zone. They estimated the concentrations of various trace elements from the rock types in each sedimentary sequence and found that the concentrations of those elements in erupted rocks from a given arc correlate with the concentration of that element being delivered in sediments via subduction. For example, Ba is enriched in rocks from the Guatemalan segment of the Central American arc, and the authors attribute that enrichment to a high concentration of barite in the biogenic oozes that make up about half of the sediment column on the incoming Cocos Plate.

One of the most interesting geochemical tracers of sediment input to arc rocks is the rare nuclide ^{10}Be. Be and its atomic neighbors Li and B are largely skipped during nucleosynthesis, and therefore their cosmic abundances are negligible (see Figure 2.12). Be has only one stable isotope, ^{9}Be. However, a small amount of ^{10}Be is made in the Earth's atmosphere by spallation of atmospheric oxygen, a process in which a high-energy cosmic ray whacks the nucleus of an oxygen atom and cleaves off a piece (see Making Connections 2.1). If that piece has four protons and six neutrons, it is a ^{10}Be nucleus. These atoms make their way to the Earth's surface, including into sediments that are about to be subducted. ^{10}Be has a half-life of 1.39 Myr, so if such atoms make their way into the source region of the arc magmas fast enough (within a few half-lives), then one might be able to find those atoms in the erupted rocks. For a convergence rate of 50 mm/yr and a slab dip

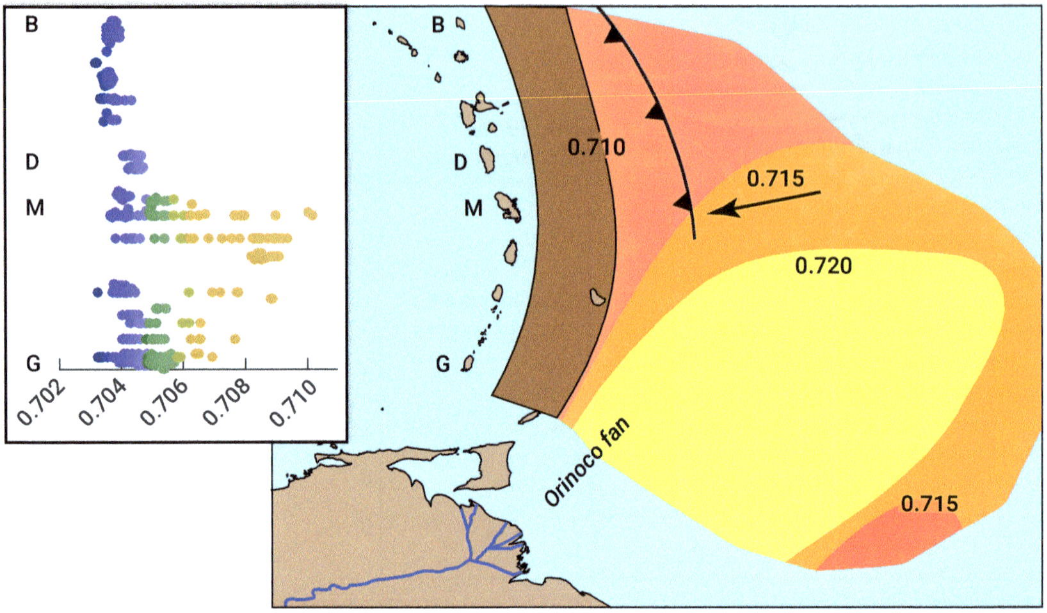

Figure 6.39 Variation with latitude of $^{87}Sr/^{86}Sr$ in volcanic rocks of the Lesser Antilles. Northern islands from Barbuda (B) to Dominica (D) have low ratios around 0.704; southern islands from Martinique (M) to Grenada (G) have ratios that range up to 0.711. Color patterns show $^{87}Sr/^{86}Sr$ in oceanic sediments; values in the southern part of the area are significantly higher than elsewhere owing to input of high-$^{87}Sr/^{86}Sr$ sediments from the Orinoco River. The subduction trench, marked by the barbed line, is largely buried by sediments from the Orinoco fan in the southern half of the area. Subduction of the Atlantic Plate is to the west (arrow), so these data indicate that Orinoco sediments are carried into the subduction zone, where their high $^{87}Sr/^{86}Sr$ ratios show up in corresponding volcanic rocks.

Source: From White & Dupré (1986). Copyright 1986 by the American Geophysical Union.

Porphyry copper deposits
Intrusions with characteristically large crystals in fine-grained groundmass (a porphyry) that contain relatively abundant copper minerals, usually as sulfides, but also as other minerals along fractures.

Supergene deposit
An ore deposit formed by the downward movement of solutions causing secondary mineralization. "Super" here means "from above," not necessarily high concentration.

of 45°, subducted sediments would take about 3 Myr, or a little more than two ^{10}Be half-lives, to reach 100 km depth—thus, about 25 percent of the originally minuscule load of ^{10}Be would remain undecayed by the time the slab reached depths of magma generation.

This sounds like an interesting way to look for subducted sediments, but the technical problems in measuring ^{10}Be are prodigious. Typical measured concentrations are on the order of a million atoms of ^{10}Be per gram of rock. A million is a big number, but given that a gram of rock contains about 10^{22} atoms, this is roughly one part in 10^{16}—one part in 10 quadrillion. This analytical job is roughly equivalent to mixing 10 red sand grains into a football stadium full of sand, and then counting the red grains.

Tera et al. (1986) compared ^{10}Be analyses from oceanic ridges, ocean islands, and rift zones with those from several arcs and found significantly higher ^{10}Be in the arc rocks. Oceanic ridges should not contain significant ^{10}Be, because their magmas should be coming out of the mantle with no surface components. As with radiogenic isotopes and trace elements, then, ^{10}Be studies indicate that at least some subducted sediment is incorporated into arc lavas.

Making Connections 6.1

Copper Deposits

Society is built on minerals and mining; just take a look around you. If what you see didn't come from a plant (wood, paper, cotton, etc.) or an animal (wool, leather, etc.), or from recycling, someone mined it. The average cell phone has over 60 different essential metals in addition to plastics (derived from mining petroleum). Green technology is dependent on rare metals like lithium and gallium. We hear a lot in the news about rare metals, but one of the biggest metals hiding in the shadows of your everyday life is copper: the average car contains about 25 kg and the average house about 200 kg. Or, to put it differently, the weight of copper in the average house is approximately equal to the weight of its inhabitants. Because copper has extremely high electrical conductivity (only silver is better), most copper (roughly 2/3) goes into wiring and transport of electricity. All those green technologies require copper, too. Even with recycling, copper demand and production have increased dramatically since around 1900, and copper consistently ranks in the world's top 10 commodities.

Where does all this copper come from? Ultimately, most comes from subduction zone magmas in the form of **porphyry copper deposits** (Figure 6.40). Exactly how this copper becomes concentrated is still under study, but in general, copper and sulfur are highly incompatible in most minerals, so they become concentrated in partial melts during melting and in residual melts during fractional crystallization. Sulfur and many transition metals like copper are so incompatible with silicates that they can form their own separate liquids. Metal-sulfide liquids are extremely dense, so why don't they sink away? Maybe many do, but vapor bubbles also nucleate on sulfide melt droplets. Experiments show that, at the right conditions, the combination of a low-density bubble plus a high-density sulfide melt droplet can actually float and accumulate.

Porphyry copper deposits are extraordinary because of what happens when they're exposed at Earth's surface. You probably remember that iron is highly soluble under reducing conditions (Fe^{2+} tends to dissolve in water), but highly insoluble under oxidizing conditions (Fe^{3+} precipitates). Copper is the exact opposite—it's highly soluble under oxidizing conditions and insoluble under reducing conditions. When a porphyry copper deposit weathers under oxidizing conditions at Earth's surface, the copper dissolves and migrates downward (Figure 6.41a). Water tables are highly reducing, so when the dissolved copper hits the water table, it precipitates out, creating a high concentration of copper sulfides called a **supergene deposit** (Figure 6.41a). In contrast, iron oxidizes near the surface to form nearly insoluble iron oxides, hydroxides, and oxyhydroxides, which create characteristic red-cemented rocks and soils (Figure 6.41a,b). Sometimes these deposits are discovered not from outcrops of copper-bearing rock, but from the Fe-rich material above them. This weathering process creates numerous spectacular mineral specimens, especially beautiful blue and green copper carbonates, oxides, sulfates, and hydroxides (Figure 6.41c,d). These overlie reduced copper-bearing sulfides deeper in the system (Figure 6.41e,f).

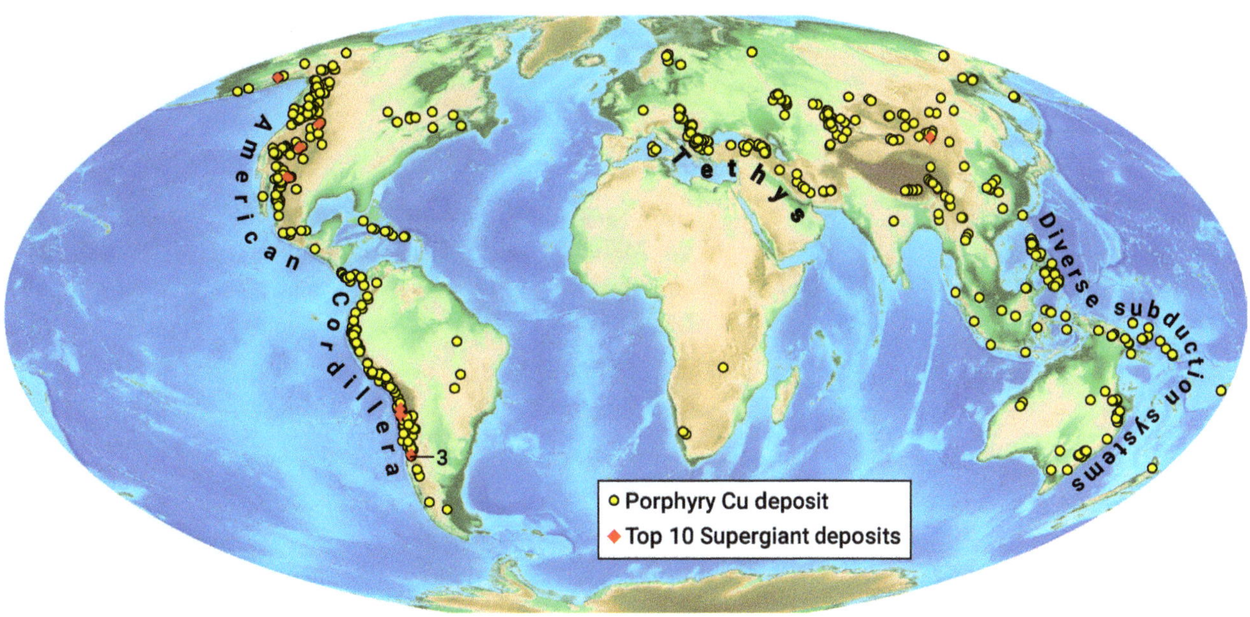

Figure 6.40 Distribution of porphyry copper deposits worldwide. Note concentration along subduction zones of western North America, South America, and the southwestern Pacific. Some deposits are not associated with subduction zones and can reflect remobilization of melts during continental collisions. The number "3" in Chile represents three supergiant deposits in that region.

Source: Modified from John et al. (2010).

continued

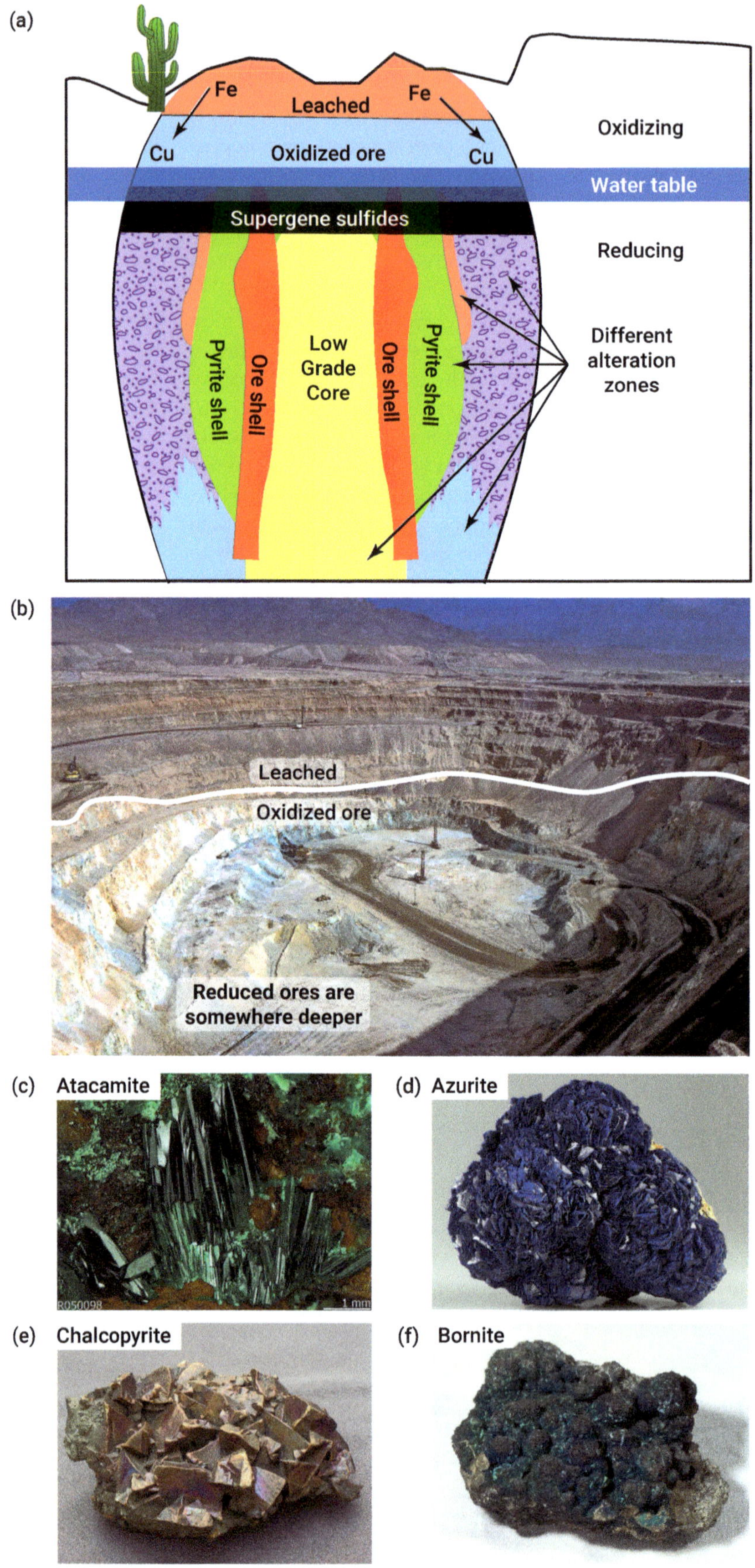

Figure 6.41 (a) Model of rock distribution in a copper sulfide-bearing porphyry and supergene enrichment processes above it. Other than the ore shell, different colors below the supergene sulfide horizon represent different types of rock alteration. Oxidized Cu migrates downward into an oxidized ore zone, while oxidized Fe remains in a leached horizon; a supergene enrichment occurs near the water table, with oxidized Cu ores above and reduced Cu ores below. Cactus is not to scale. (b) The Radomiro Tomic deposit in Chile, one of the largest copper mines on Earth. (c, d) Images of oxidized copper ore minerals: atacamite [$Cu_2(OH)_3Cl$] and azurite [$Cu_3(CO_3)_2(OH)_2$]. (e, f) Images of sulfide ores: chalcopyrite ($CuFeS_2$) and bornite (Cu_5FeS_4).

Sources: (a) Modified from Lowell & Guilbert (1970) and John et al. (2010); (b) modified from Cameron et al. (2008); (c) RRUFF database; (d) Ivar Leidus/Wikipedia; (e, f) Smithsonian Institute, National Museum of Natural History, Mineral Sciences Collection.

Summary

Key points of this chapter include the following:

- Subduction differentiates Earth from other planets and cycles water through the Earth system to create magmas.
- Subduction zones are divisible into clear geomorphic regions, including the hinge, trench, accretionary wedge, fore-arc (which may include fore-arc basins), arc (including island arcs and continental arcs), and back-arc (which may include back-arc basins). Each region contains characteristic rocks that allow identification of ancient arcs and subduction zones in the geologic record.
- The Wadati-Benioff zone of earthquakes delineates the subducting slab; earthquakes occur within the slab and do not directly outline the subduction interface.
- Seismic tomography is a method that images Earth's interior and commonly shows slabs of high seismic velocity at depth in the mantle, interpreted to be subducting plates. These plates may or may not penetrate the 660-km discontinuity, where the temperature-dependence of phase changes inhibits density-driven subduction.
- Subduction zone lithosystems include the arc lithosystem, which contains a wide range of volcanic rock compositions, especially andesite, and the HP/LT (blueschist) lithosystem, which contains unusual minerals such as glaucophane and lawsonite, stable only at high pressure and low temperature.
- The driving forces for plate motion include both viscous drag on the bottoms of plates from convecting mantle and density-driven foundering of old cold slabs.
- Earth's plates have no unique reference frame for motion, and different applications emphasize different reference frames.
- The depth of the subducted slab beneath the arc varies widely, but has a mode at about 100 km. This location does not strongly correlate with any one subduction parameter (e.g., angle of subduction, age of subducting plate, etc.), although very shallowly dipping slabs lack arcs.
- The subduction interface is thought to represent a region of intensely sheared rock, including materials from both the subducting plate, the overriding plate, and their alteration products.
- Volcanic rock geochemistry changes systematically across an arc, classically from tholeiitic closer to the trench to calc-alkaline in the heart of the arc and alkalic in the back-arc.
- Subduction creates complex thermal structures, with inverted temperature profiles, low temperatures near the core of the subducting slab, and high temperatures underneath the arc due to magma transport and convection of the mantle wedge. Dehydration of serpentinite to form olivine along the subduction interface is thought to mechanically couple the subducting slab with convecting mantle.
- Ultrahigh-pressure rocks, subducted to depths greater than 100 km, are predicted in subduction zones, but their exhumation and preservation remain a mystery.
- Subduction zone magmas arise from contributions of water, derived from dehydration of the subducting slab (likely serpentine), to the mantle wedge, catalyzing relatively low-temperature, water-present melting reactions. Direct melting of the subducting slab is rare.
- The different compressibilities of water vs. hydrous minerals give rise to different slopes in *P-T* space for water-present melting reactions (typically steep negative slopes) vs. dehydration-melting reactions (typically steep positive slopes).
- Primary arc melts may be either basaltic or andesitic, depending on how they form.
- Mineral equilibria suggest that melts beneath arcs are hot and shallow, although they may not represent the conditions at which the mantle melted.
- Sediments contribute their singular geochemical signal to arc volcanic rocks, although the mechanism or mechanisms by which they do so are not completely understood.

Notes

1. Focal mechanisms are derived for a set of seismograms and indicate the nature of slip on a fault during an earthquake.
2. Presumably until the Sun accumulates a lot of He ash, turns into a red giant, and vaporizes the Earth.
3. Graywacke (also greywacke) is commonly characterized as "dirty sandstone" and typically consists of quartz, feldspar, and lithic fragments in a clay-rich matrix—the sort of detritus that one might expect to be shed from a magmatic arc, where the lithic fragments are typically volcanic.
4. Remember from Chapter 5 that metamorphic grade generally equates to the maximum *T* that the rocks experienced.
5. Their analysis employed tasty-sounding "banana-donut kernels," which are waveforms used to model seismic wave arrivals, not snack food.
6. The melting kinetics are so sluggish that it is difficult to tell when melting begins.
7. Possibly white mice. They would not be called "geo"chemists, though—they would be xenochemists.
8. Barring rocks such as lunar anorthosites, which formed by flotation of plagioclase in the lunar magma ocean, which was basaltic.
9. There is also a clear sampling bias toward tall, photogenic, dangerous stratovolcanoes rather than low-lying, forest-covered basalts.
10. Just a quick reminder that no one has ever seen this process happen, but it seems inescapable, plus geophysical evidence suggests the presence of hydrous minerals in the cool tip of the mantle wedge.

CHAPTER 7

Volcanism and the Spectrum of Igneous Activity on Earth

Volcanoes are the most obvious, awe-inspiring, and dangerous manifestation of magmatic activity on Earth. Although some eruptions are small, quiet events that one can walk up to without being in imminent danger, the largest eruptions, which can erupt enough material to cover the state of Colorado more than 10 m deep, could well put an end to civilization. Differences in volcanic behavior arise largely from differences in composition, and as we have seen, among the planets of our solar system, Earth excels at making magmas of different compositions.

In volcanic eruptions, magma at a high temperature (typically 800–1200 °C) rises rapidly from depth to the surface. The Péclet number for magma rising at 1 m/s through a dike a few meters wide is huge—on the order of 10^6—so relatively little heat is lost to dike walls. This particularly efficient method of heat transfer, via movement of hot material, is known as **advection**. Volcanism is sometimes referred to as the Earth's way of sweating, although the analogy is imperfect, because sweating relies on evaporative cooling as well as advective movement of warm perspiration to the surface.

Advection
Here, transfer of heat by fluid (magma) flow.

7.1 The Spectrum of Volcanic Activity

Learning Objective

By the end of this section you should be able to:

7.1.1 Discuss the range of power, output, and effect on humans displayed by historic volcanic eruptions.

We begin with narratives of two eruptions that illustrate the wide range of behavior that the Earth's volcanoes can produce.

Hawai'i, 1983–2018

Much of our knowledge of basaltic eruptions comes from over a century of scientific study on the Big Island of Hawai'i (Figure 7.1). Eruptions in Hawai'i are generally non-violent and predictable, allowing relatively safe observation by scientists and the public alike.

Umbrella cloud above Mount Etna during the December 4, 2015, eruption. Etna, a stratovolcano on the island of Sicily, towers ~3400 m above the Mediterranean Sea and is one of the world's most active volcanoes. Eruptions such as this one pose a frequent hazard to air travel in Europe.

Source: Photo by Giuseppe Famiani.

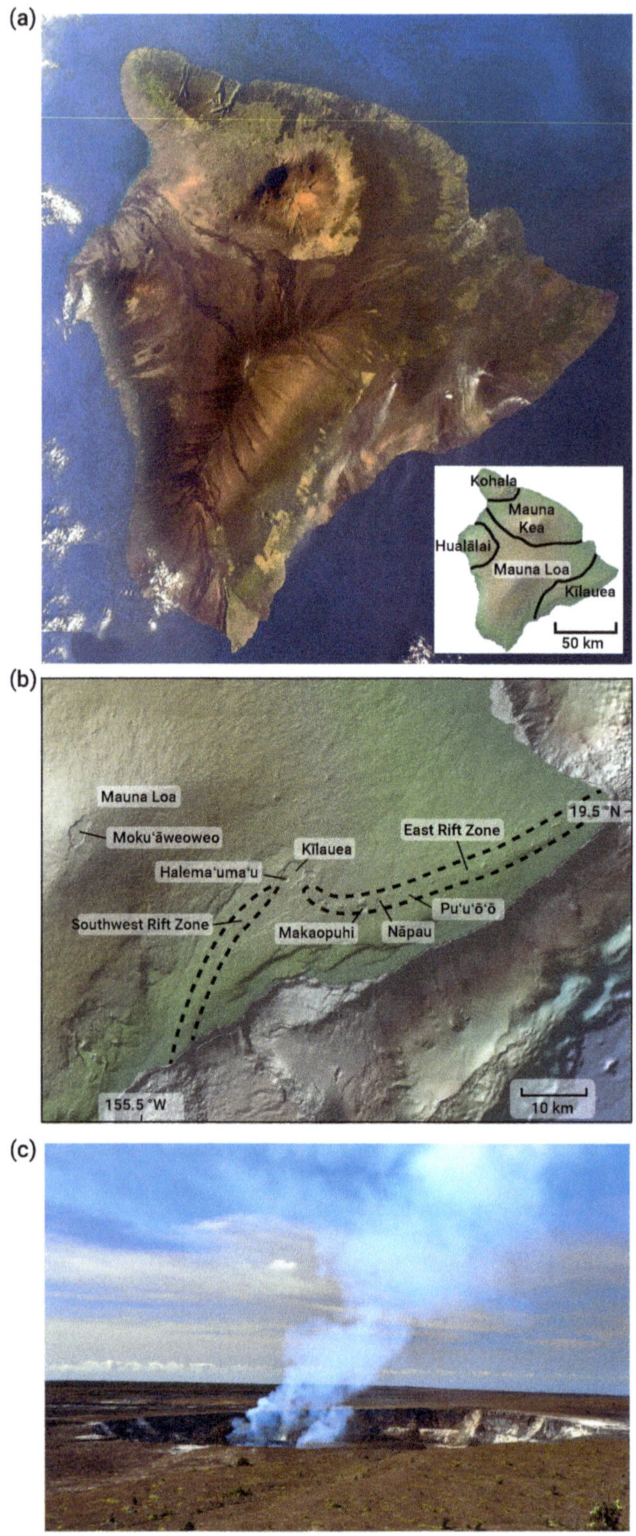

Figure 7.1 (a) The Big Island of Hawai'i consists of five overlapping shield volcanoes. Of these, Mauna Loa and Kīlauea are active, with Kīlauea in constant eruption since 1983. Photo taken from International Space Station. (b) Map of Kīlauea volcano and its **rift zones**. (c) View of smoking Halema'uma'u caldera within Kīlauea caldera as seen from the Hawaiian Volcano Observatory, which had a front-row seat on the edge of Kīlauea caldera. This picture was taken in 2017, not long before eruptions on the East Rift Zone caused summit collapse that severely damaged the observatory buildings. Gases rise from a lava lake within Halema'uma'u, which is ~1 km in diameter.

Sources: (a) NASA, Johnson Space Center; (b) map produced using GeoMapApp.; (c) AFG photo.

Rift zones
In volcanism, linear zones of crustal extension above a set of dikes that are taking advantage of that extra space.

The USGS Hawaiian Volcano Observatory (HVO) has been in operation in one form or another since 1912, building on oral histories provided by the ancient Hawaiian people.

The HVO was built on the edge of Kīlauea **caldera** (Figure 7.1c), a large depression formed via withdrawal of magma from under Kīlauea. Halema'uma'u, a smaller caldera, is nested within Kīlauea caldera. A long chain of active volcanoes, the East Rift Zone, extends 50 km east from the caldera to the eastern tip of the island. This chain is roughly parallel to the southern coast of the island, which is about 10 km to the south. The HVO had a magnificent view of Kīlauea caldera activity for over a century, but in 2018, an eruption that caused devastation along the rift zone 40 km to the east also caused partial collapse of the summit area, damaging HVO facilities beyond repair. Operations have since been relocated to Hilo.

Caldera
Steep-sided, generally ovoid depression formed by collapse after eruption or lateral withdrawal of magma.

Kīlauea caldera and the East Rift Zone have been the sites of many eruptions in the past century. Although most eruptive episodes are relatively short-lived (e.g., a year or so), an eruption that began in 1983 continued until 2018—by far the longest eruptive episode in the last two centuries. The account that follows is summarized from Wolfe (1988) and Heliker et al. (2003).

A relatively long-lived eruption occurred at Nāpau Crater (Figure 7.1b) from 1969 to 1974. Shortly thereafter, in 1975, a magnitude 7.2 earthquake struck near the coastal town of Kalapana, causing tremendous damage and two deaths. During the earthquake, parts of the island south of the East Rift Zone moved 8 m (!) south and subsided 3.5 m. Much of the damage and both deaths were a result of the ensuing tsunami.

This tectonic event, which opened up the East Rift Zone, may have set the stage for the long-lived 1983 eruption, because over the next several years, 13 intrusive episodes were detected by measurements of ground tilting. Deformation was measured using exceptionally sensitive instruments that can detect tilts of less than a microradian—the tilt you get by lifting the end of a rod 1 km long by 1 mm.

Only a few of these intrusive events resulted in an eruption, so the intrusions must have loaded the rift zone with magma. The 1983 eruption was heralded on January 2 with a series of shallow earthquakes near Makaopuhi Crater. These earthquakes tracked propagation of a dike eastward from storage areas under Makaopuhi. Tiltmeters recorded the eastward advance of the dike tip as the ground bulged away from the swelling rift zone. These geodetic measurements indicate that the dike propagated laterally at about 550 m/h (about 10 m/min) and vertically at about 70 m/h, producing a dike about 15 km long, 2.5 km high, and 3.5 m wide.

The eruption started on January 3 from a fissure in Nāpau Crater, and over the course of the next several days, magma poured from a set of fissures that stretched nearly 8 km eastward along the rift zone. This long zone of **effusion** focused down to a smaller number of vents, including one that was given the name Pu'u'ō'ō (pronounced "poo-ooo oh-oh"). By June, Pu'u'ō'ō was the main show, and quite a show it was, as lava fountains reaching 470 m in height (taller than the top of the antenna on the Empire State Building!) jetted from the vent. Fragmented, frothy magma that was thrown into the air fell back from these fountains to produce a cinder cone 225 m high and 1400 m in diameter.

Effusion
Eruption of fluid lava.

Pu'u'ō'ō got into a regular habit of sending forth fountains for a day or so, then going quiet for three to four weeks, then repeating. Fountain, rest, repeat. Fragmented material thrown up in low fountains did not have much time to cool while in the air, hit the ground hot, and coalesced with other magma blobs, forming a lava pond that overflowed in lava rivers (Figure 7.2). Lava flows, both **pāheohoe** and **a'a**, destroyed 16 houses in the sparsely populated area between Pu'u'ō'ō and the coast in 1983 and 1984.

Pāhoehoe
Basaltic lava flows with a relatively smooth, folded surface that can look like coils of rope.

A'a
Basaltic lava flows with a rough, spiky surface.

In July of 1986 the Pu'u'ō'ō system broke down as activity shifted to a new fissure 3 km to the east. As with earlier episodes, this fissure narrowed down to a single vent. A large lava pond 140 × 300 m in diameter formed over this vent, and overflows from this pond built a low, shield-shaped mound around it. The overflow eventually was roofed over by cooled magma, forming a tube that insulated the flowing lava, and a network of tube-fed flows advanced toward the ocean, reaching it in November 1986, cutting the coastal highway, and overrunning 14 houses in Kalapana. In 1989 flows broke out in a new direction

(a)

(b)

Figure 7.2 (a) Lava river from Pu'u'ō'ō (on skyline) flowing over a small lava cascade, where the steel-gray, cooled surface is momentarily broken, revealing the bright orange lava within. A small bit of vegetation that was not overrun by lava bravely hangs on. Such oases are known as **kipukas**. Photo taken July 2007. (b) Walter's Kalapana Store and Drive Inn was burned and covered by lava in June 1990 as lava flows moved relentlessly through the area. Although the flows were typically <20 cm thick as they advanced over an area, many inflated dramatically. These flows inflated so much that only the sign is left; the flow top is ~4 m above the former ground surface.

Source: USGS.

and destroyed a National Park visitor center, permanently severing the coastal road. In 1990 lava flows returned to and completely overran the battered town of Kalapana, which ceased to exist (Figure 7.2b).

Activity at the new vent gradually diminished, and the pond crusted over in June 1990, ending five-and-a-half years of lava effusion. Eruptive activity then returned to Pu'u'ō'ō, where new cycles of eruptive activity began, finally ending in 2018.

The events of the 35-year eruption of the East Rift Zone encompass many common features of basaltic eruptions, including fissure and central-vent eruptions, lava fountains, cinder cones, multiple types of lava flows, and lava tubes. Eruptions at the other end of the compositional spectrum are dramatically different.

Krakatau, 1883

The Republic of Indonesia is host to over 100 active volcanoes scattered across many islands. Many of these volcanoes, including Agung, Galanggung, and Merapi, are well known for deadly eruptions, producing vivid sunsets around the world, or affecting air traffic. Toba, at the western end of Sumatra, produced one of the largest eruptions known about 74 ka. The eruption of Tambora in 1815 was the largest eruption in modern history and threw so much material into the atmosphere that it had a significant effect on climate worldwide. But the most notorious and well-known volcano in Indonesia is Krakatau.

Before 1883, Krakatau volcano consisted of a set of islands in the Sunda Strait, a narrow passage between the large Indonesian islands of Java and Sumatra. In the 19th century, the strait was a popular trade route to the Spice Islands, and the surrounding coastlines were well-populated. The capital city of Batavia (present-day Jakarta) lies about 100 km to the east, and the city of Buitenzorg (now Bogor), the same. All was well until May 20, 1883, when the volcano awoke after 203 years of slumber.

The eruption began early in the morning with a series of loud explosions that were heard over 150 km away. These emanated from a vent, Perbuwatan, on the northern part of the main island. The volcano produced an eruption column that rose over 11 km high and dropped ash over a wide area. A week later, the situation was stable enough that a group of 86 people chartered a steamboat to visit the island, still erupting an impressive column of gas and ash. As noted by Francis (1993):

> From their boat, the party saw that the [formerly lush, tropical] island was covered with fine white dust, like snow, and that trees on the northern part of the island had been stripped bare of their foliage by falling ash. Scrambling ashore and scuffling their way through ankle-deep ash, the party found that the centre of all the excitement was the Perbuwatan crater. This was in a state of semi-continuous activity, with minor explosions taking place every five or ten minutes, showering ash 200 metres into the air, and occasionally revealing the cherry-red glow of lava in the crater. A great banner of steam rose continuously 3000 m into the air. The visiting party, who were the first and last ever to get a good look at the crater of Perbuwatan, found it to be about a kilometre in diameter and fifty metres deep, with a small pit about fifty metres deep in the center of the crater floor. It was from this pit that the steam column was escaping with a great roar. (p. 74)

Kipukas
Islands of vegetation isolated by lava flows; may serve to isolate floral or faunal populations.

After this visit, on May 27, the volcano's activity waxed and waned for several months. A second and then a third column formed, and the remaining vegetation was destroyed. The island was in near-constant eruption, gearing up for the big event on August 26 and 27. The three days leading up to August 26 were marked by increasing activity at the multiple vents on the island. Around 2 p.m. observers on a British ship reported a black cloud rising to 25 km above the volcano (see Figure 7.3), and as the afternoon progressed, the explosions became more violent, and loud enough to be heard over all of Java. The eruption continued as night fell, and ships in the vicinity, under the glare of volcanic lightning, were in trouble. As recounted by Francis (1993):

> Captain Watson [of the British ship *Charles Bal*, 20 km from the island] could not see well enough through the murk to steer away to safety, but ironically the glare from the volcano provided a terrifying lighthouse with which to check his bearings. A less pleasant night can scarcely be imagined, with a rain of hot ash falling on the ship, and the air laden with choking fumes of sulphurous gases. As Captain Watson was to write after the eruption: "The night was a fearful one: the blinding fall of sand and stone, the intense blackness above and around us, broken only by the incessant glare of various kinds of lightning and the continued explosive roars of Krakatau, made our situation a truly awful one." He did not exaggerate. To make matters worse, peculiar pinky glows of static electricity lit up the mastheads and rigging of the ship with an unearthly light, known to mariners as St Elmo's Fire. (p. 77)

Figure 7.3 Krakatau in eruption during the 1883 event.

Source: Courtesy of Leiden University Libraries, Digital Collections.

The eruptive cycle came to a climax the morning of August 27 with four colossal explosions. The largest, at 10:02 a.m., was distinctly heard as far away as Rodriguez Island in the western Indian Ocean—a distance of 4800 km. By comparison, a similar explosion in Yellowstone National Park would be heard across the entire lower 48 United States, all of Alaska, nearly to Hawai'i, and as far south as Panama. This was the loudest sound in recorded history.

As serious as these explosions were, the main loss of life from the eruption came from a series of devastating tsunamis that pounded the lowlands around the Sunda Strait. An estimated 36,000 people lost their lives, and entire villages were washed away. Additional loss of life occurred when pyroclastic flows (flows of hot, degassing pumice and volcanic ash) traveled tens of kilometers across open water to bury low-lying areas.

After the massive explosions, activity at Krakatau tapered off, ceasing on August 28. Survivors staggered about in an ash-covered landscape; ship passage was blocked by great floating rafts of pumice. When people were able to get close to the source of the calamity, they found a remarkably changed landscape. The northern two-thirds of the island was gone, taking with it the Perbuwatan vent that started the eruptive sequence. A great cliff on the remaining part of the island faced a deep submarine hole—a caldera—that formed during the eruption.

In 1927 the tip of a new volcano, Anak Krakatau (Child of Krakatau), emerged from the sea in the center of what was left of the old volcano, eventually reaching a height over 300 m above the sea. Anak Krakatau has been the site of frequent eruptions, most relatively small. However, on December 22, 2018, the southwestern part of the cone collapsed into the Sunda Strait, generating a tsunami that killed over 400 people.

Summary

The Hawai‘ian eruptions described earlier represent one end of the spectrum of volcanic activity—relatively non-violent discharge of fluid basaltic magma. This activity is quiet and predictable enough that one can sometimes walk up close to flowing lava with minimal danger if simple precautions are taken. Krakatau represents a far more dangerous, violent, and unpredictable style of volcanism—explosive discharge of gas-rich, highly viscous (stiff), energetic magma. The Krakatau eruption was one of the largest and most destructive eruptions in recorded history, but it is sobering to realize that it only ejected about 1 percent of the volume of material erupted in the largest known events.

7.2 Volcanic Explosivity Index

Learning Objectives

By the end of this section you should be able to:

7.2.1 Estimate the Volcanic Explosivity Index of volcanic eruptions from observations of their behavior and output.

7.2.2 Compare and contrast the Volcanic Explosivity Index with the magnitude scale for earthquakes, and explain both to the public.

Volcanic Explosivity Index (VEI)
A logarithmic scale of the intensity and severity of volcanic eruptions, comparable in some ways to the moment magnitude scale for earthquakes.

As recounted earlier, volcanic eruptions vary dramatically in their energy, violence, and volume of ejecta. Newhall and Self (1982) proposed a measure for comparing eruptions, now widely used, known as the **Volcanic Explosivity Index (VEI)** (Figure 7.4). Like the

Relative erupted volumes and examples	Volcanic Explosivity Index	Erupted volume, km^3	Recurrence interval (approximate)	Relative volume in animal terms
Hawai‘i, Stromboli	1	<0.0001	Daily	*Mouse*
Galeras 1993, Sinabung 2010	2	0.001	A few weeks	*Hamster*
Soufrière Hills 1995, Lassen Peak 1915	3	0.01	3 months	*Rabbit*
Pelée 1902, Eyjafjallajökull 2010	4	0.1	A few years	*Dog*
Vesuvius 79, St. Helens 1980	5	1	10 years	*Human*
Krakatau 1883, Pinatubo 1991	6	10	50–100 years	*Polar bear*
Mazama 5600 BCE, Taupo 180, Tambora 1815	7	100	500–1000 years	*African elephant*
Yellowstone 640 ka, Toba 74 ka, Taupo 25 ka	8	1000	>50,000 years	*Blue whale*

Figure 7.4 Volcanic Explosivity Index, or VEI, with approximate erupted volumes and recurrence intervals. Volumes of red cylinders are proportional to ejecta volume and roughly to animal volume, with an eruption the size of Mount St. Helens (1980), VEI 5, corresponding to a human. The largest eruptions in recorded history are elephants, but the largest in geologic history are blue whales.

Source: Data from Simkin & Siebert (2000).

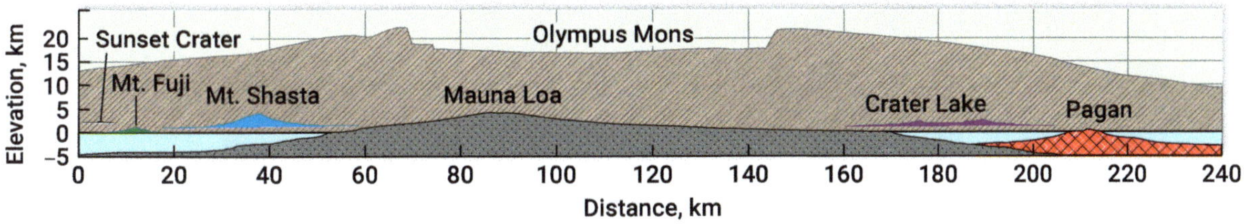

Figure 7.5 Cross sections of various volcanoes drawn to the same scale and at their appropriate elevations. On Earth, shield volcanoes such as Mauna Loa dwarf other kinds of volcanic edifices, but other planets have volcanoes that dwarf our largest shield volcanoes, such as Olympus Mons on Mars. Pagan volcano was featured in Figure 6.3.

moment magnitude scale of earthquake energy release, VEI is a logarithmic measure of an eruption's intensity. However, unlike the seismic scale, which is a quantitative measure of energy release, the VEI is a composite, semiquantitative measure based on "magnitude and/or intensity and/or destructiveness and/or (less frequently) dispersive power, violence, and energy release rate, depending on which data were available" (Newhall & Self, 1982, p. 146); these in turn are based on modern or historical records of ejecta volume, column height, qualitative descriptions by observers (e.g., "colossal"), eruption duration, and other measures. Eruptions of VEI 8 are commonly referred to as **supereruptions**.

The VEI scale ranges from 0 (modest eruptions such as the recent Kīlauea events) to 8 (the largest known prehistoric eruptions); the largest eruption in the last few centuries, the 1815 eruption of Tambora, has been assigned a VEI of 7. Each step in VEI corresponds to an approximately 10-fold increase in ejecta volume (except for the smallest steps) and an approximately 10-fold increase in recurrence interval (the mean time between eruptions of that size). With analogy to floods, if we say that the 1883 eruption of Krakatau was a "100-year eruption," then in a given year, there is approximately a 1 percent chance that a Krakatau 1883–sized eruption will occur, a 10 percent chance that a 10-year Mount St. Helens 1980–sized eruption will occur, etc.

The spectrum of volcanic activity is reflected in the wide variety of volcano shapes and sizes (Figure 7.5). These in turn are related to the composition and gas content of the magma.

Supereruptions
Infrequent eruptions that release on the order of 1000 km^3 or more of magma.

Pyroclastic
Magma fragmented and blown into the air during eruption; "fire broken."

Making Connections 7.1

Volcanism, Art, and Literature

Many of the works of the British landscape painter J. M. W. Turner are thought to have been influenced by the particularly orange and red skies above European sunsets after the Tambora eruption (Figure 7.6a). Zerefos et al. (2007) studied hundreds of paintings produced between 1500 and 1900 and found a statistically significant correlation between measured red-to-green ratios in paintings and observational measures of atmospheric dust and aerosol contents. The particularly vivid and menacing orange sky in Edvard Munch's chilling painting *The Scream* is typically attributed to dust from the Krakatau eruption (Figure 7.6b).

In popular culture, more than paintings were affected by the dreary weather of the summer of 1816 brought on by the Tambora eruption. Mary Shelley, vacationing on Lake Geneva in the gloomy, wet cold, began the novel *Frankenstein*. She noted in the introduction to the 1831 edition:

> But it proved a wet, ungenial summer, and incessant rain often confined us for days to the house … "We will each write a ghost story," said Lord Byron (Shelley, 1831).

And thus a classic, in which foul weather plays many roles, was born. Byron, living on Lake Geneva at the time, penned the poem *Darkness*, which contains these frightful lines:

> I had a dream, which was not all a dream.
> The bright sun was extinguish'd, and the stars
> Did wander darkling in the eternal space,
> Rayless, and pathless, and the icy earth
> Swung blind and blackening in the moonless air;
> Morn came and went—and came, and brought no day (Byron, 1816, p. 27)

continued

Figure 7.6 (a) *The Decline of the Carthaginian Empire* by J. M. W. Turner, 1817. Rich orange colors may have been a response to atmospheric disturbances from the Tambora eruption. (b) *The Scream* by Edvard Munch, 1893, one of four versions of this haunting painting, inspired, perhaps, by the blood-red sunsets caused by Krakatau.

Sources: (a) Tate Museum; (b) National Museum of Art, Architecture and Design/Edvard Munch.

7.3 Physical Properties of Magma

Learning Objectives

By the end of this section you should be able to:

7.3.1 Explain how changes in temperature, water content, and silica content affect the viscosity of a magma.

7.3.2 Use the known effect of solid particles on fluid viscosity to explain why lavas with more than 50 percent crystals are rarely erupted.

7.3.3 Describe how the temperature of erupted magma correlates with silica content and give typical magma temperatures.

The nature of a volcanic eruption depends almost completely on the physical properties of the magma involved, and this in turn is governed by the bulk composition of the magma. Low-silica (e.g., basaltic) magmas with low concentrations of dissolved gases produce the quietest eruptions; high-silica (rhyolitic) magmas with high concentrations of dissolved gases produce the most energetic and explosive eruptions. Position along the BADR spectrum governs how viscous (stiff) the magmatic liquid is, and the content of dissolved gases governs how much oomph a given eruption can muster. Arc volcanoes erupt the entirety of the BADR spectrum, so they exhibit most of the volcanic behaviors known on Earth.

Viscosity

What Is Viscosity? Magma **viscosity** is the dominant control on volcano behavior, and estimated values of viscosity for magmas cover a range of over 10 orders of magnitude. High viscosity means that applying a large stress to a substance results in only a small rate of deformation; you can think of viscosity as the resistance of a fluid material to deformation under applied stress, such as gravity acting on lava flowing down a slope. The SI unit for viscosity is pascal-seconds, or Pa·s. Older data commonly give viscosity in units of poise, where 1 Pa·s = 10 poise. Approximate viscosities of a few substances of geologic and culinary interest are shown in Figure 7.7. Although hot basalt has a viscosity similar to that of chilled honey, the viscosities of silicic magmas are vastly higher.

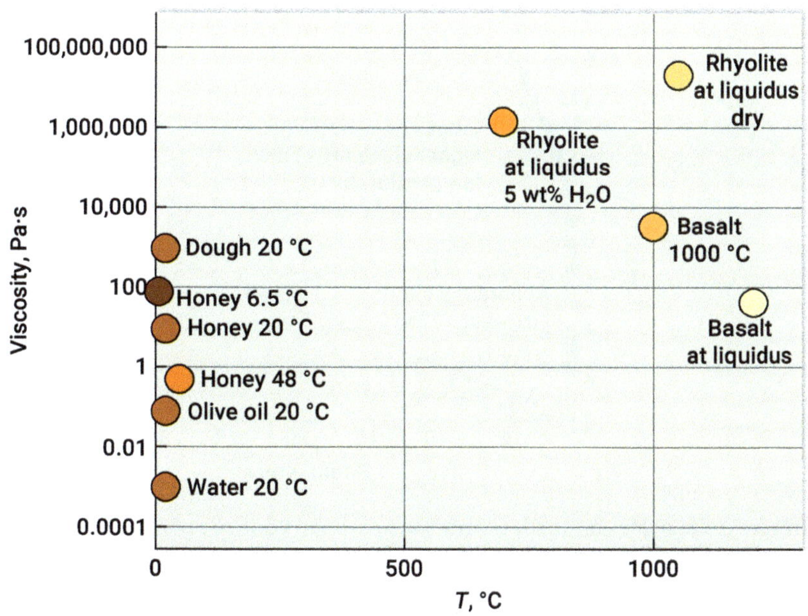

Figure 7.7 Viscosity of various materials plotted against temperature; note logarithmic scale.

Sources: Water: Kestin et al. (1978); olive oil: Diamante & Lan (2014); honey: Steffe (1996); magma viscosities estimated using Giordano et al. (2008).

Viscosity, dynamic viscosity Resistance of a fluid to flow under shear stress.

Viscosity (specifically **dynamic viscosity**) is defined conceptually as:

$$\mu = \frac{\text{stress}}{\text{strain rate}} = \frac{\sigma}{\dot{\varepsilon}} \quad \textbf{(Eq. 7.1)}$$

where μ is the dynamic viscosity, σ is shear stress, and $\dot{\varepsilon}$ is the strain rate. For such a material, stress is strictly proportional to strain rate (Figure 7.8a). A fluid exhibiting this relationship is called a **Newtonian fluid**, and magmas are broadly Newtonian. A **Bingham fluid** is like a Newtonian fluid except that it does not begin to flow until a critical yield stress is applied.

Newtonian fluid A fluid for which shear stress and strain rate are strictly proportional.

Bingham fluid A fluid in which a critical yield stress must be exceeded before it starts to flow.

Figure 7.8b shows real data for partially crystallized Hawai'ian basalt magma at different temperatures. The magma was approximately 10, 20, and 30 percent crystallized at 1170, 1150, and 1125 °C. The 1170 and 1150 °C runs show Newtonian behavior, with linear relationships between stress and strain rate that go through the origin. Viscosities are simply the slopes of the lines: $\mu = 960$ Pa·s at 1170 °C and 2460 Pa·s at 1150 °C. For the 1125 °C run, things are more complicated, but these data can be approximated as a Bingham fluid with a yield stress of approximately 10 kPa.[1] At higher strain rates, the data are crudely linear with a significantly steeper slope, corresponding to greater resistance to flow and thus a higher effective viscosity.

The data in Figure 7.8b illustrate two important points about magma viscosity. First, viscosity has a strong negative exponential dependence on temperature. Second, magma with a significant crystal content can exhibit decidedly non-Newtonian behavior owing to interactions between the crystals.

Effects of Composition, Water, and Temperature

Note from Figure 7.7 that the viscosity of water-free (anhydrous) rhyolite at its liquidus is about 20 billion times greater than the viscosity of water, and about half a million times greater than the viscosity of a water-free basalt at its liquidus. Using an empirical model for how viscosity varies with composition and T (Giordano et al., 2008), we can calculate viscosity along the BADR trend for both anhydrous and hydrous magma (Figure 7.9). Across the compositional spectrum at a given temperature, the viscosity of magma with 5 wt% H_2O is several orders of magnitude lower than that of equivalent dry magma. However, water-rich magmas exist at significantly lower temperatures, and this partially counteracts the effect.

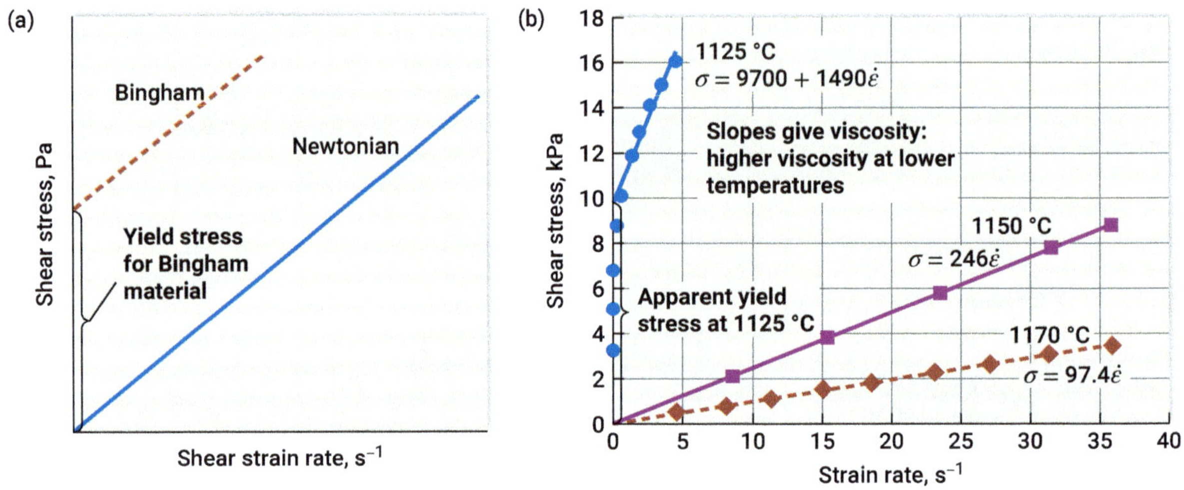

Figure 7.8 (a) For a Newtonian fluid, stress and strain rate are strictly proportional, and the slope of a plot of stress vs. strain rate is the viscosity. For a Bingham plastic fluid, a yield stress must be applied before the fluid begins to flow, after which it is pseudo-Newtonian, with strain rate proportional to stress minus yield stress. With stress in SI units of Pa and strain rate in reciprocal seconds, the units of viscosity are thus Pa/s^{-1} or Pa·s. (b) Laboratory measurements of the viscosity of Hawaiian basaltic magma at three temperatures. Plots of shear stress vs. strain rate are linear and go through the origin at the two higher temperatures, and the slope of the resulting curve is the viscosity. At the lowest temperature, the curve is distinctly nonlinear, but the remainder of the data can be approximated by a line with a non-zero intercept on the ordinate. This is Bingham behavior, and the intercept, here ~10 kPa, is an estimate of the yield stress.

Source: (b) Data from Shaw (1969).

The massive viscosity differences between magmas with compositions at the ends of the BADR spectrum result in radically different behaviors. For example, the Jeffreys equation relates the velocity of fluid flow down an inclined surface to its viscosity:

$$\mu = \frac{g\rho(\sin\alpha)d^2}{nV} \qquad \textbf{(Eq. 7.2)}$$

where μ is viscosity, g is the gravitational acceleration, ρ is density, α is the slope, d is the thickness of the flow, n is a constant (3 for broad flows and 4 for narrow, channelized flows), and V is velocity. According to this equation, the basalt from Figure 7.7 at 1200 °C would flow down a 2° slope, a typical slope on the flanks of Hawaiian volcanoes, at 6.8 m/s, or about 15 miles/hr—you would not be able to outrun that for long unless you can run a 4-minute mile pace. In contrast, the dry rhyolite in Figure 7.7 would flow at about 1.4×10^{-5} m/s, or roughly 1 m/day.

Viscosity shows a strong negative dependence on temperature. This is evident from the honey data in Figure 7.7, and is evident to anyone who has taken maple syrup or hot fudge sauce out of the refrigerator and heated it in a microwave oven. Given the range of temperatures over which igneous rocks erupt (e.g., from >1200 to <800 °C), this clearly plays a role in erupted magma viscosities.

Bond strength
For an ion, charge divided by coordination number.

Silica plays a dominant role in determining the viscosity of a magmatic liquid because of the great strength of the Si-O bond. One measure of **bond strength** is charge divided by coordination number. By this definition, Si-O bonds have a bond strength of 1, and Al-O 0.75. Other cations have lower bond strengths; for example, Mg in octahedral coordination has a bond strength of 0.33 (2/6), and Na in 10-fold coordination only 0.1. Thus, SiO_4 tetrahedra are strong, compact units, and liquids with a significant number of Si-O bonds are difficult to deform, because flow requires continuous breaking of these strong bonds.

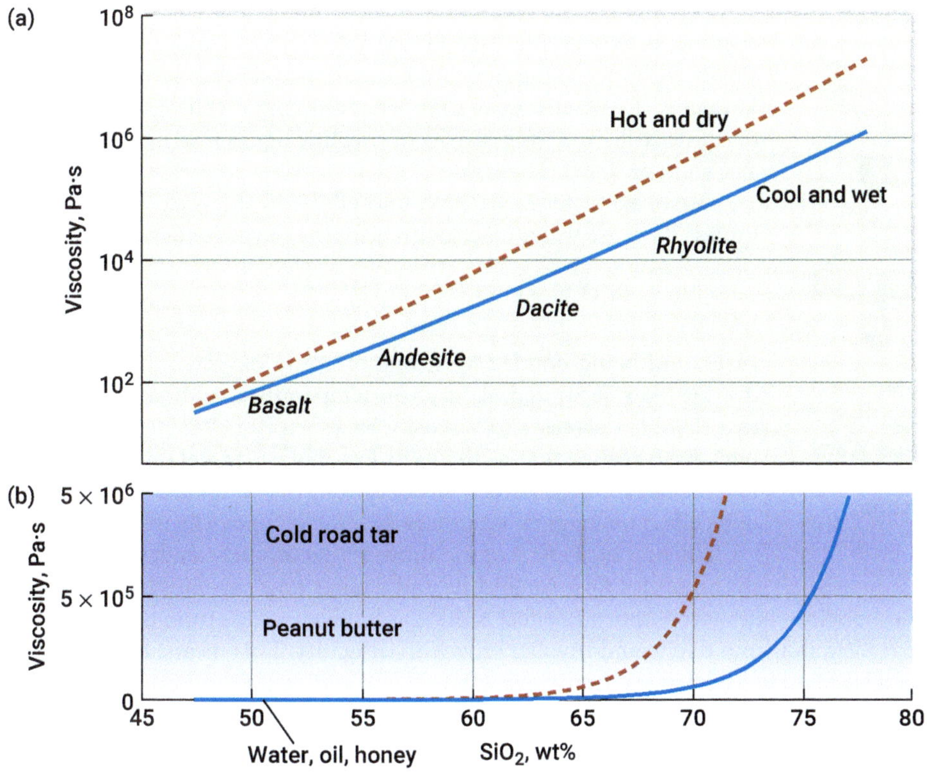

Figure 7.9 (a) Viscosity calculated along a basalt-rhyolite mixing line at liquidus temperatures for hydrous and anhydrous compositions. End-member temperatures used were 1200 and 1050 °C for anhydrous basalt and rhyolite, and 1050 and 700 °C for basalt and rhyolite with 5 wt% H_2O added. Although addition of water to a magmatic liquid greatly decreases its viscosity, it also decreases liquidus temperatures, and these two effects, as modeled here, largely cancel out. Viscosity of anhydrous rhyolite is 500,000 times greater than anhydrous basalt; adding 5 wt% water to the system decreases this ratio to about 50,000. Water-added data are plotted at anhydrous silica compositions. (b) Same data plotted on a linear scale capped at 1,000,000 Pa·s, with viscosities of everyday substances noted. With increasing SiO_2, once a magmatic liquid reaches ~65–70 wt%, its viscosity skyrockets from the realm of things that flow on short timescales (e.g., honey and peanut butter) to much more viscous materials that behave as solids over short timescales. This range marks a significant transition in the behavior of magma and resulting volcanoes.

Sources: Basalt and rhyolite compositions from Ussler & Glazner (1989) and viscosity model from Giordano et al. (2008).

Two tetrahedra can share an oxygen atom to form an Si_2O_7 group, and the oxygen they share is called a **bridging oxygen** (Figure 7.10). The Si-O-Si chain is quite strong, and large networks of such chains make a silicate liquid difficult to deform.

Bridging oxygen
An oxygen atom shared between two Si or Al tetrahedra.

Silicate liquids have structures that are related to minerals of comparable composition. For example, olivine is an orthosilicate, with SiO_4 tetrahedra isolated from one another, and olivine-rich (low-silica) liquids have SiO_4 tetrahedra that are generally isolated as well. Tetrahedra may hook up here and there, but most are isolated. As the silica content goes up, however, more and more tetrahedra are linked until, at a rhyolite composition with about 77 wt% SiO_2, the liquid has a bulk composition equivalent to a mixture of quartz, albite, and orthoclase, and essentially all the oxygen atoms are potentially bridging oxygens. This gives dry rhyolitic liquid its extreme viscosity (see Figure 7.7).

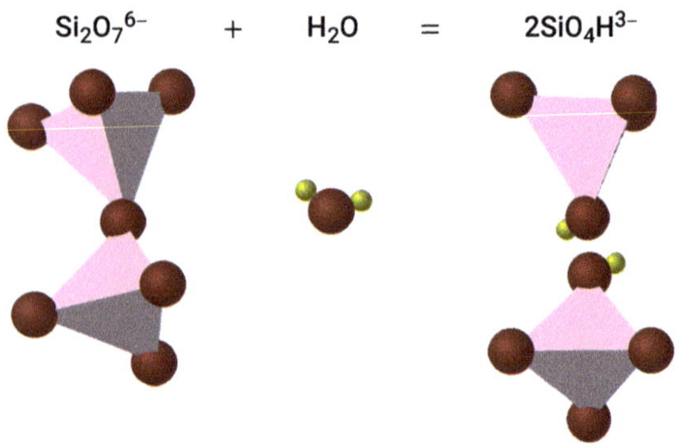

Figure 7.10 Si tetrahedra can be linked by an oxygen atom, and such bridging oxygens can form strong chains and make a highly viscous liquid. Water dissolved under pressure splits these Si-O-Si links, breaking up networks and drastically decreasing viscosity.

The increase in viscosity with silica content, going from basalt to rhyolite, is exponential (see Figure 7.9). Viscosity increases relatively slowly with increasing silica from basalt to andesite, but from andesite to dacite, the magma goes from what one might call "relatively fluid" to "quite stiff," and the viscosity of rhyolitic liquid is so high that it acts as a solid on short timescales.

Dissolved water has a profound effect on the viscosity of silicate liquids, because water reacts with bridging oxygens via the reaction:

$$Si_2O_7^{6-} + H_2O = 2SiO_4H^{3-} \qquad \textbf{(Eq. 7.3)}$$

Instead of being bonded by strong bridging oxygens, the Si tetrahedra are now separated. At a given temperature, adding 5 wt% H_2O to a rhyolite decreases its viscosity by a factor of over 10,000. It is important to realize that owing to its low molecular weight, 5 wt% H_2O means that about 25 percent of the cations in the liquid are H. In terms of its effect on magma, water clearly punches above its weight class.

The practical effect of the above is that basalts can flow quite readily, andesites flow sluggishly, and dacites and rhyolites barely flow at all. A less obvious but more important effect is that gas bubbles escape from basalt and andesite readily, but from dacite and rhyolite only with difficulty. This means that gas pressure can build up to high values in silicic magmas, driving violent eruptions.

Effect of Crystals The presence of crystals in a magma also affects its viscosity, an effect familiar to anyone who has made chocolate chip cookies. The egg-flour-sugar batter is viscous but easy to stir; add a few chocolate chips and it is still easy to stir, but keep adding them and it gets far more difficult to stir—its effective viscosity goes up rapidly as more chips are added. This increase in viscosity occurs when there are enough solid particles (chocolate chips or crystals) that they begin interfering with one another as the mixture is stirred. Estimates of how particles increase viscosity go back to Einstein (1906) (yes, *that* Einstein) for very dilute solutions, and later, Roscoe (1952) derived a relationship between relative viscosity and abundance of spheres of many sizes. This problem has been addressed over and over in geology owing to its importance in how magma and sediment slurries flow (Petford, 2003). The following empirical equation, fit to data in Costa (2005) and shown in Figure 7.11, can be used to estimate how the viscosity μ of a magma whose crystal-free viscosity is μ_0 varies with crystal fraction c:

$$\log_{10}\left(\frac{\mu}{\mu_0}\right) = \frac{9.4}{1 + e^{(7-10c)}} \qquad \textbf{(Eq. 7.4)}$$

This indicates that magma viscosities must increase dramatically as the volume percent crystals increases beyond about 50 percent. This is consistent with experimental data (e.g., Sato, 2005) and with a fundamental observation in igneous petrology: magmas with more than about 50 volume percent crystals are rarely erupted (Figure 7.11b). As noted by Marsh (1981),

> it is clear that lavas with more than 55% (vol.) phenocrysts are rare. It seems that, beyond a certain degree of crystallinity, the magma is apparently too viscous to be erupted. This cutoff in phenocryst content is sharp and may imply a sudden change in rheology from that of a fluid to essentially that of a solid. As a magma cools, its chances of erupting diminish as its crystallinity increases and approaches this transition or locking point (p. 85).

Effect of Bubbles Magmas commonly contains bubbles, and these, too, affect viscosity. However, the direction in which they change viscosity depends on a number of

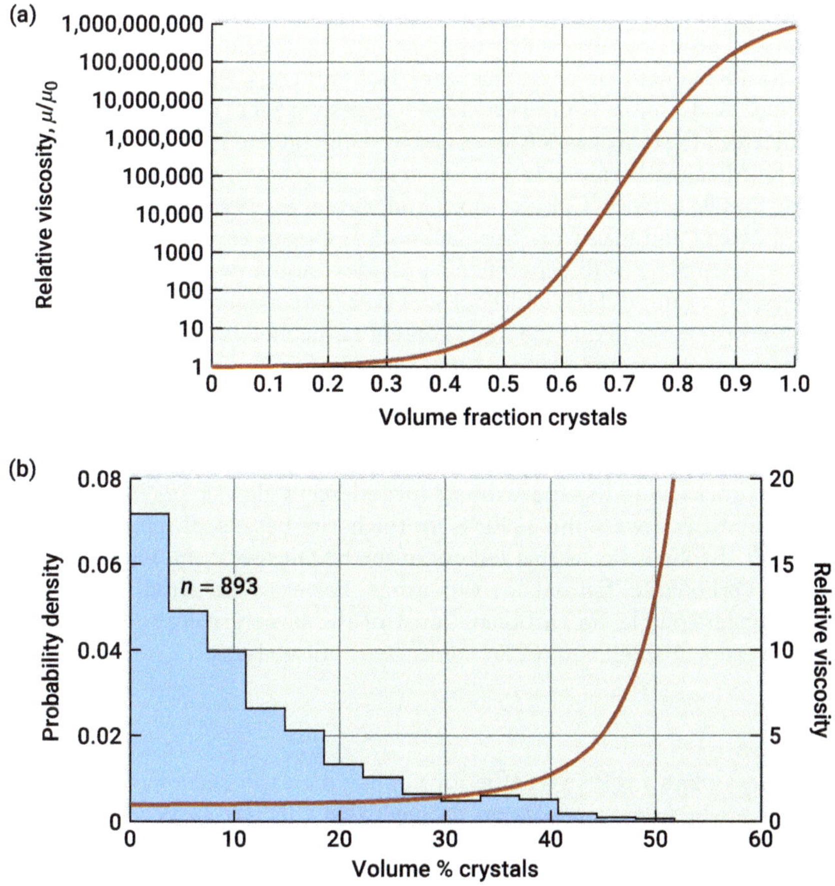

Figure 7.11 (a) Relative viscosity of a suspension of particles (e.g., crystals in magma). Relative viscosity rises slowly as volume fraction of crystals *c* rises, increasing by a factor of 10 when *c* reaches 0.5. At $c = 0.6$, it has risen by a factor of ~200, at 0.7 by ~100,000, and so on. At these high crystal contents, flow of partially molten rock requires crushing and plastic deformation of crystals. (b) Histogram of volume percent crystals in 893 volcanic rocks shows a rapid falloff with increasing crystal content, and none with >50%. Curve shows relative viscosity from (a). The curve rockets upward at ~50% crystals. This is consistent with the hypothesis that magmas more crystal-rich than this are too rigid to flow and erupt.

Source: (a) Data from Costa (2005). (b) Data from the NAVDAT dataset.

factors, including bubble size and shear rate. At low strain rates, bubbles typically reduce bulk viscosity, whereas at high strain rates, they increase it (Scaillet et al., 1998). Compared to the effects of dissolved water, temperature, and silica content, though, these effects are minor, on the order of a factor of two or three.

Hyperviscous Liquids Viscosity comes up often in discussing the behavior of magmas and volcanoes. It is common to refer to dacitic and rhyolitic magmas as being "viscous," meaning "really viscous," or "extremely viscous," but this term is not really adequate for incredibly stiff liquids such as rhyolite magmas. Here, we will use the term **hyperviscous** to refer to magmatic liquids with particularly high viscosities. The pitch (asphalt) dripping through a funnel in the famous Pitch Drop Experiment at the University of Queensland would qualify as hyperviscous; since the experiment began in 1930, nine drops have fallen, leading to a calculated viscosity on the order of 10^7 Pa·s—a viscosity comparable to that of dry silicic magma.

Hyperviscous
Having an extremely high viscosity, on the order to 10^7 Pa·s or more, such that the material seems nearly solid at human time scales.

Temperature

The temperature at which magma comes out of the ground is an important parameter in petrology, but it is surprisingly difficult to measure directly. The most obvious method is to stick a thermocouple into the lava, but this is only practical for calm, predictable eruptions of lava with a low enough viscosity that the thermocouple can actually be planted in it. A safer method is optical pyrometry, in which a telescope with a glowing filament in the field of view is trained on incandescent lava. The current going through the filament is adjusted until the filament color matches that of the lava, and the temperature is derived via a calibration curve for black body radiation. This method is best performed at night and only works for lavas in which the glowing interior is routinely exposed.

Other methods for determining the temperature of erupting lava depend on laboratory analysis of the erupted products. One widely used method involves analyzing the compositions of ilmenite and magnetite in rocks that have both minerals, which includes many dacites and rhyolites (Buddington & Lindsley, 1964; Ghiorso & Sack, 1991); the partitioning of Fe and Ti between the two phases is governed by temperature. This is an example of a geothermometer. Other geothermometers useful in igneous rocks include ones based on partitioning of Mg and Fe between Cpx and Opx (Andersen et al., 1993) and on O-isotope exchange between quartz and magnetite (see Section 5.4; Downs et al., 1981). Matching the observed phase assemblage with experiments provides another way to estimate magmatic temperatures. For example, Helz and Thornber (1987) studied crystallization of Hawai'ian basalt at 1 atm and found that the MgO content of the melt is a sensitive function of temperature, because olivine and Cpx are highly enriched in MgO and rapidly take it out of the liquid as temperature drops. Thus, the MgO content of the glass in a quenched sample provides an empirical geothermometer for rocks of that bulk composition.

Figure 7.12 summarizes observed and calculated temperatures of eruption of a broad spectrum of erupted rocks, with the approximate incandescent glow colors indicated. In general, these temperatures are around 1200 °C for the hotter basalts, dropping to about 700 °C for the cooler rhyolites. Rocks that fall out of the BADR spectrum, such as MgO-rich komatiite and carbonatite, fall outside this range. For example, Krafft and Keller (1989) found that water-soluble Na carbonate lavas of the bizarre volcano Ol Doinyo Lengai in Tanzania erupt at about 500 °C, far cooler than silicate lavas.

Density

The density of magma plays a role in whether or not magma can rise by buoyancy, and the density differences between crystals and the melt that they are suspended in controls whether those minerals will sink or float. Densities of magmas along the BADR

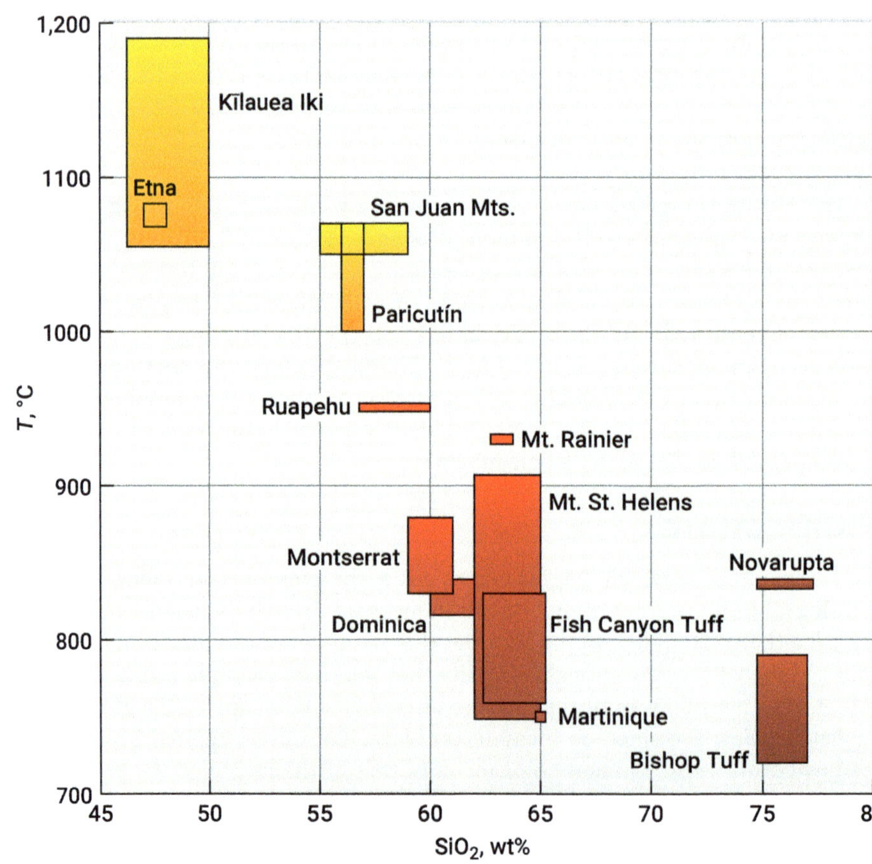

Figure 7.12 Approximate temperature values for selected magmas, measured by thermocouple or pyrometer, estimated from mineral equilibria, or estimated by experimental petrology. Each rectangle encompasses the maximum and minimum estimated temperatures and silica contents, and the approximate incandescence colors are shown.

Sources: Data from Krauskopf (1948), Wilcox (1954), Eaton & Murata (1960), Downes & Cristofolini (1973), Hildreth (1977), Archambault & Tanguy (1976), d'Arco et al. (1981), Askren et al. (1991), Donoghue et al. (1995), Venezky & Rutherford (1997), Coombs & Gardner (2001), and Lindsay et al. (2005).

trend range from about 2100 kg/m^3 for water-rich rhyolitic liquids to 2800 kg/m^3 for water-poor, crystal-rich basaltic magmas. We defer a fuller discussion of density to the next chapter.

7.4 Products of Volcanoes

Learning Objective

By the end of this section you should be able to:

7.4.1 Enumerate the various solid, liquid, and gas products emitted by volcanoes.

Volcanoes emit three main types of material: **lava**, pyroclastic material fragmented explosively during ejection from the Earth (**tephra** is a useful term that encompasses all pyroclastic material), and volcanic gases, chiefly H_2O, CO_2, and SO_2. The various types of lavas and pyroclastic rocks are discussed in Section 7.6. Many volcanoes also produce **mudflows** (also known by the Javanese term **lahars**)—highly destructive debris flows lubricated by water.

Lavas come in three varieties: runny pāhoehoe, less-runny a‘a, and stiff, slow-moving **block lava**. Tephra is named according to its size; particles less than 2 mm across are **ash**, particles between 2 and 64 mm are **lapilli**, and anything larger than 64 mm across is a **bomb** or a **block**. The distinction between bombs and blocks is based on whether the glob was soft when airborne: bombs show evidence of deformation during flight and impact, whereas blocks were solid when ejected and have angular outlines. Cow-dung bombs bring to mind a particular style of deformation during impact.

Volcanic gas compositions are measured in a number of ways. The most obvious and direct way to measure them is by direct sampling: the intrepid and suitably protected geochemist sticks a chemically inert and heat-resistant probe into the source, attaches an evacuated bottle to it, and sucks in the gases for analysis. Such sampling is generally done at **fumaroles**, vents that emit high-temperature gases (Figure 7.13).

The stew of volcanic gases varies from volcano to volcano. Typically, H_2O is the most abundant gas, followed by CO_2 and SO_2. Other significant gases are H_2S, H_2, CO, HCl, and HF. Subduction volcanoes are typically 95 mol% or more H_2O (Mount Etna in Sicily is a conspicuous exception), whereas divergent-margin and hotspot volcanoes typically have less than 90 mol% H_2O and significant amounts of CO_2 and SO_2.

Sulfur dioxide is a particularly important component of volcanic plumes because of its effects on the amount of sunlight reaching the Earth. Large eruptions can drive gases and tephra into the stratosphere, above the reach of weather systems that can flush them out. Sulfur dioxide reacts with OH^- to form an aerosol of sulfuric acid, H_2SO_4. Solid tephra particles settle out of the stratosphere efficiently, but the H_2SO_4 remains for months or years, reflecting sunlight and lowering global temperatures if abundant enough.

Lava
Molten material that flows out of the ground as a fluid.

Tephra
Term encompassing all pyroclastic material.

Mudflows, lahars
Gravity-driven flows of water-lubricated clastic material down the side of a volcano; may be hot or cold and need not be produced by an eruption.

Block lava
Viscous to hyperviscous lava that breaks into semiangular blocks during flow.

Ash
Pyroclastic particles <2 mm across (smaller than lentils).

Lapilli
Pyroclastic particles between 2 and 64 mm across (lentils to tennis balls); singular lapillus.

Bomb, block
Pyroclastic material >64 mm (tennis ball) across; bombs were molten when ejected, whereas blocks were solid.

Fumaroles
Hot volcanic gas vents.

Figure 7.13 Fumaroles in the crater of Vulcano, in Italy's Aeolian Islands, color the rocks around them yellow owing to elemental sulfur deposited by the hot gases. Sulfur may form via several mechanisms, including oxidation of H_2S by atmospheric oxygen to form $H_2O + S_8$. Clouds around the fumaroles are choking, SO_2-laden miasmas. Hikers beware! Inset shows a USGS scientist with a respirator collecting gases from a fumarole within Kīlauea caldera.

Sources: AFG photo; inset USGS.

7.5 Magma Transport in the Crust

Learning Objectives

By the end of this section you should be able to:

7.5.1 Use the physics of fluid and heat flow to explain why new lava eruptions typically start as a linear curtain of fire as a dike breaches the surface, and then choke down to one or more fire fountains as the eruption progresses.

7.5.2 Estimate the Reynolds number, and thus the likelihood that flow was turbulent, for various magma compositions and flow situations.

The typical textbook depiction of a volcano shows a conical shape sitting atop a tube that supplies the volcano with magma from a shallow tank. There are many debatable aspects of these illustrations, but perhaps the most misleading is that they give the perception that magma moves through the crust in pipes. Magma transport generally takes place in tabular fractures rather than pipes, although fractures filled with magma can freeze down in places to pipe-like conduits. These fractures may form owing to magma overpressure, which cracks the crust, or they may be preexisting fractures that invading magma exploits. Regardless, magma does not establish a conduit by pushing a cylinder of crust ahead of it, as would be required to form a pipe.

The magma that erupts from volcanoes is supplied by dikes (see Section 10.2). One might assume, then, that most eruptions would be long, linear affairs, and in fact many eruptions do begin that way, such as the Hawaiʻian eruption described in Section 7.1. However, cooling of magma generally causes fissure-type eruptions to choke down to one or more vents. This happens because for laminar (i.e., non-turbulent; see Toolbox 7.1) flow of a fluid in a planar crack, driven by excess pressure, the magma flux (volume passed per unit time) goes up as the cube of the width of the crack. Thus, if there are any variations in the width of the crack, and there always are, the magma flows more slowly in the narrow parts, which tend to freeze up. This forces the remaining magma out the non-frozen parts, where it flows even more rapidly, leading to a winner-take-all scenario where one part of the dike gets all the magma flow and the rest freezes (Figures 7.14 and 7.15). By this process, a dike that breaches the surface at several places can evolve into a single, sustained vent.

Stresses created by the topographic load of a volcano can influence dike orientations. A growing volcano weighs on the rocks below it and induces stresses that encourage them to spread under the influence of gravity. The effect of gravitational spreading can be studied with analog models, using a material such as gelatin to mimic the volcanic edifice (Fiske & Jackson, 1972; Hyndman & Alt, 1987). These experiments show that fluid injected into a ridge-like feature will form vertical dikes parallel to the ridge, and cone-like features will produce dikes that radiate from the cone's central axis. Imagine pressing down on the apex of a gelatin cone; the weight of your hand will cause the base of the cone to spread, creating tension around the circumference and radial cracks. **Radial dikes** are a common feature of the larger volcanoes (Figure 7.16) and commonly lead to eruptions on the flank of the volcano.

Radial dikes
Vertical dikes that extend radially from a magma source such as a volcanic center.

(a)

(b)

Figure 7.14 (a) Looking down on a fissure eruption from a dike breaching the surface along the East Rift Zone of Kīlauea, May 2018. Lava flows spread out, buried, and incinerated the tropical forest here on the rainier side of the island. (b) Remarkable lava fountain 450 m tall (height of the Empire State Building in New York City) at Puʻuʻōʻō, 1984. This lava fountain developed after the eruptive episode opened with a long fissure eruption.

Sources: (a) Bruce Omori/Paradise Helicopters/EPA; (b) Christina Heliker, USGS.

Figure 7.15 Speculative cross section of the rhyolite dike that fed Holocene eruptions of domes along the Inyo trend north of Mammoth Lakes, California. The Inyo domes erupted along a north-south line, and a hole slant-drilled between Obsidian Dome and Glass Creek Dome intersected the 7-m-thick dike about 640 m below the surface. Drilling near Inyo Craters, steam explosion craters where no magma erupted, also intersected a dike at depth. The dike is thought to be segmented into **en echelon** segments; dotted lines mark the general surface trace of these segments. These form a left-stepping en echelon set, consistent with a component of right-lateral movement along the north-striking normal fault system that the dike strikes parallel to.

Sources: Adapted from Eichelberger et al. (1984) and Reches & Fink (1988).

En echelon
Said of a series of overlapping, parallel cracks, dikes, etc. that form a linear trend where each element is oblique to that trend.

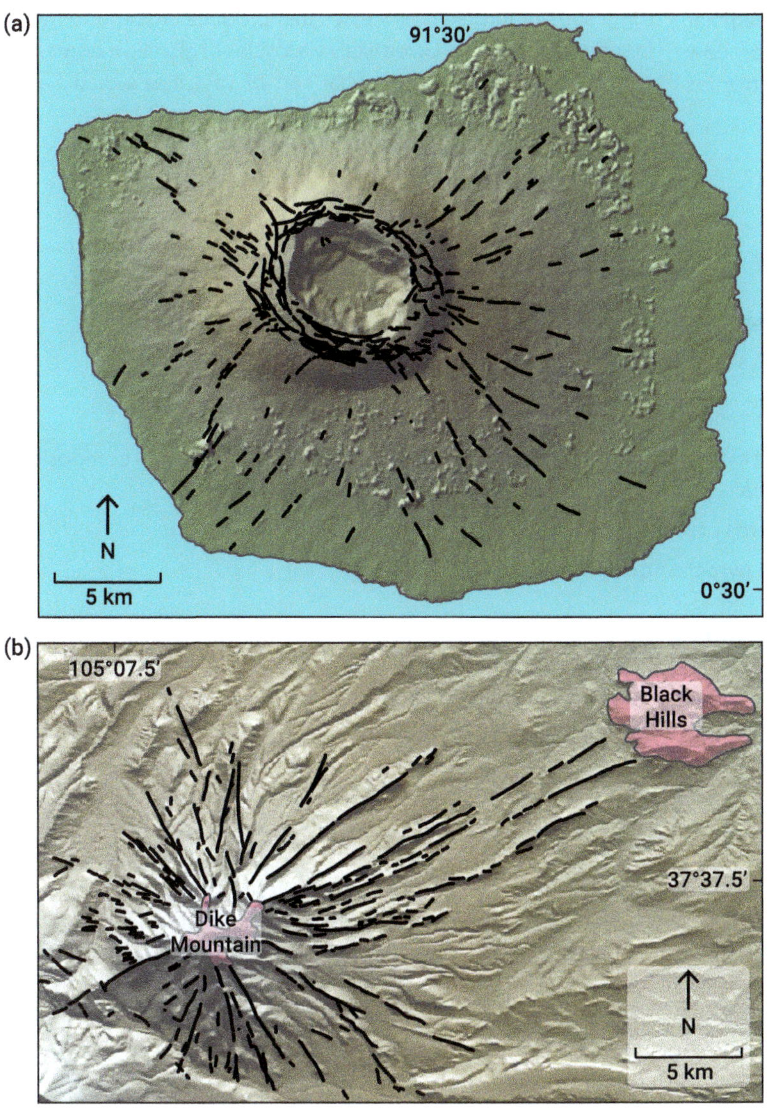

Figure 7.16 (a) Radial fissures and cones inferred to lie above radial dikes on Fernandina, Galápagos. (b) Dikes exposed in the eroded core of Dike Mountain, Spanish Peaks, Colorado. Both Dike Mountain and the Black Hills are cored by monzonite, and the dikes are dominantly the same.

Sources: (a) Modified from Chadwick & Dieterich (1995); (b) modified from Johnson (1961).

Toolbox 7.1

Laminar vs. Turbulent Flow and the Reynolds Number

Most magmas are viscous enough that they flow in a laminar manner; that is, the fluid flows in parallel layers, with no eddies and little mixing between layers. Turbulent flow, in contrast, involves chaotic motion and abundant mixing of the fluid. The Reynolds number (Re) is an important dimensionless number that is used to estimate whether a given fluid flow will be laminar or turbulent:

$$\mathrm{Re} = \frac{\rho v L}{\mu} \qquad \textbf{(Eq. 7.5)}$$

where ρ is the fluid density, v its velocity, L a length scale, and μ the viscosity. Re is the ratio of two quantities: inertial forces from the velocity and mass of the fluid (numerator) and viscous forces (denominator). When inertial forces dominate, flow is turbulent; when viscous forces dominate, it is laminar. Critical values for the transition from laminar to turbulent flow differ for different geometries; for flow between parallel plates (e.g., a dike), the critical regime lies around 2300.

Let's try this out. The viscosity of water at room temperature is about 0.001 Pa·s. For water flowing in a small stream 20-cm thick at 1 m/s (about the speed of an easy stroll), Re = $\sim 2 \times 10^5$, well into the turbulent regime. Magma, owing to its far higher viscosity, is different. For basaltic magma with a viscosity of 10 Pa·s and a density of 2500 kg/m^3, flowing through a 1-m dike at 1 m/s, Re = ~250, well below the value needed for turbulence. A velocity of 1 m/s is rather high compared to observed velocities of basaltic dike propagation, and 10 Pa·s is at the low end of typical magma viscosity (see Figure 7.9), so it is clear that magma flow is almost always laminar. Only for especially low-viscosity magmas, such as carbonatites or komatiites, is turbulent flow possible.

Granitic rocks commonly display layering defined by mafic minerals that resembles cross-bedding in sedimentary rocks, and these are commonly interpreted as having formed by the same sorts of turbulent processes that produce cross-bedding in sedimentary rocks. Given likely viscosities of granitic magmas and a length scale of 1 m, what sort of velocities would be required to achieve turbulence in a granitic magma?

7.6 A Survey of Volcanic Styles and Landforms

Learning Objectives

By the end of this section you should be able to:

- **7.6.1** Classify a volcano by observation.
- **7.6.2** Predict what kinds of products that a volcano would erupt, and the general composition of the magma.
- **7.6.3** List the hazards presented by that volcano.
- **7.6.4** Recognize a pyroclastic deposit as a fall, flow, or surge.

Hawaiian Eruptions

Hawaiian style eruptions are common among volcanoes that erupt basaltic magma. Such magmas have low viscosities, and these eruptions produce a characteristic set of landforms that owe their morphologies to this low viscosity.

Shield volcanoes
Broad, low volcanoes formed of low-viscosity basaltic magma.

Shield volcanoes, so named because they have the broad, convex-up profile of a warrior's shield lying on the ground, are by far the largest volcanoes on Earth. The Big Island, youngest of the Hawaiian chain, consists of five overlapping shield volcanoes (see Figure 7.1). Mauna Loa, one of two active volcanoes on the island, rises 9 km from the ocean floor and over 4 km above sea level. Kīlauea, the most active volcano on the island, is a smaller shield volcano on the flank of Mauna Loa at an elevation of about 1200 m. Shield volcanoes can be many tens of kilometers in radius and are composed of thousands of thin lava flows. The lavas were able to spread so far owing to three factors: (1) they are composed of low-viscosity basaltic magma, (2) lava tubes insulate the flowing magma so that it can travel tens of kilometers without significant cooling, and (3) they commonly erupt along rift zones—the surface expressions of radial dikes that vent in fissure eruptions.

Figure 7.17 (a) Pāhoehoe lava from the Kīlauea system pouring over a 15-m cliff onto a bench near the ocean. (b) Ropy pāhoehoe surface on lava near the former Waha'ula Visitor Center, which was destroyed by lava in 1989. Width of view ~2 m. (c, left) Thin pāhoehoe lava flow, ~30 cm thick, crossing emergency road constructed across recent lava flows near Kamokuna ocean entry. (c, right) Same flow two weeks later, after it had inflated to a thickness of 2–3 m.

Sources: (a) USGS; (b) AFG photo; (c) M. Patrick, courtesy of Hawaiian Volcano Observatory, USGS.

The most fluid basaltic lavas flow in channels like hot rivers (Figure 7.17) and can move at speeds ranging from a crawl to 15 km/hr (Kauahikaua et al., 1998). Where flowing rapidly or over an obstruction, the glowing interior is exposed, but otherwise the surface crusts over as it rapidly radiates energy away according to the Stefan-Boltzmann law (see Equation 2.1). As the surface crusts over, it makes a plastic skin that is dragged and folded by flow of magma in the interior. This produces the classic ropy texture of a pāhoehoe lava flow (Figure 7.17b). The front of a pāhoehoe flow consists of a crust that encloses hot, fluid magma, and magma pressure will cause breakouts that ooze fluid lava out in the subaerial equivalent of submarine pillow lavas.

Lava tube
Tunnel produced by roofing-over of a fluid lava flow; refers to both active channels and former ones.

Fluid pāhoehoe flows commonly advance in thin sheets that are only tens of centimeters thick, but as the skin at the front of the flow thickens, or if the flow runs into an obstacle, it may inflate to several meters thickness—a bladder filled with glowing magma. The flow front of a pāhoehoe flow advances more slowly than the upstream flow in lava rivers because of this fingering, inflation, and cooling; observed flow-front velocities for historic flows in Hawai'i are typically less than 10 m/hr (Rowland & Walker, 1990).

Lava rivers commonly grow a roof such that the lava ends up flowing in a **lava tube**. If the magma source shuts off or the tube is plugged, it may empty out as magma drains away, leaving behind a lava cave (Figure 7.18). Lava tubes are important in the growth of shield volcanoes and far-traveled lava flows, because they insulate the flowing magma.

A'a flows are more viscous than pāhoehoe flows and yet, perversely, they can flow quite a bit more rapidly.

Figure 7.18 Valentine Cave, a lava tube in Lava Beds National Monument, California. This is an example of a cave where the last lava to flow left behind a relatively smooth floor.

Source: Dave Bunnell/Under Earth Images/Wikimedia.

Figure 7.19 A'a flow in Royal Gardens subdivision downslope from Pu'u'ō'ō and Kupaianaha. Flow front is ~2 m tall.

Source: Courtesy of Hawaiian Volcano Observatory, USGS.

Figure 7.20 Gas plume from Halema'uma'u, inside Kīlauea caldera, in November 2008.

Source: Courtesy of Hawaiian Volcano Observatory, USGS.

Figure 7.21 Thick stack of Miocene basalt flows of the Columbia River flood basalt province exposed along the Grande Ronde River in northeastern Oregon.

Source: AFG photo.

A'a flow fronts in Hawai'i have been observed to move at speeds ranging from a crawl to over to 10 km/hr, a decent running pace, forming fast-moving, noisy, scary monsters whose glowing interiors are exposed as rough pieces of the exterior fall off the advancing front, to be overrun by the flow, tractor-tread style (Figure 7.19). Advancing basaltic lava flows can sound like a glass recycling plant on a busy day as the glassy flow surface is continually broken and reformed.

Hawaiian eruptions generate copious amounts of gas but are rarely explosive, because the low viscosity of the magma allows gas to escape easily in a sulfurous plume known as **vog** (a contraction of volcanic smog; Figure 7.20). Trade winds in Hawai'i consistently blow from the northeast (see Figure 7.1), so Kīlauea vog generally blows across the area southwest of the volcano. This desolate, uninhabited region, the Ka'ū Desert, suffers persistent acid fog and rain and is nearly lifeless. On occasion, the plume drifts to the Kona region on the western side of the island, causing discomfort for residents there and accelerating decay of metallic objects.

Flood Basalts

At irregular times, the Earth has released massive amounts of basaltic magma in flood basalt eruptions (Saunders, 2005). These are discussed in Chapter 14. Flood basalt provinces consist largely of stacks of tholeiitic basalt flows (Figure 7.21). Individual eruptions can release 1000 km^3 over relatively short time intervals (a decade or so) from dike-fed fissures tens to hundreds of kilometers long (Self et al., 2005).

There have been no flood basalt eruptions since the Miocene Columbia River event, so their behavior must be inferred from field volcanology. In 1783 and 1784 Iceland (and indeed, the entire world) suffered through a volumetrically minor but still devastating taste of what flood basalt eruption must be like. The Laki eruption began on June 8, 1783, after a week of precursory earthquakes, with violent explosive activity that settled down to a more typical fissure eruption. Basaltic magma issued from a 27-km-long vent zone, a venting dike system, for eight months. Fire fountains may have risen as high as 1.5 km (think about that—to throw a projectile that high it must leave the ground at over 600 km/hr). In all, about 15 km^3 of lava were erupted—a prodigious amount, comparable to the erupted volume of Krakatau. Although this eruption did not blow material into the stratosphere, it still had global effects owing to the huge volume of SO_2 and other gases that were belched out.

The eruption released about 120 Mt (million metric tons) of SO_2—roughly 250 times the annual output from Kīlauea in recent years, and comparable to the world's annual output from human sources—factories, power stations, automobiles, etc. (Thordarson & Self, 2003). The results were catastrophic for Iceland, in part because of the large quantity of HF emitted. Poisonous fluorine-contaminated tephra fell on fields and was ingested by animals as they grazed on dying vegetation. Seventy-five percent of Iceland's livestock perished during and after the eruption, and that led to the death of 25 percent of its human population. The Laki eruption spread an acid aerosol over Europe and produced an abnormally cold winter in 1783–1784, nearly 5 °C below normal in the eastern United States.

As bad as the Laki eruption was, true flood basalt eruptions were vastly greater in scale and effect. The major flood basalt eruptions produce over 10^6 km^3 of magma; if you spread this volume over the continental United States like cake frosting, it would be over 100 m thick. The four most prominent mass extinctions in the Phanerozoic coincide closely with four major flood basalt events (see Making Connections 14.2; Courtillot & Renne, 2003), but the particular killing mechanisms are in debate. Both CO_2 and SO_2 have been fingered as the culprits (Wignall, 2005; Self et al., 2005).

Vog
Corrosive, dense haze, rich in SO_2 and other harmful gases, produced by active volcanoes.

Columnar joints
In volcanic rocks, joints that form by thermal contraction during cooling, producing six-, five-, or seven-sided columns.

Strombolian
Eruption style produced by basaltic and andesitic magma at cinder cones, with eruption of tephra out the top and lava flows out the bottom.

Scoria
Rough, vesicular basaltic tephra.

Columnar Jointing

Lava flows are commonly cut by **columnar joints**, and these can form remarkable landforms (Figure 7.22). The joints form during cooling via thermal contraction and divide the rock into columns that typically have six, five, or seven sides, with six being the most common. The long axes of the columns are parallel to the direction of heat flow and can be used to infer how a given body cooled.

Strombolian Eruptions and Cinder Cones

Strombolian eruptions, named for the Italian volcano Stromboli, are produced by eruption of magma that is more viscous than the low-viscosity basaltic magmas characteristic of Hawaiian eruptions. In Strombolian eruptions the vent ejects incandescent blobs of magma to altitudes of tens or hundreds of meters. These fall back around the vent and pile up, producing a cinder cone (Figure 7.23).

The birth and development of Parícutin volcano (Figure 7.23a), about 300 km west of Mexico City, demonstrates how Strombolian eruptions work. The story is familiar to many from children's books.[2] The eruption took place on farmland near the village of Parícutin. In February 1943, frequent earthquakes accompanied by subterranean noises akin to thunder shook the area. On the afternoon of February 20, a farmer noticed that a small pit in a cornfield had enlarged and was now situated along a newly developing east-west fissure.[3] By midnight, a small hill about 6 m high had formed around the vent.

Over the next eight months, the cone grew to nearly 400 m in height, and about 75 percent of the total erupted volume came out as tephra during the first three years of activity. Rapid-fire explosive sequences lasting hours to days formed eruptive columns of tephra and gases several kilometers tall and ejected material at velocities up to 150 m/s. The dominant magma erupted was basaltic andesite with 55 to 60 wt% SiO_2—significantly higher in silica, and therefore more viscous, than basaltic lavas of Hawai'i.

Strombolian eruptions produce abundant lapilli, blocks, and bombs; collectively, these dark, rough products are known as **scoria**. In the absence of strong and constant winds, cinder cones typically develop symmetrical, conical profiles with a crater in the top (Figure 7.23). The crater forms as material is blown out the top, in contrast to a caldera, which forms by collapse into a vacated magma chamber. Cinder cones emit lava as well as tephra, but the lavas flow out the base of the cone, because the loose tephra that makes up the

(a) (b)

Figure 7.22 (a) Columnar jointing in the rhyolitic Bishop Tuff, eastern California, exposed in the wall of Owens River Gorge (see Figure 7.38). Columns in the fanning, tulip-like array are 1 -1.5 m across. Columns are well-developed in the upper part of the cliff and become much more irregular and larger going downward into densely welded tuff. (b) Columnar-jointed basalt at Devils Postpile National Monument, California. Columns are 20–25 m tall.

Source: AFG photos.

Figure 7.23 (a) The cinder cone of Parícutin soon after its birth in 1943, erupting tephra and gases. Lava flows issuing from the base are not evident in this photo. (b) SP Crater, north of Flagstaff, Arizona, and associated lava flow, which traveled over 8 km from the cone's center. As with other cinder cones, the lava flow exited from the base of the cone, because the cinder pile is not strong enough to contain a lava lake. Snow dusts the foreground.

Sources: (a) K. Segerstrom/NOAA/USGS; (b) Courtesy of geologist and author Wayne Ranney and photographer and glider pilot Ted Grussing.

cone is generally not strong enough to hold a lava lake. Many classic cinder cones have lava flows several kilometers long that issue from their bases (Figure 7.23b). These flows typically carry away a piece of the cone, leaving a breach, but renewed Strombolian activity can rebuild a symmetrical cone again.

Phreatic Eruptions and Maars

Interaction of rising magma with groundwater can produce conical features that superficially resemble cinder cones. If a rising dike encounters groundwater at a shallow enough level to vaporize it, but the surrounding rocks are strong enough to contain the resultant steam, then pressure can build up enough to produce a **phreatomagmatic** eruption that results in explosive excavation of a crater in the rocks above the interaction zone. **Maars** are phreatomagmatic craters, cut through older rocks, with a low rim of ejecta (Figure 7.24). They superficially resemble cinder cones but have a much lower height-to-width ratio, and the crater is excavated in older rocks, not ejecta.

Phreatomagmatic
In volcanology, pertaining to the reaction of magma with groundwater or surface water and steam and fragmental magma.

Maars
Shallow, broad craters formed by explosive reaction of magma with groundwater.

Stratovolcanoes

Cinder cones are produced by basaltic and andesitic magma with SiO_2 less than about 60 wt%. The mildly explosive Strombolian style of eruption depends on magma that is runny enough that gases can escape with ease and whip up the magma into frothy scoria. Magmas with more silica produce volcanoes that are larger, more violent, and far more dangerous. Chief among these are **stratovolcanoes** (also known as **composite volcanoes**; Figure 7.25), the tall, conical structures that many people think of when picturing a volcano. Height puts the "strato" in stratovolcano, and many stratovolcanoes do indeed reach lofty heights; magnificent Ojos del Salado on the Chile-Argentina border is 6893 m (22,615 ft) tall, and Mount Rainier rises from sea level to 4392 m above Seattle.

Stratovolcanoes, composite volcanoes
Tall, conical volcanoes formed by flows and tephra of intermediate (andesitic and dacitic) magma.

Stratovolcanoes form from magmas that are typically in the range of 60 to 70 wt% SiO_2, although lower-silica cones commonly speckle their lower flanks. These intermediate to

Figure 7.24 (a) Ubehebe Crater, a basaltic maar in Death Valley National Park, California. The maar, about 750 m in diameter, exposes tilted, brightly colored terrestrial sedimentary rocks in its walls. Basaltic lapilli blanket the surrounding countryside. (b) Hanauma Bay, a maar near Honolulu, Hawaii. A low breach in the wall of the maar allows in the sea, and thus the maar is filled with coral, fish, and tourists.

Source: AFG photos.

Figure 7.25 Stratovolcanoes from around the world. (a) Beautifully symmetrical Mount Mayon rises 2462 m above Albay Gulf in the central Philippines. (b) Volcán Popocatépetl is a lofty stratovolcano whose 5426 m summit lies but 40 km from the outskirts of Mexico City, a city of 9 million people in a metropolitan area of 22 million. (c) A cluster of stratovolcanoes on the Kamchatka Peninsula, Russia. From left to right, Ushkovsky, Klyuchevskoy (smoking), Bezymianny (inconspicuous in front of Klyuchevskoy), Tolbachik, and Zimina. Astronaut photograph taken by a member of the Expedition 38 crew, International Space Station. (d) Rugged Mount Rainier (4392 m) looms over the town of Orting, Washington, a suburb of the Seattle-Tacoma area. Orting and nearby towns were built on flat areas produced by mudflows from Mount Rainier and are at risk from future mudflows. The rugged topography of this volcano is a result of glacial erosion. Rainier stores a great deal of ice on its flanks, up to 4 km above the surrounding lowlands.

Sources: (a) Kurt Fredrickson/National Museum of Natural History/Smithsonian Institution; (b) erlucho/iStockphoto; (c) Earth Observatory/NASA; (d) Ed Ruttledge/Cascades Volcano Observatory/USGS.

Figure 7.26 Mount St. Helens in full Plinian eruption, May 18, 1980. Aerial view from the southwest; the lateral blast was to the north, on the other side of the volcano. This happened a long time ago, but it is still remarkable that there is no video of the initial stages of the eruption. The first consumer camcorders went on sale a few years later.

Source: USGS.

high silica contents are high enough that viscosities are 10^4 Pa·s, or many orders of magnitude more. At such high viscosities, dissolved gases cannot easily escape and can build to pressures high enough to drive violent eruptions with far-reaching effects, such as the 1883 eruption of Krakatau.

Stratovolcanoes erupt in a wide variety of styles, the most iconic of which are **Plinian** eruptions, highly explosive ejections of tephra and gas in the form of a tall eruption column. Their nature was brought to the public's attention in dramatic fashion by the May 18, 1980, eruption of Mount St. Helens in Washington state (Figure 7.26). The Romans of the first century CE had their own firsthand experience during the 79 CE eruption of Vesuvius; Pliny the Younger's vivid description of this event, which killed his uncle Pliny the Elder, gives us the name of this class of eruption.

Pyroclastic Falls, Flows, and Surges

Plinian eruptions are awe-inspiring, violent, ear-shattering ejections of great volumes—perhaps several cubic kilometers, or much more—of magma. Most of this material goes into the air as tephra, which is transported by three distinct mechanisms during Plinian eruptions, leading to three distinct and recognizable kinds of deposits (Figure 7.27). The term **tuff** is used to describe tephra deposits (generally those intermediate or silicic in composition), and all three of these types of deposits are different kinds of tuff.

Plinian
Explosive eruption of intermediate magma, with copious pyroclastic flow and fall deposits, producing an eruption column tens of kilometers tall.

Tuff
Generic term for a deposit of pyroclastic material.

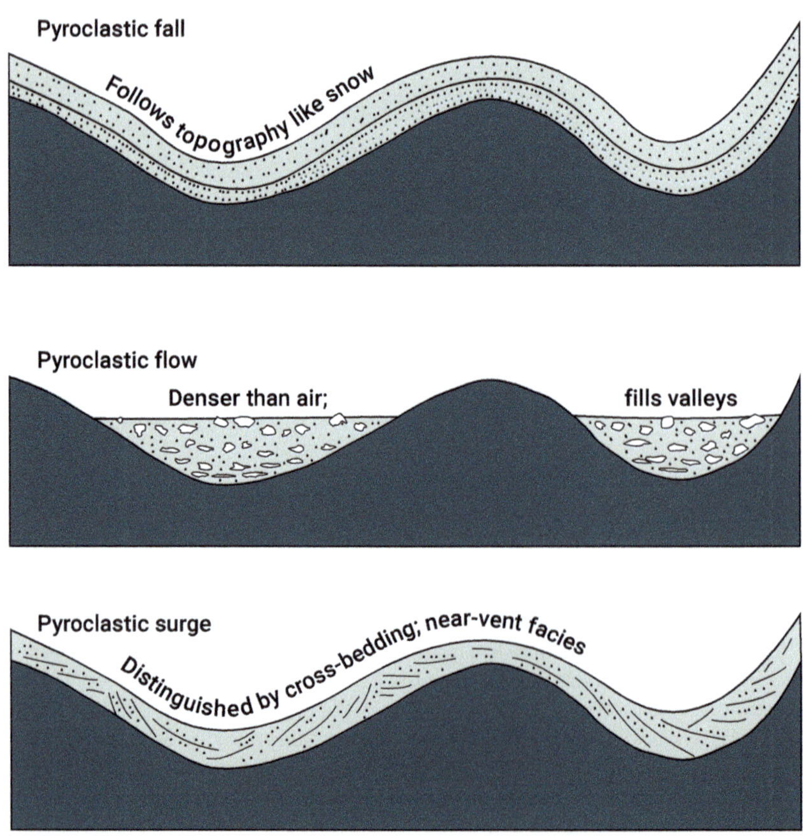

Figure 7.27 The three main types of pyroclastic deposits. Pyroclastic fall deposits are thrown into the air and fall back like snow, blanketing the countryside; they are as thick on high areas as in low ones and are typically well-bedded. Pyroclastic flow deposits are denser than air and tend to flow down valleys, although they can cross tall barriers; they are largely unbedded. Pyroclastic surge deposits show cross-bedding.

Source: Modified from Cas & Wright (1987).

Pyroclastic Falls **Pyroclastic falls** are produced by low-density pumice that is blown into the air, where it is held aloft for a while by rising air currents and wind before falling to the ground like snow. This behavior is unlike the ballistic (i.e., cannonball-like) trajectories taken by bombs and blocks in Strombolian eruptions. In large eruptions tephra leaves the vent at velocities of several hundred meters per second and can be lofted over 40 km high (Wilson, 1976).

In photos of Plinian eruptions (see Figure 7.26), most of what you see in the tephra column is destined for a pyroclastic fall deposit, accumulating somewhere downwind of the volcano. Pyroclastic fall deposits (commonly known as **air-fall tuffs**) blanket the landscape evenly, like snow; barring later redistribution, they are as thick on top of a hill as they are in a nearby valley (see Figure 7.27). Light-colored silicic pyroclastic falls even look like snow (Figure 7.28a), but unlike snow, these fluffy blankets are composed of sharp little bits of broken glass that wreak havoc on lungs, eyes, and engines and do not melt away when the weather warms up—although they do melt inside engines, coating the parts with rhyolite melt and causing rapid failure. There have been several incidents, some near-catastrophic, where jets have flown unaware into eruption clouds.

Pyroclastic fall deposits typically show well-developed bedding produced by variations in eruption intensity and weather phenomena such as wind and rain (Figure 7.29). Pyroclastic fall deposits from a given eruption typically form an elongated blanket that stretches downwind from the volcano. Figure 7.30 shows an **isopach** map of the distribution of Mount St. Helens ash. The volcanic vent lies within this blanket but is generally far from its center. Before its 1980 VEI 5 eruption, isopach mapping showed that Mount St. Helens was the most active volcano in the Cascades, because a large number of tuffs that had been identified in soil columns could be traced back to the mountain; USGS scientists Dwight Crandell and Donal Mullineaux (1978, p. C1) wrote, "Mount St. Helens has been more active and more explosive during the last 4500 years than any

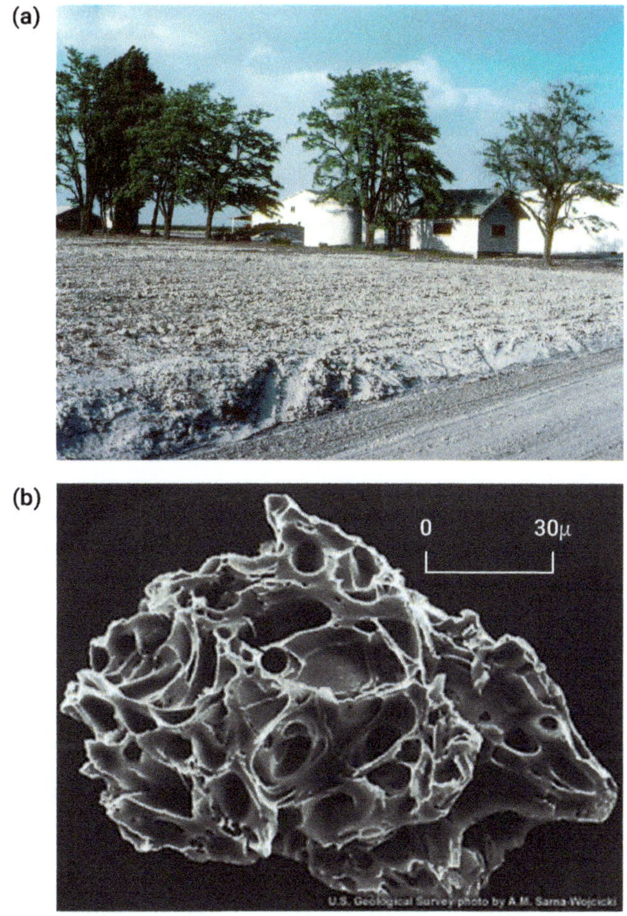

Figure 7.28 (a) Ash from the May 18, 1980, eruption of Mount St. Helens covering the ground in Connell, Washington, approximately 300 km from the volcano. It looks like snow but, unlike snow, does not melt and go away. (b) Typical particle of volcanic ash as seen by a scanning electron microscope. These sharp pieces of glass are damaging to eyes, lungs, other living tissue, and engines.

Sources: (a) Lynn Topinka, USGS; (b) A. M. Sarna-Wojcicki, USGS.

Figure 7.29 (a) Well-bedded pyroclastic fall deposits from a quarry on Lipari in the Aeolian Islands of Italy. Cliff face is ~10 m tall. (b) Andesitic pyroclastic fall deposits from Volcán Villarica, Chile. Roadcut is ~3 m tall.

Source: AFG photos.

Pyroclastic falls
Tephra that is blown into the air and falls out like snow.

Air-fall tuffs
Deposits formed by pyroclastic fall material.

Isopach
Contour connecting points of equal thickness of a rock layer.

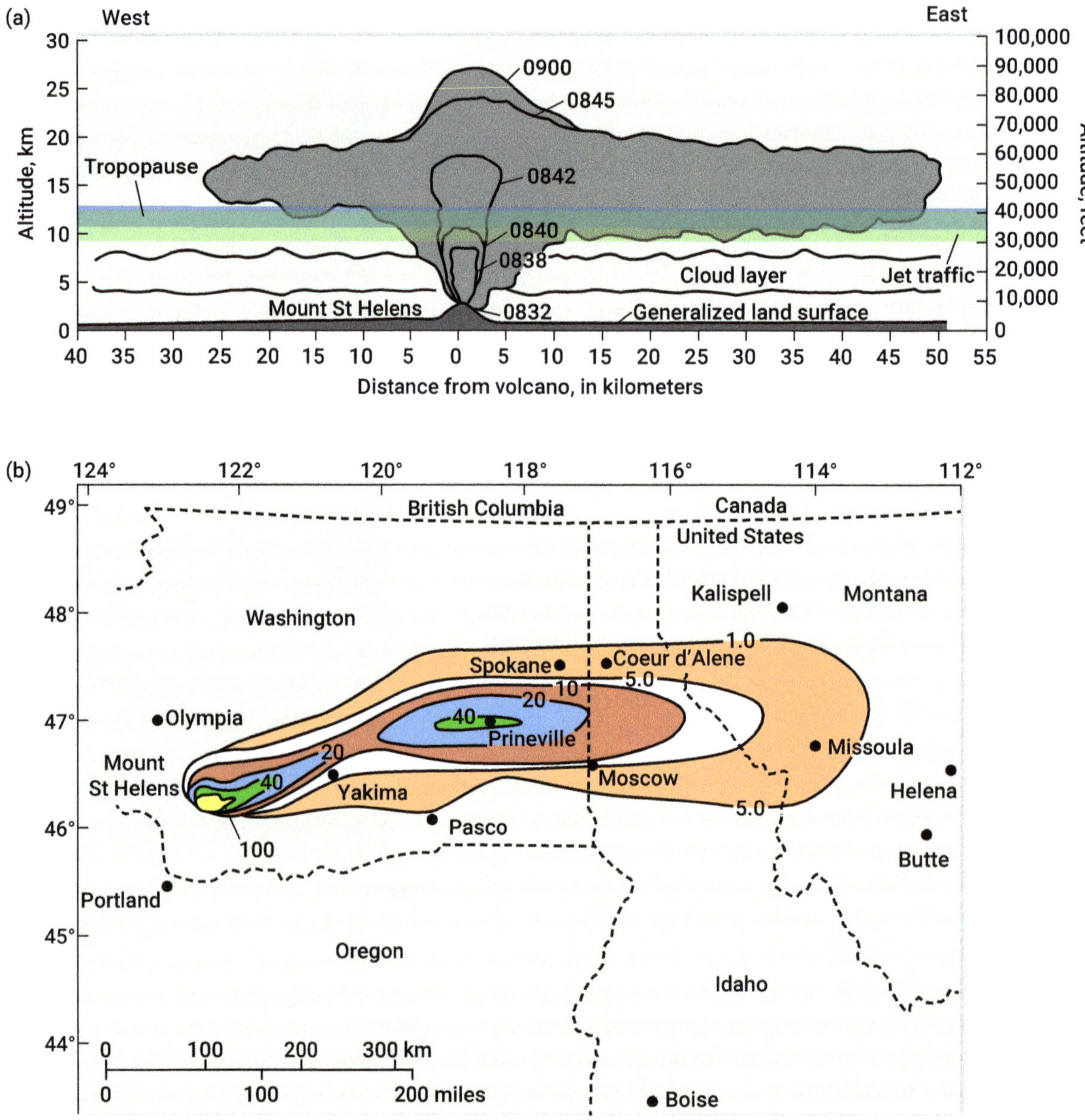

Figure 7.30 (a) Diagrammatic east-west profile of the Mount St. Helens ash plume in the half hour following onset of the May 18, 1980, Plinian eruption. The plume blasted through a low cloud layer and the altitudes at which transcontinental jets fly in the first 10 minutes and reached the stratosphere (see Figure 7.39) a few minutes later, where it spread out. (b) Isopach (layer thickness) map of pyroclastic fall material from the May 18 eruption. Elongation of the plume matches observed average wind direction in the troposphere that day well. The secondary maximum in thickness in eastern Washington is not well understood but may represent material that reached the jet stream at 10–12 km altitude before returning to Earth.

Source: From Sarna-Wojcicki et al. (1981).

other volcano in the conterminous United States." Published two years before the climactic eruption, these were prescient words, indeed.

Pyroclastic fall deposits tell you more than just which way the wind was blowing during an eruption. Peel back an air-fall tuff and you have a snapshot of what the Earth's surface was like at the moment of eruption. These two qualities and the relative ease of dating an air-fall tuff make them valuable markers in paleogeography.

Pyroclastic Flows **Pyroclastic flows** are the most devastating of short-term volcanic phenomena, and many of the well-known eruptions of historic times, such as Krakatau (1883), Tambora (1815), Valley of Ten Thousand Smokes, Alaska (1912), and Mount St. Helens (1980) were marked by expulsion of copious amounts of them. As noted by Francis (1993):

> Pyroclastic flows have some of the attributes of dinosaurs: they are huge and dangerous. But they have little of the fascination for the public that dinosaurs have, because large pyroclastic flows are rare, and have not earned the attention of the media. Unlike dinosaurs, however, pyroclastic flows are not extinct. (p. 203)

Pyroclastic flows
General term encompassing both pumice flows and block-and-ash flows.

Pyroclastic flows are not as well understood as other volcanic phenomena, because they are too dangerous to study up close. In Plinian eruptions (e.g., see Figure 7.26) the eruption column dominates the view, but the pyroclastic flows hidden beneath them are really the big show.

The terminology of pyroclastic flows and their deposits is confusing, because two distinct but gradational phenomena are lumped together. One type of pyroclastic flow, a **pumice flow**, consists of density currents of hot, incandescent, degassing pumice that are erupted at temperatures of 700 to 1000 °C. These fall back from the eruption column and flow at speeds up to 100 km/hr. Denser than air, they hug the ground and are thicker in valleys than on higher ground, producing poorly bedded deposits of pumice in an ashy matrix of ground-up pumice (see Figure 7.27). Another type forms by collapse of viscous lava domes, producing a **block-and-ash flow** of solid, angular blocks in a pumiceous matrix (Figure 7.31). A simplified classification of these deposits is given in Table 7.1.

Pumice flow
Hot density current of degassing pumice blocks and lapilli in an ashy matrix.

Block-and-ash flow
Hot density current of angular blocks, as from a collapsing dome, in an ashy matrix.

Figure 7.31 (a) Typical lightly welded pumice flow deposit of the Bishop Tuff, composed almost entirely of pumice lapilli, with faint layering at top of 1.5 m view of cliff face in quarry north of Bishop, California. (b) Pit excavated in February 2010 block-and-ash flow eruption from Soufrière Hills volcano, Montserrat, Lesser Antilles arc. Although superficially similar to (a), this 2-m pit exposes blocks of andesite, rather than pumice, in an ashy matrix. Five years after the eruption, the temperature of this deposit, 5 m below the surface, was 70 °C. (c) Only a 1-m piece of an antenna (circled) on top of the three-story control tower at the former main airport on Montserrat is left above ground after a 1997 block-and-ash flow like the one in (b) buried the eastern side of the island.

Source: AFG photos.

Table 7.1 Simplified Classification of Pyroclastic Flow Deposits

Clasts	Eruption Mechanism	Composition	Type of Pyroclastic Flow	Resulting Deposit
Vesiculated (pumice)	Column fallback	Rhyolite, dacite	Pumice flow	Ignimbrite
Non-vesiculated (lava)	Dome collapse	Rhyolite, dacite, andesite	Block-and-ash flow	Block-and-ash deposit

Source: After Wright et al. (1980)

Ignimbrite
The deposits of a pyroclastic flow; the term is generally used for pumice flows.

Welding
Post-emplacement sintering and flattening of pumice clasts in an ignimbrite.

Aspect ratio
Generally used to mean the long dimension of an object divided by the short dimension.

A solidified pyroclastic flow deposit is an **ignimbrite**, recognized in the field as a poorly bedded sheet of pumice lapilli, blocks, and ash. Because freshly emplaced pumice is frothy with gas bubbles and quite hot, the weight of the upper parts of the deposit cause compaction and **welding** in the deeper parts. Welding is a process in which pumice clasts and ground-up pumice bits in the matrix are flattened and sintered together into a denser mass, and ignimbrites are commonly referred to as welded tuffs. The degree of welding depends on how hot the deposit is, and therefore the upper and lower parts of a typical pyroclastic flow deposit are only lightly welded, because they cool rapidly. The most densely welded part of the vertical profile is typically one-quarter or one-third of the way up from the bottom, and welding is particularly prominent where thick sections of pyroclastic flows accumulate in valleys and basins. In intensely welded pyroclastic deposits, the primary texture is wiped out, and the rock turns shiny black, like obsidian (Figure 7.32).

Flattening of pumice clasts can be quite dramatic. If we define the **aspect ratio** of a pumice clast as its longest dimension divided by the dimension perpendicular to that, a golf-ball-sized pumice lapillus has an aspect ratio of 1, but in a densely welded deposit, it can be flattened to an aspect ratio of 20 or more (see Figure 7.32, inset). As this flattening involves squeezing out the gas in bubbles, it is more akin to stepping on a foam ball than stepping on a ball of clay—the clast flattens without appreciable spreading in the plane of flattening. A useful characteristic of welded ignimbrites is that the plane of flattening is usually perpendicular to gravity, thereby providing a measure of paleohorizontal—a rare thing in igneous rocks.

Ignimbrites contain lapilli and larger clasts of pumice as well as clasts of unrelated rocks picked up along the conduit during eruption. These lithic fragments are denser than the enclosing pyroclastic flow and tend to settle downward, whereas pumice clasts are generally less dense and work their way upward during flow. The result is that many

Figure 7.32 A thin Miocene ignimbrite with shiny black, densely welded zone exposed in a roadcut east of Death Valley, California. The black zone occurs in the lower third of the deposit and resembles obsidian, although relic flattened pumice clasts are visible. Inset (width of field 15 cm) shows flattened pumice clasts from below the black zone. In densely welded tuff such as this, the flattened pumice commonly turns black, and the crushed-pumice matrix darkens noticeably.

Source: AFG photos.

ignimbrites show double-grading in which lithic fragments get more abundant and denser downward and pumice clasts get more abundant and larger upward. In steeply tilted ignimbrites, such grading provides a useful indication of which way was up.

Pyroclastic flows can travel astonishing distances and surmount remarkably tall barriers. The Peach Springs Tuff was erupted 18.8 Ma from western Arizona and is found in outcrops 250 km to the west, almost to Los Angeles; the tuff of Campbell Creek was erupted in north-central Nevada 28.9 Ma and traveled over 200 km from its source; the 36-ka Campanian Ignimbrite near Naples, Italy, flowed over 1000-m ridges and across 35 km of water in the Bay of Naples; and the 22-ka Ito pyroclastic flow in Japan traveled 70 km, crossing topographic barriers over 600 m high on its journey. How do they do this?

The main driver of such far-traveled tuffs is the energy of the eruption column. During a large explosive eruption, gas and tephra can emerge from the vent at velocities up to 500 m/s (Wilson, 1980). Five hundred meters per second is about the speed of a typical bullet, and faster than the speed of sound at sea level (~350 m/s). Simple ballistic calculations show that a rock thrown straight up at 500 m/s will reach a height of nearly 13 km, ignoring the (quite significant) effects of air resistance. In addition to all this kinetic energy, an eruption column possesses abundant thermal energy. Material blasting out of the vent is a combination of gas, tephra, and lithic fragments. All of this ammo is hot, emerging at temperatures of 700 to 1000 °C, and melt emits hot gas as it froths up to pumice, adding more buoyancy to the mix.

The mix is typically denser than air as it emerges and rises via kinetic energy, but fallout of denser lithic fragments, entrainment and heating of outside air, and emission of hot gas from juvenile tephra lessens the column density as it rises, and it continues to rise buoyantly (Sparks & Wilson, 1976). What goes up must come down, and lofted ash is destined to fall downwind as a pyroclastic fall sheet. Lapilli and larger pumice blocks in the column, however, are not easily wafted by air currents and fall back down toward the vent. As these might be falling from heights of several kilometers, their potential energy is transformed back into abundant kinetic energy; upon reaching the ground, they spread out to form pyroclastic flows that race outward at high speed. It is this kinetic energy that powers pyroclastic flows long distances and over high topographic barriers. Moore and Melson (1969) made fascinating observations of this process during a small but deadly eruption of Mount Mayon (Philippines) in 1968, and their observations led to the column-fallback model of pyroclastic flow generation (Sparks & Wilson, 1976).

Block-and-ash flows are generally produced by collapse of a growing lava dome, although they can be produced by column fallback as well. As the interior of such a dome is at magmatic temperatures and likely contains a significant quantity of dissolved gas, collapse typically produces pumice as the pressure release on the interior of the dome allows gas to effervesce and escape. Block-and-ash deposits are distinguished from ignimbrites by abundant angular blocks of lava from the broken-up dome rather than being dominated by pumice (see Figure 7.31b).

The 1902 eruption of Mount Pelée in Martinique, a notorious eruption that quickly killed approximately 30,000 people in the city of St. Pierre and was the deadliest eruption of the 20th century, generated abundant block-and-ash flows. These, however, flowed down a nearby river valley and did not bury the city. The people were killed by a less well-known phenomenon, **pyroclastic surges**.

Pyroclastic surges
Turbulent, fast-moving, low-density flows of pyroclastic material and hot gases.

Pyroclastic Surges

Many explosive phenomena generate rapidly moving, outward-directed blasts of gas with a relatively low concentration of particles. These are spectacularly displayed by eruptions in shallow water, such as the 1957 eruption of Capelinhos in the Azores (Figure 7.33a), where they formed a collar around the eruption column (Moore, 1967). Pyroclastic surges are among the most destructive of volcanic phenomena; in addition to St. Pierre, many of the fatalities at Pompeii in 79 CE were caused by pyroclastic surges. Surges leave behind thin, commonly cross-bedded deposits that understate their destructiveness (Figure 7.33b).

(a) (b)

Figure 7.33 (a) Base surge (donut-shaped basal cloud surrounding eruption column) from eruption of Capelinhos, Azores, in 1957. The lighthouse in foreground, 20 m tall and dwarfed by the eruption, gives scale. (b) Pyroclastic surge deposits from phreatomagmatic eruptions near Hanauma Bay (see Figure 7.24b), island of Oahu, Hawaii people for scale. Deposits consist of basaltic pumice (dark) and broken bits of coral (white). Inset shows low-angle cross-bedding that is typical of surge deposits and absent from other pyroclastic deposits; width of field ~1 m.

Sources: (a) US Air Force/Smithsonian Institution; (b) AFG photos.

The Vesuvius eruption that buried Pompeii and Herculaneum began with around 20 hours of Plinian eruption that generated extensive pyroclastic fall deposits, locally nearly 3 m deep, to the south (Sigurdsson et al., 1982). Many people were killed by collapsing roofs during this stage of the eruption, but much of the population likely fled to safer ground. This phase was followed by a series of pyroclastic surges that swept over the doomed cities around the volcano, rapidly killing the remaining people via a combination of high temperatures (wood was charred at temperatures of at least 400 °C) and asphyxiation. The victims lay upon the relatively thin surge deposits and were buried by thin pyroclastic flow deposits that followed the surge.

Surges are distinguished in the field from fall deposits by their cross-bedding. Some are generated by overrun, trapped, and heated parcels of air that jet out of the front of an advancing pyroclastic flow. Perret (1937, p. 42) memorably described one such pyroclastic flow generating surges at Mount Pelée as "a flowing mass of incandescent material advancing with an indescribably curious rolling and puffing movement, which at the immediate front takes the form of forward-springing jets, suggesting charging lions." Few of us have stood in front of a charging lion, but the sense of menace in Perret's words is palpable.

Pyroclastic Deposits All Together

Figure 7.34a shows a typical (if such a word can be applied to such extraordinary events) cross section through an ignimbrite and its associated deposits. It is common to find well-bedded pyroclastic flow deposits below the ignimbrite; these arose from the initial phases of the Plinian column before it began to produce pyroclastic flows. In some sections a thin sequence of cross-bedded surge deposits overlies the air-fall tuff, but these are only present within a few kilometers of the caldera and are commonly missing. Above this sequence lies the main mass of the ignimbrite, which can be tens to hundreds of meters thick, and above it is a thin sequence of pyroclastic fall deposits that fell from the eruption column that produced the ground-hugging pyroclastic flows.

Lahars

As if the various pyroclastic events described above aren't bad enough, stratovolcanoes can generate **lahars**, mudflows composed of volcanic debris mobilized by water. Because stratovolcanoes are tall, they provide both gravitational potential energy and water (in the form of ice and snow). They are also subject to significant hydrothermal alteration that turns rocks such as andesite and dacite, which are full of strong, anhydrous minerals like plagioclase and Cpx, into goopy masses of weak clay. After an eruption, weakly consolidated tephra can coat the edifice. The combination of weak rocks, loose tephra, ice and snow, and topography is a dangerous combination, and stratovolcanoes routinely let loose mudflows that roar down valleys. These may be triggered by eruptions, producing hot lahars, or simply by rainfall, or by nothing tangible, producing cold lahars.

Regardless of how they are generated, lahars can wreak horrific destruction. A small eruption in 1985 of the stratovolcano Nevado del Ruiz in Colombia mobilized water in glaciers and sent a mudflow into the town of Armero, burying approximately 20,000 people; 3000 more died in other villages. In response to the Armero tragedy, the USGS and US Agency for International Development formed the Volcano Disaster Assistance

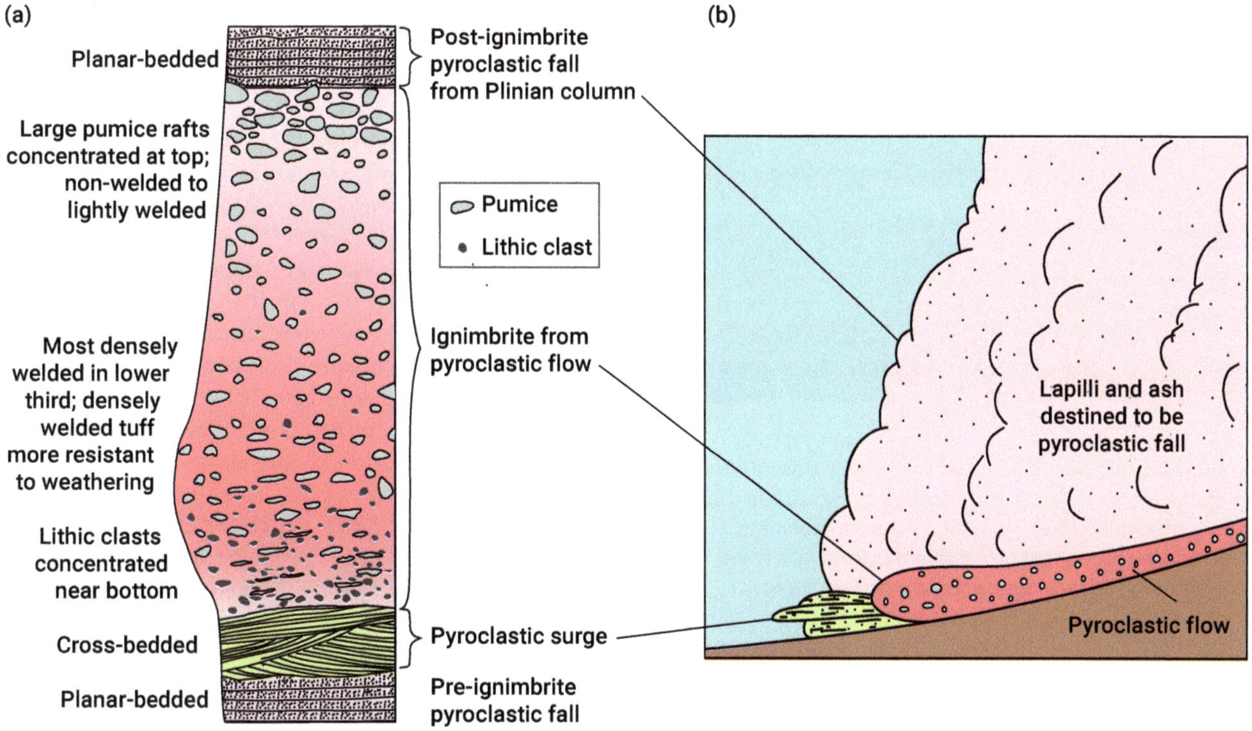

Figure 7.34 A simplified model for the genesis and zonation of pyroclastic flow deposits. (a) Idealized cross section through a pyroclastic flow deposit showing normal (larger downward) grading of dense lithic fragments and reverse (larger upward) grading of low-density pumice clasts. The main pyroclastic flow unit is sandwiched between a lower surge deposit and upper fall deposit. Left side of column indicates that densely welded tuff is more resistant to weathering than lightly welded tuff or pyroclastic surge and flow deposits. (b) Interpretation of the structure of a pyroclastic flow in action.

Source: Modified from Cas & Wright (1987).

Program (VDAP). VDAP teams have been deployed around the world to help with forecasting and monitoring eruptions, defining risk zones, and getting people out of the way. VDAP and Philippine volcanologists are credited with saving thousands of lives during the 1991 eruption of Mount Pinatubo.

Mount Rainier, which lies about 90 km southeast of Seattle, Washington, poses a significant lahar risk. Much of the flat ground around the rugged Puget Sound area is flat, because like Armero, it has been repeatedly buried by lahars. In the year 2000 approximately 150,000 people lived in areas that have been repeatedly inundated by lahars from Mount Rainier in the last 6,000 years (Sisson et al., 2001). Lahars sourced on Mount Rainier have entered the ocean at the city of Tacoma, 70 km from the volcano's summit, and may have entered the city limits of Seattle, 80 km from the summit. The volcano is cut by an extensive zone of dikes and hydrothermal alteration that weaken the edifice, making it particularly susceptible to shedding mudflows and at risk for a yet greater hazard, sector collapse.

Sector Collapse

Sometimes the weight of rock, snow, and ice on a stratovolcano exceeds the strength of the volcanic edifice, leading to **debris avalanches**. These can be triggered by eruptions or earthquakes, or have no obvious trigger, and may range in size from small landslides to wholesale collapse of an entire quadrant of the volcano. Fortunately, such **sector collapse** events are rare. The 1980 landslide that initiated the climactic blast of Mount St. Helens in 1980 was a form of sector collapse, as was the remarkably similar 1956 eruption of Bezymianny volcano in Russia. Both events involved collapse of less than

Debris avalanches
Masses of rock, soil, snow, and ice that move down the side of a mountain.

Sector collapse
Wholesale failure and collapse of a portion of a volcano; may be triggered by or trigger an eruption.

Figure 7.35 View looking south across a hummocky debris avalanche from Mount Shasta, 45 km distant. USGS description: "The hummocks are part of an exceptionally large landslide that covers an area of about 675 km^2 and has a volume of at least 45 km^3. The hummocks consist of massive blocks of lava flows and layers of unconsolidated deposits of pyroclastic flows, lahars, tephra, and sediment which were transported from Mount Shasta intact within the landslide. The flat areas between the hummocks are underlain by thoroughly mixed and shattered volcanic rocks of the landslide."

Source: S. R. Brantley, USGS.

1 km^3 of material followed by a lateral blast and a Plinian eruption. However, far larger prehistoric sector collapse deposits have been identified.

Mount Shasta is a striking Cascade stratovolcano in northern California, reaching 4322 m above sea level. Northwest of the volcano, hundreds of hills and ridges formed of andesite and pyroclastic rocks from Mount Shasta rise from a flat agricultural plain (Figure 7.35). These features rise as much as 200 m above the plain, and their origin had puzzled geologists for half a century. The 1980 Mount St. Helens eruption blasted home how they formed: they are debris from a massive sector collapse of ancestral Mount Shasta. The avalanche occurred sometime between about 380 and 300 ka and involved at least 45 km^3 of debris—the largest known Quaternary landslide on Earth.

But wait, it gets worse: even larger landslides can be generated by shield volcanoes, which are huge compared to stratovolcanoes (see Figure 7.5); like stratovolcanoes, they can grow to great height and are prone to oversteepening by eruption and intrusion. Furthermore, they produce concentrated zones of diking, such as the East Rift Zone of Kīlauea, that serve to weaken the edifice much as hydrothermal alteration along dike zones weakens other volcanoes. Bathymetric surveys around the Hawaiian Islands have found that large pieces of the islands litter the ocean floor up to 200 km from the coast (Moore et al., 1989; Figure 7.36). Several such landslides have volumes measured in the thousands of cubic kilometers—hundreds of times greater than the Shasta debris avalanche.

Discovery of these giant submarine landslides neatly explained why the islands of Oahu and Moloka'i have one side with tall cliffs facing the ocean—these are the headwalls of the landslides (see Figure 7.36). It also appears to explain why coral breccia and other wave deposits are found 326 m above sea level on Lāna'i—they were likely placed there by

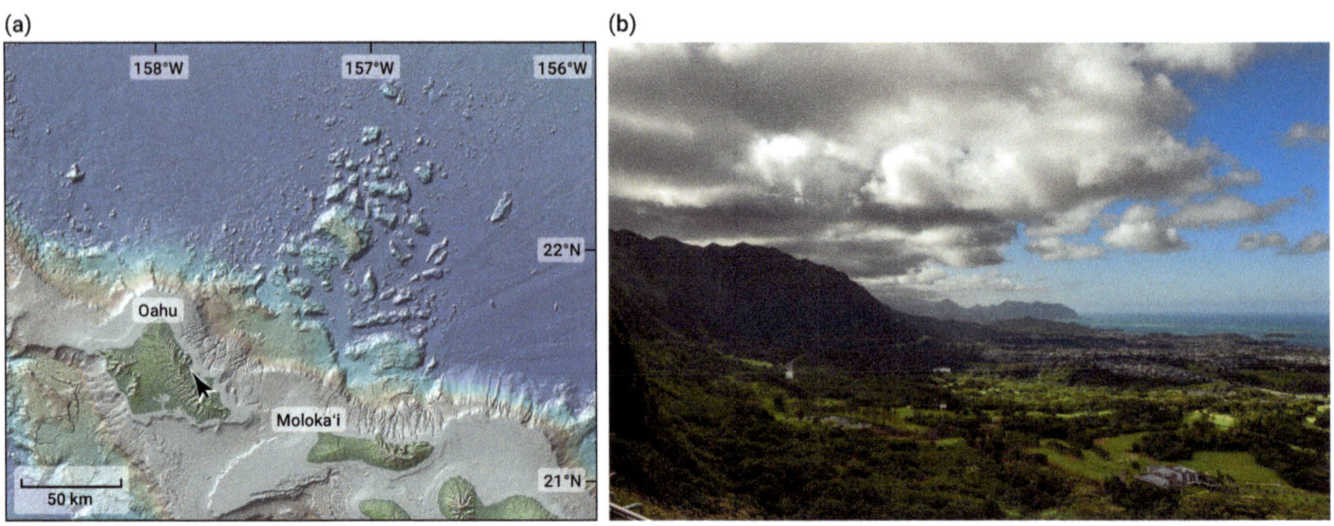

Figure 7.36 (a) Digital elevation model of bathymetry off the Hawaiian islands of Oahu and Moloka'i, showing the huge Nu'uanu (Oahu) and Wailau (Moloka'i) sector collapse landslides that deposited large chunks of the islands over the seafloor to the north. (b) Nu'uanu Pali, cliffs on the north shore of Oahu. These form part of the headwall scar of the Nu'uanu landslide. Arrow in (a) gives camera position and look direction of (b).

Sources: (a) Elevation model from GeoMapApp; (b) AFG photo.

landslide-generated tsunamis. Modeling of the effects of an Oahu-type landslide indicate that it could generate waves more than 100 m tall on Oahu and Moloka'i and, surprisingly, 70 m tall in southern California, 4000 km away (Satake et al., 2002). These results, although highly model-dependent, are certainly sobering.

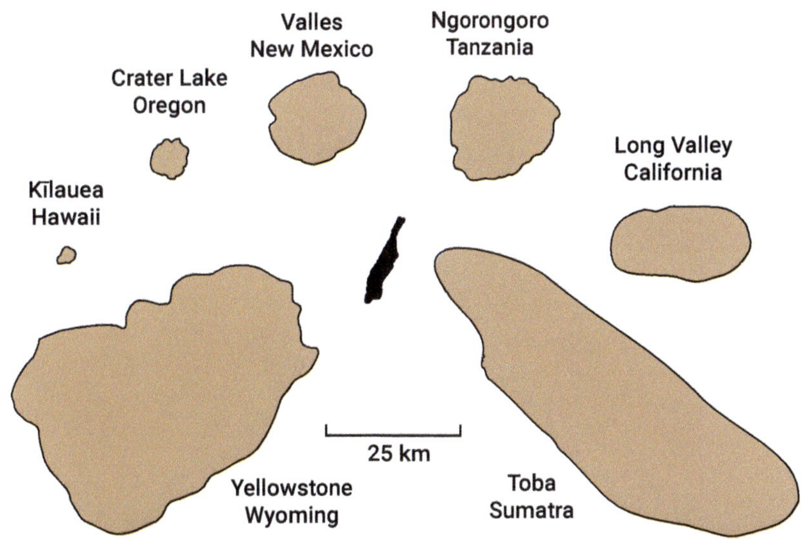

Figure 7.37 Size and shape comparison of various calderas. Curves on each outline the topographic wall of the caldera. Black figure in the middle is the outline of the island of Manhattan at the same scale.

Calderas

Size and Shape

Calderas are depressions formed when magma evacuates a magma chamber, causing the roof of the chamber to collapse. This can occur via lateral withdrawal, as at Kīlauea (see Section 7.1) and at Katmai volcano in Alaska during the massive VEI 6 eruption in 1912, or via blowing magma out the top, as in colossal silicic eruptions such as those at Mount Mazama (Crater Lake), 7600 years ago; Yellowstone, 640 ka; and Tambora in 1815. Calderas range in size from about 100 m across to a few kilometers across; the caldera of the Youngest Toba Tuff (Indonesia) is an astonishing 100 × 30 km in dimension (Figure 7.37).

The volumes of large pyroclastic eruptions are hard to visualize. The 600 km^3 of Bishop Tuff in eastern California could spread a blanket 2 m deep over the state of Colorado. The truly gargantuan eruption of the Youngest Toba Tuff, in Sumatra, 75 ka ejected an estimated 2800 km^3 of tephra (Chesner, 2012)—enough to blanket Colorado 10 m deep.

An Example: Long Valley Caldera, California

Long Valley caldera in eastern California (Figure 7.38) is one of the best-studied and best-monitored calderas in the world. The 32 × 16 km caldera depression is ringed by mountains along three-fourths of its perimeter; the relatively flat valley floor is at about 2100 m, and the mountains rise over 1200 m higher along the rim. The caldera lies in a valley between the north-trending Sierra Nevada Range and White Mountains, both of which rise to over 4200 m. The thick outflow sheet of the 767-ka Bishop Tuff rhyolite ignimbrite (Figure 7.38a) occupies this area north, east, and south of the caldera, but was blocked to the west by the Sierra Nevada.

Geology of the Long Valley caldera system has been worked out in a series of exhaustive studies, including Gilbert (1938), Bailey et al. (1976), Hildreth (1979), Wilson and Hildreth (1997), and many others. The general sequence of events, which is taken to be representative of this style of huge eruption, ran like this:

1. Eruption of basalt and andesite, then dacite, and finally high-silica rhyolite for about 4 Myr before the main eruption. The rhyolites formed an extensive set of obsidian domes, part of which is preserved as the Glass Mountain complex.
2. Eruption of the Bishop Tuff and formation of the caldera 767 ka.
3. Development of a large caldera lake and consequent sedimentation.
4. Eruption of rhyolite domes through the caldera floor from 700 to 100 ka; these are typically referred to as the **resurgent dome**. The low area between the resurgent dome and the caldera wall is called the **moat**.
5. Overflow and drainage of the caldera lake after 180 ka.
6. Migration of magmatism to a new and separate system to the west, and apparent death of the Long Valley system.

Wilson and Hildreth (1997) estimated that eruption of the entire 600-km^3 ignimbrite package took place over less than a week.

Resurgent dome
Rhyolite domes extruded within a caldera after a caldera-forming eruption.

Moat
In volcanology, the low area between the resurgent dome and the caldera wall; in some calderas the moat is filled with abundant but inconspicuous basalt flows.

(a)

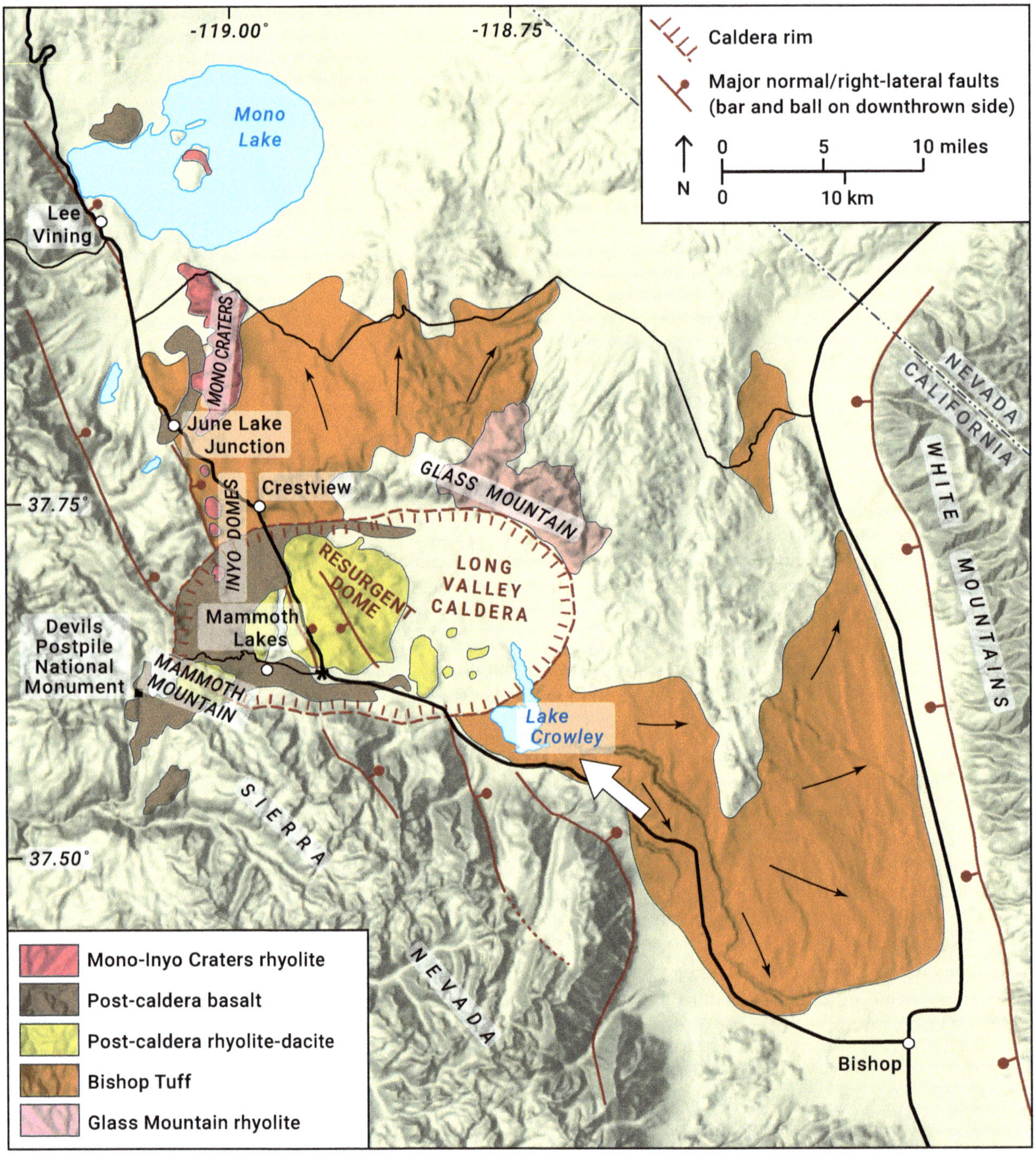

Figure 7.38 (a) Map of Long Valley caldera showing the extent of the caldera's topographic expression, setting along the eastern fault scarp of the Sierra Nevada, and thick outflow sheet of the Bishop Tuff. White arrow gives look direction of photo in (b). (b) Early morning aerial view of the caldera, looking NNW. The caldera's topographic wall is well expressed on the southwestern escarpment of Glass Mountain, a remnant of pre-caldera volcanic activity. Dashed lines indicate some of the faults in the complex right-normal oblique fault system that bounds the Sierra Nevada Range. The caldera straddles that boundary, and both the faults and topographic offset across them are subdued within the caldera. Lake Crowley, a reservoir, occupies part of the basin formerly occupied by a caldera lake. The Inyo Dome system (see Figure 7.15) parallels the fault on the far side of the caldera.

Sources: (a) Illustration by Chelsea Feeney, from Glazner et al. (2022), after USGS; (b) modified from Glazner et al. (2022).

(b)

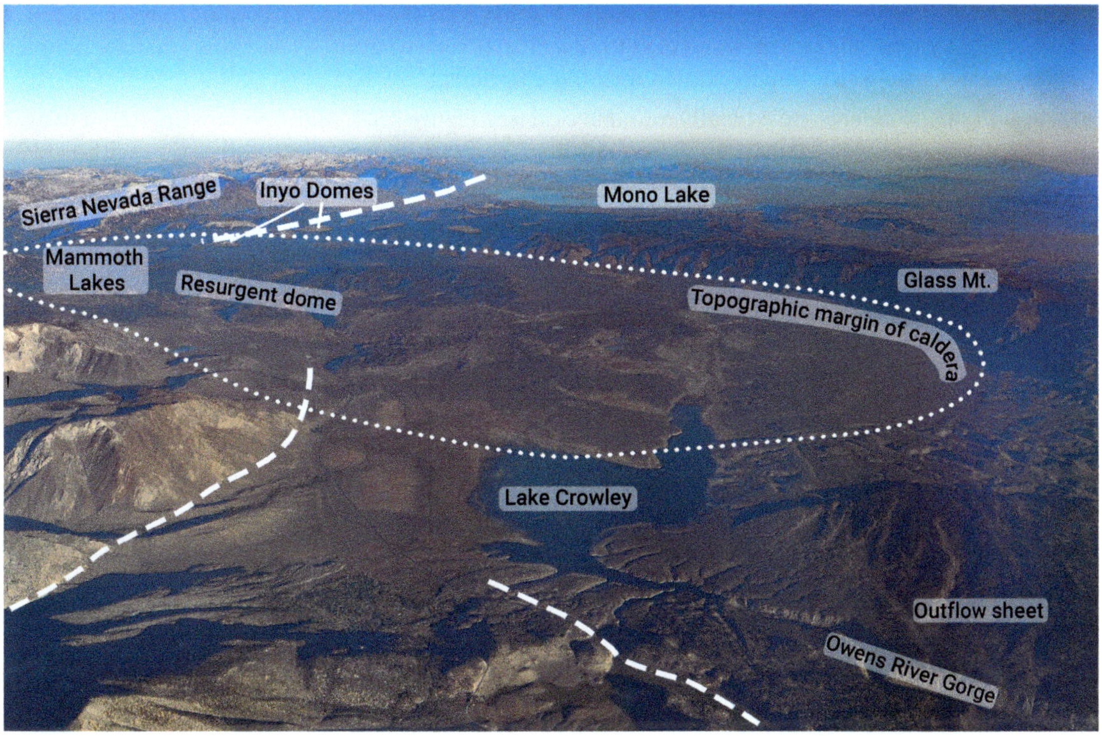

Figure 7.38 continued

How Do Caldera-Forming Supereruptions Work?

As we noted in Section 7.2, there have been no supereruptions in the Holocene, and thus none since civilization developed. As with much of geology, inferring the nature of the crime (the eruption) can only be inferred from clues left behind (pyroclastic rocks, topography, etc.). We cannot recreate such eruptions in the lab, and computer models can only take us so far.

Based on worldwide studies of calderas in various states of erosional and structural dissection, Lipman (1984) developed a generalized caldera cycle. Pre-caldera magmatism consists of a field of intermediate and silicic volcanoes. If the magma production rate is high enough, a shallow magma chamber develops, and if it is large enough and shallow enough, the roof caves in, causing catastrophic pressure release that vents much or most of the magma in a massive pyroclastic eruption. Failure of the roof around a ring fracture that outlines the magma chamber results in subsidence of the roof by as much as several kilometers; roof rocks founder into the caldera. The presumably steep-sided, cylindrical ring-fracture fault produces a kilometers-tall hole (filled with violently erupting pyroclasts) into which the walls slump, producing a topographic escarpment that retreats back from the original structural escarpment and forms the topographic feature that we identify as a caldera. Much of the erupted volume of pyroclastic rocks (perhaps half) ends up as thick intracaldera tuff; the remainder forms the outflow sheet. Intracaldera tuff in large calderas may be up to 3 km thick and display low-grade greenschist metamorphism.

In this standard model of caldera formation, such eruptions are assumed to come from liquid separated from shallow (~5 km), long-lived (1 Myr or more) magma chambers; caldera shapes are assumed to mirror the shapes of the underlying magma chambers; and unerupted material is assumed to crystallize as plutons (e.g., Bailey et al., 1976; Lipman, 1984; Lipman & Bachmann, 2015). Recent work, however, casts doubt on some of these assumptions. Particular problems include:

1. The amount of magma needed to build a shallow magma chamber and then to sustain it for 1 Myr or more against heat loss to the surface is unreasonably large. (Glazner, 2021). For example, for a sheet of magma (such as that depicted in Chapter 11,

Figure 11.9), establishing a hot geotherm and then maintaining it for 1 Myr requires input of tens of km of basaltic magma into the crust. This would greatly increase crustal thickness and lead to spectacular surface uplift—two things that are not seen in the lead-up to caldera-forming eruptions.

2. Calderas are typically round to oval (see Figure 7.37). Many plutons are round to oval, too, but many are far from such shapes, lacking any clear correlation with caldera shapes (e.g., Buddington, 1959; see Figure 10.19).
3. The common hypothesis that caldera-forming eruptions leave behind shallow plutons makes several specific predictions that can be tested using methods discussed in Chapter 9. In particular, the erupted rocks should be enriched in incompatible elements relative to the plutons, the plutons should be enriched in compatible elements relative to the volcanic rocks, and the erupted age of the ignimbrite should overlap with the magmatic lifetime of the pluton. There are few examples where this simple test is met, and many where it fails. Globally, the trace-element patterns of volcanic and plutonic rocks fail to support the hypothesis that volcanic rocks and plutonic rocks represent complements fractionated from a common parent magma.

There are many unknowns regarding the origin of silicic supereruptions and the thermal and mechanical history of caldera systems.

7.7 Links between Volcanism and Climate

Learning Objectives

By the end of this section you should be able to:

7.7.1 Contrast sulfur-rich and sulfur-poor eruptions with regard to their ability to affect global temperatures.

7.7.2 Explain why eruptions that loft material into the stratosphere have a greater effect on climate than those that do not reach as high.

Effects of Volcanism on Climate

It is easy to understand how large volcanic eruptions can influence climate by pumping large amounts of ash and gases into the atmosphere. Perhaps the most concrete illustration of this was the 1815 eruption of Tambora (Indonesia). Unlike the 1883 eruption of Krakatau, the first major eruption after the advent of a worldwide communication network (Self, 1992), little was known about the Tambora eruption outside of those affected by the eruptive event itself. However, the effects of the VEI 7 eruption, the largest in modern history, were felt worldwide. The eruption ejected about 50 km^3 of magma, most as pyroclastic flows accompanied by an eruption column that reached an altitude of over 40 km. Deaths caused by pyroclastic flows, tsunamis, starvation, and disease likely added up to over 70,000 (Oppenheimer, 2003).

The Tambora eruption had a pronounced influence on climate worldwide, with particularly severe (or perhaps just particularly well-documented) effects in Europe and North America. In New England, severe frosts occurred during every month of the year, killing crops and causing misery. Crop failures led to famine.

Troposphere
The lower 10–15 km of the atmosphere; where weather happens.

The structure of the atmosphere plays an important role in conveying the effects of an eruption to a global audience. Humans live in the **troposphere**, the lowest layer of the atmosphere (Figure 7.39). In the troposphere, temperature generally declines as one goes up in elevation. Declining temperature causes air to become denser, but increasing altitude (and thus decreasing pressure) causes it to become less dense, and how these two factors fight it out dictates the weather. Hot, humid air leads to density inversions where denser air gets above less dense air. This leads to vertical convection, a well-mixed atmosphere, and thunderstorms. Dry air is susceptible to temperature inversions, with denser air under less dense air; this leads to a stable atmosphere with little mixing and contributes to poor air quality. Rain and snow result when the atmosphere is oversaturated in water and are effective pollution scrubbers.

The top of the troposphere is the **tropopause**, a diffuse boundary with the overlying **stratosphere**. In the stratosphere, temperature increases with elevation, driven largely by absorption of ultraviolet radiation by ozone. The tropopause is generally at about 18 km near the equator, decreasing to around 8 km near the poles (see Figure 7.39). The stratosphere contains little water and is stably stratified in density, so there is essentially no weather—in particular, no precipitation to scrub it clean.

Tropopause
The boundary between the troposphere and stratosphere.

Stratosphere
The layer above the troposphere, characterized by increasing temperature with increasing altitude, dryness, and lack of precipitation.

Large volcanic eruptions typically throw ash and volcanic gases well into the stratosphere (see Figure 7.39). The latitude of a volcano is important in determining whether ejecta reach the stratosphere, because the tropopause is 10 km lower near the equator than at the poles; thus, the high-latitude 1963 eruption of Surtsey in Iceland (VEI 3) reached the stratosphere by lofting material about 15 km high, whereas ejecta from the more energetic 1981 eruption of North Pagan in the Mariana arc (VEI 4), thrown just as high, did not.

Ash particles that reach the stratosphere are not washed out by precipitation, but they do settle at rates that bring them back down to the troposphere within days to years (MacKinnon et al., 1984), depending on size. Eruptions into the stratosphere also produce **volcanic aerosols**. A particularly important aerosol is produced by sulfur dioxide (SO_2), which, through a complex series of photochemical reactions, produces tiny droplets of sulfuric acid (H_2SO_4) that can stay in the stratosphere for years. Sulfur dioxide is abundant in volcanic gases, especially those emitted from basaltic volcanoes. The combined effect of solid dust particles and volcanic aerosols is to block a small amount of short-wavelength light from reaching the Earth's surface, resulting in cooling. For major eruptions of the last few centuries, such cooling has generally lasted a few years (Self et al., 1981).

Volcanic aerosols
Fine particles and droplets emitted by volcanoes; sulfuric acid droplets are of particular importance.

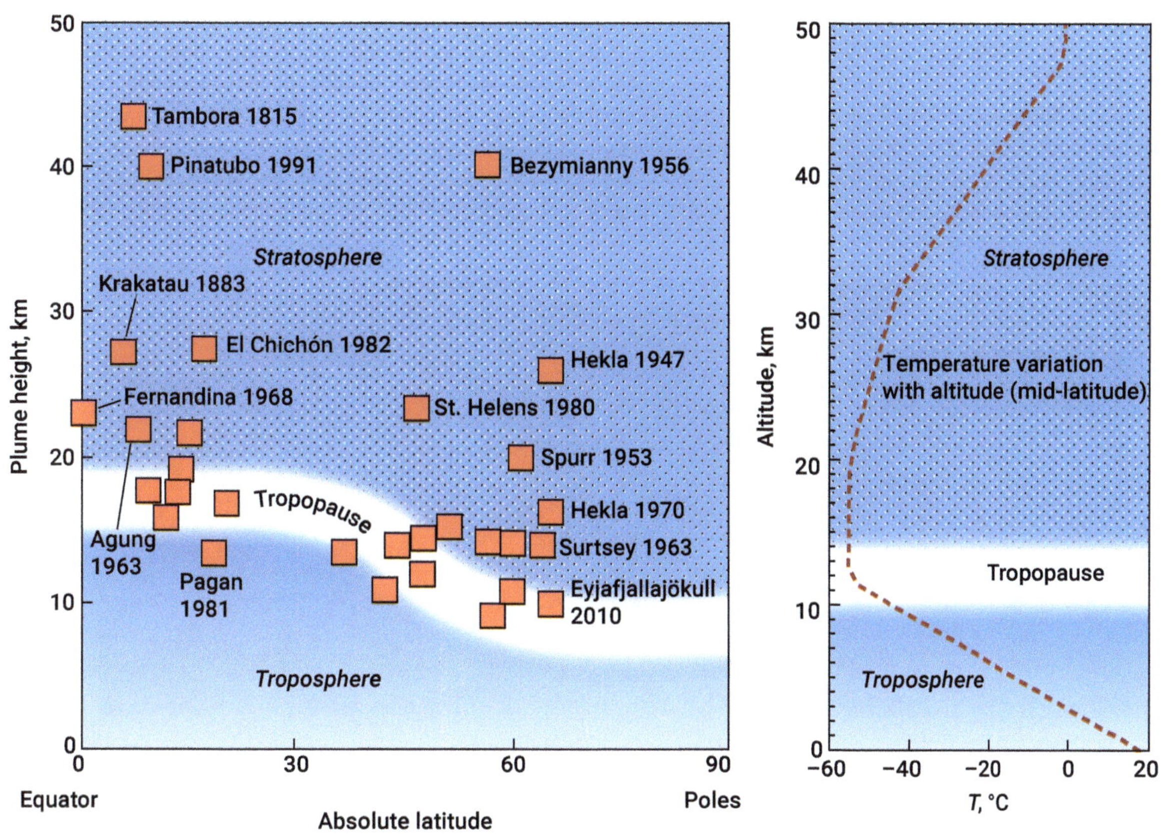

Figure 7.39 Estimated heights of plumes from selected eruptions. Whether or not a volcanic plume 10–20 km tall will reach the stratosphere depends on latitude, because the tropopause descends nearly 10 km from the equator to the poles. Dispersal into the stratosphere increases the effect of ash and gases on sunlight reaching the Earth's surface.

Source: Modified from Jakosky (1986).

Effects of Climate on Volcanism

The effect of volcanism on climate is well-established. Surprisingly, it seems that the converse is also true: climate has a detectable effect on volcanism (Rampino et al., 1979). A number of studies have found that environmental change can affect the frequency and timing of volcanic eruptions. For example, Wallmann et al. (1988) found that eruptions at Pantelleria, an island in the Mediterranean, appear to be correlated with sea-level drawdown during glacial episodes, presumably because lower sea level releases pressure on the volcanic edifice. Conversely, increased glacial ice loads are correlated with decreased eruptive activity in Iceland and California (Glazner et al., 1999; Swindles et al., 2018). These effects are small but indicate that the volcano-climate feedback goes in both directions.

Making Connections 7.2

Silicic Supereruptions and Ecosystems

As noted earlier, the VEI 7, 50-km^3 eruption of Tambora in 1815 significantly perturbed worldwide climate and had corresponding ill effects on humans that lasted several years. Scaling such an eruption up to 1000 km^3 or more, Rampino et al. (1988) suggested that such silicic supereruptions could cause "volcanic winter," a condition in which volcanic aerosols dim the sun to no brighter than its brightness on an overcast day—for years. Lack of sunlight would lead to reduced photosynthesis, reduced plant production, and consequent effects up the food chain to humans and other animals.

The exceptionally wide distributions of ashes from the Lava Creek (Yellowstone) and Youngest Toba Tuff (YTT) eruptions drive home just how much larger these events were than anything in the past several thousand years (Figure 7.40). The YTT tephra, estimated at 2800 km^3 in volume, is a meter thick in India, 2500 km away, and is found as a thin layer in Lake Malawi, 7300 km away. It is tempting to ponder the ecological devastation wrought by eruptions of this scale. The YTT was erupted 75 ka (Mark et al., 2014), and evolutionary DNA studies can be interpreted to require a serious decline in hominid numbers, down to perhaps just several thousand individuals, sometime before 60 ka. Ambrose (1998) proposed that the cause of this bottleneck was the YTT eruption.

This hypothesis received widespread attention and amplification (e.g., Rampino, 2002). It makes several predictions, including significant ecological changes and global temperature reduction in the years following the eruption. These have not been borne out. For example, Gathorne-Hardy & Harcourt-Smith (2003) noted that, on the Mentawai

(a) (c)

Figure 7.40 Supereruptions can produce remarkably thick ash deposits at great distances from the source caldera. (a, b) Distributions of recognizable (a) Lava Creek ash from Yellowstone and (b) Youngest Toba Tuff. (c) Two-meter-thick layer of Lava Creek ash interbedded with lake sediments near Shoshone in eastern California. The ash, soft enough that peopled carved out dwelling places in it during the 1920s, was deposited in Lake Tecopa, which intermittently filled a basin on one of the rivers that fed Death Valley during glacial times. Although the great thickness is partly a result of ash being washed into the lake, 2 m is still impressive given that the source in Yellowstone was 1000 km away. (d) White ash more than a meter thick from the Toba eruption in a quarry in southeastern India, 2500 km from the caldera in Sumatra.

Sources: (a) Modified from USGS; (b) modified from Costa et al. (2014); (c) AFG photo; (d) Sanjay P. K.

(b) (d)

Figure 7.40 continued

islands (350 km from Toba), nine species restricted to the islands (four primates, four squirrels, and one mouse) require that the islands have remained rainforest since before the eruption. In other words, a year or two of volcanic winter would have likely wiped out some or all these unique creatures, but they kept calm and carried on.

A second test is provided by sediment layers enclosing YTT ash in African lakes. Global volcanic winter conditions would have greatly disturbed the ecology and geochemistry of the lakes, but in Lake Malawi, neither Lane et al. (2013) nor Jackson et al. (2015) found significant faunal, geochemical, or temperature differences across the YTT interval. Further, Crick et al. (2021) found that the Toba eruption was not associated with global cooling as recorded in polar ice cores.

The consequences for a silicic supereruption would be catastrophic for civilization, but if the geologic record is a guide, such eruptions are unlikely to lead to widespread ecosystem collapse.

Summary

Key points of this chapter include the following:

- Volcanism on Earth ranges from quiet, non-explosive emission of magma to supereruptions far more violent than any that have happened since civilization began.
- The violence of an eruption depends primarily on the viscosity, volume, and gas content of the magma.
- Magma viscosity increases with increasing silica content and crystal percentage and decreases as temperature and dissolved water increase.
- Products of volcanoes include various types of lava, tephra (material blown out of a volcano), and gases.
- Magma moves through the crust primarily through planar cracks (dikes).
- Magmas, except those of exceptionally low viscosity, are too viscous to attain turbulence during flow.
- Pyroclasts form deposits classified as fall (falling from the air like snow; bedded), flow (hot, dense, gas-rich, ground-hugging flows of pumice and broken rock; non-bedded), and surge (low-density blasts; cross-bedded).
- Lahars (mudflows) are particular hazards of stratovolcanoes, where steep topography, loose rock, and abundant water in the form of snow, ice, and rain can be mobilized to form far-reaching flows.
- Shield volcanoes and stratovolcanoes are prone to sector collapse, producing huge landslides that can trigger tsunamis.
- The largest pyroclastic eruptions produce deposits 1000 or more times greater in volume than the 1980 eruption of Mount St. Helens and leave behind huge calderas.
- The largest flood-basalt eruptions are associated with mass extinctions, whereas the largest pyroclastic eruptions are not.

Notes

1. This happens to be several times less than the pressure at the bottom of an adult's foot, so if you were standing on that lava at 1125 °C, you could not count on the yield stress to keep you from sinking in. Best not to try that.
2. In particular, the excellent book *All About Volcanoes and Earthquakes* by Frederick H. Pough, which has a riveting account of the Parícutin eruption. Your first author checked this out of the library in first grade, and it clearly left an imprint.
3. At this point, your first author asked his father why the farmer didn't stick a garden hose into the pit and put the volcano out.

CHAPTER 8

Origins of Magmatic Diversity

The previous chapter emphasized the wide variety of behavior exhibited by magmas on Earth—behavior that is driven almost entirely by variations in magma composition. Given that basalt is what most planetary bodies make (see Chapter 1), how is it that the Earth also manages to produce the full BADR spectrum, as well as uncommon, bizarre compositions that lie far from it (e.g., Chapters 13 and 14)? This question has occupied igneous petrology for well over a century, in part because understanding the genesis of silicic rocks is required to understand the spectrum of volcanic behavior.

For much of the 20th century, the question of the origin of magmatic diversity was considered solved. Norman Bowen (1915a), building on work going back to Charles Darwin (1844), proposed that most igneous rocks are generated by progressive separation of crystals from an originally basaltic liquid. Bowen's 1928 book *The Evolution of the Igneous Rocks* further developed this idea of fractional crystallization, and his work influenced generations of petrological thought. Bowen argued so effectively for the dominance of fractional crystallization, and so thoroughly eviscerated those who maintained otherwise, that the problem of how the Earth generates rocks other than basalt was a non-issue; andesites, dacites, and the rest were just assumed to form from basaltic magma via fractional crystallization. The familiar Y-shaped "Bowen's Reaction Series" diagram (Bowen, 1922), a totem of this point of view, has both illuminated and confused geology students for the better part of a century.

This left an obvious problem, however—why doesn't this happen in places that are essentially all basalt, such as Hawaiʻi and along oceanic ridges? As we shall see below, separation of crystals from liquid by crystal settling is favored when the crystals are dense (e.g., olivine) and the magma has a low viscosity, two aspects of basaltic magmas. Conversely, the diversity of igneous rocks is great in arcs, places where high-silica, highly viscous magma and relatively low-density crystals (e.g., plagioclase) dominate. Both of these properties should inhibit crystal separation. Something is different about how igneous rocks evolve in subduction zones.

Several mechanisms have been proposed for turning basaltic magmas into other, more silicic compositions, or for producing other parent magmas. These include crystal-liquid separation (see Section 4.3), partial melting, magma mixing, assimilation, and liquid immiscibility. To evaluate these processes, we need to take another look at partial melting and introduce a few new concepts of geochemistry.

Although Iceland is overwhelmingly blanketed in basalt, consistent with its tectonic setting on the Mid-Atlantic Ridge, significant rhyolite is present as well. Here, the evening sun illuminates interbedded basalt and rhyolite in the mountains east of Borgarfjörður Eystri in the northeastern part of the island. Occurrences such as these raise a fundamental question: Was the rhyolite generated by fractional crystallization of basalt, by partial melting (of what?), or by some other process? In this chapter we discuss ways to tackle questions such as these.

Source: AFG photo.

8.1 Crystal-Liquid Separation during Crystallization

Learning Objectives

By the end of this section you should be able to:

8.1.1 Calculate chemical changes in magmatic liquids resulting from crystallization of one or more phases.

8.1.2 Estimate the rate of sinking or floating of a given mineral phase in a given magmatic liquid.

Evolution of Liquid Composition

In any multicomponent system undergoing crystallization, the composition of the crystals is generally different from the starting magma composition. By Baldy's law, the melt plus crystals must add up to the bulk composition of the system, so the melt must have a composition different from the magma as well. If the crystals and liquid are separated from one another at any point during crystallization, either completely or partially, then the original magma has been split into two parts, each with a composition different from the original.

Consider the binary eutectic Fo-Di, last seen in Chapter 4, and a hypothetical magma composed of bulk composition X consisting, as before, of 70 wt% Fo and 30 wt% Di (Figure 8.1). At temperatures above the liquidus for this composition, the system is all melt of composition X. As *T* is lowered, the system hits the liquidus at a little over 1800 °C, and Fo crystals begin to form; lowering *T* more causes the melt composition to fall down and to the left, along the liquidus curve, whereas the crystals remain pure Fo and fall along the right-hand compositional axis. At 1500 °C, the system consists of subequal amounts of crystals at s and melt at r. If we were to extract the melt from the system at this stage we'd have a new magma composition, s, quite different from the starting composition X. The new composition would be richer in CaO, Al_2O_3, and SiO_2 and lower in MgO than X, and as the melt migrates toward the eutectic, these trends continue.

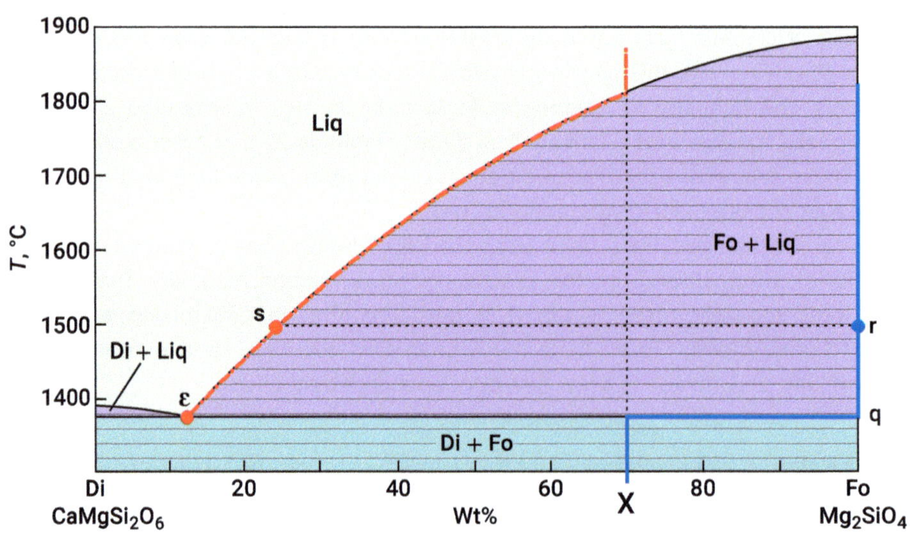

Figure 8.1 In the system Fo-Di, the composition of the liquid changes dramatically during crystallization or melting, but if the liquid is not physically separated from the crystals, then the bulk rock composition remains the same.

The steady evolution toward more silicic liquid compositions during crystallization led Bowen (1915a) to propose that the diversification of igneous rocks occurs primarily via separation of liquid from crystals during cooling.

Ways to Separate Crystals from Liquid

There are several important processes that could separate liquid from crystals, the most obvious being density sorting in the gravitational field. Magmatic liquids are less dense than the rocks they were derived from, because melting obliterates crystal structure, leading to less well-packed atoms. It has long been known that densities of glasses are several percent lower than those of their crystalline forebears (Douglas, 1907; Figure 8.2a), and extrapolating these densities to magmatic temperatures indicates that melts take up

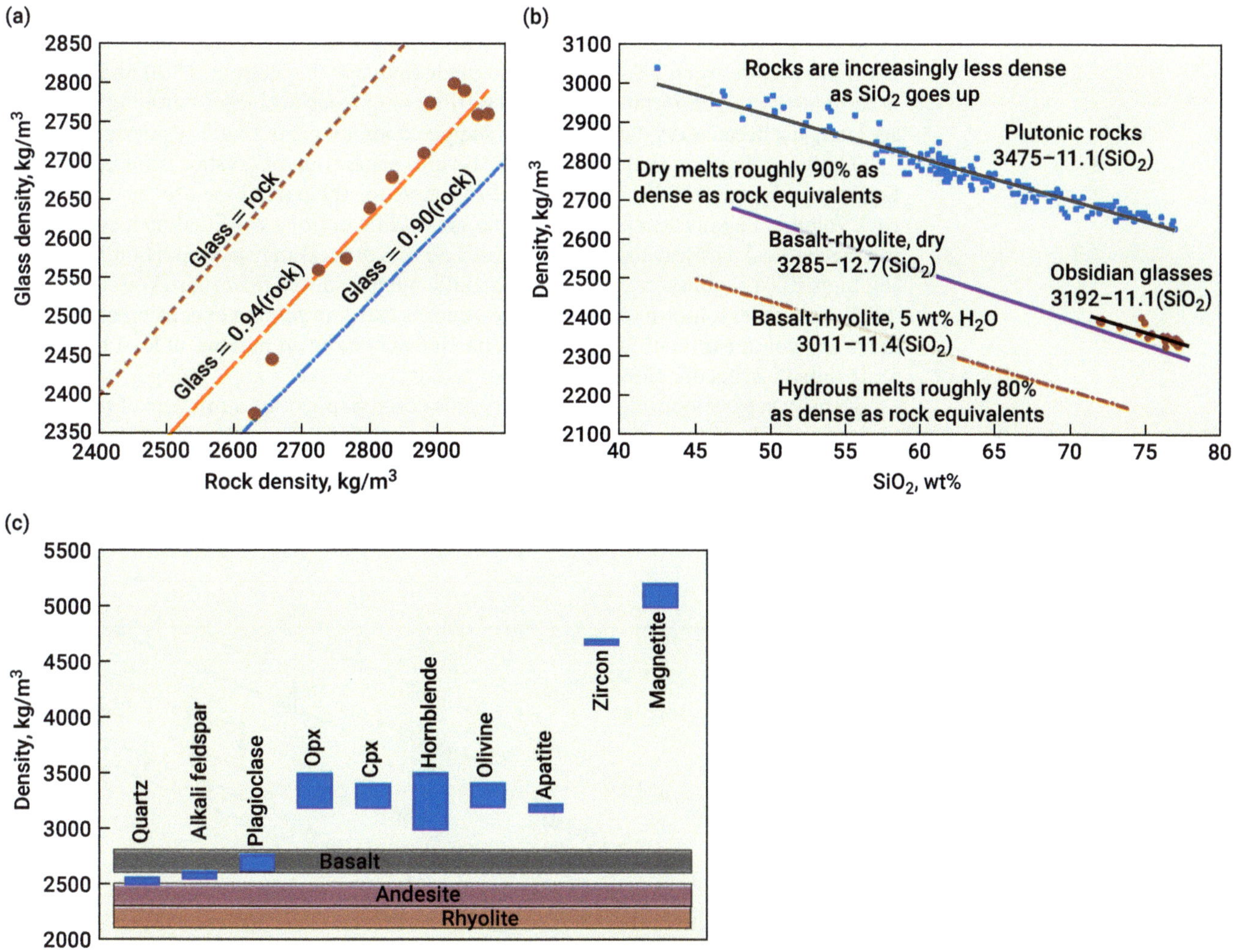

Figure 8.2 (a) Measured densities of rocks and glasses derived from them by melting at 1 atm. Glasses average about 95% of the density of the rocks they were derived from. Two low-density points that fall well below the 94% line may be abnormally low owing to foaming from volatiles released by breakdown of biotite and other hydrous minerals. (b) Densities of plutonic rocks, obsidians, and melts as a function of silica. Plutonic rocks from California range from ~3000 kg/m^3 at the gabbro end to ~2700 kg/m^3 at the granite end. Obsidian glasses are ~10% less dense, as are anhydrous melts along the BADR trend. Such melts with 5 wt% H_2O are ~20% less dense than the corresponding plutonic rocks. (c) Approximate densities of several important mineral phases and various magmatic liquids. Most minerals, especially those containing Fe, will sink in most liquids.

Sources: (a) Data from Douglas (1907). (b) Plutonic data from Bateman et al. (1984) and Baird et al. (1979), as compiled by Lee et al. (2007); obsidian data from Ambrose & Stevenson (2004); melt densities calculated along the basalt-rhyolite mixing curve of Ussler & Glazner (1989) using Glass-Density-Calc v3.2 (http://www.kaylaiacovino.com/tools-for-petrologists/).

several percent more volume than the rock that was melted to make them. Measured and calculated densities of rocks and melts similarly show that melts are significantly less dense than rocks of equivalent composition (Figure 8.2b).

The density decrease that accompanies melting is critical for two reasons. First, melts are less dense than most or all the minerals that might be suspended in them (Figure 8.2c), and such minerals will sink given enough time. Sinking rates can be estimated using **Stokes' law**:

Stokes' law
A formula for the rate of settling of spherical particles in a fluid in laminar flow.

$$v = \frac{2(\rho_c - \rho_m)gr^2}{9\mu} \quad \text{(Eq. 8.1)}$$

where v is the terminal velocity of a falling or rising particle (positive downward); ρ_c and ρ_m are the densities of crystal and melt, respectively; g is the acceleration due to gravity; and r is the particle radius. Figure 8.3 shows the calculated distances that 2-mm-diameter spheres of representative mineral species, quartz and olivine, could sink (or, for quartz in basalt, float) in a given time. These distances are surprisingly large, ranging from roughly 1 m to 1 km per decade. Such calculations were made long ago (e.g., Grout, 1918) and have been used to justify the importance of crystal settling in igneous evolution—although as we will see later, just because crystal settling *can* happen does not mean that it regularly *does*.

The second reason that density decrease aids separation of crystals from liquid is that liquid expansion provides a mechanical incentive for the partial melt to leave the melted rock volume. Expansion can drive the excess volume out of a confined melt zone if there is a mechanical pathway to lower pressure, and this process may be important in segregating high-silica rhyolitic magma from partially melted source materials (Waters & Lange, 2017). Volatile exsolution (discussed later) causes extreme volume expansion and can also drive low-volume interstitial melt out of partially crystallized magma, at least on the local scale (Sisson & Bacon, 1999).

Density separation of liquid from crystals could happen for a number of reasons. For example, olivine is typically quite a bit denser than the magmatic liquids that precipitate it; the Fo crystals could settle to the bottom of the magma chamber, with the melt decanted off the top to leave a **cumulate** rock consisting of Fo crystals—a dunite. Alternatively, the magma body could be squeezed (e.g., by tectonic forces, as at a restraining bend along

Cumulate
A term classically applied to igneous rocks thought to have formed by gravitational settling of crystals with consequent loss of magmatic liquid; see Section 14.5 for complications involving this term.

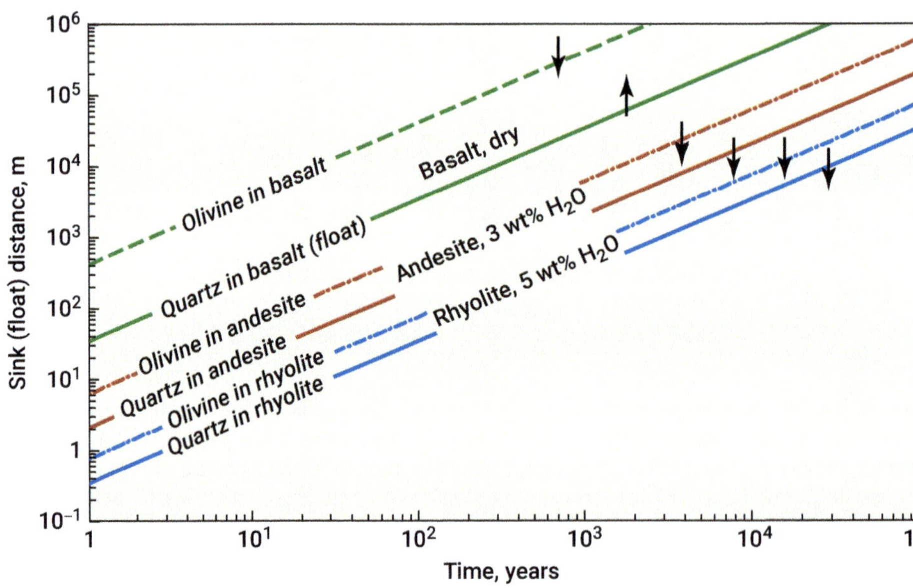

Figure 8.3 Calculated sinking or floating distances for quartz and olivine spheres, 1 mm in radius, in various magmatic liquids. Arrows give direction of movement.

a fault, or by the weight of overlying rock and magma) and the melt tapped off via fractures that are too small to admit the crystals—a process called **filter pressing**, akin to the French press method of making coffee.

Filter pressing
Separating magmatic liquid from crystals by squeezing crystal-rich magma tectonically or by forcing it through a constriction.

Although crystallization of minerals from a melt alone does not change the composition of a magma, it is clear from analytical studies that residual liquid in crystallizing magmas generally heads in a high-silica direction, commonly reaching the composition of rhyolite. Development of the electron microprobe allowed geochemists to analyze small pools of glass in the groundmasses of volcanic rocks, and it was shown long ago that the last bits of liquid in a crystallizing basalt are typically rhyolitic (e.g., Evans & Moore, 1968; Lambert et al., 1989). However, formation of several weight percent rhyolitic glass in the groundmass of a crystallizing basalt does not lead to magma diversification unless that glass can be separated to form another magma, and it is important to note that in neither of the examples just listed (Hawai'i and the Columbia River basalts) did that happen.

Crystal Fractionation and the Parameter *F*

Removal of crystals from magma changes the composition of the magma unless the crystals being removed are identical in composition to the bulk magma, and the crystals removed plus the derivative magma must add up to the original bulk composition by Baldy's law. Although, as in Chapter 4, we typically track the composition of the liquid and bulk crystals during melting and crystallization in systems such as Fo-Di, it is important to remember that nothing happens to the bulk composition of the system, which is what we measure with a whole-rock analysis, unless the crystals and melt are separated from one another. Thus, in Figure 8.1, the melt evolves during crystallization from 70 wt% Fo to 12 wt% Fo at the eutectic, *but the magma composition remains unchanged unless the melt and crystals are separated from one another.*

Now reread the previous sentence—this is really important.

Let's track these various compositions, using a parameter known as F to follow the amount of crystallization. F is simply the weight fraction of liquid in the system; thus, at and above the liquidus, $F = 1$, whereas below (but not necessarily at) the solidus, $F = 0$. Table 8.1 tracks crystallization of the bulk composition in Figure 8.1, using the methods in Section 4.1 to track the compositions and relative proportions of melt and crystals.

Figure 8.4a shows MgO in the melt and crystals as a function of F. The melt starts out at the bulk composition (45.7 wt% MgO) and evolves rapidly to lower MgO as Fo is crystallized, bottoming out at 23.3 wt% MgO when it reaches the eutectic. Conversely, the bulk crystals start out with the MgO content of pure Fo (57.3 wt%) and remain there until the liquid reaches the eutectic, at which time Di joins the crystallizing assemblage and the MgO content of the crystals goes down, bottoming out at the bulk composition when the last drop of liquid is crystallized. Note that given Figure 8.4a, at any given value of F the proportion of melt to crystals can be determined using the lever rule, using the

Table 8.1 Liquid and Solid Compositions as a Function of *F* for the Bulk Composition in Figure 8.1

		Liquid				Bulk Crystals			
T, °C	F	Wt. frac. Fo	SiO_2	MgO	CaO	Wt. frac. Fo	SiO_2	MgO	CaO
1850	1.00	0.70	46.5	45.7	7.8	1.00	42.7	57.3	0.0
1700	0.59	0.49	49.2	37.6	13.2	1.00	42.7	57.3	0.0
1600	0.47	0.36	50.9	32.5	16.6	1.00	42.7	57.3	0.0
1500	0.40	0.24	52.4	27.9	19.7	1.00	42.7	57.3	0.0
1375	0.34	0.12	54.0	23.3	22.8	1.00	42.7	57.3	0.0
1375	0.19	0.12	54.0	23.3	22.8	0.84	44.8	51.1	4.1
1375	0.00	0.12	54.0	23.3	22.8	0.70	46.5	45.7	7.8

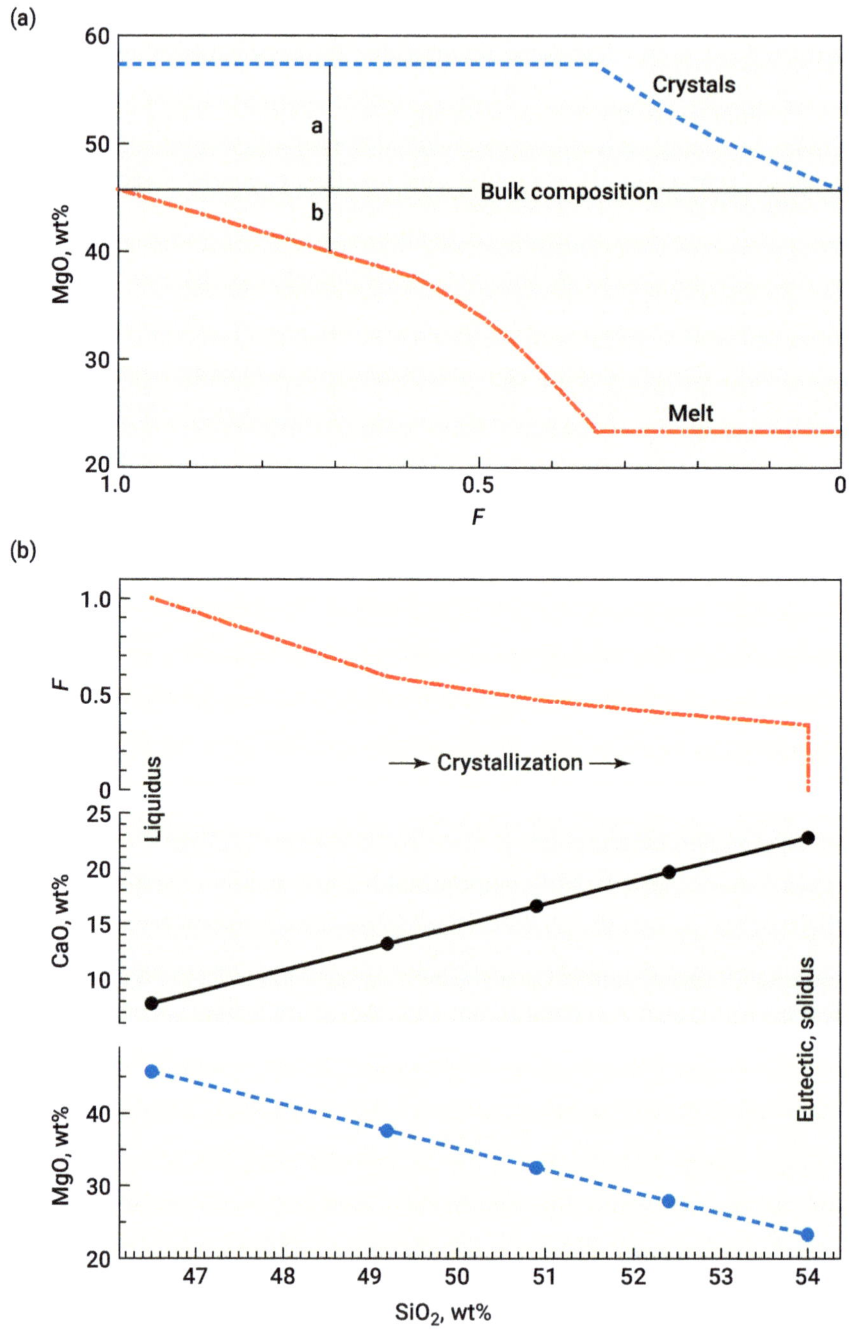

Figure 8.4 (a) Variation of wt% MgO in liquid and bulk crystals during crystallization of the magma in Figure 8.1, as a function of the weight fraction of melt, *F*. The melt starts out at the bulk composition and evolves (left to right, decreasing *F*) away from it, whereas the crystals start out as pure Fo and evolve to the bulk composition. At any value of *F*, for either oxide, the proportion of melt to crystals can be derived using the lever rule; for example, at $F = 0.7$, the proportion of melt is a/(a + b). (b) Variations of MgO and CaO with SiO_2 in lavas produced by tapping liquid from the system in Figure 8.1 at various values of *F*. Crystallization proceeds from left to right (toward higher silica); note that liquid evolution stops when the system hits the eutectic at $F = {\sim}0.34$.

bulk composition as the fulcrum. Note also that unless we separate melt from crystals, at least partially, then the bulk composition of the magma does not change in spite of the change in liquid composition.

Now let's assume that liquid and crystals are segregated by a process such as crystal settling. Assume that the magma in Figure 8.1 feeds a volcano that taps off liquid without crystals at the temperatures and *F* values in Table 8.1, producing a series of lava flows. You sample the lavas and analyze them, deriving the MgO and CaO curves in Figure 8.4b. The linear patterns are intriguing, and you might note that the lavas become lower in MgO and higher in CaO and SiO_2 with decreasing age. What are the *F* values that go with these lavas? You can't tell from the chemical values alone, because that is not something you can directly measure. You don't even know which direction *F* decreases in, although the upward decrease of MgO in the stack of lavas strongly suggests that MgO decreased as *F* decreased.

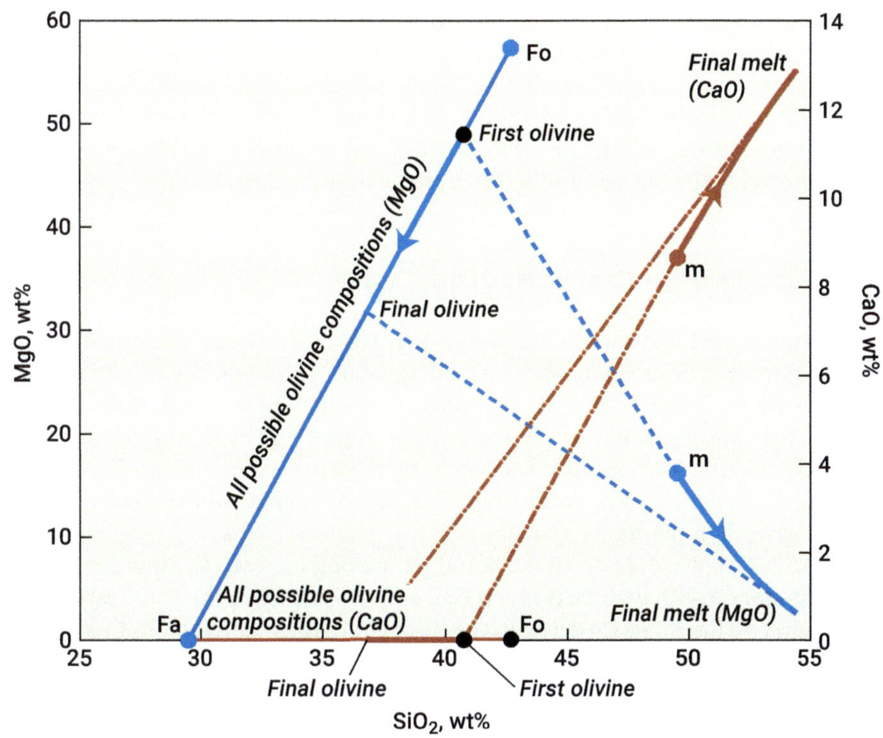

Figure 8.5 Fractionation of olivine from a basaltic liquid on plots of MgO and CaO against SiO_2. Liquid m starts to crystallize Mg-rich olivine of the indicated composition (First olivine), causing the liquid to move directly away from m on both plots in the direction of the arrows. Solid lines joining Fo and Fa are the loci of all possible olivine compositions on this diagram; for CaO, this lies along the abscissa, because CaO is not present in appreciable amounts in olivine. Removal of the initial Mg-rich olivine causes the Mg/Fe ratio of the remaining liquid to go down so that the next bit of olivine that forms is richer in Fe and poorer in MgO. As a result, the liquid path on each plot follows a curve whose tangent projects back to the Fo-Fa line. The CaO path is gently curved as well, moving away from olivine points on the abscissa.

Variations in major elements are, as always, governed by Baldy's law. Consider the case of a basaltic melt crystallizing olivine. In Figure 8.5, an Mg-rich basaltic melt (m, with the composition of a picrite) is shown on a plot of MgO vs. SiO_2, along with the olivine end-members Fo and Fa. The solid line connecting Fo and Fa is the locus of all possible olivine compositions. When olivine starts to precipitate from the melt, one of this infinite number will be chosen, but which one? We cannot tell from the diagram alone, but our knowledge of olivine-melt relations (see Figure 4.17) tells us that it will be toward the Mg-rich end of the line. Better, we can use the K_D relationship discussed in Section 4.8 to calculate the appropriate composition (see Toolbox 4.1 on modeling olivine fractionation). This is shown in Figure 8.5, and by Baldy's law, when olivine of this composition starts to form, the composition of the melt moves away from it in the direction of the arrow, toward higher SiO_2 and lower MgO.

The melt path starts out in the direction of the arrow but does not continue exactly along that path, because removing Mg-rich olivine increases the Fe/Mg ratio of the melt and causes more Fe-rich olivine to precipitate. As crystallization proceeds, the olivine moves toward Fa, with the melt moving directly away from the composition of the current olivine at every point. The olivine sweeps out a line segment, and the melt sweeps out a curve whose tangents track back to the olivine composition at that point in the magma's evolution. Keep in mind, though, that the tangent gives the composition of the olivine being removed at that particular instant, not the bulk composition of the olivine. The latter, which is the sum of all the little increments accumulated during crystallization to a given point, is found by drawing a line from the melt through the bulk composition to the olivine line (yet another expression of Baldy's law). What is the final bulk olivine composition?

The MgO melt curve in Figure 8.5 was calculated for 40 wt% olivine removal; this drives MgO down from 16.7 to 2.6 wt%, a significant amount, and SiO_2 up from 49.4 to 54.4 wt%, a less significant amount. CaO increases significantly because it is essentially absent in olivine.

Tholeiitic vs. Calc-Alkaline Trends

It is clear from experiments and observation that partially crystallized basaltic magmas contain a significant proportion of the ferromagnesian silicate minerals olivine and pyroxene. Because the Mg/Fe ratios of these phases are much higher than that of the liquid from which they are crystallizing (see Section 4.8), removing them from the liquid drives the Mg# of the liquid down rapidly, leading to significant Fe enrichment. This drives the liquid along the tholeiitic path, producing Fe-enriched magmas whose silica contents are not much higher than the starting magma.

As noted in Chapter 4, to drive the liquid path along the calc-alkaline trend requires crystallizing a phase with high Fe (to keep Fe from building up in the liquid and thus keep the Mg# high) and low silica (to drive up the silica content). Magnetite fills this role nicely. If f_{O_2} is high enough that magnetite is part of the crystallizing assemblage along with the ferromagnesian silicates, then the liquid's compositional path is calc-alkaline, appropriate for arc-related magmatism. Gill (1981, p. 274) showed that crystallization of plagioclase, Cpx, olivine, and magnetite can drive a basaltic magma along the calc-alkaline path to andesitic compositions.

Geometry of Fractional Crystallization

Oxide-oxide plots such as those in Figure 8.5 can be used to test hypotheses of magma diversification, because they are graphical tests of Baldy's law. Consider the data in Table 8.2, from Luhr and Carmichael (1980), which gives two whole-rock lava compositions along with representative olivine, plagioclase, and Cpx analyses. Our goal is to test the hypothesis that rock B was formed by separation of those three phases, in some proportions, from rock A.

Staring at the numbers in Table 8.2 is not very helpful, but plotting them is. Figure 8.6 shows three oxides from this table plotted against SiO_2. The green triangles enclose all possible mixtures of the three crystalline phases, so if rock B formed from rock A by removal of these phases, then the removed assemblage must lie on the back-projections of a vector from A to B, which must of course cross those triangles for the hypothesis to be valid. For MgO and CaO, this vector does indeed cross the compositional triangle, so that test is passed, although an exact solution is not clear from any single plot. Adding a third oxide (Al_2O_3) adds more information, and in this example, all three oxides converge on a solution given by their intersection. The dashed lines in the Al_2O_3 plot are the intersection lines from the MgO and CaO plots scaled to their proper positions on the triangle sidelines.

Table 8.2 Data for Crystal Fractionation Example

Component	SiO_2	TiO_2	Al_2O_3	FeO^t	MgO	CaO	Na_2O	K_2O
Rock A	56.5	0.8	16.7	6.5	6.0	7.7	4.0	1.1
Rock B	61.8	0.7	16.24	3.9	.98	5.3	5.2	2.5
Olivine	39.2	0.0	0.0	19.5	42.1	0.1	0.0	0.0
Plagioclase	47.5	0.0	33.9	0.4	0.0	16.8	2.4	0.0
Cpx	52.0	0.5	3.4	6.2	17.1	20.5	0.4	0.0

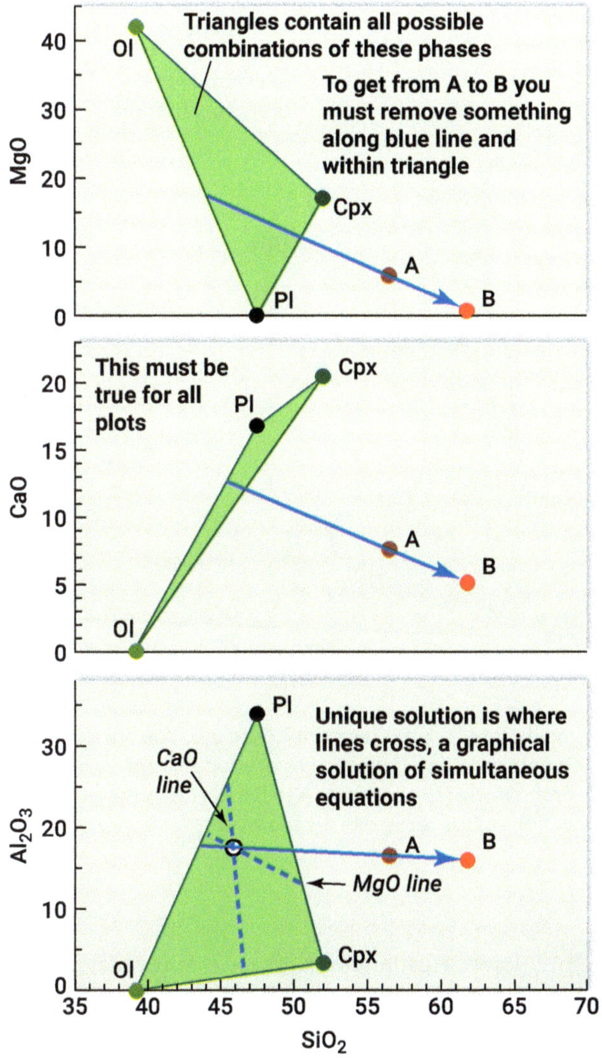

Figure 8.6 Plots showing the relationships of three oxides to SiO_2 for the crystal fractionation problem. For the MgO plot, back-projection of the line from A to B crosses the triangle connecting olivine, plagioclase, and Cpx, meaning that, *for that oxide pair*, any composition on the line within the triangle, which contains all possible combinations of those three phases, could drive A to B. Bringing in another oxide, CaO, produces another triangle and another line crossing it, and bringing in a third adds yet another triangle and line. We can transfer the crossing lines from the MgO and CaO plots to this plot; they cross within the triangle, giving the proper proportions to drive A to B. A successful hypothesis must pass all such tests.

More insight is given if we make a 3D plot with all three oxides (Figure 8.7); here, the locus of all possible mixtures is a triangle in 3D space and the back-projected vector cuts it at a point—the solution, roughly 49 wt% plagioclase, 31 wt% olivine, and 20 wt% Cpx.

The solution illustrated in Figure 8.7 is unique for these oxides but is not guaranteed to fit the others. Working with more components requires employing linear algebra to solve a least-squares problem. Such techniques are covered in a number of resources (e.g., LeMaitre, 1982; Ragland, 1989). When the entire dataset in Table 8.2 is treated, there are serious misfits for some oxides, including TiO_2 and FeO^t, whose predicted compositions in the residual magma are too high. How might one improve the model to better fit those oxides? Are any important expected mineral phases missing from the model?

Field and Petrographic Expectations

Crystal fractionation makes a number of testable predictions about the temporal and petrographic evolution of magmatic systems. For example, in the archetypal case of a magma chamber undergoing crystal fractionation under a volcano, the sequence of erupted lavas should evolve through time in the predicted direction (e.g., higher SiO_2, lower MgO,

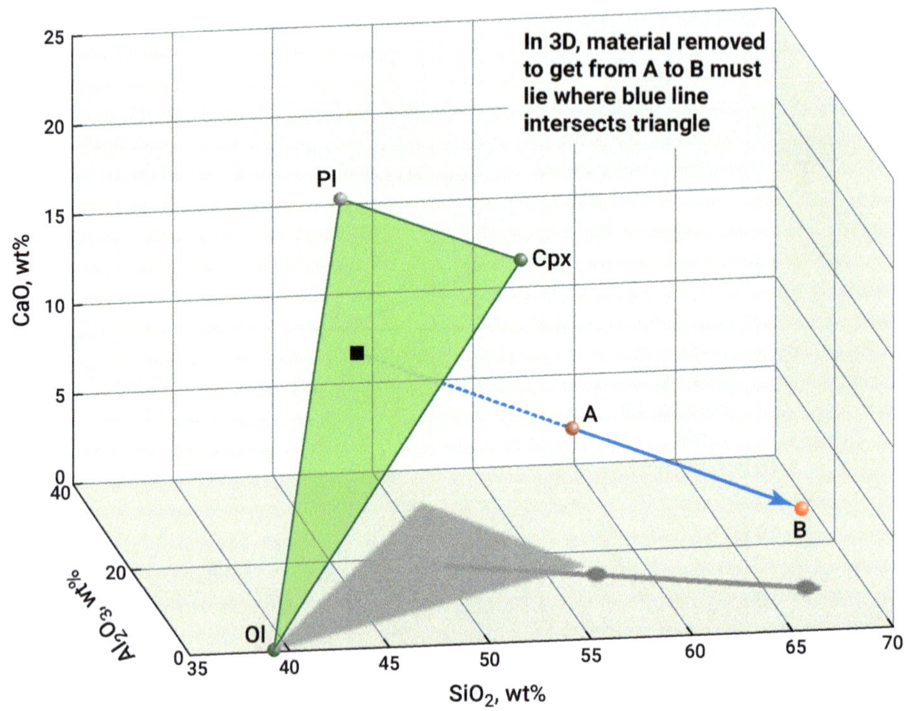

Figure 8.7 Three-dimensional plot of the oxides from Figure 8.6. You are looking into the box, and the shadow is projected onto the floor. If rock B was produced from rock A by removal of Ol, Pl, and Cpx, then back-projection of the line from A to B must cross the green triangle. This intersection gives the proportions of the three phases removed.

Figure 8.8 Zoned olivine phenocryst in basaltic magma from the Kīlauea Iki lava lake, Hawai'i. Higher birefringence (bluer color) toward rim indicates increasing Fe content, consistent with down-*T* crystallization. Width of field 2.6 mm.

Source: AFG photo, sample provided by Rosalind Helz, USGS.

lower Mg#, higher K_2O, etc.). Many magmatic systems have been interpreted in this way (e.g., Carmichael, 1964; Geist et al., 1995). Furthermore, in a magmatic system evolving via crystal fractionation, the individual mineral phases should show simple down-*T* zoning (Figure 8.8).

Lava lakes and sills have long been used as test beds for fractional crystallization. One spectacular example is the prehistoric lava lake exposed at Makaopuhi Crater on the East Rift Zone of Kīlauea (see Figure 7.1). Collapse of a younger pit crater exposed a 70-m vertical section through the older, frozen lava lake, and Moore and Evans (1967) used this section to examine crystal settling and differentiation of the basaltic magma that was ponded in the lake. They estimated that the original magma had about 7 vol% olivine phenocrysts, and the data in Figure 8.9 indicate that some of these crystals settled and piled up to as much as 20 vol% in the lower third of the column, with the basalt just above this zone clearly depleted in olivine. This differential movement of crystals and liquid changed the composition of the magma, with 13 wt% MgO in the enriched zones and only 6 wt% in the most depleted interval. However, silica, the principal index of variation along the BADR trend, only varies from about 48 to 51 wt%. Clearly, settling of olivine in this thick lava lake changed the Mg# significantly but did little to diversify the magmatic system along the BADR trend.

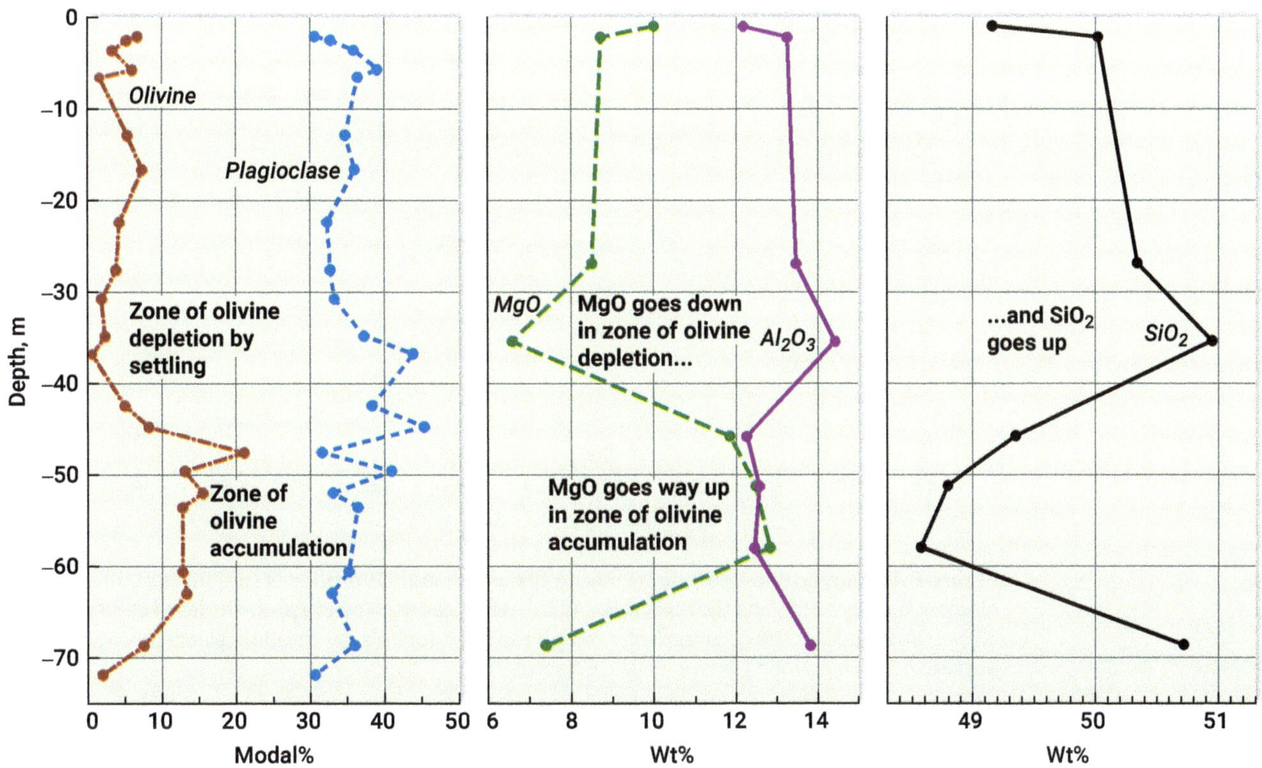

Figure 8.9 Mineralogical and geochemical variation with depth in the prehistoric Makaopuhi lava lake, Hawai'i.

Source: Data from Moore & Evans (1967).

Huge, extensive, thick gabbroic bodies known as layered mafic intrusions have long been interpreted as examples of fractional crystallization run amok. We will return to these in Chapter 14.

Crystal fractionation alone is a closed-system process; both the liquid and cumulate are subsets of the original magma, with no outside material added. Because crystallization is a natural consequence of cooling, there are no energetic barriers to fractional crystallization.

8.2 Magma Mixing, Assimilation, Partial Melting, and Other Processes

Learning Objectives

By the end of this section you should be able to:

- **8.2.1** Predict patterns on element-element diagrams that result from magma mixing or assimilation.
- **8.2.2** Describe typical field and hand sample aspects of mingled magmas.
- **8.2.3** Recognize evidence for liquid immiscibility in thin section.

Mixing, Mingling, and Mafic Magmatic Enclaves

Magma mixing is an efficient process for producing a diversity of magmas; for example, given the typical linear element-element trends shown by many BADR suites (see Section 1.2), most of the variation in such suites could be produced by mixing basaltic magma with rhyolitic magma. The geometry of magma mixing on oxide-oxide plots is

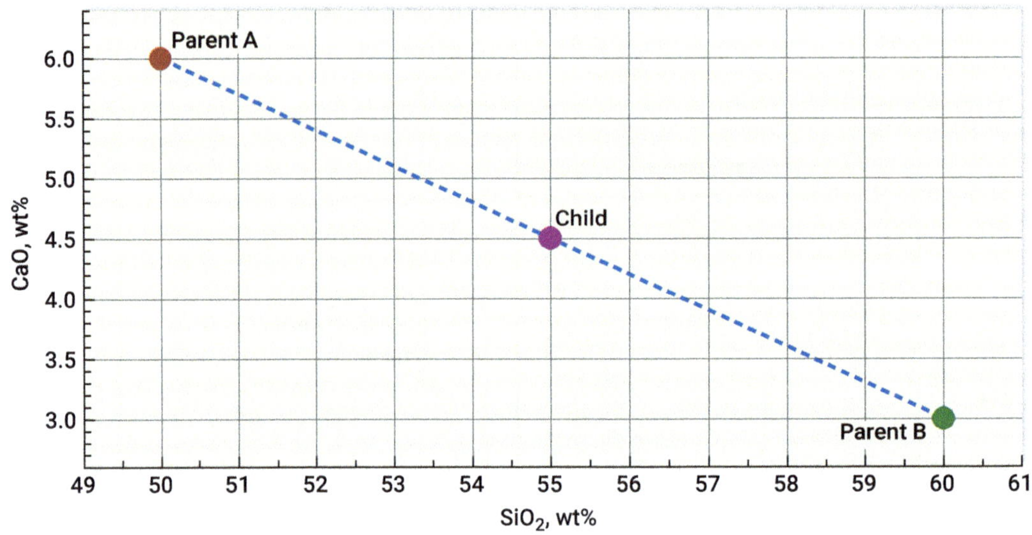

Figure 8.10 Simple two-component mixing of parent magmas produces straight lines on any element-element plot, and the lever rule can be used to determine the proportions of each parent in a given descendant. The geometry of mixing is the same for bulk assimilation. Linear mixing does not apply to combining ratios, however (see Chapter 9).

simple: mixtures lie on straight lines between the two things being mixed (Figure 8.10), with child magmas being mixtures of the two parents. Note that this straight-line relationship is not true for isotope ratios or ratios of elements (see Section 9.4 on behavior of ratios during mixing). Mixing lines are true one-dimensional lines in n-dimensional compositional space.

Magma mixing predicts field and petrographic characteristics that range from obvious to cryptic. Thorough mixing and re-equilibration of phenocrysts can result in a hybrid magma that bears little visual resemblance to its parents. Incomplete mixing of magmas is termed "mingling," and many classic areas contain lavas with clear evidence for mingling of magmas upon eruption. These include Yellowstone National Park (Wilcox, 1944), Lassen Peak in California (Clynne, 1999), Iceland (Sigurdsson & Sparks, 1981), and numerous volcanoes in Japan (Sakuyama, 1979; Koyaguchi, 1986). In these localities and countless others, lavas and pyroclastic rocks consist of streaks of contrasting composition that were sheared against each other during extrusion, resulting in a rock that commonly bears a resemblance to fudge-ripple ice cream (Figure 8.11a), especially to a hungry field geologist.

Evidence for multiple magmas inhabiting a given magmatic system simultaneously is common worldwide. Eichelberger and Izbekov (2000) recounted a number of eruptions, including the 1912 VEI 6 eruption of Katmai in Alaska, in which more than one magma type was erupted. They proposed that such multi-magma eruptions involve a dike that intersects another dike or magma chamber, giving rise to eruptions of mingled magma. Hildreth and Fierstein (2012) argued instead that the variety of magma types at Katmai arose in a shallow magma chamber that had undergone crystal fractionation. Regardless of how the various magmas arose, however, they produced spectacular mingled eruptive products.

Composite dikes contain two or more discrete compositions, typically one andesitic and one rhyolitic, whose field relations demonstrate that both were liquid at the same time (Figure 8.11b; Gibson & Walker, 1963; Wiebe, 1973; Frost & Mahood, 1987). Experiments show that if two liquids of differing viscosities are introduced into a dike, the faster, lower-viscosity fluid typically occupies the center and runs ahead of the slower, higher-viscosity fluid, which plates out onto the walls (Koyaguchi, 1985; Carrigan et al., 1994). In composite dikes, granite typically lines the walls of the dike, with diorite in the core, but the opposite also occurs.

Figure 8.11 (a) Banded lavas from Newberry volcano, Oregon, and from the 1915 eruption of Lassen Peak, California. These rocks record fluid mixing and hybridization of two magmas, one light-colored and rhyolitic and the other dark and andesitic. (b) Composite dike exposed in coastal Maine, with blobs of dark diorite in a complex granite and granodiorite matrix. Lobate margins on diorite blobs indicate that diorite magma was molten when it interacted with host granitic magma. (See Wiebe, 1973).

Source: AFG photos.

Many granitic plutons contain inclusions of diorite or gabbro that bear a strong resemblance in shape and size to pillows in pillow lavas, including fine-grained margins interpreted as chilled zones, and such field relations are generally interpreted as injection of mafic magma into cooler silicic magma (Wager & Bailey, 1953). Composite intrusions show the full range of textures, from unmixed to mingled to fully mixed.

Mafic magmatic enclaves, typically shortened to mafic enclaves, are inclusions, typically 0.1 to 1.0 m in diameter, of mafic rock in a more silicic rock, either volcanic or plutonic (Figure 8.12). They are distinguished from randomly incorporated xenoliths by their igneous compositions; smooth, globular shapes; and locally, fine-grained rims that suggest chilling. The presence of crystals from the host magma in the enclaves is strong evidence that both the enclave and host were liquid at the same time (Figure 8.12b,c).

Xenocrysts? Phenocrysts?

When two magmas mix, the crystals carried by each are generally out of equilibrium with both the other magmatic liquid and with any intermediate liquids formed by mixing. Hybrid rocks commonly retain evidence of this disequilibrium.

Many andesites show two unmistakable signs of magma mixing in their crystals: reverse-zoned, spongy plagioclase and quartz surrounded by a rim of Cpx crystals (Figure 8.13; Eichelberger, 1978). Reverse-zoned plagioclase crystals typically contain a clear core (usually andesine); a fine-grained spongy zone containing somewhat more calcic plagioclase, glass, and minute grains of oxides and Cpx; and a thin, clear rim of even more calcic plagioclase whose composition matches that of groundmass crystals (Tsuchiyama, 1985; Clynne, 1999). These crystals are interpreted as plagioclase phenocrysts from a higher-silica mixing end member that were mixed into a larger mass of less-silicic magma with which they reacted, forming a protective armor of more calcic plagioclase in equilibrium with the new host. Reaction of sodic plagioclase with melt to form more calcic plagioclase is easily explained by the An-Ab phase diagram (see Figure 8.18).

Quartz crystals in andesites are almost always surrounded by a halo of Cpx crystals, and there is typically a thin zone of rhyolitic glass separating the Cpx halo from the quartz crystal, and spherical inclusions of rhyolitic glass in the quartz (Figure 8.13b). Eichelberger (1978) interpreted this as evidence that the quartz crystals were derived from a rhyolitic magma that mixed with more mafic magma. Cpx halos are easily produced in the lab

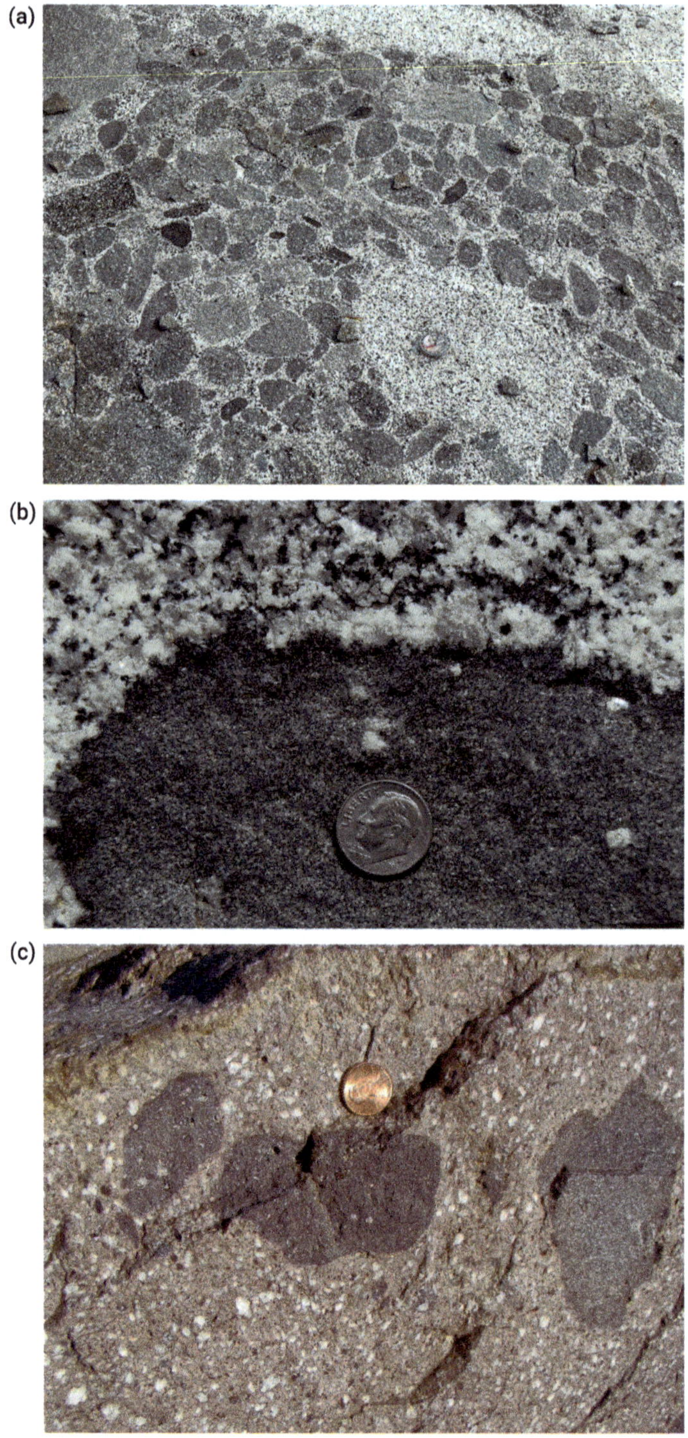

Figure 8.12 (a) Densely packed swarm of mafic magmatic enclaves in granodiorite, Yosemite National Park, California. The enclaves show a wide variety of textures and shapes and surround a central enclave of granodiorite 30 cm in diameter. (b) Fine-grained diorite enclave in El Capitan Granite, Yosemite National Park. Plagioclase phenocrysts derived from the host granite occur in the enclave, which has a dark rim that could indicate chilling against the host granite. (c) Three finely vesicular andesite enclaves in porphyritic dacite, Chaos Crags, Lassen Volcanic National Park, California. Again, the mafic enclaves contain isolated plagioclase crystals presumably derived from the host dacite, indicating mixing across the interface.

Sources: (a, b) AFG photos; (c) courtesy of Garry Hayes.

by basting quartz crystals in basaltic melt (Ussler, 1988), but are not easily explained by relevant phase diagrams (e.g., Figure 4.11).

Xenocrysts
Crystals foreign to their host magma; in mixed magmas, this term can be ambiguous.

Out-of-equilibrium crystals such as those in Figure 8.13 are typically referred to as **xenocrysts**, or foreign crystals. This terminology made sense when igneous systems were assumed to be closed and to evolve by crystal fractionation, but in a mixed-magma system, the term is confusing. In an andesitic magma produced by mixing equal parts of basaltic and rhyolitic magma, where crystals in the mixture were derived from both parents, one might have spongy reverse-zoned plagioclase and Cpx-mantled quartz derived from one parent, and clear, normally zoned calcic plagioclase, Cpx, and olivine from the other. Which are xenocrysts and which are phenocrysts? Available terminology, inherited from the 19th century, makes this issue foggy. Sometimes crystals are clearly xenocrysts (e.g., a garnet, kyanite, or staurolite crystal in a basalt would almost certainly be a truly foreign crystal), but the term should be avoided when describing features such as quartz crystals in basalt.

Mixing is an open-system process from the point of view of either of the parent magmas. Magmas generally mix with little excess energy, unlike solid solutions with different-sized cations, so there are few energetic barriers to mixing.

(a)

(b)

Figure 8.13 (a) Reverse-zoned, spongy plagioclase crystal in andesite, Coso volcanic field, California. The crystal comprises a sodic core, spongy intermediate zone, and calcic rim whose composition matches plagioclase in the groundmass. Width of field 4 mm. (b) Quartz crystal armored by a Cpx corona from the same sample as (a). The crystal contains two rounded inclusions of rhyolite glass, one partially crystallized, and a thin rhyolite glass rim separates the Cpx halo from the quartz crystal. Width of field 2 mm.

Source: AFG photos.

Assimilation

Assimilation is a process in which magma changes composition by assimilating, or dissolving, solid rock. There are significant energetic barriers to assimilation, because a magma must expend a great deal of thermal energy to eat relatively cool solid rock. Bowen (1922) used thermodynamics and the new science of phase equilibria to show that assimilation is a self-limiting process that prevents widespread digestion of wall rocks. Calculations using modern data indicate that a typical magma will congeal beyond the rheologic barrier at about 50 percent crystals after assimilation of 10 or 20 wt% rock, depending mainly upon the temperature of the material being assimilated (Glazner, 2007).

Significant assimilation should lead to widespread partially digested xenoliths trapped in magmas unless the assimilation process goes to completion. The geometry of assimilation on element-element plots is the same as that for magma mixing—straight lines between the host magma and the assimilant. However, unlike magma mixing, assimilated rocks may be sedimentary or metasedimentary, and therefore their compositions lie well outside the compositional range of igneous rocks (see Figure 1.6), and their assimilation would drive the hybrid magma off the igneous trend as well.

Although widespread assimilation is commonly invoked to explain open-system isotopic characteristics in magmatic systems, there is little evidence that bulk assimilation is a widespread process; more likely, contamination occurs by mixing with partial melts of wall rocks rather than wholesale assimilation (Glazner & Bartley, 2006). Assimilation is fundamentally an open-system process, because new material, from outside the original magma, is added.

Partial Melting

If we take any rock and melt it completely, the melt has the same composition as the rock we started with. But, as we have seen repeatedly, the first melts are commonly far from the bulk rock in composition. For example, in the system Fo-SiO_2 (see Figure 4.5), melting of Fo-rich bulk composition X results in peritectic melt π that is quartz-normative, and in general, melting begins at compositions that are more SiO_2-rich than the starting composition (see Section 4.5). We will return to this topic in Section 8.3.

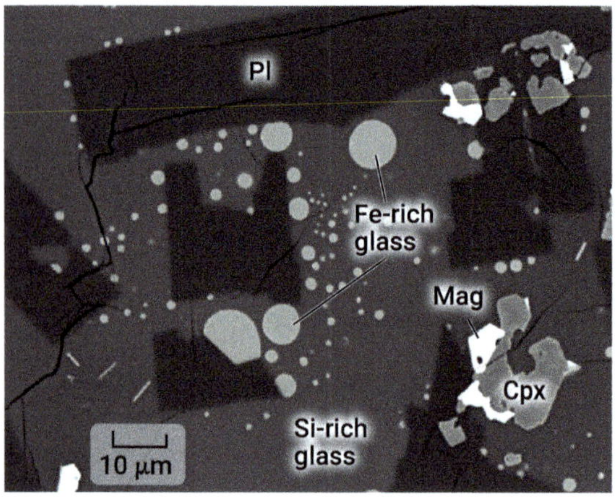

Figure 8.14 Liquid immiscibility displayed in basaltic magma in an experiment by Charlier and Grove (2012). In this backscattered electron image, brightness correlates with mean atomic number such that plagioclase (Pl) is dark gray, Cpx is medium gray, and magnetite (Mag) is bright white. Si-rich glass makes up about 50% of the imaged area, and complementary immiscible Fe-rich glass shows up as bright round globules, like drops of oil in vinegar.

Source: Courtesy of Bernard Charlier; modified from Charlier & Grove (2012), © 2012, Springer-Verlag.

Liquid Immiscibility and Other Processes

Immiscibility is familiar to anyone who has shaken vinegar-oil salad dressings. Oil and water (or vinegar, which is essentially a watery solution of acetic acid) are immiscible with one another, because water is a highly polar molecule (with a positive side and a negative side) and oil much less so. Thermodynamic stability is a balance between lowering energy and increasing entropy, and in the case of immiscible liquids, the energy cost of forcing the unlike molecules together is greater than the entropy benefit gained by mixing the molecules together. Thus, the system is more stable if the liquids are segregated from one another. Because a sphere has the lowest surface area for a given volume, immiscibility shows up as spherical globules of one liquid in another. This spherical form allows minimum contact area between the liquids, each of which thinks the other has cooties.

Microscopic examples of immiscibility are widespread, especially in the groundmasses of basalts, where coexisting globules of low-silica and high-silica glass are common (Figure 8.14; Philpotts, 1979). Although discounted by Bowen (1928), the process may be widespread (Charlier et al., 2011; VanTongeren & Mathez, 2012). Williams and Tobisch (1994) suggested that the drop-like shapes and sharp boundaries of mafic magmatic enclaves indicate that they were immiscible in the host magma. Immiscibility is a closed-system process with no energy barriers.

A number of other processes have been proposed for driving igneous magma diversification. These include chemical migration in a thermal gradient, chemical segregation driven by density differences in the gravitational field, and metasomatism. These processes may play an important role locally but will not be considered further here.

Making Connections 8.1

A Sweet Example of Partial Melting

The eutectic-like behavior of the mantle during partial melting can explain how the worldwide oceanic ridge system produces basalt of uniform composition, and partial melting of frozen sugar water can explain the way that ice pops (e.g., Popsicles®) behave in the mouth. Ice and the sugar sucrose ($C_{12}H_{22}O_{11}$) form a binary eutectic system, because neither exhibits solid solution toward the other and there are no intermediate compounds. Sucrose melts at 185 °C and ice at 0 °C, so the eutectic must lie below freezing. The experimentally determined phase diagram (Figure 8.15) shows that it lies at about −14 °C and 63 wt% sucrose. Home freezers are generally set at −18 °C (0 °F), comfortably below the eutectic. One's mouth should be around 37 °C, so popping in the pop will cause it to begin to melt.

Let's analyze what happens during fractional melting when an ice pop is heated by one's mouth. To (greatly) simplify, we'll assume that the entire pop rises uniformly in temperature; in reality, of course, the outside begins to melt before the center. The phase diagram is a standard binary eutectic, with a two-phase field of ice + sucrose below the eutectic and a one-phase field of sugar water above. The curved boundary of the ice + sugar water field gives the maximum solubility of the component H_2O at the given temperature; if any more is added, it will precipitate as ice. The curved boundary of the syrup + sucrose field gives the maximum solubility of sucrose at the given temperature; add any more, and it will precipitate as sucrose. These two curves together define the liquidus.

As the pop absorbs heat from your mouth, its temperature rises until at −14 °C it begins to melt, producing a sucrose-enriched syrup at the eutectic. This syrup (flavored with something fruity in commercial pops) is quite sweet, and sucking removes it from the partially melted matrix, in the same sense that gravitational separation can remove basaltic melt from partially molten peridotite. As more heat is supplied by blood vessels bringing in fresh, warm blood to tissues in your mouth, the temperature of the pop stays at −14 °C, but more eutectic melt is produced; the added heat supplies latent heat of melting for ice and sucrose. As you remove more and more of this by sucking, the composition of the solid matrix moves left from the bulk composition at 10 wt% sucrose toward more ice-rich compositions. Eventually, you will have sucked out all the eutectic syrup, leaving behind a spongy matrix of pure ice. At this point, the sweetness is gone, because all the sucrose has been removed, and you are left with nothing but ice on a stick at −14 °C. How disappointing. Your mouth needs to heat the ice to 0 °C before it will begin to melt again, and all that you get upon melting is water. Fractional melting effectively separated a sugar-rich fraction, leaving pure ice behind.

In the MORB analogy, the components that are enriched in basalt relative to peridotite—in particular, SiO_2, Al_2O_3, Na_2O, TiO_2, and others—represent the sugary, low-melting fraction.

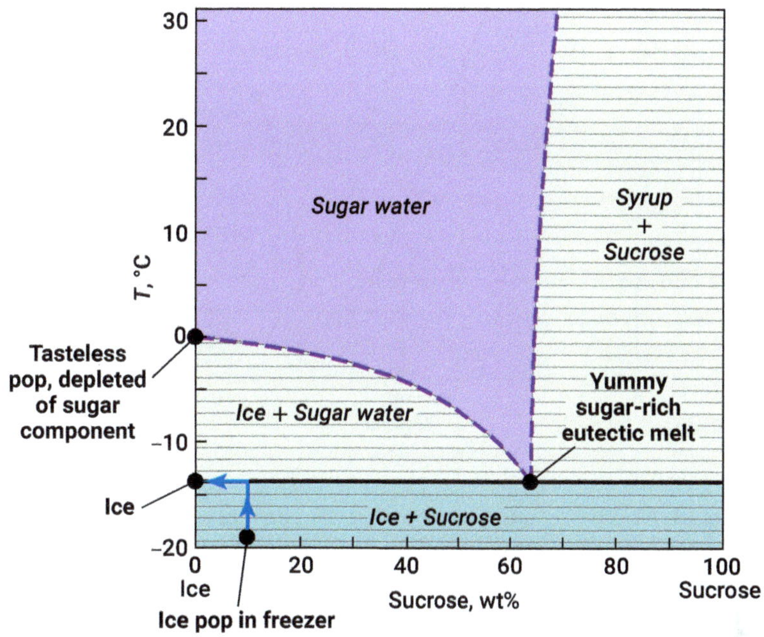

Figure 8.15 Phase diagram for the system sucrose-water. "Syrup" denotes sucrose-saturated water.

Source: Data from Young and Jones (1949).

8.3 Generation and Crystallization of Silicic Magma

Learning Objective

By the end of this section you should be able to:

8.3.1 Use enthalpy of fusion to explain why partial melts of silicate systems are commonly richer in silica and feldspar components than the rocks from which they are produced.

Making a More Silicic Magma

Although Bowen's influential works led to emphasis on down-temperature crystal fractionation as the driving force in igneous diversification, partial melting is also an efficient method for producing a range of magmas. In Section 6.7 we mentioned that if alien geochemists were to grind up the entire Earth and melt it a bit, the partial melt would have the composition of basalt. However, if they were to just skim off some of the continental crust and partially melt it (with water present), they would get rhyolitic magma. In a planetary-scale simplification, *basaltic magma is partial melt of the mantle, and rhyolitic magma is partial melt of hydrous crust*. As we will see in this section, this is true even for hydrous mafic crust.

Consider again the binary eutectic Fo-Di (see Figure 8.1). As we saw in Chapter 4, partial melting allows you to take a peridotite and separate out a basaltic liquid that differs drastically in composition from the peridotite, and such eutectic-like melting can explain the worldwide homogeneity of MORB major-element compositions. Increasing the percent melting enough to drive the liquid off the eutectic (thus melting enough to eliminate

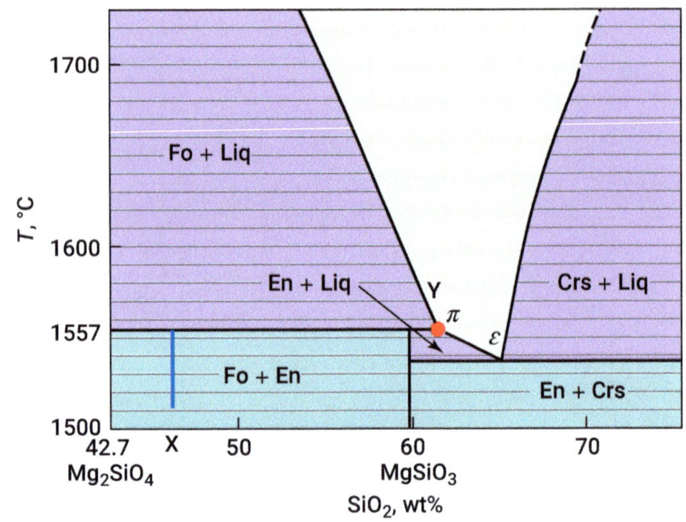

Figure 8.16 A portion of the system Fo-Si (see Figure 4.5) with the compositional axis listed as wt% SiO_2. Partially melting a model mantle with 80 wt% Fo and 20 wt% En produces a liquid π, which is quartz-normative, with ~62 wt% SiO_2—a model andesitic crust.

Di) will produce liquids that are richer and richer in the Fo component, corresponding to picrites and ultramafic lavas known as **komatiites**.

Komatiites
Volcanic rocks formed from high-temperature peridotitic magma; almost exclusively of Archean and Proterozoic age.

This is all well and good, but this example drives the liquid compositions toward more and more Fo-rich (and thus lower and lower SiO_2) compositions, and does nothing to produce high-silica rocks. Eutectic melting behavior traps the partial melt within the system. To understand how partial melting can produce high-silica rocks, we need to examine more silica-rich systems.

Peritectic melting, in which the melt composition may lie far from the bulk composition of the material being melted, is one way to diversify rocks by partial melting. We turn here to the system Fo-Si (Figure 8.16; see Figure 4.5) and our mantle peridotite composition X consisting of 80 wt% Fo and 20 wt% En. X has about 46 wt% SiO_2, a little high but within striking distance of real peridotite compositions. Raising this material to the solidus (1557 °C) produces peritectic melt π, which is quite a bit richer in SiO_2 than the starting material—in fact, it is quartz-normative and andesite-like (or as andesite-like as you're going to get in a two-component simplification of the real world). Call this new bulk composition, which has about 62 wt% SiO_2, Y. This material is a respectable model of the continental crust, which is roughly andesitic in composition (Rudnick & Gao, 2014).

Single-stage melting of our peridotitic mantle X has produced an andesitic crust Y. If we allow this material to cool and then melt it, we get eutectic melt ε, which is even more silica-rich, with about 66 wt% SiO_2.

Making Silicic Magma by Two-Stage Melting of the Mantle

The melting relationships in this simple system are reflected in the melting behavior of real rocks; single-stage melting of the mantle produces basalt, and melting hydrated basaltic rocks can produce rhyolitic magma. Sisson et al. (2005) melted hornblende-biotite gabbro samples from the Sierra Nevada batholith of California at deep-crustal pressures of 700 MPa and produced dacitic and rhyolitic melts via dehydration melting (water was supplied by dehydration of the hornblende and biotite; see Figure 6.32). Figure 8.17 illustrates the silica-rich melts produced in these experiments, along with their parent gabbros and a generic peridotite. These melts are a good chemical match for arc-related granites and granodiorites (Sisson et al., 2005; Ratajeski et al., 2005). Thus, a two-stage melting scenario involving (1) melting peridotite in a subduction zone to produce hydrous basaltic magma that crystallizes as hornblende gabbro, followed by (2) partial melting of the gabbro, can generate rhyolitic magmas and the rocks that crystallize from them, including granite.

It is not a coincidence that this rhyolitic eutectic melt is composed almost entirely of tectosilicate components—quartz and feldspar—that have the lowest enthalpies of fusion (see Table 4.1); recall that minerals with low ΔH_f pull the eutectic toward themselves (see Section 4.5). We now delve into the melting behavior of tectosilicates in detail.

Melting and Crystallization of Feldspars and Quartz

In a broad sense, the Earth's upper continental crust, the stuff we walk around on, is granitic, or more accurately granodioritic, with an average of 65 to 70 wt% SiO_2 (Rudnick & Gao, 2014). This is dramatically different from oceanic crust. Rocks at the high-silica end of the BADR spectrum, with about 77 wt% SiO_2, are composed predominantly of

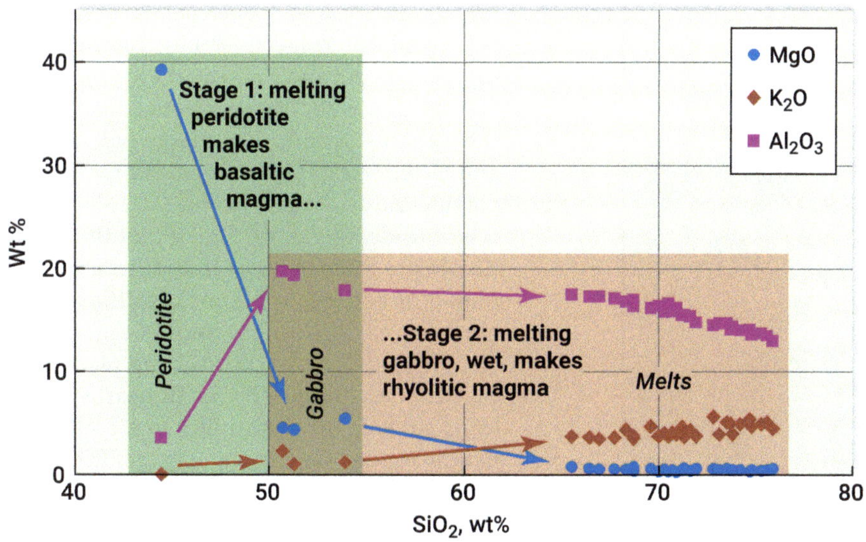

Figure 8.17 Variation in three major-element oxides vs. SiO_2, showing peridotite and its partial melt, and three gabbros and melts produced by partially melting them. Only melts produced at <900 °C, at <~25 wt% melting, are plotted. Assuming that hydrous gabbros inferred to be in the middle and lower arc crust, such as those plotted here, can be made by partially melting hydrated peridotite (see Chapter 6), then this plot illustrates a potential two-stage melting process that can produce dacitic and rhyolitic magmas.

Sources: Peridotite data from Takahashi et al. (1993); gabbro data from Sisson et al. (2005).

SiO_2, Al_2O_3, Na_2O, and K_2O (e.g., Figure 1.4), and when they crystallize, they are almost entirely quartz and one or two feldspars (one Na-rich, one K-rich). In less-silicic rocks, plagioclase is a prominent phase, and many granites contain three-feldspar assemblages of K-feldspar, albite, and plagioclase (typically oligoclase). "K-feldspar" is used here for simplicity to refer to K-rich orthoclase or microcline. The crystallization behavior of feldspars is complex, as is their subsolidus behavior.

Plagioclase

Crystal-liquid equilibria in the plagioclase system at 1 atm were worked out by Bowen (1913). He produced the well-known, fish-shaped An-Ab phase diagram of Figure 8.18. The general topology of this system is analogous to the olivine system Fo-Fa (see Figure 4.17), with the Ca end-member An playing the high-temperature role of the Mg end-member Fo. Melting and crystallization paths are analogous to those in the olivine system; early crystallizing plagioclase is An-rich, becoming more Ab-rich as cooling proceeds. The liquid is always more Ab-rich than the crystals and becomes ever more so as cooling proceeds.

Diffusion and Disequilibrium

Although the olivine and plagioclase phase diagrams share the same general form, their behaviors differ in an important way. In olivine, Mg and Fe atoms are relatively free to move and exchange places by diffusion, thus homogenizing the olivine composition, but in plagioclase, diffusion is many orders of magnitude slower.

One way to comprehend diffusion is to envision a lattice with Mg and Fe atoms jiggling around as they are parked in crystallographic sites. The atoms are moving around owing to thermal agitation, and as the temperature rises, they jiggle more and more, sometimes momentarily popping out of the lattice before settling back in. If airborne, while coming back in for a landing they'll take any unoccupied site, leaving their original site available for another atom. In this manner (sort of), diffusion disperses chemical components.

In olivine, Mg and Fe have relatively low bond strengths (2/6), and diffusion is rapid because it is relatively easy for Mg and Fe atoms to pop out of the lattice. Bond strengths

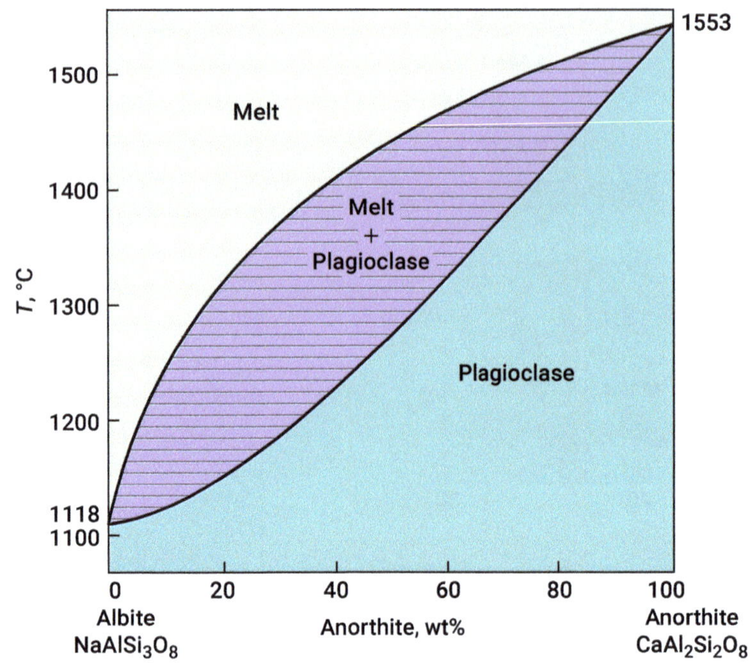

Figure 8.18 The classic 1-atm plagioclase phase diagram.

Sources: After Bowen (1913), modified by Morse (1980).

for Na and Ca in plagioclase are even weaker (~1/9 and 2/9, respectively), so you might expect diffusion of Na and Ca to be even faster. However, Na and Ca are not free agents, but must travel with their chums Si and Al as charge-balanced $NaSi^{5+}$ and $CaAl^{5+}$ units. Bond strengths for Si and Al in plagioclase are large (4/4 and 3/4, respectively), so it is considerably more difficult for thermal agitation to pry them out, and they diffuse more slowly than Mg and Fe. What makes diffusion in plagioclase really slow, however, is that getting the $NaSi^{5+}$ and $CaAl^{5+}$ units to swap places requires that all four species pop out of nearby sites at the same time. If the probability that a given pair of nearby Na and Ca atoms pop out of the lattice and exchange places is p_1 and that their mates Si and Al pop out together is p_2, both small numbers, then the probability that both pairs pop out at the same time is $p_1 \times p_2$, a much smaller number. Low probabilities lead to slow diffusion.

Diffusion rates are quantified by a temperature-dependent constant, *D*, that measures how quickly a given species diffuses (see Toolbox 8.1). *D* values for NaSi-CaAl diffusion in plagioclase are 8 to 10 orders of magnitude smaller than those for Na-K diffusion (Figure 8.19).

Slow diffusion means that it is difficult for crystallizing plagioclase to stay in equilibrium with the liquid. In equilibrium crystallization, the crystals must constantly remake

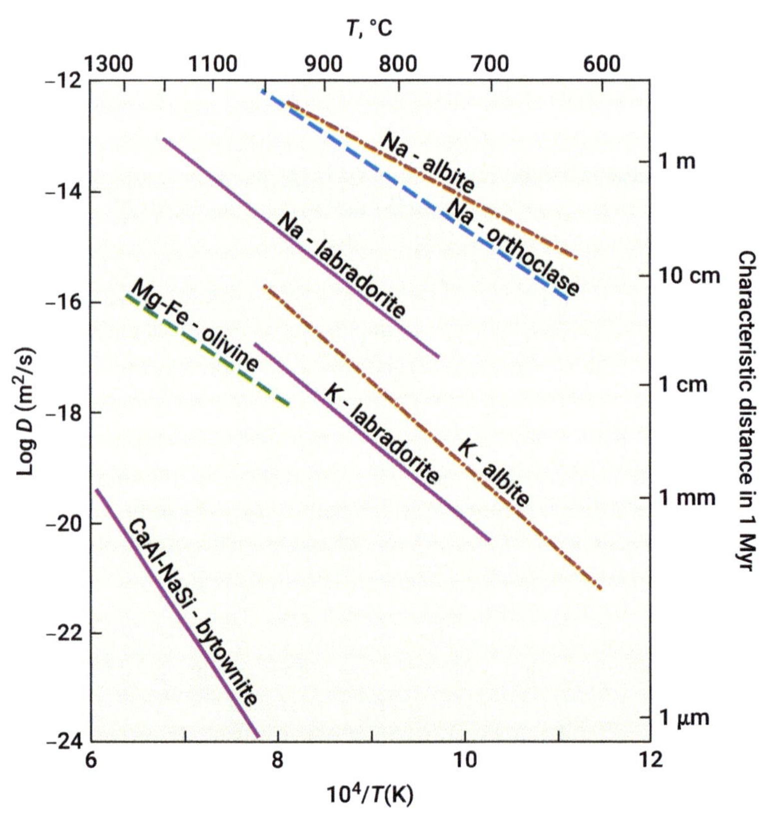

Figure 8.19 Selected diffusion constants as a function of temperature, plotted on an Arrhenius diagram. Distances on the right-hand ordinate are characteristic distances $\sqrt{Dt}$ that a species would diffuse in 1 Myr at the corresponding value of *D* (see Toolbox 8.1).

Sources: Data from Cherniak et al. (2010), Grove et al. (1984), and Chakraborty (1997).

Toolbox 8.1

Diffusion

Diffusion is a complicated topic with many levels of complexity. In its simplest form, one-dimensional diffusion of a chemical species is described by Fick's Second Law:

$$\frac{\partial C}{\partial t} = D\frac{\partial^2 C}{\partial x^2} \qquad \textbf{(Eq. 8.2)}$$

where C is concentration, t is time, D is the **diffusion coefficient** of the chemical species, and x is position. This partial differential equation has a simple message. The left side is the rate at which the concentration at a given point (x-coordinate) changes with time, and the right side is D times the second derivative of the concentration gradient—essentially the curvature of the concentration gradient. The equation says that where the concentration gradient changes rapidly over a short distance (as at a step), the concentration will change rapidly with time. It also says that the rate of change of concentration scales linearly with D.

If D is constant, then the solution to Equation 8.2 is

$$C = C_0 - C_0 erf\left(\frac{x}{2\sqrt{Dt}}\right) \qquad \textbf{(Eq. 8.3)}$$

where C_0 is the initial concentration and erf is the error function, a definite integral given by

$$erf(x) = \frac{2}{\sqrt{\pi}}\int_0^x e^{-t^2}dt \qquad \textbf{(Eq. 8.4)}$$

and typically available in programming languages and spreadsheets.

Diffusion is another thermally activated process (see Section 5.3), and the Arrhenius equation describing the temperature dependence of the diffusion coefficient is

$$D = D_0 e^{-E/RT} \qquad \textbf{(Eq. 8.5)}$$

where D is the diffusion coefficient, D_0 is the value D approaches at infinite temperature, E is the activation energy, R is the universal gas constant, and T is the absolute temperature. If a coefficient follows this relationship, then a plot of log(D) vs. $1/T$ will be a straight line. Figure 8.19 summarizes experimental data for several diffusing species of interest.

Figure 8.20 shows a theoretical diffusion profile for $D = 10^{-5}$ μm²/s, a value appropriate to Fe and Mg in olivine. Curves give the predicted profile of the diffusing component at several times, assuming that it diffuses to the right from x = 0, a boundary where the concentration of the diffusing component is held constant at an arbitrary value of 1, into the crystal, where the value is 0 (these values can be scaled to any appropriate values). This situation is relevant, for example, in the case of an Fo_{90} olivine xenocryst picked up by a more silica-rich magma that precipitates Fo_{70} olivine. In such a case, the host magma is essentially an infinite reservoir, and Fe will diffuse into, and Mg out of, the xenocryst in an attempt to bring it into Mg-Fe equilibrium with the host magma.

Given the shapes of the diffusion curves, it is difficult to say just how far into the crystal Fe has diffused after a given time interval. One common approach is to use the quantity $\sqrt{Dt}$, which has units of length, as an order-of-magnitude estimate of the effective diffusion length. The stars in Figure 8.20 show this value, which lies at a relative concentration of 0.48, for each curve; the distance this point propagates goes up by the square root of time (e.g., this concentration isopleth has migrated 10 times farther in 100 years than it had in 1 year). By this measure, Fe and Mg diffuse about 20 μm in a year and 120 μm (over a tenth of a millimeter) in 50 years. Costa and Dungan (2005) used this approach to estimate the time interval between basalt magma picking up ultramafic xenoliths (which contributed Fo-rich olivine xenocrysts) and eruption, deriving results on the order of 1 to 20 years.

A significant complication of these methods is that D is commonly a function of crystal composition, temperature, and crystallographic orientation (i.e., which direction in the crystal that diffusion is taking place). These dependencies can be accounted for if there are sufficient experimental data to calibrate them.

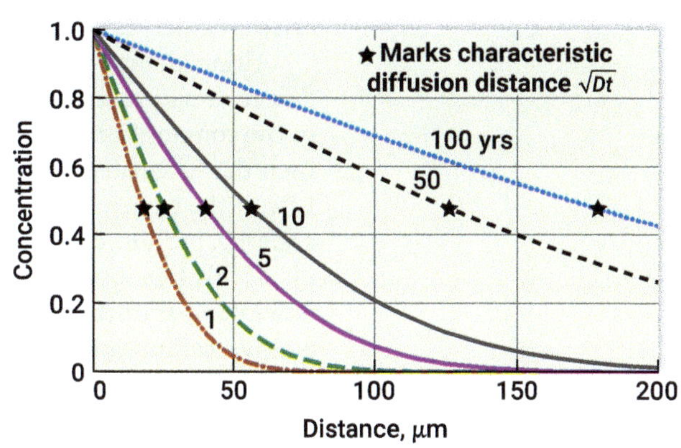

Figure 8.20 Curves showing relative concentration of Mg in an Fo-rich olivine xenocryst immersed in magma with a much lower Mg#. Concentration is scaled such that 1 is the concentration in the crystal and 0 is the concentration in the magma. Stars give $\sqrt{Dt}$ distances—essentially, an average distance—that Mg diffuses in the given time.

Figure 8.21 Zoning in plagioclase crystal seen in cross-polarized light. Differences in brightness reflect differences in optic-axis orientation, which result from differences in composition. Blue and brown tones result from dispersion (misalignment of optic axes for different colors, a common feature of triclinic minerals such as plagioclase). Width of field 3 mm. Note complexity of plagioclase zoning here compared to simplicity of olivine zoning in Figure 8.8.

Source: AFG photo.

themselves to more and more Ab-rich compositions as cooling proceeds and the liquid becomes more and more Ab-rich (see Figure 8.18). To maintain equilibrium, all the crystallized material must evolve to more Ab-rich compositions with each drop in temperature, and slow diffusion kinetics make this essentially impossible. As a result, plagioclase commonly is dramatically zoned in composition in igneous rocks (Figure 8.21). Zoning is particularly easy to see in thin sections, because plagioclase is triclinic, and thus changes in composition affect the optical properties more than in a higher-symmetry mineral. For example, in olivine, axes of the optical indicatrix must be aligned with the crystallographic axes, whereas in plagioclase, they are free to wander at will. Thus, extinction angles are zero for olivine, whereas they vary widely with composition in plagioclase.

Another consequence of slow diffusion kinetics is that plagioclase tends toward more extreme zoning than simple equilibrium crystallization predicts. Consider a bulk composition of 60 wt% An in Figure 8.18. Under equilibrium crystallization, plagioclase will start to form at about 1470 °C, and the last liquid will disappear at about 1320 °C, with the last crystals having a composition of An_{20}. Under fractional crystallization, however, crystals are removed from the liquid as they form, and so the bulk composition shifts to more and more Ab-rich compositions as cooling proceeds. In perfect fractional crystallization, a vanishingly small amount of melt and crystals make it all the way to pure Ab.

Diffusion coefficient
Here, a parameter that describes how rapidly a chemical species diffuses at a given concentration gradient.

Zoning caused by slow kinetics mimics fractional crystallization but is fundamentally different. Consider bulk composition X cooling not under true fractional crystallization, where the crystals are removed from the magma, but under conditions where each bit of plagioclase that forms is isolated from later increments by slow kinetics. In essence, each layer of new plagioclase on a crystal is chemically isolated from the rest and armors the crystal from reaction with the liquid. In this situation, the crystals will show extreme zoning, all the way to pure Ab, but the bulk composition of the magma does not change—it stays at X. All the original magmatic material is still there.

Although lack of equilibrium in natural plagioclase might be viewed as some sort of failure of nature to live up to its end of the bargain, there is a lot of information encoded in the compositional profiles of zoned crystals. If diffusion is slow relative to growth rate, then the successive layers of a zoned crystal bear a resemblance to tree rings, characteristic of the environment in which the crystal grew. Reverse-zoned plagioclase crystals in andesite, produced by magma mixing or assimilation, are a particularly important example of this (see Figure 8.13). This phenomenon, exactly the opposite of that predicted by down-temperature crystallization and Bowen's Reaction Series, was widely ignored for nearly a century until its significance was realized in the 1970s.

Solid Solutions in Ternary Systems

Figure 8.22 shows the diagram for the system diopside-plagioclase (Di-Ab-An), which makes a good model of a rock such as gabbro or diorite and a good introduction to the behavior of solid solutions in ternary phase diagrams. Although the diagram is quite simple, with nothing more than a thermal valley running down the middle from right to left, the presence of solid solution adds some interesting twists. For one, there are no ternary invariant points as in other ternary systems that we have studied. The phase rule for an isobaric (here, 1-atm) system is $F + \pi = C + 1$, and we have three components, so $F = 4 - \pi$, and thus an invariant point

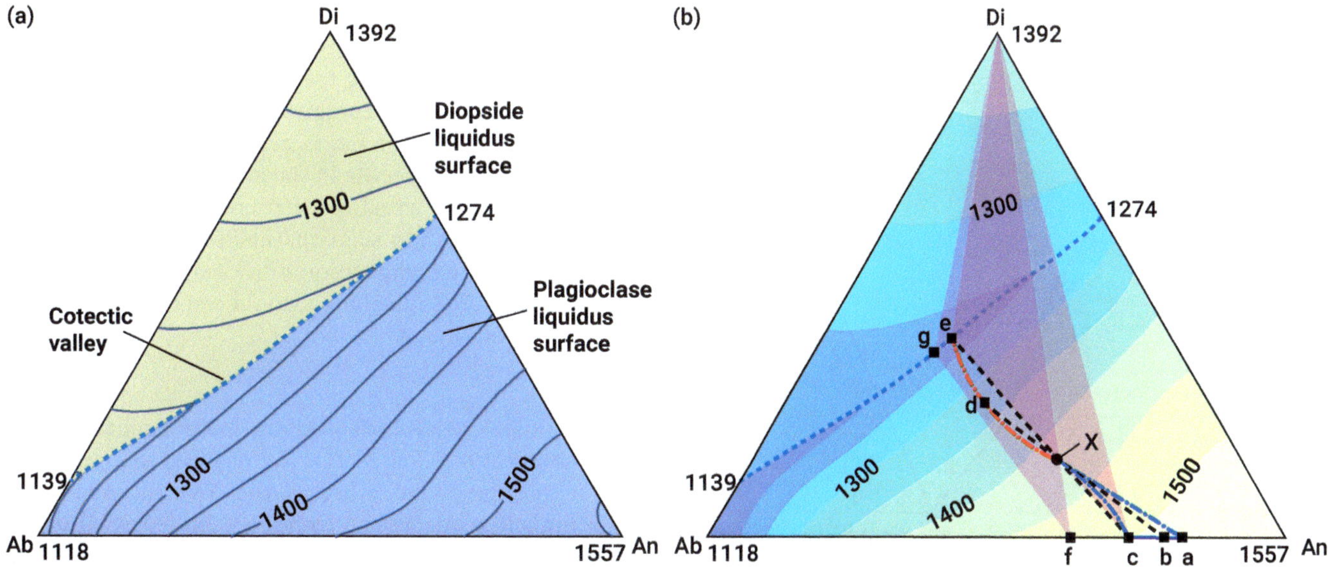

Figure 8.22 (a) Ternary system Di-Ab-An contoured for temperature. (b) Crystallization of a bulk composition in the plagioclase field, showing the approximate liquid path and the highest- and lowest-temperature three-phase triangles. Colors denote increasing temperature, dark to light.

Source: (a) Modified from Weill et al. (1980).

would require four phases. We only have two possible solid phases and one liquid, so there are no invariant points.

Let's follow crystallization of bulk composition X during cooling. There are numerous subtleties to understanding what happens in systems such as these, and Morse (1980) covers many of them. X lies above the plagioclase liquidus, so when it reaches the liquidus temperature, plagioclase will start to precipitate. What composition will these crystals be? We can't tell from the diagram alone; in going from two components to three, both plotted on a 2D piece of paper, we've given up information—in this case, compositional information. The first crystals to form will be somewhere along the Ab-An sideline, and they must be such that removing them from the liquid X will drive the residual liquid downhill on the liquidus surface. Their composition can be determined from experiments or by calculation, but not from the diagram, and Figure 8.22 shows an estimate at point *a*.

As in the binary system An-Ab, under equilibrium crystallization (unlikely, we know), the crystals must constantly reform to more Ab-rich compositions as *T* falls. Thus, the crystals move to the left, and owing to Corollary 1B, the liquid path must curve as shown. At any point along this path, a tie line between the crystals and liquid (e.g., b-d) must go through the bulk composition, and the lever rule can be used to determine the proportions of each.

Eventually, the liquid path, which has been heading in a generally diopside-ward direction, becomes rich enough in diopside that Di starts to precipitate at point e. This is the cotectic, and Di, Pl, and Liq coexist with a three-phase triangle Di-c-e linking them. The liquid path takes a sharp left-hand bend, because its path is now defined by precipitation of both Di and Pl. The Pl composition (not knowable directly from the diagram) is at c when the liquid reaches the cotectic, and by Corollary 1A, the proportions of Di and Pl being taken out are given by extrapolating the tangent to the liquid's path along the cotectic back to the tie line between Di and Pl.

As *T* drops, the three-phase triangle pivots about Di, rotating clockwise, like triangular treads on a spiral staircase. When the liquid first reaches the cotectic, the bulk

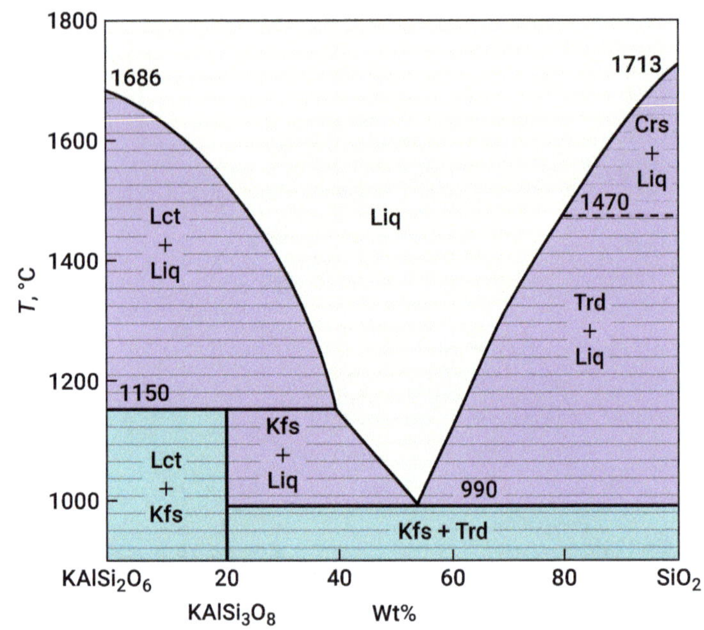

Figure 8.23 The binary system leucite-silica. Leucite and silica polymorphs react to form K-feldspar (Kfs), which melts incongruently.

Source: Modified from Schairer and Bowen (1955).

composition lies on the Pl-Liq tie line c-e, so the proportion of Di is zero; when the trailing Di-Pl edge of the triangle Di-f-g reaches the bulk composition, crystallization is complete, because at that point the proportion of Liq is zero. In this hypothetical equilibrium case, the Pl starts at a and ends at f. In a more realistic situation where reaction of Pl with Liq is too sluggish to keep up, the Pl will be zoned from the starting composition *a* to compositions considerably more sodic than it would reach under equilibrium conditions.

Alkali Feldspars

Albite melts congruently at about 1100 °C at 1 atm; the dry-melting curve has a positive slope, whereas the water-present melting curve drops precipitously in temperature to less than 700 °C at 1 GPa (see Figure 6.29; Makhluf et al., 2016). In contrast, K-feldspar (Kfs) melts incongruently at 1 atm to leucite (Lct) and a more silica-rich liquid (Figure 8.23), a peritectic relationship topologically identical to that in the system Fo-Si (see Figure 4.5). However, as shown by Tuttle and Bowen (1958), this incongruent melting is suppressed by modest (e.g., 300 MPa) pressures in the presence of water. Because water-present melting is common in crustal magmatism and leucite is rare, we will ignore the incongruent melting of K-feldspar in further discussions. It is worth noting, however, that partial melting of Kfs + Lct at low *P* will lead to production of a Si-oversaturated (Qtz-normative) liquid at the peritectic.

In plagioclase, the $NaSi^{5+}$ and $CaAl^{5+}$ units are readily mixable at igneous temperatures, as shown by the complete solid solution exhibited by the plagioclase system. In alkali feldspars, the exchangeable units are simply Na^+ and K^+. These ions have low bond strength and the same charge, so they diffuse rapidly—faster even than Mg and Fe in olivine (see Figure 8.19). However, they differ greatly in size. Na^+ has an ionic radius of about 120 pm (picometers, 10^{-12} m), whereas K^+ has an ionic radius of about 155 pm, making it 25 percent greater in radius and nearly twice as great in volume. As a consequence, as *T* falls, the Ab and Or components of alkali feldspars are more and more likely to undergo **exsolution**.

Exsolution
A process whereby a homogeneous solution splits into two phases by forming separate domains that are enriched in each component.

Consider an alkali feldspar crystal with 50 atomic percent (at%) Na and 50 at% K. Both Na-feldspar (albite) and K-feldspar (orthoclase) are stable and content when forming their own crystals, but mixing them into the same crystal introduces stress into the lattice—a lattice made for the larger K^+ ion has to collapse in a bit to host a Na^+ ion, which rattles around in the oversized site, whereas a lattice made for the smaller Na^+ ion has to stretch to accommodate a K^+ ion. At higher temperatures, where the atoms vibrate greatly, this size mismatch is not a deal-breaker, but as temperature falls, the crystal lattice tries to settle into a more orderly state and becomes less and less tolerant of distortions caused by trying to accommodate the mismatched ions.

Solvus
A curve on a binary phase diagram that separates a homogenous phase from two phases that form by exsolution as *T* drops. This concept generalizes to more components.

As temperature drops, there comes a point where the lattice can form a lower-energy phase assemblage if it self-segregates into K-rich and Na-rich domains. On the binary phase diagram, this is expressed by the **solvus**, a smooth curve that separates a region where one mixed feldspar is stable from one where the system is better off (thermodynamically) if it splits into two phases (Figure 8.24).

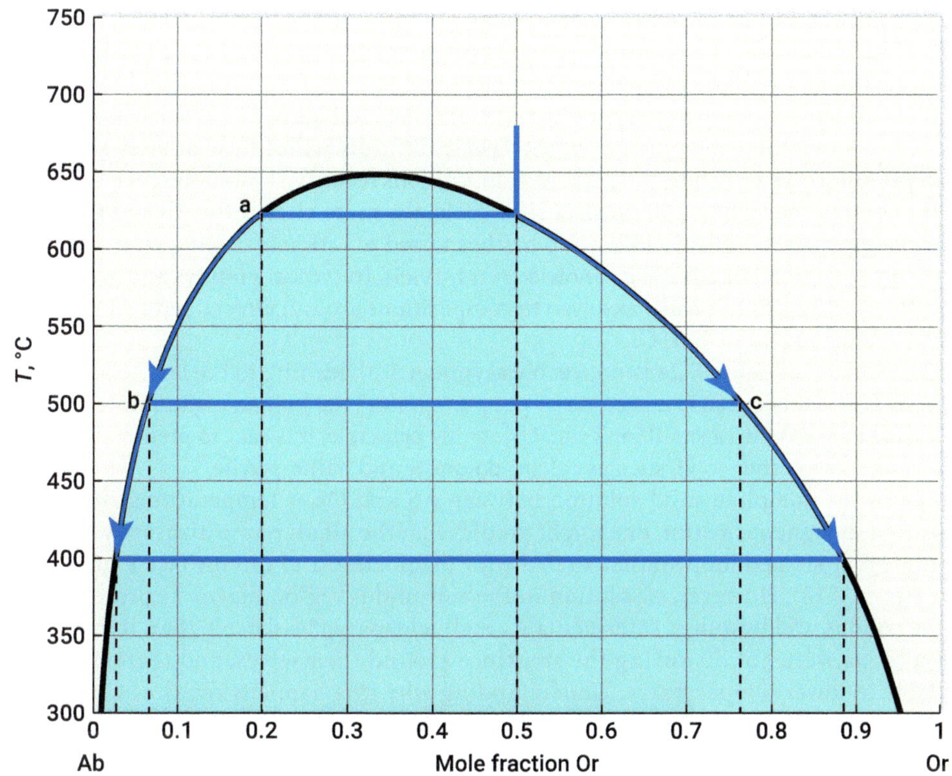

Figure 8.24 Solvus in the alkali feldspar system at 100 MPa. A single alkali feldspar with equal molar proportions of Ab and Or will hit the solvus at ~630 °C, at which time it becomes thermodynamically favorable to start segregating Ab-rich domains from Or-rich ones. Given sufficiently slow cooling, the two feldspar phases become farther and farther apart in composition. In practice, the kinetics of exsolution are so slow at temperatures below ~300 °C that the process essentially stops.

Source: Data from Waldbaum & Thompson (1969).

In Figure 8.24, a 50:50 mixed feldspar will exist as a single phase above temperatures of about 630 °C. Cooled below that, it encounters the solvus, at which time (if equilibrium prevails) small domains of Na-rich feldspar—specifically, Or_{20} (point a)—will form within the original feldspar. With further cooling, the feldspars continue their self-segregation and follow the limbs of the solvus; at 500 °C, they are at b and *c*, and at 400 °C, they are even closer to their end-member compositions. At any point the relative proportions can be determined by using the lever rule, since the Ab-rich and Or-rich domains must add up to the bulk composition. Exsolution of albite from K-rich feldspar produces perthite (Figure 8.25). Perthitic exsolution in natural feldspars ranges from thin lamellae, too fine to see with a microscope, to blebs large enough to see with the naked eye.

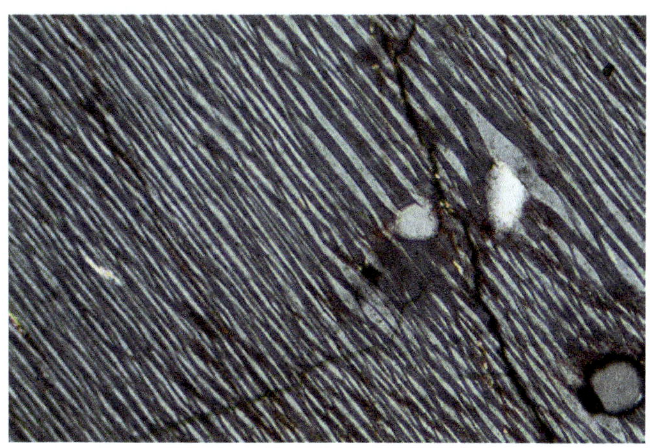

Figure 8.25 In this thin section view, linear bands of albite form perthitic laminae in a crystal of K-feldspar. Width of field 2 mm.

Source: Courtesy of Alessandro Da Mommio (www.alexstrekeisen.it).

The alkali feldspar solvus is broad enough that an intermediate alkali feldspar such as the one in Figure 8.24 would exsolve to almost pure albite and orthoclase at room temperature if not limited by reaction kinetics. However, even fast-diffusing K and Na are essentially brought to a stop as temperature falls. The data in Figure 8.19, when extrapolated to room temperature, show that even the fastest diffusers will diffuse at rates less than 1 µm per million years. Thus, the perthite compositions recorded by a cooling rock depend on the cooling rate. The more slowly a rock cools, the more cleanly the Ab and Or components segregate, and this relationship has been used to estimate cooling rates from the width of exsolution lamellae (e.g., Snow & Yund, 1985). In typical granites and granodiorites, the K-feldspar is typically exsolved to compositions around $Or_{85\text{-}90}$, indicating equilibration down to 300 to 400 °C.

The solvus in Figure 8.24 is noticeably asymmetrical, leaning to the left (toward Ab). Asymmetry occurs because it is easier to put a small ion (Na^+) into a big ion's site than to stuff a big ion (K^+) into a small ion's site. Using this principle, it is easy to predict the sense of asymmetry in other solvi, such as calcite-dolomite and halite-sylvite.

Given the complete solid solution between Ab and Or at temperatures above the solvus, in the igneous realm, one might predict that the alkali phase diagram would be a simple fish-shaped loop similar to those for plagioclase and olivine (see Figure 4.17 and Figure 8.18). However, exsolution in the subsolidus region makes its presence felt in the solidus and liquidus. Intermediate alkali feldspar is less stable than it would be if K and Na were not distorting the structure around themselves, and therefore melt is stable to lower temperatures. Understanding why this happens requires a grounding in thermodynamics (Ernst, 1976), but the result is a pinched solidus-liquidus loop with a minimum m near cp (for critical point), the crest of the solvus, as shown in Figure 8.26a.

The double-loop phase diagram means that bulk compositions on either side of m fractionate toward m with declining temperature, unlike in the plagioclase system, where liquids fractionate steadily toward Ab regardless of the starting composition. Thus, K-rich feldspar becomes more Na-rich during crystallization, and vice versa. Consider composition X in Figure 8.26a. This liquid initially precipitates high-Or feldspar a, and with cooling temperature, both the liquid and feldspar shift to more Ab-rich compositions. Under equilibrium conditions, the last drop of liquid disappears at about 770 °C, leaving just feldspar of the bulk composition X. During crystallization, the feldspar has migrated from a to X, and the liquid from X to b—almost, but not quite, to the minimum m. When this feldspar reaches the temperature of the solvus, it begins to exsolve Ab-rich feldspar at 630 °C, and from that point on, the K-rich feldspar reverses course and becomes steadily more Or-rich as Ab-rich feldspar exsolves. Thus, the evolution was first toward more sodic compositions during precipitation from the melt, then toward more potassic compositions during exsolution. A bulk composition to the left of the minimum m would undergo the opposite evolution.

If the water-saturated pressure is increased to more than about 400 MPa, an interesting thing happens: the solvus intersects the solidus (Figure 8.26b). This occurs mainly because increasing water pressure drives the solidus and liquidus temperatures down (e.g., Figure 6.29). Intersection of the solvus and solidus gives the phase diagram a eutectic (ε) (Figure 8.26b), and the diagram resembles a wide gown with asymmetrical shoulder straps and a eutectic belt around the waist. If you were to take the sides of the belt, points c and b, and pull them toward the sidelines, you would transform the diagram into a standard eutectic. Points c and b represent the maximum amount of the other component that can be stuffed into one of the end members. For example, the diagram says that at P_{H_2O} = 500 MPa, the maximum amount of Or that you can put in Ab-rich feldspar is about 22 wt%, and to put this much in, you'd want to do it at a little over 700 °C. Add any more, and you'll start to crystallize an Or-rich feldspar at *b*.

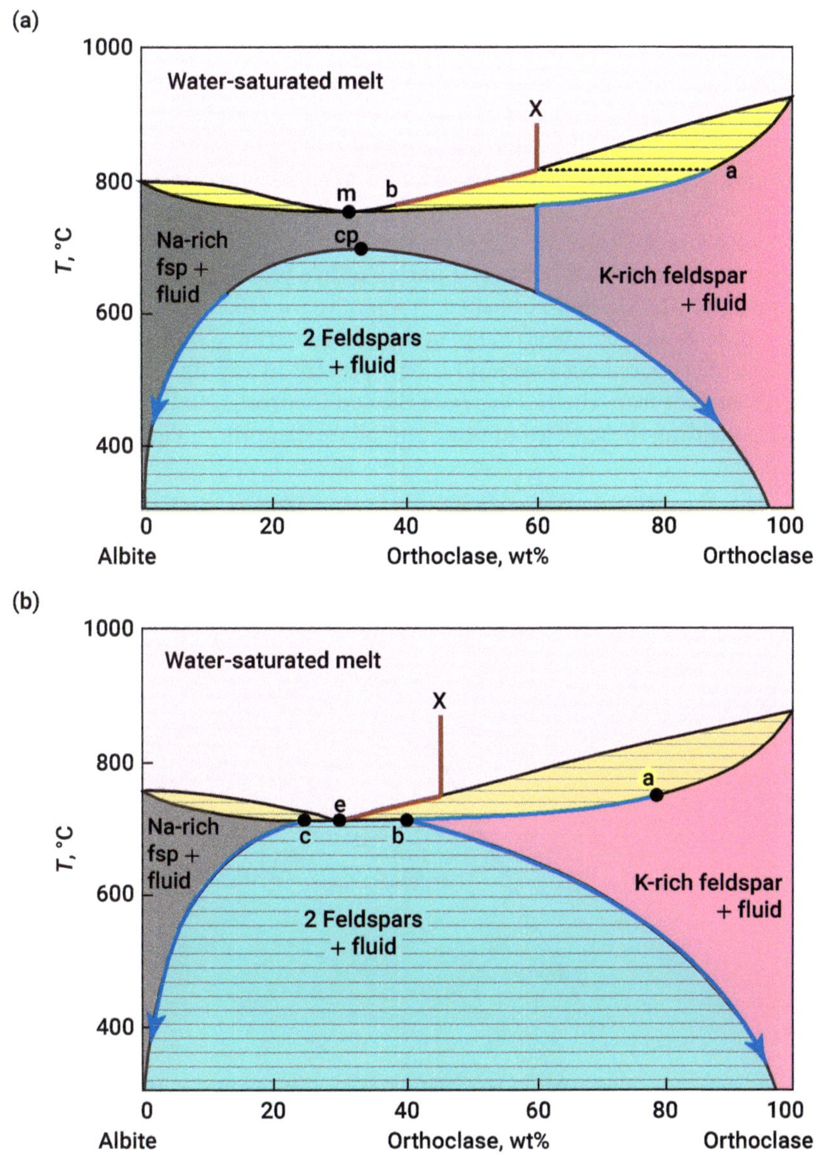

Figure 8.26 Pseudo-binary phase relations in the system Ab-Or-H_2O. At water-saturated pressures below ~350–400 MPa, the solvus and solidus are separate, with a continuous field of feldspar across the diagram between the two. Above ~400 MPa, the two meet and form a eutectic. (a) For all compositions at 300 MPa, crystallization concludes with a single feldspar, which then unmixes at the solvus; only one feldspar phase crystallizes from the melt. (b) At 500 MPa, bulk compositions not too far from the eutectic will reach the eutectic and crystallize two feldspars from the melt, whereas those farther away will fully crystallize before reaching the eutectic.

Source: Modified from Ernst (1976).

The presence of the eutectic in Figure 8.26b introduces an invariant point. Any bulk composition between c and b will first start to crystallize one feldspar along the solidus, then both c and b when the system reaches the eutectic, and then feldspars c and b will each exsolve along the solvus with further cooling. Any bulk composition outside the range c-b will only crystallize a single feldspar from the liquid (it will run out of liquid before reaching the eutectic), although that feldspar will exsolve into two at subsolidus conditions.

Granite: The System Quartz-Albite-Orthoclase-H_2O

Haplogranite
The system Si-Ab-Or-H_2O, a simplification of the high-silica granite system that leaves out important components such as CaO, MgO, and Fe oxides.

And what about granite? We can predict with confidence that each of the binaries Ab-Si, Kfs-Si, and An-Si is a simple eutectic, because there are no stable compounds that lie in composition between any of the three feldspar end members and Si. Adding water to any of these systems depresses temperatures dramatically, but eutectic behavior remains. Of particular relevance is the system Si-Ab-Or-H_2O, known as the **haplogranite** system ("haplo-" being a Greek-derived prefix meaning "simple"). Phase relations in this system with excess water were worked out over a period of five years at the Geophysical Laboratory in Washington, D. C., and published in a landmark Geological Society of America Memoir by Tuttle & Bowen (1958). This system is quaternary (four-component) owing to the presence of H_2O, but because H_2O is only present in any significant amount in the liquid, we can project from the H_2O apex of the tetrahedron onto the anhydrous base as in Figure 3.18, yielding the simple but profound phase diagram in Figure 8.27.

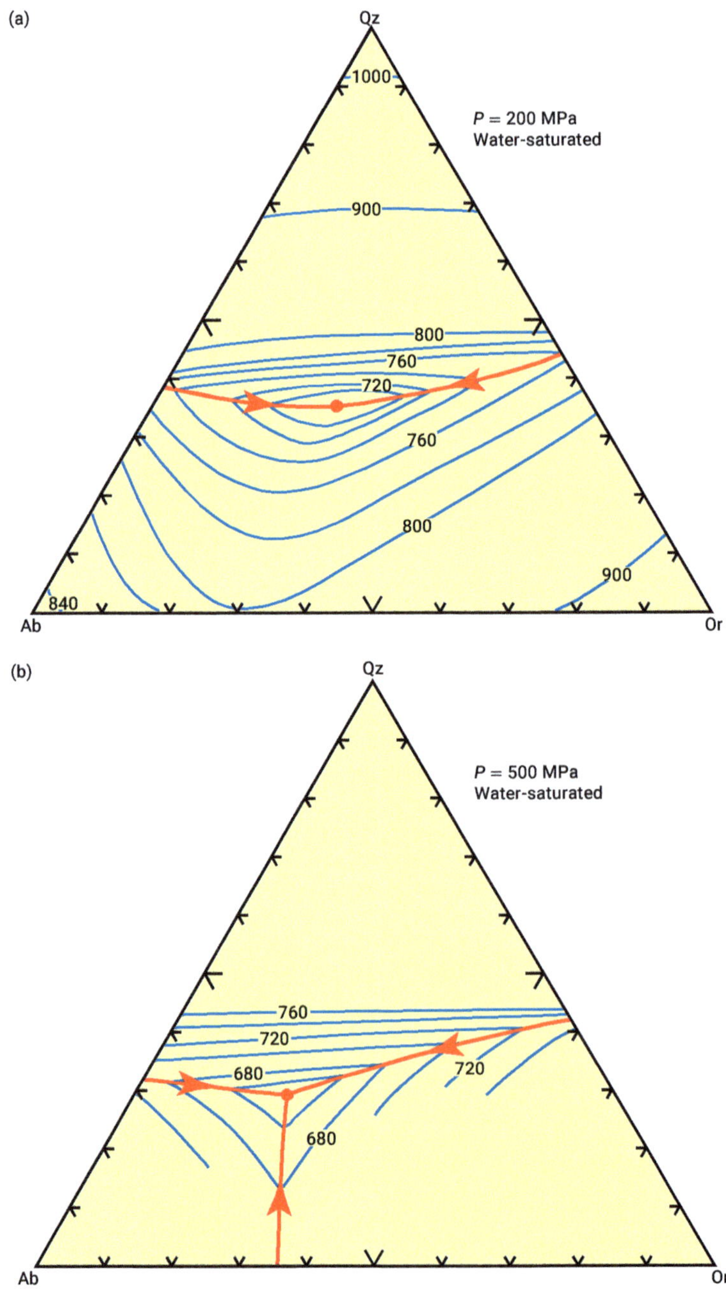

Figure 8.27 (a) Liquidus surface in the haplogranite system at 200 MPa pressure, water-saturated. The liquidus surface is divided in two by a cotectic valley that descends from eutectics in the Qz-Or and Qz-Ab systems to a minimum in the middle of the diagram. Under equilibrium crystallization, any liquid composition in the middle of the system will evolve toward, but not all the way to, the minimum. Conversely, any mixture of alkali feldspar and quartz near the middle of the system will produce a first melt that lies near, but not exactly at, the minimum. (b) Same system at $P = 500$ MPa. At this pressure, the minimum in the alkali feldspar binary has become a eutectic (see Figure 8.26b), and this eutectic plunges into the ternary to form a ternary eutectic. Again, for an arbitrary composition, melting begins, and crystallization ceases, near the eutectic.

Source: Modified from Ernst (1976).

Tuttle and Bowen's work disclosed several important features of the granite system:

1. At pressures less than about 400 MPa, the system consists of a cotectic valley between alkali feldspar and quartz that drains toward a minimum (Figure 8.27a); this is the ternary equivalent of the minimum in the binary system (Figure 8.26a).
2. Above about 400 MPa, the binary alkali feldspar system is a eutectic (Figure 8.26b), and therefore the ternary is as well, with a cotectic between Na-rich and K-rich alkali feldspar.
3. The temperature of the water-saturated minimum/eutectic drops drastically with increasing *P*, from about 1000 °C in the dry system to about 650 °C at water-saturated pressures of 500 MPa and above (Figure 8.28). This parallels the drastic drop in solidus temperature in the system Ab-H_2O (see Figure 6.29a) and occurs for the same reason.

As with the system Di-Ab-An, the paths followed by liquid and crystals upon cooling in this system cannot be told exactly from the phase diagram, because feldspar-melt tie lines are not shown. However, the general paths can be inferred from knowledge of the binary Ab-Or, and understanding the general method is important (see Toolbox 8.2). There are two key and related points:

1. For crystallization of melt compositions not near the sidelines of the system, the liquid path ends up close to the ternary minimum for pressures below about 400 MPa (see Figure 8.27a), or near the ternary eutectic for greater pressures (see Figure 8.27b).
2. For solid bulk compositions not near the sidelines, partial melting produces a melt near the minimum or near the eutectic.

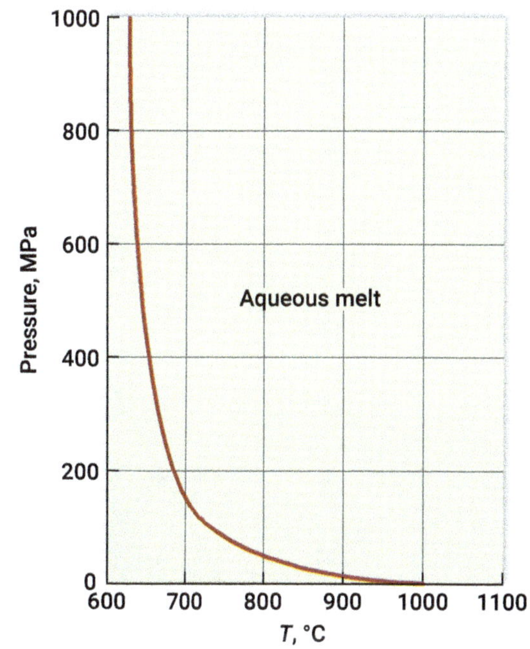

Figure 8.28 Minimum melting temperature in the haplogranite system at water saturation. The first few hundred MPa of pressure cause the temperature of melting to drop over 300 °C.

Source: Modified from Ernst (1976).

Toolbox 8.2

Crystallization and Melting in the Granite System

Magmas at the rhyolitic end of the BADR spectrum, when crystallized, are composed mostly of quartz and one or two alkali feldspars, and the system Ab-Si-Or-H_2O is relevant to their genesis and evolution. Consider a melt with composition X in Figure 8.29, undergoing equilibrium crystallization under water-saturated conditions at 200 MPa (shallow crustal depths). This melt has about 80 wt% Or and subequal amounts of Ab and Qz. It lies above the alkali feldspar liquidus, so upon cooling, that will be the first phase precipitated, at about 850 °C. What composition will precipitate? We don't know exactly, but by consulting the corresponding diagram for the binary system Ab-Or, we can see that compositions on the Or-rich side of the minimum will precipitate feldspar that is significantly more Or-rich than the melt. Given this we'll estimate that the first Or-rich feldspar is at a. Removing this from the melt drives the melt directly away from a, following the red curve. As more and more feldspar precipitates, it becomes more Ab-rich, migrating to the left, and at any given point in its evolution, the melt moves directly away from the increment of feldspar being precipitated at that moment. Eventually, the melt meets the cotectic at b. At this point, Qz starts to precipitate, and the last bit of feldspar to precipitate was at c (as given by the tangent to the melt path).

Although the last bit of feldspar to come out of the liquid is at c, the aggregate of all the feldspar precipitated to that

continued

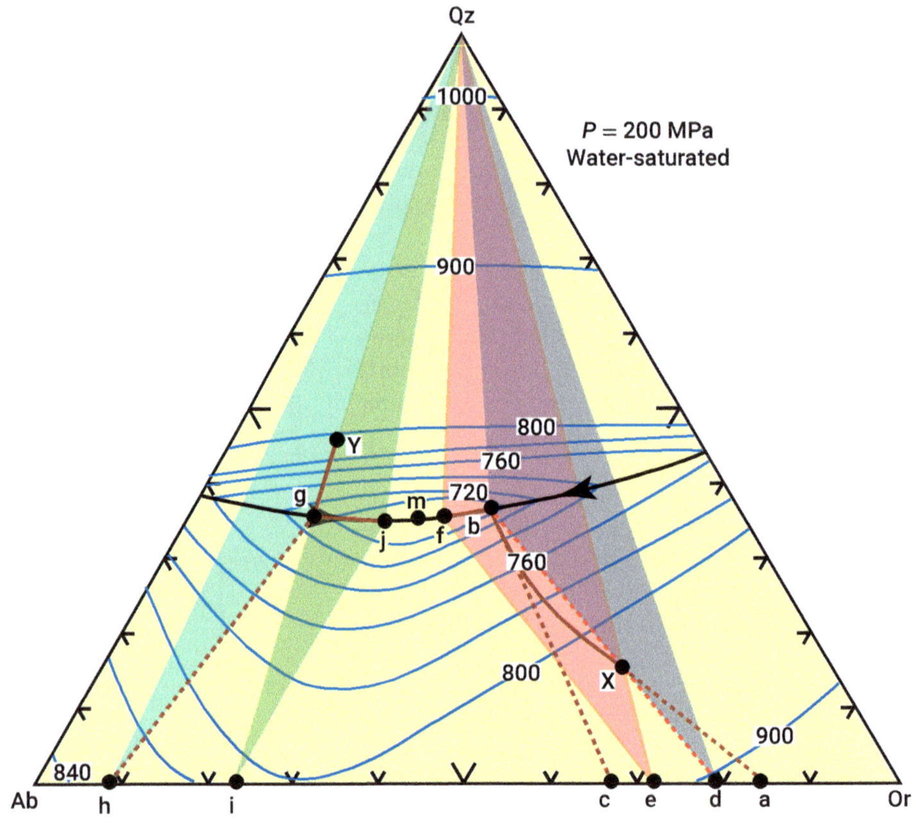

Figure 8.29 Crystallization behavior in the granite system (see text).

Source: Modified from Ernst (1976).

point must be at d. How do we know that? By Baldy's law, of course—d is what is needed so that it and melt b add up to the bulk composition X. We now have our highest-T three-phase triangle, Qz-b-d, with X on the leading edge b-d. As T falls, the melt and feldspar migrate to the left, and the triangle pivots about Qz in spiral-staircase manner. When the trailing edge Qz-e covers the bulk composition X, the percentage of melt is zero. The melt ends up at f, near but not quite at m.

Crystallization of Qz-rich composition Y is similar. Crystallization commences at about 800 °C, and the melt moves directly away from the Qz apex, because Qz has a constant composition. Upon reaching the cotectic, an Ab-rich feldspar, estimated at h, precipitates. The three-phase triangle pivots to the right, and when the trailing edge covers Y, crystallization is complete, with the last melt at j.

Partial melting of X and Y will yield first melts that lie at f and j, respectively, and it is clear that for bulk compositions in this system that are not too close to the walls, the first melt will lie near the minimum/eutectic.

An important finding from Tuttle and Bowen's study was their observation that naturally occurring granites and rhyolites—the high-silica end of the BADR spectrum—cluster in composition around the ternary minimum in the granite system. They mined (by hand) Washington's (1917) 1200-page compilation of igneous rock analyses for those with more than 80 wt% normative quartz + albite + orthoclase and showed that they cluster around the low-P ternary minimum. Figure 8.30 is a modern rendering of this task, using igneous rock analyses from circum-Pacific subduction zones and setting the cutoff to 90 wt%. The analyses are grouped around the low-P minimum (100–200 MPa). Tuttle and Bowen argued that this correspondence put to rest the long-standing idea that granite bodies form by metasomatic transformation of other rocks.

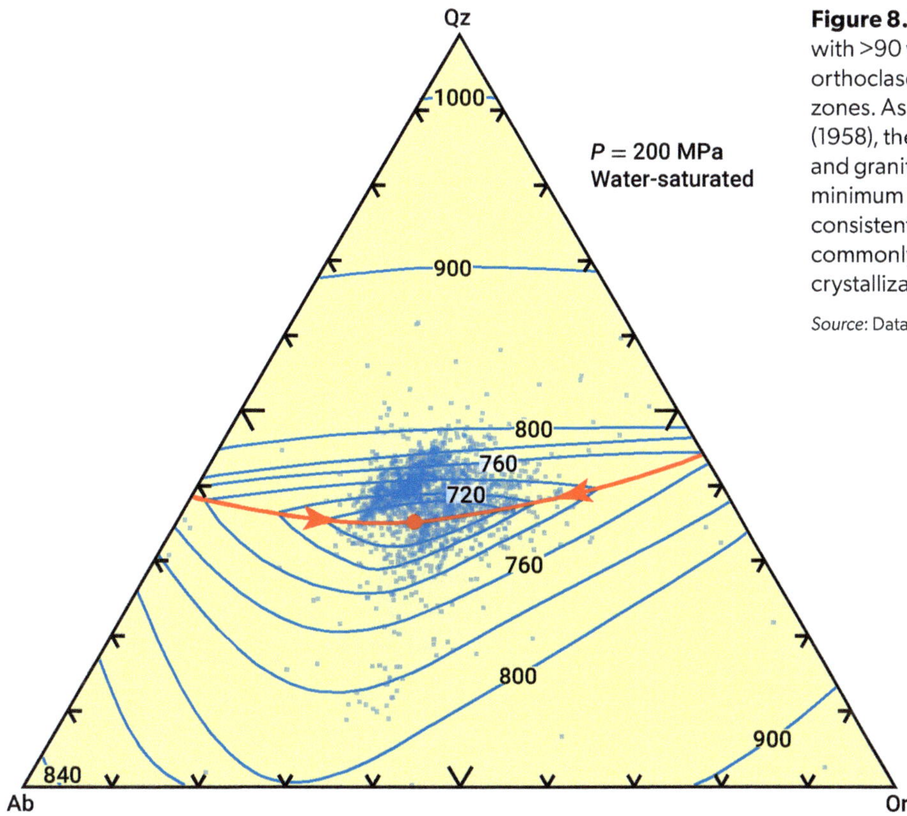

Figure 8.30 Distribution of 1811 rocks with >90 wt% normative quartz + albite + orthoclase, from circum-Pacific subduction zones. As shown by Tuttle and Bowen (1958), the compositions of natural rhyolites and granites cluster nicely about the low-*P* minimum in the system Ab-Si-Or-H_2O, consistent with the hypothesis that they commonly form by either extended fractional crystallization or low-degree partial melting.

Source: Data from Glazner et al. (2015).

Summary

Key points of this chapter include the following:

- During crystallization, the composition of the liquid may change dramatically, but no change in bulk composition occurs unless crystals are separated from liquid.
- Crystal-liquid separation tends to drive silicate liquids toward compositions richer in silica and K_2O and lower in MgO, CaO, and FeO^t.
- A number of processes can drive crystal-liquid separation, including crystal sinking, escape of buoyant liquid, volatile exsolution, and filter pressing.
- Although crystal settling can be documented in lava lakes and sills, it produces only modest changes in bulk-rock composition in those settings.
- Changes in liquid composition for a given process can be determined graphically or algebraically if relevant phase compositions are known.
- Crystal fractionation makes a number of testable predictions; in particular, for the case of a magma chamber undergoing crystal fractionation under a volcano, the sequence of erupted lavas should evolve through time in the predicted direction (e.g., higher SiO_2, lower MgO, higher K_2O, etc.).
- Crystal fractionation is a closed-system process, and therefore a suite of rocks related by this process should have invariant isotopic compositions for heavy (e.g., Sr) systems.
- Magma mixing and assimilation are open-system processes that predict variations in such isotopic systems.
- Xenocrysts that are grossly out of equilibrium with a host magma are generally signs of magma mixing or assimilation.
- Liquid immiscibility, in which silica-rich liquid coexists with Fe-rich, low-silica liquid, with one forming globules in the other, is a widespread, although generally minor, process in igneous rocks.
- Partial melting of hydrous crustal rocks, including hydrous gabbro, can produce silicic liquids; in a broad sense, rhyolitic liquid is the eutectic melt of hydrous crust.
- The alkali feldspar system Ab-Or has a broad solvus owing to the great size difference between Na and K, and alkali feldspars exsolve upon cooling below the solidus.
- Natural granitic and rhyolitic rock compositions cluster around the minimum in the water-saturated ternary system Ab-Or-Qz.

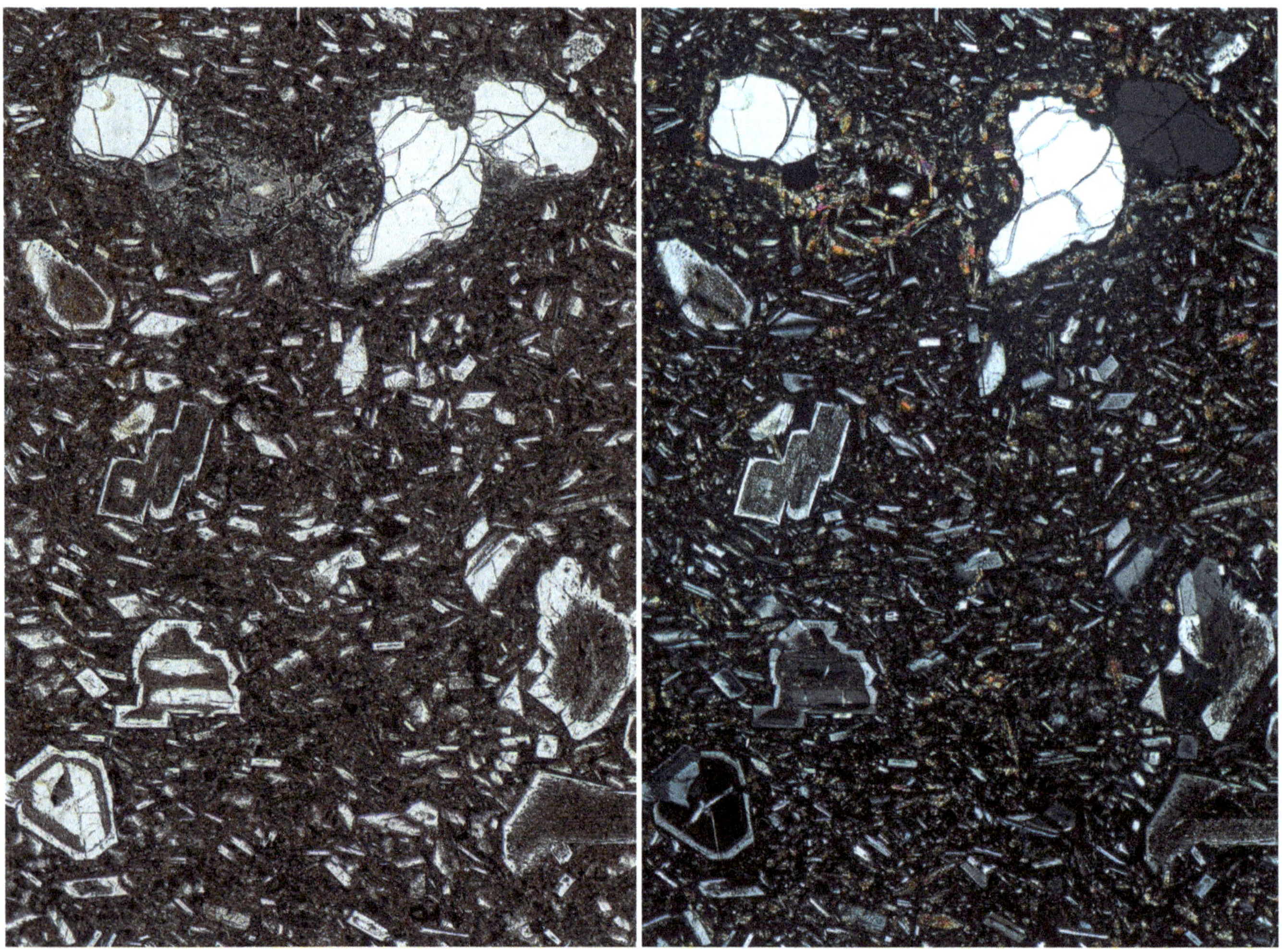

CHAPTER 9

Tools for Probing Magma Origin and Evolution

Using major element data and phase equilibria to infer how a given suite of rocks spans the range from basalt to rhyolite involves ambiguity, because crystal fractionation, magma mixing, and assimilation of the proper contaminant can all drive a mafic magma along the BADR path to rhyolitic compositions. Petrographic arguments and field data may help to resolve this ambiguity. For example, many volcanoes produce lavas that contain unambiguous evidence for magma mixing (e.g., Eichelberger, 1978; Koyaguchi, 1986), but such evidence is not always apparent, and its presence does not guarantee that mixing was the only process going on.

Fortunately, there are several other techniques available to petrologists for untangling the overlapping effects of crystal fractionation, magma mixing, and other processes. In particular, studies of trace elements and isotope systems such as $^{87}Sr/^{86}Sr$ can help to resolve these ambiguities.

9.1 Trace Elements

Learning Objectives

By the end of this section you should be able to:

9.1.1 Predict the compatibility of a trace element in a mineral using a size-charge diagram.

9.1.2 Calculate trace-element concentrations in melts under various melting and crystallization scenarios.

9.1.3 Explain why D values for a given element are generally higher for high-silica liquids than for low-silica liquids.

Magma mixing in a Miocene andesite from the Mojave Desert, California. This thin section view shows the same area, 3 mm wide, in plane-polarized (left) and cross-polarized (right) light. The three bright white crystals at the top of the image are quartz, an unexpected phenocryst phase in an andesite such as this with 59 wt% SiO_2. Larger crystals below are all plagioclase, which is reverse-zoned, with sodic cores surrounded by complex spongy zones that are in turn encased in calcic rims. Quartz crystals are surrounded by glass, which is in turn surrounded by haloes of Cpx (second-order birefringence colors). Glass is gray in plane-polarized light and black in cross-polarized light; a blob of glass containing elongated Cpx crystals joins the left and center quartz crystals. Relationships like this are common in andesites worldwide but were largely ignored before the 1970s, because they could not be understood in terms of the dominant paradigm of fractional crystallization. The current understanding of these rocks as products of open-system magma mixing and assimilation comes from isotope studies, one of the topics of this chapter.

Source: AFG photos.

Trace Elements vs. Essential Structural Constituents

So far in our discussions of phase equilibria, we have dealt exclusively with chemical components that are essential parts of the corresponding minerals—for example, Na, Ca, Al, and Si (or their corresponding oxides) in plagioclase. Chemical constituents that a crystal cannot do without are known as **stoichiometric constituents** or, a bit less cumbersomely, **essential structural constituents (ESCs)**. These are the building blocks of minerals. In contrast are **trace elements**, those elements whose concentrations are low enough (typically <~0.1 wt%, although there is no strict definition) that they do not form their own phases but instead are dissolved in minerals or melts.

Stoichiometric constituents
Elements that are an essential part of a mineral formula.

Essential structural constituents (ESCs)
Equivalent to stoichiometric constituents.

Trace elements
Elements present in (generally) low concentrations such that they do not make their own minerals but instead substitute into existing ones. Trace elements cannot be stoichiometric constituents of the mineral.

Trace elements are useful in petrology because their behavior reflects the proportions of phases that are present in a system. Thus, whereas the major-element composition of a melt being generated at an invariant point (e.g., Figure 8.1) is fixed as long as the relevant phases have not been completely melted away, trace-element compositions can vary widely with the degree of melting. Trace elements give insight into behavior that is otherwise masked by invariant points.

Compatible and Incompatible Elements

Trace elements are classified into two categories: **compatible elements** that prefer to be sited in one or more minerals in the system rather than the melt, and **incompatible elements** that are more easily accommodated by the melt than by the crystals. Compatibility is purely situational, depending on what is present in the mineral assemblage. For example, Ni is a compatible element in olivine; Ni^{2+} fits well into the octahedral Mg-Fe site and finds it more hospitable than basaltic melt. Sr, by contrast, is incompatible in olivine. Conversely, Ni is highly incompatible in plagioclase, but Sr is compatible.

Compatible elements
Trace elements that are, at equilibrium, present in higher concentration in a mineral than in coexisting melt.

Incompatible elements
Trace elements that are, at equilibrium, present in lower concentration in a mineral than in coexisting melt.

Size-Charge Relationships

Whether or not a given trace element fits in a given crystal structure, and thus whether it is compatible or not, depends on size-charge relationships. Figure 9.1 shows ionic radii plotted against charge for a number of elements of geochemical interest. Ions in red are major elements, abundant enough that they generally form their own phases, whereas ions in blue are typically trace elements. The basics of trace-element behavior can be gleaned from this diagram. For example, for ions with a 2+ charge, similarities in size of transition metals such as Mn, Fe, and Ni to Mg means that the transition metals substitute readily into Mg-bearing minerals such as olivine and Cpx—thus, they are compatible in those minerals. Similarly, Hf substitutes readily for Zr, Ta for Nb, and Ga for Al.

Conversely, elements such as Th, Ba, and the rare earth elements have sizes and charges that make them unusable by typical rock-forming minerals such as olivine, feldspars, and pyroxenes—thus, they typically are incompatible. Rb is highly incompatible in rock-forming minerals except for those that contain K (e.g., K-feldspar and biotite).

An obvious but sometimes overlooked pattern is that *most elements are incompatible in most minerals*. Although incompatible elements are quite useful, compatible elements generally give clearer information about petrologic processes.

Trace-Element Groupings

Large-Ion Lithophile Elements

Large ions with a low valence (e.g., K^+, Rb^+, Ba^{2+}, etc.; see Figure 9.1), and therefore weak bond strength, are known as large-ion lithophile elements (LILEs). Their large size makes them incompatible in most elements unless they are an ESC of a precipitating phase (e.g., K and Na in alkali feldspar). "Lithophile" means "silicate-loving."

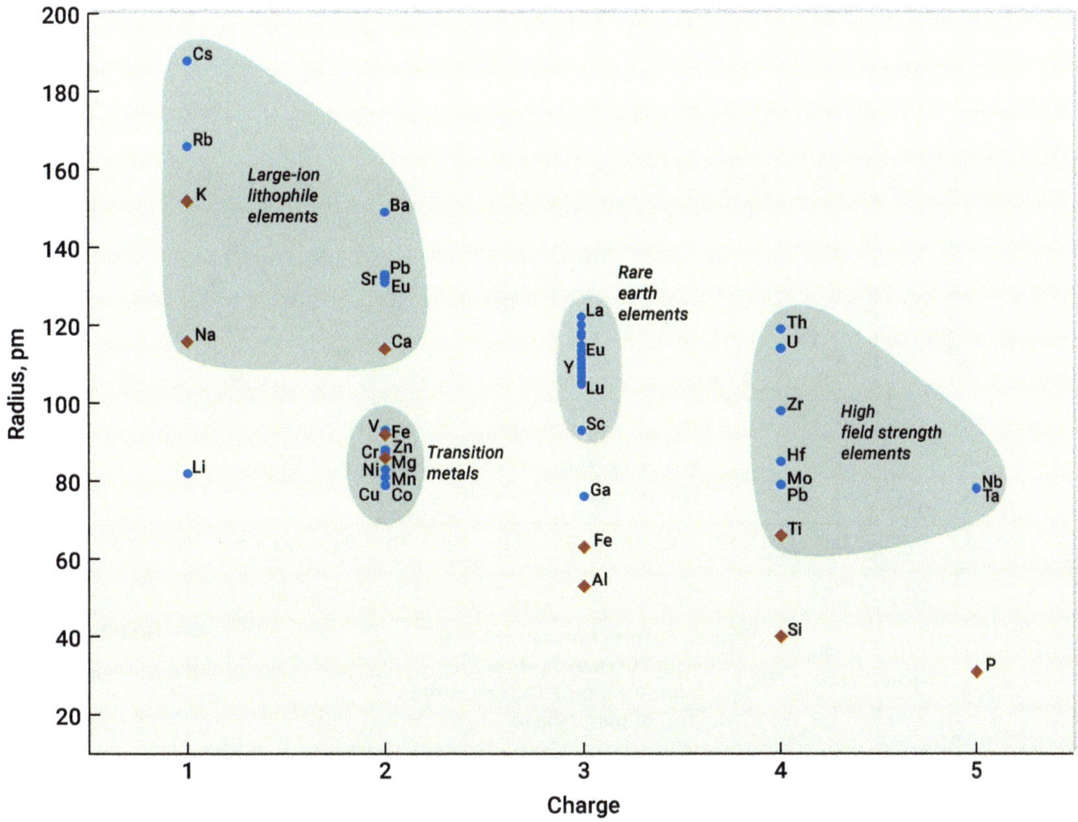

Figure 9.1 Ionic radii plotted against charge for elements of geochemical interest. Major elements are plotted in red diamonds and trace elements in blue dots (definitions are situational; see text). Elements form natural groupings based on their size-charge relationships.

Source: Radii from Veksler et al. (2006).

Transition Metals Technically, these are the elements in the "valley" in the middle of the periodic table, columns 3 to 12, but in common geochemical parlance, the term refers to the first row (Sc through Zn). They are characterized by multiple valence states. Because Fe is a transition metal, they are commonly compatible in Fe-Mg silicates such as olivine and pyroxenes.

Rare Earth Elements The rare earth elements (REEs) hold a special place in the geochemist's heart, because they behave almost—but not quite—identically, and subtle differences in their behavior can be used to infer subtle processes. REE include Sc, Y, and the lanthanides, which come after Ba in the periodic table, although the term REE is typically used to refer just to the lanthanides. All have a charge of 3+ in geologic materials except europium, which can exist partially as Eu^{2+} in magmas, and cerium, which partially converts to Ce^{4+} in oxidizing surface environments. Abundances of the REEs decline with increasing atomic number and follow the Oddo-Harkins rule (see Section 2.5), so elements with an even atomic number, such as Ce, are roughly five times more abundant than their odd counterparts, such as La. Atomic radii of the REEs decrease slightly from light to heavy, and it is this change in radius that gives the 3+ ions their slight differences in geochemical behavior. Y has 3+ valence and a radius in the range of the middle REEs. Its abundance is greater than those of the middle REEs, however, and is therefore significantly easier to measure, making it a useful proxy.

Figure 9.2a shows REE abundances for an ordinary chondrite and for a granite, plotted on a log scale because the range is nearly four orders of magnitude. The abundances

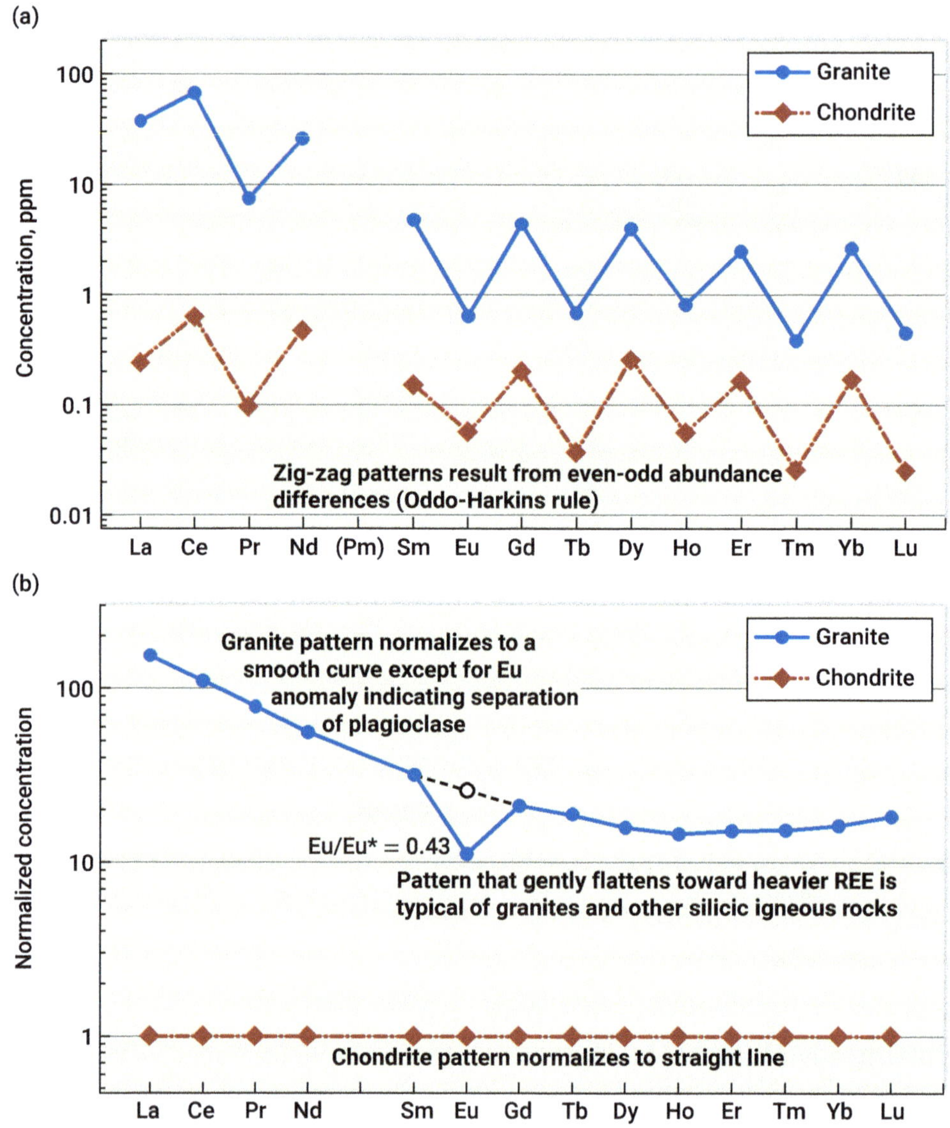

Figure 9.2 (a) REE concentrations in a granite and a chondrite. Note logarithmic scale; values range over more than a factor of 1000. Both the granite and chondrite show alternating high/low values for even/odd atomic numbers (note that odd-numbered Pm, which lies between Nd and Sm, is not shown, because it has no stable isotopes and does not occur naturally). This high/low variation in concentration is a result of the Oddo-Harkins effect (see Section 2.5). (b) Values from (a) normalized by dividing by the concentration in the chondrite. This normalization smooths out the even/odd alternation and clearly reveals a negative Eu anomaly that is not apparent in (a).

form sawtooth patterns owing to the Oddo-Harkins rule, and to remove this, it is common to divide concentrations in a sample by a known reservoir—such as chondrites, or the mantle, or average upper crust. This serves the dual purpose of taking out the sawtooth and showing the concentrations of the elements relative to the reservoir. In Figure 9.2b, this has been done by dividing both by the chondrite abundances, and the resulting patterns are called REE patterns.

High-Field-Strength Elements High-field-strength elements (HFSEs) are those with high charge and small radius, and thus high field strength (charge/radius) and bond strength (charge/coordination number). Zr^{4+} and Ti^{4+} are typical examples. They are anti-LILEs and are also typically incompatible.

Distribution Coefficients and Henry's Law

Distribution coefficient
Ratio of the concentration of a trace element in a crystal to that in coexisting melt.

The partitioning of a trace element between melt and crystals is expressed using a **distribution coefficient**, *D*, defined as

$$D_e^j = \frac{c_e^j}{c_e^{melt}} \quad \textbf{(Eq. 9.1)}$$

where the left side is the distribution coefficient for element *e* between mineral *j* and melt, and the right side refers to the concentrations of element *e* in mineral *j* and in the melt. For compatible elements, $D > 1$, and for incompatible elements, $D < 1$.

Figure 9.3 shows a compilation of measured distribution coefficients for 31 elements in three different minerals in igneous rocks, and Appendix 6 shows data for 12 more

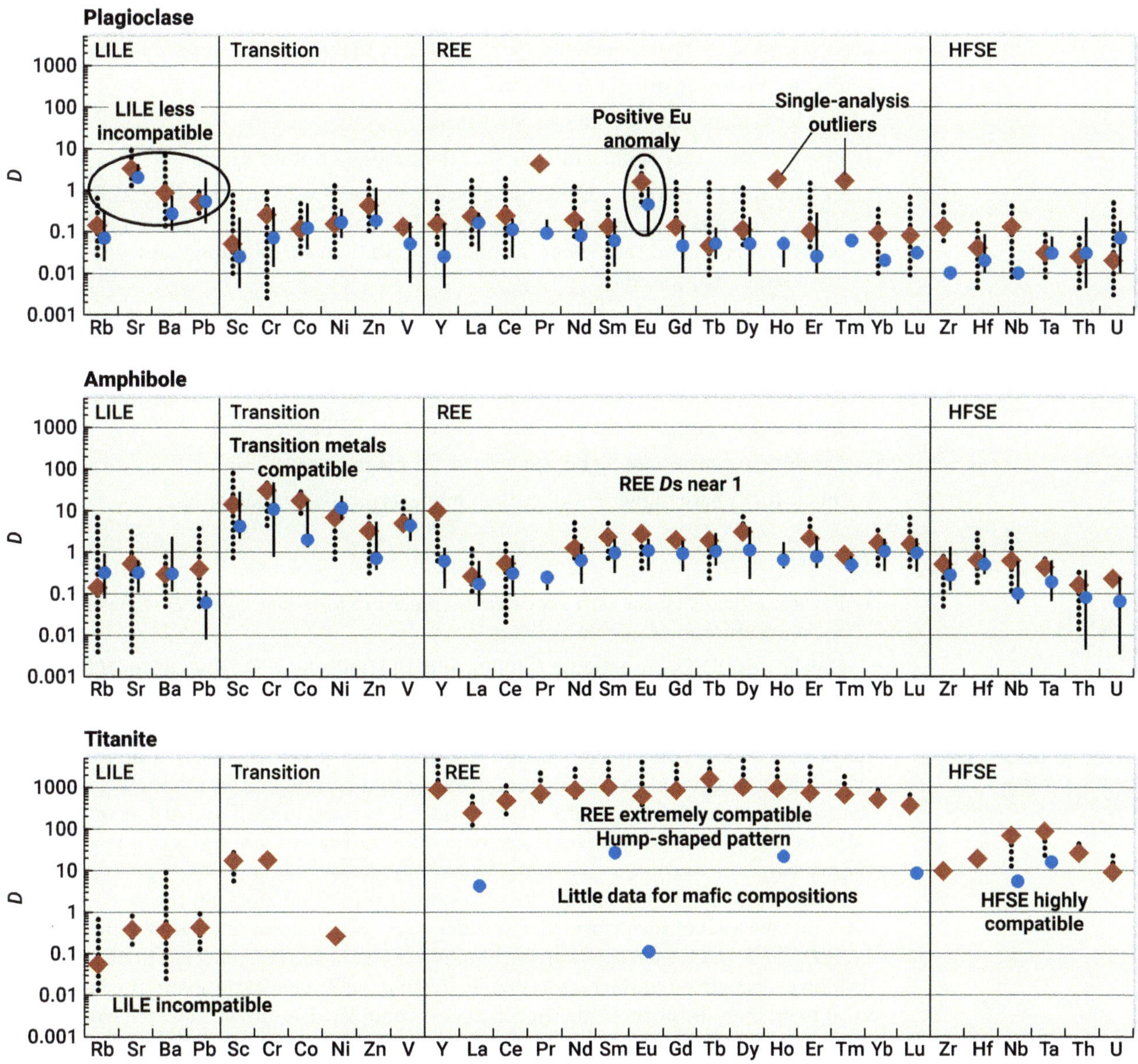

Figure 9.3 Plots of $\log_{10}$ of measured distribution coefficients for 31 elements in three minerals. Higher-silica rocks (dacites, rhyolites, and alkalic equivalents) are shown in red diamonds with dashed lines, and lower-silica rocks (basalts, andesites, and alkalic equivalents) in blue dots with solid lines. *D* values for higher-silica melts are uniformly higher than those for lower-silica melts. Dots and diamonds are medians, and lines represent 10th and 90th percentiles; dots and diamonds without lines are generally single measurements.

minerals. On these plots the elements are (1) ordered from low to high charge; (2) within lists of like charge, ordered by decreasing ionic radius; and (3) grouped into the associations noted in the previous section. From these plots, it is easy to see which elements are compatible and to see patterns within like groupings.

The data on each plot in Figure 9.3 are broken into two groups, lower-Si (chiefly basaltic and andesitic) and higher-Si (chiefly dacitic and rhyolitic). It is clear that higher-Si magmas have larger *D* values for a given element, almost without exception. This generalization has to do with the structure of silicate liquids; the high degree of order and feldspar-like structure of high-Si liquids (see Section 7.3) means that they are less hospitable to elements with odd charges and sizes than less structured, more tolerant, low-Si liquids.

Let's examine the systematics of *D* values in some representative minerals in Figure 9.3:

Plagioclase:

- *D* values for nearly all elements are <1, meaning that most elements are incompatible.
- Exceptions are Sr, and Ba and Eu in silicic rocks.
- The least-incompatible elements are LILEs.
- *D* values for Eu are significantly greater than those for other REEs.
- Pr, Ho, and Tm have $D > 1$, but these are inconsistent with surrounding REEs and are likely single analyses, as indicated by the lack of data range lines. These three REEs are odd-numbered, and thus lower in concentration than surrounding ones and were not analyzed in many studies.
- Silicic rocks have higher *D* values than mafic by roughly a factor of 2.

Amphibole:

- *D* values for transition elements are >1, and most REEs are near 1.
- *D* values in general are higher than those for plagioclase.
- Silicic rocks have higher *D* values than mafic, typically by a factor of 2.

Titanite:

- For most elements, the only reported *D* values are for silicic rocks. *D* values for mafic rocks are few and are single analyses.
- *D* values for REEs are extreme (~1000), and all HFSEs have *D* values of 10–100.

Henry's law
Here, the observation that at low concentrations, the concentration of an element in a mineral is proportional to that in coexisting melt.

The thermodynamic basis for trace-element distribution is **Henry's law**, which states that at low concentrations, the solubility of a given chemical component in a crystal is proportional to the concentration of that component in the melt. This is essentially what the definition of *D* states, and it is critical to realize that it is only applicable to components at relatively low concentration that are not ESCs of any phases in the system.

This bears repetition, because it is a common stumbling block on the journey to unraveling how a set of rocks formed. Consider a crystallizing dacite magma with 200 ppm Zr. If $D_{Zr}^{hornblende}$ is 0.5, then crystallizing amphibole will have 100 ppm Zr. If further crystallization causes zircon to start crystallizing, driving the Zr concentration in the melt down to 50 ppm, then the hornblende should have 25 ppm Zr. *The zircon, however, will have its stoichiometric allotment (67.2 wt%, or 497,000 ppm) of Zr regardless of the Zr concentration in the melt.*[1] Zr is an ESC of zircon, and its concentration in zircon is therefore not dependent on the concentration in the melt. In our example, Zr is still partitioned between melt and hornblende according to Henry's law, but not between melt and zircon, and therefore not between melt and the full crystallizing assemblage. Thus, Zr cannot be modeled as a trace element in this system. Beware!

The distribution coefficient D applies to the partitioning of an element between a single mineral phase and the melt. When more than one mineral is present, a weighted average B ("B" for bulk) must be used:

$$B_e = w_1 D_e^1 + w_2 D_e^2 + \ldots + w_j D_e^j \qquad \textbf{(Eq. 9.2)}$$

where w_j is the weight fraction of mineral j in the crystalline assemblage, and D_e^j is the distribution coefficient for element e between mineral j and melt.

Equations for Trace-Element Variations during Melting and Crystallization

If Henry's law holds, then variations in the concentration of a trace element can be calculated for various melting and crystallization scenarios. These equations are generally developed using the parameter F, the weight fraction of melt in the system (reminder: for a completely molten magma, $F = 1$, and for a solid rock, $F = 0$). Equations for various processes are given in Toolbox 9.1 on trace-element equations.

The following discussion provides equations for the evolution of trace elements in four common petrologic scenarios. Derivations of these equations are given in geochemistry texts such as Wood and Fraser (1978). For illustration, we consider a compatible element (bulk distribution coefficient $B = 10$) and an incompatible element ($B = 0.01$).

Batch Melting

Batch melting is a process in which a source material is melted a given amount, with the melt maintaining equilibrium with the crystals, and then the melt is extracted. The equation for the variation in a trace element as a function of F is

$$\frac{c_e^{melt}}{c_e^0} = \frac{1}{B + F(1 - B)} \qquad \textbf{(Eq. 9.3)}$$

where the left side is the ratio of the concentration in the melt to that in the original rock that is being melted, and B is the bulk distribution coefficient defined earlier. Equations given here and later only hold if B is constant throughout the process; more complicated procedures are used if this is not true.

Fractional Melting

In pure fractional melting, the melt is removed from the source as soon as an infinitesimal increment is formed:

$$\frac{c_e^{melt}}{c_e^0} = \frac{1}{B}(1 - F)^{\left(\frac{1}{B}-1\right)} \qquad \textbf{(Eq. 9.4)}$$

Figure 9.4a plots Equations 9.3 and 9.4 for distribution coefficients $B = 10$ and 0.01. For both the compatible and incompatible element, the initial concentration in the melt is $1/B$ (100 and 0.1, respectively), but the incompatible element's concentration in successive batches of melt drops dramatically, because nearly all of the incompatible element jumps into the melt in the first increments. Escape of tiny melt batches is unlikely in nature, so these curves represent limiting cases only. In contrast, the source rock hangs onto the compatible element until that source is almost completely melted, at which time it must give it up.

Equilibrium Crystallization

In equilibrium crystallization, the crystals remain in chemical equilibrium with the melt throughout the crystallization process:

$$\frac{c_e^{melt}}{c_e^0} = \frac{1}{B + F(1 - B)} \qquad \textbf{(Eq. 9.5)}$$

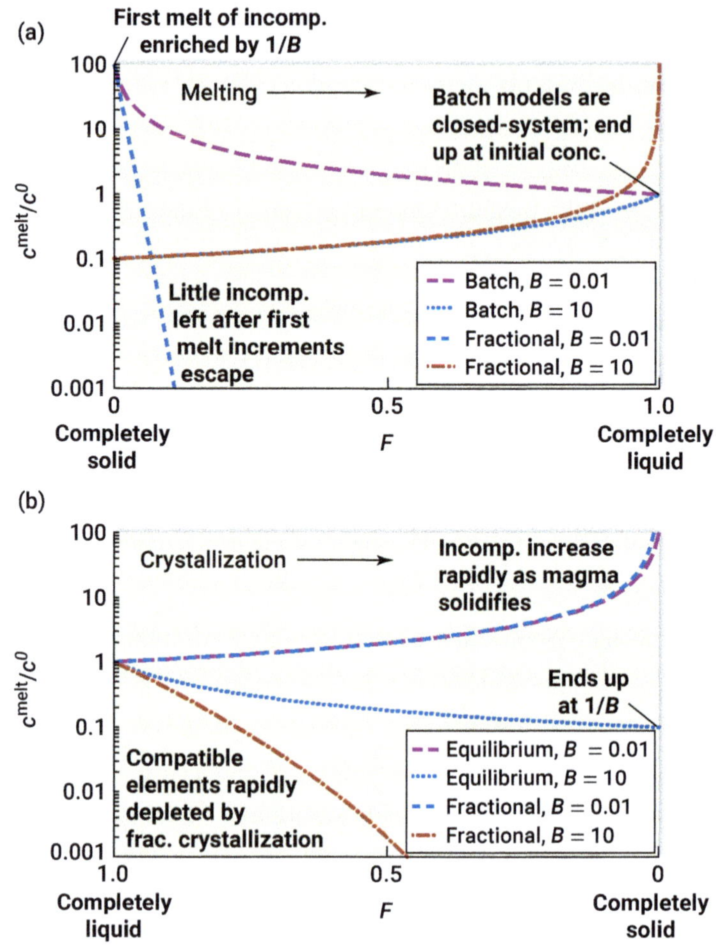

Figure 9.4 Contrasting behavior of compatible and incompatible elements under different melting and crystallization scenarios (see text). In (a), melting proceeds from completely solid ($F = 0$) on the left to completely liquid ($F = 1$) on the right; in (b), the abscissa is reversed such that fully molten magma is on the left and fully crystallized on the right.

where c_e^0 is the initial concentration of trace element *e* in the melt. Not surprisingly, the right side of this equation is the same as that for equilibrium (batch) melting; if you melt a rock to a certain *F* under closed-system, equilibrium conditions, the trace-element concentrations are the same as if you melt it all the way and then crystallize it back to the same value of *F*.

Fractional Crystallization

In fractional crystallization, the crystals are removed from chemical equilibrium with the melt the instant they are formed:

$$\frac{c_e^{melt}}{c_e^0} = F^{(B-1)} \quad \textbf{(Eq. 9.6)}$$

Figure 9.4b shows the evolution of trace-element concentrations for equilibrium and fractional crystallization. The abscissa is flipped, with $F = 1$ on the left, such that the system proceeds from entirely melt on the left to entirely crystals on the right. All the curves start at 1. The concentration of the incompatible element is essentially $1/F$ at any time. An important point is that under fractional crystallization, the concentration of the compatible element rapidly plummets, reaching values of 1/1000 the starting concentration by the time *F* reaches 0.5.

Toolbox 9.1

Some Simple Trace-Element Calculations

Let's use the equations discussed earlier to estimate trace-element concentrations during melting of the mantle. Our model mantle will be composed of 80 wt% olivine, 10 wt% Cpx, and 10 wt% Opx (we will ignore spinel). Let's look at an incompatible element, La, and a compatible element, Ni. Estimating *D* values from Appendix 6 by eye, we can calculate bulk *D* values (*B* values) from Equation 9.2:

	D_{La}	D_{Ni}
Ol	0.01	10
Cpx	0.1	5
Opx	0.01	5
Bulk (*B*)	0.019	9

Note that there are no *D* data for La in Appendix 6, but other REEs have *D* = ~0.1 for heavy REEs, decreasing to 0.01 for Nd. The true *D* for La is probably less than 0.01, but we'll use 0.01 for this exercise; the exact value makes little difference here. Batch melting Equation 9.3 can be rewritten as

$$c_e^{melt} = \frac{c_e^0}{B + F(1-B)} \quad \text{(Eq. 9.7)}$$

We will use starting concentrations of 0.7 ppm for La and 2000 ppm for Ni. Selected values calculated from Equation 9.7 are

F	c_{La}	c_{Ni}
0	36.84	222
0.1	5.98	244
0.5	1.37	400
1	0.70	2000

These values are plotted in Figure 9.5. Key points are (1) the incompatible element La is highly enriched in low-degree partial melts (small *F*), because most of it partitions into the melt; (2) as *F* increases, La becomes increasingly diluted, reaching its starting value when the source is fully melted (*F* = 1); and (3) conversely, the compatible element Ni is held in the source rock and stays at low values until large *F*.

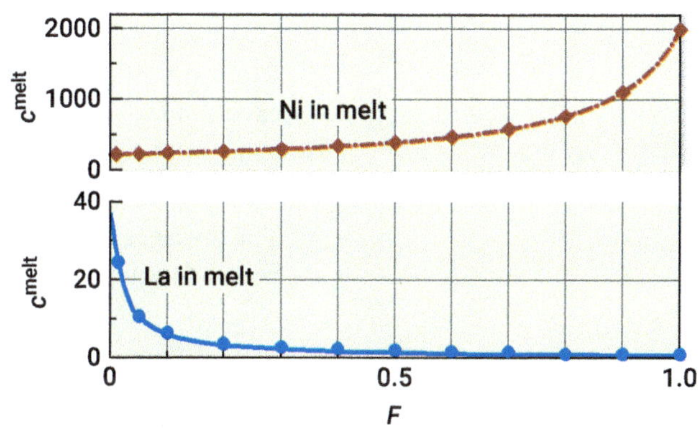

Figure 9.5 Example calculations for a compatible element (Ni) and an incompatible element (La) for batch melting of peridotite. The incompatible element jumps into the melt and is at high concentrations when *F* is low, being increasingly diluted as melting proceeds. In contrast, the compatible element is held in the source for a long time, only reaching half its concentration in the source rock when *F* = ~0.9.

9.2 Principles of Trace-Element Variation

Learning Objectives

By the end of this section you should be able to:

9.2.1 Calculate the Eu anomaly in a liquid separated from a given crystalline assemblage.

9.2.2 Predict how the following ratios in the melt will change during crystallization: incompatible/incompatible and incompatible/compatible.

9.2.3 Explain why the trace-element concentrations in rhyolites can vary by orders of magnitude even though the major-element compositions are nearly identical.

Some Rules of Thumb

Henry's law leads to a number of important principles that can be used to interpret trace-element variations in igneous rocks:

- The concentration of a highly incompatible element in liquid produced by partial melting is proportional to $1/F$. Thus, small degrees of partial melting (small F) lead to large enrichments in incompatible elements in the melt.
- Ratios of highly incompatible elements remain constant during closed-system processes such as fractional crystallization or partial melting. The individual values change, but the ratios do not.
- REEs mostly have the same charge and similar radii, so they behave similarly. However, Eu can exist in the 2+ state[2] (all the others are 3+ under magmatic conditions), so in sufficiently reducing conditions, a significant amount of Eu is Eu^{2+}. Thus, its behavior is different from the other REEs. Eu^{2+} is compatible in plagioclase, whereas the other REEs are mildly incompatible, so plagioclase crystallizing from a melt will have an excess of Eu—a europium anomaly, or Eu anomaly.
- Compatible elements are rapidly depleted in the melt by fractional crystallization, whereas incompatible elements build up as F decreases.
- The presence of a minor phase (e.g., titanite) with high D values for some elements can have a drastic effect on trace elements in the liquid.

Some Examples

Example 1: Degree of Partial Melting

If the concentration of an incompatible element in the source rock of a magma is known, then percent melting needed to generate the magma can be estimated. For example, if an incompatible element is present at 10 ppm in the source rock, then partial melt produced by 10 wt% melting ($F = 0.9$) will have 100 ppm, and 1 wt% melting will produce melt with 1000 ppm. It was noted long ago that alkali basalts from oceanic islands such as Samoa and St. Helena have much higher concentrations of incompatible elements than MORBs and their kin (Gast, 1968), and this has long been attributed to generation of alkali basalts by smaller degrees of partial melting.

Example 2: Separation and Accumulation of Plagioclase

Because the REEs have quite similar behavior in magmas except for Eu, they are sensitive indicators of magma evolution in systems where D values are significantly different between 2+ and 3+ REEs, such as plagioclase. All REEs are incompatible in plagioclase except Eu, which is less incompatible in mafic magmas and slightly compatible in silicic magmas. As a consequence, REE patterns for plagioclase typically have significantly positive Eu anomalies, and melts from which plagioclase has crystallized have corresponding negative anomalies. A classic example of this behavior occurs on the Moon, where anorthosites from the lunar highlands have large positive Eu anomalies, and basalts have correspondingly large negative Eu anomalies (Figure 9.6). These anomalies are consistent with the hypothesis that the lunar highland anorthosites formed by flotation of plagioclase in the magma ocean that made up the Moon's early history (see Figure 2.3). Removal of excess Eu in this plagioclase left behind an Eu-depleted mantle (i.e., a mantle with a negative Eu anomaly), and the lunar basalts and soils derived from them by meteorite bombardment inherited this anomaly when they formed several hundred million years later.

Example 3: Crystallization of a Sill

Sills are typically envisaged as one-shot magma injections that undergo gravity-driven fractional crystallization as they cool. A classic example is the Palisades sill, spectacularly exposed along the western side of the

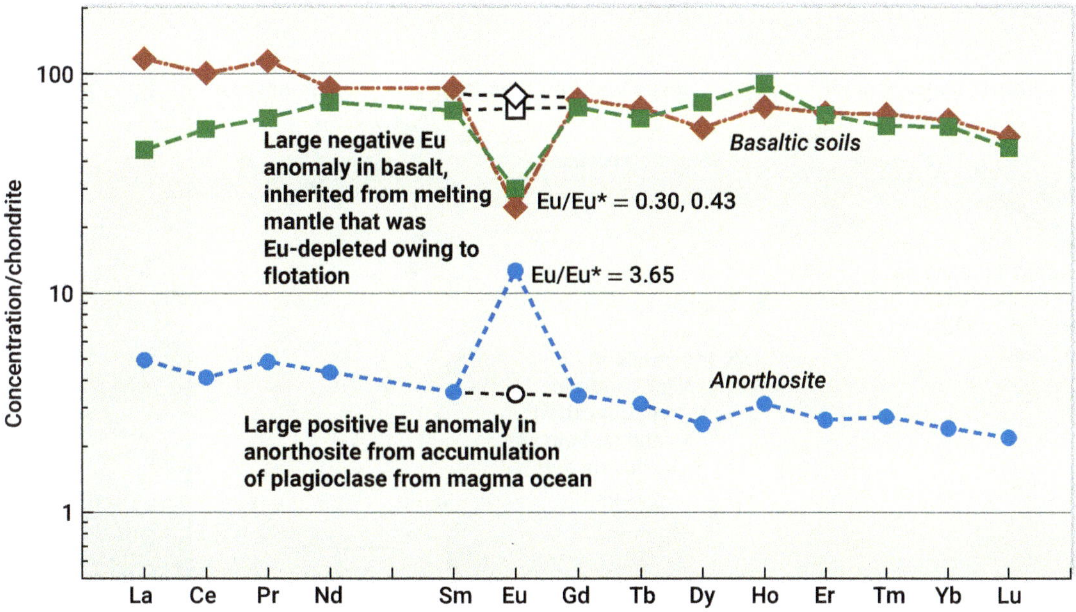

Figure 9.6 Chondrite-normalized REE patterns for lunar anorthosite and two samples of lunar soil formed by meteorite bombardment of surficial basalts. The negative anomalies in the basaltic soils complement the positive anomaly in the anorthosite. Calculated Eu anomaly values (Eu*; see Toolbox 9.2) are shown.

Source: Data from Wakita and Schmitt (1970).

Hudson River across from the island of Manhattan (Figure 9.7). This massive body of diabase, over 300 m thick here, was injected 200 Ma during the breakup of Pangea and formation of the Atlantic Ocean and associated Central Atlantic Magmatic Province (see Chapters 13 and 14). Petrologic data are more or less consistent with rapid-fire injection of several pulses that behaved more or less as a single deep magma chamber (Shirley, 1987; Husch, 1990).

Figure 9.8 shows several chemical parameters from Shirley (1987) plotted against estimated stratigraphic height for samples collected off the west end of the George Washington Bridge. There are clear, and in some cases dramatic, changes in whole-rock composition with height. In keeping with interpretation of sills elsewhere, the original magma composition is best estimated from the composition of samples chilled against the roof; these have about 7.5 wt% MgO and about 53 wt% SiO_2, good numbers for tholeiitic diabase related to continental rifting. Samples from the base, however, have much higher (albeit quite scattered) MgO and lower SiO_2 as a result of settling of olivine. MgO declines dramatically and linearly to about 10 percent of its starting value with increasing height; abruptly, about 85 percent of the way toward the top, MgO rapidly increases to its starting value of 7.5 wt%. Silica just meanders around in the low 50s, never getting very far from its starting value—behavior that is, not surprisingly, quite similar to that in the lava lake discussed in Chapter 8.

Figure 9.7 Palisades sill exposed in the escarpment on the west side of the Hudson River, New York. The local Lenape tribe called the columnar-jointed cliffs "Wee-Awk-En," meaning "rocks that look like trees," whereas Europeans thought they resembled the log fences built as defensive fortifications. To geologists, they are a reminder of the breakup of Pangea, massive magmatism that accompanied opening of the Atlantic (see Figure 10.6b), and the end-Triassic mass extinction.

Source: June Marie Sobrito/Shutterstock.

In contrast to the major elements, the incompatible element Th shows much more dramatic variation. It is enriched by a factor of four over background values in the "sandwich horizon," which is interpreted to be where the last remaining melt was concentrated as the sill reached full crystallization.

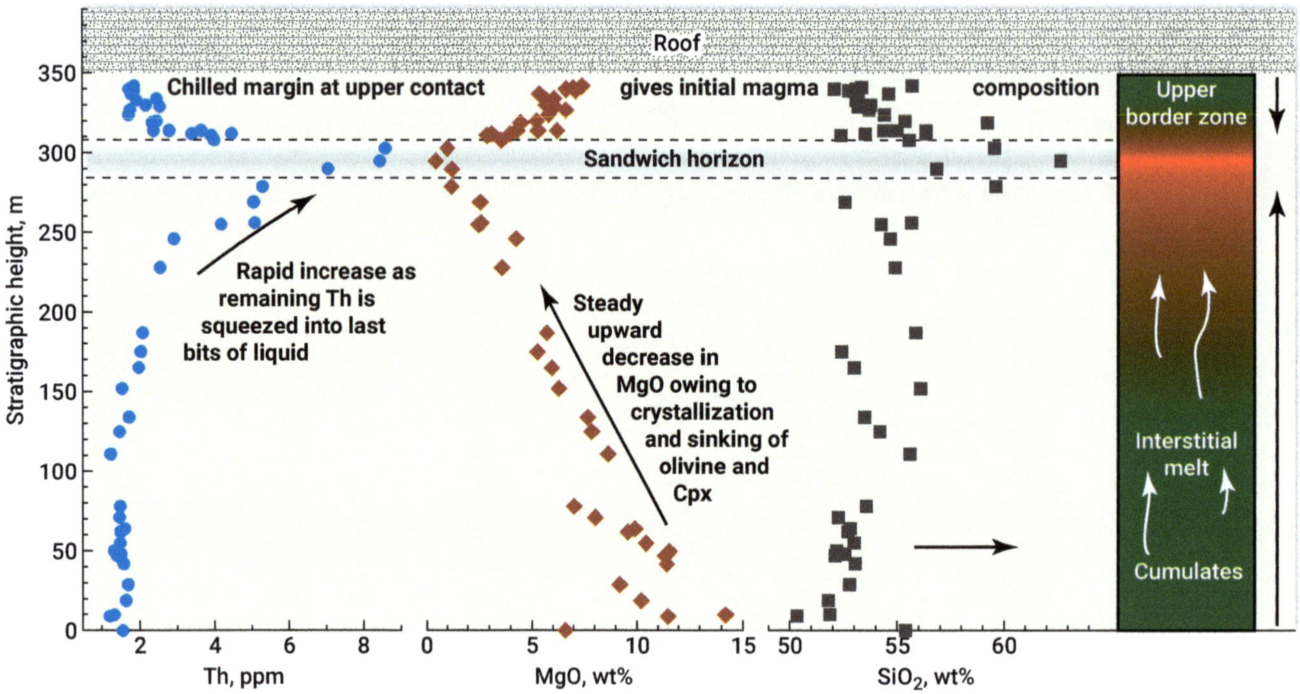

Figure 9.8 Variation in several chemical parameters with stratigraphic height in the Palisades sill. These data are consistent with highly incompatible Th being squeezed into the last-crystallizing bit of liquid and show that this happened near the top of the sill, ~300 m above the base.

Source: Data from Shirley (1987).

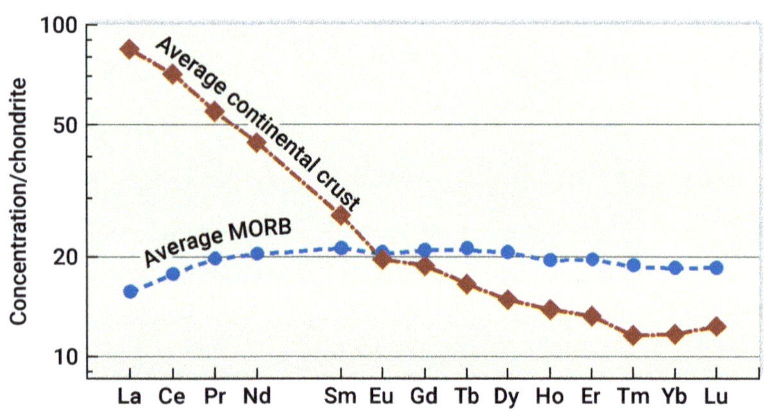

Figure 9.9 REE plots for global average MORB and average continental crust. Continental crust shows pronounced light REE enrichment and a pattern that drops steeply to the middle REEs and then flattens, whereas MORB shows a pronounced dip at the light end. This indicates derivation from mantle that has previously been depleted in light REEs relative to heavier REEs as a result of partial melting and production of continental crust. The two patterns are complementary.

Sources: Data from White & Klein (2014) and Rudnick & Gao (2014).

Example 4: Depletion of the MORB Source A thought experiment. If the Earth is broadly chondritic in composition (see Chapter 2), then if our hypothetical(?) aliens ground it up and analyzed it, it should have an overall REE pattern like that of the chondrite in Figure 9.2. Given that the core is about 30 percent of the Earth's mass and should contain almost no REEs (because REEs are lithophile), then the mantle should contain about 1.5 times chondritic abundances. And given that REEs are incompatible in mantle mineral assemblages, with light REEs more incompatible than heavy (see Figure 9.3), then REE patterns for MORBs should lie considerably above chondrites, with light REEs higher than heavy REEs. Twenty percent partial melting would give an enrichment of five times; with the mantle being 1.5 times chondrite, MORBs should have downward-sloping REE patterns at roughly 7.5 times chondrite. Right?

Wrong. MORB REE patterns have a curious convex-up pattern with significant depletion of light REEs relative to middle REEs (Figure 9.9); they look more-or-less like upside-down versions of what we predicted—they should be relatively enriched in light REEs, not relatively depleted. What is wrong with our reasoning?

Our thought experiment went wrong by leaving out a volumetrically minor but geochemically important part of the Earth—the continental crust. Since REEs are incompatible in the mantle, formation of continental crust removes REEs from the mantle and stores them in a REE-enriched place from which they cannot be effectively recycled. Thus, over Earth history, REEs and other incompatible elements have been incrementally removed from the mantle, leaving behind a progressively depleted mantle reservoir. Since light REEs are more incompatible than heavy, they have been removed in relatively greater abundance, leading to the light REE depletion seen in Figure 9.9.

Note from the crustal growth curves in Figure 2.14 that some models show steady crustal growth over Earth history, some show most crustal growth early, and some indicate that crustal mass is now decreasing via recycling of continental crust back into the mantle. There is little consensus on the specifics of separation of crust from mantle.

Toolbox 9.2

Europium Anomalies

Because REEs have such similar chemical properties, REE patterns normalized to an appropriate substance (e.g., chondritic meteorites) display patterns that are smooth except, typically, for Eu, because Eu can exist in the Eu^{2+} state. Eu^{2+} has a size and charge comparable to Sr^{2+} (see Figure 9.1), and Sr is compatible in plagioclase (see Figure 9.3), so Eu^{2+} is as well. As a consequence, melts that have separated from a plagioclase-rich source typically have a depletion in Eu—an anomaly—relative to other REEs. This is well-displayed by the lunar samples in Figure 9.6.

The Eu anomaly for a given REE pattern is defined as the ratio of Eu in the sample to the Eu value that the sample would have if the pattern were smooth (referred to as Eu*). There are different ways to calculate this, but the most logical and accepted method is given by

$$\frac{Eu}{Eu^*} = \frac{Eu_n}{\sqrt{Sm_n \times Gd_n}} \quad \text{(Eq. 9.8)}$$

where the left side is the Eu anomaly and Eu_n etc. are chondrite-normalized values. The square root of the product rather than the arithmetic mean of the neighboring REEs, Sm and Gd, is necessary owing to the log scale.

Although Eu anomalies are typically blamed on plagioclase, other phases can contribute. For example, REEs range from compatible to extremely compatible in zircon and titanite (see Figure 9.3), but both phases have lower *D* values for Eu than for other REEs. Thus, separation of these phases will impart a positive Eu anomaly on the melt or help to cancel out a negative anomaly caused by separation of plagioclase. And because titanite *D* values are gigantic for all REEs, titanite tends to gobble up all the REEs and overwhelm any plagioclase effect.

Making Connections 9.1

Obsidian Sourcing and Trade Routes

Obsidian was an important resource for Native Americans because it can be shaped into sharp tools, spear points, and arrowheads. Most obsidian worldwide is high-silica rhyolite, a composition that can arise from extended fractional crystallization of intermediate magmas or small-degree partial melts of hydrous crustal material. The major-element compositions of obsidians vary little, but trace-element compositions vary widely depending on the degree of crystallization or melting, the source composition, and whether melting or crystallization were equilibrium or fractional. Trace-element compositions thus can provide a fingerprint that characterizes individual lava flows and domes.

Obsidian fingerprinting has been used across the globe to study dispersion of material across southern Europe from prized quarry sites such as Sardinia and Lipari (Tykot, 2002) and through Washington to British Columbia from Newberry volcano in southern Oregon (Connolly et al., 2015). Eerkens et al. (2008) employed this technique to study mobility of prehistoric Owens Valley Paiute in eastern California (Figure 9.10a). An archaeological survey done in the 1980s before a highway was widened found several structures dating from two periods, about 500 to 200 and 1650 to 1450 years before present. Eerkens et al. determined trace-element compositions of the abundant obsidian chips and artifacts found there and compared them to reference compositions determined from 14 potential sources (Figure 9.10b,c). Most of the samples plotted within the fields of sources within about 200 km of the site, with none from the four sites beyond that range. They used these geochemical data and artifact sizes to infer that the populations became less mobile with time.

continued

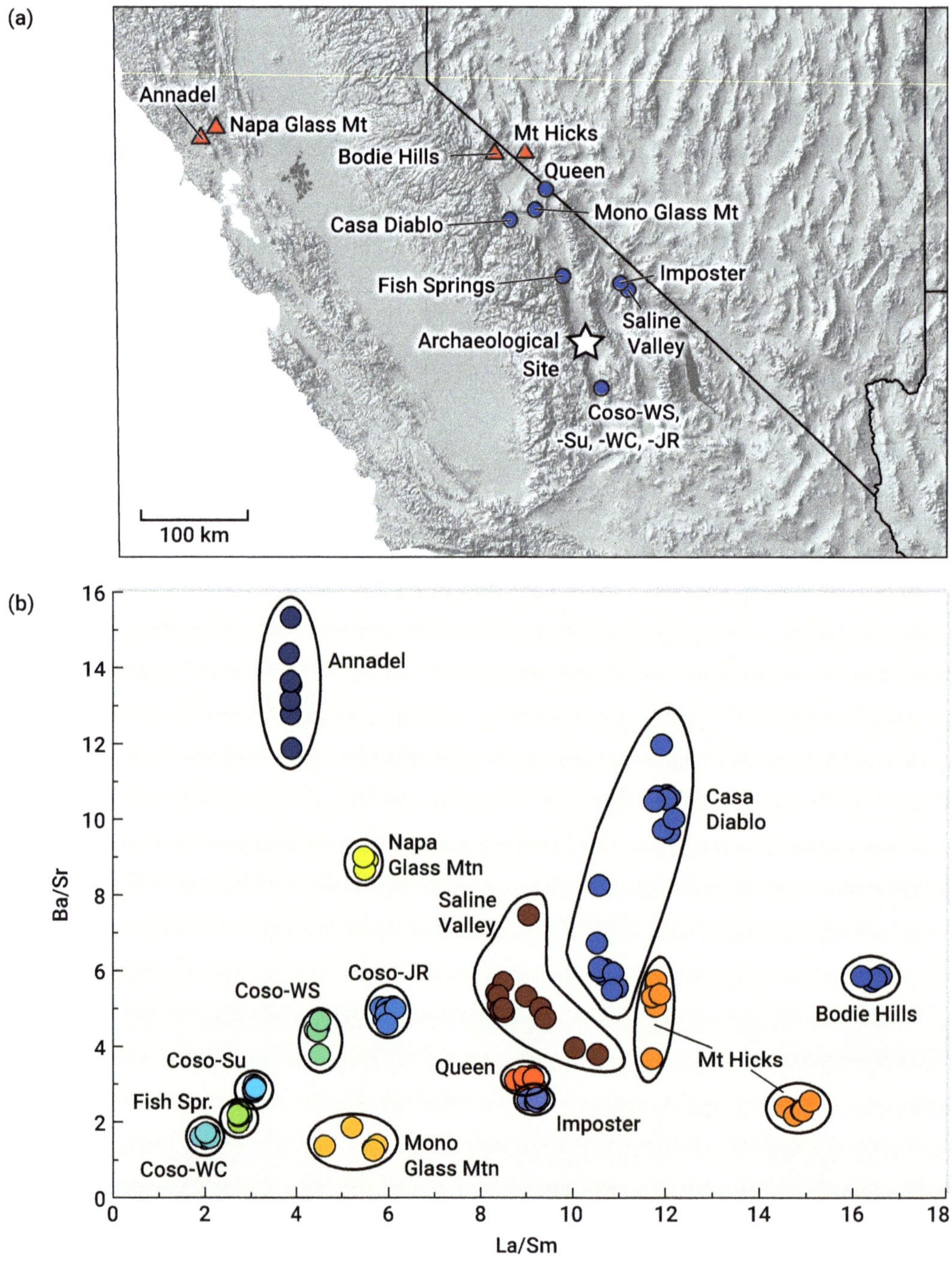

Figure 9.10 An example of using trace elements to determine the sources of archaeological samples. California and Nevada have many sources of tool-grade obsidian, all of which have the same high-silica rhyolite composition. However, different sites bear different trace-element concentrations, and these can be used to discriminate among them. (a) A portion of central California and western Nevada showing sources studied using statistical analysis to determine which elements and element ratios best separated the sites. (b) The different sources are well-separated on a plot of Ba/Sr vs. La/Sm. (c) Obsidian samples from the archaeological site cluster in several of the fields defined in (b) (blue circles in part a), but do not fall in the more distant sites (red triangles). This helps to clarify how far tool-grade material was transported and the extent of trade. The sets of points that lie above the Casa Diablo and Coso-WS fields suggest that material from unstudied sources was also present.

Source: Data courtesy of Jelmer Eerkens, from Eerkens et al. (2008).

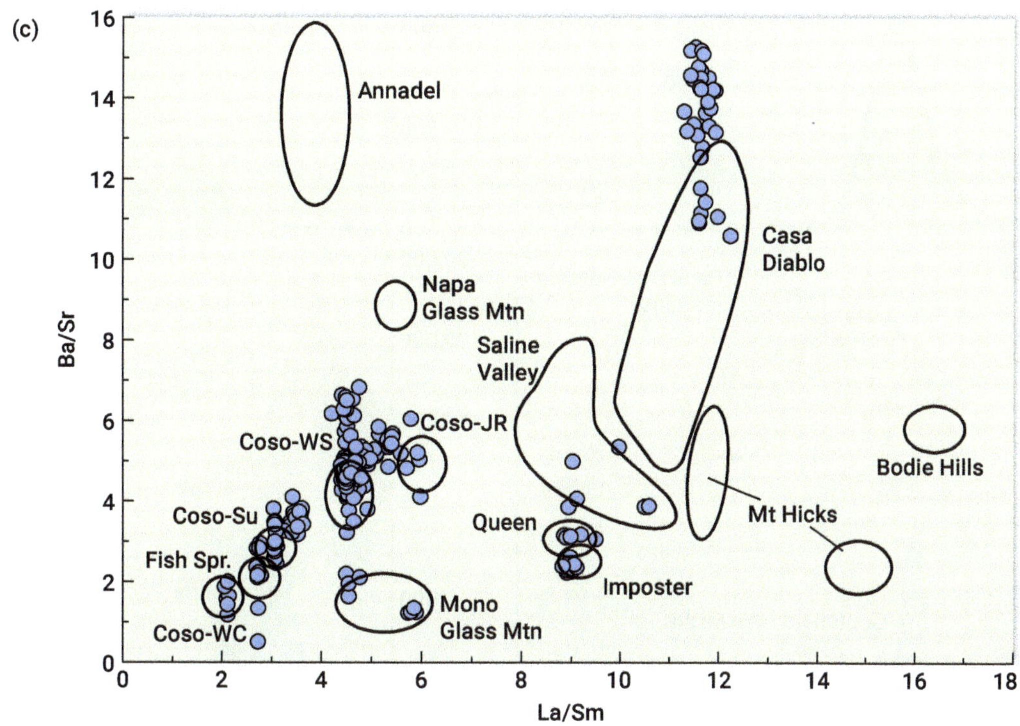

Figure 9.10 Continued

9.3 Trace Elements and Tectonic Setting

Learning Objectives

By the end of this section you should be able to:

- **9.3.1** Construct element-normalization diagrams to compare rocks from different tectonic settings.
- **9.3.2** Recognize typical characteristics of arc basalts, ocean-island basalts, and MORBs on such diagrams.

Back in Chapter 1, we noted that geologists long ago noted correlations between igneous rock compositions and geographic settings and saw that rock compositions correlate strongly with plate tectonic setting (see Figure 1.8). Aside from providing basic data for understanding how the Earth works, these correlations help to establish the provenance of terranes whose origin might be in doubt. For example, just as rocks of the ophiolite lithosystem (see Section 3.2), buried in a mountain range, tell a tale of a closed ocean basin, rocks of the arc lithosystem (see Section 6.2), spanning the range from basalt to rhyolite and gabbro to granite, might tell a story of an accreted island arc. Many of these characteristics (e.g., a wide range of SiO_2 in igneous rocks) survive metamorphism and deformation and thus can be used to untangle the geologic history of complexly deformed and metamorphosed regions.

Figure 9.11 reiterates some of the basic correlations between tectonic setting and bulk-rock geochemistry. Mid-ocean ridges and oceanic islands (see Chapter 14) show a tight range of SiO_2, overwhelmingly producing basalt, whereas arc rocks span the BADR spectrum. TiO_2, which behaves like a trace element in mantle melting unless there is a Ti-rich phase left behind during melting, tells a different story: oceanic ridges and arcs have low TiO_2, whereas ocean islands are much higher. As noted earlier, this is consistent with

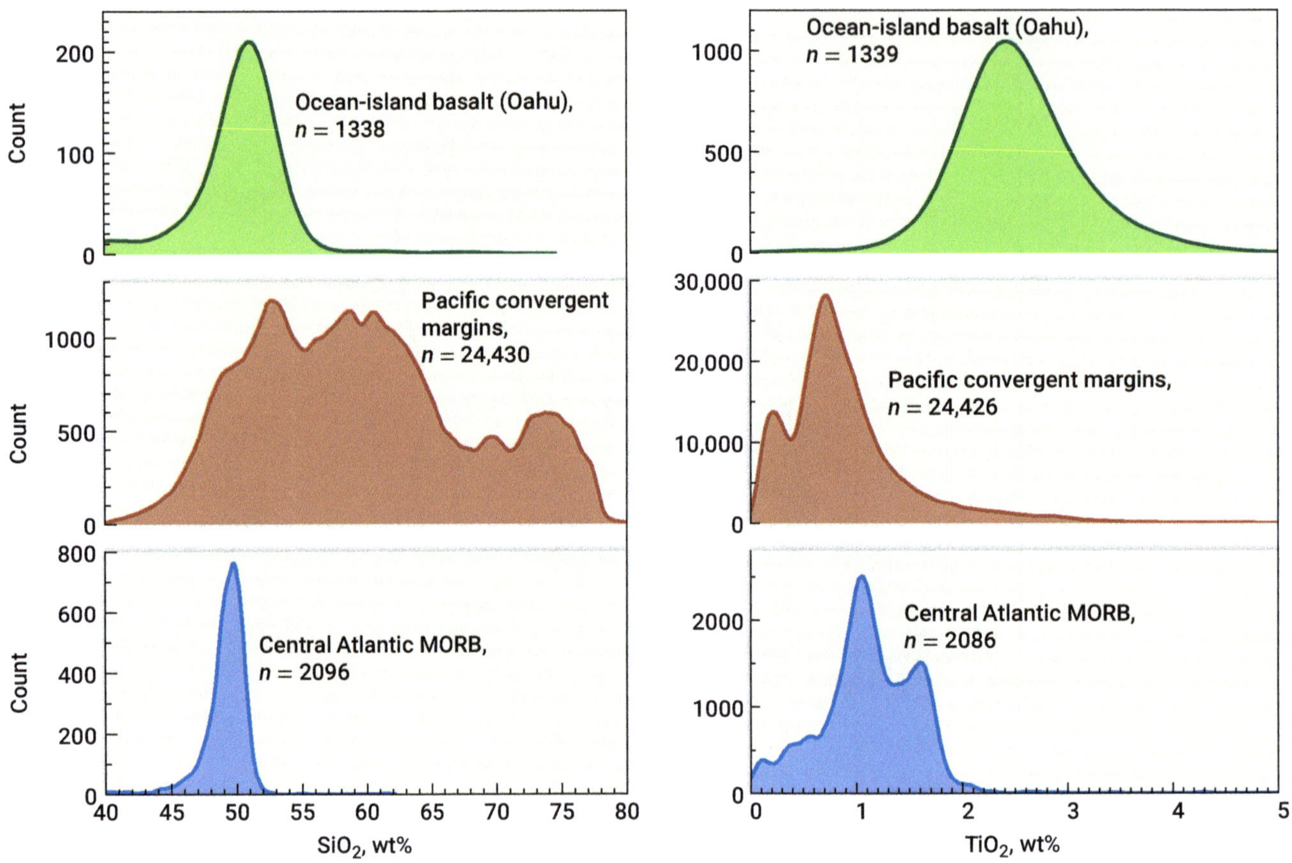

Figure 9.11 Smoothed histograms comparing distributions of SiO_2 and TiO_2 among datasets from oceanic islands (Hawaiian island of Oahu), arcs (circum-Pacific convergent margins), and mid-ocean ridges (data from central Atlantic).

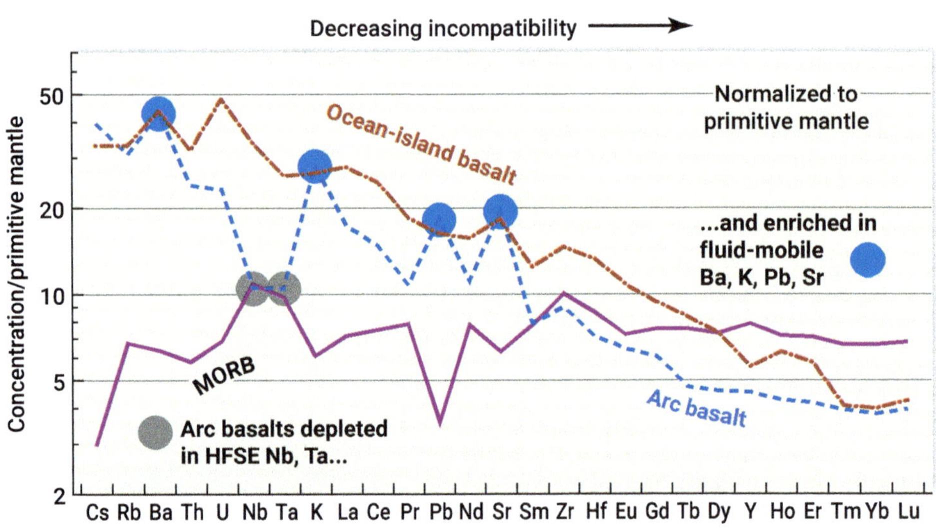

Figure 9.12 Element-normalization plot for average MORB, ocean-island basalt, and arc basalt, normalized to primitive mantle values. Key characteristics include that ocean-island basalts are generally enriched over the other types, with a smoothly decreasing pattern; MORB is quite depleted, especially of the most-incompatible elements; and arc basalts have spiky patterns, with positive spikes at fluid-mobile elements such as Ba, K, Pb, and Sr and a deep trough at Nb and Ta. Ti, not shown on this plot, shows depletion similar to that of Nb and Ta.

Sources: Data from White & Klein (2014), Kelemen et al. (2014), and Taylor & McLennan (2008); element order from White & Klein (2014); normalization data from Palme & O'Neill (2014).

ocean-island basalts being produced by lower degrees of partial melting—but the story is more complicated than that.

Trace-element patterns have long been used as fingerprints of specific tectonic settings. Specific patterns that are generally diagnostic of tectonic setting include the characteristic REE patterns of MORBs, which are usually low and convex upward (see Figure 9.9), and relative depletion of high field-strength elements (e.g., Ti, Nb, Ta) and enrichment in fluid-mobile elements (e.g., Sr, Ba, Pb) relative to other incompatible elements in arcs.

A common but imperfect tool for comparing trace-element patterns between samples is the element-normalization diagram (Figure 9.12).

These are sometimes called "spider diagrams," a misnomer that should be abandoned as that term is established for at least three other kinds of charts in common use. Such diagrams have not been standardized, but generally, they plot concentrations of incompatible elements, normalized to a standard reservoir, arranged in order of decreasing incompatibility.

Finding trace-element patterns that allow tectonic classification of a set of igneous rocks has long been a goal of geochemists. Rollinson (2013) devoted an entire chapter to various combinations of elements that have been used for this purpose. Although generally useful, the predictive power of these techniques is limited, and one should always benchmark them against datasets from known tectonic settings before use.

9.4 Isotopes and Petrogenesis

Learning Objectives

By the end of this section you should be able to:

9.4.1 Explain the various trajectories (bulk Earth, mafic crust, felsic crust, etc.) on a Sr evolution diagram.

9.4.2 Design a protocol for using $^{87}Sr/^{86}Sr$ measurements to test whether a Miocene granite body has assimilated or otherwise geochemically mixed with wall rocks of Proterozoic granitic gneiss.

9.4.3 Calculate the $^{87}Sr/^{86}Sr$ ratio in a magma formed by mixing two other magmas, given relevant data.

Sr Isotopic Evolution of the Crust

Radiogenic isotope ratios, particularly those of Sr, Nd, and Pb, are remarkably useful in studying petrogenesis, and development of these techniques in the 1970s drove a stake through the heart of many a cherished hypothesis about how magmas evolve. We have already used the $^{87}Sr/^{86}Sr$ system to determine the age of the solar system and to measure seawater contamination of MORB; here, we will use it to study magma sources, assimilation, and magma mixing.

Recall from Chapter 2 that the initial $^{87}Sr/^{86}Sr$, or $^{87}Sr/^{86}Sr_i$, of the bulk Earth is estimated at 0.699—the y-intercept of the chondrite isochron (see Figure 2.7). Recall from Section 3.8 that we can use the Rb (2.3 ppm) and Sr (7.25 ppm) abundances in chondrites to estimate the bulk $^{87}Sr/^{86}Sr$ of the Earth as about 0.756, a number that is, not surprisingly, similar to present-day $^{87}Sr/^{86}Sr$ in chondritic meteorites, but far higher than estimates of the bulk Earth's present-day $^{87}Sr/^{86}Sr$ of about 0.7045. MORBs are lower still, about 0.703 (see Figure 3.24). Why does the Earth have such a low $^{87}Sr/^{86}Sr$ relative to chondrites? The answer is most likely that Rb is more volatile than Sr (Lodders, 2003), and bodies in the solar system that accreted nearer to the Sun are more depleted in volatile elements than those that accreted farther out. Given that most meteorites seem to be derived from the asteroid belt, they should have higher Rb/Sr than the Earth.

Given the present-day bulk Earth estimate of about 0.7045, we can calculate that the bulk Earth Rb/Sr ratio is about 0.029, and that the bulk Earth's overall $^{87}Sr/^{86}Sr$ ratio has steadily increased over the past 4.56 Gyr in the manner shown in Figure 9.13.

The bulk Earth curve represents how the entire Earth has evolved, but various reservoirs in the Earth have their own evolution curves. In the grossest sense, we can divide the Earth into core, mantle, and crust and then ignore the core, because Rb and Sr are thought to be highly incompatible in it,[3] leaving the mantle and crust to compete for those elements. Now imagine for the purposes of argument (not because this is thought to be all that realistic) that the crust formed by partial melting of the mantle in one event 3.8 Ga.

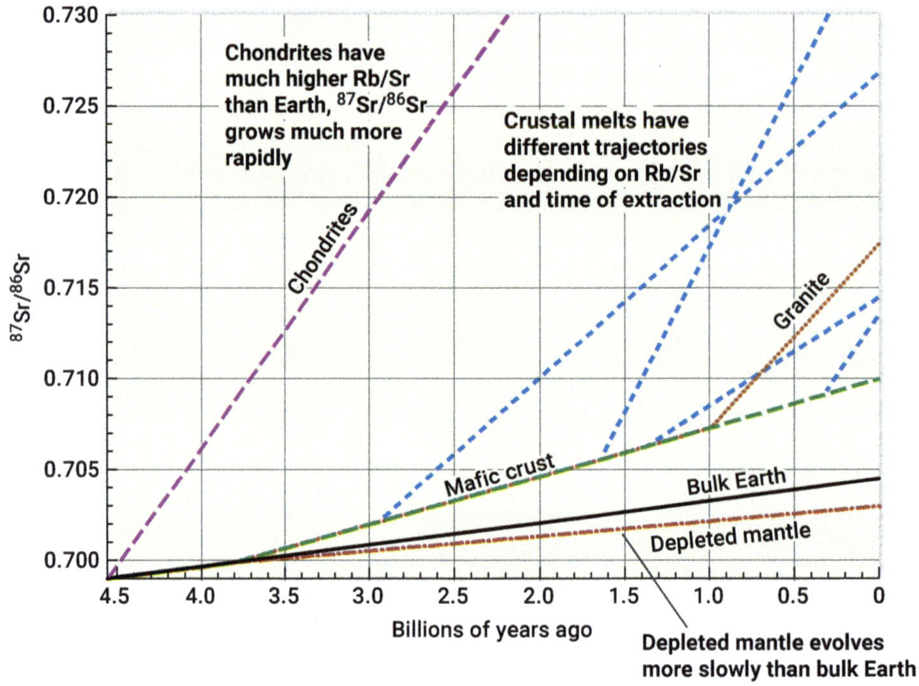

Figure 9.13 Schematic evolution of $^{87}Sr/^{86}Sr$ with time in various reservoirs in the Earth and outside it. The present-day ratio for the bulk Earth is estimated at ~0.7045, and the black curve gives its evolution from 0.699 when the Earth formed. In this simple thought experiment, there was a major event forming mafic crust 3.8 Ga, and the mafic crust, which has higher Rb/Sr because Rb is more incompatible in the mantle than Sr, evolves to a present-day ratio of 0.710. Because more Rb than Sr was extracted from the mantle during this event, the residual mantle has lower Rb/Sr and evolves along the depleted mantle curve to 0.703, a typical ratio of MORBs, which originate in this reservoir. The granite curve represents second-stage melting of the mafic crust at 1 Ga. This melt has a higher Rb/Sr than its parent for the same reason and quickly evolves to high $^{87}Sr/^{86}Sr$. Dashed blue lines show representative other second-stage melts, extracted at various times and with various Rb/Sr ratios.

How much Rb and Sr ended up in the crust during this event depends on their distribution coefficients (see Toolbox 9.3). Both are incompatible in mantle mineral assemblages (see Figure 9.3), but Rb is significantly more incompatible, because there is no comfortable place for a large, singly charged cation in mantle minerals.[4] Sr is also incompatible, but less so, because it has the same charge as and similar size to Ca, and therefore substitutes into Cpx.

As a result of their different incompatibilities, partial melts of the mantle have higher Rb/Sr than the mantle, and by Baldy's law, the residual mantle left behind after melting is left with lower Rb/Sr. As a result, the crust and mantle diverge from the bulk Earth; their slopes are given by their Rb/Sr ratios, or more specifically by the term $(^{87}Rb/^{86}Sr)\lambda$, where λ is the decay constant from Equation 2.3. The crust, with high Rb/Sr, evolves to high $^{87}Sr/^{86}Sr$, and the residual mantle, with Rb/Sr lower than the bulk Earth, evolves to $^{87}Sr/^{86}Sr$ lower than bulk Earth. This is why MORBs have such low $^{87}Sr/^{86}Sr$ (see Figure 3.24); they are produced by melting mantle that has had basalt extracted from it before.

A final note: the bulk Earth curve is not a linear curve, because it is based on the decidedly non-linear decay equation (see Equation 2.3). It sure looks linear, though. For an explanation, see Toolbox 9.4.

A key point of Earth evolution is that once elements are fractionated between continental crust and mantle, they stay that way, because continental crust is largely unsubductable.

Toolbox 9.3

Rb/Sr Ratios and Partial Melting

Calculating the Rb/Sr ratio of a partial melt of the mantle is an opportunity to test one's trace-element calculation skills. Consider the data in Table 9.1, which gives the mode and Rb and Sr contents of a hypothetical spinel peridotite. From the weight fractions of the four mineral phases and the listed *D* values, we can calculate the bulk distribution coefficient *B*; *B* for Sr is almost three times greater than that for Rb, which means that although both prefer the melt to the crystals, a little more Sr will be held back, leading to higher Rb/Sr in the melt. Using the equation for batch melting from the toolbox on trace-element equations (see Toolbox 9.1), we can calculate the Rb and Sr concentrations in the melt for any weight fraction of melting. For 5 and 20 wt% melting, the Rb/Sr weight ratio increases by 41 percent and 11 percent, respectively. Over geologic time, these lead to significant increases in $^{87}Sr/^{86}Sr$.

Table 9.1 Data and Results for Calculation of Rb/Sr in Partial Melts

	*D*s					Concentrations		
	Olivine	Cpx	Opx	Spinel	Bulk *D* (*B*)	Undepleted Upper Mantle	5 wt% Batch Melting	20 wt% Batch Melting
Wt. Fraction, *F*	0.6	0.25	0.1	0.05			0.05	0.2
Rb	0.01	0.026	0.02	0.01	0.015	0.60	9.34	2.83
Sr	0.01	0.13	0.04	0.01	0.043	20.00	220.14	85.32
Rb/Sr						0.030	0.042	0.033

Toolbox 9.4

Another Look at the Decay Equation

The plot of $^{87}Sr/^{86}Sr$ vs. time in Figure 9.13 looks like a straight line, yet it is based on the exponential decay equation $^{87}Sr = {}^{87}Sr_0 + {}^{87}Rb(e^{\lambda t} - 1)$. Why does it look straight? This is because for a small exponent (as is the case for geologic ages and decay constants, multiplied as λt), $e^{\lambda t} - 1 \approx \lambda t$.

This handy result comes from the Taylor series expansion (see your calculus book) for e^x:

$$e^x = 1 + x + \frac{x^2}{2!} + \frac{x^3}{3!} + \frac{x^4}{4!} + \ldots \quad \text{(Eq. 9.9)}$$

When x is a tiny number, then x^2 and higher powers are far tinier and can be ignored, and thus $e^x - 1 \approx x$. Let's try this. In the decay equation, the exponent x is λt. If we take the decay constant λ for ^{87}Rb from Table 2.2 (1.4×10^{-11}/yr) and multiply it by 10^9 yr, we get ~0.014. The Taylor expansion becomes

$$e^{\lambda t} = 1 + 0.014 + 0.014^2/2 + 0.014^3/6 + \ldots = 1.0141 \quad \text{(Eq. 9.10)}$$

This is quite close to the approximation 1.014 obtained by truncating the Taylor series after the *x* term.

With this result, the decay equation can be approximated as

$$\frac{^{87}Sr}{^{86}Sr} \approx \left(\frac{^{87}Sr}{^{86}Sr}\right)_0 + \left(\frac{^{87}Rb}{^{86}Sr}\lambda\right)t \quad \text{(Eq. 9.11)}$$

which is the equation of a straight line of the form $y = ax + b$ for fixed $(^{87}Sr/^{86}Sr)_0$ and $^{87}Rb/^{86}Sr$; hence the semilinear evolution of $^{87}Sr/^{86}Sr$ over the age of the Earth.

Incompatible elements such as K, Th, U, Rb, etc. are squirreled away in the continental crust, slowly drawing down the amount in the mantle. As noted earlier, it is an interesting coincidence that three of the aforementioned elements are the main producers of radioactive heat in the Earth. Except for minor recycling by delamination (see Section 9.4) and subduction erosion, the incompatible elements stay in the continental crust.

The model of crust formation 3.8 Ga (see Figure 9.13) is obviously a gross oversimplification of how the crust formed, but it is difficult to be precise here, because there is large disagreement in the community about when and how continental crust forms and grows (Condie, 2005; see Figure 2.14).

The relative incompatibilities of Rb and Sr are similar if the basaltic crust formed in our example is melted again to produce a more silicic magma. In Figure 9.13, we have extracted a granite from the prior mafic crust via hydrous partial melting (see Section 8.3), and this second-stage melt has much higher Rb/Sr than its source and therefore leaves the mafic crust evolution curve at a much steeper angle, producing a granite with a present-day $^{87}Sr/^{86}Sr$ of 0.7174. It is easy to envisage how melting events at various times, producing melts with a range of Rb/Sr ratios, can produce silicic rocks with a wide range of present-day $^{87}Sr/^{86}Sr$. What would the trajectory of a re-melt of the granite look like?

Isotopes and Open Systems

Isotope ratios such as $^{87}Sr/^{86}Sr$ provide important information about igneous diversification, because they are invariant during closed-system processes but change during open-system processes. The isotopes of Sr are assumed to be unfractionated by igneous processes, because the mass difference between ^{87}Sr and ^{86}Sr is small enough and temperatures are high enough (recall from Section 5.4 that fractionation factors decrease with increasing temperature) that the various phases in an igneous system express little preference for one or the other. That is, $^{87}Sr/^{86}Sr$ in crystals and coexisting melt, age-corrected to the time of the processes involved, should be the same. Therefore, there should be almost no variation in $^{87}Sr/^{86}Sr$ during closed-system processes such as crystal fractionation, partial melting, and liquid immiscibility. In contrast, during mixing or assimilation, isotope ratios will depend on the values in the end members involved. If they are different, then magmas formed by mixing or assimilation will have $^{87}Sr/^{86}Sr$ values intermediate between those of the end members.

Ratios of lighter isotopes such as $^{18}O/^{16}O$ can also be used to examine open-system processes, but their interpretation is more complicated, because the mass differences are large enough that the isotopes fractionate significantly at magmatic temperatures.

Case 1: Closed System, Primitive Source

Consider the data in Table 9.2. These analyses are from whole-rock samples of the Farrington pluton, a composite granite-granodiorite body in central North Carolina. Measured $^{87}Sr/^{86}Sr$ values cover a significant range, which could be a symptom of open-system processes such as magma mixing. However, if plotted on an isochron diagram (Figure 9.14a), they make a tolerably straight line, a potential isochron (see Section 2.3), whose slope indicates an age of about 500 Myr with large uncertainty. This age is consistent with the geologic history of the area (Hibbard et al., 2002), and therefore consistent with the interpretation that all sampled rocks formed about 500 Ma with the same $^{87}Sr/^{86}Sr = 0.7044$. The low initial $^{87}Sr/^{86}Sr$, similar to oceanic basalts, indicates that the granites formed from an isotopically unevolved source,

Table 9.2 Rb-Sr Data from the Farrington Pluton, North Carolina

Sample	Rb	Sr	$^{87}Rb/^{86}Sr$	$^{87}Sr/^{86}Sr$
F-11	57.8	182	0.9178	0.7097
F-15	63.8	227	0.8123	0.7094
F-32	63.8	122	1.5113	0.7144
F-41	3.0	917	0.0095	0.7049
F-42	26.9	718	0.1083	0.7030
F-62	98.0	443	0.6393	0.7080
F-64	68.8	506	0.3929	0.7070

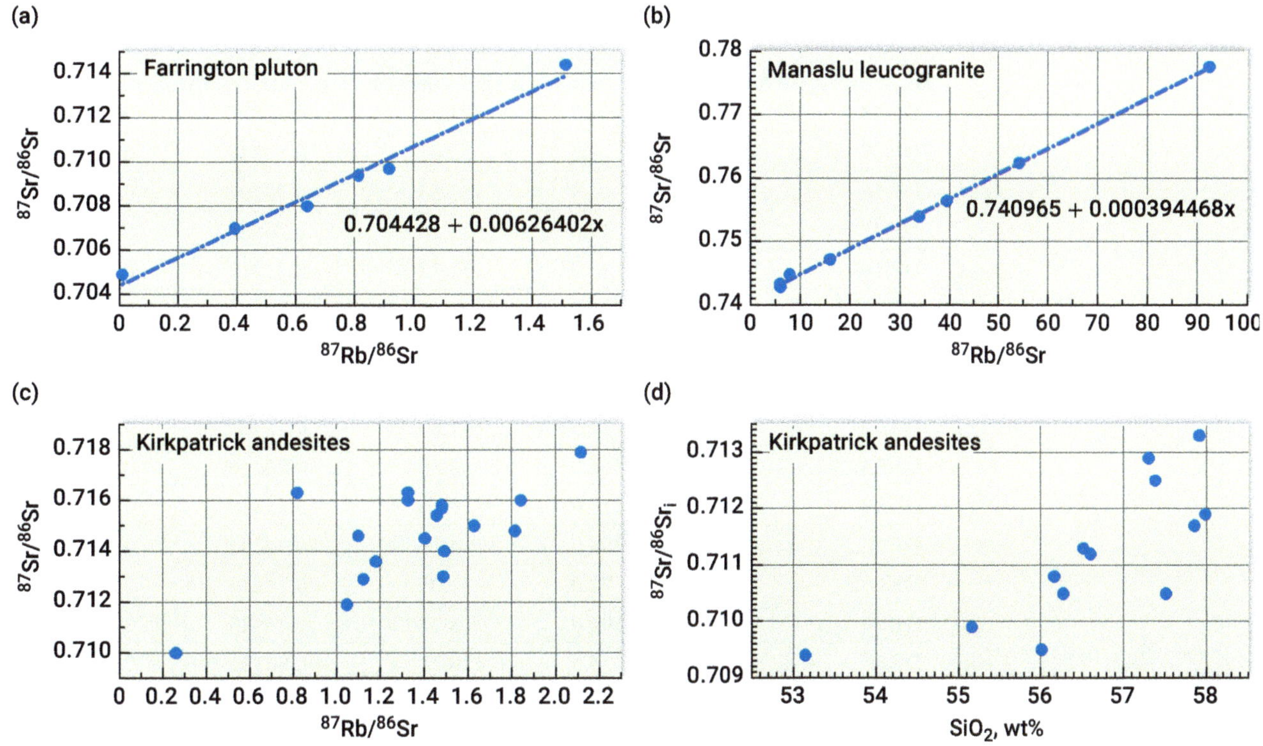

Figure 9.14 (a) Rb/Sr isochron of whole-rock granite and granodiorite samples from the Farrington pluton in North Carolina. The slope indicates crystallization ~500 Ma, and the low intercept indicates derivation from an isotopically primitive source. (b) Whole-rock data from the young Manaslu granite of the Himalaya show extremely high $^{87}Rb/^{86}Sr$ and $^{87}Sr/^{86}Sr$ and yield an isochron consistent with the age known from other methods. However, these data have an extremely high initial $^{87}Sr/^{86}Sr$, indicating that they were derived by partial melting of a source with extremely high $^{87}Sr/^{86}Sr$, such as an old silicic crust or sediments derived from same. (c) Isochron diagram for the Kirkpatrick lavas of Antarctica. The large degree of scatter shows that little age information can be gleaned. (d) Previous data for the Kirkpatrick lavas age-corrected to their known age. The scattered positive correlation with SiO_2 is consistent with open-system behavior in which basaltic magma was contaminated by mixing with a high-silica, higher-$^{87}Sr/^{86}Sr$ component to produce andesites. This pattern is solid evidence for either magma mixing or assimilation.

Sources: (a) Data from Fullagar (1971); (b) data from Deniel et al. (1987); (c) data from Faure et al. (1974).

such as by partial melting of basaltic material—in this case, from an island arc terrane accreted to eastern North America.

Case 2: Partial Melting of a Crustal Source Table 9.3 gives whole-rock data for a set of samples taken from the Manaslu leucogranite, one of a series of such young granites exposed in the high Himalaya. These data yield an isochron with an age of about 19 Myr (Figure 9.14b), consistent with the range of ages reported by other techniques (Harrison et al., 1999). However, initial $^{87}Sr/^{86}Sr$ is a whopping 0.741, far higher than the Farrington pluton example discussed earlier. These data clearly indicate that the granite was formed from a source with high $^{87}Sr/^{86}Sr$, and the most likely candidate is high-grade metasedimentary gneisses of the Tibetan Plate, above the Main Central Thrust of the Himalaya. Several melting processes have been proposed, including water-fluxed melting caused by dehydration reactions during prograde metamorphism (Le Fort et al., 1987) and shear heating by thrust faulting along the Main Central Thrust (Harrison et al., 1999).

Case 3: Assimilation or Magma Mixing Basalt and andesite lavas known as the Kirkpatrick Basalts were erupted during Jurassic rifting of Antarctica from Africa and are correlative with the widespread Karoo igneous province of southern Africa. They were erupted during a short interval about 182 Ma (Pálfy & Smith, 2000). If data from the lavas in Table 9.4 are plotted as an isochron, they yield a highly scattered array (Figure 9.14c) with

Table 9.3 Rb-Sr Data from the Manaslu Leucogranite, Nepal

Sample	Rb	Sr	$^{87}Rb/^{86}Sr$	$^{87}Sr/^{86}Sr$	$^{87}Sr/^{86}Sr_i$
DK43	326.0	8.0	121.92	0.77822	0.7452
DK45	514.5	4.5	333.81	0.83248	0.7421
DK46	346.5	6.5	150.93	0.78704	0.7462
DK47	333.0	12.0	79.26	0.76864	0.7472
DK48	380.5	10.0	113.72	0.77713	0.7463
DK49	399.5	6.5	183.98	0.79503	0.7452
DK5O	355.5	5.0	217.65	0.80201	0.7431
DK 51	387.0	66.0	17.03	0.75147	0.7469
DK52	330.0	82.5	11.60	0.74967	0.7465
DK53	293.0	146.5	5.80	0.74416	0.7426
DK54	279.0	216.0	3.76	0.74328	0.7423
DK55a	290.0	204.0	4.11	0.74292	0.7418
DK55b	274.5	72.0	11.03	0.74428	0.7413
DK56	297.0	104.0	8.27	0.74476	0.7425
DK57	347.5	4.5	218.21	0.80125	0.7422
DK58	289.0	4.0	95.49	0.77126	0.7454
DK59	270.5	65.0	12.07	0.74554	0.7423

Table 9.4 Rb-Sr and SiO_2 Data from the Kirkpatrick Lavas, Antarctica

Flow	Rb	Sr	$^{87}Rb/^{86}Sr$	$^{87}Sr/^{86}Sr$	$^{87}Sr/^{86}Sr_i$	wt% SiO_2
1	11.50	128.80	0.2572	0.7100	0.7093	53.14
2	68.60	135.00	1.4685	0.7156	0.7119	57.98
3	83.20	132.60	1.8133	0.7148	0.7102	57.51
4	69.20	94.60	2.1140	0.7179	0.7125	57.30
5	65.30	134.60	1.4021	0.7145	0.7109	56.51
6	54.90	138.60	1.1447	0.7160	0.7131	57.91
7	58.10	138.40	1.2132	0.7153	0.7122	57.38
8	67.30	130.50	1.4904	0.7140	0.7102	56.27
9	52.80	129.40	1.1792	0.7136	0.7106	56.16
10	45.20	120.40	1.0850	0.7124	0.7096	55.16
11	86.10	135.20	1.8405	0.7160	0.7113	57.85
12	65.60	127.60	1.4858	0.7130	0.7092	56.01

a poorly defined slope that is grossly consistent with the eruptive age. However, when corrected back to 182 Ma, the resulting $^{87}Sr/^{86}Sr_i$ values are still highly scattered, indicating that the system must have been open to material from outside—material with high $^{87}Sr/^{86}Sr$.

Plotting the age-corrected $^{87}Sr/^{86}Sr$ values against SiO_2 yields a scattered but clear positive correlation (Figure 9.14d). This is consistent with the hypothesis that the Kirkpatrick Basalts (which, in this dataset, are actually andesites) owe their geochemical variability at least in part to addition of silicic material with high $^{87}Sr/^{86}Sr$. We cannot tell from the data in Figure 9.14d what the nature of this component is, other than it has to be at or beyond the most contaminated sample, with $SiO_2 > 58$ wt% and initial $^{87}Sr/^{86}Sr > 0.7135$. Faure et al. (1974) used a variety of geochemical data to argue that the initial basaltic magma had $^{87}Sr/^{86}Sr = \sim 0.706$ and the contaminant $^{87}Sr/^{86}Sr = \sim 0.720$.

Toolbox 9.5

Behavior of Ratios during Mixing

Ratios of isotopes and elements mix differently than chemical parameters that are not ratios. For example, consider mixing of two magmas, A and B; A has 100 ppm Sr, and B has 300 ppm Sr. If we mix 50 g of each to make magma C, the mixture will have 200 ppm, the weighted average, of Sr. However, if A has $^{87}Sr/^{86}Sr = 0.706$ and B has $^{87}Sr/^{86}Sr = 0.708$, then the $^{87}Sr/^{86}Sr$ ratio of the resulting mixture will not be 0.707; it will be 0.7075. Do ratios violate Baldy's law? Of course not[5]; both ^{87}Sr and ^{86}Sr follow the dictates of mass balance, but their ratio does not, and the same is true for element ratios such Rb/Ba. Ratios have interesting and sometimes unpleasant mathematical properties, such as going to infinity when the denominator is small.

The arithmetic of mixing is as follows (Faure & Mensing, 2005). Consider a mixture of two components, A and B (e.g., two magmas, or a magma and a rock that is being assimilated). We can define a mixing parameter f as

$$f = \frac{A}{A+B} \quad \text{(Eq. 9.12)}$$

where m_A and m_B are the weights of the two components in the mixture. If X_A and X_B are the concentrations of element X in the two end members (weight units), the concentration of X in a mixture M is thus

$$X_M = fX_A + (1-f)X_B = f(X_A - X_B) + X_B \quad \text{(Eq. 9.13)}$$

The concentration of X in M is thus a linear function of the mixing parameter f. For an isotope ratio, however, the form of the mixing function is not linear. It becomes

$$^{87}Sr/^{86}Sr_{mix} = \frac{a}{[Sr]_{mix}} + b \quad \text{(Eq. 9.14)}$$

where

$$a = \frac{[Sr]_A[Sr]_B\left(^{87}Sr/^{86}Sr_B - {}^{87}Sr/^{86}Sr_A\right)}{[Sr]_A - [Sr]_B} \quad \text{(Eq. 9.15)}$$

and

$$b = \frac{[Sr]_A\,{}^{87}Sr/^{86}Sr_A - [Sr]_B\,{}^{87}Sr/^{86}Sr_B}{[Sr]_A - [Sr]_B} \quad \text{(Eq. 9.16)}$$

and square brackets indicate concentration. Well, that's complicated. However, the parameters a and b are constants for any given mixture, and so Equation 9.14 is just the equation of a hyperbola. If we plot the ratios against a new variable, $1/[Sr_{mix}]$, then it is transformed into a straight line.

Figure 9.15 shows mixing curves for two scenarios. In each case, component A has 200 ppm Sr, and $^{87}Sr/^{86}Sr = 0.704$. In both cases, component B has $^{87}Sr/^{86}Sr = 0.709$, but with different Sr concentrations. The mixing curves end up at $^{87}Sr/^{86}Sr = 0.709$ in both cases, but the paths they take are dramatically different.

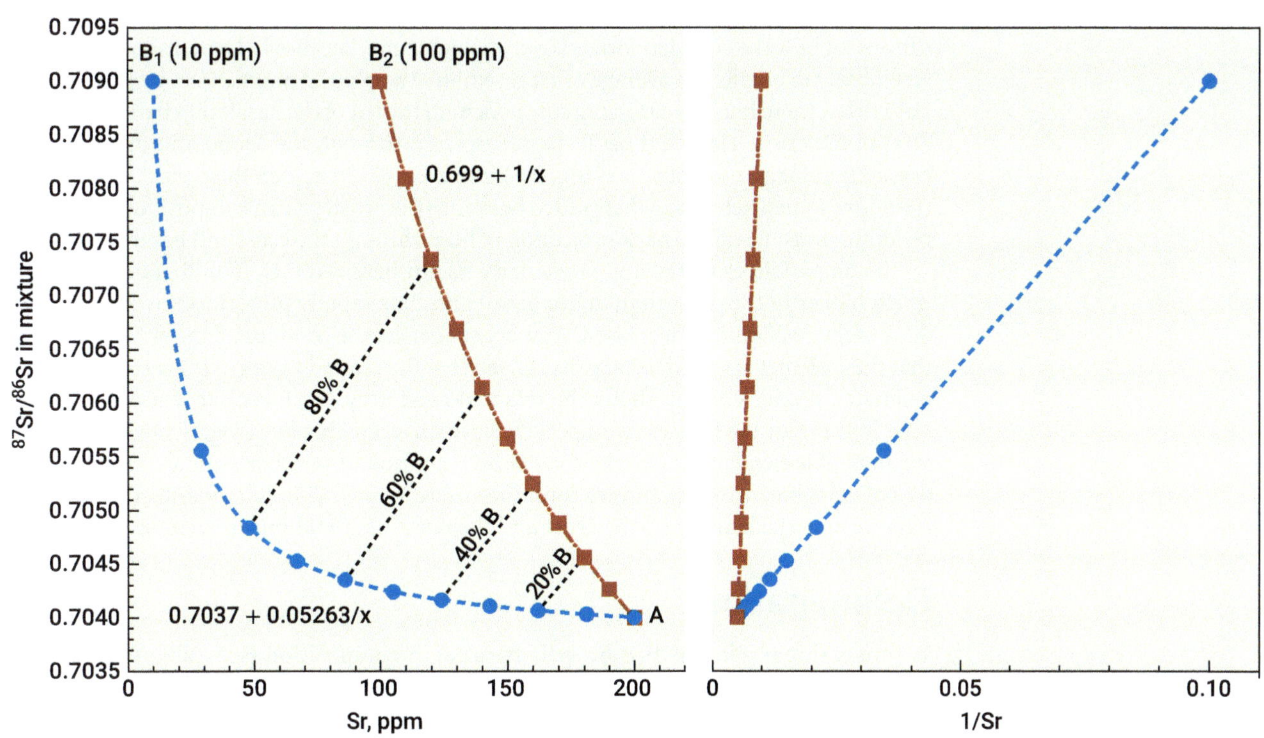

Figure 9.15 (left) Mixing curves for a sample A with 200 ppm Sr and $^{87}Sr/^{86}Sr = 0.704$ with two different B components: B_1 with $^{87}Sr/^{86}Sr = 0.709$ and 10 ppm Sr, and B_2 with $^{87}Sr/^{86}Sr = 0.709$ and 100 ppm Sr. (right) The same data with Sr recast as 1/Sr, turning the hyperbolic mixing curves into straight lines.

9.5 Mass Balance and the Crustal Paradox

Learning Objectives

By the end of this section you should be able to:

9.5.1 Critique various methods that have been used to estimate the average composition of the continental crust.

9.5.2 Illustrate the crustal paradox by plotting an incompatible element vs. SiO_2 or MgO for average arc basalt and average continental crust.

What Makes the Crust Isn't the Same as the Crust

Our current understanding of arc magmatism indicates that the primary magmas in subduction zones are predominantly basaltic, with about 50 wt% SiO_2, with an unknown, but probably fairly low, proportion of primitive andesitic magmas with about 58 wt% SiO_2 (see Chapter 6; Kelemen et al., 2014). Put another way, the net flux of material progressing upward through the Moho and adding mass to the crust has SiO_2 in the low to mid 50s. If this were the whole story of how the crust has grown over geologic time, then average crust would have similar SiO_2.

Bulk estimates of the composition of the crust, however, tell a quite different story. Early estimates came from averages of analyses of igneous rocks or of shales, which were presumed to be relatively unfractionated samples of ground-up crust (Clarke & Washington, 1924). Later estimates used averages of samples from various crustal levels combined in proportions dictated by seismic studies of crustal structure (Wedepohl, 1995; Rudnick & Gao, 2014). All these estimates tend toward andesitic compositions with about 60 wt% SiO_2, significantly higher than the basaltic flux thought to provide the primary materials for making the crust.

Figure 9.16 and Table 9.5 illustrate this paradox with estimates of the compositions of the bulk continental crust and of average continental arc basalts. Figure 9.16 has profound implications for how the crust has formed over geologic time, because if the main source of material resembles arc basalts but average crust resembles the plotted points, then by (you guessed it) Baldy's law there must be a missing component to the left of the basalt that balances the higher SiO_2 of the crust. That is, if basalt passes upward through the Moho and differentiates (by crystal fractionation, partial melting, etc.) into average crust, then there must be a more mafic residuum somewhere that, when added back into the crust, gives a composition of basalt. This residuum must lie on the back-projection of lines through the crust and basalt points.

Incompatible elements are particularly useful here, as K_2O, Th, Pb, and others are all much higher in the crust than in arc basalt and thus rapidly project to zero as the potential SiO_2 of the residuum is decreased. As a consequence, the residuum cannot have SiO_2 so low that these elements would be negative. Given that the plotted values of these elements are only estimates, no one element should be taken too seriously, but it is clear that a residuum with SiO_2 lower than the mid 40s would likely have unacceptably low (negative) values of various incompatible elements. If we take 45 wt% SiO_2 as a round-number estimate for the residuum, we get mass balance with, in very round terms, roughly one-half to one-third of the original basaltic material making up the crust and the remainder making up an unseen residuum.

Delamination

Where is this residuum? It is broadly basaltic in composition (see Table 9.5), with about 50 wt% normative plagioclase and the remainder equal parts Cpx and olivine—a good olivine gabbro. Recall from Figure 2.16 that gabbro can convert to eclogite under appropriate *P-T* conditions, and that eclogite is much denser owing to the conversion of lower-density plagioclase to higher-density garnet. There is a growing consensus that

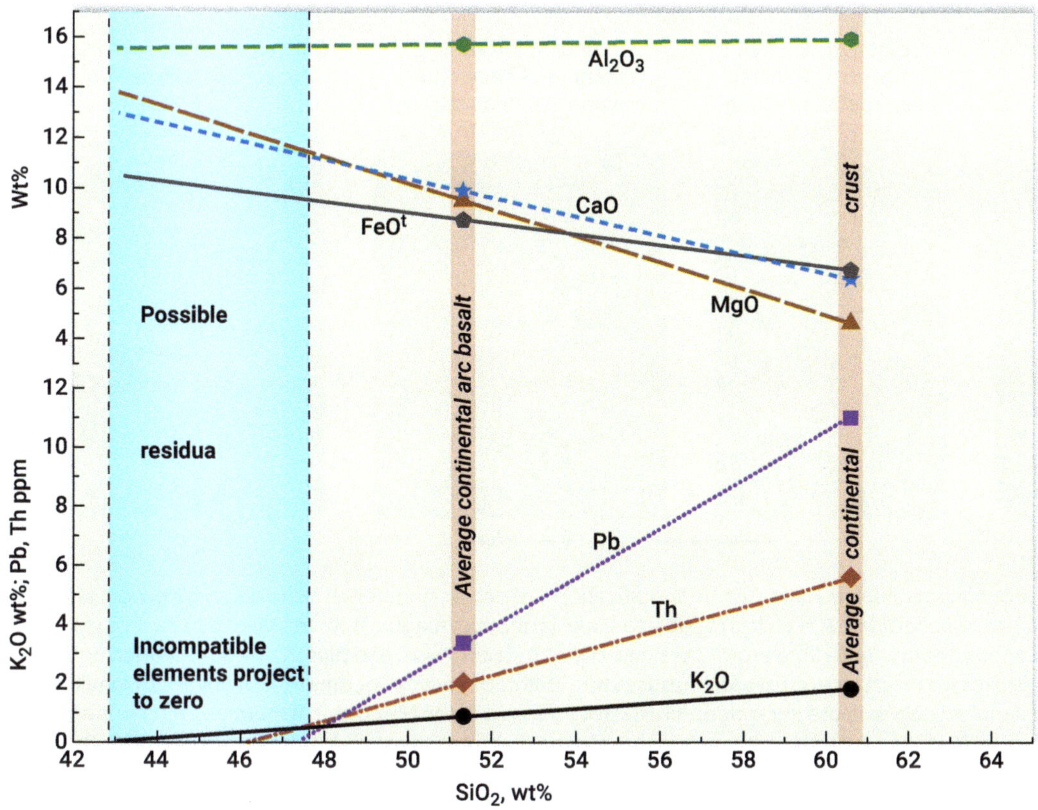

Figure 9.16 The crustal paradox. If the material passing through the Moho at arcs to build the continental crust is typical continental arc basalt with ~51 wt% SiO_2, and average continental crust has ~60 wt% SiO_2, then there must be a largely unseen reservoir of material with lower SiO_2 than arc basalt. Incompatible elements are much higher in average continental crust than in arc basalt, and a tie lie between the unseen residuum and average continental crust must go through arc basalt. As the value of these elements cannot be negative, the residuum must have ~43–47 wt% SiO_2 and be broadly gabbroic, and the lever rule indicates that this residuum should constitute the majority of the crust, because the bar for average continental arc basalt is closer to the residuum than to the bar for average continental crust. Where is this residuum?

Table 9.5 Estimates of the Average Continental Arc Basalt, Average Continental Crust, and a Residuum Needed to Mass-Balance the Crust

	Avg. Cont. Arc Basalt (Kelemen)	Avg. Cont. Crust (Rudnick & Gao)	Residuum at 45 wt% SiO_2	Residuum at 47 wt% SiO_2
SiO_2	51.3	60.6	45.0	47.0
TiO_2	1.0	0.7	1.2	1.1
Al_2O_3	15.7	15.9	15.6	15.6
FeOt	8.7	6.7	10.1	9.6
MgO	9.5	4.7	12.8	11.7
CaO	9.9	6.4	12.3	11.5
Na_2O	2.6	3.1	2.3	2.4
K_2O	0.9	1.8	0.3	0.5

Sources: Data from Kelemen et al. (2014) and Rudnick & Gao (2014).

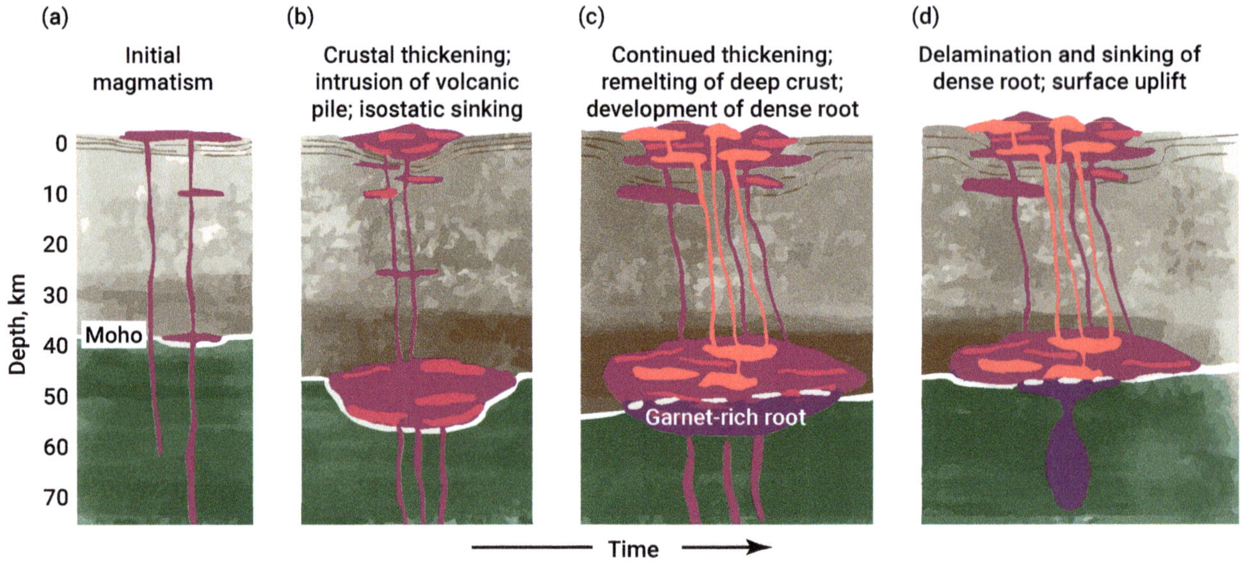

Figure 9.17 Schematic concept of delamination (or "deblobification") of dense, garnet-rich arc roots. Time progresses left to right and covers tens of millions of years; sketch is roughly to scale. (a) Input of magma from the mantle produces volcanic rocks and plutons at various depths. (b) As time progresses, the volcanic stack grows, and plutons intrude, producing a magmatic bulge that depresses the shallow crust and produces the "bathtub syncline" geometry of many arcs (Hamilton & Myers, 1974; see Chapter 10). Growth of a thick, deep-crustal root depresses the Moho, and if thermal conditions and bulk composition are suitable (c), this root may grow garnet-rich rocks at the expense of plagioclase (e.g., eclogite, garnet gabbro). Such rocks can be denser than underlying peridotite and may drip into the mantle (d), leading to shallowing of the Moho (Zandt et al., 2004). By this process, a crust that is andesitic in overall composition can be generated by input of basalt. Thin lines in upper crust are schematic crustal markers; white line is Moho; dashed white line is Moho located within rocks that were generated in the crust.

gabbroic residuum in the lower crust commonly converts to garnet-bearing, dense rock and is recycled back into the mantle by a process known as delamination (Figure 9.17; Arndt & Goldstein, 1989; Kay & Kay, 1993; Kelemen et al., 2014). Consistent with this, the lower 25 percent of a thick arc crustal section exposed in the Himalaya (Kohistan arc; Jagoutz & Schmidt, 2012) comprises garnet-bearing rocks and pyroxenites that are likely denser than upper-mantle peridotite.

The Sierra Nevada of California, a tall mountain range that lacks a deep, low-density root, provides an example of a region that may have recently undergone delamination. There, dense lower-crustal, garnet-bearing xenoliths occur in Miocene and Pliocene volcanic rocks that perforated the batholith (Dodge et al., 1986; Mukhopadhyay & Manton, 1994; Ducea & Saleeby, 1996), and a blob in the mantle with faster-than-normal P-wave velocities has been interpreted as the missing delaminated material (Jones et al., 1994). Consistent with late Cenozoic delamination, a brief burst of high-K volcanism occurred in and around the highest topography in the range at 3.5 Ma (Manley et al., 2000), and a rapid uptick in the rate of river downcutting occurred at about the same time (Stock et al., 2004; Jones et al., 2004). Taken together, these phenomena suggest that the high elevation of the range is being held up dynamically by hot, low-density mantle replacing the dense, delaminated root (Ducea & Saleeby, 1996).

9.6 "Evolution" in Igneous Systems

Learning Objective

By the end of this section you should be able to:

9.6.1 Critique the use of the term "evolution" in describing how igneous rocks form.

The word "evolution" has a long history in igneous petrology. As with Linnaean taxonomy (see Chapter 1), petrologists of a century ago borrowed the concept of evolution as ever-forward transition from a parent magma (basaltic) to the other magma types seen on Earth. Indeed, this concept was central to the most influential book ever written on igneous petrology, *The Evolution of the Igneous Rocks* (Bowen, 1928).

The concept of evolution and the assumption that rock compositions proceed in an orderly fashion from more mafic to more felsic compositions is deeply ingrained in petrology, but there are many instances where it is a confusing concept. Under the Bowen paradigm, an andesite is more evolved than a basalt, and a dacite is more evolved than an andesite. However, if the andesite was formed by mixing basalt and dacite, which is more evolved? By silica content, the dacite, but by parentage, the andesite, which is the child of the two parent magmas. As used today, "more evolved" generally means "more silica-rich."

The concept of evolution can be particularly confusing when applied to isotope ratios. "Isotopically evolved" generally means "more crustal looking" (i.e., higher $^{87}Sr/^{86}Sr$). This makes sense because the parent element, Rb, is more incompatible than the child, Sr, and is therefore relatively enriched in the crust relative to the mantle. However, crustal rocks have lower $^{143}Nd/^{144}Nd$ than the mantle, because the parent element, Sm, is more compatible than the child, Nd (see Toolbox 13.2). Thus, crustal $^{143}Nd/^{144}Nd$ ratios tend to be lower than those of the mantle, and by evolutionary terminology, these lower $^{143}Nd/^{144}Nd$ values would be considered more "evolved."

Given the confusion associated with the term, and its implication of a particular process (fractional crystallization), it is probably time to retire the term "evolution" when applied to igneous rocks.

Summary

Key points of this chapter include the following:

- Trace elements (TEs) are low-abundance constituents of minerals that substitute for essential structural constituents (ESCs; e.g., Sr substituting for Ca in plagioclase).
- A TE is classified as compatible in a given mineral if its concentration is higher in the mineral than in coexisting melt, and as incompatible if the converse is true.
- Compatibility depends on how close the size and charge of the TE are to an ESC; coupled substitutions (e.g., BaAl for KSi in alkali feldspar) are also possible.
- Most TEs are incompatible in most minerals.
- TEs can be grouped into several categories, including large-ion lithophile elements (LILEs), transition metals, rare earth elements (REEs), and high-field-strength elements (HFSEs).
- Distribution coefficients are the ratio of the concentration of a TE in a mineral to that in the melt.
- Concentration of a TE in the melt varies non-linearly with *F*, the fraction of melt in the system, and depends on whether the process is equilibrium or fractional.
- Eu anomalies in REE patterns occur because Eu is the only REE that can occur in 2+ charge, and Eu^{2+} is compatible in plagioclase.
- TEs, and in particular incompatible elements, vary with tectonic setting (e.g., MORBs have low concentrations of REEs, whereas arc basalts are enriched in fluid-mobile elements such as Ba, K, and Pb).
- Radiogenic isotope systems (e.g., $^{87}Sr/^{86}Sr$) can be used to test petrogenetic hypotheses, because closed-system fractional crystallization should not produce any changes in these ratios, whereas open-system processes such as magma mixing will if the end members have different isotopic ratios.
- Average continental crust is significantly more silica-rich (andesitic) than the typical arc basalts that are thought to be the primary material produced in subduction zones. This is a paradox.
- One solution to the paradox is delamination of a dense, garnet-rich root in arc systems.

Notes

1. We are pretending here that Hf does not exist; HfO_2 makes up 1–2 wt% of typical igneous zircon.
2. A consequence of its electron configuration; Eu^{2+} has a half-filled *f* shell, which adds a little stability.
3. Based on size-charge arguments; once again, remember that we've never actually sampled the core to measure this.
4. Some parts of the mantle contain phlogopite (Mg-rich biotite), which loves Rb ($D = \sim5$).
5. As they say, it's not just a good idea, it's the law.

CHAPTER 10

Plutons and How They Form

Intrusive bodies known as plutons are fundamental building blocks of the Earth's crust, and most of the crust that is not composed of plutons comprises recycled plutons—metamorphic and sedimentary rocks that have been processed through the rock cycle, perhaps several times, but have plutons at the roots of their family tree. Plutonic rocks commonly form tough, sparsely jointed bodies that stand up well to glacial erosion and resist propagation of faults, resulting in spectacular scenery such as that in Patagonia and in Yosemite, Kings Canyon, and North Cascades National Parks. Indeed, the boundaries of Yosemite were drawn mostly where granitic rocks give way to less scenic metamorphic and volcanic rocks.

The term "pluton" was coined by Cloos (1928) as a non-specific term for what we now call intrusive bodies; the term is generally linked to Pluto, the Roman god of the underworld and its mineral wealth. During the controversial 2006 meeting of the International Astronomical Union, during which Pluto was demoted from being a planet, a committee proposed that bodies such as Pluto be referred to as "plutons." Geologists objected, and the astronomers dropped that plan (Brumfiel, 2006).

In this chapter, we focus on the great granite, granodiorite, and tonalite plutons that are produced at subduction zones, make up the great batholiths of the world, and are parental in some sense to most of the Earth's continental crust. For simplicity, we will refer to these rocks as "granitic," realizing that their compositions may range far from true granite. Layered mafic intrusions, which form a fascinating minor branch of pluton geology, are treated in Chapter 14.

10.1 Plutons, Batholiths, and Intrusive Suites

Learning Objectives

By the end of this section you should be able to:

10.1.1 Use terms relevant to plutons correctly (e.g., pluton, stock, batholith, wall rock, screen, etc.).

10.1.2 Use terms relevant to rocks of the arc batholith lithosystem correctly (e.g., granite, granodiorite, diorite, gabbro, etc.).

Spectacular cliffs of Torres del Paine in Patagonian Chile are carved in granite sills. These were intruded into marine sedimentary rocks (dark brown) ~12.5 Ma, and contact-metamorphic assemblages in the wall rocks indicate intrusion at 2–3 km depth. Granitic rocks can be particularly resistant to glacial erosion and form some of the world's tallest cliffs (see Figure 10.1). These cliffs are ~1 km tall and the peaks rise ~2 km above glacial lakes at their base.

Source: AFG photo.

What Is a Pluton?

Over the past century, the term "pluton" has been used in multiple ways, and there is no consensus on what it means in spite of its importance in igneous petrology. Although plutons have been interpreted as the frozen remains of big magma chambers at least since Lyell (1839), several observations discussed in this chapter run counter to this simple idea. For example, many are highly composite, constructed with magmas from different sources, and many were intruded in small batches over millions of years without a pluton-sized magma chamber ever existing.

In this book we use the following definition:

> **Pluton:** A body of intrusive rock analogous to a sedimentary formation that is sufficiently large and recognizable by texture and mineralogy to form a mappable unit. It may have formed over a short or long time interval and may be composed of material from multiple sources.

Wall rocks, country rocks
The rocks into which plutonic rocks are intruded.

Screens, septa
Relatively thin panels of wall rocks within a pluton or separating two plutons.

Stock
General term for a small (on the order of a few kilometers in diameter), generally equant and isolated pluton.

The second sentence is deliberately vague enough to not be restrictive of plutonic timescales. "Pluton" is thus a generic term for any mappable intrusive body, but in geologic parlance, it is used to refer to bodies larger than, say, outcrop scale; dikes, sills, outcrop-sized blobs, and other small or thin bits of intrusive rock are typically not referred to as plutons.

The rocks intruded by a pluton are referred to as the **wall rocks** or **country rocks**, and plutons are locally separated by thin, sheet-like panels of wall rocks known as **screens** or **septa**. Plutons range from small bodies 1 or 2 km across to huge tracts of land 50 km or more in length. The term **stock** is commonly used for plutons on the smaller side, less than 100 km^2 in outcrop area.

Plutonic rocks make up much of the Earth's most dramatic landscapes. The spectacular cliffs of El Capitan and Half Dome in Yosemite Valley, California, tower more than 1 km above the valley floor, and one can easily walk to the foot of El Capitan and stare up at the vertical to overhanging face (Figure 10.1a,b), which exposes eight different plutonic rock units in intimate association (Putnam et al., 2015). Nearby Half Dome, conversely, is

(a) (b) (c)

Figure 10.1 (a) The cliff of El Capitan in Yosemite Valley, California, towers nearly 1100 m above the floor of the valley. It includes eight or more distinct, mappable plutonic units. The *Nose* route ascends the left edge, and *Dawn Wall* ascends the blank face to its right. (b) At the other end of Yosemite Valley, Half Dome rises nearly 1500 m above the valley floor; the sheer cliff is over 600 m tall. All the rock in this photo is Cretaceous granodiorite, and most of it belongs to one map unit, the Half Dome Granodiorite. Both El Capitan and Half Dome are part of the Sierra Nevada batholith. (c) Great Trango Tower (6286 m; left) and Nameless Spire (6251 m; right), composed of Late Miocene leucogranite, along the Baltoro glacier in the Pakistan Karakoram. Vertical heights of the cliffs range up to 1900 m.

Sources: (a, b) AFG photos; (c) courtesy of Mike Searle.

carved entirely in one homogenous body of granodiorite. Even taller cliffs carved in granite occur in remote areas of the Himalaya (Figure 10.1c).

What Is a Batholith?

A **batholith** is a large region, typically hundreds of kilometers in length and tens of kilometers in width, composed of numerous plutons that are related in time and space. Several massive batholiths make up large parts of the lithosphere of western North America, including the Sierra Nevada and Peninsular Ranges batholiths of California and Baja California and the Coast Ranges batholith of western Canada (Figure 10.2). Similar batholiths continue down the western margin of the Americas to Patagonia.

Batholith
A region tens of kilometers or more in lateral extent that is composed primarily of plutons.

What Is an Intrusive Suite?

Collections of plutons that are closely related in time and space are locally grouped into **intrusive suites**. One of the best-studied of these is the Tuolumne Intrusive Suite of California (Figure 10.3), which comprises four major mappable plutonic units and a great deal of internal complication (Bateman & Chappell, 1979; Coleman et al., 2012). This is one of several such intrusive suites along the crest of the Sierra Nevada Range, all of the same general age and all consisting of the same general sequence of rock types and textures; whatever the Late Cretaceous arc was doing to make these suites, it did it in the same

Intrusive suites
Groups of plutons related in time, space, outcrop pattern, and fabric; typically nested.

Figure 10.2 Mesozoic plutonic rocks in western North America. Groups of plutons that are related in time and space are commonly grouped into batholiths, and several major ones are shown.

Source: Plutonic rock outcrops were extracted from *Generalized Geological Map of the World and Linked Databases, 1995, Geological Survey of Canada Open-File Report 2915d.*

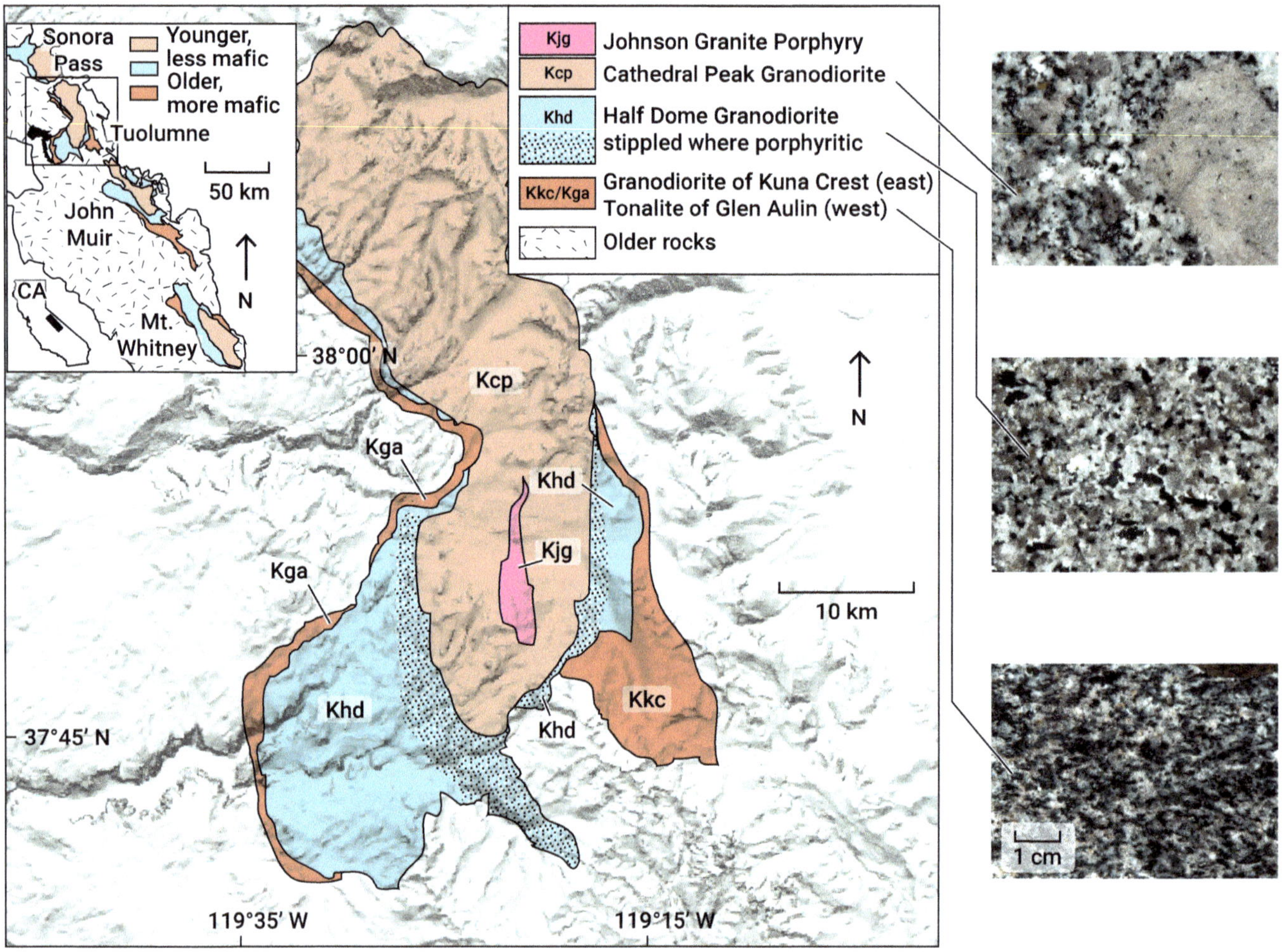

Figure 10.3 A classic example of an intrusive suite, the Tuolumne Intrusive Suite, Yosemite National Park, California. This suite comprises three major plutonic sequences that form a nested set that becomes progressively younger from the outside-in, ranging from ~94 Ma at the outer margin to ~85 Ma at the center. It is one of several such nested suites that occur along the highest part of the Sierra Nevada Range (inset); another occurs to the south of the inset map, and at least one more occurs in western Nevada, off the north end of the inset map. In each suite, the older, outer units are mafic granodiorites, succeeded inward by equigranular biotite-hornblende granodiorites that are in turn succeeded inward by biotite granodiorite with conspicuous large (>5 cm, locally 10 cm or more) K-feldspar phenocrysts. Contacts between these units are generally gradational over 10–100 m. Whatever process produced this sequence was repeatable in space and time up and down the Late Cretaceous arc over a strike length of 600 km. Box in inset shows area of main map.

Sources: Map after Huber et al. (1989); AFG photos.

repeatable manner up and down the arc at about the same time at centers that are about 150 km apart. It is possible that these are the eroded roots of stratovolcano complexes. The plutonic units of the Tuolumne Intrusive Suite were intruded continuously over several million years.

The great southeast face of El Capitan in Yosemite Valley (see Figure 10.1a) displays a dramatically different arrangement of plutonic units (Figure 10.4). El Capitan occurs at the overlap between two intrusive suites, and the cliff exposes seven mappable plutonic units, many of them dike swarms, in its more than 1 km^2 area (Putnam et al., 2015). The contrast with the neighboring Tuolumne Intrusive Suite shows some of the variety of contact relationships that plutons exhibit.

Figure 10.4 Geologic map of the kilometer-tall southeast face of El Capitan (see Figure 10.1a). The cliff exposes portions of two large granite bodies and a series of dike swarms; compositions range from mafic diorite to high-silica granite, and all, save perhaps thin cross-cutting aplite and pegmatite dikes, were intruded during a 4-Myr interval at the end of the Early Cretaceous. Much of the climbing variety on this iconic cliff arises from these different plutonic units. At the base of the cliff, just beyond the road, lies a massive rock avalanche deposit, 3600 years old, whose source has been traced to the upper part of the cliff by matching boulder compositions to the geologic map.

Source: Background photo with map overlay from Putnam et al. (2015).

Rocks of the Arc Batholith Lithosystem

Not all plutons formed above subduction zones, but the majority of the Earth's post-Archean continental plutons did, and these form much of the raw material for metamorphic processes that formed the continents (see Chapter 12). This lithosystem comprises extensive bodies composed of numerous plutons dominated by granodiorite, granite, tonalite, and related rocks, along with associated metavolcanic rocks of the arc volcanic lithosystem that ended up in the plutonic realm owing to tectonic or magmatic burial. Chief components include the following:

- *Granite, granodiorite, diorite, gabbro*: The plutonic equivalents of the BADR suite. In batholiths, the more silicic members (granodiorite and granite) are volumetrically dominant (see Figure 10.27); in volcanic regions, the opposite is true (see Figure 10.46).
- *Tonalite* (low-K granodiorite): Tonalite and other low-K variants are locally abundant.
- *Metavolcanic rocks*: Metamorphosed arc volcanic rocks that are locally common in batholiths, typically as screens bordering individual plutons.
- *Metasedimentary rocks*: Contact-metamorphosed rocks into which the plutons were intruded, also as screens.
- *Older batholithic rocks*: Many of the batholiths along the western margin of the Americas were active sporadically during intervals approaching 200 Myr, so older plutons typically make up the wall rocks of younger arc plutons.

10.2 Field Aspects of Plutonic Rocks

Learning Objectives

By the end of this section you should be able to:

10.2.1 Distinguish different types of intrusions based on textural and field relations.
10.2.2 Sketch field relationships relevant to intrusion of dikes, sills, and plutons.
10.2.3 Use terms related to dikes, sills, and plutons correctly.

Textures

Plutonic rocks by definition cool and crystallize below the Earth's surface, and consequently, they are typically coarser-grained and more fully crystallized than volcanic rocks. Differences in magma cooling and degassing history can produce dramatic differences in crystal size and abundance, yielding rocks ranging from obsidian to pumice to ash to stony rhyolite to granite, and profound geologic interpretations can sometimes be made with just a glance at a rock. The textures of extrusive and intrusive rocks are completely gradational, and the deliberately vague term **hypabyssal** is used to describe shallowly intruded rocks that do not have fully plutonic textures.

Plutonic rocks, where most crystals are large enough to see by eye, are typically described as being either fine-grained (typical crystals <1 mm across), medium-grained (1–5 mm), or coarse-grained (>5 mm). These terms can only be loose guidelines, because most igneous rocks have crystal sizes that range over several orders of magnitude and are typically log-normally distributed. Nevertheless, these terms are useful for a rough characterization of crystal size.

Hypabyssal
Intrusive rock emplaced at shallow (undefined) depth.

Sill
A dike that is parallel to foliation or bedding in host rocks, regardless of the orientation of that fabric.

Sheet
Here, used to denote relatively flat-lying dikes or sills.

Aplite
Fine- to medium-grained leucogranite, typically described as having a sugary texture.

Pegmatite
Coarse-grained to extremely coarse-grained bodies, usually of quartz and feldspars; muscovite, tourmaline, and other unusual minerals may be prominent.

Leucogranite
Light-colored granitic rock composed of quartz and feldspars with SiO_2 near 77 wt% and no more than a few percent mafic minerals; intrusive equivalent of high-silica rhyolite.

Dike swarms
Large numbers (typically hundreds or thousands) of subparallel dikes that are associated spatially and temporally.

Dikes and Sills

General Field Aspects

When the magma flux supplying a planar conduit ebbs enough that magma freezes, the frozen magma forms a dike (Figure 10.5); if the dike is intruded parallel to layering in the host rocks, it is concordant and called a **sill**. This terminology can be confusing, because the term "dike" connotes a planar, steep feature, such as the dike that Hans, the Little Dutch Boy, plugged with his finger, and "sill" connotes a planar, flat feature, such as a windowsill. However, a dike intruded parallel to steep layering or foliation is still a sill. It might be best to consider sills as special cases of dikes and leave it at that. **Sheet** is commonly used to refer to flat or gently dipping panels of magma intruded into rocks that lack a planar fabric. **Aplite** and **pegmatite** dikes are fine- to medium-grained and coarse- to extremely coarse-grained **leucogranite** dikes, respectively, that form prominent stripes across many plutons.

Dike Swarms

Dikes typically occur in sets known as **dike swarms** that may include tens of thousands of dikes and cover continent-scale areas. For example, the giant, 1.27-Ga Mackenzie dike swarm (Figure 10.6a) radiates from a source in northern Canada almost to Lake Superior, covering a distance of over 3000 km and an area roughly equivalent to that of the state of Alaska. The end-Triassic Central Atlantic Magmatic Province formed during the breakup of Pangea and radiates from a point between Florida, Guyana, and Senegal when the continents are reconstructed (Figure 10.6b), covering an area roughly the size of South America and a radial distance of over 2500 km. Dikes and sills from this swarm occur along the entire eastern seaboard of the United States; the Palisades Sill (see Section 9.2) is part of it. Not all large dike swarms are continental in scale; the Jurassic Independence dike swarm of eastern California runs parallel to the continental margin for ~600 km, and the radial dike swarm of Spanish Peaks (see Figure 7.16) only ranges for ~10 km. Large dike swarms are associated with large igneous provinces and hotspots; more about those in Chapter 14.

Figure 10.5 Dikes in the field. (a) Silicic dike (aplite) cutting granodiorite, Yosemite National Park, California. (b) Multiple dark mafic dikes, andesitic in composition, cutting granodiorite along the John Muir Trail in California. Dikes average ~30 cm thick. (c) Basalt dike, 50 cm thick, cutting layers of basaltic tephra capped by dike-fed lava flows, Reykjanes Peninsula, Iceland. This dike and the flows that it fed erupted about 800 years ago. (d) Devil's Backbone, a radial andesite dike in the stratovolcano that erupted to form Crater Lake. The dike rises from the lake to the caldera rim 390 m above and is ~10 m thick. (e) Dense swarm of Cretaceous aplite and pegmatite sheets cutting Cretaceous diorite near Pine Lake, Sierra Nevada, California. Thicker sheets in foreground are 2–3 m thick.

Source: AFG photos.

We noted in Chapter 7 that dikes typically propagate horizontally. This is certainly true for huge dike swarms and is likely common in many swarms. Ernst and Baragar (1992) showed that magma flow patterns revealed by magnetic studies indicate that the Mackenzie swarm magmas flowed vertically within about 500 km of the center point of the swarm, and horizontally at farther distances. They speculated that the 500-km radius traces out the impact zone of a mantle plume that generated the swarm (see Chapter 14).

Thickness Dike swarms are tempting targets for statistical studies, and measurements from thousands of dikes in numerous swarms around the world show that dike widths typically follow a distribution that is approximately log-normal in thickness—that is, the logarithms of thickness measurements are approximately normally distributed (Figure 10.7; e.g., Jolly & Sanderson, 1995; Delaney & Gartner, 1997; Glazner et al., 2008). Why this should be is not well understood, but it is a consistent feature of dike swarms. Further complicating description of dike swarms is the observation that individual dikes are

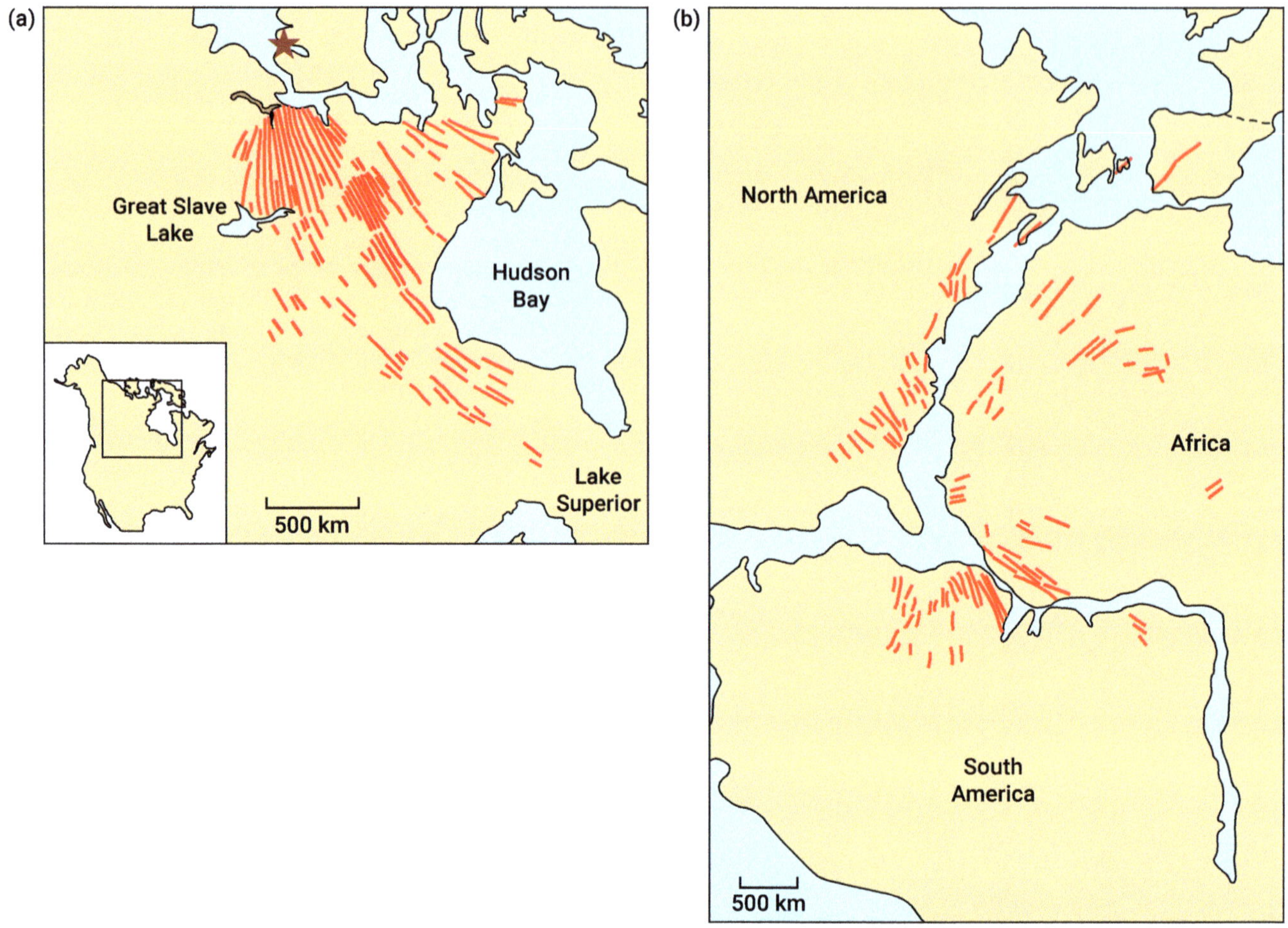

Figure 10.6 Schematic representations of two major dike swarms. (a) Map of the 1.27-Ga Mackenzie dike swarm. Red star marks the approximate focal point for this activity. (b) The Central Atlantic magmatic province (CAMP) shown on a reconstruction of the Central Atlantic region at about 200 Ma, prior to Early Jurassic continental breakup. The focal point was around what is now southern Florida.

Sources: (a) Unpublished diagram courtesy of K. L. Buchan and R. E. Ernst (Buchan & Ernst, 2004); (b) from Marzoli et al. (2014), © 2013 Elsevier B.V.

Dikelets
Here, a component of a composite dike interpreted to represent a single pulse of magma; dikelets are commonly seen to be split in half by later pulses.

typically composed of numerous injections (**dikelets**), and further complicating that is the observation that many panels within dikes are half-injections that were split by later injections (Figure 10.8a).

The highly non-normal (in a statistical sense) distribution of dike widths means that the standard and commonly reported arithmetic mean is a misleading measure of characteristic dike width, complicating description and cataloging of dike swarm characteristics. This is apparent in Figure 10.7; the arithmetic mean is over two times greater than the mode (the most common value). The practical effect is that in the field, you might walk across dozens of dikes that are about 1 m thick, and then one that is 20 m thick, and that 20-m dike contributes 20 times as much as a single 1-m dike to the mean. In such distributions, the geometric mean (n^{th} root of their product), mode, and median are more intuitive measures of dike width.

Energy of Dike Intrusion

Intrusion of dikes involves significant energy, because the wall rocks have to move to accommodate magma injection. This is a complex topic that touches on petrology, seismology, and geodesy. Julian (1983) showed that magnitude-6 earthquakes during a 1980 swarm in Long Valley caldera of eastern California could have been generated by injection of dikes on the order of a few meters thick. Using the same method, we can calculate that injecting a dike 20 m thick involves energy equivalent to an earthquake with a magnitude of about 8. As the world's great earthquakes are demonstrably tectonic in origin, rather than magmatic, one might ask if dikes that are tens of meters thick, so common in the geologic record, really record one-shot opening of tens of meters. Close examination of well-exposed dikes says that the answer is likely "no," as many thick dikes represent injection of multiple pulses of magma.

Composite Dikes In areas of excellent exposure, such as recently glaciated terrain or quarry walls (Figure 10.8), dikes are commonly found to comprise multiple dikelets whose margins are distinguished by grain-size differences (e.g., chilled margins) or thin screens of wall rock (Daly, 1914). In the Independence dike swarm, typical dikelets are about 10 cm thick, roughly 10 to 20 percent of typical dike thickness. The widespread presence of multiple dikelets in dikes indicates that dikes are preferred pathways for magma injection; that is, new injections in a given area seem to favor the pathways of prior injections rather than breaking new pathways for themselves. Dikelets commonly have a range of chemical compositions (Glazner et al., 2008a), and Benton et al. (2011) used chemical analyses to show that composite dikes of the Independence swarm in two localities formed by each new injection splitting the previous one. This produces a mirror pattern, as in Figure 10.8a, where all dikelets but the central one are halves of a single injection.

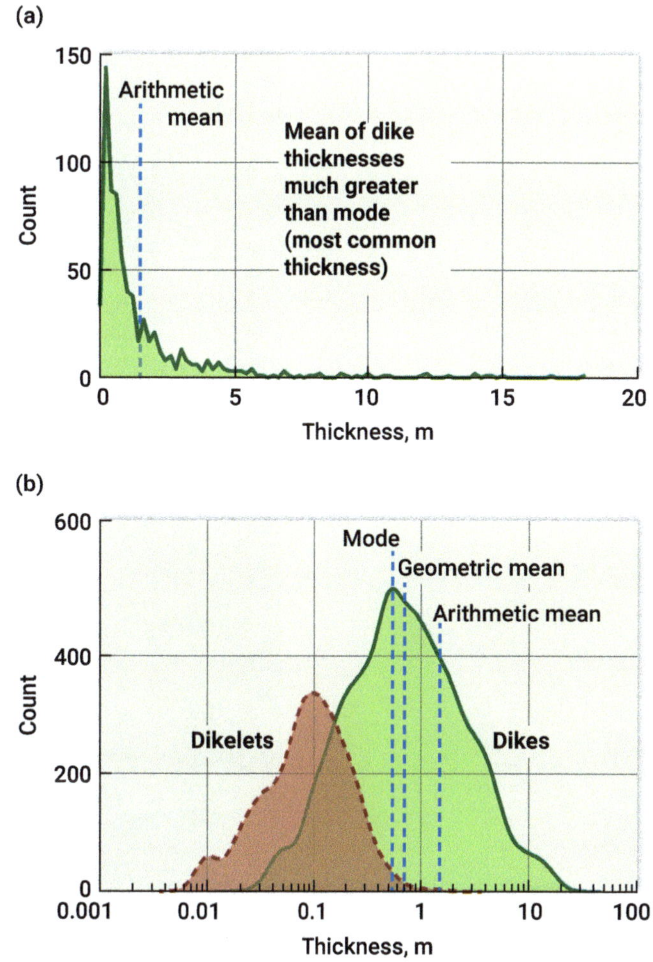

Figure 10.7 Thickness distribution of 705 dikes from the Independence swarm of eastern California. Many dike swarms worldwide have similar size distributions. (a) Histogram of thicknesses showing that most dikes are on the order of ~1 m thick, but thicknesses of individual dikes range up to nearly 20 m. The arithmetic average is difficult to discern from the distribution and is less useful than the mean of data drawn from a normal distribution. (b) Histogram of $\log_{10}$ of thickness in meters of dikes (green) and dikelets (brown). For the dikes, the mode is 0.5 m, the geometric mean is 0.7 m, and the arithmetic mean is 1.5 m—a misleading measure of typical dike width. Mode of dikelet distribution is 0.1 m.

Source: Data from Glazner et al. (2008a).

Orientation Swarms of parallel dikes owe their common orientation to something, and frequently, this is the ambient stress field; mechanical theory and analog experiments demonstrate that dikes commonly intrude perpendicular to the least compressive stress. Dikes can thus be used as paleostress indicators (e.g., Feraud & Campredon, 1983; Best, 1988), with the understanding that local stress deviations and anisotropy of the host rock can deflect dike orientations. For example, schistose rocks have a strong anisotropy that can guide injection of magma, dikes may preferentially follow preexisting fractures in a rock, and strong layers can act as a barrier to propagating dikes. We have already seen how radial stresses generated by the topographic load of a volcano can generate radial dike swarms (see Section 7.5); in contrast, far-field stresses generally control the orientations of parallel dikes.

Dikes in the Independence swarm of California strike uniformly northwest-southeast, with few exceptions. Most of these dikes were intruded about 148 Ma, but these are interspersed with both older (165-Ma) and younger (~90-Ma) parallel dikes, and in eastern California dikes as young as Miocene commonly have a northwest strike. It is probably not coincidental that these dikes were intruded parallel to concurrent subduction zones at the continental margin to the west, as these would have formed free boundaries that could balance extension caused by dike intrusion.

Giant radiating dike swarms such as those in Figure 10.6 appear parallel at the local scale but radiate at the continental scale, and their orientation is presumed to be controlled

Figure 10.8 Typical composite dikes. (a) Composite diorite dike in which the initial injection (1) was split by a second (2) whose boundaries (dashed in part for emphasis) are marked by chilled margins and color changes. Compass circle is 5 cm in diameter. (b) Cliff ~4 m tall exposing sheets of pegmatite dikes whose boundaries are locally marked by screens of the gray shale that hosts them. (c) Composite aplite dike that trapped a thin screen of granodiorite wall rock; segments that are separated in the foreground by granodiorite merge in the distance. (d) Multiply composite dike with thin, partially melted screens of granodiorite wall rock separating separate injections. Note that granodiorite screen on left terminates toward the bottom of the photo, and that a thin (3-mm) screen a few centimeters to the right of the compass body marks the boundary between two dikelets.

Source: AFG photos; locations: (a, c, d) Sierra Nevada, California; (b) Alta, Utah.

by tangential stretching of lithosphere lifted up by hotspots (Ernst & Buchan, 2001; see Chapter 14). Giant radiating dike swarms have been identified on Mars and Venus, and at least one on Mars is as laterally extensive as the Mackenzie swarm on Earth (Ernst & Buchan, 2001).

Diabasic texture
Texture characteristic of shallow basalt intrusions in which laths of plagioclase are encased in much larger crystals of pyroxene or olivine.

Poikilitic
Texture in which multiple smaller crystals of one phase are encased in a much larger crystal of another.

Dike Textures

Dikes typically have textures that are characteristic of hypabyssal rocks. A particularly common one is the texture of shallowly intruded basaltic magma, which produces a rock that is called diabase in the United States and dolerite in other English-speaking countries. In the **diabasic texture** characteristic of relatively slowly cooled basaltic magmas, laths of plagioclase are enclosed in large crystals of pyroxene or olivine, a texture quite different from extruded basalt flows unless they are particularly thick. The resulting texture is termed **poikilitic** (Figure 10.9).

Aplite and Pegmatite Dikes Most granitic rocks are cut by late-stage silicic dikes of aplite and pegmatite. In some areas, these make up several percent of the volume of the host rock (see Figure 10.5a,e and Figure 10.8b,c). Aplite dikes are generally fine-grained leucogranites with 1 percent or less mafic minerals and high-silica rhyolite compositions, with SiO_2 generally in the range 76 to 78 wt%, although recrystallized aplites may be medium- or even coarse-grained (Glazner et al., 2020).

Pegmatites are coarse-grained (typical grain size >1 cm) to exceptionally coarse-grained dikes composed mostly of quartz, albite, and K-feldspar (Figure 10.10). Many pegmatites contain large crystals of unusual minerals such as tourmaline, beryl, topaz, and spodumene (a lithium pyroxene). Pegmatites and aplites commonly occupy the same set of fractures and commonly share the same fractures, typically with pegmatite forming the core of an aplite dike (Figure 10.10c). How pegmatites form is still a subject of debate; unlike aplite dikes, pegmatites are locally of great economic importance owing to high concentrations of important metals such as Be and Li (London & Morgan, 2012).

Figure 10.9 Typical diabasic texture in a diabase sill from the Karoo province of South Africa, showing large colorful crystals of Cpx enclosing elongate laths of plagioclase. Such texture is characteristic of basaltic magma intruded at shallow depth. Width of field 2 mm.

Source: AFG photo.

General Field Aspects of Sills Flat-lying sills are naturally more difficult to see in the field than steep dikes, because they are only well-exposed on steep surfaces, such as those in Figure 10.11. Sill thicknesses vary dramatically but are more difficult to quantify than dikes. Sills up to several hundred meters in thickness are known (Button & Cawthorn, 2015), but as with dikes, sills are log-normally distributed in thickness, and especially thick ones are much less common than thinner ones. Various compilations of sill thicknesses show that modes are typically in the 10-m range (Button & Cawthorn, 2015; Coetzee & Kisters, 2016), roughly an order of magnitude greater than modes of dike thicknesses.

Saucer-Shaped Sills Some sills have upturned edges that give them the appearance of giant dinner plates (Figure 10.12). Such **saucer-shaped sills** formed by the thousands in the Karoo Basin of South Africa when basaltic magma invaded Permo-Triassic sedimentary rocks and cooled into diabase. Myriad sills connected by myriad

Saucer-shaped sills
Large sills with upturned edges.

(a) (b) (c)

Figure 10.10 (a) Typical pegmatite dike cutting granodiorite, with crystals of quartz and feldspar several centimeters across. Finger for scale. (b) Meter-scale pegmatite dikes with quartz and feldspar crystals 5 cm or more across cutting water-polished schist along the Colorado River, Grand Canyon, Arizona. Dikes and schist are Proterozoic. (c) Multiple generations of aplite dikes, Yosemite National Park. Older generations (e.g., either side of 5-cm compass circle) are recrystallized to medium grain size, comparable to that in host granodiorite, and have fuzzy margins. The youngest dike is finer-grained, has sharp margins, and is cored by pegmatite with open cavities. See Glazner et al. (2020) for more examples.

Source: AFG photos.

Figure 10.11 (a) Engineer Mountain, north of Durango, Colorado, is capped by an Oligocene or Miocene, light-colored, intermediate-to-silicic sill that was intruded into Permian sedimentary rocks. (b) The Neoproterozoic Purcell Sill exposed in Glacier National Park, Montana. The 30-m-thick diorite sill, which intrudes Neoproterozoic sedimentary rocks, is the dark stripe surrounded by bleached zones of contact metamorphism. (c) White granite sills of Miocene age intruding dark gray schistose rocks on Elba Island, Italy. Boats for scale. (d) Sill of dark gray Oligocene plagioclase-hornblende diorite in the Henry Mountains, Utah. Sill was intruded parallel to bedding in red Jurassic sandstone; a thin band of light-colored sandstone is visible above the columnar-jointed cliff, which is ~10 m tall.

Sources: (a, c, d) AFG photos; (b) courtesy of Marli Miller, geologypics.com.

Laccoliths
Magmatic blisters intruded into flat-lying sedimentary rocks, with a convex roof and a relatively flat floor; a laccolith may be faulted along part of its margin.

Figure 10.12 Aerial view looking northeast of the Golden Valley Sill, a typical saucer-shaped sill in the Karoo Basin, central South Africa. Edges of the sill dip inward, forming a flat-floored, agricultural valley. The leaf-shaped sill is ~10 km wide, 20 km long, and 100–150 m thick in outcrop, with maximum relief at the rim of 700 m. The village of Tarkastad is on the lower right.

Source: Google Earth.

dikes (Galerne et al., 2011; Muirhead et al., 2014) transported thousands of cubic kilometers of magma to shallow levels and onto the Earth's surface in the Karoo-Ferrar large igneous province (see Chapter 14)—the same system that produced the giant Ferrar sills in Antarctica. Such sills, widespread in the marine subsurface in the North Atlantic and off the northwestern Australian coast, are of great interest, because they are associated with deposits of oil and gas, forming traps and affecting hydrocarbon maturation (Polteau et al., 2008). Intrusion of massive amounts of magma into hydrocarbon-bearing sediments also likely liberated large amounts of carbon dioxide from the subsurface, possibly initiating the Paleocene-Eocene thermal maximum climatic event (Svensen et al., 2007).

Laccoliths and Lopoliths

Laccoliths are sills with a domical upper surface—blister-like bodies that make space for themselves by arching the rocks above them. The Henry Mountains of central Utah (Figure 10.13), described by G. K. Gilbert (1877) in a

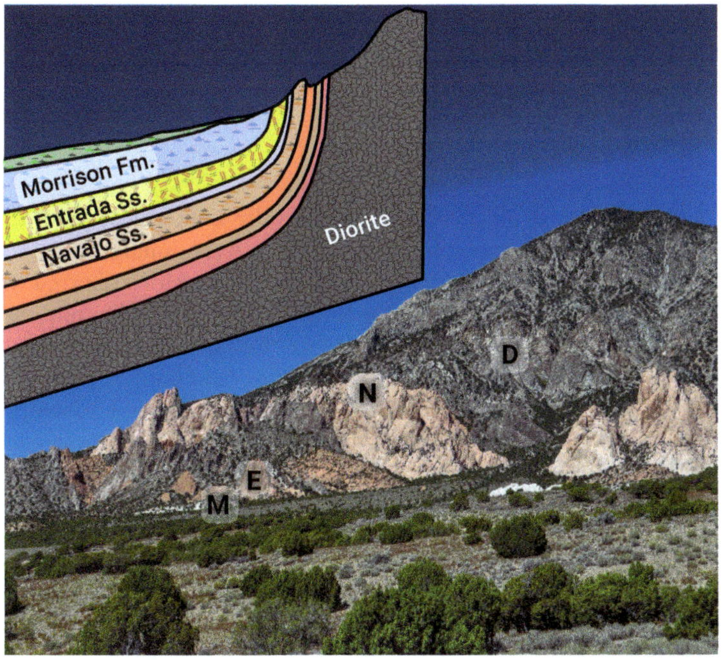

Figure 10.13 South flank of Mt. Hillers, a classic laccolith in the Henry Mountains of Utah. The range comprises six or more diorite blisters in the otherwise relatively flat surface of the Colorado Plateau, here comprising flat-lying, red and gray, Triassic to Cretaceous sedimentary rocks. Cretaceous shales under shrubbery in the foreground are underlain by massive pink and red Jurassic and Triassic sandstones at depth. These strata were originally flatly continuous across the area now occupied by the laccolith and are visible at the base of Mt. Hillers, where they have been punched up by the diorite (D), as shown by cross section. The pink Navajo Sandstone (N) of Triassic-Jurassic age forms massive cliffs and buttes in Zion, Capitol Reef, and Canyonlands National Parks; the Jurassic Entrada Sandstone (E) forms some of the most prominent arches in Arches National Park; and the Jurassic Morrison Formation (M) is famous for its dinosaur fossils.

Sources: AFG photo; cross section from Jackson & Noller (1991).

masterful monograph and by Hunt (1953) in another, are classic examples of this intrusive category. They comprise six separate diorite bulges that pushed up formerly flat-lying sandstones and shales of the Colorado Plateau in the late Oligocene and early Miocene.

Laccoliths appear to form in two stages: first a thin sill invades the enclosing strata, and then further inflation jacks up the roof to form the characteristic arched top (Corry, 1988; Jackson & Pollard, 1988; Rocchi et al., 2002). The flat base is usually inferred rather than observed, and it is possible that many laccoliths have depressed bases, as in Figure 10.14b. In the Henry Mountains, geologic and paleomagnetic studies show that early, horizontally intruded sills were rotated to steep dips by inflation of later pulses that were injected below the initial sills (Jackson & Pollard, 1988).

Lopoliths are intrusive bodies shaped somewhat like upside-down laccoliths, with flattish roofs and downwarped floors (Figure 10.14b). Downwarping of strata beneath individual plutons or entire batholiths is common (e.g., Hamilton & Myers, 1974; Glazner

Lopoliths
Magmatic blisters intruded into flat-lying sedimentary rocks, with a flat roof and a concave floor; may be faulted along part of its margin.

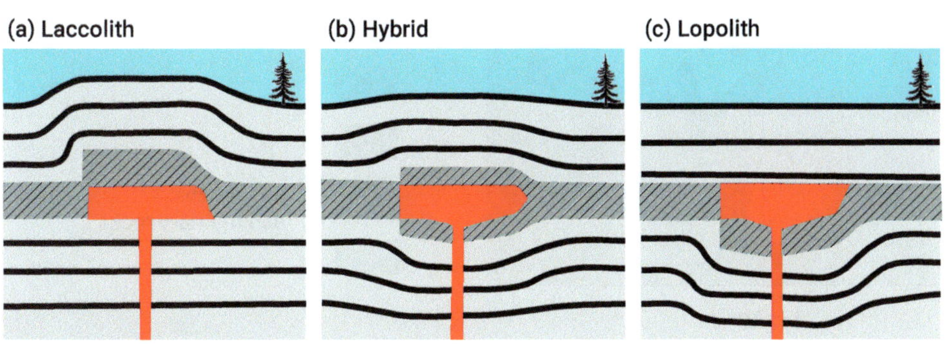

Figure 10.14 Schematic representation of emplacement of (a) laccoliths, (b) hybrid geometries, and (c) lopoliths. In the simplest sense, laccoliths raise the roof and lopoliths depress the floor, although hybrids involving both processes are common. In laccoliths, space is made by bulging up the surface and displacing the air; in lopoliths, space is made by pushing rocks below the pluton out of the way. In these schematic examples, the intrusion is shown with a fault boundary on the left and a fold on the right, as both processes occur around such intrusions. Tree indicates surface and is not necessarily to scale.

Source: Modified from Cruden (1998), Geological Society of London.

& Miller, 1997), and Cruden and McCaffrey (2001) have argued that downwarping and floor subsidence is a common space-making process for plutons. We return to this topic in Section 10.7.

Plutons

Concordant
Said of a geologic contact that is parallel to bedding or foliation.

Discordant
Said of a geologic contact that cuts across bedding or foliation.

Practical Usage of the Term
A classic pluton is generally somewhat equant in map view, a few kilometers to tens of kilometers across, and relatively homogeneous in composition, but the sizes, shapes, and petrographic variability of intrusions referred to as plutons vary dramatically. Figure 10.15 shows geologic maps of three Mesozoic plutons from a small area in eastern California. All were intruded into the same sequence of Paleozoic and Neoproterozoic sandstones, shales, and limestones, but their contact relationships vary from **concordant**, where wall-rock bedding or foliation is parallel to and wraps around the pluton (Figure 10.15b,c), to **discordant**, where wall-rock markers appear to be sliced off at the pluton contact (Figure 10.15a). The Sage Hen Flat and Papoose Flat

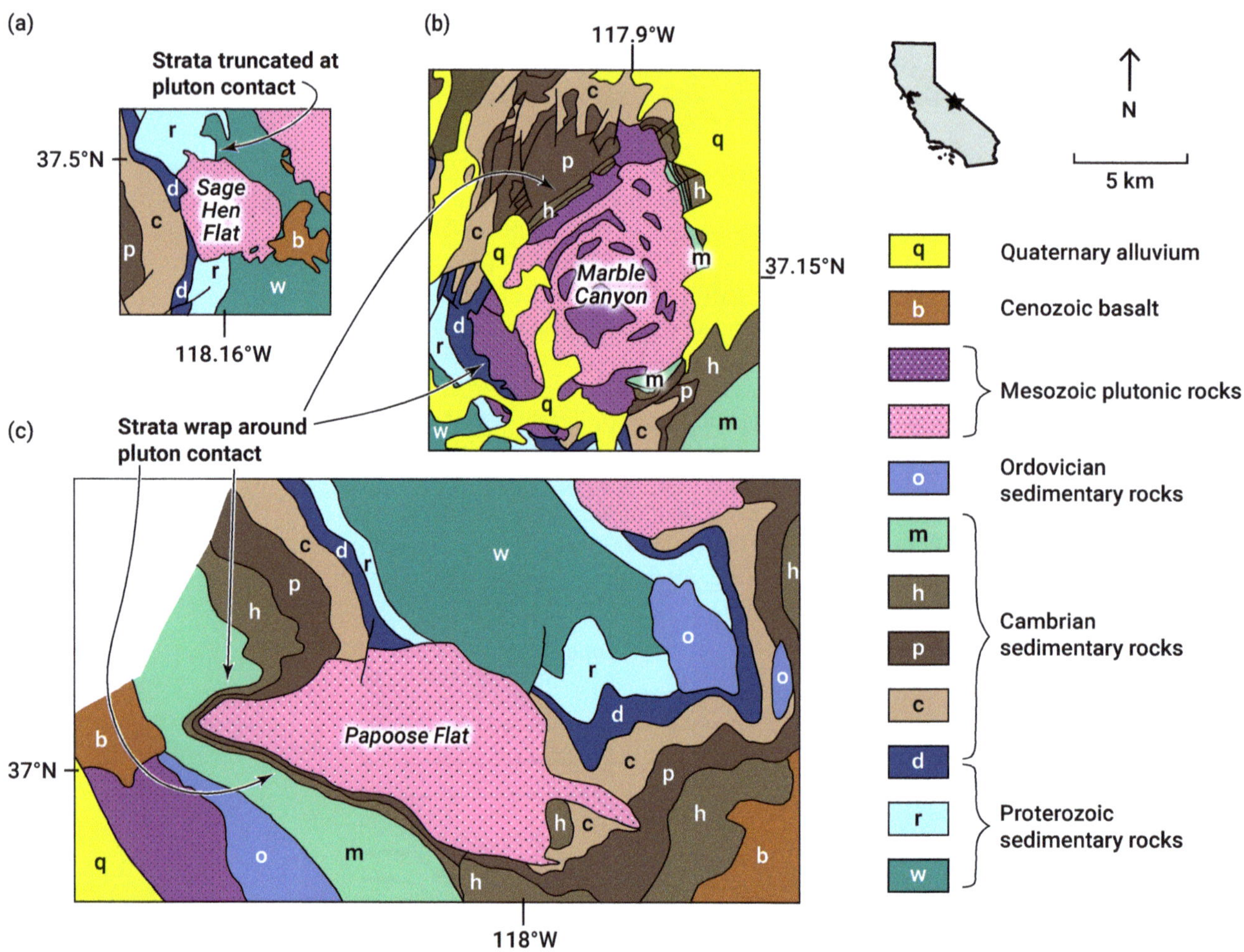

Figure 10.15 Simplified geologic maps of three plutons in the White-Inyo Range, eastern California, showing a variety of contact relations and internal complexity. The Jurassic Sage Hen Flat pluton (a) appears to sharply truncate bedding in surrounding low-grade metasedimentary wall rocks (Bilodeau & Nelson, 1993), whereas the same strata mostly wrap around the Jurassic Marble Canyon (b; Nelson, 1971) and Cretaceous Papoose Flat (c; Sylvester et al., 1978) plutons. Limestone-rich strata[1] (unit p) are thinned to 10% of their original thickness around the western end of the Papoose Flat pluton. The Sage Hen Flat and Papoose Flat plutons are homogeneous granite, whereas the Marble Canyon pluton includes a wide variety of rock types.

Sources: Adapted from Nelson (1991) and Morgan et al. (1998).

plutons consist of homogeneous granite, whereas the Marble Canyon pluton contains rocks as diverse as diorite, gabbro, monzonite, and strange rocks with more than 50 percent giant hornblende crystals.

The main thing that the three plutons in Figure 10.15 have in common is that one can draw a line around an area that contains mostly plutonic rocks and separate it on a map from an area that does not. In this sense, they are analogous to sedimentary formations, where, for example, a homogeneous package of dolomite layers might be overlain by a heterogeneous package of layers including limestone, shale, sandstone, and dolomite, with that package in turn overlain by a homogeneous set of sandstone layers. Each package could be defined as a formation, even though the middle one consists of a wide variety of rock types and is simply defined by its boundaries with other packages.

General Shape and Vertical Extent

Early concepts of plutons saw them as steep-sided, typically cylindrical plugs of plutonic rock without known floors (Figure 10.16; Daly, 1914). Daly's influential diagrams typically rendered plutons as looking something like big molar teeth with steep roots, connecting to a much larger body at depth. Metamorphic wall rocks were deemed **roof pendants**, preserved pieces of the roof of the granite body. This common term has a strong process connotation and should be avoided.

Roof pendants
Term used to describe large pieces of wall rock surrounded by plutonic rocks and assumed to be uneroded pieces of the roof.

This visual concept of plutons as the congealed remains of magma bodies that had vertical extents of tens of kilometers or more, so common in introductory geology textbooks, persists and may locally be correct. However, exposures of pluton floors are common (Figure 10.17; Rosenberg et al., 1995; Brown & McClelland, 2000; Kalakay et al., 2001; Bartley et al., 2012), and it is becoming increasingly clear from both field and geophysical data that many plutons are shaped more like pancakes a few kilometers thick than like beer cans 10 or 20 km tall (McCaffrey & Petford, 1997; Cruden et al., 2018).

Inferring the shape of a pluton below the surface from its surface expression is a difficult problem. Figure 10.18, a schematic map of a discordant pluton like that in Figure 10.15a, shows the basic ambiguity. Both honor the surface geology but make wildly different predictions about what lies at depth. Extrapolation of a pluton's boundaries to depth is generally based on the orientations of contacts and magmatic fabrics, which are commonly inferred to follow magma flow lines. Gravity data (Améglio & Vigneresse, 1999) can help to place limits on the subsurface geometry if there is a significant density difference between the pluton and the wall rocks, but these studies suffer from several ambiguities. In particular, a deep gravity low can tell you that there is a great thickness of relatively low-density rocks, such as granite, below a given area, but they cannot

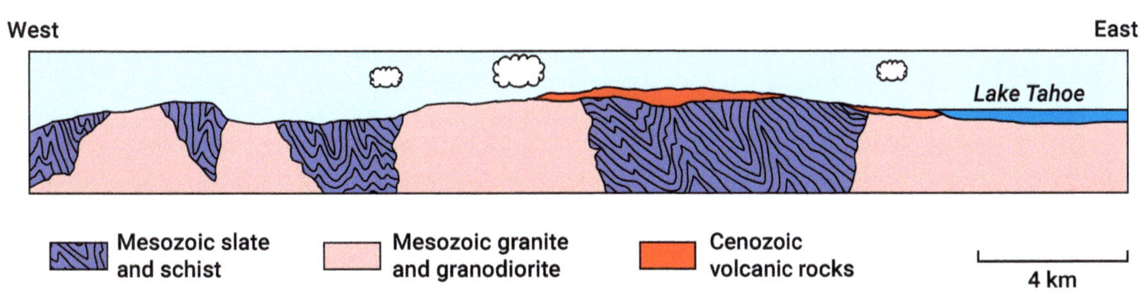

Figure 10.16 Daly's interpretation of plutonic contacts west of Lake Tahoe, California. He viewed these granite bodies as cupolas (as on the US Capitol) sticking up from a much larger body at depth that connects them, and the slates and schists as pieces of the roof that hung down into the massive magma body.

Source: Modified from Daly (1914).

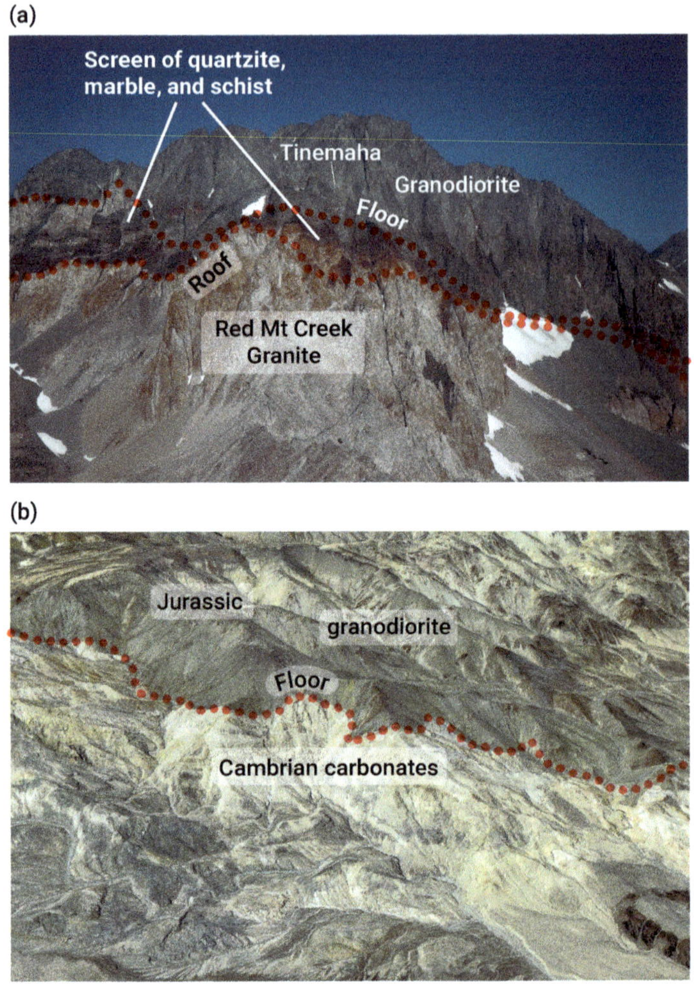

Figure 10.17 Two spectacular, although hard to access, examples of pluton floors in California. Relationships that are so well-exposed here, including the lack of xenolith blocks on the floors, are not uncommon in batholiths. (a) Aerial view from ~3700 m of the southeast face of Split Mountain (4287 m) showing the roof of a light-colored Jurassic granite pluton (below) and the floor of a dark Jurassic granodiorite (above). The two plutons are separated by a thin screen of metasedimentary rocks including dark brown quartzite and tan marble. These relationships continue for nearly 10 km to the south; for details, see Moore (1963), Coleman et al. (1995), and Bartley et al. (2012). Vertical relief in photo is ~700 m. (b) Oblique view to northwest showing the well-exposed floor of a Jurassic granodiorite pluton in the Hunter Mountain batholith, Cottonwood Mountains, Death Valley National Park. The sharply defined floor sits on bright Cambrian dolomite. Width of mid-ground ~5 km.

Sources: (a) AFG photo; (b) Google Earth, ©2019 Google.

tell you when those low-density rocks were emplaced. In the Sierra Nevada, rocks with granitic density appear to persist to about 10 km depth (Oliver et al., 1993), but rocks of granitic composition in the area range from >200 Ma to <90 Ma in age. Thus, low gravity beneath a given pluton at the surface does not require that the pluton extend to the depth of the anomaly.

Blobby vs. Elongate Plutons

Although plutons come in a wide variety of shapes and sizes in map view, there are two basic end members: blobby ones whose outlines are roughly circular in map view, and highly elongate ones. Figure 10.19 shows a selection of pluton shapes and sizes taken from the geologic map of Sequoia and Kings Canyon National Parks in California. These parks cover 3500 km^2 of exceptionally well-exposed, exceptionally rugged, high-altitude plutonic rock and were mapped over several decades by Moore and Sisson (1987). Figure 10.19 demonstrates the wide range of shapes and sizes that they demarcated as individual plutons. Clearly, most plutons there are distinctly non-circular.

One hypothesis for the blobby vs. elongate dichotomy is that they represent exposure of different levels of fundamentally mushroom-shaped—laccolithic or lopolithic—magmatic systems. Eroding down into the intrusions in Figure 10.14 would expose blobby cross sections at shallower levels, giving way to elongate cuts through the dike-like stem at deeper levels.

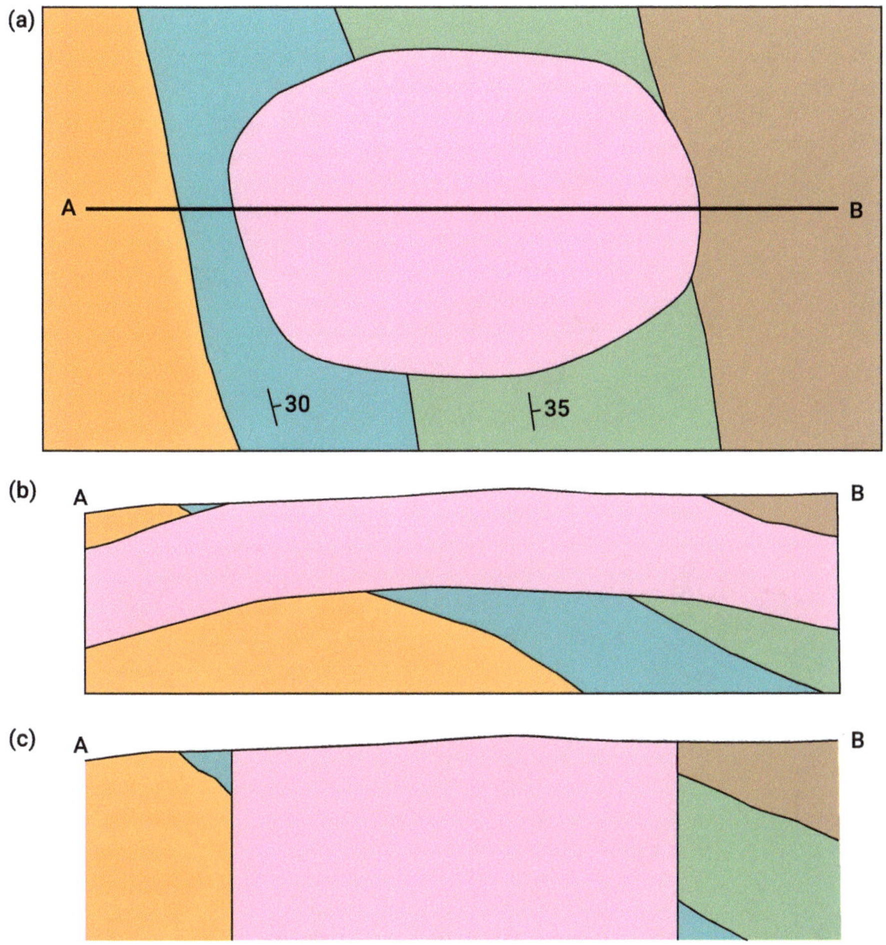

Figure 10.18 (a) Schematic geologic map of a discordant pluton with cross sections that extrapolate it to depth as (b) a relatively thin sheet or stack of sheets (pancake) and (c) a steep-sided cylinder (beer can). Careful study of the orientations of the contacts between granite and wall rocks could help to discriminate between these two hypotheses.

"Ghost Stratigraphy" in Plutons with Abundant Wall-Rock Sheets

Some elongate plutons contain a large proportion of sheet-like panels of wall rock that are parallel to the pluton's long axis. Classic examples of this are the Main Donegal pluton in Ireland (Pitcher & Berger, 1972) and the McDoogle pluton in the Sierra Nevada (Mahan et al., 2003). The latter was intruded into a shear zone with steep foliation and split off panels of the foliated rock, forming a complex, multiphase pluton that is locally 50 percent screens of wall rock ranging from a few centimeters to several tens of meters thick (Figure 10.20a).

An important feature of plutons such as the McDoogle and Main Donegal is that panels of wall rock in the pluton are found in the same order as those in the wall rocks outside the pluton, and structures in those panels (e.g., attitudes of foliation and fold axes) parallel those in the wall rocks. This "ghost stratigraphy" had long been used to argue for a granitization origin of the plutonic rocks (e.g., that the sheets of plutonic rock in Figure 10.20a were formed by metasomatic replacement of the wall rocks). Granitization has since been ruled out on isotopic and other grounds, and field relations make it clear that such plutons form by diking parallel to wall-rock foliation (e.g., Stevenson et al., 2008; Figure 10.20b). Similar relationships (panels of wall rock with congruent foliation and lineation parallel to contacts) are common in more blobby plutons, which suggests that those plutons as well started as dike-like bodies (Glazner et al., 2004).

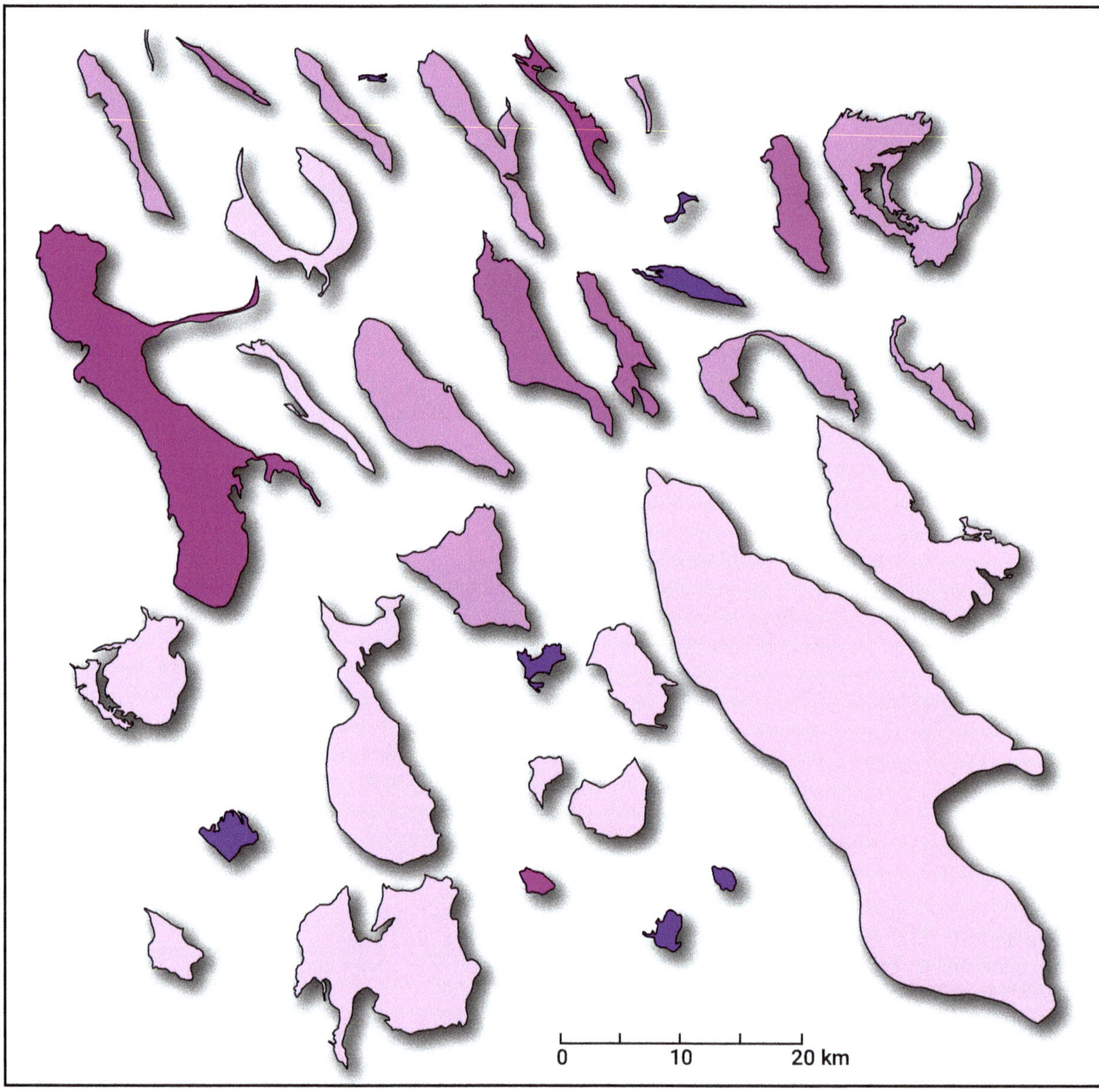

Figure 10.19 Selected plutons from the geologic map of Sequoia and Kings Canyon National Parks in California, showing the wide range of sizes and shapes encompassed. Dark purple denotes diorite and gabbro; tones range from mafic granodiorite (darker) to granite (lighter). Blobs are roughly arranged from most elongate at top to most equant at bottom.

Source: Data of Moore & Sisson (1987).

Post-kinematic, syn-kinematic, pre-kinematic After, during, or before deformation, respectively.

Timing of Intrusion and Deformation Plutons typically show sharp contacts with non-plutonic wall rocks, and the nature of the contact can locally be used to determine the timing of deformation relative to intrusion (Figure 10.21). For example, a pluton or dike that cuts folding in metamorphic or sedimentary wall rocks was presumably intruded after the deformation event that produced the folding, and an age on the pluton gives a minimum age for that deformation. Such plutons are termed **post-kinematic**. Plutons that are themselves deformed are either **syn-kinematic** (deformed along with the wall rocks) or **pre-kinematic** (intruded before the wall rocks were deformed). However, such interpretations are greatly complicated by recrystallization, which can erase much of the evidence for deformation, and it is possible that

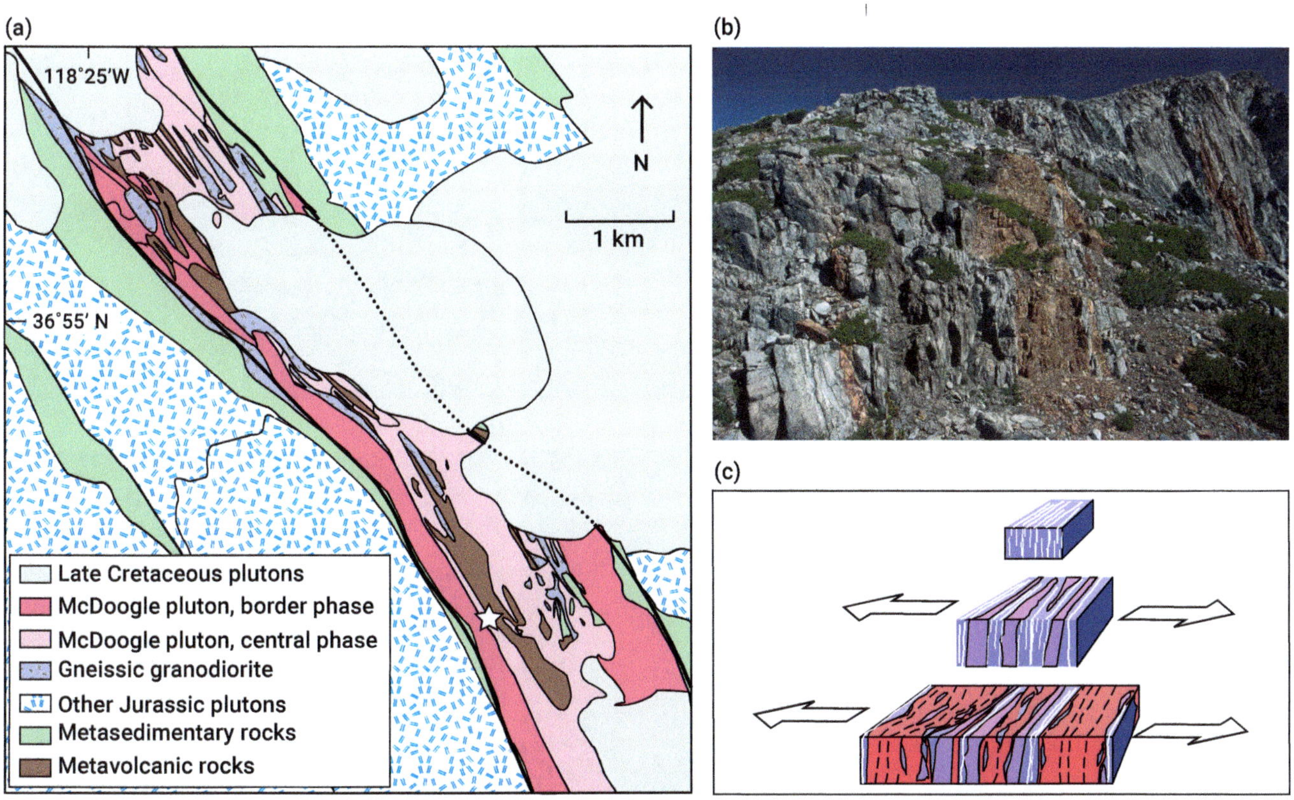

Figure 10.20 Sheeted, dike-like plutons in map view and block diagrams. (a) Simplified map of the McDoogle pluton, eastern Sierra Nevada, which comprises mafic granodiorite and quartz monzodiorite injected into foliated Jurassic wall rocks. Foliation in wall rocks is subvertical and strikes northwest, parallel to the pluton borders (heavy lines). The pluton contains abundant large screens of foliated wall rocks (schistose metavolcanic rocks and gneissic granodiorite), and each body extensively invades those wall rocks on a scale too fine to map. Star indicates location of photo in (b). (b) Photo, looking northwest, shows how sheeted gray dikes of the central phase invaded orange-weathering sheets of foliated metavolcanic wall rocks. Dikes of the pluton are uniformly coarse-grained, whether a 100-m sheet or a 2-cm dike. (c) Block diagrams illustrating emplacement of the Main Donegal Granite by injection of subvertical sheets parallel to wall-rock foliation in a shear zone. Wavy white lines in gray wall rocks schematically indicate foliation; dashed lines in granite schematically indicate internal sheets.

Sources: (a) Courtesy of Kevin H. Mahan, from Mahan et al. (2003); (b) AFG photo; (c) modified from Stevenson et al. (2008), © Carl T. E. Stevenson, Donald H. W. Hutton, Alun R. Price/Royal Society of Edinburgh 2006. Cambridge University Press, 2008.

many plutons interpreted to be post-kinematic were actually significantly deformed by faulting and folding, with the deformation features healed by recrystallization (Bartley et al., 2018; see Section 10.8).

Sharp vs. Gradational Relationships at Pluton Contacts Contact relationships of plutons with other plutons range from sharp to fully gradational, depending on the time interval between them (Figure 10.22). Plutons that are roughly the same age may show gradational contacts with one another; in the field as you approach a contact, the pluton you're walking on gradually takes on the mappable textural characteristics of the other. For example, in the Tuolumne Intrusive Suite (see Figure 10.3), most of the more than 100 km of contacts between the major units are gradational over length scales ranging from 10 m to several hundred meters (Coleman et al., 2005).

The contact between the Half Dome Granodiorite and Cathedral Peak Granodiorite in Yosemite National Park is a particularly clear example of a gradational contact. The former is recognized by euhedral hornblende phenocrysts up to 2 cm in length

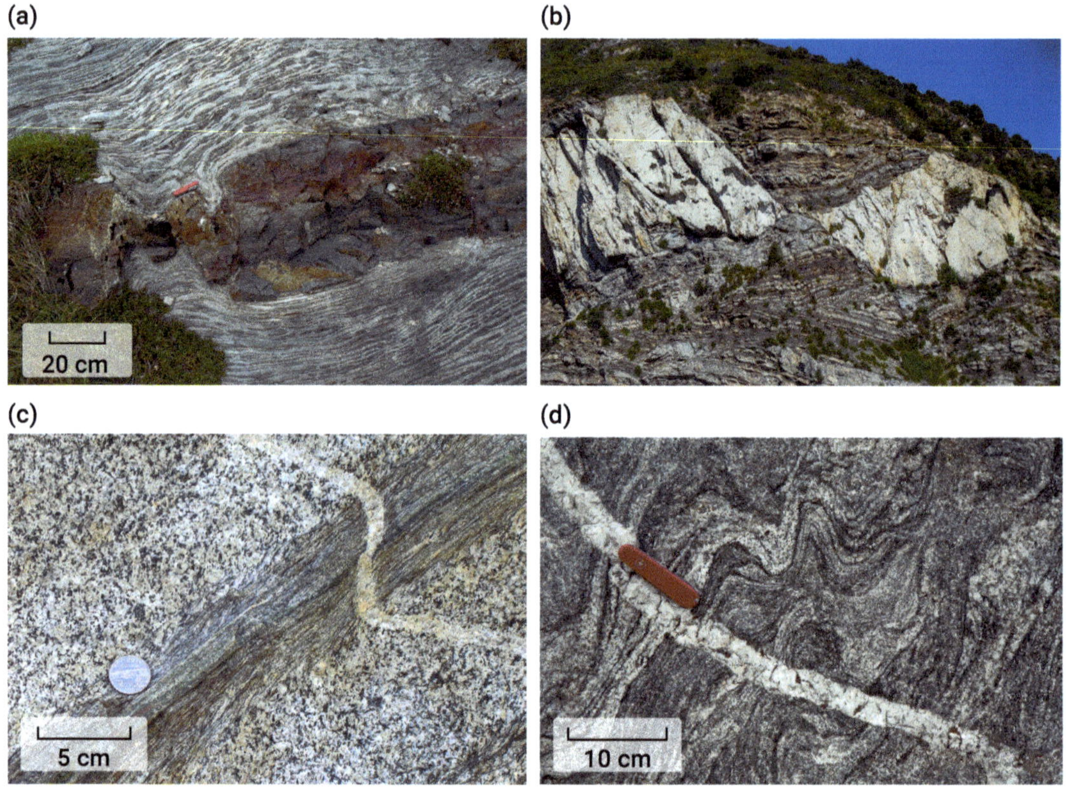

Figure 10.21 (a) Basalt dike in marble, Adamello region, Italy. Deformation caused the stronger (more competent) dike to fracture and pull apart, and the weaker marble flowed into the gap, a classic example of **boudinage**. The dike is thus pre-kinematic. (b) Deformed sills at margin of granite laccolith exposed in sea cliff on the island of Elba, Italy (see Figure 10.11). The sills were intruded after intense folding and shearing of the sedimentary wall rocks (note "<"-shaped folds under granite) and were later cut by normal faults with down-to-east (right) displacement. The 7-Ma granite is thus post-kinematic with respect to the folding and pre-kinematic with respect to the normal faulting. Sills ~20 m thick. (c) Granodiorite with a weak deformation fabric cut by a shear zone near Morgan Pass, Sierra Nevada, California. The granodiorite and shear zone are in turn cut by a thin leucogranite dike. The dike is offset by roughly the width of the shear zone, indicating a shear strain of ~1, at least in the plane of the outcrop. Shear strain across the mylonite is undetermined but likely tens to hundreds of times greater. Thus, the dike is (very late) syn-kinematic. (d) Clearly post-kinematic, undeformed pegmatite dike cutting folded gneiss.

Sources: (a, b, c) AFG photos; (d) courtesy of Marli Miller (geologypics.com).

Boudinage
A structure in deformed rocks in which a stronger layer has been pinched apart during deformation, with weaker layers flowing into the gaps. Individual fragments are commonly strung out in a line, like linked sausages; *boudin* is a French word, one translation of which is sausage.

and no phenocrysts of K-feldspar, and the latter by huge phenocrysts of K-feldspar 5 to 10 cm in length and only sparse hornblende phenocrysts (see Figure 10.3). Along the highway that bisects the park, this spectacularly exposed but curiously subtle contact is gradational over more than a kilometer (Figure 10.23; Coleman et al., 2005; Johnson & Glazner, 2010). Walking from the Half Dome Granodiorite toward the Cathedral Peak Granodiorite, one at first sees no K-feldspar phenocrysts, then ragged patches of K-feldspar about 1 cm across that encase numerous smaller minerals and are only recognizable when cleavage faces reflect the sun, then larger patches that are less packed with other crystals and easier to see, and eventually, K-feldspar crystals that are several centimeters across, euhedral, and easily recognizable. Where does one draw a contact on a geologic map in a situation like this? Bateman et al. (1983) drew it where the K-feldspar

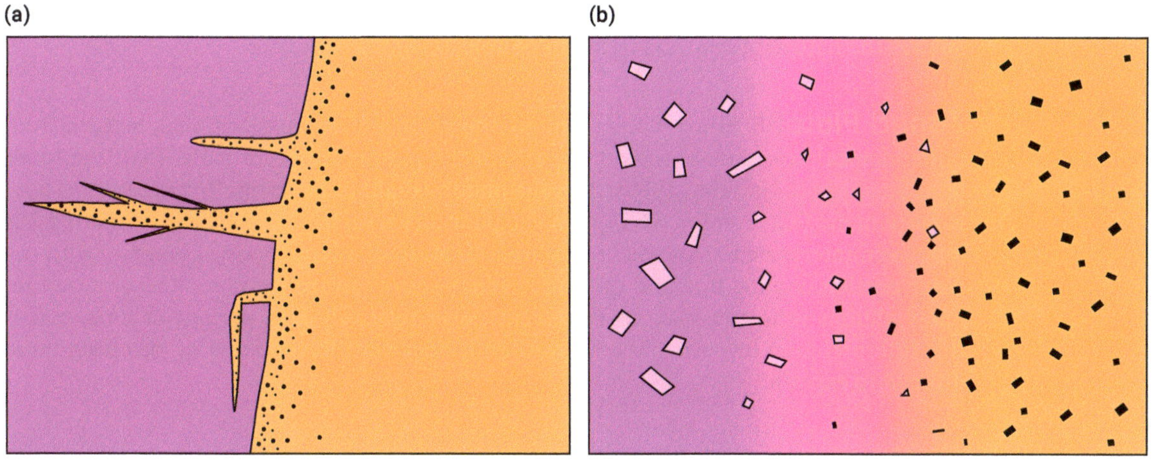

Figure 10.22 (a) Expected contact relationships when a pluton (right) cuts a significantly older pluton or other wall rocks; dikes of the pluton clearly cut the older rocks, and the invading magma forms a chilled margin (stipple). Such chilled margins are rare in rocks more silicic than basalt. (b) Gradational contact relationships commonly seen between plutons of similar age: gradational contact, with a gradational change in texture.

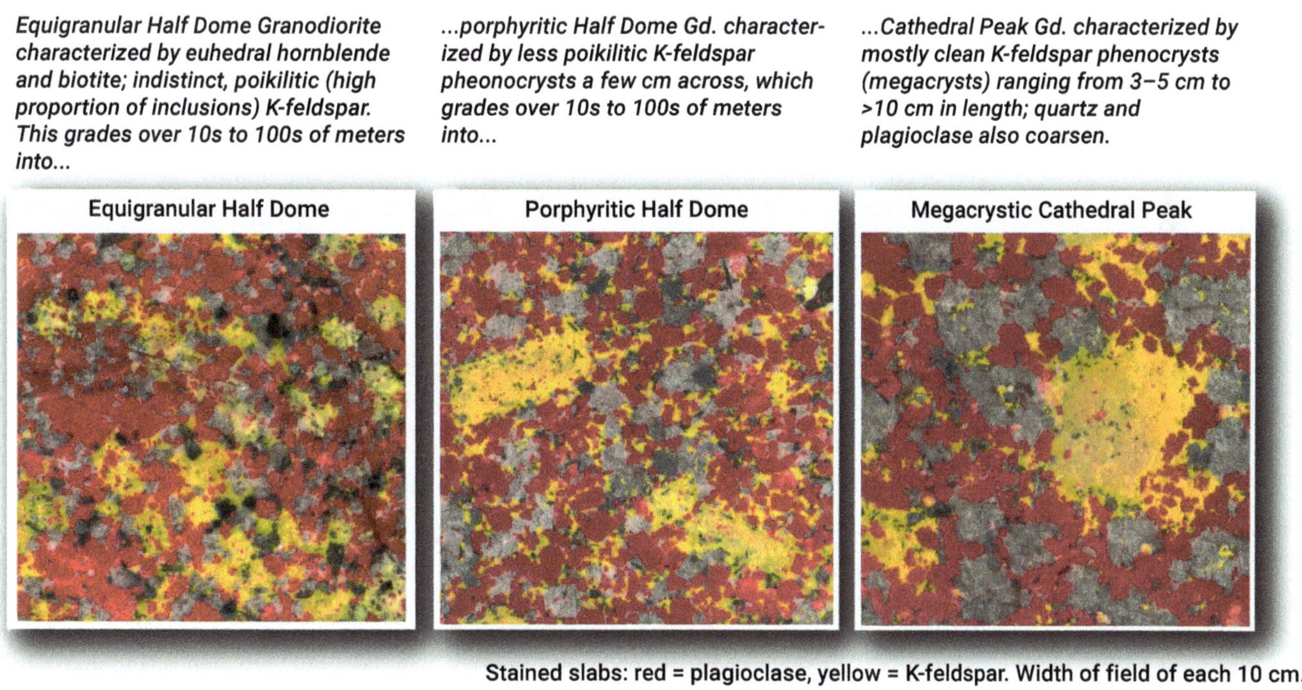

Figure 10.23 Stained slabs from a particularly broad and intriguing gradational contact between two plutons, the Half Dome and Cathedral Peak Granodiorites of Yosemite National Park (see Figure 10.3). Walking from the older Half Dome pluton into the younger Cathedral Peak pluton, the texture of the rock gradually changes, with increasing size and clarity of K-feldspar, increasing size of quartz and plagioclase, and decreasing abundance of hornblende.

Source: Breck Johnson photos.

size ramps up to 10 cm over a short distance, but clearly, this placement is arbitrary. Such is the nature of contacts between plutons of similar age.

Zoned Plutons Some blobby plutons show concentric compositional, textural, and mineralogical zonation. If the outer parts of the pluton are more mafic than the inner (e.g., Stephens & Halliday, 1980), then the pluton is said to be normally zoned; if the core is more mafic (e.g., Ayuso, 1984), then the pluton is reversely zoned. Normal zoning has long been interpreted as in situ crystal fractionation of a large body of magma, with the outer portion cooling first and driving residual liquid toward the core. Recognition that many plutons were emplaced in increments, over relatively long periods of time, makes this interpretation less straightforward than it once was. Reverse zoning has been more difficult to explain.

Enclaves and Xenoliths

The origin and evolution of mafic magmatic enclaves (see Figure 8.12) makes up an entire subfield of igneous petrology (e.g., Pabst, 1928; Didier & Barbarin, 1991), and enclaves still present many puzzles. Magmatic enclaves are ubiquitous in intermediate plutons such as granodiorites, and less common in more mafic or felsic ones. They commonly range from a few centimeters to 1 m in diameter, with 10 cm being a typical diameter in outcrop; enclaves more than a few meters across are rare. Their abundances across a given intermediate pluton range from absent over hectare-sized areas to enclave-choked regions tens or hundreds of meters across. Even given the about 50 percent abundance in regions such as that in Figure 8.12, the overall areal abundance of enclaves in granodiorite plutons is typically much less than 1 percent.

Magmatic enclaves are usually more mafic than the plutons that host them (Figure 10.24). In the Sierra Nevada batholith, enclaves typically have andesitic compositions with SiO_2 ranging from 50 to 60 wt%, whereas host granodiorites have SiO_2 of 60 wt% and above (e.g., Barbarin et al., 1989; Dorais et al., 1990). There is a curious lack of compositional overlap between the enclaves and their host plutons; the enclaves extend to lower SiO_2 than the host rocks, with a small overlap at 55 to 60 wt% SiO_2. This curiosity awaits a clear explanation.

Xenoliths are far rarer than enclaves in plutons. In the elegantly exposed Tuolumne Intrusive Suite, xenolith counts yield abundances in the range of 1 part in a million (0.0001%). Xenolith abundances are rarely quantified but appear comparably low in most other plutons. Xenoliths may form swarms, especially near contacts, but away from contacts, one can walk across bare granite for days without seeing one.

Schlieren, Ladder Structures, and Other Oddities

Schlieren
Thin, discontinuous horizons that are rich in mafic minerals.

Schlieren[2] are layers enriched in mafic minerals, and texturally, they range from wispy and indistinct to dark and dramatic (Figure 10.25). Typically, they are discontinuous, thin, dark layers that come and go. Schlieren are sparse (typically $\ll$1% of outcrop area) but widespread in intermediate and silicic plutons worldwide. The formation of schlieren is typically ascribed to processes such as magmatic currents depositing dense minerals, currents stirring cumulate minerals and lofting less-dense ones, disaggregation of mafic enclaves, and numerous others. Barbey (2009) presented a thorough discussion of the various manifestations of mafic layers in plutons.

Ladder structures are bizarre accumulations of alternating light and dark layers that typically form well-defined stripes about 1 m wide on the outcrop, with the alternating bands forming the "rungs" of the ladder (Figure 10.25c). These structures are typically planar in three dimensions and are commonly referred to as ladder dikes, but they are

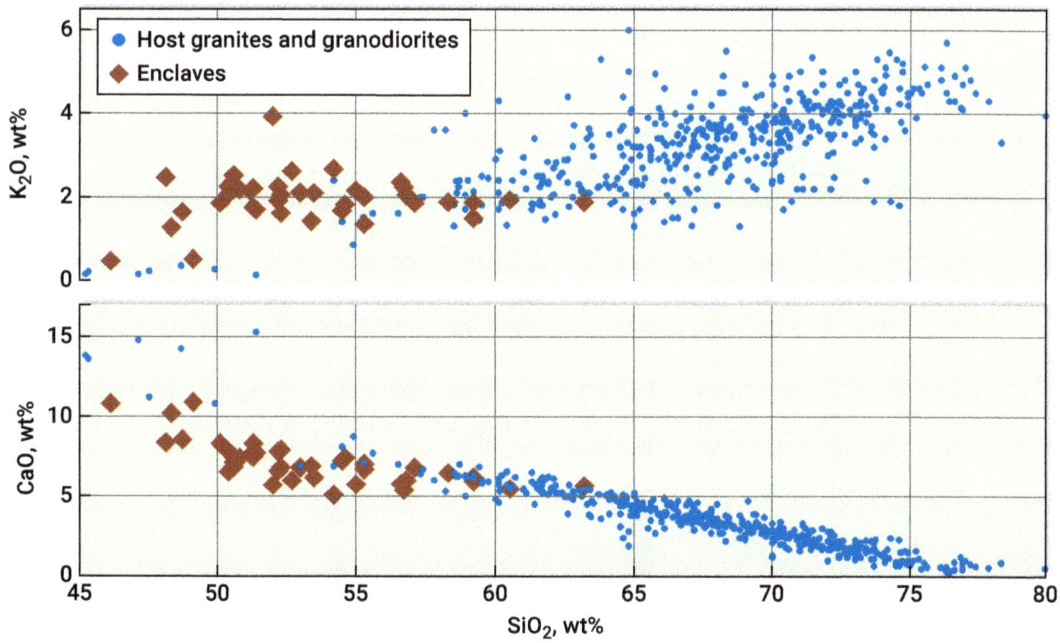

Figure 10.24 Plots of two oxides vs. SiO_2 for Sierra Nevada batholith rocks and the enclaves found in them. Enclaves extend up to about 60 wt% SiO_2, and the batholith extends down to about 55 wt% SiO_2, with a small proportion of cumulate rocks at lower values. There is little overlap between the two populations.

Sources: Data from Bateman et al. (1984) and Barbarin et al. (1989).

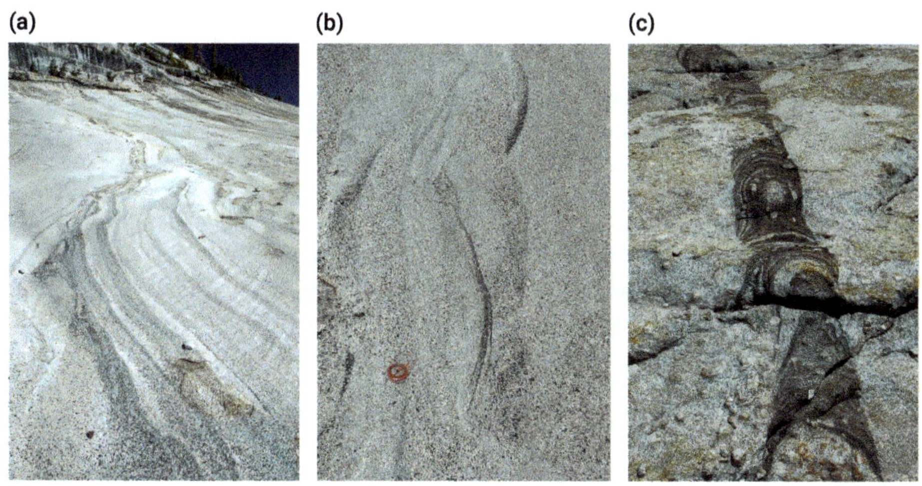

Figure 10.25 (a) Granodiorite with wavy schlieren composed of concentrations of biotite, hornblende, and magnetite, Yosemite National Park. Width of foreground ~3 m. (b) Fainter schlieren in leucogranite. Compass circle is 5 cm in diameter. (c) Ladder structure in Yosemite National Park, about 50 cm wide, composed of dark layers of biotite, hornblende, magnetite, apatite, titanite, and zircon alternating with light layers of quartz and feldspars. Here, the layers are gently curved troughs that plunge shallowly toward the viewer. In this example, both the host Cathedral Peak Granodiorite and ladder structure contain abundant 5-cm K-feldspar **megacrysts**.

Source: AFG photos.

Megacrysts
Here, unusually large crystals, typically defined as >5 cm in long dimension.

unlikely to be dikes in the traditional sense. Like other minor phenomena, they make up a minuscule but eye-catching fraction of typical granodiorite plutons worldwide. The dark layers consist of the same mafic minerals as in the host rock (e.g., biotite, hornblende, magnetite, but little plagioclase), but with anomalously high concentrations of trace phases (e.g., zircon, apatite, titanite). Monomineralic layers of titanite and magnetite occur locally.

Aid climbing
Climbing where gear such as pitons, bolts, or other equipment is used to aid upward progress.

Free climbing
Climbing in which all upward progress is done using hands, feet, and other parts of the body on natural features; equipment is used for fall protection only.

Ladder structures are typically ascribed to concentration of mafic minerals by processes such as accumulation in depressions in a magma chamber floor (Wiebe et al., 2017) or shear flow along the walls of a conduit (Weinberg et al., 2001). However, these explanations fail to explain the extremely high concentrations of trace phases, lack of plagioclase, and high concentrations of incompatible elements including REEs, which are enriched to over 1000 times chondrite in the most mafic layers. The mafic component of ladder structures and of at least some schlieren may form via liquid immiscibility (Glazner et al., 2021), but the origin of these features and their significance is still largely a mystery.

Making Connections 10.1

Rock Climbing

Thousand-meter granite cliffs make Yosemite Valley one of the world's premier rock-climbing areas. Big-wall rock climbing began in Europe (chiefly in the Dolomites) but gained prominence in North America when climbers such as John Salathé, Allen Steck, Warren Harding, Royal Robbins, Tom Frost, and Yvon Chouinard (founder of the Patagonia company) pioneered routes up the great faces of Sentinel Rock, Half Dome, and El Capitan (see Figure 10.1a,b). Early ascents took multiple days and generally relied on **aid climbing**; as an extreme example, the first ascent of the *Nose* route on El Capitan in 1958 took 45 days, and the team inserted 600 pitons and placed over 100 bolts in the rock. Times have changed; the speed record for the *Nose* as of 2024 stands at just under two hours, using removable gear that does little damage to the rock. In 1993, Lynn Hill became the first person to succeed in **free climbing** the *Nose*, a supremely difficult feat whose hardest move is rated 5.14a/b on the **Yosemite Decimal System**. Yosemite big-wall climbing remains in the public eye, as shown by the success of films such as *The Dawn Wall* (2017) and *Free Solo* (2018).

The valley is carved entirely in granite and granodiorite of two major intrusive suites, the Yosemite Valley and the slightly younger Tuolumne (see Figures 10.1, 10.3, and 10.4). One of the key ingredients for making kilometer-tall cliffs is exceptionally strong and sparsely jointed rock, and the protracted emplacement times and consequent annealing of Yosemite granite made it strong, indeed. The gigapixel base photo that Putnam et al. (2015) used for mapping was detailed enough to see 10-mm climbing ropes, and yet no brittle faults were detected offsetting any of the sharp contacts on the map (see Figure 10.4).

Climbers rely on numerous features of the rock to make their ascents. Many of these, such as ledges, flakes, huecos (holes in the rock), jugs (large, easily gripped holds), and cracks into which fingers, fists, feet, or whole bodies can be jammed, occur in many types of rock, but some are only found in granitic and other igneous rocks and are integral to some of the most iconic climbing in Yosemite. For example, mafic enclaves commonly stand in positive relief on weathered granitic rocks, providing holds (Figure 10.26a). K-feldspar megacrysts generally stick out and provide solid holds[3] (Figure 10.26b). Fine-grained aplite dikes tend to stand out and provide relief on otherwise featureless, slick granite; a system of steeply oriented aplite dikes pervades Half Dome, and one provides a relatively easy route up the west face of this formidable feature (Figure 10.26c).

Ignimbrites form another climbing surface with unique igneous features. Many ignimbrites develop weathering pockets that provide excellent, high-friction holds. Surfaces of large columnar joints in densely welded tuff yield vertical cliffs that commonly have narrow ledges formed by joints parallel to the flattening foliation or by weathered-out, highly flattened pumice lapilli. These, and cracks at the intersections of two columnar joint surfaces, provide routes up otherwise featureless cliffs (Figure 10.26d).

Figure 10.26 Various igneous features that can be utilized for rock climbing. (a) A climber on *Wicked Gravity*, a route of serious difficulty (5.12c) on El Capitan Granite near the entrance to Yosemite Valley. This climb is made possible by differential weathering, which makes the mafic enclaves stand out as "chicken heads" on the vertical surface. (b) Many of the glacier-sculpted domes in the high country of Yosemite are carved in Cathedral Peak Granodiorite. Here, a climber takes advantage of protruding K-feldspar megacrysts to climb a difficult 5.10d route. A thin aplite dike provides more opportunities for purchase. (c) Climbers on *Snake Dike*, a long route up a meter-wide aplite dike on the west flank of Half Dome at the head of Yosemite Valley. As in (a), the finer-grained dike sticks out in complex relief from the smooth Half Dome Granodiorite that hosts it. This affords a moderate (5.7) climb up the otherwise slick sunlit shoulder of Half Dome in Figure 10.1b. (d) Climbing on densely welded Bishop Tuff in Owens River Gorge (see Figure 7.38b). Routes in this area generally climb faces of large, poorly defined columnar joints in the densely welded zone (see Figure 7.22b), and joints parallel to the flattening foliation provide shallow ledges that skilled climbers can use.

Sources: (a) Steven M. Bumgardner; (b) Bryan Law; (c) Roger Putnam; (d) Cavan Images/Alamy Stock Photo.

Yosemite Decimal System
A system for rating the difficulty of a climb, with "5." followed by a number and sometimes a letter for grades 5.10 and above. 5.0–5.4 is easy; 5.5–5.9 is moderate, with good holds and few overhangs; and 5.10 and up is advanced, with smooth walls, tiny holds, and overhangs. For 5.13 and up, imagine a granite countertop, with a few chips in its surface, turned on edge, perhaps 500 m off the ground.

10.3 Origin and Evolution of Granitic Magma

Learning Objectives

By the end of this section you should be able to:

10.3.1 Explain why volcanic rocks are limited to about 50 percent maximum crystallinity.

10.3.2 Predict the crystallization sequences of dacitic magmas (using a simplified *P-T* diagram).

10.3.3 Apply Alumina Saturation Index and other simple classification schemes.

The Invisible 50 Percent

Volcanic rocks cool rapidly upon eruption and therefore provide snapshots of what the magma was like just prior to eruption, akin to laboratory experiments in which synthetic magmas are quenched (see Section 3.3). In Section 7.3, we noted the critical crystal content of about 50 vol% that separates magmas that are capable of flow from those that are semirigid, melt-soaked crystal frameworks. The dramatic increase in effective viscosity at this point is consistent with observations of volcanic rocks, in which crystal contents greater than 50 vol% are essentially unknown (see Figure 7.11). For this reason, volcanic rocks carry no record of magma crystallization beyond 50 percent—once magmas reach that point, they are too crystal-rich to erupt, and they stay put and become plutonic rocks.

In contrast, all plutonic rocks are all always fully crystallized, so natural samples provide no snapshots of magmas in the interval from 50 to 100 percent crystallization. This invisible 50 percent of crystallization determines the look and feel of plutonic rocks but has been barely explored. Before speculating about what might happen there, we need to review how crystallization proceeds in silicic magmas.

Crystallization in Granitic Systems

The great calc-alkaline batholiths generated at continental margins (see Figure 10.2) are composed largely of feldspars and quartz, typically accompanied by hornblende, biotite, and Cpx, with local abundances of other minerals such as Opx, cordierite, garnet, and muscovite and minor amounts of important accessory minerals such as zircon, titanite, and apatite. In the Sierra Nevada, bulk compositions typically range along the BADR trend from about 60 to 75 wt% SiO_2 (see Figure 10.24), and a compilation of data from circum-Pacific subduction zones shows a preponderance of compositions in this range (Figure 10.27). The origins of such magmas are still a subject

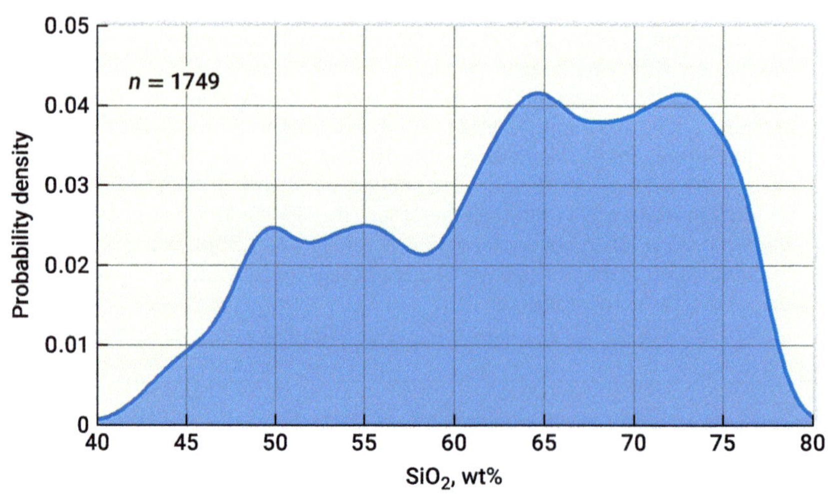

Figure 10.27 Distribution of SiO_2 in circum-Pacific (subduction-related) plutonic rocks. There is a distinct double maximum from ~63–75 wt%. In contrast, volcanic rocks are skewed toward mafic compositions; see Figure 10.46 and Glazner et al. (2015).

Source: Data from Glazner et al. (2015).

of considerable interest and research (see Chapters 6 and 8), but field and geochemical studies show that emplacement of magmas with an overall dacitic bulk composition, producing granodiorites and tonalites, is common. We turn now to the crystallization behavior of such magmas.

Crystallization of Dacitic (Granodioritic) Magma

The crystallization behavior of calc-alkaline dacitic magma is well known both from experimental work and from observations of volcanic rocks, which provide quenched samples of much longer-term natural experiments. Experimental investigation of dacitic magma evolution post-dated that of granites, because it required development of devices capable of reaching high pressures and temperatures and of the electron probe microanalyzer to determine mineral and glass compositions and proportions. A generic dacite *P-T* diagram, synthesized from several water-saturated experimental studies (e.g., Johnson & Rutherford, 1989; Scaillet & Evans, 1999; Costa et al., 2004; Holtz et al., 2005; Muir et al., 2014), is given in Figure 10.28, and a schematic view of changing mineralogy with temperature at 200 MPa in Figure 10.29. These diagrams provide guidance for understanding how dacitic magma crystallizes to form granodiorite.

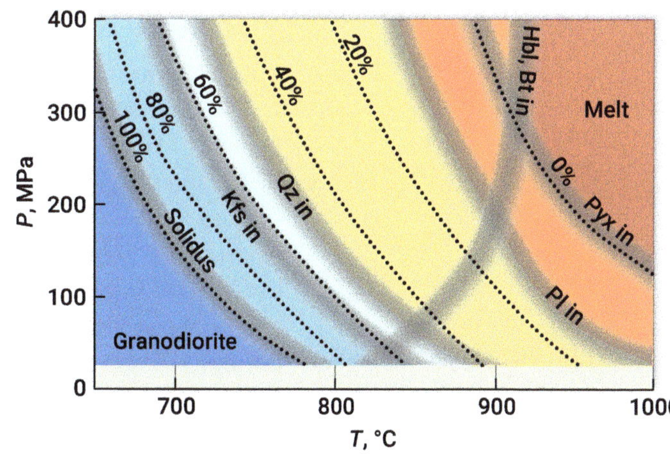

Figure 10.28 Schematic summary *P-T* diagram for water-saturated experiments on dacites at upper-crustal pressures. Gray bands show first appearance of a given phase; e.g., "Qz in" denotes the temperature at which quartz first appears upon cooling as a function of *P*. The appearance of Fe-Ti oxides is not shown, but they crystallize early, around the temperature that other mafic minerals appear. Dotted lines give crystal content. Key points are that the first 50% of crystallization consists of plagioclase, mafic silicates, and oxides, with perhaps a small amount of quartz. Crystallization of the remaining 50% produces mostly quartz, K-feldspar, and sodic plagioclase, plus key accessory minerals including zircon and titanite. Down-temperature reaction of pyroxenes to hydrous silicates is ignored.

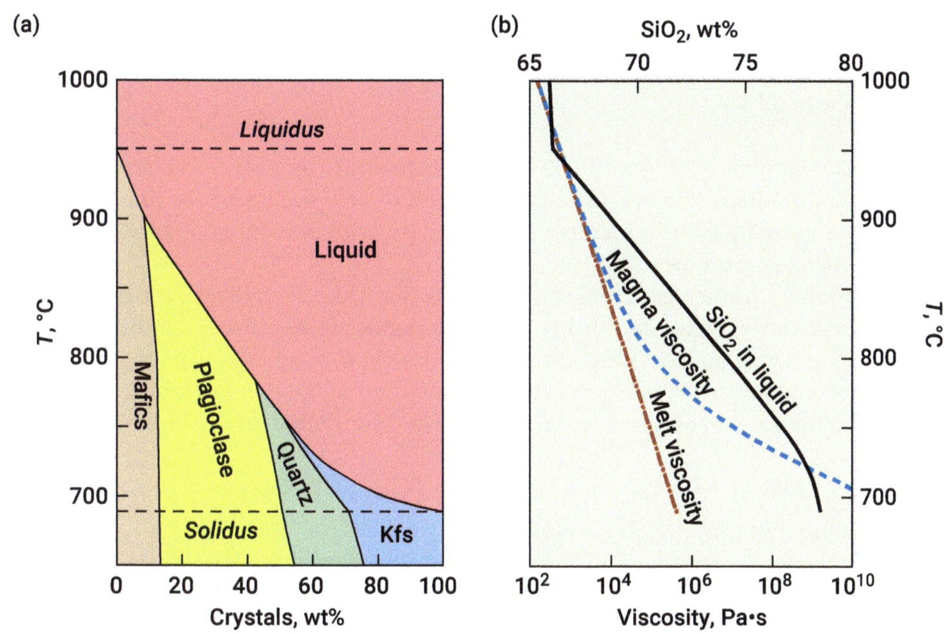

Figure 10.29 (a) Modal mineralogy as a function of temperature for the hypothetical model dacite. Key features are that the first 50% of crystallization is dominated by mafic minerals, plagioclase, and a minor amount of quartz, and that K-feldspar typically does not appear until after ~50% crystallization, at which time the magma is mechanically locked up (see Section 7.3). (b) Changes in liquid SiO_2, liquid viscosity, and magma viscosity with *T*. At the point where the magma reaches lock-up at 50% crystallization, the liquid has reached ~75 wt% SiO_2.

Sources: Liquid viscosity calculated from Giordano et al. (2008); magma viscosity from Equation 7.4.

A few important points about the *P-T* diagram:

- The solidus has a negative slope, in keeping with the melting of mostly anhydrous mineral assemblages in the presence of water.
- Liquidus *T* is 900 to 1000 °C, decreasing with increasing *P*.
- At $P > 300$ MPa (typical for intrusions) hornblende is the first phase to come in upon cooling, followed by plagioclase and biotite.
- Quartz does not appear upon crystallization until the magma is about half-crystallized, and K-feldspar (sanidine) does not appear until even higher crystal contents.
- Consistent with Section 6.6, *P-T* slopes of the hydrous phases (biotite, hornblende) are positive, whereas those of anhydrous phases (e.g., quartz, plagioclase) are negative. As a consequence, pyroxene (generally Opx), is the liquidus phase at lower *P*, whereas hornblende or biotite is on the liquidus at higher *P*.

Figure 10.29 summarizes several important points related to evolution of mineralogy and liquid composition during dacite crystallization:

- The first half of crystallization is dominated by plagioclase, mafic minerals, oxides, and minor quartz, consistent with the mineralogy of dacite lavas and tuffs. K-feldspar is minor (a few modal %) or absent at half crystallization.
- Crystallization of mafic minerals and oxides rapidly drives the liquid composition to high SiO_2, reaching 70 wt% after 20 wt% crystallization and over 75 wt% by half crystallization at about 780 °C.
- These high-silica liquid compositions are consistent with the compositions of matrix glass in crystal-rich dacites (e.g., Luhr & Melson, 1996; Bachmann et al., 2002; Sparks et al., 2008; Folkes et al., 2011).
- Melt viscosity is less than 1000 Pa·s at the liquidus but increases to about 30,000 Pa·s by half crystallization.
- At that point, the high crystal content rapidly drives the bulk viscosity of the magma to values above 10^6 Pa·s.

An important corollary of Figure 10.29 is that the mineral assemblage in a magma of granodioritic composition that is crystal-poor enough to flow should be the same as that in dacites, dominated by mafic minerals and plagioclase, with or without quartz, and with at most a few percent sanidine.

A twist found in some experimental studies is that Opx crystallizes near the liquidus, only to be replaced by biotite as *T* drops. This behavior occurs because crystallization of the anhydrous phases plagioclase, Opx, and Fe-Ti oxides removes no H_2O from the melt, causing its concentration to increase and eventually build up to the point that Opx is hydrated to biotite. Hydration can occur via a reaction such as this, written for the Mg end member:

$$\begin{array}{l} 3MgSiO_3 + [KAlSi_3O_8] + [H_2O] = KMg_3AlSi_3O_{10}(OH_2) + 3[SiO_2] \\ \text{Opx + K-feldspar in melt} + H_2O = \text{biotite} + SiO_2 \text{ in melt} \end{array} \quad \textbf{(Eq. 10.1)}$$

where components in brackets are in the melt.

The results of these studies are consistent with the reaction principle introduced by Bowen (1922) a century ago: early crystallization of pyroxene, amphibole, biotite, oxides, and intermediate plagioclase, followed by crystallization of more sodic plagioclase, quartz, and K-feldspar.

Comparison to Dacites The crystallization sequence determined in these experiments conforms to the crystallization sequence observed in dacite lavas and pyroclastic rocks. Dacites from arc volcanoes are dominated by plagioclase, mafic minerals, oxides, and various amounts of quartz. Most typically have <2 modal % K-feldspar as sanidine, even in lavas with 50 vol% crystals (de Silva et al., 1994; Sparks et al., 2008; Folkes et al., 2011).

Figure 10.30 shows a particularly crystal-rich dacite sample (65 wt% SiO_2) collected from the extruding spine of Mount St. Helens in 2005. This rock has 49 vol% phenocrysts and, consistent with experiments, has no modal sanidine (plagioclase ≫ hypersthene, hornblende, oxides; Pallister et al., 2008). In keeping with lavas worldwide, dacite lavas range up to but not beyond roughly 50 vol% phenocrysts; at higher crystal contents, the touching crystal framework makes the magma so hyperviscous that magma flow through dikes and subsequent extrusion cannot happen (see Figure 7.11).

Figure 10.30 Extremely crystal-rich dacite from a 2005 dome-forming event at Mount St. Helens contains 50 vol% crystals but is devoid of sanidine and nearly devoid of quartz, the components of which were in the melt (now glass, black) upon eruption. This is as crystal-rich as a magma can be and still be capable of flow. This image was taken in circularly polarized light.

Sources: Sample courtesy of Martin Streck; sample 315-4 of Pallister et al. (2008) and Streck et al. (2008). AFG photo.

Compositional Types of Granitic Magma

Alumina Saturation Index

The great bulk of granitic rocks in subduction-related batholiths such as those of the Cordillera (see Figure 10.2) are calc-alkaline, with a mafic mineralogy dominated by hornblende, biotite, pyroxenes, magnetite, and ilmenite. However, important provinces of granitic rocks contain noticeable quantities of Al-rich minerals, such as cordierite, garnet, and muscovite; such minerals form because the magma has an excess of Al over that needed to form feldspars.

The **alumina saturation index (ASI)** measures the excess or deficit of Al in a rock; it simply ratios the molar amount of Al to Ca, Na, and K needed for feldspars (see Toolbox 10.1). Rocks with ASI < 1 are **subaluminous**, those with ASI ~ 1 (typically within a few percent) are **metaluminous**, and those with ASI > 1 are **peraluminous**. ASI values tend to

Alumina saturation index (ASI)
A measure of whether there is excess Al in a magma beyond that needed to make feldspar; if so, then peraluminous minerals such as muscovite and garnet may form to take up the excess.

Peraluminous, metaluminous, subaluminous
Rocks with ASI significantly greater than 1, within a few percent of 1, and significantly below 1, respectively.

Toolbox 10.1

Calculating the Alumina Saturation Index

The alumina saturation index is the ratio of the number of moles of Al in a rock to the number of moles of Na, K, and Ca needed to make feldspars (Zen, 1986). Feldspar formulas can be written in oxide form as $Na_2O{\cdot}Al_2O_3{\cdot}6SiO_2$, $K_2O{\cdot}Al_2O_3{\cdot}6SiO_2$, and $CaO{\cdot}Al_2O_3{\cdot}2SiO_2$; thus, each mole of Al_2O_3 requires a mole of some combination of Na_2O, K_2O, and CaO to make feldspars. If n_i is the number of moles of an oxide, then

$$\mathrm{ASI} = \frac{n_{\mathrm{Al_2O_3}}}{n_{\mathrm{K_2O}} + n_{\mathrm{Na_2O}} + n_{\mathrm{CaO}}} \qquad \textbf{(Eq. 10.2)}$$

Thus, if w_i is the weight percentage of an oxide, then

$$\mathrm{ASI} = \frac{\dfrac{w_{\mathrm{Al_2O_3}}}{102}}{\dfrac{w_{\mathrm{K_2O}}}{94.2} + \dfrac{w_{\mathrm{Na_2O}}}{62.0} + \dfrac{w_{\mathrm{CaO}}}{56.1}} \qquad \textbf{(Eq. 10.3)}$$

where the denominators below each oxide are the respective molecular weights of the oxides. One can include a term to account for Ca taken up by apatite, but this correction is usually small. Rocks that have ASI > 1 have corundum in their CIPW norms; ASI = 1.1 corresponds to about 1.5 wt% corundum in the norm.

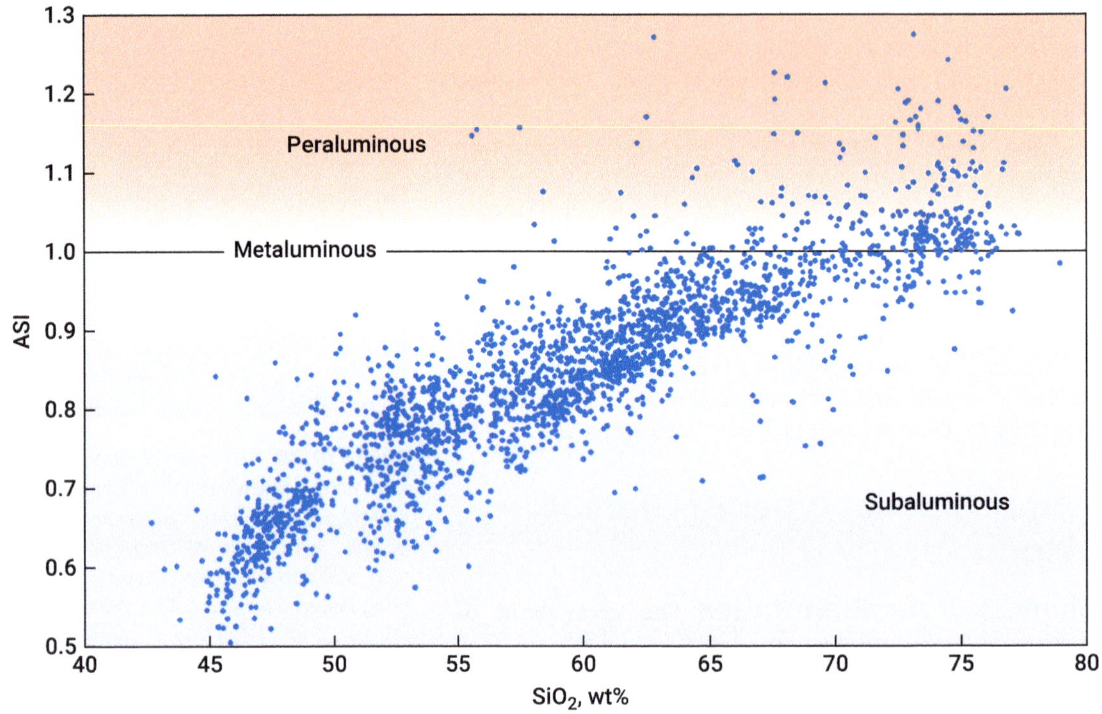

Figure 10.31 Alumina saturation index (ASI) vs. SiO_2 for rocks of the Sierra Nevada batholith. The high degree of scatter is likely due in part to mobility of alkalis during alteration.

Source: Data from Bateman et al. (1984).

scatter a lot owing to mobility of Na and K during weathering, but along the BADR trend, rocks are subaluminous at the low-silica end, reach ASI ~ 1 at about 70 wt% SiO_2, and become slightly peraluminous at higher silica values (Figure 10.31).

I-type granite
A granite born from igneous sources.

S-type granite
A granite formed via partial melting of sedimentary sources.

I-Type and S-Type Granites

Eastern Australia has been a proving ground for theories of granite genesis for many decades owing to the diversity of igneous rocks found there and keen interest from Australian geologists. Chappell and White (1974) published a short paper, less than 600 words long, that recognized two basic granite types in the Tasman Fold Belt of eastern Australia. They defined these as **I-type granite** and **S-type granite**; some of the basic distinctions between them, updated from the original paper, are given in Table 10.1. This short paper on the parentage of granites has been cited thousands of times—a kind of geo-equivalent of Watson and Crick's 1953 paper on the double-helix structure of DNA.

Chappell and White recognized that these two distinct magma types, recognizable in the field by their mineralogy, reflect different source rocks. I-type ("I" for "igneous") magmas arise from igneous precursors by one of the processes outlined in Chapter 8 (e.g.,

Table 10.1 Characteristics of I-Type and S-Type Magma Series

Characteristic	I-Type	S-Type
ASI	Generally <1.1	Generally >1.1
CIPW norm	Contains diopside or <~1 wt% corundum	Contains >1 wt% corundum
Compositional range	High to intermediate silica	Primarily high silica
Oxidation state	Relatively high; commonly magnetite-bearing	Relatively low; commonly ilmenite-bearing
$^{87}Sr/^{86}Sr$ (initial)	Magmatic, typically <0.706	High, typically >0.707
$\delta^{18}O$	Magmatic, typically <8‰	High, typically >8‰

Making Connections 10.2

Countertop Granites

In this book, we have tried to deemphasize rock classification and limit the number of rock names to the most common ones. However, there are times when even we have to say, "enough."

Q: When is a granite not a granite?

A: When it's a countertop.

That's because in the countertop industry, most rocks are categorized into either marble (soft) or granite (hard), regardless of how they form. In the case of countertop "granite," a wide range of rock types is represented, most of which are pretty far from true granite and some of which aren't even igneous rocks. That is, industry names can be far removed from geologic reality.

To be sure, some true granites are represented, meaning that they have roughly equal proportions of quartz, K-feldspar, and plagioclase; other granite cousins with varying proportions of these minerals are also common. But varieties of countertop "granite" range across the rock spectrum of felsic to mafic igneous rocks, high-grade metaigneous rocks, metamorphic rocks derived from any sort of protolith, and even some sedimentary rocks. Figure 10.32 illustrates some of the more interesting examples you may encounter. When it comes to countertops, we try not to focus on the "granite" appellation and instead appreciate the wide range of rock types represented in our kitchens. Even if they're not really granites.

Figure 10.32 Examples of rocks that are considered granites by at least some countertop distributors. (Top row) Countertop "granite" that is broadly silicic and contains common granitic minerals (quartz, K-feldspar, and plagioclase). Mount Airy actually is a granite; Baltic Brown contains spectacular ovoid K-feldspar crystals, sometimes rimmed with plagioclase; Blue Pearl is nearly 100% anorthoclase (high-*T*, Ab-rich feldspar, now exsolved into submicroscopic intergrowths of K-feldspar and plagioclase). (Middle row) Rocks farther removed from granites. Volga Blue is far too mafic to be a granite and would probably be classified as a gabbro or anorthosite. Ametista is a sillimanite gneiss (i.e., high-grade metamorphosed shale). The light partial melts are broadly granitic in composition, but the garnet and sillimanite are diagnostically metamorphic. Blue Bahia is principally a mixture of blue sodalite (a feldspathoid) and white K-feldspar, so would be classified as a feldspathoidal syenite; it's not even close to being quartz saturated. (Bottom row) Non-igneous "granites." Agata appears to be a calcareous metasediment. Garuda is a spectacular garnet-staurolite schist—another metamorphosed shale. Green Marinace is a metamorphosed conglomerate.

Sources: AFG, MJK photos; faces from By Belbury (CC BY-SA 4.0).

crystal fractionation or partial melting), whereas S-type ("S" for "sedimentary") magmas arise by partial melting of clay-rich sedimentary precursors such as shales and mudstones. S-type granites have significantly higher ASI values than I-type, and it is commonly assumed that is a consequence of higher Al in the sedimentary parent. However, this is generally not true; rather, high ASI in shaley rocks results in large part from low concentrations of Na and Ca rather than high concentrations of Al. As we shall see in Chapters 11 and 13, metashales with high ASI grow metamorphic mineral assemblages that provide useful information about P and T conditions during metamorphism.

10.4 Magma Ascent and Arrest

Learning Objectives

By the end of this section you should be able to:

10.4.1 Propose mechanisms by which magmas segregate and ascend.

10.4.2 Explain why magmas commonly stall in the crust and form plutons.

Why Does Magma Rise?

As discussed in Chapters 3 and 6, most magma generation occurs at depths of tens to hundreds of km. If a rising parcel of magma reaches the surface it becomes volcanic rock; if it freezes on the way up it becomes a pluton. What factors control ascent and freezing?

Compaction
In igneous petrology, expulsion of liquid from a crystal mush by crystal settling under the force of gravity.

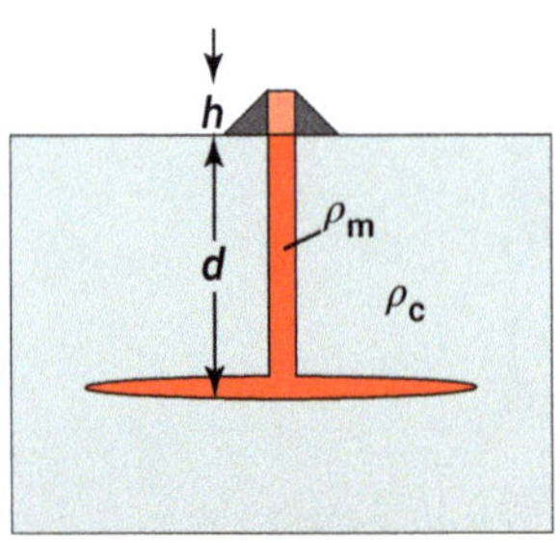

Figure 10.33 Simple buoyancy-driven model of magma ascent. Equilibrium occurs when the pressure at the base of the magma column is equal to the pressure at the same depth in the crust. Because magma is generally less dense than crustal rocks, this density difference can support a column (and thus build a volcano) that extends above the surface. The pressure at depth d in the crust is $P = \rho_c gd$, and in the magma column, $P = \rho_m gd$, which is less; to balance the columns, an additional height h is needed such that $\rho_c gd = \rho_m g(d + h)$.

Buoyancy A major factor that causes magmas to rise is buoyancy (Walker, 1989). Consider the simple model in Figure 10.33, where constant-density magma sits in constant-density crust in a system that connects a sill-like magma body to the surface via a dike. Mechanical equilibrium occurs when the pressure at the base of the magma system is equal to the pressure at an equivalent depth in the crust; for the general case where the density of the magma is less than the density of the crust, this provides a driving force for the magma to rise out of the conduit and onto the surface. Essentially, the denser crust sits on the magma body and squeezes the magma upward.

Tectonics Pressure differentials caused by regional tectonics may play a role in driving magma into the crust (Robin & Cruden, 1994; Brown & Solar, 1998). de Saint Blanquat et al. (1998) argued that subduction zone magmatism generally occurs in a transpressive tectonic regime (strike-slip along the arc coupled with shortening perpendicular to it), and that such a state of stress drives magma upward through the shear zone.

Compaction For melt to leave the partially melted volume in quantities greater than those dictated by thermal or volatile expansion, the crystal network containing the melt must undergo **compaction**. Compaction can occur via filter pressing or via the vertical stress induced by the weight of the crystals, which promotes descent of the denser crystals, ascent of the less-dense melt, and thus net upward movement of the melt. Compaction can produce a horizontal planar fabric expressed by alignment of minerals (Figure 10.34). In a purely compactional fabric, elongate minerals would be rotated toward the plane of compaction, but with their long axes randomly oriented in that fabric, whereas a fabric produced by flow should be more lineated than foliated, with long axes of elongate minerals rotated toward the flow direction.

Minimum Escape Percentage and the Wetting Angle If melt is to escape a volume of rock, it must find connected pathways out. For relatively low melt percentages, a given volume of melt can exist in completely isolated blobs or in thin, connected pathways along crystal edges, depending on wetting characteristics of the melt. If the melt and crystals have severe antipathy, the melt will attempt to minimize all contact area by beading up (effectively, standing on tiptoes) at the expense of potential energy needed to lift itself in the gravitational field (Figure 10.35a). If the melt and crystals are compatible, such that the melt tends to spread out along the crystal surface (as, e.g., alcohol or water on glass; Figure 10.35b), then the melt will attempt to maximize surface area by forming a thin interconnected network along grain edges.

These geometries are characterized by the dihedral angle θ. For melts with $\theta < 60°$ (Figure 10.35d), interconnected networks develop even at melt fractions (F) below 1 vol%. For intermediate θ ($60° < \theta < 120°$), isolated droplets form (Figure 10.35e) without interconnectivity, at least for low-melt fractions. For $\theta > 120°$, melts will attempt to separate completely from crystals to form pools. For rhyolitic partial melt in granitic rock or basaltic partial melt in olivine-rich rock, θ tends to be small (<60°), allowing (at least theoretically) for escape of melt from the source region even at melt fractions as low as a few volume percent (e.g., Waff & Bulau, 1979; Laporte & Watson, 1995).

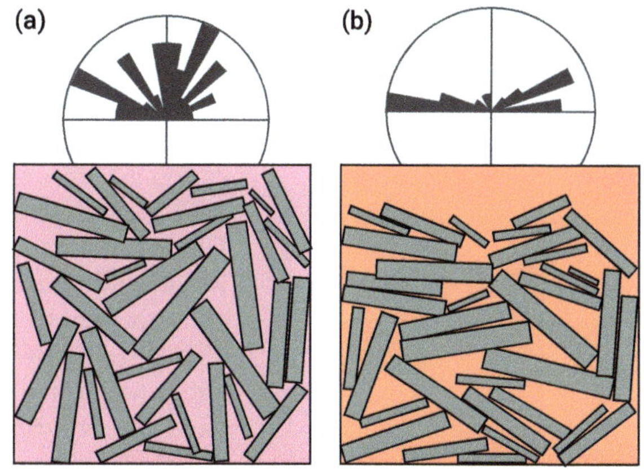

Figure 10.34 Diagrammatic illustration of compaction. (a) Lath-shaped crystals, randomly oriented and loosely packed. (b) The same crystals, packed more tightly by vertical compaction. This process, which here has produced about 10% free liquid at the top, results in significant alignment of the crystals, as shown by the rose diagrams of long axes at top.

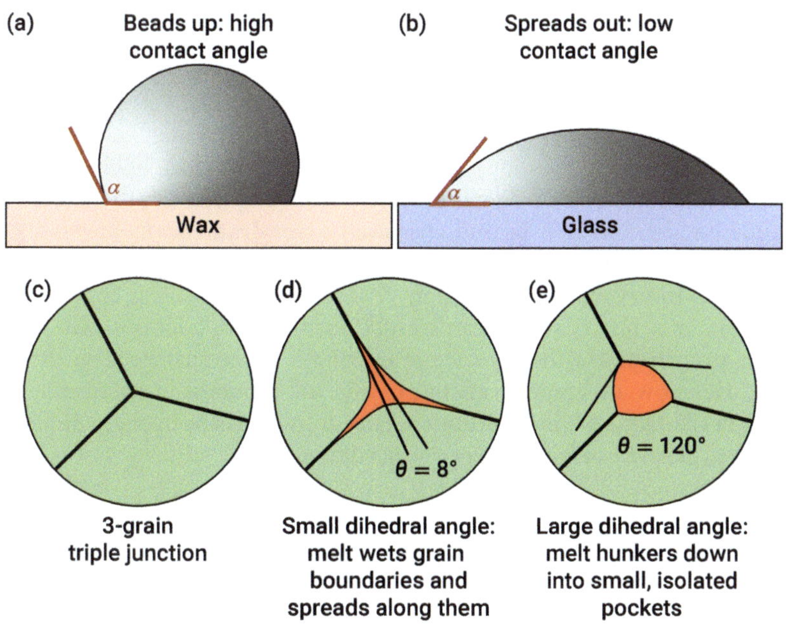

Figure 10.35 Side views of water droplets on (a) wax and (b) glass. The angle α is the contact angle; the smaller the angle, the more affinity the liquid has for the surface that it is sitting on. When the energy cost of liquid touching substrate is high, the liquid accepts higher gravitational potential energy, forming a taller droplet to lower the contact energy. (c) Schematic view of a three-grain junction (e.g., three olivine crystals) undergoing melting. If the melt tends to wet the crystals, it will extend along grain boundaries (d) and form interconnected networks at lower melt fractions than if melt-crystal contacts are high-energy, causing the melt to form isolated pockets (e). In the former case the dihedral angle is low, and in the latter, it is high. The low-angle case corresponds to silicate melts such as basaltic melt in peridotite and rhyolitic melt in granite, and the high-angle case to sulfide or metallic melts in a silicate host.

Source: (a, b) adapted from Sumner et al. (2004).

Volatile Exsolution Ultraviolent volcanic eruptions in the range of VEI 7 and above can launch tephra at speeds of several hundred meters per second to altitudes of over 40 km (Figure 7.39). The energy for such dramatic mass ejections comes from expansion of gases dissolved in the magma, and the relevant physics is borrowed from the study of cannons (Wilson, 1980).

Because the solubility of water in silicate melts is strongly pressure-dependent (see Figure 6.29), rising magma will exsolve gas if the pressure drops below the relevant solubility curve. As the magma continues to rise, the solubility of water drops further, and more and more bubbles come out, frothing the magma. Consider 1 kg of melt with 3 wt%, or 30 g, dissolved water. If the melt has a density of 2200 kg/m^3, then its volume is 454 mL—about half a liter, with the dissolved water making up about 30 mL. That water, if expanded to atmospheric pressure at 900 °C, would have a volume over 5000 times greater, a volume increase comparable to that accompanying detonation of TNT. Ejection of magma from the conduit unloads and depressurizes magma below, causing the depth at which the magma starts to exsolve to increase, driving more ejection. This runaway process will abate when the system runs out of magma or the magma supply rate from depth cannot keep up with the rate at which magma is being ejected.

Why Does Magma Stall and Freeze?

Loss of Heat Heat loss out the walls of the magma conduit is an obvious mechanism for arresting magma ascent and producing plutons. As noted in Chapter 7, flow through planar fractures is a common mechanism of magma transport in the crust, but planar fractures have a high surface area-to-volume ratio, leading to rapid heat loss and solidification. Choking-down of planar conduits to form pipe-like structures (see Section 7.5) is a heat-conserving mechanism.

Loss of Volatiles Loss of dissolved volatiles, particularly water, has a profound effect on the ability of magma to rise, for two main reasons. First, once bubbles start to form, they transport water out of the magmatic liquid, causing its viscosity to increase by orders of magnitude (see Section 7.3). Second, dissolved water dramatically lowers the liquidus and solidus temperatures of a magma (see Section 6.6), and thus loss of dissolved water causes a magma to crystallize. Increasing viscosity and increasing crystal content are death to magma movement, and volcanic eruptions that vent a lot of steam leave behind magma with a much-diminished chance of making it to the surface. The 1980 eruption of Mount St. Helens was a graphic example of this; after months of steam-rich explosions culminating in the May 18 Plinian eruption, later activity has been characterized by slow, typically non-explosive oozing of hyperviscous dacite.

Strong Layers Magma may cease to rise simply because it finds its way blocked by a rock unit that is too strong to penetrate. This mechanism is thought to determine where many sills spread out (Kavanagh et al., 2006). An important corollary is that the new sill, once cooled, can itself be a strong layer and form a barrier to later dikes. This predicts that sill complexes formed by multiple pulses will be stacked from the top down, with each new pulse spreading out below the previous one. Such downward stacking is observed in many horizontally sheeted plutons (see Section 10.7).

Neutral Buoyancy The crust is stratified in density, with less-dense rocks generally nearer the surface. If a magma is less dense than the lower crust but denser than the shallower crust, then it will encounter a depth at which its density is the same as that of the surrounding rock. A parcel of magma rising by buoyancy loses its upward buoyant force when it encounters rocks of lesser density, and this would be a favorable depth for lateral injection (Gilbert, 1877; Corry, 1988; Walker, 1989). However, most magmas have low

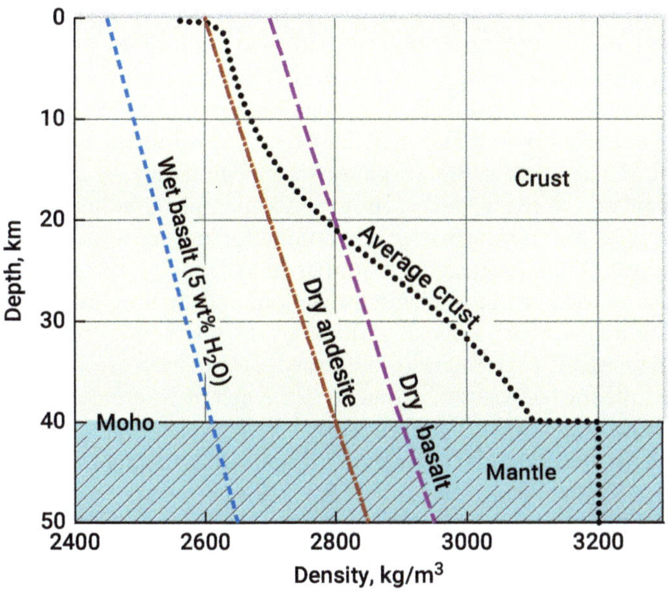

Figure 10.36 An average crustal density profile (modified to show expected drop in density in uppermost few kilometers) with predicted density profiles for columns of dry basalt (50 wt% SiO_2), andesite (55 wt% SiO_2), and basalt with 5 wt% H_2O. Densities from Figure 8.2. Only the basalt is dense enough to reach a level of neutral buoyancy, and a few percent water would negate that. More silicic lavas are too low density to encounter lower-density rocks except at the shallowest crustal levels. Effects of crystals, which would increase magma density, and vesiculation, which would decrease it, are ignored.

Sources: Modified from Christensen & Mooney (1995); average compressibility from Lange & Carmichael (1990).

enough densities that they likely never encounter crustal rocks of lower density until they are in the near-surface realm of fractured rock. Figure 10.36 compares average crustal density with expected density profiles from mafic and intermediate magmas. It is clear that most magmas, especially those with significant dissolved water, are too low-density to reach a level of neutral buoyancy. For this and other reasons, neutral buoyancy probably does not play a large role in arresting magma ascent (Menand, 2011).

10.5 Modes of Magma Ascent

Learning Objective

By the end of this section you should be able to:

10.5.1 Evaluate likelihood of different mechanisms by which magmas have been proposed to ascend.

Dikes and Sills

Dikes are the primary mechanism by which magma moves in the crust, or at least in the upper crust (Petford et al., 1993; Brown, 1994; Annen et al., 2015), and sills (see Figure 10.11) are a mechanism by which magma accumulates in the subsurface. Gently dipping sills are inferred (and locally seen) to have been fed by steep dikes (e.g., Gretener, 1969; Valentine & Krogh, 2006; Galerne et al., 2011); less commonly, sills are seen to have fed dikes, which may feed higher sills (Muirhead et al., 2014). The abrupt transition from a steep dike and horizontal extension perpendicular to the dike plane to a shallow sill and vertical extension perpendicular to the sill has preoccupied geologists for over a century.

Diapirs

Ascent of granitic magma in the form of **diapirs**—teardrop-shaped pods of magma several kilometers across—has long been invoked as a primary mechanism of magma movement. Grout (1945) conducted a series of experiments in fish tanks with various oils, clays, and syrups, producing elongated blobs of lower-density material strung out in higher-density

Diapirs
Kilometer-scale bodies of material, generally circular in map view, that have punched up through overlying layers either by buoyant ascent or owing to tectonic forcing of low-viscosity material.

material. Ramberg (1967), during a remarkable series of experiments in which layers of clays and waxes of differing densities were spun in a centrifuge at 1000 times the force of gravity, produced dramatic mushroom-shaped structures. These images influenced thought on granite emplacement, and guided field interpretation, for decades. In particular, the rising diapirs were encircled by circular synclines that formed as the buoyant material punched up through the host rocks. Blobby plutons were commonly assumed to have such features around them and were mapped to conform to this view, whether or not it was justified (see Sylvester, 1998, for an informative re-examination of such a situation).

The structures Ramberg and others produced are similar to salt diapirs, which have been extensively documented by field studies, seismic reflection, and drilling owing to their importance in localizing hydrocarbon deposits. Although the low density of halite plays a role in ascent of salt diapirs, salt's extremely low viscosity plays an even greater role. Salt moves around under the influence of stresses, primarily as a result of differential loading by strata above or by faulting.

Although many analogies have been made between salt diapirs and magma ascent in the crust (e.g., Quirk et al., 1998), there is a major difference: salt systems are essentially isothermal, whereas magmas are much hotter than the rocks they are moving through. In order for magma to move upward as a buoyant diapir, it must soften and shoulder aside the rocks above it, giving up its own heat to do so. Rising bodies of magma can give up only a limited amount of thermal energy for softening their wall rocks before they hit 50 percent crystallization, at which time rheologic lock-up occurs and ever-slowing ascent must cease (Mahon et al., 1988; see Toolbox 10.2).

Stoping

In mining terminology, a stope is an excavation from which ore has been extracted from either above or below the level of a tunnel. Daly (1903) borrowed this concept and coined

Toolbox 10.2

Energy Needed to Soften Rocks Around a Pluton

A simple model illustrates the thermal problems inherent in diapir ascent in the middle and shallow crust. Assume that a spherical pluton of radius r_p and temperature T_p is emplaced instantaneously into wall rocks of temperature T_w. The wall rocks have to be heated to a temperature at which they become ductile (T_d) in order for the diapir to rise. The thermal energy available in the pluton to soften the wall rocks is the product of the pluton volume, density, specific heat, and excess temperature, or

$$E_p = V\rho_p c_p (T_p - T_d) \quad \textbf{(Eq. 10.4)}$$

where ρ_p and c_p are the density and specific heat of the pluton, respectively, and V is the volume, given by

$$V = (4/3)\pi r_p^3 \quad \textbf{(Eq. 10.5)}$$

The thermal energy needed to heat the aureole, a spherical shell, from ambient temperature to the ductility temperature is

$$E_a = \left[(4/3)\pi\left(r_a^3 - r_p^3\right)\right]\rho_a c_a \left(T_d - T_w\right) \quad \textbf{(Eq. 10.6)}$$

where ρ_a and c_a are the density and specific heat of the aureole, respectively, and the bracketed term is its volume. Using values from Glazner and Miller (1997): r_p = 5 km, r_a = 6 km, T_p = 700 °C, T_w = 200 °C, T_d = 300 °C, ρ_p = 2500 kg/m^3, ρ_a = 2700 kg/m^3, and $c_p = c_a$ = 1000 J/(kg K), yielding calculated energies of $E_p = 5.2 \times 10^{20}$ J and $E_a = 1.0 \times 10^{20}$ J. Thus, there is enough thermal energy in the pluton to soften about five aureole volumes. The rising diapir will sweep out a cylindrical volume of softened wall rocks, and the height it can rise to is given by

$$x = \frac{E_p - E_a}{4\pi r_a^2 \rho_a c_a \left(T_d - T_w\right)} \quad \textbf{(Eq. 10.7)}$$

For these parameters, x is about 4 km if T_d = 300 °C, and about 1 km if T_d = 400 °C. These estimates ignore several important factors, such as conduction of heat to wall rocks outside the aureole, but they seem clear in showing that in the middle and upper crust, diapiric rise is limited to a few km at most.

the term **magmatic stoping** to explain contact relationships of plutonic bodies that truncate structures at their margins (see Figures 10.15 and 10.16)—in his experience, this included nearly all plutons. In the stoping hypothesis, magma makes space for itself at the level of emplacement by loosening pieces of wall rock from its roof, after which the loosened **stoped blocks** sink through the magma, presumably accumulating somewhere below the level of exposure or dissolving in the host magma.

Magmatic stoping
Hypothesis that magma bodies ascend by plucking blocks out of the roof and moving into the space thus freed up.

Stoped blocks
Xenoliths that have free-fallen through magma during stoping. This term is commonly, but mistakenly, applied as a default interpretation of any xenolith.

Stoping provides a simple explanation for contact relationships such as those in Figures 10.16 and 10.15a, seems quite reasonable, and has long been invoked as an important magma ascent process (e.g., Paterson & Fowler, 1993; Cobbing, 1999; Yoshinobu et al., 2003; Frost & Frost, 2019). However, Glazner and Bartley (2006)[4] found that the stoping hypothesis fails most of the tests that one can devise for it, including the following:

1. Plutons that rose via stoping should be packed full of xenoliths (Figure 10.37), but typical plutons have xenolith fractions $\ll$1 percent, and commonly $\ll$0.01 percent.
2. The missing stoped blocks cannot be made to disappear by assimilation, because this would leave a dramatic, recognizable imprint on the major-element, trace-element, and isotopic composition of the magma doing the assimilating.
3. Assimilation would rapidly freeze the magma.
4. Missing stoped blocks are commonly said to accumulate on unexposed pluton floors, but pluton floors are commonly exposed and are generally devoid of xenoliths (see Figure 10.17).

Although xenoliths are commonly called stoped blocks, it is clear that many xenoliths are isolated by injection of cross-cutting dikes or simply by multiple injections in a composite dike, without ever having been free-floating in magma. For example, in Figure 10.38, several fragments of granodiorite were separated from the dike walls by injection of successive dikelets, and some of these blocks were rotated significantly in the process. These high-T mafic dikes cooled relatively rapidly, producing chilled margins that delineate individual dikelets. In lower-T, water-rich granitic systems that stayed active for 1 Myr or more, recrystallization can mask much of the original intrusive structure (see Section 10.8).

Compaction and Porous Flow

Compaction, discussed in Section 10.4, is a form of porous flow, in which magma moves through a porous crystal matrix in response to a pressure gradient such as one driven by buoyancy or gravitational loading. Porous flow is akin to flow of groundwater through soil and fractured rock, but with a critical difference: the crystals through which the melt

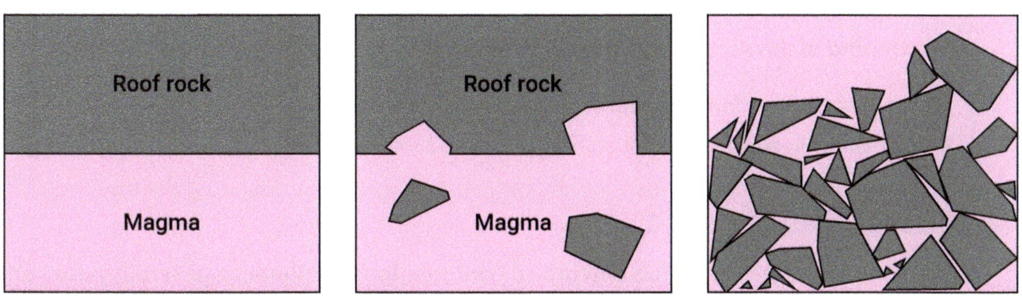

Figure 10.37 Cross section of a hypothetical squat, cylindrical body of magma rising via stoping roof rocks through itself. A general result from sedimentology is that loosely packed blocks take up roughly twice as much volume as intact rock, and thus accumulation of stoped blocks should lead to a magmatic breccia that is roughly 50% xenoliths. However, such large-volume accumulations of xenoliths are unknown, and pluton floors (see Figure 10.17) lack the expected accumulations as well.

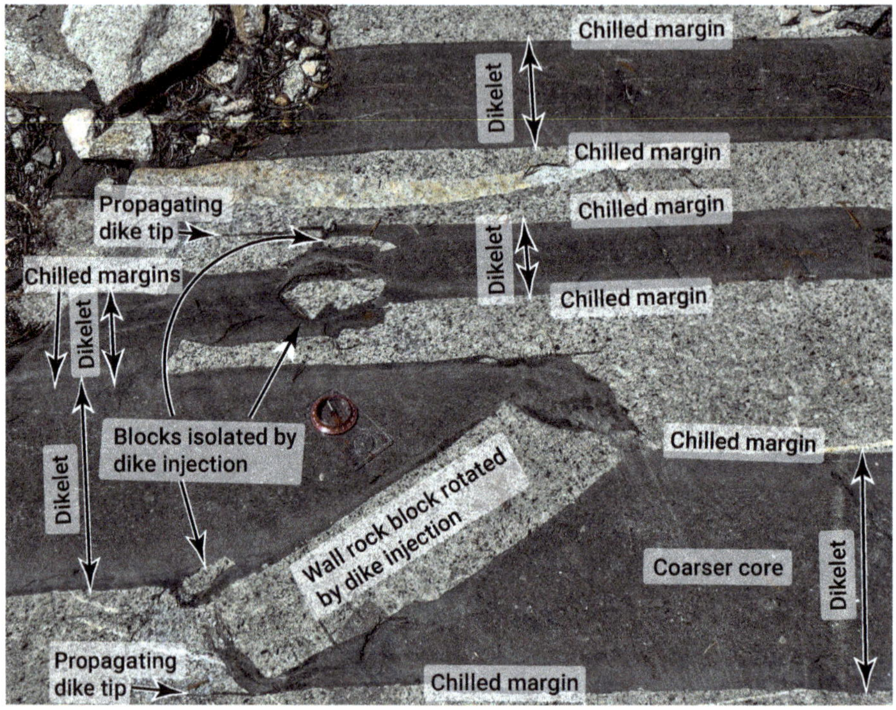

Figure 10.38 Closeup of dike injection showing separation, isolation, and rotation of wall-rock fragments by propagation of dike tips, two of which terminate in the photo. Dikelets have well-defined chilled margins, and two have merged on the left side of the photo to form a thicker dike. Large granodiorite wall-rock fragment in lower dike has rotated ~30° but is still attached to the lower wall. Oriented compass circle is 5 cm across.

Source: AFG photo.

is flowing react with the melt and may dissolve in it, opening up new channels, or be precipitated from it, closing up porosity and permeability. There is abundant evidence in mantle peridotites for replacement of peridotite by dunite during reactive porous flow (Kelemen & Dick, 1995; Kelemen et al., 1995). These important processes are limited in how far they can transport melt upward by the vertical scale of the porous matrix.

10.6 The Room Problem

Learning Objectives

By the end of this section you should be able to:

10.6.1 Explain the room problem.

10.6.2 Critically evaluate potential solutions to the room problem.

Pluton Dimensions

Room problem
The question of how space is made in the crust for plutonic bodies.

Interpretation of the shape and downward continuation of a pluton is generally based on what you can see at the surface coupled with prevailing wisdom about what happens at depth. Cross sections drawn through plutons (see Figure 10.16) generally show massive volumes of plutonic material taking the place of rocks, of whatever variety, that formerly occupied that volume. In many batholiths, the rocks at the surface are entirely granite and granodiorite, literally for as far as the eye can see (e.g., Figure 10.1b). Where did all the rock that used to be there go? This is the **room problem**.

The room problem is particularly acute for large magma bodies because of the relatively short cooling times involved. For example, to insert a cylinder of magma 20 km in diameter and 10 km deep (~3000 km^3, the volume of the largest supereruptions) into the shallow crust fast enough to keep most of it in a dominantly molten state requires assembling the magma body on timescales of about 500 kyr (e.g., Gelman et al., 2013). This in turn requires moving that same volume of relatively cold wall rock somewhere else in that same time period, a formidable problem in crustal remodeling. For example, if only half of the 10-km thickness of such a magma body were accommodated by surface uplift, that would require uplift of 5 km at a sustained rate of 1 cm/yr. Areas on Earth that have risen at such rates near volcanoes do so for periods measured in only years or decades (Pritchard & Simons, 2004; Singer et al., 2014). Long-term (≥1 Myr) rapid uplift elsewhere (e.g., in the Himalaya; Lavé & Avouac, 2001; Zeitler et al., 2014; Guevara et al., 2022) is forced by tectonics and river incision, not magmatism.

As noted in Section 10.2, the same observable map pattern at the surface can lead to dramatically different interpretations about what happens at depth, and thus to estimates of pluton volume (see Figure 10.18). The dimensions of plutons (particularly their vertical extents) remain unclear, even in areas with extensive geochemical, geochronologic, and geophysical data, such as the southern Rocky Mountains; compare the kilometer-scale pancakes of Tappa et al. (2011) and Mills and Coleman (2013) with the 20-km-tall beer cans of Lipman and Bachmann (2015).

Some Unworkable Solutions to the Room Problem

The room problem has occupied those who study plutons since the igneous nature of granite and other rocks was established in the 19th century. An early hypothesis for taking care of the room problem was granitization, in which granite is formed by metasomatic replacement of preexisting rocks by hot fluids. In essence, hot fluids bring in chemical components that transform the preexisting rock to granite, and carry the exchanged components out, to be deposited somewhere else. Field relations of wall rocks truncated at a pluton contact, such as those in Figure 10.15a,c, led many geologists to granitization as a solution (e.g., Reynolds, 1946; Bateman, 1965). The debate between those who favored an igneous, intrusive origin for granite versus those who favored a metasomatic origin raged for decades; it was not until Tuttle and Bowen's (1958) experimental study in the granite-water system that granitization was largely abandoned. Although the granitization controversy raged before radiogenic isotope methods were developed, granitization would predict that the granite should have isotopic ratios similar to the rock that was granitized, unless the granitizing fluid swamped out the original signature. Studies of $^{87}Sr/^{86}Sr$ in the 1970s demonstrated striking mismatches between many plutons and their wall rocks, further nailing the granitization coffin shut.

Stoping is often invoked as a space-making process for plutonic rocks, but stoping actually dramatically diminishes the volume available for free magma, replacing it with block-choked, immobile magmatic breccia (see Figure 10.37). Stoping could make space for a pluton to rise into its final resting place, at the expense of taking up roughly twice as much space deeper in the body, but this seems not to happen on Earth, because the predicted magmatic breccias are nowhere found.

Assimilation of wall rocks has been invoked to remove preexisting rocks from a rising magma's path (e.g., Pitcher & Berger, 1972), but this can be ruled out on thermal and chemical grounds (Bowen, 1922; Glazner & Bartley, 2006; Glazner, 2007).

Workable Solutions

Releasing Bends in Faults

One possible solution to the room problem involves emplacing magma into folds or faults, such as at releasing bends in strike-slip faults or into extensional fault zones. Field relations have long suggested this (e.g., Hutton & Reavy, 1992; Tikoff & Teyssier, 1992; Vines & Law, 2000; Acocella et al., 2000; Kalakay et al., 2001).

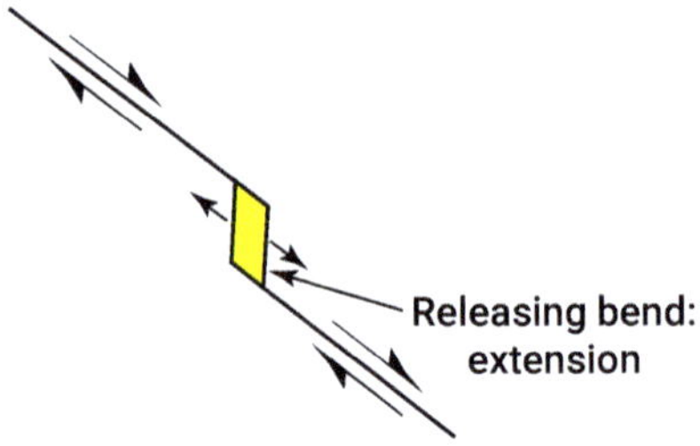

Figure 10.39 Schematic map view of a strike-slip fault showing how crustal voids can form at releasing bends (e.g., right steps in a right-lateral fault system). Such crustal expansion can be balanced by distant shortening (as at a subduction zone) or more locally at a restraining bend (e.g., a left step in a right-lateral fault). Similar systems can form in gently dipping thrust faults and extensional faults.

Space thus made can be balanced somewhere else; for example, space made at a releasing bend in a strike-slip fault can be balanced by uplift at a restraining bend or transferred to a plate boundary (Figure 10.39). Emplacement of sheeted dike complexes at oceanic ridges is an example of tectonic space-making, where the space needed by the injected magma is compensated for by thrust faulting at a distant subduction zone. This hypothesis has been criticized on the grounds that faults make space slowly (mm/yr), and thus any plutons emplaced in this manner would consist of recognizable sheeted dikes (Paterson & Fowler, 1993). However, recognition of slow, incremental emplacement (discussed later) has largely negated this criticism.

Laccoliths A second workaround is emplacement as laccoliths, sills, and lopoliths; these mechanisms largely avoid the room problem, because space is made by some combination of lifting the roof and dropping the floor. Lifting the Earth's surface just pushes air aside, an easy task that we accomplish every day with our bodies; dropping the floor of a pluton requires pushing the rocks below aside. This is easier than moving shallow rocks because deeper rocks are hotter for several reasons: (1) the geothermal gradient, (2) because they have had magma passing through them, and (3) because they have a hot magma body sitting on them. Thus, rocks under a pluton are naturally more ductile than those above it. Depressing the floor of a pluton requires lifting the Earth's surface a bit around it in order to balance volume.

However, even the mechanisms described here are difficult to envision if large plutons are emplaced rapidly. Lifting the Earth's surface 5 km in 100 kyr to emplace a 5-km laccolith as a large body of magma requires making a localized mountain range as tall as the Andes in a geologically brief interval, and such an event would produce an unmistakable sedimentary and geomorphic record of erosion, drastic river rearrangement, and other events. No such events have yet been recognized in the geologic record on Earth. The problem disappears, for any pluton geometry, if magma is emplaced incrementally, over a much longer timescale than needed to produce a large magma body.

Sinking Plutons

Many introductory textbooks show plutons, even in the shallow crust, as rising diapiric blobs that shoulder the wall rocks aside and push them upward like a salt diapir. This concept, although appealing and visually striking, is uncommon. Wall rocks do bend up around laccoliths, but field studies rarely confirm the predictions of a rising diapir. More common are curious anti-diapir contact relations in which the wall rocks, rather than bowing up over the pluton, bow down under it (Figure 10.40). Wall rocks dipping under plutonic rocks occur at a variety of scales ranging from small plutons (Marble Canyon pluton in Figure 10.15; Glazner & Miller, 1997; Sylvester, 1998) to large batholiths (Hamilton & Myers, 1974; Glazner et al., 2003). Wall-rock strata bend down approaching the pluton or batholith such that the youngest rocks are closest to the plutonic rocks. These relationships are common enough, especially around batholiths, that they demand explanation.

One hypothesis for why strata dip down beneath pluton is return flow, a conjectured process in which the wall rocks descend to fill the space left behind by diapirically rising magma—in essence, they are sucked down into that space in a narrow channel adjacent to the rising magma (e.g., Tobisch et al., 1986). This hypothesis depends on a diapiric ascent mechanism and predicts a narrow, high-strain shear zone adjacent to the pluton that shows pluton-up movement.

An alternative hypothesis is that the downturned strata simply indicate descent of the plutonic rocks relative to the wall rocks. This could happen for at least two different reasons. First, magma that is near neutral buoyancy when emplaced can become denser than the wall

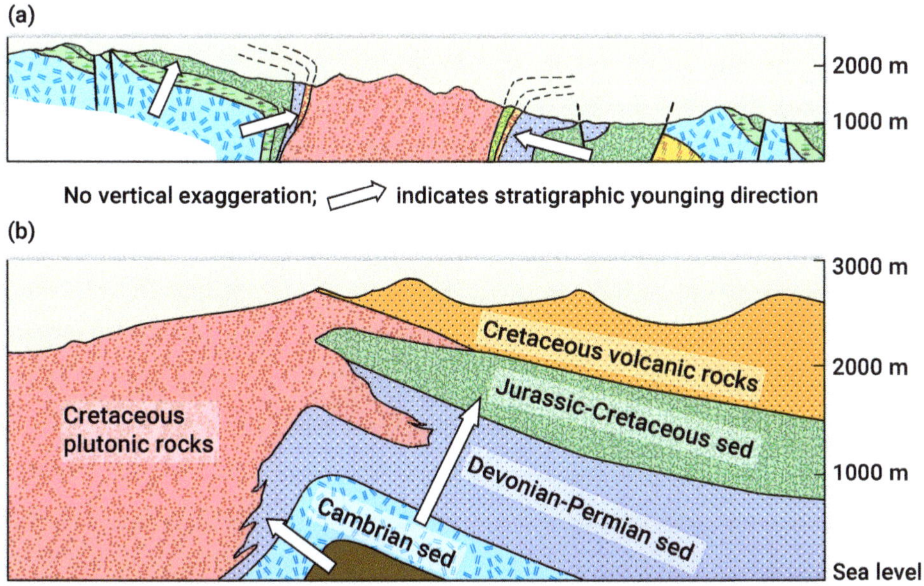

Figure 10.40 (a) Cross section through the Jurassic Marble Canyon pluton of the Inyo Range in California (see Figure 10.15b). Neoproterozoic and Cambrian wall-rock strata dive down adjacent to the pluton margin in a tight rim anticline, leaving youngest strata against the pluton, a relationship opposite to that predicted by diapiric ascent. (b) Cross section of the Boulder batholith in Montana. Strata ranging from Cambrian to Cretaceous are again folded down adjacent to the plutonic rocks, resulting in "bathtub syncline" geometry.

Sources: (a) from Glazner & Miller (1997), based on geologic map by Nelson (1971); (b) from Hamilton & Myers (1974).

rocks upon cooling owing to the significant density increase that accompanies crystallization (see Figure 8.2; Glazner, 1994; Glazner & Miller, 1997). Second, magmatic processes commonly build up a column of magmatic rocks several kilometers thick, and such a pile must sink under its own weight in an attempt to reach isostatic equilibrium. This process likely accounts for the "bathtub syncline" geometry of many batholiths—plutonic rocks are emplaced above wall rocks that they heat up, soften, push aside, and settle into as more plutonic and volcanic rocks pile up above them (Hamilton & Myers, 1967; Glazner et al., 2003).

10.7 Incremental Emplacement

Learning Objectives

By the end of this section you should be able to:

10.7.1 Critically evaluate evidence for incremental emplacement of plutons.

10.7.2 Interpret results of U-Pb geochronology in terms of concordia diagrams.

Duration of Pluton Assembly

As already noted, early concepts of plutons as frozen diapirs or other large bodies of magma required that the magma body invade and occupy its final resting place at a geologically rapid rate, because magmas cool rapidly. Glazner et al. (2004) calculated the conductive cooling time for a body of magma 5 km deep, 20 km wide, and infinite in the third dimension (a 2D model) emplaced instantaneously with its top 15 km below the surface. Instantaneous emplacement of such a huge body is impossible, and purely conductive cooling is unlikely, but these parameters were chosen to estimate the maximum time that such cooling might take. Even with these choices, however, the entire pluton cooled below the point at which it was 50 percent crystallized in about 500 kyr. They interpreted this as a maximum emplacement duration for a large pluton. More sophisticated thermal models have been run (see summary in Blundy & Annen, 2016), but all indicate emplacement durations of less than 1 Myr, even for the largest plutons, if they were emplaced as one large batch of magma.

Incremental Emplacement and Composite Plutons

Concordia diagram
A plot of $^{206}Pb/^{238}U$ vs. $^{207}Pb/^{235}U$, used to interpret U-Pb data in terms of age and as a check that the two chronometers are in sync.

When the precision of geochronologic analyses was poor, geochronology and thermal modeling were not in conflict, because geochronology could not resolve the times (<1 Myr) required to test the thermal models. However, development of U-Pb zircon dating techniques reset this argument, because it is now possible to determine ages with precisions much less than 100 kyr (Schoene, 2014; see Toolbox 10.3).

Toolbox 10.3

Uranium-Lead Zircon Geochronology

There are many different isotopic geochronometers in use, but the most valuable for plutonic studies is the U-Pb system applied to the mineral zircon ($ZrSiO_4$). This system is used primarily to determine crystallization ages of igneous rocks, and there are several reasons why it excels at this task:

1. There are two separate clocks running: decay of ^{238}U to ^{206}Pb and decay of ^{235}U to ^{207}Pb. ^{238}U has a half-life of 4.468 Gyr (so the Earth now has about half of the ^{238}U that it was born with), and ^{235}U has a half-life of 704 Myr (so the Earth only has about 1% of its original ^{235}U; see Table 2.2). Thus, there are two separate, non-linear clocks keeping time, and if the system has been disturbed, then these two clocks will not agree. This provides an internal check on the quality of the result.
2. Zircon, with its Zr^{4+} ion, readily takes up the similarly sized U^{4+} ion (see Figure 8.1) but has no place for Pb^{2+}. As a result, zircon crystals form with an abundance of parent U but essentially no child Pb.
3. Pb atoms that form by decay of U are not particularly happy in the zircon lattice, but diffusion of Pb in zircon is so slow that it nevertheless is largely held onto even at magmatic temperatures.
4. Zircon crystals are highly resistant to chemical weathering and mechanical abrasion. Resistance to weathering helps to keep the parent and child isotopes in place, and resistance to abrasion means that zircon crystals can survive multiple cycles of erosion and deposition in sedimentary rocks. The oldest material known on Earth is not a rock but zircon crystals that formed more than 4 Ga that are found in metasedimentary rocks that were laid down about 3 Ga (Cavosie et al., 2007). One can only speculate about how many cycles of erosion and deposition filled that billion-year gap.

Geochronologic treatment of U-Pb data is different from that of Rb-Sr data. Lead has four stable isotopes: non-radiogenic ^{204}Pb (1.4% of Pb on Earth); ^{206}Pb (24.1%) and ^{207}Pb (22.1%), which form by decay of U; and ^{208}Pb (52.4%), which forms by decay of ^{232}Th. In an ideal zircon with no initial Pb, the only Pb present will be that formed by decay of U (^{206}Pb and ^{207}Pb; we are ignoring Th), so there is no common stable isotope to ratio measured values to, like ^{86}Sr in the Rb-Sr system (see Section 2.3). Instead of an isochron diagram, U-Pb data are analyzed with a **concordia diagram** (Figure 10.41). A zircon forming today, with no Pb, will plot at the origin on this diagram, because the $^{206}Pb/^{238}U$ and $^{207}Pb/^{235}U$ ratios are both zero, but as time passes, both ratios will increase, sweeping out the concordia curve. In a well-behaved zircon crystal, both ratios will plot on concordia, giving a well-determined estimate of the age at which the zircon crystallized.

U-Pb dating of zircon is, of course, far more complicated and exacting than this brief introduction can relate. Dating begins with field work and thoughtful choice of samples, followed by:

1. Rock crushing or disaggregation
2. Separation of zircons and other dense minerals using a water table that works something like a gold miner's pan
3. Magnetic separation

4. Processing through heavy liquids (dense organometallic liquids that are, unfortunately, both volatile and toxic)
5. Hand picking of choice zircon crystals
6. Thermal treatment to anneal radiation damage
7. Acid treatment to dissolve away damaged and U-rich domains that are most susceptible to chronologic disturbance
8. Dissolution in hydrofluoric acid at high pressure
9. Passage of the dissolved material over several ion-exchange columns to separate U and Pb
10. Analysis on a thermal ionization mass spectrometer

With these painstaking techniques, U-Pb ages can now be determined with a precision of better than 0.1 percent—a remarkable, if hard-won, achievement.

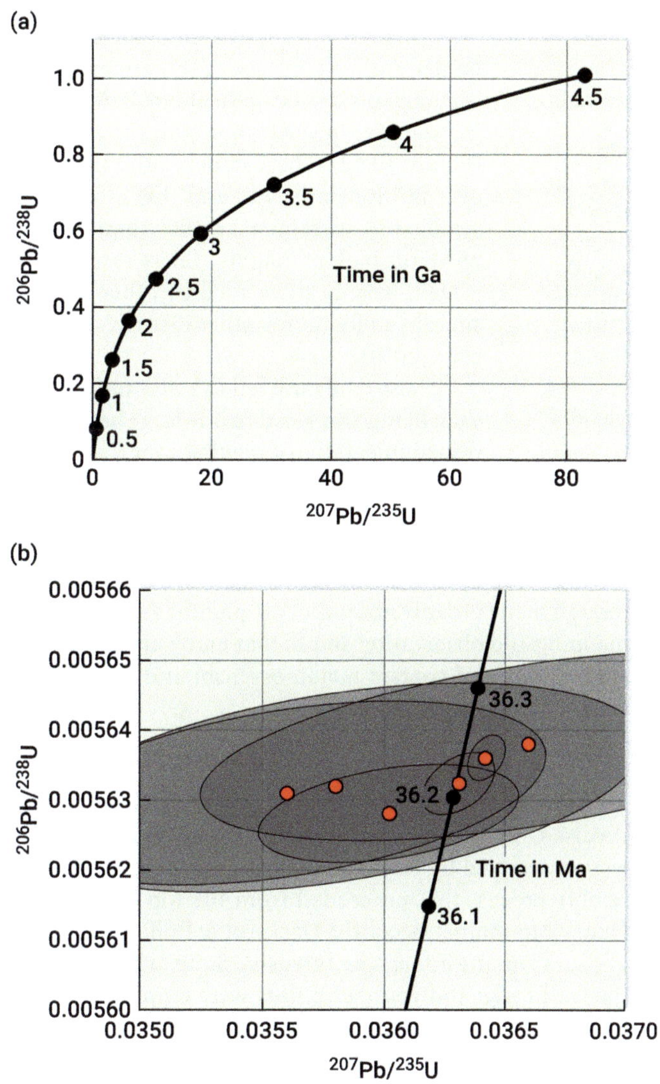

Figure 10.41 The concordia diagram for U-Pb dating. A given zircon crystal, if undisturbed and lacking any "common" (non-radiogenic) Pb, will crawl up the curve as time passes. If the system has been undisturbed, the ages derived from decay of ^{238}U and ^{235}U should agree and thus lie on the curve. (a) Full diagram; perfect U-Pb systems with no initial Pb would plot along the line at the age at which the zircons crystallized. (b) An example of high-precision data showing points clustered around concordia. These data, from an ignimbrite in the Organ Mountains, New Mexico, are interpreted as indicating crystallization at 36.215 Ma ± 0.016 Myr—a precision of 16 kyr.

Source: (b) From Rioux et al. (2016); see that paper for explanation of error ellipses.

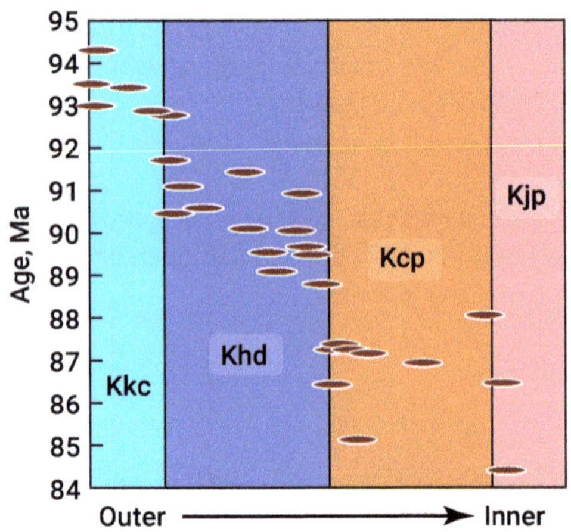

Figure 10.42 Summary of precise U-Pb zircon ages from the Tuolumne Intrusive Suite, plotted against position relative to mapped pluton boundaries (see Figure 10.3). Ellipses give approximate precision of ages (height) and location (width). The three main units of the suite were emplaced over nearly 10 Myr, with the Half Dome Granodiorite (Khd) and Cathedral Peak Granodiorite (Kcp) taking 3–4 Myr each. Kkc = granodiorite of Kuna Crest; Kjp = Johnson Granite Porphyry.

Sources: Ages from Coleman et al. (2004) and summary of Paterson et al. (2016).

Incremental emplacement Construction of a plutonic body in small increments rather than via intrusion of the entire body all at once.

Modern U-Pb dating applied to the classic Tuolumne Intrusive Suite led to surprising results. As of the turn of the century, numerous ages on granites and granodiorites in Yosemite had been collected by a variety of methods ranging back to the early days of K-Ar dating in the 1950s (Curtis et al., 1958). These ages spanned about 10 Myr, a range highly inconsistent with the Bateman and Chappell (1979) interpretation that the three main plutons of the suite were intruded as large magma bodies in rapid succession. Thermal models indicated that this would have happened within 1 Myr or so, much less than the time span indicated by the various dating attempts (Glazner et al., 2004). Thus, the wide range of ages was assumed to be a result of analytical problems.

Coleman et al. (2004) attempted to better nail down the Yosemite ages by determining precise U-Pb zircon ages on the major units. They found that they do indeed span about 10 Myr, with a consistent pattern of becoming younger inward from both the east and west margins toward the center (Figure 10.42). These data require that the intrusive bodies were assembled over millions of years, and this in turn requires that the plutons were assembled in increments rather than brought to the site of emplacement as one large blob of magma. These results, highly controversial when first presented, have been verified by other labs, and similar large age ranges have since been found for other intrusive suites (e.g., Matzel et al., 2006; Memeti et al., 2010).

These data have led to a new understanding of how plutons grow: rather than being emplaced in one large batch of magma, many or most grow by repeated injections of magma over time spans longer than the cooling history of even a large batch of magma—a process known as **incremental emplacement**, which produces composite plutons.

A persistent and significant criticism of the hypothesis that most plutons are highly composite is the lack of evidence for boundaries between the increments. Although many broadly granodioritic plutons are at least superficially homogeneous, careful mapping and observation shows that many are extremely complex internally (e.g., Bartley et al., 2018), and contact metamorphism and annealing (see Section 10.8) may erase evidence of incremental assembly.

Top-Down Assembly

Precise U-Pb geochronology on numerous plutons and intrusive suites indicate that they were assembled by emplacement of new pulses of magma beneath earlier ones in an assembly process that proceeded from the top down. This has already been noted for the Henry Mountains laccoliths (Jackson & Pollard, 1988) and is a reasonable interpretation of data from the Tuolumne Intrusive Suite, where the earliest pulse was emplaced against the wall rocks and younger pulses were emplaced at successively deeper levels, with the whole package arched in laccolithic manner (Coleman et al., 2012). Top-down assembly is consistent with focusing of magma emplacement at strong layers and rheologic contrasts (Menand, 2011).

10.8 Annealing and Recrystallization in Plutons

Learning Objectives

By the end of this section you should be able to:

10.8.1 Articulate the arguments for textural and chemical re-equilibration of plutons to subsolidus metamorphic conditions.

10.8.2 Propose a mechanism by which plutonic rocks re-equilibrate at metamorphic conditions.

Evidence for the Metamorphic Life of Plutons

Igneous or Metamorphic?

When magma erupts and is quickly cooled, its texture is a reliable recorder of what the magma was like just prior to eruption. For example, there is little question that a basalt with phenocrysts of olivine and plagioclase set in a fine-grained or glassy matrix consisted of melt with crystals of olivine and plagioclase just prior to eruption. For rapidly cooled magmas, little change happens during the quench period other than rapid crystallization of myriad microscopic crystals from the melt. However, plutonic rocks, which cool much more slowly, are more difficult to interpret. Kilometer-scale bodies of magma can cool over periods of time that are long enough for significant textural changes to occur; there is every gradation from rapidly cooled hypabyssal rocks whose textures mimic those of volcanic rocks to coarse-grained plutons that cooled over several million years and have highly modified textures that reflect this long passage through metamorphic conditions.

Textural modification is a controversial topic in petrology that can inflame the passions of even the most docile petrologist. This topic is, in essence, a question of whether plutons are igneous or metamorphic—a dichotomy that is as misleading as it is simple. Plutons are intruded as magmas and therefore are by definition igneous, but many undergo a long (from one to several Myr) history at temperatures corresponding to the amphibolite and greenschist facies. They are thus subjected to a long metamorphic history after their igneous birth. Which are they, igneous or metamorphic?

The viewpoint taken here is that plutons are *both* igneous and metamorphic—or, more accurately, metaigneous. This duality is central to the character of any plutonic rock. It is built into the magmatic systems that produce plutons; the "look and feel" of a plutonic rock such as a granite is igneous by definition, yet it reflects a long metamorphic history.

Cooling Rates and Textural Memory

Volcanic rocks can cool at rates of hundreds of degrees per minute upon eruption, effectively quenching a record of their high-T state in which melt congeals as glass. Relatively small intrusive bodies intruded at shallow depth, such as subvolcanic domes and shallow laccoliths, cool more slowly such that melt fully crystallizes to a fine-grained groundmass, but phenocrystic texture (e.g., millimeter-scale plagioclase crystals in an aphanitic matrix) in such rocks is still a reliable indicator of what the magma was like upon intrusion. Large plutonic systems cool rapidly through the nominal solidus (600–700 °C) but then can take a million years or more to cool below 300 °C (see Section 11.7), producing a coarse-grained rock that might bear little textural resemblance to a volcanic rock produced from the same magma. This long, slow cooling can have a profound effect on texture and on mineral compositions. Phenocrystic texture in such a rock (e.g., K-feldspar crystals several centimeters long in a medium-grained matrix; discussed later) can be highly misleading.

Information from Mineral Compositions

Some of the most compelling evidence for metamorphic modification of plutonic textures comes from mineral compositions that indicate formation at temperatures well below the nominal solidus (~650 °C) of granitic rocks. Giant K-feldspar crystals (megacrysts; see Figures 10.3 and 10.43) are common in granodiorites around the world, and their large size and euhedral shapes have led many to the conclusion that they form early in the magma's history when there was plenty of room to grow (Vernon & Paterson, 2008; Moore & Sisson, 2008; Holness et al., 2018). However, this is contradicted by phase equilibria studies and abundant other evidence that they grow late in the magma's crystallization history (e.g., Figures 10.28 and 10.29). The compositions of these giant crystals indicate crystallization at temperatures below 600 °C; in particular, they are too Or-rich and An-poor to have crystallized at the higher temperatures (e.g., >700 °C) needed for a melt-rich magma (see Figure 8.24; Glazner & Johnson, 2013). Their compositions are consistent with the limbs of the binary Ab-Or solvus at temperatures on the order of 400 to 550 °C. Temperatures computed

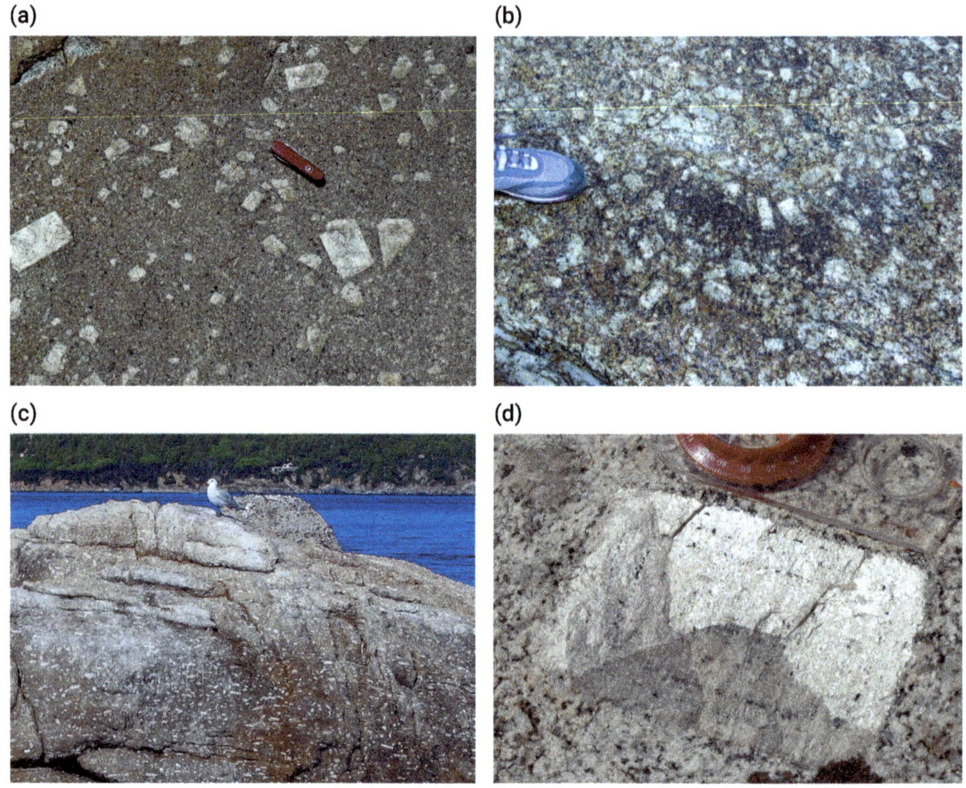

Figure 10.43 K-feldspar megacrysts. (a) Glacially polished surface in the Cathedral Peak Granodiorite, Yosemite National Park, California. Slight weathering makes the crystals show up well. Knife is 8 cm long. (b) Wet beach exposure of megacrystic granite of the S-type Peninsula pluton near Cape Town, South Africa. (c) A small island of megacrystic, 7-Ma granite near Elba, Italy; seagull for scale. (d) Cleavage surface on 10-cm megacryst in the Cathedral Peak Granodiorite showing Carlsbad twinning, which is ubiquitous in K-feldspar megacrysts. Upper part and left and right wings are reflecting the sun, whereas the darker wedge-shaped area in the bottom half is in the other twin, which is not oriented to reflect. Zonal arrangement of sparse minerals, chiefly hornblende here, is also ubiquitous.

Source: AFG photos.

using the two-feldspar thermometer (Putirka, 2008) typically give values of 750 to 800 °C for dacite lavas, consistent with experimental studies (see Figure 10.28), whereas granites give much lower temperatures, below 600 °C.

Quartz compositions in these same rocks also indicate low temperatures of crystallization and recrystallization. The Ti content of quartz is a function of temperature, because there is a miscibility gap between quartz and rutile (TiO_2) just as there is between albite and orthoclase. Ackerson et al. (2018) found low Ti concentrations in quartz crystals from the Tuolumne Intrusive Suite and estimated crystallization temperatures of about 475 to 550 °C, well below the nominal solidus and consistent with the K-feldspar data. Similarly, magnetite crystals in granites are typically almost devoid of Ti, as contrasted with magnetite in volcanic rocks of comparable composition, which contain several percent of the ulvöspinel[5] (Fe_2TiO_4) component. Although the position of the magnetite-ulvöspinel solvus is not well known and is complicated by magnetic effects (compare Ghiorso, 1997, with Lilova et al., 2012), it is clear that magnetite that is nearly pure Fe_3O_4 must form at temperatures well below the nominal granite solidus.

Phase equilibria of dacitic magmas, petrography of dacites, and mineral compositional data thus require that the textures and mineral compositions of dacitic magmas undergo thorough reconstruction during passage through the invisible last 50 percent of crystallization (see Section 10.3) on their way to becoming granodiorites.

Consequences of Incremental Emplacement

Thermal Cycling and Repeated Hot-Water Baths Data summarized earlier indicate that many large plutons were emplaced in increments over several million years, typically in top-down order. A natural consequence of this is that T must fluctuate up and down repeatedly, a process known as thermal cycling. Experimental work demonstrates that thermal cycling has a profound effect on crystal size distribution. As T rises, crystals dissolve, and the proportion dissolved from small crystals is larger than that dissolved from large ones; as T falls, the newly dissolved material has to find a place to re-precipitate—musical chairs for molecules—and the large crystals have much more area, and thus many more sites to attach to, than the small ones. The net effect is that thermal cycling transfers mass from small crystals (many of which cease to exist) to large ones in one manifestation of an important process known as **dissolution-reprecipitation**—10 important syllables that we can abbreviate as DR.

Dissolution-reprecipitation Mineral replacement process by which slightly unstable material dissolves in a fluid and reprecipitates in a more stable form.

The DR process is important in systems other than rocks. The evolution of snow from aggregates of fluffy, feathery powder through granular snow to névé (snow that has been partially melted, recrystallized, and compacted) to firn (an even denser product made from névé that has survived from previous years) proceeds largely via DR driven by daily variations in T. In the kitchen, thermal cycling of your freezer causes ice in ice cream to grow via DR, producing an undesirable crunchy feel (Donhowe & Hartel, 1996). This is DR with real economic consequences, enough to make intrepid food scientists in cold-weather gear watch ice crystals grow while sitting in a freezer for hours (Caillet et al., 2003).

The ability of thermal conduction to move heat through a growing pluton is limited by the slow rate at which heat diffuses through rock. However, in a crystallizing magma, there is commonly an agent ready and willing to move heat rapidly—water. Subduction-generated magmas are quite water-rich, containing, by various estimates, at least 5 wt% and commonly more (e.g., Carmichael, 2002; Wallace, 2005). Owing to the low molecular weight of water, this is a huge volume, most of which must depart when the rock crystallizes, because typical crystallized equivalents (e.g., granodiorite) have only about 0.5 wt% H_2O. And 5 wt% H_2O dissolved in the magma equates to about 10 percent of the magma's volume; thus, a pluton 2 km thick emits enough water upon crystallization to fill a swimming pool with the area of the pluton to a depth of 200 m (Figure 10.44).

Most of this water must escape the crystallizing magma to be released at the surface as hot springs vapor or fixed into the wall rocks in the form of hydrous metamorphic minerals formed during metamorphism. As low-density water vapor must travel upward to find its way out, rocks above the crystallizing magma, including rocks formed by cooling of earlier increments of magma, must be bathed repeatedly in waves of hot water released by crystallizing magma pulses below. This is speculative, but essentially required by what we know of magma systems. The effects of this bath are likely profound.

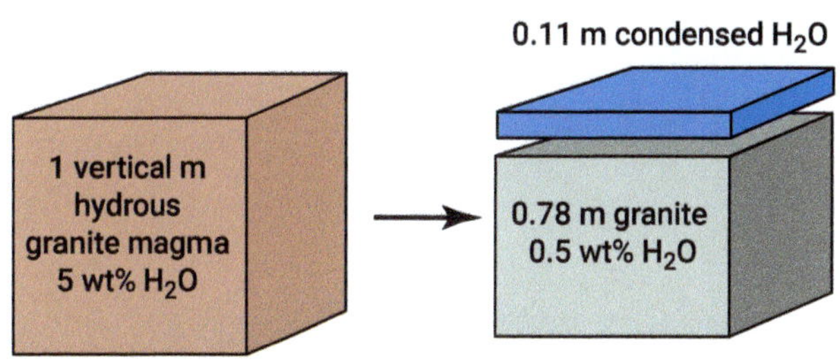

Figure 10.44 Schematic illustration of the volume of water contained in a moderately hydrous granitic magma, where 5 wt% of a cubic meter of magma with a density of 2200 kg/m^3 (see Figure 8.2) is 110 kg, which is 0.11 m^3 of liquid water. The remaining part of the magma, 2090 kg, crystallizes to a rock with a density of ~2750 kg/m^3, with a volume of 0.76 m^3. Thus, a given thickness of hydrous magma ends up as rock only ~75% as thick, and hot water vapor equivalent in condensed form to ~10% of the original thickness must escape to the surface through the rocks above.

Self-Metamorphism In the traditional model of pluton emplacement, hot magma comes in, cools, and that's that. In the incremental model, a pulse of magma comes in and cools partially or completely, and then another pulse comes in, and so on, possibly for millions of years. These surges of magma release pulses of heat and water that travel upward and baste the rocks above the active site of magma injection. In top-down systems, much of what is being basted is earlier parts of the incrementally emplaced pluton itself. The rocks into which the pluton was emplaced (e.g., shale) are clearly contact-metamorphic wall rocks, and we expect them to be metamorphosed (see Chapter 11). It is equally clear, but generally ignored, that the earlier pulses of magma in an incrementally emplaced pluton end up as the contact-metamorphosed wall rocks of later pulses. Thus, incrementally emplaced plutons are self-metamorphic, and their textures and mineral compositions reflect this in both their mineral equilibria, noted earlier, and in their textures.

Textural Healing

Although classic plutonic texture with its interlocking large crystals is by definition igneous, several observations indicate that crystals in plutonic rocks are commonly recrystallized, changing their compositions and rearranging their grain boundaries in the process. These include the following:

1. K-feldspar megacrysts in granites and granodiorites worldwide commonly show evidence of having been broken by aplite dikes and then healed across the fracture (Figure 10.45a; Pitcher, 1993, p. 70ff.; Johnson & Glazner, 2010). Typically, the healed fracture can be traced across the crystal by a trail of quartz blebs. Brittle fracture of a crystallizing magma can only occur when the magma is rich enough in crystals to provide a rigid framework—perhaps 70 percent crystalline or more (Hibbard & Watters, 1985). These observations thus indicate significant recrystallization of these large crystals at a late stage of the magma's crystallization history.
2. Plutonic K-feldspar crystals have much higher K and lower Ca (i.e., more Or component and less An component) than volcanic sanidine (Glazner & Johnson, 2013). Recalling from Figure 8.24 that K-rich feldspar becomes more sodic with down-T crystallization, the only way for the crystals to become more potassic is by exsolution along the solvus to relatively low temperatures.
3. It is common in granitic rocks to find brittle fractures that can be easily traced where they cut markers such as mafic enclaves or dikes, only to disappear where they cut only the granitic host (Bartley et al., 2018). Figure 10.45b shows a particularly clear example of this. As with fractured megacrysts, the crystallizing magma cannot brittlely

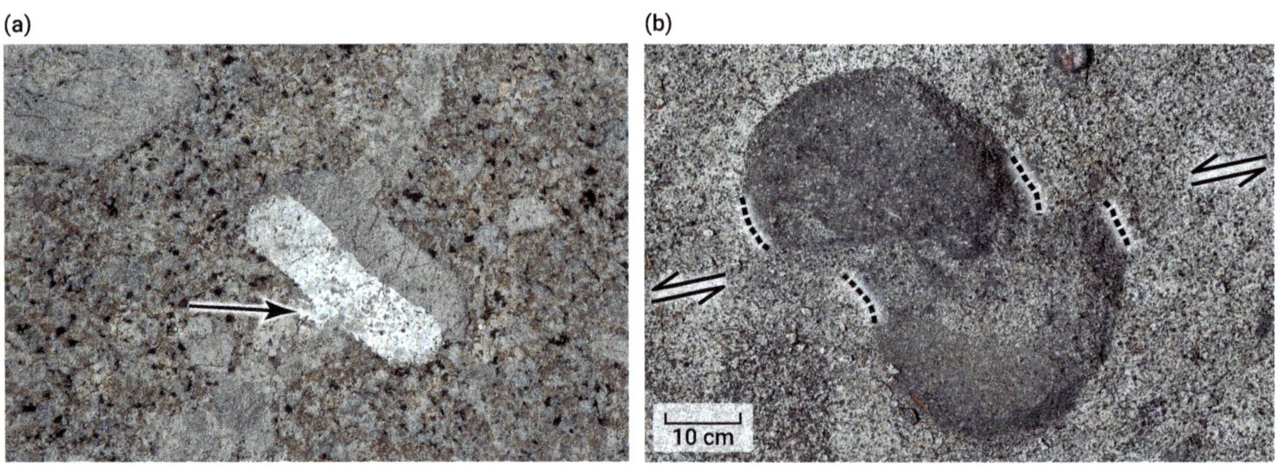

Figure 10.45 (a) A K-feldspar megacryst in the Cathedral Peak Granodiorite, Yosemite National Park. This crystal was cut by a leucogranite dike but has healed across the break, as demonstrated by sunlight reflecting off a large cleavage face on the lower half of the twinned crystal. That side of the crystal apparently annexed some of the K-feldspar in the dike, which is now in crystallographic continuity with it and reflects brightly (arrow). Compass edge is 5 cm long. See Glazner & Johnson (2013). (b) Enclave that was cut by a fault with 10 cm left-lateral separation. The fault clearly cuts the enclave but is not visible in the granodiorite on either side, which must have been solid enough during faulting to transmit shear stresses to the enclave. A likely explanation for disappearance of the fault in the granodiorite is textural healing. Such relationships are common in these granodiorite plutons. See Bartley et al. (2018) for more examples.

Source: AFG photos.

fracture until it is largely crystallized, so features such as this record complete or near-complete healing of crystals cut by a brittle fault. None of the phases in the rock (K-feldspar, quartz, plagioclase, biotite, hornblende) show any evidence at naked-eye or electron-microscope scale for having been deformed (Bartley et al., 2018).

4. Mafic and ultramafic rocks locally show strong evidence for volume-for-volume replacement of layered plutonic rocks by another rock type. For example, dunite bodies locally form dike-like bodies in peridotites and pyroxenites, but veins of chromite in the dunite trace out former layering in the host rock, indicating that the dunite replaced the host. McBirney and Hunter (1995) presented evidence that anorthosite in the Skaergaard Intrusion in Greenland (see Section 14.5) formed by replacement of layered gabbro, such that islands of gabbro are preserved with the orientation of their layering intact. They argued that many of the classic textures in the Skaergaard are not primary igneous features, but rather resulted from recrystallization or metasomatism.

10.9 Relationships between Plutonic and Volcanic Rocks

Learning Objectives

By the end of this section you should be able to:

10.9.1 Articulate compositional differences between volcanic and plutonic rocks.

10.9.2 Argue for and against mechanisms to produce compositional differences between volcanic and plutonic rocks.

Compositional Distribution

The relationships between plutonic and volcanic rocks have occupied petrologists for centuries. In the color frontispiece to his 1839 masterwork *Elements of Geology*, Lyell showed dikes from a magma chamber, part of a composite pluton, feeding an erupting volcano. This illustration, and the hundreds or thousands of similar ones that have graced textbooks and papers in the following centuries, have firmly cemented a link between the two great classes of igneous rocks, but the details remain relatively unknown.

It has been pointed out for over a century that at the Earth's surface, volcanic rocks are generally more mafic than plutonic rocks (Daly, 1914). In compilations of igneous rock compositions from western North America and from circum-Pacific subduction zones, plutonic rocks average roughly 5 wt% higher SiO_2 than volcanic rocks (Glazner et al., 2015). In these compilations, volcanic rocks are far more abundant and younger than plutonic rocks; these biases may bias the resulting averages, but it is clear by observation that low-silica rocks are much more common in the volcanic record than in the plutonic. This distinction is clear in Figure 10.46, which compares the distributions of SiO_2 in volcanic and plutonic rocks from western North America. The mode for the volcanic rocks is near 50 wt% SiO_2, whereas that for plutonic rocks is above 70 wt% SiO_2. Circum-Pacific subduction zones show a similar pattern.

The reasons for this strong dichotomy between volcanic and plutonic rocks are generally attributed to the physical properties of the magmas. In particular, the far higher viscosities of silicic magmas make them far slower to ascend (unless explosive decompression occurs), and thus more likely to freeze before reaching the surface.

Compositional Comparison

The volcanic-plutonic connection has long been viewed through the lens of the Bowen paradigm that crystal fractionation drives parental magmas to higher silica. This has led to the commonplace assumption that typical plutons are the crystallized remains

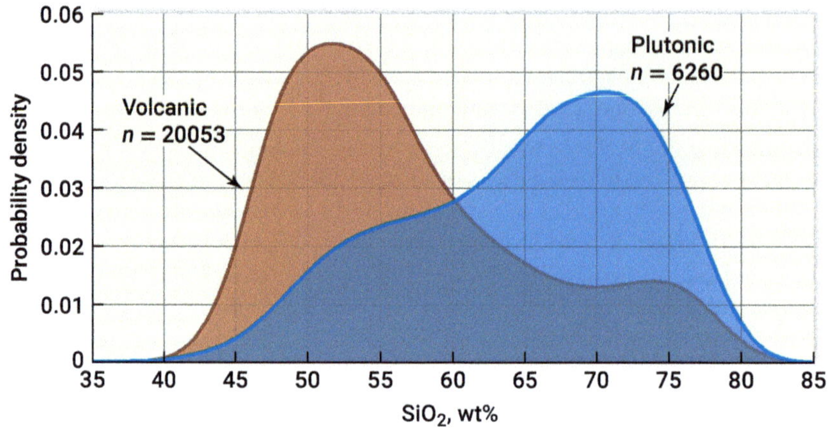

Figure 10.46 Comparison of SiO_2 contents of volcanic and plutonic rocks from western North America. These probability density plots show that that analyzed plutonic rocks are far more silica-rich than analyzed volcanic rocks, consistent with field observations going back at least to Daly (1914). Note that although the area under each curve is unity, there are over three times as many volcanic as plutonic analyses in the database.

Source: Data from the NAVDAT dataset (Glazner et al., 2015).

of magma that lost a crystal-poor magmatic fraction to eruption; thus, plutons and the erupted, melt-rich magmas that escaped from them are complementary parts which, when summed, add up to the original magma. This revitalized old concept, commonly referred to as the mush model (Hildreth, 2004; Bachmann & Bergantz, 2004), can be tested using the concepts developed in Chapter 9.

One test relies on the trace-element principle that liquid separated from a crystallizing magma should be enriched in incompatible elements and depleted in compatible elements, whereas the corresponding melt-depleted mush should have complementary enrichments and depletions. For example, Sr and Ba, compatible in plagioclase and K-feldspar, should be significantly enriched in plutons relative to volcanic rocks derived from them by separation of liquid. Glazner et al. (2015) tested this hypothesis using a large dataset of analyses from circum-Pacific subduction zones and found no evidence for such a relationship. In general, volcanic and plutonic rocks in this dataset show the same trace-element patterns, and Glazner et al. argued that plutonic rocks typically are simply magmas that froze in the crust before erupting rather than the frozen remains of magma chambers that fractionated to produce more-silicic volcanic rocks. This controversial topic is critical to understanding crustal magmatism.

Summary

- Plutons are recognizable, mappable intrusive bodies analogous to sedimentary formations. They may be homogeneous or heterogeneous and may have accumulated over timescales of several million years.
- Plutonic rocks form some of the world's most spectacular landforms owing to their strength and consequent resistance to erosion.
- Many plutons, especially those emplaced in the upper and middle crust, were constructed from myriad dikes.
- Dikes of meter-scale-and-greater thickness are typically composite, formed from multiple parallel injections.
- Massive dike swarms can transport magma thousands of kilometers from its source.
- Many shallow plutons were emplaced as laccoliths that made space for themselves by lifting their roof rocks.
- Some elongate plutons, rich in parallel wall-rock screens, are essentially dike swarms.

- Contacts between similar-aged plutons are commonly gradational over length scales ranging from meters to hundreds of meters.
- The dominant granodiorite plutons of arc batholiths crystallize from dacitic magmas whose first 50 percent of crystallization consist of mafic minerals and plagioclase; quartz typically does not begin to crystallize until about 50 percent crystallization, and K-feldspar does not join the crystallizing assemblage until the magma is well past 50 percent crystallization and the corresponding lock-up point.
- Magma ascent in the upper and middle crust is predominantly by diking. Diapirs and stoping are not viable ascent mechanisms.
- Precise U-Pb zircon geochronologic studies have shown that many plutons were emplaced in small increments over timescales ranging from 0.1 to 5 million years. Recognition of slow, incremental emplacement largely solves the room problem (accounting for the space needed for the intruded magma).
- Plutons are metaigneous rocks in which early intruded increments are the contact metamorphic wall rocks of later increments.
- Evidence for metamorphic recrystallization of granitic rocks includes formation of K-feldspar megacrysts from interstitial, late-crystallized K-feldspar; K-feldspar megacrysts that have healed after being cut by late-stage dikes; healing of minerals across brittle faults; low-temperature compositions of many minerals, including highly potassic K-feldspar and Ti-free magnetite; low temperatures recorded by two-feldspar geothermometry; and ubiquity of nearly pure albite along grain boundaries.

Notes

1. These strata form the Poleta Formation, familiar to the thousands of students who have learned geologic mapping in the Poleta folds area.
2. Another English geo-word taken from German, meaning "streaks."
3. Until they cleave along one of their perfect cleavages, but that almost never happens.
4. This paper hit a nerve, eliciting three commentary responses from a total of 11 authors (Clarke & Erdmann, 2008; Paterson et al., 2008; Yoshinobu & Barnes, 2008). The comments and the reply by Glazner and Bartley (2008) provide interesting fodder for discussion.
5. Another good name for a heavy metal band.

CHAPTER 11

Contact Metamorphism, Hydrothermal Systems, and Thermochronology

Hot magma rises into cooler rocks before it erupts or freezes. If magma makes it to the surface, most of its heat is dissipated to the atmosphere by convection and radiation, but if it freezes in the crust, its excess heat is transferred to its wall rocks, with effects that range from subtle to profound. Heat transfer may be by conduction, which is quite slow, or by convection of hydrothermal fluids, which can be an order of magnitude faster and have the side effect of producing geothermal energy systems, hot springs, and ore deposits.

Rocks are terrible conductors of heat—or, to put a positive spin on it, excellent insulators. Granite would make an acceptable, if heavy, picnic cooler. Bricks are essentially rocks with some air in them, and the combination of low thermal conductivity and high melting temperature makes them useful as refractory insulation in ovens.

The amount of thermal energy carried by magma is surprisingly high. Consider a medium-sized cylindrical pluton, 5 km in radius and 2 km thick, intruded at a temperature of 800 °C and cooling to ambient conditions of 100 °C. A little calculation shows that in so cooling, the body will release on the order of 10^{20} J (100 exajoules), or enough energy to power the United States for several decades. Our energy problems are solved! Well, not really, because the energy is released over a huge area (how do we collect it?), over a time span of a million years or so (we'd have to be patient), and the second law of thermodynamics reminds us that only a fraction of this is available for useful work. But that energy can have a profound effect on the rocks surrounding the pluton.

The wall rocks of plutons are subjected to contact metamorphism induced by heat from invading magma and from circulating fluids derived both from crystallizing magma and from surrounding rocks. In this west-looking aerial view across the eastern Sierra Nevada, light-colored Triassic granite in the foreground intrudes dark brown metamudstone and metasandstone that form the ridge crest. Cretaceous granite, granodiorite, and diorite can be seen in the distance. The metasedimentary rocks were deposited in the Paleozoic, folded and faulted by Mesozoic shortening, and then intruded by plutons. They are now tightly folded into steep packages that strike northwest, parallel to the Sierra Nevada batholith and the elongation direction of many plutons. These rocks have been intensely contact metamorphosed, and skarn mineralization of marbles behind the brown ridge was a major source of tungsten in the twentieth century. (See Section 11.5 and Figure 11.26.) We don't know why the rocks in the upper center are blue and are not going there to find out.

Source: AFG photo.

11.1 Contact Metamorphism

Learning Objective

By the end of this section you should be able to:

11.1.1 Predict the changes in texture and mineralogy that a rock undergoes during contact metamorphism depending on proximity to the heat source.

A Few Examples

Contact metamorphism
Metamorphism driven by proximity to an igneous body. Includes both thermal effects and fluid-rock interaction.

Contact aureole
The region adjacent to an igneous body whose mineralogy, geochemistry, and/or texture have been influenced by the intrusion (i.e., the contact metamorphic rocks around an intrusion).

Slates
Fine-grained platy rocks composed mostly of chlorite, various clay minerals, and quartz.

Hornfels
Literally a stone that resembles animal horn: hard, fine-grained, and usually dark gray or black with little layering and no tendency to split along planes. Hornfels forms adjacent to plutons as rocks thoroughly recrystallize.

Porphyroblasts
Relatively large crystals that grow in a metamorphic rock. Similar to the igneous term "phenocryst" but applies to metamorphic rocks.

Granoblastic
A texture in metamorphic rocks in which grains are approximately equidimensional and grain boundaries meet at triple junctions with angles near 120° in 2D cuts.

Contact metamorphism is the set of processes that changes the character of wall rocks around a pluton. As the name implies, it generally occurs near pluton contacts, and the area affected by contact metamorphism is the **contact aureole**, by nature a zone with fuzzy boundaries. Metamorphism in the contact aureole occurs as a result of diverse interacting factors, including heat flowing from the pluton; hot, water-rich fluids permeating and flowing within the wall rocks; and reaction of wall rocks with fluids and magma.

Around many plutons, changes in the wall rocks that are visible to the naked eye can be used to map zones of metamorphism. In the Santa Rosa Range of north-central Nevada, contact metamorphism affected Triassic low-grade metasedimentary rocks (chiefly **slates**) that were intruded by Cretaceous granitic plutons (Figure 11.1), producing abundant **hornfels**. The mapped aureoles extend up to a few kilometers from contacts with the granites. Compton (1960) used the first appearance of **porphyroblasts** of cordierite $[Al_3(Mg,Fe)_2AlSi_5O_{18}{\cdot}nH_2O]$, andalusite, biotite, or staurolite $[(Mg,Fe)_2Al_9Si_4O_{23}(OH)]$ to define the outermost contact zone. These crystals, typically millimeter-scale, give the rock a spotted appearance (Figure 11.1, inset). The contact between the outer and inner aureole was based on matrix grain size. Here and in other examples, contacts were defined on the basis of features visible in the field rather than by laboratory analysis.

Textural Changes and Textural Equilibrium

Hornfels is characterized by **granoblastic** texture in which the minerals are equant and of fairly uniform size (Figure 11.2). The texture in such a rock is a compromise between the crystal habit the relevant mineral prefers (e.g., quartz typically grows in a prismatic habit with tapered ends) and the need to pack the crystals together with no open space—that is, with all faces touching other faces. Clearly, most crystal shapes cannot fill space without leaving gaps, so alternate crystal shapes are required.

One solution would be for the crystals to take on cubic shapes, because cubes can be packed to fully fill the space available. However, cubes, with eight sharp corners and 12 sharp edges, are not the most stable shape that a crystal can form. Consider the schematic 2D atomic lattice in Figure 11.3. Atoms in the interior have all their bonds linked to other atoms and have eight neighbors. Those on faces, however, have only five neighbors, and one of their bonds is dangling, without a partner. This atom is in a higher-energy environment than the interior atom. The 90° corner atom has two dangling bonds and only three neighbors. The other side of the lattice has been beveled to produce a 135° corner, and an atom parked there has four neighbors, a better state of affairs than at the 90° corner. In a general sense, atoms in the interior of a crystal have the lowest free energy (i.e., are most stable), followed at successively higher energy by faces, obtuse edges, acute edges, and corners (see Figure 11.3).

Spheres have no corners or edges but cannot fill space without gaps. What compromise shapes do the crystals take on? Studies of annealed metals, foams, human cells, and other space-filling, surface-area-minimizing materials show that individual units commonly assume subspherical, polyhedral shapes (Figure 11.4a). Matzke (1946) used a binocular microscope to optically isolate and sketch individual bubbles in a long-lived soap foam, counting the number of faces of each and the number of sides on each face. He found that

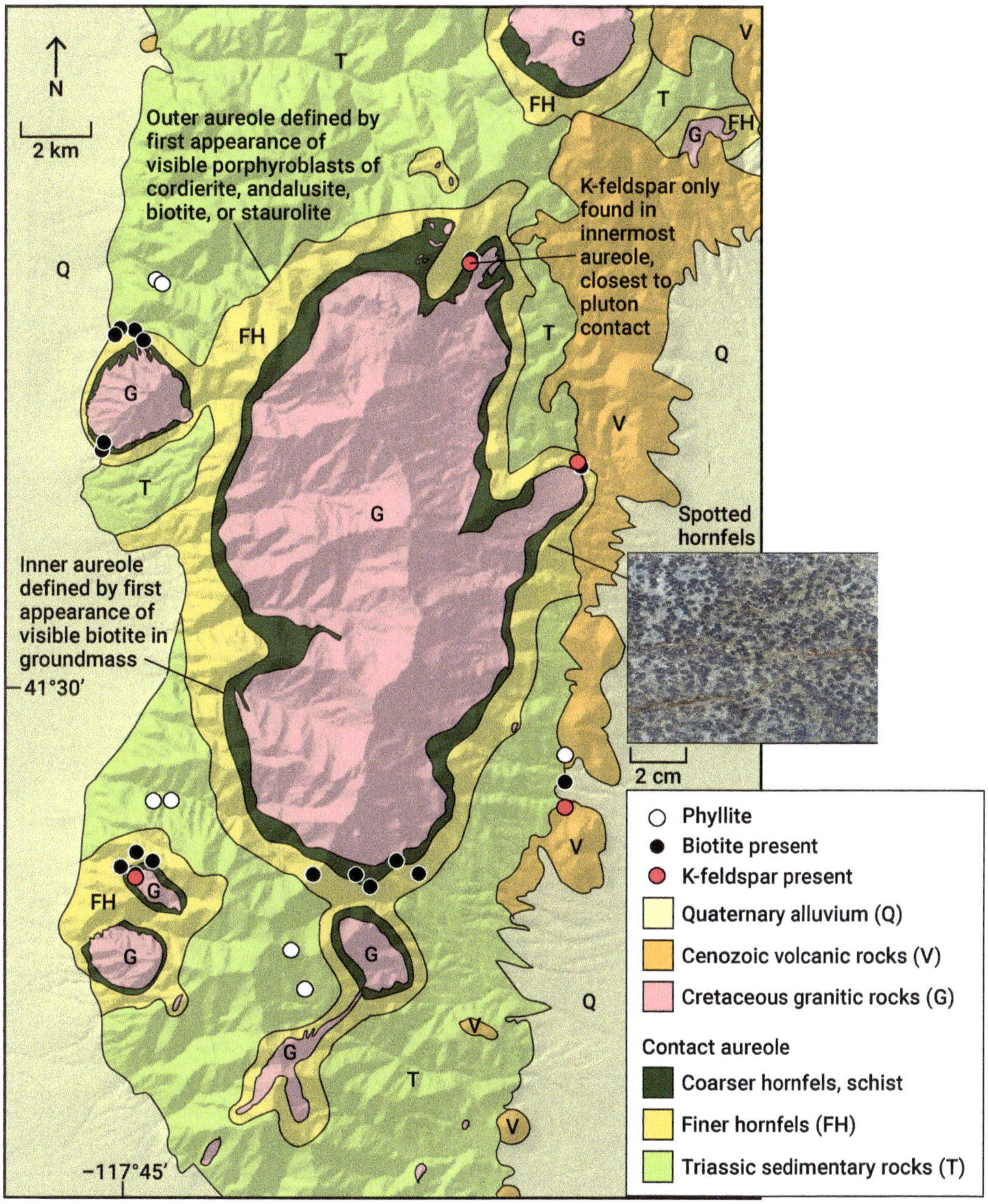

Figure 11.1 Geologic map of the Santa Rosa Range of north-central Nevada, simplified from Compton (1960). Here, Cretaceous granites intruded Mesozoic metasedimentary rocks (phyllite and slate derived from mudstone, sandstone, and carbonates), producing a contact aureole whose extent was mapped in the field on the basis of appearance of key index minerals, especially biotite, staurolite, cordierite, and andalusite. Dots give locations of samples that were examined petrographically in thin section. Inset shows typical sample of spotted hornfels with cordierite porphyroblasts.

Sources: Map adapted from Compton (1960); inset: MJK photo.

bubbles with 12 to 15 faces dominate the array, with five being the dominant number of sides on a face (Figure 11.4b). Lead pellets compressed in a steel cylinder, cells of human fat tissue, and crystals of brass show similar distributions (Marvin, 1939; Smith, 1964). Annealed rocks consisting largely of a single mineral, such as quartzite and marble, have fundamentally the same texture.

A thin section of a granoblastic rock is a 2D cut through a packed array of crystals such as those in Figure 11.4, and a characteristic of such arrays is that three-grain junctions

Figure 11.2 Quartz crystals in quartzite showing granoblastic texture. Note that crystals at most triple junctions meet at ~120° angles. Field of view 2 mm.

Source: Alessandro Da Mommio (http://www.alexstrekeisen.it).

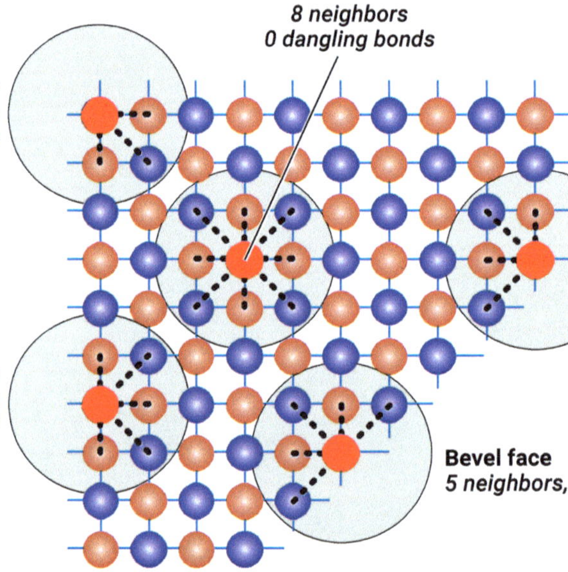

Figure 11.3 Schematic 2D atomic lattice showing the different neighborhoods and bond satisfaction for atoms found in various positions. In this example, interior atoms are the most stable and 90° corners the least; beveling the corner increases its stability.

tend to meet at angles close to 120°. A material with a dominance of 120° grain boundaries, such as that in Figure 11.2, is said to have foam texture. Development of this texture, as with most geochemical processes, is an energy-minimization strategy (Figure 11.5).

Grain Size

Changes in mineral size and shape are commonly evident around plutons even when there are no obvious mineral reactions. For example, limestone, dolomite, and chert commonly coarsen from submillimeter grain size far from the contact to marble or quartzite with grain sizes in the mm range adjacent to the pluton (Figure 11.6).

Many famous marbles used for sculpture, such as the Carrara marble of northern Italy (Michelangelo's *David*) and Parian marble from the Greek island of Paros (*Winged*

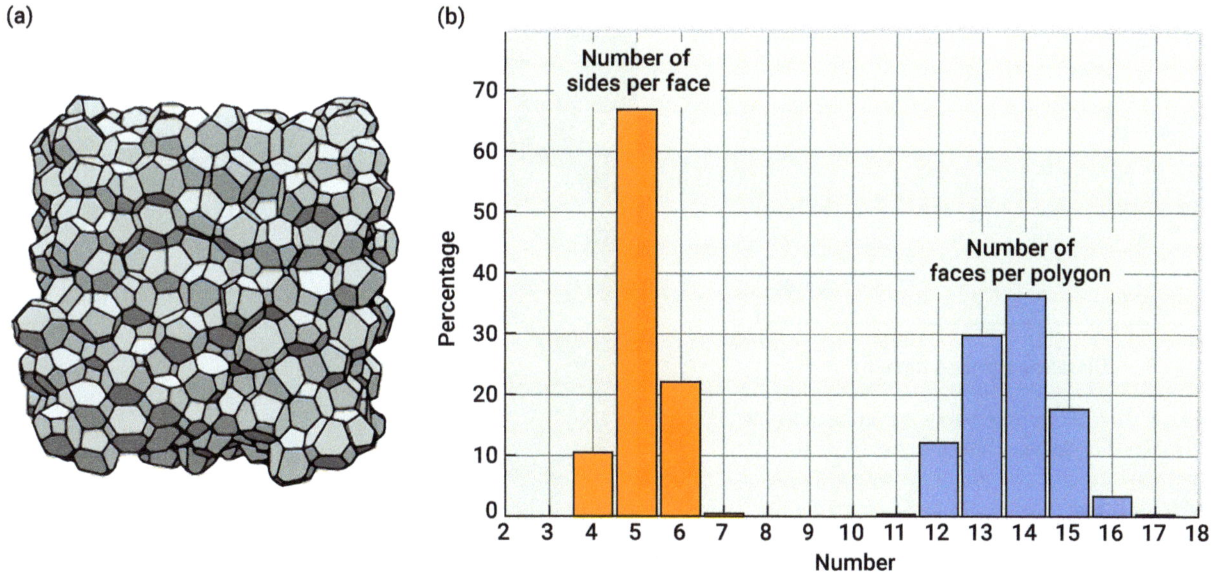

Figure 11.4 (a) Mathematical simulation of foam structure. (b) Distribution of faces per polyhedron and edges per face in soap foam. Pentagonal faces and 14-sided polyhedra are the most common. The simulation in (a) and crystals in granoblastic rocks such as marble (see Figure 11.7) have a similar distribution.

Sources: (a) Modified from Kraynik et al. (2003); (b) Data from Matzke (1946).

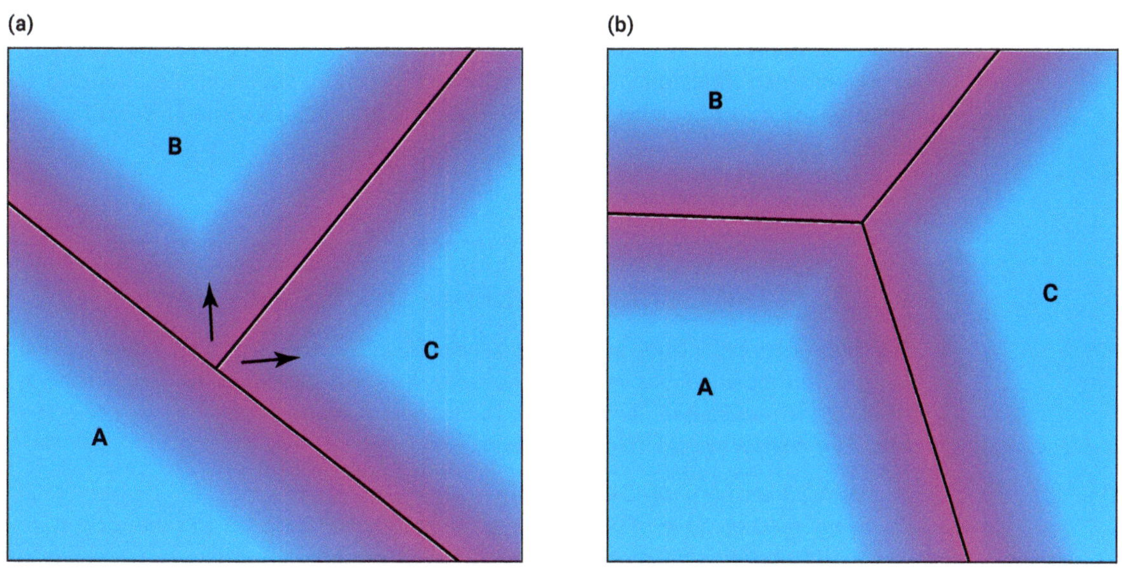

Figure 11.5 Explanation of development of 120° triple junctions among grains of the same mineral. (a) Three crystals, A, B, and C, meet at a triple junction with two 90° and one 180° angle (the straight face of A). Magenta colors indicate higher-energy parts of the crystal near surfaces where atoms are not fully bonded; corners on B and C are higher energy than faces on any of the crystals. As a result, corners on B and C retreat (arrows), and A forms a corner to fill the space. (b) Resulting 120° angles have a lower total energy than the configuration in (a). The total length of high-energy grain boundaries is also reduced.

Source: Modified from Mason (2013).

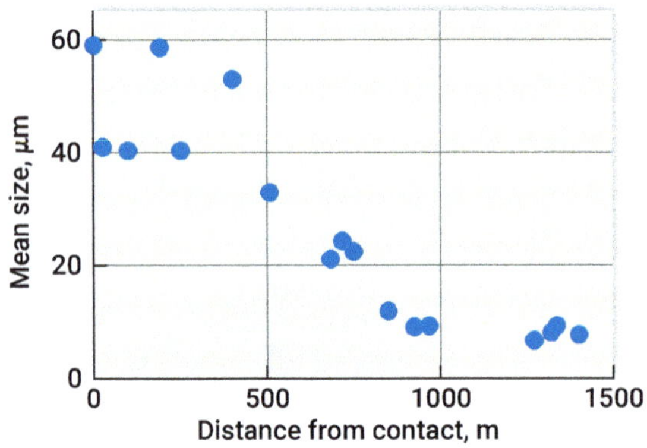

Figure 11.6 Size of quartz crystals in contact aureole around quartz diorite near Kyoto, Japan. Crystals in this metachert increase in diameter by roughly a factor of 10 approaching the contact. Values are the geometric mean of grain diameters measured in thin section.

Source: Modified from Okudaira et al. (2013).

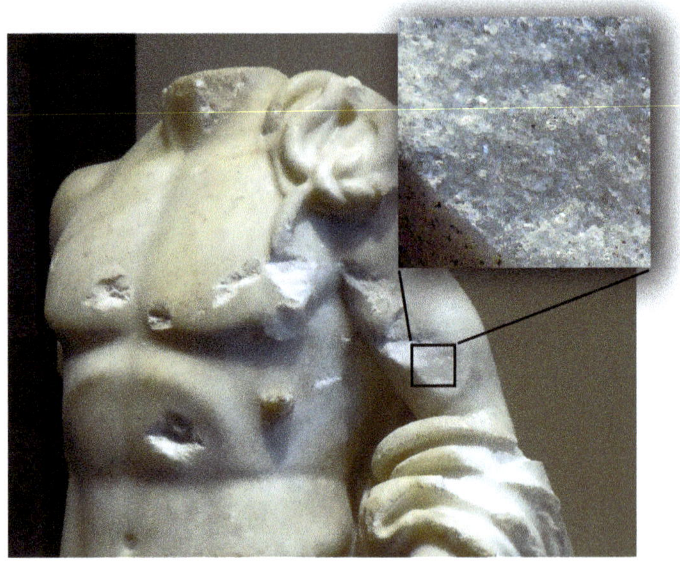

Figure 11.7 Beat-up statue of Mercury, second century CE, in the Temple of Romulus, Roman Forum. Inset (width 20 mm) shows a broken part of Mercury's left arm, with coarse calcite crystals ~1 mm across that underlie the smoothly carved exterior (Parian marbles typically have a grain size closer to 0.2–0.5 mm; Renfrew & Peacey, 1968). The limestone that was metamorphosed to form this marble probably had a grain size on the order of 10 µm or so, 100 times smaller.

Source: AFG photos.

Victory of Samothrace) are composed of relatively coarse-grained calcite marbles derived from fine-grained limestone. The long, slow annealing of such marbles, coupled with the relative softness of calcite, makes them especially crack-free and suitable for sculpture (Figure 11.7).

Rocks of the Contact Metamorphic Lithosystem

This lithosystem can include any rocks that have been invaded by plutons; indeed, many plutons have been contact metamorphosed by younger plutons or by younger increments of themselves (see Section 10.8). The list here comprises rocks that are characteristic of the contact metamorphic lithosystem.

- *Marble*: metamorphosed limestone or dolomite.
- *Skarn*: reaction zone of calc-silicate minerals between a pluton and the rocks it intruded; economically important minerals may be present.
- *Hornfels*: hard, fine-grained, unfoliated rock.
- *Calc-silicates*: silicates with high calcium contents (e.g., diopside, actinolite, titanite, etc.) formed by reaction of magma or fluids with carbonates.
- *Quartz veins*: concordant (parallel to rock textures) or discordant, continuous or discontinuous veins filled with quartz.
- *Monomineralic rocks*: rocks that consist of a single mineral. In addition to marble, rocks can be made entirely of garnet, diopside, epidote, magnetite, hornblende, etc.
- *Ore deposits*: Deposits of economically valuable minerals (ores). Can include economic concentrations of Cu, Co, Au, W, Zn, Pb, Sn, and other elements.

11.2 Thermal Evolution of Pluton Aureoles

Learning Objectives

By the end of this section you should be able to:

11.2.1 Predict qualitatively the temperature evolution around plutons depending on style of emplacement (one-pulse vs. incremental).

11.2.2 Apply solutions of the heat-flow equation to predict temperatures around a cooling dike.

Physics of Heat Flow

Heat flow in igneous and metamorphic systems is critically important to how such systems behave. Because it is difficult and expensive to take the Earth's temperature (e.g., by drilling boreholes), much of our understanding of what happens when heat is transferred from hot rocks or magma to cooler rocks comes from mathematical models of heat flow. Thus, it is worth delving into how such problems are solved.

When a pluton is emplaced, it loses heat to the wall rocks around it. How hot the wall rocks become depends upon several factors, including the starting temperature of the wall rock (which depends on the geothermal gradient, and varies with depth), the temperature of the invading magma (mafic is normally hotter than felsic), the initial crystallinity of the magma (more crystalline releases less heat), the depth of emplacement, and whether cooling is solely by conduction, aided by convection of fluids in the wall rocks or magma, or affected by other heat-transfer processes such as radiation or advection.

Heat conduction is governed by Fourier's law (see Section 2.7), which has the same mathematical form as Fick's law of chemical diffusion. To review, Fourier's law states that the heat flux q (flow of heat per unit area per unit time) at a point is directly proportional to the thermal gradient dT/dz at that point:

$$q = -k\frac{dT}{dz} \quad \text{(Eq. 11.1)}$$

where T is temperature and z is the spatial dimension. The proportionality constant k is the thermal conductivity, and the negative sign means heat flows from hot materials to cold materials. Fourier's law is applicable in steady-state cases, where the thermal gradient does not change with time. We can use this equation to estimate the heat flux if, for example, we know the temperature at two closely spaced points along with the thermal conductivity. Surface heat flux (see Figure 2.19) is an important quantity that can be measured by drilling holes in rocks or sticking probes in sediment, and then measuring both the temperature gradient and the thermal conductivity of the material.

When temperature is changing with time, a different equation applies. The equation for time-dependent heat conduction in a 1D system is

$$\frac{\partial T}{\partial t} = \kappa\frac{\partial^2 T}{\partial z^2} \quad \text{(Eq. 11.2)}$$

where t is time, z is the spatial dimension, and κ (kappa) is the thermal diffusivity, which is related to thermal conductivity (k) by the relation $\kappa = \dfrac{k}{\rho c}$ where ρ is density and c is specific heat (heat capacity). Remember to be careful to distinguish k from κ! They're related, but represent different properties.

Although Equation 11.2 is a partial differential equation, the physical significance of the terms in it are easy to understand. The left side, $\dfrac{\partial T}{\partial t}$, is simply the rate at which temperature changes at a given point (z-coordinate) in the system. Remember that the partial

Specific heat
The amount of thermal energy needed to increase the temperature in a unit mass of rock by a unit amount. Units are usually J/g·K or kJ/kg·K.

derivative represents the derivative $\frac{dT}{dt}$ at a slice through the model where the spatial coordinate z is held constant, and that there may be a different value of $\frac{dT}{dt}$ for each value of z. The right side, the second derivative of temperature with respect to z, is the curvature of the temperature profile. Where the temperature profile is sharply curved, temperature varies quickly; where it is not curved (straight line), it does not vary with time at all.

The heat-flow equation, a second-order partial differential equation, generally cannot be integrated into a simple function that gives us temperature as a function of space and time. Carslaw & Jaeger (1959) provided solutions for a large number of cases (e.g., a sphere, an infinite sheet, etc.), but solutions for real-life problems involve infinite series, integrals that cannot be solved analytically, or both. Lovering (1935; see also Jaeger, 1964) gave the solution for intrusion of a dike, and it is worth following this example to see how such solutions are handled (see Toolbox 11.1).

Numerical Model—One Pulse

In real-world applications, differential equations such as the heat-flow equation are solved with numerical methods (e.g., Furlong et al., 1991) rather than by simplifying the problem into one of the cases for which an analytical solution is known (you don't hit Mars with a rocket by assuming that its orbit is a circle!). These methods are conceptually simple, but there are numerous subtleties involved in obtaining accurate results, and even highly accurate solutions to the problem can be incorrect if the model does not properly reflect reality.

Consider instantaneous emplacement of a 2-km-thick sill whose top is 5 km below the surface (instantaneous emplacement of such a thick sill is unlikely, but, conceptually, it's a place to start). We wish to model how temperature varies with depth and time as a result of this intrusion. This is a 3D problem, but if the sill is a thin plate (Figure 11.8), we can approximate temperature evolution with a 1D model through the center of the plate. To model this numerically, we need relevant physical parameters and two types of information: **initial conditions** and **boundary conditions**. Initial conditions specify the temperature structure at the start of the model, and boundary conditions specify what happens at the boundaries of the model. Numerical methods are then used to move the initial T conditions forward in time, subject to the boundary conditions.

Initial conditions
The starting conditions for a numerical model; here, the initial temperature distribution.

Boundary conditions
How the model space interacts with the world outside; in a thermal model, typical boundary conditions are constant temperature, constant temperature gradient, constant heat flux, or insulated (zero heat flux).

Let's consider a model that goes from the surface to a depth of 30 km, with the following parameters:

- Geothermal gradient: 20 °C/km
- Mean surface temperature: 15 °C
- Initial magma temperature: 1000 °C
- Magma solidus temperature: 700 °C

There are several other important parameters that must be specified as well, such as the heat capacities of the rock and magma and the latent heat of the magma. The model presented here uses typical values (e.g., Furlong et al., 1991).

First, the initial conditions. Before the magma is intruded, the temperature at 30 km depth is 15 + 20 × 30 = 615 °C. Instantaneous intrusion of the sill splits the Earth at 5 km depth, so at the moment the sill comes in, the temperature just above the sill at 5 km depth is 15 + 20 × 5 = 115 °C, the same as the temperature just below the sill at 7 km depth. The initial conditions are thus the following temperature profile: 15 °C at the surface, increasing by 20 °C/km to 115 °C at 5 km, where the magma at 1000 °C covers the next 2 km. Below that, the 20 °C/km profile takes off again from a starting point of 115 °C at 7 km, reaching 615 °C at 32 km.

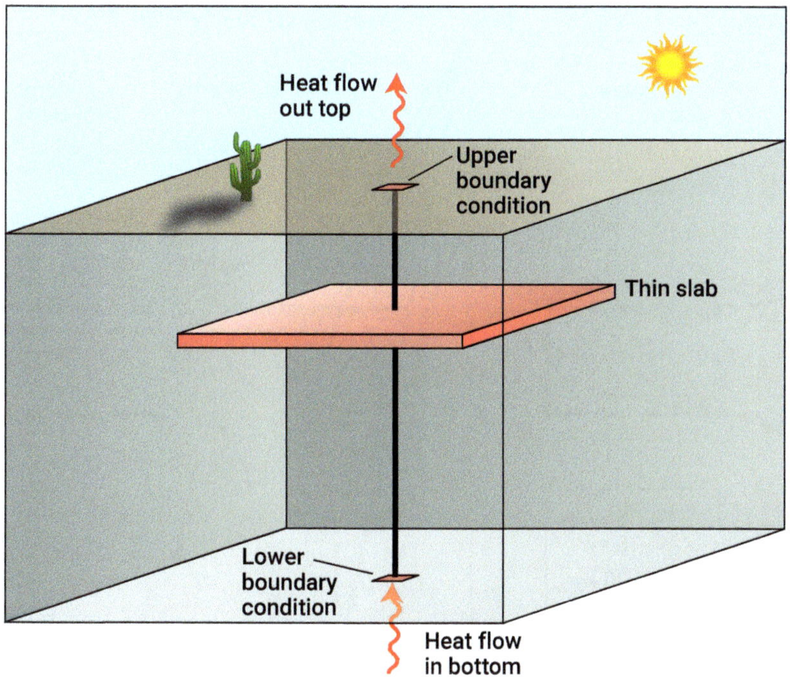

Figure 11.8 Schematic 3D view of a 1D thermal model of an intrusion. Although the problem is inherently 3D, if the boundaries of the slab are reasonably far from its center, the thermal evolution of the center can be approximated with a 1D model. The thermal evolution near the edges, though, cannot. Boundary conditions generally set a constant upper temperature at the surface, and either a constant temperature or constant heat flow at the bottom of the model. (Cactus and sun not to scale.)

Next, the boundary conditions. The upper boundary condition is given by setting the surface temperature to 15 °C, maintained there by the mean annual temperature of the atmosphere. The lower boundary condition is less certain. One common approach is to fix it at its initial temperature (in this case, 615 °C); another is to set the geothermal gradient at the lower boundary to a fixed value (in this case, 20 °C/km). If the model goes deep enough that this point is far away from the action, then the choice makes little difference to the calculations. For our models, we used a constant temperature gradient.

Once the initial and boundary conditions have been set, solving for the temperature evolution just requires a lot of arithmetic (see Toolbox 11.2). The results in Figure 11.9a show the evolution of the temperature profile over 1 Myr following intrusion. Significant points include the following:

- The center of the sill cools from 1000 °C at the time of intrusion to below the nominal solidus, 700 °C, in less than 100,000 years, a relatively short time.
- The upper 15 km of the rock column was below 300 °C (sub-greenschist facies) before the sill was intruded, but after intrusion, much of the region below 3 km depth resides in the greenschist and amphibolite facies (see Figure 5.10), for a million years or more. Remember from Chapter 5 that temperature is a key determinant of metamorphic grade.
- The geothermal gradient at the surface, 20 °C/km at the start, rises rapidly to a peak of 90 °C/km at 200 kyr after intrusion, after which it declines asymptotically back to the starting value (see Figure 11.9c). This peak value is quite high compared to regional heat flow estimates (see Figure 2.19).

Clearly, intrusion of the sill has a dramatic effect on the thermal history of its wall rocks.

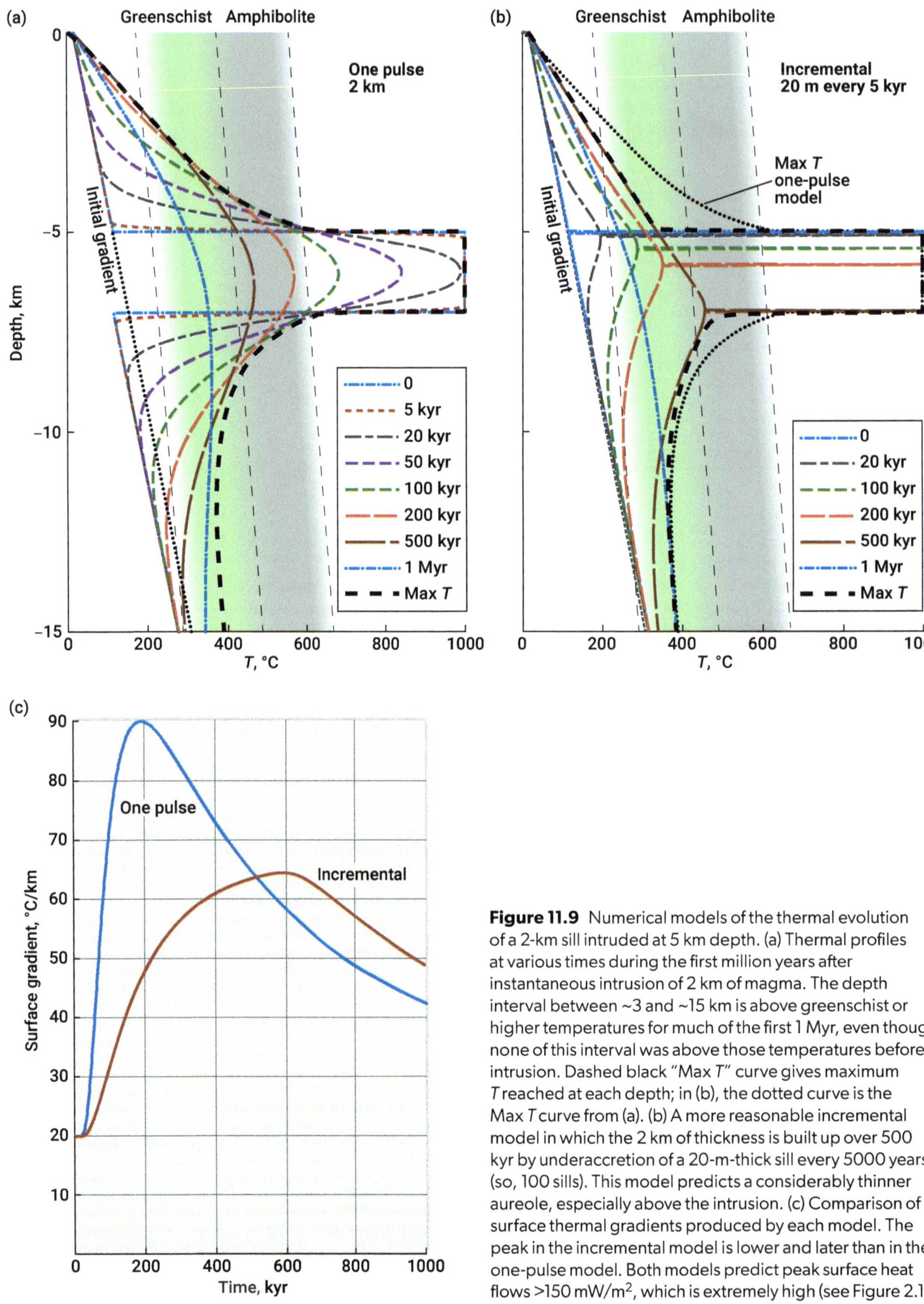

Figure 11.9 Numerical models of the thermal evolution of a 2-km sill intruded at 5 km depth. (a) Thermal profiles at various times during the first million years after instantaneous intrusion of 2 km of magma. The depth interval between ~3 and ~15 km is above greenschist or higher temperatures for much of the first 1 Myr, even though none of this interval was above those temperatures before intrusion. Dashed black "Max *T*" curve gives maximum *T* reached at each depth; in (b), the dotted curve is the Max *T* curve from (a). (b) A more reasonable incremental model in which the 2 km of thickness is built up over 500 kyr by underaccretion of a 20-m-thick sill every 5000 years (so, 100 sills). This model predicts a considerably thinner aureole, especially above the intrusion. (c) Comparison of surface thermal gradients produced by each model. The peak in the incremental model is lower and later than in the one-pulse model. Both models predict peak surface heat flows >150 mW/m^2, which is extremely high (see Figure 2.19).

Toolbox 11.1

Heat Flow Out of a Dike

The following analysis is from Lovering (1935) and Jaeger (1964); students should examine the Lovering paper to see the beautiful curves on graph paper—drawn by hand, lettered by hand, and painstakingly computed by hand—50 years before spreadsheets became a thing. There are several important aspects of these models that should be kept in mind:

- Their treatment is 1D (i.e., the dike is considered to be an infinite panel of rock, so cooling occurs only perpendicular to its surface and symmetrically about the dike center, and one need compute only half of the entire solution).
- The dike is assumed to be intruded instantaneously in one pulse, with no flow in the dike after intrusion.
- Latent heat of crystallization (see Section 4.4) is ignored.

The solution is given in terms of dimensionless quantities, which are real-world coordinates (distance, time) scaled by parameters of the problem (dike width, thermal diffusivity) such that the resulting quotient is dimensionless. The dimensionless quantities are:

- Length: $\lambda = x/d$, where x is the horizontal coordinate (distance from the dike center) and d is the half-width of the dike.
- Time: $\tau = \kappa t/d^2$, where t is time and κ is thermal diffusivity. The situation when $\tau = 1$ is the "**heat flow rule of thumb**."
- Temperature: $T^* = (T - T_0)/(T_m - T_0)$, where T is temperature, T_m is the magma temperature, and T_0 is the temperature of the wall rocks before intrusion.

You might ask, why use dimensionless numbers instead of real distances, times, and temperatures? The answer is that dimensionless numbers provide a general solution—a set of curves—that can be scaled to any relevant problem, as we will see.

The solution for the time and distance dependence of temperature—the solution to the heat-flow equation—in this case is given by (Furlong et al., 1991)

$$T^*(\lambda, \tau) = \frac{1}{2}\left\{ \operatorname{erf}\left(\frac{\lambda + 1}{2\sqrt{\tau}}\right) - \operatorname{erf}\left(\frac{\lambda - 1}{2\sqrt{\tau}}\right) \right\} \quad \textbf{(Eq. 11.3)}$$

and as we saw in Section 8.3, the error function erf is given by a definite integral whose value cannot be calculated analytically but is readily available through numerical approximation in programming languages and spreadsheets. Although the equation looks intimidating, it is easy to implement.

A set of curves for this solution is given in Figure 11.10a. The system starts at the time of dike intrusion with $\tau = 0$ and evolves to $\tau = 20$ (and on to $\tau = \infty$, where $T^* = 0$). But what does this mean? You can see that the center of the dike is at about half its initial T above the wall rock T when $\tau = 1$, but how do we interpret this? The full solution is given if we know the magma temperature T_m, T_0, κ, and d. Suppose we want to model a 2-m-thick dike intruded at 1000 °C into 100 °C wall rocks that have a thermal diffusivity (κ) of 32 m^2/yr (1 mm^2/s). In this case, T_m = 1000 °C, T_0 = 100 °C, κ = 32 m^2/yr, and d = 1 m (half-width, remember?). In this case, τ = 1 corresponds to t = (1 m)^2/(32 m^2/yr) = 1/32 yr, or about 11 days—a 2-m-thick dike should freeze in a matter of days. What if our dike half-width is 10 m instead of 1 m? Then τ = 1 corresponds to 100/32, or about 3 years—a 20-m-thick dike takes years to freeze. What if the thermal diffusivity around our dike is twice as large? Then cooling occurs twice as fast. The curves are the same, they're just scaled for different values of τ, which can accommodate any values of κ, t, or d.

Be careful not to take the curves in Figure 11.10 too seriously—models are meant to guide thinking, not imprison it. For example, this model ignores latent heat. Lovering suggested handling this either by increasing the width of the dike such that the total thermal energy accounts for the latent heat, or by increasing the initial T to do the same. Alternatively, one can solve the equations using numerical methods (see Toolbox 11.2).

One parting thought concerns T at the dike margin (λ = 1). Notice in Figure 11.10a that the first few curves (τ = 0.1, 0.4) pass through T^* = 0.5, which is the midpoint between magma T and wall rock T, until the system cools enough that the dike margin falls below this T, which happens at about τ = 0.75. This value of T^* (0.5) is the maximum T reached by the wall rocks under the assumptions of the model—a one-pulse dike with no heat of crystallization that cools solely via thermal conduction.

Let's again dimensionalize the numbers by setting d = 5 m, κ = 32 m^2/yr, T_m = 1200 °C, and T_0 = 100 °C (Figure 11.10b). This is a thick dike, but the contact T according to this model is the same as it would be for a thin dike—this T persists longer, for months, but never gets above 650 °C, an approximation of the water-present solidus for juicy wall rocks.

This result is in sharp contrast with numerous dikes around the world that show evidence for pervasive melting at their contacts, extending many meters into the wall rocks (e.g., Al-Rawi & Carmichael, 1967; Kitchen, 1989). Why is the prediction of the thermal model so far off the mark? (Hint: see Section 7.1.)

continued

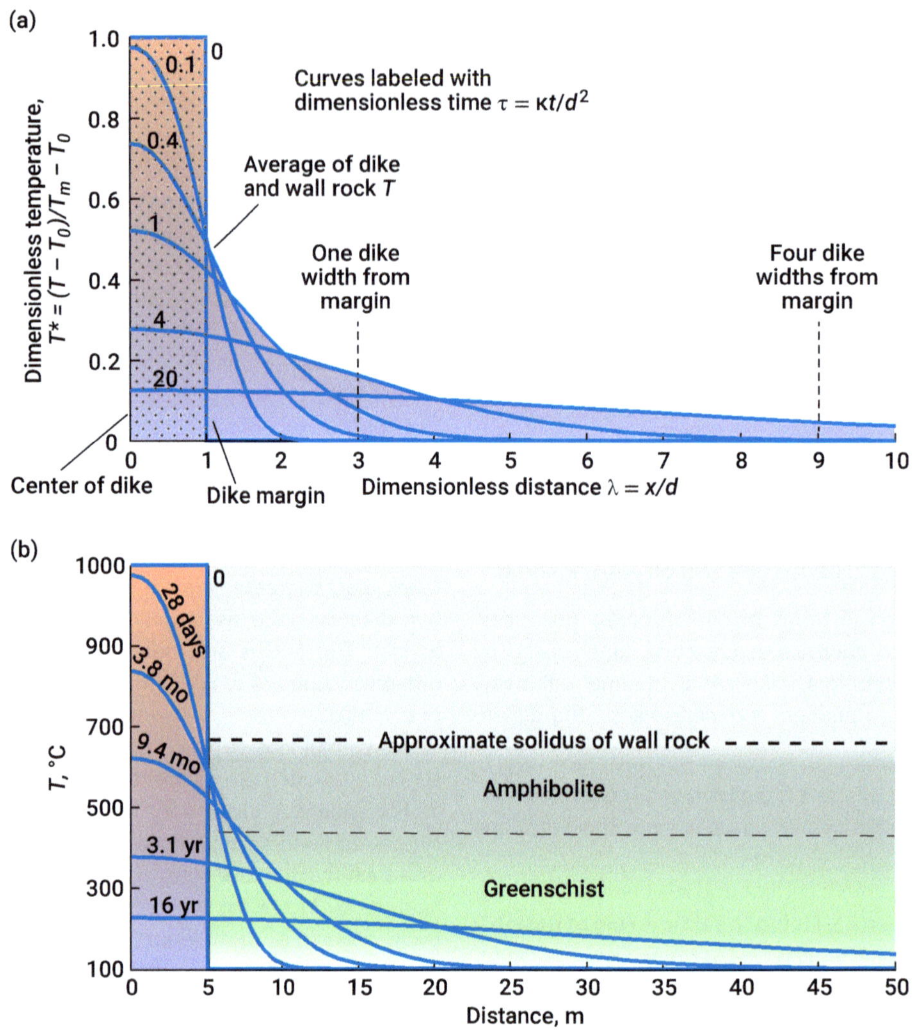

Figure 11.10 Dike cooling, dimensionless and dimensioned. (a) Cooling of a dike calculated from the error function solution of Lovering (1935), and expressed in dimensionless parameters. This solution can be scaled to a dike of any width and an intrusion of any temperature. Note that initial cooling is quite rapid; the center of the dike cools to half its original temperature by the time that dimensionless time $\tau = 1$; it takes four times longer to reach about 25% of the original temperature, and five times that ($\tau = 20$) to cut the temperature in half again. (b) Solution for heat flow around a one-pulse dike scaled to relevant values: dike thickness = 10 m (half-width 5 m), magma $T = 1000$ °C, wall rock $T = 100$ °C, and thermal diffusivity = 32 m^2/yr. Curves are labeled for time in days, months (mo), and years (yr). Note that wall rocks attain temperatures in the greenschist facies and, within a few meters of the dike margin, amphibolite facies, but only for a brief time. They never reach the solidus of the wall rocks, even though many dikes show evidence of significant melting at their margins (see text).

Heat flow rule of thumb
Relationship between time (t) and distance (d) when $\tau = 1$. For a specified value of κ (thermal diffusivity), this can be used to estimate timescales of heat diffusion ($t = d^2/\kappa$, given a length scale, d), or heat penetration distances ($d = \sqrt{\kappa t}$, given a timescale, t).

Numerical Model—Incremental

Figure 11.9b shows another model that produces a 2-km-thick sill, but here, the body was intruded incrementally, by top-down insertion of one hundred 20-m-thick sills every 5000 years. The key difference is that the thermal effects are muted; only a small portion of the region above the sill reaches greenschist conditions, and the peak surface heat flow is only two-thirds of the value for the one-pulse case. We can expect thinner aureoles around incrementally emplaced plutons. In at least some cases, this provides a better match to observed aureole dimensions than assuming that all the magma came in at once (Chen & Nabelek, 2017).

Toolbox 11.2

Numerical Heat Flow Solution in a Spreadsheet

Numerical solutions to differential equations are so widely used, and so important, that it is worth taking time to demystify them so that you, the student, can be an informed consumer of science that involves numerical solutions. Consider the time-dependent, 1D heat-flow equation presented in Equation 11.2. The right-hand side is κ times the second derivative of temperature with respect to distance. Recalling from calculus that the derivative of a curve can be approximated by the slope of a small segment of it, consider the curve in Figure 11.11.

Three points, *a*, *b*, and *c*, are separated by equal Δx intervals. The first derivative of segment *a-b* can be approximated by

$$\frac{\Delta T_{ba}}{\Delta x} \qquad \textbf{(Eq. 11.4)}$$

where ΔT_{ba} is $T_b - T_a$, etc. That's easy. Approximation of the second derivative is less easy but straightforward; it is the derivative of the first derivative, and is thus, at point *b*,

$$\frac{\left(\frac{\Delta T_{cb}}{\Delta x} - \frac{\Delta T_{ba}}{\Delta x}\right)}{\Delta x} = \frac{\Delta T_{cb} - \Delta T_{ba}}{(\Delta x)^2} \qquad \textbf{(Eq. 11.5)}$$

If we expand the ΔT terms, the numerator becomes $(T_c - T_b) - (T_b - T_a) = T_c - 2T_b + T_a$, and thus an approximation of the second derivative is

$$\frac{(T_c - 2T_b + T_a)}{(\Delta x)^2} \qquad \textbf{(Eq. 11.6)}$$

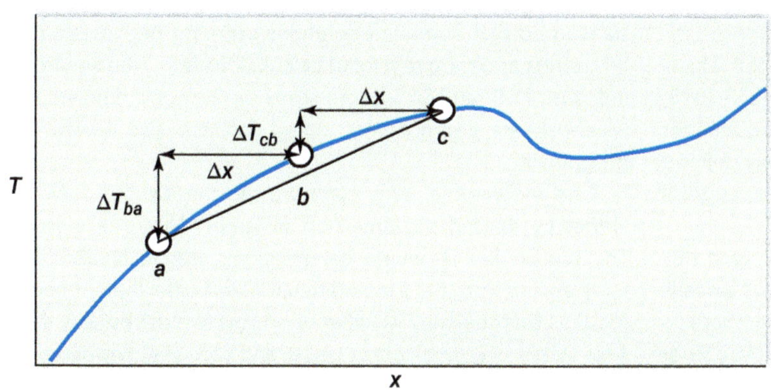

Figure 11.11 A curve with three marked points. Approximation of the first derivatives of segments *a-b* and *b-c*, and the second derivative of segment *a-c*, are discussed in the text. Note that if the points were spread farther apart they would fail to capture the shape of the curve properly—in particular, the curve goes through an inflection point just to the right of point *c*, changing from convex-up to convex-down.

The derivatives here are spatial derivatives (i.e., how *T* varies in space—in this case, along a line). The left side of Equation 11.2 is a time derivative—how *T* varies with time. An approximation of the time derivative is

$$\frac{\Delta T}{\Delta t} \qquad \textbf{(Eq. 11.7)}$$

where the numerator is the change in *T* (what we want to calculate) over a time step of length Δt.

An approximation for the time-dependent heat-flow equation (see Equation 11.2) is thus

$$\frac{\Delta T}{\Delta t} = \kappa \frac{(T_c - 2T_b + T_a)}{(\Delta x)^2} \qquad \textbf{(Eq. 11.8)}$$

Expanding ΔT, moving Δt to the right side, and making a change in notation yields

$$T_{t=n+1} = T_{t=n} + (\Delta t)\kappa \frac{(T_{x+1} - 2T_x + T_{x-1})}{(\Delta x)^2} \qquad \textbf{(Eq. 11.9)}$$

where *n* represents a particular time step, *n* + 1 is the next one, and *x* + 1 and *x* – 1 are the grid points just beyond and just before the point of interest *x*. Everything on the right side of the equation is calculated at the present time step, and the left side gives the *T* distribution at the next.

That's almost all there is to it; the rest is just spreadsheet gymnastics. We need to (1) choose values for κ, Δt, and Δx; (2) select initial conditions; (3) select boundary conditions; and then (4) program the cells to propagate the solution.

The spreadsheet in Figure 11.12 shows the beginning of the solution. Values for k (abbreviation for κ, thermal diffusivity), dt (Δt), and dx (Δx) are in the named[1] blue cells in row 3. The initial conditions are defined by the problem we're solving. Let's look at a 1-m dike at 1000 °C instantaneously emplaced into wall rocks at 0 °C. We need to model only half of the dike, because the solution is symmetrical about the centerline; we'll model this as an insulated boundary condition, because no heat flows across the centerline—it flows symmetrically out both sides. So, the initial conditions are 1000 for cells within the dike and 0 for those outside it. These are in column B, which is time *t* = 0.

continued

	A	B	C	D	E	F	G	H	I	J	K
1	**Numerical simulation of the temperature field around a dike**										
2											
3		k, m2/day	***0.10***			dt, day	***1.00***			dx, m	***0.50***
4		***Initial cond↓***				Stability 2kdt/dx^2		0.80			
5	***time (days)→ x (m)↓***	0.0	1.0	2.0	3.0	4.0	5.0	6.0	7.0		
6	0.0	***1000***								***←Boundary condition***	
7	0.1	***1000***									
8	0.2	***1000***									
9	0.3	***1000***									
10	0.4	***1000***									
11	0.5	***1000***									
12	0.6	***0***									
13	0.7	***0***									
14	0.8	***0***									
15	0.9	***0***									
16	1.0	***0***									
17	1.1	***0***									
18	1.2	***0***									
19	1.3	***0***									
20											
21											

Figure 11.12 The beginnings of a spreadsheet solution to the heat-flow equation. Boxed blue values in row 3 are constants, and bold italic values in column B are the initial conditions.

Now the fun part. What goes in column C, the next time step (in this case, one day later)? The answer is given by Equation 11.9. As an example, the formula in cell C13 is =B13+dt*k*(B14-2*B13+B12)/dx^2, where dt, k (actually κ – thermal diffusivity, NOT conductivity – but styling in spreadsheets is often limited), and dx are named cells from row 3 containing the corresponding values. Note that the temperature in this cell calculated at the first time step (1 day, column C) is based entirely on the values from the previous time step, time 0 (column B).

As you see, the value calculated this way for cell C13 is . . . 0, because the values used from column B (B12, B13, and B14) are all zeros. Not very exciting. But do the same for the point at $x = 0.6$ (cell C12) by dragging our newly defined formula up, and the value is 400. This point is on the edge of the dike, and it's feeling the heat.

We can fill the rest of the values in column C with this formula to estimate T at the first time step, but this fails for cell C6 at the top of the column, because there is no appropriate cell above and to the left of it. We're at the boundary of the model and need to implement the insulated boundary condition. Insulated means no heat flow, and no heat flow means that the thermal gradient is zero (i.e., temperatures in the border cell and the ones below it, row 6 and row 7, should be the same). So, we simply set the top cell (row 6) to the value in the cell below (row 7) and propagate the formulas through the spreadsheet. In Figure 11.13, the solution has been run forward seven days, and the high T of the dike is warming up the wall rocks, at least nearby.

That's all there is to it for this simplest sort of solution, although our solution covers only a small part of space and time—we've gone forward only a week. We can run it further forward in time by dragging the formulas to the right, but be warned that numerical solutions done properly use small time steps and require a *lot* of calculations (e.g., the solutions in Figure 11.9 ran 1500 depth steps for 200,000 time-steps). Spreadsheets are not well suited for this—you'll want to program it.

One more topic: Why can't we run the solution forward further by making the time-step *dt* larger? Why not a year? The reason is that numerical solutions require small steps for stability. For the forward-difference model here, theory requires that the stability criterion in row 4 must be less than 1 for stability. Try making *dt* larger and see what happens; all sorts of interesting behavior occurs when the stability ratio is right around 1 (best seen in plots of a column, representing model temperature distribution). There are more complicated but far more elegant and efficient numerical methods for solutions than the brute-force (clunky) method employed here.

	A	B	C	D	E	F	G	H	I	J	K
1	**Numerical simulation of the temperature field around a dike**										
2											
3		k, m2/day	*0.10*			dt, day	*1.00*			dx, m	*0.50*
4		*Initial cond↓*				Stability	2kdt/dx^2	0.80			
5	*time (days)→ x (m)↓*	0.0	1.0	2.0	3.0	4.0	5.0	6.0	7.0		
6	0.0	*1000*	1000	1000	1000	1000	990	980	965	*←Boundary condition*	
7	0.1	*1000*	1000	1000	1000	1000	990	980	965		
8	0.2	*1000*	1000	1000	1000	974	964	944	929		
9	0.3	*1000*	1000	1000	936	923	887	872	850		
10	0.4	*1000*	1000	840	840	782	772	745	734		
11	0.5	*1000*	600	680	600	616	590	591	580		
12	0.6	*0*	400	320	400	384	410	409	420		
13	0.7	*0*	0	160	160	218	228	255	266		
14	0.8	*0*	0	0	64	77	113	128	150		
15	0.9	*0*	0	0	0	26	36	56	69		
16	1.0	*0*	0	0	0	0	10	16	27		
17	1.1	*0*	0	0	0	0	0	4	7		
18	1.2	*0*	0	0	0	0	0	0	2		
19	1.3	*0*	0	0	0	0	0	0	0		
20											
21											

Figure 11.13 The formulas have been propagated and the cells colored using conditional formatting, with a hot color (red) used for hot cells grading to a cool color (blue) for the wall rocks.

11.3 Isograds and the Metamorphic Evolution of Pluton Aureoles

Learning Objectives

By the end of this section you should be able to:

11.3.1 Use a mineral paragenesis diagram to predict the sequences of mineral assemblages you might observe in contact metamorphosed mudstones and metabasalts.

11.3.2 Use chemographic diagrams to illustrate mineral reactions.

Metamorphism of Mudstones

The heat dumped into the wall rocks by our pluton is bound to have a significant effect on them, but what we see depends on what type of rock is being cooked. If the wall rocks are sandstone composed entirely of quartz, little will happen. If the pressure is low enough (<100 MPa or so) and the magma hot enough (1000 °C or so), then the quartz might turn to tridymite (see Figure 3.13), but that's about it. The quartz crystals will likely rearrange their grain boundaries and develop a coarser granoblastic texture (see Figure 11.2), but no new minerals will grow. A fine-grained sandstone turns into a (somewhat) coarser-grained quartzite. If the wall rocks are limestone or dolomite, and they do not chemically react with the magma, they will likely coarsen, develop a granoblastic texture, and turn into marble. But again, new minerals will not form. Yes, marble and quartzite have important commercial value, and the kinetics of grain coarsening are fundamental to aspects of structural geology and materials science. But, from a mineralogical perspective, nothing happens. If instead

Table 11.1 Minerals in the Contact Aureole of Plutons in the Santa Rosa Range

Mineral	Abbreviation	Formula
Quartz	Qz	SiO_2
Albite	Ab	$NaAlSi_3O_8$
Muscovite	Ms	$KAl_3Si_3O_{10}(OH)_2$
Chlorite	Chl	$(Mg,Fe,Al)_6(Si,Al)_4O_{10}(OH)_8$
Carbonates	Cal, Dol	$CaCO_3$, $CaMg(CO_3)_2$
Fe-Ti oxides	Ox	Fe_3O_4, $FeTiO_3$
Cordierite	Crd	$Al_3(Mg,Fe)_2AlSi_5O_{18}{\cdot}nH_2O$
Biotite	Bt	$K(Fe,Mg)_3(AlSi_3O_{10})(OH)$
Andalusite, sillimanite	And, Sil	Al_2SiO_5
Staurolite	St	$(Mg,Fe)_2Al_9Si_4O_{23}(OH)$
Plagioclase	Pl	$NaAlSi_3O_8$-$CaAl_2Si_2O_8$
K-feldspar	Kfs	$KAlSi_3O_8$

the wall rocks are chemically more interesting, new minerals can develop, and these can be used to quantify conditions and processes of metamorphism.

Shale is chemically interesting because its high alumina saturation index (see Section 10.5) means that excess Al is available to form minerals whose appearances depend on both pressure and temperature. Several of these occur in the Santa Rosa aureole (see Figure 11.1). As the chemical compositions of many of these denizens of the contact metamorphic world are complicated and unfamiliar, the key minerals in the Santa Rosa Range aureole and their simplified compositions are given in Table 11.1.

The first six rows in Table 11.1 are the low-grade (greenschist) metamorphic mineral assemblage to which the shaley sedimentary rocks were metamorphosed, before intrusion of Cretaceous magmas. The rest of the minerals were produced by contact metamorphism. Observations in the field and thin section reveal the following mineral assemblages, all with Qz, Ms, Pl, and oxides:

- Cordierite-andalusite-biotite, Crd-And-Bt
- Cordierite-biotite, Crd-Bt
- Staurolite-andalusite-biotite, St-And-Bt
- Andalusite-biotite, And-Bt

The outermost mapped boundary of the aureole is marked by the appearance of Bt (pseudomorphing Chl in the host rocks) and porphyroblasts of And. Crd occurs near the outer margin of the aureole, but always with And or Bt (or both).

OK, now what? The map pattern makes it clear that the changes visible in the aureole are related to proximity to the plutonic rocks, but what does that mean? Can we use this information to place any limits on the *P* and *T* conditions of metamorphism?

Metamorphic Facies and Metamorphic Zones

Let's start with the metamorphic facies diagram, first introduced in Figure 5.10 and important enough to reproduce again here (Figure 11.14). Some observations:

- Before the pluton intruded, shaley rocks contained Ab, Ms, Chl, and Qz—a classic lower-greenschist-facies assemblage (see later).
- The facies diagram says nothing directly about Bt, an aureole mineral common to most samples.

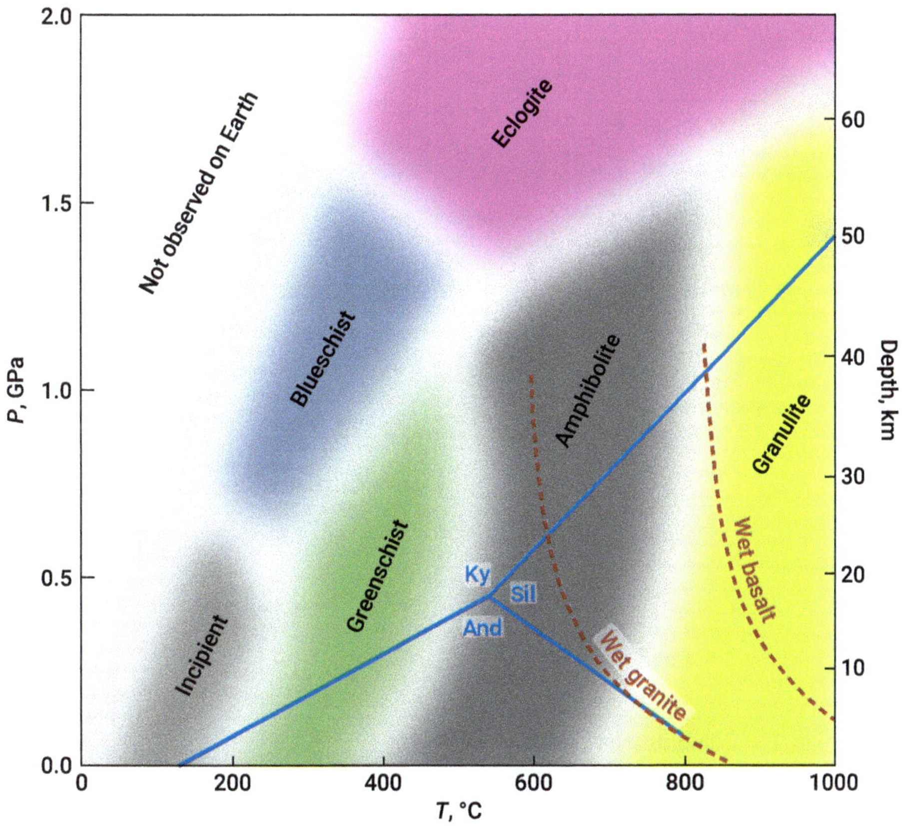

Figure 11.14 The five principal metamorphic facies (plus "incipient"). Key reactions include the boundaries among the Al_2SiO_5 polymorphs (kyanite, andalusite, and sillimanite), and the red curves for water-present ("wet") melting of granite and basalt.

Source: Adapted and simplified from Spear (1995).

- The presence of And rather than Ky as the dominant aluminosilicate means that the *P-T* trajectory followed by the rocks passed below the aluminosilicate triple point where Ky, And, and Sil meet; thus, metamorphism happened at depths less than about 15 km.
- There is no sign that the aureole rocks melted, and so the maximum *T* reached was below the wet granite solidus (if free water was present), or somewhat higher if melting was by dehydration of one of the hydrous minerals.

Back in Chapter 5, we introduced the paragenesis diagram (see Figure 5.5). Figure 11.15 shows paragenesis diagrams for shales and basalts as they are progressively metamorphosed along a relatively low-*P* trajectory—low enough to miss the Ky field. As peak *T* goes up, key **index minerals**—Bt, Crd, And, and so forth—generally appear, each defining an **isograd**. The rock volumes between isograd surfaces are **metamorphic zones**; thus, metashales generally go through the Bt zone, Crd zone, And zone, etc., as *T* increases along a low-*P* path, as in contact metamorphism around a shallow pluton. The Bt zone lies in the greenschist facies, the Crd and St zones form in the lowest amphibolite facies, the And zone in the mid-amphibolite facies, etc.

Several details of the paragenesis diagrams in Figure 11.15 bear elaboration. First, many low-grade metabasalts display two Pl phases, one relatively pure Ab and another of intermediate composition (typically ~An_{20}). This occurs because the Pl series has low-*T* mixing gaps that are not relevant at igneous temperatures but do become evident in rocks held at relatively low *T* for millions of years. This particular compositional gap for sodic plagioclase is known as the **peristerite gap**. Second, the exact bulk composition of a rock matters; for

Index minerals
Minerals whose appearance marks an increase in metamorphic grade.

Isograd
A surface (line on a map) marking the boundary of a metamorphic zone.

Metamorphic zones
Rock volumes (areas on a map) bounded by isograds.

Peristerite gap
A compositional gap in plagioclase between nearly pure albite and oligoclase that is locally seen in metamorphic rocks.

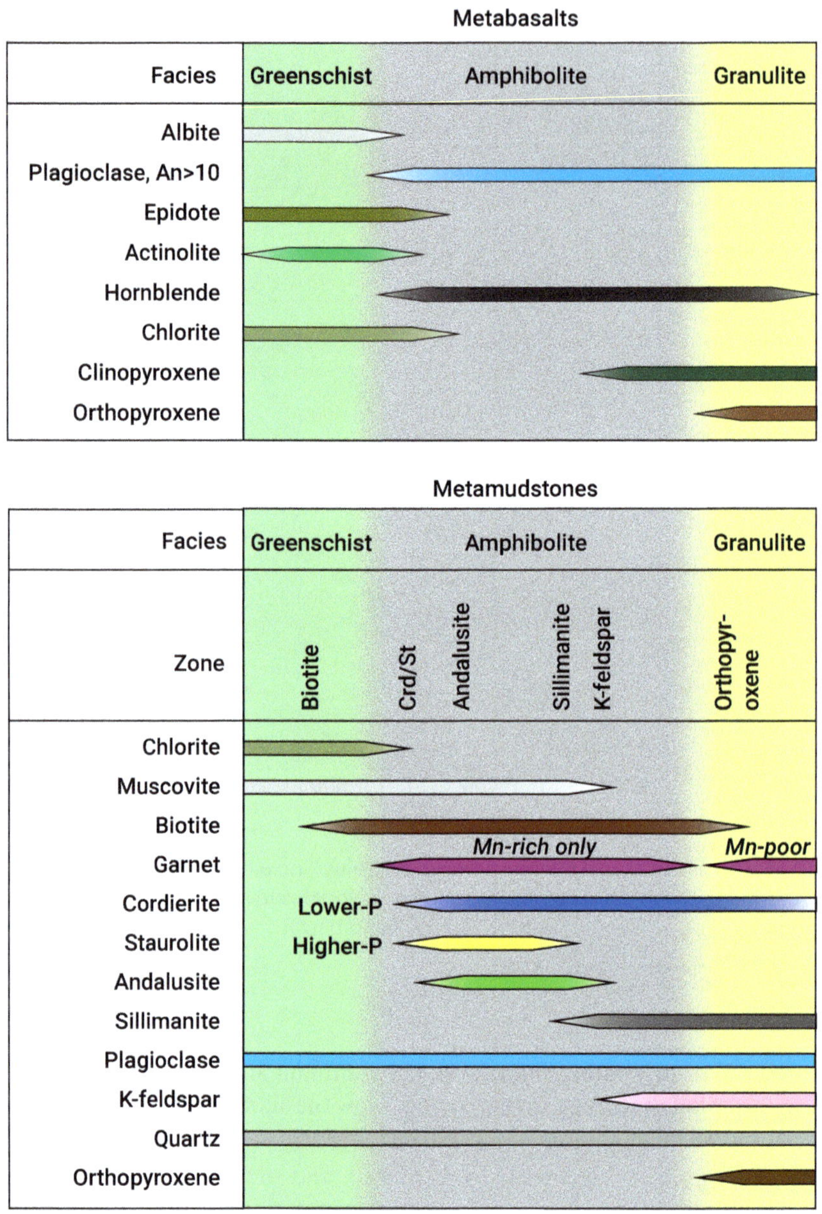

Figure 11.15 Mineral parageneses for low-pressure metamorphic rocks. Exact mineral assemblage depends on bulk rock composition (e.g., garnet is not normally common, but can form in Mn-rich rocks), and presence of cordierite vs. staurolite depends on pressure.

Sources: Modified from Ernst (1969) and Pattison & Vogl (2005).

example, low-grade garnets typically form only in high-Mn rocks (spessartine garnet is $Mn_3Al_2Si_3O_{12}$). Third, pressure matters. Low-*P* contact aureoles (<~10 km) form Crd in the amphibolite facies, whereas higher-*P* contact aureoles (>~10 km depth) form St (e.g., Pattison & Vogl, 2005). In the Santa Rosa contact aureole, St and Crd do not occur together. Compton (1960) identified St in two areas on the center west and center east sides of the field area; this observation suggests the center of the contact aureole exposes deeper rocks than to the north and south. Last, you might think from the aluminosilicate phase diagram that a rock should have either And or Sil, but not both. From an equilibrium thermodynamic perspective, that's true, and And and Sil should not overlap, as shown. But an aluminosilicate, once formed, can persist even if the rock enters the stability field of a different aluminosilicate.

It is important to appreciate the difference between a facies, which defines a region of *P-T* space, and a zone, which is defined by a particular mineral assemblage and indicates particular *P-T* conditions for a particular bulk composition. It is worth repeating that a rock layer made of pure Qz or Cal can go through the *P-T* ranges encompassed by Figure 11.15 and undergo no mineralogical changes at all while its muddy neighbor in the next bed goes wild, changing from fine-grained Qz and clay to a flashy assemblage that includes Ms, St, and Grt (see "Garuda[2]" in Figure 10.32). It all depends on bulk composition.

Reactions and Projections

The minerals listed earlier and their complicated formulas make analysis appear daunting, so as we have done in previous chapters, let's make this clearer with a chemographic diagram simplified by projection. A common projection for Al-rich shaley rocks is the AKF projection, handy for rocks with abundant K_2O but little CaO.

A common formulation for the A-, K-, and F-components for an AKF diagram is

$$A = Al_2O_3 + Fe_2O_3\text{-}(K_2O + Na_2O)$$

$$K = K_2O$$

$$F = FeO + MgO + MnO$$

where the oxides are expressed in moles, not wt%; the simplest way to do this is to divide wt% values by the respective oxide weights, giving moles per 100 g. The choice of components may seem a little arbitrary, but the AKF diagram is not a proper projection, unlike those in Section 5.2; it is just a graphical way to display some of the mineral compositions, compatibilities, and reactions deduced for the aureole.

Figure 11.16 is an AKF diagram with 20 whole-rock analyses from the Santa Rosa Range aureole plotted along with the locations of end-member minerals (real mineral compositions have solid solution and produce bands, but we're simplifying here, too). We can use the observed mineral associations and this diagram to work out some of the reactions that accompanied metamorphism.

The metashales started out with the greenschist facies assemblage Qz + Ab + Ms + Chl + carbonates (Cal and/or Dol). As noted earlier, the outermost aureole is marked by first appearance of porphyroblasts, and these are typically Bt and And. Crd is found near this outermost contact, but only with Bt and And. From this, Compton (1960) surmised that the following reaction took place:

$$\text{Ms} + \text{Chl} + \text{Qz} = \text{Crd} + \text{And} + \text{Bt} + H_2O \qquad \textbf{(Eq. 11.10)}$$

We leave balancing this reaction to the student as a character-building exercise. We can see it, more or less, on the AKF diagram (Figure 11.16a); remembering that we are projecting from Qz and water, the Ms-Chl tie line is replaced by the Crd-And-Bt triangle, and the bulk rock compositions lie comfortably near both reactants and products, satisfying Baldy's law. The presence of St in some rocks, rather than Crd, could reflect the following reaction at higher *P* (Figure 11.16b):

$$\text{Ms} + \text{Chl} = \text{St} + \text{Bt} + \text{Qz} + H_2O \qquad \textbf{(Eq. 11.11)}$$

Closer to the plutonic contact, Kfs makes an appearance. It can form by dehydration of either of the micas (the only K-bearing phases in the metashales), and the general reactions can be seen on the AKF diagram (Figure 11.16c), unbalanced, diagrammatically, and ignoring H_2O and Qz, as

$$\text{Ms} = \text{Kfs} + \text{Sil} \qquad \textbf{(Eq. 11.12)}$$

$$\text{Bt} = \text{Kfs} + \text{Fe-Ti Oxide} \qquad \textbf{(Eq. 11.13)}$$

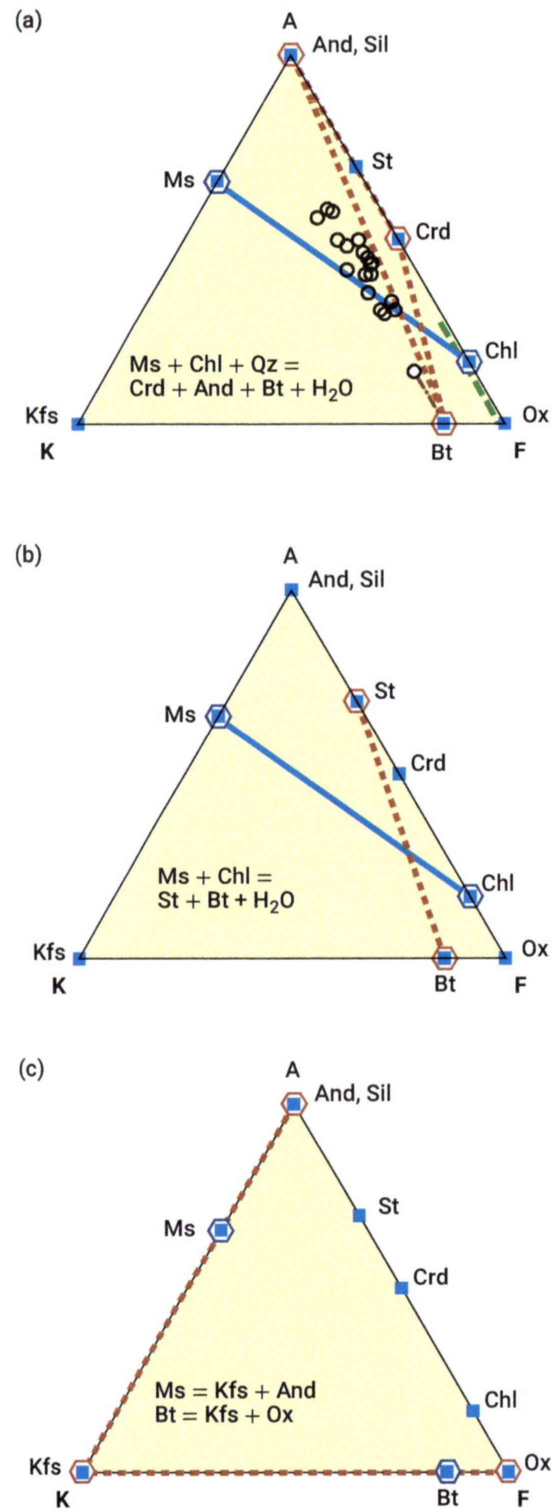

Figure 11.16 AKF ternary composition plots showing locations of common minerals and reaction relations relevant to contact metamorphism of shales in the Santa Rosa Range, northern Nevada. (a) Cordierite + biotite + andalusite assemblages likely form from chlorite + muscovite. Whole-rock compositions are shown as circles. (b) Staurolite + biotite assemblages also likely form from chlorite + muscovite (but at higher pressure). (c) Muscovite breaks down to form sillimanite and K-feldspar, whereas biotite breaks down to form K-feldspar and oxides.

Source: Modified from Pattison & Vogl (2005).

Calc-Silicate Rocks

Metamorphism around the Alta Stock Rocks that consist of carbonates and quartz (a common assemblage in marine sedimentary rocks worldwide) are particularly interesting when contact metamorphosed, because they form a wide variety of calc-silicate minerals. A classic example of this is the aureole around the Cenozoic Alta stock near Salt Lake City, Utah (Figure 11.17). This body of granodiorite was intruded about 35 Ma into Paleozoic sedimentary rocks including limestone, dolomite, sandstone, and, of most interest, limestone and dolomite with siliceous nodules and interbeds. These mixed rocks, consisting of fine-grained carbonate and silicate grains and chert nodules, were ripe for reaction when the Alta and neighboring Cottonwood stocks were intruded.

The mapped aureole of the Alta stock is wide (2–3 km, comparable to the width of the pluton itself). The outermost zone of new mineral growth is shown on Figure 11.17 as the talc isograd. Inside this boundary (closer to the pluton), talc is found in rocks of appropriate composition. Moving toward the pluton, one crosses more isograds: tremolite, forsterite, and finally, close to the pluton contact, periclase. These mineral assemblages are clear in thin section (Figure 11.18), and the pertinent reactions are easily visualized with a chemographic diagram.

Chemographic Analysis of the Alta Aureole The main minerals found in the aureole are listed in Table 11.2. There are many other possible minerals in this chemical system, including diopside ($CaMgSi_2O_6$), enstatite ($MgSiO_3$), serpentine

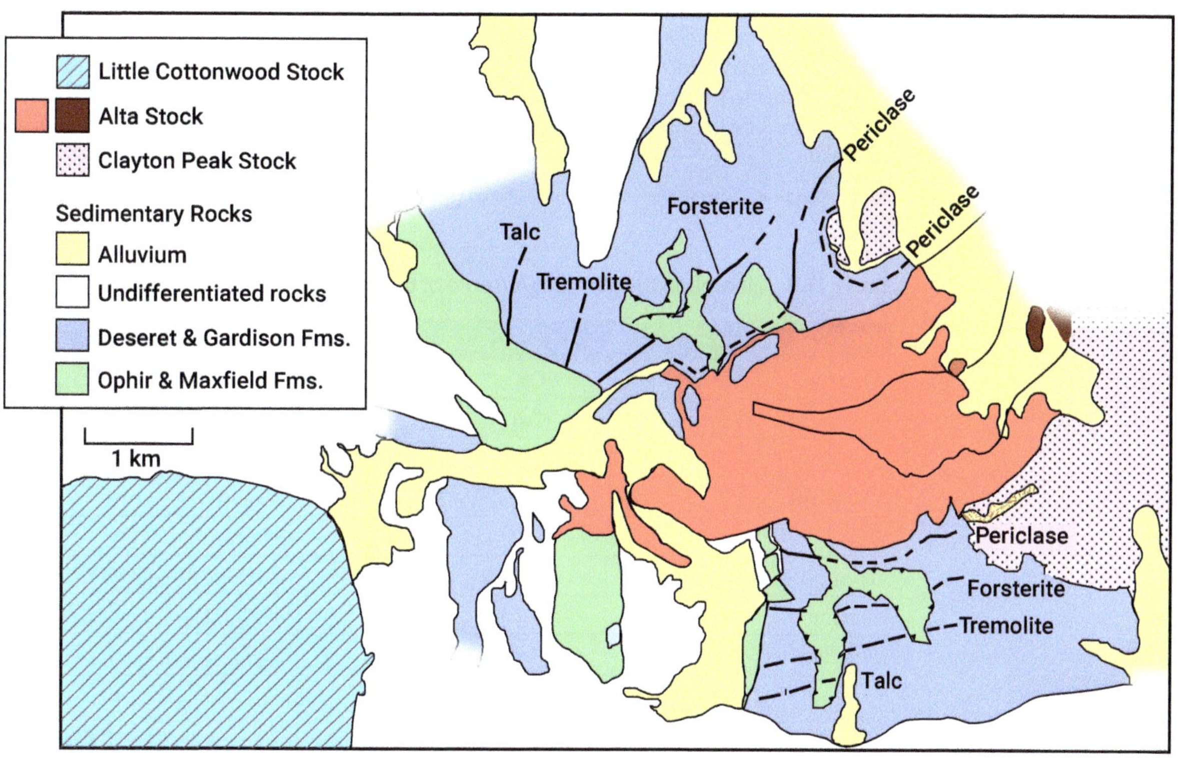

Figure 11.17 Map of the Alta region, Utah, showing distribution of rocks and contact metamorphic isograds around the Alta stock. The talc, tremolite, forsterite, and periclase isograds are common in contact metamorphosed carbonates and represent increasing temperature toward the pluton.

Source: Modified from Stearns et al. (2020).

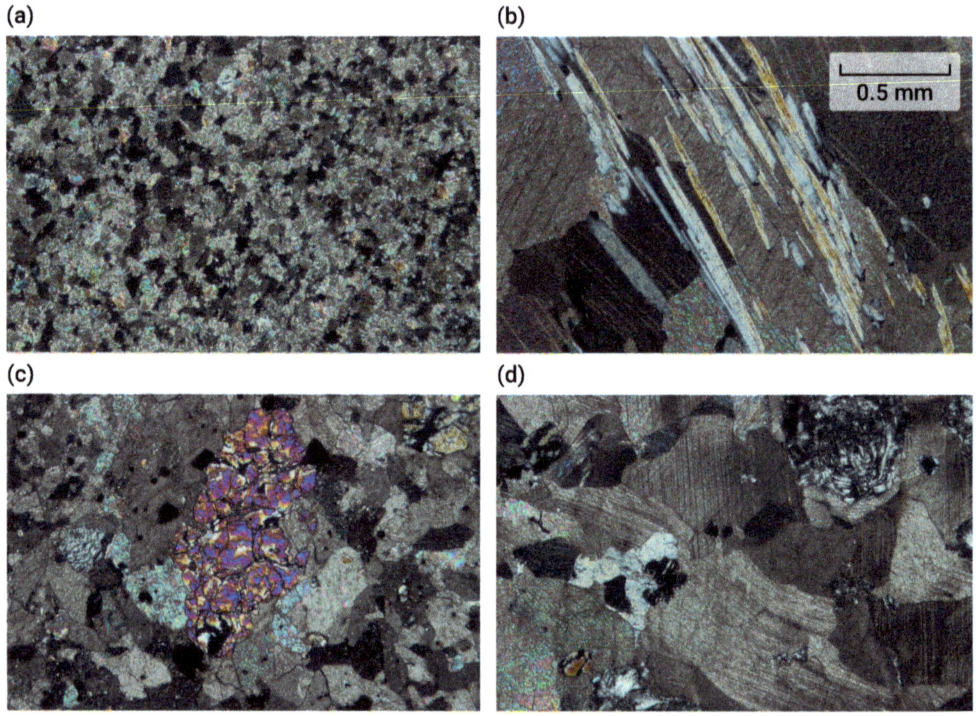

Figure 11.18 Thin section photomicrographs (crossed polarizers) from the (a) talc, (b) tremolite, (c) forsterite, and (d) periclase zones in the contact aureole of the Alta stock. Note the increase in calcite grain size going from the talc zone to the higher-grade zones. Mean grain diameter increases by a factor of more than 10. (a) Talc forms patches with second-order birefringence (much lower than calcite). (b) Tremolite needles cross-cut matrix carbonates. (c) Forsterite is the brightly colored crystal in the middle. (d) Swirly gray patches in the upper right region of the photo are brucite that formed after periclase. Periclase is highly unstable and reacts rapidly with water to produce brucite. This distinctive texture of brucite indicates pseudomorphing after periclase.

Sources: AFG photos; thin sections courtesy of Cameron Davidson.

Table 11.2 Minerals in the Contact Aureole of the Alta Stock

Mineral	Abbreviation	Formula
Quartz	Qz	SiO_2
Calcite	Cal	$CaCO_3$
Dolomite	Dol	$CaMg(CO_3)_2$
Forsterite	Fo	Mg_2SiO_4
Talc	Tlc	$Mg_3Si_4O_{10}(OH)_2$
Tremolite	Tr	$Ca_2Mg_5Si_8O_{22}(OH)_2$
Brucite	Brc	$Mg(OH)_2$
Periclase	Per	MgO

($Mg_3Si_2O_5(OH)_4$), etc., but they do not occur at Alta either because they were not stable at those *P-T* conditions or because the bulk compositions were not right.

Applying the phase rule to this system involves a couple twists relating to *P* and fluid composition that we have not yet discussed. Five chemical components are required to describe these minerals: SiO_2, MgO, CaO, CO_2, and H_2O (so $C=5$). By the mineralogical phase rule (see Section 5.2; $\pi = C + 2 - F$), we would expect that a random bulk composition at random P and T (divariant) should therefore consist of five phases. One of these is a mixed H_2O-CO_2 fluid phase that does not stick around (except perhaps in tiny fluid inclusions in the minerals), so we might expect four minerals to be preserved in a typical rock. However, fluid pressure doesn't have to be equal to rock pressure. Rather, the pressure of the fluid phase might be considered as another intensive variable ($P_{fluid} \neq P$). Thus, there are in general actually *three* independent intensive variables in this system (P, T, and P_{fluid}), not two, and the system is trivariant ($\pi = C + 3 - F = 3$). This doesn't change the number of phases that might be present in a rock (it's still 5, including a fluid phase), but different rocks at the same T and rock P could, in principle, have different assemblages of solid minerals.

Although this rationale makes sense theoretically, it ignores a few commonsense factors. First, for rocks at depths greater than ~5 km, P_{fluid} is probably close to P of the rock (so we revert to only 2 degrees of freedom – P and T). Second, X_{CO_2} (the molar ratio of CO_2 to total $H_2O + CO_2$) may not be variable, and instead might be fixed by external conditions (e.g., mixing of fluids derived from the pluton and host rock). Any time we fix a composition, we reduce the number of degrees of freedom by 1. Altogether, this means that the degrees of freedom may be only 1 (because $P_{fluid} = P$ and we fixed X_{CO_2}). In this case, most rocks should consist of only three minerals (not four) plus a fugitive fluid that has fled the scene.

Plotting five chemical components requires a 4D, tetrahedron-like thing,[3] and once again, our 3D brains are ill-suited for this task. However, if we assume that H_2O and CO_2 are present in excess, we can project from them and plot the remaining components in a triangle (Figure 11.19). The original sedimentary rocks consisted primarily of just three minerals, Cal, Dol, and Qz (Figure 11.19a), so all possible bulk compositions lie in the left-hand sub-triangle. We can immediately tell that we should not find any metamorphic assemblages consisting only of phases in the right-hand subtriangle (e.g., Qz + Tr + Tlc), because such an assemblage would have a bulk composition outside the range encompassed by the sedimentary rocks. One or more of these right-side phases could occur in a rock, but they would have to coexist with calcite or wollastonite (or both).

In the unmetamorphosed sedimentary rocks, Cal and Dol are dominant over Qz, so bulk compositions lie toward the lower part of the shaded area, near the Cal-Dol join (recall from Chapter 4 that a join is a line segment in composition space that joins two compositions). The silicate part of the rocks comprises dispersed Qz in the limestone and dolomite, and chert nodules in dolomite. Formation of Tlc, defining the talc isograd, can form via only one reaction from this combination of minerals: Qz + Dol = Tlc + Cal (note that in reactions inferred from the chemographic diagram, we ignore the fluid phase; this one has to consume water and produce carbon dioxide). This reaction is evident theoretically from the crossing relationship on a chemographic diagram (Figure 11.19b), and Tlc + Cal reaction rims on chert nodules in dolomite (Figure 11.20) are smoking-gun observational evidence for this reaction. The geographic area between the talc and tremolite isograds is the talc zone.

The tremolite zone is recognized by the assemblage Tr + Dol + Cal. There are two crossing reactions that can produce Tr: Tlc + Cal = Tr + Dol and Dol + Qz = Tr + Cal (Figure 11.19c). Which one took place? Cook and Bowman (2000) argued that the second reaction was largely responsible for Tr formation, based in part on the absence of Qz in these rocks, indicating that the second reaction went to completion. Moving inward, the forsterite zone is marked by the reaction Tr + Dol = Fo + Cal, and the periclase zone by Dol = Per + Cal. Periclase in the aureole has been completely hydrated during cooling to Brc—a **retrograde** reaction (or **retrogression**).

It is important to emphasize again that a rock will show an isograd reaction only if its bulk composition is appropriate. The isograds above are all based on rocks whose bulk compositions are largely dolomite with a little quartz either dispersed in the dolomite or collected into chert nodules, and such bulk compositions will show all the isograds. A rock layer consisting only of quartz, however, will not show any of the reactions—it will just be plain old quartz (albeit coarser, recrystallized quartz) in the aureole. Similarly, a rock

Retrogression/retrograde
Back-reaction or re-equilibration of minerals during cooling and exhumation from the maximum metamorphic conditions. For example, retrogression of periclase as it cools and reacts with water typically forms retrograde brucite.

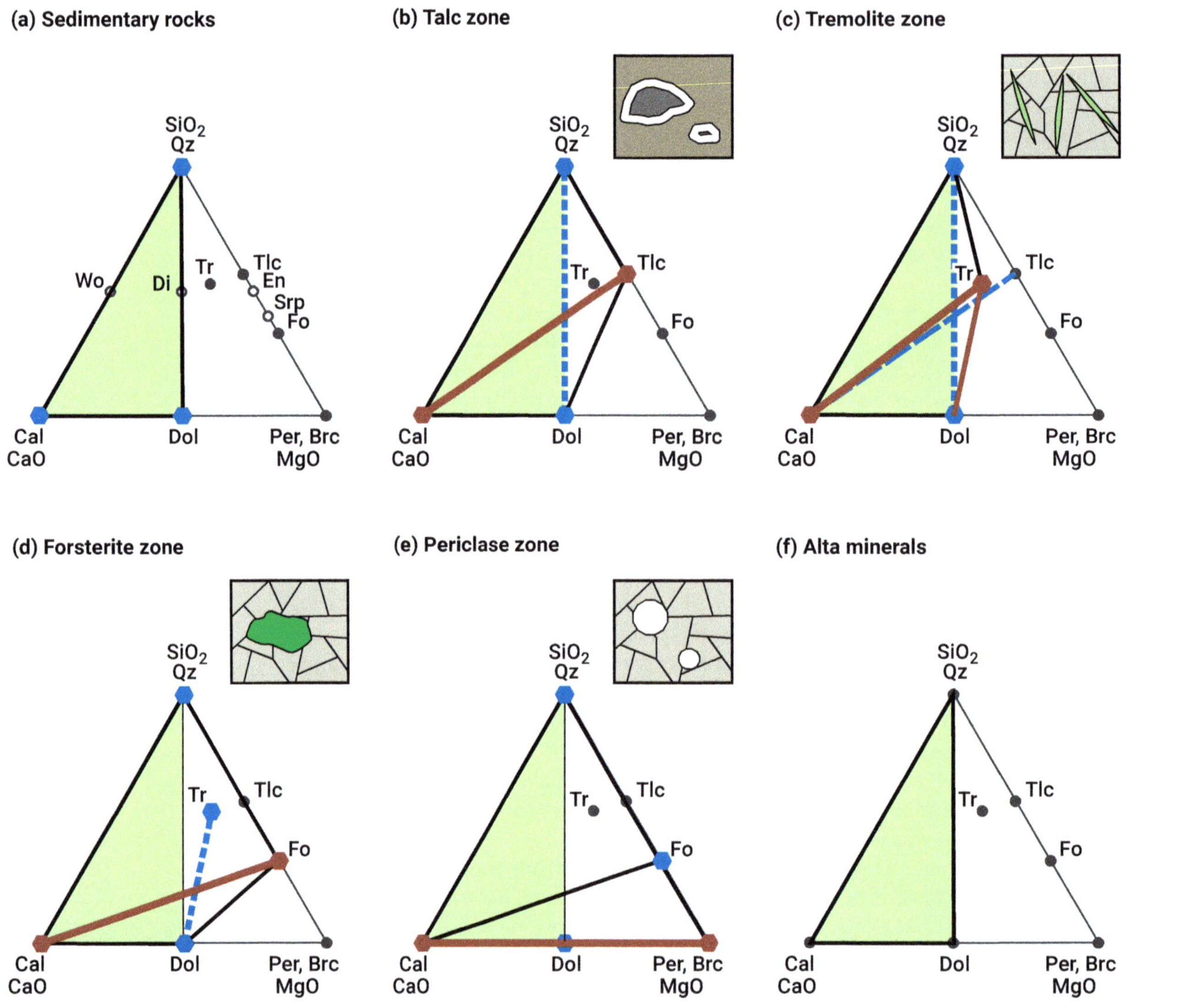

Figure 11.19 SiO_2-CaO-MgO composition diagrams, showing mineral compatibilities at the Alta contact aureole, arranged according to mineral zones. The triangle for siliceous dolomites is shaded in green. (a) Unmetamorphosed rocks with possible minerals for a large range of bulk compositions. (b–e) Talc, tremolite, forsterite, and periclase zones, with observed mineral compatibilities. Small diagrams show schematic textures in outcrop (b) and thin section (c–e). Blue lines connect reactant minerals; red lines connect product minerals. (f) The minerals that are observed in the Alta rocks, corresponding to their restricted bulk composition.

consisting only of calcite will remain calcite throughout the aureole, and a rock consisting solely of dolomite will change only at periclase isograd and, even then, only for certain fluid compositions.

P-T-X Conditions of the Alta Aureole It makes sense that the assemblages closer to the pluton reached higher peak metamorphic temperatures, but how can we estimate these temperatures? Cook and Bowman (1994) used Cal-Dol thermometry as a guide to peak conditions. This technique relies on the solvus between Cal and Dol (Figure 11.21a). The solubility of Dol (or Mg) in Cal depends on temperature, and they assumed that the amount of Mg in the Cal reflects the peak temperature (see Section 5.3). From Cal compositions in rocks where Dol was present, they showed that temperatures increase toward the contact, as expected, approaching about 600 °C adjacent to the pluton

(Figure 11.21b). Three samples from the Per zone yield much lower temperatures than the rest, probably as a result of partial re-equilibration (retrogression) during cooling.

The reactions inferred here, written out with volatiles and balanced, are

$$3\text{Dol} + 4\text{Qz} + H_2O = \text{Tlc} + 3\text{Cal} + 3CO_2 \quad \textbf{(Eq. 11.14)}$$

$$5\text{Dol} + 8\text{Qz} + H_2O = \text{Tr} + 3\text{Cal} + 7CO_2 \quad \textbf{(Eq. 11.15)}$$

$$\text{Tr} + 11\text{Dol} = 8\text{Fo} + 13\text{Cal} + 9CO_2 + H_2O \quad \textbf{(Eq. 11.16)}$$

$$\text{Dol} = \text{Per} + \text{Cal} + CO_2 \quad \textbf{(Eq. 11.17)}$$

Figure 11.20 Talc zone: brownish chert nodule in gray dolomite showing a reaction rim of talc and calcite ~1 cm thick. The slippery feel of talc is evident when rubbing a finger over the rim.

Source: AFG photo.

Because these reactions involve one or both volatile components, the composition of the volatile phase (X_{CO_2}) affects the location of the reaction in *P-T* space. The phase rule once again comes in handy here. For an *n*-component system, a reaction involving *n* + 1 phases is univariant, meaning that it plots as a line on a *P-T* diagram. For example, in the two-component system Ab-H_2O (see Figure 6.29), the melting reaction Ab + vapor = melt is univariant and defines the water-present solidus. However, the first three reactions above (Equations 11.14–11.16) have only five phases (remember that CO_2 and H_2O are combined in the fluid phase), so in a five-component system, the reactions are divariant and form surfaces in P-T-X_{CO_2} coordinates. These reactions will, however, plot as lines if one of the intensive variables is fixed.

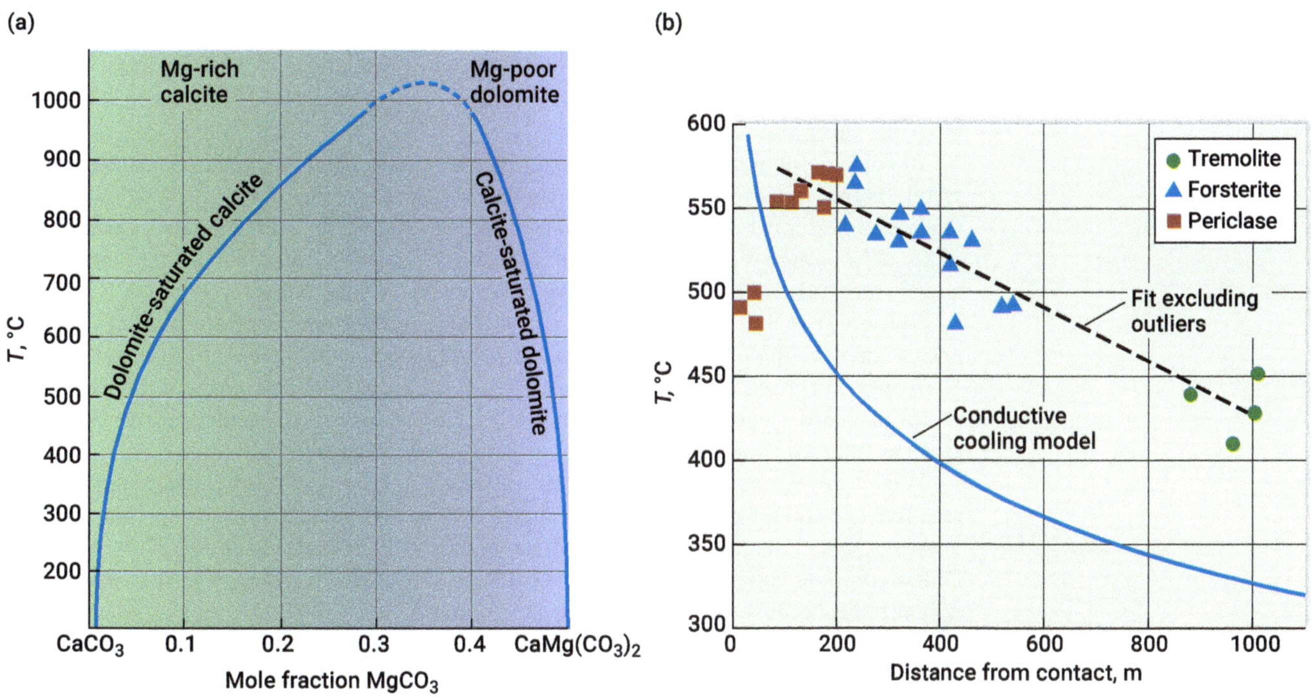

Figure 11.21 (a) Calcite-dolomite solvus. (b) Temperature vs. distance from the contact as calculated using the Mg content of calcite that coexists with dolomite. A reference conductive cooling model is also shown.

Sources: (a) Modified from Anovitz & Essene (1987); (b) From Cook et al. (1997).

Customarily, we fix P and look at reactions in T-X_{CO_2}. Figure 11.22 shows such a diagram at $P = 150$ MPa from Cook & Bowman (2000). Construction of such diagrams involves significant calculation using internally consistent thermodynamic databases, which is a complicated topic. There are univariant curves along which five phases are in equilibrium, and they meet at invariant points (labeled a–e) with six phases. Using these curves, we can work out a T-X_{CO_2} history that is consistent with observed mineral assemblages and inferred temperatures. Note that the history given here is not unique; it is just one possible path.

Recall that the fluid phase is a mixture of H_2O and CO_2, and X_{CO_2} will change as reactions proceed, because all the reactions involve these components on one side or the other. Formation of Tlc requires that X_{CO_2} is less than the value at invariant point *a*. Consider a bulk composition at 300 °C with $X_{CO_2} = 0.2$. When this assemblage reaches ~300 °C, Tlc starts to form via the reaction in Equation 11.14 (Dol + Qz + H_2O = Tlc + Cal + CO_2). One might expect that the reaction would go to completion at this temperature, but things are more complicated. The reaction consumes H_2O and produces CO_2, so X_{CO_2} goes up as the reaction proceeds. Adding heat to the system causes the system to crawl to the right, up in temperature, along the reaction curve. As long as Dol, Qz, Tlc, and Cal coexist, X_{CO_2} is fixed by the temperature. We say that the fluid composition is **buffered** by the solid assemblage along the univariant reaction curve. However, if one of the reactants, say Qz, is used up, the univariant relationship is destroyed, and the assemblage could leave the curve as temperature rises. In this case, the path would proceed vertically upward in T, reaching the Tlc + Cal curve and producing Tr.

Buffer
Here, to limit the composition of a phase or an intensive variable such as T via a chemical reaction. Solid-solid reactions often buffer the compositions of fluids.

If all the reactants remain when the system reaches point a (in fact, not all the Qz was used up), the system is stuck there with rising thermal input, because this is an invariant point, with six phases present. However, as long as heat keeps flowing into the system, it cannot stay there forever, and eventually, one of the reactants is eliminated by reaction. What is the reaction? Observation shows that Tlc is present going in and Tr going out, so the net reaction must use up Tlc to produce Tr via the pair of Tr-producing reactions that meet at point *a*. At the Tr isograd in the field, the assemblage Tr + Cal + Dol and the observed sequence Dol + Qz followed by Tr + Cal + Dol indicates that most Tr is produced by reaction 11.15; thus, Tlc must be the limiting reactant, and when it is used up, the system crawls up reaction 11.15. This reaction consumes H_2O and produces CO_2, thus driving X_{CO_2} up as the fluid composition remains buffered along the reaction curve.

At some point along reaction 11.15, Qz is used up (as shown by field relations), and the system is free to leave the curve with increasing temperature. The next place on its path that we can infer is Fo-forming reaction 11.16. Although this reaction covers a wide range of X_{CO_2}, the path in Figure 11.22 has been drawn to cross the reaction at a temperature estimated from Cal-Dol thermometry. This temperature occurs at low X_{CO_2}, which is buffered by reaction 11.16 and increases with temperature, because the reaction produces 9 moles of CO_2 for every mole of H_2O (Equation 11.16). The system can leave the curve for reaction 11.16 once Tr is used up. From that point, the fluid must go to even more water-rich compositions in order to cross the Per-forming reaction at appropriate temperatures (~600 °C, again as inferred from Cal-Dol thermometry; Figure 11.21b).

Why don't we see diopside? After all, at some point around 450 °C, the system crosses reaction Tr + Cal + Qz = Di + H_2O + CO_2. However, the rocks do not display this reaction, because Qz was the limiting reactant for reaction 11.14 and was used up. Thus, although Di is a common component of Qz-rich calc-silicate rocks elsewhere, it is absent from the Alta aureole, because the rocks are too Qz-poor.

This approximate P-T-X_{CO_2} path tells us something about fluid flow in the Alta aureole. The Tlc- and Tr-forming reactions are hydration reactions that consume H_2O to form hydrous minerals and liberate CO_2, driving X_{CO_2} up. Closer to the pluton contact, though, X_{CO_2} was far lower, and the fluid was largely H_2O. Based on these observations and oxygen isotopes, Cook & Bowman (1994) inferred that water-rich fluids flowed in a down-temperature direction, away from the pluton, during metamorphism.

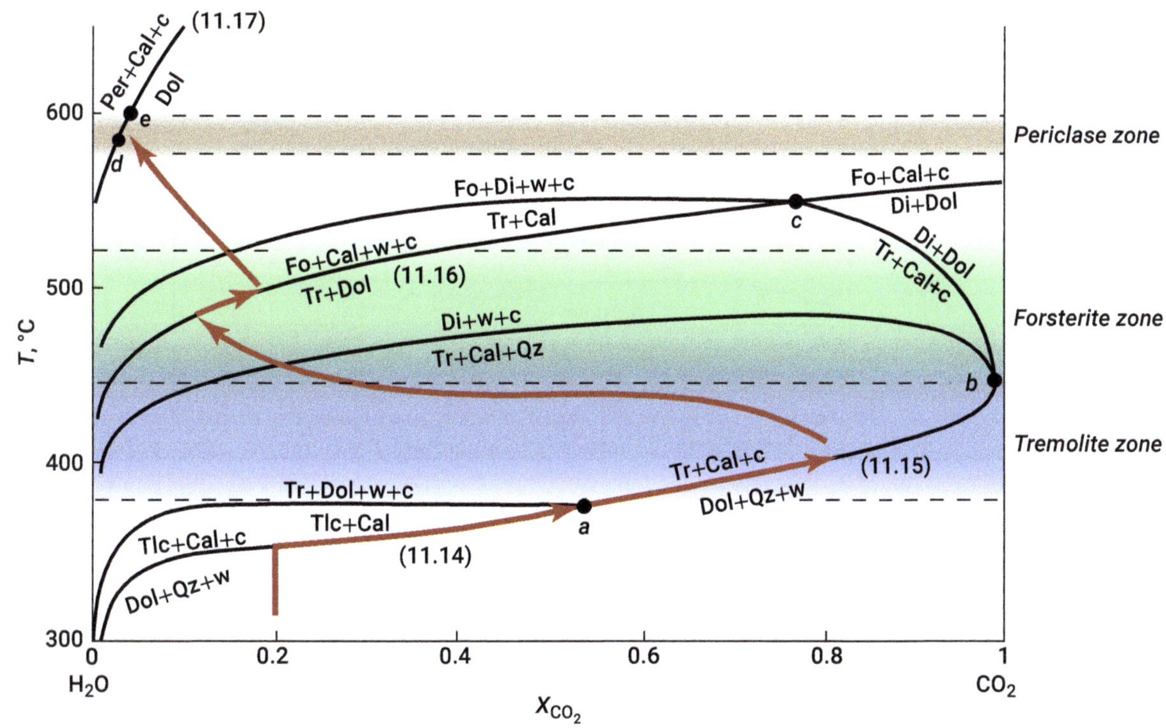

Figure 11.22 Temperature vs. composition diagram for siliceous dolomites at a fixed pressure of 150 MPa. "w" and "c" are abbreviations for H_2O and CO_2. Rocks in the Alta aureole experienced only reactions described in Equations 11.14–11.17. Dots mark invariant points where three or more reactions intersect. Red curve marks the path that rocks from the Alta contact aureole might have followed (see text); they evolve to high X_{CO_2} as reactions produce CO_2, then toward low X_{CO_2} as they interact with water closer to the pluton.

Source: Modified from Cook & Bowman (2000).

Aureoles of Unusual Size

Although the outer boundary of an aureole cannot be precisely defined, it makes sense that aureole thickness, as defined by some parameter (e.g., first appearance of And or a particular maximum temperature reached) should correlate with pluton size. Thermal modeling of conductive cooling does indeed show that all else being equal (e.g., magma temperature), the thickness of the aureole should scale with pluton dimensions (see Equation 11.3; see Figure 11.10), such that the aureole around a 2-km-thick sill should be twice as thick as that around a 1-km-thick sill. However, relative aureole thicknesses (aureole thickness divided by estimated pluton thickness) in a compilation by Annen (2017) vary by more than an order of magnitude. This is evident in Figure 11.1, where the mapped thicknesses of aureoles around different pluton exposures vary dramatically within one mountain range.

There are several reasons why apparent aureole thickness might not correlate well with pluton size. Aureole thickness should be measured perpendicular to the plutonic contact, but this direction can be difficult to determine. In addition, the subsurface exposure of a pluton might be much larger than what breaks through to the surface. Hidden plutons can lead to orphaned aureoles with no obvious parents (Figure 11.23). On the map in Figure 11.1, Compton (1960) mapped several zones of contact metamorphism without surface pluton exposure and inferred that they are underlain at shallow depth by plutons that did not quite make it to the surface (or, to put it more accurately, the surface has not made it down to them yet).

There are many other reasons for aureoles of unusual size. The sheet-like Manaslu granite of Nepal (see Section 9.3) is several kilometers thick, but its upper aureole is only

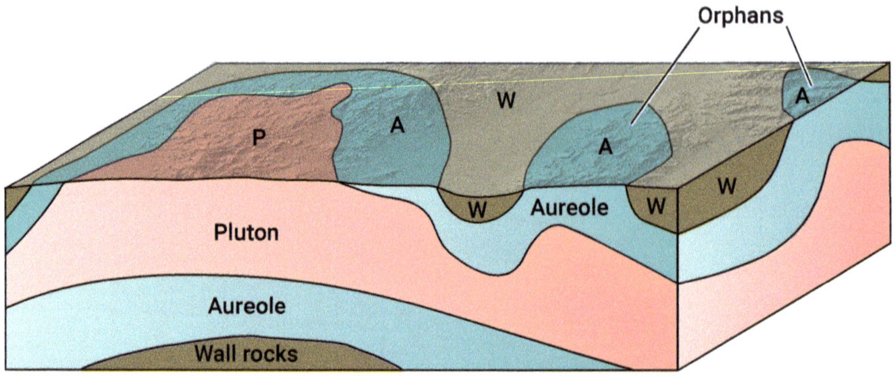

Figure 11.23 Schematic diagram illustrating how plutons that do not quite breach the surface can give rise to orphan or unusually wide aureoles. P = pluton, A = aureole, W = wall rocks.

about 100 m thick. Annen et al. (2006) explained the skimpy upper aureole by incremental emplacement in which heat was supplied over a long period of time rather than in one large blast, such that the thermal effects were confined to a small region adjacent to the intrusion. Widespread extensional thinning in the region does, however, complicate interpretations.

In contrast to Manaslu, the spectacular aureole of the Alta stock extends 2 km from the contact, even though the pluton outcrop area is only a few kilometers across. The Alta aureole is wider than thermal models predict (see Figure 11.21b) even if the pluton were emplaced all at once rather than incrementally. Cook and Bowman (1994) proposed that heat was carried farther than expected via advective flow of hot fluids outward from the pluton—implying that the entire thermal event took place even more rapidly than one-pulse emplacement would indicate. However, Stearns et al. (2020) dated metamorphic titanite in the aureole and found that it is significantly younger than the Alta stock itself. They explained these younger ages and the width of the aureole as a result of fluid flow from the younger, larger, adjacent Little Cottonwood pluton through the Alta stock over several million years. Thus, the contact aureole, so carefully analyzed earlier, appears to be composite, reflecting thermal and fluid inputs from at least two plutons.

Predictions of aureole thickness based on conductive cooling models are invalid if convection of hydrothermal fluids occurs in and around the pluton (see Section 11.6). The inner aureole around such a body will be thinner and potentially much richer in mineral variety and minerals of economic value, while the outer aureole (the region that sees elevated heat) can be larger.

11.4 Incremental Emplacement of Large Plutons

Learning Objective

By the end of this section you should be able to:

11.4.1 Compare, using sketches, batch emplacement of a pluton with incremental emplacement, and evaluate the viability of the following processes in each scenario: crystal fractionation, stoping, and crystal settling.

A great deal of effort and countless CPU cycles have been spent modeling the effects of pluton emplacement, both one-pulse and incremental, on the thermal and mechanical evolution of aureoles (e.g., Bowers et al., 1990; Annen, 2017; Chen & Nabelek, 2017). Figure 11.24 shows a simplified thermal model of emplacement of a batholithic

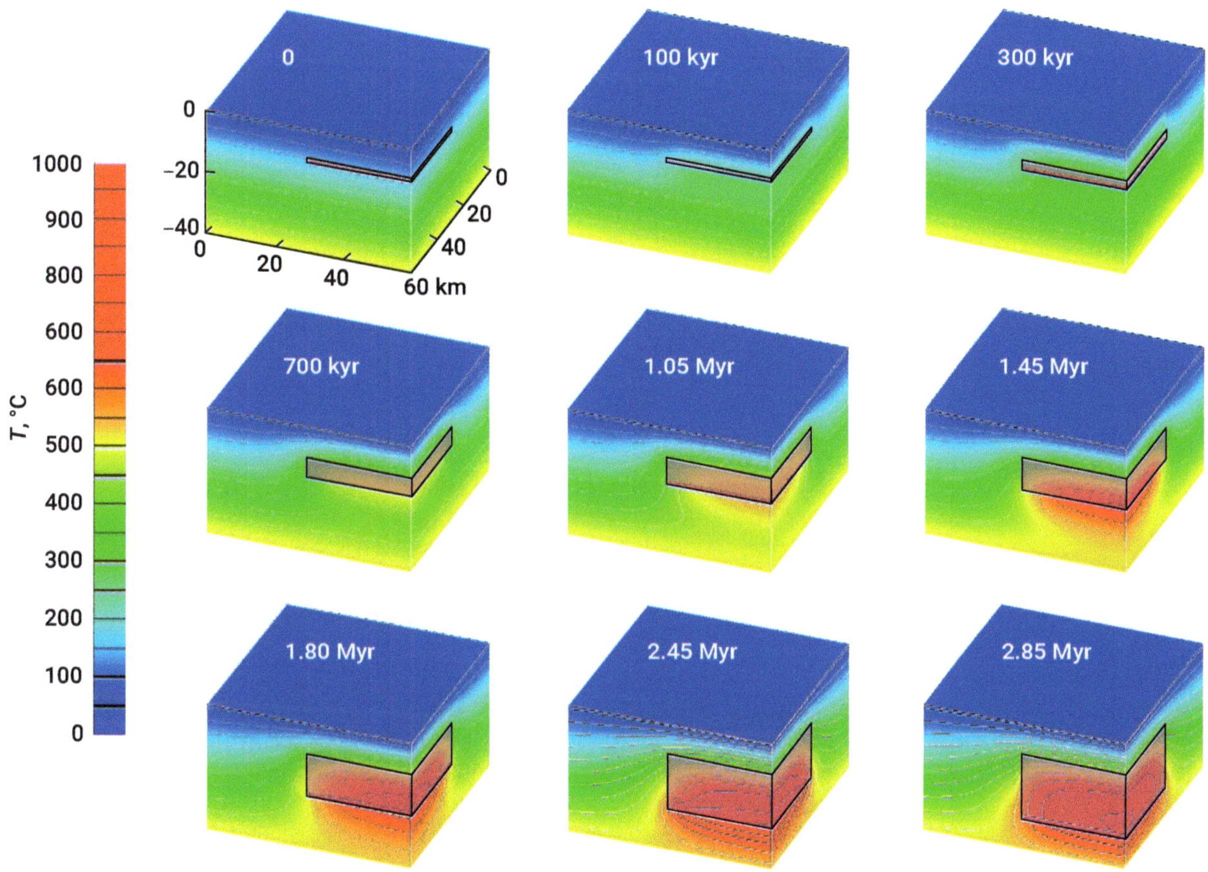

Figure 11.24 Models of incremental emplacement of a sill every 150 kyr for a total period of 2.85 Myr. Each sill is 1000 m thick, and sills are emplaced from the top down. At first, sills cool quickly, but after four or five emplacements, a hot zone develops and, after 1.5 Myr, melt persists between intrusions. This hot zone grows downward, and wall rocks melt around it. Metamorphism occurs not only in the country rock, but also in previously emplaced sills.

(30 × 30 × 20 km) pluton over 2.85 Myr via top-down injection of twenty 1000-m sills every 150 kyr. As with the model in Figure 11.9, this scenario is unrealistic (e.g., instantaneous emplacement of 1000-m-thick sills is unlikely), yet it illuminates some fundamental principles. These include the following:

- Early increments cool below the solidus well before the next increment is intruded.
- Within 1 Myr, a hot zone forms, and after about 1.5 Myr, this hot zone stays partially molten, such that the next increments are emplaced into a mushy zone. This mushy zone expands downward with time owing to all the earlier increments that have dumped their heat into the system and because the zone of injection is moving downward into hotter rocks.
- By the end of the injection period, the zone above the solidus is smaller than the body of intruded rock, but much larger than a single injection, and shaped something like an oblate, rounded box.
- The partially molten zone at the end of the model includes wall rocks below the pluton.
- Isotherms are lifted around the intrusive body such that temperatures corresponding to greenschist and amphibolite grade occur several kilometers higher than before intrusion started.
- Nearly the entire pluton is at amphibolite grade or higher.

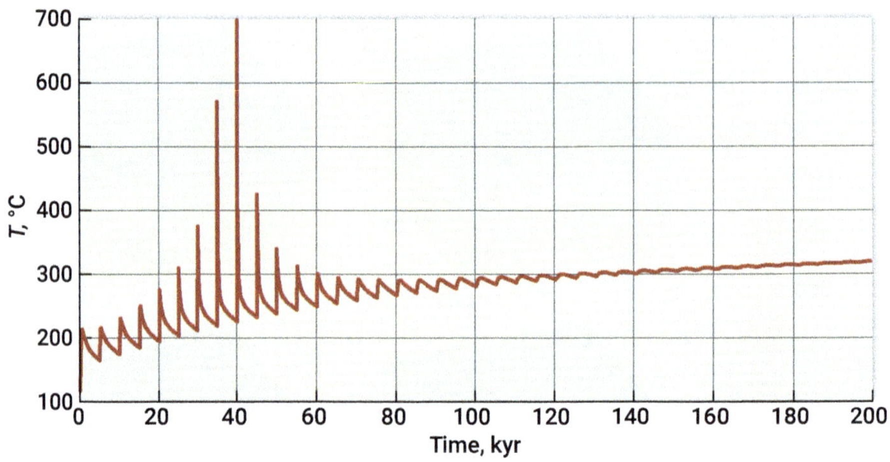

Figure 11.25 Temperature-time history at a level of 5200 m for an incrementally emplaced pluton based on the model illustrated in Figure 11.9b (20-m-thick sills emplaced every 5000 years for 500 kyr). Before 40 kyr, this level represents wall rock, but as the pluton grows, this region fills with sills. This depth witnesses large temperature swings until ~100 kyr, after which temperature gradually rises with small thermal oscillations.

An important but commonly overlooked feature of pluton emplacement is the effect that pluton heat has on the pluton itself. Figure 11.25 shows the evolution of T with time at a depth of 5200 m in the incremental emplacement model of Figure 11.9b. This is 200 m deep in the eventual pluton, and that depth consisted of wall rocks until the pluton had built down to that level 40 kyr after the start of the model. Thus, T fluctuations before 40 kyr indicate T cycling of the wall rocks as the pluton built downward, and fluctuations after 40 kyr represent T variations within the newly constructed part of the pluton. As noted in Chapter 10, numerous experimental studies have shown that T fluctuations have a profound effect on crystal growth, causing small crystals to dissolve and reprecipitate onto larger ones, ever enlarging crystal sizes via dissolution-reprecipitation (see Section 10.8). Figure 11.25 shows that an incrementally emplaced pluton could be subjected to numerous T cycles, perhaps thousands, during its lifetime.

The models in Figures 11.9b and 11.24 reveal a fundamental truth: *early emplaced batches of magma in an incrementally emplaced pluton are the contact-metamorphic wall rocks of later batches.* This fact holds even in bodies emplaced in single huge batches, because the outer, rapidly solidified parts of the pluton would then be flooded by massive amounts of heat and fluids released by the slower-cooling interior. Plutons contact-metamorphose themselves.

11.5 **Skarns: Reaction between Magma and Wall Rocks**

Learning Objective

By the end of this section you should be able to:

11.5.1 Balance chemical reactions occurring in carbonate-rich contact aureoles, identifying chemical components that must be derived from the magma.

Silicate Magma vs. Carbonate Wall Rocks

In the previous sections, we have discussed textural changes (e.g., development of granoblastic texture) and reactions induced in wall rocks by plutonic heat and circulating fluids, noting in particular that carbonate rocks containing silicate minerals produce a

Table 11.3 Some Reactions Important in Skarn Formation

Magmatic Components	+	Wall Rock Components	=	Reactant Mineral(s)	+	Gas
SiO_2	+	$CaCO_3$ *calcite*	=	$CaSiO_3$ *wollastonite*	+	CO_2
$2SiO_2$	+	$CaMg(CO_3)_2$ *dolomite*	=	$CaMgSi_2O_6$ *diopside*	+	$2CO_2$
$3SiO_2 + Al_2O_3$	+	$3CaCO_3$ *calcite*	=	$Ca_3Al_2Si_3O_{12}$ *grossular*	+	$3CO_2$
$8SiO_2$	+	$5CaMg(CO_3)_2 + H_2O$ *dolomite + H_2O*	=	$Ca_2Mg_5Si_8O_{22}(OH)_2 + 3CaCO_3$ *tremolite + calcite*	+	$7CO_2$
$4SiO_2$	+	$3CaMg(CO_3)_2 + H_2O$ *dolomite + H_2O*	=	$Mg_3Si_4O_{10}(OH)_2 + 3CaCO_3$ *talc + calcite*	+	$3CO_2$

rich variety of contact metamorphic minerals in calc-silicate assemblages. When the wall rocks are carbonates and the intruding magma chemically reacts with them, a **skarn** develops. Skarns are reaction zones that locally exhibit large grain size (e.g., baseball-sized garnets) and thick monomineralic zones. They commonly also host economic deposits of Fe, W, Cu, Zn, Mo, Sn, Au, and other metals.

Skarn
A reaction zone of calc-silicate minerals between a pluton and the rocks it intruded.

Skarns develop owing to the fundamental incompatibility of hot silicate magma with carbonate rocks. Typically, invading magma is prevented from significant reaction with wall rock carbonates by formation of a chilled margin or relatively impermeable reaction zone. Locally, however, there can be significant reaction. Skarns resulting from reaction of magma with limestone are characterized by Ca-rich minerals such as grossular-andradite garnets, wollastonite, epidote, and tremolite. Skarns developed in dolomite typically develop Mg-rich minerals such as serpentine, diopside, olivine, phlogopite, and spinel. Magma is a ready source of SiO_2, and magmatic heat destabilizes calcite and dolomite, whose CaO and MgO combine with the magma to form typical skarn minerals via reactions including those in Table 11.3.

Some Skarn Examples

The Pine Creek skarn of eastern California, located in the Sierra Nevada about 30 km south of Long Valley caldera (see Section 7.6), exemplifies skarns of economic significance. The skarn developed along contacts between Paleozoic metasedimentary rocks and Mesozoic plutonic rocks and provided tungsten, an essential component of hardened steel, hardened drill bits, and old-fashioned incandescent lights, from World War I until the 1990s. Tungsten, extracted from scheelite ($CaWO_4$), was a critical metal during World War II and the Korean and Vietnam Wars, and much geologic exploration in eastern California was driven by the need for more of it. Paul Bateman and colleagues at the USGS mapped five 15-minute geologic quadrangles in the area, and Bateman (1965) produced a report detailing the geology of the tungsten deposits.

The several kilometers of tunnels and shafts in the Pine Creek mines reveal how the skarn developed along the contact between intruding granitic rocks and marble that comprises part of a steeply dipping panel of hornfels and marble derived from shale, sandstone, and limestone (Figure 11.26). Skarns developed along a complex and interfingering contact between silicate magma and marble, and only parts of the skarn contain enough scheelite to have been worth mining (Figure 11.27).

Although skarns are typically dominated by a few minerals, some are more diverse. The Crestmore quarry in southern California, mined out of a small hill of marble east of Los Angeles, has yielded an astounding variety of minerals—166 at last count, including nine that were discovered there. Quarrying for marble and abundant core drilling has exposed the interior of the system, showing that the calc-silicate skarn is intimately interfingered with intruding tonalite along a contact zone ranging from centimeters to tens of

Figure 11.26 View northwest from the summit of Mount Tom, showing Broken Finger Peak (above the word "Triassic") and the Pine Creek mine entrance. The western side of the ridge consists of a steeply dipping screen of dark brown hornfels and dark gray marble sandwiched between Triassic granitoids on the east and Cretaceous granitoids on the west. The hornfels-granite contact is barren of mineralization, but the marble-granite contact is richly decorated with skarns containing scheelite ($CaWO_4$). The entrance to the tunnel and shaft complex of the Pine Creek Mine, which ceased operations in 1991, is marked by buildings in this 2007 photo. Most of the mine buildings have since been removed.

Source: AFG photo.

meters thick. Burnham (1959) mapped the skarn and enumerated the various reactions between the invading magma and carbonate wall rocks. The main minerals in the skarn are arranged in zones around the tonalite (Figure 11.28). Passing outward from the plutonic rocks, these are a garnet zone consisting mostly of brown grossular; a zone consisting entirely of green vesuvianite, $Ca_{10}(Mg,Fe)_2Al_4(SiO_4)_5(Si_2O_7)_2(OH,F)_4$; and a monticellite zone consisting mostly of that mineral, a Ca-bearing olivine ($CaMgSiO_4$). Outside the monticellite zone, the marble was largely unaffected by the silicate magma.

Origin of Monomineralic Zones

The common occurrence of monomineralic zones in skarns (Figure 11.29) is puzzling given that the mineralogical phase rule (see Section 5.2) predicts that for an arbitrary bulk composition at arbitrary *P-T* conditions, the number of phases should equal the number of chemical components. At Crestmore, there were certainly at least four or five major components (e.g., CaO, MgO, SiO_2, Al_2O_3, H_2O), so there should be three or four minerals present in a typical sample.

An explanation for such mineral segregations was worked out by Korzhinskii (1950) and expanded upon by Thompson (1959), who introduced the concept of local equilibrium, illustrated by the following thought experiment. Consider a rod consisting of large volumes of periclase (Per, MgO) and quartz (Qz), separated by a zone

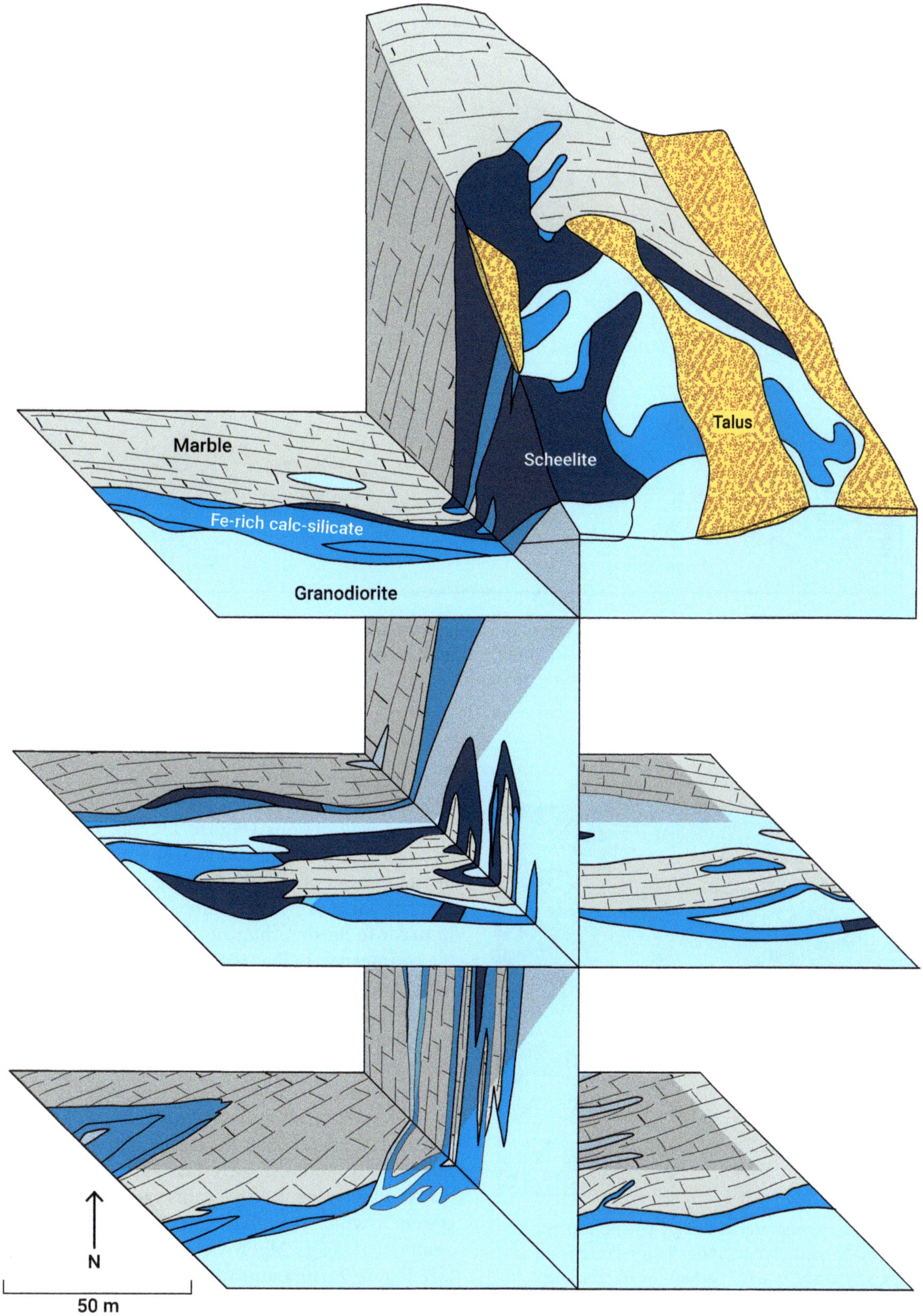

Figure 11.27 Cross sections of Pine Creek mine showing complex structural relationships between economic scheelite deposits, marble, calc-silicates, and Cretaceous granodiorite.

Source: Modified from Bateman (1965), who drew this diagram based on data from tunnel walls and drill core.

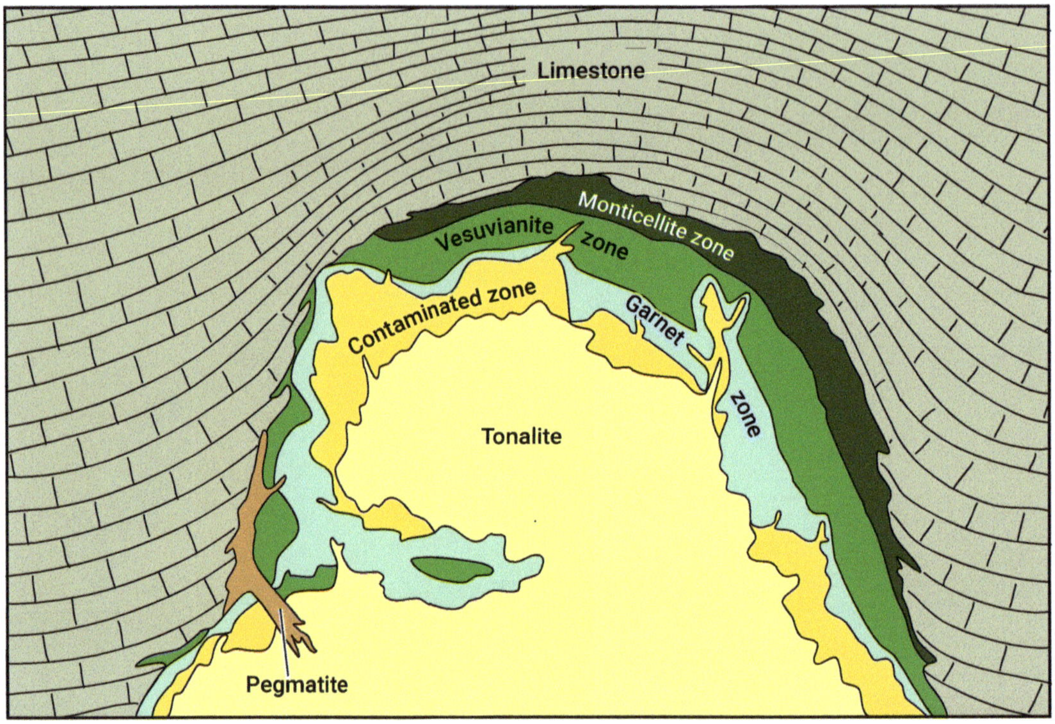

Figure 11.28 Classic cross section of the Crestmore Quarry, showing distribution of mineralogic zones. The "contaminated" zone represents mixtures of tonalite and skarn. Garnets in the garnet zone are Ca-rich grossular-andradite. Vesuvianite is a complex Ca-Mg-Al silicate. Monticellite is a Ca-Mg olivine-structured mineral.

Source: Modified from Burnham (1959).

Figure 11.29 Typical monomineralic zones developed in a skarn. A zone of red-brown garnet (under fingers) is surrounded by coarse-grained epidote, which is in turn surrounded by coarse calcite. These rocks originated as Paleozoic carbonate rocks that were engulfed by Cretaceous granodiorite. Armstrong Canyon, eastern Sierra Nevada, California.

Source: AFG photo.

in which powdered Per and Qz are mixed in smoothly varying proportions (Figure 11.30a). If held at a *T* high enough that reactions can proceed in reasonable amounts of time, these phases, which are not stable together, will react to form zones of forsterite (Fo) and enstatite (En) in order to harden their borders against their incompatible counterparts (Figure 11.30b). However, these reaction zones themselves contain incompatibilities in two places, where the Per + Fo and Fo + En zones meet and where the Fo + En and En + Qz zones meet, because at these interfaces Per + En and Fo + Qz, respectively, are in contact. At these interfaces, pure zones of Fo and En will form (Figure 11.30c), and given time, these will grow to eliminate the two-phase zones, resulting in a succession of monomineralic zones (Figure 11.30d).

An interesting feature of this process is that it can take a system in which the concentrations of one or more chemical components (e.g., MgO) vary smoothly across the system (see Figure 11.30a) and turn it into one in which concentrations are a step function (e.g., 100 wt% MgO in Per, 57.3 wt% in Fo, 40.2 wt% in En, and 0 wt% in Qz). This is the opposite of the more normal situation in which diffusion tends to iron out chemical gradients.

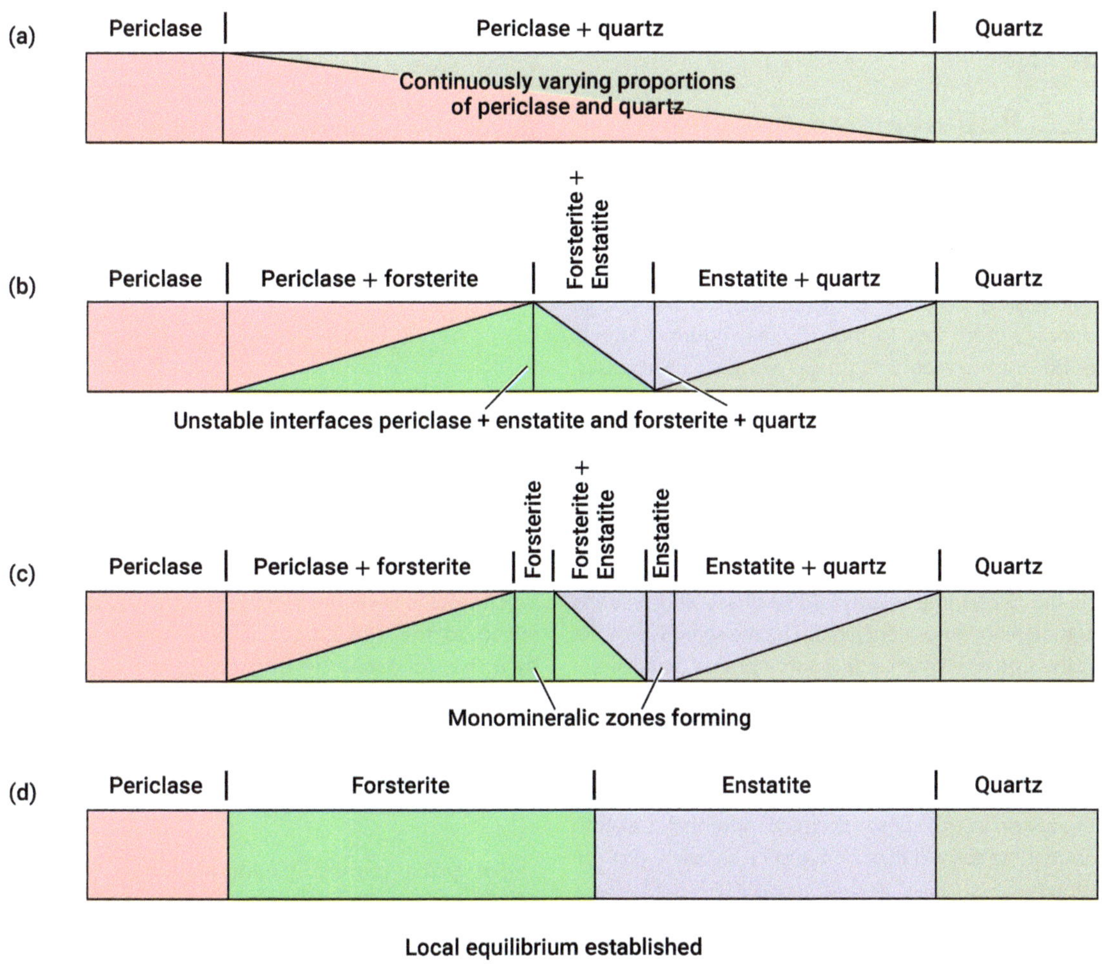

Figure 11.30 Thought experiment illustrating how monomineralic zones can form through local equilibrium. In all cases, a rod of solid material is constructed such that 100% periclase (MgO) is stable at one end and quartz is stable at the other. (a) Periclase and quartz are mixed in continuously varying proportions along the rod, then heated. (b) Higher MgO mixtures toward the left form olivine, while higher SiO_2 mixtures toward the right form enstatite. Mineral proportions vary continuously across these zones. However, there are two unstable interfaces where periclase and enstatite are in contact and where forsterite and quartz are in contact. (c) Further reaction and expansion of these unstable interfaces starts to produce monomineralic zones of forsterite and enstatite. (d) Expansion of the monomineralic zones ultimately stabilizes to a final configuration with only monomineralic zones.

Source: Modified from Thompson (1959).

The key to this process is for one or both of the components to be mobile and able to migrate to zones where they are needed by reaction. Korzhinskii (1950) referred to these as perfectly mobile components—components whose activities (essentially, thermodynamic concentrations; see Toolbox 11.3) are fixed outside the system of interest and that are able to migrate freely. In our example, the activities of both MgO and SiO_2 are set outside the zone of reaction (we're assuming infinite rods of each), and as long as at least one can migrate freely, the reactions will proceed. Skarns are reaction zones between reservoirs of carbonate rock and silicate magma, and the development of (nearly) monomineralic zones implies fixed activity and significant mobility of one or more of the major chemical components in the system.

Toolbox 11.3

Chemical Potential and Activity

In Section 3.4, we introduced the concept of Gibbs free energy *G* and noted that at a given *P* and *T*, a system at equilibrium will have the minimum value of *G*. Thus, systems do their best to minimize *G* by moving components around (e.g., moving SiO_2 from Trd to β-Qz when crossing that phase boundary from low to high *P*; see Figure 3.13). In solutions, be they gases, magmatic liquids, or crystals, a related critical concept is that of **chemical potential**, which is defined as

$$\mu_i = \left(\frac{\partial G}{\partial n_i}\right)_{P,T,n_{j\neq i}} \quad \textbf{(Eq. 11.18)}$$

where *G* is the Gibbs free energy of the phase and n_i is the number of moles of component *i* in it. The meaning is simple: How does the free energy change if some of that component is moved into or out of the phase, everything else being held constant (*P*, *T*, and all other chemical components $n_{j\neq i}$)? If μ_i is higher in one phase than another, then the system can lower *G* by moving that component from the phase with the higher value to the lower. In this sense, chemical potential is akin to pressure in that matter will flow from high μ to low μ if given the opportunity.

The activity of a component in a solution (a_i) is defined by the equation

$$\mu_i = \mu_i^\circ + RT\ln(a_i) \quad \textbf{(Eq. 11.19)}$$

where μ_i° is the chemical potential of component *i* in its **standard state**, which commonly is defined as the pure component at *P* and *T* of interest. For example, if we are studying the NaCl-KCl system (halite-sylvite) at 1 bar and 500 °C, the standard state for NaCl component would be pure halite at 1 bar and 500 °C. The relationship described by Equation 11.19 was originally derived from the ideal gas law.

One way to think about activity is as a measure of the effective pressure (a material's partial pressure) relative to the pressure if the phase were pure. In an ideal mixture of gases, partial pressure (relative to total pressure) of a component scales with the proportion of that gas component (i.e., its mole fraction relative to all gases present). Thus, it is commonly stated that an ideal solution is one where $a_i = X_i$, where *a* and *X* are the activity and mole fraction, respectively, of component *i*. This expression of ideality holds for minerals when there is only one crystallographic site on which different elements mix. For example (glossing over some structural complications), in orthopyroxene, $(Fe,Mg)_1SiO_3$, Fe and Mg can be considered to mix on only one site, and the activity of enstatite ($MgSiO_3$) is given by the mole fraction of Mg:

$$a_{En} = X_{Mg} = Mg/(Mg+Fe) \quad \textbf{(Eq. 11.20)}$$

where Mg and Fe are atomic proportions. This gives rise to a simple 1:1 relationship between activity and composition (Figure 11.31a). However, most minerals have more than one crystallographic site on which mixing occurs. For example, in olivine, with composition $(Fe,Mg)_2SiO_4$, Fe and Mg mix on two sites. In these cases, statistical mechanics shows that the ideal activity is the mole fraction raised to the site multiplicity, a value typically denoted by α. So, for forsterite (Mg_2SiO_4), the activity is given by the mole fraction of Mg raised to the power $\alpha = 2$ (Figure 11.31b):

$$a_{Fo} = \left[X_{Mg}\right]^2 = \left[Mg/(Mg+Fe)\right]^2 \quad \textbf{(Eq. 11.21)}$$

For typical garnets $[(Fe,Mg,Mn,Ca)_3Al_2Si_3O_{12}]$, the ideal activity of grossular would be the mole fraction of Ca raised to the power of three:

$$a_{Grs} = \left[X_{Ca}\right]^3 = \left[Ca/(Fe+Mg+Mn+Ca)\right]^3 \quad \textbf{(Eq. 11.22)}$$

And so on. The activity of enstatite also includes a site multiplicity, but it is unity, so we normally omit it. You might be wondering, too, about the other sites in these minerals. We can ignore any site that is completely filled with a single element, because the mole fraction (*X*) is unity. For example, the tetrahedral site in olivine is normally filled with Si, or nearly so (i.e., $X_{Si} = 1.0$). Similarly, we do not (normally) have to consider O, because it is typically the only anion. We could include those sites in the activity expression, but with *X*'s equal to 1, they would not affect the activity. However, as we move on to more chemically complicated minerals where mixing does occur on multiple sites (e.g., think of a typical mica or amphibole), the ideal activity expressions rapidly become much more complicated.

Three other considerations are important:

1. Few real materials mix ideally, the alkali feldspars (see Section 8.3) being an important case in point. We call

such solutions "non-ideal," and derive expressions to relate activity to composition. In such systems, activity-composition relationships can be complicated (Figure 11.31c). The cause of non-ideality in this system is the size mismatch between Na and K.

2. Activities of components are only meaningful when referenced to a specific standard state. For example, using a_{SiO_2} for the activity of SiO_2 in a silicate melt is meaningless without an explicit statement of the standard state (e.g., β-quartz at P and T). The activity calculated relative to Trd at P and T would be different, and the activity calculated relative to Trd at 1 atm and T could be really different. Choice of standard state does not affect the final results of a calculation as long as consistency is maintained.
3. The ideal relationship $a_i = X_i$ for single-site mixing holds only when the standard state is the P and T of interest. For example, consider pure pyrope ($Mg_3Al_2Si_3O_{12}$) garnet at 1 GPa and 800 °C. The activity of the $Mg_3Al_2Si_3O_{12}$ component is 1 relative to a standard state of pure pyrope garnet at those conditions (1 GPa and 800 °C), but 330,000 relative to a standard state of pure pyrope garnet at 1 atm and 800 °C.

Key point: always be clear about standard states when discussing activities.

Chemical potential
Change in Gibbs free energy per mole when a small amount of a component is added to a phase, with all other variables held constant; at equilibrium, the chemical potential of a given component is the same in all phases.

Standard state
Here, a reference state for activity, typically a pure phase (e.g., mineral) composed of that component at a specified P and T.

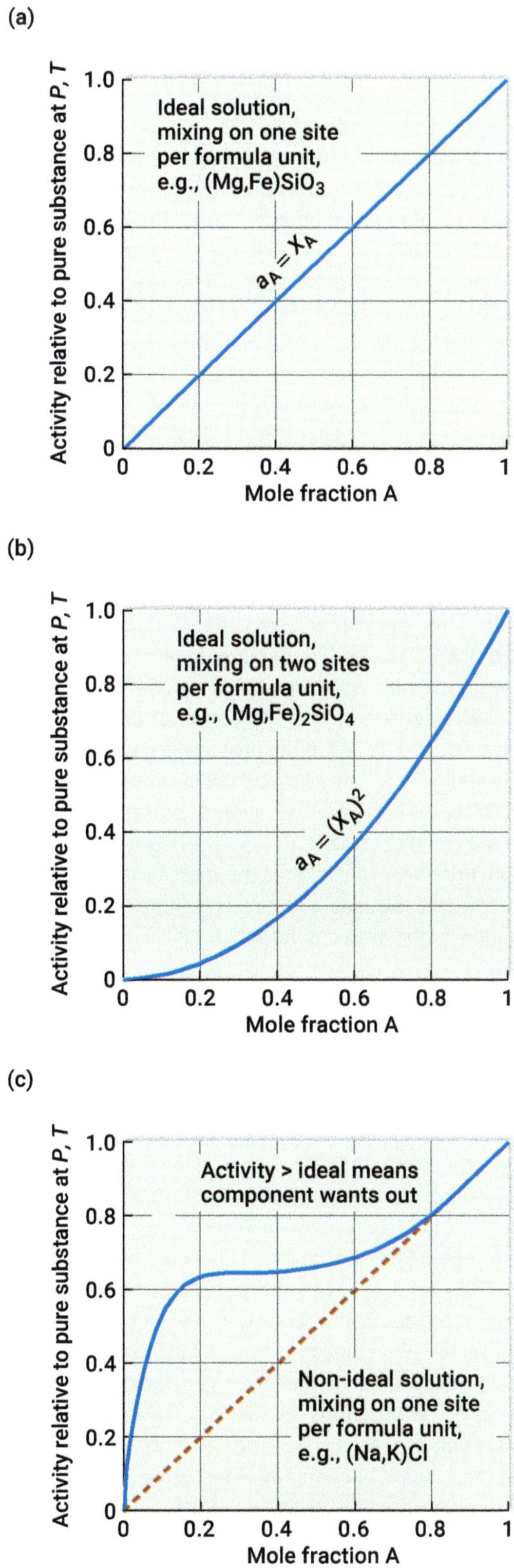

Figure 11.31 Activity-composition relationships for different types of mixing. (a) Ideal solutions with mixing on a single site show a simple 1:1 relationship. (b) Ideal solutions with mixing on two sites (e.g., forsterite in olivine) show a simple parabolic relationship. (c) Non-ideal solutions can show quite complicated relationships, as in this case for alkali feldspar.

Making Connections 11.1

Cement Production and Anthropogenic CO_2

Concrete is a remarkable substance: it is strong, cheap, easy to cast into any shape, non-flammable and non-conducting, and made from abundant and widespread raw materials. It can even be used to 3D-print structures. Concrete forms the foundations of houses, bridges, and buildings, and is made into dams, roads, aqueducts, and other structures of modern civilization. About 30 billion metric tons of concrete are currently produced every year. That's enough to pave over the island of Manhattan 20 m deep, every year. Elhacham et al. (2020) estimated that concrete makes up over half of all human-produced material and that the total mass of concrete produced will exceed the Earth's dry biomass in the 2030s (Figure 11.32a). Production of concrete is following an exponential growth curve, with a doubling time of about 15 years.

Concrete is a conglomerate-like mixture of sand and gravel bound with cement. Humankind has spent millennia finding the right mixture of materials to produce a strong, lasting binder that can withstand the elements, seawater, seismic shaking, repeated pounding by traffic, and other assaults, and in so doing developed a process that is analogous to skarn metamorphism followed by retrograde hydration. The ancient Romans were quite good at making concrete, producing architectural marvels such as the Pantheon in Rome, which sports an unreinforced concrete dome 43 m in diameter that was built nearly 2000 years ago (Figure 11.32b).

Cements are designed to harden (set) via chemical reactions that occur when water is added to cement powder. In its most basic form, CaO (lime) is mixed with water to form portlandite, $Ca(OH)_2$, which is called slaked lime in the building trade. This can be used as mortar to hold bricks together, or as plaster to put on a wall. Over time, portlandite reacts with CO_2 in the atmosphere to form calcite via the reaction

$$Ca(OH)_2 + CO_2 = CaCO_3 + H_2O \qquad \textbf{(Eq. 11.23)}$$

The resulting product is only weakly cemented and is not useful as a structural building material.

Today's main cement type, Portland cement, is descended from various successful home brews that were found both by accident and by design. Lime is obtained by heating calcite from limestone to more than 750 °C, driving the decomposition (calcination) reaction to the right:

$$CaCO_3 = CaO + CO_2 \qquad \textbf{(Eq. 11.24)}$$

Production of lime is energy-intensive, and so various cheap additives have been tried over the millennia to stretch it out, much as cheap additives are added to processed foods to bulk them up. Roman architects found that mixing Italy's abundant fine volcanic ash with slaked lime produced a much stronger cement than slaked lime alone. In England and France, builders found that lime produced by calcining limestone that contained silica and clay produced strong cement.

The formula for Portland cement, refined after extensive experimentation, was patented in 1827. It is produced by firing limestone and 20 to 25 percent clay to 1450 °C in huge rotating kilns—massive, inclined steel pipes with firebrick lining, as long as a football field, into which the crushed limestone-clay mixture is poured. The kiln rotates slowly, and the mixture flows toward the lower end, where a roaring gas flame heats the mixture to between 1400 and 1500 °C. As the mixture makes its way inexorably toward searing hell, several prograde reactions that parallel reactions during contact metamorphism of shaley limestones occur. The resulting product comes out the low end of the kiln as a slag-like substance known as clinker. Clinker is mixed with a small amount of gypsum, ground up, and sold as Portland cement.

A main ingredient in clinker is the high-*T* reaction product Ca_3SiO_5, which produces a key setting (hardening) reaction via hydration when water is added to the Portland cement powder:

$$2Ca_3SiO_5 + 7H_2O = Ca_3Si_2O_7 \cdot 4H_2O + 3Ca(OH)_2 \qquad \textbf{(Eq. 11.25)}$$

This is but one of several such reactions, and other components such as Al, Fe, and S play key roles as well. This reaction is significantly exothermic, and setting of a large concrete structure can be sped up by years if a system of cooling pipes is built in.

As is evident from Equation 11.24, calcining limestone releases CO_2 to the atmosphere. Current data indicate that cement production accounts for about 8 percent of anthropogenic CO_2 release, with about half of this from calcination and the other half from producing the energy needed to drive reactions in the kiln. Given the importance of concrete to civilization and its huge carbon footprint, the search is on for alternative materials and methods of production that are more environmentally friendly (Jacoby, 2020).

(a)

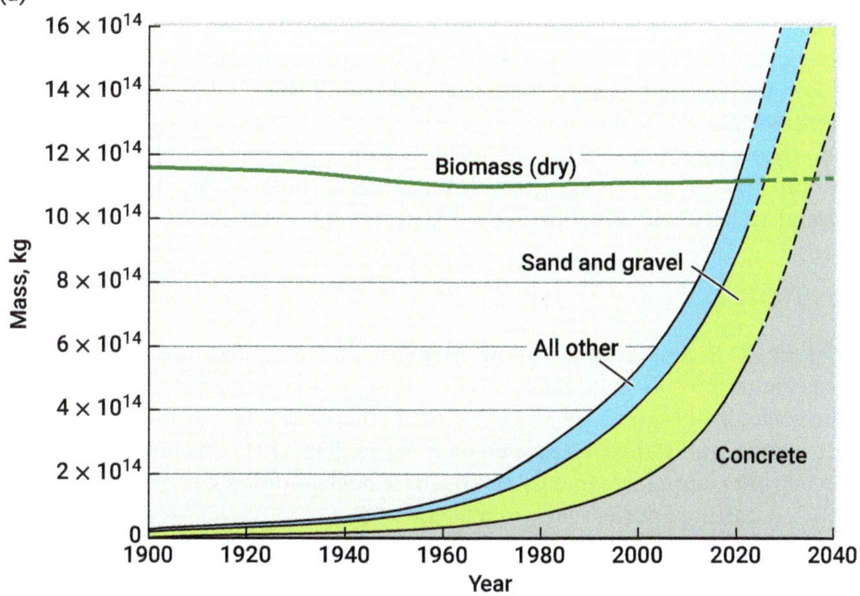

(b)

Figure 11.32 (a) The current mass of all human-produced material to date is about the same as the Earth's biomass minus the water living things contain, and in the 2030s, concrete alone will surpass the dry biomass. Dashed lines are extrapolations. (b) Looking up into the concrete dome of the Pantheon in Rome. The dome is 43 m in diameter, and the 7.6-m opening (oculus) at the apex is 43 m above the floor. This marvel of concrete engineering was built nearly 2000 years ago.

Sources: (a) Modified from Elhacham et al. (2020); (b) Yana Marudova/Unsplash.

11.6 Hydrothermal Systems around Intrusions

Learning Objectives

By the end of this section you should be able to:

11.6.1 Develop and illustrate a conceptual model that explains why heat flow is so high in geothermal areas.

11.6.2 Use stable isotopes of water to identify sources of hydrothermal fluids and processes affecting them.

If the rocks surrounding the pluton are sufficiently permeable and there is any water around, a cooling magma body in the upper parts of the crust will set up a hydrothermal convection system. These hot-water pumps are important for several reasons, including their efficiency in carrying heat away from the magma, their ability to concentrate elements of low abundance but high value (e.g., gold), and their ability to focus magmatic heat into extractable zones of geothermal energy. Surface evidence of hydrothermal activity, commonly obvious and spectacular, includes fumaroles, hot springs, geysers, boiling mud pots, and areas of hot, chalky ground. Yellowstone National Park is a wonderland of hydrothermal activity, and we will use it to illustrate some of these phenomena.

Yellowstone

Heat Budget

Estimates of heat flow at Yellowstone National Park within the main caldera (see Figure 7.37) average about 2 W/m^2, roughly 40 times greater than typical heat flow in the geologically calm eastern half of North America and 20 times more than that in the active Basin and Range. Integrated over the caldera, that adds up to about 5 GW, a huge amount (but comparable to only two or three nuclear or large hydroelectric generating stations). Recalling Fourier's law from Section 2.7:

$$q = -k\frac{dT}{dz} \qquad \textbf{(Eq. 11.26)}$$

We can use heat flow and an estimate of thermal conductivity (k) to estimate geothermal gradient, *assuming that heat flow is by conduction alone*. Using $k = 3$ W/m/°C, we get $\frac{dT}{dz} = 0.7$ °C/m, or a whopping 700 °C/km. Given that the water-present solidus of crustal rocks is typically less than 700 °C, this would require the presence of partially molten rocks 1 km below the surface. However, drilling shows that temperatures are likely not much above 300 °C at 1-km depth, and this means that our assumption of heat transfer by conduction must be wrong. Instead, another process, ascent of hot water, is transferring heat much more effectively. As noted by Lowenstern & Hurwitz (2008), "the enormous hydrothermal system is able to act as a wick, transferring heat advectively from a deep but very large magma source" via hot, rising, hydrous fluids.

Types of Hydrothermal Activity

There are two main types of hydrothermal activity at Yellowstone, hot water and vapor (steam) dominated (Fournier, 1989). Hot-water systems account for much of the activity in the western half of the park, where the boiling curve under hydrostatic conditions defines the maximum T at a given depth. Under these conditions, the weight of water defines P in the water-filled crack system, and if a given spot heats above the boiling T at that P, the water boils, sending up steam bubbles that carry heat away and maintain T at the boiling point.

At the surface, hot-water systems produce boiling hot springs and geysers that generate neutral to mildly alkaline water. These waters contain significant chloride (Cl^-) and trace amounts of silica that, over decades and millennia, produce terraces and mounds of **siliceous sinter** and **travertine** (Figure 11.33). Old Faithful has built up a gigantic mound of siliceous sinter. Bubbling mud pots occur where such water comes close to the surface, boils a bit, then flows laterally and to lower ground, where it emerges as a boiling spring. In Yellowstone, these waters come from sources at about 180 to 270 °C; given the boiling point vs. depth curve, this temperature corresponds to hydrostatic depths of 100 to 550 m.

Siliceous sinter
Deposits of opaline or amorphous silica around hot springs and geysers.

Travertine
$CaCO_3$ deposited by spring-fed, and commonly heated, waters.

In vapor-dominated systems, which are prevalent in the high plateau in the eastern part of the park, steam and other gases fill open fractures but are largely trapped beneath an impervious cap. These systems exist at pressures below the boiling curve (if not, the steam would condense to water). When these vapors are discharged at the surface, they produce fumaroles, acidic boiling pools, and mud pots, without significant discharge of chloride-rich hot water. Acidic waters form when H_2S (the gas with the rotten-egg smell,

Figure 11.33 Surface phenomena associated with geothermal activity. (a) Boiling mud pots, fumaroles, and hydrothermally altered basaltic **hyaloclastite** (granular material formed by fragmentation as lava enters water) at the Námafjall geothermal area, Iceland. Red and tan rock has been highly altered to clays by acidic thermal waters; gray material in foreground is colored by pyrite and other sulfides. (b) Thermal pools at Hot Creek, Long Valley caldera, eastern California. The striking pale blue pools owe their color to various factors, chiefly that they are too hot (>70 °C) for bacteria and algae to grow, as they do in the adjacent cooler, darker, greenish pool. Blue pools are lined with white calcite and clay minerals that reflect sunlight, and finely divided minerals in the water scatter the blue end of the visible spectrum, giving rise to the pretty aqua color. This area, once popular for bathing, was closed in 2006 after the plumbing system changed and fumaroles and geysers shifted to previously safe areas. (c) Travertine at Minerva Terrace, Mammoth Hot Springs, Yellowstone National Park. Pools are fed by water that is saturated in $CaCO_3$; as it spills over the edges of the pools, a little is deposited, leading to dripstone rims. (d) White Dome Geyser in Yellowstone National Park, a 4-m-tall, cone-type geyser built of siliceous sinter. This geyser erupts irregularly at intervals ranging from 15 minutes to 3 hours.

Source: AFG photos.

very toxic and noticeable) in steam reacts with surface O_2 to form H_2SO_4 (sulfuric acid); this strong acid is quite effective in breaking down feldspars and glass in Yellowstone rhyolites. Such hydrothermal alteration produces abundant alteration minerals, including clays, zeolites, micas, and hydrothermal (low-T) K-feldspar cut by veins of opal, chalcedony, quartz, calcite, fluorite, pyrite, and iron oxides. Intense hydrothermal alteration leaves behind barren rocks composed of quartz, clays, various sulfates, and a little native sulfur.

Hyaloclastite
Granular material formed by fragmentation as lava enters water.

Sources of Water in Hydrothermal Systems

The Meteoric Water Line

Hydrothermal systems and geothermal energy systems depend on water. In some geothermal areas, the amount of water available controls how much energy can be extracted from the ground. All power plants reinject water back into the system after energy has been extracted from it, and some even inject outside water, such as treated sewage, as a hedge against water losses. Because of its central importance, the source of the working fluid in hydrothermal systems is of great interest.

Geochemists quickly solved the question of where hydrothermal waters come from once oxygen and hydrogen isotopes were employed. Recall from Section 5.4 that we used oxygen isotopes to test the hypothesis that seawater is the main source of water responsible for hydration of MORB at MORs. Similar studies on land using oxygen and hydrogen isotopes show that precipitation, not magma, is the source of most water in hydrothermal systems.

Hydrogen in the hydrosphere is nearly all (99.985%) ordinary ^{1}H with a single electron orbiting a nucleus made of a single proton. The remainder is ^{2}H, also known as deuterium, with a single neutron in its nucleus, and a trace amount of radioactive ^{3}H, called tritium, with two neutrons. Fractionation of isotopes from one another by mass-dependent processes (see Section 5.4) depends on the ratios of the masses. With masses that differ by a factor of two, ^{2}H and ^{1}H are strongly fractionated from one another during processes such as evaporation. As with oxygen, relative amounts of these two isotopes are characterized by a delta value:

$$\delta^2\text{H} = \delta\text{D} = 1000\left(\frac{R_{\text{sample}}}{R_{\text{standard}}} - 1\right) \qquad \textbf{(Eq. 11.27)}$$

where the ratios are ^{2}H/^{1}H on an atomic basis and the standard is VSMOW. The symbol δD has traditionally been used for this measure (D for deuterium), but we will go with δ^2H for consistency with oxygen.

Globally, δ^2H values show a huge range compared to oxygen, varying from near zero in the tropics and subtropics (e.g., along the Gulf Coast) to less than –200‰ in northern Canada and Greenland, and less than –400‰ in Antarctica. Regardless, the continent-scale pattern of variation is nearly the same as that for δ^{18}O (see Figure 5.14), with a large rain-shadow effect on the eastern side of the Cordillera. δ^2H and δ^{18}O are strongly correlated with each other in precipitation, a relationship known as the meteoric water line (Craig, 1961; Figure 11.34) and given approximately by the equation

$$\delta^2\text{H} = 10 + 8\delta^{18}\text{O} \qquad \textbf{(Eq. 11.28)}$$

Principal Source of Geothermal Water—Precipitation When geothermal waters are examined with both oxygen and hydrogen isotopes, an interesting pattern emerges: Data from a given area smear from the meteoric water line to the right (higher δ^{18}O)—that is, δ^2H values from a given geothermal system are relatively constant, whereas δ^{18}O values range from the local position on the meteoric water line to much higher values (Figure 11.35). Why does δ^{18}O change a lot and δ^2H only a little? The answer for the horizontal arrays is mass balance; O makes up roughly 50 percent of the atoms in rocks, whereas H is a much smaller component of rocks but makes up two-thirds of the atoms in a water molecule. When water and rocks interact and exchange isotopes, the amount of H in the water vastly exceeds H in the rock, so the water δ^2H will not change much (although the rock δ^2H can). In contrast, the O concentrations are more evenly matched, and the isotopic composition of the water shifts toward the heavier rock values.

The dominance of meteoric water in geothermal fluids is clear from Figure 11.35, because the values scatter from the meteoric line toward higher values (closer to magmatic) rather than the other way around. Having δ^2H information makes the case much stronger, because these values match the local meteoric water rather than being anchored at magmatic values, which tend to range from –60 to –90 for δ^2H and +5 to +6 for δ^{18}O. However, the δ^2H values show an interesting pattern: water from chloride-rich boiling pools has δ^2H consistently in the range of local meteoric water (in other words, the values smear to the right along a horizontal trend), whereas water from vapor-dominated systems takes off on trajectories with a positive slope, leading to much higher δ^2H. This happens because

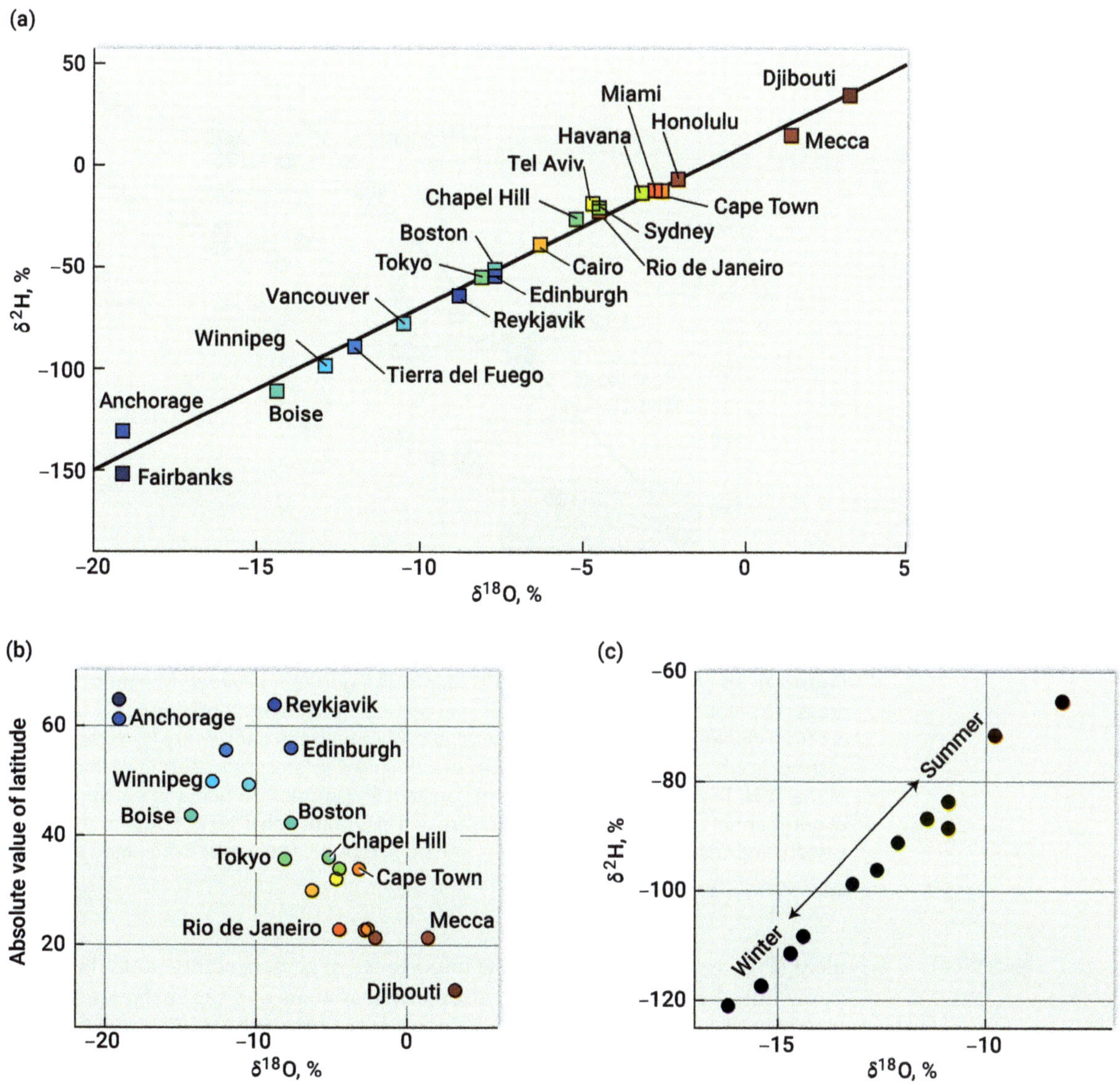

Figure 11.34 (a) Correlation between $\delta^{18}O$ and $\delta^{2}H$—the meteoric water line—for precipitation in selected world cities. City symbols are color-coded by latitude. Low-latitude tropical air masses have much higher $\delta^{18}O$ and $\delta^{2}H$ than those at high latitudes. (b) Correlation between the absolute value of latitude and $\delta^{18}O$. Three labeled points in this compilation fall off of the main latitude-$\delta^{18}O$ correlation: Boise, Reykjavik, and Edinburgh. Boise is at about the latitude of Boston but has low $\delta^{18}O$ and $\delta^{2}H$, characteristic of a much more northerly latitude, such as central or northern Canada. The rain shadow effect (see Figure 5.14) lowers Boise's isotope values. In contrast, Reykjavik and Edinburgh have much higher $\delta^{18}O$ and $\delta^{2}H$ than the correlation predicts; Reykjavik's values predict a latitude comparable to that of Boston, which is over 20° farther south. Why might Reykjavik and Edinburgh deviate from the global correlation in a way that makes them look warmer than they should? (c) Seasonal isotope data from Boise, showing local correlation between $\delta^{18}O$ and $\delta^{2}H$. High vs. low values correspond with summer (warmer) vs. winter (colder) precipitation, respectively. Most precipitation in Boise falls during the winter, so average values are quite low.

Sources: Data from the IAEA Global network of isotopes in precipitation and from https://wateriso.utah.edu/waterisotopes/pages/spatial_db/SPATIAL_DB.html.

boiling fractionates hydrogen isotopes readily, with lighter ^{1}H going into the steam phase, leaving behind water and steam condensate that is enriched in ^{2}H.

Geysers

Geysers are striking phenomena in which boiling water is periodically ejected from a hot-spring pool or an open vent (see Figure 11.33d). Old Faithful in Yellowstone is perhaps the world's most iconic geyser, erupting roughly every 90 minutes or so to heights of 40 m or

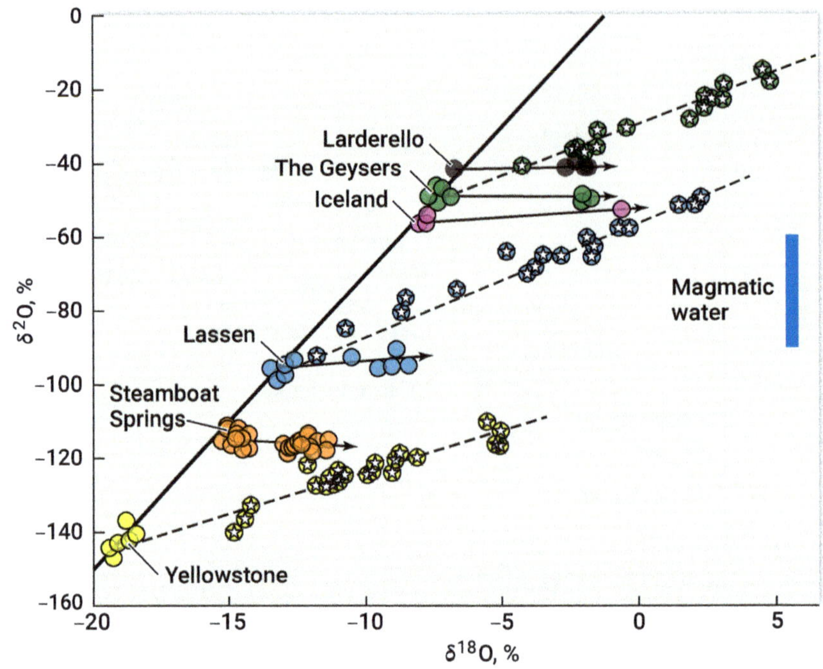

Figure 11.35 Variation of $\delta^{18}O$ and $\delta^{2}H$ at selected geothermal areas. Minimum values for all areas are pinned at the meteoric water line, consistent with a meteoric source. Solid symbols are from chloride-rich boiling pools, whereas those with star centers are from vapor-dominated systems in which much more boiling occurs; boiling depletes the waters in ^{1}H and thus leads to higher $\delta^{2}H$. The smear to higher $\delta^{18}O$ at constant $\delta^{2}H$ in chloride-rich pools arises from reaction of geothermal waters with rocks, which are rich in O with high $\delta^{18}O$ but poor in H. The Geysers, Lassen, and Yellowstone have data for both boiling pools and vapor-dominated systems.

Source: Modified from White (2013).

more (Figure 11.36a). The scientific importance of geysers comes about because they are exotic links between subsurface hydrothermal systems and the surface, and their behavior is a guide to how water and other fluids move in the subsurface and how such systems respond to earthquakes, Earth tides, and atmospheric conditions (Hurwitz et al., 2014).

Hurwitz and Manga (2017) noted that there are only about 1000 geysers known worldwide, about half of which are in Yellowstone National Park. Iceland, New Zealand, and the Kamchatka Peninsula in eastern Russia host a large proportion of the rest. Geysers occur in locations that span the tectonic spectrum: hotspot (Yellowstone), subduction zone (New Zealand, Chile, Russia), oceanic ridge (Iceland), and continental rift (Kenya).

The current understanding of how geysers operate (Hurwitz & Manga, 2017) is based on a system with one or more enclosed cavities that are laterally offset from and connected to a throttled opening to the surface (Figure 11.36b). The mixture of steam bubbles and liquid water in the bubble trap is highly compressible. After an eruption, heated, circulating meteoric water recharges the system, P and T build, and water in the system is heated to the boiling point (or just below) at the relevant altitude. The throttle keeps the system from freely expelling the expanding water-steam mixture, much as the paper shell around a firecracker holds in gases from burning gunpowder—until it bursts. In geysers, the driving force for eruption seems to be when the system reaches a critical point where most of the water is at the boiling point. Because boiling point increases with P (see Figure 5.3), any eruption of water decreases P on the remaining column. This drop in P causes some of the remaining water to boil, which forces more water out. This water loss causes a larger drop in P and more boiling, ultimately leading to a runaway process that empties enough of the system for the eruption to cease and the cycle to begin again. Initial phases of these eruptions are typically liquid-dominated, shifting to vapor-dominated as pressure in the reservoir decreases.

Figure 11.36 (a) Old Faithful in full roar on a clear winter day, with snow covering sinter. (b) A model for how typical geysers work in which two systems – a recharge cavity (bubble trap) and a connected vent with a throttle (constriction) – interact to produce periodic eruptions. After an eruption, the system recharges with heated meteoric water, some of which boils into bubbles which collect in the bubble trap, producing a highly compressible water-steam mixture. Fluids leak out through the throttle as the system fills and heats up. Once much of the water is at the boiling point, further leakage of material depressurizes the column enough that more water boils. This forces more water out, and further depressurization proceeds in a runaway process that uses *PV* energy stored in the bubble trap to drive a roaring eruption.

Sources: (a) Jacob W. Frank, National Park Service; (b) modified from Hurwitz & Manga (2017).

Making Connections 11.2

Hydrothermal Systems and DNA Sequencing

Hot springs present unique challenges to life, and yet life manages to hang on even in springs with near-boiling temperatures or highly acidic waters. One of the most striking geologic sights on Earth, Grand Prismatic Spring in Yellowstone National Park (Figure 11.37), is an example. The center of the pool, where deep blue, slightly alkaline water wells up at more than 85 °C, is largely devoid of life. This water, only about 8 °C below the boiling point at this altitude, cools as it flows radially outward, eventually becoming cool enough (~74 °C) for the cyanobacteria *Synechococcus* to survive after having made a few evolutionary adjustments. *Synechococcus* photosynthesizes using chlorophyll, but the normal green color is masked by high concentrations of carotenoid pigments (the same compounds that color carrots). These shield the thermally stressed organism from intense, high-altitude sunlight and accompanying ultraviolet light and give the bright yellow band its color; the green band is yellow bacteria seen through blue water. Moving outward, the orange band, where water has cooled to about 65 °C, contains both *Synechococcus* and other bacteria that use

continued

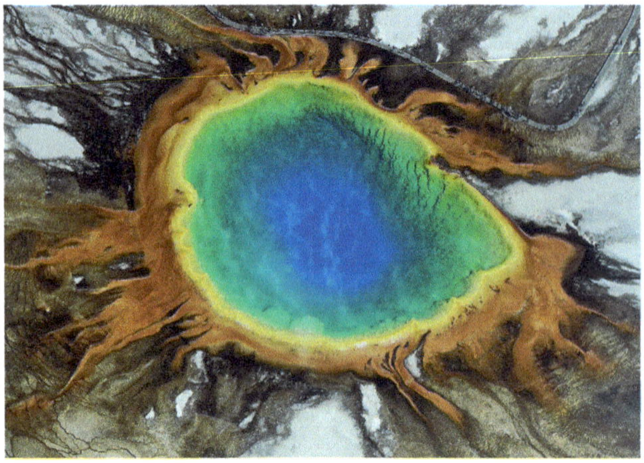

Figure 11.37 Aerial view of Grand Prismatic Spring in Yellowstone National Park; people on boardwalk at top give scale. The blue water in the center is the hottest, and as it cools, different bacteria with different colors colonize the floor—yellow bacteria toward the center and orange bacteria at the edge. The green color represents optical mixing of blue water and yellow bacteria.

Source: Carsten Steger (CC BY-SA 4.0).

different types of chlorophyll and different carotenoids. From the orange band, the water spills onto a broad, low, pale brown siliceous sinter terrace. Research on these unusual bacteria helped advance DNA testing today.

The story starts in the 1960s, when microbiologist Thomas Hudson of Indiana University was interested in bacteria in hot springs, especially gooey pink bacterial filaments growing in 85 °C water in Octopus Spring, a few kilometers northeast of Grand Prismatic. At that time, 85 °C was well above the assumed maximum thermal limit for life. In 1965, he and undergraduate Hudson Freeze collected bacteria from nearby Mushroom Pool. A year later, they had isolated and cultured a bacterium that they named *Thermus aquaticus*, and in 1976, a key enzyme, *Taq* polymerase, was isolated.

Fast forward to the early 1980s, when a novel technique of DNA manipulation known as polymerase chain reaction, or PCR, was being developed. PCR offered a way to create billions of copies of a given DNA sequence, a process known as DNA amplification. This technique is central to molecular genetics today, used for procedures ranging from analysis of ancient DNA (think *Jurassic Park*, only real), to identification of pathogens, to screening for cancer cells, to determining one's genetic ancestors, to forensics (think *CSI: City of Your Choice*). PCR consists of three steps:

1. The DNA sample is heated to 95 °C for 30 to 60 seconds to snip the hydrogen bonds that link the two strands of the DNA.
2. Molecules called primers, which act as keys to identify the region of interest on the split DNA, are introduced into the reaction vessel, which has been cooled to between 50 and 65 °C. These primers stick to the ends of the target section.
3. The reaction vessel is heated to 72 °C, and a polymerase enzyme goes to work. The polymerase finds the short segments where the DNA has been completed by the primers and moves along the strand, completing it by adding complementary DNA nucleotides as needed (e.g., if it finds an A, it adds a T; if it finds a G it adds a C; etc.). This process takes from seconds to minutes depending on the length of the section to be extended.

At the end of this cycle, the original DNA strand has been copied (1 strand has become 2). Repeating the process doubles this to 4 strands, another repeat makes 8, and *n* repeats produces 2^n strands. This is exponential growth; 30 cycles produces about 1 billion copies, and 40 cycles about 1 trillion.

T. aquaticus, the **thermophile** organism isolated from Mushroom Pool, is key here. Its DNA polymerase, *Taq* polymerase ("*Taq*" from the parent organism's name), can tolerate high reaction temperatures without breaking down. Earlier DNA polymerases used for PCR (e.g., from *Escherichia coli*) were destroyed at the high temperatures (72 °C) employed during the extension step; using *Taq* polymerase, a hardier version that functions in its parent thermophile, means that the reaction vessel can remain closed during DNA amplification, greatly increasing the efficiency and speed of the process. PCR is now a multibillion dollar per year industry, built in part on a microorganism from a Yellowstone hot spring.

Thermophile
A microorganism that survives at temperatures >40 °C.

11.7 Thermochronology and the Meaning of Isotopically Determined Ages

Learning Objectives

By the end of this section you should be able to:

11.7.1 Compare and contrast the concepts of geochronology and thermochronology.

11.7.2 Propose an analytical scheme for determining the *T-t* history of a pluton from its initial crystallization to its exposure at Earth's surface.

Geochronometers

In Chapter 2, we introduced the Rb-Sr isochron as a method for determining the age of the solar system, and in Chapter 10, we introduced U-Pb zircon geochronology as a means of determining intrusive ages of plutons. These and other methods of geochronology depend upon a key principle—that the components involved (e.g., Rb and Sr for the former, and U and Pb for the latter) remain in the system, undisturbed, since the time that the geochronologic clock started.

For the Rb-Sr system in the Tieschitz meteorite (see Section 2.3), we assumed the components of the meteorite were assembled from solar nebula material that had a uniform $^{87}Sr/^{86}Sr$ ratio, and that the system remained undisturbed during the next 4.6 billion years, until its components were disassembled and analyzed in a laboratory in Paris (Minster & Allègre, 1979). Linearity of the resulting isochron is consistent with these assumptions, but by no means proves them. For the U-Pb system, we assumed that the zircons crystallized with U of uniform isotopic composition (inherited from the melt from which they crystallized), that no Pb was incorporated into the crystal, and that no U or Pb leaked out since the crystal formed. Consistency of the two geochronometers $^{238}U/^{206}Pb$ and $^{235}U/^{207}Pb$ is consistent with these assumptions, but again, does not prove them.

One of the earliest geochronologic systems to be widely applied is the K-Ar system, which relies on decay of ^{40}K to ^{40}Ar.[4] As with transformation of ^{87}Rb to ^{87}Sr by β-decay (see Section 2.3), parent and child have virtually the same mass, but the main decay scheme is electron capture, in which a proton combines with an electron and transforms into a neutron. This leaves the atomic mass essentially the same, but causes the atomic number to decrease by one. The K-Ar system was developed in the early days of mass spectrometers because it is relatively easy to separate and measure Ar and K concentrations in minerals and rocks. The modern implementation of the technique uses a variation known as the $^{40}Ar/^{39}Ar$ method ("argon forty-thirty-nine" or "argon-argon" method; see Toolbox 11.4). For simplicity, we will refer to both methods as the K-Ar system, and to ages determined by either as Ar ages.

The K-Ar system has an obvious problem as a geochronometer—Ar is a noble gas. Although it can be forced into forming a limited range of compounds under duress, it has no place being in feldspar, biotite, muscovite, and other K-bearing minerals. Its parent K is an essential structural constituent of those minerals, but once it decays, its offspring Ar wants out in the worst way, to be free to roam the atmosphere, unattached to other atoms. The K-Ar system is contrary to the U-Pb zircon system; in the latter, child nuclide Pb has less incentive and little ability to leak out.

The fugitive nature of Ar is both a bug and a feature of the K-Ar system. For volcanic rocks, which cool rapidly, Ar stays put, because diffusion at surface temperatures is so slow that leakage is negligible. Thus, Ar ages on unaltered volcanic rocks and their minerals can be both precise and accurate estimates of eruption ages. For more slowly cooled plutonic and metamorphic rocks, however, leakage of child Ar is significant. Although this may seem like a problem, with careful work one can use this leakage to determine points along the cooling path of the rock.

Cooling Ages

Closure Temperature

Consider the one-pulse intrusive model in Figure 11.9a. Let's say intrusion occurred at 100 Ma, and in the intervening time, tectonics and erosion have exposed the middle of the body. You collect a sample and determine an age—what age do you get?

Well, it depends. Based on current understanding, a U-Pb zircon age will give you the intrusive age—in this case, 100 Ma—or, more precisely, the age at which zircon crystallized in the magma. Commonly, that's close to the intrusion age, although older ages from earlier-crystallized zircons or inherited zircon crystals may be common and must be screened for. An Ar age, however, is different, because leakage of Ar during cooling

(Figure 11.38) violates the requirement that the decay product stay put. The center of the pluton took about 10 Myr to cool below greenschist facies. What age does the K-Ar system record?

Again, it depends—in this case on which mineral is being dated—because Ar diffuses at different rates in different minerals. Consider a hypothetical situation in which Ar leaks out rapidly above a certain temperature, but once it cools below that temperature, the gate closes, and newly formed Ar is trapped in the crystals. We will call this temperature the **closure temperature**, T_c. In such a case, an Ar age would record the time at which the pluton (or more specifically that particular part of the pluton) cooled below T_c.

Closure temperature (T_c)
The temperature at which loss of the radiogenic isotope (^{206}Pb, ^{40}Ar, etc.) becomes sufficiently slow that it is essentially wholly retained. Above T_c, the radiogenic isotope is (mostly) lost, and the isotopic clock is not counting; below T_c, the radiogenic isotope is (mostly) retained, and the age follows the normal age equation.

It turns out that real geologic systems behave in a manner somewhat like this. Consider a spherical K-bearing crystal of radius *r* (crystals are not generally spherical, but this is a convenient starting point mathematically).[5] As Ar forms, it diffuses at a rate that depends exponentially on T (see Section 8.3) in yet another Arrhenius relation:

$$D = D_0 e^{-E/RT} \qquad \textbf{(Eq. 11.29)}$$

If the diffusion rate is slow compared to the radius of the crystal, then most of the Ar stays in the crystal; if it is fast, then most leaks out. "Slow" is a relative term, of course. We can use the characteristic diffusion distance $\delta = \sqrt{Dt}$ (e.g., Figure 8.21) to estimate how far the Ar will travel, but this depends on the time involved. We have assumed (and will shortly show) that Ar diffusion is negligible at surface temperatures, so we will use good round numbers, such as 1 and 0.1 Ma, as rough estimates of the time it takes to cool from magmatic temperatures, where Ar diffusion is rapid, to near-surface temperatures, where it is extremely slow (see Figure 11.38).

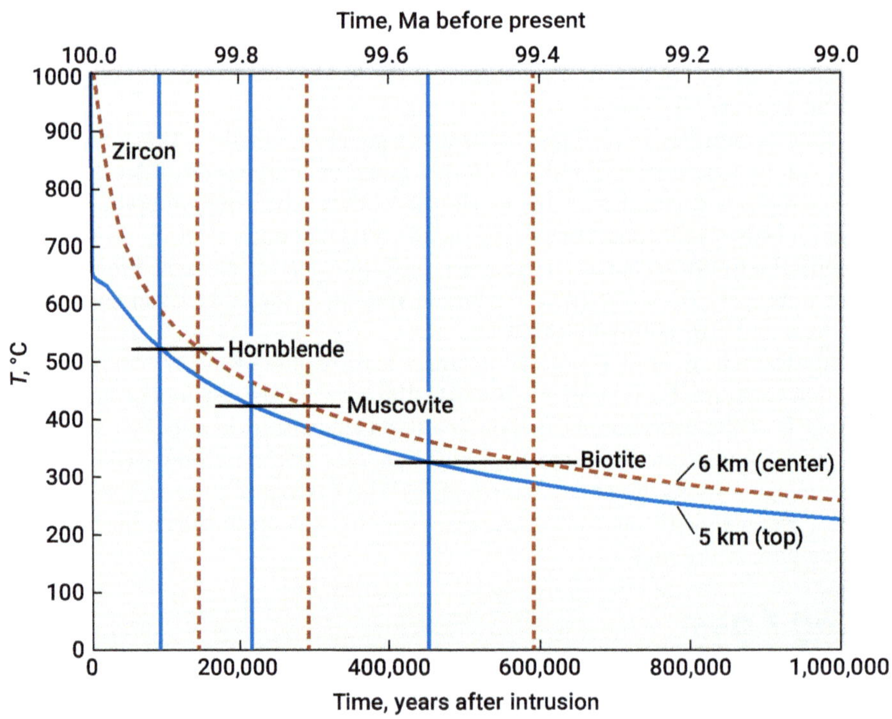

Figure 11.38 Conductive cooling profile for a sheet-like pluton, 2 km thick with its top at 5 km. Curves show cooling of the center (6 km) and top (5 km). Dotted lines give the ages that would be recorded by the K-Ar system for hornblende, muscovite, and biotite, given the depicted closure temperatures (horizontal lines). Solid vertical lines give ages that would be recorded by the top of the intrusion.

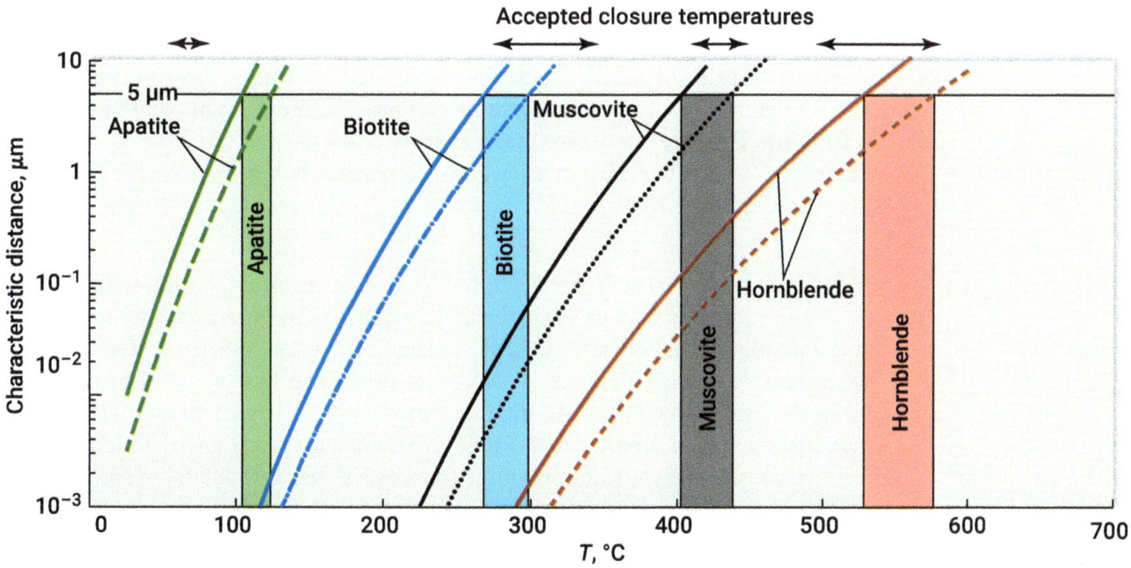

Figure 11.39 Average diffusion distances for decay products in various minerals: Ar in hornblende, biotite, and muscovite, and He in apatite. Each system has a different Arrhenius relation for the *T* dependence of *D*. Curves show characteristic distance $\delta = \sqrt{Dt}$ (over times of 1 Myr (solid lines) and 0.1 Myr (dotted lines). If we arbitrarily select 5 μm as an acceptable distance for closure (in other words, the maximum distance we'll let something diffuse and still call the system "closed"), then we get *T* estimates of ~530–580 °C for hornblende, 400–440 °C for muscovite, 270–300 °C for biotite, and 100–120 °C for apatite. Double-headed arrows give commonly used closure temperature bounds calculated by the method of Dodson (1973); the simple characteristic distance method used here gives comparable results for all but apatite.

Armed with this estimate and Arrhenius parameters (D_0 and E) for the minerals of interest, we can calculate characteristic distance δ as a function of T (Figure 11.39). This figure shows experimentally determined curves for Ar diffusion in biotite, muscovite, and hornblende, along with He diffusion in apatite, where He forms by α-decay of U and Th. Each curve has the same exponential shape, but the T ranges at which δ increases rapidly from the μm range (significantly smaller than crystals chosen for analysis) to the tens to hundreds of μm range (comparable to the dimensions of crystals chosen for analysis) vary greatly.

Consider biotite. Figure 11.39 predicts that Ar will diffuse tens of millimeters in 1 Myr at temperatures above 600 °C. By 275 °C, however, the characteristic distance has decreased to less than 10 μm, a factor of more than a million. By this temperature, the Ar is essentially locked into the crystal over typical timescales of cooling; the gate is closed. Similar stories can be told for the other minerals, with significant differences in T_c.

The key point of Figure 11.39 is that each mineral will lock in its decay product at a different T as the system cools, and thus geochronometers record the time when the rock cooled through the closure temperature for the relevant system. This is when the isotopic clock starts. In broad terms, using accepted closure temperatures, an Ar date on hornblende will record when the rock cooled through about 525 °C, muscovite through about 420 °C, and biotite through about 300 °C. The U-He clock in apatite does not start until the rock cools below about 75 °C. Note that given a geothermal gradient of 20 °C/km, the apatite He gate is continuously open for any rock deeper than 4 km. Thus, the He system in apatite can be used to estimate when a piece of rock was brought through that isotherm by tectonics and erosion.

To more fully answer the question at the start of this section, the age you get depends upon which system you use for dating. U-Pb dates on zircon and monazite have closure temperatures in the magmatic range (~1000 °C), and thus presumably date the age of crystallization of those minerals (and these ages, in turn, are typically assumed to closely approximate the time of intrusion). Ar dates on hornblende, muscovite, and biotite record the time at which the rock passed through those successively lower closure temperatures, and other systems can provide more points on the temperature-time (*T-t*) path. Cooling curves for many plutons show a rapid fall from zircon U-Pb through hornblende Ar, then a lag of several million years before the system cools through biotite Ar (Figure 11.40).

Zircon Saturation Temperature The extremely high closure temperature of Pb in zircon is higher than the *T* at which many intermediate and silicic magmas are intruded (see Figure 10.37). However, Zr, the key essential structural constituent of zircon, is a trace element, so zircon is rarely the first mineral to crystallize from a magma. Magmas commonly intrude before zircon begins crystallizing, and thus the age recorded by a zircon crystal reflects when the crystal grew, which could be tens or hundreds of thousands of years after the magma was intruded. Given the high precision of U-Pb zircon age determinations (see Section 10.7), this distinction is important (e.g., Samperton et al., 2015).

Fortunately, Watson and Harrison (1983) performed experiments, later updated by Boehnke et al. (2013), that allow empirical estimation of the *T* at which zircon would start to crystallize in a given magma (see Toolbox 11.5). This *T* is typically 700 to 800 °C for granites and granodiorites (Miller et al., 2003), well below the nominal closure *T* of about 1000 °C. This is why igneous zircons date crystallization, not closure.

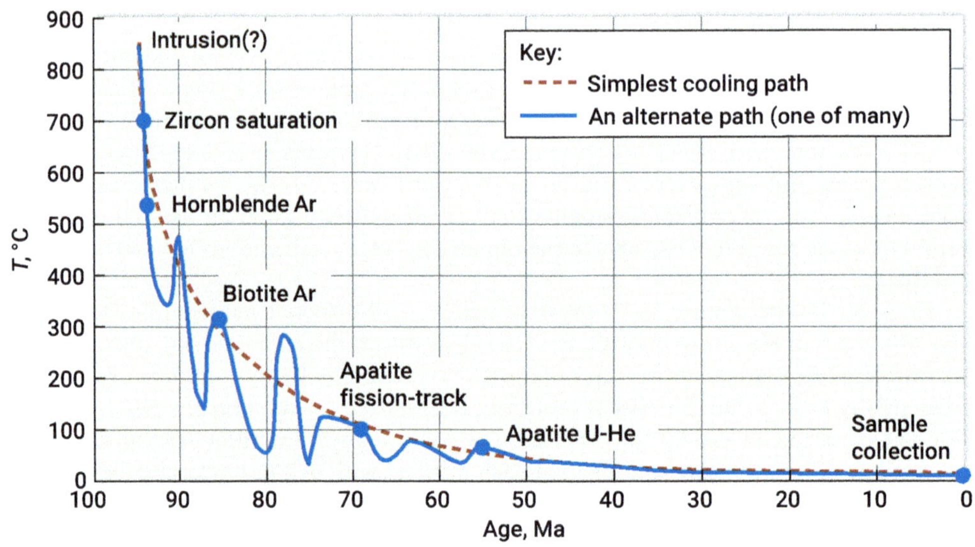

Figure 11.40 Composite *T-t* cooling curve for the outermost unit of the Tuolumne Intrusive Suite (see Figure 10.3), showing rapid cooling through hornblende Ar temperatures followed by slower cooling through biotite Ar and lower closure temperature systems. The two curves show possible paths through the data; a simple monotonic cooling curve and a more complicated, pulsing one. A given point (e.g., the biotite Ar age at 86 Ma) tells you that the rock likely did not exceed the closure temperature after that time, but gives no information about what happened before.

Sources: Data combined from sources in different areas (Dumitru, 1990; House et al., 2001; Davis et al., 2012), so the resulting curve is only schematic.

P-T-t Evolution of Contact Metamorphic Aureoles

The pressure-temperature-time (*P-T-t*) evolution of a point in the contact aureole near a big blob of magma that suddenly appears next door might seem simple—it heats up, then cools down—but there are interesting subtleties and complications. Consider the one-pulse intrusion of a 2-km-thick sill, modeled in Section 11.2 and shown in Figure 11.9a. We can easily track the *T-t* evolution of the points in the aureole, and Figure 11.41a shows how T varies with t for points 100 m above and 100 m below the sill (dashed curves). Sure enough, they heat up and then cool down, but their paths are highly asymmetrical—extremely rapid heating to peak temperatures of about 600 °C roughly 10 kyr after intrusion, followed by cooling below 400 °C in a few hundred thousand years, and asymptotic cooling back to ambient levels over millions of years. Pressure for the point above the sill is static (Figure 11.41b), whereas the point below is instantaneously depressed 2 km deeper, where it remains for the duration.

Figure 11.41a also shows *T-t* evolution of the same points for the incremental model of Figure 11.9b, wherein a 20-m-thick sill is injected every 5000 years (solid curves). This is considerably more interesting for two reasons. First, the lower point is continually buried to greater and greater depth (and thus higher and higher P) during the 500 kyr of pluton assembly (see Figure 11.41b). Second, the *T-t* path of the lower point cycles up and down over a 50 °C range owing to its proximity to the base of the pluton, where we decreed that all new injections of magma occur (remember that this may or may not happen in nature, but we made it happen in the model). The path is a sawtooth with 1000 teeth, one for each injection. As noted in Chapter 10, T cycling can have a profound effect on crystal size and rock texture, as can the pulses of hot water released by the injection pulses or mobilized in hydrothermal systems. The point above the pluton also shows T cycling, but because the injection site moves farther and farther away from the top of the pluton, the sawtooth signal quickly dies out. Note that the maximum temperatures reached by these points are lower than for the one-pulse model, and for the point above the sill, much lower.

Now consider the *P-t* paths taken by these two points in the aureole. As with the one-pulse model, the upper point is fixed at 4900 m depth, so its P does not change. The lower point, however, evolves to higher and higher P until the last injection, at which time it stays put and cools off (see Figure 11.41b).

We can combine data for these models into *P-T* plots (Figure 11.41c). We can also add relative time marks to these modeled *P-T* paths; they start at time 0 and end 0.5 Ma later. If the rocks we're looking at are metasedimentary, let's say schists derived from our beloved shales, then we can add two more *P-T* points to the paths, because we know that (1) they formed at the Earth's surface and (2) you collected them at the Earth's surface. Thus, we can add $P = 0$ and $T = 0$ (within rounding error) to the curves twice.

What else can we add? Let's say that the shales are Paleozoic, deposited at 500 Ma; that puts time on our first (0,0) point. Since we're making this up, let's say that the pluton gives a U-Pb zircon age of 100 Ma, and that you collected the schist from an outcrop that was covered by 20-Ma sedimentary rocks (meaning that it was at the surface by 20 Ma). Then the origin also marks arrival of the rocks at the surface at 20 Ma and their presence there today. So, we know where they were at 500 Ma, 100 Ma, and 99.5 Ma (the end of our thermal model). Somehow they got from the surface to about 5 km depth over the 400-Myr interval between deposition and intrusion; this might have been a smooth path, as shown, or they may have gone deeper, been brought back up, gone deeper again . . . 400 Myr is a long time. The same is true for the exhumation path after intrusion ceased; the rocks made it to the surface between 99 and 20 Ma, but that path could have been complicated with intervening periods of burial, heating, etc.

This example is a **forward model**—predicting what will happen in a given situation. Geologists are typically faced with **inverse problems** in which something happened—a pluton was intruded, rocks were metamorphosed, etc.—and the task is to determine when and how this occurred. General metamorphic assemblages (see Figure 11.16), including

Forward model
Prediction of how a system will behave given specific boundary and initial conditions and equations governing its physical and/or chemical behavior.

Inverse problems
Identification of what processes occurred to produce a set of observations. Most geologic questions are inverse problems—we seek to understand past processes that formed the rock record.

Plateau age
The weighted average age for the flat portion of an Ar release spectrum. Plateau ages commonly correspond with the later increments of gas released at the highest temperatures.

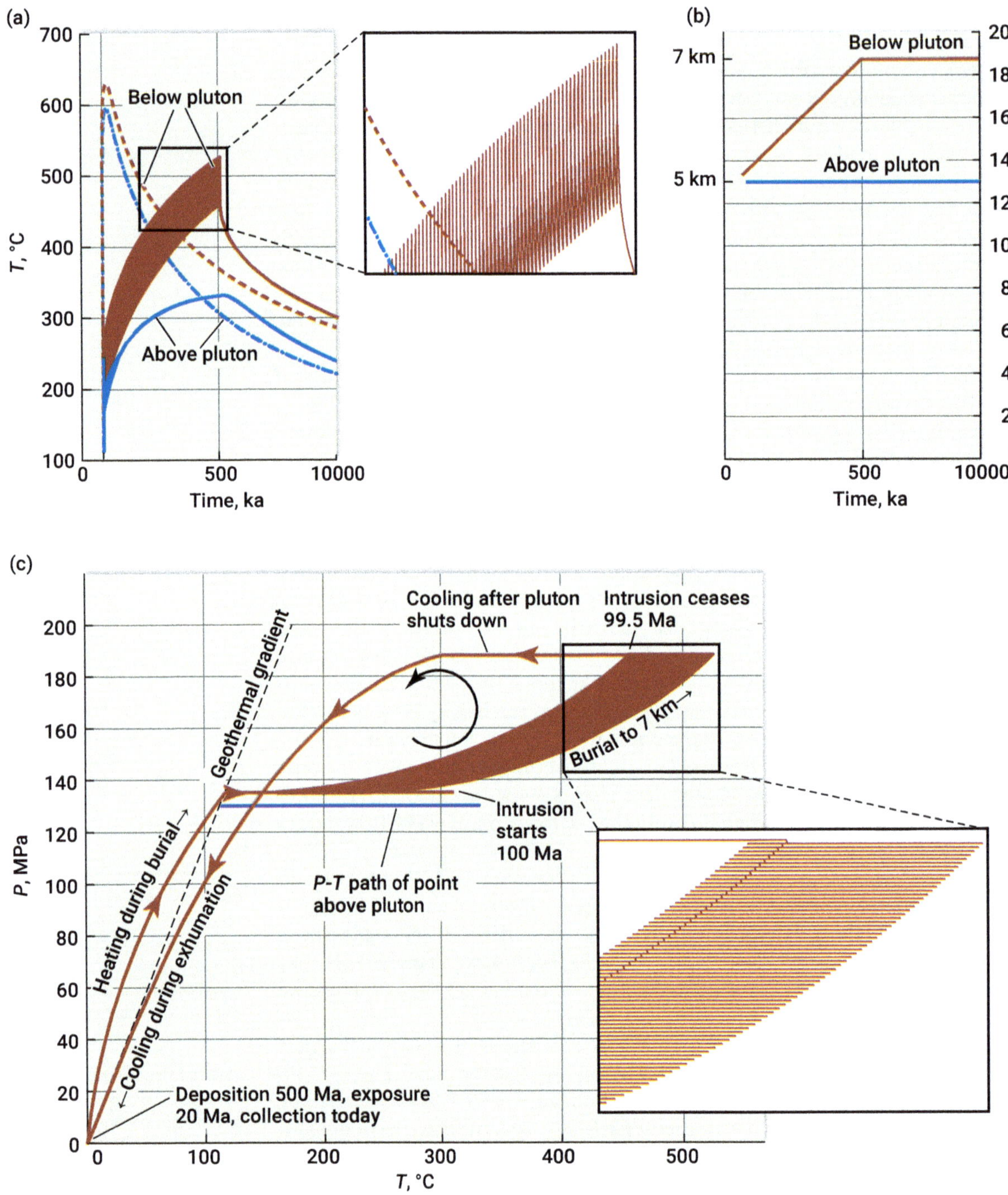

Figure 11.41 Models of the effects of one-pulse vs. incremental pluton emplacement on thermal histories and *P-T* paths for rocks located immediately above and immediately below a pluton. (a) Temperature vs. time. The one-pulse models (dashed lines) show extremely rapid heating followed by more protracted cooling. The incremental model (solid lines) shows a broadly similar shape, but slower heating and large temperature oscillations (inset). (b) Pressure vs. time for incremental model. The point below the pluton gradually increases in pressure as magma is intruded above it. The one-pulse intrusion (not shown) would simply impose two different pressures. (c) *P-T* paths for incremental model. The point above the pluton simply heats and cools at constant pressure. Below the pluton, temperatures oscillate as pressure and average temperature increase, creating a sawtooth *P-T* path with 1000 horizontal teeth (inset). Post-intrusion cooling at first occurs isobarically, then with decreasing pressure as erosion exhumes the rock. Erosional path of point above pluton omitted for clarity; where might it plot in panel c?

observations as simple as the identity of the Al_2SiO_5 polymorph, if present, can tell you general *P-T* conditions; studies of metamorphic reactions can narrow down *P* and *T* further; and thermochronology and basic field geology (e.g., presence of schist under 20-Ma sedimentary rocks, as above) can put time marks on the *P-T* path. In this way, the geologic and tectonic evolution of a given area can be worked out.

Total gas age
The weighted average of all increments of gas released when measuring $^{40}Ar/^{39}Ar$ ages.

Toolbox 11.4

K-Ar and $^{40}Ar/^{39}Ar$ Dating

The K-Ar method is simple in principle, and K-Ar ages on igneous rocks started appearing in the 1950s. However, numerous complications make obtaining precise ages difficult. Natural Ar is nearly all ^{40}Ar (99.6%), nearly all produced over the Earth's lifetime by decay of ^{40}K; the rest consists of ^{36}Ar (0.334%) and ^{38}Ar (0.063%).

A purified sample of a K-bearing substance (e.g., biotite, K-feldspar, volcanic glass) is split into two aliquots. K concentration is measured in one by several possible techniques, and ^{40}K is calculated from that based on the known ratio of ^{40}K to total K. Ar is released by fusing the other aliquot in a vacuum, and a known quantity of "spike"—in this case, ^{38}Ar—is mixed with it. The gas is run through a simple mass spectrometer (many labs originally used homemade ones) that measures the ratios of the three Ar isotopes. Contaminating Ar is mathematically stripped out of the analysis by assuming that all ^{36}Ar is atmospheric, and the resulting $^{38}Ar/^{40}Ar$ ratio is converted to an age using the calculated ^{40}K concentration. The decay equation is more complicated than that for the Rb-Sr system, because ^{40}K undergoes branching decay; nearly 90% of ^{40}K atoms β-decay to ^{40}Ca, which is not terribly useful for dating, because it is so common in rocks and minerals.

The K-Ar system suffers from several issues that reduce its precision and accuracy: (1) splitting the sample for separate K and Ar analyses introduces unavoidable error, because the two aliquots will never be truly identical in K and ^{40}Ar concentrations; (2) correction for contaminating Ar can be significant; and (3) there is no way to assess how much ^{40}Ar loss has occurred after the rock or mineral formed.

To address these problems, the clever $^{40}Ar/^{39}Ar$ dating technique was developed. A purified sample is placed in a nuclear reactor[6] for several days and irradiated with neutrons to convert ^{39}K to ^{39}Ar (the ^{39}K atoms are shot with neutrons and cough up protons). ^{39}Ar will eventually β-decay back to ^{39}K, but the half-life is 269 years, far longer than the laboratory work takes (we hope!), and so its concentration can be assumed to be constant. The sample is then placed in a high-vacuum furnace at the inlet of the mass spectrometer and slowly heated. Ar is released in stages of increasing temperature, and the $^{40}Ar/^{39}Ar$ ratio, after stripping out atmospheric contamination (based on $^{36}Ar/^{40}Ar$), yields an age for each stage.

The $^{40}Ar/^{39}Ar$ technique offers several advantages over K-Ar dating: (1) sample splitting error is eliminated, (2) correction for atmospheric Ar is more precise, and (3) the way in which Ar is released as the sample is heated reveals information about Ar loss. Consider Figure 11.42, which shows measured age profiles for four muscovite crystals from Proterozoic metamorphic rocks in New Mexico (Heizler et al., 1997). All three are messy at the low-*T* (left) end of the spectrum when the first bits of Ar are being released; this is the most loosely held Ar, typically found near fractures where the path to freedom is nearby. This early gas typically gives ages that are younger than the true age, because some of the gas has been lost. After that, the samples show different behavior—they "plateau" out at an age that represents the most retentive parts of the crystals and can be interpreted using Ar loss theory.

Sample A shows a beautiful flat spectrum with a long series of increments whose ages are just under 1400 Ma. The weighted average of the increments in this plateau is called the **plateau age**. Sample B climbs to a plateau by about 20 percent of gas release and follows A with some wiggles. Sample C climbs for the first 30 percent of gas release to the plateau of A, consistent with a similar age but having suffered some Ar loss. Sample D, a deformed muscovite crystal, displays a less attractive spectrum—it is a climber that just barely seems to reach a plateau in the last 25 percent of gas release. It is easy to see how deformation would enhance Ar loss. Despite these complexities, the plateau age for this sample is similar to the others, which are not noticeably deformed.

K-Ar dating would tell a different story. The weighted average of all the increments for a given sample is called the **total gas age** and is essentially the same as a K-Ar age (minus the sample splitting error). These range from 1388 Ma for A and 1395 Ma for B to 1379 Ma for C and 1306 Ma for D. Clearly, the $^{40}Ar/^{39}Ar$ method is able to see through Ar loss better than the K-Ar method. We might not know how to interpret that age (yet), but at least we can begin to make sense of this complicated system.

continued

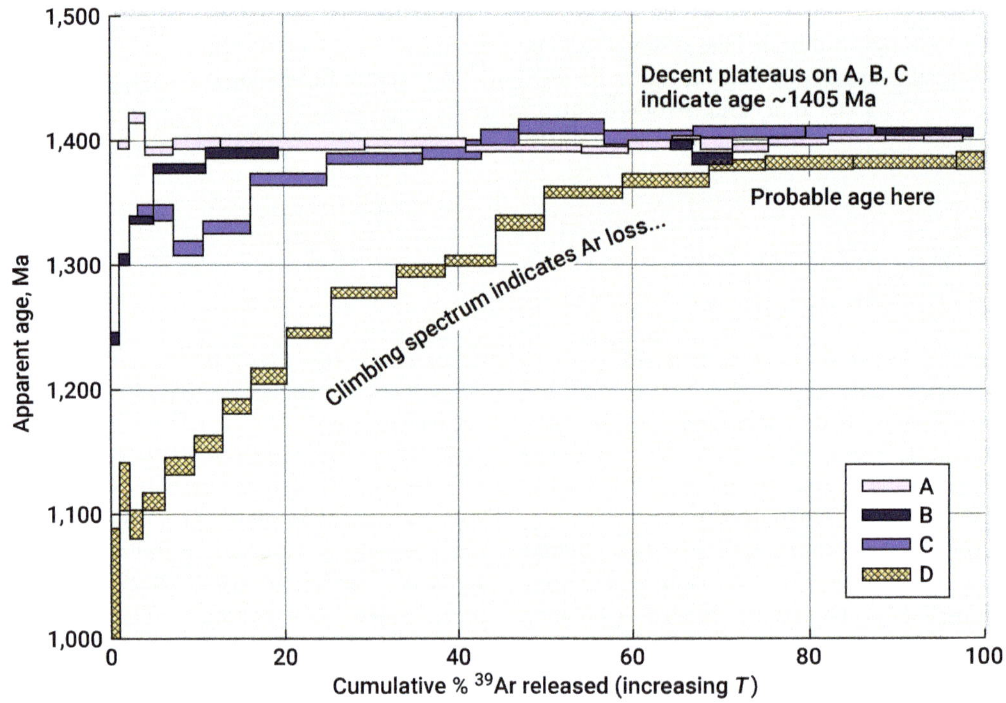

Figure 11.42 Ar-release spectra from muscovite crystals from the Manzano Mountains of New Mexico. Samples A, B, and C are from relatively undeformed crystals from pegmatite, schist, and peraluminous granite, respectively; sample D is a deformed crystal from a shear zone in the granite. The undeformed crystals settle down to relatively clear plateaus after about 50% of the gas has been released, whereas the deformed crystal's spectrum climbs until nearly 80% gas release, after which it settles down to a plateau similar in age to that of its undeformed counterpart. Total-gas ages, equivalent to K-Ar ages, are proportional to the areas under the curves, so in this case, the Ar-Ar method is able to see through partial Ar loss and get a better estimate of the true age.

Source: From Heizler et al. (1997), 1997 Published by Elsevier B.V.

Toolbox 11.5

Zircon Saturation Temperature

We have repeatedly seen how important U-Pb ages on zircon are to determining ages of geologic events. Proper interpretation of these ages depends on knowing at what point in the magma's history zircon started to crystallize. You should recall from introductory chemistry that the solubility product constant, typically denoted K_{sp}, is used to estimate when a given material will precipitate from a solution. Silicate liquids are solutions, and zircon is a precipitate, so in a general sense, there should be a product that expresses when zircon will precipitate. For the reaction

$$ZrO_2 + SiO_2 = ZrSiO_4 \quad \textbf{(Eq. 11.30)}$$

where the components on the left are in the liquid and the right side is zircon, we can write

$$K_{sp} = [ZrO_2][SiO_2] \quad \textbf{(Eq. 11.31)}$$

where bracketed quantities are concentrations of the species in the magmatic liquid.

For simple solutions (the kind you deal with in chemistry classes), K_{sp} provides a reasonable estimate of whether a given compound will precipitate, and tables give relevant values. Silicate liquids are far from simple solutions, and their chemical behavior is difficult to predict. Watson & Harrison (1983) conducted a series of experiments in magmas of different bulk compositions and Zr concentrations and came up with an empirically based model for predicting the T at which zircon will begin to crystallize from a melt of a given composition. As recalibrated by Boehnke et al. (2013), this expression is

$$T(\mathrm{K}) = 10108/(13.44 + 1.16M - \ln[\mathrm{Zr,melt}]) \quad \textbf{(Eq. 11.32)}$$

where the left side is Zr concentration in the melt (ppm) at zircon saturation, *T* is in kelvins (very important!), and

$$M = \frac{Na + K + 2Ca}{Al \times Si} \quad \textbf{(Eq. 11.33)}$$

where the element terms on the right side are cation fractions (e.g., moles Na per total cation moles, etc.). These can be tricky to get right; the user is advised to reproduce *M* values from the paper before proceeding.

Zircon saturation temperatures can be put to several uses (Hanchar & Watson, 2003). For example, aplite dikes (see Figures 10.5 and 10.10) form from late-stage silicic melt that enters fractures in a cooling granitic pluton. Such plutons nearly always have begun crystallizing zircon by the time aplite dikes segregate, so the Zr concentration in the aplite is a measure of *T*, assuming that the aplite liquids did not bring hitchhiking zircons into the fractures with them. This by itself is not all that informative, because it just provides an undated temperature on the cooling path, but combined with a model of how the compositions of high-silica rhyolite liquids vary with *T* and *P* (see Section 8.6), it provides a possible way to estimate the *P* of intrusion (Putnam et al., 2015).

Summary

Key points of this chapter include the following:

- Contact metamorphism encompasses the processes that occur in rocks adjacent to a pluton (the contact aureole) owing to heat and fluids supplied by the intruded magma. This includes both textural changes (typically grain coarsening) and mineral reactions.
- Rocks in a contact aureole that are composed predominantly of a single mineral such as calcite or quartz undergo recrystallization and coarsening, developing a foam texture characterized by 120° triple junctions in thin section. Marble and quartzite are examples.
- Heat flow around a pluton can be modeled with the time-dependent heat-flow equation if heat flow is by conduction only.
- Maximum temperatures reached in the contact aureole depend strongly on whether the pluton was intruded in one pulse or incrementally, and whether hydrothermal circulation occurred.
- Wall rocks of simple mineralogy, such as limestone or sandstone, show few mineralogic changes during contact metamorphism, but shaley rocks or those containing both carbonate and silicate layers (e.g., cherty limestones) undergo dramatic changes that can be used to estimate *P* and *T*.
- In incrementally emplaced plutons, early increments are the contact-metamorphic wall rocks of later increments. Plutonic texture reflects these metamorphic processes.
- Reaction of silicate magma with carbonate wall rocks produces skarns which may host ore deposits.
- Convection of fluids around plutons greatly increases the rate of cooling and produces geothermal phenomena such as fumaroles and geysers.
- Oxygen and hydrogen isotopes demonstrate that most hydrothermal fluids are meteoric in origin—that is, derived from precipitation.
- Many systems used for geochronology, such as the K-Ar system, leak decay products at high temperatures such that their isotopic clocks do not start until the system cools below a critical closure temperature.
- The variation in closure temperature among different isotopic systems can be exploited to determine time-temperature points along the cooling path. This is the basis of thermochronology.

Notes

1. Named cells are given a name to use in formulas; thus, cell C3 here is named k (for kappa, not thermal conductivity). Naming a cell aids readability and has the advantage of not having to anchor the location with $ when the formula is propagated.
2. This happens to be a photo of the first author's kitchen countertop.
3. Impress your friends; this is a *4-simplex* or *pentachoron*.
4. ^{40}K actually decays to two different child isotopes, ^{40}Ar and ^{40}Ca, about 10 percent of the time to the former. This is again a statistical process. Decay to ^{40}Ca is by β-decay.
5. This assumption has a long tradition in various fields, including stand-up physics comedy routines involving a joke with the punchline "Consider a spherical cow . . ."
6. Historical note: when your first author was in graduate school at UCLA, there was a small reactor in a laboratory behind the snack bar in the engineering building, and we would take samples there, behind the candy and hot dogs, to be irradiated. Seriously. The reactor was decommissioned and removed in 1984, prior to the Los Angeles Olympics.

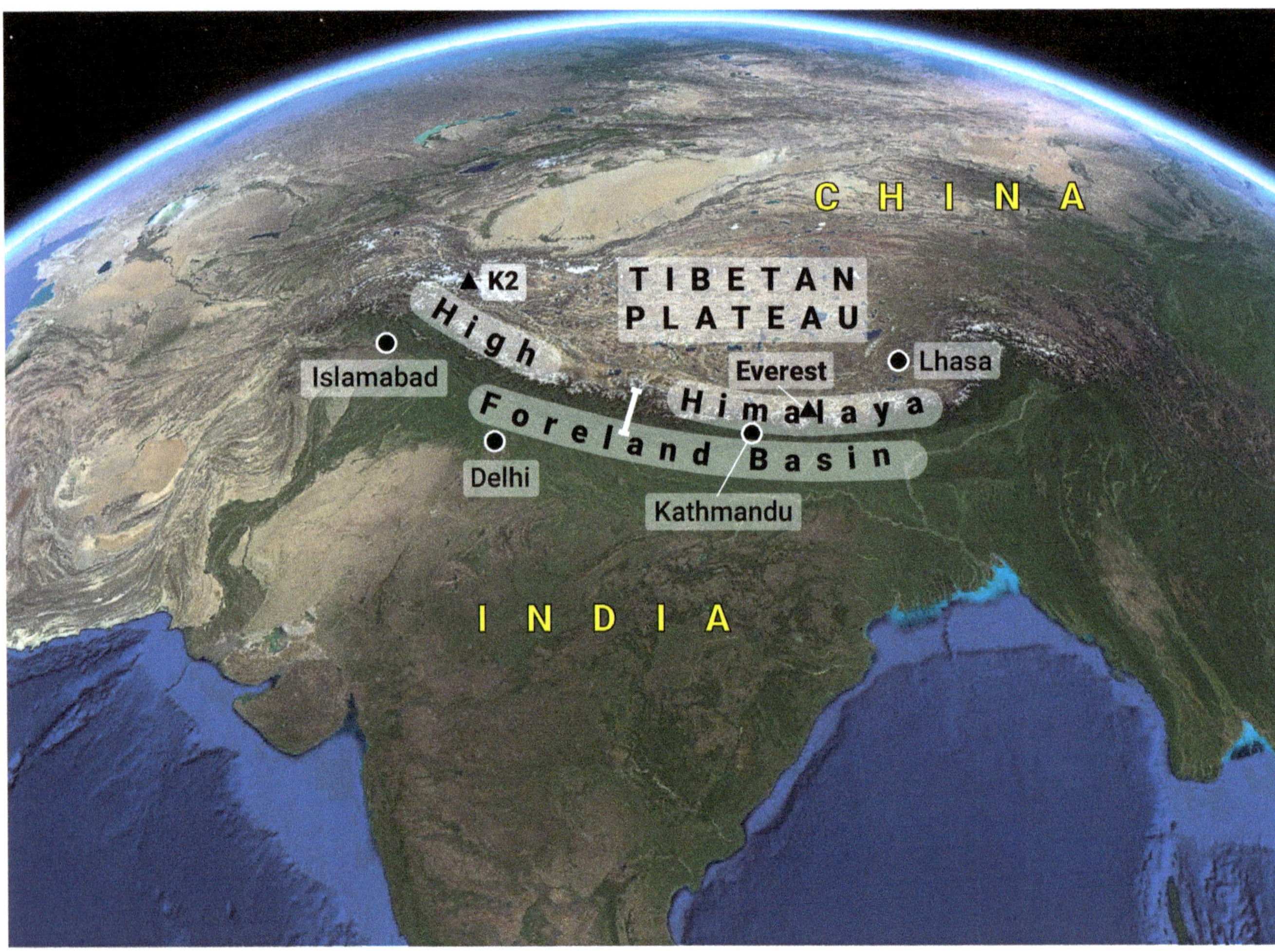
CHINA
TIBETAN PLATEAU
K2
High Himalaya
Islamabad
Everest
Lhasa
Foreland Basin
Delhi
Kathmandu
INDIA

CHAPTER 12

Making Continents: Continental Collision, Accretion, and Regional Metamorphism

Subduction of oceanic lithosphere with normal crustal thickness produces magmatic arcs, HP/LT metamorphism, and other lithosystems that serve to distill the mantle into lower-density crustal material. If the downgoing slab contains crust of abnormal thickness, such as a continent, island arc, or oceanic plateau, then normal subduction is inhibited. Partially subducted material becomes welded to the overlying plate, producing a thickened zone of crustal material known as an **orogenic belt** or **orogen**. Collisional events produce the regional metamorphic and ultrahigh-pressure (UHP) lithosystems, and these are responsible for processing and stabilizing much of the continental crust.

Orogenic belt, orogen
An elongate region that has undergone lithospheric shortening.

12.1 Plate Motions, Orogens, and Metamorphism

Learning Objectives

By the end of this section you should be able to:

12.1.1 Construct a series of sketches illustrating the tectonic events that lead up to the formation of orogens.

12.1.2 Illustrate the main tectonic components of a continent-continent collision zone and where metamorphism is likely to occur.

12.1.3 Illustrate the likely evolution of continents in terms of age progressions and terrane types.

The Himalaya, a dramatic and ongoing example of continental collision. In this Google Earth view looking north, the foreground comprises the flat plains of India, no more than a few hundred meters above sea level. These give way to the snow-covered High Himalaya whose apex is Mount Everest, the tallest mountain on Earth at 8849 m. The Himalaya encompass parts of India, Nepal, Bhutan, Pakistan, and southwestern China. Immediately north of the Himalaya lies Tibet, most of which is a high plateau above 4000 m. The Himalaya and Tibetan Plateau owe their high elevation to underthrusting of Indian continental lithosphere beneath southern Asia. The white bar cutting across the central Himalaya between Delhi and Kathmandu shows the location of a structural cross section discussed in Figure 12.2.

Source: GoogleLandsat/CopernicusData SIO, NOAA, US Navy, NGA, GEBCOIBCAOTMap Mobility, annotations by authors.

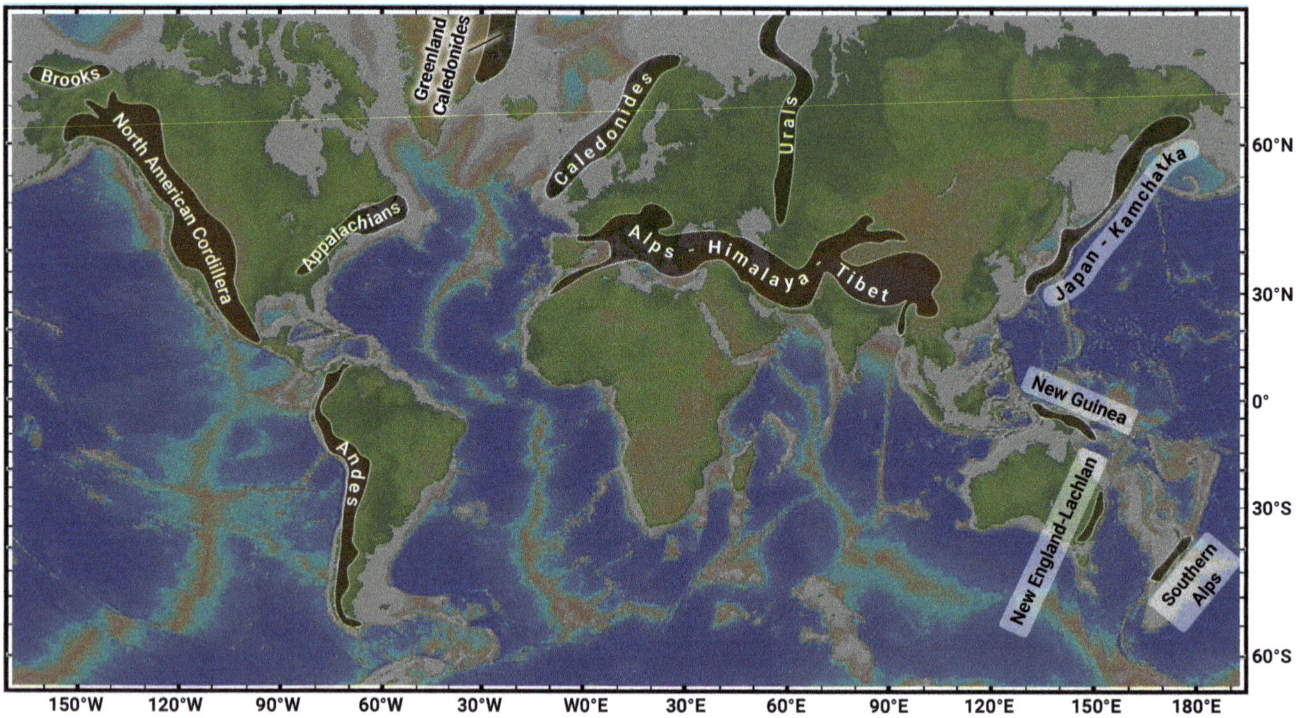

Figure 12.1 Map showing locations of some of the major orogens. Many orogens have recently experienced or are undergoing active shortening, leading to high topography (e.g., Alps-Himalaya-Tibet, New Guinea, Southern Alps). Other orogens (e.g., Appalachians, Caledonides, Urals) are hundreds of millions of years old, and why they still stand (relatively) tall is not well understood.

Source: GeoMapApp base map, annotations by authors.

Orogenic Belts and Crustal Thickening

Many of the world's great mountain chains, including the Alps, Caledonides, Appalachians, and Himalaya (Figure 12.1), are orogens formed by continental collision. The Cordillera of North and South America are long-lived orogens formed along continental subduction zones, although evidence for collision there is less obvious and more controversial. Orogens are characterized by significant crustal shortening and consequent crustal thickening (Figure 12.2). The thickest crust on Earth, 85 to 90 km, is in the Himalaya (Kind et al., 2002), and much of the crust in the central Andes is over 50 km thick (Rivadeneyra-Vera et al., 2019). The Appalachian Mountains in eastern North America, Caledonide ranges in the British Isles and Norway, and the mountains of eastern Greenland are the eroded roots of a complex set of orogenic belts that formed in several stages during Paleozoic ocean closures that formed Pangea. They formed a continuous chain in Pangea that was dispersed by the opening of the Atlantic.

Continental collision
Collision of continental material on a downgoing plate with a continental mass on the overriding plate.

Accretion
Suturing of an island arc or continental mass on a downgoing plate to a continental mass on the overriding plate.

Collision and Accretion

Collision as a Consequence of Subduction

Continents are low-density passengers on lithospheric plates, and when they or island arcs, oceanic plateaus (see Chapter 14), or other pieces of low-density crustal material are carried to a subduction zone, they jam it, leading to **continental collision** or **accretion** (Dewey & Bird, 1970). Figure 12.3 demonstrates two of several possible scenarios for what happens when lithosphere that resists subduction meets a subduction zone; for numerous other scenarios, see Moores and Twiss (1995, p. 213).

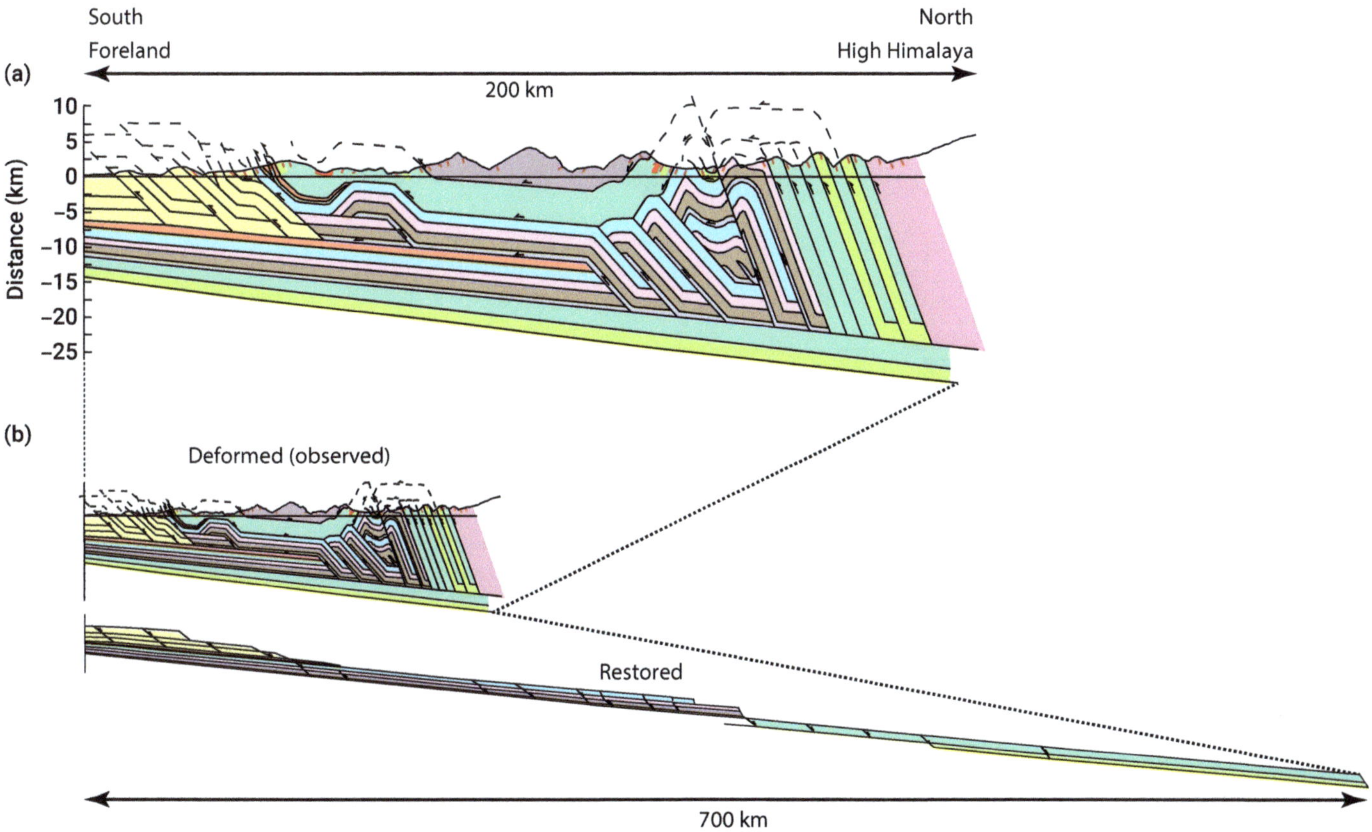

Figure 12.2 Cross section through western Nepal showing extreme tectonic shortening and deep burial (20 km) of shallow crustal rocks under the master thrust now exposed in the High Himalaya. (a) Present-day deformed section, roughly 200 km wide. Small black arrows show that successively higher sheets have been transported southward (to the left) relative to lower sheets owing to northward underthrusting of the Indian Plate. Active thrusts have shallow dips, but stacking rotates thrust sheets so that many now dip steeply to the north (right). (b) When the deformed section is restored, it is 700 km wide, requiring roughly 500 km of shortening for this section. Additional shortening occurred farther north in the High Himalaya and Tibetan Himalaya.

Source: Modified from Robinson et al. (2006); no vertical exaggeration.

Figure 12.3a depicts subduction of continent-bearing oceanic lithosphere beneath continental lithosphere. Subduction closes the oceanic basin separating the two continental pieces, and eventually, the continental mass on the subducting plate reaches the trench. At this point, subduction largely stops, although intense deformation of the continental lithosphere may continue for tens of millions of years. Classic examples include the collision between India and Asia to produce the Himalaya, collision between Africa and ancestral North America to produce the Appalachians, and collision between Europe and Adria (a small plate associated with North Africa) to produce the Alps.

Figure 12.3b shows what can happen when ocean-ocean subduction brings a continental mass to the subduction zone under an island arc. The continental mass again jams the subduction zone, but in this case, the direction (**polarity**) of subduction can reverse—from under the island arc (toward the right) to under the continent (toward the left). This polarity switch converted the former passive margin of the continent, like that off the eastern coast of the United States today, to an active subduction margin, like that off the Pacific Northwest or the western margin of South America. Regardless of whether subduction reverses polarity or simply ceases, collision of the continent and island arc serves to enlarge the continent by accretion. The attached island arc is known as an **accreted terrane**.

Polarity
Here, the direction in which a subducted plate is descending, generally expressed as the facing direction from the arc to the incoming plate. For example, the Cascade arc (see Figure 6.5) is west-facing, whereas the Mariana arc (see Figure 6.3) is east-facing.

Accreted terrane
A terrane (e.g., an island arc) that has been attached to a continent.

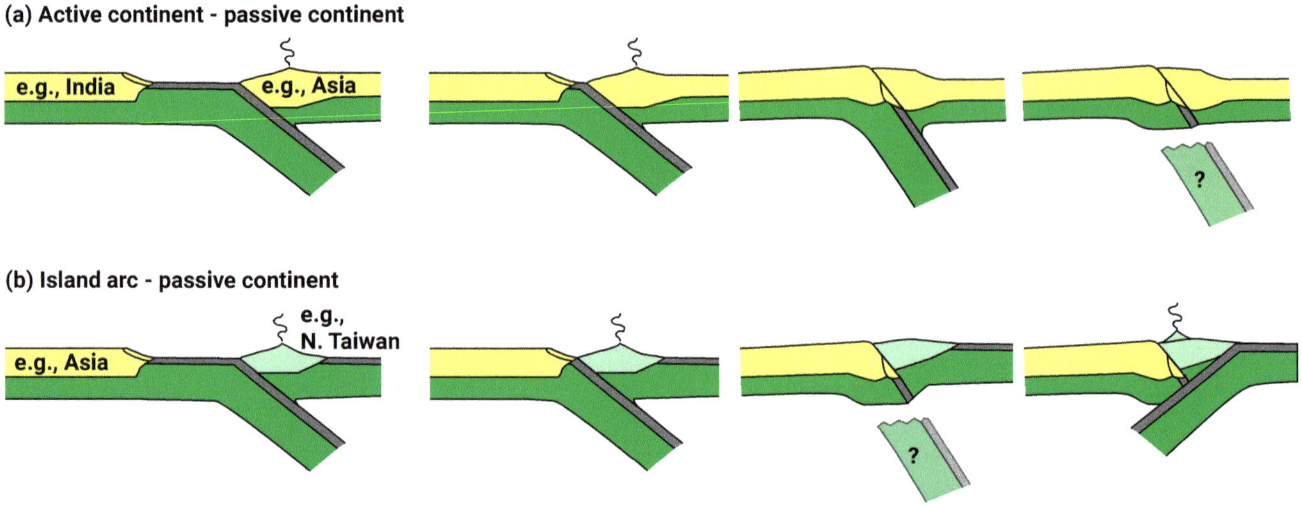

Figure 12.3 Schematic plate tectonic diagrams of collisions. Note that the boundary at the bottom of the lithosphere is drawn as sharp for convenience; in reality, the lithosphere-asthenosphere boundary is gradational. (a) Collision between two continents jams the subduction zone. The fate of the subducted oceanic lithosphere is not known, but it could detach (as shown) by a postulated process called slab breakoff, or gradually become incorporated into the upper mantle. Cessation of subduction at one boundary requires rearranging plate motions elsewhere (not shown). (b) Collision between an arc and a continent may cause subduction polarity to reverse.

Source: Modified from Moores & Twiss (1995).

Collision by Transform Faulting Although most of Earth's collisional belts are formed by partial subduction of a buoyant piece of lithosphere, transform faulting can also accrete buoyant material to a continent, as in southeastern Alaska and southwestern Yukon. The western margin of North America is largely a transform boundary where the Pacific Plate slides northwest relative to the North American Plate. The San Andreas Fault in California and the Fairweather-Queen Charlotte fault system of western Canada take up this motion along much of the margin except between northern California and British Columbia, where subduction of the small Juan de Fuca Plate produces the Cascades chain of volcanoes. The plate boundary makes a sharp turn westward in the armpit of Alaska; to the west, the Pacific Plate is being subducted beneath Alaska and the Bering Sea, producing the Aleutian chain.

The Yakutat block is an oceanic plateau on the Pacific Plate that once passively slid northwest along the transform system but now finds itself being rammed into the armpit (Figure 12.4). Ouch! Underthrusting has formed the lofty St. Elias Mountains, which include many of the tallest peaks in North America, including the second-highest, Yukon's Mount Logan (5959 m). Mount St. Elias, on the US-Canada border, is 5489 m tall and yet lies only 10 km from the ocean. Alaska is largely formed of accreted terranes such as the Yakutat block.

California and adjacent Mexico provide interesting examples of how pieces of continents can switch allegiance as they are moved around. Complicated plate interactions along the western margin of North America (see Section 13.1) detached a thin sliver, Baja California, from northern Mexico about 6 Ma, opening a narrow ocean basin behind it. This process transferred Baja California, formerly part of North America, to the Pacific Plate, where it and other parts of coastal California are being carried northward along the San Andreas transform fault. The dramatic topography around Los Angeles results from the large, left-stepping "Big Bend" in the right-lateral San Andreas Fault, which causes a Yakutat-style collision in which the lithosphere carrying Los Angeles is being thrust under the 3000-m-tall San Gabriel Mountains. Although it

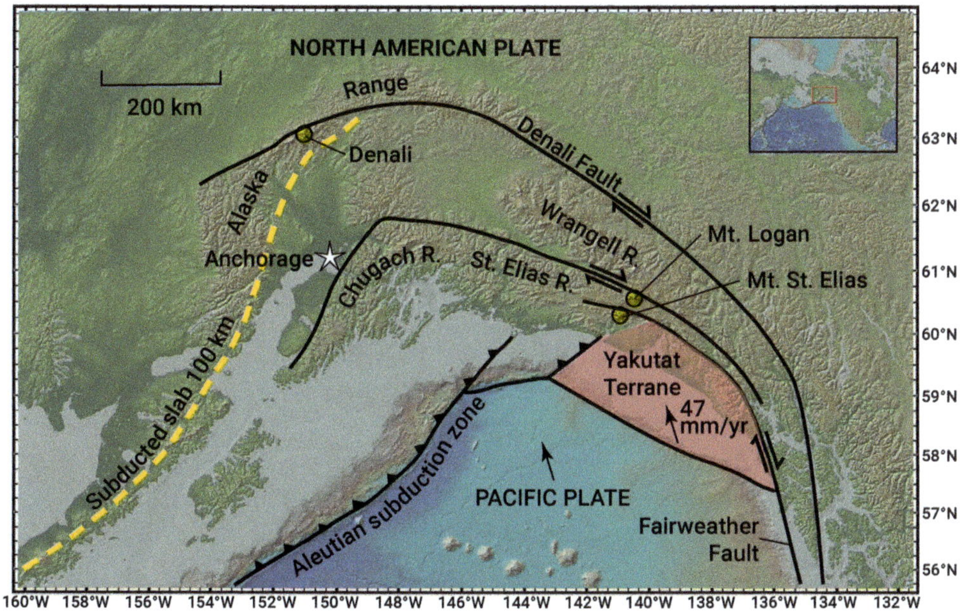

Figure 12.4 Map of northwestern North America showing relative movement of the Pacific Plate. The Yakutat terrane slides laterally along the Fairweather Fault and is being shoved northwestward under Alaska. Arrows and velocity on Pacific Plate and Yakutat block are motion relative to North America.

Source: Modified from Reeder (2016).

is commonly said that motion along the San Andreas will make Los Angeles and San Francisco neighbors in 15 Myr or so, more likely additional parts of California north of the Big Bend will be transferred to the Pacific Plate as well, and the metropolitan areas will be carried northward together to enjoy the same fate as the Yakutat block, enlarging Alaska yet again.

Regional Metamorphism Results from Collision and Accretion

Although the widespread and abundant nature of high-grade metamorphic rocks on the continents has been known for centuries, their origin was largely a mystery until the development of plate tectonics theory. Before then, mountain belts were generally explained as the culminating products of geosynclines, crustal downwarps that accumulated thick sequences of volcanic and sedimentary rocks and, somehow, promoted orogeny and mountain building. Acceptance of plate tectonics led to abandonment of the geosyncline concept, and to recognition that most mountain belts are the products of collision and accretion, aside from mountain chains such as the Andes that are generated largely by subduction magmatism.

Within this framework, then, **regional metamorphism** does not usually represent burial by sediment but rather burial beneath thrust sheets. The widespread igneous and metamorphic terranes that comprise continents were largely formed by these processes. Whether modern-style plate tectonics operated in the Archean is a topic of debate (e.g., Brown et al., 2020), so the origin of older terranes is less certain.

A critical aspect of regional metamorphic terranes is that they commonly contain abundant high-grade metamorphic rocks derived from volcanic and sedimentary rocks that were deposited at the surface. Such **supracrustal** rocks tell a story of deposition, burial to midcrustal or greater depths, and return to the surface for your inspection.

Regional metamorphism
General term for metamorphism of large areas, as contrasted with metamorphism near igneous bodies (contact metamorphism) or along shear zones (dynamic metamorphism).

Supracrustal
Literally, on top of the crust; pertains to sedimentary and volcanic rocks that were deposited at Earth's surface.

Rocks of the Regional Metamorphic Lithosystem

These lithosystems are dominated by metasedimentary and metaigneous rocks of the medium-pressure/medium temperature (MP/MT) series, along with eclogite-facies rocks of the HP/LT and UHP series, which are commonly hydrated and retrograded to MP/MT facies during ascent. Common rocks include the following:

- *Slate*: Very low-grade, incipiently crystallized metamudstone. Cleaves readily (slaty cleavage).
- *Phyllite*: Low-grade, finely recrystallized metamudstone. Has a foliation, but individual grains may be barely visible. Typically has a silky luster from fine-grained muscovite and chlorite.
- *Schist*: Coarsely crystalline metamorphic rock with strong planar fabric (foliation or schistosity). Often derived from metamudstone and consequently rich in muscovite, garnet, quartz, and key indicator minerals such as staurolite or kyanite.
- *Gneiss*: Layered and high-grade metamorphic rock; if the protolith can be determined, the gneiss may be classified as follows:
    - *Orthogneiss*: Gneiss derived from an igneous parent, with typical igneous mineralogy.
    - *Paragneiss*: Gneiss derived from a sedimentary parent, typically Al-rich with abundant muscovite and garnet, although marbles and calc-silicates can also be paragneisses.
- *Greenschist*: Low-grade metabasalt with mineralogy characteristic of the greenschist facies (chlorite, epidote, albite, and sometimes actinolite).
- *Amphibolite*: Intermediate-grade metabasalt, with mineralogy characteristic of the amphibolite facies (hornblende and plagioclase).
- *Granulite*: High-grade metabasalt with mineralogy characteristic of the granulite facies (orthopyroxene and commonly clinopyroxene and garnet). Often generalized to any rock with granulite facies mineralogy, including felsic orthogneisses and paragneisses.
- *Migmatite*: Gneissic rock with quartz-feldspar bands formed by partial melting.
- *Eclogite*: Metabasalt with mineralogy consistent with HP/LT and UHP (eclogite-facies) metamorphism. Often generalized to any rock with eclogite-facies mineralogy.
- *Marble*: Metamorphosed limestone or dolomite dominated by carbonate.
- *Calc-silicate*: Metamorphosed mixture of limestone or dolomite plus sand, silt, or clay. Contains abundant Ca-rich silicates such as actinolite, epidote, garnet, diopside, wollastonite, etc.
- *Quartzite*: Metamorphosed sandstone.
- *Ultrahigh-temperature (UHT) metamorphism*: Metamorphism at $T > 900$ °C. Mineral associations such as sapphirine + quartz and orthopyroxene + sillimanite + quartz are common indicators of UHT metamorphism.

Cratons and Continental Construction

Basement
Metamorphic and igneous rocks that form the foundation of the crust, below any sedimentary rock cover.

Geologic maps of North America, and indeed the world, show a preponderance of sedimentary rocks at the surface, and fully 90 percent of the United States on the 2005 Geologic Map of North America is so depicted. However, scratch this surface cover and you will find metamorphic and igneous **basement** beneath. You might have to scratch a little vigorously in some places; for example, Paleozoic sedimentary rocks in Grand Canyon National Park are about 1.5 km thick, and equivalent rocks in the Michigan basin reach

Rocks of the UHP (Ultrahigh-Pressure) Lithosystem

The UHP lithosystem is dominated by continental rocks that have experienced HP/LT metamorphism. Common rocks include the following:

- *Felsic orthogneisses*: Commonly most abundant but, for reasons we do not fully understand, rarely preserve evidence for UHP metamorphism.
- *Mafic eclogite*: Commonly hosted in orthogneiss and represents original basaltic sills and dikes. These rocks are commonly how UHP terranes are recognized.
- *Paragneisses*: Represent metasediments (quartzites, schists, marbles, and calc-silicates) and demonstrate that some rocks originated at Earth's surface.
- *Garnet peridotite* and other ultramafic rocks: Thought to represent part of the mantle wedge above the subducting continental crust.
- *Whiteschist*: A rare rock consisting mainly of quartz and muscovite, but often containing high-pressure talc, kyanite, Mg-rich garnet, and relicts of coesite.

5 km in thickness. Here and elsewhere in much of North America, these sedimentary rocks were deposited on Precambrian basement—Proterozoic and, locally, Archean metamorphic and igneous rocks. Sedimentary cover is less extensive in Canada, half of which comprises Precambrian basement. This vast area of exposed basement, centered on Hudson Bay, is known as the Canadian **shield**.

Regions of igneous and high-grade metamorphic rocks that have been tectonically stable for periods on the order of hundreds of millions of years are termed **cratons**. Extensive geochronologic studies have established that the North American craton comprises belts of similar-aged rocks that, in a broad sense, become younger outward from a core of Archean (>2.5 Ga) rocks that runs through the north-central United States and northern and central Canada (Figure 12.5). The eastern and western margins of the continent comprise island-arc and other terranes that were accreted to the cratonal core during the Phanerozoic. In the southwestern United States, island arcs were accreted to the continent between 1.8 and 1.6 Ga, metamorphosed at temperatures and pressures of about 600 to 750 °C and 700 MPa (corresponding to the amphibolite facies at ~25 km depth; see Figure 5.10), then intruded by less-deformed granites about 1.4 Ga (Williams & Karlstrom, 1996). Elsewhere across the North American core, granulite-facies rocks commonly reached *P-T* conditions of about 800 °C and 1.0 GPa, although *P* ranges widely from about 0.6 GPa to at least 1.6 GPa (e.g., Pattison et al., 2003; Dragovic et al., 2016; Dumond et al., 2018). The record becomes progressively more difficult to read going into the early Proterozoic and Archean, and is less well known going north into the Arctic, but it seems clear that orogenic processes (especially accretion) have more than doubled the area of the North American continent over the past few billion years.

Shield
Large, low-relief area of basement rocks in a craton.

Craton
Generally low-relief core of a continent that has been tectonically stable for prolonged periods, on the order of hundreds of millions of years.

Examples of Continental Collision

Here, we present examples of landscapes and rocks from two well-studied and accessible orogens—the European Alps and North American Appalachians. These orogens have served as development and testing grounds for quantitative methods to investigate metamorphic and tectonic processes.

The Alps are one of the most accessible and best-studied examples of continental collision owing to a confluence of deep glacial erosion, stunning exposure, proximity to dozens of major universities, and a long tradition of mountaineering coupled with geologic study. Many basic concepts in tectonics were developed based on Alpine rocks, including the formation of **nappes** (Figure 12.6a) and thrust faults (Figure 12.6b).

Nappe
Refers to both large-scale recumbent folds and large-scale thrust sheets.

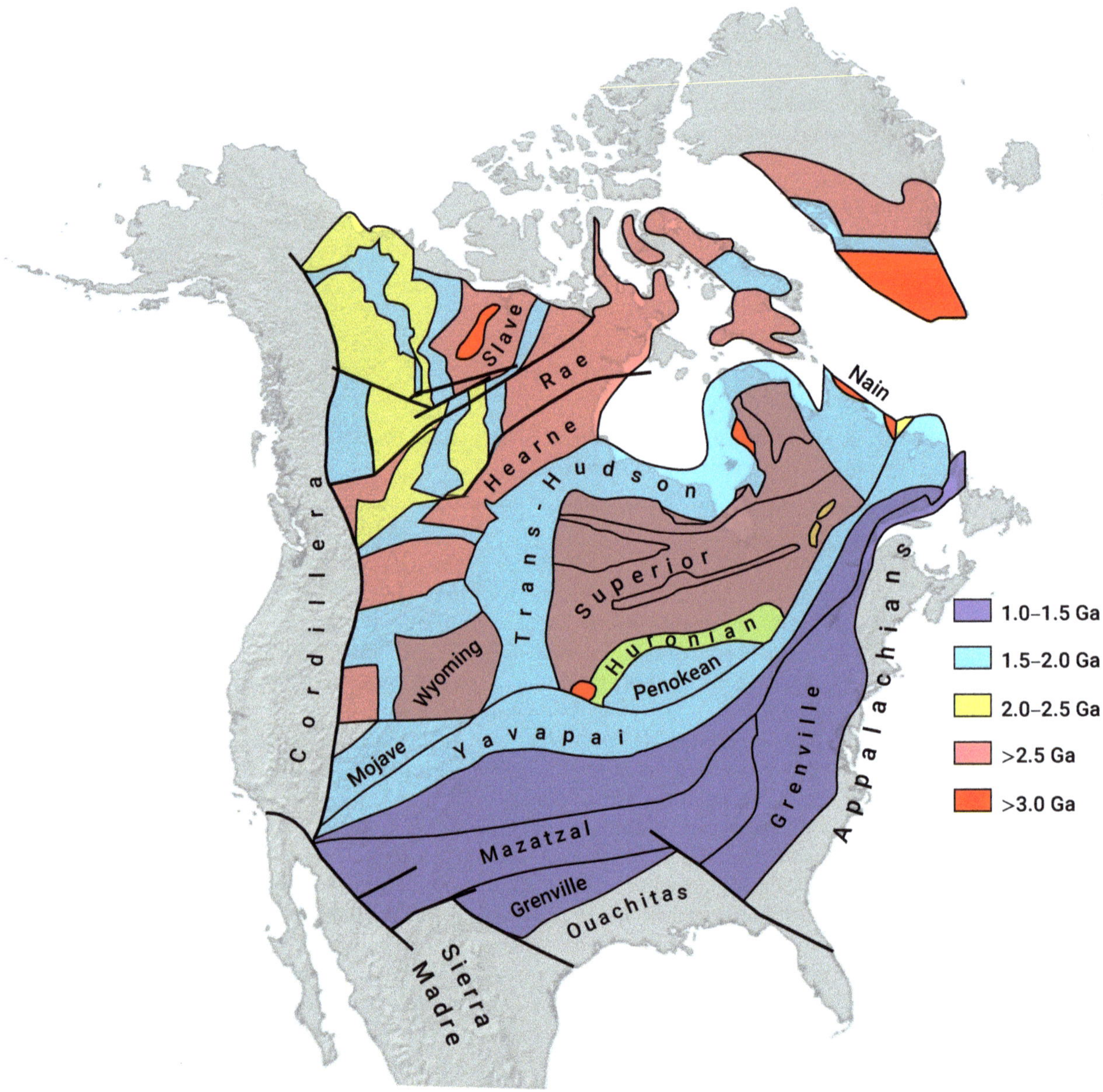

Figure 12.5 Ages of belts of basement rocks in North America. Names are specific tectonic provinces and/or orogens.

Sources: Modified from Hoffman (1988) and Thurston (2011).

The Alps formed in a general sense from closure of an ocean basin between Europe and Africa, starting around 60 Ma and culminating in a collision around 30 Ma. In between those times, oceanic lithosphere and thinned continental lithosphere of Europe were subducted southward. Representatives of many of these rocks were spit back out from both the subduction zone and the core of the orogen, leading to a wide variety of metamorphic rocks, depending on the protolith (deep marine rocks, shallow marine rocks, altered oceanic crust and mantle, continental rocks) and conditions of metamorphism (HP/LT, UHP, MP/MT). In general, the western Alps record the early subduction history better, preserving many blueschist- and eclogite-facies rocks, whereas the central Alps, with many greenschist- and amphibolite-facies rocks, record the latest collision better. Diverse rock types and metamorphic histories along the orogen record different parts of the tectonic

Figure 12.6 (a) Striking folds in the Dent de Morcles ("Morcles' tooth"), a 2969-m peak in the Swiss Alps, record the Alps-forming collision of Africa with Europe. Large-scale folds that are lying on their sides (recumbent), as in this photo, are locally known as nappes. These folds are part of the underside of the Morcles nappe in Switzerland. Here, Cretaceous and younger strata are complexly folded. (b) The Glarus Thrust in the Swiss Alps emplaced dark Permo-Triassic rocks over pale rocks ranging in age from Jurassic to Paleogene. This region is now preserved as a UNESCO World Heritage Site. Inset is a sketch of the thrust from Escher von der Linth, who first recognized the unusual older-on-younger relationship. Note the abundance of snow in the sketch from July 1812, compared to the photograph from July 2018; atmospheric CO_2 concentration in 1812 was only about two-thirds of today's.

Sources: (a) Bernhard Edmaier/Science Photo Library; (b) Christian Heine/Wikimedia; (inset) Hans C. Escher von der Linth/Wikimedia.

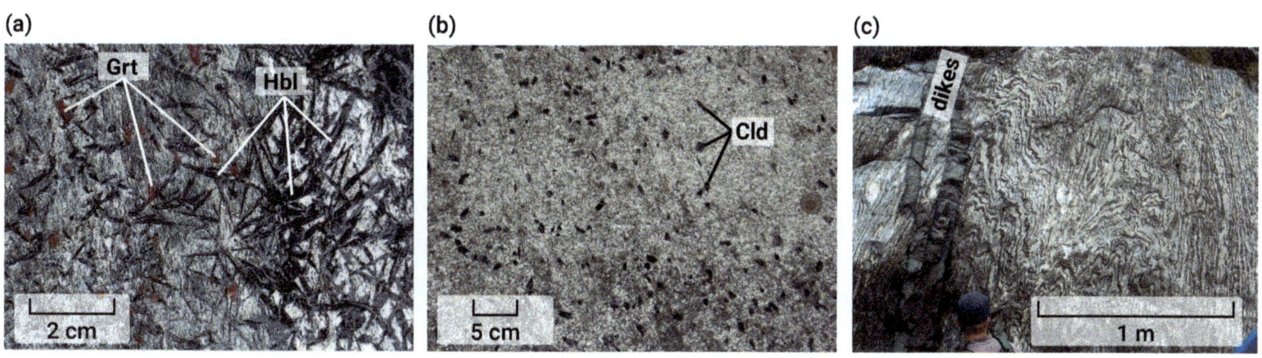

Figure 12.7 Photogenic metamorphic rocks from the Alps. (a) Schist locally known as Garbenschiefer with laths of hornblende and shiny garnets in foliation defined by muscovite, paragonite (Na version of muscovite), and chlorite, from Tauern Window, Austria. (b) This paving stone from the Sanctuary of Oropa, Biella, Italy, is made of eclogite-facies metamudstone with abundant large, blue-black chloritoid crystals. (c) Mafic dikes with chilled margins cutting intensely folded, thinly interbedded marbles (white) and calc-silicates (dark), Adamello, Italy.

Source: MJK photos.

history (Figure 12.7). This overall pattern of orogenesis revealed in the Alps—subduction to collision—is recorded in numerous orogens worldwide spanning billions of years.

The tectonic and metamorphic history of the Appalachian Mountains of eastern North America is much more complicated than the Alps and is significantly less well-known owing to greater age and poorer exposure (Figure 12.8a). Metamorphic rocks are nonetheless quite spectacular and are key to unraveling the tectonic history (Figure 12.8b,c). The Appalachians evolved through a series of Paleozoic collisions, the Taconic (~450 Ma), Acadian (375–400 Ma), and Alleghanian (~300 Ma) orogenies, involving

Figure 12.8 (a) View of Mount Moosilauke, New Hampshire, from Black Mountain, New Hampshire, showing typical vegetated Appalachian terrain, with limited exposure compared to the Alps (see Figure 12.6). Here, low-*P*, high-*T* quartzites and schists on Mount Moosilauke were thrust over MP/MT quartzites on foreground Black Mountain ~400 Ma. (b) This isoclinal fold on Mount Monadnock, New Hampshire, deforms Devonian low-*P*, high-*T* schists and quartzites of the New England Appalachians. Person for scale. (c) Metamorphosed graded bedding on Mount Washington, New Hampshire. Light-colored, fine-grained layers under knife are metasandstones. Darker, coarsely crystalline layers contain relict sillimanite, which formed after earlier andalusite (*T*-first *P-T* path; see Section 12.3). The original stratigraphic topping direction is to the left.

Sources: (a, c) MJK photos; (b) Edward E. Meyer/Wikimedia.

Diamictites
Poorly sorted sediment with grain sizes ranging from clay to boulder. Many diamictites represent glacial deposits, including tills (tillite) and diamictites containing dropstones (ice-rafted pebbles, cobbles, and boulders) in fine-grained marine sediments.

island arcs (Taconic), continental slivers (Acadian), and the main mass of Gondwanaland (Alleghanian). Although multiple subduction zones must have been active, only scraps of blueschists and eclogites avoided retrogression and are preserved for study. Collision in all three orogenies drove regional metamorphism up to amphibolite facies, overprinting earlier metamorphism formed during subduction and earlier collisional events. What makes the Appalachians difficult to understand (but also interesting!) is that the series of collisions variably overprinted prior metamorphism.

Making Connections 12.1

Collisions and Climate

The Earth's climate is fickle, ever-changing, and controlled in large part by the amount of CO_2 in the atmosphere. Carbon dioxide is fed into the atmosphere from volcanoes, natural oxidation of organic matter during decomposition, metamorphic decarbonation reactions, and anthropogenic activities such as burning fossil fuels and making cement. It is removed from the atmosphere by photosynthesis, burial of organic material, and weathering of Ca-bearing silicate minerals to make limestone.

Distinctive rocks known as **diamictites** are found scattered throughout Australia, India, Africa, Antarctica and South America (Figure 12.9). Such deposits of pebbles, cobbles, and boulders in a muddy matrix are commonly lithified glacial till, and associated scratched cobbles and bedrock surfaces confirm a glacial origin for many of them. Recognition of these deposits, all of roughly Pennsylvanian-Permian age (~300 Ma), on widely dispersed continents was one of the clues that led to the idea that those continents had once been joined together into a supercontinent we now call **Gondwanaland** (Wegener, 1912).

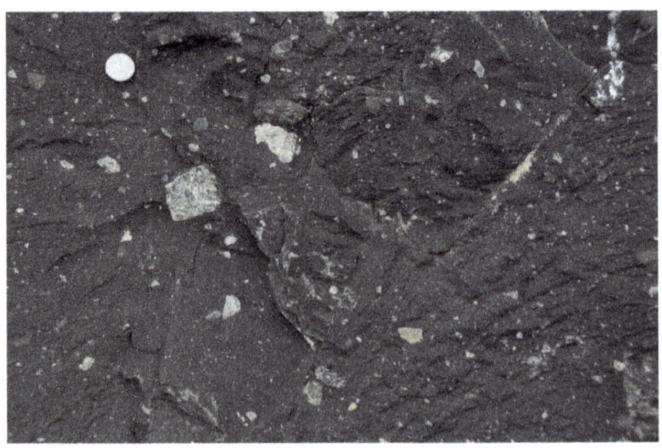

Figure 12.9 Diamictite, the famous Dwyka Tillite, from South Africa. This and correlative units provided some of the earliest data that led to the concept of continental drift, and they also are evidence for widespread low-latitude glaciation in the late Paleozoic, a glaciation that can be tied to continental collision. Coin in upper left is 2 cm in diameter.

Source: AFG photo.

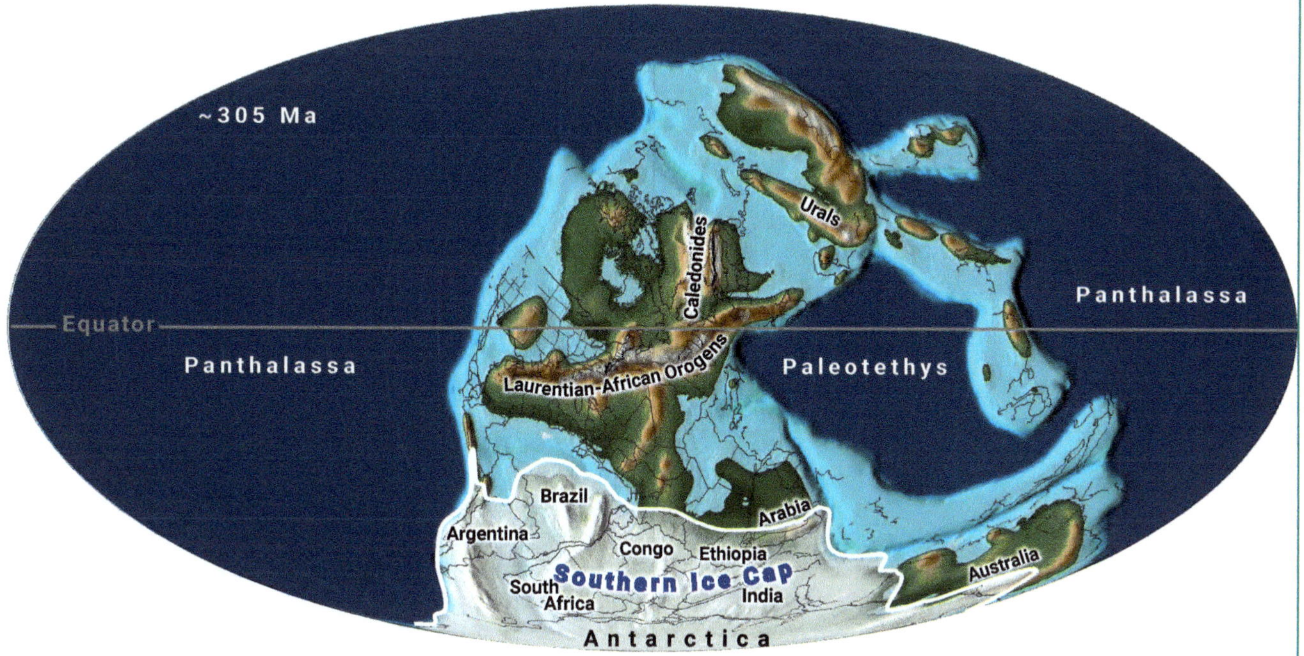

Figure 12.10 Plate configuration at 300 Ma, showing numerous orogenic belts near the Equator after formation of Pangea. Panthalassa and Paleotethys are names of ancient oceans.

Source: Modified from Scotese (2021).

Gathering the continents into Gondwanaland brings these far-flung glacial deposits into continuity. Once plates are restored to their 300-Ma positions, Gondwanaland covers the South Pole, but the northern limit of the glacial deposits ranged well north of the pole, to at least 45° S latitude (Figure 12.10). Widespread glaciation at such low latitudes (i.e., so close to the equator) indicates a cold climate, but what caused it? One process that led to CO_2 drawdown in the Paleozoic was colonization of dry land by vascular plants (Algeo et al., 1995), but this began in earnest significantly before Pennsylvanian glaciation.

However, another factor was at work: closure of proto-Atlantic oceans by a series of arc-continent and continent-continent collisions that welded Gondwanaland to eastern North America and Europe (see Figure 12.10). This suture generated a mountain range over 4000 km long, running east-west, located near the equator. Tall mountains at tropical latitudes provided a perfect environment for intense chemical weathering. Macdonald et al. (2019) argued that the abundance of ophiolitic rocks in this suture zone led to even greater CO_2 drawdown, because basaltic rocks in the ophiolites are significantly higher in Ca than typical granitic rocks. Erosion of calcium-rich rocks provides more Ca^{2+} and, ultimately, reduces atmospheric CO_2 by precipitating $CaCO_3$ (calcite and aragonite) and $CaMg(CO_3)_2$ (dolomite).

12.2 Continental Geotherms

Gondwanaland
A supercontinent formed during the latest Proterozoic from the assembly of India, South America, Africa, Arabia, Australia, and Antarctica.

Learning Objectives

By the end of this section you should be able to:

12.2.1 Explain the geochemical processes that cause the principal heat-producing elements to be concentrated in the upper crust.

12.2.2 Apply the heat-flow equation to show how heat generation will produce a steady-state curved thermal profile.

Concentration of Heat-Producing Elements in the Upper Crust

Present-Day Heat Production Heat production in the crust plays a critical role in metamorphism. As we saw in Section 2.4, the only elements on Earth that still produce significant thermal energy from radioactive decay are K, U, and Th. All three of these are incompatible elements, and such elements are preferentially partitioned into the melt during melting events. Because melts are less dense than the rocks they were generated from, they tend to rise. Melting of the mantle produces basaltic melt that rises to form oceanic and continental crust, in the process moving K, U, and Th out of the mantle and upwards. Hydrous melting of basaltic rocks in the lower crust (see Section 8.3) produces more silicic melts, which rise and carry the heat-producing elements higher. The result of these progressive distillations has been to concentrate the heat-producing elements in the crust, and particularly in the upper crust.

This upward segregation is evident in geochemical estimates of the compositions of various Earth reservoirs (Figure 12.11). K, U, and Th concentrations, and total heat production, are nearly twice as high in the upper crust than in the middle crust, and more than five times higher than in the lower crust. Values for the oceanic crust are lower still, and values for the mantle barely show up. These estimates accord with data from the field; for example, average heat production for metamorphic rocks from the Canadian Shield, taken from earthchem.org, is around 0.2 pW/kg, between values for the middle and lower crust, whereas that for upper-crustal, K-rich granites from the Sierra Nevada batholith (see Section 10.1) is 1–2 pW/kg, literally off the chart.

Surface heat flow varies by roughly a factor of 10 across North America, from around 15 to 150 mW/m^2 (see Figure 2.19). A curious and important finding from surveys of heat flow in plutonic rocks is that in many areas, there is a strong linear correlation between surface heat flow and surface heat production. One model that explains this relationship

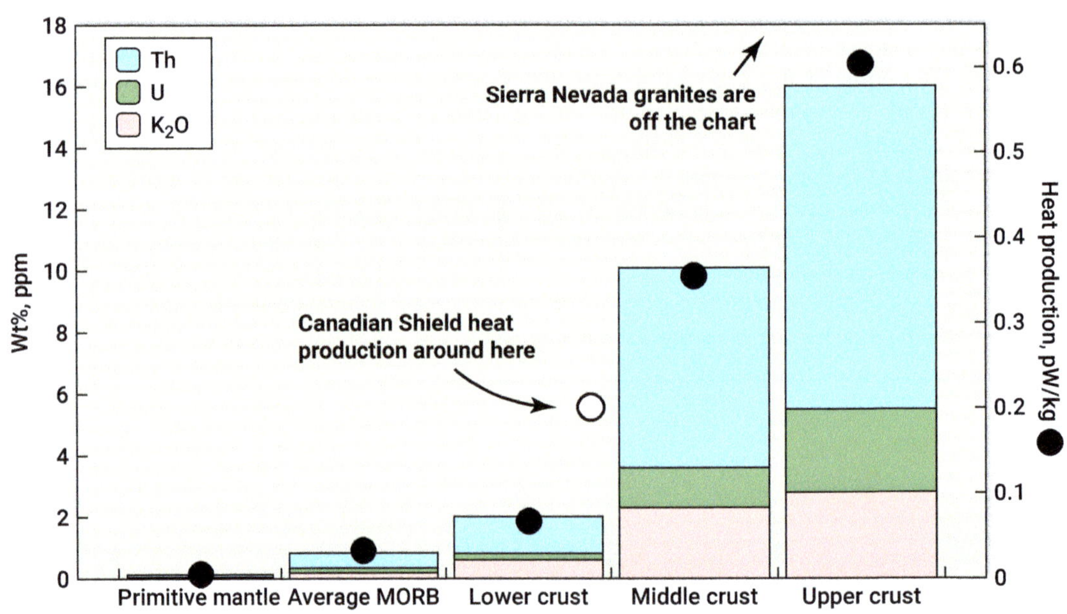

Figure 12.11 Concentrations of heat-producing elements (colored bars) and associated heat production (blacks dots) from various reservoirs. There is a steady increase from the mantle to oceanic crust to continental crust, and from deep to shallow in continental crust. This is a consequence of the incompatible nature of K, U, and Th.

Sources: Data from Rudnick & Gao (2014), Palme & O'Neil (2014), and White & Klein (2014).

and accounts for the data in Figure 12.11 is where the abundance of heat-producing elements decreases exponentially downward (Lachenbruch, 1970). This model is non-unique, but it is a starting point and captures the essential point that heat production in the crust decreases dramatically with depth.

Effect of Internal Heat Production on the Geotherm

In Section 11.2, we introduced the 1D heat-flow equation:

$$\frac{\partial T}{\partial t} = \kappa \frac{\partial^2 T}{\partial z^2} \qquad \textbf{(Eq. 12.1)}$$

Reviewing from that section, recall that the left side shows how T varies with time at a given point (z value) in the model, and that the right side, the second derivative of the temperature profile, is how the slope $\frac{dT}{dz}$ changes at that point—essentially the curvature of the profile. A memory aid from calculus reminds us how the sign of a second derivative gives us whether the curve is concave-up (+) or concave-down (–):

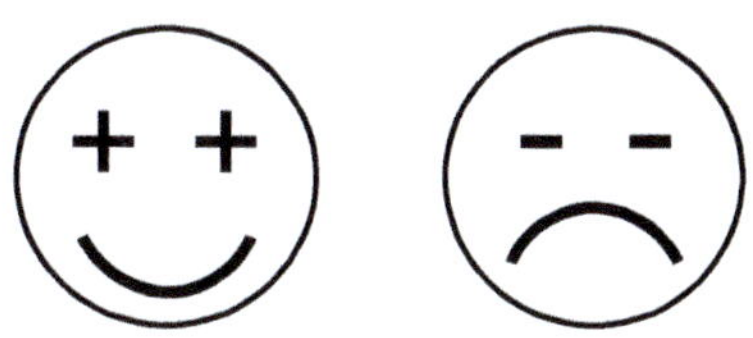

Here, the mouths represent the general shape of the temperature profile (T vs. z). Imagine that you take a steel bar out of a hot oven and clamp it between two ice blocks. Heat will flow out of the ends, producing a temperature profile that is high in the middle and 0 °C at both ends. For such a concave-down profile, the sign is negative, corresponding to the unhappy face, and therefore $\frac{\partial^2 T}{\partial z^2} < 0$. From the heat-flow equation, then, $\frac{\partial T}{\partial t}$ must be negative (κ is always positive), and thus temperatures in the concave-down part of the profile will decrease with time and the profile will evolve to a flatter shape.

If there is internal heat generation, however, things are different. Heat generation adds a second term to the differential equation:

$$\frac{\partial T}{\partial t} = \kappa \frac{\partial^2 T}{\partial z^2} + \frac{A(z)}{c} \qquad \textbf{(Eq. 12.2)}$$

where $A(z)$ is the rate of heat generation (W/kg) as a function of z and c is specific heat (see Section 2.7). By dimensional analysis, this term must have units of temperature per time, the same as the left side of the equation (e.g., °C/s).

From our example here, we saw that if the temperature profile is concave-downward, then T will decrease with time, because the right side of the equation is negative. However, heat generation can zero this out and produce a steady-state, concave-down profile. Alternatively, if a thermal profile is straight such that $\frac{\partial^2 T}{\partial z^2} = 0$, internal heat generation ($A > 0$) will cause $\frac{\partial T}{\partial t}$ to be positive, and thus the temperature will increase.

A Reference Continental Geotherm

Exponentially decreasing heat production leads to a simple steady-state solution of the heat-flow equation (Lachenbruch, 1970):

$$T = T_s + \frac{Q^* z}{k} + \frac{A_0 D^2}{k}(1 - e^{-z/D}) \quad \textbf{(Eq. 12.3)}$$

where T is temperature, T_s is the mean annual surface temperature, Q^* is heat flow from the mantle into the base of the crust, k is thermal conductivity, A_0 is surface heat production, z is depth, and D is a measure of length that governs how quickly heat production dies out going downward. Values used in modeling here are $Q^* = 0.025$ W/m^2, $k = 2.5$ W/m/°C, $A_0 = 4.3$ μW/m^3, and $D = 15$ km. With these choices and assuming a surface temperature of 0 °C, the steady-state geotherm is

$$T = 0.01z + 387(1 - e^{-z/15000}) \quad \textbf{(Eq. 12.4)}$$

where z is in meters. The geothermal gradient in °C/m is then given by the derivative with respect to depth:

$$\frac{dT}{dz} = 0.01 + 0.0258e^{(-z/15000)} \quad \textbf{(Eq. 12.5)}$$

We can interpret this equation as follows: The background temperature gradient in the absence of heat-producing elements is independent of depth and is given by the first term, 0.01 °C/m or 10 °C/km; heat production increases the gradient to about 35 °C/km at the surface, decreasing to 20 °C/km at a depth of about 15 km (moderate depth in the crust), and 11 °C/km, nearly down to the background rate, at 50 km.

As we will investigate later, collision perturbs the geotherm, but given enough time, continental crust will approach a steady-state thickness (on the order of 40 km), heat flow, and temperature structure. Knowledge of the equilibrium temperature profile provides a reference against which thermal perturbations caused by tectonic events can be compared.

Heat Production in the Precambrian

In most rocks today, U produces slightly more heat than Th, with K a distant third. But the half-lives of ^{40}K, ^{238}U, and (especially) ^{235}U are much shorter than ^{232}Th, so in the past, U and K produced much more heat. In fact, if we go back 2 or 3 Gyr, heat production from U and K were about two and four times higher, respectively, and total heat production (globally) was about two times higher. Some work suggests that U, Th, and K were not as enriched in crust over mantle during the Archean as they are today (Gard et al., 2019), implying that heat production in the crust was rather similar, and that formation and reworking of crust have gradually enriched it in heat-producing elements. But data for the Archean are sparse, and if concentrations were comparable, equilibrium temperatures in the deep crust (~30 km depth) would increase by over 300 °C, raising the predicted steady-state metamorphic grade from amphibolite-facies to UHT granulite-facies. Mantle heat flux during the Archean must have been higher, too, and increased heat flow helps explain the preponderance of granulites in ancient cratonic cores, especially UHT granulites (e.g., Kelsey & Hand, 2015).

Toolbox 12.1

Thermal Energy from Radioactive Decay

In Chapter 2, we learned that although radioactive decay of short-lived isotopes such as ^{26}Al contributed a great deal of heat to the early Earth, the only radioisotopes left in high enough abundances to produce significant thermal energy now are ^{238}U, ^{235}U, ^{232}Th, and ^{40}K. The U and Th isotopes decay first by α-decay (producing He nuclei) and then by a complicated and long chain of α- and β-decays, eventually ending up at stable isotopes of Pb. For example, ^{238}U undergoes α-decay with a half-life of 4.468 Gyr, producing ^{234}Th, which has a half-life of only 24 days and forms ^{234}Pa (protactinium); ^{234}Pa in turn β-decays to ^{234}U, with a half-life of about one minute; and so on through 11 more steps down to ^{206}Pb, which is stable. ^{40}K, in contrast, undergoes single step but branching decay to ^{40}Ar and ^{40}Ca (see Section 11.8).

Each of these steps releases energy, and the total energy emitted by a single atom of ^{238}U going through the entire decay chain to ^{206}Pb can be constructed by adding up the energies of all the intermediate steps. This comes to 51.8 million electron-volts per atom, or 8.3×10^{-12} J/atom. In 1 g of ^{238}U, there are $N/238$ atoms, where N is Avogadro's number, so decay of 1 kg of ^{238}U would produce $(8.3 \times 10^{-12}$ J/atom$) \times (6.02 \times 10^{23}$ atom/mol$) \times (0.238$ kg/mol$) = 2.1 \times 10^{13}$ J/kg! That is a big number, but this energy is released slowly. Recall from Section 2.3 that the decay constant is defined as $\lambda = \ln(2)/t_{1/2}$; from Table 2.2, this is 1.55×10^{-10}/yr. If λ is small, this quantity gives the proportion of a given amount of the isotope that decays in a year. This is the last bit of information that we need to calculate the rate of heat generation for ^{238}U; multiplying the rate in J/kg given earlier by λ, we get 3240 J/kg/yr.

The rate of heat generation by these isotopes in rocks, A, is a function of the abundances of the parent elements and the relative abundances of the radioisotopes (i.e., not all natural U is ^{238}U). For K, U, and Th, these values are in Table 12.1.

Table 12.1 Heat Production of Potassium, Uranium, and Thorium

Element	Heat Production A, W/kg	Example Rock	H, W/kg in Example
K	3.43×10^{-9}	3.34 wt% (K_2O)	9.51×10^{-11}
U	9.83×10^{-5}	8.3 ppm	8.16×10^{-10}
Th	2.64×10^{-5}	19.6 ppm	5.17×10^{-10}
Total			1.43×10^{-9}

To calculate the rate of heat production of a given rock, these values must be scaled to their abundance in the rock and then summed, and this requires paying attention to units. K concentrations in rocks are generally reported as wt% K_2O, while U and Th concentrations are reported as ppm (see Toolbox 1.1). To convert K_2O to K, we must multiply by the appropriate gravimetric factor (0.830; see Appendix 9), and the resulting number, 2.82, gives wt% K in the rock, which is the number of grams of K per 100 g of the rock. Dividing this by 100 gives grams of K per gram of the rock. Values for U and Th are micrograms per gram of the rock, so dividing them by 10^6 gives grams of the element per gram of the rock. Taken together, total heat production H is given by

$$H = A_K \times (\text{wt\%K}_2\text{O}) \times 0.83/100 + A_U \times (\text{ppm U})/10^6 + A_{Th} \times (\text{ppm Th})/10^6 \text{ W/kg} \quad \textbf{(Eq. 12.6)}$$

or

$$H = \begin{bmatrix} 2.85 \times (\text{wt\%K}_2\text{O}) + 9.83 \\ \times (\text{ppm U}) + 2.64 \times (\text{ppm Th}) \end{bmatrix} \times 10^{-11} \text{ W/kg} \quad \textbf{(Eq. 12.7)}$$

12.3 *P-T* Predictions from Thermal Models

Learning Objectives

By the end of this section you should be able to:

12.3.1 Predict how the geometry, heat production, and erosion history of a thrust belt will affect the *P-T* paths of resulting metamorphic rocks.

12.3.2 Predict how *P-T* paths may have changed over billions of years as Earth cooled.

12.3.3 Predict the metamorphic field gradient (distribution of *P-T* conditions preserved by rocks in the field) likely to be observed in different tectonic settings.

12.3.4 Predict qualitatively the long-term evolution of crustal strength during and after orogenesis.

What Types of *P-T* Paths Might Rocks in Collisional Orogens Follow?

Rocks in regional metamorphic terranes start somewhere (in the case of supracrustal rocks, at Earth's surface), are taken to some higher *P* and *T* owing to burial and tectonic events, and then return to the surface, where they are available for collection and analysis. Such rocks contain information about the *P-T* paths that they followed, and a significant preoccupation of metamorphic petrologists lies in using this information for tectonic reconstructions.

Let's try a few thought experiments. Let's say we are in an area where the geothermal gradient is 15°C/km, and mud and sand have accumulated in a basin to significant depth. We leave it to the student to sketch out the following scenarios in order to get a feel for the assumptions that must be made in each.

Thought Experiment 1: Rapid Burial Followed by Erosion

In our first thought experiment, crustal shortening caused by continental collision causes the basin to be rapidly (say, in less than a million years) overrun by a thrust fault, burying the muds and sands under a 20-km-thick thrust plate. After burial, erosion slowly (say, over tens of millions of years) removes the thrust plate, exposing the rocks to the surface again. What might we expect for the *P-T* path?

Consider a point initially at 5 km depth, on the ambient geotherm. Burial will increase the pressure by about 530 MPa ($P = \rho g h$ where $\rho = 2700$ kg/m^3, a nice value for upper-crustal rocks). The point moves steeply up on the *P-T* diagram (Figure 12.12a). After burial these cold rocks will warm up for two reasons: (1) hot rocks at the base of the thrust sheet were placed on top of them, and (2) heat from deeper in the Earth will eventually return to leaking upward, once the negative geothermal gradient caused by placing hot rocks over cold ones is ironed out. Our buried point is now 25 km deep; given enough time, it will sit at 25 km depth and warm back up to the initial geotherm. Right?

Nope, wrong, for two reasons. First, erosion is an important feature of mountain belts, and excess topography produced by orogeny causes dramatic increases in erosion. Our buried point will be carried toward the surface as erosion proceeds (or, to view it differently, the surface will erode down toward the point), and so pressure will decrease. The

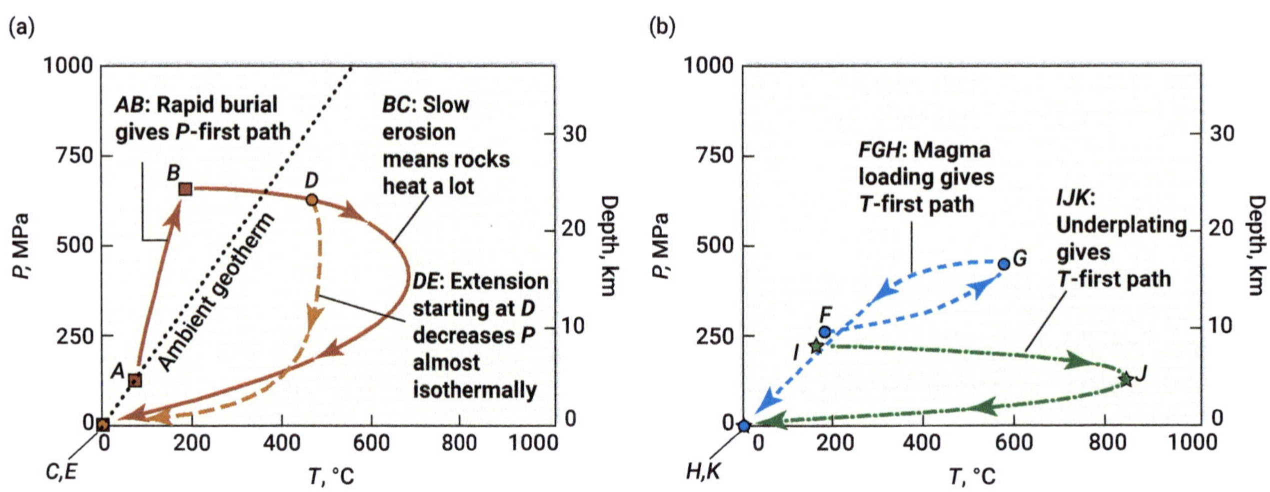

Figure 12.12 Thought experiment for likely *P-T* histories of rocks in response to different tectonic scenarios. (a) Overthrust scenarios with *P*-first *P-T* paths: In path *A-B-C*, erosion exhumes rocks slowly so that they reach relatively high temperature; In path *A-B-D-E*, extension starting at *D* exhumes rocks rapidly and short-circuits heating. (b) Contact metamorphic scenarios illustrating expected *T*-first *P-T* paths. Path *F-G-H* represents magma loading (burial under plutons). Path *I-J-K* represents magma underplating.

other reason this simple experiment is incorrect is that upper-crustal rocks of the lower plate are enriched in heat-producing elements, and burying these insulates them and allows significant excess heat to build up. We will explore later how much the rocks heat up.

Given these considerations, we can predict that our buried point, which started at *A*, will jump in pressure rapidly to *B*, after which it will heat up and pressure will drop as erosion unroofs it, eventually bringing it to the surface for collection at *C* (see Figure 12.12a). On a *P-T* plot with the pressure axis increasing upward (a petrologist's perspective), our rock will experience a **clockwise *P-T* path**. Note that the same *P-T* evolution will describe an anti-clockwise path if *P* is drawn to increase downward (a geophysicist's perspective). Although terminology based on analog clocks is widespread, to decrease confusion we will refer to a *P-T* path such as that in Figure 12.12a as a ***P*-first *P-T* path**, to indicate that pressure increases rapidly at first, with temperature lagging behind.

Clockwise *P-T* path
A path that describes a clockwise loop on a *P-T* diagram when the pressure axis is drawn to increase upward.

***P*-first *P-T* path**
A path on a *P-T* diagram where pressure increases significantly before temperature does; equivalent to a clockwise *P-T* path, but invariant with respect to which way the axes are drawn.

Thought Experiment 2: Rapid Burial Followed by Extension

Erosion is not the only process that can exhume rocks; extension is important, too. Extension can remove material faster than erosion and short-circuit the heating portion of the *P-T* path. Rocks decrease in pressure more nearly isothermally as extension begins (point *D* in Figure 12.12a), and they may reach relatively low *P* at moderately high *T*. Low-*P* minerals like andalusite or cordierite can be signifiers of this kind of process.

Thought Experiment 3: Burial Under a Plutonic Pile

Now let's consider the effect of intrusion of granite magma at 900 °C on the same pile of sedimentary rocks. Our reference point is now at 10 km depth (point *F* in Figure 12.12b), but still on the ambient geotherm, until a 10-km stack of horizontal sheets of plutonic rocks is intruded above it. Intrusion occurs over 10 Myr, starting at 9 km depth, and averaging 1 mm/yr of magmatic inflation. Consistent with Section 10.9, we will keep intruding new sheets of magma 1 km above our target rock, as rising magma meets the mechanical barrier provided by earlier sheets that have solidified.

Our reference point heats up owing to intrusion of hot magma 1 km above. How much it heats up depends on several unspecified parameters (e.g., how thick each sheet is), but intrusion of 10 km of magma should perturb things quite a bit and increase pressure by about 270 MPa, to point *G*, for a granite density of 2700 kg/m^3. However, as intrusion occurs over 10 Myr and the consequent crustal thickening probably puffs up the surface, erosion will knock down the total pressure increase to some extent. Initial temperature increase should follow intrusion after a (geologically) short delay, whereas pressure increases steadily over 10 Myr, so this is an example of a ***T*-first *P-T* path** (see Figure 12.12b) that loops back on itself. Eventually, erosion and faulting bring the rock to the surface (point *H*).

***T*-first *P-T* path**
A path on a *P-T* diagram where temperature increases significantly before pressure does; equivalent to an anti-clockwise *P-T* path, but invariant with respect to which way the axes are drawn.

Thought Experiment 4: Underplating by Magma

In this scenario, we consider the *P-T* path of a point 1 km above the upper magma sheet in the previous scenario (point *I* in Figure 12.12b, so at a bit lower pressure). This point is subjected to most of the heat (and fluids) being emitted by the magma stack growing below, so it heats up, but it is not subjected to greater pressure. It follows a horizontal path to higher temperature (point *J*), then cools down, moving toward the surface (point *K*) as erosion works on the excess topography caused by the intrusions. The path is clockwise on the axes as drawn, but clearly represents an event in which increasing temperature was the initial event.

Calculating *P-T* Paths

The Physical Model

Using numerical methods to predict *P-T* paths involves a number of significant assumptions involving the distribution of heat-producing elements, the rate and progress of erosion, thickness of the thrust plate, etc. The exponential distribution of heat-producing elements was discussed earlier. A reasonable model of erosion also depends upon the exponential function. We will assume that the rate of erosion is proportional to excess topography generated during orogeny, and that this in turn is proportional to the excess crustal thickness added by collision. Thus, in our model, if 20 km of crustal thickness was added during collision and the initial erosion rate is E_0, then the rate of erosion when this has been taken down to 10 km of excess is $E_0/2$, and at 5 km of excess $E_0/4$, etc. The erosion rate is then simply

$$E(t) = E_0 e^{(-E_0 t / y_{xs})} \qquad \textbf{(Eq. 12.8)}$$

where $E(t)$ is the erosion rate at time t and y_{xs} is the excess crustal thickness added by shortening.

As with the thermal models presented in Section 11.2, the programming is not difficult but requires a lot of bookkeeping. In these models, our spatial reference point is the Earth's surface, and so unlike the models in Section 11.2, heat is being transported not only by conduction but also by advection, because sanding down the surface by erosion means that, from the surface's point of view, hot rock is being advectively transported toward it.

Two thermal models are presented here. We assume that a section of crust whose ambient geotherm is given by Equation 12.4 is cut by a thrust fault that instantaneously emplaces a 20-km-thick thrust sheet on top of adjoining crust—in effect, doubling the upper 20 km of the crust (Figures 12.13). As we are assuming instantaneous emplacement, the upper portion of the geotherm is doubled, producing a saw-toothed (Z-shaped) profile. Is this reasonable? No, but Oxburgh and Turcotte (1974), who pioneered this sort of investigation, showed that the Péclet number (see Section 3.6) for thrust faulting is on the order of tens. As this is significantly greater than one, it at least means that the thermal structure is being dragged along by faulting faster than it can re-equilibrate. So, instantaneous emplacement is not a bad starting point.

This doubled crust with the Z-shaped geotherm is the starting point for our model (see Figure 12.13). The model uses a depth step of 200 m and a time step of 500 yr. Upper and lower boundary conditions are a surface temperature of 0 °C and a bottom heat flux of 0.025 W/m^2, and models were run for 100 Myr.

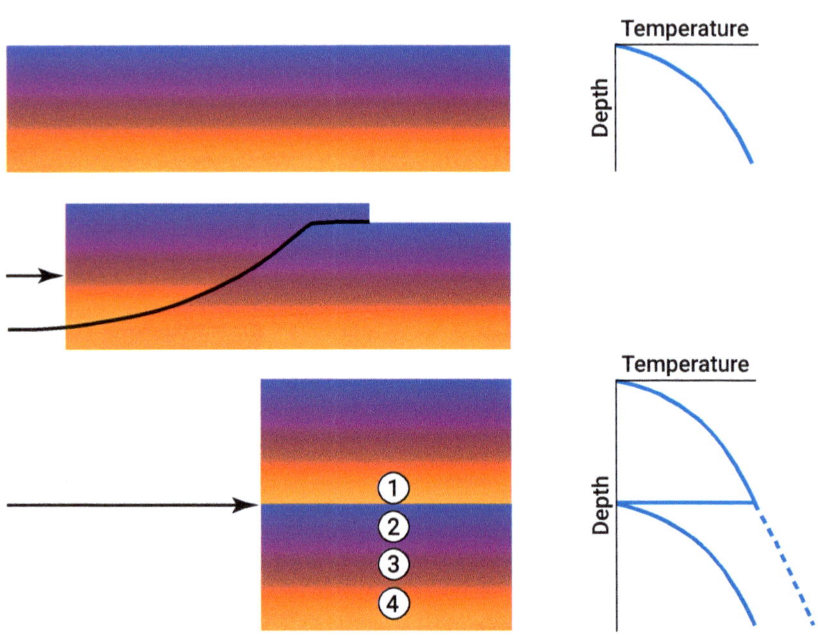

Figure 12.13 Schematic illustration of how a Z-shaped geotherm results from rapid emplacement of a thrust sheet. This is the limiting case for instantaneous emplacement. Numbers indicate watch points tracked during calculations.

Model Results Model results are shown in Figure 12.14, which tracks the *P-T* evolution of the four watch points over 100 Myr, with points plotted every 10 Myr. The two models use the same parameters except that the initial erosion rate is 0.1 mm/yr for one and 1 mm/yr for the other.

Significant aspects of these models include:

- Initial thrust emplacement pushes all the points in the lower plate into the HP/LT regime (blueschist and eclogite facies).
- Point 1, in the upper plate, initially cools significantly because it is emplaced on top of the cool surface of the lower plate, but then rapidly heats back up.
- Maximum temperatures reached are 150 to 200 °C higher for the slower-eroding (hereafter "slower") model.
- The upper plate is effectively eroded away for the faster model, with points nearly returning to their starting depths and

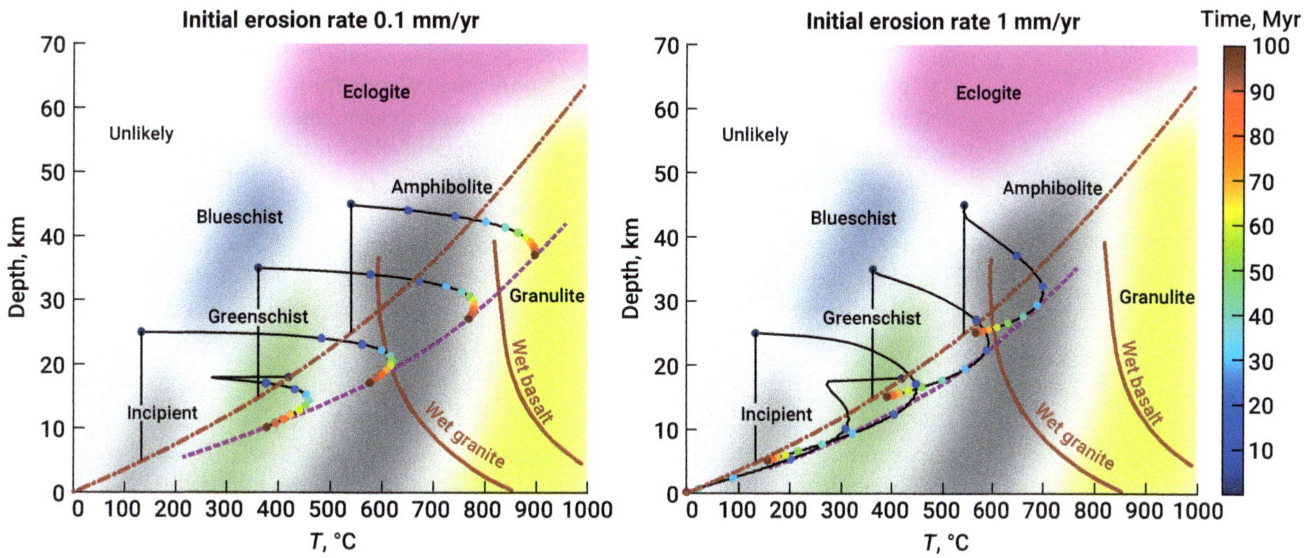

Figure 12.14 *P-T* paths from two models that differ only in erosion rate. In both models, a 20-km thrust plate is emplaced instantaneously. Surface heat production is high and declines downward with length scale 15 km. Dotted lines give ambient pre-thrust geotherm, from Equation 12.4. The left panel shows a slow erosion rate of 0.1 mm/yr, and the right panel a higher rate of 1 mm/yr. The four paths correspond to the four watch points in Figure 12.13, and dots along the paths are drawn every 10 Myr. All points in the lower plate (2, 3, and 4) initially jump in depth by 20 km, and then erosion brings them part or all the way back to the surface during the model time of 100 Myr. Point 1 initially cools down owing to being thrust over cool surface rocks, but all points are warming up by 10 Myr after emplacement owing to burial of heat-producing elements. The key point of this comparison is that erosion rate has a strong effect on maximum temperatures reached. Dashed line connects points of maximum *T* on the curves and corresponds to a metamorphic field gradient.

temperatures, but less than half the upper plate is eroded away in 100 Myr for the slower model.

- Anything below point 2 in the slower model crosses the wet granite solidus; thus, appropriate rock compositions will partially melt and become **migmatite**.
- Heating paths are considerably faster than cooling paths (note spacing of 10-Myr points on the paths).

Migmatite
Partially molten gneissic rock.

Parameter Space and the Purpose of Modeling

Only one parameter, the erosion rate, was varied for the models in Figure 12.14. One can also vary surface heat production, the heat production length scale, and the thickness of the upper plate. In the jargon of numerical modeling, examining input variables is referred to as "exploring parameter space" (or something similar). In these models, the purpose of such exploration is generally not to find a combination of parameters that will match a particular field example, but rather to see what variables are most important in controlling the outcome. The models in Figure 12.14 clearly show that erosion rate has a large effect on the model results. Similarly, surface heat production and the length scale for exponential decay of heat production have large effects. Even with just these few parameters to adjust, the number of possible paths through *P-T* space is infinite, and any given point in *P-T* space can be reached with many different combinations of parameters. Finding a set of parameters that matches observational data does *not* mean that the chosen parameters are the correct ones!

Metamorphic Field Gradients

What Parts of the *P-T* Path Do Metamorphic Rocks Preserve?

For the *P*-first *P-T* paths in Figure 12.14, maximum *T* is reached after maximum *P*. Suppose that a column of rocks that underwent one of these histories was laid bare at the Earth's surface for collection (note that for multiple levels to be exposed, the terrane must have been tilted, or various levels brought to the surface by faulting), and that you used thermobarometry (discussed later) to determine *P-T* conditions of formation. Recalling from Section 5.2 that metamorphic rocks generally record mineral assemblages that were developed at or close to maximum *T*, rocks collected from various levels within terranes represented by the models should record the maximum *T* that the given level reached. The dashed lines connect points of maximum *T* on the *P-T* paths and produce a **metamorphic field gradient**.

Metamorphic field gradient The distribution of *P-T* conditions recorded by rocks as exposed in the field.

An important result from thermal modeling is that the metamorphic gradient observed in the field does not in general correspond to the *P-T* path that any given rock parcel followed, but rather is a summary of the maximum temperatures that the rocks experienced and the pressures at which those temperatures were reached.

Some Classic Metamorphic Field Gradients Metamorphic field gradients are generally subdivided into three categories according to their *P/T* distributions. HP/LT distributions with low thermal gradients ("cold") commonly fall within blueschist and eclogite facies and represent subduction metamorphism. Low-pressure/high-temperature (LP/HT) distributions with high thermal gradients ("hot") typically form in association with plutons and represent contact metamorphism. MP/MT distributions commonly form during collisional metamorphism as continental crust evolves toward an equilibrium geotherm. Several specialized names have been applied to these distributions over the years,[1] but only "Barrovian" (after pioneering petrologist George Barrow; Figure 12.15) is still in common use. MP/MT field gradients are the most widespread and

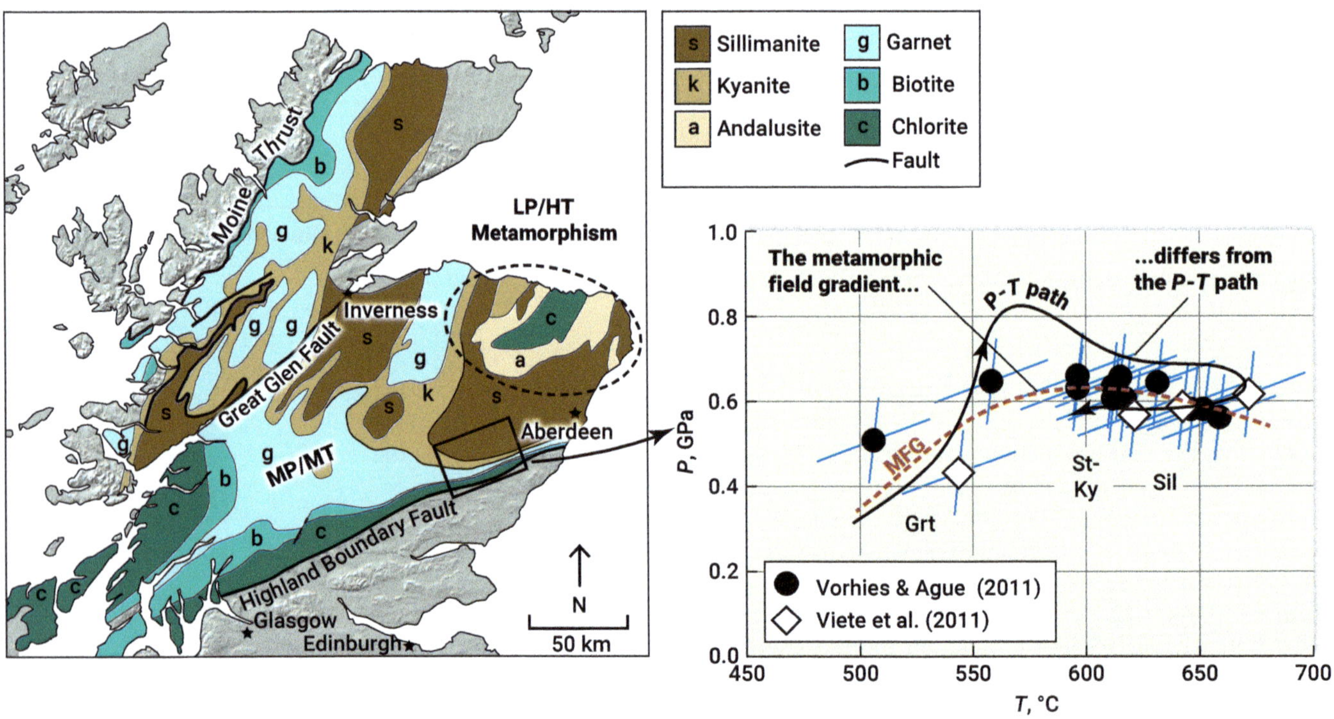

Figure 12.15 Metamorphic map of the Scottish Highlands showing distribution of mineral zones developed in metamudstones. Most of the region is characterized by MP/MT ("Barrovian") metamorphism, but northeast Scotland has LP/HT rocks. Box with arrow shows region where *P-T* conditions have been calculated and direction of increasing metamorphic grade. *P-T* diagram on the right shows the metamorphic field gradient (MFG) from southeast to northwest within the Barrovian zone. A *P*-first *P-T* path calculated for sillimanite-zone rocks crosses the metamorphic field gradient, as thermal models predict.

Sources: Base map modified from Winter (2010); diagram data from Vorhies & Ague (2011) and Viete et al. (2011).

are found in orogenic belts worldwide. The most common mineral isograds (biotite, garnet, staurolite, etc.) correspond with MP/MT field gradients. The concave downward shape displayed by the metamorphic field gradient in Figure 12.15 (but not by our thermal modeling) is observed in orogens worldwide.

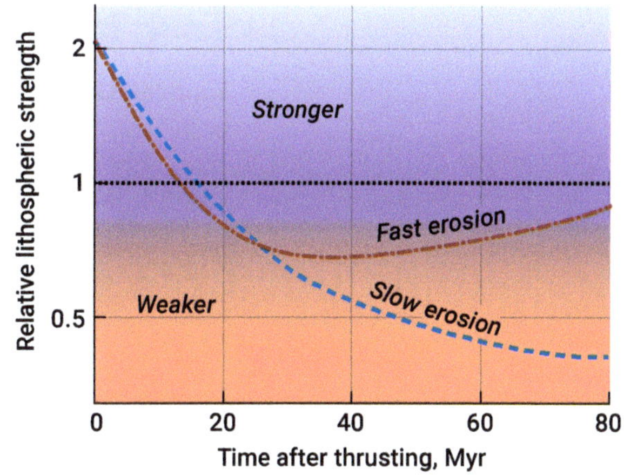

Figure 12.16 Representative models showing how lithospheric strength varies after a thrust event stacks up the crust. Relative lithospheric strength is the strength of the column in tension divided by that of unmodified crust. The two models depicted show fast and slow erosion. For a variety of model parameters, the strength of the column starts out significantly higher than its pre-thrusting value but falls within some tens of millions of years to values 50–75% that of unmodified crust. This weakening, coupled with excess topography, can explain the common observation that extension (i.e., spreading of the crustal column) commonly follows crustal shortening.

Source: Modified from Glazner & Bartley (1985).

Lithospheric Strength Changes and Post-Thickening Extension

As thermal models of thrust belts with radiogenic heat production show, the middle crust can get quite hot as long as upper-crustal rocks of the lower plate remain buried. In Figure 12.14, the watch points end up 100 to 150 °C above the pre-thrust geotherm. This typically happens 20 to 50 Myr after thrusting, depending on model parameters, and it has a significant effect on lithospheric strength, because rocks soften dramatically as T increases and the deformation mechanism changes from brittle failure to ductile flow (see Figure 2.21). Once in the ductile regime, rock strength decreases exponentially with temperature. Thus, we can predict that a section of crust following the thermal evolution in Figure 12.14 will be significantly weaker some tens of millions of years after thrusting than it was before thrusting.

However, there is another factor at work. Recall from Figure 2.21 that the upper crust is strong, even though it is made of relatively weak quartz-rich rocks, because it is relatively cool. In our simple model, thrusting doubles this cool, strong upper crust, so initially, the lithosphere could be stronger than it was before the thrusting event. Glazner and Bartley (1985) used thermal modeling like that in Section 12.3 to show that lithospheric strength is 1.5 to 2 times greater initially, but declines with midcrustal heating, dropping back to ambient values by 20 Myr and to values as little as 50 to 75 percent of ambient by 40 Myr (Figure 12.16). If temperatures exceed about 700 °C, partial melting further reduces rock strength, possibly as much as an order of magnitude for melt volumes of as little as 10 percent (Rosenberg & Handy, 2005). Rock strength also depends on the degree to which solid grains interact, and uncertainties in these interactions represent another area of active research.

The weakness of the lower crust has been invoked to explain various syn-collisional processes, including both the formation of extensional faults and the growth of plateaus, as the lower crust oozes laterally (e.g., Clark & Royden, 2000; Beaumont et al., 2001). This result may also help explain why so many thrust belts later undergo extension. For example, in Nevada and adjacent parts of the Basin and Range, Late Cretaceous thrusting ended about 60 Ma and was followed by widespread extension that began about 30 Myr later. Collision in the Norwegian Caledonides terminated with extensional faulting in the early Devonian. Although the Himalaya have been forming since about 60 Ma, they have already exhibited extensional faulting since 20 to 25 Ma.

In a global sense, burial of heat-producing elements during thrust faulting and subsequent lithospheric weakening contributes to the Wilson cycle—continental collision buries heat-producing elements and produces mountains and excess crustal thickness, but subsequent weakening caused by radiogenic heating facilitates extension, which may open new ocean basins.

12.4 Metamorphic Reactions

Learning Objectives

By the end of this section you should be able to:

12.4.1 Illustrate how a reaction affects stable mineral assemblages on binary and ternary composition diagrams.

12.4.2 Use stable mineral assemblages in the context of a petrogenetic grid to quantify *P-T* conditions of metamorphism.

12.4.3 Explain why aluminum is a key constituent of most chemical geobarometers.

12.4.4 Quantify *P-T* conditions from chemical or other observations through the use of simple analytical expressions.

In Section 11.3, we introduced index minerals, metamorphic zones, and isograds, all concepts that can be used to estimate *P-T* conditions of metamorphism. In this section, we will show how to plot mineral assemblages chemically, and how assemblage stability relates to reactions and *P-T* conditions.

Reactions in the System $NaAlSiO_4$-SiO_2

Consider the system with components $NaAlSiO_4$-SiO_2 or Nph-Qz. There are four possible stable phases in this binary system:

- Quartz (Qz) SiO_2
- Albite (Ab) $NaAlSi_3O_8$
- Jadeite (Jd) $NaAlSi_2O_6$
- Nepheline (Nph) $NaAlSiO_4$

You might think that three components (SiO_2, Al_2O_3, Na_2O) are needed to describe these phases, but they can be represented on a line with just two components, because Jd and Ab are linear combinations of Nph and Qz. Figure 12.17 shows a *P-T* phase diagram with relevant chemographic illustrations (lines with minerals along them), showing which minerals are stable. The two reaction lines (1 and 2) divide *P-T* space into three fields. Reactions are given by the labels (e.g., reaction 2 is Ab = Jd + Qz crossing from lower to higher *P*).

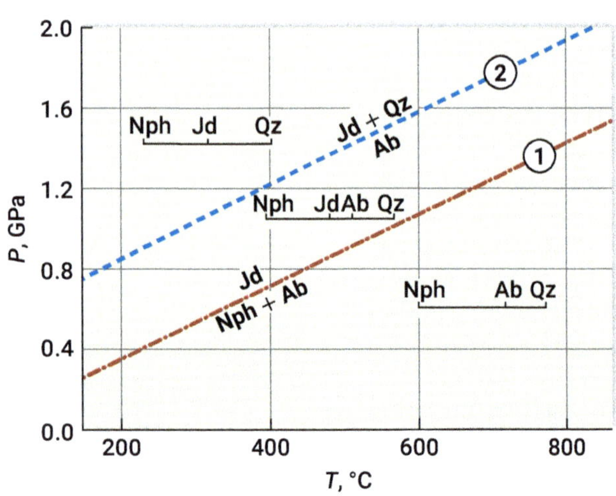

Figure 12.17 Pressure vs. temperature diagram showing the locations of two *P*-dependent reactions that divide *P-T* space into three areas. Horizontal lines show possible stable mineral assemblages.

The phase rule (see Section 3.4) says that if you are in an arbitrary place in *P-T* space, a two-component system should contain two phases. Thus, if we choose a random composition and then throw a dart at the phase diagram to choose a random *P* and *T*, we will generally land in one of the divariant fields (divariant meaning that one can tweak two intensive parameters—here, *P* and *T*—independently, and maintain the same phase assemblage; see Section 3.4). It's unlikely that we would hit one of the univariant lines (a reaction boundary). The chemographic diagrams tell us the stable phase assemblages in each field. In the lowest-*P* field, they are Nph + Ab and Ab + Qz; in the intermediate-*P* field, they are Nph + Jd, Jd + Ab, and Ab + Qz; and finally, at the highest-*P* field, Nph + Jd and Jd + Qz. Even this simple diagram helps us interpret rocks. For example, the assemblage Nph + Ab occurs in alkaline igneous rocks and is stable only in the lowest-*P* field (most igneous rocks crystallize at relatively low *P*), whereas Jd + Qz occurs in blueschists and eclogites (associated with subduction zones) and is stable only in the highest-*P* field.

There are several possible reactions among these phases:

$$\text{Nph} + \text{Ab} = \text{Jd} \qquad \textbf{(Eq. 12.9)}$$

$$\text{Ab} = \text{Jd} + \text{Qz} \qquad \textbf{(Eq. 12.10)}$$

$$\text{Nph} + \text{Qz} = \text{Ab} \qquad \textbf{(Eq. 12.11)}$$

$$\text{Nph} + \text{Qz} = \text{Jd} \qquad \textbf{(Eq. 12.12)}$$

The first two are represented by reaction lines on the *P-T* diagram in Figure 12.17. The other two are not shown, because Nph and Qz are never stable together and react under any conditions to form Ab or Jd. Recalling the principle that Al goes from fourfold (tetrahedral) coordination to sixfold (octahedral) coordination with increasing pressure (see Section 3.3), the two reactions represent progressive conversion of Al^{IV} in Nph and Ab to Al^{VI} in Jd.

It is clear that Jd is favored by high pressure, and we could define an isograd marking increasing *P* by its first appearance. However, appearance of Jd is compositionally dependent—bulk compositions between Nph and Ab will produce Jd when they cross reaction 1, but compositions between Ab and Qz will not. Instead, they produce Jd only after they cross reaction 2. Isograds are generally defined by mineral assemblages,[2] and in this system, the assemblage Jd + Qz is stable only above reaction 2. If you were working in rocks where Qz was ubiquitous, a common situation, then first appearance of Jd would be diagnostic of reaction 2. If you were working in very alkaline (uncommon) rocks, the first appearance of Jd would be diagnostic of reaction 1. The dependence of reaction histories and mineral assemblages on bulk compositions means that investigation of different rock types in an area can help define *P-T* conditions or a *P-T* history better than investigation of a single type of rock.

The System KASH

Geologic Map of a Four-Component Area

We now consider a four-component system, K_2O-Al_2O_3-SiO_2-H_2O, in an exercise adapted from Best (2003). Systems of chemical components are sometimes referred to by their initials, so this one is KASH, or perhaps KA$H. There are hundreds of minerals that consist of these components, but this is a thought experiment, so we will concern ourselves with just the subset of muscovite (Ms), K-feldspar (Kfs), quartz (Qz), corundum (Crn), and an aluminosilicate polymorph, either kyanite, andalusite, or sillimanite (Als). This is a four-component system, and the mineralogic phase rule predicts that we should usually find four phases (unless we happen to hit a reaction boundary or invariant point). One of these phases is (was) water vapor, so there should be three solid phases in a rock of arbitrary bulk composition. The following triangular projection of the four-component tetrahedron from H_2O onto the K_2O-Al_2O_3-SiO_2 plane (Figure 12.18) will serve as our chemographic diagram.

Note that the minerals Ms and Als are plotted halfway between the relevant apices. This tells us that the components are being plotted on a molar basis (i.e., Als is halfway between Al_2O_3 and SiO_2 owing to reaction stoichiometry):

$$Al_2O_3 + SiO_2 = Al_2SiO_5 \qquad \textbf{(Eq. 12.13)}$$

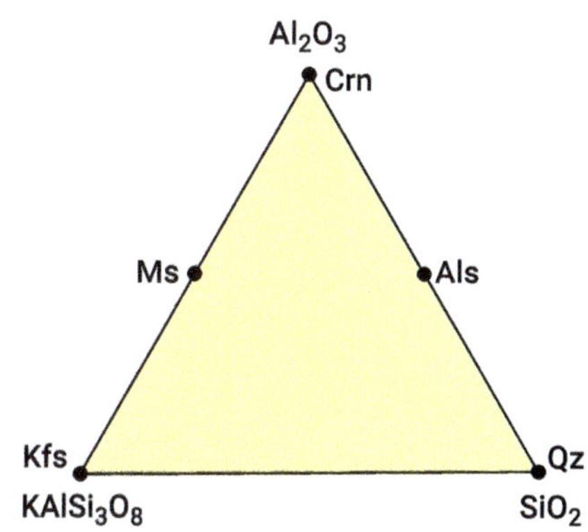

Figure 12.18 Ternary composition diagram showing plotting positions of minerals in the Al_2O_3-$KAlSi_3O_8$-SiO_2 system.

Suppose you have made the following map of the Trendy Idaho Creek area (Figure 12.19), which comprises folded metasedimentary strata that you mapped as units A, B, C, and D. The bulk compositions of these strata lie in the system KASH.

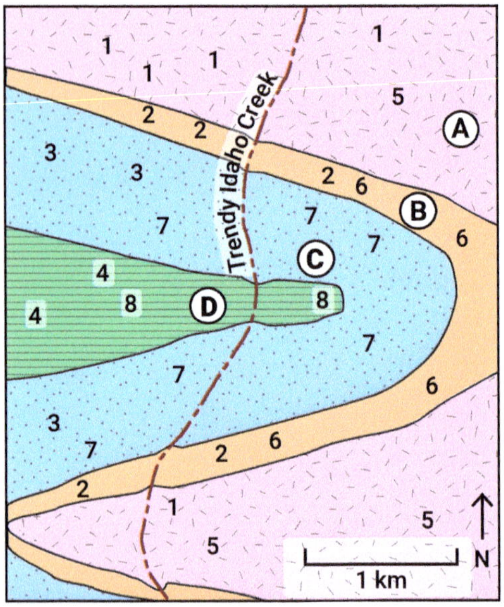

Figure 12.19 Geologic map of the Trendy Idaho Creek area, showing folded rock units (A–D) and distributions of observed mineral assemblages (1–8). (Mental exercise: Is the fold that runs east-west across the center of the map an antiform or synform?)

Thin section examination reveals the following mineral assemblages at the indicated spots on the map:

1. Ms + Crn + Als
2. Ms
3. Ms + Als + Qz
4. Kfs + Ms + Qz
5. Kfs + Crn + Als
6. Kfs + Crn
7. Kfs + Als
8. Kfs + Als + Qz

Sillimanite is the observed aluminosilicate phase.

Although the distribution of mineral assemblages might at first appear chaotic, notice that within each map unit, mineral assemblages change systematically from west to east. In some map units, these changes occur at similar geographic locations. For example, unit D, in the center of the prominent fold, and its flanking unit C both show shifts in assemblage (from Kfs + Ms + Qz to Kfs + Als + Qz and from Ms + Als + Qz to Kfs + Als, respectively) in about the same region. In fact, you can draw a boundary on the map that separates these assemblages. Although these spatially coherent changes could be fortuitous, we might hypothesize that the bulk compositions of units C and D allow them to capture a change in assemblage associations because of a reaction. If so, then units A and B must have sufficiently different bulk compositions that they do not see this particular reaction. As we follow units A and B eastward, they instead show systematic changes in assemblages (from Ms + Crn + Als to Kfs + Crn + Als, and from Ms to Kfs + Crn) in similar geographic areas, east of the changes in units C and D. Perhaps their bulk compositions allow them to see a reaction that C and D do not. If each map unit comprises strata of similar composition, the simplest way to explain these observations is via a gradient in peak metamorphic conditions, reflected in two reactions. The first (in the west) is captured by units C and D, and the second (in the east) is captured by units A and B.

Although the mineral assemblages may look formidable, they are amenable to graphical analysis. Start with the west side of the area. Clearly, Ms is stable there, as it is found in each map unit. Conversely, Ms is absent in the eastern half of the area. Assemblages 1, 3, and 4 occur in the west and require that mineral compatibility triangles there are arranged as in western area of Figure 12.20. These compatibilities show that the bulk compositions of units A, B, C, and D must plot somewhere within the indicated triangles (B plots at Ms).

Mineral assemblages change toward the east. For example, in unit C, assemblage 3 (Ms + Als + Qz) is replaced by assemblage 7 (Kfs + Als); the tie line between Ms and Qz is broken, replaced by one from Kfs to Als that flips the compatibility triangles. This is called a "tie-line flip" or a "crossing tie-line reaction" (see Section 5.2). Chemographic relations show that the reaction that changes the mineral assemblages must be

$$\text{Ms} + \text{Qz} = \text{Kfs} + \text{Als} \qquad \textbf{(Eq. 12.14)}$$

This simplified and unbalanced reaction omits water, because we are projecting from it.

From observations in the west, we know that compositions A, C, and D must plot somewhere within their compatibility triangles. Mineral assemblages from the central area help to narrow things down: unit C must plot on the Kfs-Als tie line, above the broken (dotted) Ms-Qz tie line, and D must lie within the triangle defined by the overlap region of triangles Kfs-Ms-Qz and Kfs-Als-Qz.

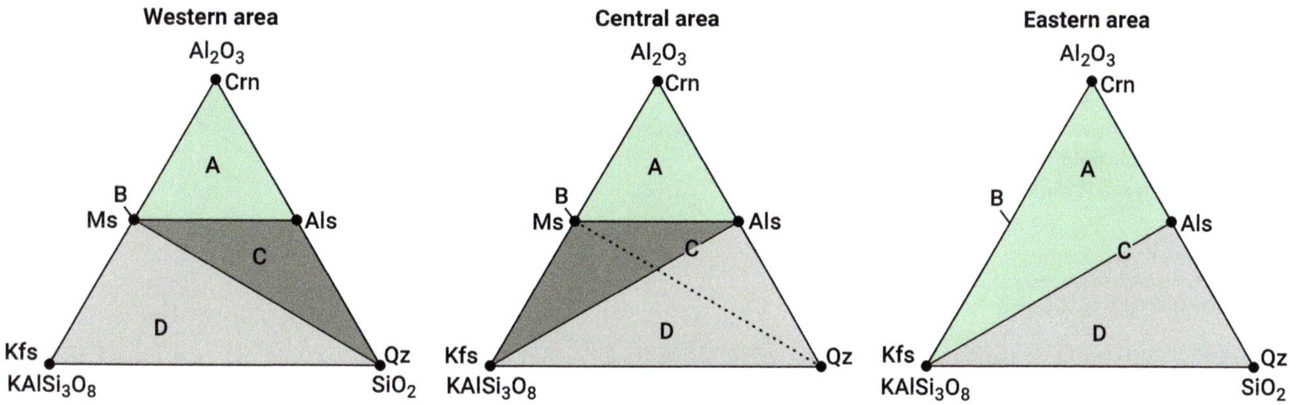

Figure 12.20 Mineral compatibilities for the western, central, and eastern sides of the Trendy Idaho Creek area. All minerals are present across the region except for muscovite, which is absent in the east.

Mineral compatibilities change again in the east, as shown by assemblages 5, 6, and 7. Kfs + Crn replaces Ms, which is no longer stable, so does not appear on the diagrams. Now, a new set of just two compatibility triangles is present (Kfs-Als-Qz and Kfs-Als-Crn).

The terminal stability of a mineral occurs at higher temperature than a reaction involving multiple minerals. For example, the breakdown of muscovite alone:

$$\text{Ms} = \text{Kfs} + \text{Crn} \quad \textbf{(Eq. 12.15)}$$

occurs at higher temperature than the breakdown of muscovite plus quartz together (Ms + Qz = Kfs + Als). This means that peak temperatures increased to the east. In principle, if we have experimental data for these reactions, we can begin to quantify peak *P-T* conditions in the field area, simply by looking at mineral assemblages.

Petrogenetic Grid

We can gain some insight into the numerical *P-T* conditions of metamorphism by comparing observed mineral assemblages to experimental data and thermodynamic calculations using a **petrogenetic grid**. Petrogenetic grids show univariant curves that mark reactions in the relevant chemical system or in subsets of it. Figure 12.21 is a simple petrogenetic grid. Of particular interest to the Trendy Idaho Creek area are the reactions that control the crossover reaction Ms + Qz = Kfs + Als and the (terminal) dehydration reaction Ms = Kfs + Crn.

Petrogenetic grid
A *P-T* diagram showing reactions that control appearance and breakdown of phases in a given chemical system.

Recalling that the Trendy Idaho mineral assemblages can be approximated in the KASH system, let's see how the KASH petrogenetic grid might be used to infer temperature and pressure. Note that the grid contains all the minerals we considered in our system, plus water vapor (V, which we projected from in our chemographic diagrams) and silicate liquid (L). Here are some of pertinent results from mapping and thin section analysis:

- Sillimanite (Sil) is the only aluminosilicate phase present.
- There is no evidence of melting in any of the samples.
- The assemblage Ms + Qz is stable in the western part of the area, but . . .
- . . . is replaced by Kfs + Sil in the central area, and . . .
- . . . Ms is replaced by Kfs + Crn in the eastern part of the area.

With these observations, and in the context of a KASH petrogenetic grid (Figure 12.21), we can narrow down the *P-T* conditions of metamorphism. Peak metamorphism occurred in the Sil wedge, on the high-*T* sides of the Ky-Sil and And-Sil boundaries. The western area contains the assemblage Ms + Qz (not Kfs + Als) and the assemblage Ms +

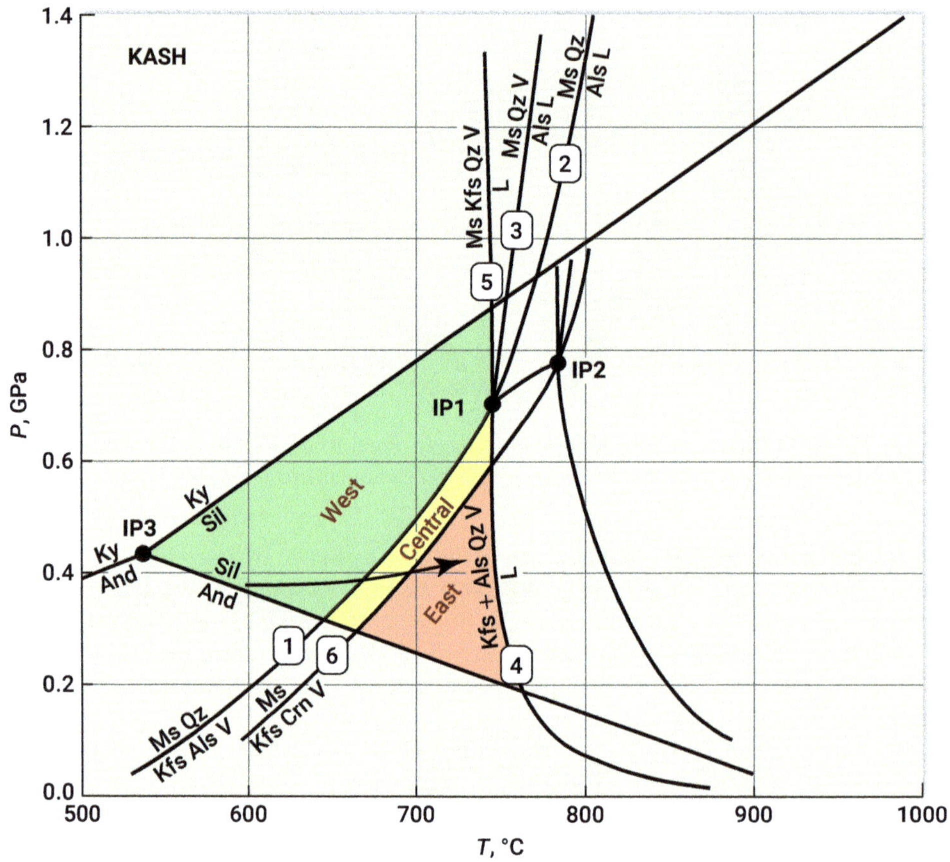

Figure 12.21 Petrogenetic grid for KASH chemical system, shaded with assemblage information from the Trendy Idaho region. Arrow shows one possible metamorphic field gradient. Black dots are invariant points.

Source: Modified from Spear (1995) to include quartz-absent reactions (cluster of reactions at higher temperature).

Kfs + Qz (not liquid), which require conditions on the low-*T* sides of reactions 1 and 5. The central area contains the assemblage Kfs + Sil as well as pure Ms (in unit B), so conditions there peaked on the high-*T* side of reaction 1 but the low-*T* side of reaction 6. The lack of partial melt means that peak *T* in the central area did not exceed that of reaction 4, and pressure was below invariant point 1 (~600 MPa). In the eastern area, the occurrence of Kfs + Crn rather than Ms and the assemblage Kfs + Sil + Qz rather than liquid means that peak *T* was below reaction 4 but above reaction 6.

Given these constraints, peak metamorphic conditions fall in the indicated regions on Figure 12.21, at relatively low *P* and high *T*. If these rocks experienced a *T*-first *P*-*T* path, as in Figure 12.12, a metamorphic field gradient might resemble the arrow in Figure 12.21. To test that hypothesis, we might look farther west for prograde andalusite and farther east for melt textures or, perhaps, a plutonic heat source. Of course, just because rocks peaked at low *P*/high *T* doesn't mean they necessarily followed a *T*-first *P*-*T* path. For example, we might find kyanite farther west, or relicts of kyanite in our rocks, suggesting a higher-*P* history. Although a good working hypothesis helps drive science by predicting behavior (maybe these rocks experienced a *T*-first *P*-*T* path driven by a heat source to the east), it's best to consider alternatives until definitive observations arise.

The System KFMASH and the AFM Diagram

We now consider a six-component system: K_2O-FeO-MgO-Al_2O_3-SiO_2-H_2O, or KFMASH. This system represents the minimum number of chemical components needed to describe the common minerals in a metamudstone and allows us to visualize associations among (Fe,Mg)-bearing minerals like garnet, biotite, chlorite, staurolite, etc. The difficulty with a six-component system is that few humans can visualize 6D composition space. Fortunately, metamudstones commonly contain quartz and evolve water, so we can project from SiO_2 and H_2O directly, reducing the number of components to four. To reduce the system to a convenient three-component system that we can graph on a ternary diagram, we project from either muscovite at $T < \sim 700$ °C or K-feldspar at higher T to remove K_2O, leaving behind Al_2O_3, FeO, and MgO to form the **AFM diagram** (Thompson, 1957; Figure 12.22).

AFM diagram
A ternary composition diagram representing the KFMASH system as projected from quartz, water, and either muscovite or K-feldspar. Not the same as an igneous AFM diagram, which is simply a composition plot.

The numerators of the F and M components are easily calculated from the number of moles of FeO and MgO in a mineral [n(FeO) and n(MgO)]. The A component is trickier, because unlike other ternary diagrams that simply plot compositions, the formal projection from muscovite and K-feldspar affects the numerator value of A:

$$\text{Numerator of A with muscovite} = n(Al_2O_3) - 3n(K_2O) \quad \textbf{(Eq. 12.16)}$$

$$\text{Numerator of A with K–feldspar} = n(Al_2O_3) - n(K_2O) \quad \textbf{(Eq. 12.17)}$$

where $n(Al_2O_3)$ is the number of moles of Al_2O_3, etc. (so you have to convert from weight percent to molar quantities). Once these are calculated, proportions are determined normally in proportion to FeO and MgO (e.g., when projecting from muscovite):

$$A = \left[n(Al_2O_3) - 3n(K_2O)\right] / \left[n(FeO) + n(MgO) + n(Al_2O_3) - 3n(K_2O)\right] \quad \textbf{(Eq. 12.18)}$$

$$F = n(FeO) / \left[n(FeO) + n(MgO) + n(Al_2O_3) - 3n(K_2O)\right] \quad \textbf{(Eq. 12.19)}$$

$$M = n(MgO) / \left[n(FeO) + n(MgO) + n(Al_2O_3) - 3n(K_2O)\right] \quad \textbf{(Eq. 12.20)}$$

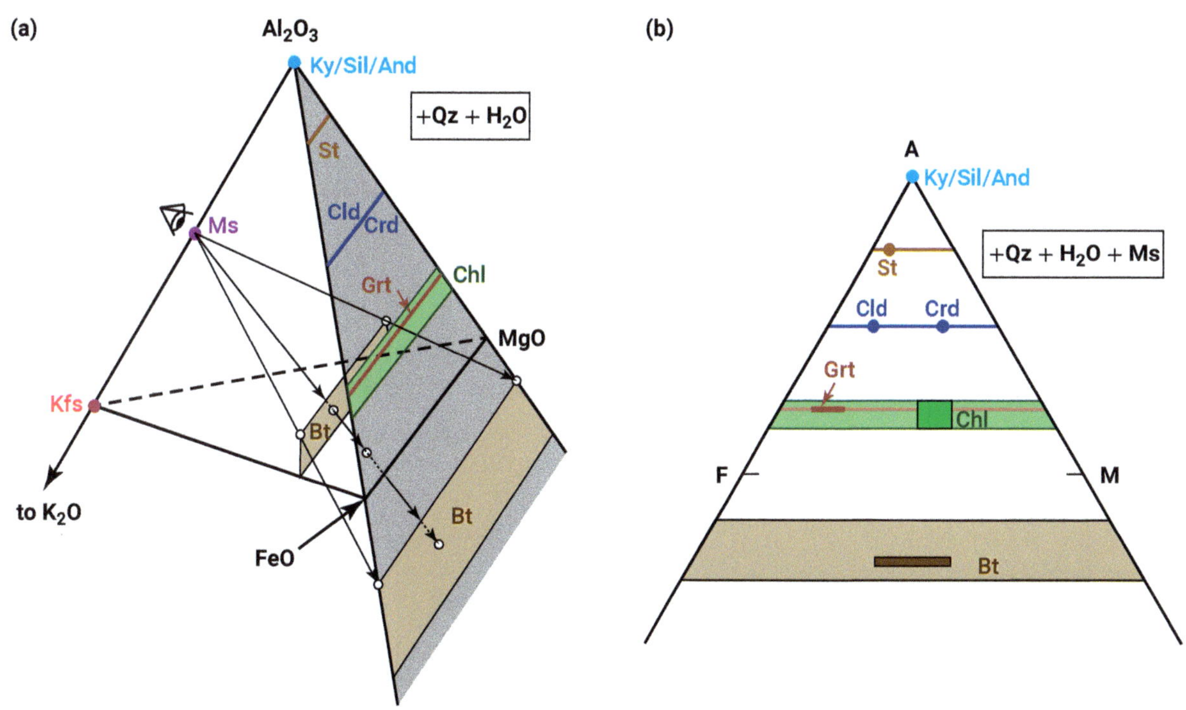

Figure 12.22 (a) Al_2O_3-K_2O-FeO-MgO tetrahedron, showing plotting positions of different minerals. Muscovite and K-feldspar plot along the Al_2O_3-K_2O edge, and biotite plots in the interior. All other common minerals (including chlorite) plot on the Al_2O_3-FeO-MgO plane. (b) AFM projection from muscovite. More typical plotting positions of AFM phases are shown with darker shading and dots.

Most minerals in metamudstones have no K_2O, so the normalization simplifies to just FeO, MgO, and Al_2O_3. For example, a chlorite with 4.5 moles of FeO, 4.5 moles of MgO, and 3 moles of Al_2O_3 would plot at A = 0.25, F = 0.375, M = 0.375, whereas a garnet with 2.7 moles of FeO, 0.3 moles of MgO, and 1 mole of Al_2O_3 would plot at A = 0.25, F = 0.675, M = 0.075. Biotite is the only common mineral besides muscovite and K-feldspar that contains Al_2O_3 and K_2O, and its plotting position in a muscovite projection seems quite strange: it appears to have "negative Al." This doesn't mean we're creating antimatter (which would be lethal), it's simply an artifact of projections.[3] Any initial discomfort with biotite's negative A-values fades with practice.

Most presentations of AFM diagrams illustrate minerals with wide ranges of possible Fe-Mg contents. Although this is accurate across all possible rocks, mineral compositions are much more restricted in most metamudstones. In order of decreasing Fe/(Fe + Mg), they partition as

$$\text{Grt}(\text{typically } 0.8-0.9) > \text{St} \sim \text{Cld}(\text{slightly less than Grt}) > \text{Bt} \sim \text{Chl}(\text{typically } 0.5-0.6) > \text{Crd}$$

What's the value in an AFM diagram? Alone, not much, just as a single KASH diagram is of limited value. But remember how the sequence of mineral associations and reactions in the KASH system allowed us to say something about the *P-T* conditions of rock formation in the Trendy Idaho problem? We used that example to develop the concept, but that kind of variation in bulk rock compositions is rare. In contrast, metamudstones are as common as, well, mud. Consequently, their sequence of AFM diagrams, mineral associations, and reactions—the KFMASH petrogenetic grid—can be used to interpret *P-T* conditions of rock formation in numerous orogens (e.g., Spear & Cheney, 1989). In fact, each isograd represents the first appearance of a characteristic mineral or (more correctly) mineral association.

This means that if you know the stability of mineral associations, you can infer likely *P-T* conditions in the field, simply from observations of mineral assemblages. For example, in typical metamudstones, the staurolite isograd represents the first stable association of staurolite plus biotite.[4] According to the KFMASH petrogenetic grid, this assemblage (isograd) does not become stable until about 575 °C. So, commonly, when you see staurolite in the field with biotite, it means the rocks were at least 575 °C. The sequence of mineral isograds at lower pressure also differs from intermediate pressures, so the sequence of mineral isograds can also indicate *P* (e.g., Carmichael, 1978; Pattison & Vogl, 2005). If you know your AFM diagrams, you can tell roughly where you are in *P-T* space. More refined *P-T* estimates usually require much more (expensive) processing of samples and instrumental measurements.

Thermobarometry

Chemical Thermobarometry: Calculating *P* and *T* from Mineral Chemistry

An important petrologic application of thermodynamics lies in estimating the maximum *P-T* conditions that a rock reached and the *P-T* path rocks followed both before and after reaching that condition. Several reactions serve as important thermobarometers.

Example: The Garnet-Biotite Geothermometer and GASP Geobarometer

A reaction that has a small change in volume (ΔV) but a large change in enthalpy (ΔH) has a steep slope on a *P-T* plot and makes a good geothermometer. A reaction with a small ΔH and a large ΔV makes a good geobarometer.

Fe^{2+} and Mg^{2+} have similar sizes and exchange readily with one another, and their exchange forms the basis of many geothermometers. In this exercise, we will use what is perhaps the most common of these, the garnet-biotite geothermometer (Ferry & Spear, 1978).

Differential partitioning of Fe^{2+} and Mg^{2+} between garnet and biotite reflects disparity between the cubic site in garnet, which prefers the larger cation (Fe^{2+}), versus the octahedral site in biotite, which prefers the smaller cation (Mg^{2+}). Many geobarometers are based on changes in aluminum coordination from Al^{IV} (tetrahedral) to Al^{VI} (octahedral), as discussed in Section 3.3. We will use the breakdown of anorthite, in which all Al is in tetrahedral coordination, to quartz plus grossular and kyanite, both of which have Al in octahedral coordination, for our geobarometer. The examples here are taken from Spear (1995), who developed these thermobarometers in the context of thermodynamics.

The garnet-biotite geothermometer reaction is

$$\underset{\text{almandine}}{Fe_3Al_2Si_3O_{12}} + \underset{\text{phlogopite}}{KMg_3AlSi_3O_{10}(OH)_2} = \underset{\text{pyrope}}{Mg_3Al_2Si_3O_{12}} + \underset{\text{annite}}{KFe_3AlSi_3O_{10}(OH)_2} \quad \textbf{(Eq. 12.21)}$$

Here, almandine and pyrope refer to the components in a solid solution of garnet that contains both Fe and Mg in its cubic site (as well as other elements), whereas phlogopite and annite refer to the Mg and Fe components in a solid solution of biotite. Garnet plus biotite commonly coexist in metamorphic rocks from upper greenschist to lower granulite facies conditions, so this reaction is widely applicable.

Using thermodynamic data, we can write T as a function of P and the equilibrium constant K:

$$T = \frac{52100 + 2.39P}{19.5 - R\ln K} - 273 \quad \textbf{(Eq. 12.22)}$$

where T is in °C and P is in MPa. The equilibrium constant for this reaction is

$$K = \frac{\left(X_{Prp}^{Grt}\right)^3 \left(X_{Ann}^{Bt}\right)^3}{\left(X_{Phl}^{Bt}\right)^3 \left(X_{Alm}^{Grt}\right)^3} \quad \textbf{(Eq. 12.23)}$$

Here, the X quantities are mole fractions of individual garnet and biotite components; for example, X_{prp}^{Grt} is the mole fraction of Mg (pyrope component) in the cubic site of garnet, etc.:

$$X_{Prp}^{Grt} = \frac{Mg}{Ca + Fe + Mg + Mn} \quad \textbf{(Eq. 12.24)}$$

where Ca, Fe, Mg, and Mn are the main cations that occupy the cubic site of garnet. When these ratios are written out and denominators are canceled, the expression for K simplifies to

$$K = \left(\frac{(Mg/Fe)^{Grt}}{(Mg/Fe)^{Bt}}\right)^3 \quad \textbf{(Eq. 12.25)}$$

The cube exponents come from the stoichiometry of garnet and biotite (3 Fe or Mg atoms per formula unit for each end member component); a fuller explanation requires study of solid-state thermodynamics,[5]—for example, in the book *Elementary Thermodynamics for Geologists* by Wood and Fraser (1978). But, briefly, the equilibrium constant represents a form of entropy (the configurational entropy), and its absolute magnitude depends on the randomness of mixing among cations. The more sites there are in a mineral over which mixing occurs, the greater the randomness and higher the entropy. The fact that cations mix randomly over three sites per formula unit in both garnet and biotite leads to the cubic exponents. If cations mixed randomly over a different number of sites (e.g., over 1 alkali site in feldspar, 2 octahedral sites in olivine, etc.), the exponent would be different (1 for feldspar and 2 for olivine). Major and minor element compositions needed to determine these mole fractions are measurable with an electron probe.

Figure 12.23a shows Equation 12.22 plotted on a *P-T* diagram, with lines giving the variation of *P* with *T* for a given value of *K*. Note that for a given value of *K*, slopes are steep, consistent with the small ΔV of the reaction. This reaction therefore makes a good geothermometer—we do not need to know *P* very well to estimate *T*.

The GASP (garnet-aluminosilicate-silica-plagioclase) geobarometer is based on the reaction

$$\underset{\text{anorthite}}{3CaAl_2Si_2O_8} = \underset{\text{grossular}}{Ca_3Al_2Si_3O_{12}} + \underset{\text{kyanite}}{2Al_2SiO_5} + \underset{\text{quartz}}{SiO_2} \quad \textbf{(Eq. 12.26)}$$

This reaction has a relatively large negative ΔV owing to the conversion of Al^{IV} in plagioclase to Al^{VI} in garnet and kyanite. Using thermodynamic data, equilibrium for this reaction is given by the equation

$$T = \frac{48360 + 66.1P}{151 + R\ln K} - 273 \quad \textbf{(Eq. 12.27)}$$

where *T* is in °C and *P* is in MPa. This reaction, too, has been studied experimentally (e.g., Koziol & Newton, 1988). *K* is defined as

$$K = \frac{\left(X_{grs}^{Grt}\right)^3}{\left(X_{an}^{Pl}\right)^3} \quad \textbf{(Eq. 12.28)}$$

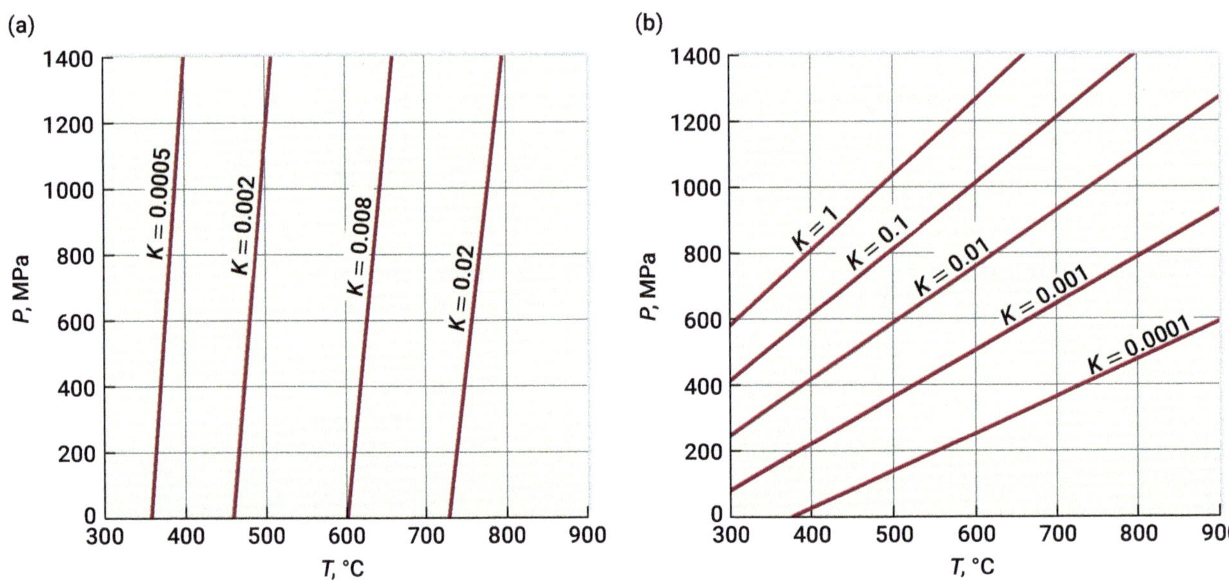

Figure 12.23 (a) Isopleths (lines of constant value) of the equilibrium constant for the garnet-biotite thermometer, showing steep slopes; *T* is thus relatively insensitive to *P*. (b) Isopleths of the equilibrium constant for the GASP barometer showing shallower slopes.

Source: Modified from Spear (1995).

As discussed for the garnet-biotite thermometer, the cubic exponent for the grossular component comes from the site multiplicity in garnet (3 cubic sites per formula unit). The cubic exponent on the plagioclase term comes from the stoichiometry of the reaction (the reaction involves 3 moles of anorthite).

As before:

$$X_{Grs}^{Grt} = \frac{Ca}{Ca + Fe + Mg + Mn} \qquad \textbf{(Eq. 12.29)}$$

and

$$X_{An}^{Pl} = \frac{Ca}{Ca + Na + K} \qquad \textbf{(Eq. 12.30)}$$

Figure 12.23b shows a P-T plot contoured for K for this reaction. Note that the lines of constant K are significantly less steep (T is more P-dependent) than those for the garnet-biotite reaction: GASP allows us to estimate P fairly well (to within 100 MPa) even if we do not know T very well. In Toolbox 12.2 we show how to use mineral compositions to calculate P-T conditions using the geothermometer and geobarometer just discussed.

Toolbox 12.2

Estimating P and T for a Schist

Application of the garnet-biotite geothermometer and GASP geobarometer is computationally straightforward, and we will apply them to data from a garnet-kyanite schist of the Crabtree terrane in central North Carolina. This Al-rich metamudstone is Neoproterozoic to early Paleozoic in age and contains abundant muscovite.

The first step is to evaluate the texture of the rock carefully (Figure 12.24), looking to see that the minerals appear to be in textural equilibrium and that there are not obvious problems such as multiple compositional populations of a given mineral. We also need to check for chemical zoning and infer (using petrologic criteria) which compositions best approximate equilibrium. These are essential, although complicated and interpretive, aspects of thermobarometry. In our example, we will assume that the garnet rim, matrix biotite, and matrix plagioclase were in equilibrium. Compositions of these minerals, determined with an electron probe, are in Table 12.2, and a worksheet for you is in Table 12.3.

Equations 12.22 and 12.27 relate T and P to the value of K, which in turn is a function of the mineral compositions for the relevant reaction—specifically, cation ratios. These are easily calculated from electron probe analyses.

The T(Grt-Bt) and T(Grt-Pl) lines give temperatures corresponding to two pressures for the two thermobarometric equations. Note that the Grt-Bt temperature varies little over the range of pressure (signifying a good geothermometer),

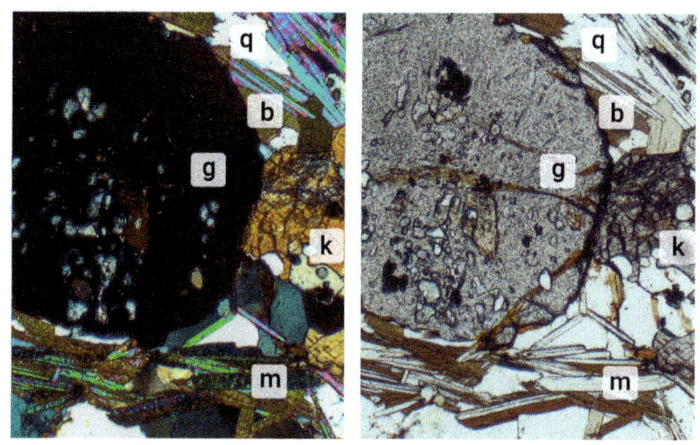

Figure 12.24 Photomicrographs of the Crabtree terrane schist in crossed (left) and plane-polarized (right) light. Garnet (g) and kyanite (k) porphyroblasts sit in a matrix of muscovite (m), biotite (b), and quartz (q). Garnet has abundant inclusions of quartz and other minerals. The right and upper sides of the garnet crystal have what appear to be planar faces, whereas the lower side, where inclusions are concentrated, is more rounded and scalloped. These observations indicate that the upper/right part of the crystal is more likely to be in equilibrium with other minerals than the lower part. Analyses of the garnet rim and adjacent biotite crystals should provide a good estimate of peak T. Width of each field 4 mm.

Source: AFG photos.

continued

Table 12.2 Electron Probe Analyses of Minerals in Kyanite Schist from the Crabtree Terrane, North Carolina

Oxide	Biotite wt%	Garnet wt%	Plagioclase wt%
Al_2O_3	19.87	21.60	25.28
FeO	18.70	33.84	0.05
MnO	0.04	0.58	0.00
MgO	9.64	3.70	0.00
CaO	0.01	3.60	7.08
Na_2O	0.18	0.02	7.64
K_2O	8.67	0.01	0.09

Source: Electron probe data collected by J. Alex Speer, in Stoddard & Blake (1994).

Table 12.3 Worksheet for Calculating P and T (partially completed—fill in gray cells)

Oxide	Biotite moles/100 g	Garnet rim moles/100 g	Plagioclase moles/100 g	Oxide Weight
Al_2O_3	0.195			101.96
FeO	0.260			71.85
MnO	0.001			70.94
MgO	0.239			40.30
CaO	0.000			56.08
Na_2O	0.003			61.98
K_2O	0.092			94.20
Mg/Fe	0.919			
K(Grt-Bt)				[Use Eq. 12.25]
Ca/(Ca+Fe+Mn+Mg) in Grt		0.101		[Use Eq. 12.29]
Ca/(Ca+Na+K) in Pl				[Use Eq. 12.30]
K(Grt-Pl)				[Use Eq. 12.28]
	***T* Calculations**			
	T(Grt-Bt), °C			[Use Eq. 12.22]
	T(Grt-Pl), °C			[Use Eq. 12.27]
	ΔT			
	P, MPa	**500**	← Vary this to drive ΔT to zero	

whereas the Grt-Plag temperature varies over a much greater range. As both equations are linear (e.g., *T* is a linear function of *P*), they describe lines on a *P-T* plot. Figure 12.25 is the line for the Grt-Bt thermometer.

The corresponding line for the GASP barometer intersects this Grt-Bt thermometer line somewhere; your job is to find that intersection point. If the mineral compositions perfectly represent equilibrium, the intersection point represents one *P-T* condition on the rock's *P-T* path, likely near the maximum *T*. Note that the goal seek function in a spreadsheet can be used to determine this automatically by finding a *P* at which the difference in temperatures is zero.

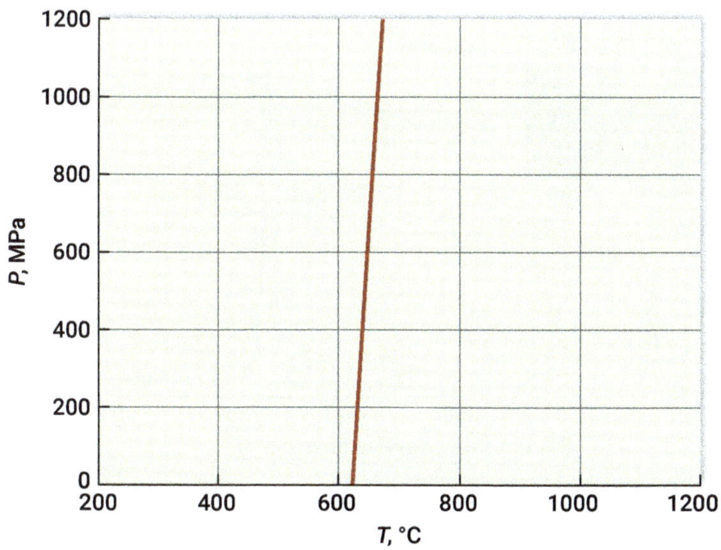

Figure 12.25 Line for the garnet-biotite geothermometer, from data and calculations in Tables 12.2 and 12.3 and Equation 12.22. In order to estimate a singular T and P for this sample, the corresponding line for the GASP geobarometer (Equation 12.27) needs to be plotted.

Raman Spectroscopy

The thermobarometric methods in the previous section rely on chemical equilibrium among minerals, as quantified using thermodynamics. In recent years, wholly different approaches have been developed using Raman spectroscopy. Briefly, Raman spectroscopy measures the frequency of molecular vibrations using a laser with a beam diameter on the order of 1 to 2 μm. The shapes and positions of Raman peaks depend on how ordered a crystal is (more ordered crystals yield sharper Raman peaks), bond length (shorter, stronger bonds resonate at higher frequencies), mineral composition (e.g., jadeite vs. diopside), mineral structure (e.g., kyanite vs. sillimanite), pressure, temperature, and atomic ordering (e.g., the distribution of Al and Si atoms in feldspar). In Raman spectroscopy of carbonaceous material (Beyssac et al., 2002), peak shapes and areas of graphite change systematically with T, but not with P, as carbon atoms order themselves more regularly at higher temperature (Figure 12.26a). This approach is now widely applied to low- to moderate-grade metamorphic rocks, where thermodynamically based thermometers are generally lacking. Raman spectroscopy can also be used for geobarometry. Commonly, a quartz inclusion that was trapped in a garnet at elevated P will attempt to expand more than the garnet as the rock is exhumed to Earth's surface. Garnet resists this expansion, and stress builds up between the inclusion and the host garnet. In garnet, which is normally optically isotropic, this stress flexes the mineral into anisotropy that can be seen under cross-polarized light (Figure 12.26b). Raman peak positions shift under stress (Figure 12.26b), and knowing the physical properties of minerals, we can use the magnitude of the present-day stress to back-calculate the possible P-T conditions under which the inclusion was trapped (Rosenfeld & Chase, 1961). This gives rise to a line on a P-T diagram, which can be steeply sloped for some inclusion-host pairs (a good thermometer) and shallowly sloped for others (a good barometer).

Many thermobarometric comparisons using chemical (thermodynamics) and Raman measurements yield similar results (e.g., Endo et al., 2012), but some spectacular discrepancies between Raman-based barometry and thermodynamics have also been discovered (e.g., Castro & Spear, 2017). Explanations for these discrepancies are varied and represent an area of active research.

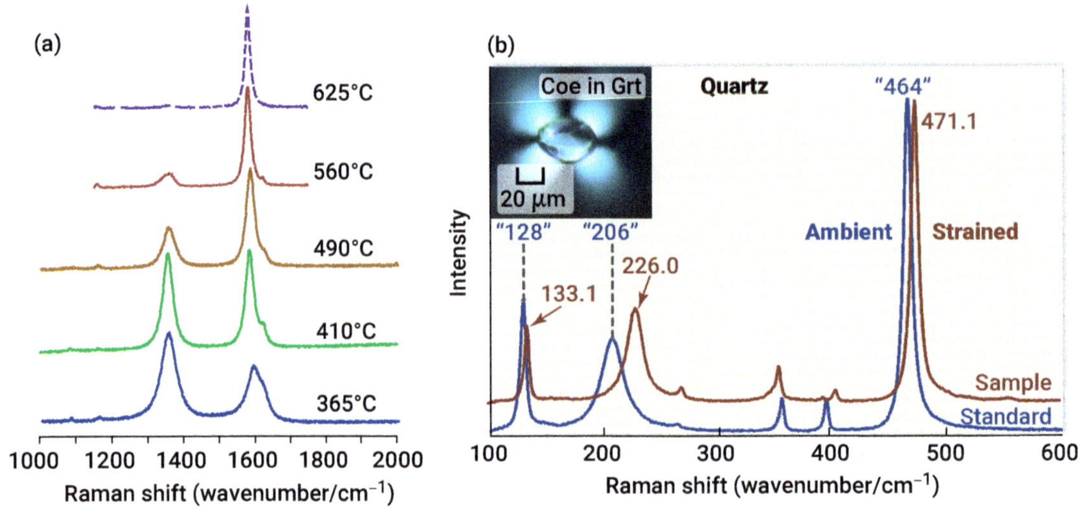

Figure 12.26 (a) Raman-based thermometry of carbonaceous rocks, showing how Raman peaks sharpen and change relative heights (and areas) with increasing temperature. Wavenumber is reciprocal wavelength, and therefore a measure of vibrational frequency. (b) Inclusions that are under stress exhibit strain-induced shifts to Raman peak positions; in this case, an increase in wavenumber indicates compressive stress. The magnitude of this shift can be converted to either a thermometric or barometric line. (Inset) Photomicrograph of a coesite inclusion in garnet under cross-polarized light. Normally, garnet is isotropic (black) in crossed polars, but stress around the inclusion warps the garnet structure so it is no longer cubic, inducing light fringes (birefringence).

Sources: (a) Modified from Beyssac et al. (2016) and Simoes et al. (2007); (b) modified from Kohn et al. (2023); (inset) from Campomenosi et al. (2018).

12.5 Mineral Assemblage Diagrams

Learning Objectives

By the end of this section you should be able to:

12.5.1 Distinguish between the value of a mineral assemblage diagram vs. a petrogenetic grid for inferring *P-T* conditions and *P-T* paths.

12.5.2 Explain why retrograde reactions are disfavored, but how they can be used to estimate *P-T* paths.

What Can We Do with Mineral Assemblage Diagrams?

If you think back to our mapping project at Trendy Idaho Creek, you may remember that each rock recorded only a subset of possible reactions in the KASH system. So, although the petrogenetic grid allowed us to see all possible reactions and use them to estimate *P-T* conditions, it did not quite represent what any particular rock might show.

Mineral assemblage diagrams
Diagrams, usually *P-T* diagrams, that show thermodynamically predicted mineral assemblages for a specific rock composition. Also called a pseudosection or bulk-composition-specific phase diagram.

Mineral assemblage diagrams (**MADs**) can help. MADs, also commonly referred to as "pseudosections," are predictions of mineral assemblages and mineral compositions determined using principles of equilibrium thermodynamics. For a specified rock composition, the equilibrium state corresponding to minimum Gibbs free energy at some *P-T* (and fluid) condition corresponds to a particular assemblage of minerals with particular compositions. The MAD maps out these mineral assemblages and compositions across *P-T*-composition space. Exactly what a MAD shows depends critically on the thermodynamic underpinnings, and different thermodynamic databases and mineral solution models can yield different MADs for the same rock. Sorting out these inconsistencies can be . . . maddening.

One reason to construct a MAD is to better illustrate the series of mineral assemblages that a specific sample would be expected to develop. A second is because most rocks are chemically complicated. Petrogenetic grids may be relatively straightforward to interpret when the number of chemical components is small, say five or less. But to describe the chemical complexity of minerals in many rocks can require many more components. For example, if you want to study a rock that contains an amphibole, you're signing up for at least seven chemical components (Na_2O, CaO, FeO, MgO, Al_2O_3, SiO_2, H_2O) and, depending on what else is in the rock, maybe a couple more (Fe_2O_3 and MnO). In these cases, petrogenetic grids become extremely complicated, and extracting relevant information about assemblages from the confusion of reactions challenges even the most flexible of minds. Although MADs are restricted to a particular rock composition, they clear out the clutter.

Yet another reason we construct MADs is because, as soon as we define a bulk composition, we can simultaneously predict not only what minerals are present (the assemblage), but also mineral modes (abundances) and compositions. These predictions are valuable because minerals commonly have chemical zoning that reflects earlier *P-T* conditions, and textures can tell us whether a mineral is growing or dissolving. These chemical and textural characteristics help define *P-T* paths, which are valuable for interpreting tectonic evolution.

Example of the Utility of a MAD

One example, among hundreds of published MADs, models eclogite assemblages from the Dora Maira region of the Alps (Groppo et al., 2007; Figure 12.27). This rock contains a peak assemblage of coesite, muscovite with high Si-content (3.55 atoms per formula unit, compared to the ideal 3), Na-rich clinopyroxene (>50% jadeite component, called omphacite), and Ca-rich garnet (>30% grossular). These minerals have reacted to different compositions during exhumation, especially lower-Si muscovite and even amphibolite-facies assemblages. Modeling this rock shows that the peak assemblage (Grt + Ms + Coe + Cpx) is stable only at the highest *P-T* conditions. Modeling of muscovite, clinopyroxene, and garnet compositions plus thermobarometry indicates peak *P-T* at 3.7–3.8 GPa and 725–750 °C. Modeling retrograde compositions indicates cooling during exhumation. These kinds of models and thermobarometric data now form the basis for most published *P-T* paths and corresponding tectonic models.

Retrograde Overprints

In general, any change in *P-T* conditions can change mineral compositions and assemblages. Along their prograde *P-T* paths, most rocks experience dehydration reactions. For example, breakdown of chlorite plus quartz to form garnet is one of the largest generators of water during subduction. Because the porosity of rocks is extremely low, this water escapes, commonly along fractures and shear zones, typically leaving little water in a rock for these reactions to operate in reverse. So, garnet will not normally back-react to chlorite plus quartz unless some source of water is available. Of course, deeper rocks might still be dehydrating, and that water could infiltrate a rock to produce hydrous minerals. But outside of shear zones, retrograde overprinting is commonly minor—many metamorphic rocks preserve assemblages and compositions representative of their maximum temperatures for hundreds of Myr.

Nonetheless, retrograde overprints can provide key information about exhumation histories, especially for UHP and UHT assemblages that are particularly unstable during exhumation and cooling. Radial fractures around quartz inclusions indicate they were once coesite and experienced UHP conditions. In high-*P* rocks, formation of intergrowths and coronas of low-*P* minerals like orthopyroxene, cordierite, or spinel (Figure 12.28a) indicates that they evolved to low-*P* conditions while still hot, which requires rapid exhumation, likely via tectonic extension. Retrograde reactions in UHT rocks are often used to model their *P-T* paths during cooling.

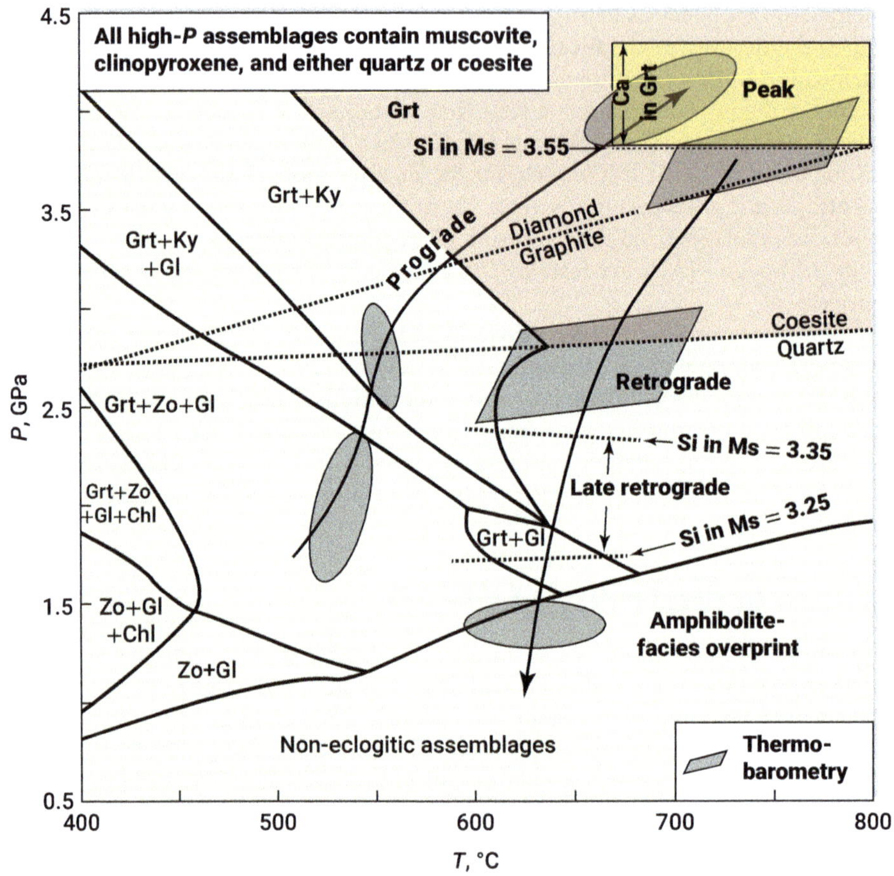

Figure 12.27 Example of mineral assemblage diagram for rocks from the Dora Maira UHP terrane. Observed mineral assemblage (garnet + clinopyroxene + muscovite + coesite) is restricted to the upper right corner of the diagram (region labeled simply "Grt"). Thermobarometry (gray polygons) and compositions (lines and shading) help to estimate peak *P-T* conditions and *P-T* path. Minerals are chemically zoned, and different compositions are associated with different textural associations. Separate *P-T* conditions can be calculated by linking together different associations and compositions that are thought to have equilibrated together. Connecting these independently-calculated *P-T* conditions delineates the overall *P-T* path. The ellipses show range of *P-T* conditions consistent with garnet compositions (lowest *P*, *T* = garnet core; intermediate *P*, *T* = garnet mantle; highest *P*, *T* = garnet rim) as modeled in a different rock from the same region.

Sources: Modified from Groppo et al. (2007, 2019).

Even if mineral assemblages do not change, mineral compositions can re-equilibrate during cooling. The garnet-biotite thermometer tells us that as temperature increases, garnet and biotite compositions should become more similar. But the reverse is also true: as temperature decreases, garnet and biotite compositions should attempt to separate (garnet becomes more Fe-rich, while biotite becomes more Mg-rich). Slow diffusion in garnet often limits this process, so garnet commonly develops diffusion profiles on crystal rims that are enriched in Fe and depleted in Mg (Figure 12.28b,c). In favorable cases, the shapes of these diffusion profiles can be used to determine cooling histories in rocks, because faster cooling means less diffusion and thus narrower composition profiles.

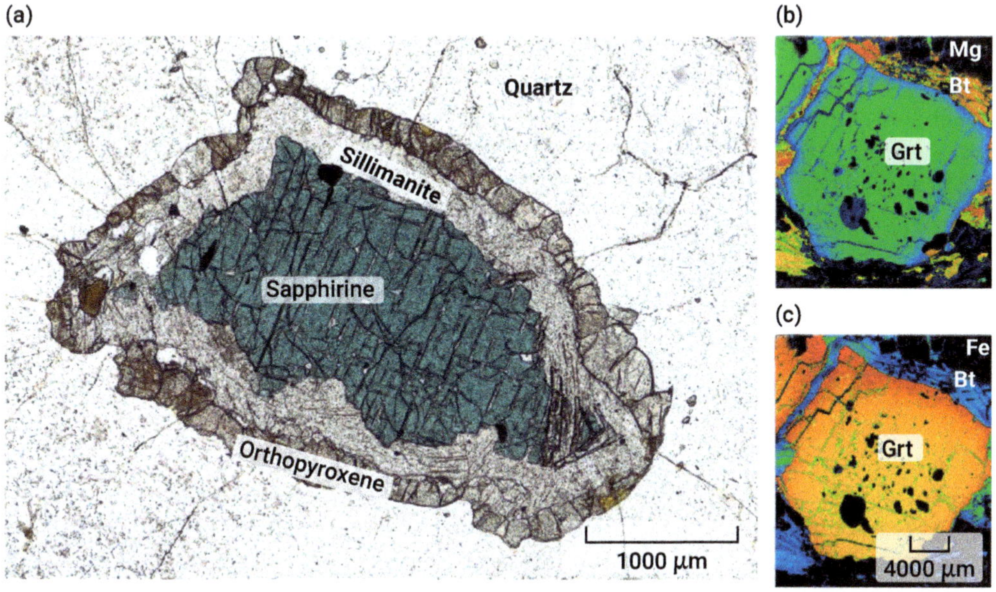

Figure 12.28 Examples of retrograde reactions. (a) Coronas of sillimanite and orthopyroxene between sapphirine [$(Mg,Al)_8(Si,Al)_6O_{20}$] and quartz, showing destabilization of a characteristic UHT assemblage (sapphirine + quartz) during cooling. Monomineralic bands are an example of local equilibrium (see Figure 11.30). From the Napier Complex, Antarctica. (b, c) Composition maps of a garnet crystal surrounded by biotite. "Hotter" colors (orange to red) indicate higher concentrations; "cooler" colors (green to blue) indicate lower concentrations. Rim of garnet is depleted in Mg, and enriched in Fe, especially adjacent to biotite. These chemical changes reflect reactions occurring as temperature decreased. These maps were made on an electron probe microanalyzer.

Sources: (a) Courtesy of Martin Hand; (b, c) from Zou et al. (2021), Spring Nature BV.

12.6 Petrochronology and Fabrics: Putting Time on the *P-T* Path

Learning Objectives

By the end of this section you should be able to:

12.6.1 Formulate a hypothesis for how the temperature-time history of a rock might be collected and used to limit possible tectonic histories.

12.6.2 Predict the fabrics that may be preserved in a porphyroblast depending on the timing of its growth relative to deformation.

What Is Petrochronology?

Petrochronology combines petrology with geochronology by using petrologic features to interpret ages (Engi et al., 2017). Although the basic purpose of geochronology may be to determine ages, applications of geochronology, especially in metamorphic petrology, necessarily require placing that date in the context of a geologic process. As we saw earlier (see Figure 12.14), a metamorphic *P-T* path can easily span tens of millions of years, whereas in some rocks, we can measure the age of a crystal, or domain within a crystal, to within 100 kyr or less, a small percentage of the total *P-T* path. In that context, how might

Petrochronology
Geochronology in the context of petrology. Links time (rates and durations) with rock-forming processes and *P-T* conditions.

we interpret an age that is, say, 30 Ma? If the age represents the early part of the *P-T* path, peak metamorphism could be much younger, and the metamorphic and tectonic cycle proceeded relatively rapidly. Conversely, if it represents late-stage retrogression, burial and heating could be much older, and the metamorphic cycle proceeded relatively slowly.

These considerations emphasize that although *P-T* paths alone do help define tectonic processes (is the shape *P*-first or *T*-first?), adding time defines rates of heating, burial, cooling, and exhumation. The rates of burial and heating can tell us plate convergence rates in subduction systems, or thickening rates in orogens, whereas the rate of exhumation and cooling can tell us whether an orogen exhumed via erosion (relatively slowly) or via extensional faulting (relatively rapidly).

Dating Events on a *P-T* Path

To put the "petro" in petrochronology, most petrologists either link an age to diagnostic inclusions of minerals inside the dated mineral, or they use the chemistry of the dated mineral. That is, we no longer measure only an age (just "date things"); we also simultaneously determine the mineral assemblage or measure the chemistry of the material we dated (or both). For example, we might measure the major element composition of a garnet that we date using Sm-Nd geochronology, or the Zr content of titanite that we date using U-Pb geochronology, or the REE composition of monazite that we date using Th-Pb geochronology. We then use those compositional data (e.g., major elements in garnet, Zr in titanite, REE composition of monazite) to interpret the *P-T* conditions of formation of the material we dated.

The mineral inclusion method focuses on diagnostic minerals or their compositions (or both) trapped within the dated mineral. For example, titanite crystals from UHP rocks of the western Alps contain inclusions of omphacite (Na-rich clinopyroxene) of varying composition (Rubatto & Hermann, 2001; Figure 12.29). The compositions of omphacite inclusions link specific domains of titanite to pressure, and the ages of those domains then date those portions of the *P-T* path. These results indicate unusually rapid exhumation of UHP rocks, even faster than the rate of Alpine subduction (see Figure 12.29). Other examples link ages of zircons to inclusions of UHP minerals like coesite or diamond (e.g., Katayama et al., 2001; Kaneko et al., 2003; Liu & Liou, 2011).

Otherwise, petrologists employ two main methods to correlate mineral compositions with *P-T* paths. Trace-element thermometry represents the most direct method. Rutile, titanite, and zircon all contain sufficient U to permit U-Pb dating, and concentrations of certain trace elements in them are *T*-sensitive and have been calibrated as thermometers. For example, one of the most widely employed trace-element thermometers, "Zr-in-rutile," is represented by the reaction

$$\begin{aligned} ZrSiO_4 &= ZrO_2 + SiO_2 \\ \text{Zircon} &= \text{Zr-rutile} + \alpha\text{-Quartz} \end{aligned} \qquad \textbf{(Eq. 12.31)}$$

Here, zircon and α-quartz represent (nearly pure) minerals commonly present in rocks, whereas Zr-rutile represents zirconium oxide with the rutile structure (i.e., a zirconium oxide chemical component in rutile, not the mineral baddeleyite). The Zr content of rutile represents the amount of this component dissolved in a crystal and is highly sensitive to temperature. The Ti content of zircon and the Zr content of titanite have also been calibrated as thermometers (Watson et al., 2006; Hayden et al., 2008). For each of these minerals, we can simultaneously measure U-Pb age and trace-element concentration (Zr in rutile or titanite, Ti in zircon) and link ages directly with temperature and the *P-T* path.

An example of this approach for the Zr-in-titanite thermometer shows the temperature-time history for the uppermost high-grade metamorphic rocks of the central Himalaya (Figure 12.30; Kohn & Corrie, 2011). These rocks represent some of the deepest levels of the Himalayan orogen and reveal that rocks achieved high temperatures in the partial melting field for surrounding rocks as early as about 40 Ma and remained there

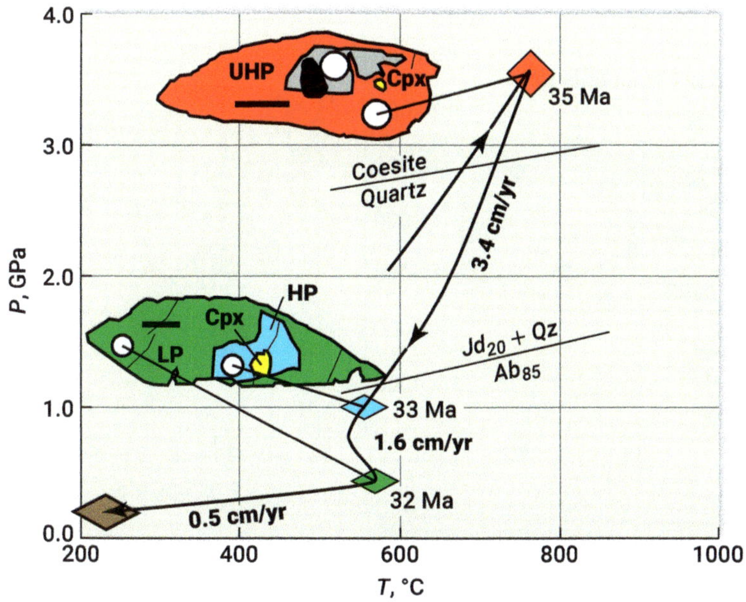

Figure 12.29 Titanite data and *P-T*-time history from Dora Maira massif calc-silicates, western Italian Alps. Sketches show titanite textures evident in backscattered electron images and locations of some analytical spots (dots). Several ages were averaged to derive the pooled age listed for each stage on the *P-T* path. Titanite contains inherited cores and omphacite (Na-rich Cpx) inclusions with compositions distinctive of ultrahigh-pressure (UHP) and later high-pressure (HP) conditions. LP = low pressure; Jd_{20} = 20 mol% jadeite component in Cpx, Ab_{85} = 85 mol% albite component in plagioclase. Scale bars are 50 μm.

Sources: Modified from Kohn (2017), additional data from Rubatto & Hermann (2001).

for about 20 Myr. Partial melting greatly reduces the strength of rocks (e.g., Rosenberg & Handy, 2005), so this part of the crust should have been quite weak. We do not fully understand how these rocks could have managed to stay so hot for so long in an active orogen. Other geochronologic measurements indicate the rocks then rapidly cooled, which is ascribed to a combination of thrusting over cooler rocks and extensional exhumation.

A different, less direct petrochronologic approach links the composition of the dated mineral to the growth of other minerals that can in turn be linked to the *P-T* path. The most common example links REE concentrations and patterns in monazite or zircon to the abundance of garnet. Garnet preferentially takes up Y and heavy REE (HREE) (see Section 9.1). This means that garnet scavenges Y and HREE, and as the volume of garnet increases, Y and HREE concentrations decrease in any coexisting monazite or zircon. Conversely, if garnet dissolves, Y and HREE concentrations increase in monazite or zircon. Thermodynamic models of whole-rock reactions predict both the *P-T* path that a specific rock experienced and the volume of garnet along that path. By correlating Y or HREE concentrations and patterns of monazite or zircon to garnet volume, and garnet volume to the *P-T* path, monazite and zircon ages can be tied to the *P-T* path.

Ding et al. (2021) provided an example of this approach for the same structural level of the Himalaya as studied by Kohn & Corrie (2011), but about 750 km to the east (or one-third the length of the Himalaya). Different portions of monazite crystals show decreasing Y content from about 46 to about 41 Ma, corresponding with garnet growth, followed by increasing Y contents, corresponding with garnet dissolution (Figure 12.30b). Garnet growth and dissolution can be linked to a *P-T* path to calculate the temperature-time history of these rocks (Figure 12.30c), which is remarkably similar to rocks from the central Himalaya. These data point to similar tectonic processes along strike.

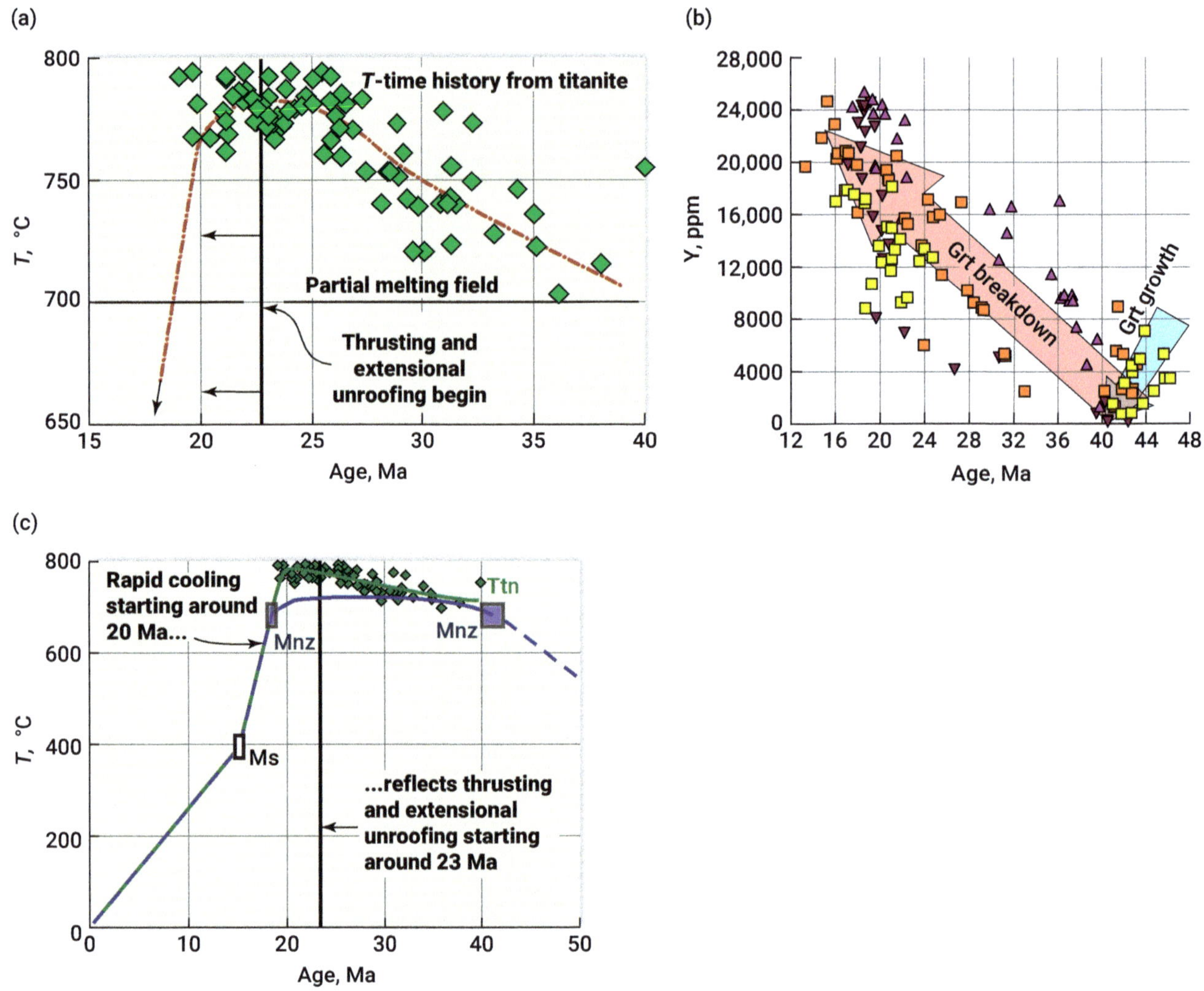

Figure 12.30 (a) Temperature-time history for high-grade rocks in the central Himalaya, based on a combination of Zr-in-titanite thermometry and U-Pb dating. Titanite crystals contain multiple domains of different compositions that develop during progressive metamorphism of these rocks. Analysis of these different domains provides discrete *T-t* points. (b) Raw data for Y (yttrium) content in monazite vs. age, showing an early period of likely decreasing concentrations (corresponding with garnet growth) followed by a protracted period of increasing concentrations (garnet breakdown). Different symbols represent data from different rocks of similar bulk composition from the same small area. (c) Comparison of temperature-time history for eastern Himalaya based on petrochronology of monazite with titanite data from central Himalaya. Rapid cooling commencing ~20 Ma likely reflects thrusting on the Main Central Thrust and extension on the South Tibetan detachment system.

Sources: (a) Modified from Kohn & Corrie (2011); (b) from Ding et al. (2021); (c) data from Kohn & Corrie (2011) and Ding et al. (2021).

Fabric
The texture of a rock, which can include its primary depositional characteristics (e.g., stratigraphic layering), but in metamorphic rocks more commonly refers to the preferred arrangement of linear or planar features, especially from elongate or platy minerals.

Metamorphic Fabric

What Are Metamorphic Fabrics? Most rocks do not contain randomly oriented minerals; rather, any platy or elongate minerals are commonly aligned in some way. Very generally, this alignment is referred to as **fabric**. Although linear elements dominate a few fabrics to form **lineations**, most rocks contain at least some platy minerals, usually micas, and develop planar fabrics. In metamorphic rocks, these planar fabrics are referred to generically as **foliations** (after the Latin word *folium/a*, meaning leaf/leaves) or **schistosities** in schists and some gneisses. They can also be called **cleavages**, because foliated rocks often split or cleave (this is a different use of the word cleavage than in mineralogy). In structural geology, planar fabrics are labeled "S" (for schistosity); S_0 is reserved for original depositional layering, and all later fabrics are labeled in order of formation—S_1 is the oldest identifiable schistosity, S_2 the next oldest, etc. Often earlier fabrics are preserved in large metamorphic crystals called porphyroblasts.

Lineations
Linear fabrics, usually in a metamorphic rock.

Foliations
Planar fabrics, usually in a metamorphic rock.

Schistosities
Foliations formed by micas and other sheet silicates in schists. Labeled "S" in structural geology (e.g., S_1, S_2, etc.).

Cleavages
In rocks, planes of weakness along which they cleave or split (e.g., along foliations).

Why Study Fabrics?

Fabrics are useful for two main reasons. First, they tell us about relative sequences of deformation events. Deformation is how a rock responds to differential stress, and differential stresses tend to align minerals. By studying different sequences of fabrics in a rock, we can learn about the different deformation events that a rock has experienced. In favorable cases, fabrics can be linked to major structural evolution in mountain belts. For example, many orogens show an early phase of tight folding and thrusting, which creates a strong fabric in metasedimentary rocks and igneous rocks with abundant micas, then a later stage of upright folding or warping, which creates an overprinting fabric. Simply by looking at one rock in the right location, you can identify these two orogen-scale deformations.

The second main reason to study fabrics is because, as metamorphic crystals grow, they interact with fabrics to produce textures that link them to these deformation phases. Metamorphic porphyroblasts can grow either before, during, or after a deformation event (called "pre-," "syn-," and "post-" deformation porphyroblasts, respectively). These textures allow us to construct a relative sequence of timings. If a metamorphic mineral grows after deformation (post-deformation), it can overprint the fabric, showing up as a "cookie-cutter" outline against the backdrop of textural features in the matrix, or it can preserve inclusions that align with matrix minerals (Figure 12.31a,b). In this context, petrologists often talk about an "internal foliation" (the foliation inside the crystal) and the "external foliation" (the foliation in the matrix). At the opposite end of the spectrum, if a metamorphic mineral grows before a phase of deformation (pre-deformation), it can act as a rigid body around which a fabric develops (Figure 12.31b,c). Commonly, we see quartz pressure shadows (small, triangular wedges of quartz that form on two sides of a crystal; Figure 12.31b,c), but sometimes rigid crystals break, and other minerals fill in the fractures. If there is an internal fabric in the porphyroblast because of an earlier deformation, that fabric is at an angle to the matrix foliation. Lastly, a porphyroblast that grows during deformation (syn-deformation) may show spiral inclusion trails if shear deformation caused the rigid crystal to rotate. The internal trails are at an angle to the external foliation in the core of the crystal but become parallel to the external foliation at the edge of the crystal (Figure 12.31d). Of course, a rock may experience multiple deformation events and multiple episodes of crystal growth, so a porphyroblast could be post-deformational with respect to an early fabric, syn-deformational with respect to an intermediate one, and pre-deformational with respect to a third. Sorting through all these complexities can be challenging!

In the past, structural geologists and metamorphic petrologists approached the interaction between metamorphic mineral growth and fabric development somewhat differently. Here, we have implicitly assumed that deformation is an "event," limited in time, whereas growth of a metamorphic porphyroblast may be protracted. Classically, structural geologists have adopted the reverse view—that metamorphic porphyroblasts grow during an "event," limited in time, whereas deformation is protracted. Fortunately, geologists do talk to each other, and these end-member perspectives have faded away with the recognition that both deformation and metamorphic mineral growth can be protracted. Nonetheless, structural geologists do talk about pre-, syn-, and post-metamorphic fabric development. This can become confusing, because if a metamorphic mineral forms before a fabric, the fabric is "post-metamorphic," but the mineral is "pre-deformational." Although "syn" is always equivalent, be careful when reading about "pre" and "post."

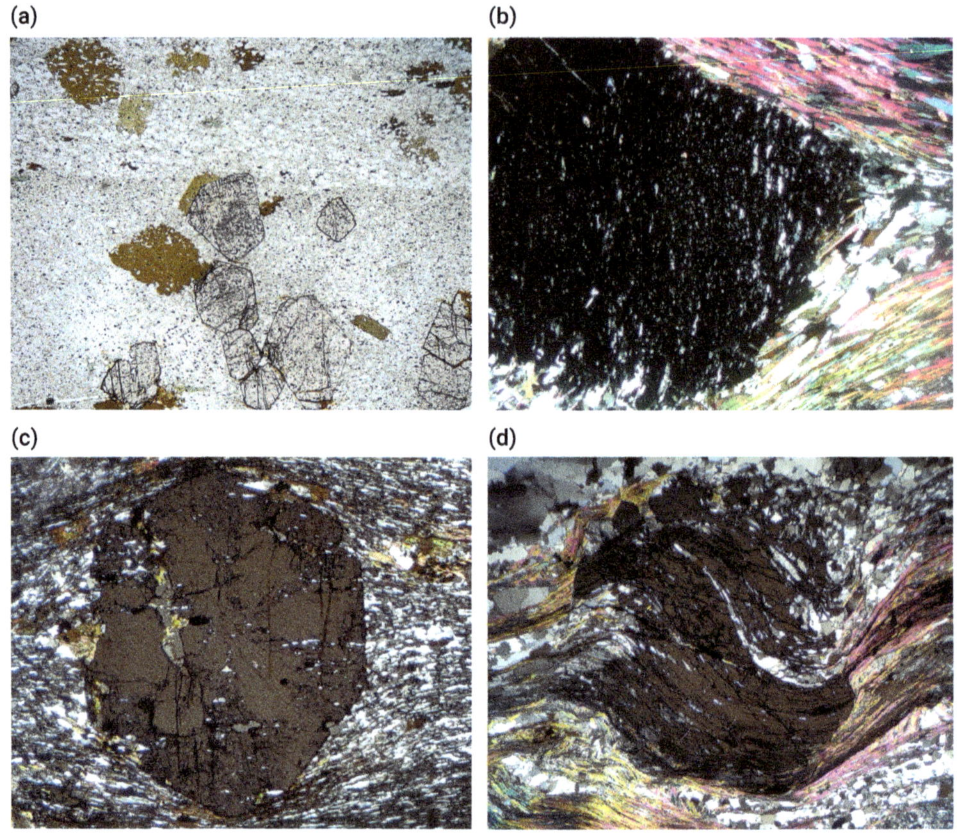

Figure 12.31 (a) Photomicrograph of post-deformational garnet and biotite. Biotite crystals are randomly oriented, and both garnet and biotite overprint the matrix fabric, which is running approximately horizontally across this image. (b) Photomicrograph of garnet in crossed polars (black) with white quartz inclusion trails, oriented nearly vertically. The garnet postdates an early fabric (represented by the aligned inclusions), but before a second fabric represented by inclined micas in the matrix. A quartz pressure shadow occurs to the right. (c) Photomicrograph of a pre-deformational garnet in partially crossed polars. Foliation warps around a core that shows no preferred alignment of inclusions. Possible alignment of inclusions in garnet rims may indicate some growth after deformation. Quartz pressure shadows occur to the right and left. (d) Photomicrograph of a syn-deformational garnet in partially crossed polars. Inclusion trails change orientation smoothly from upper left to lower right across the crystal, forming a sigmoidal pattern.

Source: MJK photos.

Why do we care when a mineral grows relative to deformation? Partly, it's so that we can understand how deep in an orogen deformation was occurring. More generally, however, deformation proceeds faster than isotherms can relax (e.g., at a high Péclet number; see Section 3.7), so deformation perturbs the thermal structure of the crust. The timing of deformation relative to *P-T* evolution allows us to interrelate deformation and thermal structure on the scale of an entire orogenic thickness.

An Example that Links Deformation and Metamorphism

Selverstone (1985) presented one of the first fully quantitative examples linking deformation explicitly to metamorphism (Figure 12.32). Her work showed that after initial formation of a strong fabric and growth of lawsonite ($CaAl_2Si_2O_7(OH)_2{\cdot}H_2O$, a mineral stable only at HP/LT conditions), garnets grew syn-deformationally along a *P-T* path that showed two periods of burial. The rocks were subsequently sheared and then crenulated (earlier schistosity

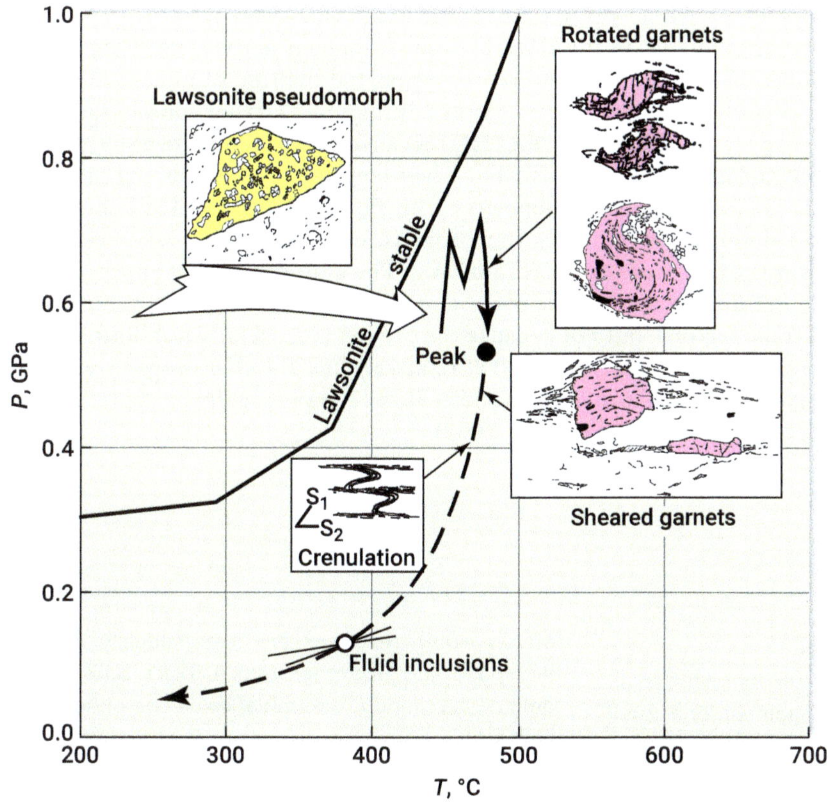

Figure 12.32 Relationships between fabric development, metamorphic mineral growth, and *P-T* evolution for the upper Schiefferhülle, Tauern Window, Alps. Pseudomorphs after lawsonite indicate initial metamorphism under HP/LT (lawsonite-stable) conditions; thermodynamic modeling of chemical zoning in syn-deformational garnet yields the M-shaped *P-T* path while S_1 was forming. S_2 shearing created a crenulation fabric and sheared garnets. The densities and compositions of different sets of late-stage fluid inclusions help to define a point on the retrograde *P-T* path.

Source: Modified from Selverstone (1985).

was kinked). The earlier fabrics formed during thickening, whereas later fabrics formed during extensional thinning. Her calculations of pressures and temperatures quantified the conditions under which these deformations occurred and simultaneously linked deformation, metamorphism, tectonics, and thermal evolution in an orogen. This kind of endeavor underpins much metamorphic research today.

12.7 Igneous Activity in Collisional Orogens

Learning Objective

By the end of this section you should be able to:

12.7.1 Qualitatively predict the amount of melt produced during collisional orogenesis depending on rock types and thermal structure.

What about igneous rocks—surely they form during collisions, right? Well, sometimes yes, sometimes no. After subduction ceases, melting generally occurs in response to heating of the crust because of thickening (radioactive heating) and thermal relaxation. These are slow processes. Some orogens, especially the Alps, show little igneous activity, whereas in the Himalaya, most igneous activity postdates initial collision by over 30 Myr. Melts often represent partial melting of mica-rich rocks, either by water-saturated melting (see reaction 5 in Figure 12.21) or muscovite dehydration melting (see reaction 2 in Figure 12.21), because these are the lowest-T melting reactions, and therefore are reached first. They tend to produce leucogranites that may host economically important deposits of Li, Be, and other strategic metals. Igneous rocks can dominate in older, hotter orogens, especially UHT terranes, where high temperatures drive more extensive melting. The abundance of melt reflects a combination of P-T history and source rock. The Himalaya have relatively abundant partial melts in some regions, but not because the orogen is especially hot. Rather, metamudstones are abundant, and these melt at relatively low T. Other orogens with abundant melts (e.g., the Acadian orogen in eastern New England) really do represent highly elevated geotherms (e.g., Spear et al., 2002).

Summary

Key points of this chapter include the following:

- Orogens represent the culmination of the collision between a continent and either an arc, a continental fragment, or a continent, and typically result in thick crust and widespread regional metamorphism. Ultimately, collisions form basement rocks in shields that resist further deformation.
- Erosion of orogens strongly influences atmospheric CO_2 concentrations over multimillion-year timescales.
- The temperature structure of the crust (its geotherm) depends on heat-producing elements, mantle heat flux, and movement of rocks via thickening, thinning, and erosion, plus movement of fluids and melts.
- Collisions are predicted to drive P-first P-T paths, in which loading is followed by a temperature increase as the geotherm evolves, then exhumation and cooling as the crust thins and/or erodes. Intrusions and/or magma loading are predicted to drive T-first P-T paths, dominated by heating (contact metamorphism).
- The distribution of metamorphic assemblages and P-T conditions as observed in the field (the metamorphic field gradient) results from complex thermal evolution and does not normally represent the P-T path followed by any one parcel of rock.
- Post-orogenic thermal evolution predicts widespread crustal weakening within a few tens of millions of years, allowing extensional collapse of thickened crust.
- Metamorphic reactions are systematically distributed in P-T space and, together with a rock's bulk composition, define what mineral assemblages can be observed across a field area.
- The (metamorphic) AFM diagram readily illustrates the main mineral assemblages expected in metamorphosed mudstones.
- Thermobarometry quantifies pressures and temperatures. Classical methods apply thermodynamic equations of equilibrium using mineral compositions and balanced chemical reactions. Spectroscopic methods are now also used.
- Mineral assemblage diagrams (MADs) use thermodynamics to predict mineral assemblages and mineral compositions for a specific rock composition. These are commonly used to infer P-T paths from metamorphic rocks, based on sequences of mineral assemblages and mineral compositions (from zoned minerals).
- Petrochronology links P-T paths with time, usually by dating minerals that can be linked either to characteristic assemblages (e.g., through inclusions or chemical composition) or to temperature through direct thermometry.
- Rock fabrics link metamorphism and P-T-t paths to deformation and integrate metamorphic petrology and structural geology.
- Igneous activity in collisional orogens commonly represents relatively low-T partial melting of metamudstones. Leucogranites are common products of orogenesis.

Notes

1. These include Sanbagawa or Franciscan for HP/LT (named after subduction complexes in Japan and California) and Abukuma or Buchan for LP/HT (named after localities in Japan and Scotland.
2. Even if they're commonly referred to a single mineral. For example, the "staurolite isograd" normally refers to an association of staurolite + biotite + muscovite + quartz (+ assumed H_2O). The *P-T* conditions of this assemblage are well-defined. Otherwise, depending on rock composition, staurolite can occur with other minerals over an enormous region of *P-T* space.
3. Projections require a projection point (in this case, muscovite) and a projection plane (the plane formed by Al_2O_3, FeO, and MgO). What we're really doing is plotting compositions in an AKFM tetrahedron, where biotite plots in the interior. If you put your eye on muscovite (on the edge of the tetrahedron, between K_2O and Al_2O_3) and look at biotite (inside), it projects below the FeO-MgO base.
4. Why "plus biotite"? Because in less common rocks with very high Al_2O_3 contents, staurolite can be stable over a wide range of temperatures. In these rocks, the occurrence of staurolite is less remarkable or useful, except as an indicator of rock composition.
5. Which you definitely should do.

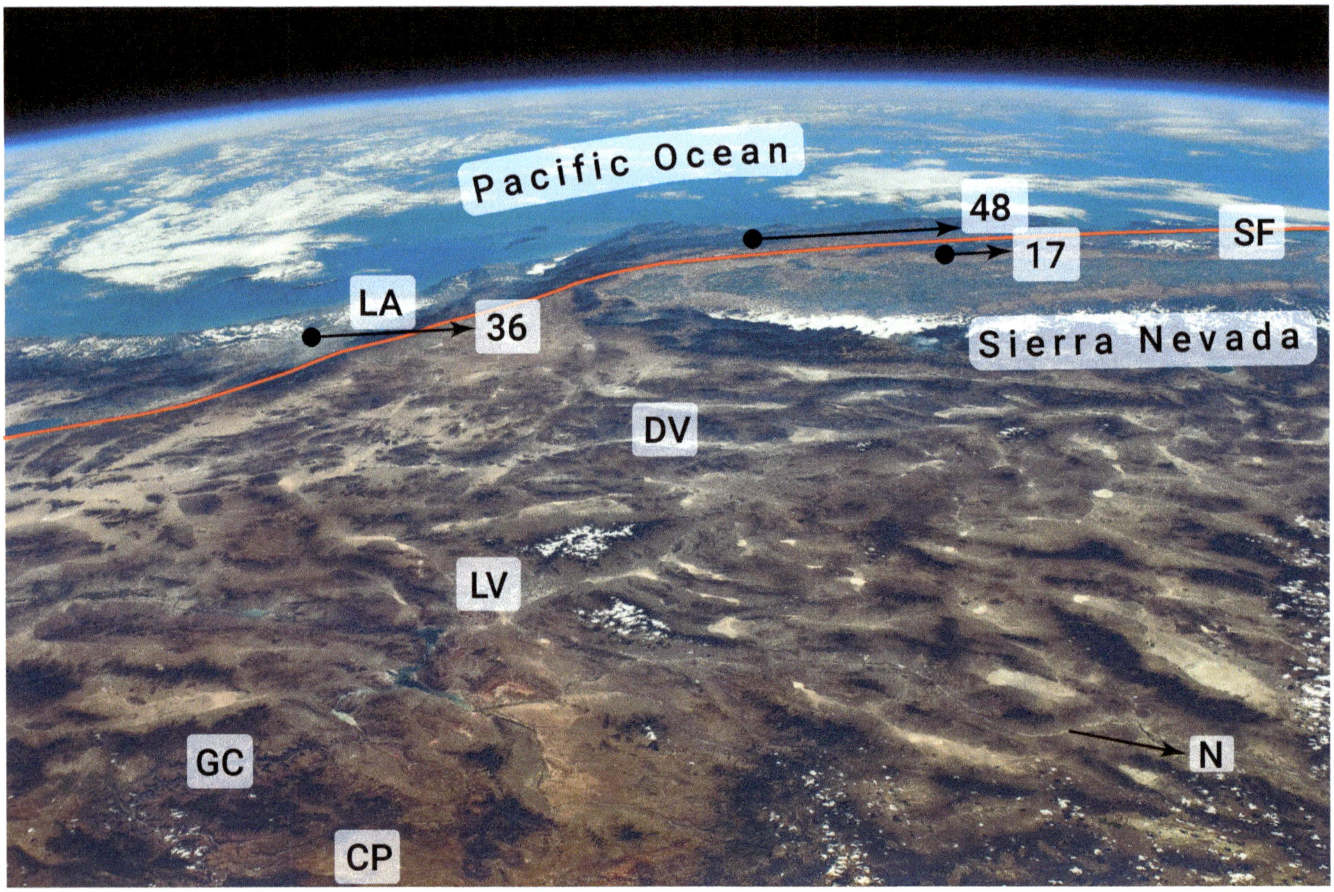

Pacific Ocean
48
17
SF
LA
36
Sierra Nevada
DV
LV
GC
N
CP

CHAPTER 13

Continental Rifting and Magmatism

The inevitable course of plate tectonics means that continents formed through collisions must eventually split apart. Protracted continental rifting produces ocean basins, returning us to Chapter 3 and the mid-ocean ridge system. Pangea, laboriously assembled over 150 Myr by numerous collisions (see Chapter 12), eventually rifted apart to form the Atlantic Ocean, leaving behind fault-bounded rift basins (e.g., the Newark and Durham basins) filled with coarse continental sedimentary detritus, diabase dikes and sills, and the fossil remains of dinosaurs and other beasts. Some continental rifts split off thin ribbons of continental lithosphere, such as Baja California, that are fated to be re-accreted to a continent in the distant future. Some continental rifts, such as the East African Rift system, struggle along for tens of millions of years without developing a new ocean basin.

Continental rifting gives rise to the continental rift lithosystem, with distinctive igneous rocks, associated fault-controlled sedimentary rocks, and metamorphic rocks exhumed by normal faults. Continental rifting is responsible for breathtaking scenery at places such as Death Valley, the eastern escarpment of the Sierra Nevada, the East African Rift, and Lake Baikal (Figure 13.1), and much of our understanding of human evolution comes from deposits in rift valleys (see Making Connections 13.1).

View westward across the southwestern Basin and Range province to the California coast from the International Space Station. The Basin and Range is bounded here on the west by the Sierra Nevada Range and the mountains north of Los Angeles (LA). The San Andreas Fault, which separates the North American and Pacific Plates, is the red line shown running across the photo from San Francisco (SF) to LA. The western edge of the Colorado Plateau (CP), deeply incised by the Colorado River in the Grand Canyon (GC), occupies the lower-left corner. This arid area, in the rain shadow of California mountains, is mostly internally drained and dotted by dry lakes that mark the locations of perennial lakes during the Last Glacial Maximum. Arrows show GPS-derived velocities (mm/yr) of representative points relative to the Colorado Plateau. Difference in vectors in central California across the San Andreas Fault shows ~30 mm/yr of right-lateral slip at the plate boundary; mismatch between vector in LA basin and the plate boundary results in California south of the fault (Pacific Plate) being thrust under the Basin and Range (North American Plate), a form of continental collision that forms the mountains north of Los Angeles. LV = Las Vegas; DV = Death Valley.

Source: NASA, annotations by authors.

(a)

(b)

Figure 13.1 (a) View northwest across Death Valley, a classic example of Basin and Range–style continental rifting. The viewpoint, Dante's View, lies on the crest of the Black Mountains at 1670 m elevation; Badwater, the lowest point in North America, is a mile below, 86 m below sea level. The Panamint Range across the valley tops out at 3368 m, over two miles above the valley floor. The Black Mountains are composed of high-grade Proterozoic gneisses that were intruded by Miocene plutons and then exhumed from the middle crust by faulting. Miocene and younger rift-related volcanic rocks blanket the range to the north. This view encompasses many of the elements of the continental rift lithosystem: high topography and narrow, sediment-filled basins, with syn-extensional volcanic rocks tilted by normal faulting in fault contact with high-grade metamorphic rocks exhumed by normal faulting. (b) The 500-m-tall western escarpment of the Gregory Rift in northern Tanzania, as viewed from the slopes of the bizarre volcano Ol Doinyo Lengai, which spews out lava composed largely of water-soluble, Na-rich carbonates. The escarpment exposes layers of lava and tephra, represented by nearly horizontal stripes in the cliff wall, derived from extensive volcanic activity in and near the rift.

Sources: (a) AFG photo; (b) from Scoon (2018), Springer Nature BV.

13.1 Continental Rifts

Learning Objectives

By the end of this section you should be able to:

13.1.1 Compare rates of extension in oceanic spreading centers to continental rifts.

13.1.2 Summarize the wide range of magmatism found in Earth's major continental rifts.

Although of singular importance for understanding continental lithospheric structure and chemistry, as well as Earth's long-term carbon budget, continental rifts are small by most measures. For example, as we saw in Section 3.1, the ocean ridge system produces new lithosphere paved with basalt at a rate of about 8000 m^2 (about the area of one or two football fields) per day. Continental rifts produce new lithospheric area, too, but at a much smaller rate; spreading in the Basin and Range and the East African Rift add up to about 30 m^2 (about the area of two soccer goals) per day, together accounting for less than 0.5 percent of the area produced daily by oceanic ridges. Likewise, the volume of magma produced in continental rifts is a fraction of a percent of Earth's total magma output. And whereas western North America hosts by far the largest abundance of **metamorphic core complexes** on Earth, these represent only a few percent of exposed metamorphic rocks on the continent, and a fraction of a percent globally.

Metamorphic core complexes
Geographically limited suites of metamorphic and plutonic rocks exposed by extension. Typically, low-angle faults and shear zones juxtapose unmetamorphosed or hydrothermally altered sedimentary and volcanic rocks above the fault against high-grade metamorphic rocks below.

Magma chemistry in continental rifts can be bewilderingly complex in comparison to other magmatic systems. For example, MORBs are homogeneous, despite large differences in spreading rates, structural style of extension, ridge depth, and other parameters, and most arcs show a well-defined array of compositions stretching from basalt to rhyolite. In contrast, the magmatic products of continental rifting are typically richer in the alkalis Na and K, incompatible elements, and high-field-strength elements (HFSE; Figure 13.2). However, to add to the confusion, not all rifts are distinguished by alkaline magmatism—Cenozoic igneous rocks in the Basin and Range province, Earth's largest active continental rift system, are compositionally much like arc rocks in spite of many having been erupted and intruded over 1000 km from the plate margin.

In this chapter, we discuss several examples of continental rifting, causes of rift-related magmatism and controls on its composition, the importance of subcontinental mantle, and metamorphic rocks in extensional terranes. Alkaline rocks are weird and economically important components of both rift and intraplate magmatism, and we split our discussion of them between this chapter and the next. We begin with a discussion of three

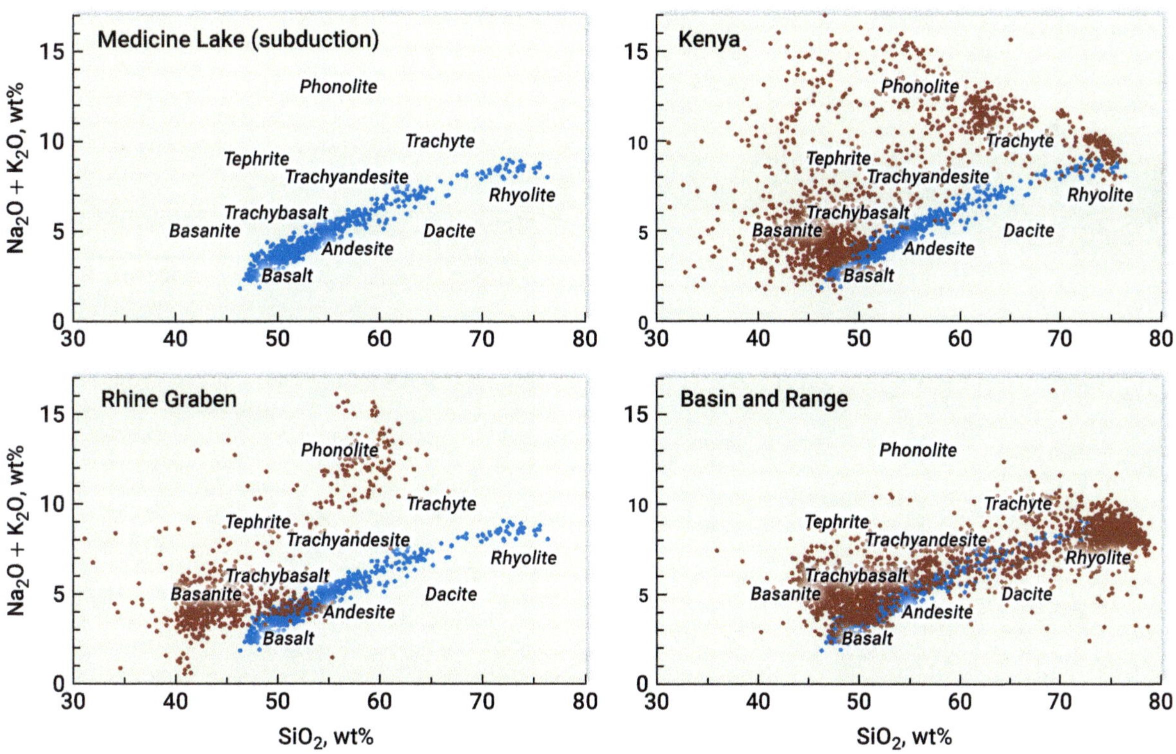

Figure 13.2 Three rifts compared on a TAS plot with arc rocks from Medicine Lake (see Figure 1.4) in blue for comparison. All are bimodal to some extent and are dominated by basaltic rocks and their more alkalic cousins (e.g., basanite). Kenya contains a significant proportion of rhyolite as well as phonolite; Rhine Graben has only phonolite and no rhyolite. The Basin and Range, although somewhat bimodal, contains a significant proportion of andesite and dacite and is compositionally like a subduction assemblage.

Sources: Data from GEOROC (georoc.eu) precompiled datasets except Medicine Lake, extracted from EarthChem.

areas that illustrate the diversity of features in continental rifts and focus on the Basin and Range province of western North America.

East African Rift

The East African Rift system (Figure 13.3a) is the southern leg of the rift-rift-rift Afar triple junction whose other arms are oceanic rifts under the Red Sea and Gulf of Aden. This triple junction separates the African, Arabian, and Somali Plates. The rift system splits the Ethiopian Plateau and continues south along the eastern margin of the East African Plateau into Kenya and Tanzania. A separate branch, the Western Rift, runs down the western side of the East African Plateau, along the western borders of Uganda and Tanzania, and through the small nations of Rwanda and Burundi. Large, elongate, fault-bounded lakes occur in both branches of the rift, although semicircular Lake Victoria,[1] the putative source of the Nile, occupies a position on the East African Plateau between the rift branches. Africa east of the western branch of the rift, along with the adjacent portion of the Indian Ocean, comprise the Somali Plate (Bird, 2003).

The rift system has been active since at least the earliest Miocene (~24 Ma) but has only accomplished 35 to 40 km total extension (Hendrie et al., 1994). GPS data and plate movements show that the rift is spreading east-west at 3 to 6 mm/yr in the Ethiopia-Tanzania

Natrocarbonatite
Magma consisting largely of Na and K carbonates.

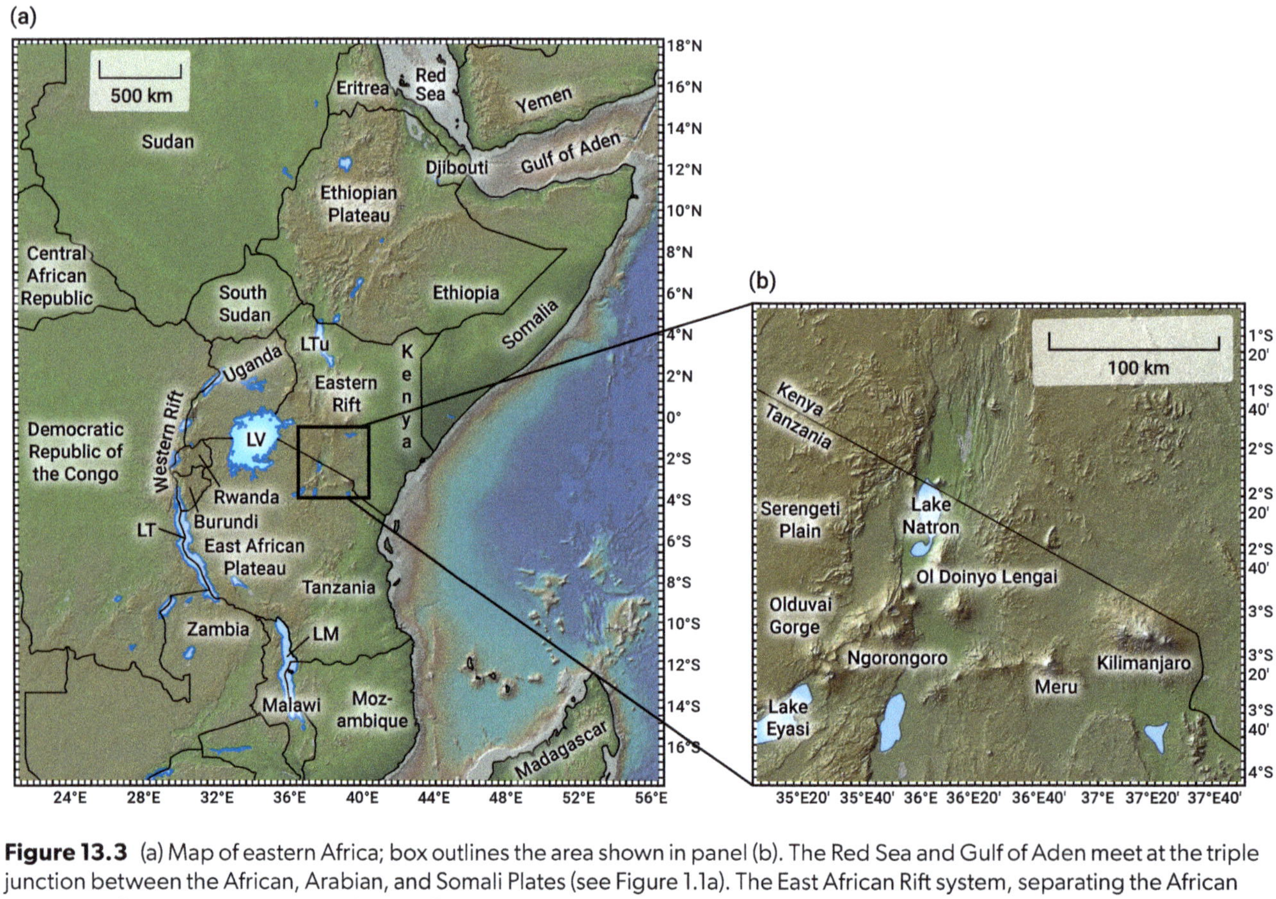

Figure 13.3 (a) Map of eastern Africa; box outlines the area shown in panel (b). The Red Sea and Gulf of Aden meet at the triple junction between the African, Arabian, and Somali Plates (see Figure 1.1a). The East African Rift system, separating the African and Somali Plates, runs southwest through Ethiopia and then south, splitting into western and eastern branches. Lake Victoria (LV) occupies the plateau between these branches. (b) Rift topography and volcanoes in the Gregory Rift (eastern branch of the East African Rift system) around the Tanzania-Kenya border. Stepped rift topography controls the locations of lakes, and alkalic volcanoes and the active **natrocarbonatite** volcano Ol Doinyo Lengai are concentrated around the rift. LTu = Lake Turkana; LT = Lake Tanganyika; LM = Lake Malawi.

Source: Maps produced using GeoMapApp.

segment, which is 10 to 20 times slower than the average mid-ocean ridge.

Volcanic rocks associated with the East African Rift are typically highly alkalic. On a TAS diagram (see Figure 13.2; see also Figure 1.3), they lie well above the linear trend defined by typical subduction-related volcanic suites and extend to far lower silica contents (<45 wt% SiO_2). At a given silica content, rocks from the rift (Figures 13.4 and 13.5) are typically classified as basanites and trachybasalts (vs. basalts in arcs), trachyandesites and phonolites (vs. andesites), and trachytes (vs. dacites). The rare Si-undersaturated minerals leucite ($KAlSi_2O_6$), nepheline ($Na_3KAl_4Si_4O_{16}$), and melilite, a group with the approximate formula $(Ca,Na)_2(Al,Mg,Fe)(Si,Al)_2O_7$, are common in these more alkalic lavas.

Figure 13.4 One of the world's most iconic volcanoes, Mount Kilimanjaro (5,895 m; 19,341 ft), rises over 5 km above Moshi, Tanzania. Rocks erupted from Kilimanjaro include **nephelinites**, alkali basalts, and phonolites.

Source: Muhammad Mahdi Khan/Wikipedia (GNU Free Documentation License, Version 1.2).

Nephelinites
Alkaline volcanic or hypabyssal rocks consisting largely of nepheline and augite and lacking olivine and plagioclase.

(a)

(b)

Figure 13.5 (a) Arguably Earth's strangest volcano, Ol Doinyo Lengai rises above the basin of Lake Natron in northern Tanzania. This volcano is unique in erupting natrocarbonatite lava that crystallizes largely to the rare Na-K-Ca carbonate minerals nyerereite [$Na_2Ca(CO_3)_2$] and gregoryite [$(Na_2,K_2,Ca)CO_3$]. The lavas are largely water-soluble, so lava and tephra, which are black when they first erupt, rapidly weather to gray, partially dissolve, and wash into the lake basin when it rains. (b) Carbonatite lava flow, a few days old, from Ol Doinyo Lengai. Natrocarbonatite lava has a viscosity on the order of 0.1–1 Pa·s, comparable to olive oil or honey at room temperature. The lava resembles watery mud in appearance and flow characteristics, although it erupts at a temperature of ~500 °C, glowing a faint red at night. Some magmas froth like soda as they erupt, because they contain so much CO_2 gas. Black lava turns white after a day or two of reaction with water vapor in the air. Spatter cones in mid-ground are 2–4 m tall.

Sources: (a) Clem23/Wikimedia; (b) Tobias Fischer, University of New Mexico.

Rhine Graben

The Rhine Graben rift system (Figure 13.6) cuts Europe north of the Alps and localizes the course of the Rhine River after it exits Switzerland. The upper Rhine Graben follows the French-German border northward to Frankfurt; the lower Rhine Graben, separated from the upper by a highland area known as the Rhenish Massif, runs northwestward through northern Germany and the Netherlands and on to the North Sea. These grabens and others west of the Alps form lowlands running from the Mediterranean to the North Sea.

The Rhine Graben system began to develop in the Eocene as a series of isolated basins that had coalesced by the Oligocene into a continuous rift. Although scattered magmatism in and around the Rhine Graben system commenced as early as 70 Ma (Schmitt et al., 2007), most volcanism occurred between 45 and 5 Ma, with scattered volcanism in the last 5 Myr in the Rhenish Massif (Wilson & Downes, 2006). Tectonism and volcanism thus began during oceanic convergence and continued through the main phase of continental collision and beyond.

Figure 13.6 Location of major rifts in Europe. Laacher See is a former explosive crater that has filled in with water (a maar). BG = Besse Graben; LRG = Lower Rhine Graben; RG = Rhône Graben; URG = Upper Rhine Graben.

Source: Map produced using GeoMapApp.

Volcanic rocks of the Rhine Graben are chemically similar to those of the East African Rift (see Figure 13.2): highly alkalic, with a predominance of low-silica, high-alkali mafic rocks (basanites on the TAS diagram) and phonolites. Unlike the East African Rift, high-silica rhyolitic compositions are rare. Phonolite magma erupted spectacularly in a VEI 6 event about 13 ka at Laacher See volcano in the Rhenish Massif, and the surrounding farmland is punctured by maars and calderas (Figure 13.7), some of which produced VEI 5 eruptions in the Pleistocene. The consequences of a future VEI 5 or greater eruption for Europe would be ruinous.

Figure 13.7 Crater lakes (maars) formed by volcanic explosions in the Eifel region, Germany.

Source: Martin Schildgen/Wikipedia (GNU Free Documentation License, Version 1.2).

Basin and Range

The Basin and Range province of western North America (Figure 13.8) is the largest and most active area of continental extension on Earth today. Given its importance in understanding the relationships between plate tectonics and magmatism (e.g., Atwater, 1970; Dickinson & Snyder, 1979) and how large-scale crustal extension works (Wernicke & Burchfiel, 1982), coupled with easy access, it is one of the best-studied regions on Earth—and yet many important questions remain about how the abundant magmatism there was generated, how it relates to plate tectonics, and how locally exposed metamorphic rocks reflect tectonic history.

The Basin and Range extends south from Oregon and Idaho through Nevada (Figure 13.9), eastern California, and Utah, then through Arizona, New Mexico, and northern Mexico. In the following discussion, we will focus on the northern part of the Basin and

Figure 13.8 Map of the southwestern United States and northwestern Mexico, showing extent of the Basin and Range province.

Source: Illustration by Chelsea Feeney, from Glazner et al. (2022).

Figure 13.9 Jet flights from central and eastern United States to Nevada and California pass over the Basin and Range. This view, looking north out the airplane window over eastern Nevada, shows typical Basin and Range topography, with ranges (dark with vegetation except where recently burned, as in center foreground) separated by basins that locally contain playas, as in Lake and Cave Valleys. Ely, Nevada, a former Pony Express station and mining town, lies in one of the basins. Extensional unroofing of the Snake Range and Schell Creek Range has locally exposed greenschist- to amphibolite-facies metamorphic rocks in metamorphic core complexes.

Source: AFG photo.

Range that stretches from the Sierra Nevada to central Utah. Most of this region is characterized by internal drainage, with generally short drainage systems that end at what are now dry lakes (**playas**). This northern section is referred to as the **Great Basin**.

Playas
The dry, vegetation-free floors of geomorphic basins, covered with clay or evaporite salts.

Great Basin
The northern part of the Basin and Range province; drainages currently terminate inside the continent and do not reach the ocean.

Geologic History

The geologic history of the Great Basin ties together many of the petrologic and plate tectonic processes covered so far in this book. The oldest rocks in the province are about 1.7 Ga metamorphic rocks; these and younger Proterozoic plutonic rocks are the basement upon which thick sections of marine sedimentary rocks were deposited in the late Proterozoic and Paleozoic. Fault-bounded ranges expose thick, tilted sections of these strata, locally several kilometers thick, as well as underlying basement. All are overlain by volcanic deposits, largely Oligocene and Miocene, that are hundreds of meters to a few kilometers thick.

A brief geologic history of the Great Basin includes the following.

1. The supercontinent Rodinia started rifting about 800 Ma to form a new passive margin that ran, in present geography, from northern Utah southwestward to southern California. The conjugate piece of Rodinia comprises one or more of Antarctica, Australia, and China (e.g., Hoffman, 1991).
2. Thick sequences of marine rocks were deposited from the late Proterozoic to the end of the Paleozoic. These range from thin, near-shore facies (sandstone, shale) in the east to thick (several kilometer) sections of carbonates in central Nevada to thinner, deep-water facies including shale and chert in western Nevada.
3. During the Late Paleozoic the passive margin, which had existed for roughly half a billion years, was converted to a subduction margin. The oldest plutons from this arc are Permo-Triassic (~240–270 Ma; Cecil et al., 2019); the arc was more fully developed along eastern California and western Nevada by the Jurassic, about 180 Ma.

4. Magmatism related to subduction of the Farallon Plate (and probably others, now consumed) continued through the Cretaceous; this arc produced much of the Sierra Nevada and Peninsular Ranges batholiths (see Figure 10.2). Arc volcanism shut down abruptly 70 to 80 Ma.
5. Magmatism returned to the Great Basin about 40 Ma as silicic volcanism, largely pyroclastic flows, swept south across Nevada, reaching southern Nevada about 15 Ma. A second wave of volcanism, sweeping northward across the Mojave Desert and Arizona, reached southern Nevada about the same time.
6. In the Oligocene, about 30 Ma, a spreading ridge between the Pacific and Farallon Plates reached the subduction zone, producing a strike-slip (transform) boundary—the San Andreas Fault—between the Pacific and North American Plates, and two triple junctions that migrated north and south.
7. Volcanism became increasingly mafic with time, with Pliocene and younger volcanism largely basaltic.

Figure 13.10 shows four frames from an animation of Pacific-Farallon-North American Plate interactions over the past 38 Myr.[2] These were based on the plate-circuit

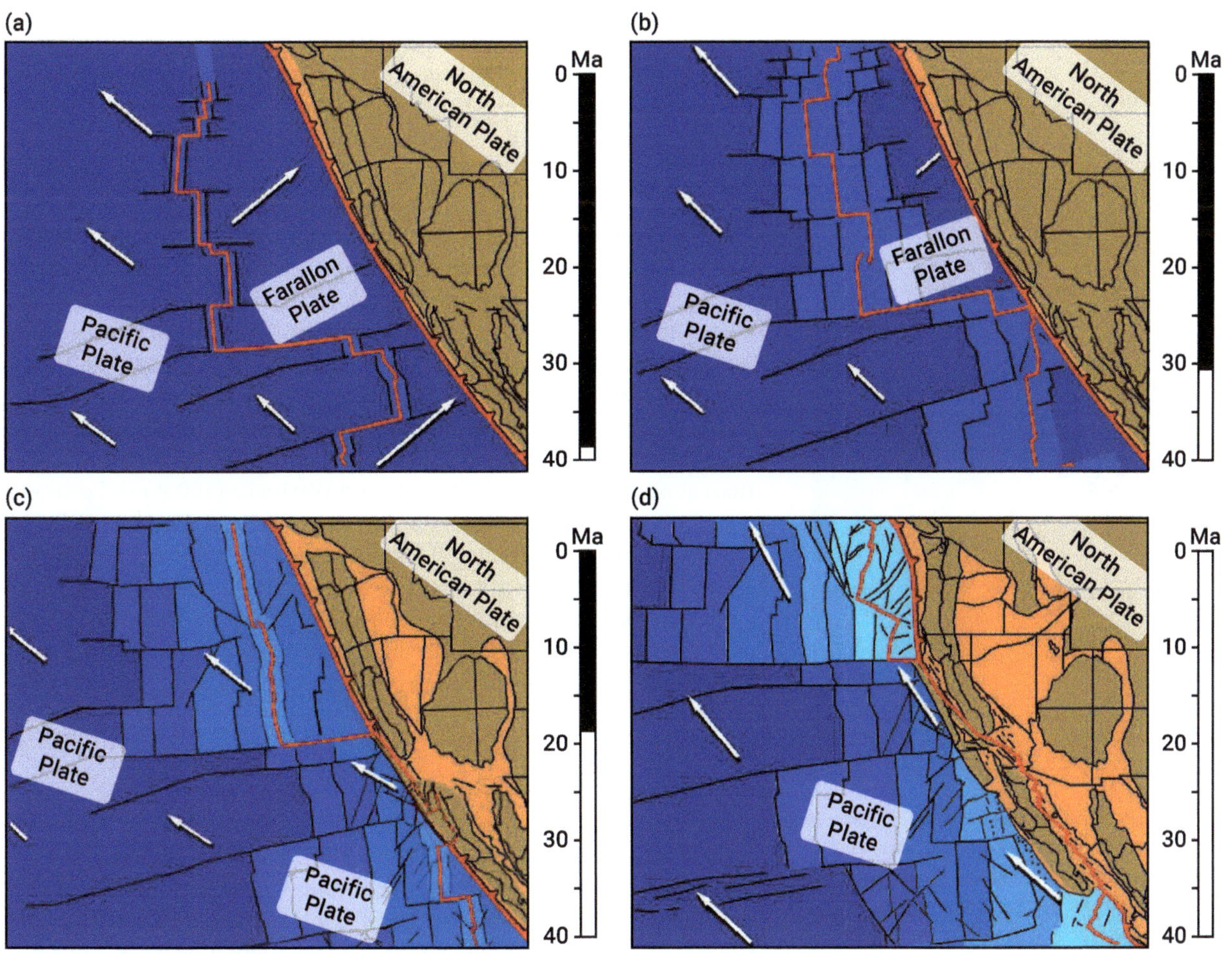

Figure 13.10 Reconstruction of the extensional history of the Basin and Range in the context of plate motions. (a) 38 Ma: The Farallon Plate simply subducts beneath North America, but spreading ridges are approaching the plate boundary. (b) ~30 Ma: The Pacific-Farallon spreading center first encounters the subduction zone, splitting the Farallon Plate into a northern segment (Juan de Fuca Plate) and a southern segment (Cocos Plate). The San Andreas strike-slip fault system initiates. (c) 18 Ma: Major rifts and other extensional systems have formed (orange regions). The San Andreas Fault system lengthens, and rifting starts to separate Baja California from continental North America. (d) 0 Ma: Modern configuration with widespread extension, a lengthy San Andreas Fault system, and a new ocean basin—the Gulf of California.

Source: From Atwater & Stock (1998), courtesy of Tanya Atwater.

reconstruction of Atwater and Stock (1998), with extension restored. At 38 Ma, subduction occurred along the west coast of North America, but at about 30 Ma, the Pacific and North American Plates came into contact as the Pacific-Farallon spreading ridge met the trench at the latitude of what is now the US-Mexico border. By 18 Ma, two triple junctions—the Mendocino in the north and the Rivera in the south—had formed and were migrating away from each other, leaving a lengthening transform boundary between them. Extension was underway by this time and continues to the present day. About 6 Ma, Baja California split off from mainland Mexico, forming a new, narrow ocean basin: the Gulf of California (e.g., Oskin & Stock, 2003). There, new lithosphere is being formed along short, northeast-trending spreading centers linked together by northwest-striking transform faults.

The Mendocino triple junction, now off northern California, marks the southern end of the Cascade arc, and the Rivera triple junction, off central Mexico, corresponds to the northern end of the Trans-Mexican arc. Recognition of the general history shown in Figure 13.10 by Atwater (1970) led directly to a number of deductions about the magmatic history of western North America (e.g., Christiansen & Lipman, 1972; Coney & Reynolds, 1977; Dickinson & Snyder, 1979). For example, these reconstructions seem to require that subduction was continuous along the western margin of North America at 40 Ma but was shut off as the transform boundary developed. Logically, the arc must have progressively shut down from south to north as the Mendocino triple junction migrated northward (e.g., Dickinson & Snyder, 1979). However, these general predictions are not supported by space-time patterns of magmatism (Glazner, 2022). Such disparities between predictions from reconstructed plate interactions and direct geologic observations are not uncommon.

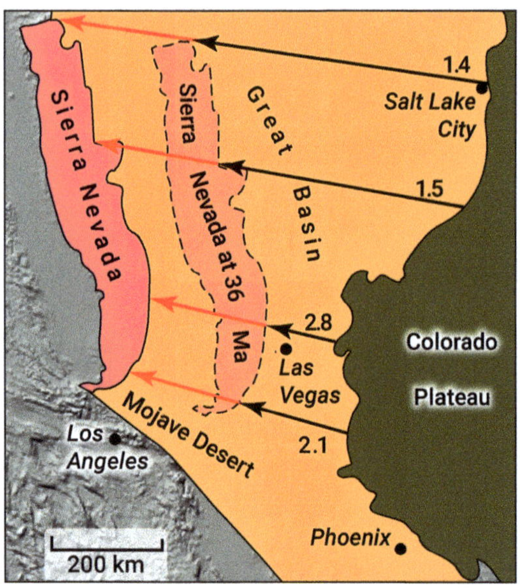

Figure 13.11 The Great Basin is bounded by the rigid Sierra Nevada block on the west and the rigid Colorado Plateau block on the east. McQuarrie and Wernicke (2005) used geologic constraints to restore the Sierra Nevada to its position relative to the plateau at 36 Ma. Extension since then has moved the block a net distance of about 220 km (red arrows). For the paths indicated, this movement produced stretching factors (final line length divided by initial line length; see Section 13.2), ranging from a maximum of 2.8 near Las Vegas to 1.4 across northern Nevada and Utah near Salt Lake City.

Source: Data from McQuarrie & Wernicke (2005).

Extension

Lithospheric extension began in the middle Cenozoic, with the main pulses of extension roughly following magmatism and sweeping both southward and northward toward southern Nevada (Glazner & Bartley, 1984; Axen et al., 1993). Extension dramatically remade the lithosphere, slicing the crust with both steeply and shallowly dipping faults and bringing deep rocks up to the surface.

Although the Basin and Range is the largest active extensional province on Earth, the rate of extension across it is significantly lower than at oceanic ridges. McQuarrie and Wernicke (2005) used geologic correlations to restore extension across the Great Basin and found an area increase of about 153,000 km^2 over the past 36 Myr (Figure 13.11). In their reconstruction, the relatively rigid Sierra Nevada and Colorado Plateau blocks separated from one another by about 220 km over the past 36 Myr, a long-term average of about 6 mm/yr. At present, the Sierra Nevada block is moving about 13 mm/yr toward the northwest, away from the Colorado Plateau.

Magmatism

The great volumes of volcanic rocks in the Basin and Range were documented in the 19th-century geologic surveys of the region led by John Wesley Powell, Ferdinand V. Hayden, Clarence King, and others. It was recognized long ago (e.g., Lindgren, 1915) that Cenozoic magmatism in the Great Basin generally comprises thick stacks of intermediate volcanic rocks topped by basalts erupted in the past 5 million years or so. Intermediate rocks are generally card-carrying members of the BADR association, although many are somewhat more alkalic than those of the Cascade arc (compare Basin and Range with Medicine Lake in Figure 13.2).

Such rocks are generally found above subduction zones, and there was subduction going on at least in the earlier part of the Cenozoic (see Figure 13.10), but two things make the occurrence of such calc-alkaline rocks in the Basin and Range, and further inland in the Colorado Plateau and Rio Grande Rift, highly unusual. First, such rocks occur much farther from the plate boundary than in any known active or fossil subduction zone. For example, even after extension is restored, massive outpourings of intermediate calc-alkaline rocks occurred over 1000 km from the trench in Colorado and New Mexico; in sharp contrast, magmatism in the Andean arc rarely reaches even half that distance inland (see Figure 6.24; Ramos & Folguera, 2009). Second, the magmatic record does not correspond to predictions of the plate tectonic history (see Figure 13.10). For example, in the corridor from New Mexico and Arizona through eastern California and southern Nevada, magmatism was absent when subduction was occurring and only began after it ceased, opposite the deduction made from the reconstructions in Figure 13.10. The correspondence between plate tectonics and magmatism in western North America remains a puzzle, and several important pieces have not yet been found (Glazner, 2022).

Making Connections 13.1

Hominids and East African Rift Valley

The East African Rift (Figure 13.12a) is renowned for its fossils of ancient primates ranging from *Proconsul* (~20 Ma) to early examples of *Homo sapiens* (possibly 350 ka). Indeed, much of our understanding of hominid evolution is based on fossils from that region. Possibly the most famous fossil is Lucy,[3] a relatively complete skeleton of *Australopithecus afarensis* from Ethiopia discovered in 1974 and dated to about 3.2 Ma. What is so spectacular about fossils of Lucy and other *Austraolpithecus* fossils (Figure 13.12b) is that they indicate she walked upright but had a relatively small skull. Two principal features of humans are that we have relatively large brains and that we walk upright. From fossils like Lucy, we know that our ancestors started walking upright millions of years before our brains grew up. But Lucy is by no means the only big discovery in the East African Rift. Other major hominid finds there include *Ardipithecus* (4–6 Ma), *Kenyanthropus* (~3 Ma) from Kenya, and both *Paranthropus* (1–2.5 Ma) and our genus *Homo* (0–2 Ma), who lived together in Ethiopia, Kenya, Tanzania, and Malawi.

Why are so many hominid fossils found in the Rift Valley? Three factors are important—exposure, preservation, and (paleo)ecology. We know our ancestors and cousins have been roaming the entire continent of Africa for millions of years, because there are other important fossil localities in South Africa, Chad, and Morocco. Tectonic activity and accompanying erosion along the Rift Valley have provided an important service by exposing strata and the fossils they contain. For example, Olduvai Gorge in Tanzania (Figure 13.12c) exposes fossils of *Homo habilis* (one of the earliest human species, ~1.8 Ma), *Paranthropus boisei* (~1.8 Ma), *Homo erectus* (~1.4 Ma), and *Homo sapiens*. And it's not only bones and teeth that we find there—the basic evolution of stone tool technology can also be traced through the Olduvai deposits, from simple tools with chipped edges to sophisticated flakes whose shape and size were controlled by heating in fires.

Preservation matters, too. Sedimentary strata are abundant, but relatively few preserve vertebrate fossils. Volcanogenic sediments, especially finer-grained varieties, tend to preserve fossils better than other types of sediments. Partly, it's a matter of depositional energy. Systems that deposit coarse-grained sediments tend to crush and scatter bones and teeth, unlike quieter settings such as floodplains and lakes. Geochemistry also matters, and sediments that include an ash component tend to preserve fossils better than sediments derived from other sources. Ash beds can also directly preserve footprints of ancient hominids. The 1976 discovery of hominid trackways in the 3.7-Ma ash beds at Laetoli, Tanzania (Figure 13.12d), unequivocally demonstrated that our ancestors or their cousins walked upright by then. If Rift Valley volcanoes weren't spewing ash, we wouldn't have these records.

Finally, fossils of a particular animal (hominid or otherwise) will not be found in an area if that animal did not want to live there. Hominids are thought to have lived in a landscape characterized by open spaces but also wooded areas—savannas or mixed grassland and woods, for example. An upright stance helps in the open, but there is safety and food near trees. Perhaps not surprisingly, topography invites ecosystem diversity that in turn promotes faunal diversity, and the regions on Earth that have the greatest diversity of mammals are often topographically diverse (Figure 13.12a). The diverse topography and ecosystems in East Africa during the last several million years may have been more appealing to hominids, leading to denser populations and a greater likelihood that we might find their fossils there.

continued

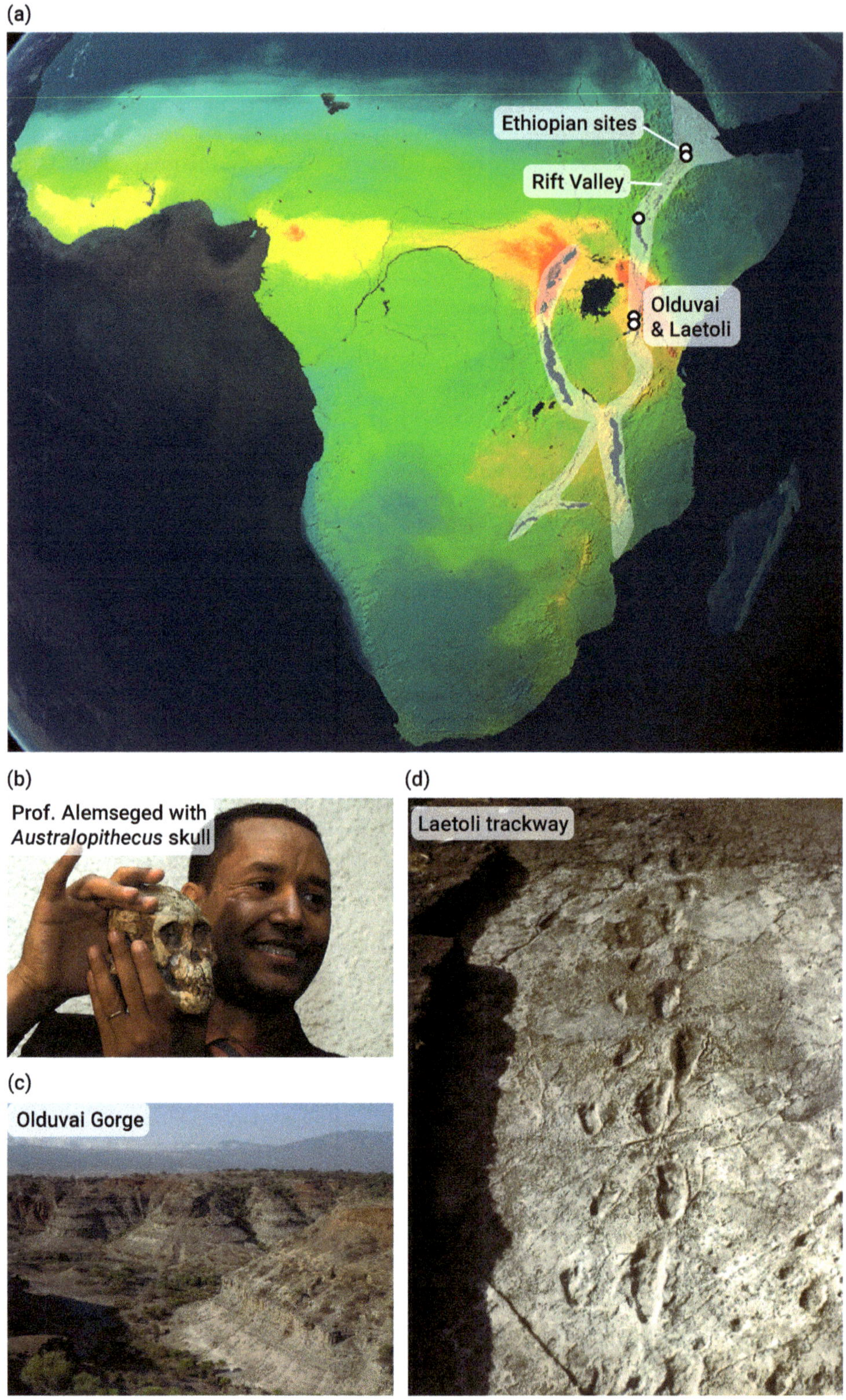

Figure 13.12 (a) Map of Africa, shaded for mammal species diversity, showing the region of the East African Rift (white shading) and a few key fossil localities (dots). Hotter colors (red and orange) represent greater diversity. (b) Paleoanthropologist Dr. Zeresenay Alemseged with fossil skull of juvenile *Australopithecus afarensis* that his research group discovered in Dikika, Ethiopia. (c) Photo of Olduvai Gorge with rift volcanoes in the background. (d) Adult and child hominid tracks at Laetoli, probably of *Australopithecus*, showing that hominids already walked upright by 3.7 Ma.

Sources: (a) Clinton Jenkins, biodiversitymapping.org; (b) Andrew Heavens, via Natural History Museum of London; (c) Yiming Wang, Max Planck Institute; (d) Images of Africa Photobank/Alamy Stock Photo.

Rocks of the Continental Rift Lithosystem

This lithosystem is dominated by alkalic volcanic rocks and associated rift-related, terrestrial sedimentary rocks. Extension can expose metamorphic rocks formed through earlier processes. Common rocks include the following:

- *Bimodal basalt-rhyolite associations*, locally with abundant *andesite* and *dacite*; typically more alkalic than comparable rocks in subduction settings.
- Widespread *tholeiitic basalt* (see Section 4.6).
- Alkalic bimodal associations of alkali basalt, basanite, and nephelinite with trachyte, phonolite, and rhyolite.
- *Mantle xenoliths* of peridotite and other ultramafic rocks, carried to the surface by basalts and more exotic alkalic mafic magmas.
- Rare, highly potassic, low-silica rocks bearing one or more of leucite, kalsilite ($KAlSiO_4$), melilite [$(Ca,Na)_2(Al,Mg)(Al,Si)SiO_7$], phlogopite, and perovskite ($CaTiO_3$, the mineral, not the structure) along with olivine, clinopyroxene, and magnetite; these weird and exceedingly uncommon rocks include *katungite*, *ugandite*, and *melilitite*.
- *Natrocarbonatite* and *nephelinite* lavas at Ol Doinyo Lengai in Tanzania, the only known active carbonatite volcano; these water-soluble carbonatite lavas are predominantly composed of minerals rich in Na_2CO_3.
- In eroded rifts, plutonic equivalents of the volcanic rocks above including *syenite* and *nepheline syenite*.
- *Intrusive calcicarbonatite-ijolite* complexes; the latter, dominated by nepheline and clinopyroxene, is the intrusive equivalent of nephelinite.
- Sedimentary rocks in the continental rift lithosystem include the following:
    - Basin-fill facies ranging from *conglomerate* through *sandstone* to *shale* and *evaporites*.
    - Fault-scarp *breccias* and *megabreccias* produced by collapse of steep topography along faulted mountain fronts; these are important components of many basins that are bounded by normal faults.
    - *Lacustrine sedimentary rocks*, which provide important environmental information.
    - *Fluvial sedimentary rocks*, which provide records of axial river flow.
- *Metamorphic rocks*, exposed in metamorphic core complexes. Typically, these consist of rocks metamorphosed at MP/MT conditions in the greenschist and amphibolite facies, although HP/LT and LP/HT rocks are also possible.

13.2 Fault Styles and Magma Emplacement in Continental Rifting

Learning Objectives

By the end of this section you should be able to:

13.2.1 Make sketches showing why the classic horst-and-graben model of extensional faulting fails to explain many features of continental rifts.

13.2.2 Sketch alternative fault geometries that are more consistent with observations of continental rifts.

High-Angle, Low-Angle, and Rotated Normal Fault Systems

The textbook view of a rift is horst-and-graben structure, in which high-standing blocks (horsts) are flanked by normal faults dipping 60° under valleys (grabens). As with many textbook views, this one is oversimplified. For example, seismic imaging shows that normal faults are typically **listric faults**, curving to shallower dips at depth rather than maintaining a steep dip into the middle and lower crust. Faulting around grabens is typically asymmetrical, with a master fault on one side facing a significantly smaller one on the other. These are termed **half-grabens**.

Listric faults
Curved normal faults that flatten at depth; extension causes strata near the fault to dip toward the fault, producing an anticline.

Half-grabens
Asymmetric grabens with significantly more displacement on one side than the other.

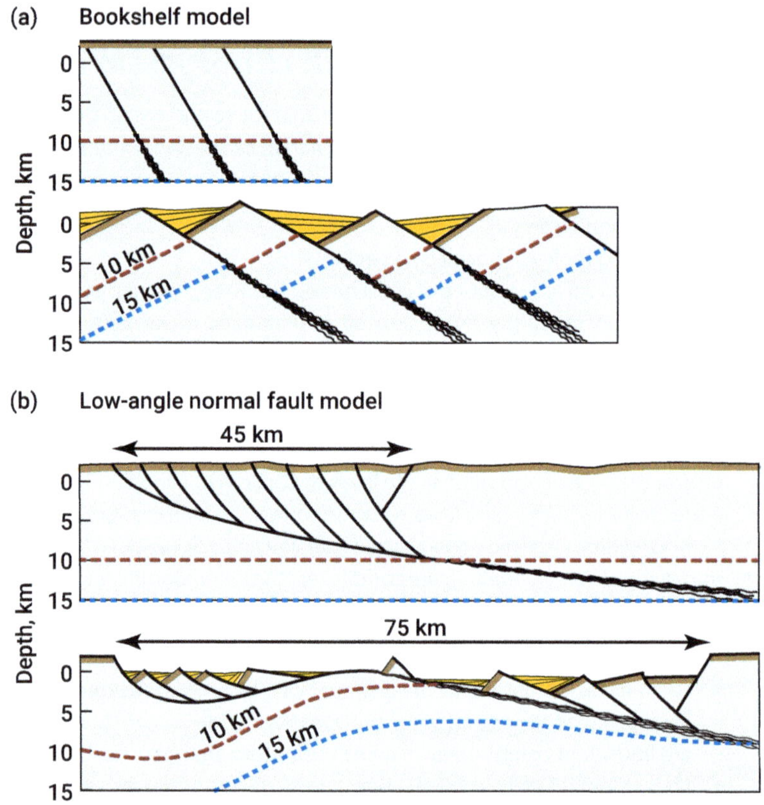

Figure 13.13 Two interpretations of the nature of extensional faulting; dashed lines indicate reference depths in the unfaulted blocks. In each example, $\beta \sim 2$, and brittle faults in the upper crust transition to ductile behavior and production of cataclasite and mylonite in the middle crust (see "Fault Rocks" section that follows). (a) Bookshelf model: dominantly unidirectional tilting produces a series of half-grabens and brings midcrustal rocks significantly closer to the surface. (b) Low-angle normal fault model: bookshelf faults as in (a) root into a master detachment fault, producing a series of half-grabens. In addition to tilting, isostatic rebound caused by unloading of the lower plate can exhume midcrustal rocks.

Source: Modified from Singleton & Mosher (2012),

Stretching factor β
The width of an extended region divided by its original width (McKenzie, 1978).

Detachment faults
Large-scale, low-angle faults separating an upper plate from a lower plate; when unqualified, the term generally refers to extensional faults in contrast to thrust faults.

Dynamic metamorphism
Transformation of a rock into cataclasite or mylonite via movement along faults.

Fault gouge, fault breccia
Finely comminuted, unconsolidated rock material; fault gouge with a substantial proportion of clasts is fault breccia. May be foliated if it contains significant layer silicates.

Two current interpretations of extensional fault style are shown in Figure 13.13. The magnitude of extension can be quantified by the total distance by which points on either side of a rift have diverged, and also by the **stretching factor β**, which is defined as the ratio of the final width of the extended area to its starting width (McKenzie, 1978). In Figure 13.13, $\beta \sim 2$ for both panels.

Crustal extension is commonly accompanied by significant tilting of beds in the hanging wall; this may be accomplished by bookshelf-style faulting (Figure 13.13a), slip on listric faults, or isostatic rebound of the footwall, which may have had many kilometers of rocks pulled off it by faulting (Figure 13.13b). Large extensional faults with tens of kilometers of displacement, such as those on the east side of Death Valley (see Figure 13.1a), commonly dip at low angles; these **detachment faults** place shallow-crustal rocks on mid-crustal rocks brought up in the footwall, as in Figure 13.13b. These systems commonly constitute metamorphic core complexes.

Does Magma Travel along Faults?

The worldwide association of extension with magmatism is undeniable. On the continents, non-subduction volcanism is rare except in rift zones, such as the Basin and Range and East African Rift, and along hotspot tracks (see Figure 1.1; see Chapter 14). Given that crustal magma transport is dominantly through fractures, and that faults are fractures, it might seem like normal faults must act as conduits for magma transport. However, theory and observation contradict the notion that dikes follow normal faults. If this were true, then in eroded volcanic areas, it would be common to find dikes intruded along fault zones, with different units on either side and evidence for shear deformation of the dike's wall rocks, and perhaps of the dike itself. However, such relationships are rare.

In the classic normal fault system (see Figure 13.13a), the least compressive stress is horizontal, the greatest compressive stress is vertical, and faults should dip about 60°

(Figure 13.14). Field studies of extended terranes do commonly find high-angle normal faults. However, mechanical theory predicts that dikes should form perpendicular to the least compressive stress, and thus vertically (90° dip) in the shallow crust, and this, too, is borne out by field observation (see Figure 10.5). Thus, field observation and theory confirm that dikes do not commonly follow normal faults to the surface. Rather, they find or make their own paths along steep fractures.

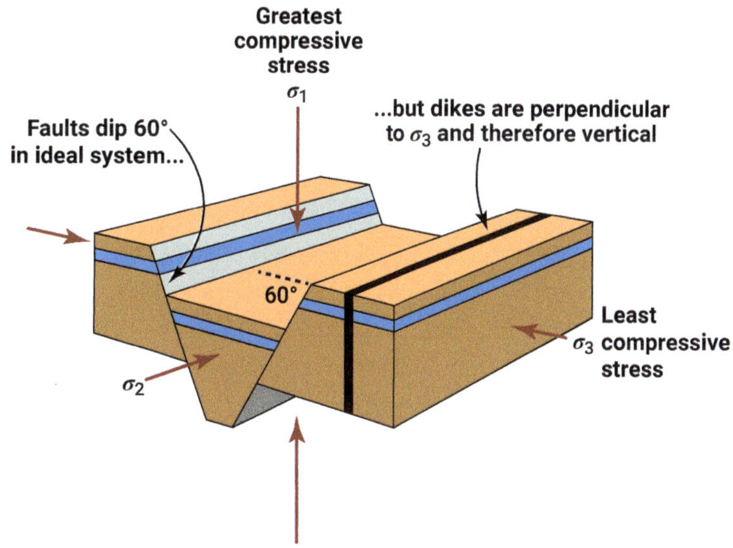

Figure 13.14 Idealized diagram of extensional structures, illustrating geometry with respect to stress field. Although faults should dip 60° with respect to Earth's surface, dikes should intrude along vertical fractures. So, magmas in extensional systems rarely intrude along faults, consistent with field observation.

Fault Rocks: Cataclasite and Mylonite

Movement along faults grinds down the rocks on either side, a process known as **dynamic metamorphism**. Within a few kilometers of the surface, this produces unconsolidated **fault gouge** and **fault breccia**. At deeper levels, confining pressure cements the ground-up material into **cataclasite**, and at greater depth still, the resulting rock is foliated **mylonite** (Figure 13.15).

The transition from generation of non-foliated cataclasite to foliated mylonite is generally attributed to the change in deformation mechanism from brittle failure to ductile flow (Sibson, 1977). The resistance of a rock to brittle failure generally increases with pressure, and thus depth, up to the point where temperatures are high enough that ductile flow takes over. The crossover point is a function of temperature, rock type, and strain rate but generally lies in the middle crust, typically 10 to 20 km depth, at 250 to 350 °C (see Figure 2.21). Figure 13.16 is a generalized cross section of a fault showing the regimes where gouge, cataclasite, and mylonite develop.

Cataclasite
A low-*T* rock formed through brittle fracturing, shattering, or granulation of its constituent minerals. Essentially a cemented breccia.

Mylonite
An intensely deformed rock in which grain size reduction has occurred ductilely. Typically formed at high temperature.

(a)

(b)

Figure 13.15 Examples of fault rocks. (a) Cataclasite developed along a fault in granite from the Marble Mountains, California. Stained red by Fe oxides, multiple generations of cataclasite production are indicated by the clast of cataclasite below the red pencil cap. If unconsolidated, this would be fault breccia (abundant clasts in fault gouge). (b) Mylonite showing asymmetrical fabric features that indicate sense of shear. In this example, the asymmetrical tails on the large clast below the pencil indicate top-to-right sense of shear.

Sources: (a) AFG photo; (b) courtesy of Marli Miller, geologypics.com.

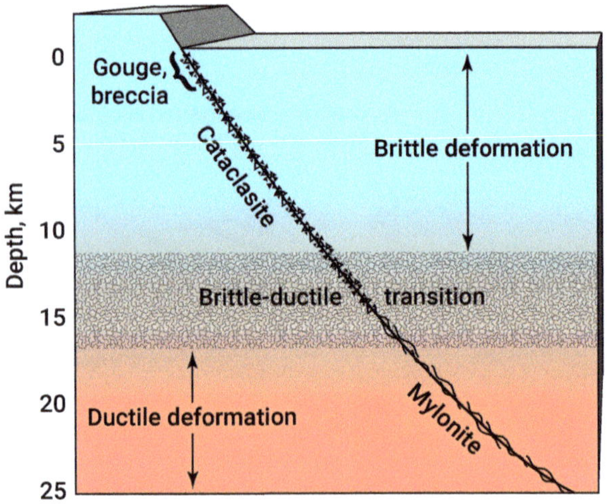

Figure 13.16 Cross section of a fault showing development of gouge and breccia at shallow depths, cataclasite at greater depth, and mylonite below the midcrustal brittle-ductile transition.

Source: Adapted from Fossen (2016).

13.3 **Rifting and Melting**

Learning Objectives

By the end of this section you should be able to:

13.3.1 Compare and contrast models of lithospheric thinning.

13.3.2 Interpret tomographic images in terms of seismic wave velocity, rock type, and thermal structure.

Spreading, Upwelling, and Stretching

In Section 3.1, we discussed formation of magma at oceanic ridges by upwelling of the mantle under oceanic ridges. This globe-encircling system is the most prolific magma producer on Earth. Mantle upwelling and plate divergence proceed in concert at an average rate of about 45 mm/yr. In contrast, continental rifting is more sluggish. The Great Basin is currently spreading at 12 to 15 mm/yr, whereas the rate across the East African Rift system is only 4 to 5 mm/yr.

In Figure 13.17, the distance between points A and B has doubled, and thus the stretching factor $\beta = 2$. This highly schematic representation of lithospheric stretching illustrates several important points:

1. Stretching results in significant crustal thinning. Mass balance requires that the crustal area between points A and B be the same before and after stretching, so for

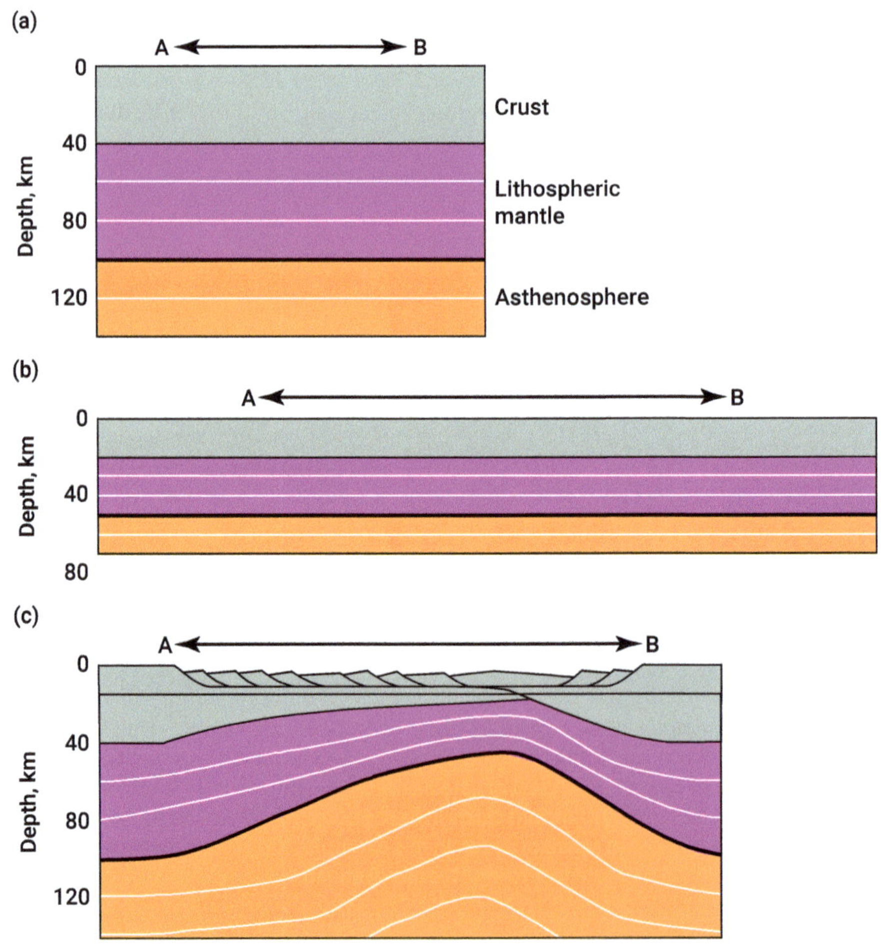

Figure 13.17 Schematic illustration of lifting of isotherms as a result of rifting and lithospheric thinning (note that the gradational LAB is here depicted with a sharp line for simplicity). (a) In this highly schematic diagram, the lithospheric mantle extends to 100 km depth beneath a crust 40 km thick. (b) The entire column is instantaneously stretched to double its initial width, which requires thinning it vertically by 50% and thus doubling the geothermal gradient. (c) A more realistic rift opens up over a finite period of time. The crust stretches in a manner like those in Figure 13.13, and the distance between future rift shoulders A and B in (a) is doubled, leading to an average stretching factor $\beta = 2$. Shallow isotherms and the LAB are lifted ~20 km; compare to the situation at oceanic ridges in Figure 3.21.

$\beta = 2$, the average crust in the rift is half of its original thickness. If stretching is heterogeneous (as is likely), some sections of crust will be thinned even more.

2. Stretching brings isotherms closer together and thus increases the geothermal gradient. Instantaneous, homogeneous stretching with $\beta = 2$, as in Figure 13.17b, results in doubling of the geothermal gradient. For a more realistic geometry with extension occurring over a finite period of time, this packing of the isotherms will be less pronounced (Figure 13.17c).
3. Lithospheric stretching raises the lithosphere-asthenosphere boundary (LAB) and thus thins the lithosphere as well as the crust.

Although perfectly homogeneous stretching will not change the temperature of a rock, it does lower its confining pressure in a manner analogous to mantle upwelling. This upwelling can catalyze both decompression melting in the mantle and partial melting of the crust owing to the increased geothermal gradient.

Thermal erosion
Transformation of lithosphere to asthenosphere through heating (thermal weakening); thermal erosion contrasts with physical erosion, which (in this context) would be the physical removal or replacement of stiff material with weaker material.

Receiver function analysis
Interpretation of the seismic record at a given seismic station (receiver) in terms of velocity interfaces along the seismic ray path.

Thinning of the Lithosphere

In Section 2.8, we discussed the definition of "lithosphere"—the all-important, relatively rigid outer shell of the Earth whose pieces are commonly equated with plates. Recall that the base of the lithosphere (the LAB) is typically a gradational boundary: upper mantle that is weak enough to shear under plate tectonic stresses at plate-tectonic rates is labeled asthenosphere, whereas upper mantle (and overlying crust) that is too stiff to shear so readily is labeled lithosphere. The strength of rocks depends exponentially on temperature. So, magma ascending into the lithosphere warms and weakens it and can gradually transform it into asthenosphere. This process of **thermal erosion** thins the lithosphere.

The location of a boundary with as fuzzy a definition as the LAB is difficult to pin down, especially when we can't observe it directly (e.g., by drilling a really deep hole). Instead, we rely on proxy methods that identify boundaries that are presumed to correlate with a change in mechanical strength. One technique relies on **receiver function analysis** of seismic waves from distant earthquakes (see Toolbox 13.1) to map out seismic interfaces in the crust and upper mantle. In the western United States, receiver function analysis commonly shows an interface with a relatively sharp downward decrease in S-wave velocity at depths of 50 to 100 km. S-wave velocity depends directly on shear strength, so this velocity contrast is commonly interpreted as the LAB. The interface is seismically sharp enough that it likely does not reflect temperature alone, but also encompasses changes in bulk composition or the presence of partial melt (Hopper & Fischer, 2018). Partial melt is a particularly effective way to decrease both the velocity of S-waves and a rock's shear strength.

In the western United States, the LAB is typically at depths of 80 to 100 km under the Colorado Plateau and Great Plains, shallowing to less than 80 km, and locally less than 70 km, under much of the Great Basin (Figure 13.18). This is broadly consistent with extensional thinning of the lithosphere: for stretching factors of β ~1.4–2.8, as in Figure 13.11, 100-km-thick lithosphere would be thinned to $100/\beta$ km, or roughly 70 to less than 40 km, by the mechanisms in Figure 13.17.

An additional way to examine the LAB is to use S-wave tomography. This computationally intense method uses Rayleigh surface waves to model S-wave velocity at depth.

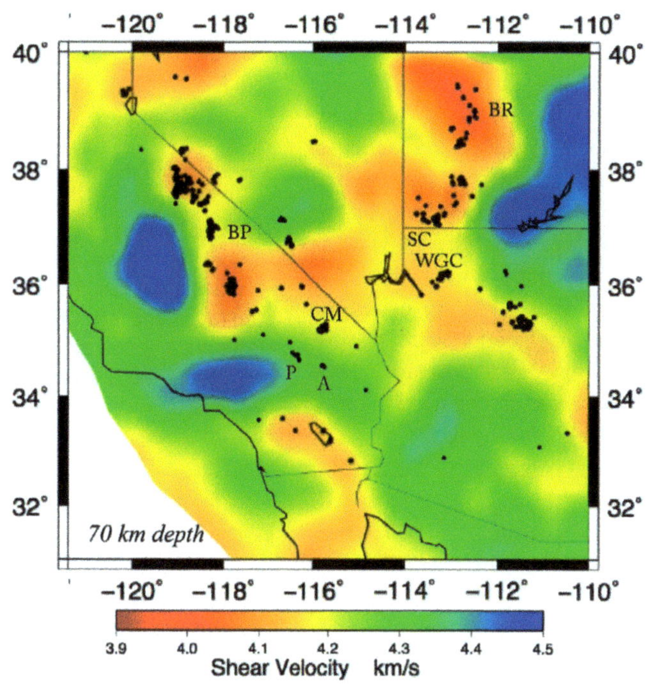

Figure 13.18 Tomographic image showing S-wave velocity at 70 km depth in the southern Great Basin. Black dots show locations of young (<1 Ma) volcanism, which mostly (but not entirely) correlate with areas with low shear velocity. Volcanic fields are Big Pine (BP), Cima (CM), Pisgah (P), Amboy (A), Black Rock (BR), Snow Canyon (SC), and Western Grand Canyon (WGC).

Sources: From Plank & Forsyth (2016), © 2016. American Geophysical Union. All Rights Reserved; from data in Rau & Forsyth (2011).

Toolbox 13.1

Receiver Function Analysis

Students in introductory geology courses are typically taught how to determine the distance to an earthquake by finding the time difference between the P- and S-wave arrivals. The seismic record of an earthquake at a given seismometer contains far more information than just these arrivals, and receiver function analysis (Figure 13.19) is one way to mine this rich record.

Consider a P-wave from a distant earthquake that comes into a seismometer at a relatively steep angle and encounters a velocity interface in the crust. The P-wave will refract (just as light does at a velocity interface such as the surface of a lens), and its energy will also split, with some partitioned into an S-wave. This secondary S-wave travels more slowly than the P-wave and reaches the receiver later than the P-wave. With an estimate of P- and S-wave velocities in the crust in hand, the time difference between the direct P-wave and secondary S-wave gives an estimate of the depth of the interface. A P-wave may also reflect downward off the surface and back upward off the crustal interface (giving rise to P- and S-waves at each bounce), and with a sufficiently clear record, these bounces can be used to refine the interface depth.

A further piece of information in receiver function analysis comes from the polarity of the waves. Seismic velocities in the crust generally increase downward, so an ascending P-wave will generally cross interfaces into slower and slower material. However, a wave passing through a low-velocity zone will pass from slower to faster material, and the polarity of the waves will flip. This makes receiver function analysis well-suited for detecting zones of partially molten rock (e.g., Chu et al., 2010).

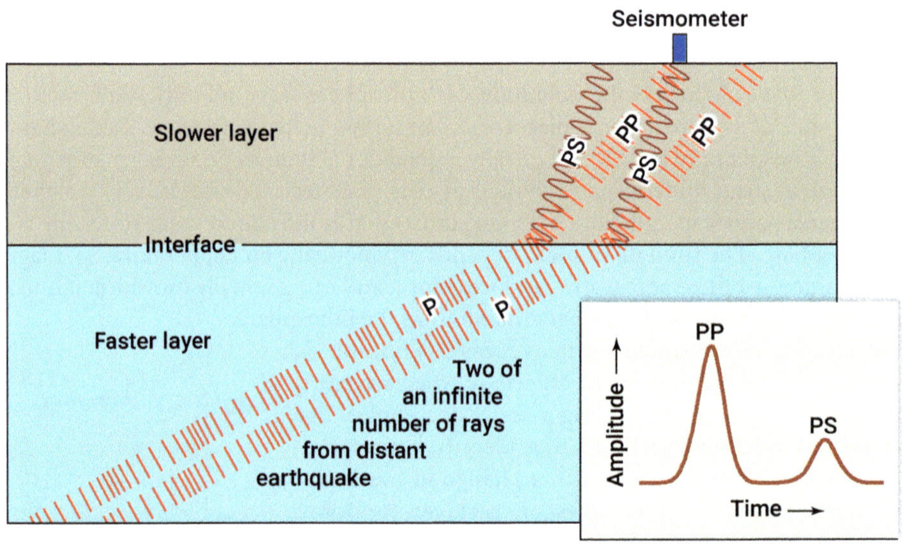

Figure 13.19 Receiver function analysis, simplified. A seismometer sitting above a sharp velocity interface in the crust or upper mantle will receive more than just direct P- and S-wave arrivals. Here, the first arrival is a P-wave that refracted at the interface (here, PP for the ray on the left). This original P-wave also produces an S-wave (here, PS for the ray on the left), but it travels at a slower speed, bends more sharply, and misses the receiver. The PS arrival represents refraction of a parallel ray (the P-wave on the right) at the same interface, and the difference in arrival time of the PP and PS phases gives an estimate of the depth of the interface.

This works because long-period Rayleigh waves, with exceptionally long wavelengths of tens to hundreds of kilometers, "feel" deep into the crust and upper mantle, and thus velocity variations at depth affect their overall velocity. Rau & Forsyth (2011) found that S-wave velocity at 70 km depth varies substantially under the Great Basin (Figure 13.18). Most of the area in Figure 13.18 is slower than a reference Earth value for S-wave velocity at 70 km depth (4.5 km/s; see Figure 2.15), but several zones show particularly slow S-wave transmission. Most areas of volcanism less than 1 Myr old lie within these especially slow zones, consistent with the presence of at least some partial melt in the upper mantle under the volcanic fields. A number of studies (e.g., Mordick & Glazner, 2006; Putirka & Platt, 2012; Till et al., 2013; Plank & Forsyth, 2016) have used igneous thermobarometry to estimate the *P-T* conditions at which basalts last equilibrated with peridotite (i.e., conditions at which basaltic liquids left their peridotite nurseries and headed for the surface). Depths of last equilibration generally range from about 40 to about 75 km, in some cases within 5 to 10 km of the Moho (crust-mantle boundary).

13.4 Subcontinental Lithospheric Mantle

Learning Objectives

By the end of this section you should be able to:

13.4.1 Interpret Sr and Nd isotope trends in terms of different mantle reservoir sources.

13.4.2 Use geochemical observations to argue for or against crustal contamination vs. enriched mantle sources.

Isolation from the Convecting Mantle

MORBs are derived from upper-mantle peridotite by partial melting under mid-ocean ridges. This result is consistent with experimental results on melting peridotite (see Figure 3.16) and by seismic studies of mid-ocean ridge systems. The low $^{87}Sr/^{86}Sr$ (see Figure 3.24) and light REE depletion (see Figure 9.9) of MORBs further indicate that the mantle from which MORB is derived has been depleted in incompatible elements over a significant portion of Earth history. As noted in Section 9.2, the missing incompatible elements (in these cases, Rb and light REE) are permanently sequestered in the continental crust aside from relatively minor percentages that may be returned during subduction of sediments or continental slivers or by delamination (see Section 9.5). The upper mantle is convectively cycled through the oceanic ridge system, then back into the mantle during subduction. Unlike mantle under the continents, oceanic mantle is subject to melt extraction again if it ends up back at an oceanic ridge. You might think of oceanic upper mantle as a social butterfly, leaving home and visiting oceanic ridges (and other parts of the mantle) over and over.

Continental crust enjoys a much stronger bond and longer relationship with its underlying mantle. Continents are largely unsubductable, so mantle lithosphere under them remains isolated from the convecting mantle. Compared to the behavior of oceanic upper mantle, the mantle under continents is more of a wallflower. This **subcontinental lithospheric mantle** (SCLM) does not undergo the multiple cycles of partial melting that oceanic upper mantle experiences, and as a result, it not only retains incompatible elements and volatiles that would be lost during melting but accumulates more owing to fluxing by subduction-derived fluids. Rb, U, and Th are particularly important incompatible elements owing to their use in isotope geochemistry (see Table 2.2). As a result, whereas MORBs have $^{87}Sr/^{86}Sr$ lower than bulk Earth (see Section 3.8), SCLM can have $^{87}Sr/^{86}Sr$ significantly higher than bulk Earth.

Subcontinental lithospheric mantle (SCLM)
The stiff part of the mantle beneath the continents. Because it is attached to buoyant continental crust, this part of the mantle is especially resistant to subduction and recycling.

In discussing SCLM, it is useful to bring in the Sm-Nd isotope system (see Toolbox 13.2). This is another radiogenic system but with a key difference: the parent/child (Sm/Nd) ratio is generally lower in the crust and continental lithosphere than in the mantle, the opposite of the situation for Rb/Sr. As a consequence, SCLM typically has a *lower* radiogenic isotope ratio for Nd ($^{143}Nd/^{144}Nd$) than the mantle or MORB.

SCLM is geochemically distinct from suboceanic lithospheric mantle for two main reasons. First, mantle attached to a continent is not subject to repeated cycles of melting and depletion in incompatible elements and volatiles such as H_2O. Second, subduction enriches the mantle under arcs and continents in a variety of geochemically important components. Subduction brings hydrated minerals, oceanic crust (which is enriched in incompatible elements compared to the oceanic mantle from which it was derived), and sediment shed off the continents not only into the source regions of arc magmas, but beyond.

Toolbox 13.2

The Sm-Nd Isotope System

In this book, we have used the Rb-Sr isotope system in many different ways. This system was one of the first isotope systems to be explored in geochemistry, because it is analytically simpler than others. Parent/child (Rb/Sr) ratios vary greatly in nature owing to the relative sizes and charges of Rb^+ and Sr^{2+}. As a consequence, $^{87}Sr/^{86}Sr$ has a wide range in natural samples (e.g., from 0.702 to >0.8), and thus the relatively low precision of the first mass spectrometers was sufficient to see meaningful differences. Rb and Sr are also easy to separate in the laboratory. Even though the half-life of ^{87}Rb is 10 times the age of the Earth (see Table 2.2), Rb/Sr varies enough that measurable differences in $^{87}Sr/^{86}Sr$ accumulate rapidly.

The Rb/Sr system is compromised, however, by the geochemical mobility of Rb and Sr. In particular, Rb, a singly charged LILE, is subject to movement by hydrothermal alteration and weathering. This is particularly problematic when trying to age-correct measured $^{87}Sr/^{86}Sr$, and thus the Rb/Sr method should only be applied to pristine samples.

The Sm/Nd system complements the Rb/Sr system. The isotope ^{147}Sm α-decays to ^{143}Sm with a half-life of 106 Gyr. Although this is only twice as long as the half-life of ^{87}Rb, changes in $^{143}Nd/^{144}Nd$ accumulate far more slowly, because natural variations in the parent/child (Sm/Nd) ratio are tiny. Both are REE, and separation by geochemical processes is minimal. Whereas the Rb/Sr ratio in rocks ranges over several orders of magnitude, the Sm/Nd ratio rarely varies by even a factor of two.

Long half-life and small parent/child variations mean that changes in $^{143}Nd/^{144}Nd$ are tiny (e.g., in the western United States, $^{87}Sr/^{86}Sr$ varies from 0.703 to >0.712, or about 10%, whereas $^{143}Nd/^{144}Nd$ varies from 0.5112 to 0.5133, or about 0.5%). Measuring such tiny changes required great improvements in mass spectrometer sensitivity and stability, and development of new procedures for chemically separating Sm from Nd.

Because changes in $^{143}Nd/^{144}Nd$ are minuscule, ratios are commonly reported in ε (epsilon) notation, similar to the δ notation used for oxygen isotopes (see Section 5.4, Equation 5.6), but a factor of 10 more sensitive. Also, the reference value is a theoretical chondritic reservoir (**CHUR**) corrected to the time of interest rather than seawater.

CHUR
A theoretical reservoir with a chondritic Sm/Nd ratio; as with VSMOW and δ notation, CHUR is used as a baseline for ε notation when discussing variations in $^{143}Nd/^{144}Nd$.

Ocean-island basalts (OIBs)
Basalts from ocean islands that are not associated with spreading centers or subduction zones. The geochemistry of OIBs more closely approaches primitive mantle or bulk silicate Earth than does that of MORBs.

Depleted mantle
Mantle that has been depleted in incompatible elements, especially large-ion lithophile elements (K, Rb, Cs, Sr, Ba) and light REE (La, Ce, Pr, Nd), which are preferentially sequestered in continental crust.

As a consequence, basaltic rocks erupted on the continents have much greater variability in radiogenic isotope ratios than those from the oceans.

The Pacific Northwest provides an interesting example of the contrast between relatively young oceanic mantle and SCLM. Figure 13.20 shows a cross section through the northwestern United States, illustrating crustal and lithospheric thicknesses. Particularly noticeable is that crust west of about 116°W latitude is thinner than that to the east. Crust to the west contains abundant, relatively young (<300 Ma) accreted oceanic terranes, whereas crust to the east contains ancient components, including the Wyoming Craton, where some rocks are as old as 3.5 Ga. The boundary between young and old crust is exposed along the Western Idaho shear zone (and other structures to the north and south) and presents as a stark difference in $^{87}Sr/^{86}Sr$ for recent volcanic rocks—lower values within the young oceanic terranes to the west (as low as 0.703), and higher ones in the older cratonal rocks to the east (up to 0.720). Although SCLM is normally quite thick (e.g., ~200 km at 106°W longitude), it is thin between about 116 and 110°W longitude beneath the Snake River Plain. Passage of the Yellowstone hotspot over the last 15 Myr is thought to have eroded the lower part, yielding similar lithospheric thicknesses even though the crust to the east is much thicker.

Depleted and Primitive Mantle

The isotope systematics of MORBs, **ocean-island basalts** (**OIBs**), and continental basalts illustrate the differences between SCLM and oceanic mantle. MORBs generally form a tight array that lies at the low end of measured $^{87}Sr/^{86}Sr$ and high end of $^{143}Nd/^{144}Nd$ ratios (Figure 13.21a). This location on the isotopic array ($^{87}Sr/^{86}Sr = 0.7026$, $^{143}Nd/^{144}Nd = 0.51311$; Salters & Stracke, 2004) is known as **depleted mantle** (or depleted MORB mantle). The corresponding point for the **bulk silicate Earth** (i.e., bulk Earth minus the core),

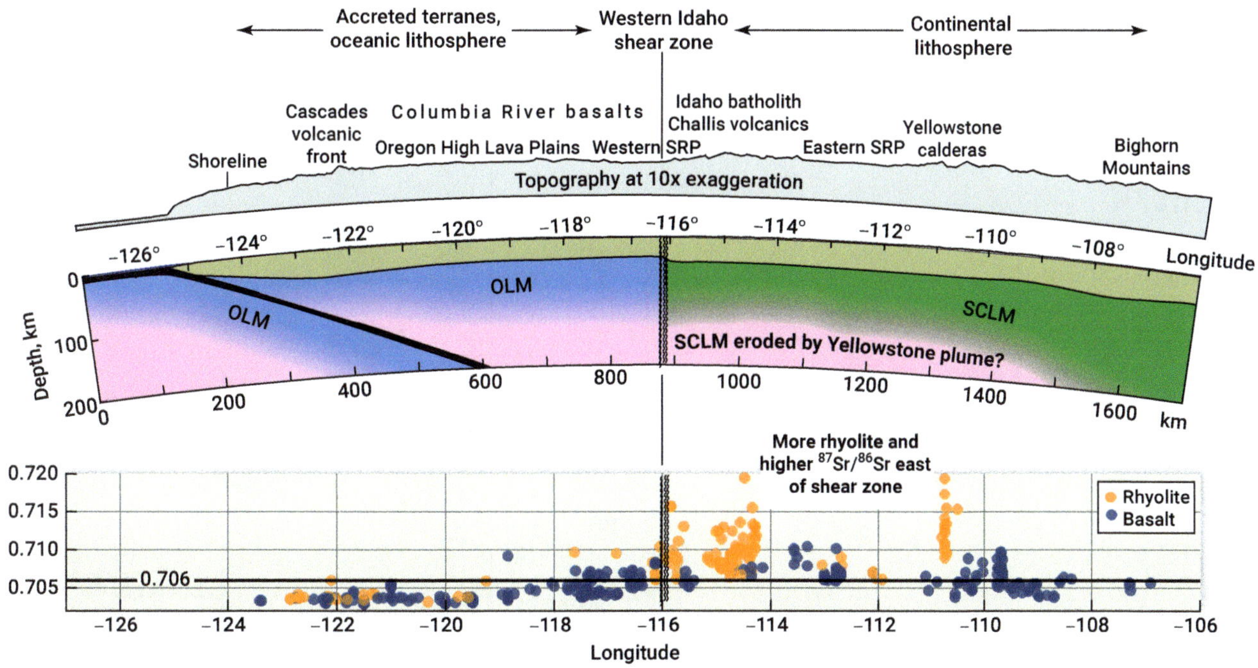

Figure 13.20 Interpretive east-west cross section from the Oregon coast to north-central Wyoming at latitude 44°N, showing the Cascadia subduction zone and the inferred lithospheric boundary coinciding with the mapped western Idaho shear zone. Lithospheric thickness increases to >200 km toward the continental interior but farther west it is 100–150 km thick. $^{87}Sr/^{86}Sr$ in young volcanic rocks increases significantly crossing the western Idaho shear zone eastward, especially in rhyolites. Topography along the cross section is drawn true to scale, showing that the Earth's surface is essentially a smooth sphere; topography above, in gray, is exaggerated by a factor of 10. Sr-isotope data, from earthchem.org, were projected onto the cross section from a swath between 40–47°N. OLM = oceanic lithospheric mantle; SCLM = sub-continental lithospheric mantle; SRP = Snake River Plain.

Sources: Crustal thickness from Stanciu et al. (2016); LAB from Yuan & Romanowicz (2010); slab geometry from McCrory et al. (2012).

interpreted as representing primitive, unprocessed mantle, lies toward higher $^{87}Sr/^{86}Sr$ (0.7045) and lower $^{143}Nd/^{144}Nd$ (0.51263), and OIBs generally lie in between these two points. This general trend, known as the **mantle array** (Figure 13.21b), suggests that OIBs are derived from both depleted mantle and from primitive mantle that reflects the Earth's overall isotopic composition.

Bulk silicate Earth, BSE
The estimated overall composition of the Earth after subtracting the core, as estimated from meteorites and models of solar system formation. Because the mantle dominates the silicate budget, primitive (unprocessed) mantle is thought to have a composition similar to BSE.

Mantle array
Refers to the nearly linear array of oceanic basalt compositions on a plot of Nd vs. Sr isotopes. The term has been generalized to other isotope plots.

Enriched Mantle or Crustal Contamination?

In contrast to the tight arrays shown by MORB and OIB, basalts from the continents scatter over a much larger area of the Sr-Nd isotope diagram (Figure 13.21c). There are two obvious explanations for why continental basalts have such scatter: either the mantle source has a highly variable composition, or continental crust contaminates melts as they traverse the crust. For brevity, we will use $^{87}Sr/^{86}Sr$ in the following discussion, but similar arguments can be made for $^{143}Nd/^{144}Nd$ (with "low" substituting for "high," etc.) and other radiogenic isotope systems.

Crustal Contamination and Open Magmatic Systems

Rising magmas can assimilate crustal rock or mix with partial melts from the crust (see Section 8.2). We know this because of visible evidence of partially digested xenoliths and geochemical indications of mixing between uncontaminated magma and its contaminant. The Kirkpatrick lavas of Antarctica have already been presented as an example of contamination of basaltic magma by crustal material with relatively high $^{87}Sr/^{86}Sr$ (see Figure 9.14c,d). Recall that continental crust is enriched in SiO_2 and has a higher Rb/Sr ratio, which over

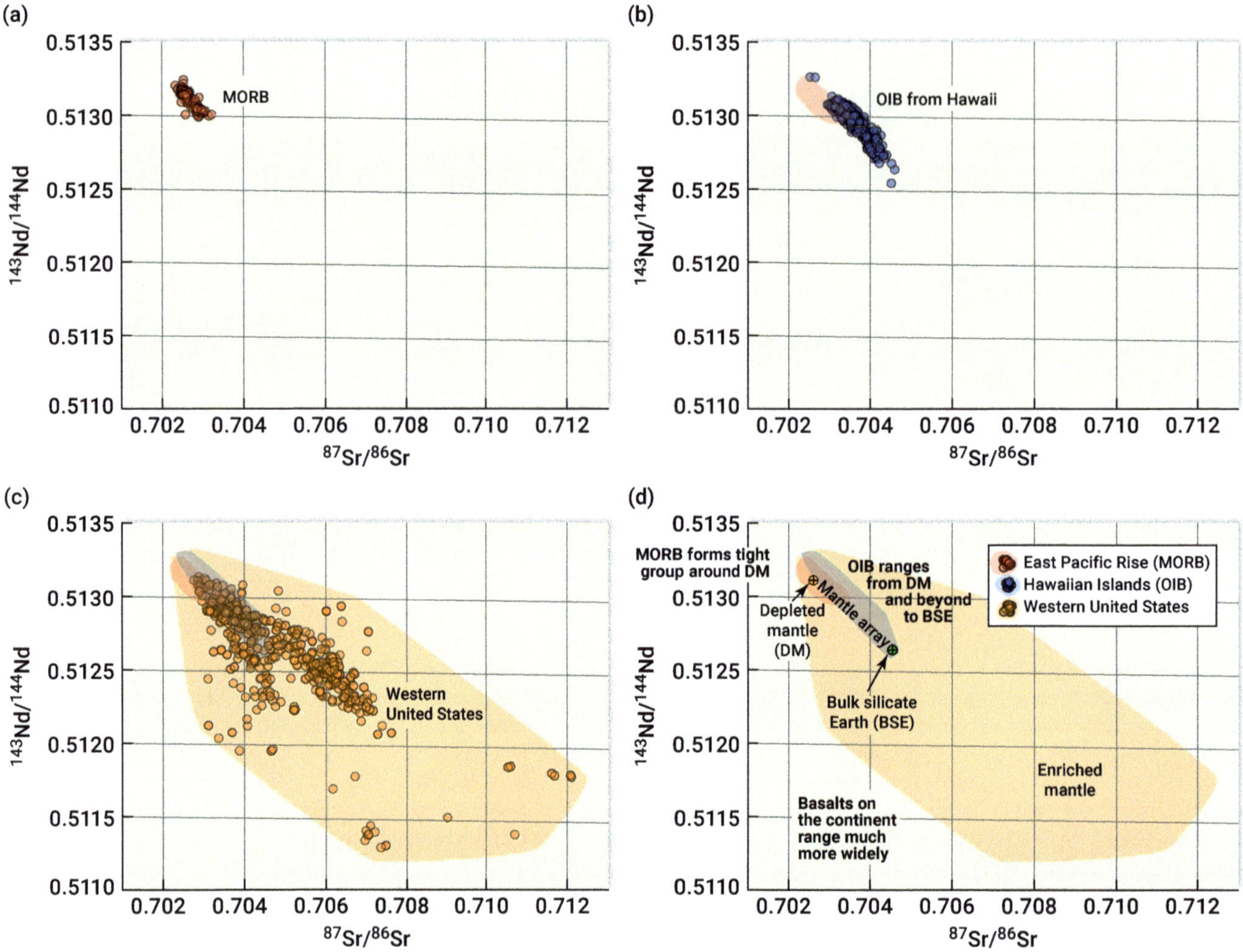

Figure 13.21 Nd and Sr isotope compositions and trends for basalts from different tectonic settings. (a) Typical mid-ocean ridge basalt from the East Pacific Rise, showing highly restricted range of compositions. (b) Typical ocean island basalt from Hawai'i, showing limited trend toward lower $^{143}Nd/^{144}Nd$ and higher $^{87}Sr/^{86}Sr$. (c) Western United States, showing wide range extending to very low $^{143}Nd/^{144}Nd$ and very high $^{87}Sr/^{86}Sr$. (d) Summary identifying different mantle sources for magmas in different tectonic settings.

time decays to increase $^{87}Sr/^{86}Sr$ faster than in the mantle. Consequently, a tell-tale sign of crustal contamination is a positive correlation of $^{87}Sr/^{86}Sr$ values (corrected back to the emplacement age of the suite) with SiO_2.

Enriched mantle
Mantle that is enriched in incompatible elements, especially large-ion lithophile elements and light REE. Characterized by relatively high Rb/Sr and $^{87}Sr/^{86}Sr$, and low Sm/Nd and $^{143}Nd/^{144}Nd$ ratios.

Enriched Mantle and the "0.706 Line"

Although it is tempting to attribute the scatter of isotopic compositions in Figure 13.21c to crustal contamination, several compelling lines of evidence indicate that at least some, and in some cases most, of the range in compositions comes from the mantle itself—**enriched mantle** with higher Rb/Sr, and lower Sm/Nd, than primitive or depleted mantle. Such evidence includes:

1. High initial $^{87}Sr/^{86}Sr$ (e.g., >0.705) in primitive basalts, diorites, and gabbros that show no petrographic evidence for contamination.
2. Regionally coherent patterns of high $^{87}Sr/^{86}Sr$ in basalts.
3. High $^{87}Sr/^{86}Sr$ in basalts that carry peridotite and pyroxenite xenoliths from the mantle (discussed later); the physical presence of mantle xenoliths indicates that the host basalts rose rapidly and thus had little time to interact with the crust.
4. High $^{87}Sr/^{86}Sr$ in mantle xenoliths carried in basalts with similarly high $^{87}Sr/^{86}Sr$.

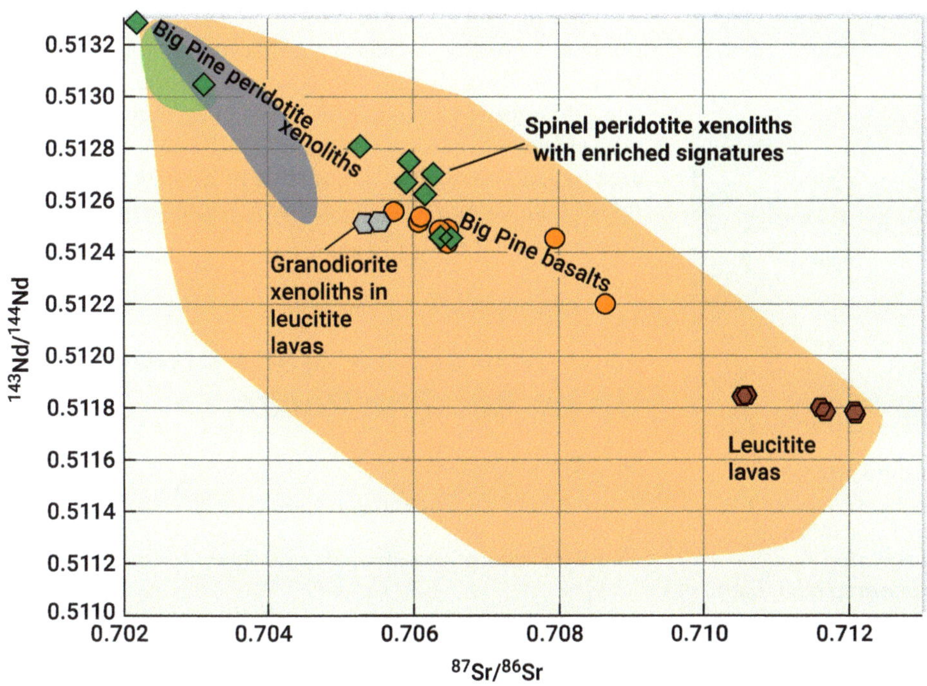

Figure 13.22 Isotopic data for basalts and xenoliths from the Big Pine volcanic field, eastern California, showing a wide range of compositions that cannot be explained simply by crustal contamination. Xenoliths indicate that both depleted and enriched mantle are present. Leucitite lavas from a small exposure east of the Sierra Nevada have some of the most extreme isotopic values found among volcanic rocks in western North America.

In the early days of Sr-isotope geochemistry, Hedge (1966) and Leeman (1970) applied the new Rb/Sr technique to volcanic rocks in the western United States and found consistent regional variations in $^{87}Sr/^{86}Sr$, with high values (>0.706) in east-central California. Kistler & Peterman (1973) found that initial $^{87}Sr/^{86}Sr$ in Mesozoic plutons generally match those in the basalts, and this was later extended to $^{143}Nd/^{144}Nd$ (DePaolo, 1981; Farmer & DePaolo, 1983).

Across the Basin and Range, initial $^{87}Sr/^{86}Sr$ generally increases eastward, and a curve separating values largely less than 0.706 to the west from those largely greater than 0.706 to the east is known as the 0.706 line ("seven-oh-six line"). The 0.706 line is commonly equated with the western edge of North American basement. This is likely true for the cross section in Figure 13.20, where the tectonic history is well known, rock exposures are good, and the tectonic boundary is sharp and steep, but is less certain for much of western North America. Data scatter in some areas is quite large, meaning magmas must either tap different sources or experience variable contamination (or both), so the "line" may be more properly described as a band. A further complicating factor is deformation. For example, in southern California, Triassic plutons with a depleted isotopic signature were apparently thrust eastward over Proterozoic lithosphere in the Mojave Desert, and later Jurassic plutons in the same region show an enriched signature (Miller et al., 1995).

Test Case: Eastern California Lavas and Plutons The Quaternary Big Pine volcanic field in eastern California lies within the domain of $^{87}Sr/^{86}Sr > 0.706$ and is unusual because several lava flows in the field contain spinel peridotite xenoliths. Host lavas and most of their xenoliths have high $^{87}Sr/^{86}Sr$ and low $^{143}Nd/^{144}Nd$—the enriched mantle signature, consistent with the hypothesis that the enriched signature reflects the mantle source rather than contamination (Figure 13.22; Beard & Glazner, 1995). Two of the analyzed xenoliths have low $^{87}Sr/^{86}Sr$ and high $^{143}Nd/^{144}Nd$, suggesting that depleted mantle is also present in the source region of the basalts. Leucite-bearing lavas (leucitites) crop out near the Big Pine field; these have bizarre isotopic compositions and confirm that some isotopically weird stuff lies under eastern California. More on such rocks in Section 13.6.

The geochemistry of Cretaceous plutonic rocks in the eastern Sierra Nevada also implicates enriched mantle. Plutons there show a large range in SiO_2 contents (gabbro to granite) but have an essentially constant, enriched isotopic signature with initial $^{87}Sr/^{86}Sr$

Figure 13.23 The eastern escarpment of the Sierra Nevada near Big Pine, California, exposes a thick series of diorite sheets capped by granite (this area lies just south of Split Mountain, shown in Figure 10.17a). Pleistocene lavas of the Big Pine volcanic field poured out of cinder cones that erupted through the alluvial fans shed off the range; eight cinder cones, some deeply eroded, are visible in this photo. The lavas, peridotite xenoliths that they carry, diorite, and granite all have the isotopic signature of enriched mantle.

Source: AFG photo.

~ 0.706 (Figure 13.23). Recalling that crustal contamination predicts a positive correlation between $^{87}Sr/^{86}Sr$ and SiO_2, the uniformly high $^{87}Sr/^{86}Sr$ independent of SiO_2 thus appears to arise from the mantle and not from interaction with old continental crust. Apparently, granitic rocks in this area were distilled out of the mantle in the Late Cretaceous, rather than being recycled products from older continental crust (Coleman et al., 1992).

13.5 Mantle Samples Brought Up by Alkaline Rocks

Learning Objective

By the end of this section you should be able to:

13.5.1 Interpret origins of mantle xenoliths based on mineralogy.

Alkaline rocks commonly bring up samples of the mantle when they erupt. These ultramafic xenoliths are by their nature unusual because they ended up on the surface rather than staying in the mantle where they belong; how representative they are of the mantle is not known. Mantle xenoliths range up to some tens of centimeters in maximum dimension, and to transport them to the surface requires two things: that the magma rose rapidly enough to entrain the xenoliths, and that there was a conduit with an aperture of at least xenolith dimensions that ran from the mantle to the surface (although there is no requirement that the entire conduit was open at the same time). Typical estimates using Stokes' law (see Section 8.1) and 20- to 40-cm-diameter xenoliths yield minimum ascent velocities on the order of 0.5 m/s, or about 1 km/hr. Thus, minimum transit times from the upper mantle should be no longer than two to three days, and perhaps much faster—kimberlites (see Section 13.6) may hit speeds of 200 km/hr during ascent (Russell et al., 2019).

Mantle xenoliths (Figure 13.24) are common in lavas worldwide, in all tectonic settings, although they are most prevalent in alkaline host rocks and almost unknown in tholeiitic (low-alkali) basalts. The reason for this dichotomy is a mystery. Classification of mantle xenoliths is based on modal mineralogy, fabric, chemical composition, and presence of minor phases such as mica or amphibole (Wilshire et al., 1988; Pearson et al., 2014). Observed xenolith compositions fall over most of the Ol-Opx-Cpx classification triangle (see Figure 3.17).

There are two main types of ultramafic xenoliths (Table 13.1). Xenoliths of the Cr-diopside type contain bright-green, Cr-rich Cpx and spinel along with Mg-rich olivine and orthopyroxene (see Figure 13.24). These tend to fall toward the Ol apex of the classification triangle, in the peridotite and dunite fields. Xenoliths of the Al-augite type contain Al- and Ti-rich Cpx, and comparatively Fe-rich olivine and pyroxenes. These fall along or near the Ol-Cpx side of the classification triangle (clinopyroxenites and wehrlites). In both suites, spinel is the main Al-bearing mineral, although garnet, indicating derivation from greater depth (see Figure 3.9), sometimes occurs.

Pyroxene-rich xenoliths (wehrlites, clinopyroxenites, websterites, orthopyroxenites; see Figure 3.17) commonly occur as veins in Cr-diopside peridotite hosts. Pyroxenites are chemically much closer in composition to basalt than is peridotite (see Appendix 10), whereas harzburgite (orthopyroxene plus olivine) and dunite (nearly pure olivine) can form by extracting melt from peridotite (see Figure 3.18). Pyroxenite veins of the Al-augite group are thus interpreted as basaltic veins that didn't quite make it out of the mantle before they crystallized (Wilshire & Shervais, 1975).

Table 13.1 Features of Principal Xenolith Groups

	Cr-Diopside Group	Al-Augite Group
Rock types	**Peridotite**, dunite, websterite, wehrlite	**Clinopyroxenite**, wehrlite, dunite, peridotite
Texture	Metamorphic tectonite layers	Branching and cross-cutting igneous layers (e.g., Cpx-rich veins cutting Cr-diopside peridotite)
Mineral preferred orientation	Common	Uncommon
Mineralogy	**Olivine, Opx, Cpx,** spinel, pargasite (Na-rich hornblende), phlogopite (Mg-rich biotite)	**Cpx, Olivine,** spinel, Opx, kaersutite (Na-Ti-rich oxy-hornblende), phlogopite

Bold indicates dominant.
Source: From Wilshire & Shervais (1975).

Of importance for understanding SCLM are relatively rare xenoliths that contain hydrous, alkali-rich minerals. Chief among these minerals are Mg-rich biotite (phlogopite) and amphibole, either hornblende or kaersutite (an Na- and Ti-rich hornblende). These minerals show that SCLM is enriched both in water and alkalis. Because both constituents promote melting at lower temperature, they indicate that SCLM has been fertilized. The source of the fertilizing alkali-rich fluids is debated, but dewatering of subducting slabs likely contributes. Otherwise, speculation involves interaction with compositionally exotic melts discussed below.

Figure 13.24 Cr-diopside group peridotite xenolith in Pleistocene basanite (alkali-rich basalt) lava from San Carlos, Arizona. This is the most common variety of ultramafic xenolith in western US localities. The peridotite comprises mostly olivine with bright-green, Cr-rich Cpx, dark brown Opx, and black spinel.

Source: James St. John/Flickr.

13.6 Alkaline Magmatism

Learning Objectives

By the end of this section you should be able to:

13.6.1 Use geochemistry and mineralogy to distinguish among diverse alkaline rocks.

13.6.2 Justify the value of investigating alkaline rocks when they are so rare.

Interesting and Odd

Alkaline igneous rocks, those richer in K and Na than the usual BADR array, have long fascinated petrologists. There are many reasons for this, among them that their unusual compositions lead to crystallization of uncommon minerals, and some alkaline rocks are sources of rare and important commodities such as REE, Nb, and diamonds. A result of this fascination is that their nomenclature is a tangled mess of rock names based on place, and some of these names are unique to a single, remote locality. For example, the name "kakortokite" (an especially pretty variety of nepheline syenite) seems to have only been applied to the small, rarely visited area in southern Greenland where it was first described. In the East African Rift, names applied to more-or-less basaltic rocks include basalt, basanite, tephrite, and other names on the TAS diagram, but also hawaiite, lamproite, absarokite, mugearite, benmoreite, leucitite, nephelinite, shoshonite, ankaratrite, foidite, and more. These names are meaningful to those working on alkaline rocks, but to few others, and we will keep the nomenclature simple here.

Those interested in traveling the alkaline path have many (albeit conflicting) resources for sorting out the terminology, including LeMaitre (2002) and citations therein. According to O'Brien et al. (2005), alkaline magmas make up less than 1 percent of all igneous rocks but account for nearly half of igneous rock names. Many are strangely appropriate to certain poetic forms, as in this Tanka poem celebrating (?) alkaline rock names:

Italite, orbite,
Orvietite, phonolite;
Jacupirangite,
Benmoreite and leucitite;
Just a few alkaline rocks.

We split our discussion of alkaline rocks between this chapter and the next, because such rocks are prominent in continental rifts but also occur in a wide variety of tectonic environments including subduction zones, cratons, and ocean islands. Here, we cover alkaline rocks associated with rifting, and others whose associations are less clear.

Lamprophyres and Lamproites

Alkaline rocks range across much more of the compositional spectrum than those of the BADR suite, and as a consequence, many uncommon minerals, some of which are quite valuable, are found in them. This section and the next two briefly discuss four entangled, important branches of the alkaline rock family. Figure 13.25 introduces the general geochemical characteristics of these rocks. Compared to typical continental crust and the BADR suite, they are low to extremely low (zero) in SiO_2, high to extremely high in MgO, low to extremely high in K_2O, and high in TiO_2. They are also greatly enriched in incompatible elements, including a number of economically valuable ones.

Lamprophyres
Rare, ultrapotassic mafic to ultramafic igneous rocks, usually exposed as dikes or diatremes, with phenocrysts of Cpx and either biotite or hornblende, but lacking phenocrystic plagioclase.

Diatreme
Funnel- or carrot-shaped pipe filled with breccia composed of fragmented magma and wall rock; may reach downward 1 km or more. Diatremes are likely expressed as maars at the surface and terminate downward in dikes.

Lamproites
K_2O- and MgO-rich, low-silica, hypabyssal or extrusive rocks with high concentrations of incompatible elements, high Fe^{3+}/Fe^{2+}, and exotic mineralogy.

Lamprophyres

Lamprophyres are mafic to ultramafic, porphyritic rocks that have conspicuous phenocrysts of mafic minerals but lack phenocrystic plagioclase. As a group, they overlap geochemically with alkalic basalts but extend to much lower SiO_2, higher MgO, and higher incompatible and HFSE elements (see Figure 13.25).

The classic lamprophyre is a dike with phenocrysts of biotite or hornblende and Cpx, and a prominent mass of such is Ship Rock (Figure 13.26), an eroded **diatreme** in northwestern New Mexico. Three massive radial dikes emanated from the 500-m-tall, 27-Ma monument, known to the Navajo people as *Tsé Bit'a'í*, the "rock with wings." The rock comprising Ship Rock (minette) contains phenocrysts of Mg-rich (phlogopitic) biotite and Mg-rich Cpx in a groundmass of the same with sanidine and apatite; plagioclase is absent.

Lamproites

Lamproites are similar to lamprophyres but extend to extremely high K_2O, with $K_2O \gg Na_2O$. They are characterized by an unusual mineral assemblage that typically includes one or more of Ti-rich, Al-poor phlogopite; Ti-rich, K-rich, hornblende-like amphibole (richterite); Fe-rich leucite; and Fe-rich sanidine. According to LeMaitre (2002, p. 16), "Minor and common accessory phases include priderite, wadeite, apatite, perovskite, magnesiochromite, titanian magnesiochromite and magnesian titaniferous magnetite with less commonly, but characteristically, jeppeite, armalcolite, shcherbakovite, ilmenite and enstatite."[4] Williams et al. (1982, p. 204) stated, "There is a good deal of mineralogical variation in the suite and a correspondingly complex and rather outlandish terminology has become accepted."

Lamproites have silica contents broadly centered on those of basaltic rocks (Figure 13.25a), but with exceedingly high concentrations of incompatible elements. LILEs and HFSEs are on the order of 100 times their concentration in primitive mantle and 10 times that in continental arc basalt (Figure 13.25b). Extreme concentrations of elements

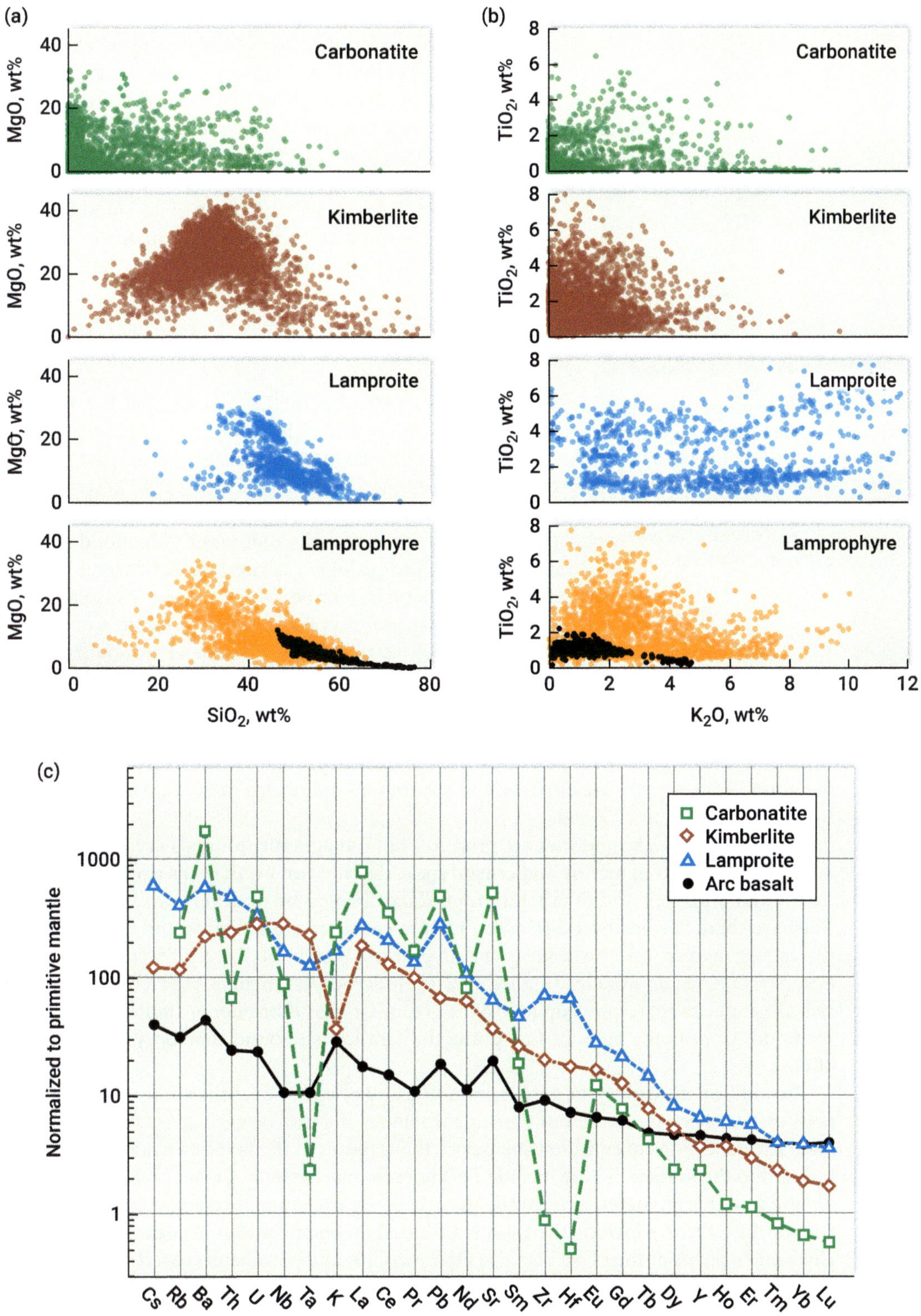

Figure 13.25 (a) Silica vs. MgO for carbonatites, kimberlites, lamproites, and lamprophyres (colored symbols) compared to a typical subduction-related suite (Medicine Lake; black symbols in bottom panel). The alkaline rocks reach exceedingly low SiO_2 and high MgO compared to arc rocks, which we use as a proxy for continental crust. (b) A high-field-strength oxide (TiO_2) vs. an incompatible oxide (K_2O). Again, both are exceedingly high compared to the continental crust. (c) Mean values for carbonatite, kimberlite, and lamproite, normalized to primitive mantle, as compared to average continental arc basalt. These volumetrically tiny components of the crust reach extreme values of many incompatible elements.

Sources: Data for averages from precompiled datasets from GEOROC (georoc.eu); arc basalt from Kelemen et al. (2014).

Figure 13.26 Ship Rock, showing dikes that radiate out from it. The summit lies nearly 500 m above the plain, which is underlain by Cretaceous shales. Ship Rock likely represents the neck of a diatreme, emplaced about 1 km below a now-eroded maar. Eroded igneous bodies such as this are scattered across the Four Corners region, providing positive relief on the otherwise flat plateau; for another kind of igneous landform on the Colorado Plateau, see Figure 10.13.

Source: Venti Views/Unsplash.

such as Ba, Nb, and Ti, coupled with a high Fe^{3+}/Fe^{2+} ratio, leads to the unusual mineralogy of many lamproites.

Lamproite localities are scattered across the globe, typically outside of cratons (Bergman, 1987). North American examples include the Leucite Hills, Wyoming, and several localities in Montana; the Navajo volcanic field of the Four Corners region; a few localities in the Sierra Nevada; and Hills Pond, Kansas. The discovery of diamond-bearing lamproites in Australia and Arkansas has made the unusual mineralogy of these uncommon rocks of more than passing interest.

Kimberlites

General Characteristics and Occurrence

Kimberlites are volatile-rich, low silica, MgO-rich rocks with abundant olivine (~50 modal %), carbonate (>5 wt% CO_2), and water (>5 wt% H_2O) and up to several wt% K_2O (see Figure 13.25). Kimberlites are typically found as part of the breccia that fills diatremes ("diamond pipes") along with high-*P* eclogite, garnet peridotite, and locally, diamond xenocrysts. Kimberlites can occur as isolated intrusions but are more typically found in clusters of several to dozens, such as the Kimberley cluster in South Africa, from which kimberlites get their name. They are more common in and adjacent to Archean cratons, although they have perforated numerous other areas, too, such as in a belt stretching from Tennessee to Quebec (Figure 13.27a). Kimberlites excavate mantle xenoliths from greater depth than other alkaline rocks, and diamond-bearing pipes are of particular economic importance in South Africa, Lesotho, Botswana, Siberia, and Canada.

Kimberlites
Low-silica, high-MgO, serpentinized breccias, rich in olivine, carbonate, and H_2O. They are composed of deeply sourced magma carrying a large proportion of xenocrysts entrained from mantle and crustal wall rocks.

Macrocrysts
Term used for larger (generally 0.5–10 mm) crystals in kimberlite; includes both phenocrysts and xenocrysts.

Mineralogically, kimberlites are mixtures of (1) magmatic minerals crystallized from melt, (2) xenocrysts of mantle and crustal rocks derived from wall rocks ripped up during ascent and eruption, and (3) alteration products formed by reaction with various fluids. They are characterized by a distinctive texture consisting of a diverse and unusual set of crystals termed **macrocrysts** set in a fine-grained matrix of serpentinized olivine, calcite, garnet, Cpx, and oxides (Figure 13.27b). Macrocrysts include Mg-rich olivine; Mg-rich ilmenite; pyrope; phlogopite; bright-green, Cr-rich diopside; enstatite; and locally, diamond. Commonly rims of fine-grained alteration surround macrocrysts, especially olivine.

In addition to xenocrysts, garnet peridotite and eclogite xenoliths are common in kimberlites, and there are numerous thermobarometers that can be used on such rocks to estimate *P-T* conditions of equilibration before those pieces of the mantle were ripped up and brought to the surface. These include Fe-Mg exchange between garnet and either Cpx or olivine as a thermometer, akin to the garnet-biotite exchange thermometer discussed in Section 12.4. Others include thermometers based on compositions of coexisting Opx and Cpx and on partitioning of Na between Opx and Cpx, and barometers based on the Al content of Opx and exchange of Ca between olivine and Cpx.

Figure 13.28 shows a collection of estimated *P-T* conditions from garnet peridotite xenoliths in kimberlites from southern Africa, plotted on the reference continental and oceanic geotherms from Figure 2.20. The *P-T* estimates for the African samples nicely correspond with the continental geotherm, and such data are one way in which geophysical estimates of geotherms can be tested. The intersection of the geotherm with mantle adiabats is one way to define the thickness of the lithosphere, here quite thick (>200 km). For this *P-T* dataset, about half of the points predict origins in the diamond stability field.

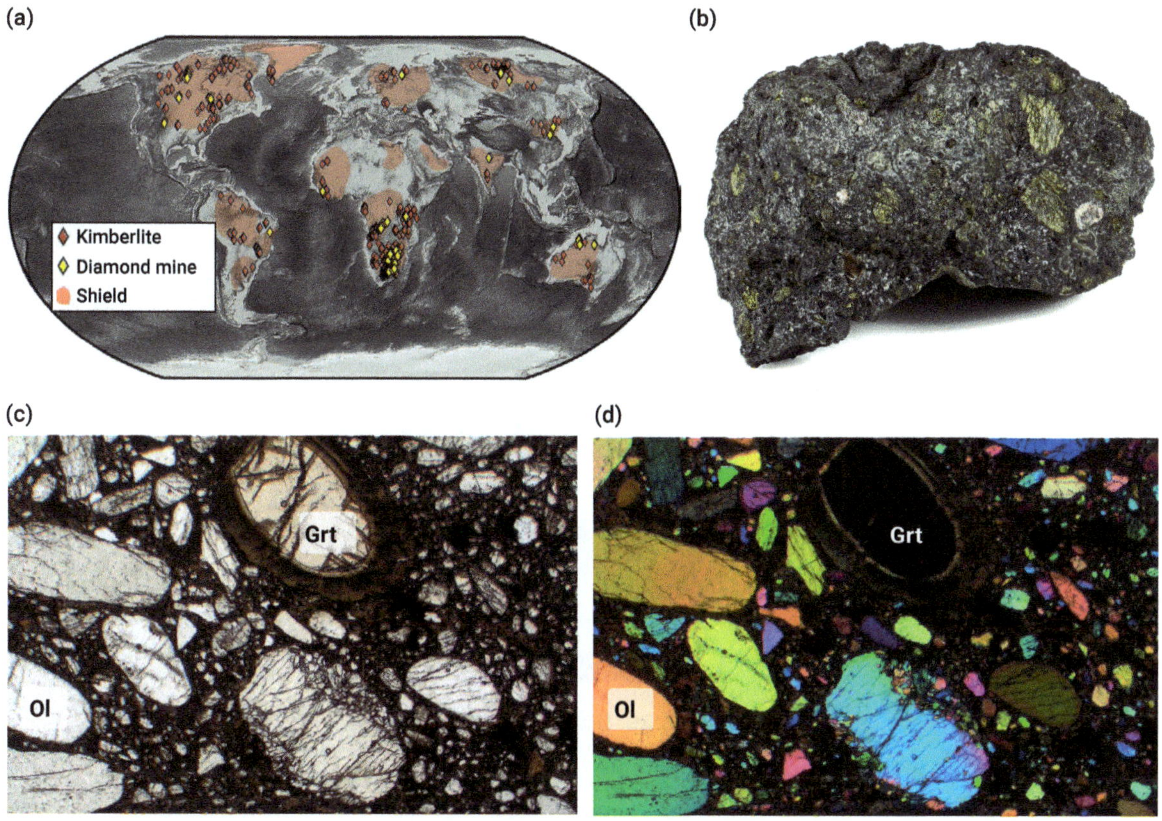

Figure 13.27 (a) Map showing the distribution of known kimberlites and kimberlites with diamond mines. Nearly all are on continental shields. (b) South African kimberlite with centimeter-scale macrocrysts of olivine, smaller diamonds, and a serpentinized matrix. (c, d) Thin section photomicrographs of Udachnaya kimberlite in plane and crossed polarized light, respectively, showing macrocrysts of olivine and garnet.

Sources: (a) From Giuliani & Pearson (2019), Andrea Giuliani, Mineralogical Society of America; (b) Bjoern Wylezich/Shutterstock; (c, d) Alessandro Da Mommio (www.alexstrekeisen.it).

Diamonds Research on kimberlite diamonds and their inclusions is extensive (e.g., see review in Smit et al., 2022). Inclusions in some diamonds indicate that the diamonds came from deeper in the mantle than their kimberlite host melts, so the inclusions carry information about deeper mantle mineralogy. Examples of unusual inclusions are $CaSiO_3$- and $MgSiO_3$-rich phases interpreted to represent former lower-mantle perovskite-group minerals now inverted to lower-P structures, ferropericlase [(Fe,Mg)O], and iron-rich metal and sulfide inclusions that are interpreted to represent former melts. The latter inclusions are found preferentially in so-called CLIPPIR (Cullinan-Like, Inclusion-Poor, relatively Pure, Irregularly shaped, and Resorbed) diamonds (Figure 13.29; Smith et al., 2016). CLIPPIR diamonds are the largest and most flawless ever found and include the Cullinan, Constellation, and Koh-i-Noor Diamonds. Blue diamonds (e.g., the Hope Diamond at the Smithsonian Institution) also appear to originate from the lower mantle, and their compositions and inclusions indicate derivation, at least in part, from subducted oceanic lithosphere (Smith et al., 2018, 2021). In particular, the blue of the Hope Diamond is caused by relatively high concentrations of boron, and Smith et al. (2018) proposed that this came from serpentine in oceanic lithosphere that was subducted to and below the transition zone. The deeper they came from, the higher the price!

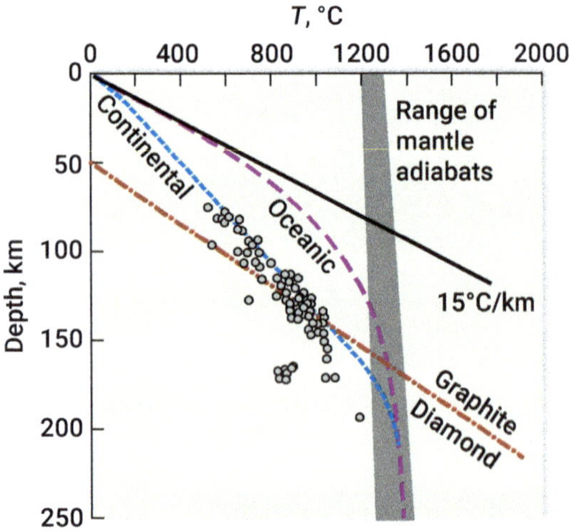

Figure 13.28 *P-T* estimates for garnet peridotites from South Africa, plotted with the geotherms from Figure 2.20. The points mostly plot along the continental geotherm, providing independent confirmation of the geophysically estimated curve. About half the samples come from depths within the diamond stability field.

Sources: Graphite-diamond curve from Day (2012); data compilation and mantle adiabats from Pearson et al. (2014).

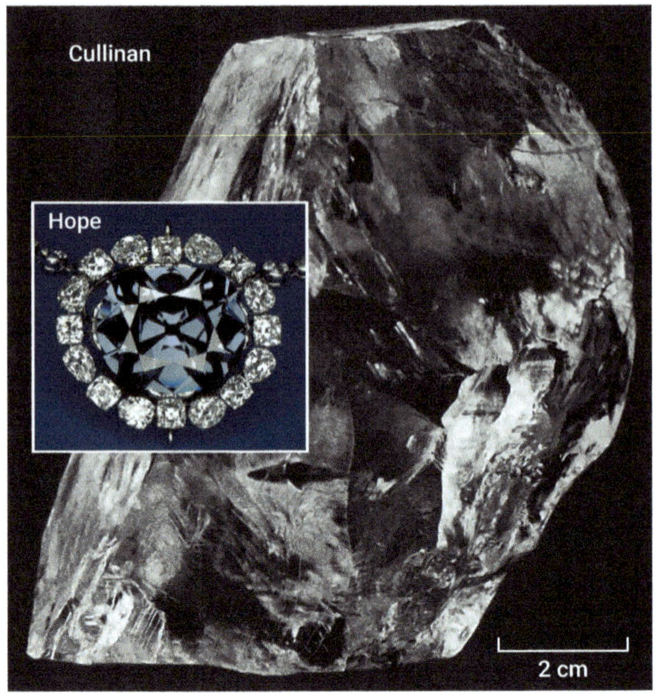

Figure 13.29 Diamonds from the deeper mantle. The Cullinan Diamond, found in a South African mine in 1905, is the largest gem-quality diamond ever found. It was the size of an adult fist and weighed 621 g (1.3 lbs) before being cleaved into pieces that produced nine major gems and a large number of smaller ones. (Inset) The walnut-sized Hope Diamond is famous for its striking blue color, caused by boron, that may have come from serpentinite that was delivered to transition-zone depths by deeply subducted oceanic lithosphere.

Sources: Cullinan image from Plate 1, The Cullinan (1908) by Joseph Asscher, principal cutter; Hope diamond courtesy of Smithsonian Institution.

Carbonatites

Carbonatites
Carbonate-rich intrusive and volcanic rocks, commonly associated with highly alkaline silicate rocks.

Calciocarbonatite
Magma consisting largely of calcium carbonate.

General Characteristics and Mineralogy

Carbonatites, igneous rocks composed predominantly of calcite, dolomite, and other carbonates, form an unusual branch of the igneous rock family: weird cousins that don't really fit in with everyone else, have lots of odd names, and generally wreak havoc when they show up. Much of the framework for understanding these odd associations came from early work by Brøgger (1921) at Fen, Norway, and von Eckermann (1948) at Alnö, Sweden, where carbonatites, nepheline syenites, and other alkaline silicate rocks intruded Precambrian metamorphic rocks ~550 Ma. In both localities, carbonate rocks and their silicate associates comprise varying proportions of calcite, dolomite, alkali feldspar, various pyroxenes including Cpx with a significant acmite ($NaFe^{3+}Si_2O_6$) component, nepheline, melanite (Ti-bearing garnet), olivine, biotite, melilite, apatite, Fe-Ti oxides, and cancrinite (a complex Na-Ca feldspathoid with carbonate).

The odd minerals that occur in carbonatites and accompanying silicate rocks have proven to be an irresistible motivation for composing new rock names, a sampling of which includes alvikite, tinguaite, alnöite, beforsite, ouachitite, sövite, melteigite, urtite, jacupirangite, vibetoide, cromaltite, juvite, malignite, pulaskite, damkjernite, and uncompaghrite. However, rocks at Alnö and many other intrusive carbonatites are dominated by **calciocarbonatite** (in older terminology, sövite), nepheline syenite, and ijolite (ee-uh-lite), a distinctive plutonic rock, composed largely of nepheline and acmite-rich augite, that is common enough in carbonatite complexes to deserve its own unusual name.

Carbonatite complexes typically show widespread metasomatic alteration of their host rocks. This process of **fenitization** produces a number of dramatic chemical changes, primarily dumping Na, K, or both into the host rocks (e.g., Elliott et al., 2018) and producing rocks, fenites, that can be almost entirely alkali feldspar. For decades, this was a puzzle, because the carbonatites themselves have little Na or K. Observation of natrocarbonatite lavas at Ol Doinyo Lengai (Dawson, 1962) raised the question of whether primary carbonatite magmas are alkali-rich and convert to calcite and dolomite carbonatite after reaction with groundwater and other agents of alteration.

Fenitization
Addition of Na and K to host rocks around carbonatite intrusions. Crystallization of carbonatite magmas evolves highly alkali- and H_2O-rich fluids that react with the surrounding rock to produce an alkali-rich rock called fenite.

The odd chemistry and mineralogy of carbonatites makes them economic targets. They host important deposits of niobium, used in jet engines and superconducting magnets (e.g., in MRI scanners) and the REEs, used across a wide range of technologies (e.g., 10–15 kg of lanthanum per electric vehicle battery). The world's largest source of niobium, in Brazil, is hosted by carbonatites of the Barreiro Complex, and the largest source of REEs, at Bayan Obo, China, has also been interpreted as a carbonatite. Light REEs are particularly valuable, and carbonatites typically are enriched several hundred times chondrite at the light end (see Figure 13.25c).

Origin of Carbonatite Magmas

The origin of carbonatitic magmas is complex. There are three main hypotheses: (1) as low-degree partial melts of carbonate-bearing (carbonated) peridotite; (2) as residual liquids produced by extensive crystallization of silica-undersaturated, alkali-rich magmas; and (3) by liquid immiscibility, with carbonatite liquid separating from silicate liquid.

Partial Melting At 1 atm, calcite and dolomite do not melt; they decompose to oxides plus CO_2, a key step in making cement (see Equation 11.24). The existence of carbonate liquids (potential carbonatitic magmas) was not shown until Wyllie & Tuttle (1959) made molten carbonate at 100 MPa in the system $CaO\text{-}CO_2\text{-}H_2O$, and since then, a large number of experimental studies have shown conditions under which carbonatitic magmas may exist. A recent summary of generalized *P-T* relations for CO_2- and H_2O-bearing peridotite is given in Figure 13.30. Starting material for the experiments on which this diagram is based was "damp" peridotite (~0.5 wt% H_2O in the starting material; this makes amphibole at pressures below 3 GPa) with a few weight percent CO_2. The solidus rises to about 1075 °C at 2 GPa. In the absence of CO_2, it continues at about 1100 °C to 3 GPa, then curves back sharply as amphibole breaks down, as discussed in Figure 6.32. However, in the presence of CO_2, the solidus curves back sharply at lower *P*, 2 GPa, owing to melting of dolomite-bearing peridotite to form carbonatitic liquids. The region of dolomite-bearing peridotite is limited on the low- and high-*P* sides by the reactions

$$\underset{\text{Opx}}{4MgSiO_3} + \underset{\text{Dol}}{CaMg(CO_3)_2} = \underset{\text{Ol}}{2Mg_2SiO_4} + \underset{\text{Cpx}}{CaMgSi_2O_6} \qquad \textbf{(Eq. 13.1)}$$

and

$$\underset{\text{Dol}}{CaMg(CO_3)_2} + \underset{\text{Opx}}{2MgSiO_3} = \underset{\text{Cpx}}{CaMgSi_2O_6} + \underset{\text{Mgs}}{2MgCO_3} \qquad \textbf{(Eq. 13.2)}$$

We use Opx for En, Ol for Fo, and Cpx for Di, in recognition of Fe in the starting compositions used in the experiments.

For this system, in the interval from 2 to 3 GPa, melting of carbonated peridotite begins at less than 1000 °C to produce a carbonatitic melt. Water in the system is tied up in amphibole and does not participate in the melting reaction until *T* is raised to about 1075 °C, at which time amphibole dehydrates. At $P < 2$ GPa or $P > 3$ GPa, dolomite is not present to participate in relatively low-*T*, carbonate-driven melting, whereas at $P > 3$ GPa, coincidentally, amphibole has broken down and freed up water that joins with CO_2 to facilitate melting.

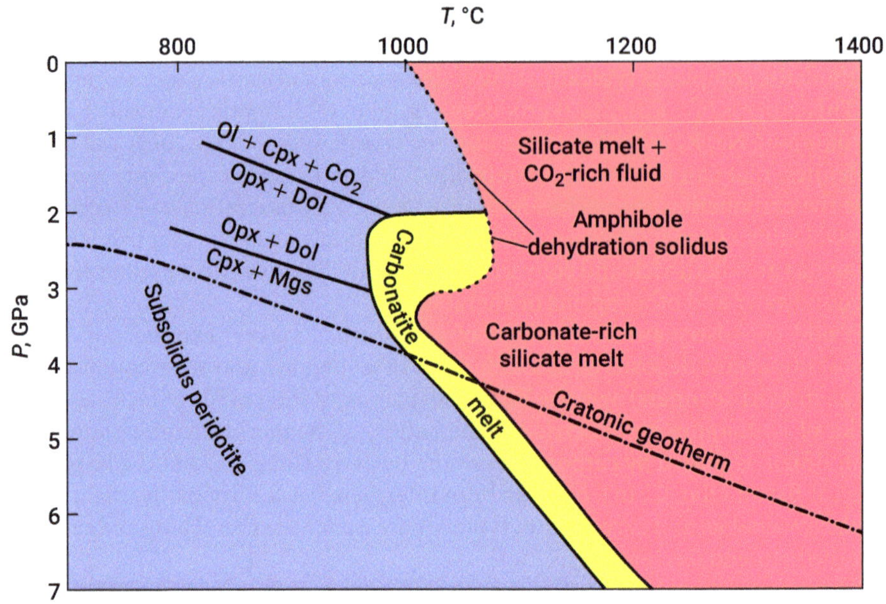

Figure 13.30 Summary of melting relations in fertile peridotite with ~0.5 wt% H_2O and 2–3 wt% CO_2. Subsolidus reactions given by Equations 13.1 and 13.2 are shown; these help to delimit the *P* range in which carbonatitic melts are produced. Relations below 1 GPa ignore the water-present solidus and are schematic only.

Source: Modified from Yaxley et al. (2021).

Carbonatitic melts produced in this way are Mg-rich, with Ca typically making up less than half of total Ca + Mg. Natrocarbonatite lavas have viscosities on the order of only 0.1 to 1 Pa·S at about 500 °C (Keller & Krafft, 1990)—like the viscosity of warm honey—and the viscosity of carbonatite melts at about 1000 °C should be much lower still. Hunter & McKenzie (1989) showed experimentally that carbonate melts have low dihedral angles (see Figure 10.35) and should effectively wet silicate grain boundaries. Combined with their ultralow viscosity and low density, these melts should rise as intergranular films, with abundant opportunities to react with enclosing silicates. Such reactions may drive melts to more Ca-rich carbonate compositions (Yaxley et al., 2022).

Fractional Crystallization Crystallization of non-carbonate phases such as olivine and Cpx will drive up the concentration of carbonate in the magmatic liquid, just as crystallization of these anhydrous phases drives up the concentration of H_2O. Extensive crystallization can lead to a calcite-rich carbonatitic liquid that, if separated, is carbonatite magma. It was long thought that fractional crystallization could not produce true natrocarbonatitic liquids owing to a thermal divide in the system Na_2CO_3-$CaCO_3$ at 0.1 GPa (upper-crustal pressures; Figure 13.31). For a $CaCO_3$-rich composition (e.g., 70 wt% $CaCO_3$), the eutectic between nyerereite and calcite traps crystallizing liquid and prevents it from becoming more Na-rich, whether crystallization occurs in a closed system or is fractional, with liquid separated from crystals.

However, Weidendorfer et al. (2017), showed that for natural compositions, the double-eutectic diagram with a thermal divide becomes a peritectic in which, with decreasing *T*, calcite and liquid react to form nyerereite via the reaction

$$CaCO_3 + Na_2CO_3(\text{Liquid}) = Na_2Ca(CO_3)_2 \qquad \textbf{(Eq. 13.3)}$$

Under closed-system crystallization, a calcite-rich liquid on the $CaCO_3$ side of the diagram will cease crystallization at the peritectic temperature, but under fractional

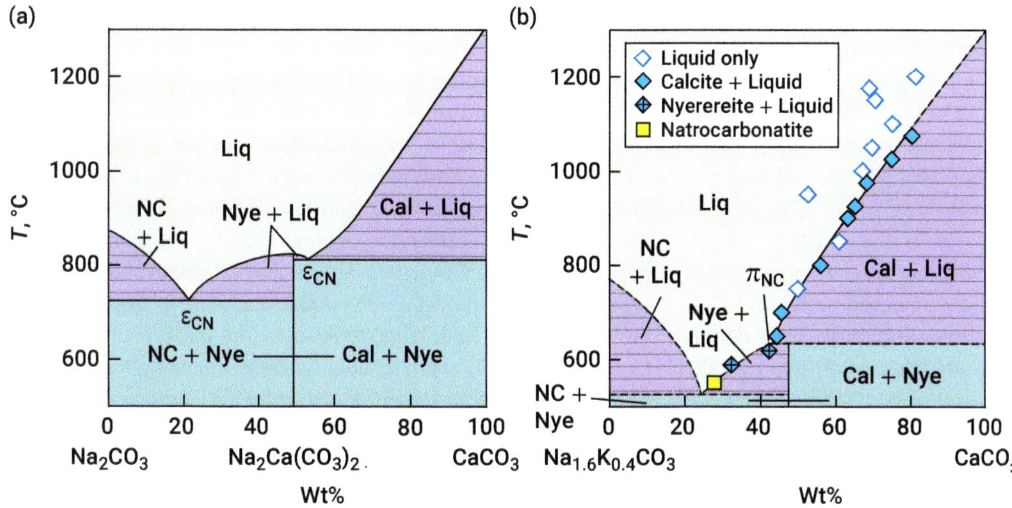

Figure 13.31 (a) Phase relations in the system Na_2CO_3-$CaCO_3$ at 0.1 GPa. The system is a double eutectic, and liquids on either side are trapped at the corresponding eutectic. (b) Pseudo-binary section of the pseudo-ternary Na_2CO_3-K_2CO_3-$CaCO_3$ at 0.1 GPa. This section is cut at a Na/K ratio of 4 (think of a vertical cut through a three-dimensional structure such as Figure 4.10 that goes through one of the apices). These experiments used natural natrocarbonatite lava seeded with calcite and other phases to examine saturation of those phases. The presence of the peritectic (π_{NC}) allows liquids to progress to highly sodic compositions, approaching those in natural natrocarbonatite lavas, under fractional crystallization.

Source: (a) Modified from Cooper et al. (1975); (b) modified from Weidendorfer et al. (2017).

crystallization, the peritectic allows the liquid to move to higher and higher Na contents with decreasing *T*, a situation comparable to that in the system Mg_2SiO_4-SiO_2 (see Figure 4.5), with calcite playing the role of forsterite.

Liquid Immiscibility Given the dramatic structural differences between silicate melts, whose properties are largely determined by silicate tetrahedra and their polymerization, and carbonate melts, with ionically bonded planar CO_3^{2-} ions, it is not surprising that petrographic evidence for immiscibility is widespread (e.g., Treiman & Essene, 1985; Church & Jones, 1995). Experimental data confirm the existence of a silicate-carbonate immiscibility gap (e.g., Hamilton et al., 1979; Lee & Wyllie, 1994).

Limnic
Relating to a body of low-salinity water such as a lake.

Making Connections 13.2

Shaken, Not Stirred—Limnic Eruptions and Their Fatal Consequences

You know what happens if you shake up a can of carbonated beverage, then pop the top—a frothy, sticky mess spurts out, splattering the immediate vicinity. Lakes may burp out CO_2 as well in events knowns as **limnic** eruptions. The water in most lakes undergoes vertical mixing at least once a year, but a small fraction does not, owing to stable density stratification. If located in volcanic regions, such lakes can accumulate high concentrations of CO_2 in their deep waters. At some point, a physical trigger—intrusion of magma, earthquake, landslide, etc.—can cause some of the CO_2-charged water to rise. As it does, CO_2 bubbles nucleate, which causes the bubbling mass to rise faster, forming more and more bubbles. The rising column of water sucks up deep CO_2-rich lake waters, and the whole mass explodes at the surface, much as your

continued

beverage did when you released the pressure. However, CO_2 is denser than air. So, while the shaken can may be an amusing joke on your friends, a limnic eruption can have deadly consequences. The expelled CO_2 can form a dense cloud that spreads out and rolls down slopes, hugging the land surface and suffocating animal life.

Tragically, this happened *twice* in the western African country of Cameroon in the 1980s, in maar-filling lakes. In 1984, Lake Manoun catastrophically released a cloud of CO_2 gas that suffocated 37 people in low-lying areas. Their deaths remained a mystery until two years later, when nearby Lake Nyos released an even larger cloud of gas, killing over 1700 people and 3500 head of cattle (Figure 13.32a), the death zone extending some 25 km from the lake. The exact trigger is not known, but the area is an active alkaline volcanic field. The eruption stirred up the lakebed, dispersing oxidized iron so that the water now appears red-orange (Figure 13.32b). At both lakes, engineers have installed degassing systems to prevent future catastrophes (Figure 13.32c). No pumps are needed, because CO_2 gas pressure at the bottom of the lake drives the fountain of gas-rich water. Much farther east, in the East African Rift, Lake Kivu, surrounded by millions of inhabitants, also has CO_2-charged deep water. Fortunately, the risk of a limnic eruption there seems less than for the Cameroon lakes.

Figure 13.32 (a) Cattle killed by CO_2 gas cloud from Lake Nyos in 1986. (b) Lake Nyos today; red color is from iron that was disturbed from the lake bottom during the eruption. (c) Degassing system at Lake Nyos spouts CO_2-charged water.

Sources: (a) Peter Turnley/Corbis/VCG via Getty Images; (b) Professor George Kling; (c) Bill Evans, USGS.

13.7 Controls on the Compositions of Alkaline Magmas

Learning Objective

By the end of this section you should be able to:

13.7.1 Predict how changes in percent melting, pressure, and volatiles affect the alkaline character of magmas generated by melting of peridotite.

Degree of Partial Melting

Alkaline magmas have weird trace-element compositions to go along with their weird major-element compositions. One simple explanation for high concentrations of the incompatible elements in mafic alkaline rocks is that they were produced by smaller degrees of partial melting than more typical basalts (e.g., Gast, 1968). Nephelinites and other highly alkaline mafic rocks typically are enriched in highly incompatible elements by factors of 100 to 200 over primitive mantle (see Figure 13.25c), and this could be explained by extraction of melt when the weight fraction of magmatic liquid F was 0.005 to 0.01 (0.5–1 percent). It is possible that low degrees of partial melting of primitive mantle contribute to high incompatible element concentrations in alkali basalts, but other aspects of the geochemistry of alkaline rocks argue against a primitive mantle source.

It is also not known if it is physically possible to extract melt when its proportion is only 1 wt% or less, even when the dihedral angle is small (see Section 10.4). Although small degrees of partial melting may be important, it seems that some other enrichment processes must be in play.

Depth of Generation

From a major-element perspective, the mafic ends of alkaline magmatic systems are distinguished by lower silica and higher K_2O than mafic rocks in arc or ocean-ridge systems (see Appendix 10; see also Figure 13.2). In the terminology of the basalt tetrahedron (see Figure 4.15), these rocks are nepheline-normative alkali basalts and their more alkalic relatives, rather than olivine tholeiites or quartz tholeiites. We saw in Chapter 4 that melting in the system Fo-Di-Qz at low pressure begins at a quartz-normative peritectic that lies in the Fo-Di-Qz triangle, but that increasing pressure drives this invariant point into the Fo-Di-En triangle, where it becomes olivine-normative. Thus, higher *P* drives the beginning of melting to lower silica contents, at least in this simple three-component system.

To address alkalis, Green and Falloon (1998) modeled mantle melting in the system Nph-Fo-Qz (Figure 13.33). This system includes an alkali phase, either Ab or Nph at low to moderate pressure, but lacks Cpx, an important phase in the mantle. At 1 atm, the

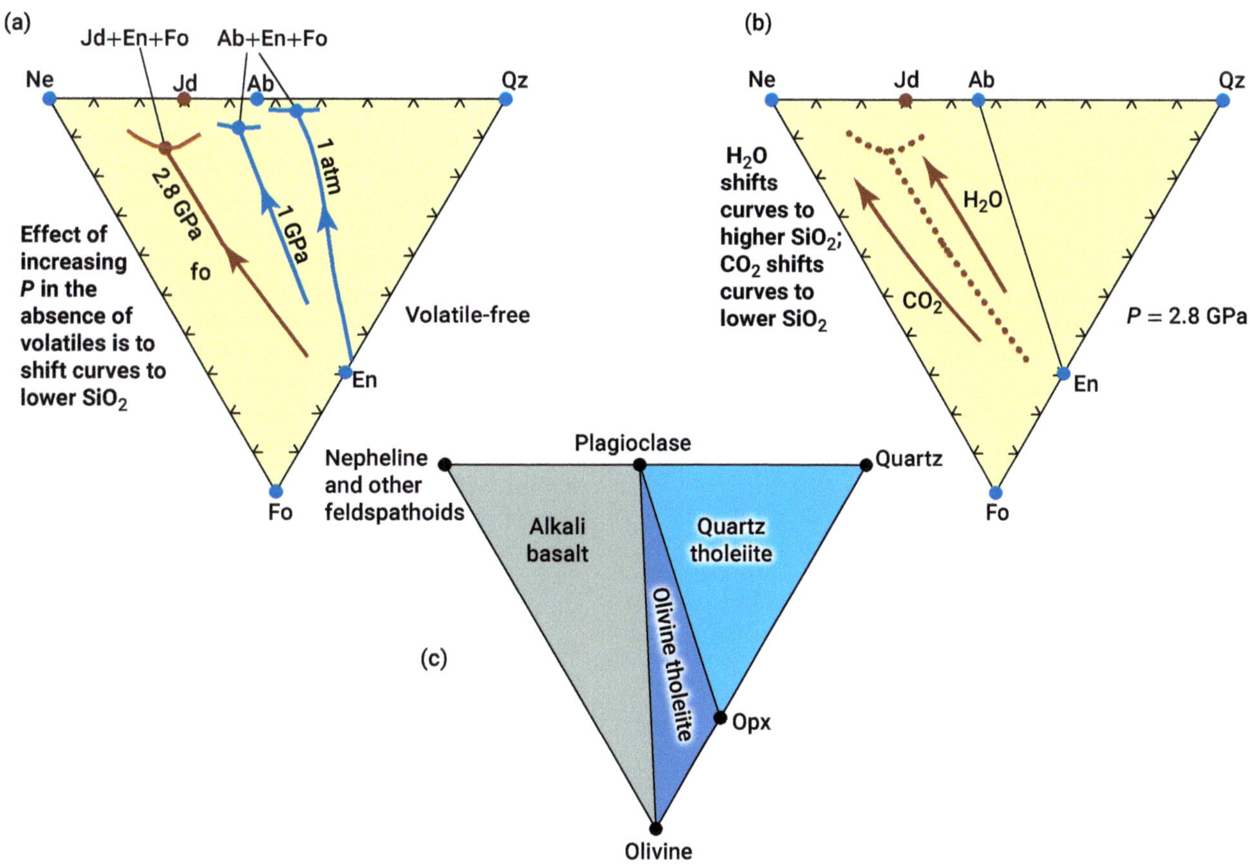

Figure 13.33 Melting relations in the Fo-Ne-Qz ternary. (a) At low *P*, melting occurs outside the Ol-Opx-Pl triangle, and melts are quartz-normative. With increasing *P*, melting shifts away from quartz toward more alkalic compositions. (b) H_2O and CO_2 also shift melting toward and away from quartz, respectively. (c) The base of the basalt tetrahedron (see Figure 4.15) for reference.

Source: After Green & Falloon (1998).

reaction curve between the fields of Fo and En (1-atm curve with arrow in Figure 13.33a) lies well to the right of the Ab-En join, toward Qz. This curve terminates at a peritectic invariant point involving Fo, Ab, and En; any bulk composition lying in the Fo-Ab-En triangle will begin melting there, and a bulk composition rich in Fo and En, with a little Ab, would make a reasonable model mantle composition in this system. This point actually lies *outside* the Fo-En-Ab triangle, on the side away from Fo. So, when assemblages of En + Ab + Fo melt at 1 atm, the reaction must be a crossover, En + Ab = Fo + Liq. Any liquid in equilibrium with Fo + En (i.e., our model basalts) must lie on the reaction curve, on the Qz side of the Ab-En join. Thus, low-*P* melting produces quartz-normative (relatively SiO_2-rich) quartz tholeiites.

At a pressure of 1 GPa, however, things are different. Increasing *P* shifts the Fo + En curve to the left, along with the Fo + Ab + En invariant point. Both now lie *within* the Fo-Ab-En triangle; the curve is now a cotectic rather than a reaction curve, and the invariant point is a eutectic instead of a peritectic. At this pressure, liquids are olivine-normative (relatively SiO_2-poor). At 2.8 GPa, corresponding to a depth of about 90 km, Ab is no longer stable, having been replaced by Jd + Qz (see Figure 13.17a; see also Figure 12.17). The Fo + En curve has shifted still farther away from Qz and toward Ne, and liquids produced at the Jd + En + Fo invariant point or along the Fo + En curve are nepheline-normative (extremely SiO_2-poor) alkali basalts.

This relationship between *P* and the silica content of low-degree partial melts holds in more compositionally complex systems. Figure 13.34 shows melt compositions for a fertile peridotite (fertile in the sense of not having had basaltic partial melt already extracted) as a function of the weight fraction of liquid (*F*; see Section 8.1) at pressures ranging from 1 to 5 GPa (i.e., at depths ranging from the base of typical crust to ~160 km). These curves, from Tomlinson and Holland (2021), were calculated from a thermodynamic model that was tested against experimental results. At $F = 0.02$ (2 wt% melt), silica drops precipitously from more than 54 wt% for melting at 1 GPa to about 46 wt% at 5 GPa. SiO_2 content is generally low (<50 wt%), but does not depend so much on pressure at higher liquid fractions.

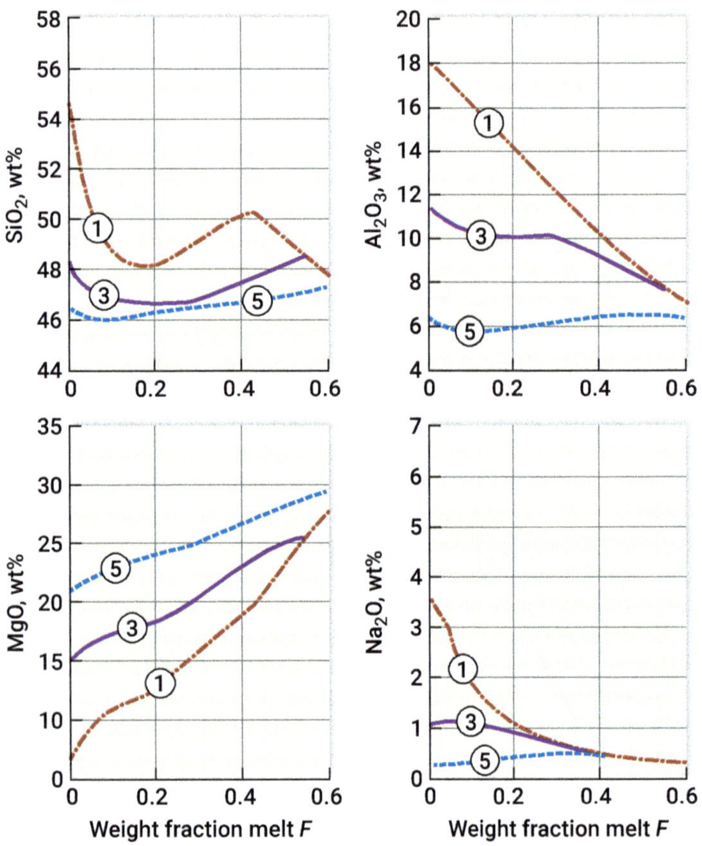

Figure 13.34 Models of peridotite melting, showing changes in melt composition with weight fraction of melt, contoured for pressure at 1, 3, and 5 GPa. Melting at higher *P* should lead to lower SiO_2 contents in melts, but should also suppress high Na_2O; thus melting normal mantle at high pressure should not produce low-SiO_2, high-Na_2O alkali basalts.

Source: From Tomlinson & Holland (2021).

In simple systems, higher *P* by itself seemed to explain high alkalis of alkalic rocks (see Figure 13.33), but predictions for realistic rock compositions show that increasing *P* actually drives down the Na_2O contents of partial melts (see Figure 13.34), so much so that melting at depths greater than about 100 km (>~3 GPa) should produce Na_2O-*poor* magmas. This can be rationalized because at crustal pressures, Na is largely incompatible in mantle minerals except for plagioclase, which melts out in the first several percent melting (leading to high Na_2O contents). However, as *P* increases, Na becomes increasingly compatible as the jadeite component in Cpx, which is less susceptible to melting. As we saw in prior chapters, the change in Al coordination from fourfold in plagioclase to sixfold in pyroxene affects mineralogy and mineral reactions, leading in this case to changes in melt chemistry. Overall, it seems that higher-*P* melting alone has difficulty explaining magmas that are both highly alkaline, but also low-SiO_2.

CO_2 and H_2O

We have repeatedly seen the critical role that H_2O plays in igneous petrology—in particular, its role in reducing viscosity (see Figure 7.9) and in decreasing melting temperatures (see Figure 6.29). Water accomplishes these feats by prying apart polymerized SiO_4 tetrahedra at bridging oxygens (see Figure 7.10). Breaking down bridging oxygens increases the entropy (S) of the liquid, because highly polymerized silicate liquids have relatively ordered structures—tetrahedra are chained to other tetrahedra by strong Si-O-Si bonds rather than isolated and bound by weaker bonds.

Increasing entropy lowers the free energy G owing to the relationship $G = H - TS$ (H is enthalpy, or heat content), which you will learn when you take a course in thermodynamics. Larger S means lower G, so the more SiO_2 in a liquid, the more dissolved H_2O lowers its free energy. The net effect of this is to both lower the melting temperature (because the free energy of the liquid is decreased by dissolved H_2O, whereas the free energies of any solid phases are unaffected) and to shift liquid compositions toward those whose free energy is lowered the most and thus favored by the thermodynamic gods—highly polymerized, SiO_2-rich liquids. As a result, although anhydrous pressure tends to shift liquids to lower-silica compositions (see Figure 13.33a), addition of H_2O counters this and shifts them back toward more silica-rich compositions (see Figure 13.33b).

Carbon dioxide behaves differently. It is far less soluble in silicate melts than H_2O (Figure 13.35), and whereas H_2O is more soluble on a weight basis in more silicic melts, CO_2 is more soluble in low-silica melts, particularly in alkalic melts such as leucitites. Like anhydrous pressure, CO_2 has little effect on polymerization and shifts melt compositions the opposite way, to lower-silica compositions (see Figure 13.17b).

Source Composition

The simplest explanation for why some areas produce alkalic magmas and others do not is that the composition of the mantle being melted differs from place to place. Cpx-rich veins in mantle xenoliths (see Section 13.5) provide evidence for prior diking

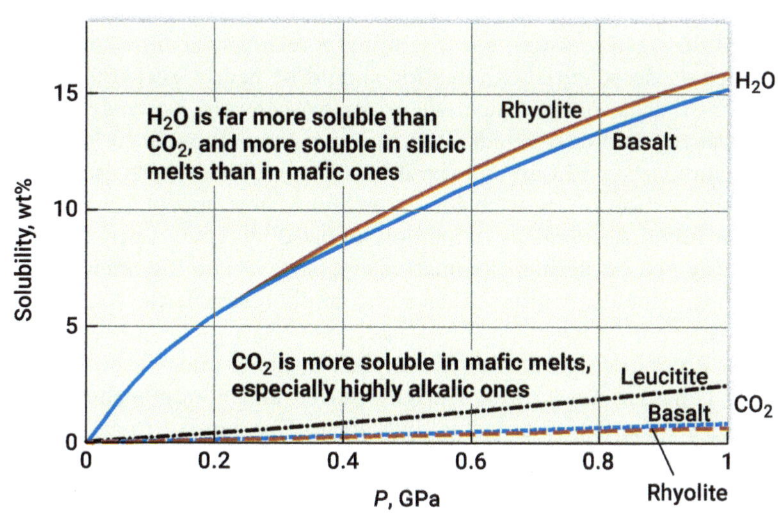

Figure 13.35 Solubilities of H_2O and CO_2 in different magmas and pressures. H_2O is much more soluble than CO_2 and, unlike CO_2, is more soluble in more silicic magmas.

Source: Modified from Duan (2014).

of typical mantle peridotite by basaltic magmas. These magmas bring in incompatible elements and, when the peridotite plus enriched vein is melted, could produce alkalic melts. Similarly, prior intrusion of carbonatitic magmas into peridotite could metasomatize the mantle. Spatially distinct domains of metasomatically enriched mantle—a "plum pudding" mantle (see Section 14.2)—could explain regional variations in magma composition.

How might these basalts and carbonatites intrude the upper mantle? There are several possible mechanisms. For example, rifting to form a new ocean basin stretches the lithosphere, and thinned lithosphere that is transitional from continental to oceanic should be invaded by basaltic dikes that modify the bulk composition of a given volume; given time, oceanic lithospheric mantle along new continental margins could be converted to enriched mantle. Other mechanisms include subduction, which releases volatiles and magmas that pass through and metasomatize the mantle wedge, and delamination (see Section 9.5), which can send crustal material back into the mantle to enrich it. But perhaps the most effective mechanism for enrichment in rift settings lies in the intrusion of small-degree partial melts of carbonatite or other alkalic magmas from deeper sources that rise and freeze in cooler, shallower mantle. Remelting of peridotite that has been repeatedly fertilized with such material could, in a general sense, explain the highly alkalic magmas found in rift zones.

13.8 **Metamorphic Rocks in Extensional Terranes**

Learning Objective

By the end of this section you should be able to:

13.8.1 Predict exhumation *P-T-t* paths in an extensional setting.

Extension by itself does not necessarily lead to increases in temperature that might drive metamorphism (see Figure 13.17), but extension can be important for exposing previously metamorphosed rocks at the surface. The rates and mechanisms of extension can also lead to distinctive post-peak metamorphic *P-T* paths. Very generally, rapid removal of rock should decrease pressure, but if faulting is relatively shallow and metamorphic rocks are relatively deep, initial exhumation should be nearly isothermal. However, extension steepens the geotherm, which will evolve back toward its steady-state condition. Starting around the mid-crust (~0.5 GPa), rising rocks will begin to "feel" the surface temperature boundary condition and should cool rapidly and nearly isobarically. Thus, the exhumation portion of an extensionally exhumed metamorphic rock might be expected to show initial isothermal exhumation followed by rapid, nearly isobaric cooling. Broadly speaking, the isothermal exhumation and isobaric cooling resemble a *P*-first *P-T* path, although the isobaric cooling, taken alone, might resemble the cooling portion of a *T*-first *P-T* path.

The Priest River metamorphic core complex, in Idaho and Washington, illustrates this expected behavior. Rocks were initially buried and metamorphosed in the Late Cretaceous during the Sevier orogeny, then abruptly exhumed to at least midcrustal levels in the early Tertiary (Stevens et al., 2015, 2016). Metamorphic *P-T* paths show nearly isothermal exhumation from at least 1.0 to about 0.6 GPa at a temperature of nearly 800 °C (Figure 13.36). Exhumation caused these partially molten rocks to melt a little more. But as they approached the mid-crust, the rocks started cooling nearly

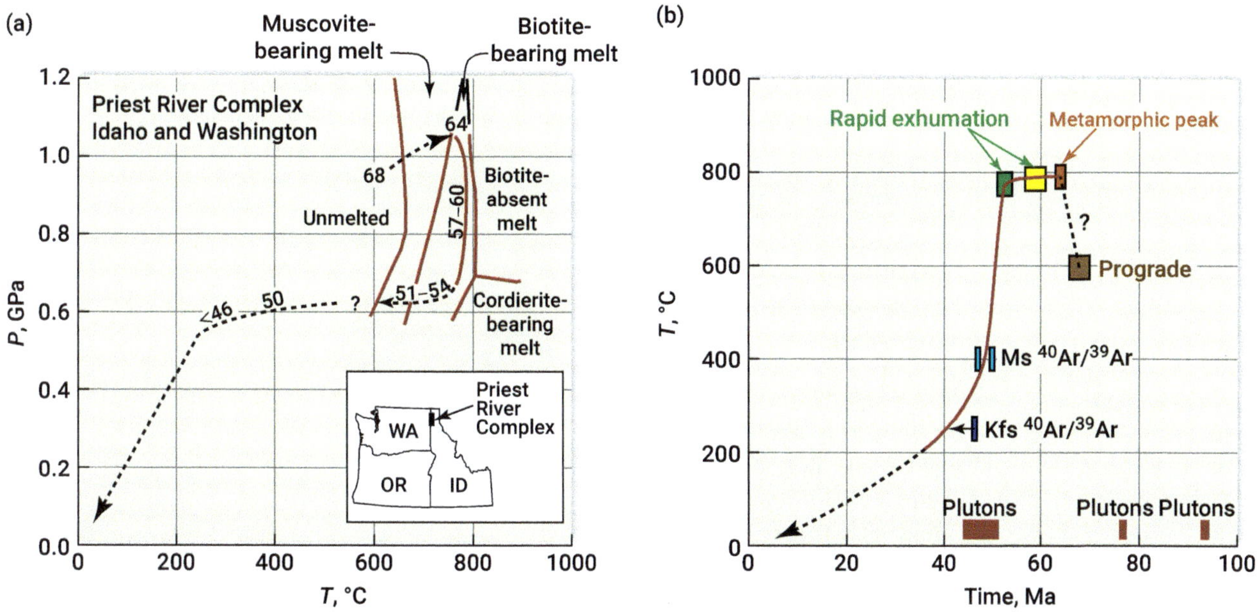

Figure 13.36 (a) Pressure-temperature diagram, showing nearly isothermal decrease in pressure at a temperatures of nearly 800 °C followed by cooling at roughly constant pressure of ~0.6 GPa. Dashed lines show more speculative parts of path. The different mineral assemblage fields help estimate the *P-T* path. Rapid exhumation via extensional faulting should yield nearly isothermal decrease in pressure, as observed. Inset shows location of the Priest River Complex in Idaho and Washington. (b) Temperature-time diagram, showing protracted high temperatures of nearly 800 °C followed by extremely rapid cooling at ~50 Ma. Rapid cooling is predicted to follow rapid exhumation, as observed. The timing of pluton emplacement does not obviously correlate with rapid exhumation.

Sources: (a) Modified from Stevens et al. (2015); (b) based on data reported in Stevens et al. (2015, 2016).

isobarically, and melts crystallized. Mineral assemblage diagrams and monazite petrochronology delineate this high-temperature history. But rapid cooling didn't stop there. Muscovite and K-feldspar $^{40}Ar/^{39}Ar$ ages from nearby (Doughty & Price, 1999) indicate extremely rapid cooling—possibly a few hundred °C/Myr—down to temperatures of at least 400 °C, and possibly as low as 250 °C. Such cooling histories are common throughout the northern metamorphic core complexes (e.g., Foster et al., 2007; Stevens et al., 2017).

Exactly what drives core complex extension and exhumation? Mechanical heterogeneities clearly localize extensional shear zones along weak horizons, and these structures bound complexes such that low-grade or unmetamorphosed rocks are juxtaposed atop a high-grade metamorphic core. However, regional extensional stress regimes related to changes in plate convergence or changes in the geometry of the subducting plate began much later in the tectonic history of North America than exhumation, at least for the northern core complexes. Similarly, the timing of igneous activity does not seem to correlate with extension and rapid exhumation, so softening and collapse of the crust does not seem likely. One intriguing possibility invokes mechanical contrasts between young, accreted terranes and old cratonal basement (Stevens et al., 2017). Depending on margin geometry, convergence can initiate upward flow of rocks, leading to extension and exhumation. This mechanism might explain why North American core complexes have a spatial distribution that parallels the 0.706 line separating older vs. younger portions of the North American continent.

13.9 A Few More Alkalic Poems

We leave this chapter with a few more poems made from the names of alkaline rocks, knowing that more such names are coming in Chapter 14. First a few haiku:

Monchiquite, minette,
Uncompaghgrite, theralite;
Alkaline plutons

Melilitolite,
Marienbergite, urtite,
Nephelinolite

Another tanka:

I see you, leucite,
phlogopite, and diopside.
Why are you wyomingite
And not cedricite?
Is phlogopite essential?

And, finally, a limerick:

There once was a rock called beforsite
Carbonatite made of doLOmite
Rock names are so scattered
That students feel battered
To most, well it just seems malignite

Summary

- Continental rifting is an important process, but it forms new lithosphere at a rate less than 1 percent of that of oceanic ridges.
- The Basin and Range of western North America and the East African Rift system are the most prominent active rifts on Earth.
- Continental rifting produces a distinctive lithosystem consisting of (1) volcanic rocks (typically alkalic and commonly bimodal); (2) fault-basin sedimentary assemblages that include conglomerates, sandstones, shales, lacustrine deposits, and evaporites with local fault breccias and megabreccias; and (3) high-grade metamorphic rocks exposed in the footwalls of low-angle normal faults.
- Magmatism is clearly associated with lithospheric extension on the continents, but magmas rarely follow faults (i.e., it is rare to find dikes along faults).
- Fault rocks (fault breccias, cataclasites, and mylonites) are commonly exposed in the footwalls of normal faults.
- Stretching of the lithosphere promotes melting by compressing the geotherm (bringing hot rocks closer to the surface) and promoting decompression melting.
- Isotopic data (Sr and Nd) indicate that significant volumes of subcontinental mantle have been isolated from the convective mantle for hundreds of millions to billions of years. This is enriched mantle.
- Along continental margins, accreted oceanic lithosphere can be recognized by the oceanic isotopic characteristics (low $^{87}Sr/^{86}Sr$, high $^{143}Nd/^{144}Nd$) of associated igneous rocks.
- Mantle xenoliths in alkali basalts and more alkalic volcanic rocks are typically spinel peridotites, although garnet peridotites are locally brought up. Some of these show evidence for mantle metasomatism in the form of clinopyroxenite veins and unusual accessory minerals.
- There sure are a lot of names for weird, obscure, and volumetrically insignificant alkalic rocks.
- Lamprophyres, lamproites, kimberlites, and carbonatites are volumetrically insignificant alkalic rocks of great economic importance. Commonly rift-related, they are characterized by odd geochemistry (typically very low SiO_2, high MgO, and extreme concentrations of many incompatible elements) and carry unusual minerals, including diamonds.
- There are several potential mechanisms by which carbonatitic melts can form, including partial melting of dolomite-bearing peridotite, extensive fractional crystallization, and liquid immiscibility.
- Large-scale lithospheric extension commonly brings up high-grade metamorphic rocks in the footwalls of normal faults, as, for example, in metamorphic core complexes.

Notes

1. Many of these lakes were (re)named by colonial explorers for British royalty, although names relevant to local cultures have now sometimes been reassigned (e.g., Turkana).
2. *Northeast Pacific and Western North American Plate Tectonic History* animation: Tanya Atwater, Educational Multimedia Visualization Center, UC Santa Barbara.
3. When the fossils were first discovered, the Beatles' song "Lucy in the Sky with Diamonds" was playing in camp (hey, it was the 70s!), giving rise to her popular name.
4. Your authors had to look up many of these minerals but did know that armalcolite was originally recognized in lunar samples and named after astronauts Neil Armstrong, Buzz Aldrin, and Michael Collins.

CHAPTER 14

Intraplate Magmatism, Hotspots, and Large Igneous Provinces

Until this point, we have dealt mainly with lithosystems that fit into the plate tectonic framework that governs major Earth systems. Most of the following make sense and are understood at some level:

- Magmatism at oceanic ridges and at subduction zones
- Hydration and metamorphism of oceanic lithosphere at ridges
- HP/LT metamorphism in subduction zones
- Thermal (LP/HT) metamorphism of rocks around plutons
- Regional (MP/MT) metamorphism and deformation during continental collision
- Magmatism during continental rifting

These processes occur at active or incipient plate boundaries. However, important classes of magmatism and associated metamorphism occur seemingly without regard to plate boundaries. These may result from well-understood processes, such as impacts, or from more mysterious processes, such as those that produce chains of volcanic islands. Such intraplate magmatism is the topic of this chapter.

14.1 Hotspots and Their Tracks

Learning Objectives

By the end of this section you should be able to:

14.1.1 Predict the geographic distribution of volcanoes for a hotspot track under arbitrary boundary conditions (plate motion and hotspot motion).

14.1.2 Compare the geochemical character of magmas produced by continental hotspots with those produced by oceanic hotspots.

14.1.3 Characterize the differences in magma output during hotspot initiation vs. steady-state hotspot magmatism.

Galápagos Islands as seen from the European Space Agency's Copernicus Sentinel-2 multispectral satellite. The basaltic islands form a hotspot system. The largest is Isabela, with Fernandina to its west (left), San Salvador to its east, and Santa Cruz at the right edge of the image. Six calderas are visible, five of them on Isabela.

Source: Contains modified Copernicus Sentinel data (2020), processed by the European Space Agency.

Aseismic ridges
Long, linear, seismically quiet, volcanically inactive seafloor ridges.

Hotspot
Postulated source of magma that forms volcanic chains, islands, or seamounts far from plate boundaries; may or may not be sourced from the deep mantle or be much hotter than surrounding mantle.

Plume
Postulated relatively deep-seated, hot, buoyant mantle that, as it ascends, can give rise to hotspots.

The origins of isolated seamounts, ocean islands, and **aseismic ridges** have long intrigued geologists. As plate tectonic theory developed in the 1960s, many topographic and geologic features, such as mid-ocean ridges, earthquakes, trenches, and mountain ranges, fell into place. How, then, to explain a string of volcanoes in the middle of the ocean, far away from any plate margins? These anomalies were recognized early in the development of plate-tectonic theory. Wilson (1963) and Morgan (1971) proposed that they are **hotspot** tracks formed from a deep mantle **plume**, and that plate motion over hotspots creates lines of volcanoes (Figure 14.1). Multiple hotspots, distributed in both space and time, could potentially explain the diversity of islands and volcanic chains spread out over Earth's oceans, as well as some enigmatic magmatic trends on the continents. In the following section, we discuss a few classic hotspots and their geologic histories, partly to illustrate their common characteristics, but also to highlight deficiencies in our understanding of how these curious features form.

Reference Frames and Paleomagnetism

How do we define a reference frame for measuring the motion between different features on the Earth? With difficulty. Any point on the Earth is rotating with the Earth's spin, moving with its lithospheric plate relative to other plates, and moving up or down owing to

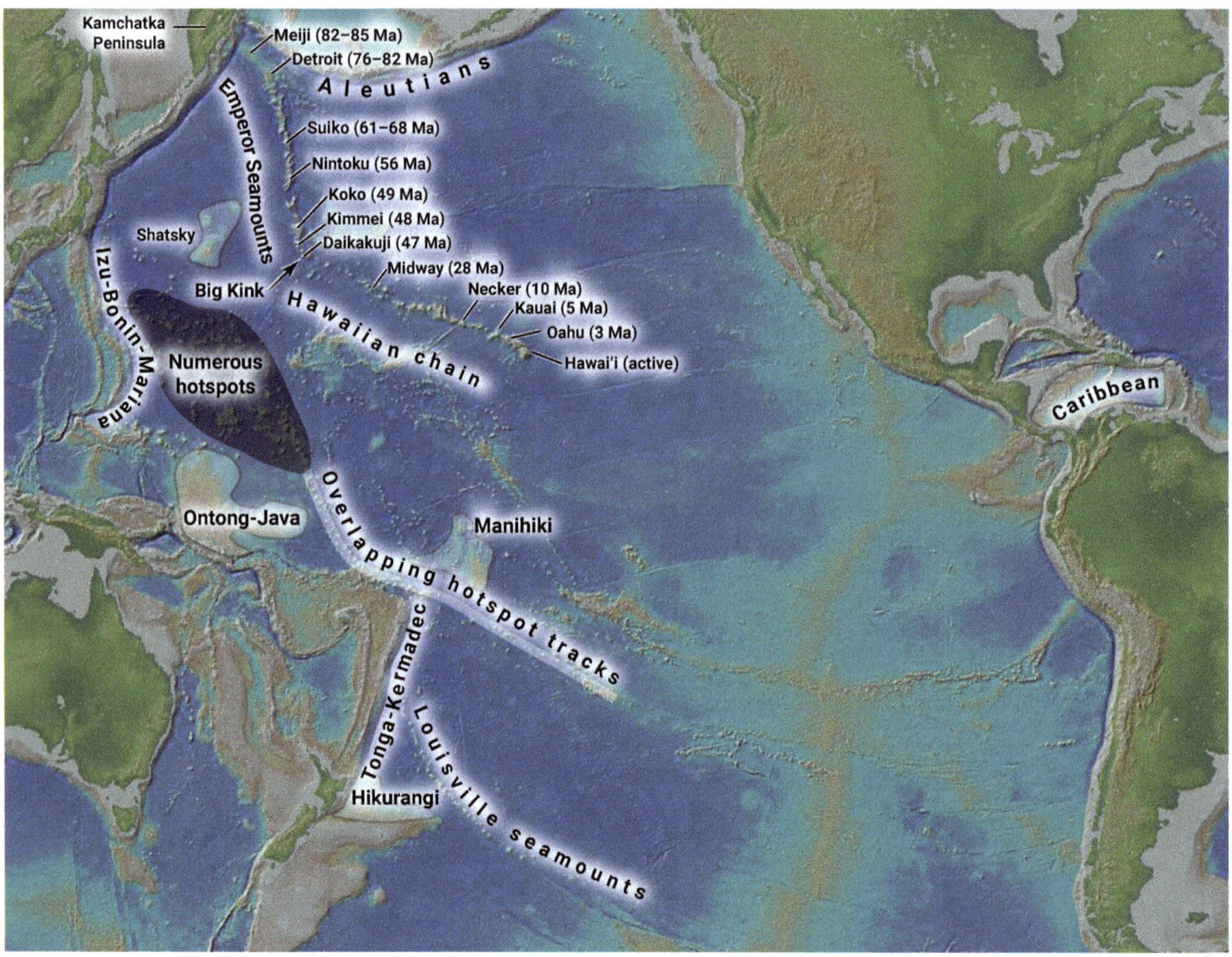

Figure 14.1 Bathymetric and topographic map of the Pacific and surrounding continents, highlighting the track of the Hawaiian-Emperor seamount chain with ages along the track. Also shown are major subduction systems, other hotspot tracks, and oceanic plateaus that may be related to hotspot activity (Shatsky, Ontong-Java, Manihiki, and Hikurangi).

Source: Map produced using GeoMapApp.

tectonics, isostasy (e.g., glacial unloading), and other effects. These various motions mean little unless tied to a measurable reference frame. Earth's orientation in space is defined relative to distant stellar objects in the ICRF (International Celestial Reference Frame), and positions of points on Earth's surface are defined relative to the ITRF (International Terrestrial Reference Frame), which gives the locations of points on the surface along with their measured velocities relative to the rotating reference frame defined by the ICRF.

Defining relative motion of plates is relatively easy, and GPS measurements tied to the ITRF provide a way to monitor these motions in real time. Long-term monitoring of plate motions relative to the ITRF provides one way to define absolute plate motions. Other reference frames for plate motion were discussed in Section 6.3 and Figure 6.19. The no-net-rotation reference frame, in which the sum of motions of all the plates relative to the spinning Earth is zero, is typically used as a proxy for absolute plate motion, under the assumption that the entire lithosphere is riding along with the rest of the Earth rather than moving at a different rate. Long ago, Morgan (1972) conjectured that hotspots are fixed relative to one another and locked to the planet's spin, and thus provide a global reference frame for measuring plate motion. How do we test this conjecture?

One test is provided by paleomagnetic measurements (see Toolbox 14.1). If the fixed-hotspot hypothesis is true, the latitude of rocks formed when that part of the plate passed over the hotspot (the **paleolatitude**) should be the same. The Hawaiian-Emperor seamount chain provides an excellent **proving ground** for this hypothesis. If the Hawaiian hotspot has been fixed in latitude over its history, the paleolatitude of all the islands and seamounts formed over the hotspot should be the same as its current latitude at the south end of the island of Hawai'i. As noted below, the Hawaiian chain fails this test, but some other hotspot collections pass; this is a symptom of the uncertainty that pervades the study of hotspots.

The fixed-hotspot reference frame also predicts significantly different absolute plate motions than the no-net-rotation reference frame (see Figure 6.19a,c). Although both predict plate motions that are consistent with the tracks of the Hawaiian and Yellowstone hotspots, they predict significantly different motions for Africa, Asia, and South America.

Paleolatitude
The latitude of a point on the Earth's surface when a given unit of interest (e.g., a lava flow) was formed.

Proving ground
A physical or conceptual place for testing theories, ideas, equipment, tactics, etc.; a proving ground is *not* a place where theories are proven.

Hawaiian-Emperor Seamount Chain

The Big Kahuna When people talk about hotspot tracks, they mention Hawai'i first, and for good reason. The Hawaiian-Emperor seamount chain (HESC), with a staggering 6000-km length decorated every 50 to 100 km with a basaltic seamount or island, is the clearest in terms of numbers of volcanoes and consistency of age and geographic distributions (see Figure 14.1). A pronounced kink approximately halfway along the track (creatively referred to here as the "Big Kink") adds variety. The generic explanation for a hotspot track—a stationary magma source that regularly cooks up volcanoes as a plate moves over it—is broadly apt for the HESC. There is definitely a quasi-continuous magma source, and movement of the Pacific Plate certainly could generate the chain of volcanoes. But deviations of the HESC from this simple predictive model have helped refine our understanding of hotspots and sparked many controversies.

Some Questions about the HESC

1. Where and when did it start? Good question! We don't know. The oldest seamounts are Meiji (~83 Ma) and Detroit (76–82 Ma) in the extreme northwest, so the hotspot was certainly active before 80 Ma. Older portions of the seamount chain could have been subducted beneath Kamchatka.
2. Is the hotspot stationary relative to . . . something? No, it has moved, but the amount is unclear, even down to how that movement should be measured. Paleomagnetic evidence shows that volcanic rocks at Detroit Seamount were erupted at about 34°N latitude, but Hawai'i today is at about 19.5°N latitude (Figure 14.2), indicating that the hotspot has moved southward relative to the magnetic axis, which is presumably coincident with the spin axis when averaged over tens of thousands of years. If Earth's

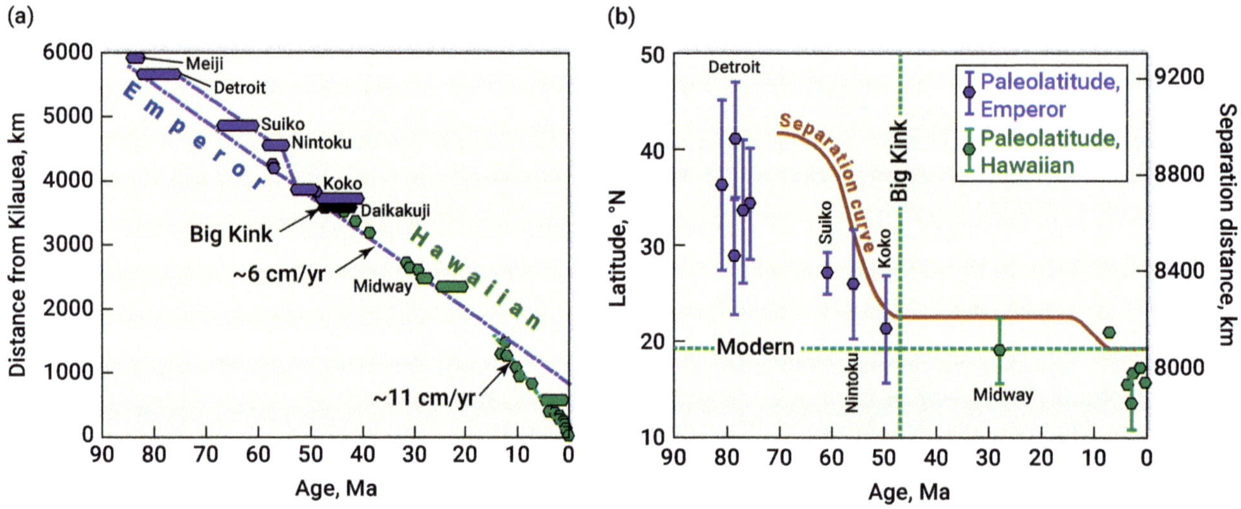

Figure 14.2 (a) Distance from Hawai'i along the hotspot track vs. seamount and island age showing a nearly constant velocity between ~80 to ~20 Ma and a near doubling between 10 and 0 Ma. An alternate velocity record has been proposed for Detroit, Suiko, and Nintoku Seamounts (bent line). (b) Evidence for hotspot wander: paleolatitude vs. age shows a decrease to ~47 Ma (the Big Kink), then a similar latitude as modern Hawai'i. Right axis gives separation between the HESC and Louisville seamount chains (red solid line), which decreases to the approximate time of the Big Kink, then is similar to modern. The modern inclination does not correspond with Hawai'i's latitude owing to minor distortion of the theoretical magnetic dipole field.

Source: (a) Data from O'Connor et al. (2013).

spin axis has remained constant since Detroit Seamount formed, this latitudinal shift would represent an about 1600 km southward drift of the hotspot. There are several complications, however. First, Earth's spin axis might have changed (although this would predict that seamounts in other oceans should show latitudinal shifts, and many do not; Doubrovine & Tarduno, 2008). Second, modern paleomagnetic results for Hawai'i predict a different latitude than its present position (see Figure 14.2), owing to irregularities in the geomagnetic field, so quantifying the magnitude of drift has an accuracy limit of several degrees. Arguably the best evidence for movement is the distance between the HESC and other Pacific hotspots. If hotspots were fixed, these distances should be constant, regardless of Earth's spin axis. However, the distance between the HESC and the Louisville seamount chain in the southern Pacific has decreased by nearly 1000 km in the last 70 Myr (see Figure 14.2; Jiang et al., 2021). Most of this drift occurred before about 50 Ma, when the Big Kink formed.

3. What caused the Big Kink? Good question! See Jiang et al. (2021) for a discussion. Some argue that the bend represents a change in hotspot motion as it shifted from drifting south to remaining more stationary (e.g., Tarduno et al., 2003; Bono et al., 2019; see Figure 14.2). Others argue that Pacific Plate motion changed abruptly at about 50 Ma as plates reorganized along the Aleutian chain (e.g., Hu et al., 2022) and subduction initiated along the western Izu-Bonin-Mariana and Tonga-Kermadec Trenches, shifting Pacific Plate motion westward. Regardless of mechanism, any change in plate motion must coincide with changes in mantle circulation at depth. If hotspot magma sources do originate deep in the mantle, their surface volcanic expression must reflect not only plate motion but their potentially tortuous path through the mantle. In that context, it should not be surprising that the distance between the HESC and the Louisville seamount chain shows a rate change at about the time of the Big Kink.

Yellowstone

The Yellowstone hotspot, as well known as the HESC, is at the end of an archetypal continental hotspot track (Figure 14.3). It is generally interpreted to have migrated from northwest Nevada to modern-day Yellowstone National Park over the last 16 to 17 Myr (e.g.,

Pierce & Morgan, 1992), or, to put it in the hotspot frame, the North American Plate is interpreted to have moved southwest over it during that time. The latter half of this path closely follows the modern topographic depression of the Snake River Plain, whereas the earlier path inhabits topographically diverse and rugged terrain along the northern edge of the Basin and Range. The standard interpretation for the Yellowstone hotspot (Pierce & Morgan, 1992) has the following characteristics.

Initiation Initiation of silicic Yellowstone magmatism is commonly assigned to magmatic centers in northwestern Nevada (see Figure 14.3). However, we infer this mostly because High Rock and McDermitt are the oldest caldera systems that we know about along a line of calderas whose ages increase westward. Numerous older magmatic centers, especially in accreted terranes to the west, have been interpreted to represent an oceanic plateau associated with the Yellowstone hotspot that formed as early as 56 Ma (Camp & Wells, 2021). The complexity of tectonics in this area (accretion, changes in trench position, Basin and Range extension) does not currently permit clear identification of a pre-17 Ma track, and most studies assume initiation about 16.5 Ma.

Link to Columbia River Basalts The Columbia River Basalts represent a relatively recent outpouring of flood basalt. The age of this large igneous province ranges from about 17.2 to about 6 Ma, but by far the largest magmatic flux (~200,000 km^3, or >90%) occurred between 16.6 and 16.0 Ma (Kasbohm & Schoene, 2018). This time period overlaps the earliest identified Yellowstone calderas (see Figure 14.3; Benson et al., 2017).

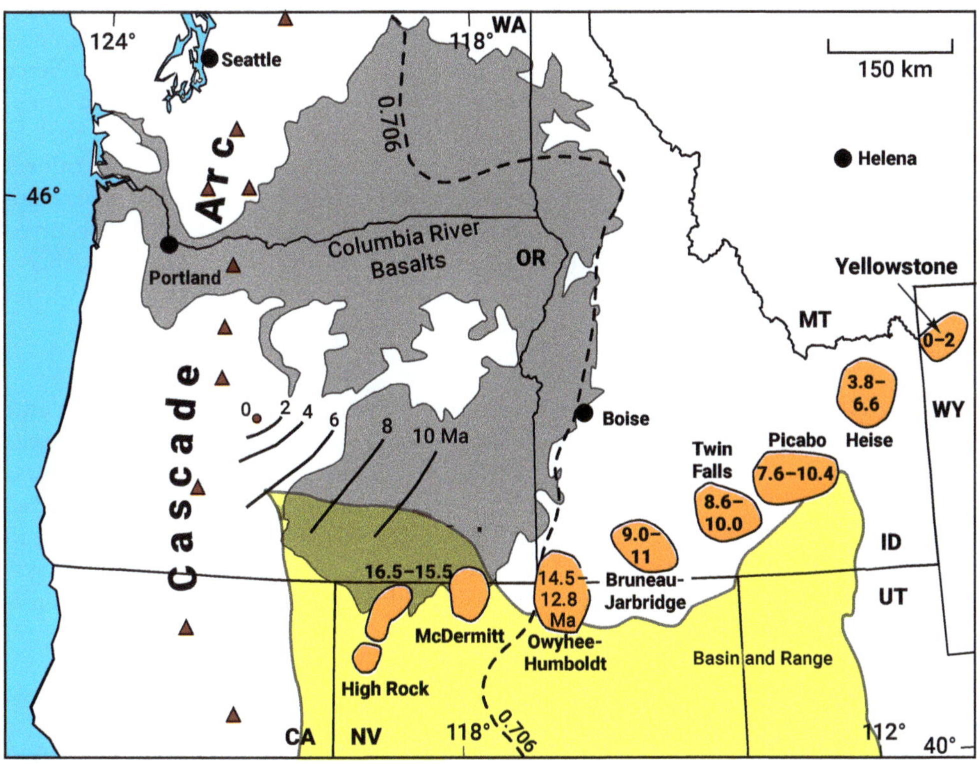

Figure 14.3 Track of the Yellowstone hotspot from northwestern Nevada through southern Idaho to northwestern Wyoming over the last ~16.5 Myr. Major silicic eruptive centers are labeled with range of ages for each in Ma. Columbia River Basalts (gray) erupted mostly between 16.6 and 16.0 Ma, coeval with the first silicic volcanic centers. Age contours that decrease westward represent silicic magmatism in Oregon that terminates near the Cascade arc. Yellow overlay denotes the Basin and Range province; dashed line is the 0.706 line (see Section 13.4).

Source: Modified from Coble & Mahood (2015).

The combined output of these early silicic volcanoes is about ~3000 km^3, which is large by modern eruption standards, but only a few percent of the volume of the flood basalts.

Progression After about 16 Ma, volcanism moved eastward across southern Idaho to form major silicic centers that most recently erupted at Yellowstone (see Figure 14.3). Ages broadly decrease eastward along the track and equate to a migration rate about 3 to 4 cm/yr, but correction for regional extension reduces this to between 2 and 2.5 cm/yr over the last 10 Myr. Faulting in the region also appears to have migrated in concert with the Yellowstone trend at a rate of about 2 cm/yr (Pierce & Morgan, 2009). Complicating the Yellowstone hotspot picture, silicic systems not only migrated eastward across Idaho but also northwestward across Oregon at rates of about 3 to about 1 cm/yr, to reach the Cascades today (see Figure 14.3; Jordan et al., 2004). This is but one of dozens of ill-understood volcanic migrations that occurred in western North America during the Cenozoic (Glazner, 2022).

Interaction with Lithosphere The Yellowstone hotspot track started west of the 0.706 line (dashed line in Figure 14.3), in lithosphere composed of accreted oceanic terranes, and crossed into Precambrian lithosphere in southwestern Idaho (see Figure 14.3). The character of silicic volcanism changed dramatically as the magmatic centers progressed from young accreted to old cratonal lithosphere (Nash et al., 2006). For example, in silicic volcanism along the track, the $^{143}Nd/^{144}Nd$ ratio dropped dramatically, from values characteristic of oceanic lithosphere (0.5130; see Figure 13.21) to values characteristic of enriched subcontinental lithospheric mantle (0.5121), and $^{87}Sr/^{86}Sr$ increased (see Figure 13.20).

Modern Yellowstone Volcanic System Yellowstone today is topographically high (~2500 meters), with a ~400-km diameter topographic swell around it (Pierce & Morgan, 2009). Renowned for its abundant geysers, mud pots, hot springs, and fumaroles, Yellowstone National Park was created in 1872 as America's first national park to preserve these volcanic features and surrounding wilderness. The Yellowstone caldera is wholly contained within the park and represents the remnants of the last supereruption at about 630 ka (Lava Creek eruption). Two older supereruptions have also occurred at Yellowstone: Huckleberry Ridge (2.1 Ma) and Mesa Falls (1.3 Ma). The Huckleberry Ridge and Lava Creek eruptions (~2500 and ~1000 km^3, respectively) were blue whales (see Figure 7.4), dwarfing any eruptions in recorded history.

Integration Most models of hotspot initiation invoke impingement of a large-volume plume head at the base of the lithosphere with massive volumes of decompression melts derived from rising mantle. The Columbia River Basalts could represent this event (unless the hotspot initiated earlier, in which case the flood basalts represent something else; Camp & Wells, 2021). Movement of North American lithosphere across the plume tail then gave rise to smaller-volume melting of continental crust to form silicic magma systems and their calderas, represented by the early High Rock and McDermitt fields transitioning east-northeast to Yellowstone today.

Great Meteor

The standard interpretation of the Great Meteor hotspot is that it tracked southeastward across Canada and New England, then out into the Atlantic Ocean. There, it shifted northward to form the Corner Seamounts, then crossed the Mid-Atlantic Ridge to form the Great Meteor and Atlantis Seamounts (Figure 14.4). If this interpretation is correct, it is one of the oldest active hotspots, having initiated by at least 125 Ma, and possibly even before 200 Ma (Heaman & Kjarsgaard, 2000). However, Great Meteor has numerous puzzling characteristics. First, most of its igneous activity, minuscule and compositionally

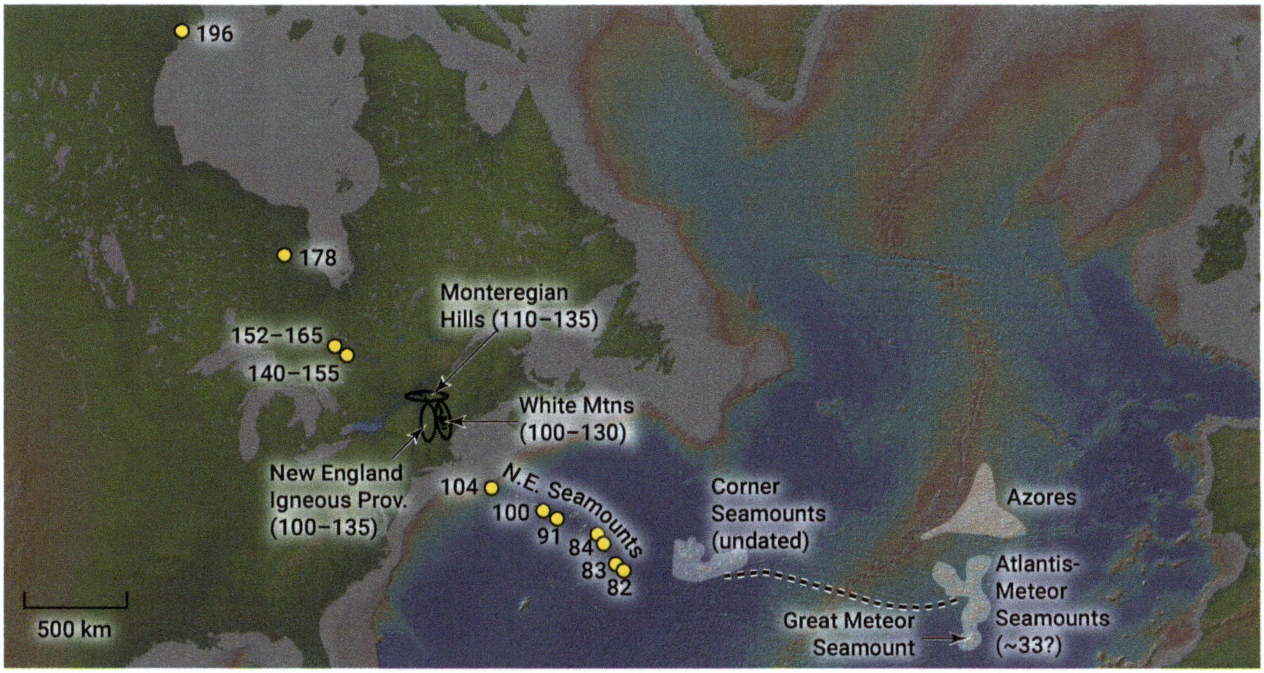

Figure 14.4 Track of the Great Meteor hotspot, possibly from far northern Canada through Ontario, Quebec, and New England through the New England seamount chain and across the Mid-Atlantic Ridge. Other than the New England Seamounts, the track is relatively speculative.

Sources: Ages summarized from Heaman et al. (2004) and Boemmels et al. (2021).

bizarre, is inferred from age progression of kimberlites across central Canada, switching over to highly alkalic magmas in Quebec and New Hampshire. Second, magmatism in Quebec and New Hampshire lasted substantially longer than other hotspots—tens of millions of years. Third, although the New England Seamounts show a systematic progression of younger ages toward the southeast, the relationship to the undated Corner Seamounts and Meteor Seamounts is obscure. Probably, we should refer to this hotspot as the "New England" hotspot (based on the New England seamount chain) rather than the "Great Meteor" hotspot, whose connection is problematic.

We included Great Meteor in this section because it illustrates common difficulties in interpreting hotspots. Many hotspot tracks have magmatic gaps (e.g., Sleep, 2002) and unusual bends, and are not always well dated. Great Meteor is particularly difficult to trace because of its interaction with a spreading center. It does not have a clear start or an end, and other than the New England seamount chain, evidence for a hotspot track is somewhat unconvincing. Conversely, not every hotspot has to follow the patterns of Hawai'i or Yellowstone, which have consistently erupted large volumes of magma. A less productive source might develop the sparse record that is inferred for Great Meteor.

Rocks of the Oceanic Hotspot Lithosystem

These systems are dominated by variably metamorphosed mafic volcanic and intrusive rocks, upper-mantle ultramafic rocks, and deep-water sedimentary rocks. Chief components include pillow basalts, gabbros, and basaltic dikes. These are similar protoliths to those seen in the HP/LT lithosystem (see Section 6.2), but associated metamorphism is greenschist- to amphibolite-facies, not HP/LT.

Toolbox 14.1

Paleomagnetism

The Earth's magnetic field can be approximated by a dipole field, like that of a bar magnet (Figure 14.5). In such a field the inclination (dip) of the magnetic lines varies with latitude according to the formula

$$I = \tan^{-1}(2\tan(\lambda)) \qquad \textbf{(Eq. 14.1)}$$

where I is the inclination of the field and λ is latitude. Inclination is 0° at the magnetic equator, and 90° at the magnetic poles.

Cooling magma picks up Earth's magnetic field when appropriate minerals, typically magnetite, cool below the **Curie point**. The orientation of the field locked into the rock at this point can be measured, and this gives a measure of how far the rock was from the magnetic pole at the time it cooled. Basaltic lava flows typically carry a strong record of the ambient field when they were erupted and cooled, although that record can be corrupted by later events, such as hydrothermal alteration or lightning strikes.

In 2023, the geomagnetic north pole was located at 80.8°N, 72.7°W, northwest of Greenland, and was moving northwestward by almost 0.1°/yr. A lava flow records magnetic paleolatitude at the time the flow cooled, and thus determining actual paleolatitude requires collecting data from a sufficient number of lava flows that were erupted over a long enough period of time—on the order of 10 kyr—that **secular variation** can be averaged out, but not over so long a period of time that plate movement caused paleolatitude to vary.

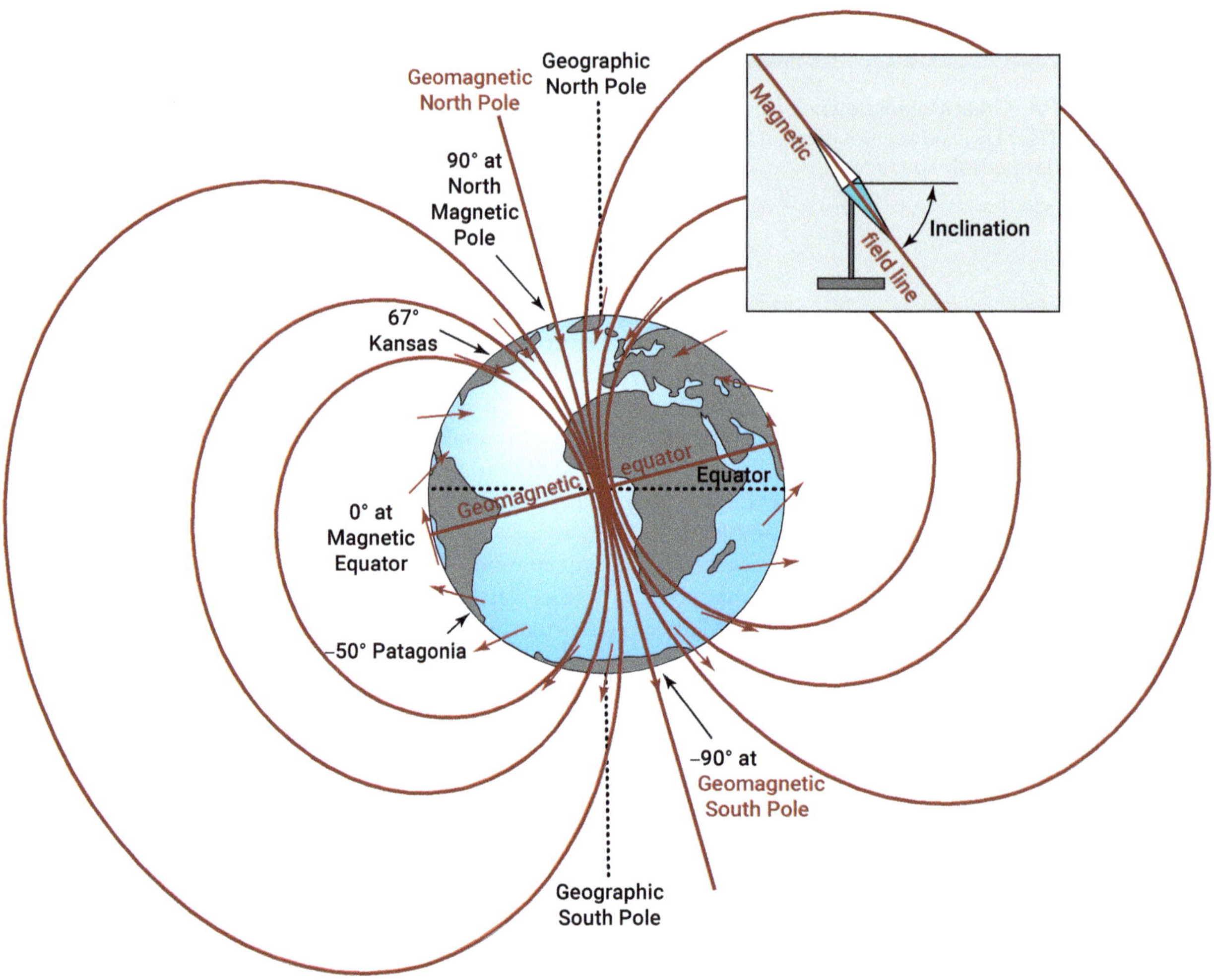

Figure 14.5 The inclination (angle relative to horizontal) of the magnetic field varies with latitude following Equation 14.1. Inset shows a magnetic needle parallel to the local field, with the north-seeking end shaded. The angle below (+) or above (–) horizontal is the inclination. In this idealized view of the general case, the magnetic pole (where the inclination is 90°) is not coincident with the geographic pole. The inclination of the field thus gives you geomagnetic paleolatitude rather than true paleolatitude.

14.2 Mantle Plumes

Learning Objectives

By the end of this section you should be able to:

14.2.1 Draw a labeled whole-Earth cross section, illustrating potential sources of plumes.

14.2.2 Propose features you would look for, both geologic and geochemical, to identify an ancient plume.

14.2.3 Explain how mantle tomography contributes to the debate about plume origins.

Plumes vs. Plates

The existence of hotspot tracks of varying consistency is not in doubt, but what produces them is. Although a large proportion of the geologic community, and most textbooks, present the existence of narrow mantle plumes coming from a thin layer known as **D″**, which lies just above the **core-mantle boundary** (**CMB**), as settled science, a substantial community of skeptics remains. This debate, typically framed as "plumes vs. plates" (e.g., Foulger et al., 2005), is complicated and crosses into advanced topics in seismology, geochemistry, and several other fields. A brief review of the basic issues follows.

In the plume hypothesis:

1. Mantle plumes arise from a deep source, likely D″, although a shallower source such as the mantle transition zone is another option; these are not exclusive of one another.
2. The upper mantle actively convects and is generally well mixed.
3. The lower mantle actively convects.
4. Subducted slabs penetrate the 660-km discontinuity, descend into the lower mantle, and may end up fertilizing D″.
5. Plumes and associated flood basalts can lead to continental breakup.

In the plate hypothesis:

1. Hotspot tracks and other intraplate magmatism result from extension of the lithosphere.
2. The amount and type of magmatism depend on the fertility (see Section 6.6), and thus composition, of the upper mantle where extension occurs.
3. Subducted slabs stall out at the 660- or 1000-km discontinuities and rarely reach the lower mantle.
4. Convection in the lower mantle is sluggish.
5. Convection in the upper mantle is driven by sinking plates.
6. The upper mantle is heterogeneous—the "plum pudding" model.

As this last term is widespread but few people today seem to know what plum pudding is, here is a general recipe: make a mantle of cake batter (butter, flour, eggs, etc.) and toss in chopped-up plums, dates, raisins, currants, and carrots. Bake and top with a crust of butter, sugar, and rum. The plums, raisins, and so on represent compositional variability in the mantle, scattered somewhat randomly throughout it.

The Plume Hypothesis

The mantle's lower boundary, the CMB, is a profound compositional break in the Earth, far more stark than the mantle's upper boundary, the Moho, because it separates molten Fe-Ni alloy from the silicate mantle. Although there is great uncertainty in the

Curie point
Temperature above which ferromagnetism disappears owing to thermal agitation; for natural magnetite, the Curie point, which varies with composition, is ~570 °C. Named for Pierre, the lesser-known Curie.

Secular variation
Movement of the geomagnetic poles over time; the geomagnetic poles seem to average out to the spin axis on time scales of 10^4 years.

D″
(dee-double-prime) A designation, from an archaic seismologic classification of mantle layers, for the lowermost few hundred kilometers of the mantle; a thermal and likely compositional boundary layer that is not identical to the core-mantle boundary owing to its considerable thickness.

Core-mantle boundary (CMB)
The profound chemical and thermal boundary between the core and mantle.

temperature profile of the lower mantle and core, it is likely that the outer core is several hundred to 1000 °C hotter than simple extrapolation of T in the lower mantle to the CMB would predict (Figure 14.6; Frost et al., 2022). That mismatch implies the existence of an overheated thermal boundary layer in the lowermost mantle. This overheated layer, for which we will use the D″ designation, is less dense than the rock above and will rise buoyantly if able. The plume hypothesis is predicated on the assumption that this material rises through the mantle as narrow, tubular jets.

The Griffiths-Campbell Experiments Much of the general conception of mantle plumes is based on a set of experiments done by Griffiths and Campbell (1990). They injected hot (80 °C), dyed glucose syrup into clear syrup at ambient laboratory temperature (20 °C). These temperatures make the heated syrup almost 300 times less viscous than the ambient syrup. As the heated material was pumped in, it inflated into a sphere above the entrance hole, and as the buoyant sphere inflated, it began to rise atop a lengthening conduit (Figure 14.7), producing a structure that looks like a mushroom with a cap and stalk. As the head of the dyed syrup approached the free surface, it spread laterally, with rising material still feeding and enlarging it. The volume of the head grew faster than the rate at which heated syrup was being supplied, because it heated and entrained the unheated syrup through which it rose. These elegant experiments and images from them have guided much of the interpretation of plumes.

Caveat about Models, Experimental and Otherwise It is important to keep in mind that compelling images produced by a laboratory experiment or numerical model are not confirmation that the process being modeled actually occurs in the real world. Ramberg's clay diapir experiments (see Section 10.4) loomed over interpretation

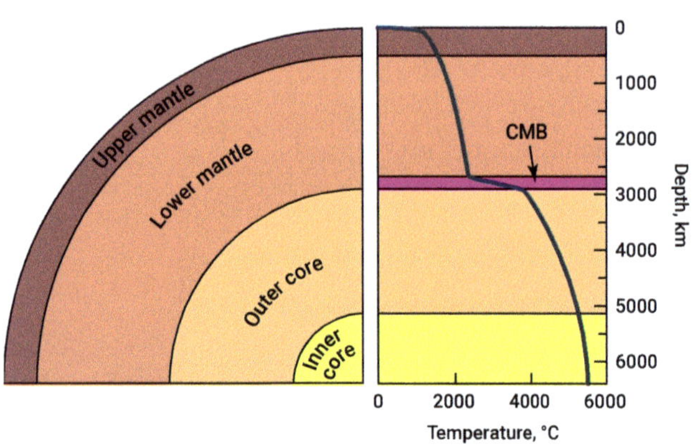

Figure 14.6 Schematic illustration of the postulated thermal boundary layer in the lowermost mantle. Extrapolating the theoretical mantle geotherm downward results in a CMB temperature over 1000 °C lower (~2100–2600 °C) than the predicted temperature of the outermost mantle (3600–4000 °C). This discrepancy likely occurs within a thin thermal boundary layer in the lowermost mantle where the temperature rises several hundred degrees over a distance of ~100 km.

Source: Created using estimated geotherm from Andrault et al. (2016).

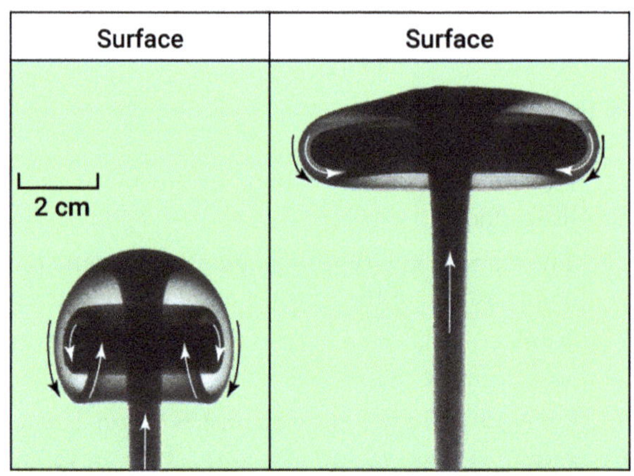

Figure 14.7 Images from heated syrup experiments of Griffiths and Campbell. Darker regions represent hotter syrup; lighter regions are cooler. White arrows show flow lines within the plume; black arrows show flow lines adjacent to plume. Thin tail represents source of syrup being pumped into the system. (Left) Approximately halfway through plume rise, the plume head is relatively equant. (Right) As plume approaches surface, it flattens. Continued addition of hot syrup, pumped in from below, expands the plume head laterally.

Source: Photos from Griffiths & Campbell (1990). © 1990 Published by Elsevier B.V.

of plutons for decades even though the experiments lacked a critical feature—the exponential dependence of viscosity on temperature. Clay with a constant viscosity is a poor analog for magma that transforms from a highly viscous liquid to a brittle solid as it cools, and although the diapir model for magma ascent in the middle and upper crust was commonly invoked in the 20th century, it has largely been abandoned owing to thermal shortcomings, conflicts with field data, and better understanding of the mechanics of salt diapirs.

The syrup models in Figure 14.7 similarly lack numerous critical features of the mantle, including vertical changes in viscosity and thermal conductivity as well as phase changes such as the olivine-spinel and spinel-perovskite transitions (see Section 3.3). An important difference between the syrup experiments and the Earth is that the hot syrup plume was actively pumped into the tank rather than allowed to form by buoyancy differences. The experiments are valuable, but reiterating the previous statement, compelling images are not compelling evidence that the modeled process actually occurs. On the other hand, among geologists, the famous Hubbert and Rubey (1959) beer-can experiment has wide acceptance, or at least an enthusiastic fan club, probably because it involves beer.

Predictions of the Plume Hypothesis

The plume hypothesis makes numerous predictions, including the following, many of which are testable:

1. Plumes take about 100 Myr to reach the base of the lithosphere from D″ (Campbell, 2005), an average ascent rate (~30 mm/yr) that is comparable to plate-spreading rates.
2. Plumes are composed of mantle with a high potential temperature (see Section 3.7) and should, at least initially, erupt high-MgO basalts (picrites; see Section 6.7) whose compositions reflect this. (Recall from Figure 13.34 that MgO increases as F increases.)
3. When the hot plume head reaches the lithosphere, it spreads out into a pancake, heats the surrounding mantle, and thermally erodes the lithosphere as rising material from the plume tail continues to feed it.
4. This excess thermal energy produces massive amounts of magma that forms flood basalts and large igneous provinces.
5. The rate of magma production is highest when the plume head first hits the lithosphere and declines dramatically after that.
6. Buoyancy of the plume causes domal uplift of the lithosphere and surface.
7. If the plume rises under continental lithosphere, thermal weakening coupled with uplift can lead to continental rifting.
8. Movement of lithospheric plates over hotspots causes hotspot tracks.

The Plate Hypothesis

The plume hypothesis has been criticized in many prominent forums (e.g., Anderson, 2000; Hamilton, 2003; Foulger et al., 2005; and papers cited therein), and defended in many others (e.g., Ernst & Buchan, 2001; Campbell, 2005). Foulger (2010) presented a lengthy critique of the plume hypothesis and compared it with the plate hypothesis.

The plate hypothesis is something of an all-purpose explanation for magmatism in that the volume and geochemical character of a given magmatic province is a function of how many pieces of fruit (in the plum pudding model) a given extensional system taps, and what kinds of fruit they are. Extension is the process that drives magmatism in the plate hypothesis, but this is inferred in many cases (e.g., Hawai'i) rather than observed. This flexibility makes the plate hypothesis inherently difficult to test. A comparison of these two hypotheses is given in Figure 14.8 and Table 14.1.

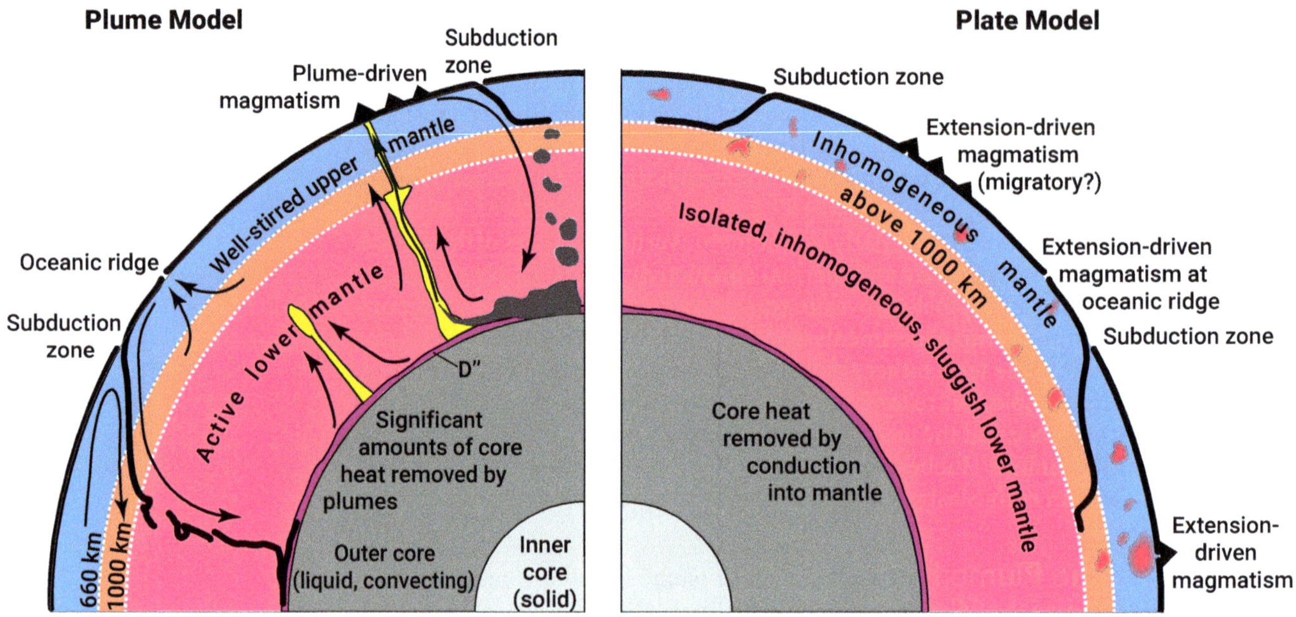

Figure 14.8 Schematic comparison of the plume hypothesis and alternative plate hypothesis for the origin of hotspot tracks and other varieties of intraplate magmatism. Red blobs in plate model represent fertile heterogeneities ("plums"). See text for discussion.

Source: Modified from Anderson (2005) and Koppers et al. (2021).

Table 14.1 Comparison of Plume and Plate Hypotheses

Feature	Plume Model	Plate Model
Fundamental process	Deep mantle plumes rising from the core-mantle boundary	Extension within the lithosphere and upper mantle
Driving force	Thermal buoyancy of hot, buoyant plume material	Plate tectonic stresses, including rifting, mantle upwelling near plate boundaries, and interactions with subducting slabs; also topographic stresses induced by volcano loading
Magmatism pattern	Linear tracks or hotspots; large igneous provinces (LIPs) with high eruption rates	Divergent boundaries (mid-ocean ridges); hotspots with lower eruption rates; scattered intraplate volcanism
Geochemical signature	Distinctive isotopic and trace-element signatures due to deep mantle origin	Varied signatures depending on local mantle composition and interaction with surrounding lithosphere
Lithospheric interaction	Plume melts through or weakens the overlying lithosphere	Plate movements cause extensional stress, thinning, and rifting of the lithosphere, allowing melts to rise
Evidence	Hotspot tracks; linear age progressions; seismic tomography; ocean-island and LIP geochemistry	Seismic and GPS data on stress states; observations of intraplate rifting and mantle flows
Strengths	Accounts for hotspot tracks, LIPs, and some unique geochemical features	Accounts for any intraplate volcanism, consistent with plate tectonics, and accounts for diverse geochemistry
Weaknesses	Relies on deep mantle processes for which there is limited direct evidence; deep mantle tomography is subject to significant interpretation; struggles to explain some intraplate volcanism and range of isotopic compositions	Can't explain hotspot tracks or LIPs as easily; requires complex interactions between various plate tectonic processes

14.3 Mantle Reservoirs

Learning Objectives

By the end of this section you should be able to:

14.3.1 Illustrate, using isotope plots, how different mantle reservoirs are identified.

14.3.2 Predict the isotopic compositions of different basalts, depending on their source.

How Heterogeneous Is the Mantle?

What does the mantle look like chemically? Is it homogeneous because mantle convection has thoroughly mixed it, or heterogeneous because subduction and delamination keep dumping lithospheric junk into it? Are hotspot sources different from MORB sources?

It is perhaps unsurprising that our understanding of mantle chemistry is highly inferential, because we have few direct observables. Seismic tomography provides evidence that the mantle is heterogeneous, but velocity variations can arise from variations in temperature, or composition, or which mineralogic phases are present for a given composition, and sorting these out is difficult. A more direct way to examine chemical heterogeneities in the mantle is to examine materials that come out of the mantle—xenoliths and basalts. But xenoliths are rare, and sample not only the uppermost mantle, but only a point in the uppermost mantle, and perhaps even not a representative point in the mantle. After all, the xenolith had to be caught up in magma to be carried to the surface for us to examine, and places that melt in the mantle may not be representative of places that don't melt. Consequently, geochemists have generally relied on basalt geochemistry to study variations in the mantle.

The use of basalt geochemistry to map out mantle heterogeneity carries its own difficulties. Remember that melting is commonly eutectic-like, fixing major-element compositions within a small range. Different magmas can be more and less alkalic, but from a major-element perspective, one basalt is much like another, and their compositions do not tell us much about what was melting (other than "mantle"). Major-element compositions also depend strongly on post-melting processes such as fractional crystallization. Trace elements are the next logical step, and although different magma sources result in different trace-element concentrations and patterns, these depend on the depth and degree of melting. These are useful things to know, but they mask the fundamental question of this section, so geochemists instead turn to a third geochemical tracer—isotope ratios—that *do not depend on the degree of partial melting or fractional crystallization*. As we saw in Sections 3.8 and 13.4, MORBs have homogeneous Nd and Sr-isotope ratios. MORBs are also homogeneous in many other isotopic systems, indicating that at least one source of basalts in the mantle is homogeneous. But how many other components are there? And what do they represent?

Compositional Variation in Oceanic Islands

Tahiti, Hawai'i, Samoa, Bora Bora, Kauai ... these names might conjure up visions of a tropical vacation for most people, but if you are a petrologist, you will undoubtedly think of alkaline rocks such as tahitiite, hawaiite, and kauaiite. Hotspot-related oceanic islands are overwhelmingly basaltic, but alkaline rocks and their consequent name explosion are typically part of the package, and some islands produce extremely alkalic lavas. In keeping with the general pattern, the more unusual the rock, the more studied and oversampled it is—especially on beautiful tropical isles.

TAS plots of data from four ocean island groups—Samoa, the Society Islands (which include Tahiti and Bora Bora), the Canary Islands, and the Hawaiian Islands—illustrate this variability (Figure 14.9). All are overwhelmingly basaltic, although the Canary Islands dataset contains a significant proportion of trachytes and phonolites with about 60 wt% SiO_2.

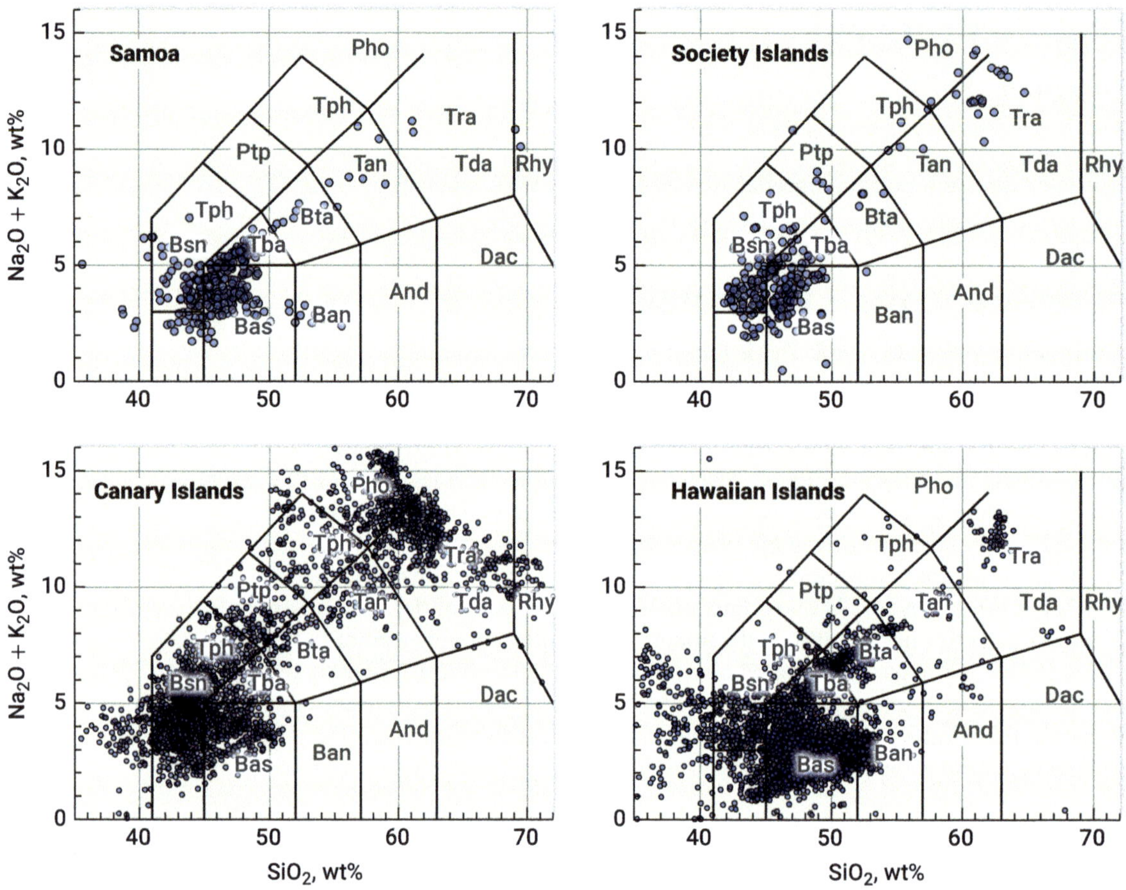

Figure 14.9 TAS plots of data from four ocean island groups: three from the Pacific Ocean (Samoa, Society Islands, Hawaiian Islands) and one from the Atlantic (Canary Islands). Note that, as always, small volume alkalic rocks are generally oversampled.

Source: Data from GEOROC (georoc.eu) compilations, screened for whole-rock analyses of volcanic rocks.

All groups have a concentration of low- to high-alkali basalts and then scatter along a high-alkali trend toward trachytes and phonolites. Recall that arc rocks tend to scatter along or just under the dividing line that separates basalt, andesite, and dacite from their cousins trachybasalt, trachyandesite, and trachydacite (e.g., Figures 1.3 and 13.2). The Galápagos Islands, not shown, are predominantly low-alkali basalts that plot in the middle of the basalt field, in the same place as low-alkali Hawaiian basalts.

Hawaiʻi is likely the best-studied of all the OIB systems and provides at least one template for how such hotspot systems work. The general evolution there is given in Table 14.2.

The massive volcanoes Mauna Loa and Mauna Kea (see Figure 7.1a) on the island of Hawaiʻi illustrate two of these stages. Mauna Loa is active, with significant eruptions at the summit caldera in 1984 and 2022. Mauna Kea's last confirmed eruption was approximately

Table 14.2 Stages of Hawaiian Volcanism

Stage	Lava Composition	Relative Volume
Pre-shield	Alkali basalt	A few percent
Shield	Tholeiitic basalt	>95%
Post-shield	Alkalic basalts—trachybasalt ranging to trachyte and nephelinite	A few percent
Post-erosional or rejuvenated	Low-silica alkalic—basanite; depleted source	≪1%

4500 years ago, and the summit is home to 13 (at last count) expensive telescope facilities that take advantage of the relatively low air pollution, low light pollution, high altitude (4200 m, 13,800 ft), and volcanic quietude.

Mauna Kea lies 40 km north of Mauna Loa; although not exactly "downstream" along the northwest-trending hotspot track (see Figure 14.1), Mauna Kea is assumed to be in a more advanced stage of the volcano life cycle than Mauna Loa, and this is reflected in the morphology of the volcanoes (Figure 14.10). Mauna Loa's profile is smooth except for the summit caldera, but Mauna Kea's summit is dotted with cinder cones from the alkalic post-shield phase of volcanism. These cones and associated flows contain both mantle and crustal (gabbroic) xenoliths (Jackson et al., 1982), in keeping with a general and unexplained observation: tholeiitic lavas worldwide are essentially devoid of mantle xenoliths, whereas alkalic lavas are fertile hunting grounds for them.

Figure 14.10 Aerial view looking southwest across the snow-dusted summits of Mauna Kea (foreground) and Mauna Loa. These huge volcanoes rise 9 km above the ocean floor (see Figure 7.5). Mauna Kea is at a more advanced stage of evolution than Mauna Loa and is marked by numerous cinder cones from the post-shield stage of alkalic volcanism.

Source: International Gemini Observatory/NSF's NOIRLab/AURA/J. Pollard.

Isotopic Discriminators of Mantle Sources

The main discriminators we use to identify different mantle sources are isotopes of Nd, Sr, and Pb, with additional discrimination from isotopes of O and He. Pb isotope geochemistry is quite complicated, and we will only skim the surface here (see Toolbox 14.2). On isotope plots of Nd vs. Sr (Figure 14.11), MORB has a relatively limited range, but different OIBs fan out in a squid-like fashion toward three different end members. These end members are called EM1 ("ee-em-one" for Enriched Mantle 1), EM2 (Enriched Mantle 2), and HIMU (high-μ, where $\mu = {}^{238}U/{}^{204}Pb$; Toolbox 14.2). It may seem like we should need only three reservoirs to explain Nd and Sr isotopes.

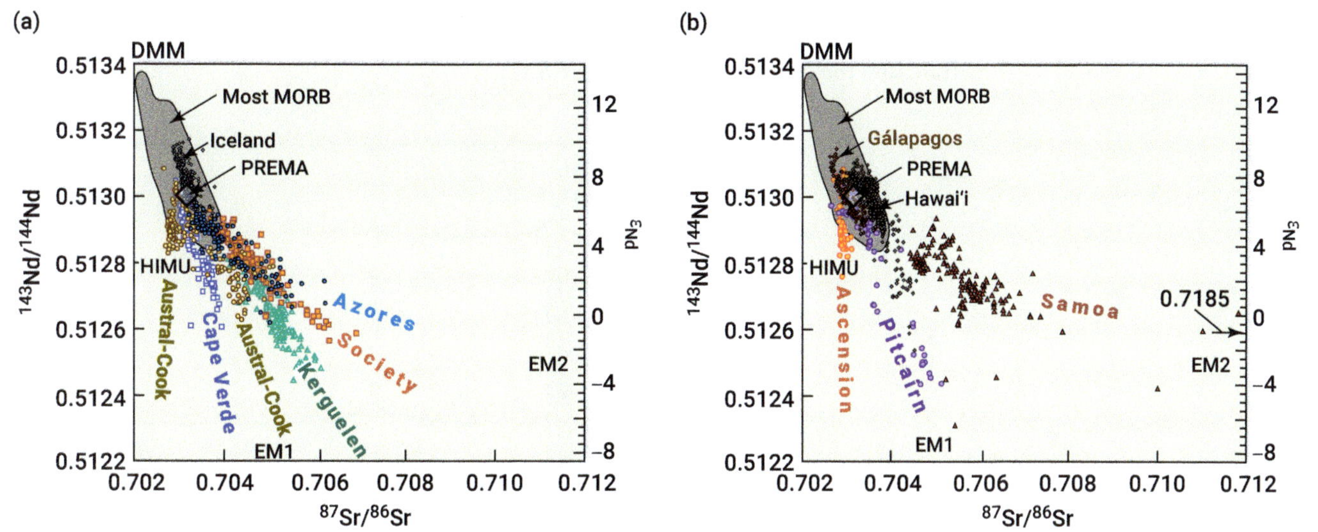

Figure 14.11 Isotope plot (the "squid") of Nd vs. Sr showing five primary isotopic reservoirs that contribute to OIBs: depleted MORB mantle (DMM), prevalent mantle (PREMA), high-μ (HIMU), and Enriched Mantle 1 and 2 (EM1 and EM2, respectively). (a) Data sets from five distinctive ocean island chains showing how isotopes fan out from nearly vertical arrays at HIMU to more nearly horizontal arrays toward EM2. Iceland straddles a hotspot and has more MORB-like compositions. (b) Datasets from three additional ocean island chains showing similar spread. Galápagos and Hawai'i are more MORB-like, although Hawai'i does extend toward EM1.

Source: Modified from White (2010).

That's true, and we identify more reservoirs because some of these end members, especially HIMU, are defined more clearly with Pb isotopes, which represent more dimensions in isotope space. Trends of OIBs converge toward a common region that goes by a variety of monikers; we will use PREMA (for PREvalent MAntle; Zindler & Hart, 1986).

Broadly speaking, the five reservoirs in Figure 14.11 (DMM, PREMA, HIMU, EM1, and EM2) are thought to represent different components of the mantle that are either primordial (leftover from the original differentiation of the Earth) or recycled and replenished through processes of plate tectonics. DMM represents one end member of mantle from which MORB is extracted. Clearly, a single DMM cannot be the only source of MORB, or the isotopes would show no spread. But many MORBs have much higher $^{143}Nd/^{144}Nd$ and lower $^{87}Sr/^{86}Sr$ than other basalts, so DMM must be a major component in MORB formation. EM1 and EM2 are thought to represent pelagic sediment (sediment composed of marine organisms, clays, and chemical precipitates) and sediment eroded from continents, respectively. The source of HIMU is harder to pin down because it is difficult to create rocks with high-μ (i.e., rocks with a particularly high U/Pb ratio), but it may represent subducted oceanic crust.

That leaves PREMA. Perhaps its most distinctive characteristic is that PREMA has high $^3He/^4He$ (see Toolbox 14.2). Basalts with high $^3He/^4He$ are thought to tap relatively undegassed, primitive regions of the mantle. MORB has a restricted range of relatively low $^3He/^4He$ (Figure 14.12), and this is taken to indicate a source that is both highly mixed and substantially degassed by prior melting. In contrast, OIBs show much more variable $^3He/^4He$, extending from low, MORB-like values to very high values. If this part of the mantle experienced some degassing, it was to a lesser extent than DMM.

So, what kind of picture emerges from these isotope studies? First, MORB is homogeneous in multiple isotopic systems, which seems to indicate a relatively homogeneous source in the upper mantle. This interpretation must be tempered, however, with a recognition that there is significant isotopic variability to MORB compositions, and that the higher the degree of melting, the more homogenized the melt becomes. So, the upper mantle may be more geochemically heterogeneous than MORB might at first indicate. The large ranges in isotopic compositions of OIB indicate geochemically isolated reservoirs that different islands tap to different degrees. Some of these sources likely reside in the deep mantle. Exactly where sources like EM1, EM2, and HIMU reside remains an open question. Inasmuch as EM1, EM2, and HIMU are thought to represent subducted components, they could reside wherever subducted components end up. Seismic tomography suggests that some subducted slabs stall out in the transition zone, whereas others penetrate into the deep mantle (see Figure 6.17).

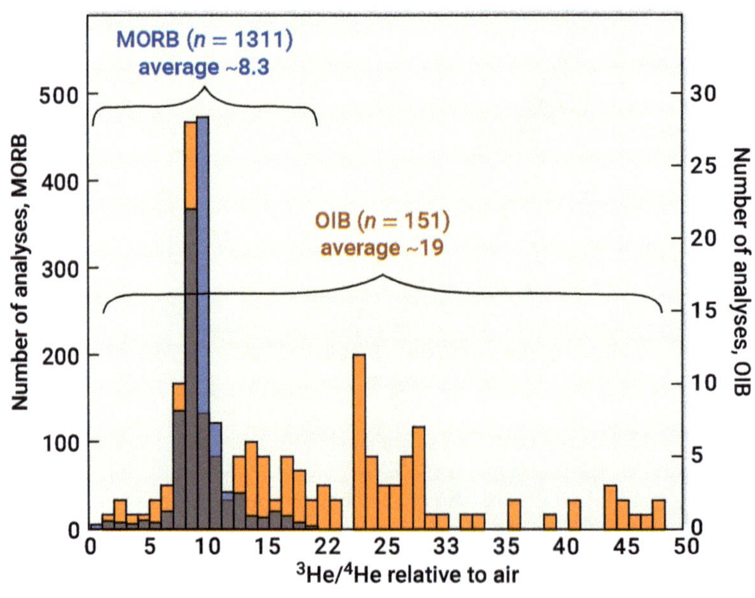

Figure 14.12 Helium isotope data, normalized to ratio in air, for MORB and OIB. MORB has a narrow range of ratios, consistent with it tapping a relatively homogeneous mantle source. OIB spreads out to much higher ratios, consistent with it tapping not only a MORB source, but also relatively undegassed, old mantle.

Source: Data from EarthChem.

Toolbox 14.2

A Brief Introduction to Pb and He Isotope Geochemistry

The isotope geochemistry of Pb is complicated because it is a mixed family—three of its stable isotopes are radiogenic children of three different parents: $^{232}Th \rightarrow {}^{208}Pb$, $^{235}U \rightarrow {}^{207}Pb$, and $^{238}U \rightarrow {}^{206}Pb$. Each of these systems is a long decay chain. The fourth stable isotope, ^{204}Pb, is non-radiogenic and thus plays the normalizing role that ^{86}Sr plays in the Rb/Sr system. The complicated nature of this system means that it can be used to solve geochemical problems in diverse creative ways; e.g., as we saw in Section 10.7, decay of two different U isotopes to two different Pb isotopes gives two independent clocks in U-Pb geochronology.

Although Pb isotopes form a complicated system, old materials with high U/Pb ratios progressively develop higher ratios of $^{206}Pb/^{204}Pb$ and $^{207}Pb/^{204}Pb$ over time, just as materials with high Rb/Sr develop high $^{87}Sr/^{86}Sr$ over time. The variable μ (mu), which is the $^{238}U/^{204}Pb$ ratio, is commonly used in Pb-isotope geochemistry.

The isotope geochemistry of He is simple. The Earth's primordial $^{3}He/^{4}He$ ratio is estimated as about 3×10^{-4}, but production of α particles (^{4}He nuclei) by decay of ^{232}Th, ^{235}U, ^{238}U, and other heavy radionuclides has diluted the $^{3}He/^{4}He$ ratio of the atmosphere by over two orders of magnitude, to about 1.3×10^{-6} (Porcelli & Ballentine, 2002). The ratio of any Earth reservoir with U or Th, such as the crust, is also decreasing with time. The rate of decrease is proportional to (U + Th)/He. He is extremely incompatible, because it does not bond to anything, and thus it is more strongly partitioned into melts than even the strongly incompatible elements U and Th. Regions that have previously melted consequently will have higher (U + Th)/He and develop lower $^{3}He/^{4}He$ over time.

Making Connections 14.1

OIBs and Polynesian Migrations

The human colonization of Polynesia represents one of the largest geographic expansions of humans in history. Although not technically an empire in the sense of an area under single political rule, the area populated by linguistically and culturally related Polynesian peoples, in excess of 35,000,000 km^2, dwarfs the Roman and Ottoman Empires (~5,000,000 km^2), is two-and-a-half times the size of the Spanish Empire (~14,000,000 km^2), and exceeds even the vast Mongol Empire (~24,000,000 km^2; Taagepera, 1979, 1997). Expansion into Polynesia occurred relatively quickly, too. With the development of advanced sailing technologies, people originating from what is now Taiwan spread southeastward, reaching Samoa about 3000 years ago (e.g., Gray et al., 2009; Skoglund et al., 2016). After a pause of nearly 2000 years, they then spread rapidly across the Pacific to settle in far-flung Rapa Nui (Easter Island), Hawai'i, and New Zealand (Figure 14.13). Curiously, however, outlier communities about 500 years old are found back west of Samoa. How did all these communities originate? What physical evidence exists for these migrations? And what kinds of interactions occurred between island chains?

Like all humans, Polynesians carried many materials with them when they traveled. Some are found at archeological sites, including stone tools for pounding taro root, working wood, hammering and polishing, and other tasks (see Figure 14.13, inset). Most tools were made from basalt, but basalts are fairly consistent in their mineralogy and major-element composition. How can we tell where the tools came from? Geochemistry to the rescue! Areas of interest include oceanic arcs around the Pacific Rim that encircle a vast area of the Pacific Ocean dotted by islands. Way back in Section 1.4 we learned about the "andesite line" that separates andesite-spewing arc volcanoes from basalts of the Pacific basin (see Figure 14.13). The geochemistry of arc basalts differs dramatically from that of OIBs (see Figures 9.11 and 9.12). A particularly clear discriminator is the concentration of HFSEs such as Ti, Nb, and Zr. These are generally incompatible during melting of the mantle, and so, as discussed in Section 13.6, their concentrations in a melt depend on the degree of melting. This could be a problem, but we can minimize this effect by looking at the ratio of an HFSE to another incompatible element, because their concentrations will vary in sympathy with the degree of partial melting, and thus their ratio will remain the same (see Section 9.2). A commonly used ratio, one of many, is Nb/La. Figure 14.14a compares

continued

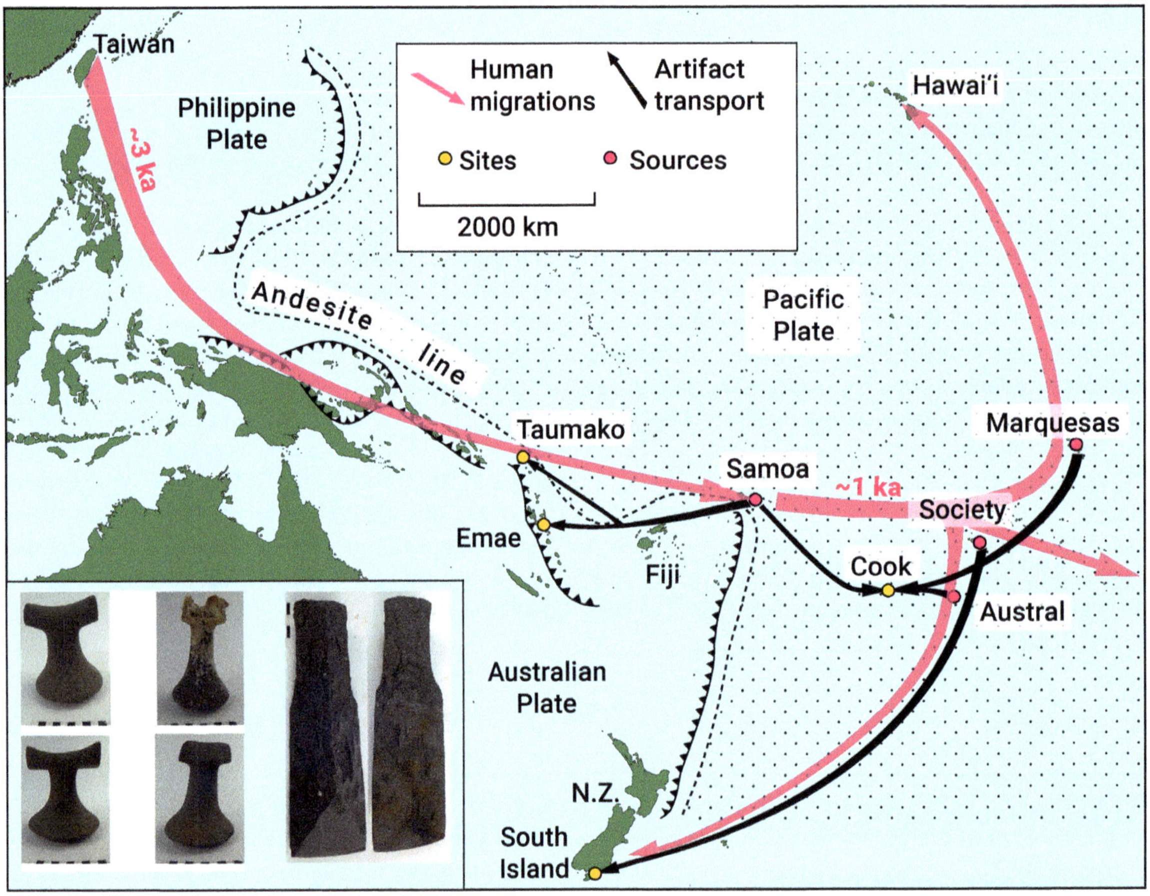

Figure 14.13 Map of the central and western Pacific, showing the location of Polynesian islands and migration and transport routes. Initial sailors left Taiwan and settled Samoa and neighboring island groups before voyaging to Hawai'i, Rapa Nui, and New Zealand. Dot pattern marks the area of the Pacific Basin outside the andesite line. Inset photos show examples of Polynesian tools (scale in cm).

Sources: Map based on Weisler et al. (2016), Ramsay et al. (2021), and Hermann et al. (2023); inset from Richards & Günther (2019), © 2019 Michelle J. Richards, Jasmin Günther.

this ratio for southwestern Pacific island arcs with that for southwestern Pacific ocean islands; it is far higher for the intraplate islands than for the island arcs.

Isotopic data are highly variable among ocean islands (see Figure 14.11) and allow further discrimination among island groups. By comparing trace elements and isotopic compositions, artifacts can commonly be traced back to sources in specific island groups, or even to specific quarries. In the case of the western Polynesian outliers (islands of Taumato and Emae), several artifacts, based on their high Nb/La ratios, clearly did not come from local volcanic arcs, but rather from intraplate oceanic islands (see Figure 14.14a). Within that group, isotopic data allow them to be traced with some confidence to Samoa, over 2000 km to the east (Figure 14.14b).

These data indicate that some Polynesians moved back westward 2000 years after occupation of Samoa. Geochemistry also indicates interaction between widely distributed communities within the Polynesian triangle. For example, materials from Mangaia Island, in the geographically central Cook Islands, in part come from the nearby Austral Islands, but also the distant Marquesas Islands and (again) Samoa (see Figure 14.14b). Artifacts from the South Island of New Zealand are interpreted to come from the Society Islands, although possibly also the Austral Islands. That Polynesians managed to travel such vast distances over empty oceans stands as a tribute to their fortitude, creativity, and persistence. That they transported materials from island to island for thousands of years stands as a testament to a human need for social connection and senses of history, tradition, and place.

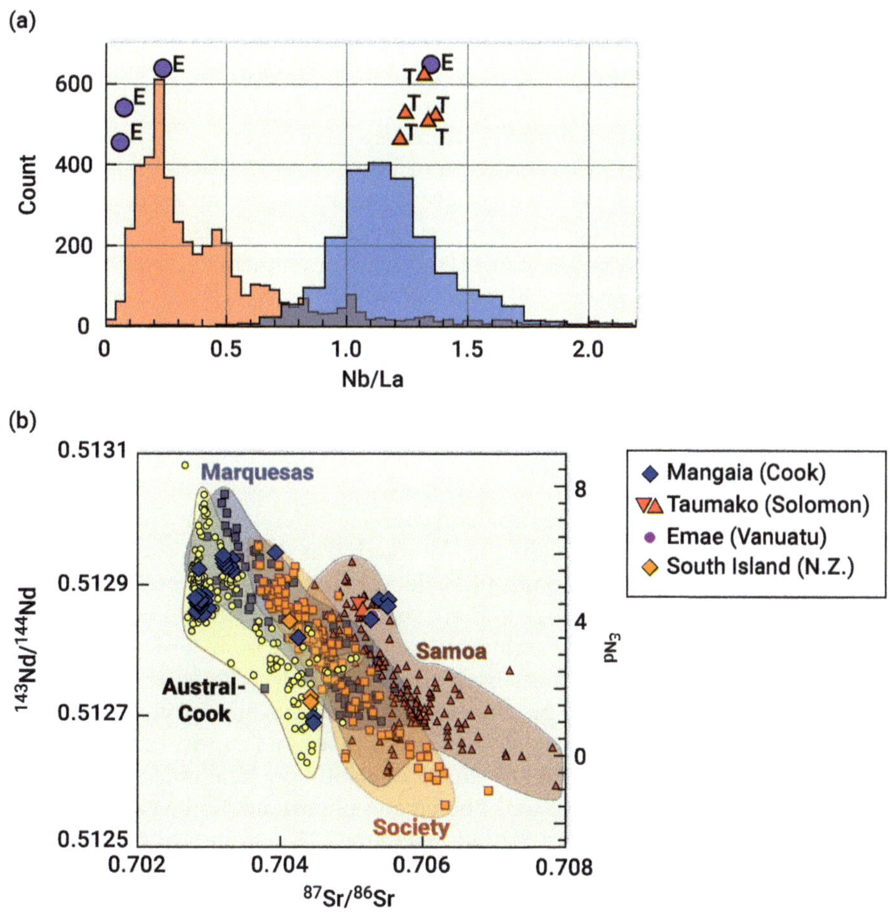

Figure 14.14 (a) Comparison of Nb/La in arc basalts and ocean islands, showing the relative depletion in HFSEs (here, Nb) in arcs. In the southwestern Pacific, this provides a good geochemical source tracer. Basalt artifact data are plotted for two western sites (vertical placement is arbitrary): Emae (E) and Taumako (T), both of which are in island arc settings. Three of the Emae artifacts have low Nb/La, consistent with their arc setting, but one does not; in contrast, all five Taumako artifacts have high Nb/La, indicating that they were brought from ocean islands. Isotopic data in (b) indicate that the source was likely Samoa. (b) Nd- and Sr-isotope compositions of Polynesian artifacts compared to reference data from several island groups.

Sources: (a) Artifact data from Hermann et al. (2023); (b) reference data from White (2010); artifact data from Weisler et al. (2016), Ramsay et al. (2021), Hermann et al. (2023).

14.4 Large Igneous Provinces

Learning Objectives

By the end of this section you should be able to:

14.4.1 Predict the flux history of magmatic output for a large igneous province.

14.4.2 Predict the evolution of magmatism after large igneous province formation.

14.4.3 Use sketches to illustrate the impact of oceanic plateaus on convergent tectonics.

Large igneous provinces (LIPs)
Intraplate magmatic province with area >100,000 km^2 and volume >100,000 km^3 with large magmatic fluxes (e.g., >75% erupting in <5 Myr; Kerr, 2014).

Large igneous provinces (**LIPs**) are regions on Earth with abundant igneous rocks that were erupted during a geologically brief interval. That is, they have unusually large magmatic fluxes. We have talked previously about other igneous provinces, such as the New

England igneous province in Vermont and New Hampshire, which represents dozens of plutons with an area of a few thousand square kilometers. But because it developed over several tens of millions of years, the magmatic flux was relatively low. Contrast that with one of the classic LIPs like the Deccan Traps in India (500,000 km^2; 600,000 km^3), which formed in 1 Myr, or the massive Ontong-Java Plateau (~2,000,000 km^2; ~44,000,000 km^3), much of which formed within a few Myr.

To produce so much melt so fast, the source must be unusually hot, fertile, or volatile-rich. There is little evidence to suggest that LIP magmas are especially rich in volatiles, so the source is usually thought to represent unusually hot and fertile ("thermochemical") plumes with a deep mantle origin. If so, LIPs form when a superheated, melt-rich plume head reaches the base of the lithosphere. However, to reach the surface, the magmas must traverse chemically and physically complex lithosphere, so the chemical and physical characteristics that might implicate a plume can be equivocal. We separate the following sections into flood basalts, which form on continental lithosphere, and oceanic plateaus, which form on oceanic lithosphere.

Flood Basalts

Flood basalts have two main characteristics—they are mostly basaltic in composition, and they flood the landscape. Although this definition may sound trite, embedded in it are two important points. First, flood basalts really are mafic. Some are virtually 100 percent basalt, and although andesite can also occur, the volume of material more silicic than andesite in a flood basalt province is tiny. The more felsic rocks are important because they provide the most easily datable material (e.g., feldspars or zircon amenable to $^{40}Ar/^{39}Ar$ or U-Pb dating), but they are not a major magmatic component. Second, to flood a landscape, basalts must be both voluminous and runny. Individual basalt flows in a LIP can easily exceed 1000 km^3 in volume and flow hundreds of kilometers. In the next section, we talk about two well-known and important flood basalt LIPs—the Columbia River Basalts in the United States (see Figure 7.21) and the Deccan Traps[1] in India (Figure 14.15). The even more voluminous Siberian Traps are discussed later in the context of climate change and extinctions (see Making Connections 14.2).

Columbia River Basalt Group

The Columbia River Basalt Group (Figure 14.16) represents a minimum volume of 210,000 km^3 (Reidel et al., 2013), mostly erupted within 0.5 Myr (Kasbohm & Schoene, 2018; Cahoon et al., 2020). Six major members span nearly

Figure 14.15 (a) Enormously thick stack of basalt flows in the Deccan Traps, Maharashtra, India. (b) Prayer room carved out of basalt in the massive Ajanta Caves, a set of 29 Buddhist cave monuments on the Deccan Plateau northeast of Mumbai. Excavation and carving of basalt to form the rooms at this UNESCO World Heritage Site began around 4000 years ago.

Sources: (a) Nicholas (Nichalp)/Wikimedia Commons; (b) Dey.sandip/Wikimedia.

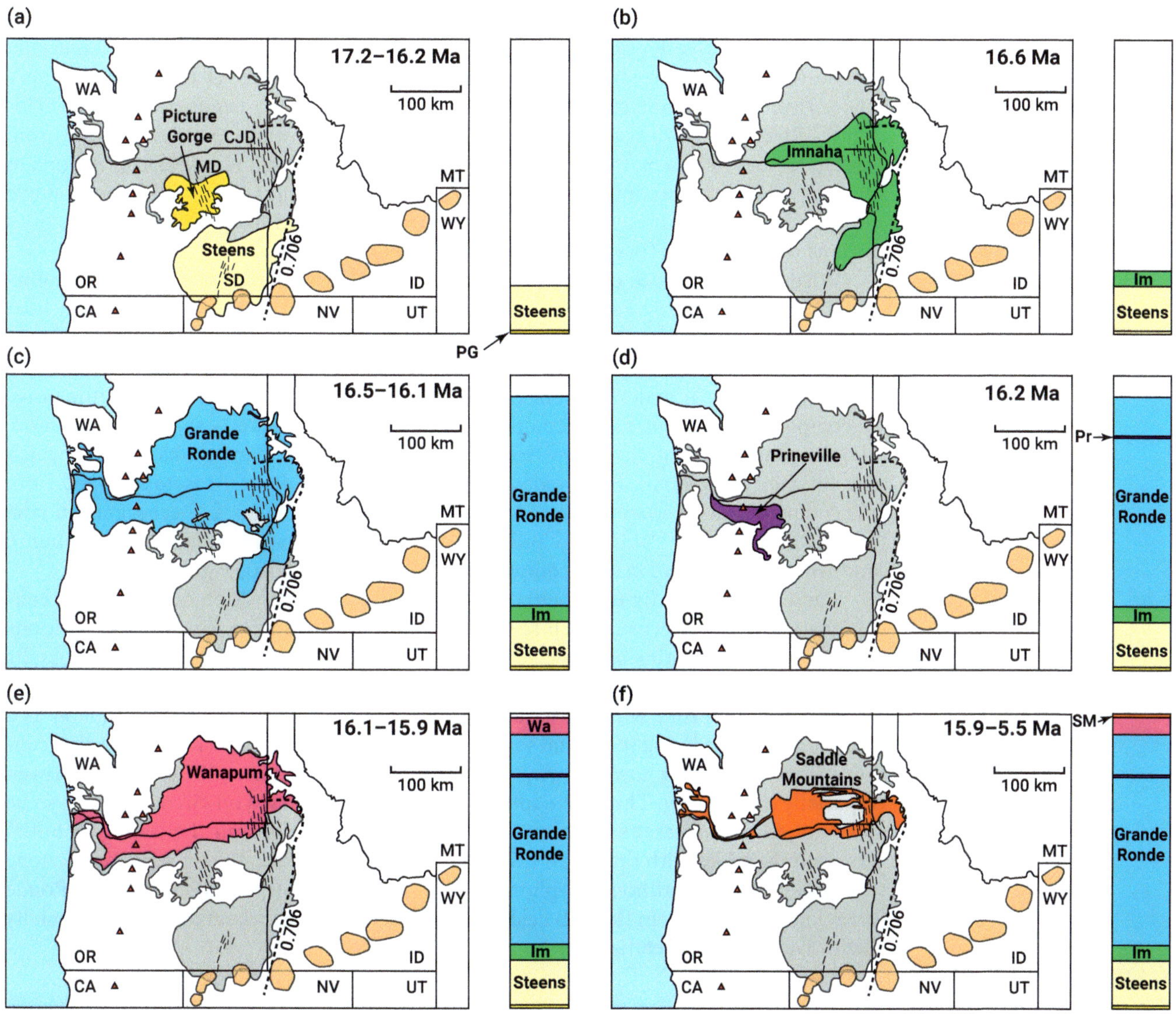

Figure 14.16 Age progression of Columbia River basalts. Columns to the right of the maps show relative percentage of magma accumulated for each major unit. Groups of short dashed lines indicate dike swarms interpreted to represent vent areas (SD = Steens dike swarm; MD = Monument dike swarm; CJD = Chief Joseph dike swarm). (a, b) The Picture Gorge (PG), Steens, and Imnaha (Im) basalts are oldest; the Steens is the largest of these early basalts. (c, d) The intermediate-age Grande Ronde is by far the largest and most rapidly erupted; the tiny Prineville (Pr) interfingers with it. (e, f) The youngest Wanapum (Wa) and Saddle Mountains (SM) basalts are relatively small. Dashed line is the 0.706 line, generally delineating the transition from accreted terranes on the west and old continental crust to the east. The blobs trending from northwestern Nevada across southern Idaho to northwestern Wyoming are the major volcanic fields of the Yellowstone hotspot.

Sources: Adapted from Reidel et al. (2013), Coble & Mahood (2015), and Soderberg & Wolff (2023).

12 Myr of eruption. The Picture Gorge started earliest but is small by flood basalt standards (2400 km^3); it lasted a relatively long time, about 1 Myr (17.2–16.1 Ma), and thus represents low-flux magmatism. The Steens basalt is the first big eruption sequence (17.0 to 16.4 Ma; >30,000 km^3), and it overlaps in age with the moderately sized Imnaha (~11,000 km^3; ~16.6 Ma). The later Grand Ronde basalt is the monster of the bunch, with about 149,000 km^3 erupted between 16.5 and 16.1 Ma, overlapping in age with the tiny Prineville (590 km^3; ~16.2 Ma). The moderately sized Wanapum followed with about 12,000 km^3 erupted between 16.1 and 15.9 Ma. The small (~2400 km^3) Saddle Mountains completed the sequence and extends the chronology from 15.9 to 5.5 Ma.

Vent sites have been identified in eastern Oregon, west-central Idaho, and southeastern Washington (see Figure 14.16), and flows spread out in all directions, but especially westward down the ancestral Columbia River, reaching the Pacific Ocean. Virtually every major river in the region exposes these basalts, and spectacular exposures along the Columbia River and nearby canyons give rise to the name. Today, we can drive the route from vent to the ocean in a matter of hours, but it might have taken a few weeks for magma to flow the same distance, and about 10 years of quasi-constant magma injection for a single flow to form fully (Self et al., 1998). Individual vents erupted magma for durations of as little as weeks to years (Biasi & Karlstrom, 2021).

What was the source of the Columbia River basalts, and why did so much magma erupt so quickly? Many geochemists assume that a plume was responsible, and at least one database classes the Columbia River basalts with the Yellowstone-Snake River Plain volcanic province, although they are geographically distinct. But pinning down sources is difficult. Looking at major-element compositions (Figure 14.17a), rocks show little range in composition, plotting nearly completely as basalt and andesite. A small subset is more alkaline, but many of these are from the youngest Saddle Mountains basalts, which are volumetrically small. From oldest to youngest units, Nd and Sr isotopes trend from the compositional overlap region of MORB and OIB (Picture Gorge, Steens, and Imnaha units) toward lower $^{143}Nd/^{144}Nd$ and higher $^{87}Sr/^{86}Sr$ (Grande Ronde, then Wanapum), and still further toward crustal compositions (Saddle Mountains; Figure 14.17b).

These trends broadly reflect either changing source composition or progressive contamination with old crust, which seems necessary to explain the extreme compositions of the Saddle Mountains basalts (see Figure 14.17b). Because the Columbia River basalts erupted on the edge of a major extensional province (potential magma source 1), east of an active subduction zone whose geometry was changing (potential source 2), arguably as a mantle plume was rising (potential source 3) through continental crust (confounding contaminant), identifying how and why so much magma erupted in so little time has proven especially challenging. The plume interpretation is mostly based on circumstance—some (but not all) flood basalts are associated with hotspot tracks, and the timing of initial eruption coincides with the oldest silicic volcanic centers associated with the Yellowstone hotspot. Otherwise, although a plume is certainly possible (the massive Grande Ronde eruptions fall squarely in the OIB field in Figure 14.17b), the geochemistry is not especially definitive, and interpretations vary.

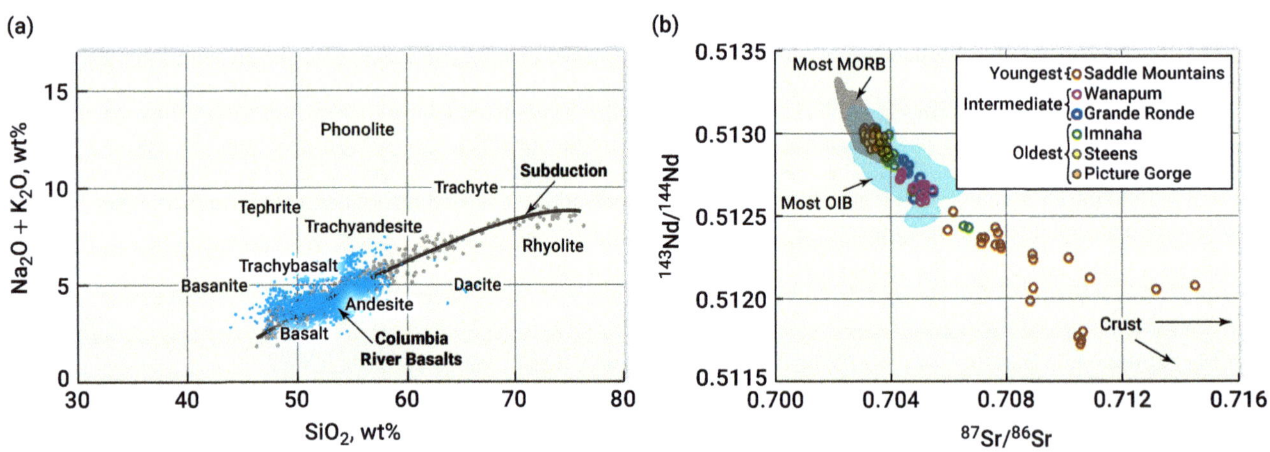

Figure 14.17 (a) TAS diagram showing that the Columbia River basalts follow the subduction trend in the basalt and andesite fields. A few are more alkaline (mostly from the youngest Saddle Mountains unit). (b) Neodymium and strontium isotope data for the Columbia River basalts, split out according to major members. A general trend is evident from oldest to youngest, starting in the field of overlapping MORB and OIB, transitioning toward compositions more similar to crust.

Sources: Data from GEOROC Compilation: Continental Flood Basalts (doi.org/10.25625/WSTPOX), and from Stracke (2021).

Deccan Traps Moving up in scale, the Deccan Traps in India (Figure 14.18) erupted a volume of over 600,000 km^3 (Richards et al., 2017) between 66.5 and 65.5 Ma (Schoene et al., 2019; Sprain et al., 2019). The volume of some individual flows reaches 10,000 km^3, as big as or larger than entire members of the Columbia River basalts, and some crossed the entire subcontinent from west to east, a distance up to 1000 km, to empty into the Bay of Bengal. The greatest flux of magma occurred about 66.0 Ma, right as an asteroid struck the opposite side of the Earth, spelling the demise of non-avian dinosaurs and many other organisms during the Cretaceous-Paleogene (K-Pg) extinction. A lively debate centers on whether the asteroid impact could have catalyzed volcanism (Richards et al., 2017), and whether climate change induced by the Deccan Traps or the asteroid killed off the dinosaurs; conversely, perhaps volcanism softened the climatic impact and helped some organisms survive (e.g., Chiarenza et al., 2020). The rate of magmatism varies through the sequence, either in fits and starts with major pulses before and after the impact (based on U-Pb geochronology; Schoene et al., 2019), or with increasing flux after the impact (based on $^{40}Ar/^{39}Ar$ geochronology; Sprain et al., 2019). Like the Columbia River basalts, earlier eruptions (Kalsubai and Lonavada Groups) were clearly subsidiary to later eruptions (Wai Group). The main chronologic question is whether the Wai Group spans the K-Pg extinction boundary (Schoene et al., 2019) or starts at the boundary (Sprain et al., 2019; Figure 14.18b).

Interpretations for the origin of the Deccan Traps commonly invoke a plume head, centered near the densest accumulation of feeder dikes (see Figure 14.18b). The Seychelles microcontinent (Mascarene Ridge), far to the southwest, also contains Deccan-like magmas, so it is interpreted to have rifted from India soon after eruptions. The trace of the hotspot trends nearly due south along an aseismic ridge, then bends west, hops the Carlsberg Spreading Center (meaning that spreading across the ridge has now separated the continuous track into two segments), and continues southwestward to the island of Reunion (see Figure 14.18a). In one hypothesis, the 66-Ma seismic shock of the asteroid at Chicxulub, Mexico, triggered the massive release of magmas representing the Wai Group. This hypothesis presupposes magmas were waiting in storage, which would still be consistent with a plume model.

Geochemically, Deccan magmas resemble Columbia River basalts, except that there is a much greater abundance of basalt, but also a greater range of compositions (Figure 14.19a),

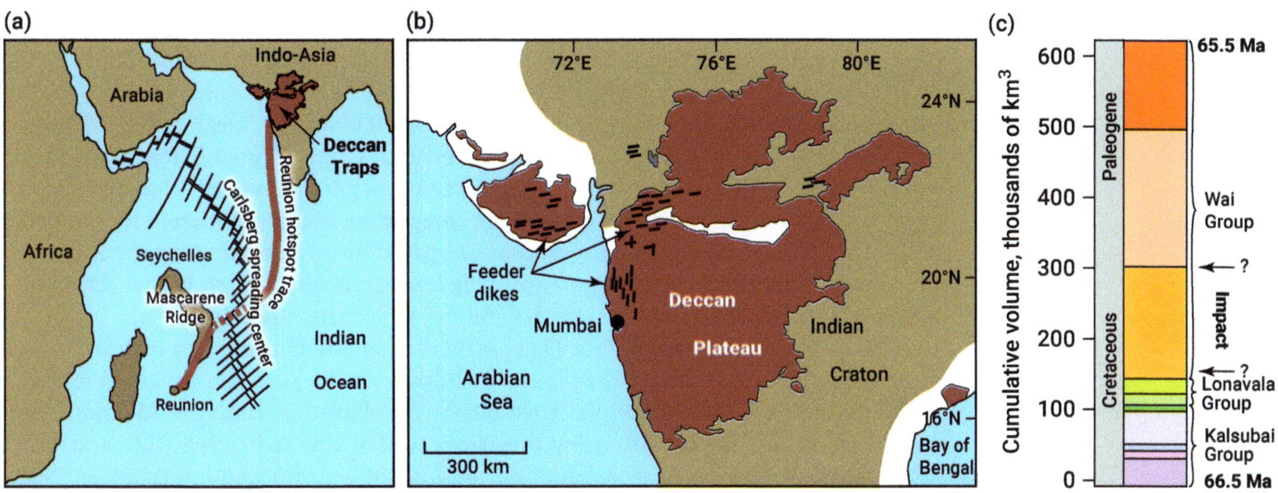

Figure 14.18 (a) Regional map of western Indian Ocean, showing location of Deccan Traps, Reunion hotspot trace, Seychelles, the Mascarene Ridge, and Reunion Island. (b) Map of central India, showing extent of Deccan Traps. Note the exposure in the east along the coast of the Bay of Bengal. Feeder dikes were located as much as 1000 km away. (c) Column showing volume of different units. The Wai Group dominates. The timing of asteroid impact is debated as either early in the Wai Group or partway through.

Sources: (b) Based on Mittal et al. (2021); (c) based on Schoene et al. (2019); © Blair Schoene, Michael P. Eddy, C. Brenhin Keller, and Kyle M. Samperton 2021.

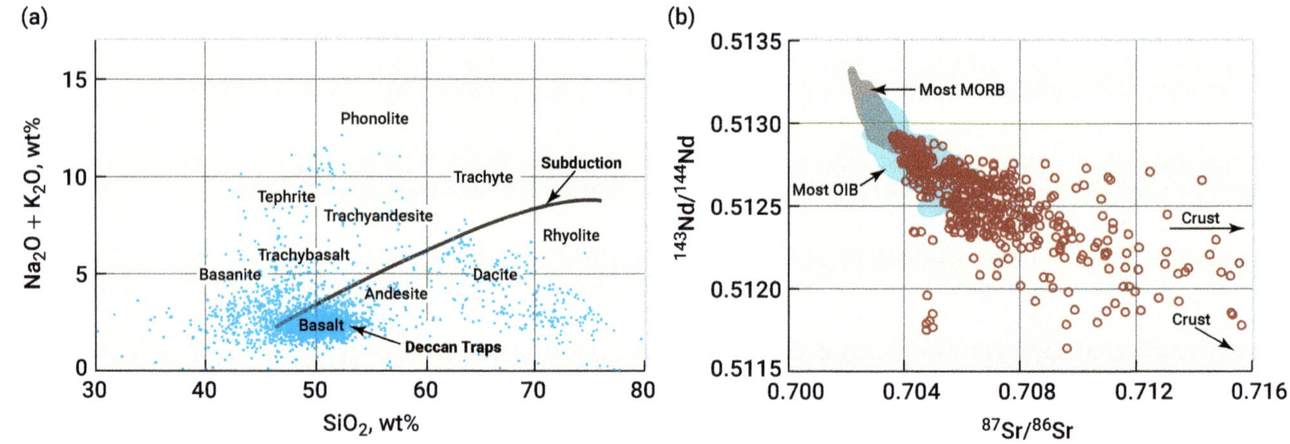

Figure 14.19 (a) TAS diagram shows the Deccan Traps are mostly basalt, although they scatter across extremely alkaline to odd, high-silica, low-alkali compositions, possibly owing to alteration. (b) Age-corrected Nd- and Sr-isotope data for the Deccan Traps span from the field of overlapping MORB and OIB toward continental crust, similar to Columbia River basalts.

Source: Data from GEOROC Compilation: Continental Flood Basalts (doi.org/10.25625/WSTPOX).

including highly alkalic and highly silicic rocks. Isotopes again show compositions spanning the intersection between MORB and OIB toward crust (Figure 14.19b). Some of this spread represents crustal contamination. The degree to which contamination affects the densest cluster of data is not immediately clear, but the general overlap with the OIB field (plus other geochemical characteristics) is similar to OIBs, which many geochemists interpret are the products of plumes.

Oceanic Plateaus

Ontong-Java Moving farther up in scale to the very top, the Ontong-Java Plateau, in the western Pacific (see Figure 14.1), erupted a volume of at least 44,000,000 km^3 in only a few million years around 120 Ma (Kerr, 2014). But even that immense volume may be substantially underestimated. Possibly, the Ontong-Java Plateau was once part of a larger plateau (Figure 14.20; Taylor, 2006) that rifted apart, driving the Manihiki and Hikurangi Plateaus to the east and south (see Figure 14.1). Putting them back together, these three plateaus could represent 60,000,000 to 90,000,000 km^3 (Neal et al., 2019), forming the largest known LIP. Oceanic plateaus generally lack magnetic lineations, because widespread volcanism overprints the signature of magnetic reversals that oceanic crust develops at spreading ridges. This overprint complicates reconstruction of former plateau positions and ages. The original positions are based on surrounding magnetics of unaffected oceanic crust plus direct geochronology, which is often much harder for mafic vs. felsic rocks, because mafic rocks possess fewer datable minerals.

Until around 25 Ma, the Pacific Plate, with its attendant Ontong-Java Plateau, was subducting southwestward when the plateau collided with the Solomon arc and jammed the subduction system. This collision thrust bits of the plateau up onto the Solomon Islands and switched the subduction facing direction, so that the Australian Plate and various microplates to the south now subduct northeastward beneath Ontong-Java (e.g., Neal et al., 2019). This collisional orogen allows investigation of an oceanic plateau from samples taken on land, which are much less expensive to collect than deep-sea samples.

Geochemically, Ontong-Java shows a restricted range in major-element compositions (basically basalt; Figure 14.21a), even more restricted than for the Deccan Traps (see Figure 14.19). Isotopically, Nd and Sr isotopes are also more limited, and many analyses occupy the overlap region between MORB and OIB (Figure 14.21b), although with some

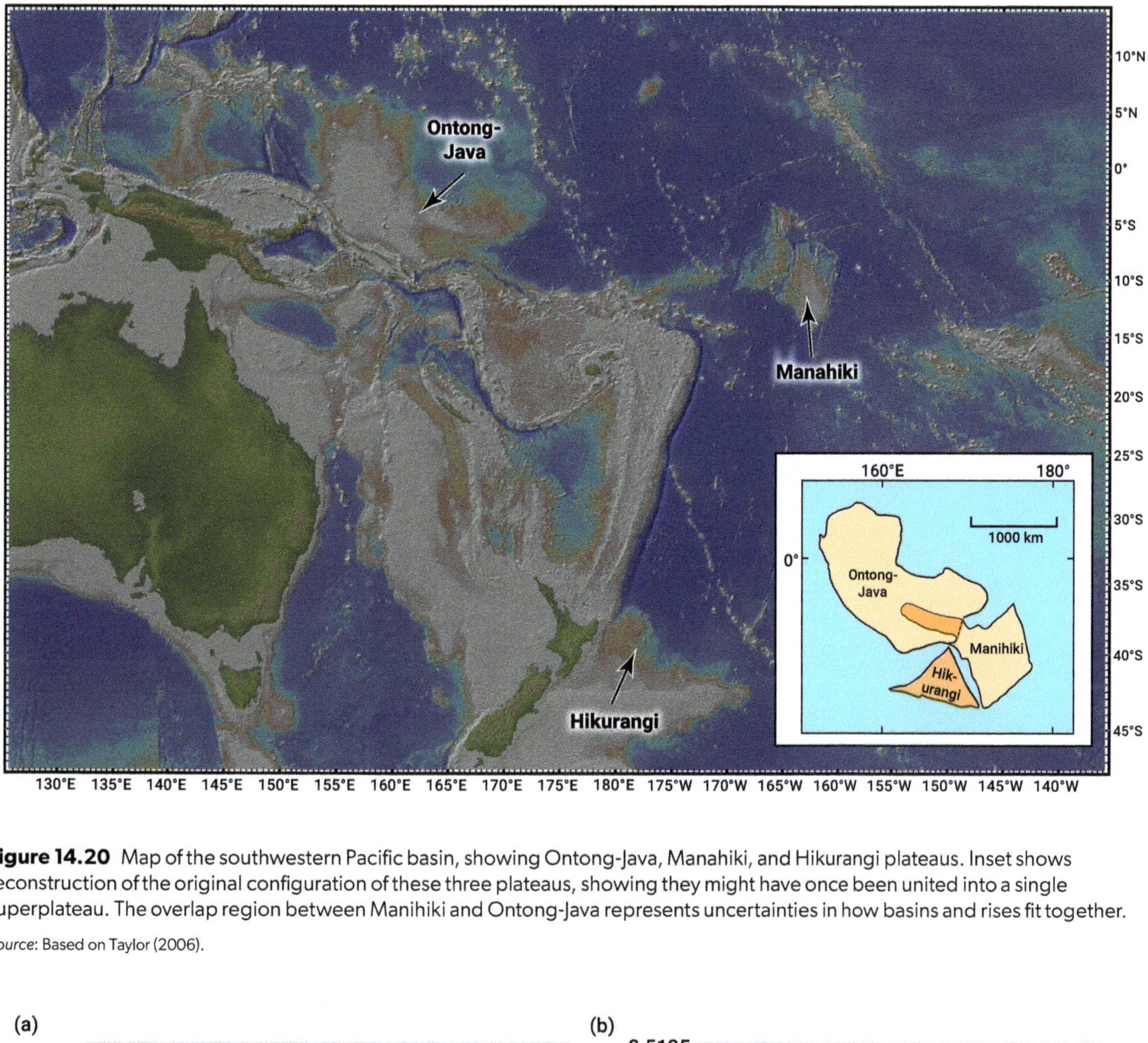

Figure 14.20 Map of the southwestern Pacific basin, showing Ontong-Java, Manahiki, and Hikurangi plateaus. Inset shows reconstruction of the original configuration of these three plateaus, showing they might have once been united into a single superplateau. The overlap region between Manihiki and Ontong-Java represents uncertainties in how basins and rises fit together.

Source: Based on Taylor (2006).

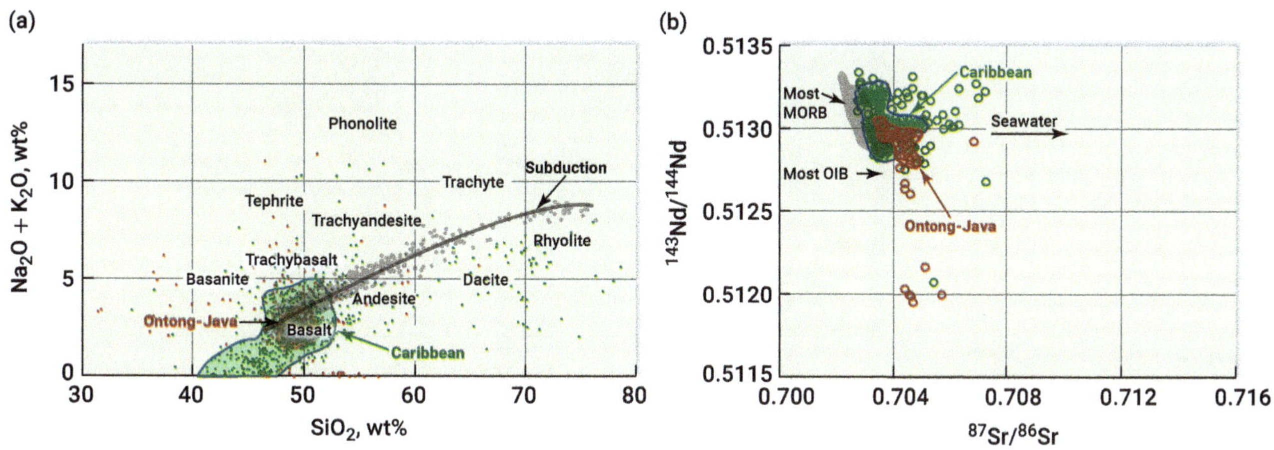

Figure 14.21 Geochemical data for the Ontong-Java and Caribbean Plateaus. (a) TAS diagram, showing that basalt and even more mafic rocks dominate. (b) Age-corrected Nd- and Sr-isotope data, showing a highly restricted range, generally at the overlap region between MORB and OIB. Seawater alteration is common in these submarine basalts and pulls compositions to the right (toward more radiogenic Sr-isotope compositions).

Source: Data from GEOROC Compilation: Oceanic Plateaus (doi.org/10.25625/JRZIJF).

scatter toward seawater compositions. As summarized in Neal et al. (2019), material from the thickest part of the plateau was deposited over 1 km below sea level, although subaerial deposits have been found elsewhere. Geodynamically, formation of a submarine plateau is difficult to accomplish, because the mantle source was likely hot and buoyant, and the crust was also buoyant relative to mantle. Dense residues or stalled crystallized basalts might offset the inherent buoyancy of mantle and crust to reduce elevations, and observed fast shear-wave velocities below the plateau could indicate dense eclogite. Notoriously, the Ontong-Java, Hikurangi, and Manihiki Plateaus all lack later-forming seamount chains, which has caused some to question whether the plume hypothesis applies to them. However, both the Ontong-Java and Hikurangi Plateaus abut subduction zones, so a hotspot track could have been present, but subducted.

Caribbean The Caribbean Plateau formed within the Farallon Plate (east of the Pacific Plate) between about 85 and 90 Ma and survived multiple tectonic assassination attempts (Kerr, 2014; Figure 14.22). The tectonic configuration prior to about 100 Ma in the current region of Central America was complex and remains uncertain, but in one model, the Farallon Plate was overriding plates to the east when a plume head impinged upon it. The buoyancy contrast between young, hot lithosphere above the plume head and old, cold lithosphere to the west initiated subduction of the Farallon Plate beneath the new plateau (see Figure 14.22). This process effectively split out the plateau from the Farallon Plate to form the Caribbean Plate. Alternatively (Burke, 1988), the Farallon Plate was subducting eastward to form a proto-Caribbean arc until the (already formed) plateau

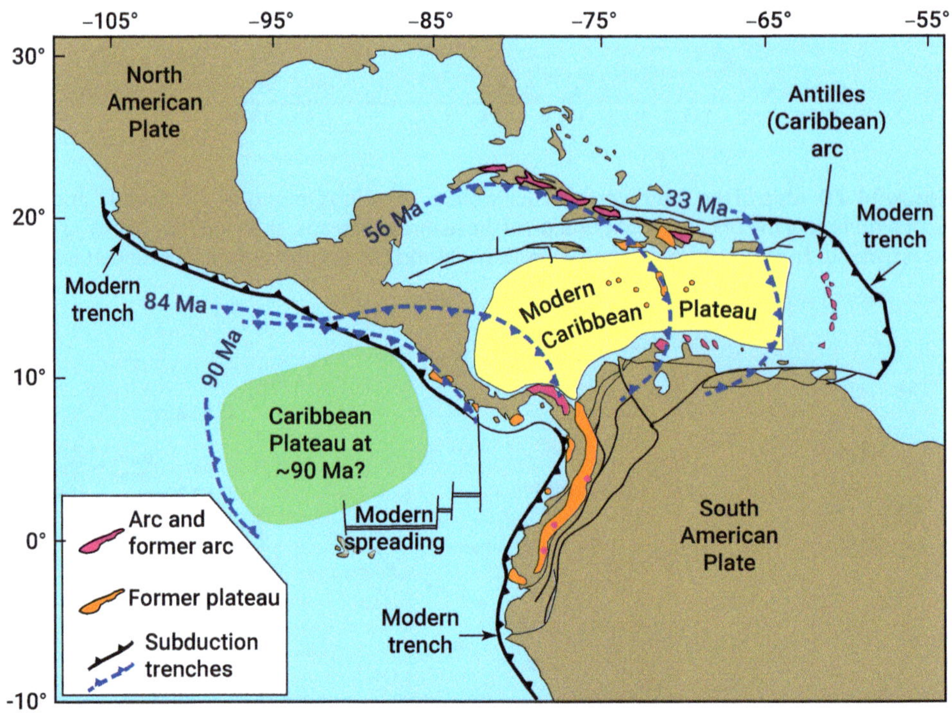

Figure 14.22 Highly simplified reconstruction of tectonic evolution of the Caribbean Plateau. The plateau first formed ~85–90 Ma within the Farallon Plate. Subduction initiated to the west of the plateau for reasons we do not fully understand, carving a piece out of the Farallon Plate to form the Caribbean Plate. The Caribbean Plate has since migrated eastward (blue set of proposed trench locations, labeled with age in Ma) and is bounded to the west and east today by the Central American and Caribbean (Antilles) subduction systems (black trench symbols). Remnants of former arcs and the plateau are scattered around the margins of the modern plate.

Sources: Modified from Burke (1988), Boschman et al. (2019), and Riel et al. (2023).

collided with the arc. The resistance of the plateau to subduction caused a switch in subduction polarity (so that the newly formed Atlantic Plate started subducting westward beneath the plateau), and also initiated a separate subduction zone within the Farallon Plate to form a new distinct plate—the Caribbean Plate. Regardless of its formation, the Caribbean Plate then migrated eastward to its present position, shaving off bits to the north and south along the way (see Figure 14.22). Unlike the Ontong-Java Plateau, subaerial deposition of plateau rocks is well-documented (Buchs et al., 2018).

Geochemically, the Caribbean LIP looks much like other LIPs with many major element and isotopic data plotting in the basalt field (see Figure 14.21a) and in the overlap region of MORB and OIB (see Figure 14.21b). Like the Ontong-Java Plateau, isotopic data also scatter toward seawater compositions. Unusually, rocks with especially high MgO and low SiO_2 and $Na_2O + K_2O$ (komatiites and picrites) occur within uplifted fragments of the plateau. Komatiite and picrite magmas are especially dense, and their melting temperatures are especially high (~1600 °C or higher); in fact, most komatiites formed during the Archean, when Earth's interior was hotter. The young komatiites and picrites of the Caribbean LIP generally support an origin of the plateau from an unusually hot plume.

Making Connections 14.2

The Siberian Traps and the End-Permian Extinction

Eruptions of some flood basalts correlate with climate and extinction events. The Siberian Traps (Figure 14.23) are the largest known eruptions of continental flood basalts and represent one of the strongest cases for an eruption-climate-extinction connection. Because so much of the province is now buried, the total volume is not known precisely, but estimates range from at least 3,000,000 to possibly 16,000,000 km^3 (Dobretsov, 2005; Dal Corso et al., 2022). This is up to 25 times the size of the Deccan Traps, 75 times the size of the Columbia River basalts, and by some estimates,

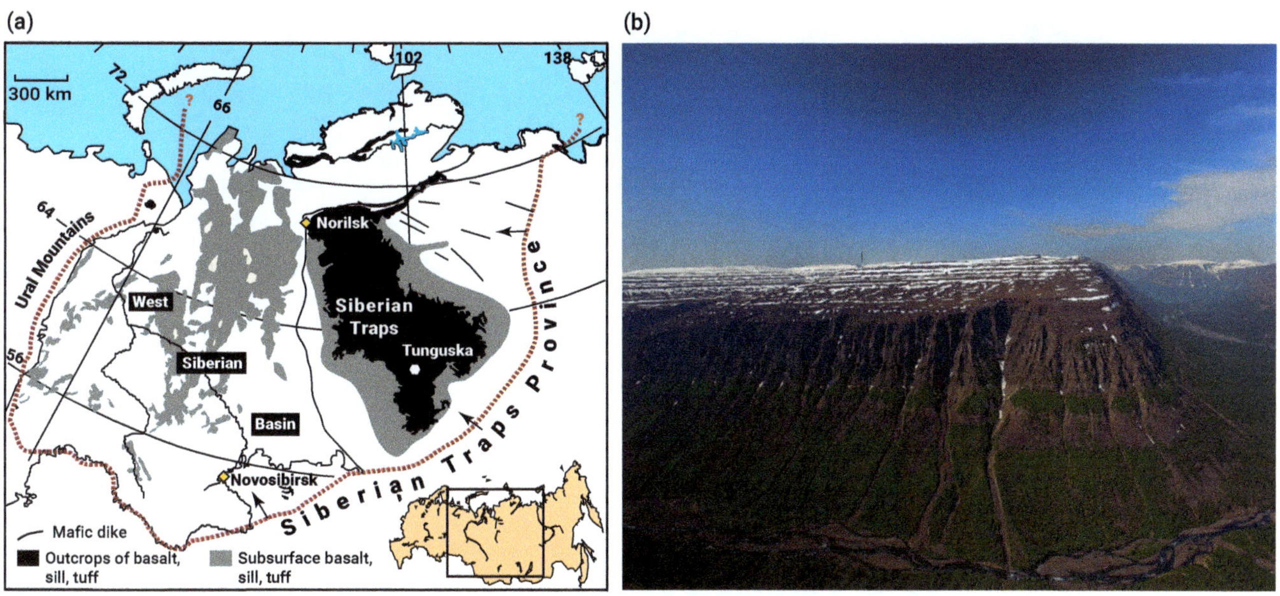

Figure 14.23 (a) Map showing extent of the Siberian Traps (inset shows main map overlain on a map of Russia). Although rocks are exposed over a huge region (black), the subsurface area (dark gray), as determined from boreholes and geophysical imaging, is many times larger. The location of the 1908 Tunguska impact (see Figure 14.37) is also shown. The West Siberian Basin is the source of most of Russia's oil and natural gas reserves. (b) Photo of the Siberian Traps showing widespread thick sequences of repetitive flows.

Sources: (a) Modified from Ernst & Bell (2010); (b) from Wignall (2011), courtesy of Paul Wignall, Springer Nature BV.

continued

enough to bury both the United States and Canada under nearly a kilometer of basalt! The Siberian Traps erupted and intruded between 252.3 to 251.4 Ma, precisely spanning the end-Permian extinction at 251.9 Ma (Burgess & Bowring, 2015).

Erwin (1996) and Dal Corso et al. (2022) linked eruption of the Siberian Traps with major disruptions of Earth geochemistry and with extinctions. The end-Permian represents the most significant extinction in Earth's history, spelling doom for roughly 90 percent of all marine species, 70 percent of terrestrial vertebrate families, and even 50 percent of insect families, which are otherwise remarkably resistant to extinction. It's no wonder that the end-Permian extinctions represent *the* major boundary between the Paleozoic and Mesozoic eras. The Earth's environment went haywire, with dramatic increases in temperature, acidified oceans, global contamination from heavy metals and acid rain, and major perturbations to the carbon cycle. The death of so many land plants and carbonate-secreting marine organisms collapsed peat and reef systems to produce the so-called early Triassic "coal gap" and "reef gap" that lasted for over 5 Myr (Retallack et al., 1996; Flügel & Kiessling, 2002; Figure 14.24). A superabundance of fungal spores, algae, and bacteria suggests Earth became a "slime world," generally inhospitable to most multicellular life (e.g., Visscher et al., 1996; Mays et al., 2021). Pollen during this time is unusually misshapen, indicating extreme environmental stress (e.g., Visscher et al., 2004). Life is tenacious, but still, it is a wonder that anything survived.

What could have caused such catastrophic ecosystem change? Surely not eruptions alone, right? Although massive eruptions can indeed modify global climate, strange circumstances surrounding the Siberian Traps created special climatic impact. Unlike other flood basalts, the Siberian Traps erupted through abundant coals and evaporites. Heat from the intrusions burned the coals and reacted with salts to produce especially high levels of CO_2 and halogens (acid rain; see Erwin, 1996; Dal Corso et al., 2022). Persistent high levels of CO_2 led to high temperatures and ocean acidification, stressing corals and other carbonate-secreting organisms (Knoll et al., 2007), while halogens created acid rain and degraded Earth's ozone layer (Black et al., 2012), leading to higher mutation rates (Visscher et al., 2004). The combination of timing and special geologic setting implicates the Siberian Traps as the cause of this extinction.

Figure 14.24 (a) Fossil of a cynodont. These animals took a major hit during the end-Permian extinction that coincided with eruption of the Siberian Traps, but the line managed to survive and, eventually, evolve into mammals. Still, their demise opened opportunities for dinosaurs to dominate during the Mesozoic, at least until the Cretaceous-Paleogene extinction and eruption of the Deccan Traps, when dinosaurs went extinct and mammals rose to dominance. (b) El Capitan in Guadelupe Mountains National Park, Texas. The prominent cliffs represent Permian reefs, ecosystems that completely collapsed at the end of the Paleozoic during eruption of the Siberian Traps.

Sources: (a) Roger Smith, Iziko Museums of South Africa Natural History Collections; (b) Ricraider/Wikimedia.

14.5 Layered Mafic Intrusions

Learning Objectives

By the end of this section you should be able to:

14.5.1 Sketch the classic big-tank model of how a layered mafic intrusion crystallizes.

14.5.2 Predict magma evolution for the big-tank model of magma crystallization.

14.5.3 Compare and contrast mechanisms for creating layering and concentrations of crystals.

Scattered around the world are eye-catching mafic intrusions that exhibit dramatic layering defined by variations in the proportions and sizes of the constituent minerals. Although uncommon, these **layered mafic intrusions** (**LMIs**) have played a starring role in petrology textbooks for the past century because of their assumed utility in understanding processes that occur in large magma chambers. LMIs have long been assumed to form from a body of magma that was emplaced all at once and then crystallized as heat was lost—the classic big-tank model of magma crystallization. They are also of great economic importance because they are the world's main sources of Pt, Pd, and other **platinum-group elements** (**PGEs**). LMIs of particular petrologic and economic interest include the Bushveld Complex of South Africa, the Stillwater Complex of Montana, and the Skaergaard Intrusion of eastern Greenland. The first two are major producers of PGEs. Despite the disproportionate attention paid to many of them, the origins of these enigmatic complexes remain obscure.

Layered mafic intrusions (LMIs)
Mafic plutons characterized by banding defined by dramatic variations in mineral proportions, sizes, and compositions.

Platinum-group elements (PGEs)
A group of six metals: Ru, Rh, and Pd, which lie just before Ag in row 5 of the periodic table, and Os, Ir, and Pt, which lie below the previous three and just before Au in row 6.

PGEs are chemically unreactive, have extremely high melting points, and make fine jewelry, but their main utility lies in catalyzing chemical reactions. Most Pd is used for automobile catalytic converters, which convert carbon monoxide and other harmful compounds to less harmful forms. Most Pt is used for jewelry and catalytic converters; a tiny but important (to us) bit is used in experimental petrology (see Section 3.3).

Terminology and the Implications of the Term "Layered"

Rock Names

Like most branches of petrology, the terminology of LMIs can be intimidating. We will use a simplified and somewhat loose nomenclature, in keeping with Appendix 3. Important rock types include norite (gabbro with Opx > Cpx), peridotite, anorthosite, gabbro, orthopyroxenite, and chromitite (rock composed largely of chromite, a spinel with nominal formula $FeCr_2O_4$).

Caution about Use of Sedimentary Terminology

Because LMIs are layered, it is common to refer to paleovertical sequences through them as magmatic "stratigraphy." However, this terminology implies that the layers were deposited in sequence from bottom to top as are sediments in a stratigraphic sequence, or at least in sequence from the floor up and roof down. This has been the dominant interpretation of LMIs over the past century, but there are many contradictory pieces of evidence, discussed later, that call into question a simple sedimentary interpretation. Hence, stratigraphic nomenclature should be avoided, as it presupposes a process that may not be applicable. Modal layering (i.e., layering defined by differences in mineral proportions) is typically interpreted as having formed by deposition of crystals from magma, but many processes can produce layering, so a depositional origin should not be assumed. "Banded intrusion" is perhaps a less genetic and therefore more appropriate term for these bodies, but the term "layered mafic intrusion" is deeply entrenched.

Bushveld Complex

We begin with the largest known LMI, the Bushveld Complex of South Africa (Figure 14.25), which lies in the northern part of the country, near the capital of Pretoria and west of Kruger National Park with its lions, cheetahs, elephants, leopards, and other non-geologic things of interest. The geology of the Bushveld, summarized in Kinnaird (2005) and Cawthorn (2015), is of critical economic importance, because it produces large fractions of world production of Pt, Pd, Cr, and other valuable metals. The tabular complex is exposed over 350 km east-west and 200 km north-south, about the area of, and similar in map view to, West Virginia. It is composed of layered mafic and ultramafic rocks and overlying felsic volcanic and intrusive rocks, all emplaced about 2.05 Ga.

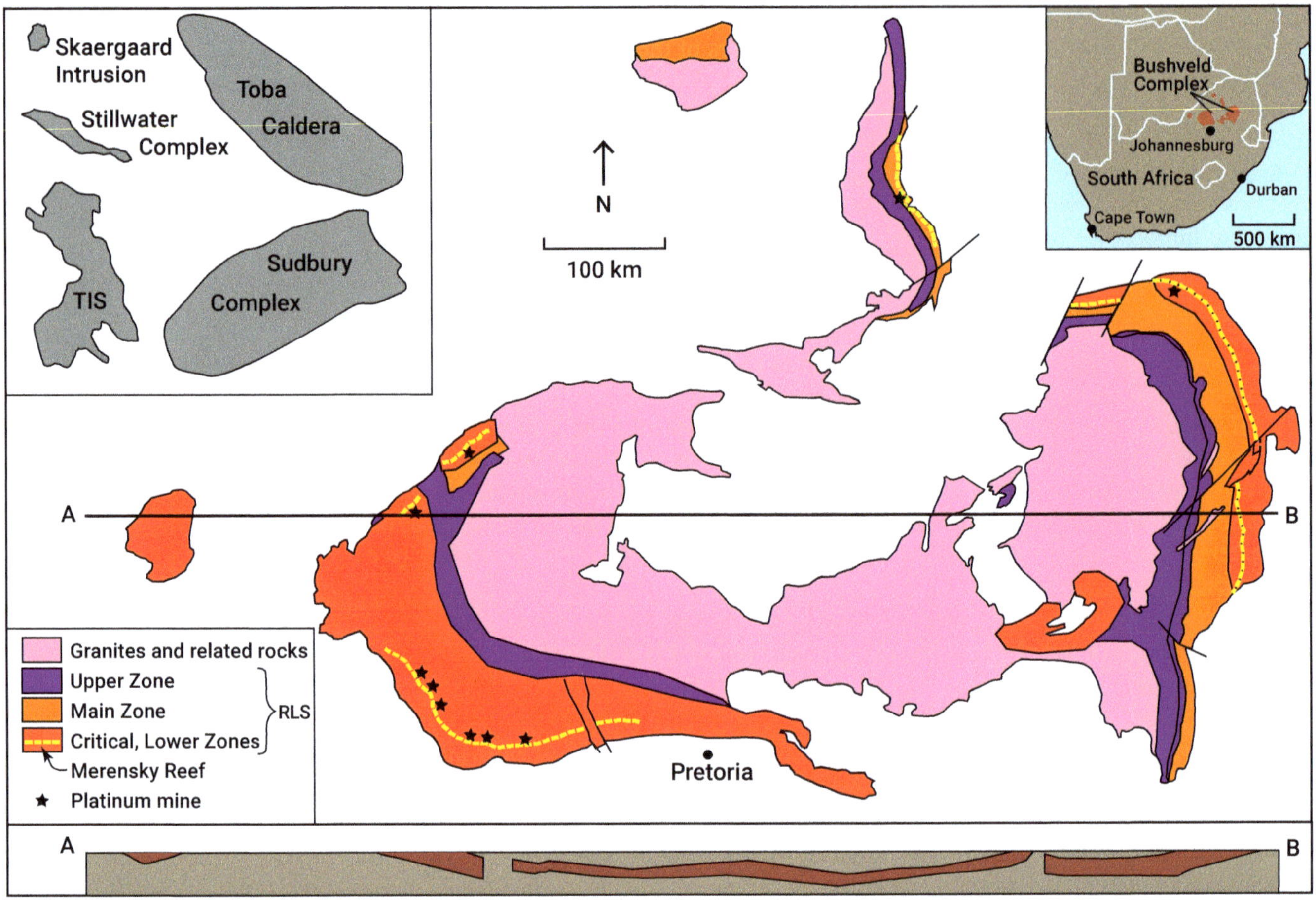

Figure 14.25 Generalized geologic map of the Bushveld Complex of South Africa. Lower panel shows a gravity-based estimate of the Rustenburg Layered Suite (RLS) in the subsurface. Inset (right) shows location on larger map of South Africa. Inset (left) shows various other large geologic features at the same scale, for comparison. TIS = Tuolumne Intrusive Suite.

Sources: Simplified from Taylor et al. (2005); lower panel from Cawthorn (2015).

Mafic and ultramafic rocks make up the Rustenburg Layered Suite (RLS), the main target of economic interest. With a thickness of around 8 km, the RLS has an aspect ratio of about 40 and is thus similar in shape to, but relatively thinner than, a children's book such as *The Cat in the Hat* laid on its side, or perhaps a thick, warped serving platter (Figure 14.25, lower panel).

A generalization of layering in the RLS is shown in Figure 14.26a. From the bottom up, the RLS is divided into a Lower Zone of ultramafic rocks; a Critical Zone (ultramafic rocks, anorthosites, and norites) that contains most of the chromite and PGEs; a Main Zone of gabbro, anorthosite, norite, and pyroxenite; and an Upper Zone of similar rock types. The An# (mole% An) of plagioclase and Mg# (mole% Mg relative to Mg + Fe) of Opx vary systematically with position in the layered sequence, decreasing upward with a few kinks.

A common feature of LMIs is that important economic mineralization is restricted to thin, planar zones known as **reefs**. In the Bushveld Complex, most PGE production comes from the celebrated Merensky Reef, which lies near the top of the 1500-m-thick Critical Zone and above several important chromitite layers (Figure 14.27a). The Merensky Reef, remarkably, comprises only 50 cm of the 8 km thickness of the RLS, and yet produces over half of the world's PGEs.

Reefs
In ore deposits, metalliferous mineral deposits; the term is typically used for those of particular economic value.

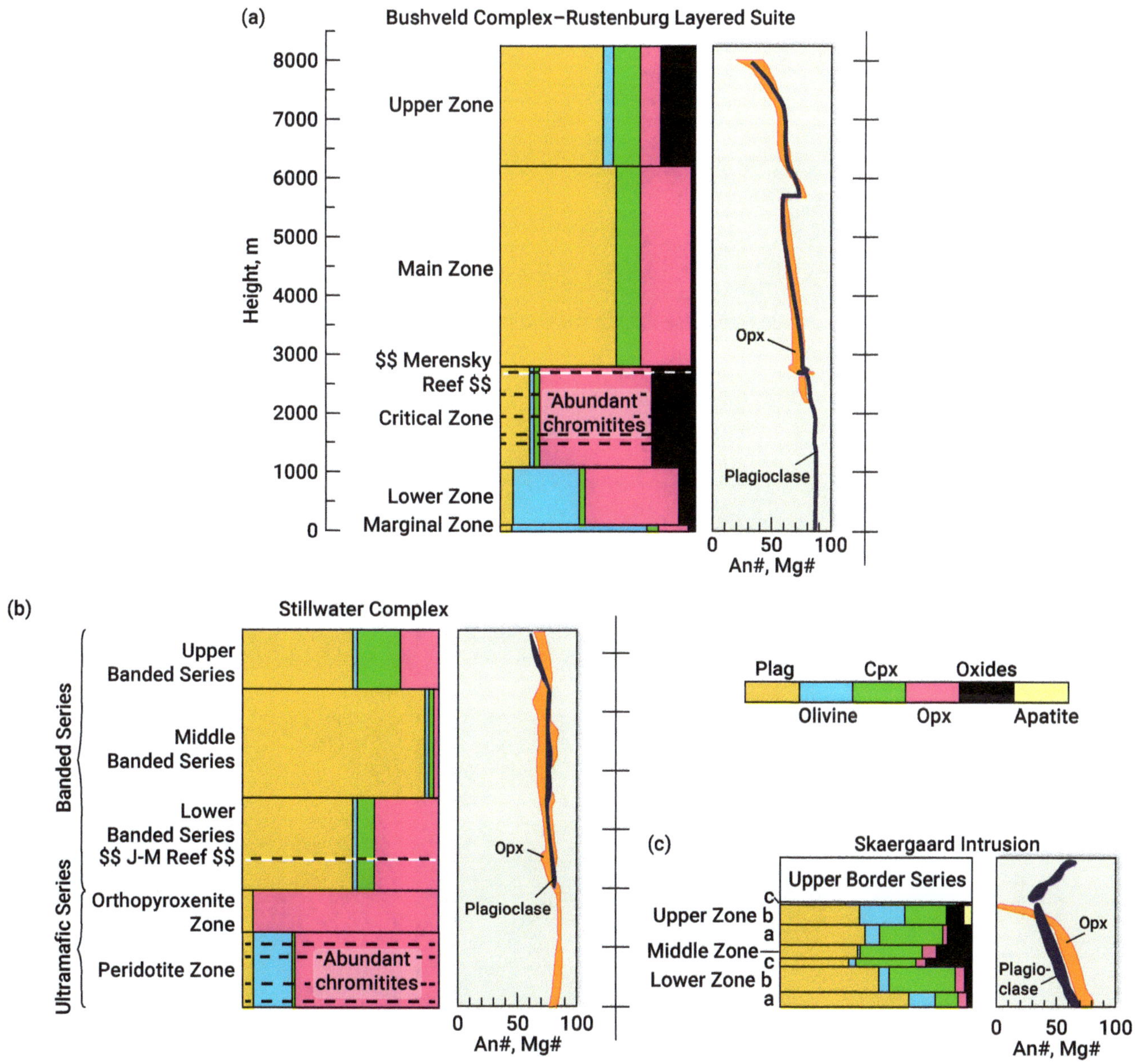

Figure 14.26 Generalized variations in modal mineralogy with vertical position of the Bushveld, Stillwater, and Skaergaard LMIs, drawn at the same scale. Blocks indicate averages of generally very spiky data for each named unit. (a) Rustenburg Layered Suite of the massive Bushveld Complex, South Africa. Lower units are largely Opx and olivine; the Critical Zone unit contains numerous chromitite layers and the PGE-rich Merensky Reef (white-black dashed line). (b) The Stillwater Complex of Montana resembles the Bushveld in having Opx-rich lower units overlain by more plagioclase-rich gabbros and norites. Chromitites and the PGE-rich J-M Reef (white-black dashed line) occur in the lower of two ultramafic units. (c) In contrast, the much smaller Skaergaard Intrusion of Greenland is largely gabbroic throughout. Plagioclase becomes monotonically more sodic passing upward in the layered series, reversing this trend in the Upper Border Series. Opx and olivine (not shown) become dramatically more Fe-rich, reaching Mg-free ferrosilite and fayalite at the top of the layered series. See Figure 14.29 for banding nomenclature.

Sources: (a) Modified from Boudreau (2019, p. 21); (b) modified from Boudreau (2019, p. 26) with kind permission of Alan Boudreau; (c) modified from Namur & Humphreys (2018), Salmonsen & Tegner (2013).

The Bushveld Complex comprises an astonishing volume of magma, perhaps as much as 1,000,000 km^3 (Cawthorn, 2015), with the RLS making up approximately one-third of that. A million cubic kilometers is a difficult volume to comprehend, but if spread out over the contiguous United States, it would make a frosting over 100 m thick (but see Making Connections 14.2 for an even greater magmatic event). The Bushveld Complex is a large igneous province (see Section 14.4), indeed.

Figure 14.27 A small sampling of the curious field aspects of layered intrusions. (a) Black seams of chromitite ~20 cm thick in anorthosite in the Bushveld Complex, South Africa. (b) A particularly sulfide-rich piece of ore from the J-M Reef (Stillwater Complex, Montana) containing black serpentinized olivine and altered pyroxene with brassy sulfides pyrrhotite, pentlandite, and chalcopyrite. At 2024 prices, this hand sample might contain roughly $100 (minus the cost of extraction) worth of Pt and Pd. (c) "Inch-scale" layering in steeply tilted anorthosite, Stillwater Complex. Black layers, generally doublets, are composed of altered Opx. (d) Multiple mafic-felsic couplets in Upper Zone A and (e) strongly banded gabbro, Skaergaard Intrusion, Greenland; hammer is 50 cm long. (f) Strongly layered granodiorite, Rock Creek, eastern Sierra Nevada, showing that modal banding of the same style and scale as in the Skaergaard also occurs in typical calc-alkaline granitic rocks.

Sources: (a) Courtesy of Sarah Penniston-Dorland; (b, c, f) AFG photos; (d, e) with kind permission of Alan Boudreau.

Stillwater Complex

The Neoarchean (2700 Ma) Stillwater Complex in southwestern Montana, near Yellowstone National Park, extends 45 km along strike and has a maximum exposed thickness of 6.5 km of layered mafic and ultramafic rocks (Figure 14.28). The geology of the Stillwater Complex was reviewed by Boudreau (2016). It is sandwiched between truly ancient

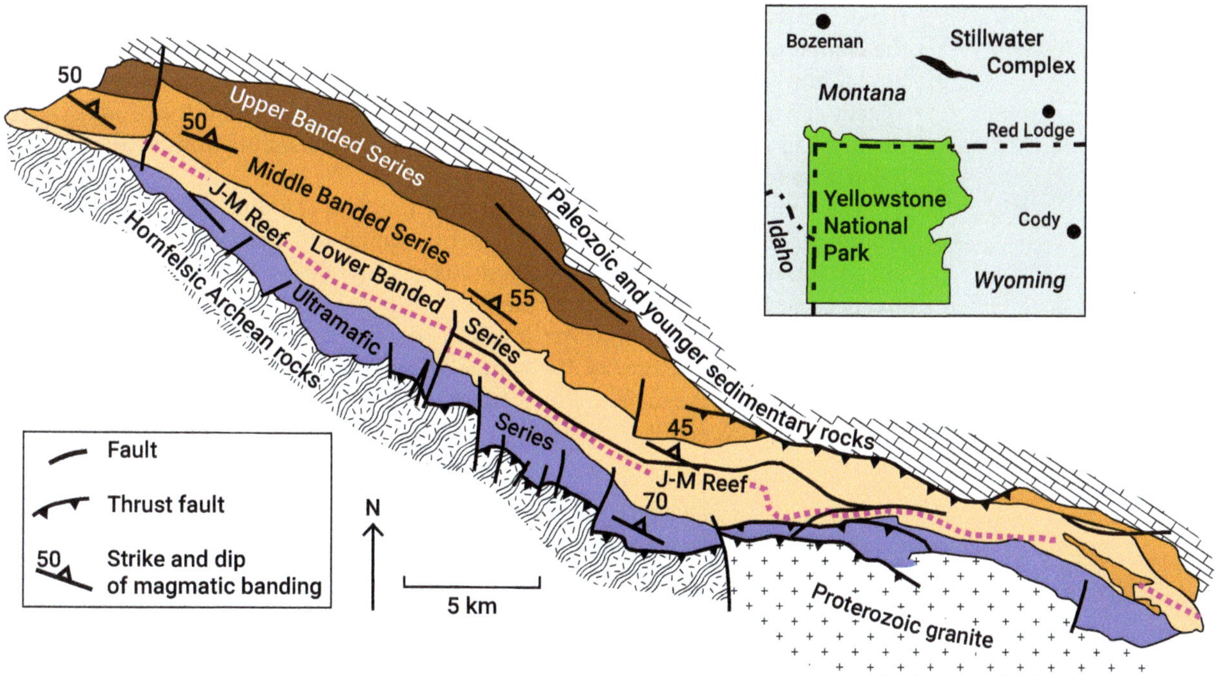

Figure 14.28 Geologic map and schematic cross section of the Stillwater Complex. The Stillwater sits on the edge of the Beartooth Mountains, a block of largely Neoarchean rocks. The complex has been steeply tilted to the northeast along reverse faults that bound the range, and it is possible that the complex and J-M Reef extend under the valley to the northeast.

Source: Modified from Boudreau (2016).

(Paleoarchean, 3270 Ma) metasedimentary rocks and Paleozoic sedimentary rocks that are nearly 3 Gyr younger, and lies in a basement uplift that has tilted it steeply to the northeast. Between surface exposures, mine workings, and drill core, the internal structure of the complex is well documented.

The Stillwater Complex comprises a lower Ultramafic Series and a Banded Series composed of interlayered gabbro, norite, and anorthosite (see Figure 14.26b). Compositions of olivine, pyroxenes, and plagioclase vary with distance from the bottom of the sequence (Boudreau, 2016), similar to but less consistent than variations in the Skaergaard Intrusion (discussed later). Production of PGEs is from an interval known as the J-M Reef (see Figure 14.27b), which lies near the base of the Banded Series and is 1 to 3 m thick. The ore zone was discovered by surface exploration but, owing to the steep dip of the rock units, is extracted by underground mining. Sulfides, which typically make up a few percent of the J-M Reef, contain roughly 1500 ppm Pt and 5500 ppm Pd along with significant Ni and Cu.

A particularly eye-catching feature of the Stillwater Complex is a zone characterized by "inch-scale" banded anorthosites (see Figure 14.27c). These occur in the lower part of the Banded Series within a noritic unit that lies above the J-M Reef. Banding, which may have been horizontal before the complex was steeply tilted, is defined by thin bands of Opx with a bimodal spacing of either about 1 cm or 3 to 6 cm. Thin and thick anorthosite bands typically alternate with Opx, giving an overall appearance of double Opx stripes.

Skaergaard Intrusion

Field Aspects

The Skaergaard LMI has received more petrologic scrutiny than any other in spite of its remote location, high latitude, ice cover, short field season, and polar bears. The intrusion is exposed in fjords on the eastern coast of Greenland, at roughly the latitude of the northern edge of Iceland. It was discovered by Lawrence Wager, a geologist and member of a 1930–1931 expedition to survey Greenland as a possible stop on aerial routes across the North Atlantic.[2] Wager and Deer (1939) published a massive treatise on the complex, which they interpreted as the remains of a deep magma chamber, initially liquid, in which crystals formed, settled to the bottom, and accumulated in layers as the system crystallized. This basic interpretation is widely accepted today, and cartoons of the inferred magma chamber and processes within adorn numerous textbooks and journal articles (e.g., Irvine et al., 1998).

The complex is semicircular, about 10 km across (Figure 14.29), and was emplaced around 56 Ma during opening of the North Atlantic and eruption of associated flood basalts. It is thus not truly intraplate magmatism, as the Bushveld and Stillwater occurrences are, because it formed during continental rifting and lies on the edge of the North American continent. Like the Stillwater, it was intruded near the contact between Precambrian gneiss and supracrustal rocks. Its shape is interpreted to be box-like, with its top initially about 3 km below the surface. Most authors assume that it was filled to a depth of about 4 to 5 km in one event, after which crystallization ensued.

Banding The Skaergaard Intrusion consists primarily of banded gabbro with interlayered anorthosite (see Figure 14.27d,e). Southward tilting of the complex coupled with deep glacial erosion and over 1 km of vertical relief have exposed a column of banded rocks about 3 km thick (see Figure 14.26c). Wager and Deer (1939) subdivided the intrusion into four units. About 70 percent of the exposed thickness comprises the Layered Series, a sequence of gabbros, banded on the decimeter to meter scale, in which plagioclase becomes more Na-rich, and olivine more Mg-rich, moving upward. The Layered Series is subdivided into Lower, Middle, and Upper Zones, and these are further subdivided into

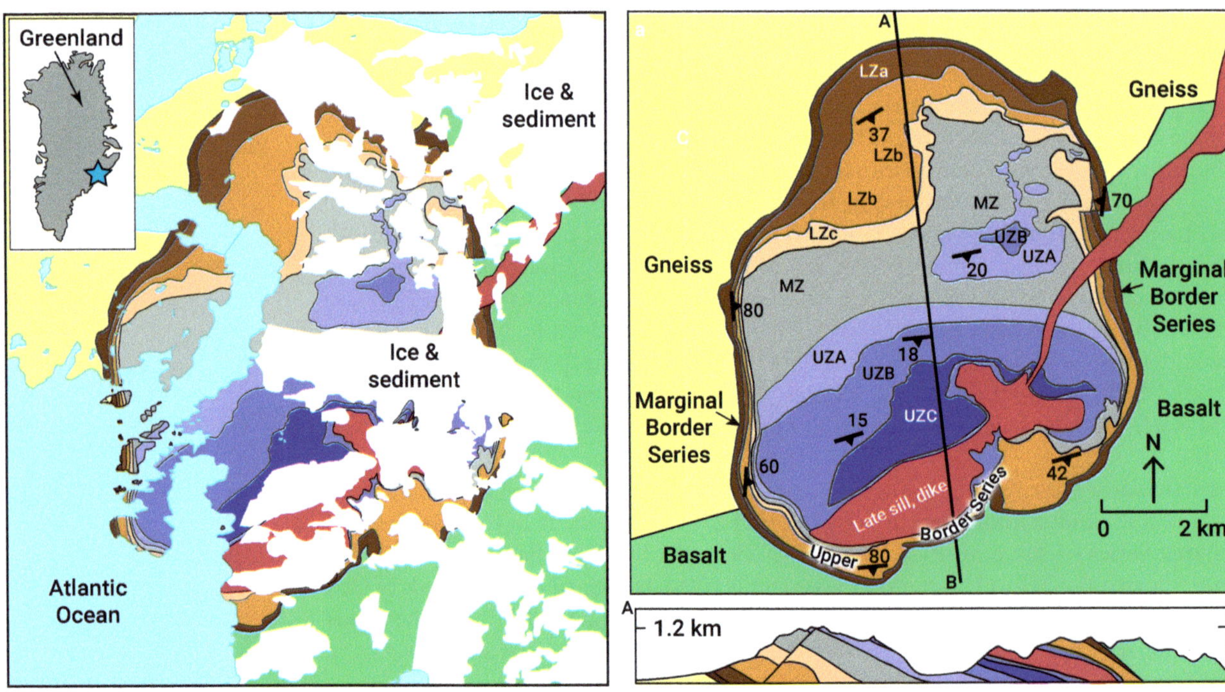

Figure 14.29 Geologic map and cross section of the Skaergaard Intrusion. The left panel shows the geologic map with ice and water, and the right panel is an interpretation of the geology with ice and water cover removed, with cross section below. Strike-and-dip symbols give orientation of magmatic banding. The Layered Series comprises a Lower Zone, which is subdivided into units a, b, and c; a Middle Zone; and an Upper Zone, subdivided into units A, B, and C. The Marginal and Upper Border Series comprise a thin collar of banded rocks whose banding is roughly parallel to the outer contact of the intrusion.

Sources: Base maps modified from Kurt Hollocher (muse.union.edu/hollochk), which was modified from McBirney (1989); cross section modified from McBirney (2007, p. 233).

subzones. The thinner, complementary Upper Border Series, which is exposed along the southern margin of the intrusion, reverses this sequence, and the two meet at a thin sandwich horizon like that in the Palisades Sill (see Figure 9.8). Layering in these units, which typically dips about 20° south, abuts a thin, steep zone of yet more banded gabbros, the Marginal Border Series, along contacts with host gneisses (see Figure 14.29). See McBirney (1989) for details.

Compositional Variation Minerals in the gabbros show smooth patterns in composition passing upward in the Layered Series (see Figure 14.26c). The An# in Pl decreases smoothly from 75 at the deepest levels to 30 at the top of the Layered Series, and the Mg# in Cpx decreases over the same interval from 70 all the way down to 0, with most of this decrease in the upper part of the Layered Series. Olivine becomes more Fe-rich as well, reaching pure fayalite at the top. The thinner Upper Border Series mimics these trends in reverse. These remarkable patterns demand an explanation, but what is it?

How Do LMIs Form?

The Standard Model The Skaergaard LMI is shaped approximately like a thick book, about 11 km long, around 8 km wide, and about 4 km thick—more *Moby Dick* (or, perhaps more fittingly, *Infinite Jest*) than *The Cat in the Hat*. A volume-weighted average of the units that comprise it yields a bulk composition of high-Ti tholeiite, similar to East Greenland flood basalts that were erupted during opening of the North Atlantic (Nielsen, 2004). A model for intrusion and crystallization of the Skaergaard LMI, summarized from multiple sources, is as follows (some version of this scenario has been applied to many LMIs; a key feature shared by all is a starting point of a large, crystal-poor tank of magma):

1. Approximately 300 km^3 of basaltic magma intruded into the upper crust to form a deep chamber filled with crystal-poor magma. The chamber was roofed with Eocene basalts about 3 km deep.
2. Crystallization proceeded inward from the bottom, top, and sides, but settling of crystals built up the chamber floor faster than the sides or roof.
3. As crystallization proceeded, the composition of the residual liquid changed to lower and lower An# and Mg# owing to crystallization of calcic Pl and Mg-rich Ol, Cpx, and Opx.
4. Convection in the magma chamber kept the residual liquid mixed, and minerals crystallizing from it changed composition sympathetically.
5. Slumps of crystallized mush from the sides and drips from the ceiling added material to the floor (Layered Series).
6. Turbidity currents of crystal-rich slurries created cross-bedding, trough structures, and other sedimentary features in the banded rocks.
7. Final solidification occurred at the Sandwich Horizon, with the most evolved rock and mineral chemistries.

This developmental sequence, advanced by Wager and Deer (1939), has been expanded upon by many others and taught widely in the intervening decades for Skaergaard and other LMIs. The model is intuitively appealing, but the big-tank model for LMIs has sprung a number of leaks and is in need of a thorough cleanup. This is a situation akin to the big-tank assumption that dominated interpretation of granitic plutons until precise U-Pb zircon geochronology showed that many granitic plutons could not have existed as big tanks (see Section 10.8). U-Pb geochronology is significantly more difficult in LMIs than in granites owing to the scarcity of zircon in mafic rocks; in such low-Zr rocks, baddeleyite (ZrO_2) is extremely scarce and difficult to find and separate, and zircon is typically only found in small granite segregations that were the last bits of liquid to crystallize. As a consequence, it is far harder to date LMIs precisely than it is to date granitic rocks.

The Bushveld Complex has a volume about 2000 times greater than that of the Skaergaard, but even this West Virginia–sized dinner plate is viewed by many as having been filled with essentially crystal-free magma (e.g., Latypov et al., 2022).

Some Problems with the Standard Model

There are several reasons to question the standard model, including the following:

1. How does a large, crystal-poor magma chamber form? The standard model for the Skaergaard requires an empty cavity about 4 km deep into which crystal-poor magma was instantaneously emplaced. This is easy to draw or to model in a computer, but difficult to accomplish in the real Earth. Realistic thermal models of magma emplacement into upper crustal rocks indicate that establishing a crystal-poor magma chamber of a given volume requires intruding at least twice that volume of magma, most of which freezes before things are warmed up enough that the final magma chamber can be established (Gelman et al., 2013; Glazner, 2021). In the Skaergaard and other LMIs, there is little evidence for this precursory magma, which should have crystallized to simple gabbro.
2. Emplacing a magma body 4 km thick requires displacing the rocks above and/or below 4 km vertically—and in the Skaergaard, this movement occurred under a relatively thin roof (~3 km). This is not difficult to do on million-year timescales, because erosion off the top and isostatic subsidence below can accommodate kilometer-scale displacements; in granitic plutons, relaxation of the big-tank model solved the so-called "room problem" (see Section 10.7). However, displacing rocks several kilometers fast enough to pour in crystal-free magma without freezing most of it requires moving things vertically at extreme rates.
3. High-precision U-Pb geochronology indicates pulsed intrusion of the Stillwater and Bushveld over intervals of about 3 Myr and about 5 Myr, respectively (Mungall et al., 2016; Wall et al., 2018; Scoates et al., 2021). Importantly, data indicate out-of-sequence intrusion; that is, the ages do not become monotonically younger upward in the sequences toward a sandwich horizon. These controversial results are consistent with time spans over which large granitic plutons accumulate and with a less catastrophic view of formation of LMIs. In contrast, two ages for the main Skaergaard body are consistent with, but do not require, much faster crystallization of that body (Wotzlaw et al., 2012).
4. It is difficult to explain the sequences of minerals found in layers. For example, in Figure 14.27d, layers are defined by, from left to right, a sprinkling of just dark minerals, a layer of just plagioclase, a bimodal layer that is half dark minerals and half plagioclase, another bimodal layer that is about 25 percent dark minerals and the rest plagioclase, another half-and-half layer, and another defined just by dark minerals. These are not normal crystallization sequences.
5. Interpretation of features that resemble cross-bedding and graded bedding as being formed by sedimentary processes runs afoul of the physics of how fluids move and cannot explain formation of similar features on the walls and roof of the Skaergaard. This topic is important enough to deserve its own section.

Caveats about Treating Magmas as Sedimentary Systems

Problems with Crystal Settling as a Dominant Mechanism

It has been common to view many plutonic bodies, especially LMIs, as the products of sedimentation of crystals from a magma chamber. Sinking of dense materials in a fluid is a process with which we are all familiar—as, for example, when we walk through the swash zone at a beach. However, the idea that layering in LMIs represents simple sequential settling of phases from a crystallizing magma was in trouble from its earliest conception. Obvious issues include:

1. The lower parts of the Skaergaard have more plagioclase in them than the upper parts (see Figure 14.26c). This is a problem because it was recognized early on that plagioclase would likely have floated in the relatively Fe-rich, dense basaltic liquid that the Skaergaard crystallized from (McBirney & Noyes, 1979; Krättli & Schmidt, 2021; see Section 8.1).
2. In layers produced by particle settling, such as turbidites, particles of a given density are larger at the base of the layer than at the surface, producing size-graded layering. This is uncommon in igneous layers, and inverse grading, with larger particles at the top, is common, especially in granodiorites (Wilshire, 1969; Solgadi & Sawyer, 2008).
3. The vertical sequence of minerals in a given layer commonly correlates with density (e.g., Fe-Ti oxides, olivine, pyroxene, and hornblende at the bottom; feldspars and quartz at the top) but not with calculated sinking rates.
4. Layering like that in the main mass of the Skaergaard is also found on formerly vertical walls and, upside-down, on roofs (Campbell, 1978). It is difficult to see how settling in the gravity field can explain similar layering on surfaces of such different orientations.

Hydraulic Considerations Formation of cross-bedding in sedimentary rocks requires turbulence, as does graded bedding produced by turbidity currents and similar processes. The physics of fluids predicts that turbulence can occur if the Reynolds number (Re; see Section 7.5), the ratio of inertial forces provided by fluid movement to viscous resistance by the fluid itself, exceeds a critical value on the order of 1000. McBirney & Noyes (1979) calculated that Re for Skaergaard magma would have been less than 1, more than a thousand times below the critical value for the transition to turbulence. To put it differently, for the magma to have reached the turbulent regime would have required flow velocities on the order of several tens of meters per second (about 100 mph), an absurd result. The high viscosities of silicate liquids, even fluid basalts, generally prevent the turbulence required to make many of the structures that are common in water- or wind-laid sediments (Glazner, 2014).

Crystal Cementation Clastic sediments are passively carried by water or air, fluids with which they scarcely react. Piles of such particles are not cemented and can avalanche, producing deposits such as turbidity currents. Crystals in magma, in contrast, are precipitated by and generally react with the host silicate liquid, making them more akin to evaporite deposits than to clastic deposits (Boudreau, 2019, p. 13). Evaporite minerals in modern saline lakes and playas comprise mats of tightly interlocked crystals (Casas & Lowenstein, 1989; Warren, 2016). Precipitation has important consequences for sedimentary interpretations. As crystallization proceeds, new material is added to existing crystals, cementing them together and giving the interlocking texture that distinguishes igneous from clastic rocks (Figure 14.30). Adhesion and intergrowth of crystals during crystallization is a familiar process to anyone who has made or eaten rock candy, which is coarse-grained sucrose precipitated from water. Such precipitation yields a solid, even tooth-breaking intergrowth rather than loose crystals. Similarly, metals produced by casting from melt form tightly interlocked crystal aggregates rather than loose crystal piles, a feature that contributes much to their strength.

Bowen (1928, p. 167) noted that "if a mass of discrete crystals of a soluble salt lie in the bottom of a beaker of saturated solution the mass will gradually be converted into a solid cake of the salt." Thus, even in the absence of additional precipitation, crystals tend to grow onto one another. It is likely that igneous systems behave similarly, and that

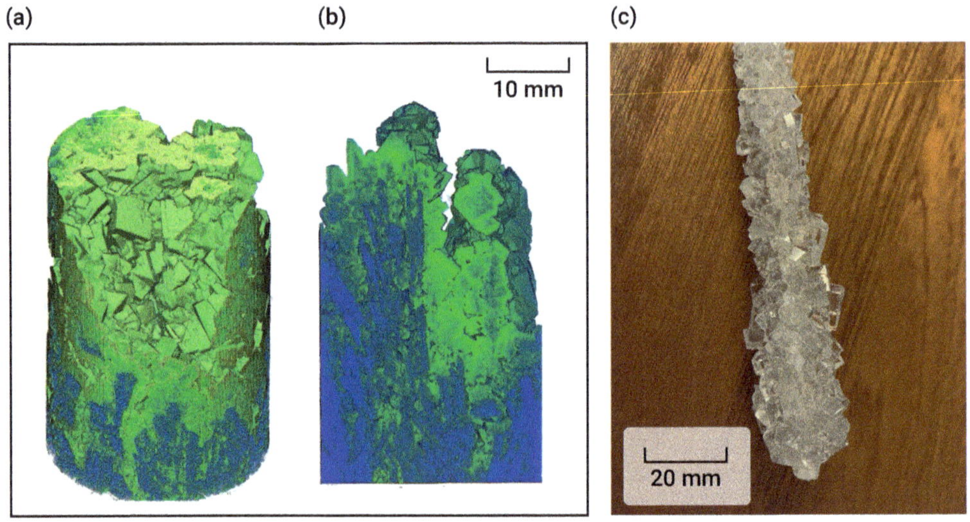

Figure 14.30 X-ray computed tomographic images showing that crystallization from a fluid generally leads to a tightly cemented crystal mass rather than a loose pile of crystals: (a) oblique view with top truncated; (b) cross section showing topography of upper surface. In this experiment, a water solution saturated in chalcanthite ($CuSO_4 \cdot 5H_2O$) and epsomite ($MgSO_4$) at 50 °C was poured into a 25-mm centrifuge tube and cooled at 10 °C/hr. Bladed crystals of epsomite (blue in these false-color images) grew up from the bottom of the tube and are intergrown with pseudo-cubic alpersite ($Mg_{0.6}Cu_{0.4}SO_4 \cdot 7H_2O$; green), which shows significant Cu-Mg zoning with brighter Cu-rich rims. Crystals are tightly intergrown. (c) A piece of rock candy made by suspending a wooden skewer covered in granulated sugar into a supersaturated solution of sugar in water. Crystals nucleate on the seed crystals and grow into a tightly interlocked mass rather than raining out of the solution and forming loose accumulations on the floor of the container.

Source: AFG photos, images in (a) and (b) from experiments by AFG.

cemented crystal mush in igneous systems is quite unlike the accumulations of clastic sediments to which it is commonly compared.

Sequential Crystallization The particles in a clastic sedimentary system are all present when sedimentation occurs. Crystallizing magmas are different because they precipitate different phases at different temperatures, and this concept applies to all magma types, not just layered mafic intrusions. For example, we saw in Figure 10.28 that a typical dacitic magma, in cooling on its way to becoming a granodiorite pluton, starts crystallization with one or more mafic phases (e.g., hornblende, biotite, magnetite, Cpx) followed by plagioclase, and that quartz and K-feldspar do not begin to precipitate until the magma is near 50 percent crystallinity. Thus, by the time appreciable amounts of quartz and K-feldspar have grown, the magma is far past the rheologic lock-up point (see Figure 7.11) and tightly cemented by additional growth onto already touching crystals. Sequential precipitation (i.e., that not all the mineral phases in a magma are present throughout the magma's crystallization history) is typically ignored in studies that consider the formation of layering by sedimentary processes (e.g., Clarke & Clarke, 1998; Solgadi & Sawyer, 2008). The granodioritic layers in these studies (identical to those in Figure 14.27f) contain all the phases in the rock, including significant amounts of K-feldspar, and could not have flowed down a magma chamber wall or through a dike carrying that crystalline assemblage.

Cumulus Terminology Wager et al. (1960) introduced the term "cumulate" for igneous rocks formed by crystal accumulation. The crystals so accumulated were termed

cumulus crystals, surrounded by **intercumulus liquid**, with plagioclase as an exemplar of cumulate processes. Since then, various authors have proposed alternate definitions, including:

1. Crystal settling in a manner analogous to the way sedimentary particles settle out of a water column (Jackson, 1961).
2. Formation of a framework of touching mineral crystals concentrated through fractional crystallization, a definition which allows for crystal flotation as well as sinking (Irvine, 1982).
3. Concentration of crystals and/or loss of melt such that rocks do not represent the original magma composition (Vernon & Collins, 2011).

Although these definitions are simple, they do not directly guide the identification of such rocks.

Numerous authors (e.g., Campbell, 1978; McBirney & Hunter, 1995; McBirney, 2009; Boudreau, 2019) identified features of layered intrusions, including many discussed earlier, that contradict a simple cumulate interpretation. In spite of this, the cumulate interpretation of layered mafic rocks and, indeed, many layered granitic rocks has persisted. However, the strong genetic connotation of the terminology forces an interpretation that is commonly only speculative. We concur with Boudreau (2019, p. 10) that "using cumulus terminology, which traditionally has interpretive implications, has pernicious effects on the understanding of the processes that formed the rocks."

Cumulus crystals
Crystals that accumulate preferentially relative to liquid; assumed to reflect processes that separate crystals and liquids via crystal flotation, crystal settling, compaction, etc.

Intercumulus liquid
Magmatic liquid trapped between cumulus crystals.

Metamorphic differentiation (metamorphic segregation)
Development of banding or mineral segregations from a uniform protolith during metamorphism.

Alternative Methods for Forming Layers in Plutonic Rocks

Metamorphic Differentiation

Many classes of rocks exhibit banding that was not produced by crystal sedimentation. **Metamorphic differentiation** (or **metamorphic segregation**) refers to banding, common in gneisses and migmatites, that is defined by differences in mineral proportions (Figure 14.31). These are typically alternating mafic and felsic layers, although banding may be defined by differences in proportion of micas or other minerals. Banding in metamorphic rocks may be a relic of the structure of the protolith (e.g., sedimentary layering), but in many cases, it clearly cuts across such depositional layering or was developed in a non-layered rock such as basalt (e.g., Bowes & Park, 1966; Williams, 1990).

Deformation

Deformation of metamorphic rocks can produce banding by a number of different mechanisms. For example, Moore et al. (2020) describe how mafic-felsic banding was produced in a high-grade shear zone in Norway via selective dissolution and reprecipitation of components. Williams (1990) described cleavages in low-grade metasedimentary and metavolcanic rocks defined by mica-rich layers alternating with ones rich in quartz or carbonate minerals.

(a)

(b)

Figure 14.31 Examples of banding in rocks produced by various non-sedimentary processes. (a) Orthogneiss in Prague, Czech Republic, showing folded gneissic banding, with mafic and felsic minerals segregated into bands. Width of field ~1 m. (b) Migmatite from India showing that partial melt is segregated into discrete bands rather than soaking through the rock to produce a more uniform texture.

Sources: (a) Huhulenik/Wikimedia (CC BY 3.0); (b) MJK photo.

Liesegang Banding and Geochemical Self-Organization Liesegang banding is an eye-catching phenomenon in which reacting chemicals produce bands of precipitate at ever-increasing distances from one another and at ever-increasing time intervals. The experiment in Figure 14.32 shows that bands developed when concentrated ammonia (aqueous NH_4OH solution) reacted with $MgSO_4$. The sulfate was dissolved in gelatin dyed with red food coloring, and the clear ammonia was poured on top. Over the next hour, as ammonia diffused into the gelatin (with a gravity assist), a band of $Mg(OH)_2$ 9 mm thick precipitated at the interface. Thinner rings then formed at successively greater spacings and doubling time intervals.

Liesegang banding or related phenomena are common in many rock types. A particularly beautiful and familiar example is "picture sandstone" from the Triassic Chinle formation in southern Utah (Figure 14.33a), which has been made into countless coasters, bookends, and other trinkets. Banding is defined by variations in the concentration of Fe oxides, and the patterns are far more complicated than those exhibited in a test tube or petri dish. They typically show cutoffs reminiscent of erosional features in sedimentary rocks, trough-like structures, and concentrations of Fe oxides at discontinuities where one set of bands meets another. The mechanism by which Liesegang bands form has been under debate since their discovery by Liesegang (1896).

Liesegang bands in sandstones and other rocks have received a great deal of geologic attention over the last century (e.g., Carl & Amstutz, 1958; Fu et al., 1994; Shahabpour, 1998; Dietrich & Jacob, 2017). In recent years, the possible role of microbes in mediating Fe-oxide cementation (Kettler et al., 2015) and similarities between oxide structures on Earth and features on Mars (Chan et al., 2005) have brought new interest to the phenomenon in Earth science.

Ortoleva's (1994) book *Geochemical Self-Organization* showcases a large variety of processes that can drive geochemical patterning in addition to Liesegang banding, with

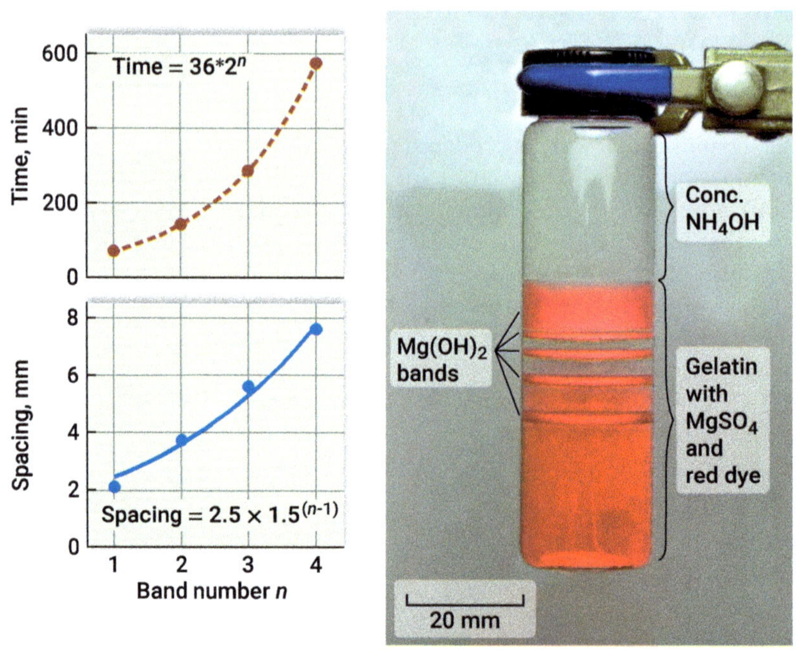

Figure 14.32 Experimental production of Liesegang bands in gelatin. Although one might expect NH_4OH and $MgSO_4$ to simply react and form an ever-thickening layer of $Mg(OH)_2$ at their interface, well-defined bands instead form at ever-increasing distances from the interface. This phenomenon is still not well-understood.

Source: Experiments by AFG; AFG photo.

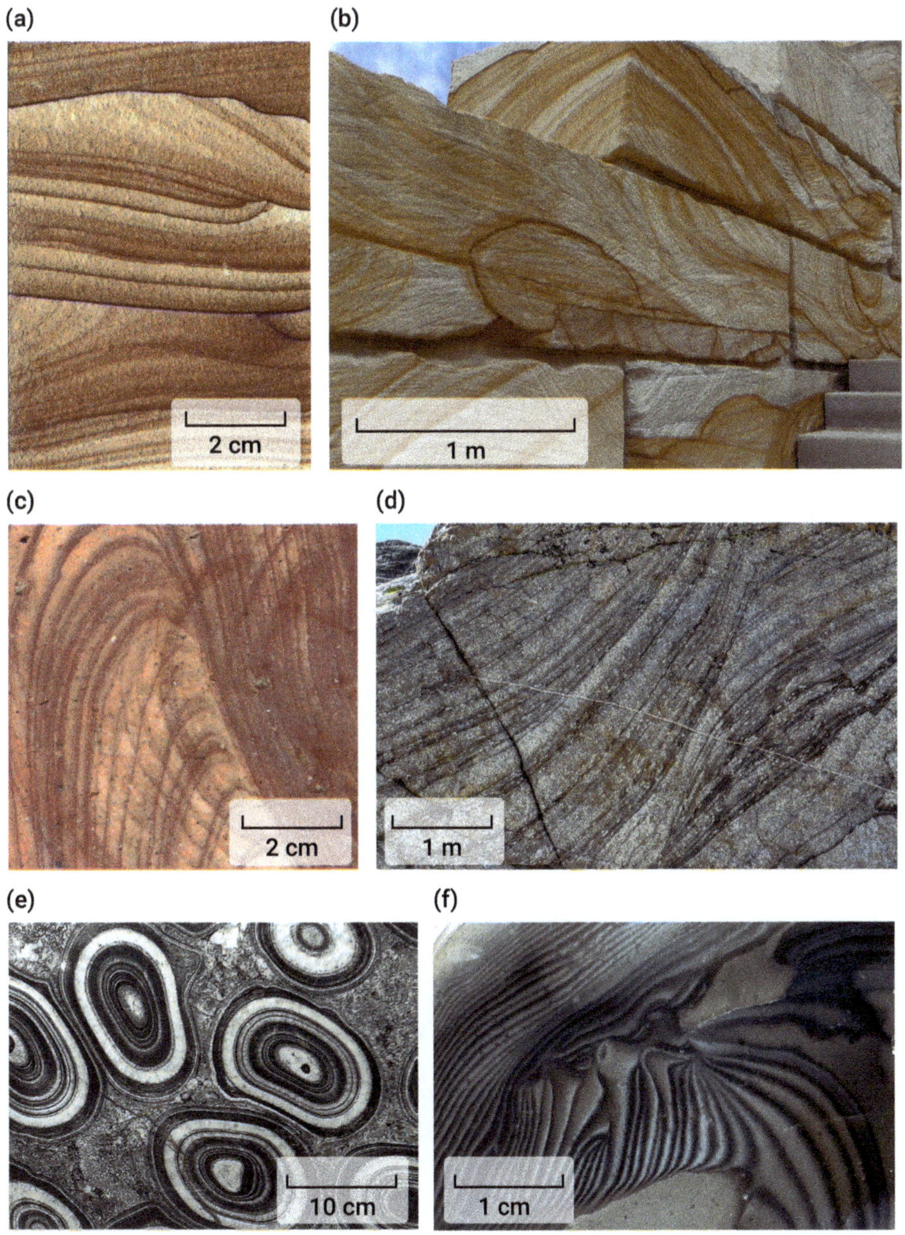

Figure 14.33 Examples of geochemical self-organization leading to Liesegang-like banding, and superficially similar banding in plutonic rocks. (a) Banding in picture sandstone from the Triassic Chinle Formation near Kanab, Utah. Bands occur in sets whose boundaries are typically marked by a darker concentration of oxides along the "unconformity." Banding superficially resembles sedimentary cross-bedding, but cannot be because it formed by reacting pore fluids after deposition and cementation. (b) Blocks of Triassic Hawkesbury Sandstone at Barangaroo Reserve, Sydney, Australia. Hewn blocks display the three-dimensional nature of the banding. (c) Fe-oxide bands in a trivet cut from "placerite," a decorative stone mined from a 31 Ma rhyolite ignimbrite east of Placerville, California. This material is unusual in showing band sets that meet at "unconformities," as in (a) and (b), as well as a set of bands that cuts across the others. (d) Strongly layered granodiorite of the marginal Tuolumne Intrusive Suite near Saddlebag Lake, California, showing large-scale, non-overprinting banding that resembles Fe-oxide banding in sandstones and other porous rocks. How these form is unknown, but arguments above rule out a standard crystal sedimentation origin. (e) Orbicular diorite from Finland. (f) Polished slab of banded "Polish flint" in Jurassic limestone, eastern Poland.

Source: (a-d, f) AFG photos; (e) MJK photo; (f) sample provided by Stephen Challener.

applications to oscillatory zoning in crystals, metamorphic differentiation, patterning in agates, structure of sedimentary basins, and other phenomena. Boudreau & McBirney (1997) and Boudreau (2011) have summarized a number of chemical processes that could serve to produce the sorts of banding and crystal size variations seen in LMIs and other banded plutonic rocks.

Thermal Gradients Experimental crystallization of magma in a strong thermal gradient can produce dramatic differentiation, with mafic components segregated into the hot end of the experimental charge and felsic components into the cool end. In one experiment, Huang et al. (2009) put andesite powder and water in a sealed capsule and held it for 66 days, with one end at 950 °C and the other at 350 °C. After quenching, the charge was found to contain all melt at the hot end, all crystals at the cool end, and a smoothly varying change in mineralogy from hot to cold, with successive appearance of Fe-Ti oxides, hornblende, plagioclase, biotite, quartz, and K-feldspar—essentially the order expected during cooling of an intermediate magma (see Figure 10.28), and roughly the order seen in mafic-felsic bands such as those in Figure 14.27e,f. Lesher and Walker (1988) have proposed that crystal-melt separation in a thermal gradient can explain some of the layering features seen in LMIs, and Lundstrom (2009, 2016) has proposed that chemical segregation in a thermal gradient can explain much of the crust's chemical zonation.

Summary Crystal settling has long been the paradigm under which LMIs have been interpreted, but evidence cited here, especially the difficulties in making a huge, crystal-free magma chamber, fluid dynamic arguments against turbulent sedimentary processes, and the likelihood of crystal cementation, require that this paradigm be re-evaluated. There are many alternatives to crystal sedimentation for producing segregation and banding in chemical systems.

14.6 Impacts

Learning Objectives

By the end of this section you should be able to:

- **14.6.1** Propose features you would look for across scales ranging from hundreds of kilometers to micrometers to identify an impact site.
- **14.6.2** Apply the frequency vs. energy diagram to predict the likelihood of impact events.

Cratering History of the Inner Planets

In Chapter 2, we learned that Earth and other planets formed by gravitational accretion of material in the solar nebula. Formation of Earth by accretion likely led to its being a big, glowing blob of magma early in its history, and if the giant-impactor hypothesis of Moon formation (see Figure 2.3) is correct, then the Moon was molten, too. Accretion continued after cooling and solidification, and **bolide** impacts produced impact melt and dramatic shock metamorphism.

Bolide
In geology, a large, crater-forming impactor; in astronomy, an exceptionally bright meteor (fireball).

The Moon and Mercury, airless worlds without plate tectonics, carry a long-term record of these impacts in their scarred faces (Figure 14.34), and the density of craters on a given surface is a measure of that surface's age. Before the Apollo missions, the relative ages of various lunar surfaces had been worked out by quantifying the numbers of craters per unit area, and dating of returned samples placed these relative ages on the timescale. The heavily crated anorthosite of the lunar highlands (see Section 9.2) makes bright patches on the Moon's surface that formed more than 4 Ga, whereas the less-cratered **lunar maria** are dark plains of basaltic melt that formed 3 to 4 Ga and filled huge, older impact craters. Cratering studies indicate that the rate of cratering

Lunar maria
Relatively dark basalt "seas" that fill impact basins on the Moon's near side; singular mare.

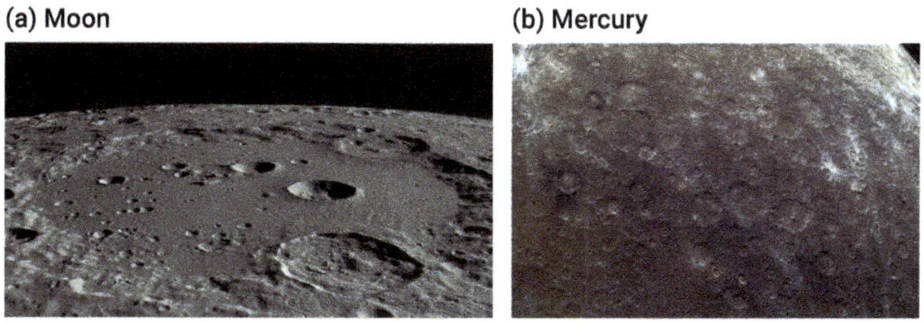

Figure 14.34 (a) Clavius Crater, which is over 200 km in diameter and fills most of this photograph, is located near the Moon's south pole on the near side. It illustrates the intense cratering that characterizes the lunar highlands. Based on the density of craters in Clavius, its age is estimated as ~3.85 Ga. Science fiction fans might recognize Clavius as the site of the lunar base in the 1968 film *2001: A Space Odyssey*; the alien monolith was found buried nearby. (b) The surface of Mercury is also densely cratered.

Source: (a) NASA/USGS; (b) NASA.

was more than 100 times higher prior to 4 Ga, and has been relatively steady for the past 3 Gyr.

Earth's Cratering History

Earth presents a distinctly smoother face to the universe than the Moon, because most of its surface is covered either by rocks younger than 3 Ga or by water, and because erosion is effective at erasing craters. Nevertheless, approximately 190 impact craters have been identified (Figure 14.35). Only a few of these have been identified in the oceans; most are concentrated in continental areas where Precambrian, and particularly Archean, basement is at or near the surface. Earth's largest-known impact structures are Vredefort in South Africa (2.02 Ga), Chicxulub in Mexico (66 Ma), and Sudbury in Ontario (1.85 Ga); all are about 150 km in diameter. The best-known impact structure may be Barringer Crater in Arizona (Figure 14.36a), formed 48 ka by impact of a 50-m iron meteorite. Barringer is, however, dwarfed by the larger structures.

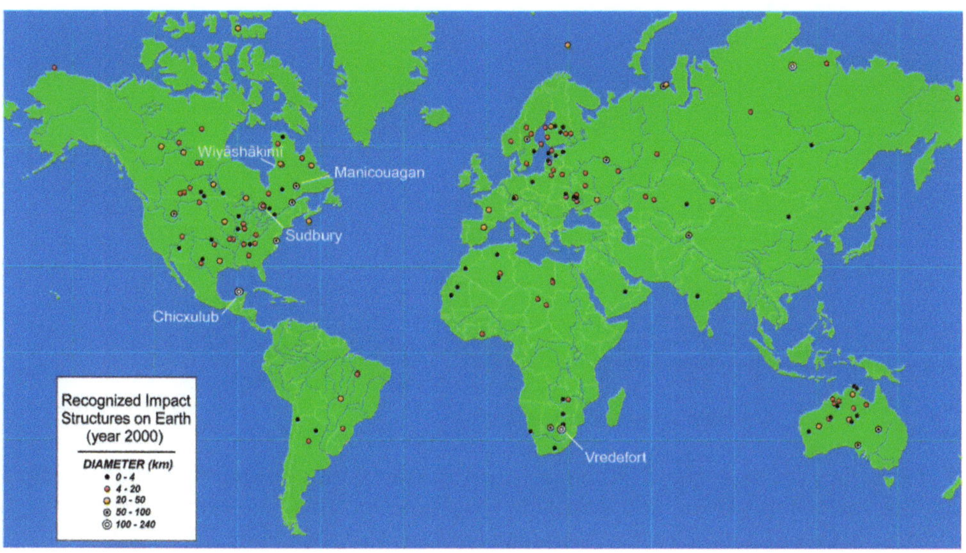

Figure 14.35 Map of recognized impact structures on Earth.

Source: Lunar and Planetary Institute.

(a) Barringer Crater, Arizona (b) Lac Wiyâshâkimî, Quebec

Figure 14.36 (a) Well-known Barringer Crater (also known as Meteor Crater) in Arizona, about 1 km in diameter, and (b) the far larger Lac Wiyâshâkimî (Clearwater Lakes) pair of impact scars east of Hudson Bay, Canada. The white blob in the smaller of the two lakes is the outline of Barringer Crater to scale. Barringer Crater formed ~50 ka from an iron meteorite estimated to be ~50 m in diameter. The Lac Wiyâshâkimî craters (Figure 14.36b), long thought to be from impact of a binary asteroid, instead apparently formed 180 Myr apart, ~290 and 470 Ma (Schmieder et al., 2015).

Sources: (a) Courtesy of Marli Miller (geologypics.com); (b) The Earth Observatory/NASA.

Although supereruptions threaten civilization, impacts have far higher potential for extinction-level events. The Chicxulub impact structure in Yucatan, although barely evident at the surface, caused the dinosaur-ending K-Pg extinction, coincidentally (or not) accompanied by flood basalt eruptions in India. The 1.85 Ga Sudbury impact event was even larger. The statistics of impacts are like other random events, such as floods; if a so-called 100-year event happens this year, one cannot count on 100 years of safety before the next one. Impacts are **stochastic** events that should follow a **Poisson distribution**, although specific events can perturb the randomness. One such non-random sequence of impacts was a rain of chondritic meteorites that peppered Earth in the Ordovician, 470 Ma (Korochantseva et al., 2007); the eastern Lac Wiyâshâkimî crater (Figure 14.36) was likely formed in this event. This swarm of Earth-crossing L-chondrites (a low-Fe subset of the chondrites), many of which show evidence for shock metamorphism, most likely formed as a result of an impact on the L-chondrite parent asteroid. Figure 14.37 is an estimate of size-frequency statistics for impacts.

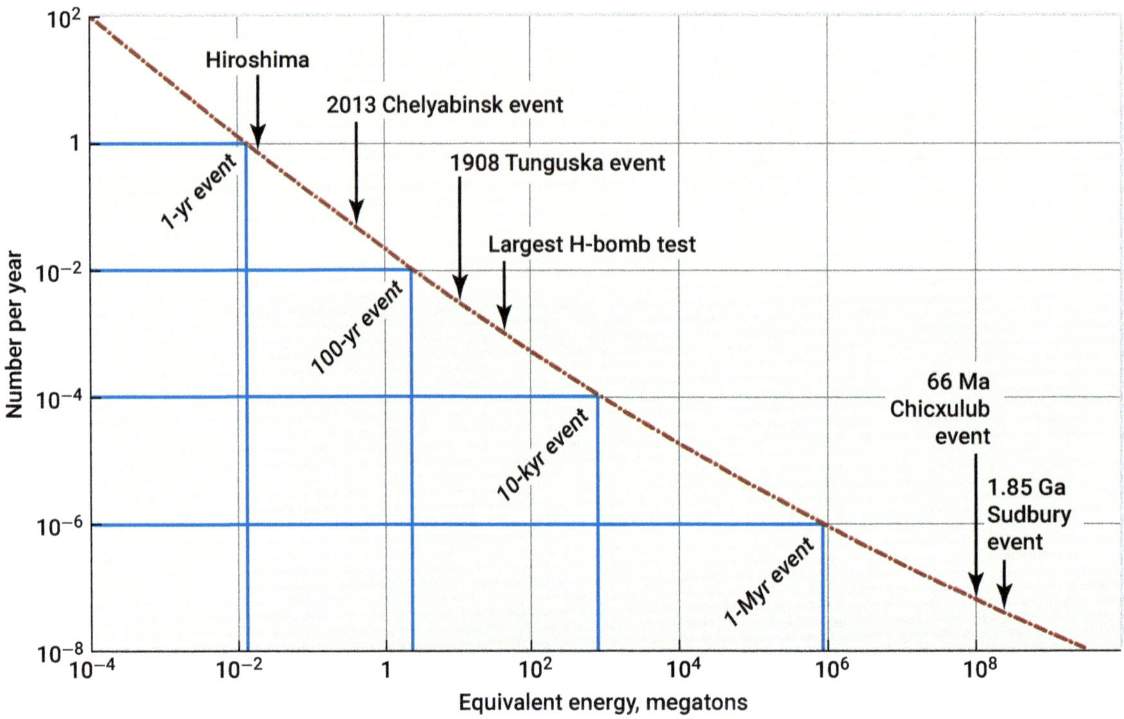

Figure 14.37 Approximate frequency of impacts of a given energy. The ordinate gives the number of impacts of the corresponding energy expected in a given year; thus, 10^{-4}/yr means that in 1 Myr, 100 such events would be expected. Curve extrapolated to Sudbury. Energy is expressed in megatons of TNT, a unit typically used for hydrogen bombs; 1 Mt of TNT = 4.2×10^{15} J. Note that energies ascribed to the Chicxulub and Sudbury impact events are equivalent to about one million of the largest H-bombs ever detonated.

Source: Modified from Shoemaker (1983).

Impact Melting and Shock Metamorphism

Impact Melting **Hypervelocity impacts** on solid planetary bodies induce both melting and metamorphism in **target rocks** unfortunate enough to be hit. Cratering events have been extensively studied in the field, in hydrodynamic computer models, and in the laboratory, where experiments involve unique equipment, such as firing projectiles at target rocks at velocities up to 5 km/s through welded-together naval gun barrels (Yang & Ahrens, 1998). Another way to put on the squeeze is to use the Sandia Z Machine, a power generator capable of storing about 20 MJ of electrical energy (equivalent to the potential energy of a 1-ton weight lifted 2 km) and use magnetic force to get small projectiles moving at up to 50 km/s; this can produce shock pressures up to 2000 GPa(!) (Knudson, 2012). A general view of a generic impact is given in Figure 14.38; transient shock pressures reach 50 GPa (*P* equivalent to a depth of ~1500 km), ejecta are blown out of the crater, and impact melt forms. How much melt forms depends on the energy of the impact.

In a rough approximation, it takes about 1 MJ/kg to heat a rock from surface temperatures to the liquidus and melt it; a little arithmetic shows that 1 Mt of energy is enough to melt about 1.5×10^6 m^3, roughly the volume of a cube with an edge length of a football field, if all the impact energy were devoted to melting. However, a considerable amount of energy is expended in fracturing the target, frictional losses (think of how a short stretch of gravel can absorb the energy of a runaway truck by fracture and friction in the gravel), and kinetic energy of ejecta. A generic 1-Myr event, with an energy of 10^6 Mt, could produce about 1500 km^3 of melt, roughly comparable to the largest caldera-forming eruptions, and a Chicxulub-sized event (10^8 Mt) could produce 100 times that. The largest event in Figure 14.37 produced the Sudbury impact structure, whose volume is estimated at some tens of thousands of cubic kilometers. The estimated impact energy (2×10^8 Mt) could produce roughly 10 times that (see Toolbox 14.3), so at least in that example, most of the impact energy went into pulverizing the target, kinetic energy of ejecta, and other processes.

Stochastic
Randomly determined.

Poisson distribution
Statistical distribution describing the probability of a given number of independent stochastic events happening in a given interval of time or space.

Hypervelocity impacts
Collisions of material moving at several kilometers per second with a planetary body.

Target rocks
Any rocky material on a planetary body affected by a bolide strike.

Shock Metamorphism The pressures produced during shock metamorphism are, well, shockingly high (Figure 14.39). Smoking-gun evidence of metamorphism from a hypervelocity impact, taken from French (1998), includes:

1. **Shatter cones** (Figure 14.40a), which form at shock pressures of 1 to 5 GPa.
2. Stishovite, a high-*P* polymorph of SiO_2; other high-*P* polymorphs such as diamond and coesite commonly occur in impact rocks but are not diagnostic. Stishovite can

Shatter cones
Curved, striated, conical fracture surfaces, typically tens of centimeters long, that are characteristic of relatively low shock pressures (1–5 GPa).

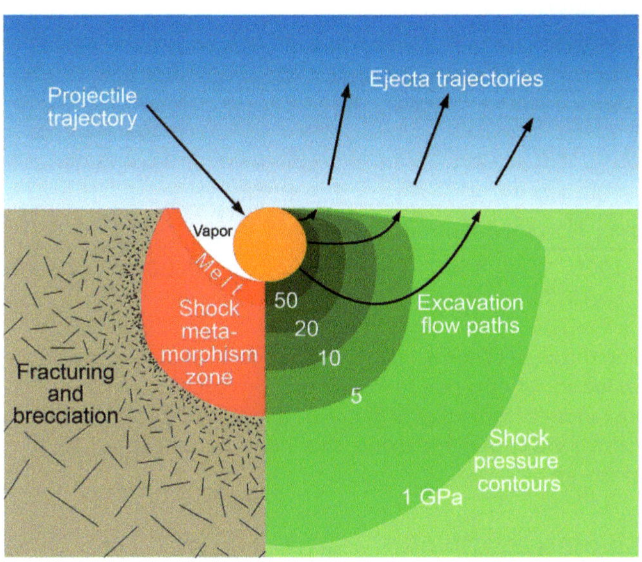

Figure 14.38 Schematic illustration of melt generation and shock metamorphism, including contours of pressure, from a large impact.

Sources: Illustration courtesy of David Kring, after French (1998) and Melosh (1996).

Diaplectic glasses
Amorphous material in which long-term crystallographic order has been destroyed by a shock wave; unlike normal glasses, diaplectic glasses can be annealed back into something approximating the original crystal.

form at shock pressures of 12 to 15 GPa, below its equilibrium transition *P* (see Figure 14.39), and was identified at Barringer Crater in 1962, a year after S. M. Stishov, a Soviet physicist, first synthesized it.

3. Planar fractures and deformation bands in quartz and other minerals (Figure 14.40b), formed at about 7 to 35 GPa.
4. Formation of **diaplectic glasses** from quartz and feldspars at 35 to 45 GPa.

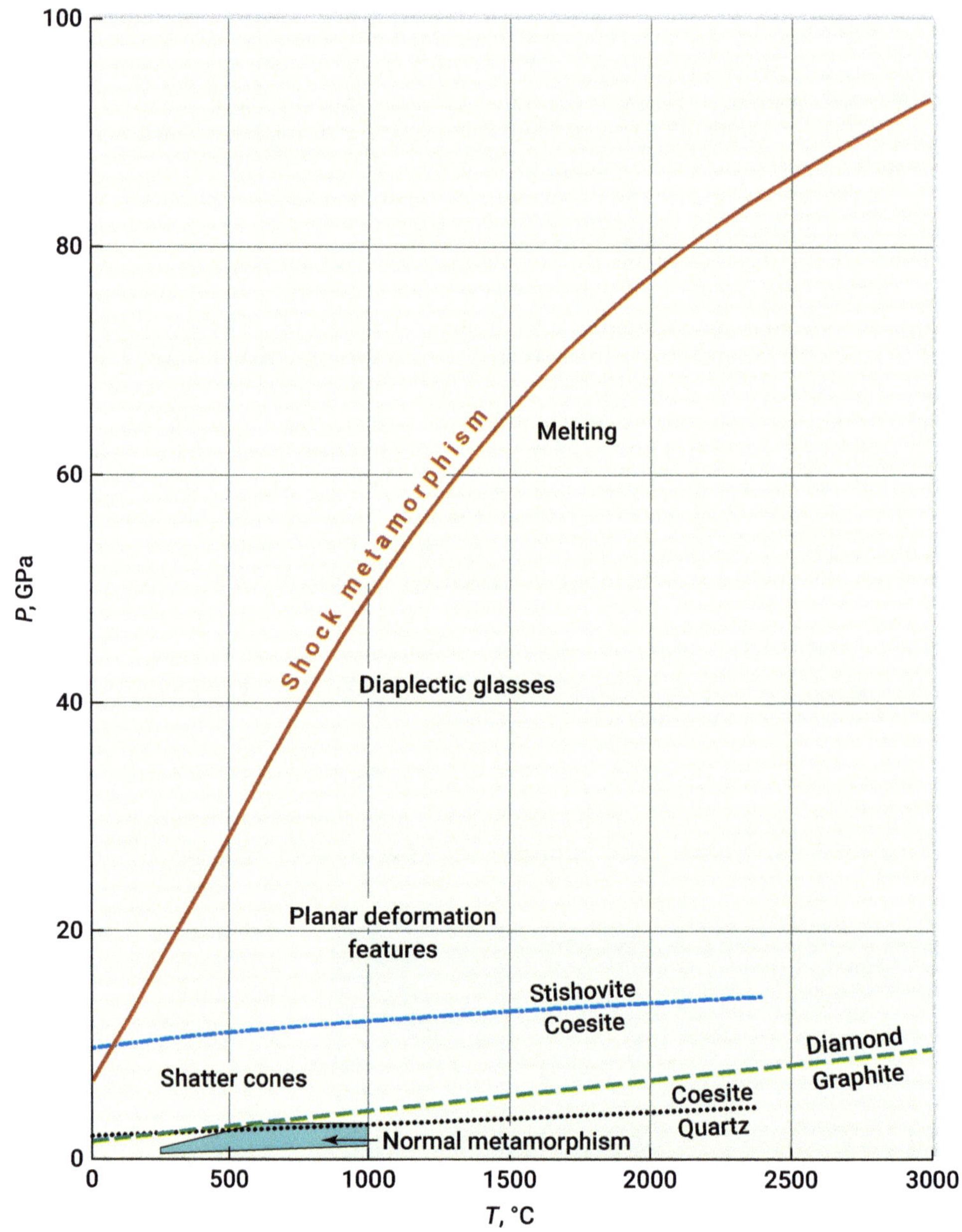

Figure 14.39 *P-T* diagram showing the zone of normal crustal metamorphism (squashed down into the bottom left corner), curves for equilibrium polymorphic reactions such as quartz = coesite, and approximate pressure ranges of formation of features such as shatter cones. Curve labeled "Shock metamorphism" shows the approximate post-shock temperatures reached in crustal rocks for the given shock pressure.

Source: Modified from French (1998).

Figure 14.40 Textural evidence for impacts and shock metamorphism. (a) Shatter cone near the Sudbury impact structure. (b) Shocked quartz from the Chicxulub impact under cross-polarized light, showing characteristic planar deformation features (lines).

Source: (a) Photo courtesy of David Kring, from Kring (2023); (b) David Kring.

Toolbox 14.3

Estimating the Energy of an Impact

Imagine that a spherical asteroid with a diameter of 1 km is headed for a head-on impact with Earth at 10 km/s (Figure 14.41). How much energy will the impact release? Let's assume that this is an inelastic collision—the asteroid sticks to the Earth rather than the two objects bouncing off each other like billiard balls. If we assume that the asteroid is chondritic in composition, then a round-number estimate of density is 3000 kg/m^3, and thus its mass is roughly 1.6 × 10^{12} kg. Kinetic energy is $(\frac{1}{2})mv^2$, so that works out to 7.8 × 10^{19} J. Wow, that sounds like a lot! It is, hundreds of times larger than the largest hydrogen bomb ever detonated, but well below Chicxulub in size (see Figure 14.37). But wait a minute . . . if we're on the asteroid, we see Earth approaching at 10 km/s. Earth's mass is about 6 × 10^{24} kg, so that means its kinetic energy is 3 × 10^{32} J. That's a disturbingly greater amount, way, way off the chart! Which is correct?

Neither, although one is much closer than the other. The proper way to approach this problem is by conservation of momentum (*mv*, mass times velocity). If we take Earth as fixed, then its velocity is zero and conservation of momentum is expressed by

$$m_e(0\ \text{km/s}) + m_a(10^4\ \text{m/s}) = (m_e + m_a)v_{e+a} \qquad \textbf{(Eq. 14.2)}$$

where m_e and m_a are the masses of Earth and the asteroid, respectively, and v_{e+a} is the velocity of the stuck-together Earth + asteroid after the collision (relative to the pre-collision Earth). This velocity is thus

$$v_{e+a} = \left(\frac{m_a}{m_e + m_a}\right)v_0 = 2.7 \times 10^{-9}\ \text{m/s} \qquad \textbf{(Eq. 14.3)}$$

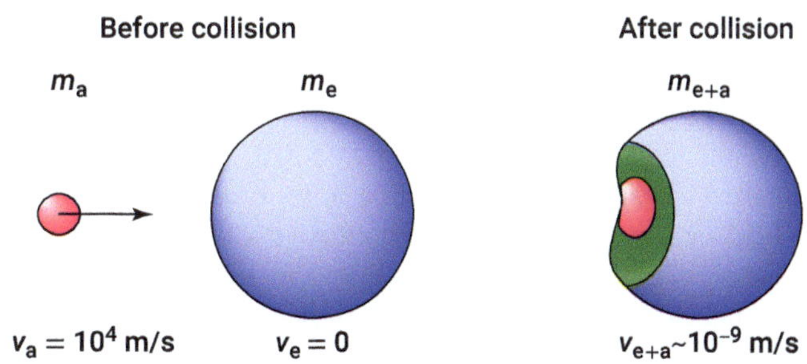

Figure 14.41 Collision of two planetary bodies. Note that for clarity the asteroid is shown vastly larger than it should be; for comparison, the Chicxulub impactor, relative to a standard 12-inch globe, would be about the size of a grain of sand.

This is not very fast, about 10 cm/yr, and equates to a kinetic energy of around 2 × 10^7 J, or about 3×10^{-11} percent of the asteroid's kinetic energy. The kinetic energies before and after the collision are thus

$$E_a = \frac{1}{2}m_a v_0^2 \qquad \textbf{(Eq. 14.4)}$$

and

$$E_{e+a} = \frac{1}{2}(m_e + m_a)v_{e+a}^2 \qquad \textbf{(Eq. 14.5)}$$

continued

Their difference, after a bit of algebra, works out to

$$\Delta E = \frac{1}{2} m_a v_0^2 \left[1 - \left(\frac{m_a}{m_e + m_a} \right) \right] \quad \textbf{(Eq. 14.6)}$$

The last term is just the ratio of the asteroid's mass to that of the Earth + asteroid. This is very close to zero and is why using the (small) impactor's kinetic energy, although technically incorrect, is a good approximation of the energy release. Nearly all the kinetic energy is used to accomplish work—shock metamorphism, melting, kinetic energy of ejected material, etc.

What if we do this calculation from the point of view of the asteroid, where the huge Earth is speeding toward it? In that case, then we can just switch the subscripts "e" and "a" in the equations above, turning Equation 14.6) into

$$\Delta E = \frac{1}{2} m_e v_0^2 \left[1 - \left(\frac{m_e}{m_e + m_a} \right) \right] \quad \textbf{(Eq. 14.7)}$$

Now the term outside the brackets is quite large, but the bracketed term is very close to zero, and ΔE works out to the same value obtained by Equation 14.6.

Summary

- Magmatism far from plate boundaries (intraplate magmatism) is an important part of the Earth's magmatic output.
- Migrating foci of intraplate magmatism are known as hotspots.
- Volcanic rocks lock in the orientation of the Earth's magnetic field when they cool below the Curie point (the temperature above which a mineral cannot retain a magnetic signal) for the relevant mineral, which is typically magnetite. This field gives the geomagnetic paleolatitude.
- Earth's geomagnetic field can be approximated as a dipole (i.e., a bar magnet); the axis of the field is not necessarily coincident with Earth's spin axis, but averages out to the spin axis on timescales of tens of thousands of years.
- Paleomagnetic studies show that many hotspots have not remained stationary relative to one another.
- Hotspot magmatism is typically mafic (e.g., Hawaii and many other Pacific hotspots) or bimodal (Yellowstone), and the mafic rocks may be alkalic.
- The predominant hypothesis for the origin of hotspots (plume hypothesis) is that they arise from plumes of buoyant, hot mantle that arise from the CMB or mantle transition zone.
- An alternative hypothesis (plate hypothesis) is that plumes arise from extension-driven melting of fertile zones in a heterogeneous upper mantle, and that migratory behavior arises from migration of extension.
- Studies of radiogenic isotopes (primarily Sr, Nd, and Pb) demonstrate that the source regions of MORBs are significantly more homogeneous than those of ocean islands.
- Important mantle reservoirs and their typical abbreviations are DMM (depleted MORB mantle), PREMA (prevalent, or dominant, mantle), EM1 (enriched mantle 1, likely enriched by subduction of pelagic sediment), EM2 (enriched mantle 2, likely enriched by subduction of eroded continental material), and HIMU (high $^{238}U/^{204}Pb$; possibly subducted oceanic crust).
- Large igneous provinces are defined by production of extremely large volumes of magma (>100,000 km^3) over timescales of 5 Myr or less. They are dominantly basaltic.
- Layered mafic intrusions are mafic plutons characterized by dramatic banding that is defined by variations in mineral proportions, sizes, and compositions. Some are hosts of important deposits of platinum-group elements, Cr, and other valuable commodities.
- These intrusions have classically been interpreted as the frozen remains of a large tank of crystal-poor magma that developed banding primarily via crystal settling.
- Significant problems with the classical interpretation include (1) the difficulty of emplacing a huge volume of crystal-poor magma in the shallow crust without crystallizing much of it, (2) the difficulty of explaining observed crystallization sequences, (3) cementation of crystals in crystallizing magmas, and (4) fluid-dynamic arguments against developing cross-bedding and other sedimentary features with which banding is commonly compared.
- Numerous other processes can and do produce banding in igneous rocks. These include metamorphic differentiation, partial melting, thermal gradients, and a number of self-organizing geochemical processes, including the Liesegang phenomenon.
- Bolide impacts can produce large volumes of impact melt and a characteristic array of features known as shock metamorphism.

Notes

1. "Traps" comes from a Swedish word meaning "stairs." The step-like nature of repeated basalt flows gives rise to the name.
2. In addition to arctic exploration, Wager was an accomplished mountaineer. He was selected for the 1933 British assault on Mount Everest, the first since the 1924 expedition in which Mallory and Irvine perished near the summit. Wager and his partner made it within about 1000 feet of the summit before turning back. In doing so, they set an altitude record for climbing without supplemental oxygen, a record that stood for 45 years.

CHAPTER 15

What's Next?

15.1 Whence Petrology?

The material in textbooks is commonly presented as a settled body of knowledge that one must learn in order to move on to more advanced topics. But as we have emphasized in the preceding chapters, this is not true for petrology and plate tectonics, even at the undergraduate level. Geology is a young science, petrology a younger branch of it, and plate tectonics younger still. There is much we just do not yet know, and this final chapter summarizes some of the major issues that deserve further work. It is our hope that some of the young minds studying this book might be inspired to solve some of these problems.

15.2 Some Outstanding Problems in Petrology and Plate Tectonics

The questions presented here are grouped loosely and range from whole-Earth issues, such as the driving forces for plate tectonics, to more specific questions such as the origin of ultrahigh-temperature metamorphism.

Plate Tectonics

- We have excellent documentation of how plates have moved relative to one another (plate kinematics), but little knowledge of what drives the plates (plate dynamics). What are the driving forces for plate tectonics?
- When did modern plate tectonics, with subduction and seafloor spreading, begin?
- If there was a time before plate tectonics, how did heat escape from the Earth in that period?
- If there was a time before plate tectonics, what processes formed early continents?
- How and when did Earth differentiate into the geochemical reservoirs (continental crust, oceanic crust, upper mantle, etc.) that we see today?
- Why do the majority of transform faults strike approximately east-west?
- Why do arcs in the western Pacific commonly exhibit back-arc spreading, whereas those along the eastern Pacific are largely associated with back-arc shortening?

The iconic "Earthrise" photo taken by Apollo 8 astronaut Bill Anders on December 24, 1968, on the first expedition in which astronauts circled the Moon. This image brought home the stark contrast between the geologically dead, gray, airless, lifeless Moon and the blue oceans, dynamic atmosphere, and life of the geologically active Earth. In this photo, north is to the right, with Antarctica under dense clouds on the left, the deserts of northwestern Africa visible in the lower right, and South America under clouds at the top. In the book *100 Photographs that Changed the World* (Editors of Life, 2003), mountaineer and wilderness photographer Galen Rowell called this "the most influential environmental photograph ever taken."

Source: NASA.

Subduction

- Is the primary magma in subduction zones basalt or andesite, or is it something else?
- If basalts and andesites are the main material that rises through the Moho in arcs, how did continents become significantly more silicic than that?
- Is delamination of dense mafic material an important process in developing a more silicic continental crust?
- What is the source of the material in arc magmas? For example, in a subduction-related pluton or lava flow, where did the K, Na, Mg, P, etc. come from?
- How does subduction initiate?
- What is the main driver of magma formation in arcs—introduction of H_2O or asthenospheric rise?
- Is H_2O essential to forming arc magmas, or does it just increase the volume of magma produced?
- Why are volcanic arcs typically sharply bounded on the trench side and diffuse on the back-arc side?
- How much surficial H_2O and CO_2 is subducted and then recycled to the atmosphere by magmatism?
- Gas fluxes are commonly estimated from observations at volcanoes. Is there a significant diffuse flux into the atmosphere of CO_2 and other important volatiles from intrusive processes?
- How much surficial H_2O and CO_2 is subducted into the deeper mantle?
- If significant H_2O and CO_2 are deeply subducted, are they ever returned to the surface?
- What physical processes bring high-*P* and ultrahigh-*P* rocks back to the surface?
- Are exhumed subduction-zone rocks representative of the *P*-*T* structure of subduction zones, or are they a biased sample set?
- What role does serpentine play in lubricating subduction interfaces and driving melting?
- How fast are metamorphic rocks subducted and exhumed?
- Are subducted rocks exhumed in a single stage, or does it take multiple stages?

Oceanic Ridges and Ophiolites

- How do ophiolites form?
- Are ophiolites random samples of oceanic crust, or are they preserved because they are specific to a particular tectonic setting?
- How do melts form under oceanic ridges and collect to erupt at the ridges?
- How widespread are ecosystems powered by serpentinization reactions?

Hotspots and Other Intraplate Processes

- What drives hotspot magmatism?
- What are plumes, and are there any that originate from the core–mantle boundary?
- Can the plume vs. plate models be reconciled, or can we rule out one or the other?
- What caused the kinks in the Hawaiian-Emperor and other Pacific hotspot tracks?

Collisions

- Why are some ancient collision belts (e.g., Appalachians, Urals) still topographically high?
- How fast do rocks cycle through the orogenic cycle of burial, heating, metamorphism, cooling, and exhumation?

- How do geotherms evolve in collision zones, and how well do simple thermal models capture this behavior?
- How do collisions and volcanism influence Earth's climate (e.g., through the weathering cycle, CO_2 input, etc.)?
- How much does incipient partial melting weaken rocks, and what role does partially molten rock play in the geodynamics of collision zones?
- How does the transition from subduction to collision occur?
- How and why does the transition from collision to extension occur?

Magmatism

- Classical magma chambers have proven hard to find. Are typical volcanic eruptions (VEI < 7) sourced from such magma chambers or from dike swarms?
- Where in the crust are magma chambers located?
- How good is our ability to geophysically detect eruptible volumes of magma in the crust?
- What is the life cycle of a silicic supereruption, and can we tell if any are imminent now?
- What are the environmental effects of silicic supereruptions, and can any be tied to mass extinctions?
- What are the principal mechanisms of heat loss from intruded magma?
- Can climate change affect volcanism on 100- or 1000-year timescales?
- What is the global tsunami risk of sector collapse of oceanic volcanoes (e.g., Hawai'i, La Palma)? Do these generate megatsunamis, or do the landslides form as a series of smaller events?
- Why are mantle xenoliths common in alkali basalts but almost unknown in tholeiitic basalts?
- How representative are mantle xenoliths of average upper mantle?
- Were any large plutons emplaced in the traditional manner as large magma blobs, or are all such plutons emplaced incrementally?
- How widespread is downward movement of plutons during and after emplacement (as, e.g., by magma loading and isostatic sinking)?
- By what mechanisms and pathways do small-degree partial melts aggregate to form dikes?
- Are diapirs viable magmatic ascent mechanisms in the crust? In the upper mantle?
- Where and when did the crystals in a given igneous rock nucleate?
- What happens, texturally and chemically, during the last 50 percent of crystallization of a plutonic rock?
- Regarding plutons, to what extent do magmas differentiate at the observed level of emplacement?
- What are the links between migmatites and silicic magmatism?
- Are high-silica magmas (e.g., granites, rhyolites) primarily formed by fractional crystallization or by partial melting?
- Volcanic rocks with negative Eu anomalies are common. Where are the source rocks with positive Eu anomalies that should have been left behind?
- What is the significance of mafic magmatic enclaves in plutons?
- What forms the banding in layered mafic intrusions and other banded plutons?
- Do layered mafic intrusions form all at once or in increments, as many more silicic plutons do?
- Are layered mafic intrusions widespread in the crust?
- What is the significance of nested plutonic suites in the Sierra Nevada, western Nevada, and elsewhere? Are they the roots of former arc volcanoes?

Metamorphism

- How fast do crystals grow during metamorphism?
- How closely do metamorphic crystals approach chemical equilibrium?
- To what degree do metamorphic crystals retain their original isotopic and chemical compositions?
- What drives ultrahigh-T metamorphism—unusual radiogenic heating, extension, or something else?
- How much does differential stress bias estimates of metamorphic pressures?
- How do new methods for estimating P and T of metamorphism, such as Raman spectroscopy, compare with traditional thermodynamic methods?
- Are fluids necessary for whole-rock metamorphic equilibrium to occur? What happens during fluid-absent metamorphism?
- To what degree and over what length scales do neighboring rocks communicate chemically during metamorphism?
- How effectively and rapidly do fluids move geochemically important solutes?
- Over what timescales do metamorphic fluids flow?

Tectonics

- The recent (past 30 Myr) plate tectonic history of western North America seems fairly well known, and yet space-time patterns of magmatism fail to match what should have happened. Why?
- What drives large intraplate earthquakes?
- The Andes seem like a good analog for western North America in the early Cenozoic, and yet magmatism has occurred 1000 km farther inland in North America (after removing extension) than in the Andes. Why?
- Is the Indo-Asian collision a good analog for past continent-continent collisions? Why or why not? Are there better analogs?

Miscellaneous

- How much of Earth's water is stored in the mantle?
- Is there a way to classify igneous and metamorphic rocks that provides more insight into their importance and how they form?
- How are ore deposits linked to magmatism and tectonics?

15.3 Epilogue

Petrology has been a "hard" science, based in chemistry and physics more than description, for about a century, and modern concepts of plate tectonics have been around for about half that time. In this book, we have tried to illustrate how these two branches of geology combine into a coherent, albeit evolving, view of how the outer part of the solid Earth works. Perhaps more importantly, we have tried to show how the solid Earth interacts with other critical Earth systems, including the climate, oceans, atmosphere, hydrosphere, cryosphere, and biosphere; how petrologic processes can bring about catastrophes and mass extinctions but also keep the Earth in a habitable state; and how petrology and plate tectonics provide us with many of life's requirements and life's joys, including energy, metals, building materials, mountains, scenery, topography, coastlines, and beautiful tropical islands.

We hope that this book challenges you to make your own discoveries about petrology and plate tectonics.

APPENDIX 1

Selected Mineral Abbreviations

This is a subset of the abbreviations in Whitney and Evans (2010). An asterisk (*) indicates a group name.

Ab	Albite	En	Enstatite
Act	Actinolite	Ep	Epidote
Adr	Andradite	Fa	Fayalite
Afs	Alkali feldspar	Fo	Forsterite
Alm	Almandine	Fs	Ferrosilite
Als	Aluminosilicate*	Fsp	Feldspar*
Amp	Amphibole*	Gln	Glaucophane
An	Anorthite	Gr	Graphite
And	Andalusite	Grs	Grossular
Ann	Annite	Grt	Garnet*
Ano	Anorthoclase	Hbl	Hornblende
Ap	Apatite	Hd	Hedenbergite
Arg	Aragonite	Hem	Hematite
Atg	Antigorite	Hyp	Hypersthene
Aug	Augite	Ilm	Ilmenite
Brc	Brucite	Jd	Jadeite
Bt	Biotite*	Kfs	K-feldspar
Cal	Calcite	Kln	Kaolinite
Cam	Clinoamphibole*	Kls	Kalsilite
Chr	Chromite	Ky	Kyanite
Cld	Chloritoid	Lct	Leucite
Coe	Coesite	Liq	Liquid
Cpx	Clinopyroxene*	Lws	Lawsonite
Crd	Cordierite	Mag	Magnetite
Crn	Corundum	Mc	Microcline
Crs	Cristobalite	Mgs	Magnesite
Ctl	Chrysotile	Mll	Melilite
Czo	Clinozoisite	Mnt	Montmorillonite
Di	Diopside	Mnz	Monazite
Dia	Diamond	Ms	Muscovite
Dol	Dolomite	Nph	Nepheline

Ol	Olivine*
Omp	Omphacite
Opl	Opal
Opq	Opaque mineral
Or	Orthoclase
Per	Periclase
Pg	Paragonite
Pgt	Pigeonite
Ph	Phengite
Phl	Phlogopite
Pl	Plagioclase*
Prh	Prehnite
Prl	Pyrophyllite
Prp	Pyrope
Prv	Perovskite
Py	Pyrite
Qz	Quartz
Rt	Rutile
Sa	Sanidine
Sch	Scheelite
Ser	Sericite
Sil	Sillimanite
Sme	Smectite
Spd	Spodumene
Spl	Spinel
Sps	Spessartine
Srp	Serpentine*
St	Staurolite
Sti	Stishovite
Thr	Thorite
Tlc	Talc
Tur	Tourmaline
Tr	Tremolite
Trd	Tridymite
Tro	Troilite
Tpz	Topaz
Ttn	Titanite (sphene)
Usp	Ulvöpinel
Wo	Wollastonite
Wus	Wüstite
Zeo	Zeolite*
Zrn	Zircon

APPENDIX 2

Review of Rock-Forming Minerals

A2.1 Chemical Exchanges and Solid Solutions

The following review covers the major minerals relevant to most igneous and metamorphic rocks except for SiO_2 polymorphs, which are described in Section 3.4 of the main text. To understand the compositional behavior of minerals, it is helpful to commit to memory some patterns of solid solution behavior related to chemical substitutions in crystals (Table A2.1).

As you probably know, there is complete solid solution between fayalite (Fe_2SiO_4) and forsterite (Mg_2SiO_4) end members in olivine, representing exchange between Fe^{2+} and Mg^{2+} in the crystal. With few exceptions, mineral end members linked by this particular exchange, Fe-Mg, exhibit complete solid solution behavior. So, for example, there is complete solid solution between Fe and Mg end members in the micas (annite-phlogopite), amphiboles (ferroactinolite-tremolite), pyroxenes (hedenbergite-diopside), garnets (almandine-pyrope), etc. Other exchanges show varying degrees of solid solution compatibility. Among the alkali cations, Na and Ca show substantial solid solution behavior (often complete, or nearly so, at higher temperatures), although substitution must be charge-balanced with another cation—for example, $Na^{1+}Si^{4+}$ for $Ca^{2+}Al^{3+}$ or $Na^{1+}Al^{3+}$ for $Ca^{2+}Mg^{2+}$. Such minerals generally show extensive solid solution at higher temperatures, but more limited solid solution at lower temperatures. In contrast, Ca^{2+} and K^{1+} never mix well.

Such relations are especially evident in the feldspars and micas. At low temperatures, Na- and K-rich compositions separate along a solvus (also known as a miscibility gap) into distinctly different minerals (e.g., albite vs. K-feldspar, paragonite vs. muscovite). The other major exchange observed in many mineral groups is Ca^{2+} for either Mg^{2+} or Fe^{2+}. Generally, Ca^{2+} and Mg^{2+} do not mix well (limited solid solution with a large solvus), whereas Ca^{2+} and Fe^{2+} mix a bit better, but still not well (limited solid solution with a smaller solvus). Compositions of coexisting minerals from the same groups but different compositions are sometimes used to estimate temperatures in rocks, a concept referred to as solvus thermometry. Tschermak exchange [AlAl for (Mg,Fe)Si] leads to increased chemical diversity in mica, chlorite, pyroxene, and amphibole.

Last, learning about compositional exchanges can help you learn mineral compositions, too. Your authors don't actually memorize the compositions of all the different varieties of pyroxenes, amphiboles, feldspars, etc.; they learn one and derive the others by knowing how they are linked via compositional exchange. For example, annite [$KFe_3AlSi_3O_{10}(OH)_2$] and phlogopite [$KMg_3AlSi_3O_{10}(OH)_2$] are linked through three Fe-Mg exchanges. It's easier to learn one end member (say, phlogopite) and derive the other (subtract 3 Mg and add 3 Fe to get annite) than it is to try to memorize both annite and phlogopite.

Table A2.1 Common Chemical Exchanges in Minerals and the Degree of Solid Solution

Exchange	Occurs In	Degree of Solid Solution
Fe-Mg	Diverse silicates, carbonates, oxides	Complete
Ca-Mg	Pyroxenes, amphiboles, carbonates	Highly limited, usually a large miscibility gap
Ca-Fe	Pyroxenes, amphiboles, carbonates	Highly limited but smaller miscibility gap than for Ca-Mg
NaSi-CaAl, NaAl-Ca(Mg,Fe)	Plagioclase, pyroxenes, amphiboles	Complete at high *T*, can show miscibility gaps at low *T*
Na-K	Alkali feldspars, micas, amphiboles	Complete at high *T*, miscibility gaps at low *T*
KSi-CaAl	Feldspars, micas	Almost none, large miscibility gaps
Ca-Fe-Mg-Mn	Garnets	Complete (different crystallographic site than for other silicates)

A2.2 Feldspars

Three main end members:

- K-feldspar (microcline, orthoclase, sanidine, adularia) = $KAlSi_3O_8$
- Albite = $NaAlSi_3O_8$
- Anorthite = $CaAl_2Si_2O_8$

Feldspars are the most abundant mineral in Earth's crust and are found in a wide range of mafic through felsic igneous rocks, their metamorphic equivalents, and diverse clastic metasedimentary rocks (metasandstones, metashales, etc.). Feldspars are divided into two main classes based on their solid solution behavior. The alkali feldspars form a solid solution between the end-members albite and K-feldspar, whereas the plagioclase feldspars form a solid solution between the end-members albite and anorthite (Figure A2.1a). Following principles for alkali cation substitutions, Ca and K end members hardly

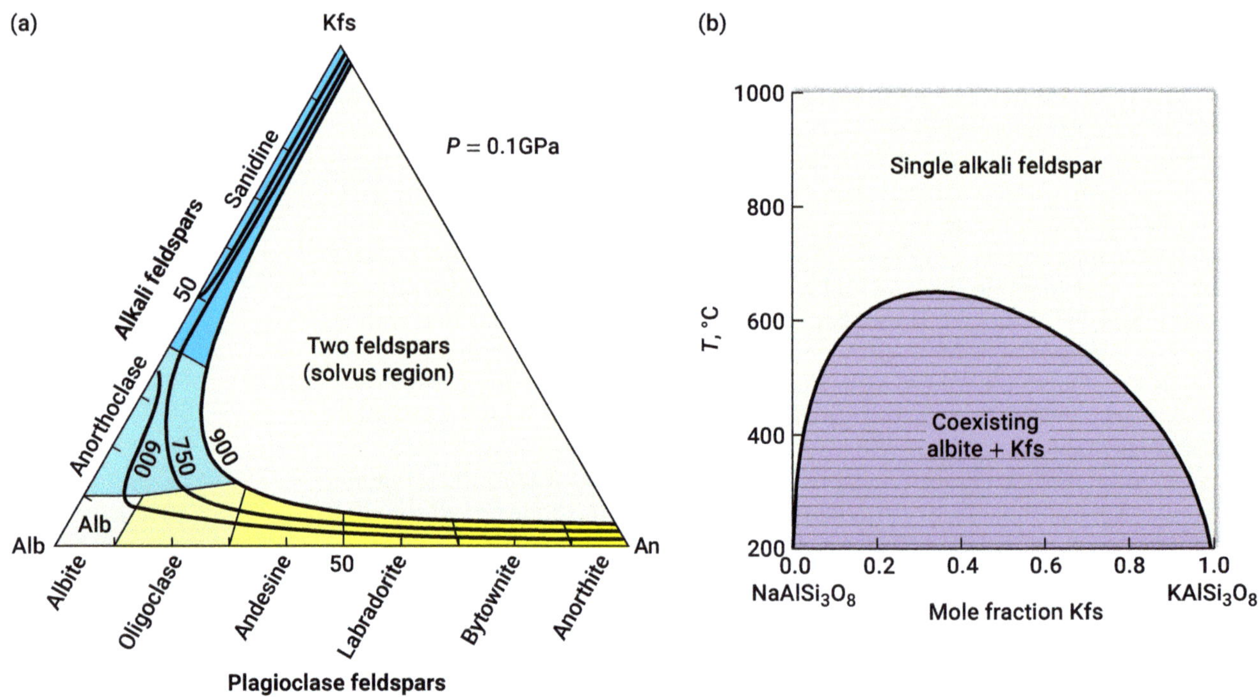

Figure A2.1 Compositional variation in feldspars. (a) Ternary composition diagram showing boundaries on names. Curves give degree of ternary solid solution at different temperatures (°C). (b) Solvus in the alkali feldspar system.

Source: Modified from Spear (1995).

mix at all, Na and K end members mix moderately well, and Na and Ca end members mix the best. Thus, at all temperatures, anorthite and K-feldspar show a pronounced solvus, although at very high temperature (at ~900 °C and higher), a ternary feldspar with substantial concentrations of all three components can form (anorthoclase; Figure A2.1a; "Blue Pearl" in Figure 10.32). At moderate temperatures (>~700 °C), albite and K-feldspar show complete solid solution, but at lower temperatures, a pronounced solvus forms between Na-rich and K-rich feldspars (Figure A2.1b). Formation of perthite (Na-rich feldspar lamellae in K-rich feldspar host) and antiperthite (K-rich feldspar lamellae in Na-rich feldspar host) reflects exsolution (separation of one phase from another) along this solvus. High-resolution electron imaging of plagioclase feldspars also shows solvi at subsolidus temperatures, but relations are complicated, and macroscopically, a plagioclase crystal may appear compositionally homogeneous. The play of colors in labradorite (a variety of moderately calcic plagioclase) reflects exsolution at a submicron scale.

A2.3 **White Micas**

Three main end members:

- Muscovite = $KAl_2AlSi_3O_{10}(OH)_2$ (super common)
- Paragonite = $NaAl_2AlSi_3O_{10}(OH)_2$ (rare)
- Margarite = $CaAl_2Al_2Si_2O_{10}(OH)_2$ (super rare)

White micas, especially muscovite, are common in felsic igneous and metamorphic rocks. Like the feldspars, end members are distinguished by different alkali cations (Na, K, and Ca) and generally follow the same chemical behavior: Na and Ca mix the best, Na and K next, and Ca and K the least (Figure A2.2). However, unlike the feldspars, white micas are not stable at high temperature, so they are always separated by solvi. Sometimes talc $[(Fe,Mg)_3Si_4O_{10}(OH)_2]$ is called a white mica, because it is structurally similar and appears nearly colorless in hand sample. Technically, however, it's not a mica, because it lacks an alkali cation. Also, normally, muscovite has a bit of extra Mg and Fe substituting for Al (reverse Tschermak exchange), so a typical composition might be more like $K(Fe,Mg)_{0.2}Al_{1.8}Al_{0.8}Si_{3.2}O_{10}(OH)_2$.

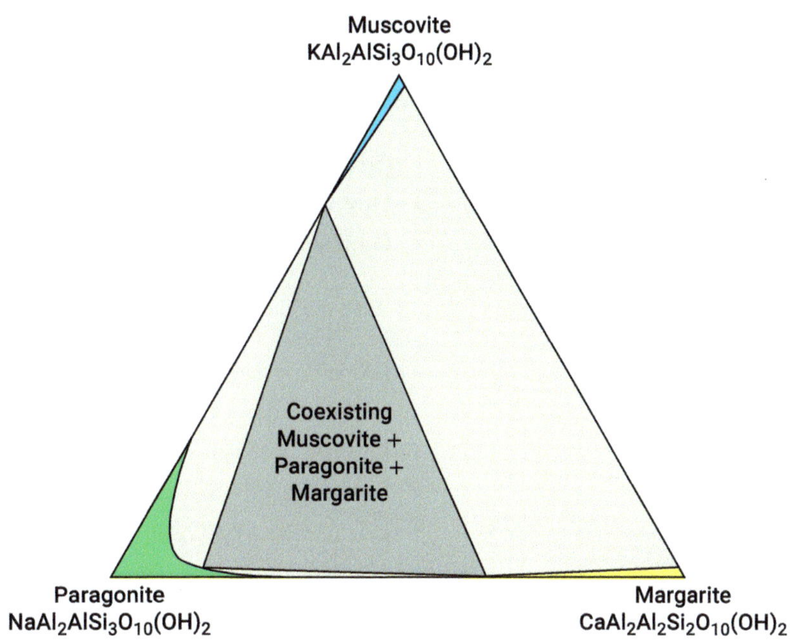

Figure A2.2 Compositional variation in white micas muscovite, paragonite, and margarite. The fields for muscovite and margarite are exaggerated for illustration as the solubility of K in margarite and Ca in muscovite is extremely low.

Source: Based on MJK unpublished data.

A2.4 **Olivine**

Two main end members:

- Forsterite = Mg_2SiO_4
- Fayalite = Fe_2SiO_4

Olivine is the most abundant mineral of the upper mantle, and its chemistry strongly controls the compositions of basalts. It is also found in calc-silicate rocks. Compositionally, olivine is nearly entirely described by a solid solution between the forsterite and fayalite end members, although most mantle rocks are highly Mg-enriched (~90% forsterite component). Tiny amounts of Ca, Mn, and Ni can substitute for Fe and Mg and are sometimes used as petrogenetic tracers.

A2.5 **Biopyriboles (Biotite, Pyroxene, Amphibole)**

Biotite, pyroxenes, and amphiboles (and talc) bear structural and chemical similarities, and so are sometimes lumped together into a single group: the "biopyriboles" (get it? BIOtite, PYRoxene, and amphIBOLE). These distinct minerals also often appear similar in hand sample (dark little specks), and so can be easily mistaken for each other. If you're unsure what that little black mineral is, cover your bases, and call it a biopyribole! Biotite, pyroxenes, and amphiboles are extremely common across a wide range of igneous and metamorphic rock compositions and *P-T* conditions.

Biotite and talc

Here we list biotite and talc together because of their structural similarities, but they are separate minerals that can coexist:

- Biotite = $K(Fe,Mg)_3AlSi_3O_{10}(OH)_2$
- Talc = $(Fe,Mg)_3Si_4O_{10}(OH)_2$

Ther are two main end members for biotite:

- Phlogopite = $KMg_3AlSi_3O_{10}(OH)_2$
- Annite = $KFe_3AlSi_3O_{10}(OH)_2$

Biotite and talc are a bit like olivine, in that Fe-Mg exchange largely dominates their chemistry. Normally, however, biotite has some extra Al, so a typical composition might be more like $K(Fe,Mg)_{2.5}Al_{0.5}Al_{1.5}Si_{2.5}O_{10}(OH)_2$. Talc often forms in Mg-rich rocks (e.g., altered ultramafic rocks), so it is commonly highly enriched in the Mg-endmember.

Pyroxenes

The three most common pyroxenes are calcic clinopyroxene (that's nearly always what we mean when we say or write "Cpx"), orthopyroxene, and sodic clinopyroxene (omphacite):

- Cpx = $Ca(Fe,Mg)Si_2O_6$
- Opx = $(Fe,Mg)_2Si_2O_6$
- Omphacite (solid solution between Cpx and jadeite) = $(Na,Ca)(Fe,Mg,Al)Si_2O_6$

Five main end members:

- Diopside = $CaMgSi_2O_6$
- Hedenbergite = $CaFeSi_2O_6$
- Enstatite = $Mg_2Si_2O_6$
- Ferrosilite = $Fe_2Si_2O_6$
- Jadeite = $NaAlSi_2O_6$

Pyroxenes comprise orthopyroxene (Fe- and Mg-rich), calcic and sodic clinopyroxenes (Ca- and Na-rich), and a compositionally intermediate, high-temperature clinopyroxene called pigeonite. When petrologists say "clinopyroxene" (or "Cpx"), they nearly always mean a calcic clinopyroxene, because it is more common than sodic pyroxene or pigeonite. Recalling that Ca doesn't mix well with Fe and Mg, it's unsurprising that there

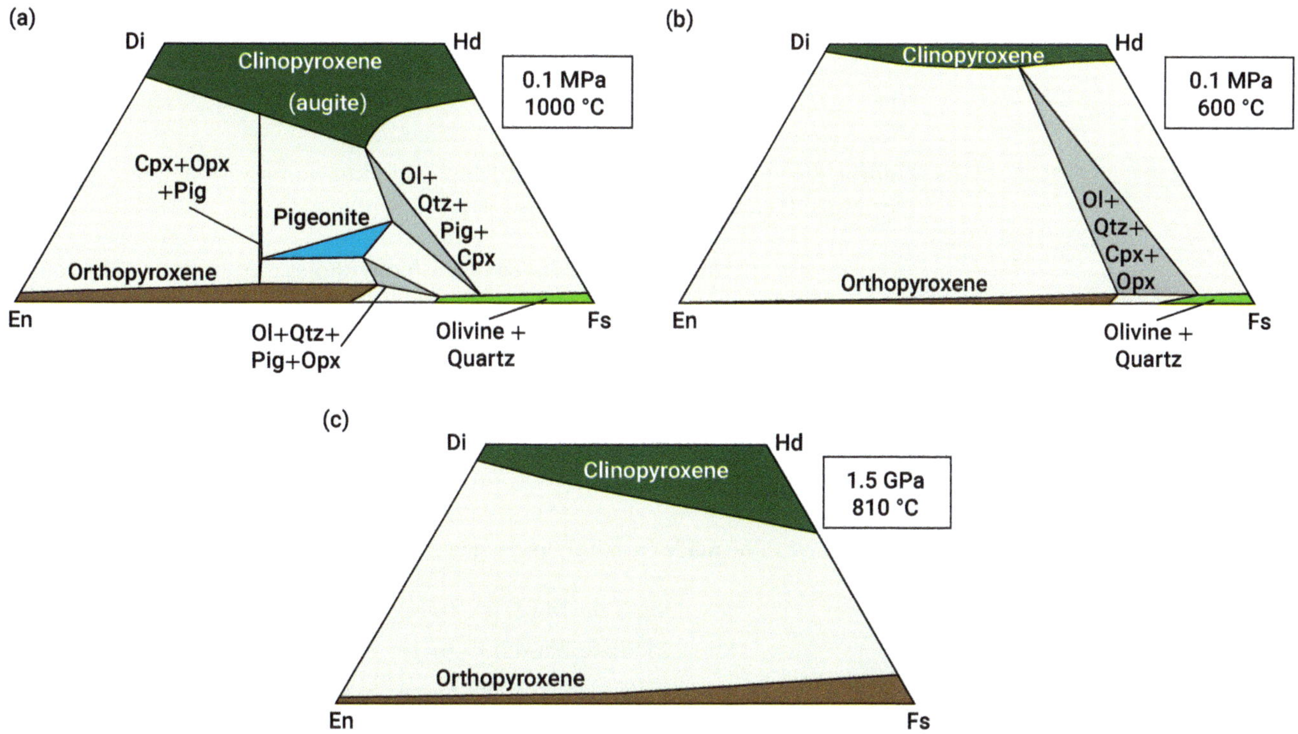

Figure A2.3 Compositional variation in pyroxenes expressed on the pyroxene quadrilateral (composition field defined by end-members enstatite, ferrosilite, diopside, and hedenbergite). (a) Pyroxene relationships at low *P* and high *T*. Pigeonite is stable, but ferrosilite is unstable with respect to olivine (fayalite) plus quartz. (b) Pyroxene relationships at low *P* and low *T*. Pigeonite is no longer stable. (c) Pyroxene relationships at high *P* and intermediate *T*. Neither pigeonite nor olivine are stable. Note the larger miscibility gap (separation of compositions) for Mg-rich pyroxenes (diopside and enstatite) compared to Fe-rich pyroxenes (hedenbergite and ferrosilite).

Source: Based on Spear (1995).

are miscibility gaps among calcic clinopyroxene, pigeonite, and orthopyroxene (Figure A2.3). Consequently, many rocks contain two pyroxenes (e.g., the average upper-mantle peridotite, granulite-facies rocks, etc.). Pigeonite is stable only at high temperature and is therefore typically found in volcanic rocks (Figure A2.3a), whereas sodic clinopyroxene is restricted either to unusually Na-rich igneous rocks or to high-*P* metamorphic rocks. Omphacite is a solid solution between calcic clinopyroxene and jadeite and is an important indicator of eclogite-facies metamorphism. Calcic clinopyroxene that is a bit deficient in Ca, augite, is common in volcanic rocks.

Amphiboles

There are numerous amphiboles with numerous end members, such as:

- Tremolite = $Ca_2Mg_5Si_8O_{22}(OH)_2$
- Ferroactinolite = $Ca_2Fe_5Si_8O_{22}(OH)_2$
- Anthophyllite (orthoamphibole) and cummingtonite (clinoamphibole) = $Mg_7Si_8O_{22}(OH)_2$
- Ferroanthophyllite (orthoamphibole) and grunerite (clinoamphibole) = $Fe_7Si_8O_{22}(OH)_2$
- Glaucophane = $Na_2Mg_3Al_2Si_8O_{22}(OH)_2$
- Ferroglaucophane = $Na_2Fe_3Al_2Si_8O_{22}(OH)_2$

Two amphiboles have more intermediate compositions:

- Hornblende = a wide range, but a typical composition might be $[Na_{0.25}][Na_{0.2}Ca_{1.8}][(Fe,Mg)_{3.8}Fe^{3+}{}_{0.4}Al_{0.8}][Al_{1.25}Si_{6.75}]O_{22}(OH)_2$ where square brackets separate different sites. Clearly, it's messy.
- Gedrite (orthoamphibole) = a range, but a typical composition might be $[Na_{0.5}][(Fe,Mg)_2][(Fe,Mg)_4Al][Al_{1.5}Si_{6.5}]O_{22}(OH)_2$.

Amphiboles are the most complicated of the biopyriboles. Structurally, they combine either talc or a mica plus two pyroxenes, and so inherit the chemical complexities of both. For example, here's how to make tremolite from talc and diopside:

$$\begin{aligned}&1\,\text{Mg-Talc: } Mg_3Si_4O_{10}(OH)_2\\&+2\,\text{Diopside: } CaMgSi_2O_6(= Ca_2Mg_2Si_4O_{12})\\&=1\,\text{Tremolite: } Ca_2Mg_5Si_8O_{22}(OH)_2\end{aligned}$$

Glaucophane can be made from talc and jadeite:

$$\begin{aligned}&1\,\text{Mg-Talc: } Mg_3Si_4O_{10}(OH)_2\\&+2\,\text{Jadeite: } NaAlSi_2O_6(= Na_2Al_2Si_4O_{12})\\&=1\,\text{Glaucophane: } Na_2Mg_3Al_2Si_8O_{22}(OH)_2\end{aligned}$$

And anthophyllite (an amphibole in the orthorhombic crystal system) can be made from talc and enstatite:

$$\begin{aligned}&1\,\text{Mg-Talc: } Mg_3Si_4O_{10}(OH)_2\\&+2\text{Enstatite: } Mg_2Si_2O_6(= Mg_4Si_4O_{12})\\&=1\,\text{Anthophyllite: } Mg_7Si_8O_{22}(OH)_2\end{aligned}$$

Like biotite, talc, pyroxenes, and olivine, amphiboles exhibit complete solid solution between Fe and Mg end members. However, also like the clino- and orthopyroxenes, they exhibit miscibility gaps between calcic and non-calcic amphiboles (Figure A2.4a). Another miscibility gap exists between Al+Na-rich amphiboles and Al+Na-poor amphiboles: hornblende vs. actinolite (two calcic clinoamphiboles), and gedrite vs. anthophyllite (two orthoamphiboles; Figure A2.4b). Two other amphiboles—cummingtonite and anthophyllite—have nearly identical compositions but different structures. Cummingtonite

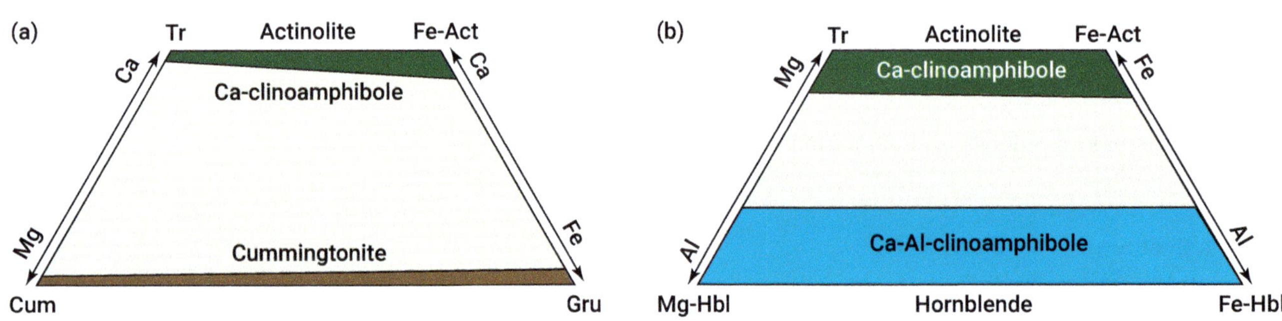

Figure A2.4 Schematic diagrams illustrating compositional variation among a small subset of amphiboles. (a) Calcic and non-calcic amphiboles exhibit a miscibility gap, similar to calcic and non-calcic pyroxenes. (b) The calcic clinoamphiboles may also have a miscibility gap between hornblende and actinolite.

Source: Based on Spear (1995).

is a clinoamphibole, whereas anthophyllite is an orthoamphibole. This would be a little like having enstatite vs. clinoenstatite (which does exist but is very rare) coexisting in a rock. In metamorphosed, hydrothermally altered volcanic rocks, it's possible to have four different stable amphiboles (the clinoamphiboles hornblende and cummingtonite, plus the orthoamphiboles anthophyllite and gedrite).

Like sodic clinopyroxenes, sodic amphiboles form either in highly Na-rich igneous rocks or in high pressure rocks. Blueschists derive their name from blue glaucophane. A miscibility gap probably exists between glaucophane and actinolite.

A2.6 Garnets

Five main end members:

- Pyrope = $Mg_3Al_2Si_3O_{12}$
- Almandine = $Fe_3Al_2Si_3O_{12}$
- Spessartine = $Mn_3Al_2Si_3O_{12}$
- Grossular = $Ca_3Al_2Si_3O_{12}$
- Andradite = $Ca_3Fe^{3+}{}_2Si_3O_{12}$

Garnets are common in deeper rocks of the upper mantle (garnet peridotites) and in diverse metamorphic rocks of the mid to lower crust. Structurally, garnets are bit different from other Fe-Mg minerals, because most cation mixing occurs on a much larger site (a cubic site) than in most silicates (an octahedral site). Consequently, garnets not only greatly prefer Fe over Mg (because Fe has a larger ionic radius), they also take substantial Ca and Mn. Thus, typical garnets have compositions expressed as $(Mg,Fe,Mn,Ca)_3Al_2Si_3O_{12}$. In contact aureoles, skarns, and some metacarbonates, ferric iron can substitute for Al, giving rise to andradite garnets, which can include substantial grossular component.

A2.7 Chlorite

Four main end members:

- Clinochlore = $Mg_5Al_2Si_3O_{10}(OH)_8$
- Chamosite = $Fe_5Al_2Si_3O_{10}(OH)_8$
- Mg-Al chlorite = $Mg_{4.5}Al_3Si_{2.5}O_{10}(OH)_8$
- Fe-Al chlorite = $Fe_{4.5}Al_3Si_{2.5}O_{10}(OH)_8$

Chlorite is a common, low-grade metamorphic mineral across a wide range of bulk compositions. It exhibits complete solid solution between Fe and Mg end members, although most chlorite is intermediate in Fe/Mg. Chlorite is commonly referenced to clinochlore and chamosite end members, but most exhibit higher Al contents. For example, a composition of $(Fe,Mg)_{4.5}Al_3Si_{2.5}O_{10}(OH)_8$ is typical of metamorphosed mudstones and felsic metaigneous rocks.

A2.8 Serpentine

Three polymorphs and two main compositional end members:

- Antigorite, chrysotile, lizardite = $Mg_3Si_2O_5(OH)_4$
- Fe-Antigorite, -chrysotile, -lizardite = $Fe_3Si_2O_5(OH)_4$

Serpentine is common in hydrothermally altered mafic and ultramafic rocks, and antigorite arguably controls the frictional strength and temperature structure of subduction zones. Although serpentine shows complete solid solution between Mg and Fe end members, it usually forms in Mg-rich rocks, and so is a nearly pure Mg end member. Chrysotile and lizardite are restricted to low temperatures (<~350 °C), whereas antigorite is stable at higher temperatures.

A2.9 Diagnostic Metamorphic Minerals

Four minerals:

- Cordierite = $(Fe,Mg)_2Al_4Si_5O_{18}{\cdot}H_2O$
- Staurolite = $(Mg,Fe)_2Al_9Si_4O_{23}(OH)$
- Chloritoid = $(Fe,Mg)_2Al_4Si_2O_{10}(OH)_4$
- Lawsonite = $CaAl_2Si_2O_7(OH)_2{\cdot}H_2O$

These minerals are not especially common, but the first three are important index minerals in metamorphosed aluminous rocks, whereas lawsonite is diagnostic of the blueschist facies and forms only under high-*P*, low-*T* conditions. Chloritoid, staurolite, and cordierite show complete solid solution between Fe and Mg end members, although chloritoid and staurolite are commonly quite Fe-rich (similar to garnet), while cordierite is commonly Mg-rich.

A2.10 Aluminosilicates

Three polymorphs:

- Kyanite, andalusite, sillimanite = Al_2SiO_5

Aluminosilicates are common in moderate- to high-grade metashales. We use the stable aluminosilicate in metamorphic rocks to help delineate *P-T* conditions. Compositionally, they are all commonly nearly pure.

A2.11 Epidote

Four main end members:

- Zoisite = $Ca_2Al_3Si_3O_{12}(OH)$
- Clinozoisite = $Ca_2Al_3Si_3O_{12}(OH)$
- Pistacite = $Ca_2(Fe^{3+},Al)_3Si_3O_{12}(OH)$
- Allanite = $(REE,Ca)_2(Fe^{3+},Fe^{2+},Al)_3Si_3O_{12}(OH)$, where REE = rare earth elements

Epidote group minerals form across a wide range of relatively calcic igneous and metamorphic rocks (e.g., calc-silicates, metabasalts, and certain high-pressure plutons). Epidote is also a common alteration product of plagioclase and forms along fractures in igneous rocks. Most epidote is a solid solution between clinozoisite and pistacite end members with a typical composition of about $Ca_2Fe^{3+}{}_{0.5}Al_{2.5}Si_3O_{12}(OH)$. Allanite is stabilized primarily by REE, but it also takes substantial Th and U and can be used as a geochronometer. Zoisite has virtually the same composition as clinozoisite but belongs to the orthorhombic crystal system. Zoisite is moderately common in high-pressure blueschists and eclogite.

A2.12 **Oxides**

Four oxides are particularly common:

- Rutile = TiO_2
- Ilmenite = $FeTiO_3$
- Magnetite = Fe_3O_4
- Spinel = $(Fe,Mg)(Al,Cr)_2O_4$

Rarer oxides include:

- Hematite = Fe_2O_3
- Corundum = Al_2O_3

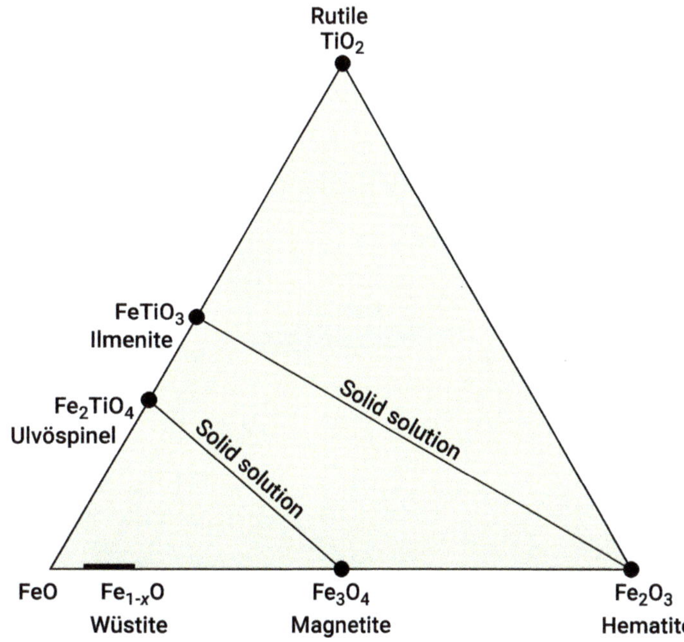

Figure A2.5 Compositional triangle showing Fe-Ti oxide phases and solid solution bands.

Most rocks contain oxides in addition to silicates. Ilmenite and rutile are mainly stabilized by Ti (see Figure A2.5). Magnetite tends to form in moderately oxidized rocks (because most Fe is Fe^{3+}), whereas hematite forms only in highly oxidized rocks (because all Fe is Fe^{3+}), and corundum only in highly aluminous rocks (e.g., metabauxites or highly altered rocks). Spinel is a common mineral in the upper mantle and is stabilized by Al, although it also can contain substantial Cr^{3+} (to form chrome-spinel). The Zr content of rutile is a commonly applied thermometer in high-pressure rocks, and rutile can sometimes be used as a U-Pb geochronometer.

A2.13 **Carbonates**

Two common end members:

- Calcite = $CaCO_3$
- Dolomite = $CaMg(CO_3)_2$

Two compositionally intermediate carbonates:

- Ankerite = $Ca(Mg,Fe)(CO_3)_2$
- Magnesite-siderite = $(Mg,Fe)CO_3$

Carbonates are common in marbles, calc-silicates, and many metaigneous rocks (because of seafloor alteration) and form carbonatites. Compositionally, carbonates follow systematics comparable to pyroxenes and amphiboles—(nearly) complete solid solution between Fe and Mg end members (ankerite is an exception) with miscibility gaps between calcite and dolomite, and between dolomite and magnesite-siderite solid solutions (Figure A2.6a). Compositions of coexisting calcite and dolomite have been used to estimate metamorphic temperatures (Figure 11.21; Figure A2.6b).

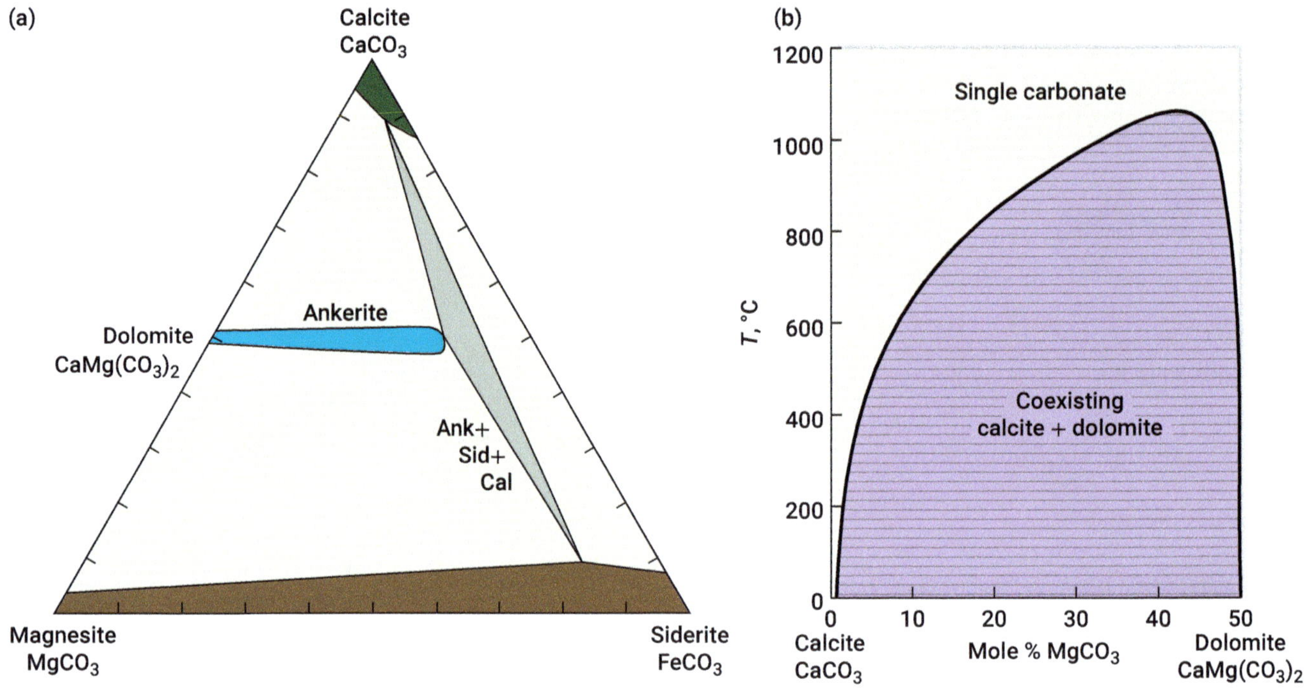

Figure A2.6 (a) Compositional variation among carbonates at 550 °C showing pronounced solvi. (b) *T-X* phase diagram showing solvus between calcite and dolomite.

Source: Based on Spear (1995).

A2.14 **Accessories**

Nearly limitless possibilities, but here are some common examples:

- Zircon = $ZrSiO_4$
- Apatite = $Ca_5(PO_4)_3(OH, F)$
- Tourmaline = $Na(Fe,Mg)_3Al_6(Si_6O_{18})(BO_3)_3(OH)_4$
- Monazite = $(REE)PO_4$
- Pyrite = FeS_2
- Titanite = $CaTiSiO_5$

Accessory minerals are ubiquitous in igneous and metamorphic rocks, although they are rarely abundant. Rather, they are stabilized by the trace elements Zr (zircon), P (apatite), REE (monazite), B (tourmaline), and S (a sulfide, often pyrite). Titanite requires both high Ti and high Ca. Zircon and monazite are particularly important, because they take up substantial U and Th, but very little Pb, so they make good geochronometers. Titanite can be used as a geochronometer (it also takes up U), although correction for common Pb is important. The Zr content of titanite and the Ti content of zircon are used as thermometers. Many other accessory minerals are possible—it all depends on what other elements are enriched in a rock.

APPENDIX 3

Description and Classification of Igneous and Metamorphic Rocks

A3.1 Igneous Rocks

The volume edited by LeMaitre (2002) is an exceedingly detailed look at classification of igneous rocks and provides a thorough glossary of rock names.

Describing igneous rocks requires one to interpret observations in some kind of intellectual framework. These interpretations can range from solid ("this is quartz") to speculative and model-driven ("this rock is a cumulate" or "this is igneous cross-bedding"). It is critical to avoid using terminology that forces a particular origin on a rock or feature when that origin is in doubt.

Terms Describing Igneous Rocks

Terms Describing Field Occurrence

- **Volcanic, extrusive:** Formed from erupted magma.
- **Pyroclastic:** Magma fragmented and blown into the air upon eruption.
- **Plutonic, intrusive:** Formed from magma that cooled and solidified in the Earth.
- **Hypabyssal:** Intruded at shallow depths; roughly the upper kilometer.

Terms Describing General Mineralogical and Chemical Composition

- **Color Index:** For plutonic rocks, the color index (CI) is defined as the volume percent of mafic minerals in the rock; in practice, this is the area percent, because we generally see rock surfaces or slices. This is a useful term for describing rocks in the field and correlates strongly with bulk composition, but it is notoriously difficult to estimate accurately. Glazner and Walker (2020) describe a digital aid for estimating color index with a mobile device.
- **Leucocratic, felsic, silicic:** These terms are not strictly synonymous, but generally they are applied to rocks which are light colored (leucocratic), and hence rich in light-colored minerals (e.g., feldspars, quartz, or feldspathoids). Felsic is a portmanteau of feldspar or feldspathoid and silica. Felsic rocks rich in quartz and feldspars are by inference rich in SiO_2 (silicic), a term generally applied to rocks with >63 wt% SiO_2, if

the composition is known. The prefix "leuco-" means light-colored; for example, leucocytes are white blood cells.

- **Mafic, melanocratic:** Again, these are not synonymous but are applied to rocks which are dark colored (melanocratic), and hence rich in dark-colored minerals that are rich in FeO and MgO (e.g., pyroxenes, amphiboles, olivine). Mafic is a portmanteau of magnesium and iron (Fe). The prefix "melano-" means dark; melanoma is a skin cancer characterized by dark lesions.
- **Ultramafic:** Applies to rocks with very high color indices, on the order of 90 or more.
- **Feldspathoidal:** Rocks which contain a feldspathoid mineral (e.g., leucite, nepheline).

Terms Describing Chemical Composition

Although the previously described terms carry with them certain aspects of the chemical composition of a given igneous rock, there exist numerous other more well-defined chemical parameters with which you should be familiar. However, these terms have not been systematized, so the numbers below are only suggestions. "Acid" and "basic" are archaic terms.

- **Silicic, acid:** Rocks with >63 wt% SiO_2. Generally, 78 wt% SiO_2 is the upper limit of silica for igneous rocks that have not been hydrothermally altered.
- **Intermediate:** Rocks with 52–63 wt% SiO_2.
- **Mafic, basic:** Rocks with 45–52 wt% SiO_2.
- **Ultramafic, ultrabasic:** Rocks with <45 wt% SiO_2.
- **Silica saturation:** The abundance of SiO_2 relative to other oxides:
  - **Oversaturated** magma will, in theory, precipitate quartz upon crystallization and have quartz in the norm.
  - **Undersaturated** magma will precipitate a mineral that reacts with quartz (e.g., a feldspathoid or Mg-rich olivine) upon crystallization and will have olivine or nepheline (or, in extremely alkalic rocks, weird non-mineral things) in the norm.
  - **Saturated** magmas will precipitate a more silica-rich mineral (e.g., orthopyroxene or albite) upon crystallization but will not have any free silica left to precipitate quartz. These typically have Opx in the norm.
- **Alumina saturation:** Refers to the amount of Al_2O_3 in a rock relative to the alkalis (Na_2O, K_2O) and CaO. This is covered in Toolbox 10.1. The definitions for this important index, where oxides are in molar amounts, are as follows:
  - **Peraluminous:** $(CaO + Na_2O + K_2O) < Al_2O_3$. Contains normative corundum.
  - **Metaluminous:** $(Na_2O + K_2O) < Al_2O_3 < (CaO + Na_2O + K_2O)$. Contains prominent anorthite in the norm.
  - **Subaluminous:** $Al_2O_3 = (Na_2O + K_2O)$. Contains a small amount of normative anorthite.
  - **Peralkaline:** $Al_2O_3 < (Na_2O + K_2O)$. Contains acmite, Na-metasilicate, or kalsilite in the norm.

Classifying Volcanic Rocks by Chemical Composition

The total alkali-silica (TAS) diagram (Figure A3.1; Le Bas et al., 1986) is generally used to assign names to volcanic rocks if a chemical analysis is available. This diagram splits out a basaltic andesite field; suggested silica cutoffs for the BADR trend are given. There are multiple methods for further splitting these fields (particularly the alkalic ones); see LeMaitre (2002) for details.

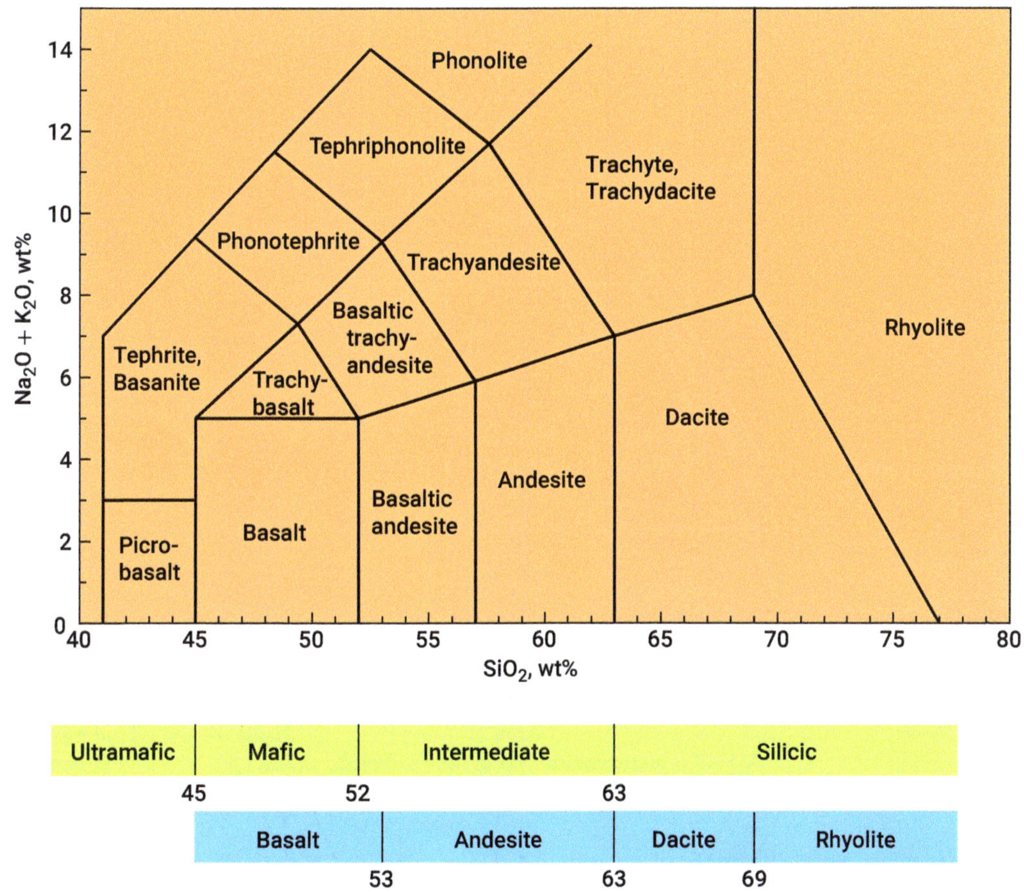

Figure A3.1 Total alkali-silica (TAS) diagram with recommended nomenclature, and silica divisions for rock classes. Suggested silica divisions for the BADR trend are at the bottom. None of the sharp lines on this diagram should be taken too literally, as rocks do not respect these divisions.

Source: Modified from Le Bas et al. (1986).

Classifying Plutonic Rocks by Mineralogical Composition

Standard IUGS Method for Classification of Plutonic Rocks

The traditional method for naming quartz-bearing plutonic rocks consists of estimating the modal (volume) percentages of the minerals quartz (Q), alkali feldspar (A), and plagioclase (P), and then plotting the point on the upper triangle in Figure A3.2 to determine a name. For rocks with a feldspathoid instead of quartz, the lower triangle is used ("foid" is shorthand for whatever feldspathoid is present, typically nepheline or leucite). This system was devised half a century ago by a large committee of the International Union of Geological Sciences (IUGS; Streckeisen, 1976).

Although this system is widely used, Glazner et al. (2019) pointed out a number of significant problems with it. These include the following:

1. A given plutonic body typically scatters across multiple fields, making naming arbitrary.
2. It lacks important information about the rock's mineralogy—in particular, the abundance of mafic minerals.

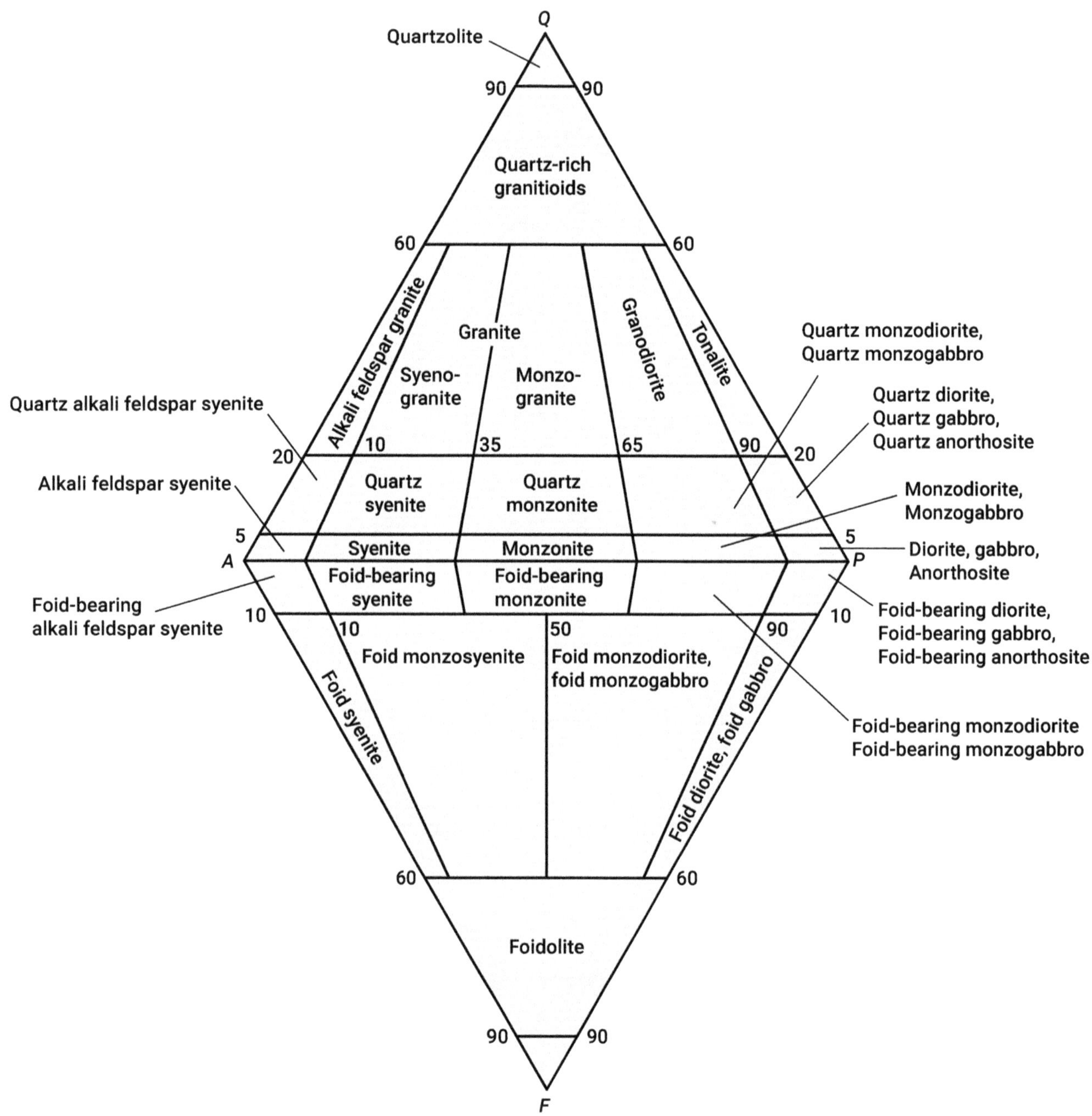

Figure A3.2 Traditional IUGS diamond for classifying plutonic rocks. See Figure A3.3 for a simpler scheme.

Source: Modified from Streckeisen (1976).

3. Geochemical trends were not considered when making field boundaries, so a sequence of rocks related by fractional crystallization or partial melting typically cuts across many boundaries at low angles, radiating from the P apex.
4. It is difficult to remember.
5. The quantitative data (modal percentages) used to define a name are then discarded.

Point #1 can be particularly problematic; a quartz-bearing pluton with about 90 percent plagioclase might have as many as 10 different names applied to it depending on minor variations in mineralogy.

Glazner et al. (2019) proposed a simplified classification that remedies many of these problems. In particular, it is far simpler, more forgiving of the natural variability that plutonic bodies exhibit, retains the quantitative data used in deriving the name, and gives the color index in the name.

Simpler Method for Classification of Plutonic Rocks The following procedures involve the same observations needed to classify a rock with the IUGS method, and thus the full IUGS name can always be applied. All percentages are modal (vol%).

QAP: Rocks with quartz + feldspars >10 percent (upper half of IUGS diamond)

1. Estimate the proportions of quartz (Q), alkali feldspar (A), plagioclase (P), and mafic minerals (M), and the identities of mafic minerals and accessories. (*Example: 20,20,50,10; biotite, hornblende, titanite.*)
2. Assign a root name based on where the QAP estimate, normalized to 100, falls on the upper triangle in Figure A3.3. (*Example: granodiorite.*)
3. Prefix the rock name with unnormalized QAP (e.g., 20,20,50). The proportion of mafic minerals (color index) is implicit in these numbers as 100 minus their sum. (*Example: 20,20,50 granodiorite; color index is thus 10.*)
4. Prefix the resulting name with relevant mafic minerals, using defined abbreviations if desired. The prefix should list these in increasing abundance, so that the most abundant is closest to the root name. (*Example: Hbl-Bt 20,20,50 granodiorite.*)
5. Important accessory minerals can be denoted by, for example, titanite-bearing or Ttn-bearing. (*Example: Ttn-bearing hbl-bio 20,20,50 granodiorite.*)

Note that these steps are identical to the IUGS method, except that there are far fewer bins in which to put the rocks, and the bins simply indicate broad QAP proportions. For example, granites and granodiorites are quartz-rich and distinguished broadly by alkali feldspar > plagioclase or by plagioclase > alkali feldspar, respectively. Syenites are rich in alkali feldspar and poor in plagioclase + quartz. Diorites are rich in plagioclase and poor in alkali feldspar + quartz, and so on.

QAF: Rocks with feldspathoid(s) + feldspars >10 percent (lower half of IUGS diamond)

Classification of feldspathoid-bearing rocks is the same as with quartz-bearing rocks, except that the identity of the feldspathoid replaces "foid" (e.g., nepheline syenite rather than foid syenite).

OOC and POOC: Ultramafic and gabbroic rocks

- **Ultramafic rocks:** Olivine-pyroxene rocks with <10 percent felsic minerals are named via the olivine-orthopyroxene-clinopyroxene (OOC) triangle. The same simplification principles apply to these rocks: estimate mineral proportions, and then name the rock with these numbers and the simplified, blurred boundary classification in Figure A3.4a. Preface the name with other important minerals such as garnet or spinel; their proportion is 100 minus the sum of OOC.
- **Gabbroic rocks:** Gabbroic rocks can be classified using a tetrahedron with apices of plagioclase, olivine, orthopyroxene, and clinopyroxene (POOC; Figure A3.4b). Note that sharp planar boundaries here are actually gradational bands. The base of the tetrahedron is the OOC triangle of Figure A3.4a; the apices are anorthosite, dunite, orthopyroxenite, and clinopyroxenite. Troctolite lies along the plagioclase-olivine edge, and the interior is gabbro or norite

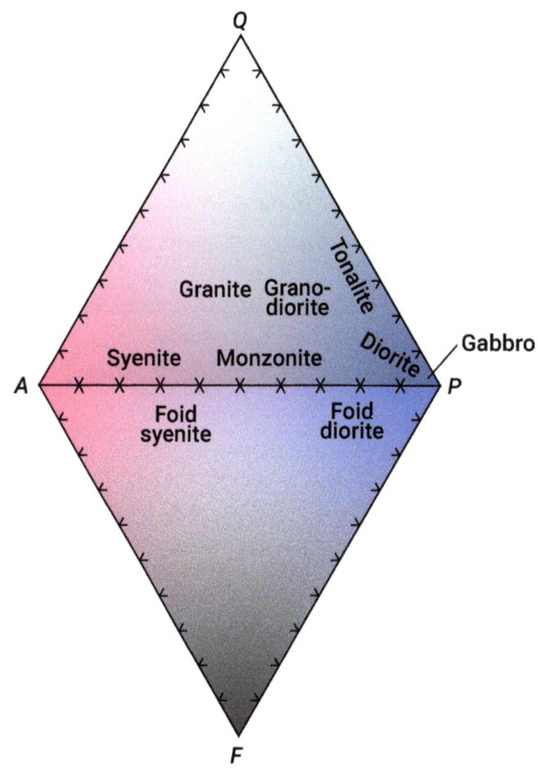

Figure A3.3 Simplified classification. Boundaries are fuzzy, fields overlap, and names are redundant of the numeric values, which can be converted into formal IUGS names if desired.

Source: From Glazner et al. (2019); © 2019 The Geological Society of America.

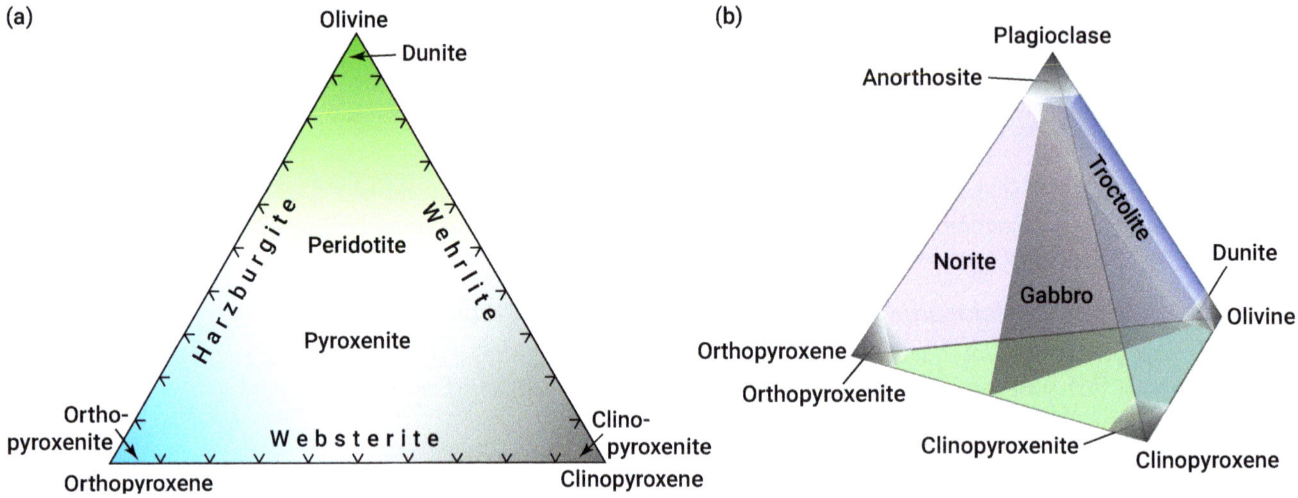

Figure A3.4 Simplified classification of mafic and ultramafic rocks using (a) the OOC (olivine-Opx-Cpx) triangle and (b) the POOC (plagioclase-olivine-Opx-Cpx) tetrahedron. Note that boundaries in the tetrahedron are drawn as planes for drafting convenience but should be considered fuzzy suggestions only.

depending on whether the dominant pyroxene is clinopyroxene or orthopyroxene. Hybrid names such as gabbronorite and qualified names such as olivine gabbro are unnecessary (but can be used if desired) when the defining mineralogy is given in the name (e.g., *50,10,20,20 gabbro*).

Table A3.1 gives typical major and accessory minerals for plutonic and compositionally equivalent volcanic rocks.

Table A3.1 Common Mineral Associations in Selected Igneous Rocks

Plutonic Rocks			Volcanic Rocks		
Name	**Major Minerals**	**Accessory Minerals**	**Accessory Minerals**	**Major Minerals**	**Name**
Granite	K-feldspar, sodic plagioclase, quartz, biotite, hornblende	Apatite, magnetite, titanite, zircon, allanite, ilmenite, muscovite, garnet, monazite	Apatite, fayalite, garnet, ilmenite, magnetite, titanite, zircon	Sanidine, sodic plagioclase, quartz, anorthoclase, biotite, hornblende	**Rhyolite**
Granodiorite	Plagioclase, biotite, hornblende, Cpx, K-feldspar, quartz	Apatite, zircon, titanite	Apatite, garnet, ilmenite, magnetite, K-feldspar	Biotite, quartz, Cpx hornblende, Opx, plagioclase	**Dacite**
Diorite	Cpx, biotite, hornblende, plagioclase, quartz	Allanite, apatite, garnet, ilmenite, magnetite, titanite, zircon	Apatite, quartz (xenocrystic)	Cpx, hornblende, Opx, olivine, plagioclase, magnetite, ilmenite	**Andesite**
Gabbro	Cpx, Opx, plagioclase, olivine	Apatite, biotite, chromite, hornblende, ilmenite, magnetite, K-feldspar, quartz, titanite	Apatite, quartz (xenocrystic)	Plagioclase, olivine, Cpx, Opx, magnetite, ilmenite	**Basalt**
Ultramafic rocks	Olivine, Cpx, Opx, serpentine	Apatite, chromite, ilmenite, magnetite, phlogopite, plagioclase, spinel, garnet, diamond	Cr-spinel	Olivine, Cpx	**Komatiite**

Common Textures and Structures in Igneous Rocks

There is no unanimous agreement on the meaning of the terms "texture" and "structure." In general, however, texture relates to small-scale features, and especially to the size, shape, and manner of joining (articulation) of the individual grains. Structure is mainly reserved for larger features and implies that some force other than crystallization was active in determining the arrangement of the individual elements. Some small-scale features involving the juxtaposition of different textural elements may be regarded as structures. The term "fabric" is a general term including both textures and small-scale structures. Some of the more common terms are given in the following lists.

Textures

Terms Describing the Crystallinity of the Rock

- **Holocrystalline:** Denotes a rock consisting entirely of crystals.
- **Hypocrystalline:** Denotes a rock containing both crystals and glass.
- **Glassy:** Denotes a rock, or part of a rock, that is composed entirely of glass.

Terms Describing the Size of Crystals in a Rock

- **Phaneritic:** Indicates that individual crystals in a rock are large enough to be distinguished by the unaided eye.
    - **Coarse-grained:** Crystals average >5 mm.
    - **Medium-grained:** Crystals average 1–5 mm.
    - **Fine-grained:** Crystals average <1 mm.
- **Pegmatitic:** Indicates a rock of overall igneous character that is abnormally coarse-grained, with the average crystal >1 cm across.
- **Microcrystalline:** Indicates a holocrystalline rock in which the individual crystals are too small to be distinguished clearly by the unaided eye. With the aid of low-power magnification, however, the granular texture can be seen.
- **Cryptocrystalline:** Indicates a holocrystalline rock in which the individual crystals are too small to be seen, even microscopically. The rock can be seen to be crystalline, however, by its effect on polarized light.
- **Aphanitic:** Denotes a texture in which crystals are too small to be seen with the naked eye. It is a useful field term and is not restricted to igneous rocks. Aphanitic rocks may be either cryptocrystalline or microcrystalline.
- **Equigranular:** Denotes a crystalline rock in which individual crystals are all of generally similar size.
- **Inequigranular:** Denotes a rock in which the grains vary considerably in size.
- **Seriate:** Denotes a rock in which there is a continuous gradation in grain size from largest to smallest.
- **Porphyritic:** Denotes a rock in which at least two distinctly different sizes of crystal are present: larger phenocrysts surrounded by a finer groundmass or matrix. The phenocrysts may be abundant or sparse, and they may be large or small. However, they are distinctly larger than the groundmass crystals, and intermediate size gradation is lacking.
- **Vitrophyric:** Porphyritic with a groundmass consisting largely of glass.
- **Aphyric:** Crystal-free (glassy).

Terms Describing the Form of Individual Grains in a Rock

- **Euhedral:** Denotes individual crystals perfectly bounded by their own crystal faces.
- **Subhedral:** Denotes individual crystals with external faces partly or imperfectly developed.
- **Anhedral:** Denotes shapeless crystal grains with no external faces developed on them.

The following terms refer to **crystal habit**:

- **Equant:** Roughly equidimensional.
- **Prismatic:** Bounded by planes that form an open-ended tube (like a segment of a pencil).
- **Acicular:** Needle-like.
- **Fibrous:** Shaped like fibers.
- **Tabular:** Plate-like.
- **Platy:** Forming thin plates, like mica.

Terms Describing the Form and Relationship of Grains in the Rock as a Whole

- **Hypidiomorphic:** Denotes a rock consisting predominantly of subhedral grains, the commonest texture of phaneritic igneous rocks. It is typical of granites and is sometimes referred to as **granitic texture** when referring to felsic igneous rocks.
- **Allotriomorphic:** Denotes a rock in which most of the grains are anhedral. A common example of this texture occurs in aplites: fine- or medium-grained, equigranular rocks consisting largely of quartz and alkali feldspars. The term **aplitic texture** can be used for such rocks.
- **Ophitic:** Denotes a rock in which large, anhedral pyroxene grains enclose euhedral, tabular plagioclase grains. This texture is common in diabase and is sometimes referred to as diabasic texture.
- **Poikilitic:** Denotes a rock in which grains of various minerals in random orientation are completely enclosed within a larger crystal of another mineral. Ophitic texture is a special case of poikilitic texture.
- **Felty:** Denotes a rock in which microlites (tiny needle-like crystals) of euhedral feldspar are interwoven in an irregular fashion. (Note: felty and trachytic are generally petrographic terms, not field terms.)
- **Trachytic:** Denotes a rock in which the microlites are arranged in a subparallel manner, probably as a result of flow in the magma.

Structures

Plutonic Rocks

- **Xenolith:** A foreign fragment of rock.
- **Enclave:** Generally used to refer to mafic inclusions that were partially molten, and thus soft, when incorporated into the host.
- **Inclusions:** These are crystals or rocks which differ in mineralogy or texture from the surrounding groundmass. They may have crystallized from the same magma as the rock in which they are enclosed, or they may be foreign to it.
- **Schlieren:** Streaky collections of mafic minerals in felsic plutonic rocks. Singular is schliere.
- **Orbicular structure:** Spheroidal or irregularly rounded bodies consisting of concentric zones of alternating light and dark minerals which are the same as those of the rock surrounding the spheroids. The structure occurs locally in a variety of medium-grained plutonic rocks, but most orbicular rocks are granites, gabbros, and norites.

- **Miarolitic cavities:** Found in granitic rocks, these are openings of small to large size and/or irregular shape into which the crystals of the rock project. This important texture indicates crystallization at shallow depth. The crystals exposed against the opening are bounded by crystal faces. The cavities were formed late in the crystallization history by exsolution of a fluid phase.
- **Massive structure:** Rocks which lack any of the previously described structures, and which are not layered or banded.

Volcanic Rocks

- **Inclusions, xenoliths, and xenocrysts:** Foreign rocks and crystals are common in volcanic rocks. Many xenoliths are similar to the enclosing rocks as they represent earlier consolidated portions of similar magma torn from the walls of the volcanic vent, or pieces of a lava flow crust that were rolled back into the flow interior.
- **Vesicular:** Holes or cavities in the rock which were bubbles formed by exsolution of vapor from the magma on eruption.
- **Eutaxitic:** A texture found in ignimbrites in which flattened pumice shards, typically glassy, are aligned by flow compaction.
- **Amygdaloidal:** Vesicles partly or completely filled with secondary minerals, typically quartz, opal, chalcedony, calcite, or zeolites.
- **Vuggy:** Irregularly shaped pockets, commonly lined with euhedral crystals. It is analogous to miarolitic fabric in phaneritic rocks.
- **Dyktitaxitic:** Common texture in many basalts in which interstitial gas cavities have crystals projecting into them.
- **Pumiceous:** A frothy vesicular structure consisting of small, closely spaced cavities separated by thin partitions of glass. Pumice is most common in rhyolitic to dacitic rocks.
- **Scoriaceous:** Said of highly vesiculated, rough-surfaced mafic lavas.
- **Spherulitic:** Radially arranged fibers of alkali feldspar and quartz in a glassy or sometimes felsic aphanitic matrix.
- **Perlitic:** Closely spaced sets of concentric cracks (like onion skin) in glassy rocks. There is usually a distinct uncracked central kernel.
- **Massive:** Volcanic rocks which lack other structures and are not layered, banded, or closely jointed are described as massive.

A3.2 Metamorphic Rocks

Basic Nomenclature

Metamorphic rock names are fairly simple. There are five main names to describe overall rock texture, which are then modified to describe mineralogy or overall bulk composition:

- **Slate:** A cleavable rock whose grain size is too small to see, even with a hand lens.
- **Phyllite:** A foliated rock whose grain size can be seen only with a hand lens. These typically have a silky luster.
- **Schist:** A foliated rock whose grains can be seen without a hand lens. A rock's schistosity is normally defined by preferred alignment of sheet silicates (usually micas) in layers, although amphiboles can sometimes form a schistosity.
- **Gneiss:** A catchall term for high-grade metamorphic rocks that are more resistant to erosion (and collection!) than schists, phyllites, or slates. Commonly, gneisses show alternating bands of light (felsic) and dark (mafic) minerals. The darker bands contain higher concentrations of micas or amphiboles and are commonly foliated, whereas the lighter bands are richer in quartz and feldspar.

  - **Orthogneisses** represent metamorphosed igneous rocks; felsic rocks (metagranites, metadiorites, etc.) can be quite homogeneous texturally, with only a weakly developed foliation.
  - **Augen gneisses** are metagranites that have experienced grain-size reduction of plagioclase and quartz and contain eye-shaped grains of K-feldspar.
  - **Paragneisses** are high-grade metasedimentary rocks.
  - **Migmatites** (metamorphic rocks that have started to melt) are usually classified as gneisses; the light bands represent felsic partial melts, whereas the darker bands are the leftover ("restitic") solid part of the rock.
- **Granofels** and **hornfels:** Non-foliated rocks with homogeneous, random distributions of minerals. Hornfelses are very fine-grained and form in contact aureoles. Granofelses form in a variety of settings, including subduction zones and continental collisions, and represent rocks that somehow escaped significant deformation, often because they contain deformation-resistant minerals. A garnet-rich metachert without a distinctive texture would be classed as a granofels. Many granulites that anneal texturally at high temperature exhibit granofelsic texture.

Once you've decided on the rock name, you can characterize it further with compositional or mineralogical modifiers. For example, you might describe your rock as a quartzofeldspathic (quartz- and feldspar-rich) garnet-biotite-muscovite schist. This kind of name explains what the mineralogy is. Although some geologic naming conventions list minerals in order of decreasing abundance, metamorphic petrologists more commonly focus on presence of either distinctively large crystals (called porphyroblasts) or minerals that convey useful information about the rock's metamorphic conditions. For example, a garnet-staurolite-biotite schist probably contains other more abundant minerals (muscovite, quartz, and plagioclase), but the garnet-staurolite-biotite association has temperature significance, so most petrologists would list them first and assume you know that other minerals are likely present.

Other ways of naming metamorphic rocks focus on the whole-rock composition of the protolith. The term "metabasite" refers to a metamorphosed rock of basaltic composition, so a garnet metabasite would tell a petrologist that the rock is broadly mafic and contains garnet. Another common term is "pelite" (an archaic name for a mudstone). Metapelites are extremely common schists, so one of our rocks above might be described as a garnet-staurolite-biotite metapelite or metapelitic schist. Obviously, a bulk rock composition implies a level of interpretation beyond mere mineralogy (how do you know it's a metabasalt and not a metabasaltic andesite?), so if the goal is purely descriptive, these terms should be avoided. The advantage of characterizing bulk composition, even approximately, is that with a little information about metamorphic grade and whole-rock composition, a petrologist can predict what minerals are likely present.

Common whole-rock compositional modifiers include:

- **Basite:** Basalt (*sensu lato*).
- **Pelite:** Mudstone (the Greek word *pelos* means mud).
- **Psammite:** Sandstone.
- **Marl:** Mixture of carbonate and mud (mixture of a limestone and a pelite). When metamorphosed, forms calc-silicates.
- **Calc-pelite:** Mixture of carbonate and mud, but muddier than a marl.

Some wholly metamorphic names that also imply whole-rock compositions include:

- **Quartzite:** Metamorphosed quartz-rich sandstone.
- **Marble:** Metamorphosed limestone or dolomite.

Textures

Rock textures parallel rock names, but it is useful to define them directly:

- **Foliation:** A parallel arrangement or distribution of minerals, including compositional layering, slaty cleavage, and schistosity. Often used synonymously with schistosity.
- **Schistosity:** The planar orientation of minerals (usually micas) producing a tendency for the rock to split.
- **Slaty cleavage:** The finer-grained equivalent of schistosity. Slates split because micro-crystals of sheet silicates are preferentially arranged in planes.
- **Gneissosity/gneissic layering:** The alternation of mineralogically distinct layers such as micaceous and quarto-feldspathic layers.
- **Lineation:** The parallel alignment of linear elements in a rock. This includes aligned prismatic minerals (e.g., hornblende or sillimanite) as well as elongate grain aggregates (e.g., stretched pebbles).
- **Crenulation cleavage:** Microfolding (crenulation) of a preexisting schistosity to form a new planar fabric.

Mineral Textures

Crystal forms have numerous names, but three of the most common refer to how well crystal faces are developed. Many metamorphic petrologists have adopted the igneous terms "euhedral," "subhedral," and "anhedral" for crystals that range from well-formed with good crystal faces to poorly formed crystals. There is a parallel set of three terms for metamorphic crystals:

- **Idioblastic:** A metamorphic mineral with well-developed crystal faces (equivalent to euhedral)
- **Subidioblastic:** A metamorphic mineral with some trace of crystal form (equivalent to subhedral)
- **Xenoblastic:** A metamorphic mineral with no trace of crystal form (equivalent to anhedral)

Here, "blastic" comes from a Greek word meaning bud or shoot and refers to growth of minerals in a solid rock.

Different minerals show different tendencies to form good crystal faces. These can be ordered into a **crystalloblastic series** (from idioblastic to xenoblastic) that broadly parallels mineral groups and structures:

- Oxides (rutile, ilmenite, and magnetite), also titanite
- Orthosilicates (garnet, kyanite, staurolite, andalusite, zircon), apatite, and tourmaline
- Epidote group (epidote, (clino)zoisite, lawsonite)
- Chain silicates (pyroxenes, amphiboles)
- Sheet silicates (micas, chlorite, talc)
- Carbonates (calcite, dolomite)
- Scapolite, cordierite, feldspar
- Quartz

Other common metamorphic textural terms include:

- **Porphyroblast:** A metamorphic mineral that is large compared to the groundmass or matrix. Common porphyroblasts are garnet, staurolite, kyanite, and andalusite.

- **Poikiloblast:** A large crystal with many inclusions. If there are many inclusions distributed uniformly, these can be described as sieve textured. Common poikiloblasts are garnet, staurolite, andalusite, and cordierite.
- **Pseudomorph:** Secondary mineral or assemblage of minerals that have replaced an earlier mineral but retained its original shape. Pseudomorphs provide information about earlier mineral assemblages. Common pseudomorphs include chlorite after garnet or white mica + epidote + chlorite pseudomorphs after lawsonite.
- **Pressure shadow:** A region in a rock protected from deformation by the presence of a relatively rigid object. Shadows occur where a foliation wraps around a porphyroblast in a schist. Quartz pressure shadows adjacent to garnet are common, because quartz solubility is P-dependent. So, it dissolves on the side of the garnet that experiences maximum compressive stress and reprecipitates nearby where stress is lower.
- **Corona:** A texture in which a core phase is surrounded by another mineral and indicates a reaction between the core phase and matrix minerals. Hornblende coronas around garnet from Gore Mountain, New York, or cordierite coronas around garnets are examples.
- **Symplectite:** Intergrowth of two or more minerals as a result of either deformation or reactions that produce both phases. Examples include plagioclase + amphibole intergrowths after omphacite in altered eclogite.

Deformed Rocks

The texture of many rocks reflects the deformation that they have experienced. Shearing generally turns large grains into smaller ones, but the mechanisms depend on the strength of the original crystal, the temperature and strain rate of shearing, and the mineralogy of the matrix. The following terms are all interpretive, because they not only describe the appearance of minerals in the rock, they also imply a mechanism by which the rock was deformed:

- **Cataclasite:** A low-T rock formed through brittle fracturing, shattering, or granulation of its constituent minerals. Essentially a cemented breccia.
- **Phyllonite:** A rock that resembles a phyllite but has been formed through intense deformation and grain size reduction. Silky films of mica or chlorite are smeared out in the schistosity. Typically formed at intermediate temperature.
- **Mylonite:** An extremely intensely deformed rock in which grain size reduction has occurred ductilely, with minimal brittle fracturing and cataclasis. Typically formed at high temperature. Brittle fault zones transition to ductile mylonite zones with increasing depth and temperature.
- **Pseudotachyllite:** A rock in which heating during shearing has caused melting. Generally appears as a very dark glass, although it commonly recrystallizes to fine-grained minerals. Tachyllite is an old name for basaltic glass, which it resembles. Pseudotachyllites indicate former brittle fault zones.
- **Porphyroclast** (not porphyroblast)**:** A large relict mineral in a matrix of finer-grained material whose grain size has been reduced through strain.
- **Sigma, delta grains:** Sheared grains with tails of finer-grained crystals that curve into the matrix in a shape that resembles the Greek letters sigma (σ) or delta (δ). The shapes of these grains and their tails indicate sense of shear.

APPENDIX 4

Periodic Table of the Elements

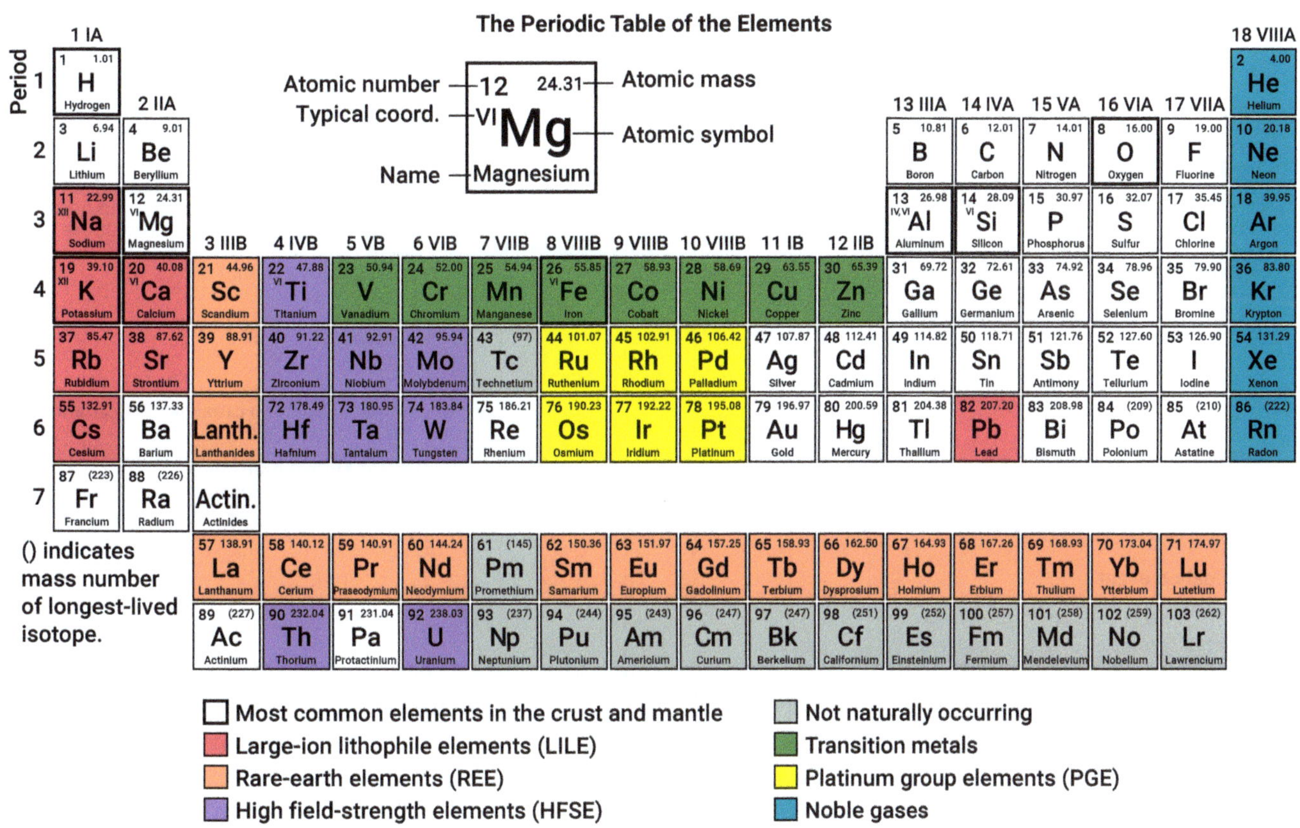

Figure A4.1 Periodic table of the elements, with specific geochemical groupings indicated by shading. Roman numerals give typical coordination numbers for major cations. Note that although the rare earth elements (REE) include Sc and Y, REE as used in practice typically refers to just the lanthanide elements (atomic numbers 57-71). For elements with no naturally occurring isotopes, an atomic mass in parentheses gives the mass number of the longest-lived isotope.

APPENDIX 5

Geologic Time Scale

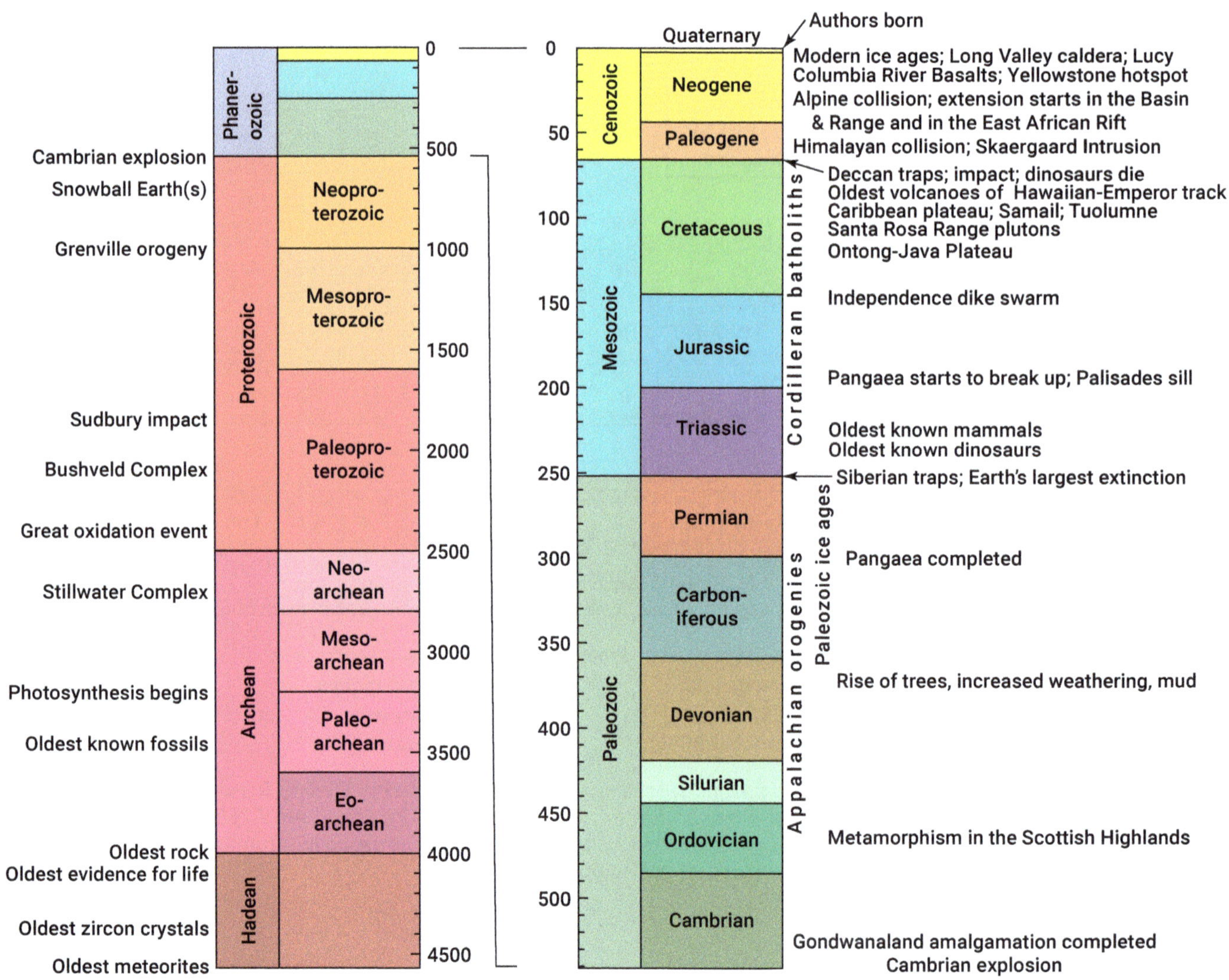

Figure A5.1 Simplified geologic time scale with selected events marked.

APPENDIX 6

Mineral/Melt Distribution Coefficients

The following graphs summarize mineral/melt distribution coefficients from data compiled in the Geochemical Earth Reference Model (GERM) Partition Coefficients Database (https://kdd.earthref.org/KdD/). These compilations are meant to provide guidelines, and readers are encouraged to compile data themselves for elements and minerals of interest.

Data were compiled for 15 minerals and processed as follows:

1. Data from simple systems (e.g., olivine in the system forsterite-jadeite) were eliminated.
2. Bulk compositions were assigned to high-Si (dacite and equivalent or more silicic), low-Si (andesite and equivalent or less silicic), and ultramafic (peridotite and equivalent; carbonatite). Only data from the high- and low-Si groups are presented.
3. Titanite data for silicic rocks are from Bachmann et al. (2005) and Ackerson (2011).
4. Thirty-one elements were assigned to trace-element groups (LILE, etc.) and then ranked from low to high charge and, within charge groups, low to high atomic number.

Data are plotted for high-Si compositions as orange dots with dashed range lines, and for low-Si compositions as blue dots with solid range lines. Plotted points are medians, and range lines end at 10th and 90th percentiles. Points without lines typically indicate single analyses.

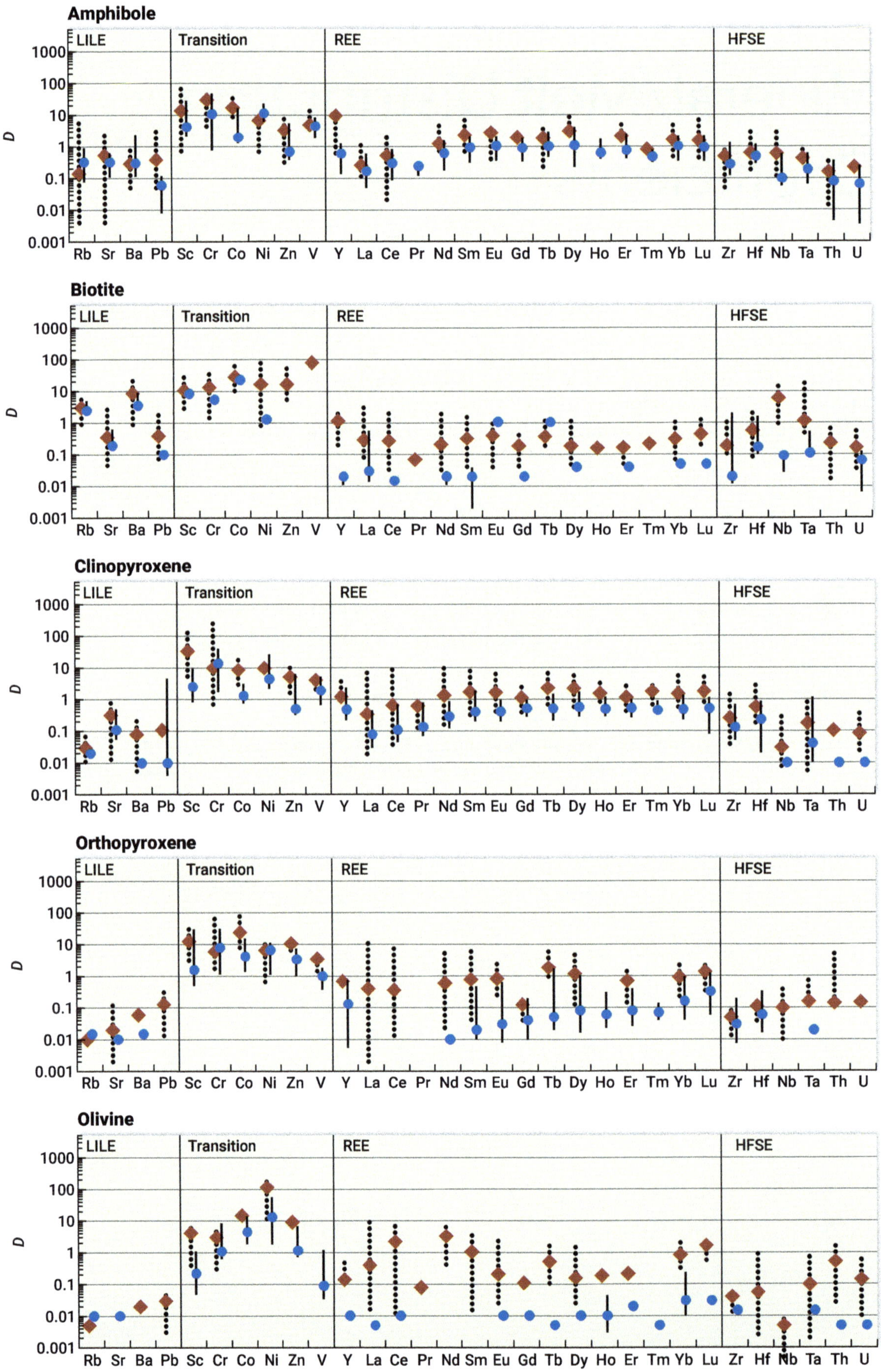

Amphibole
Biotite
Clinopyroxene
Orthopyroxene
Olivine
LILE
Transition
REE
HFSE
D
1000
100
10
1
0.1
0.01
0.001
Rb Sr Ba Pb Sc Cr Co Ni Zn V Y La Ce Pr Nd Sm Eu Gd Tb Dy Ho Er Tm Yb Lu Zr Hf Nb Ta Th U

Plagioclase

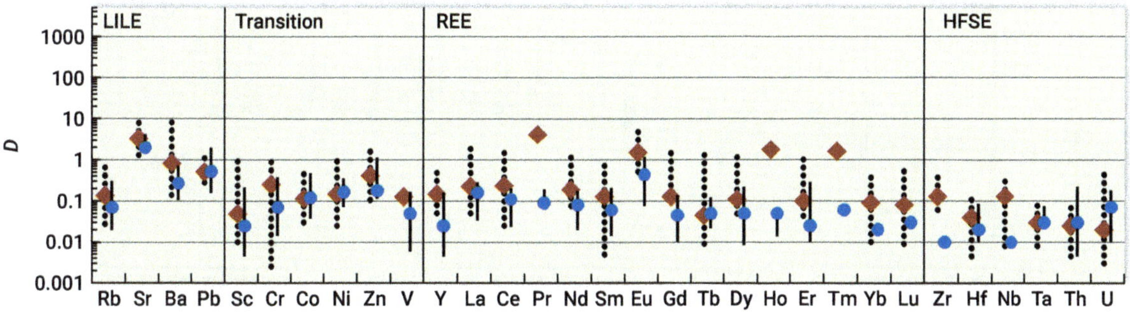

K-feldspar

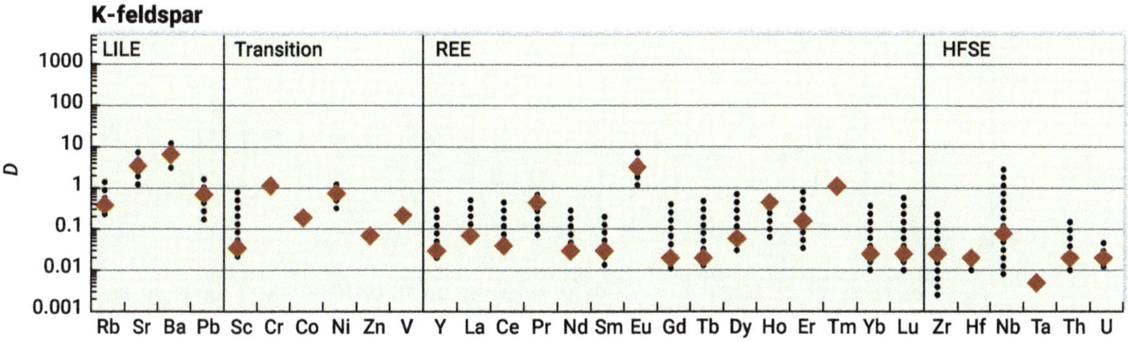

Titanite

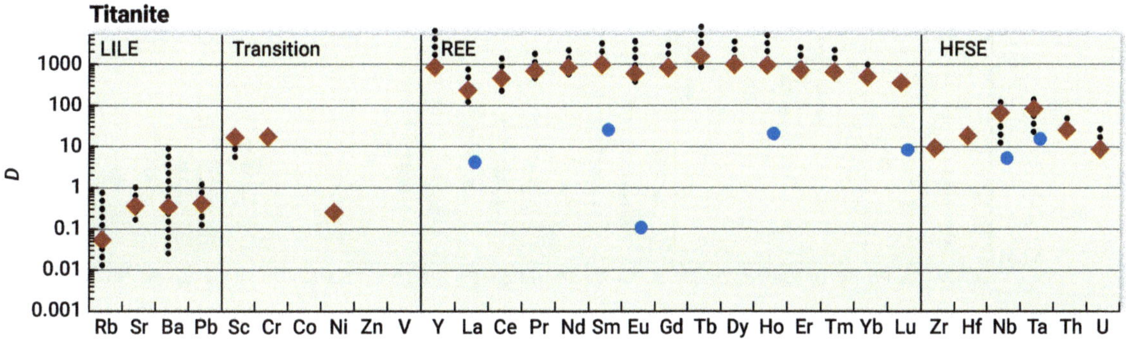

Zircon

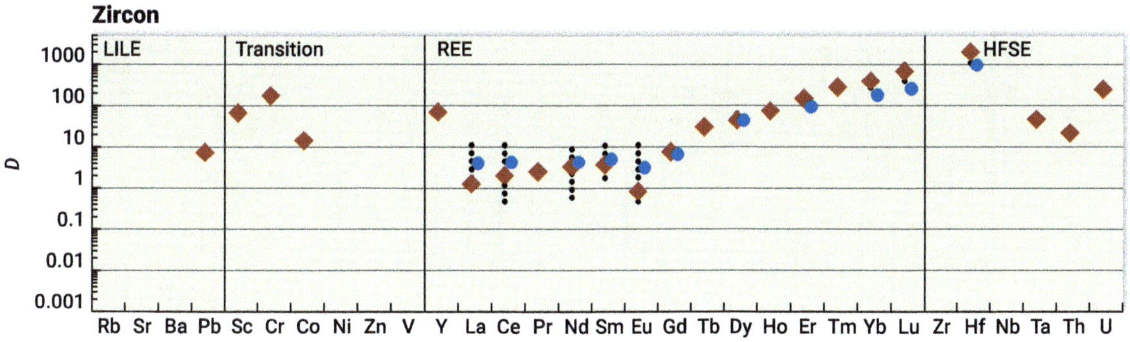

Allanite

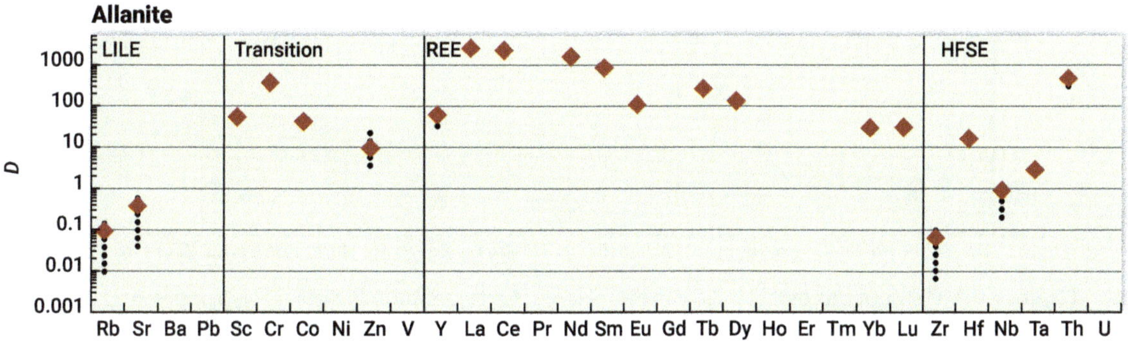

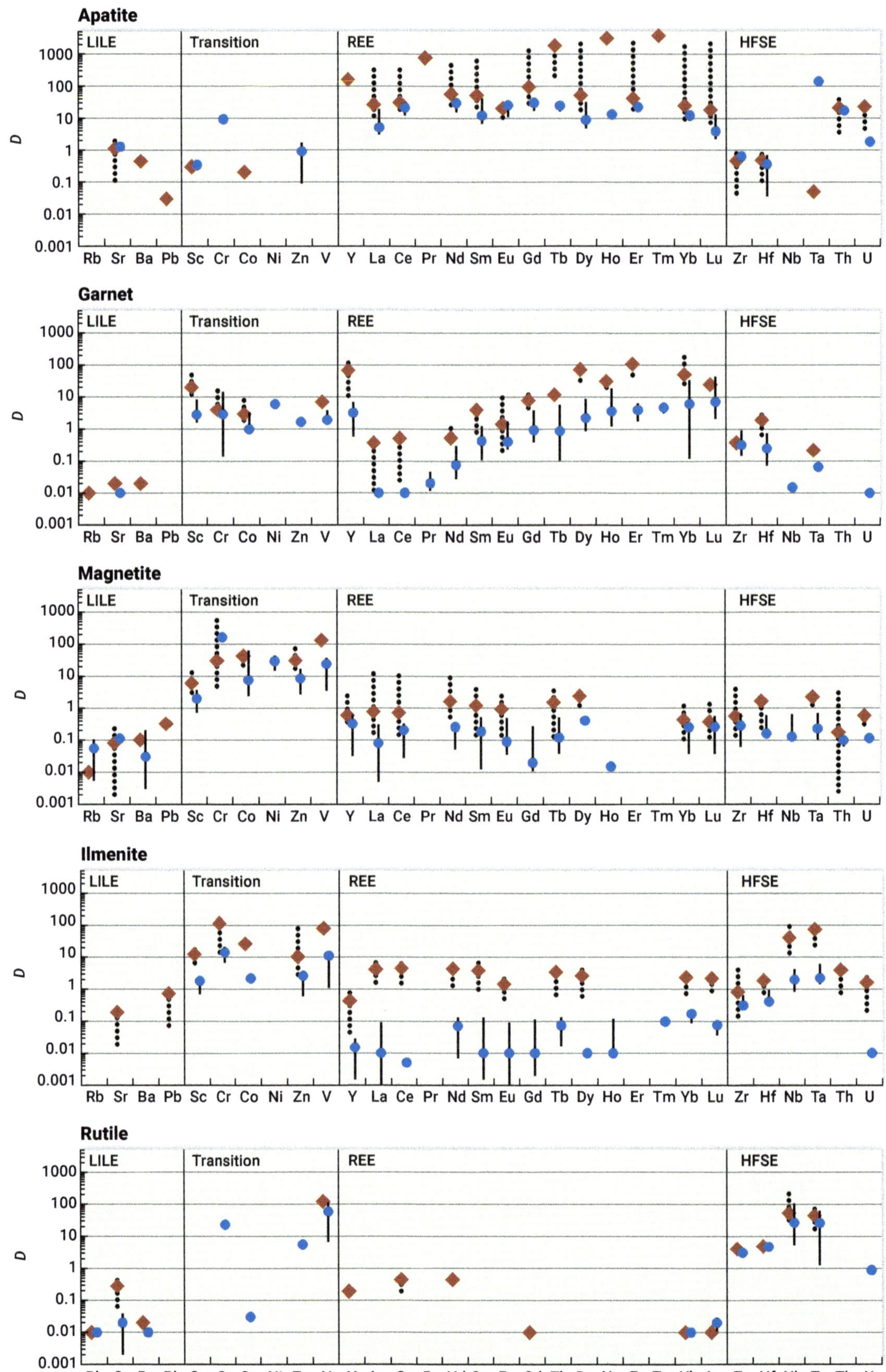

Figure A6.1 Distribution coefficients for mafic minerals, feldspars and key accessory minerals, and additional accessory minerals and Fe-Ti oxides. LILE = large-ion lithophile elements; Transition = Period 4 transition metals; REE = rare earth elements; HFSE = high-field-strength elements.

APPENDIX 7

Mineral Formulas and Formula Weights

Mineral	Molecular Weight (g/mol)	Formula
Oxides, hydroxides		
Brucite	58.32	$Mg(OH)_2$
Spinel group		
Chromite	223.84	$FeCr_2O_4$
Hercynite	173.81	$FeAl_2O_4$
Magnetite	231.54	Fe_3O_4
Spinel	142.27	$MgAl_2O_4$
Ulvöspinel	223.59	Fe_2TiO_4
Hematite group		
Corundum	101.96	Al_2O_3
Hematite	159.69	Fe_2O_3
Ilmenite	151.75	$FeTiO_3$
Periclase	40.3	MgO
Rutile	79.9	TiO_2
Orthosilicates		
Garnet group		
Almandine	497.75	$Fe_3Al_2Si_3O_{12}$
Andradite	508.18	$Ca_3Fe_2Si_3O_{12}$
Grossular	450.45	$Ca_3Al_2Si_3O_{12}$
Pyrope	403.13	$Mg_3Al_2Si_3O_{12}$
Spessartine	495.03	$Mn_3Al_2Si_3O_{12}$
Aluminosilicates		
Andalusite	162.05	Al_2SiO_5
Kyanite	162.05	Al_2SiO_5
Sillimanite	162.05	Al_2SiO_5

Mineral	Molecular Weight (g/mol)	Formula
Chloritoid (Fe)	503.82	$Fe_2Al_4Si_2O_{10}(OH)_4$
Chloritoid (Mg)	440.72	$Mg_2Al_4Si_2O_{10}(OH)_4$
Olivines		
Fayalite	203.78	Fe_2SiO_4
Forsterite	140.69	Mg_2SiO_4
Staurolite (Fe)	851.87	$Fe_2Al_9Si_4O_{23}(OH)$
Staurolite (Mg)	788.77	$Mg_2Al_9Si_4O_{23}(OH)$
Titanite	196.06	$CaTiSiO_5$
Zircon	183.3	$ZrSiO_4$
Chain silicates		
Pyroxenes, pyroxenoids		
Acmite	231.01	$NaFeSi_2O_6$
Diopside	216.55	$CaMgSi_2O_6$
Enstatite	100.39	$MgSiO_3$
Ferrosilite	131.93	$FeSiO_3$
Hedenbergite	248.1	$CaFeSi_2O_6$
Jadeite	202.14	$NaAlSi_2O_6$
Wollastonite	116.16	$CaSiO_3$
Amphiboles		
Ferroactinolite	970.08	$Ca_2Fe_5Si_8O_{22}(OH)_2$
Ferroedenite	991.97	$NaCa_2Fe_5AlSi_7O_{22}(OH)_2$
Ferroglaucophane	878.17	$Na_2Fe_3Al_2Si_8O_{22}(OH)_2$
Cummingtonite	780.82	$Mg_7Si_8O_{22}(OH)_2$
Edenite	834.26	$NaCa_2Mg_5AlSi_7O_{22}(OH)_2$
Glaucophane	783.55	$Na_2Mg_3Al_2Si_8O_{22}(OH)_2$
Grunerite	1001.62	$Fe_7Si_8O_{22}(OH)_2$
Hastingsite	990.87	$NaCa_2Fe_4FeAl_2Si_6O_{22}(OH)_2$
Magnesiohastingsite	864.7	$NaCa_2Mg_4FeAl_2Si_6O_{22}(OH)_2$
Pargasite	835.83	$NaCa_2Mg_4AlAl_2Si_6O_{22}(OH)_2$
Tremolite	812.37	$Ca_2Mg_5Si_8O_{22}(OH)_2$
Tschermakite	815.52	$Ca_2Mg_3Al_2Al_2Si_6O_{22}(OH)_2$
Sapphirine	344.62	$Mg_2Al_4SiO_{10}$
Sheet silicates		
Biotites		
Annite	511.89	$KFe_3AlSi_3O_{10}(OH)_2$

Mineral	Molecular Weight (g/mol)	Formula
Phlogopite	417.26	$KMg_3AlSi_3O_{10}(OH)_2$
Chlorites		
Chamosite	713.55	$Fe_5Al(AlSi_3O_{10})(OH)_8$
Clinochlore	555.8	$Mg_5Al(AlSi_3O_{10})(OH)_8$
White micas		
Margarite	398.19	$CaAl_2Al_2Si_2O_{10}(OH)_2$
Muscovite	398.31	$KAl_3Si_3O_{10}(OH)_2$
Paragonite	382.2	$NaAl_3Si_3O_{10}(OH)_2$
Prehnite	395.38	$Ca_2Al_2Si_3O_{10}(OH)_2$
Serpentine (antigorite, chrysotile, lizardite)	554.23	$Mg_6Si_4O_{10}(OH)_8$
Talc	379.27	$Mg_3Si_4O_{10}(OH)_2$
Framework silicates		
Feldspars		
Albite	262.22	$NaAlSi_3O_8$
Anorthite	278.21	$CaAl_2Si_2O_8$
Celsian	375.47	$BaAl_2Si_2O_8$
Microcline	278.33	$KAlSi_3O_8$
Orthoclase	278.33	$KAlSi_3O_8$
Sanidine	278.33	$KAlSi_3O_8$
Feldspathoids		
Kalsilite	158.16	$KAlSiO_4$
Leucite	218.25	$KAlSi_2O_6$
Nepheline	142.05	$NaAlSiO_4$
Nepheline (Na,K)	584.33	$Na_3KAl_4Si_4O_{16}$
Sodalite	969.22	$6NaAlSiO_4 \cdot 2NaCl$
Perovskite	135.98	$CaTiO_3$
Quartz	60.08	SiO_2
Other silicates		
Cordierite (Fe)	648.04	$Fe_2Al_4Si_5O_{18}$
Cordierite(Mg)	584.96	$Mg_2Al_4Si_5O_{18}$
Epidote	483.23	$Ca_2FeAl_2Si_3O_{12}(OH)$
Lawsonite	314.24	$CaAl_2Si_2O_7(OH)_2 \cdot H_2O$
Pumpellyite	943.09	$Ca_4MgAl_5Si_6O_{23}(OH)_3 \cdot 2H_2O$
Zoisite	454.36	$Ca_2Al_3Si_3O_{12}(OH)$

Mineral	Molecular Weight (g/mol)	Formula
Other groups		
Carbonates		
Calcite	100.09	$CaCO_3$
Dolomite	184.4	$CaMg(CO_3)_2$
Magnesite	84.31	$MgCO_3$
Siderite	115.86	$FeCO_3$
Sulfates		
Anhydrite	136.14	$CaSO_4$
Baryte	233.4	$BaSO_4$
Gypsum	172.17	$CaSO_4 \cdot 2H_2O$
Phosphates		
Chlorapatite	520.77	$Ca_5(PO_4)_3Cl$
Fluorapatite	504.31	$Ca_5(PO_4)_3F$
Hydroxylapatite	502.32	$Ca_5(PO_4)_3OH$
Monazite	235.09	$CePO_4$
Halite	58.44	NaCl

APPENDIX 8

Mineral Oxide Compositions

Group	Mineral	SiO_2	TiO_2	Al_2O_3	Fe_2O_3	FeO	MnO	MgO	CaO	Na_2O	K_2O	H_2O	Other	Oxide
Olivine	Forsterite	42.71						57.29						
Olivine	Fayalite	29.49				70.51								
	Zircon	32.78											67.22	ZrO_2
	Titanite	30.65	40.75						28.60					
Garnet	Pyrope	44.71		25.29				29.99						
Garnet	Almandine	36.21		20.48		43.30								
Garnet	Spessartine	36.41		20.60			42.99							
Garnet	Grossular	40.02		22.64					37.35					
Garnet	Andradite	35.47			31.42				33.11					
Aluminosilicate	Sillimanite, kyanite, andalusite	37.08		62.92										
	Staurolite	28.21		53.86		16.87						1.06		
	Chloritoid	23.85		40.48		28.52						7.15		
	Zoisite	39.67		33.66					24.68			1.98		
	Epidote	37.30		21.10	16.52				23.21			1.86		
	Cordierite(Mg)	51.36		34.86				13.78						
	Cordierite(Fe)	46.36		31.47		22.17								
Opx	Enstatite	59.85						40.15						
Opx	Ferrosilite	45.54				54.46								
Cpx	Diopside	55.49						18.61	25.90					
Cpx	Hedenbergite	48.44				28.96			22.60					
Cpx	Jadeite	59.45		25.22						15.33				
Pyroxenoid	Wollastonite	51.72							48.28					
Amphibole	Tremolite	59.17						24.81	13.81			2.22		
Amphibole	Ferroactinolite	49.55				37.03			11.56			1.86		
Amphibole	Edenite	50.42		6.11				24.16	13.44	3.71		2.16		
Amphibole	Ferroedenite	42.40		5.14		36.21			11.31	3.12		1.82		
Amphibole	Pargasite	43.13		18.30				19.29	13.42	3.71		2.16		
Amphibole	Ferropargasite	37.47		15.90		29.87			11.66	3.22		1.87		
Amphibole	Magnesiohastingsite	41.69		11.79	9.23			18.64	12.97	3.58		2.08		
Amphibole	Hastingsite	36.38		10.29	8.06	29.00			11.32	3.13		1.82		
Amphibole	Glaucophane	61.35		13.01				15.43		7.91		2.30		
Amphibole	Ferroglaucophane	54.74		11.61		24.54				7.06		2.05		
Mica	Muscovite	45.25		38.40							11.82	4.52		
Mica	Paragonite	47.16		40.02						8.11		4.71		
Biotite	Phlogopite	43.20		12.22				28.98			11.29	4.32		
Biotite	Annite	35.21		9.96		42.11					9.20	3.52		
Sheet silicate	Pyrophyllite	66.70		28.30								5.00		
Sheet silicate	Talc	63.37						31.88				4.75		
Chlorite	Clinochlore	32.43		18.35				36.26				12.97		
Chlorite	Amesite													
Serpentine	Antigorite	43.36						43.63				13.00		
Clay	Kaolinite	46.55		39.50								13.96		
Feldspar	Orthoclase	64.76		18.32							16.92			
Feldspar	Albite	68.74		19.44						11.82				
Feldspar	Anorthite	43.19		36.65					20.16					
Feldspar	Celsian	32.01		27.16									40.84	BaO
Silica	Quartz, coesite, tridymite, cristobalite	100.00												
Feldspathoid	Nepheline (Na,K)	41.13		34.90						15.91	8.06			
Feldspathoid	Kalsilite	37.99		32.23							29.78			
Feldspathoid	Leucite	55.06		23.36							21.58			
Oxide	Periclase							100.00						
Oxide	Corundum			100.00										
Oxide	Hematite				100.00									
Oxide	Rutile		100.00											
Fe-Tioxide	Imenite		52.65			47.35								
	Perovskite		58.76						41.24					
Spinel	Spinel			71.67				28.33						
Spinel	Hercynite			58.66		41.34								
Spinel, Fe-Ti oxide	Magnetite				68.97	31.03								
Spinel	Ulvöspinel		35.73			64.27								
Spinel	Chromite					32.10							67.90	Cr_2O_3
Hydroxide	Brucite							69.11				30.89		
Carbonate	Calcite								56.03				43.97	CO_2
Carbonate	Dolomite							21.86	30.41				47.73	CO_2
Apatite	Hydroxylapatite								55.82			1.79	42.39	P_2O_5
Apatite	Fluorapatite								55.60				42.2, 3.77	P_2O_5, F
	Monazite												30.18, 69.81	P_2O_5,Ce_2O_3

Figure A8.1 Oxide compositions (wt%) of selected minerals. Each column is shaded such that higher values are darker. Data from LeMaitre (1982).

APPENDIX 9

Oxide Molecular Weights and Gravimetric Factors

To convert wt% oxide to wt% metal, multiply by the factor, which is the ratio of the metal mass to the mass of the oxide. To convert wt% metal to wt% oxide, divide by the factor.

Oxide	Molecular Weight (g/mol)	Gravimetric Factor
Al_2O_3	101.96	0.5293
B_2O_3	69.62	0.3106
BaO	153.34	0.8957
CaO	56.08	0.7147
CO_2	44.01	0.2729
Cr_2O_3	151.99	0.6842
Fe_2O_3	159.69	0.6994
FeO	71.85	0.7773
H_2O	18.02	0.1119
HfO_2	210.49	0.8480
K_2O	94.20	0.8301
La_2O_3	325.81	0.8527
Li_2O	29.88	0.4646
MgO	40.30	0.6030
MnO	70.94	0.7745
Na_2O	61.98	0.7419
Nb_2O_5	265.81	0.6990
NiO	74.71	0.7858
P_2O_5	141.94	0.4364
Rb_2O	186.94	0.9144
SiO_2	60.08	0.4674
SO_3	80.06	0.4005
SrO	103.62	0.8456
Ta_2O_5	441.89	0.8190
ThO_2	264.04	0.8788
TiO_2	79.90	0.5995
U_3O_8	842.08	0.8480
UO_2	270.03	0.8815
Y_2O_3	225.81	0.7874
ZrO_2	123.22	0.7403

APPENDIX 10

Plots of Average Igneous Rock Compositions

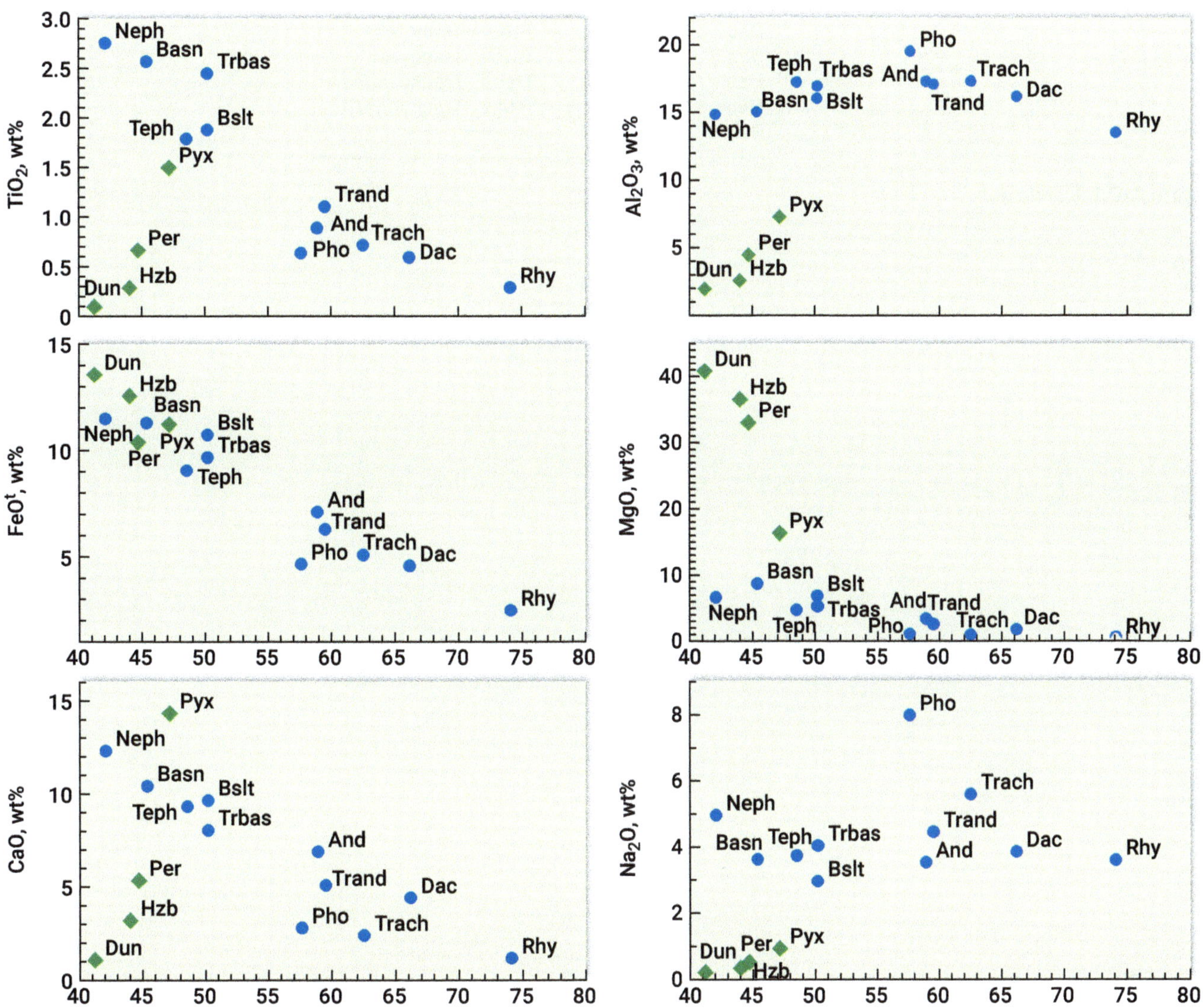

Figure A10.1 Various oxides plotted against silica for average igneous rock compositions.

Source: From the compilation of LeMaitre (1976).

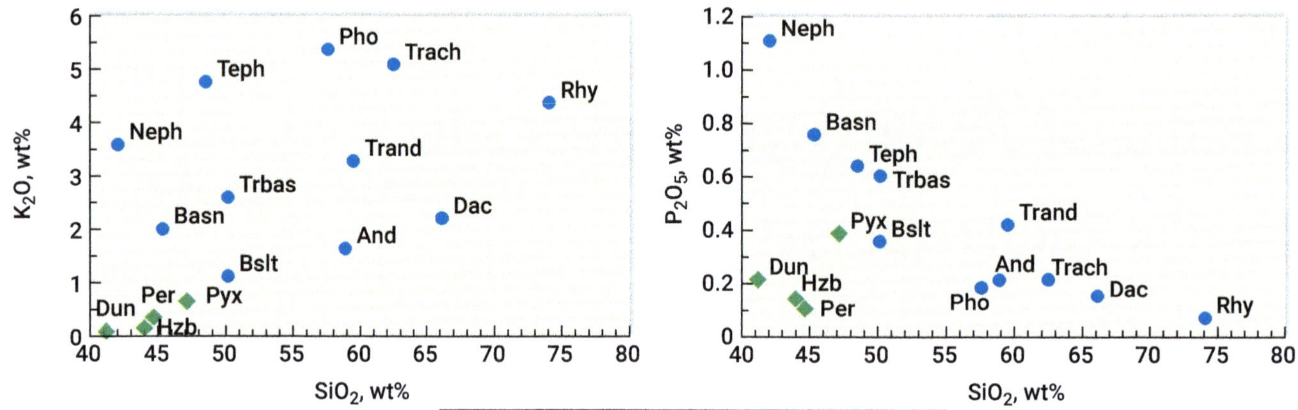

And	Andesite	Pho	Phonolite
Basn	Basanite	Pyx	Pyroxenite
Bslt	Basalt	Rhy	Rhyolite
Dac	Dacite	Teph	Tephrite
Dun	Dunite	Trach	Trachyte
Hzb	Harzburgite	Trand	Trachyandesite
Neph	Nephelinite	Trbas	Trachybasalt
Per	Peridotite		

Figure A10.1 *(continued)*

Glossary

A'a—Basaltic lava flows with a rough, spiky surface.

Abyssal—Deeper parts of the ocean, >2000–3000 m water depth.

Accreted terrane—A terrane (e.g., an island arc) that has been attached to a continent.

Accretion—Suturing of an island arc or continental mass on a downgoing plate to a continental mass on the overriding plate.

Accretionary wedges—The wedges of sediment and metamorphic rocks scraped off the subducting slab; the part of the slab that is not subducted.

Adiabatic—An adiabatic process is one in which heat does not enter or leave the system.

Advection—Here, transfer of heat by fluid (magma) flow.

AFM diagram—A ternary composition diagram representing the KFMASH system as projected from quartz, water, and either muscovite or K-feldspar. Not the same as an igneous AFM diagram, which is simply a composition plot.

Aid climbing—Climbing where gear such as pitons, bolts, or other equipment is used to aid upward progress.

Air-fall tuffs—Deposits formed by pyroclastic fall material.

Alteration—Chemical changes in a rock relative to a higher-temperature reference condition. Usually refers to the addition of water or other components, such as hydration of basalts or high-T anhydrous minerals (e.g., olivine) to produce hydrous amphiboles or sheet silicates (e.g., serpentine).

Alumina saturation index (ASI)—A measure of whether there is excess Al in a magma beyond that needed to make feldspar; if so, then peraluminous minerals, such as muscovite and garnet, may form to take up the excess.

Anhydrous—Water-free.

Aphelion, perihelion—Points in an orbit farthest and closest, respectively, to the body around which orbiting takes place; plurals end in "ia."

Aplite—Fine- to medium-grained leucogranite, typically described as having a sugary texture.

Arrhenius equation, plot—Mathematical expression that describes the rate of a process (e.g., reaction or diffusion) as an exponential function of activation energy and absolute temperature; a graph that plots the logarithm of a rate vs. reciprocal temperature.

Asbestiform—A textural term meaning fibrous or able to be split into fibers.

Asbestos—A commercial product made from fibrous (asbestiform) minerals.

Aseismic ridges—Long, linear, seismically quiet, volcanically inactive seafloor ridges.

Ash—Pyroclastic particles <2 mm across (smaller than lentils).

Aspect ratio—Generally used to mean the long dimension of an object divided by the short dimension.

Asthenosphere—Relatively weak zone in the upper mantle, softened by high temperatures and capable of flow at plate tectonic strain rates.

Austenite—Also γ-Fe; medium-T Fe polymorph that can dissolve up to 2 wt% C.

Back-arc—The part of a subduction zone that is behind the arc and farthest from the trench; the back-arc region can contain basins floored by actively spreading oceanic crust.

Back-arc basins—Basins in the back-arc; often a site of active spreading.

Baldy's law—Some of it plus the rest of it equals all of it.

Basement—Metamorphic and igneous rocks that form the foundation of the crust, below any sedimentary rock cover.

Basin and Range province—A region of active crustal extension, volcanism, and high heat flow in western North America characterized by north-trending mountain ranges separated by valleys. The Basin and Range encompasses most of Nevada, eastern California, western Utah, western and southern Arizona, and part of northern Mexico.

Batholith—A region tens of kilometers or more in lateral extent that is composed primarily of plutons.

Bimodal—Here, a statistical distribution with two maxima (modes).

Bingham fluid—A fluid in which a critical yield stress must be exceeded before it starts to flow.

Black smokers—Plumes of fluid in deep-sea hydrothermal systems, colored dark gray or black by precipitating sulfides.

Block-and-ash flow—Hot density current of angular blocks, as from a collapsing dome, in an ashy matrix.

Block lava—Viscous to hyperviscous lava that breaks into semi-angular blocks during flow.

Bolide—In geology, a large, crater-forming impactor; in astronomy, an exceptionally bright meteor (fireball).

Bomb, block—Pyroclastic material >64 mm (tennis ball) across; bombs were molten when ejected, whereas blocks were solid.

Bond strength—For an ion, charge divided by coordination number.

Boudinage—A structure in deformed rocks in which a stronger layer has been pinched apart during deformation, with weaker layers flowing into the gaps. Individual fragments are commonly strung out in a line, like linked sausages; *boudin* is a French word, one translation of which is sausage.

Boundary conditions—How the model space interacts with the world outside; in a thermal model, typical boundary conditions are constant temperature, constant temperature gradient, constant heat flux, or insulated (zero heat flux).

Bridging oxygen—An oxygen atom shared between two Si or Al tetrahedra.

Buffer—Here, to limit the composition of a phase or an intensive variable such as T via a chemical reaction. Solid-solid reactions often buffer the compositions of fluids.

Bulk silicate Earth, BSE—The estimated overall composition of the Earth after subtracting the core, as estimated from meteorites and models of solar system formation. Because the mantle dominates the silicate budget, primitive (unprocessed) mantle is thought to have a composition similar to BSE.

Calciocarbonatite—Magma consisting largely of calcium carbonate.

Caldera—Steep-sided, generally ovoid depression formed by collapse after eruption or lateral withdrawal of magma.

Carbonate compensation depth—Depth in the ocean at which the rate of calcite dissolution equals the rate at which calcite particles (e.g., foram shells) are supplied from above; typically, 4–5 km. Below this depth calcite does not accumulate, and thus the abyssal plains are covered in siliceous mud rather than calcite mud.

Carbonate-silicate cycle—The weathering of silicates, transport to oceans, and precipitation of carbonates, to reduce CO_2 concentrations in the atmosphere, followed by subduction of carbonates and reaction with silicates to form CO_2, which escapes to the atmosphere.

Carbonatites—Carbonate-rich intrusive and volcanic rocks, commonly associated with highly alkaline silicate rocks.

Cataclasite—A low-T rock formed through brittle fracturing, shattering, or granulation of its constituent minerals. Essentially a cemented breccia.

Cementite—Iron carbide, Fe_3C; has been found in a meteorite and given the name cohenite after its discoverer, so it is technically a mineral, but a vanishingly rare one.

Chemical potential—Change in Gibbs free energy per mole when a small amount of a component is added to a phase, with all other variables held constant; at equilibrium, the chemical potential of a given component is the same in all phases.

Chemographic diagrams—Graphical representations of a chemical reaction.

Chondrites—Primitive meteorites composed of relatively unprocessed material from the solar nebula.

Chondrules—Formerly molten, spherical globules of olivine and pyroxene in chondritic meteorites.

CHUR—A theoretical reservoir with a chondritic Sm/Nd ratio; as with VSMOW and δ notation, CHUR is used as a baseline for ε notation when discussing variations in $^{143}Nd/^{144}Nd$.

CIPW norm—An algorithm for converting the oxide composition of an igneous rock into a standard set of anhydrous minerals that describes a possible equilibrium assemblage for that bulk composition.

Clapeyron equation—Used to predict the slope of a reaction boundary on a P-T diagram in terms of entropy and volume changes.

Cleavages—In rocks, planes of weakness along which they cleave or split (e.g., along foliations).

Clockwise P-T path—A path that describes a clockwise loop on a P-T diagram when the pressure axis is drawn to increase upward.

Closure temperature (T_c)—The temperature at which loss of the radiogenic isotope (^{206}Pb, ^{40}Ar, etc.) becomes sufficiently slow that it is essentially wholly retained. Above T_c, the radiogenic isotope is (mostly) lost, and the isotopic clock is not counting; below T_c, the radiogenic isotope is (mostly) retained, and the age follows the normal age equation.

Columnar joints—In volcanic rocks, joints that form by thermal contraction during cooling, producing six-, five-, or seven-sided columns.

Compaction—In igneous petrology, expulsion of liquid from a crystal mush by crystal settling under the force of gravity.

Compatible elements—Trace elements that are, at equilibrium, present in higher concentration in a mineral than in coexisting melt.

Component—One of a set of the smallest number of chemical constituents needed to describe the compositions of all the phases in a system.

Concordant—Said of a geologic contact that is parallel to bedding or foliation.

Concordia diagram—A plot of $^{206}Pb/^{238}U$ vs. $^{207}Pb/^{235}U$, used to interpret U-Pb data in terms of age and as a check that the two chronometers are in sync.

Congruent melting—Melting of a phase in which the composition of the liquid formed is the same as the composition of the phase being melted.

Contact aureole—The region adjacent to an igneous body whose mineralogy, geochemistry, and/or texture have been influenced by the intrusion (i.e., the contact metamorphic rocks around an intrusion).

Contact metamorphism—Metamorphism driven by proximity to an igneous body. Includes both thermal effects and fluid-rock interaction.

Continental collision—Collision of continental material on a downgoing plate with a continental mass on the overriding plate.

Cordillera—A system of generally parallel mountain ranges and basins; generally used for those along the western margins of the Americas, from Alaska to Patagonia.

Core infall—Postulated event during Earth formation in which immiscible globules of metallic Fe-Ni solution sank to form the core.

Core-mantle boundary (CMB)—The profound chemical and thermal boundary between the core and mantle.

Coupling depth—The point along the top of the subducting plate where it binds with the overriding mantle, driving circulation of the mantle wedge. Dehydration of weak serpentine along the interface to form strong olivine is thought to control this depth.

Craton—Generally low-relief core of a continent that has been tectonically stable for prolonged periods, on the order of hundreds of millions of years.

Critical point—The point on a liquid → gas transition curve on a P-T diagram at which the distinction between liquid and gas disappears.

Crossing reaction—A reaction where multiple products are made by combining multiple reactants; on a chemographic diagram, the geometric figure (line segment, triangle, etc.) defined by the products crosses the figure formed by the reactants.

Cumulate—A term classically applied to igneous rocks thought to have formed by gravitational settling of crystals with consequent loss of magmatic liquid; see Section 14.5 for complications involving this term.

Cumulus crystals—Crystals that accumulate preferentially relative to liquid; assumed to reflect processes that separate crystals and liquids via crystal flotation, crystal settling, compaction, etc.

Curie point—Temperature above which ferromagnetism disappears owing to thermal agitation; for natural magnetite, the Curie point, which varies with composition, is ~570 °C. Named for Pierre, the lesser-known Curie.

D″—(dee-double-prime) A designation, from an archaic seismologic classification of mantle layers, for the lowermost few hundred kilometers of the mantle; a thermal and likely compositional boundary layer that is not identical to the core-mantle-boundary owing to its considerable thickness.

Debris avalanches—Masses of rock, soil, snow, and ice that move down the side of a mountain.

Dehydration melting—The breakdown of a hydrous mineral to form a melt in the absence of free water.

Depleted mantle—Mantle that has been depleted in incompatible elements, especially large-ion lithophile elements (K, Rb, Cs, Sr, Ba) and light REE (La, Ce, Pr, Nd), which are preferentially sequestered in continental crust.

Detachment faults—Large-scale, low-angle faults separating an upper plate from a lower plate; when unqualified, the term generally refers to extensional faults in contrast to thrust faults.

Diabasic texture—Texture characteristic of shallow basalt intrusions in which laths of plagioclase are encased in much larger crystals of pyroxene or olivine.

Diamictites—Poorly sorted sediment with grains sizes ranging from clay to boulder. Many diamictites represent glacial deposits, including tills (tillite) and diamictites containing dropstones (ice-rafted pebbles, cobbles, and boulders) in fine-grained marine sediments.

Diapirs—Kilometer-scale bodies of material, generally circular in map view, that have punched up through overlying layers either by buoyant ascent or owing to tectonic forcing of low-viscosity material.

Diaplectic glasses—Amorphous material in which long-term crystallographic order has been destroyed by a shock wave; unlike normal glasses, diaplectic glasses can be annealed back into something approximating the original crystal.

Diatreme—Funnel- or carrot-shaped pipe filled with breccia composed of fragmented magma and wall rock; may reach downward 1 km or more. Diatremes are likely expressed as maars at the surface and terminate downward in dikes.

Dike—A tabular, thin (generally meter-scale) intrusive body. Term is used for both active magma pathways and the frozen remains thereof.

Dikelets—A component of a composite dike interpreted to represent a single pulse of magma; dikelets are commonly seen to be split in half by later pulses.

Dike swarms—Large numbers (typically hundreds or thousands) of subparallel dikes that are associated spatially and temporally.

Diffusion coefficient—Here, a parameter that describes how rapidly a chemical species diffuses in a given concentration gradient.

Discordant—Said of a geologic contact that cuts across bedding or foliation.

Dissolution-reprecipitation—Mineral replacement process by which slightly unstable material dissolves in a fluid and reprecipitates in a more stable form.

Distribution coefficient—Ratio of the concentration of a trace element in a crystal to that in coexisting melt.

Divariant assemblage—A phase assemblage in which you can make small independent changes in either or both of two intensive variables and maintain the same phase assemblage. On a *P-T* diagram, a divariant assemblage occupies an area between two reaction curves.

Dynamic metamorphism—Transformation of a rock into cataclasite or mylonite via movement along faults.

Effusion—Eruption of fluid lava.

En echelon—Said of a series of overlapping, parallel cracks, dikes, etc. that form a linear trend where each element is oblique to that trend.

Enriched mantle—Mantle that is enriched in incompatible elements, especially large-ion lithophile elements and light REE. Characterized by relatively high Rb/Sr and $^{87}Sr/^{86}Sr$, and low Sm/Nd and $^{143}Nd/^{144}Nd$ ratios.

Essential structural constituents (ESCs)—Equivalent to stoichiometric constituents.

Eutectic—In eutectic melting, two or more solid phases melt to a liquid whose composition is a mixture of the compositions of the phases, and this melting occurs at a fixed temperature and with fixed composition regardless of the proportions of the solid phases. Greek derivation, meaning "easily melting."

Exsolution—A process whereby a homogeneous solution splits into two phases by forming separate domains that are enriched in each component.

Exsolve—Form by coming out of solution from another phase.

Extensive variable—One whose value depends on the amount (extent) of the system or phase considered (e.g., mass).

Fabric—The texture of a rock, which can include its primary depositional characteristics (e.g., stratigraphic layering), but in metamorphic rocks more commonly refers to the preferred arrangement of linear or planar features, especially from elongate or platy minerals.

Facies—The textural and/or mineralogical character of a rock, indicative of the conditions under which it formed.

Fault gouge, fault breccia—Finely comminuted, unconsolidated rock material; fault gouge with a substantial proportion of clasts is fault breccia. May be foliated if it contains significant layer silicates.

Felsic—Denoting a rock with a high abundance of feldspars and quartz or, in alkalic rocks, feldspathoids such as leucite or nepheline.

Fenitization—Addition of Na and K to host rocks around carbonatite intrusions. Crystallization of carbonatite magmas evolves highly alkali- and H_2O-rich fluids that react with the surrounding rock to produce an alkali-rich rock called fenite.

Ferrite—Also α-Fe; low-T Fe polymorph with low C solubility.

Fertility—The degree to which a rock is able to produce melt after a given thermal input.

Filter pressing—Separating magmatic liquid from crystals by squeezing crystal-rich magma tectonically or by forcing it through a constriction.

Foliations—Planar fabrics, usually in a metamorphic rock.

Forbidden zones—In this context, regions on a phase diagram in which no phase composition may plot at equilibrium.

Fore-arc—The region of a subduction zone between the arc and trench; includes the accretionary wedge and any fore-arc basins.

Formation water—Pore water in sediments or sedimentary rocks (i.e., in a sedimentary rock formation).

Forward model—Prediction of how a system will behave given specific boundary and initial conditions and equations governing its physical and/or chemical behavior.

Fractional crystallization, crystal fractionation—The process of separating magmatic liquid from crystals, or vice versa, resulting in liquid-enriched and liquid-depleted volumes of different composition.

Fractional melting—A melting process in which melt is extracted as it is formed, rather than being allowed to continuously react with residual crystals.

Fractionate—Separate similar elements, chemicals, or isotopes from one another.

Free climbing—Climbing in which all upward progress is done using hands, feet, and other parts of the body on natural features; equipment is used for fall protection only.

Fumaroles—Hot volcanic gas vents.

Geobaric gradient—Rate at which pressure increases with depth.

Geotherm—Variation of temperature with depth.

Geothermal gradient—Rate at which temperature increases with depth; typically given in °C/km.

Geothermometer, geobarometer—A method, generally based on geochemical measurements of phase compositions, that can estimate T or P at a given point in a rock or magma's life cycle.

Gibbs free energy—The amount of chemical energy available to do work in a system at constant P and T.

Gondwanaland—A supercontinent formed during the latest Proterozoic from the assembly of India, South America, Africa, Arabia, Australia, and Antarctica.

Granoblastic—A texture in metamorphic rocks in which grains are approximately equidimensional and grain boundaries meet at triple junctions with angles near 120° in two-dimensional cuts.

Great Basin—The northern part of the Basin and Range province; drainages currently terminate inside the continent and do not reach the ocean.

Half-grabens—Asymmetric grabens with significantly more displacement on one side than the other.

Hand sample—Said of a rock specimen that fits conveniently in one's hand for close examination.

Haplogranite—The system Si-Ab-Or-H_2O, a simplification of the high-silica granite system that leaves out important components, such as CaO, MgO, and Fe oxides.

Heat capacity—The amount of heat needed to produce a given increase in temperature for a given quantity of material; typical units are J °C^{-1} kg^{-1}.

Heat flow—Flux of heat through a surface (e.g., surface of the Earth or wall of a dike), in units of energy per unit time per unit area (e.g., W/m^2).

Heat flow rule of thumb—Relationship between time (t) and distance (d) when $\tau = 1$. For a specified value of κ (thermal diffusivity), this can be used to estimate timescales of heat diffusion ($t = d^2/\kappa$, given a length scale, d), or heat penetration distances ($d = \sqrt{\kappa t}$, given a timescale, t).

Henry's law—Here, the observation that at low concentrations, the concentration of an element in a mineral is proportional to that in coexisting melt.

Hornfels—Literally a stone that resembles animal horn: hard, fine-grained, and usually dark gray or black with little layering and no tendency to split along planes. Hornfels form adjacent to plutons, as rocks thoroughly recrystallize.

Hotspot—Postulated source of magma that forms volcanic chains, islands, or seamounts far from plate boundaries; may or may not be sourced from the deep mantle or be much hotter than surrounding mantle.

Hyaloclastite—Granular material formed by fragmentation as lava enters water.

Hydration—Addition of water as a structural component in minerals, either as H_2O or OH^- (or both). Hydration of olivine (Mg_2SiO_4) and enstatite ($MgSiO_3$) can form serpentine [$Mg_3Si_2O_5(OH)_4$]; hydration of anhydrite ($CaSO_4$) can produce gypsum ($CaSO_4{\cdot}2H_2O$). Glass can be hydrated simply by dissolving H_2O into it.

Hydrothermal system—Circulation of water through fractures or rocks, driven by temperature and buoyancy gradients. Hot, buoyant water rises above MORs and other heat sources and is replaced by cooler water, driving a flow system.

Hypabyssal—Intrusive rock emplaced at shallow (undefined) depth.

Hypervelocity impacts—Collisions of material moving at several kilometers per second with a planetary body.

Hyperviscous—Having an extremely high viscosity, on the order to 10^7 Pa·s or more, such that the material seems nearly solid at human timescales.

Ignimbrite—The deposits of a pyroclastic flow; the term is generally used for pumice flows.

Incompatible elements—Trace elements that are, at equilibrium, present in lower concentration in a mineral than in coexisting melt.

Incongruent melting—Melting of a phase in which the composition of the melt differs from the composition of the phase being melted.

Incremental emplacement—Construction of a plutonic body in small increments rather than via intrusion of the entire body all at once.

Index minerals—Minerals whose appearance marks an increase in metamorphic grade.

Initial conditions—The starting conditions for a numerical model; here, the initial temperature distribution.

Intensive variable—One whose value is the same in all parts of a system at equilibrium and does not depend on the amount of the system considered (e.g., temperature).

Intercumulus liquid—Magmatic liquid trapped between cumulus crystals.

Interior reaction—A reaction where a single product is made by combining multiple reactants; on a chemographic diagram the product lies within the volume formed by the reactants.

Intrusive suites—Groups of plutons related in time, space, outcrop pattern, and fabric; typically nested.

Inverse problems—Identification of what processes occurred to produce a set of observations. Most geologic questions are inverse problems—we seek to understand past processes that formed the rock record.

Iron peak elements—The quintet of elements centered on Fe (Cr, Mn, Fe, Co, and Ni); these elements have the highest nuclear binding energy and thus the most stable nuclear configurations.

Island arc—A chain of volcanoes associated with subduction of oceanic lithosphere; may or may not be arcuate.

Isochron—A linear array on an isotope diagram whose slope defines an age. The ordinate is the radiogenic isotope and the abscissa the parent isotope, both normalized to a stable isotope of the radiogenic element.

Isograd—A surface (line on a map) marking the boundary of a metamorphic zone.

Isopach—Contour connecting points of equal thickness of a rock layer.

Isopleths—Contours of constant value; most isopleths are for composition, but they can also be for modal abundance, seismic properties, etc.

Isotopic signature—The characteristic isotope composition of a major reservoir (e.g., the isotopic signature of the mantle or ocean).

I-type granite—A granite born from igneous sources.

Juvenile water—Water that is (theoretically) sourced from Earth's origin, so it hasn't participated in Earth's hydrologic cycle previously. More generally, water that comes from the mantle or from crystallization of magmas. Clearly, this is a problematic term, because Earth's mantle and its magmas do participate in the hydrologic cycle.

Kilonova—Explosive event produced by the merger of two neutron stars or a neutron star and black hole.

Kimberlites—Low-silica, high-MgO, serpentinized breccias, rich in olivine, carbonate, and H_2O. They are composed of deeply sourced magma carrying a large proportion of xenocrysts entrained from mantle and crustal wall rocks.

Kipukas—Islands of land isolated by lava flows; may serve to isolate floral or faunal populations.

Komatiite—Volcanic rocks formed from high-temperature peridotitic magma; almost exclusively of Archean and Proterozoic age.

Laccolith—Magmatic blisters intruded into flat-lying sedimentary rocks, with a convex roof and a relatively flat floor; may be faulted along part of its margin.

Lahars—Mudflows composed of loose material on the flank of a volcano.

Lamproites—K_2O- and MgO-rich, low-silica, hypabyssal or extrusive rocks with high concentrations of incompatible elements, high Fe^{3+}/Fe^{2+}, and exotic mineralogy.

Lamprophyres—Rare, ultrapotassic mafic to ultramafic igneous rocks, usually exposed as dikes or diatremes, with phenocrysts of Cpx and either biotite or hornblende, but lacking phenocrystic plagioclase.

Lapilli—Pyroclastic particles between 2 and 64 mm across (lentils to tennis balls); singular lapillus.

Lapse rate—The rate at which temperature decreases with altitude.

Large igneous provinces (LIPs)—Intraplate magmatic province with area >100,000 km^2 and volume >100,000 km^3 with large magmatic fluxes (e.g., >75% erupting in <5 Myr; Kerr, 2014).

Lava—Molten material that flows out of the ground as a fluid.

Lava tube—Tunnel produced by roofing-over of a fluid lava flow; refers to both active channels and former ones.

Layered mafic intrusions (LMIs)—Mafic plutons characterized by banding defined by dramatic variations in mineral proportions, sizes, and compositions.

Leucogranite—Light-colored granitic rock composed of quartz and feldspars with SiO_2 near 77 wt% and no more than a few percent mafic minerals; intrusive equivalent of high-silica rhyolite.

Lever rule—Graphical method for determining the proportions of phases in a system from a chemographic diagram.

Limnic—Relating to a body of low-salinity water such as a lake.

Lineations—Linear fabrics, usually in a metamorphic rock.

Liquidus—For a given bulk composition at a given pressure, the temperature at which crystallization begins upon cooling.

Listric faults—Curved normal faults that flatten at depth; extension causes strata near the fault to dip toward the fault, producing an anticline.

Lithostatic—Confining pressure given by the weight of the overlying rocks per unit area.

Lithosphere—Outer, rigid shell of the Earth, with a sharp upper boundary (the surface) and a gradational lower boundary that is typically equated to an isotherm such as 1200 °C; typical lithosphere is ~100 km thick.

Lithosystem—A set of rock types and related structures, such as faults or fabrics, that are collectively diagnostic of a given environment of formation.

Lopoliths—Magmatic blisters intruded into flat-lying sedimentary rocks, with a flat roof and a concave floor; may be faulted along part of its margin.

Lower mantle—Mantle below 660 km.

Lunar maria—Relatively dark basalt "seas" that fill impact basins on the Moon's near side; singular mare.

Maars—Shallow, broad craters formed by explosive reaction of magma with groundwater.

Macrocrysts—Term used for larger (generally 0.5–10 mm) crystals in kimberlite; includes both phenocrysts and xenocrysts.

Mafic—Denoting a rock with a high abundance of ferromagnesian minerals such as pyroxenes and amphiboles.

Magma—Naturally occurring, fully or partially molten rock material generated within a planetary body, consisting of melt with or without crystals and bubbles and containing a high enough proportion of melt to be capable of intrusion and extrusion.

Mantle array—Refers to the nearly linear array of oceanic basalt compositions on a plot of Nd vs. Sr isotopes. The term has been generalized to other isotope plots.

Mantle wedge—A wedge of mantle toward the arc-side of a subduction zone with sufficiently high temperature in its core to flow readily.

Magmatic stoping—Hypothesis that magma bodies ascend by plucking blocks out of the roof and moving into the space thus freed up.

Megacrysts—Here, unusually large crystals, typically defined as >5 cm in long dimension.

Mélange—A tectonic mixture of high-*P* rocks, such as eclogite, in a sheared, weak matrix of metasedimentary rocks and serpentinite.

Metamorphic core complexes—Geographically limited suites of metamorphic and plutonic rocks exposed by extension. Typically, low-angle faults and shear zones juxtapose unmetamorphosed or hydrothermally altered sedimentary and volcanic rocks above the fault against high-grade metamorphic rocks below.

Metamorphic differentiation (metamorphic segregation)—Development of banding or mineral segregations from a uniform protolith during metamorphism.

Metamorphic facies—A metamorphic mineral assemblage, developed in a rock of basaltic bulk composition, characteristic of a *P-T* region in the crust and upper mantle. Alternatively, the *P-T* region in which that mineral assemblage is stable in a metabasalt. The facies are named for metabasaltic rocks (e.g., greenschist, amphibolite), but relevant *P-T* conditions can be applied to other rock compositions (i.e., a metashale might be described as a schist metamorphosed under amphibolite-facies conditions).

Metamorphic field gradient—The distribution of *P-T* conditions recorded by rocks as exposed in the field.

Metamorphic grade—A qualitative characterization of metamorphic temperature, which often also correlates with pressure. A high-grade metamorphic rock forms at elevated temperature, whereas a low-grade metamorphic rock forms at lower temperature.

Metamorphic zones—Rock volumes (areas on a map) bounded by isograds.

Metamorphism—Literally a change (meta) in shape or form (morph). Usually implies growth of minerals in response to a different pressure or temperature than the rock's original formation conditions (e.g., formation of a metamorphic schist from a sedimentary shale). Deformation can also form metamorphic rocks without changing mineral assemblage (e.g., shearing can transform a granite into a gneiss).

Metasomatism—Literally a change (meta) in body (soma). Usually refers to changes in whole-rock compositions because of fluid flow, which can either add or remove chemical constituents. Hydration is a form of metasomatism.

Meteoric water—Water that comes from the sky in any form—snow, sleet, rain, etc.

Meteoroids, meteors, meteorites—Respectively, pieces of interplanetary matter, the incandescent streaks those pieces leave in the sky, and the debris that hits a planet if the meteoroid survives passage through the atmosphere.

Mid-ocean ridges (oceanic ridges)—Linear ridges in the ocean basins across which seafloor spreading takes place and new oceanic lithosphere is generated.

Migmatite—Partially molten gneissic rock.

Mineral assemblage diagrams—Diagrams, usually *P-T* diagrams, that show thermodynamically predicted mineral assemblages for a specific rock composition. Also called a pseudosection or bulk-composition-specific phase diagram.

Mineral facies—A collection of minerals characteristic of a given set of *P* and *T* conditions for a given bulk composition.

Moat—In volcanology, the low area between the resurgent dome and the caldera wall; in some calderas the moat is filled with abundant but inconspicuous basalt flows.

Mode—Volume percentages of minerals in a rock.

Mohorovičić discontinuity (Moho)—Downward jump in seismic velocity from 6–7 km/s to ~8 km/s; typically interpreted as the crust-mantle interface.

Mudflows, lahars—Gravity-driven flows of water-lubricated clastic material down the side of a volcano; may be hot or cold and need not be produced by an eruption.

Mylonite—An intensely deformed rock in which grain size reduction has occurred ductilely. Typically formed at high temperature.

Nappe—Large-scale recumbent fold, generally associated with thrust faulting.

Natrocarbonatite—Magma consisting largely of Na and K carbonates.

Nebula—A cloud of gas and dust in space; gases are predominantly H and He.

Negative feedback—A process whose effects counteract or reverse that process (e.g., weathering has a negative feedback on atmospheric CO_2 concentration).

Nephelinites—Alkaline volcanic or hypabyssal rocks consisting largely of nepheline and Cpx and lacking olivine and plagioclase.

Neutron star—Dense ball of neutrons left after a supernova explosion; may have a mass comparable to the Sun in a volume 20 km across.

Newtonian fluid—A fluid for which shear stress and strain rate are strictly proportional.

Norm—A rock composition recalculated in terms of compatible end-member minerals, typically using the CIPW algorithm.

Nuclide—A specific isotopic form of an element, defined by the numbers of protons and neutrons in the nucleus.

Ocean-island basalts (OIBs)—Basalts from ocean islands that are not associated with spreading centers or subduction zones. The geochemistry of OIBs more closely approaches primitive mantle or bulk silicate Earth than does that of MORBs.

Ophiolite—Association of deep-sea sediments (typically chert), pillow lavas, basaltic dikes, gabbro, and peridotite; interpreted to be a piece of oceanic crust and upper mantle.

Orogenic belt, orogen—An elongate region that has undergone lithospheric shortening.

Oxygen fugacity—The partial pressure of oxygen (O_2) in a system corrected for deviations from the ideal gas law; typically written as f_{O_2}.

Pāhoehoe—Basaltic lava flows with a relatively smooth, folded surface that can look like coils of rope.

Paleolatitude—The latitude of a point on the Earth's surface when a given unit of interest (e.g., a lava flow) was formed.

Paragenesis diagram—A diagram that shows the appearance and disappearance of key minerals across a range of *P-T* conditions from low-grade to high-grade.

Pearlite—The product of eutectic exsolution of austenite to form intergrown plates of cementite and ferrite.

Péclet number—A dimensionless number giving the ratio of the rate at which material moves to the rate at which heat is transported by diffusion.

Pegmatite—Coarse-grained to extremely coarse-grained bodies, usually of quartz and feldspars; muscovite, tourmaline, and other unusual minerals may be prominent.

Peraluminous, metaluminous, subaluminous—Rocks with ASI significantly >1, within a few percent of 1, and significantly below 1, respectively.

Peristerite gap—A compositional gap in plagioclase between nearly pure albite and oligoclase that is locally seen in metamorphic rocks.

Peritectic melting—Melting in which a phase or phase assemblage melts to a liquid of a different composition plus a different phase or phase assemblage.

Petrochronology—Geochronology in the context of petrology. Links time (rates and durations) with rock-forming processes and *P-T* conditions.

Petrogenetic grid—A *P-T* diagram showing reactions that control appearance and breakdown of phases in a given chemical system.

***P*-first *P-T* path**—A path on a *P-T* diagram where pressure increases significantly before temperature does; equivalent to a clockwise *P-T* path, but invariant with respect to which way the axes are drawn.

Phase—A particular form and composition of matter that is at least theoretically separable from other phases.

Phase diagrams—Diagrams that show stable phase assemblages for a given set of variables such as bulk composition, *P*, and *T*.

Phase equilibria—The nature of equilibrium phase assemblages that occur in a chemical system under specified conditions of *P*, *T*, and other intensive variables.

Phase rule—A thermodynamic formula relating the numbers of components, phases, and degrees of freedom in a system.

Phengite—Muscovitic mica, common in metamorphic rocks, with excess Si (Si > Al) that is charge-balanced by cations such as Mg.

Phreatomagmatic—In volcanology, pertaining to the reaction of magma with groundwater or surface water, producing steam and fragmental magma.

Picrite—An unusually Mg-rich basalt with abundant phenocrysts of olivine.

Planetesimals—Solid objects forming in a nebular disk owing to gravitational forces; essentially baby planets.

Plateau age—The weighted average age for the flat portion of an Ar release spectrum. Plateau ages commonly correspond with the later increments of gas released at the highest temperatures.

Platinum-group elements (PGEs)—A group of six metals: Ru, Rh, and Pd, which lie just before Ag in row 5 of the periodic table, and Os, Ir, and Pt, which lie below the previous three and just before Au in row 6.

Playas—The dry, vegetation-free floors of geomorphic basins, covered with clay or evaporite salts.

Plinian—Explosive eruption of intermediate magma, with copious pyroclastic flow and fall deposits, producing an eruption column tens of kilometers tall.

Plume—Postulated relatively deep-seated, hot, buoyant mantle that, as it ascends, can give rise to hotspots.

Pluton—A body of intrusive rock analogous to a sedimentary formation that is sufficiently large and recognizable by texture and mineralogy to form a mappable unit. It may have formed over a short or long time interval and may be composed of material from multiple sources.

Poikilitic—Texture in which multiple smaller crystals of one phase are encased in a much larger crystal of another.

Poisson distribution—Statistical distribution describing the probability of a given number of independent stochastic events happening in a given interval of time or space.

Polarity—Here, the direction in which a subducted plate is descending, generally expressed as the facing direction from the arc to the incoming plate. For example, the Cascade arc (see Figure 6.5) is west-facing, whereas the Mariana arc (see Figure 6.3) is east-facing.

Polybaric melting—Mantle melting that occurs at successively shallower depths as a parcel of mantle rises; the resulting magma is a mixture of melts derived over a range of pressures.

Porphyroblasts—Relatively large crystals that grow in a metamorphic rock. Similar to the igneous term "phenocryst" but applies to metamorphic rocks.

Porphyry copper deposits—Intrusions with characteristically large crystals in fine-grained groundmass (a porphyry) that contain relatively abundant copper minerals, usually as sulfides, but also as other minerals along fractures.

Post-kinematic, syn-kinematic, pre-kinematic—After, during, or before deformation, respectively.

Potential temperature—The temperature that a rising parcel of mantle material would attain if brought to the surface adiabatically.

Proving ground—A physical or conceptual place for testing theories, ideas, equipment, tactics, etc.; a proving ground is *not* a place where theories are proven.

Pumice flow—Hot density current of degassing pumice blocks and lapilli in an ashy matrix.

Pyroclastic—Magma fragmented and blown into the air during eruption; "fire broken."

Pyroclastic falls—Tephra that is blown into the air and falls out like snow.

Pyroclastic flows—General term encompassing both pumice flows and block-and-ash flows.

Pyroclastic surges—Turbulent, fast-moving, low-density flows of pyroclastic material and hot gases.

Radial dikes—Vertical dikes that extend radially from a magma source such as a volcanic center.

Radiogenic—Produced by radioactive decay.

Radionuclides—Radioactive nuclides.

Receiver function analysis—Interpretation of the seismic record at a given seismic station (receiver) in terms of velocity interfaces along the seismic ray path.

Red giant—A giant star with a relatively low-temperature outer shell that glows red.

Reefs—In ore deposits, metalliferous mineral deposits; the term is typically used for those of particular economic value.

Regional metamorphism—General term for metamorphism of large areas, as contrasted with metamorphism near igneous bodies (contact metamorphism) or along shear zones (dynamic metamorphism).

Residence time—In a volume of something (e.g., water, air, glacial ice, oceanic lithosphere) subject to steady-state inflow and outflow, the average time that a unit of the substance spends in that volume. Residence time is given by volume divided by inflow rate (e.g., the residence time of water in an overflowing 200-liter bathtub being fed at 20 liters per minute is 10 minutes). The concept can be generalized to area.

Resurgent dome—Rhyolite domes extruded within a caldera after a caldera-forming eruption.

Retrogression/retrograde—Back-reaction or re-equilibration of minerals during cooling and exhumation from the maximum metamorphic conditions. For example, retrogression of periclase as it cools and reacts with water typically forms retrograde brucite.

Rift zones—In volcanism, linear zones of crustal extension above a set of dikes that are taking advantage of that extra space.

Roof pendants—Term used to describe large pieces of wall rock surrounded by plutonic rocks and assumed to be uneroded pieces of the roof.

Room problem—The question of how space is made in the crust for plutonic bodies.

r-process, s-process—Processes that inject neutrons into nuclei; rapid (r) occurs during supernova explosions or neutron star collisions, whereas slow (s) occurs mainly late in the life of low-mass stars such as the Sun.

Saucer-shaped sills—Large sills with upturned edges.

Schistosities—Foliations formed by micas and other sheet silicates in schists. Labeled "S" in structural geology (e.g., S1, S2, etc.).

Schlieren—Thin, discontinuous horizons that are rich in mafic minerals.

Scoria—Rough, vesicular basaltic tephra.

Screens, septa—Relatively thin panels of wall rocks within a pluton or separating two plutons.

Seamount—Isolated, commonly flat-topped submarine mountain; generally an extinct volcano.

Sector collapse—Wholesale failure and collapse of a portion of a volcano; may be triggered by or trigger an eruption.

Secular variation—Movement of the geomagnetic poles over time; the geomagnetic poles seem to average out to the spin axis on timescales of 10^4 years.

Seismic tomography—A mathematical technique that uses arrival times of a large number of earthquakes at a large number of stations to estimate velocity variations at depth.

Serpentine endemic—A plant that grows exclusively on serpentine soils.

Serpentine soils—Soils developed on serpentinite and other ultramafic rocks (e.g., peridotites).

Serpentinization—Formation of serpentine by hydration of olivine, pyroxene, and/or glass.

Shatter cones—Curved, striated, conical fracture surfaces, typically tens of centimeters long, that are characteristic of relatively low shock pressures (1–5 GPa).

Sheet—Here, used to denote relatively flat-lying dikes or sills.

Sheeted dikes—Sets of parallel dikes in which the walls of each dike are older dikes.

Shield—Large, low-relief area of basement rocks in a craton.

Shield volcanoes—Broad, low volcanoes formed of low-viscosity basaltic magma.

Siliceous sinter—Deposits of opaline or amorphous silica around hot springs and geysers.

Sill—A dike that is parallel to foliation or bedding in host rocks, regardless of the orientation of that fabric.

Skarn—A reaction zone of calc-silicate minerals between a pluton and the rocks it intruded.

Slates—Fine-grained platy rocks composed mostly of chlorite, various clay minerals, and quartz.

Solidus—For a given bulk composition at a given pressure, the temperature at which melting begins upon heating.

Solvus—A curve on a binary phase diagram that separates a homogenous phase from two phases that form by exsolution as T drops. This concept generalizes to more components.

Specific heat—The amount of thermal energy needed to increase the temperature in a unit mass of rock by one unit of temperature. Units are usually J/g·K or kJ/kg·K.

Standard state—Here, a reference state for activity, typically a pure phase (e.g., mineral) composed of that component at a specified P and T.

Stochastic—Randomly determined.

Stock—General term for a small (on the order of a few kilometers in diameter), generally equant and isolated pluton.

Stoichiometric constituents—Elements that are an essential part of a mineral formula.

Stokes' law—A formula for the rate of settling of spherical particles in a fluid in laminar flow.

Stoped blocks—Xenoliths that have free-fallen through magma during stoping. This term is commonly, but mistakenly, applied as a default interpretation of any xenolith.

Stratosphere—The layer above the troposphere, characterized by increasing temperature with increasing altitude, dryness, and lack of precipitation.

Stratovolcanoes, composite volcanoes—Tall, conical volcanoes formed by flows and tephra of intermediate (andesitic and dacitic) magma.

Stretching factor β—The width of an extended region divided by its original width (McKenzie, 1978).

Strombolian—Eruption style produced by basaltic and andesitic magma at cinder cones, with eruption of tephra out the top and lava flows out the bottom.

S-type granite—A granite formed via partial melting of sedimentary sources.

Subaerial—Literally below the air, so on the Earth's surface.

Subcontinental lithospheric mantle (SCLM)—The stiff part of the mantle beneath the continents. Because it is attached to buoyant continental crust, this part of the mantle is especially resistant to subduction and recycling.

Submarine—Literally below the ocean, so in water.

Supercritical fluid—A fluid at *P-T* conditions above the critical point at which the gas-liquid distinction disappears.

Supracrustal—Literally, on top of the crust; pertains to sedimentary and volcanic rocks that were deposited at Earth's surface.

Supereruptions—Infrequent eruptions that release on the order of 1000 km^3 or more of tephra in one eruptive cycle. No supereruptions have (yet) occurred since civilization arose.

Supergene deposit—An ore deposit formed by the downward movement of solutions causing secondary mineralization. "Super" here means "from above," not necessarily high concentration.

Supernova—A massive star that has undergone explosive death, either by the onset of carbon burning in a white dwarf (Type I) or by contraction and rebound of the depleted core of a massive star (Type II).

Target rocks—Any rocky material on a planetary body affected by a bolide strike.

Tephra—Term encompassing all pyroclastic material.

Ternary system—A system in which three components are needed to describe the compositions of all phases present.

Terrane—A region, generally bounded by faults, with specific geologic characteristics that differ from those of adjacent regions; *terrain* refers to a parcel of land and its physical characteristics.

***T*-first *P-T* path**—A path on a *P-T* diagram where temperature increases significantly before pressure does; equivalent to an anti-clockwise *P-T* path, but invariant with respect to which way the axes are drawn.

Thermal boundary layer—Region parallel to a boundary over which the variation in temperature is significant.

Thermal conductivity—A measure of how well a material conducts heat; units of W/m/°C.

Thermal diffusivity—A coefficient that describes how rapidly a temperature perturbation moves through a material by conduction.

Thermal divide—On a phase diagram, a ridge on the liquidus surface equivalent to a topographic divide; a cooling liquid will flow away from the divide along the liquidus surface.

Thermal erosion—Transformation of lithosphere to asthenosphere through heating (thermal weakening); thermal erosion contrasts with physical erosion, which (in this context) would be the physical removal or replacement of stiff material with weaker material.

Thermobarometer, geothermobarometer—Generic term for an estimate of temperature and pressure based on equilibrium thermodynamics of a chemical reaction.

Thermophile—A microorganism that survives at temperatures >40 °C.

Tie lines, joins—On a binary *T*-X phase diagram, isothermal lines connecting the phases in equilibrium in a two-phase field. This concept generalizes to higher dimensions (e.g., a tie plane in a three-component phase diagram), and to *P*-X and other diagrams.

Total gas age—The weighted average of all increments of gas released when measuring $^{40}Ar/^{39}Ar$ ages.

Trace elements—Elements present in (generally) low concentrations such that they do not make their own minerals but instead substitute into existing ones. Trace elements cannot be stoichiometric constituents of the mineral.

Transition zone—Mantle between 410 and 660 km.

Travertine—$CaCO_3$ deposited by spring-fed, and commonly heated, waters.

Tributary reaction point—In a ternary system, a reaction point where, going down *T*, two curves that bound liquidus fields meet and a third emerges. At such a point, the reaction going down *T* is always the crystalline phase that lies between the tributaries + liquid = the other two phases.

Tropopause—The boundary between the troposphere and stratosphere.

Troposphere—The lower 10–15 km of the atmosphere; where weather happens.

Tuff—Generic term for a deposit of pyroclastic material.

Univariant assemblages—Phase assemblages in which you can make small changes in only one intensive variable and maintain the same phase assemblage; other intensive variables must adjust to the change made. On a *P* -*T* diagram, a univariant assemblage lies along a reaction curve.

Upper mantle—Mantle above 410 km.

Variation diagram—A plot of one chemical parameter of rock composition against another; typically, SiO_2 is plotted on the abscissa.

Viscosity, dynamic viscosity—Resistance of a fluid to flow under shear stress.

Vog—Corrosive, dense haze, rich in SO_2 and other harmful gases, produced by active volcanoes.

Volcanic aerosols—Fine particles and droplets emitted by volcanoes; sulfuric acid droplets are of particular importance.

Volcanic Explosivity Index (VEI)—A logarithmic scale of the intensity and severity of volcanic eruptions, comparable in some ways to the moment magnitude scale for earthquakes.

Volcanic front—The abrupt appearance of volcanoes away from the trench in the direction of subduction.

Volcanogenic massive sulfide (VMS) deposits—Deposits of massive sulfides, usually Fe-, Zn-, and Cu-rich, associated with volcanic rocks and interpreted to reflect concentration in localized hydrothermal systems.

Wadati-Benioff zones—The curved planes of seismicity that dip away from a trench and delineate the subducting plate.

Wall rocks, country rocks—The rocks into which plutonic rocks are intruded.

Welding—Post-emplacement sintering and flattening of pumice clasts in an ignimbrite.

White dwarf—Very dense stellar core remnant; may have the mass of the Sun in a volume comparable to the Earth.

White smoker—Plumes of fluid in deep-sea hydrothermal systems, colored white by precipitating sulfates and silica.

Whole-rock analysis—A chemical analysis of a bulk rock sample rather than of one or more of its component parts (minerals or glass).

Wilson cycle—Cycle of seafloor spreading, continental collision and amalgamation, and subsequent rifting to form new ocean basins.

Xenocrysts—Crystals foreign to their host magma; in mixed magmas, this term can be ambiguous.

Yosemite Decimal System—A system for rating the difficulty of a climb, with "5." followed by a number and sometimes a letter for grades 5.10 and above. 5.0–5.4 is relatively easy; 5.5–5.9 is moderate, with good holds and few overhangs; and 5.10 and up is advanced, with smooth walls, tiny holds, and overhangs. For 5.13 and up, imagine a granite countertop, with a few chips in its surface, turned on edge, perhaps 500 m off the ground.

References

Ackerson, M. R. (2011). Trace element partitioning between titanite and groundmass in silicic volcanic systems. MS thesis, Department of Geological Sciences, University of North Carolina, Chapel Hill, 69 p.

Ackerson, M. R., Mysen, B. O., Tailby, N. D. & Watson, E. B. (2018). Low-temperature crystallization of granites and the implications for crustal magmatism. *Nature* **559**, 94–97.

Acocella, V., Rossetti, F., Faccenna, C., Funiciello, R. & Lazzarotto, A. (2000). Strike-slip faulting and pluton emplacement in southern Tuscany: the Campiglia Marittima case. *Italian Journal of Geosciences* **119**, 517–528.

Aharonson, O., Zuber, M. T. & Rothman, D. H. (2001). Statistics of Mars' topography from the Mars Orbiter Laser Altimeter: slopes, correlations, and physical models. *Journal of Geophysical Research E: Planets* **106**, 23723–23735.

Al-Rawi, Y. & Carmichael, I. S. E. (1967). A note on the natural fusion of granite. *American Mineralogist* **52**, 1806–1814.

Algeo, T. J., Berner, R. A., Maynard, J. B. & Scheckler, S. E. (1995). Late Devonian oceanic anoxic events and biotic crises: "rooted" in the evolution of vascular land plants? *GSA Today* **5**, 45–66.

Ambrose, S. H. (1998). Late Pleistocene human population bottlenecks, volcanic winter, and differentiation of modern humans. *Journal of Human Evolution* **34**, 623–651.

Ambrose, W. R. & Stevenson, C. M. (2004). Obsidian density, connate water and hydration dating. *Mediterranean Archaeology and Archaeometry* **4**, 5–16.

Améglio, L. & Vigneresse, J. L. (1999). Geophysical imaging of the shape of granitic intrusions at depth: a review. *Geological Society Special Publication* **168**, 39–54.

Andersen, D. J., Lindsley, D. H. & Davidson, P. M. (1993). QUILF: A pascal program to assess equilibria among Fe-Mg-Mn-Ti oxides, pyroxenes, olivine, and quartz. *Computers & Geosciences* **19**, 1333–1350.

Anderson, A. T., Jr., Clayton, R. N. & Mayeda, T. K. (1971). Oxygen isotope thermometry of mafic igneous rocks. *Journal of Geology* **79**, 715–729.

Anderson, D. L. (2000). The thermal state of the upper mantle: no role for mantle plumes. *Geophysical Research Letters* **27**, 3623–3626.

Anderson, D. L. (2005). Scoring hotspots: The plume and plate paradigms. *Geological Society of America Special Paper* **388**, 31-54.

Anderson, D. L. & Hart, R. S. (1976). An Earth model based on free oscillations and body waves. *Journal of Geophysical Research* **81**, 1461–1475.

Andrault, D., Monteux, J., Le Bars, M. & Samuel, H. (2016). The deep Earth may not be cooling down. *Earth and Planetary Science Letters* **443**, 195–203.

Angiboust, S., Langdon, R., Agard, P., Waters, D. & Chopin, C. (2012). Eclogitization of the Monviso ophiolite (W. Alps) and implications on subduction dynamics. *Journal of Metamorphic Geology* **30**, 37–61.

Annen, C. (2017). Factors affecting the thickness of thermal aureoles. *Frontiers in Earth Science* **5**, 1–13.

Annen, C., Blundy, J. D., Leuthold, J. & Sparks, R. S. J. (2015). Construction and evolution of igneous bodies: towards an integrated perspective of crustal magmatism. *Lithos* **230**, 206–221.

Annen, C., Scaillet, B. & Sparks, R. S. J. (2006). Thermal constraints on the emplacement rate of a large intrusive complex: the Manaslu leucogranite, Nepal Himalaya. *Journal of Petrology* **47**, 71–95.

Anovitz, L. M. & Essene, E. J. (1987). Phase equilibria in the system $CaCO_3$-$MgCO_3$-$FeCO_3$. *Journal of Petrology* **28**, 389–415.

Archambault, C. & Tanguy, J. C. (1976). Comparative temperature measurements on Mount Etna lavas: problems and techniques. *Journal of Volcanology and Geothermal Research* **1**, 113–125.

Archer, D. (2005). Fate of fossil fuel CO_2 in geologic time. *Journal of Geophysical Research: Oceans* **110**, 1–6.

Arculus, R. J. (2003). Use and abuse of the terms calcalkaline and calcalkalic. *Journal of Petrology* **44**, 929–935.

Arndt, N. T. & Goldstein, S. L. (1989). An open boundary between lower continental crust and mantle: its role in crust formation and crustal recycling. *Tectonophysics* **161**, 201–212.

Askren, D. R. R., Whitney, J. A. & Roden, M. F. (1991). Petrology and geochemistry of the Huerto Andesite, San Juan volcanic field, Colorado. *Contributions to Mineralogy and Petrology* **107**, 373–386.

Asscher, J. (1908). *The Cullinan.* Joseph Asscher & Co., 16 plates.

Atwater, T. (1970). Implications of plate tectonics for the Cenozoic tectonic evolution of western North America. *Geological Society of America Bulletin* **81**, 3513–3536.

Atwater, T. & Stock, J. (1998). Pacific-North America plate tectonics of the Neogene southwestern United States: an update. *International Geology Review* **40**, 375–402.

Axen, G. J., Taylor, W. J. & Bartley, J. M. (1993). Space-time patterns and tectonic controls of Tertiary extension and magmatism in the Great Basin of the western United States. *Geological Society of America Bulletin* **105**, 56–76.

Ayuso, R. A. (1984). Field relations, crystallization, and petrology of reversely zoned granitic plutons in the Bottle Lake Complex, Maine. *US Geological Survey Professional Paper* **1320**, 58 p.

Bachmann, O. & Bergantz, G. W. (2004). On the origin of crystal-poor rhyolites: extracted from batholithic crystal mushes. *Journal of Petrology* **45**, 1565–1582.

Bachmann, O., Dungan, M. A. & Bussy, F. (2005). Insights into shallow magmatic processes in large silicic magma bodies: the trace element record in the Fish Canyon magma body, Colorado. *Contributions to Mineralogy and Petrology* **149**, 338–349.

Bachmann, O., Dungan, M. A. & Lipman, P. W. (2002). The Fish Canyon magma body, San Juan volcanic field, Colorado: rejuvenation and eruption of an upper-crustal batholith. *Journal of Petrology* **43**, 1469–1503.

Bailey, R. A., Dalrymple, G. B. & Lanphere, M. A. (1976). Volcanism, structure, and geochronology of Long Valley Caldera, Mono County, California. *Journal of Geophysical Research* **81**, 725–744.

Baird, A. K., Baird, K. W. & Welday, E. E. (1979). Batholithic rocks of the northern Peninsular and Transverse Ranges, southern California: chemical composition and variation. In: Abbott, P. L. & Todd, V. R. (eds) *Mesozoic Crystalline Rocks: Peninsular Range Batholith and Pegmatites, Point Sal Ophiolite*. San Diego State University, 111–132.

Barazangi, M. & Isacks, B. L. (1976). Spatial distribution of earthquakes and subduction of the Nazca plate beneath South America. *Geology* **4**, 686–692.

Barbarin, B., Dodge, F. C. W., Kistler, R. W. & Bateman, P. C. (1989). Mafic inclusions, aggregates, and dikes in granitoid rocks, central Sierra Nevada Batholith, California—analytic data. *US Geological Survey Bulletin* **1899**, 27 p.

Barbey, P. (2009). Layering and schlieren in granitoids: a record of interactions between magma emplacement, crystallization and deformation in growing plutons. *Geologica Belgica* **12**, 109–133.

Barrow, G. (1893). On an intrusion of muscovite-biotite gneiss in the southeastern Highlands of Scotland and its accompanying metamorphism. *Quarterly Journal of the Geological Society of London* **49**, 330–358.

Bartley, J. M., Glazner, A. F. & Coleman, D. S. (2018). Dike intrusion and deformation during growth of the Half Dome pluton, Yosemite National Park, California. *Geosphere* **14**, 1283–1297.

Bartley, J. M., Glazner, A. F. & Mahan, K. H. (2012). Formation of pluton roofs, floors, and walls by crack opening at Split Mountain, Sierra Nevada, California. *Geosphere* **8**, 1086–1103.

Bateman, P. C. (1965). Geology and tungsten mineralization of the Bishop district, California. *US Geological Survey Professional Paper* **470**, 208 p.

Bateman, P. C. & Chappell, B. W. (1979). Crystallization, fractionation, and solidification of the Tuolumne intrusive series, Yosemite National Park, California. *Geological Society of America Bulletin* **90**, 465–482.

Bateman, P. C., Dodge, F. C. W. & Bruggman, P. E. (1984). Major oxide analyses, CIPW norms, modes, and bulk specific gravities of plutonic rocks from the Mariposa 1° x 2° sheet, central Sierra Nevada, California. *US Geological Survey Open-File Report* **OF 84-0162**, 59 p.

Bateman, P. C., Kistler, R. W., Peck, D. L. & Busacca, A. J. (1983). Geologic map of the Tuolumne Meadows quadrangle, Yosemite National Park, California. *US Geological Survey Map* **GQ-1570**, scale 1:62,500.

Beard, B. L. & Glazner, A. F. (1995). Trace element and Sr and Nd isotopic composition of mantle xenoliths from the Big Pine volcanic field, California. *Journal of Geophysical Research* **100**, 4169–4179.

Beaumont, C., Jamieson, R. A., Nguyen, M. H. & Lee, B. (2001). Himalayan tectonics explained by extrusion of a low-viscosity crustal channel coupled to focused surface denudation. *Nature* **414**, 738–742.

Benioff, H. (1949). Seismic evidence for the fault origin of oceanic deeps. *Geological Society of America Bulletin* **60**, 1837–1856.

Benson, T. R., Mahood, G. A. & Grove, M. (2017). Geology and $^{40}Ar/^{39}Ar$ geochronology of the middle Miocene McDermitt volcanic field, Oregon and Nevada: silicic volcanism associated with propagating flood basalt dikes at initiation of the Yellowstone hotspot. *Geological Society of America Bulletin* **129**, 1027–1051.

Benton, T. J., Gaynor, S. & Glazner, A. F. (2011). Testing composite dike intrusion models with chemical analysis. *Abstracts with Programs—Geological Society of America* **43**, 68.

Bergman, S. C. (1987). Lamproites and other potassium-rich igneous rocks: a review of their occurrence, mineralogy and geochemistry. *Geological Society of London Special Publication* **30**, 103–190.

Berner, R. A., Lasaga, A. C. & Garrels, R. M. (1983). The carbonate-silicate geochemical cycle and its effect on atmospheric carbon dioxide of the past 100 million years. *American Journal of Science* **283**, 641–683.

Best, M. G. (1988). Early Miocene change in direction of least principal stress, southwestern United States: conflicting inferences from dikes and metamorphic core-detachment fault terranes. *Tectonics* 7, 249–259.

Best, M. G. (2003). *Igneous and Metamorphic Petrology*. Blackwell Publishing, 729 p.

Beyssac, O., Cox, S. C., Vry, J. & Herman, F. (2016). Peak metamorphic temperature and thermal history of the Southern Alps (New Zealand). *Tectonophysics* **676**, 229–249.

Beyssac, O., Goffe, B., Chopin, C. & Rouzaud, J. N. (2002). Raman spectra of carbonaceous material in metasediments: a new geothermometer. *Journal of Metamorphic Geology* **20**, 859–871.

Biasi, J. & Karlstrom, L. (2021). Timescales of magma transport in the Columbia River flood basalts, determined by paleomagnetic data. *Earth and Planetary Science Letters* **576**, doi.org/10.1016/j.epsl.2021.117169.

Bilodeau, B. J. & Nelson, C. A. (1993). Geology of the Sage Hen Flat pluton, White Mountains, California. *Geological Society of America Map and Chart Series* **MCH077**.

Bina, C. R. & Helffrich, G. (1994). Phase transition Clapeyron slopes and transition zone seismic discontinuity topography. *Journal of Geophysical Research* **99**, 15853–15860.

Birch, F. (1960). The velocity of compressional waves in rocks to 10 kilobars. *Journal of Geophysical Research* **56**, 1083–1102.

Bird, P. (2003). An updated digital model of plate boundaries. *Geochemistry, Geophysics, Geosystems* **4**, doi.org/10.1029/2001GC000252.

Black, B. A., Elkins-Tanton, L. T., Rowe, M. C. & Peate, I. U. (2012). Magnitude and consequences of volatile release from the Siberian Traps. *Earth and Planetary Science Letters* **317**, 363–373.

Blackwell, D. & Richards, M. C. (2004). The 2004 geothermal map of North America, explanation of resources and applications. *Geothermal Resource Council Transactions* **28**, 317–320.

Blundy, J. D. & Annen, C. J. (2016). Crustal magmatic systems from the perspective of heat transfer. *Elements* **12** 115–120.

Blundy, J. D., Cashman, K. & Humphreys, M. (2006). Magma heating by decompression-driven crystallization beneath andesite volcanoes. *Nature* **443**, 76–80.

Bodnar, R. J., Azbej, T., Becker, S. P., Cannatelli, C. & Fall, A. (2013). Whole Earth geohydrologic cycle, from the clouds to the core: the distribution of water in the dynamic Earth system. *Geological Society of America Special Paper* **500**, 431–461.

Boehnke, P., Watson, E. B., Trail, D., Harrison, T. M. & Schmitt, A. K. (2013). Zircon saturation re-revisited. *Chemical Geology* **351**, 324–334.

Boemmels, J. R. C., Crespi, J. M., Webb, L. E. & Fosdick, J. C. (2021). $^{40}Ar/^{39}Ar$ and LA-ICP-MS U–Pb geochronology for the New England portion of the Early Cretaceous New England-Quebec igneous province: Implications for the postrift evolution of the eastern North American Margin. *American Journal of Science* **321**, 365–391.

Boettcher, A. L. & Wyllie, P. J. (1968). Melting of granite with excess water to 30 kilobars pressure. *Journal of Geology* **76**, 235–244.

Bonatti, E., Brunelli, D., Fabretti, P., Ligi, M., Asunta Portaro, R. & Seyler, M. (2001). Steady-state creation of crust-free lithosphere at cold spots in mid-ocean ridges. *Geology* **29**, 979–982.

Bonatti, E., Ligi, M., Carrara, G., Gasperini, L., Turko, N., Perfiliev, S., Peyve, A. & Sciuto, P. F. (1996). Diffuse impact of the Mid-Atlantic Ridge with the Romanche Transform: an ultracold ridge-transform intersection. *Journal of Geophysical Research* **101**, 8043–8054.

Bono, R. K., Tarduno, J. A. & Bunge, H. P. (2019). Hotspot motion caused the Hawaiian-Emperor Bend and LLSVPs are not fixed. *Nature Communications* **10**, doi.org/10.1038/s41467-019-11314-6.

Boschman, L. M., van der Wiel, E., Flores, K. E., Langereis, C. G. & van Hinsbergen, D. J. J. (2019). The Caribbean and Farallon plates connected: constraints from stratigraphy and paleomagnetism of the Nicoya Peninsula, Costa Rica. *Journal of Geophysical Research* **124**, 6243–6266.

Boudreau, A. (2011). The evolution of texture and layering in layered intrusions. *International Geology Review* **53**, 330–353.

Boudreau, A. E. (2016). The Stillwater Complex, Montana—overview and the significance of volatiles. *Mineralogical Magazine* **80**, 585–637.

Boudreau, A. (2019). *Hydromagmatic Processes and Platinum-Group Element Deposits in Layered Intrusions.* Cambridge University Press, 275 p.

Boudreau, A. E. & McBirney, A. R. (1997). The Skaergaard layered series. Part III. Non-dynamic layering. *Journal of Petrology* **38**, 1003–1020.

Bowen, N. L. (1913). The melting phenomena of the plagioclase feldspars. *American Journal of Science* **35**, 577–599.

Bowen, N. L. (1914). The ternary system; diopside-forsterite-silica. *American Journal of Science* **38**, 207–264.

Bowen, N. L. (1915a). Crystallization-differentiation in silicate liquids. *American Journal of Science* **39**, 175–191.

Bowen, N. L. (1915b). The crystallization of haplobasaltic, haplodioritic, and related magmas. *American Journal of Science* **40**, 161–185.

Bowen, N. L. (1922). The reaction principle in petrogenesis. *Journal of Geology* **30**, 177–198.

Bowen, N. L. (1928). *The Evolution of the Igneous Rocks.* Princeton University Press, 333 p.

Bowen, N. L. & Andersen, O. (1914). The binary system MgO-SiO_2. *American Journal of Science* **37**, 487–500.

Bowen, N. L. & Schairer, J. F. (1935). The system MgO-FeO-SiO_2. *American Journal of Science*, **29**, 151–217.

Bowen, N. L. & Tuttle, O. F. (1949). The system MgO–SiO_2–H_2O. *Geological Society of America Bulletin* **60**, 439–460.

Bowers, J. R., Kerrick, D. M. & Furlong, K. P. (1990). Conduction model for the thermal evolution of the Cupsuptic aureole, Maine. *American Journal of Science* **290**, 644–665.

Bowes, D. R. & Park, R. G. (1966). Metamorphic segregation banding in the Loch Kerry basite sheet from the Lewisian of Gairloch, Ross-shire, Scotland. *Journal of Petrology* **7**, 306–334.

Bown, J. W. & White, R. S. (1994). Variation with spreading rate of oceanic crustal thickness and geochemistry. *Earth and Planetary Science Letters* **121**, 435–449.

Boyd, F. R. & England, J. L. (1960). Apparatus for phase-equilibrium measurements at pressures up to 50 kilobars and temperatures up to 1750 °C. *Journal of Geophysical Research* **65**, 741–748.

Brace, W. F. & Kohlstedt, D. L. (1980). Limits on lithospheric stress imposed by laboratory experiments. *Journal of Geophysical Research* **85**, 6248–6252.

Brady, K. U., Kruckeberg, A. R. & Bradshaw, H. D. (2005). Evolutionary ecology of plant adaptation to serpentine soils. *Annual Review of Ecology, Evolution, and Systematics* **36**, 243–266.

Brøgger, W. C. (1921). Die Eruptivgesteine des Kristianiagebietes: IV. Das Fengebiet in Telemark, Norwegen. *Norske Vindenskapsakademi i Oslo Skrifter I Mat-Nat Klasse* 1920, 9, 408 pp.

Bromiley, G. D. & Pawley, A. R. (2003). The stability of antigorite in the systems MgO-SiO_2-H_2O (MSH) and MgO-Al_2O_3-SiO_2-H_2O (MASH): the effects of Al^{3+} substitution on high-pressure stability. *American Mineralogist* **88**, 99–108.

Brown, E. H. & McClelland, W. C. (2000). Pluton emplacement by sheeting and vertical ballooning in part of the southeast Coast Plutonic Complex, British Columbia. *Geological Society of America Bulletin* **112**, 708–719.

Brown, M. (1994). The generation, segregation, ascent and emplacement of granite magma. *Earth Science Reviews* **36**, 83–130.

Brown, M., Johnson, T. & Gardiner, N. J. (2020). Plate tectonics and the Archean Earth. *Annual Review of Earth and Planetary Sciences* **48**, 291–320.

Brown, M. & Solar, G. S. (1998). Granite ascent and emplacement during contractional deformation in convergent orogens. *Journal of Structural Geology* **20**, 1365–1393.

Brumfiel, G. (2006). Plutons, planets and dwarves. *Nature*, doi.org/10.1038/news060821-4.

Buchan, K. L. & Ernst, R. E. (2004). Diabase dyke swarms and related units in Canada and adjacent regions. *Geological Survey of Canada Map* **2022A**, with map 1:5,000,000 and accompanying notes.

Buchs, D. M., Kerr, A. C., Brims, J. C., Zapata-Villada, J. P., Correa-Restrepo, T. & Rodríguez, G. (2018). Evidence for subaerial development of the Caribbean oceanic plateau in the Late Cretaceous and palaeo-environmental implications. *Earth and Planetary Science Letters* **499**, 62–73.

Buddington, A. F. (1959). Granite emplacement with special reference to North America. *Geological Society of America Bulletin* **70**, 671–747.

Buddington, A. F. & Lindsley, D. H. (1964). Iron-titanium oxide minerals and synthetic equivalents. *Journal of Petrology* **5**, 310–357.

Burgess, S. D. & Bowring, S. A. (2015). High-precision geochronology confirms voluminous magmatism before, during, and after Earth's most severe extinction. *Science Advances*, doi.org/10.1126/sciadv.1500470.

Burke, K. (1988). Tectonic evolution of the Caribbean. *Annual Review of Earth and Planetary Sciences* **16**, 201–230.

Burnham, C. W. (1959). Contact metamorphism of magnesian limestones at Crestmore, California. *Geological Society of America Bulletin* **70**, 879–920.

Burnham, C. W. & Davis, N. F. (1974). The role of H_2O in silicate melts: II, thermodynamic and phase relations in the system $NaAlSi_3O_8$-H_2O to 10 kilobars, 700 degrees to 1100 degrees C. *American Journal of Science* **274**, 902–940.

Burnham, C. W., Holloway, J. R. & Davis, N. F. (1969). Thermodynamic properties of water to 1,000 °C and 10,000 bars. *Geological Society of America Special Paper* **132**, 96 p.

Burov, E. B. & Watts, A. B. (2006). The long-term strength of continental lithosphere: "jelly sandwich" or "crème brûlée"? *GSA Today* **16**, 4–10.

Button, A. & Cawthorn, R. G. (2015). Distribution of mafic sills in the Transvaal Supergroup, northeastern South Africa. *Journal of the Geological Society* **172**, 357–367.

Byron, G. G. B. (1816). *The Prisoner of Chillon and Other Poems.* John Murray, 60 p.

Cahoon, E. B., Streck, M. J., Koppers, A. A. P. & Miggins, D. P. (2020). Reshuffling the Columbia River Basalt chronology—Picture Gorge Basalt, the earliest- and longest-erupting formation. *Geology* **48**, 348–352.

Caillet, A., Cogné, C., Andrieu, J., Laurent, P. & Rivoire, A. (2003). Characterization of ice cream structure by direct optical microscopy. Influence of freezing parameters. *Food Science and Technology* **36**, 743–749.

Cameron, E. M., Leybourne, M. I., Palacios, C. & Reich, M. (2008). Economic geology models 1. Geochemical exploration and metallogenic studies, northern Chile. *Geoscience Canada* **35**, 97–108.

Camp, V. E. & Wells, R. E. (2021). The case for a long-lived and robust Yellowstone hotspot. *GSA Today* **31**, 4–10.

Campbell, I. H. (1978). Some problems with the cumulus theory. *Lithos* **11**, 311–323.

Campbell, I. H. (2005). Large igneous provinces and the mantle plume hypothesis. *Elements* **1**, 265–269.

Campomenosi, N., Mazzucchelli, M. L., Mihailova, B., Scambelluri, M., Angel, R. J., Nestola, F., Reali, A. & Alvaro, M. (2018). How geometry and anisotropy affect residual strain in host-inclusion systems: coupling experimental and numerical approaches. *American Mineralogist* **103**, 2032–2035.

Cann, J. R. (1970). Upward movement of granitic magma. *Geological Magazine* **107**, 335–340.

Cann, J. R. (1974). A model for oceanic crustal structure developed. *Geophysical Journal of the Royal Astronomical Society* **39**, 169–187.

Canup, R. M. & Asphaug, E. (2001). Origin of the Moon in a giant impact near the end of the Earth's formation. *Nature* **412**, 708–712.

Carey, S. W. (1975). The expanding Earth—an essay review. *Earth Science Reviews* **11**, 105–143.

Carl, J. D. & Amstutz, G. C. (1958). Three-dimensional Liesegang rings by diffusion in a colloidal matrix, and their significance for the interpretation of geological phenomena. *Geological Society of America Bulletin* **69**, 1467–1468.

Carlson, W. D. & Rosenfeld, J. L. (1981). Optical determination of topotactic aragonite-calcite growth kinetics: metamorphic implications. *Journal of Geology* **89**, 615–638.

Carmichael, D. M. (1978). Metamorphic bathozones and bathograds: a measure of the depth of post-metamorphic uplift and erosion on the regional scale. *American Journal of Science* **178**, 769–797.

Carmichael, I. S. E. (1964). The petrology of Thingmuli, a Tertiary volcano in Eastern Iceland. *Journal of Petrology* **5**, 435–460.

Carmichael, I. S. E. (2002). The andesite aqueduct: perspectives on the evolution of intermediate magmatism in west-central (105–99°W) Mexico. *Contributions to Mineralogy and Petrology* **143**, 641–663.

Carmichael, I. S. E., Turner, F. J. & Verhoogen, J. (1974). *Igneous Petrology.* McGraw-Hill, 739 p.

Carrigan, C. R., Dmowska, R. & Holton, J. R. (1994). Two-component magma transport and the origin of composite intrusions and lava flows. In: Ryan, M. P. (ed) *Magmatic Systems.* Academic Press, 319–354.

Carslaw, H. S. & Jaeger, J. C. (1959). *Conduction of Heat in Solids.* Oxford University Press, 510 p.

Cas, R. A. F. & Wright, J. V. (1987). *Volcanic Successions, Modern and Ancient: A Geological Approach to Processes, Products, and Successions.* Allen & Unwin, 528 p.

Casas, E. & Lowenstein, T. K. (1989). Diagenesis of saline pan halite: comparison of petrographic features of modern, Quaternary and Permian halites. *Journal of Sedimentary Petrology* **59**, 724–739.

Castro, A. E. & Spear, F. S. (2017). Reaction overstepping and re-evaluation of peak P–T conditions of the blueschist unit Sifnos, Greece: implications for the Cyclades subduction zone. *International Geology Review* **59**, 548–562.

Cavosie, A. J., Valley, J. W. & Wilde, S. A. (2007). The oldest terrestrial mineral record: a review of 4400 to 4000 Ma detrital zircons from Jack Hills, Western Australia. *Developments in Precambrian Geology* **15**, 91–111.

Cawthorn, R. G. (2015). The Bushveld Complex, South Africa. In: Charlier, B., Namur, O., Latypov, R. & Tegner, C. (eds) *Layered Intrusions*. Springer, 517–587.

Cecil, M. R., Ferrer, M. A., Riggs, N. R., Marsaglia, K., Kylander-Clark, A., Ducea, M. N. & Stone, P. (2019). Early arc development recorded in Permian-Triassic plutons of the northern Mojave Desert region, California, USA. *Geological Society of America Bulletin* **131**, 749–765.

Chacko, T., Cole, D. R. & Horita, J. (2001). Equilibrium oxygen, hydrogen and carbon isotope fractionation factors applicable to geologic systems. *Reviews in Mineralogy and Geochemistry* **43**, 1–81.

Chacko, T. & Deines, P. (2008). Theoretical calculation of oxygen isotope fractionation factors in carbonate systems. *Geochimica et Cosmochimica Acta* **72**, 3642–3660.

Chadwick, W. W. & Dieterich, J. H. (1995). Mechanical modeling of circumferential and radial dike intrusion on Galapagos volcanoes. *Journal of Volcanology and Geothermal Research* **66**, 37–52.

Chakraborty, S. (1997). Rates and mechanisms of Fe-Mg interdiffusion in olivine at 980°–1300°C. *Journal of Geophysical Research* **102**, 12317–12331.

Chan, M. A., Bowen, B. B., Parry, W. T., Ormö, J. & Komatsu, G. (2005). Red rock and red planet diagenesis: comparisons of Earth and Mars concretions. *GSA Today* **15**, 4–10.

Chappell, B. & White, A. (1974). Two contrasting granite types. *Pacific Geology* **8**, 173–174.

Charlier, B. & Grove, T. L. (2012). Experiments on liquid immiscibility along tholeiitic liquid lines of descent. *Contributions to Mineralogy and Petrology* **164**, 27–44.

Charlier, B., Namur, O., Toplis, M. J., Schiano, P., Cluzel, N., Higgins, M. D. & Auwera, J. V. (2011). Large-scale silicate liquid immiscibility during differentiation of tholeiitic basalt to granite and the origin of the Daly gap. *Geology* **39**, 907–910.

Chayes, F. (1964). A petrographic distinction between Cenozoic volcanics in and around the open oceans. *Journal of Geophysical Research* **69**, 1573–1588.

Chen, Y. & Nabelek, P. I. (2017). The influences of incremental pluton growth on magma crystallinity and aureole rheology: numerical modeling of growth of the Papoose Flat pluton, California. *Contributions to Mineralogy and Petrology* **172**, 1–16.

Cherniak, D. J., Zhang, Y. & Cherniak, D. (2010). Cation diffusion in feldspars. *Reviews in Mineralogy and Geochemistry* **72**, 691–733.

Chesner, C. A. (2012). The Toba Caldera Complex. *Quaternary International* **258**, 5–18.

Chiarenza, A. A., Farnsworth, A., Mannion, P. D., Lunt, D. J., Valdes, P. J., Morgan, J. V. & Allison, P. A. (2020). Asteroid impact, not volcanism, caused the end-Cretaceous dinosaur extinction. *Proceedings of the National Academy of Sciences* **117**, 17084–17093.

Chopin, C. (1984). Coesite and pure pyrope in high-grade blueschists of the Western Alps: a first record and some consequences. *Contributions to Mineralogy and Petrology* **86**, 107–118.

Christensen, N. I. & Mooney, W. D. (1995). Seismic velocity structure and composition of the continental crust: a global view. *Journal of Geophysical Research* **100**, 9761–9788.

Christiansen, R. L. & Lipman, P. W. (1972). Cenozoic volcanism and plate-tectonic evolution of the western United States. II. Late Cenozoic. *Philosophical Transactions of the Royal Society of London. Series A, Mathematical and Physical Sciences* **271**, 249–284.

Chu, R., Helmberger, D. V, Sun, D., Jackson, J. M. & Zhu, L. (2010). Mushy magma beneath Yellowstone. *Geophysical Research Letters* **37**, doi.org/10.1029/2009GL041656.

Church, A. A. & Jones, A. P. (1995). Silicate-carbonate immiscibility at Oldoinyo Lengai. *Journal of Petrology* **36**, 869–889.

Clague, D. A. & Straley, P. F. (1977). Petrologic nature of the oceanic Moho. *Geology* **5**, 133–136.

Clark, M. K. & Royden, L. H. (2000). Topographic ooze: building the eastern margin of Tibet by lower crustal flow. *Geology* **28**, 703.

Clarke, D. B. & Clarke, G. K. C. (1998). Layered granodiorites at Chebucto Head, South Mountain Batholith, Nova Scotia. *Journal of Structural Geology* **20**, 1305–1324.

Clarke, D. B. & Erdmann, S. (2008). Is stoping a volumetrically significant pluton emplacement process?: Comment. *Geological Society of America Bulletin* **120**, 1072–1074.

Clarke, F. W. & Washington, H. S. (1924). *The Composition of the Earth's Crust*. US Government Printing Office, 117 p.

Cloos, H. (1928). Zur Terminologie der Plutone. *Fennia* **50**(2), 1–10.

Cloos, M. (1982). Flow melanges: numerical modeling and geologic constraints on their origin in the Franciscan subduction complex, California. *Geological Society of America Bulletin* **93**, 330–345.

Clynne, M. A. (1999). A complex magma mixing origin for rocks erupted in 1915, Lassen Peak, California. *Journal of Petrology* **40**, 105–132.

Coats, R. R. (1962). Magma type and crustal structure in the Aleutian arc. In: MacDonald, G. A. & Kuno, H. (eds) *Crust of the Pacific Basin: A.G.U. Geophysical Monograph* **6**, 92–109.

Cobbing, E. J. (1999). The Coastal Batholith and other aspects of Andean magmatism in Peru. *Geological Society of London Special Publication* **168**, 111–122.

Coble, M. A. & Mahood, G. A. (2015). Geology of the High Rock caldera complex, northwest Nevada, and implications for intense rhyolitic volcanism associated with flood basalt magmatism and the initiation of the Snake River Plain–Yellowstone trend. *Geosphere* **12**, 58–113.

Coetzee, A. & Kisters, A. (2016). The 3D geometry of regional-scale dolerite saucer complexes and their feeders in the Secunda Complex, Karoo Basin. *Journal of Volcanology and Geothermal Research* **317**, 66–79.

Cole, D. R., Horita, J., Polyakov, V. B., Valley, J. W., Spicuzza, M. J. & Coffey, D. W. (2004). An experimental and theoretical determination of oxygen isotope fractionation in the system magnetite-H_2O from 300 to 800°C. *Geochimica et Cosmochimica Acta* **68**, 3569–3585.

Coleman, D. S., Bartley, J. M., Glazner, A. F. & Law, R. D. (2005). Incremental assembly and emplacement of Mesozoic plutons in the Sierra Nevada and White and Inyo Ranges, California. *Geological Society of America Field Forum Field Trip Guide (Rethinking the Assembly and Evolution of Plutons: Field Tests and Perspectives, 7–14 October 2005)*. Geological Society of America, 55 p.

Coleman, D. S., Bartley, J. M., Glazner, A. F. & Pardue, M. J. (2012). Is chemical zonation in plutonic rocks driven by changes in source magma composition or shallow-crustal differentiation? *Geosphere* **8**, 1568–1587.

Coleman, D. S., Frost, T. P. & Glazner, A. F. (1992). Evidence from the Lamarck Granodiorite for rapid Late Cretaceous crust formation in California. *Science* **258**, 1924–1926.

Coleman, D. S., Glazner, A. F., Miller, J. S., Bradford, K. J., Frost, T. P., Joye, J. L. & Bachl, C. A. (1995). Exposure of a Late Cretaceous layered mafic-felsic magma system in the central Sierra Nevada batholith, California. *Contributions to Mineralogy and Petrology* **120**, 129–136.

Coleman, D. S., Gray, W. & Glazner, A. F. (2004). Rethinking the emplacement and evolution of zoned plutons: geochronologic evidence for incremental assembly of the Tuolumne Intrusive Suite, California. *Geology* **32**, 433–436.

Compton, R. R. (1960). Contact metamorphism in the Santa Rosa Range, Nevada. *Geological Society of America Bulletin* **71**, 1383–1416.

Condie, K. C. (1997). *Plate Tectonics and Crustal Evolution*. Butterworth Heinemann, 282 p.

Condie, K. C. (2005). *Earth as an Evolving Planetary System*. Elsevier Academic Press, 447 p.

Coney, P. J. & Reynolds, S. J. (1977). Cordilleran Benioff zones. *Nature* **270**, 403–406.

Connolly, T. J., Skinner, C. E. & Baxter, P. W. (2015). Ancient trade routes for Obsidian Cliffs and Newberry Volcano toolstone in the Pacific Northwest. In: Ozbun, T. L. & Adams, R. L. (eds) *Toolstone Geography of the Pacific Northwest*. Archaeology Press, Simon Fraser University, 180–192.

Cook, S. J. & Bowman, J. R. (1994). Contact metamorphism surrounding the Alta stock: thermal constraints and evidence of advective heat transport from calcite + dolomite geothermometry. *American Mineralogist* **79**, 513–525.

Cook, S. J. & Bowman, J. R. (2000). Mineralogical evidence for fluid-rock interaction accompanying prograde contact metamorphism of siliceous dolomites: Alta Stock aureole, Utah, USA. *Journal of Petrology* **41**, 739–757.

Cook, S. J., Bowman, J. R. & Forster, C. B. (1997). Contact metamorphism surrounding the Alta stock: finite element model simulation of heat- and $^{18}O/^{16}O$ mass-transport during prograde metamorphism. *American Journal of Science* **297**, 1–55.

Coombs, M. L. & Gardner, J. E. (2001). Shallow-storage conditions for the rhyolite of the 1912 eruption at Novarupta, Alaska. *Geology* **29**, 775–778.

Cooper, A. F., Gittins, J. & Tuttle, O. F. (1975). The system Na_2CO_3-K_2CO_3-$CaCO_3$ at 1 kilobar and its significance in carbonatite petrogenesis. *American Journal of Science*, **275**, 534–560.

Corliss, J. B., Dymond, J., Gordon, L. I., Edmond, J. M., von Herzen, R. P., Ballard, R. D., Green, K., Williams, D., Bainbridge, A., Crane, K. & van Andel, T. H. (1979). Submarine thermal springs on the Galápagos Rift. *Science* **203**, 1073–1083.

Corry, C. E. (1988). Laccoliths: mechanics of emplacement and growth. *Geological Society of America Special Paper* **220**, 110 p.

Costa, A. (2005). Viscosity of high crystal content melts: dependence on solid fraction. *Geophysical Research Letters* **32**, doi.org/10.1029/2005GL024303.

Costa, A., Smith, V. C., Macedonio, G. & Matthews, N. E. (2014). The magnitude and impact of the Youngest Toba Tuff super-eruption. *Frontiers in Earth Science* **2**, 1–8.

Costa, F. & Dungan, M. (2005). Short time scales of magmatic assimilation from diffusion modeling of multiple elements in olivine. *Geology*, **33**, 837–840.

Costa, F., Scaillet, B. & Pichavant, M. (2004). Petrological and experimental constraints on the pre-eruption conditions of Holocene dacite from Volcán San Pedro (36° S, Chilean Andes) and the importance of sulphur in silicic subduction-related magmas. *Journal of Petrology* **45**, 855–881.

Cottrell, E., Birner, S. K., Brounce, M., Davis, F. A., Waters, L. E. & Kelley, K. A. (2021). Oxygen fugacity across tectonic settings. In: Moretti, R. & Neuville, D. R. (eds) *Magma Redox Geochemistry* **266**, 33–61.

Courtillot, V. E. & Renne, P. R. (2003). On the ages of flood basalt events. *Comptes Rendus—Geoscience* **335**, 113–140.

Craig, H. (1961). Isotopic variations in meteoric waters. *Science* **133**, 1702–1703.

Crandell, D. R. & Mullineaux, D. R. (1978). Potential hazards from future eruptions of Mount St. Helens Volcano, Washington. *US Geological Survey Bulletin* **1383**, 26 p.

Crick, L., Burke, A., Hutchison, W., Kohno, M., Moore, K.A., Savarino, J., Doyle, E. A., Mahony, S., Kipfstuhl, S., Rae, J.W. & Steele, R. C. (2021). New insights into the ~74 ka Toba eruption from sulfur isotopes of polar ice cores. *Climate of the Past* **17**, 2119–2137.

Cross, W., Iddings, J. P., Pirsson, L. V. & Washington, H. S. (1902). A quantitative chemico-mineralogical classification and nomenclature of igneous rocks. *Journal of Geology* **10**, 555–690.

Cruden, A. R. (1998). On the emplacement of tabular granites. *Journal of the Geological Society of London* **155**, 853–862.

Cruden, A. R. & McCaffrey, K. J. W. (2001). Growth of plutons by floor subsidence: implications for rates of emplacement, intrusion spacing and melt-extraction mechanisms. *Physics and Chemistry of the Earth* **26**, 303–315.

Cruden, A. R., McCaffrey, K. J. & Bunger, A. P. (2018). Geometric scaling of tabular igneous intrusions: Implications for emplacement and growth. In: Breitkreuz, C., Rocchi, S. (eds) *Physical Geology of Shallow Magmatic Systems: Dykes, Sills and Laccoliths*. Springer, 11–38.

Curtis, G. H., Evernden, J. F. & Lipson, J. (1958). Age determination of some granitic rocks in California by the potassium-argon method. *Special Report—California Division of Mines and Geology* **54**, 1–16.

d'Arco, P., Maury, R. C. & Westercamp, D. (1981). Geothermometry and geobarometry of a cummingtonite-bearing dacite from Martinique, Lesser Antilles. *Contributions to Mineralogy and Petrology* **77**, 177–184.

Dal Corso, J., Song, H., Callegaro, S., Chu, D., Sun, Y., Hilton, J., Grasby, S. E., Joachimski, M. M. & Wignall, P. B. (2022).

Environmental crises at the Permian–Triassic mass extinction. *Nature Reviews Earth and Environment* **3**, 197–214.

Daly, R. A. (1903). The mechanics of igneous intrusion. *American Journal of Science* **15**, 269–298.

Daly, R. A. (1914). *Igneous Rocks and Their Origin*. McGraw-Hill, 563 p.

Darwin, C. (1844). *Geological Observations on the Volcanic Islands, Visited During the Voyage of H.M.S. Beagle, Together with Some Brief Notices on the Geology of Australia and the Cape of Good Hope*. Smith, Elder, 175 p.

Davis, J. W., Coleman, D. S., Gracely, J. T., Gaschnig, R. & Stearns, M. (2012). Magma accumulation rates and thermal histories of plutons of the Sierra Nevada batholith, CA. *Contributions to Mineralogy and Petrology* **163**, 449–465.

Dawson, J. B. (1962). Sodium carbonate lavas from Oldoinyo Lengai, Tanganyika. *Nature* **195**, 1075–1076.

Day, H. W. (2012). A revised diamond-graphite transition curve. *American Mineralogist* **97**, 52–62.

de Saint Blanquat, M., Tikoff, B., Teyssier, C. & Vigneresse, J. L. (1998). Transpressional kinematics and magmatic arcs. *Geological Society of London Special Publication* **135**, 327–340.

de Silva, S. L., Self, S., Francis, P. W., Drake, R. E. & Ramirez R. C. (1994). Effusive silicic volcanism in the Central Andes: the Chao Dacite and other young lavas of the Altiplano-Puna volcanic complex. *Journal of Geophysical Research* **99**, 17805–17825.

Defant, M. J. & Drummond, M. S. (1993). Mount St. Helens: potential example of the partial melting of the subducted lithosphere in a volcanic arc. *Geology* **21**, 547–550.

Delaney, P. T. & Gartner, A. E. (1997). Physical processes of shallow mafic dike emplacement near the San Rafael Swell, Utah. *Geological Society of America Bulletin* **109**, 1177–1192.

Deniel, C., Vidal, P., Fernandez, A., Fort, P. & Peucat, J.-J. (1987). Isotopic study of the Manaslu granite (Himalaya, Nepal): inferences on the age and source of Himalayan leucogranites. *Contributions to Mineralogy and Petrology* **96**, 78–92.

DePaolo, D. J. (1981). A neodymium and strontium isotopic study of the Mesozoic calc-alkaline granitic batholiths of the Sierra Nevada and Peninsular ranges, California. *Journal of Geophysical Research* **86**, 10470–10488.

DePaolo, D. J. & Wasserburg, G. J. (1979). Petrogenetic mixing models and Nd-Sr isotopic patterns. *Geochimica et Cosmochimica Acta* **43**, 615–627.

Deschamps, F., Guillot, S., Godard, M., Chauvel, C., Andreani, M. & Hattori, K. (2013). In situ characterization of serpentinites from forearc mantle wedges: timing of serpentinization and behavior of fluid-mobile elements in subduction zones. *Chemical Geology* **269**, 262–277.

Dewey, J. F. & Bird, J. M. (1970). Plate tectonics and geosynclines. *Tectonophysics* **10**, 625–638.

Diamante, L. M. & Lan, T. (2014). Absolute viscosities of vegetable oils at different temperatures and shear rate range of 64.5 to 4835 s^{-1}. *Journal of Food Processing* **2014**, 1–6.

Dickinson, W. R. & Snyder, W. S. (1979). Geometry of subducted slabs related to San Andreas transform. *Journal of Geology* **87**, 609–627.

Didier, J. & Barbarin, B. (1991). The different types of enclaves in granites—nomenclature. *Enclaves and Granite Petrology, Developments in Petrology* **13**, 19–23.

Dietrich, S. & Jacob, K. H. (2017). Understanding Earth: the self-organization concept and its geological significance; on the example of Liesegang-structures and electric fields. In: Müller, S., Plath, P., Radons, G., Fuchs, A. (eds) *Complexity and Synergetics*. Springer, doi.org/10.1007/978-3-319-64334-2_9.

Ding, H., Kohn, M. J. & Zhang, Z. (2021). Long-lived (ca. 22–24 Myr) partial melts in the eastern Himalaya: petrochronologic constraints and tectonic implications. *Earth and Planetary Science Letters* **558**, doi.org/10.1016/j.epsl.2021.116764.

Dobretsov, N. L. (2005). Large igneous provinces of Asia (250 Ma): Siberian and Emeishan traps (plateau basalts) and associated granitoids. *Geologiya i Geofizika* **46**, 870–890.

Dodge, F. C. W., Calk, L. C. & Kistler, R. W. (1986). Lower crustal xenoliths, Chinese Peak lava flow, central Sierra Nevada. *Journal of Petrology* **27**, 1277–1304.

Dodson, M. H. (1973). Closure temperature in cooling geochronological and petrological systems. *Contributions to Mineralogy and Petrology* **40**, 259–274.

Donhowe, D. P. & Hartel, R. W. (1996). Recrystallization of ice cream during controlled accelerated storage. *International Dairy Journal* **6**, 1191–1208.

Donoghue, S. L., Gamble, J. A. & Stewart, R. B. (1995). Magma mingling in an andesite pyroclastic flow of the Pourahu Member, Ruapehu volcano, New Zealand. *Journal of Volcanology and Geothermal Research* **68**, 177–191.

Dorais, M. J., Whitney, J. A. & Roden, M. F. (1990). Origin of mafic enclaves in the Dinkey Creek pluton, central Sierra Nevada batholith, California. *Journal of Petrology* **31**, 853–881.

Doubrovine, P. V. & Tarduno, J. A. (2008). A revised kinematic model for the relative motion between Pacific oceanic plates and North America since the Late Cretaceous. *Journal of Geophysical Research* **113**, 1–20.

Doughty, P. T. & Price, R. A. (1999). Tectonic evolution of the Priest River complex, northern Idaho and Washington: A reappraisal of the Newport fault with new insights on metamorphic core complex formation. *Tectonics* **18**, 375–393.

Douglas, J. A. (1907). On changes of physical constants which take place in certain minerals and igneous rocks, on the passage from the crystalline to the glassy state; with a short note on eutectic mixtures. *Quarterly Journal of the Geological Society* **63**, 145–161.

Downes, M. J. & Cristofolini, R. (1973). Some experimental studies on the 1971 Lavas from Etna [and discussion]. *Philosophical Transactions of the Royal Society of London. Series A, Mathematical and Physical Sciences* **274**, 55–62.

Downs, W. F., Touysinhthiphonexay, Y. & Deines, P. (1981). A direct determination of the oxygen isotope fractionation between quartz and magnetite at 600 and 800°C and 5 kbar. *Geochimica et Cosmochimica Acta* **45**, 2065–2072.

Dragovic, B., Guevara, V. E., Caddick, M. J., Baxter, E. F. & Kylander-Clark, A. R. C. (2016). A pulse of cryptic granulite-facies metamorphism in the Archean Wyoming Craton revealed by Sm–Nd garnet and U–Pb monazite geochronology. *Precambrian Research* **283**, 24–49.

Duan, X. (2014). A general model for predicting the solubility behavior of H_2O–CO_2 fluids in silicate melts over a wide range of pressure, temperature and compositions. *Geochimica et Cosmochimica Acta* **125**, 582–609.

Ducea, M. N. & Saleeby, J. B. (1996). Buoyancy sources for a large, unrooted mountain range, the Sierra Nevada, California: evidence from xenolith thermobarometry. *Journal of Geophysical Research* **101**, 8229–8244.

Duffield, W. A. (1972). A naturally occurring model of global plate tectonics. *Journal of Geophysical Research* **77**, 2543–2555.

Dumitru, T. A. (1990). Subnormal Cenozoic geothermal gradients in the extinct Sierra Nevada magmatic arc: consequences of Laramide and post-Laramide shallow-angle subduction. *Journal of Geophysical Research* **95**, 4925–4941.

Dumond, G., Williams, M. L. & Regan, S. P. (2018). The Athabasca granulite terrane and evidence for dynamic behavior of lower continental crust. *Annual Review of Earth and Planetary Sciences* **46**, 353–386.

Eaton, J. P. & Murata, K. J. (1960). How volcanoes grow. *Science* **132**, 925–938.

Ebelman, J. J. (1845). Sur les produits de la decomposition des especes minerales de la famille des silicates. *Annales des Mines* **7**, 3–66.

Editors of Life. (2003). *100 Photographs That Changed The World.* Life Books, 176 p.

Eerkens, J. W., Spurling, A. M. & Gras, M. A. (2008). Measuring prehistoric mobility strategies based on obsidian geochemical and technological signatures in the Owens Valley, California. *Journal of Archaeological Science* **35**, 668–680.

Eichelberger, J. C. (1978). Andesitic volcanism and crustal evolution. Nature **275**, 21–27.

Eichelberger, J. C. & Izbekov, P. E. (2000). Eruption of andesite triggered by dyke injection: contrasting cases at Karymsky Volcano, Kamchatka and Mt Katmai, Alaska. *Philosophical Transactions of the Royal Society. Series A: Mathematical, Physical and Engineering Sciences* **358**, 1465–1485.

Eichelberger, J. C., Lysne, P. C. & Younker, L. W. (1984). Research drilling at Inyo Domes, Long Valley Caldera, California. *Eos, Transactions American Geophysical Union* **65**, 721–725.

Eiler, J. M. (2001). Oxygen isotope variations of basaltic lavas and upper mantle rocks. *Reviews in Mineralogy and Geochemistry* **43**, 319–364.

Einstein, A. (1906). A new determination of molecular dimensions (Eine neue Bestimmung der Moleküldimensionen). *Annalen der Physik* **4**, 289–306.

Elhacham, E., Ben-Uri, L., Grozovski, J., Bar-On, Y. M. & Milo, R. (2020). Global human-made mass exceeds all living biomass. *Nature* **588**, 442–444.

Elliott, H. A. L., Wall, F., Chakhmouradian, A. R., Siegfried, P. R., Dahlgren, S., Weatherley, S., Finch, A. A., Marks, M. A. W., Dowman, E. & Deady, E. (2018). Fenites associated with carbonatite complexes: a review. *Ore Geology Reviews* **93**, 38–59.

Emiliani, C. (1957). Temperature and age analysis of deep-sea cores. *Science* **125**, 383–387.

Endo, S., Wallis, S. R., Tsuboi, M., Torres De Leon, R. & Solari, L. A. (2012). Metamorphic evolution of lawsonite eclogites from the southern Motagua fault zone, Guatemala: insights from phase equilibria and Raman spectroscopy. *Journal of Metamorphic Geology* **30**, 143–164.

Engi, M., Lanari, P. & Kohn, M. J. (2017). Significant ages—an introduction to petrochronology. *Reviews in Mineralogy and Geochemistry* **83**, 1–12.

England, P. C. & Katz, R. F. (2010). Melting above the anhydrous solidus controls the location of volcanic arcs. *Nature* **467**, 700–703.

Environmental Protection Agency. (2005). *Guidance for Developing Ecological Soil Screening Levels. OSWER Directive.* US Environmental Protection Agency.

Ernst, R. E. & Baragar, W. R. A. (1992). Evidence from magnetic fabric for the flow pattern of magma in the Mackenzie giant radiating dyke swarm. *Nature* **356**, 511–513.

Ernst, R. E. & Bell, K. (2010). Large igneous provinces (LIPs) and carbonatites. *Mineralogy and Petrology* **98**, 55–76.

Ernst, R. E. & Buchan, K. L. (2001). Mantle plumes: Their identification through time. *Geological Society of America Special Paper* **352**, 593 p.

Ernst, W. G. (1964). Petrochemical study of coexisting minerals from low-grade schists, eastern Shikoku, Japan. *Geochimica et Cosmochimica Acta* **28**, 1631–1668.

Ernst, W. G. (1969). *Earth Materials.* Prentice-Hall, 149 p.

Ernst, W. G. (1976). *Petrologic Phase Equilibria.* W. H. Freeman and Co., 333 p.

Erwin, D. H. (1996). The mother of mass extinctions. *Scientific American* **7**, 72–78.

Eskola, P. (1915). On the relations between the chemical and mineralogical composition in the metamorphic rocks of the Orijärvi region. *Bulletin de la Commission géologique de Finlande* **44**, 109–145.

Evans, B. W. & Moore, J. G. (1968). Mineralogy as a function of depth in the prehistoric Makaopuhi tholeiitic lava lake, Hawaii. *Contributions to Mineralogy and Petrology* **17**, 85–115.

Farmer, G. L. & DePaolo, D. J. (1983). Origin of Mesozoic and Tertiary granite in the western United States and implications for pre-Mesozoic crustal structure. 1. Nd and Sr isotopic studies in the geocline of the northern Great Basin. *Journal of Geophysical Research* **88**, 3379–3401.

Faul, U. H. (2001). Melt retention and segregation beneath mid-ocean ridges. *Nature* **410**, 920–923.

Faure, G., Bowman, J. R., Elliot, D. H. & Jones, L. M. (1974). Strontium isotope composition and petrogenesis of the Kirkpatrick basalt, Queen Alexandra Range, Antarctica. *Contributions to Mineralogy and Petrology* **48**, 153–169.

Faure, G. & Mensing, T. M. (2005). *Isotopes: Principles and Applications.* Wiley, 897 p.

Feraud, G. & Campredon, R. (1983). Geochronological and structural study of Tertiary and Quaternary dikes in southern France and Sardinia: an example of the utilization of dike swarms as paleostress indicators. *Tectonophysics* **98**, 297–325.

Ferry, J. M. & Spear, F. S. (1978). Experimental calibration of the partitioning of Fe and Mg between biotite and garnet. *Contributions to Mineralogy and Petrology* **66**, 113–117.

Fiske, R. S. & Jackson, E. D. (1972). Orientation and growth of Hawaiian volcanic rifts: the effect of regional structure and gravitational stresses. *Proceedings of the Royal Society of London. A. Mathematical and Physical Sciences* **329**, 299–326.

Flügel, E. & Kiessling, W. (2002). A new look at ancient reefs. *SEPM Special Publication* **72**, 3–9.

Folkes, C. B., de Silva, S. L., Wright, H. M. & Cas, R. A. F. (2011). Geochemical homogeneity of a long-lived, large silicic system: evidence from the Cerro Galán caldera, NW Argentina. *Bulletin of Volcanology* **73**, 1455–1486.

Fossen, H. (2016). *Structural Geology*. Cambridge University Press, 510 p.

Foster, D. A., Doughty, P. T., Kalakay, T. J., Fanning, C. M., Coyner, S., Grice, W. C. & Vogl, J. (2007). Kinematics and timing of exhumation of metamorphic core complexes along the Lewis and Clark fault zone, northern Rocky Mountains, USA. *Geological Society of America Special Paper* **434**, 207–232.

Foulger, G. R. (2010). *Plates vs. Plumes: A Geological Controversy*. Wiley-Blackwell, 364 p.

Foulger, G. R., Natland, J. H., Presnall, D. C. & Anderson, D. L. (eds). (2005). Plates, plumes and paradigms. *Geological Society of America Special Paper* **388**, 881 p.

Fournier, R. O. (1989). Geochemistry and dynamics of the Yellowstone National Park hydrothermal system. *Annual Review of Earth and Planetary Sciences* **17**, 13–53.

Francheteau, J. Needham, H. D., Choukroune, P., Juteau, T., Seguret, M., Ballard, R. D., Fox, P. J., Normark, W., Carranza, A., Cordoba, D. & Guerrero, J. (1979). Massive deep-sea sulphide ore deposits discovered on the East Pacific Rise. *Nature* **277**, 523–528.

Francis, P. (1993). *Volcanoes: A Planetary Perspective*. Clarendon Press, 443 p.

French, B. M. (1998). *Traces of Catastrophe: A Handbook of Shock-Metamorphic Effects in Terrestrial Meteorite Impact Structures*. LPI Contribution No. 954, 120 p.

Frisch, W. (2016). Paired metamorphic belts. In: *Encyclopedia of Marine Geosciences*. Springer, doi.org/10.1007/978-94-007-6238-1.

Frost, B. R. & Frost, C. D. (2019). *Essentials of Igneous and Metamorphic Petrology*. Cambridge University Press, 362 p.

Frost, D. A., Avery, M. S., Buffett, B. A., Chidester, B. A., Deng, J., Dortman, S. M., Li, Z., Liu, L., Lv, M. & Martin, J. F. (2022). Multidisciplinary constraints on the thermal-chemical boundary between Earth's core and mantle. *Geochemistry, Geophysics, Geosystems* **23**, doi.org/10.1029/2021GC009764.

Frost, T. P. & Mahood, G. A. (1987). Field, chemical, and physical constraints on mafic-felsic magma interaction in the Lamarck Granodiorite, Sierra Nevada, California (USA). *Geological Society of America Bulletin* **99**, 272–291.

Fryer, P., Wheat, C. G. & Mottl, M. J. (1999). Mariana blueschist mud volcanism: implications for conditions within the subduction zone. *Geology* **27**, 103–106.

Fu, L., Milliken, K. L. & Sharp Jr, J. M. (1994). Porosity and permeability variations in fractured and Liesegang-banded Breathitt sandstones (Middle Pennsylvanian), eastern Kentucky: diagenetic controls and implications for modeling dual-porosity systems. *Journal of Hydrology* **154**, 351–381.

Fukao, Y. & Obayashi, M. (2013). Subducted slabs stagnant above, penetrating through, and trapped below the 660 km discontinuity. *Journal of Geophysical Research* **118**, 5920–5938.

Fullagar, P. D. (1971). Age and origin of plutonic intrusions in the Piedmont of the southeastern Appalachians. *Geological Society of America Bulletin* **82**, 2845–2862.

Furlong, K. P., Hanson, R. B. & Bowers, J. R. (1991). Modelling thermal regimes. *Reviews in Mineralogy* **26**, 437–505.

Fyfe, W. S. (1978). The evolution of the earth's crust: modern plate tectonics to ancient hot spot tectonics? *Chemical Geology* **23**, 89–114.

Galerne, C. Y., Galland, O., Neumann, E. R. & Planke, S. (2011). 3D relationships between sills and their feeders: evidence from the Golden Valley sill complex (Karoo Basin) and experimental modelling. *Journal of Volcanology and Geothermal Research* **202**, 189–199.

Galili, N., Shemesh, A., Yam, R., Brailovsky, I., Sela-Adler, M., Schuster, E. M., Collom, C., Bekker, A., Planavsky, N., Macdonald, F. A. & Préat, A. (2019). The geologic history of seawater oxygen isotopes from marine iron oxides. *Science* **365**, 469–473.

Gao, X. & Wang, K. (2014). Strength of stick-slip and creeping subduction megathrusts from heat flow observations. *Science* **345**, 1038–1041.

Garcia-Casco, A., Torres-Roldan, R. L., Millan, G., Monie, P. & Schneider, J. (2002). Oscillatory zoning in eclogitic garnet and amphibole, northern serpentinite mélange, Cuba: a record of tectonic instability during subduction? *Journal of Metamorphic Geology* **20**, 581–598.

Gard, M., Hasterok, D., Hand, M. & Cox, G. (2019). Variations in continental heat production from 4 Ga to the present: evidence from geochemical data. *Lithos* **342–343**, 391–406.

Gast, P. W. (1960). Limitations on the composition of the upper mantle. *Journal of Geophysical Research* **65**, 1287–1297.

Gast, P. W. (1968). Trace element fractionation and the origin of tholeiitic and alkaline magma types. *Geochimica et Cosmochimica Acta* **32**, 1057–1086.

Gathorne-Hardy, F. J. & Harcourt-Smith, W. E. H. (2003). The super-eruption of Toba, did it cause a human bottleneck? *Journal of Human Evolution* **45**, 227–230.

Geist, D., Howard, K. A. & Larson, P. (1995). The generation of oceanic rhyolites by crystal fractionation: the basalt-rhyolite association at Volcán Alcedo, Galápagos archipelago. *Journal of Petrology* **36**, 965–982.

Gelman, S. E., Gutierrez, F. J. & Bachmann, O. (2013). On the longevity of large upper crustal silicic magma reservoirs. *Geology* **41**, 759–762.

Ghiorso, M. S. (1997). Thermodynamic analysis of the effect of magnetic ordering on miscibility gaps in the FeTi cubic and rhombohedral oxide minerals and the FeTi oxide geothermometer. *Physics and Chemistry of Minerals* **25**, 28–38.

Ghiorso, M. S. & Sack, R. O. (1991). Fe-Ti oxide geothermometry: thermodynamic formulation and the estimation of intensive variables in silicic magmas. *Contributions to Mineralogy and Petrology* **108**, 485–510.

Gibson, I. L. & Walker, G. P. L. (1963). Some composite rhyolite/basalt lavas and related composite dykes in eastern Iceland. *Proceedings of the Geologists' Association* **74**, 301–318.

Gilbert, G. K. (1877). Geology of the Henry Mountains. *Government Printing Office*, 160 p.

Giletti, B. J. (1986). Diffusion effects on oxygen isotope temperatures of slowly cooled igneous and metamorphic rocks. *Earth and Planetary Science Letters* 77, 218–228.

Gill, J. (1981). *Orogenic Andesites and Plate Tectonics.* Springer-Verlag, 392 p.

Gilotti, J. A. (2013). The realm of ultrahigh-pressure metamorphism. *Elements* 9, 255–260.

Giordano, D., Russell, J. K. & Dingwell, D. B. (2008). Viscosity of magmatic liquids: a model. *Earth and Planetary Science Letters* **271**, 123–134.

Giuliani, A. & Pearson, D. G. (2019). Kimberlites: from deep Earth to diamond mines. *Elements* **15**, 377–380.

Glazner, A. F. (1994). Foundering of mafic plutons and density stratification of continental crust. *Geology* **22**, 435–438.

Glazner, A. F. (2007). Thermal limitations on incorporation of wall rock into magma. *Geology* **35**, 319–322.

Glazner, A. F. (2014). Magmatic life at low Reynolds number. *Geology* **42**, 935–938.

Glazner, A. F. (2019). The ascent of water-rich magma and decompression heating: a thermodynamic analysis. *American Mineralogist* **104**, 890–896.

Glazner, A. F. (2021). Thermal constraints on the longevity, depth, and vertical extent of magmatic systems. *Geochemistry, Geophysics, Geosystems* **22**, doi.org/10.1029/2020GC009459.

Glazner, A. F. (2022). Cenozoic magmatism and plate tectonics in western North America: have we got it wrong? *Geological Society of America Special Paper* **553**, 95–108, https://doi.org/10.1130/2021.2553(09).

Glazner, A. F. & Bartley, J. M. (1984). Timing and tectonic setting of Tertiary low-angle normal faulting and associated magmatism in the southwestern United States. *Tectonics* **3**, 385–396.

Glazner, A. F. & Bartley, J. M. (1985). Evolution of lithospheric strength after thrusting. *Geology* **13**, 42–45.

Glazner, A. F. & Bartley, J. M. (2006). Is stoping a volumetrically significant pluton emplacement process? *Geological Society of America Bulletin* **118**, 1185–1195.

Glazner, A. F. & Bartley, J. M. (2008). Reply to comments on "Is stoping a volumetrically significant pluton emplacement process?" *Geological Society of America Bulletin* **120**, 1082–1087.

Glazner, A. F., Bartley, J. M. & Coleman, D. S. (2016). We need a new definition for 'magma'. *EOS*, https://eos.org/opinions/we-need-a-new-definition-for-magma.

Glazner, A. F., Bartley, J. M. & Coleman, D. S. (2019). A more informative way to name plutonic rocks. *GSA Today* **29**, doi.org/10.1130/GSATG384A.1.

Glazner, A. F., Bartley, J. M., Coleman, D. S., Gray, W. & Taylor, R. Z. (2004). Are plutons assembled over millions of years by amalgamation from small magma chambers? *GSA Today* **14**, 4–11.

Glazner, A. F., Bartley, J. M., Coleman, D. S. & Lindgren, K. (2020). Aplite diking and infiltration: a differentiation mechanism restricted to plutonic rocks. *Contributions to Mineralogy and Petrology* **175**, doi.org/10.1007/s00410-020-01677-1.

Glazner, A. F., Bartley, J. M., Hamilton, W. B. & Carl, B. S. (2003). Making space for batholiths by extrusion of sub-batholithic crust. *International Geology Review* **45**, 959–967.

Glazner, A. F., Bartley, J. M. & Law, B. S. (2021). Immiscibility and the origin of ladder structures, mafic layering, and schlieren in plutons. *Geology* **49**, 86–90.

Glazner, A. F., Carl, B. S., Coleman, D. S., Miller, J. S. & Bartley, J. M. (2008a). Chemical variability and the composite nature of dikes from the Jurassic Independence dike swarm, eastern California. *Geological Society of America Special Paper* **438**, 455–480.

Glazner, A. F., Coleman, D. S. & Bartley, J. M. (2008b). The tenuous connection between high-silica rhyolites and granodiorite plutons. *Geology* **36**, 183–186.

Glazner, A. F., Coleman, D. S. & Mills, R. D. (2015). The volcanic-plutonic connection. In: Breitkreuz, C. & Rocchi, S. (eds) *Physical Geology of Shallow Magmatic Systems: Dykes, Sills, and Laccoliths.* Springer International Publishing, 1–22.

Glazner, A. F. & Johnson, B. R. (2013). Late crystallization of K-feldspar and the paradox of megacrystic granites. *Contributions to Mineralogy and Petrology* **166**, doi.org/10.1007/s00410-013-0914-1.

Glazner, A. F., Manley, C. R., Marron, J. S. & Rojstaczer, S. (1999). Fire or ice: anticorrelation of volcanism and glaciation in California over the past 800,000 years. *Geophysical Research Letters* **26**, 1759–1762.

Glazner, A. F. & Miller, D. M. (1997). Late-stage sinking of plutons. *Geology* **25**, 1099–1102.

Glazner, A. F., Sylvester, A. G., & Sharp, R. P. (2022). *Geology Underfoot in Death Valley and Eastern California.* Mountain Press, 318 p.

Glazner, A. F. & Walker, J. D. (2020). StraboTools: A mobile app for quantifying fabric in geology. *GSA Today* **30**, 1–8.

Goldschmidt, V. M. (1911). Die kontaktmetamorphose im Kristianiagebiet. *Geologiska Föreningen i Stockholm Förhandlingar* **34**, 812–819.

Gounelle, M. & Meynet, G. (2012). Solar system genealogy revealed by extinct short-lived radionuclides in meteorites. *Astronomy and Astrophysics* **545**, 1–9.

Gray, R. D., Drummond, A. J. & Greenhill, S. J. (2009). Language phylogenies reveal expansion pulses and pauses in Pacific settlement. *Science* **323**, 479–483.

Green, D. H. & Falloon, T. J. (1998). Pyrolite: a Ringwood concept and its current expression. In: Jackson, I. N. S. (ed), *The Earth's Mantle: Composition, Structure, and Evolution.* Cambridge University Press, 311–380.

Green, T. H. & Ringwood, A. E. (1966). Origin of the calc-alkaline igneous rock suite. *Earth and Planetary Science Letters* **1**, 307–316.

Gregory, R. T. & Taylor, H. P. (1981). An oxygen isotope profile in a section of Cretaceous oceanic crust, Samail Ophiolite, Oman: evidence for $\delta^{18}O$ buffering of the oceans by deep (>5 km) seawater-hydrothermal circulation at mid-ocean ridges. *Journal of Geophysical Research* **86**, 2737–2755.

Greig, J. W. & Barth, T. F. W. (1938). The system $Na_2O{\cdot}Al_2O_3{\cdot}2SiO_2$ (nephelite, carnegieite)-$Na_2O{\cdot}Al_2O_3{\cdot}6SiO_2$ (albite). *American Journal of Science* **35-A**, 93–112.

Gretener, P. E. (1969). On the mechanics of the intrusion of sills. *Canadian Journal of Earth Sciences* **6**, 1415–1419.

Griffiths, R. W. & Campbell, I. H. (1990). Stirring and structure in mantle starting plumes. *Earth and Planetary Science Letters* **99**, 66–78.

Groppo, C., Ferrando, S., Gilio, M., Botta, S., Nosenzo, F., Balestro, G., Festa, A. & Rolfo, F. (2019). What's in the sandwich? New P–T constraints for the (U)HP nappe stack of southern Dora-Maira Massif (Western Alps). *European Journal of Mineralogy* **31**, 665–683.

Groppo, C., Lombardo, B., Castelli, D. & Compagnoni, R. (2007). Exhumation history of the UHPM Brossasco-Isasca Unit, Dora-Maira Massif, as inferred from a phengite-amphibole eclogite. *International Geology Review* **49**, 142–168.

Grout, F. F. (1918). Two-phase convection in igneous magmas. *Journal of Geology* **26**, 481–499.

Grout, F. F. (1945). Scale models of structures related to batholiths. *American Journal of Science* **243**, 260–284.

Grove, T. L., Baker, M. B. & Kinzler, R. J. (1984). Coupled CaAl-NaSi diffusion in plagioclase feldspar: experiments and applications to cooling rate speedometry. *Geochimica et Cosmochimica Acta* **48**, 2113–2121.

Guevara, V. E., Smye, A. J., Caddick, M. J., Searle, M. P., Olsen, T., Whalen, L., Kylander-Clark, A. R. C., Jercinovic, M. J. & Waters, D. J. (2022). A modern pulse of ultrafast exhumation and diachronous crustal melting in the Nanga Parbat Massif. *Science Advances* **8**, doi.org/10.1126/sciadv.abm2689.

Gunter, M. E., Belluso, E. & Mottana, A. (2007). Amphiboles: environmental and health concerns. *Reviews in Mineralogy and Geochemistry* **67**, 453–516.

Hacker, B. R. (2008). H_2O subduction beyond arcs. *Geochemistry, Geophysics, Geosystems* **9**, doi.org/10.1029/2007GC001707.

Hacker, B. R., Andersen, T. B., Johnston, S., Kylander-Clark, A. R. C., Peterman, E. M., Walsh, E. O. & Young, D. (2010). High-temperature deformation during continental-margin subduction & exhumation: the ultrahigh-pressure Western Gneiss Region of Norway. *Tectonophysics* **480**, 149–171.

Hamilton, D. L., Freestone, I. C., Dawson, J. B. & Donaldson, C. H. (1979). Origin of carbonatites by liquid immiscibility. *Nature* **279**, 52–54.

Hamilton, W. B. (2003). An alternative Earth. *GSA Today* **13**, 4–12.

Hamilton, W. B. (2007). Driving mechanism and 3-D circulation of plate tectonics. *Geological Society of America Special Paper* **433**, 1–25.

Hamilton, W. B. & Myers, W. B. (1967). The nature of batholiths. *US Geological Survey Professional Paper* **554–C**, 30 p.

Hamilton, W. & Myers, W. B. (1974). Nature of the Boulder batholith of Montana. *Geological Society of America Bulletin* **85**, 365–378.

Hanchar, J. M. & Watson, E. B. (2003). Zircon saturation thermometry. *Reviews in Mineralogy and Geochemistry* **53**, 89–112.

Harrison, T. M. (2009). The Hadean crust: evidence from >4 Ga zircons. *Annual Review of Earth and Planetary Science* **37**, 479–505.

Harrison, T. M., Grove, M., McKeegan, K. D., Coath, C. D., Lovera, O. M. & Le Fort, P. (1999). Origin and episodic emplacement of the Manaslu intrusive complex, central Himalaya. *Journal of Petrology* **40**, 3–19.

Hasterok, D., Halpin, J. A., Collins, A. S., Hand, M., Kreemer, C., Gard, M. G. & Glorie, S. (2022). New maps of global geological provinces and tectonic plates. *Earth-Science Reviews* **231**, doi.org/10.1016/j.earscirev.2022.104069.

Hayden, L. A., Watson, E. B. & Wark, D. A. (2008). A thermobarometer for sphene (titanite). *Contributions to Mineralogy and Petrology* **155**, 529–540.

Heaman, L. M. & Kjarsgaard, B. A. (2000). Timing of eastern North American kimberlite magmatism: continental extension of the Great Meteor hotspot track? *Earth and Planetary Science Letters* **178**, 253–268.

Heaman, L. M., Kjarsgaard, B. A. & Creaser, R. A. (2004). The temporal evolution of North American kimberlites. *Lithos* **76**, 377–397.

Hedge, C. E. (1966). Variations in radiogenic strontium found in volcanic rocks. *Journal of Geophysical Research* **71**, 6119–6126.

Heizler, M. T., Ralser, S. & Karlstrom, K. E. (1997). Late Proterozoic (Grenville?) deformation in central New Mexico determined from single-crystal muscovite $^{40}Ar/^{39}Ar$ age spectra. *Precambrian Research* **84**, 1–15.

Heliker, C., Swanson, D. A., Takahashi, T. J. & Norton, G. A. (2003). The PuʻuʻŌʻō-Kūpaianaha eruption of Kīlauea Volcano, Hawaiʻi: the first 20 years. *US Geological Survey Professional Paper* **1676**, 206 p.

Helz, R. T. & Thornber, C. R. (1987). Geothermometry of Kilauea Iki lava lake, Hawaii. *Bulletin of Volcanology* **49**, 651–668.

Hendrie, D. B., Kusznir, N. J., Morley, C. K. & Ebinger, C. J. (1994). Cenozoic extension in northern Kenya: a quantitative model of rift basin development in the Turkana region. *Tectonophysics* **236**, 409–438.

Hermann, A., Gutiérrez, P., Chauvel, C., Maury, R., Liorzou, C., Willie, E., Phillip, I., Forkel, R., Rzymski, C. & Bedford, S. (2023). Artifact geochemistry demonstrates long-distance voyaging in the Polynesian Outliers. *Science Advances* **9**, doi.org/10.1126/sciadv.adf4487.

Hermann, J. & Rubatto, D. (2014). Subduction of continental crust to mantle depth: geochemistry of ultrahigh-pressure rocks. In: Turekian, K. K. & Holland, H. D. (eds) *Treatise on Geochemistry*, 2nd edition, Vol. 4 Elsevier, 309–340.

Hess, H. H. (1955). Serpentines, orogeny, and epeirogeny. *Geological Society of America Special Paper* **62**, 391–408.

Hibbard, J. P., Stoddard, E. F., Secor, D. T. & Dennis, A. J. (2002). The Carolina zone: overview of Neoproterozoic to early Paleozoic peri-Gondwanan terranes along the eastern flank of the southern Appalachians. *Earth-Science Reviews* **57**, 299–339.

Hibbard, M. J. & Watters, R. J. (1985). Fracturing and diking in incompletely crystallized granitic plutons. *Lithos* **18**, 1–12.

Hildreth, E. W. (1977). The magma chamber of the Bishop Tuff: gradients in temperature, pressure, and composition. PhD thesis, University of California at Berkeley.

Hildreth, W. (1979). The Bishop Tuff: evidence for the origin of compositional zonation in silicic magma chambers. *Geological Society of America Special Paper* **180**, 43–75.

Hildreth, W. (1981). Gradients in silicic magma chambers: implications for lithospheric magmatism. *Journal of Geophysical Research* **86**, 10153–10192.

Hildreth, W. (2004). Volcanological perspectives on Long Valley, Mammoth Mountain, and Mono Craters: several contiguous but discrete systems. *Journal of Volcanology and Geothermal Research* **136**, 169–198.

Hildreth, W. & Fierstein, J. (2012). The Novarupta-Katmai eruption of 1912—largest eruption of the twentieth century: centennial perspectives. *US Geological Survey Professional Paper* **1791**, 259 p.

Hilton, R. G. & West, A. J. (2020). Mountains, erosion and the carbon cycle. *Nature Reviews Earth & Environment* **1**, 284–299.

Hoffman, P. F. (1988). United plates of America, the birth of a craton: early Proterozoic assembly and growth of Laurentia. *Annual Review of Earth and Planetary Science* **16**, 543–603.

Hoffman, P. F. (1991). Did the breakout of Laurentia turn Gondwanaland inside-out? *Science* **252**, 1409–1412.

Hoffman, P. F. & Schrag, D. P. (2000). Snowball Earth. *Scientific American* **282**, 68–75.

Hofmeister, A. M. (1999). Mantle values of thermal conductivity and the geotherm from phonon lifetimes. *Science* **283**, 1699–1706.

Holness, M. B., Clemens, J. D. & Vernon, R. H. (2018). How deceptive are microstructures in granitic rocks? Answers from integrated physical theory, phase equilibrium, and direct observations. *Contributions to Mineralogy and Petrology* **173**, doi.org/10.1007/s00410-018-1488-8.

Holtz, F., Sato, H., Lewis, J., Behrens, H. & Nakada, S. (2005). Experimental petrology of the 1991–1995 Unzen dacite, Japan. Part I: phase relations, phase composition and pre-eruptive conditions. *Journal of Petrology* **46**, 319–337.

Hopper, E. & Fischer, K. M. (2018). The changing face of the lithosphere-asthenosphere boundary: imaging continental scale patterns in upper mantle structure across the contiguous US with Sp converted waves. *Geochemistry, Geophysics, Geosystems* **19**, 2593–2614.

House, M. A., Wernicke, B. P. & Farley, K. A. (2001). Paleo-geomorphology of the Sierra Nevada, California, from (U-Th)/He ages in apatite. *American Journal of Science* **301**, 77–102.

Hu, G. & Clayton, R. N. (2003). Oxygen isotope salt effects at high pressure and high temperature and the calibration of oxygen isotope geothermometers. *Geochimica et Cosmochimica Acta* **67**, 3227–3246.

Hu, J., Gurnis, M., Rudi, J., Stadler, G. & Müller, R. D. (2022). Dynamics of the abrupt change in Pacific Plate motion around 50 million years ago. *Nature Geoscience* **15**, 74–78.

Huang, F., Lundstrom, C. C., Glessner, J., Ianno, A., Boudreau, A., Li, J., Ferre, E. C., Marshak, S. & DeFrates, J. (2009). Chemical and isotopic fractionation of wet andesite in a temperature gradient: experiments and models suggesting a new mechanism of magma differentiation. *Geochimica et Cosmochimica Acta* **73**, 729–749.

Hubbert, M. K. & Rubey, W. W. (1959). Mechanics of fluid-filled porous solids and its application to overthrust faulting. *Geological Society of America Bulletin* **70**, 115–166.

Huber, N. K., Bateman, P. C. & Wahrhaftig, C. (1989). Geologic map of Yosemite National Park and vicinity, California. *US Geological Survey Map* **I-1874**, scale 1:125,000.

Humphris, S. E. & Thompson, G. (1978). Hydrothermal alteration of oceanic basalts by seawater. *Geochimica et Cosmochimica Acta* **42**, 107–125.

Hunt, C. B. (1953). Geology and geography of the Henry Mountains region, Utah: a survey and restudy of one of the classic areas in geology. *US Geological Survey Professional Paper* **228**, 244 p.

Hunter, R. H. & McKenzie, D. (1989). The equilibrium geometry of carbonate melts in rocks of mantle composition. *Earth and Planetary Science Letters* **92**, 347–356.

Hurley, P. M. & Rand, J. R. (1969). Pre-drift continental nuclei. *Science* **164**,1229–1242.

Hurwitz, S. & Manga, M. (2017). The fascinating and complex dynamics of geyser eruptions. *Annual Review of Earth and Planetary Sciences* **45**, 31–59.

Hurwitz, S., Sohn, R. A., Luttrell, K. & Manga, M. (2014). Triggering and modulation of geyser eruptions in Yellowstone National Park by earthquakes, earth tides, and weather. *Journal of Geophysical Research* **119**, 1718–1737.

Husch, J. M. (1990). Palisades sill: origin of the olivine zone by separate magmatic injection rather than gravity settling. *Geology* **18**, 699–702.

Hutton, D. H. W. & Reavy, R. J. (1992). Strike-slip tectonics and granite petrogenesis. *Tectonics* **11**, 960–967.

Hyndman, D. W. & Alt, D. (1987). Radial dikes, laccoliths, and gelatin models. *Journal of Geology* **95**, 763–774.

Iliadis, C. (2015). *Nuclear Physics of Stars*. Wiley-VCH Verlag GmbH & Co., 672 p.

Ingram, B. L. & DePaolo, D. J. (2022). *Sr Isotopes in Seawater: Stratigraphy, Paleo-Tectonics, Paleoclimate, and Paleoceanography*. Cambridge University Press, 75 p.

Irvine, T. N. (1982). Terminology for layered intrusions. *Journal of Petrology* **23**, 127–162.

Irvine, T. N., Andersen, J. C. Ø. & Brooks, C. K. (1998). Included blocks (and blocks within blocks) in the Skaergaard intrusion: geologic relations and the origins of rhythmic modally graded layers. *Geological Society of America Bulletin* **110**, 1398–1447.

Jackson, E. D. (1961). Primary textures and mineral associations in the ultramafic zone of the Stillwater Complex, Montana. *US Geological Survey Professional Paper* **358**, 106 p.

Jackson, E. D., Beeson, M. H. & Clague, D. A. (1982). Xenoliths in volcanic rocks from Mauna Kea Volcano, Hawaii. *US Geological Survey Open-File Report* **82–201**, 19 p.

Jackson, L. J., Stone, J. R., Cohen, A. S. & Yost, C. L. (2015). High-resolution paleoecological records from Lake Malawi show no significant cooling associated with the Mount Toba supereruption at ca. 75 ka. *Geology* **43**, 823–826.

Jackson, M. D. & Noller, J. S. (1991). Geologic map of the Copper Creek Benches quadrangle, Garfield County, Utah. *Utah Geological and Mineral Survey Open-File Report* **209**, 47 p.

Jackson, M. D. & Pollard, D. D. (1988). The laccolith-stock controversy: new results from the southern Henry Mountains, Utah. *Geological Society of America Bulletin* **100**, 117–139.

Jacoby, M. (2020). Shrinking concrete's giant carbon footprint. *C&EN Global Enterprise* **98**, 26–30.

Jaeger, J. C. (1964). Thermal effects of intrusions. *Reviews of Geophysics* **2**, 443–466.

Jagoutz, O. & Schmidt, M. W. (2012). The formation and bulk composition of modern juvenile continental crust: the Kohistan Arc. *Chemical Geology* **298–299**, 79–96.

Jakosky, B. M. (1986). Volcanoes, the stratosphere, and climate. *Journal of Volcanology and Geothermal Research* **28**, 247–255.

Jiang, Z., Li, S., Liu, Q., Zhang, J. & Zhou, Z. (2021). The trials and tribulations of the Hawaii hot spot model. *Earth-Science Reviews* **215**, doi.org/10.1016/j.earscirev.2021.103544.

John, D. A., Ayuso, R. A., Barton, M. D., Blakely, R. J., Bodnar, R. J., Dilles, J. H., Gray, F., Graybeal, F. T., Mars, J. C., McPhee, D. K. & Seal, R. R. (2010). Porphyry copper deposit model. *US Geological Survey Scientific Investigations Report* **2010-5070-B**, 169 p.

Johnson, B. R. & Glazner, A. F. (2010). Formation of K-feldspar megacrysts in granodioritic plutons by thermal cycling and late-stage textural coarsening. *Contributions to Mineralogy and Petrology* **159**, 599–619.

Johnson, M. C. & Rutherford, M. J. (1989). Experimentally determined conditions in the Fish Canyon Tuff, Colorado, magma chamber. *Journal of Petrology* **30**, 711–737.

Johnson, R. B. (1961). Patterns and origin of radial dike swarms associated with West Spanish Peak and Dike Mountain, south-central Colorado. *Geological Society of America Bulletin* **72**, 579–589.

Johnston, A. D. & Draper, D. S. (1992). Near-liquidus phase relations of an anhydrous high-magnesia basalt from the Aleutian Islands: implications for arc magma genesis and ascent. *Journal of Volcanology and Geothermal Research* **52**, 27–41.

Jolly, R. J. H. & Sanderson, D. J. (1995). Variation in the form and distribution of dykes in the Mull swarm, Scotland. *Journal of Structural Geology* **17**, 1543–1556.

Jones, C. H., Farmer, G. L. & Unruh, J. (2004). Tectonics of Pliocene removal of lithosphere of the Sierra Nevada, California. *Geological Society of America Bulletin*, **116**, 1408–1422.

Jones, C. H., Kanamori, H. & Roecker, S. W. (1994). Missing roots and mantle "drips": regional Pn and teleseismic arrival times in the southern Sierra Nevada and vicinity, California. *Journal of Geophysical Research* **99**, 4567–4601.

Jordan, B. T., Grunder, A. L., Duncan, R. A. & Deino, A. L. (2004). Geochronology of age-progressive volcanism of the Oregon High Lava Plains: implications for the plume interpretation of Yellowstone. *Journal of Geophysical Research* **109**, doi.org/10.1029/2003JB002776.

Julian, B. R. (1983). Evidence for dyke intrusion earthquake mechanisms near Long Valley caldera, California. *Nature* **303**, 323–325.

Kalakay, T. J., John, B. E. & Lageson, D. R. (2001). Fault-controlled pluton emplacement in the Sevier fold-and-thrust belt of southwest Montana, USA. *Journal of Structural Geology* **23**, 1151–1165.

Kaneko, Y., Katayama, I., Yamamoto, H., Misawa, K., Ishikawa, M., Rehman, H. U., Kausar, A. B. & Shiraishi, K. (2003). Timing of Himalayan ultrahigh-pressure metamorphism: sinking rate and subduction angle of the Indian continental crust beneath Asia. *Journal of Metamorphic Geology* **21**, 589–599.

Karson, J. A., Frueh-Green, G. L., Kelley, D. S., Williams, E. A., Yoerger, D. R. & Jakuba, M. (2006). Detachment shear zone of the Atlantis Massif core complex, Mid-Atlantic Ridge, 30°N. *Geochemistry, Geophysics, Geosystems* **7**, doi.org/10.1029/2005GC001109.

Kasbohm, M. & Schoene, B. (2018). Rapid eruption of the Columbia River flood basalt and correlation with the mid-Miocene climate optimum. *Science Advances* **4**, doi.org/10.1126/sciadv.aat8223.

Kasting, J. F. (2019). The Goldilocks planet? How silicate weathering maintains Earth "just right." *Elements* **15**, 235–240.

Katayama, I., Maruyama, S., Parkinson, C. D., Terada, K. & Sano, Y. (2001). Ion micro-probe U-Pb zircon geochronology of peak and retrograde stages of ultrahigh-pressure metamorphic rocks from the Kokchetav Massif, northern Kazakhstan. *Earth and Planetary Science Letters* **188**, 185–198.

Kauahikaua, J., Cashman, K. V. & Thornber, C. R. (1998). Observations on basaltic lava streams in tubes from Kilauea Volcano, island of Hawai'i. *Journal of Geophysical Research* **103**, 27303.

Kavanagh, J. L., Menand, T. & Sparks, R. S. J. (2006). An experimental investigation of sill formation and propagation in layered elastic media. *Earth and Planetary Science Letters* **245**, 799–813.

Kawahata, H., Nohara, M., Ishizuka, H., Hasebe, S. & Chiba, H. (2001). Sr isotope geochemistry and hydrothermal alteration of the Oman ophiolite. *Journal of Geophysical Research* **106**, 11083–11099.

Kay, R. W. (1978). Aleutian magnesian andesites: melts from subducted Pacific ocean crust. *Journal of Volcanology and Geothermal Research* **4**, 117–132.

Kay, R. W. & Kay, S. M. (1993). Delamination and delamination magmatism. *Tectonophysics* **219**, 177–189.

Kelemen, P. B. & Dick, H. J. B. (1995). Focused melt flow and localized deformation in the upper mantle: juxtaposition of replacive dunite and ductile shear zones in the Josephine Peridotite, SW Oregon. *Journal of Geophysical Research* **100**, 423–438.

Kelemen, P. B., Hanghøj, K. & Greene, A. R. (2014). One view of the geochemistry of subduction-related magmatic arcs, with an emphasis on primitive andesite and lower crust. In: Turekian, K. K. & Holland, H. D. (eds) *Treatise on Geochemistry: Second Edition*, Vol 3. Elsevier, 749–805.

Kelemen, P. B., Whitehead, J. A., Aharonov, E. & Jordahl, K. A. (1995). Experiments on flow focusing in soluble porous media, with applications to melt extraction from the mantle. *Journal of Geophysical Research* **100**, 475–496.

Keller, J. & Krafft, M. (1990). Effusive natrocarbonatite activity of Oldoinyo Lengai, June 1988. *Bulletin of Volcanology* **52**, 629–645.

Kelley, D. S., Karson, J. A., Fruh-Green, G. L., Yoerger, D. R., Shank, T. M., Butterfield, D. A., Hayes, J. M., Schrenk, M. O., Olson, E. J., Proskurowski, G. & Jakuba, M. (2005). A serpentinite-hosted ecosystem: the Lost City hydrothermal field. *Science* **307**, 1428–1434.

Kelsey, D. E. & Hand, M. (2015). On ultrahigh temperature crustal metamorphism: Phase equilibria, trace element thermometry, bulk composition, heat sources, timescales and tectonic settings. *Geoscience Frontiers* **6**, 311–356.

Kennett, B. L. N. (1991). IASPEI 1991 seismological tables. *Terra Nova* **3**, 122 p.

Kerr, A. C. (2014). Oceanic Plateaus. In: Turekian, K. K. & Holland, H. D. (eds) *Treatise on Geochemistry: Second Edition*, Vol 4. Elsevier, 631–667.

Kerswell, B. C., Kohn, M. J. & Gerya, T. V. (2021). Backarc lithospheric thickness and serpentine stability control slab-mantle coupling depths in subduction zones. *Geochemistry, Geophysics, Geosystems* **22**, 1–27.

Kestin, J., Sokolov, M. & Wakeham, W. A. (1978). Viscosity of liquid water in the range −8 °C to 150 °C. *Journal of Physical and Chemical Reference Data* **7**, 941–948.

Kettler, R. M., Loope, D. B., Weber, K. A. & Niles, P. B. (2015). Life and Liesegang: outcrop-scale microbially induced diagenetic structures and geochemical self-organization phenomena produced by oxidation of reduced iron. *Astrobiology* **15**, 616–636.

Kind, R., Yuan, X., Saul, J., Nelson, D., Sobolev, S. V., Mechie, J., Zhao, W., Kosarev, G., Ni, J., Achauer, J. & Jiang, M. (2002). Seismic images of crust and upper mantle beneath Tibet: evidence for Eurasian plate subduction. *Science* **298**, 1219–1221.

Kinnaird, J. A. (2005). The Bushveld Large Igneous Province. www.largeigneousprovinces.org/sites/default/files/BushveldLIP.pdf.

Kistler, R. W. & Peterman, Z. E. (1973). Variations in Sr, Rb, K, Na, and initial $^{87}Sr/^{86}Sr$ in Mesozoic granitic rocks and intruded wall rocks in central California. *Geological Society of America Bulletin* **84**, 3489–3512.

Kitchen, D. E. (1989). The disequilibrium partial melting and assimilation of Caledonian granite by Tertiary basalt at Barnesmore, Co. Donegal. *Geological Magazine* **126**, 397–405.

Kleine, T. & Rudge, J. F. (2011). Chronometry of meteorites and the formation of the Earth and Moon. *Elements* **7**, 41–46.

Knapp, A. B. & Kassianidou, V. (2008). The archaeology of Late Bronze Age copper production. Politiko Phorades on Cyprus. *Anatolian Metal IV*, 135–147.

Knoll, A. H., Bambach, R. K., Payne, J. L., Pruss, S. & Fischer, W. W. (2007). Paleophysiology and end-Permian mass extinction. *Earth and Planetary Science Letters* **256**, 295–313.

Knudson, M. D. (2012). Megaamps, megagauss, and megabars: using the Sandia Z Machine to perform extreme material dynamics experiments. *AIP Conference Proceedings* **1426**, 35–42.

Kohlstedt, D. L., Evans, B. & Mackwell, S. J. (1995). Strength of the lithosphere: constraints imposed by laboratory experiments. *Journal of Geophysical Research* **100**, 17517–17602.

Kohn, M. J. (2014). Himalayan metamorphism and its tectonic implications. *Annual Review of Earth and Planetary Sciences*, **42**, 381–419.

Kohn, M. J. (2017). Titanite petrochronology. *Reviews in Mineralogy and Geochemistry* **83**, 419–441.

Kohn, M. J., Castro, A. E., Kerswell, B., Ranero, C. R. & Spear, F. S. (2018). Shear heating reconciles thermal models with the metamorphic rock record of subduction. *Proceedings of the National Academy of Sciences* **115**, 11706–11711.

Kohn, M. J. & Corrie, S. L. (2011). Preserved Zr-temperatures and U-Pb ages in high-grade metamorphic titanite: evidence for a static hot channel in the Himalayan orogen. *Earth and Planetary Science Letters* **311**, 136–143.

Kohn, M. J., Mazzucchelli, M. & Alvaro, M. (2023). Elastic thermobarometry. *Annual Review of Earth and Planetary Sciences*, **51**, 331–366.

Kohn, M. J. & Valley, J. W. (1998). Effects of cation substitutions in garnet and pyroxene on equilibrium oxygen isotope fractionations. *Journal of Metamorphic Geology* **16**, 625–639.

Koppers, A. A. P., Becker, T. W., Jackson, M. G., Konrad, K., Müller, R. D., Romanowicz, B., Steinberger, B. & Whittaker, J. M. (2021). Mantle plumes and their role in Earth processes. *Nature Reviews Earth and Environment* **2**, 382–401.

Korenaga, J. (2018). Crustal evolution and mantle dynamics through Earth history. *Philosophical Transactions of the Royal Society. Series A: Mathematical, Physical and Engineering Sciences* **376**, doi.org/10.1098/rsta.2017.0408.

Korochantseva, E. V., Trieloff, M., Lorenz, C. A., Buykin, A. I., Ivanova, M. A., Schwarz, W. H., Hopp, J. & Jessberger, E. K. (2007). L-chondrite asteroid breakup tied to Ordovician meteorite shower by multiple isochron ^{40}Ar-^{39}Ar dating. *Meteoritics & Planetary Science*, **42**, 113–130.

Korzhinskii, D. S. (1950). Equilibrium factors in metasomatism. *Izvestiya Akademii Nauk SSSR, Seriya Geologicheskaya* **3**, 21–49.

Koyaguchi, T. (1985). Magma mixing in a conduit. *Journal of Volcanology and Geothermal Research* **25**, 365–369.

Koyaguchi, T. (1986). Textural and compositional evidence for magma mixing and its mechanism, Abu volcano group, southwestern Japan. *Contributions to Mineralogy and Petrology* **93**, 33–45.

Koziol, A. M. & Newton, R. C. (1988). Redetermination of the anorthite breakdown reaction and improvement of the plagioclase-garnet-Al_2SiO_5-quartz geobarometer. *American Mineralogist* **73**, 216–223.

Krafft, M. & Keller, J. (1989). Temperature measurements in carbonatite lava lakes and flows from Oldoinyo Lengai, Tanzania. *Science* **245**, 168–170.

Krättli, G. & Schmidt, M. W. (2021). Experimental settling, floatation and compaction of plagioclase in basaltic melt and a revision of melt density. *Contributions to Mineralogy and Petrology* **176**, doi.org/10.1007/s00410-021-01785-6.

Krauskopf, K. B. (1948). Lava movement at Parícutin volcano, Mexico. *Geological Society of America Bulletin* **59**, 1267–1284.

Kraynik, A. M., Reinelt, D. A. & van Swol, F. (2003). Structure of random monodisperse foam. *Physical Review E* **67**, doi.org/10.1103/PhysRevE.67.031403.

Kring, David A. (2023). *Astronaut Training for Operations in Impact-Cratered Terrains.* LPI Contribution No. 3001, 570 p.

Kruckeberg, A. R. (2002). *Geology and Plant Life: The Effects of Landforms and Rock Type on Plants.* University of Washington Press, 362 p.

Kuno, H. (1959). Origin of Cenozoic petrographic provinces of Japan and surrounding areas. *Bulletin of Volcanology* **20**, 37–76.

Kushiro, I. (1972). Determination of liquidus relations in synthetic silicate systems with electron probe analysis: the system forsterite-diopside-silica at 1 atmosphere. *American Mineralogist* **57**, 1260–1271.

Lachenbruch, A. H. (1970). Crustal temperature and heat production: implications of the linear heat-flow relation. *Journal of Geophysical Research* **75**, 3291–3300.

Lambert, R. S. J., Marsh, I. K. & Chamberlain, V. E. (1989). The occurrence of interstitial granite-glass in all formations of the Columbia River Basalt Group and its petrogenetic implications. *Geological Society of America Special Paper* **239**, 321–332.

Lane, C. S., Chorn, B. T. & Johnson, T. C. (2013). Ash from the Toba supereruption in Lake Malawi shows no volcanic winter in East Africa at 75 ka. *Proceedings of the National Academy of Sciences* **110**, 8025–8029.

Lange, R. A. & Carmichael, I. S. E. (1990). Thermodynamic properties of silicate liquids with emphasis on density, thermal expansion and compressibility. *Reviews in Mineralogy* **24**, 25–64.

Laporte, D. & Watson, E. B. (1995). Experimental and theoretical constraints on melt distribution in crustal sources: the effect of crystalline anisotropy on melt interconnectivity. *Chemical Geology* **124**, 161–184.

Latypov, R., Chistyakova, S., Barnes, S. J., Godel, B., Delaney, G. W., Cleary, P. W., Radermacher, V. J., Campbell, I. & Jakata, K. (2022). Chromitite layers indicate the existence of large, long-lived, and entirely molten magma chambers. *Scientific Reports* **12**, doi.org/10.1038/s41598-022-08110-6.

Lavé, J. & Avouac, J. P. (2001). Fluvial incision and tectonic uplift across the Himalayas of central Nepal. *Journal of Geophysical Research* **106**, 26561–26591.

Le Bas, M. J., Le Maitre, R. W., Streckeisen, A. & Zanettin, B. (1986). A chemical classification of volcanic rocks based on the total alkali-silica diagram. *Journal of Petrology* **27**, 745–750.

Le Fort, P., Cuney, M., Deniel, C., France-Lanord, C., Sheppard, S. M. F., Upreti, B. N., & Vidal, P. (1987). Crustal generation of the Himalayan leucogranites. *Tectonophysics*, **134**, 39–57.

Lee, C.-T. A., Morton, D. M., Kistler, R. W. & Baird, A. K. (2007). Petrology and tectonics of Phanerozoic continent formation: from island arcs to accretion and continental arc magmatism. *Earth and Planetary Science Letters* **263**, 370–387.

Lee, T., Papanastassiou, D. A. & Wasserburg, G. J. (1976). Demonstration of ^{26}Mg excess in Allende and evidence for ^{26}Al. *Geophysical Research Letters* **3**, 41–44.

Lee, W. & Wyllie, P. J. (1994). Experimental data bearing on liquid immiscibility, crystal fractionation, and the origin of calciocarbonatites and natrocarbonatites. *International Geology Review* **36**, 797–819.

Leeman, W. P. (1970). The isotopic composition of strontium in late-Cenozoic basalts from the Basin-Range province, western United States. *Geochimica et Cosmochimica Acta* **34**, 857–872.

LeMaitre, R. W. (1976). The chemical variability of some common igneous rocks. *Journal of Petrology* **17**, 589–637.

LeMaitre, R. W. (1982). *Numerical Petrology: Statistical Interpretation of Geochemical Data*. Elsevier, 291 p.

LeMaitre, R. W. (2002). *Igneous Rocks: A Classification and Glossary of Terms*. Cambridge University Press, 236 p.

Lesher, C. E. & Walker, D. (1988). Cumulate maturation and melt migration in a temperature gradient. *Journal of Geophysical Research* **93**, 10295–10311.

Liesegang, R. E. (1896). Über einige Eigenschaften von Gallerten. *Naturwissenschaftliche Wochenschrift* **11**, 353–362.

Ligi, M., Cuffaro, M., Chierici, F. & Calafato, A. (2008). Three-dimensional passive mantle flow beneath mid-ocean ridges: an analytical approach. *Geophysical Journal International* **175**, 783–805.

Lilova, K. I., Pearce, C. I., Gorski, C., Rosso, K. M. & Navrotsky, A. (2012). Thermodynamics of the magnetite-ulvöspinel (Fe_3O_4-Fe_2TiO_4) solid solution. *American Mineralogist* **97**, 1330–1338.

Lindgren, W. (1915). The igneous geology of the Cordilleras and its problems. In: *Problems of American Geology*. Yale University Press, 234–286.

Lindsay, J. M., Trumbull, R. B. & Siebel, W. (2005). Geochemistry and petrogenesis of late Pleistocene to Recent volcanism in Southern Dominica, Lesser Antilles. *Journal of Volcanology and Geothermal Research* **148**, 253–294.

Liou, J. G. (1971). P-T Stabilities of laumontite, wairakite, lawsonite, and related minerals in the system $CaAl_2Si_2O_8$-SiO_2-H_2O. *Journal of Petrology* **12**, 379–411.

Lipman, P. W. (1984). The roots of ash flow calderas in western North America: windows into the tops of granitic batholiths. *Journal of Geophysical Research* **89**, 8801–8841.

Lipman, P. W. & Bachmann, O. (2015). Ignimbrites to batholiths: integrating perspectives from geological, geophysical, and geochronological data. *Geosphere* **11**, doi:10.1130/GES01091.1.

Lipman, P. W., Prostka, H. J. & Christiansen, R. L. (1972). Cenozoic volcanism and plate-tectonic evolution of the western United States. I. Early and middle Cenozoic. *Philosophical Transactions of the Royal Society of London. Series A, Mathematical and Physical Sciences* **271**, 217–248.

Liu, F. I. & Liou, J. G. (2011). Zircon as the best mineral for P-T-time history of UHP metamorphism: a review on mineral inclusions and U-Pb SHRIMP ages of zircons from the Dabie-Sulu UHP rocks. *Journal of Asian Earth Sciences* **40**, 1–39.

Lodders, K. (2003). Solar system abundances and condensation temperatures of the elements. *Astrophysical Journal* **591**, 1220–1247.

London, D. & Morgan, G. B. VI (2012). The pegmatite puzzle. *Elements* **8**, 263–268.

Lorenz, R. D., Turtle, E. P., Stiles, B., Le Gall, A., Hayes, A., Aharonson, O., Wood, C. A., Stofan, E. & Kirk, R. (2011). Hypsometry of Titan. *Icarus* **211**, 699–706.

Love, S. G. & Brownlee, D. E. (1993). A direct measurement of the terrestrial mass accretion rate of cosmic dust. *Science* **262**, 550–553.

Lovering, T. S. (1935). Theory of heat conduction applied to geological problems. *Geological Society of America Bulletin* **46**, 69–94.

Lowell, J. D. & Guilbert, J. M. (1970). Lateral and vertical alteration-mineralization zoning in porphyry ore deposits. *Economic Geology* **65**, 373–408.

Lowell, R. P. & Rona, P. A. (2002). Seafloor hydrothermal systems driven by the serpentinization of peridotite. *Geophysical Research Letters* **29**, doi.org/10.1029/2001GL014411.

Lowell, R. P., Rona, P. A. & Von Herzen, R. P. (1995). Seafloor hydrothermal systems. *Journal of Geophysical Research* **100**, 327–352.

Lowenstern, J. B. & Hurwitz, S. (2008). Monitoring a supervolcano in repose: heat and volatile flux at the Yellowstone Caldera. *Elements* **4**, 35–40.

Luhr, J. F. & Carmichael, I. S. E. (1980). The Colima volcanic complex, Mexico I. Post-caldera andesites from Volcán Colima. *Contributions to Mineralogy and Petrology* **71**, 343–372.

Luhr, J. F. & Melson, W. G. (1996). Mineral and glass compositions in June 15, 1991, pumices: evidence for dynamic disequilibrium in the dacite of Mount Pinatubo. In: Newhall, C. G. & Punongbayan, R. S. (eds) *Fire and Mud: Eruptions and Lahars of Mount Pinatubo, Philippines.* US Geological Survey, 733–750, https://pubs.usgs.gov/pinatubo/.

Lundstrom, C. C. (2009). Hypothesis for the origin of convergent margin granitoids and Earth's continental crust by thermal migration zone refining. *Geochimica et Cosmochimica Acta* **73**, 5709–5729.

Lundstrom, C. C. (2016). The role of thermal migration and low-temperature melt in granitoid formation: can granite form without rhyolitic melt? *International Geology Review* **58**, 371–388.

Lyell, C. S. (1839). *Elements of Geology.* John Murray, 316 p.

Macdonald, F. A., Swanson-Hysell, N. L., Park, Y., Lisiecki, L. & Jagoutz, O. (2019). Arc-continent collisions in the tropics set Earth's climate state. *Science* **364**, 181–184.

MacKinnon, I. D. R., Gooding, J. L., McKay, D. S. & Clanton, U. S. (1984). The El Chichón stratospheric cloud: solid particulates and settling rates. *Journal of Volcanology and Geothermal Research* **23**, 125–146.

Mahan, K. H., Bartley, J. M., Coleman, D. S., Glazner, A. F. & Carl, B. S. (2003). Sheeted intrusion of the synkinematic McDoogle pluton, Sierra Nevada, California. *Geological Society of America Bulletin* **115**, 1570–1582.

Mahon, K. I., Harrison, T. M. & Drew, D. A. (1988). Ascent of a granitoid diapir in a temperature varying medium. *Journal of Geophysical Research* **93**, 1174–1188.

Makhluf, A. R., Newton, R. C. & Manning, C. E. (2016). Hydrous albite magmas at lower crustal pressure: new results on liquidus H_2O content, solubility, and H_2O activity in the system $NaAlSi_3O_8$–H_2O–NaCl at 1.0 GPa. *Contributions to Mineralogy and Petrology* **171**, doi.org/10.1007/s00410-016-1286-0.

Manley, C. R., Glazner, A. F. & Farmer, G. L. (2000). Timing of volcanism in the Sierra Nevada of California: evidence for Pliocene delamination of the batholithic root. *Geology* **28**, 811–814.

Mark, D. F., Petraglia, M., Smith, V. C., Morgan, L. E., Barfod, D. N., Ellis, B. S., Pearce, N. J., Pal, J. N. & Korisettar, R. (2014). A high-precision $^{40}Ar/^{39}Ar$ age for the Young Toba Tuff and dating of ultra-distal tephra: forcing of Quaternary climate and implications for hominin occupation of India. *Quaternary Geochronology* **21**, 90–103.

Marks, N., Zierenberg, R. A. & Schiffman, P. (2015). Strontium and oxygen isotopic profiles through 3 km of hydrothermally altered oceanic crust in the Reykjanes geothermal system, Iceland. *Chemical Geology* **412**, 34–47.

Marsh, B. D. (1981). On the crystallinity, probability of occurrence, and rheology of lava and magma. *Contributions to Mineralogy and Petrology* **78**, 85–98.

Marshall, P. (1911). *Oceania. Handbuch der Regionalen Geologie.* C. Winter, 36 p.

Marvin, J. W. (1939). The shape of compressed lead shot and its relation to cell shape. *American Journal of Botany* **26**, 280–288.

Marzoli, A., Jourdan, F., Bussy, F., Chiaradia, M. & Costa, F. (2014). Petrogenesis of tholeiitic basalts from the central Atlantic magmatic province as revealed by mineral major and trace elements and Sr isotopes. *Lithos* **188**, 44–59.

Mason R. (2013). *Petrology of the Metamorphic Rocks.* Unwin Hyman, 230 p.

Matzel, J. E. P., Bowring, S. A. & Miller, R. B. (2006). Time scales of pluton construction at differing crustal levels: examples from the Mount Stuart and Tenpeak Intrusions, North Cascades, Washington. *Geological Society of America Bulletin* **118**, 1412–1430.

Matzke, E. B. (1946). The three-dimensional shape of bubbles in foam—an analysis of the role of surface forces in three-dimensional cell shape. *American Journal of Botany* **33**, 58–80.

Mays, C., McLoughlin, S., Frank, T. D., Fielding, C. R., Slater, S. M. & Vajda, V. (2021). Lethal microbial blooms delayed freshwater ecosystem recovery following the end-Permian extinction. *Nature Communications* **12**, doi.org/10.1038/s41467-021-25711-3.

McArthur, J. M., Howarth, R. J. & Bailey, T. R. (2001). Strontium isotope stratigraphy: LOWESS version 3: best fit to the marine Sr-isotope curve for 0-509 Ma and accompanying look-up table for deriving numerical age. *Journal of Geology* **109**, 155–170.

McBirney, A. R. (1989). *Geological Map of the Skaergaard Intrusion, East Greenland, 1:20,000.* Department of Geology, University of Oregon, 1 sheet.

McBirney, A. R. (2007). *Igneous Petrology.* Jones and Bartlett, 550 p.

McBirney, A. R. (2009). Factors governing the textural development of Skaergaard gabbros: a review. *Lithos* **111**, 1–5.

McBirney, A. R. & Hunter, R. H. (1995). The cumulate paradigm reconsidered. *Journal of Geology* **103**, 114–122.

McBirney, A. R. & Noyes, R. M. (1979). Crystallization and layering of the Skaergaard intrusion. *Journal of Petrology* **20**, 487–554.

McCaffrey, K. J. W. & Petford, N. (1997). Are granitic intrusions scale invariant? *Journal of the Geological Society* **154**, 1–4.

McCrory, P. A., Blair, J. L., Waldhauser, F. & Oppenheimer, D. H. (2012). Juan de Fuca slab geometry and its relation to Wadati-Benioff zone seismicity. *Journal of Geophysical Research* **117**, 1–24.

McCulloch, M. T., Gregory, R. T., Wasserburg, G. J. & Taylor Jr, H. P. (1981). Sm-Nd, Rb-Sr, and $^{18}O/^{16}O$ isotopic systematics in an oceanic crustal section: evidence from the Samail Ophiolite. *Journal of Geophysical Research* **86**, 2721–2735.

McDonough, W. F. & Sun, S. S. (1995). The composition of the Earth. *Chemical Geology* **120**, 223–253.

McKenzie, D. (1978). Some remarks on the development of sedimentary basins. *Earth and Planetary Science Letters* **40**, 25–32.

McKenzie, D. P. & Bickle, M. J. (1988). The volume and composition of melt generated by extension of the lithosphere. *Journal of Petrology* **29**, 625–679.

McQuarrie, N. & Wernicke, B. P. (2005). An animated tectonic reconstruction of southwestern North America since 36 Ma. *Geosphere* **1**, 147–172.

McSween, H. Y., De Sanctis, M. C. & Prettyman, T. H. (2014). Unique, antique Vesta. *Elements* **10**, 39–44.

Meeker, G. P., Lowers, H. A., Swayze, G. A., Van Gosen, B. S., Sutley, S. J. & Brownfield, I. K. (2006). Mineralogy and morphology of amphiboles observed in soils and rocks in El Dorado Hills CA. *US Geological Survey Open-File Report* 2006-1362, 55 p.

Melosh, H. J. (1996). *Impact Cratering: A Geologic Process.* Oxford University Press, 245 p.

Memeti, V., Paterson, S., Matzel, J., Mundil, R. & Okaya, D. (2010). Magmatic lobes as "snapshots" of magma chamber growth and evolution in large, composite batholiths: an example from the Tuolumne intrusion, Sierra Nevada, California. *Geological Society of America Bulletin* **122**, 1912–1931.

Menand, T. (2011). Physical controls and depth of emplacement of igneous bodies: a review. *Tectonophysics* **500**, 11–19.

Miller, C. F., McDowell, S. M. & Mapes, R. W. (2003). Hot and cold granites? Implications of zircon saturation temperatures and preservation of inheritance. *Geology* **31**, 529–532.

Miller, J. S., Glazner, A. F., Walker, J. D. & Martin, M. W. (1995). Geochronologic and isotopic evidence for Triassic-Jurassic emplacement of the eugeoclinal allochthon in the Mojave Desert region, California. *Geological Society of America Bulletin* **107**, 1441–1457.

Mills, R. D. & Coleman, D. S. (2013). Temporal and chemical connections between plutons and ignimbrites from the Mount Princeton magmatic center. *Contributions to Mineralogy and Petrology* **165**, 961–980.

Minster, J. F. & Allègre, C. J. (1979). ^{87}Rb-^{87}Sr chronology of H chondrites: constraint and speculations on the early evolution of their parent body. *Earth and Planetary Science Letters* **42**, 333–347.

Mittal, T., Richards, M. A. & Fendley, I. M. (2021). The magmatic architecture of continental flood basalts I: Observations from the Deccan Traps. *Journal of Geophysical Research: Solid Earth* **126**, doi.org/10.1029/2021JB021807.

Miyashiro, A. (1961). Evolution of metamorphic belts. *Journal of Petrology* **2**, 277–311.

Molnar, P. H. (2015). *Plate Tectonics: A Very Short Introduction.* Oxford University Press, 136 p.

Moore, J., Beinlich, A., Piazolo, S., Austrheim, H. & Putnis, A. (2020). Metamorphic differentiation via enhanced dissolution along high permeability zones. *Journal of Petrology* **61**, doi.org/10.1093/petrology/egaa096.

Moore, J. G. (1963). Geology of the Mount Pinchot quadrangle, southern Sierra Nevada, California. *US Geological Survey Bulletin* **1130**, 152 p.

Moore, J. G. (1967). Base surge in recent volcanic eruptions. *Bulletin of Volcanology* **30**, 337–363.

Moore, J. G., Clague, D. A., Holcomb, R. T., Lipman, P. W., Normark, W. R. & Torresan, M. E. (1989). Prodigious submarine landslides on the Hawaiian Ridge. *Journal of Geophysical Research* **94**, 17465–17484.

Moore, J. G. & Evans, B. W. (1967). The role of olivine in the crystallization of the prehistoric Makaopuhi tholeiitic lava lake, Hawaii. *Contributions to Mineralogy and Petrology* **15**, 202–223.

Moore, J. G. & Melson, W. G. (1969). Nuées ardentes of the 1968 eruption of Mayon Volcano, Philippines. *Bulletin of Volcanology* **33**, 600–620.

Moore, J. G. & Sisson, T. W. (1987). Preliminary geologic map of Sequoia and Kings Canyon National Parks, California. *US Geological Survey Open-File Report* **87-651**, map scale 1:125,000.

Moore, J. G. & Sisson, T. W. (2008). Igneous phenocrystic origin of K-feldspar megacrysts in granitic rocks from the Sierra Nevada batholith. *Geosphere* **4**, 387–400.

Moores, E. M. & Twiss, R. J. (1995). *Tectonics.* W. H. Freeman & Co., 415 p.

Morbidelli, A., Chambers, J., Lunine, J. I., Petit, J. M., Robert, F., Valsecchi, G. B. & Cyr, K. E. (2000). Source regions and timescales for the delivery of water to the Earth. *Meteoritics and Planetary Science* **35**, 1309–1320.

Mordick, B. E. & Glazner, A. F. (2006). Clinopyroxene barometry of basalts from the Coso and Big Pine volcanic fields, California. *Contributions to Mineralogy and Petrology* **152**, 111–124.

Moretti, R. & Neuville, D. (2021). Redox equilibria: from basic concepts to the magmatic realm. *Geophysical Monograph* **266**, doi.org/10.1002/9781119473206.ch1.

Morgan, S. S., Law, R. D. & Nyman, M. W. (1998). Laccolith-like emplacement model for the Papoose Flat pluton based on porphyroblast-matrix analysis. *Geological Society of America Bulletin* **110**, 96–110.

Morgan, W. J. (1971). Convection plumes in the lower mantle. *Nature* **230**, 42–43.

Morgan, W. J. (1972). Deep mantle convection plumes and plate motions. *American Association of Petroleum Geologists Bulletin* **56**, 203–213.

Morse, S. A. (1980). *Basalts and Phase Diagrams.* Springer-Verlag, 493 p.

Muhs, D. R., Bettis, E. A., III, Been, J. & McGeehin, J. P. (2001). Impact of climate and parent material on chemical weathering in loess-derived soils of the Mississippi River Valley. *Soil Science Society of America Journal* **65**, 1761–1777.

Muir, D. D., Blundy, J. D., Rust, A. C. & Hickey, J. (2014). Experimental constraints on dacite pre-eruptive magma storage conditions beneath Uturuncu Volcano. *Journal of Petrology* **55**, 749–767.

Muirhead, J. D., Airoldi, G., White, J. D. L. & Rowland, J. V. (2014). Cracking the lid: sill-fed dikes are the likely feeders of flood basalt eruptions. *Earth and Planetary Science Letters* **406**, 187–197.

Mukhopadhyay, B. & Manton, W. I. (1994). Upper-mantle fragments from beneath the Sierra Nevada batholith: partial fusion, fractional crystallization, and metasomatism in a subduction-related ancient lithosphere. *Journal of Petrology* **35**, 1417–1450.

Müller, R. D., Sdrolias, M., Gaina, C. & Roest, W. R. (2008). Age, spreading rates, and spreading asymmetry of the world's ocean crust. *Geochemistry, Geophysics, Geosystems* **9**, 1–19.

Mungall, J. E., Kamo, S. L. & McQuade, S. (2016). U-Pb geochronology documents out-of-sequence emplacement of ultramafic layers in the Bushveld Igneous Complex of South Africa. *Nature Communications* **7**, doi.org/10.1038/ncomms13385.

Namur, O. & Humphreys, M. C. S. (2018). Trace element constraints on the differentiation and crystal mush solidification in the Skaergaard intrusion, Greenland. *Journal of Petrology* **59**, 387–418.

Nash, B. P., Perkins, M. E., Christensen, J. N., Lee, D. C. & Halliday, A. N. (2006). The Yellowstone hotspot in space and time: Nd and Hf isotopes in silicic magmas. *Earth and Planetary Science Letters* **247**, 143–156.

Neal, C. R., Coffin, M. F. & Sager, W. W. (2019). Contributions of scientific ocean drilling to understanding the emplacement of submarine large igneous provinces and their effects on the environment. *Oceanography* **32**, 176–192.

Nelson, C. A. (1971). Geologic map of the Waucoba Spring quadrangle, Inyo County, California. *US Geological Survey Map* **GQ-921**, *scale 1:62,500.*

Nelson, C. A. (1991). Geologic history of the White-Inyo Range. In: Hall, C. A. & Ernst, W. J. (eds) *Natural History of the White-Inyo Range, Eastern California*. University of California Press, 42–74.

Newhall, C. G. & Self, S. (1982). The volcanic explosivity index (VEI): an estimate of explosive magnitude for historical volcanism. *Journal of Geophysical Research* **87**, 1231–1238.

Nicholls, J. & Stout, M. Z. (1982). Heat effects of assimilation, crystallization, and vesiculation in magmas. *Contributions to Mineralogy and Petrology* **81**, 328–339.

Nielsen, T. F. D. (2004). The shape and volume of the Skaergaard intrusion, Greenland: implications for mass balance and bulk composition. *Journal of Petrology* **45**, 507–530.

Nutman, A. P. (2006). Antiquity of the oceans and continents. *Elements* **2**, 223–227.

O'Brien, H. E., Peltonen, P. & Vartiainen, H. (2005). Kimberlites, carbonatites, and alkaline rocks. In: Lehtinen, M., Nurmi, P. A., Rämö, O. T. (eds) *Precambrian Geology of Finland—Key to the Evolution of the Fennoscandian Shield*. Elsevier, 605–644.

O'Connor, J. M., Steinberger, B., Regelous, M., Koppers, A. A. P., Wijbrans, J. R., Haase, K. M., Stoffers, P., Jokat, W. & Garbe-Schönberg, D. (2013). Constraints on past plate and mantle motion from new ages for the Hawaiian-Emperor Seamount Chain. *Geochemistry, Geophysics, Geosystems* **14**, 4564–4584.

O'Hanley, D. S. (1992). Solution to the volume problem in serpentinization. *Geology* **20**, 705–708.

Ochs, F. A. & Lange, R. A. (1997). The partial molar volume, thermal expansivity and compressibility of H_2O in $NaAlSi_3O_8$ liquid: new measurements and an internally consistent model. *Contributions to Mineralogy and Petrology* **129**, 155–165.

Ohring, M. (1995). *Engineering Materials Science*. Academic Press, 827 p.

Okudaira, T., Bando, H. & Yoshida, K. (2013). Grain-boundary diffusion rates inferred from grain-size variations of quartz in metacherts from a contact aureole. *American Mineralogist* **98**, 680–688.

Oldham, R. D. (1906). The constitution of the interior of the Earth, as revealed by earthquakes. *Quarterly Journal of the Geological Society of London* **62**, 456–475.

Oliver, H. W., Moore, J. G. & Sikora, R. F. (1993). Internal structure of the Sierra Nevada batholith based on specific gravity and gravity measurements. *Geophysical Research Letters* **20**, 2179–2182.

Oppenheimer, C. (2003). Climatic, environmental and human consequences of the largest known historic eruption: Tambora volcano (Indonesia) 1815. *Progress in Physical Geography* **27**, 230–259.

Ortoleva, P. J. (1994). *Geochemical Self-Organization*. Oxford University Press, 411 p.

Oskin, M. & Stock, J. (2003). Pacific-North America plate motion and opening of the Upper Delfín basin, northern Gulf of California, Mexico. *Geological Society of America Bulletin* **115**, 1173–1190.

Oxburgh, E. R. & Turcotte, D. L. (1974). Thermal gradients and regional metamorphism in overthrust terrains with special reference to the Eastern Alps. *Schweizerische Mineralogische und Petrographische Mitteilungen* **54**, 641–662.

Oze, C., Fendorf, S., Bird, D. K. & Coleman, R. G. (2004). Chromium geochemistry of serpentine soils. *International Geology Review* **46**, 97–126.

Pabst, A. (1928). Observations on inclusions in the granitic rocks of the Sierra Nevada. *University of California Publications in Geology* **17**, 325–386.

Pálfy, J. & Smith, P. L. (2000). Synchrony between Early Jurassic extinction, oceanic anoxic event, and the Karoo-Ferrar flood basalt volcanism. *Geology* **28**, 747–750.

Pallister, J. S., Thornber, C. R., Cashman, K. V, Clynne, M. A., Lowers, H. A., Mandeville, C. W., Brownfield, I. K. & Meeker, G. P. (2008). Petrology of the 2004–2006 Mount St. Helens lava dome: implications for magmatic plumbing and eruption triggering. *US Geological Survey Professional Paper* **1750**, 647–702.

Palme, H. & O'Neill, H. (2014). Cosmochemical estimates of mantle composition. In: Turekian, K. K. & Holland, H. D. (eds) *Treatise on Geochemistry: Second Edition*, Vol 3. Elsevier, 1–39.

Palmer, M. R. & Edmond, J. M. (1989). The strontium isotope budget of the modern ocean. *Earth and Planetary Science Letters* **92**, 11–26.

Paterson, S. R. & Fowler, T. K., Jr. (1993). Extensional pluton-emplacement models: do they work for large plutonic complexes? *Geology* **21**, 781–784.

Paterson, S., Memeti, V., Mundil, R. & Zak, J. (2016). Repeated, multiscale, magmatic erosion and recycling in an upper-crustal pluton: implications for magma chamber dynamics and magma volume estimates. *American Mineralogist* **101**, 2176–2198.

Paterson, S. R., Pignotta, G. S., Farris, D., Memeti, V., Miller, R. B., Vernon, R. H. & Žák, J. (2008). Is stoping a volumetrically significant pluton emplacement process?: Discussion. *Geological Society of America Bulletin* **120**, 1075–1079.

Patterson, C. C., Tilton, G. R. & Inghram, M. G. (1955). Age of the Earth. *Science* **121**, 69–75.

Pattison, D. R. M., Chacko, T., Farquhar, J. & McFarlane, C. R. M. (2003). Temperatures of granulite-facies metamorphism: constraints from experimental phase equilibria and thermobarometry corrected for retrograde exchange. *Journal of Petrology* **44**, 867–900.

Pattison, D. R. M. & Vogl, J. J. (2005). Contrasting sequences of metapelitic mineral-assemblages in the aureole of the tilted Nelson Batholith, British Columbia: implications for phase equilibria and pressure determination in andalusite-sillimanite-type settings. *Canadian Mineralogist* **43**, 51–88.

Peacock, M. A. (1931). Classification of igneous rock series. *Journal of Geology* **39**, 54–67.

Peacock, S. M., Rushmer, T. & Thompson, A. B. (1994). Partial melting of subducting oceanic crust. *Earth and Planetary Science Letters* **121**, 227–244.

Pearson, D. G., Canil, D. & Shirey, S. B. (2014). Mantle samples included in volcanic rocks: xenoliths and diamonds. In: Turekian, K. K. & Holland, H. D. (eds) *Treatise on Geochemistry: Second Edition*, Vol. 3. Elsevier, 169–253.

Pearson, D. G. & Wittig, N. (2014). The formation and evolution of cratonic mantle lithosphere–evidence from mantle xenoliths. In: Turekian, K. K. & Holland, H. D. (eds) *Treatise on Geochemistry: Second Edition*, Vol. 3. Elsevier, 255–292.

Penniston-Dorland, S. C., Kohn, M. J. & Manning, C. E. (2015). The global range of subduction zone thermal structures from exhumed blueschists and eclogites: rocks are hotter than models. *Earth and Planetary Science Letters* **428**, 243–254.

Perret, F. A. (1937). *The Eruption of Mt. Pelée, 1929–1932*. Carnegie Institution of Washington, 126 p.

Petford, N. (2003). Rheology of granitic magmas during ascent and emplacement. *Annual Review of Earth and Planetary Sciences* **31**, 399–427.

Petford, N., Kerr, R. C. & Lister, J. R. (1993). Dike transport of granitoid magmas. *Geology* **21**, 845–848.

Philpotts, A. R. (1979). Silicate liquid immiscibility in tholeiitic basalts. *Journal of Petrology* **20**, 99–118.

Pierce, K. L. & Morgan, L. A. (1992). The track of the Yellowstone hotspot: volcanism, faulting, and uplift. *Geological Society of America Memoir* **179**, 1–53.

Pierce, K. L. & Morgan, L. A. (2009). Is the track of the Yellowstone hotspot driven by a deep mantle plume? *Journal of Volcanology and Geothermal Research* **188**, 1–25.

Pitcher, W. S. (1993). *The Nature and Origin of Granite*. Chapman & Hall, 387 p.

Pitcher, W. S. & Berger, A. R. (1972). *The Geology of Donegal: A Study of Granite Emplacement and Unroofing*. Wiley-Interscience, 435 p.

Plank, T. & Forsyth, D. W. (2016). Thermal structure and melting conditions in the mantle beneath the Basin and Range province from seismology and petrology. *Geochemistry, Geophysics, Geosystems* **17**, 1312–1338.

Plank, T. & Langmuir, C. H. (1993). Tracing trace elements from sediment input to volcanic output at subduction zones. *Nature* **362**, 739–743.

Plümper, O., King, H. E., Geisler, T., Liu, Y., Pabst, S., Savov, I. P., Rost, D. & Zack, T. (2017). Subduction zone forearc serpentinites as incubators for deep microbial life. *Proceedings of the National Academy of Sciences* **114**, 4324–4329.

Polteau, S., Mazzini, A., Galland, O., Planke, S. & Malthe-Sørenssen, A. (2008). Saucer-shaped intrusions: occurrences, emplacement and implications. *Earth and Planetary Science Letters* **266**, 195–204.

Porcelli, D. & Ballentine, C. J. (2002). Models for the distribution of terrestrial noble gases and the evolution of the atmosphere. *Reviews in Mineralogy and Geochemistry* **47**, 411–480.

Pritchard, M. E. & Simons, M. (2004). Surveying volcanic arcs with satellite radar interferometry: the Central Andes, Kamchatka, and beyond. *GSA Today* **14**, 4–11.

Putirka, K. D. (2008). Thermometers and barometers for volcanic systems. *Reviews in Mineralogy and Geochemistry* **69**, 61–120.

Putirka, K. & Platt, B. (2012). Basin and Range volcanism as a passive response to extensional tectonics. *Geosphere* **8**, 1274–1285.

Putnam, R., Glazner, A. F., Coleman, D. S., Kylander-Clark, A. R. C., Pavelsky, T. & Abbot, M. I. (2015). Plutonism in three dimensions: Field and geochemical relations on the southeast face of El Capitan, Yosemite National Park, California. *Geosphere* **11**, doi.org/10.1130 /GES01133.1

Quirk, D. G., D'Lemos, R. S., Mulligan, S. & Rabti, B. M. R. (1998). Insights into the collection and emplacement of granitic magma based on 3D seismic images of normal fault-related salt structures. *Terra Nova* **10**, 268–279.

Ragland P. C. (1989). *Basic Analytical Petrology*. Oxford University Press, 369 p.

Ramberg, H. (1967). Model experimentation of the effect of gravity on tectonic processes. *Geophysical Journal International* **14**, 307–329.

Ramos, V. A. & Folguera, A. (2009). Andean flat-slab subduction through time. *Geological Society Special Publication* **327**, 31–54.

Rampino, M. R. (2002). Supereruptions as a threat to civilizations on Earth-like planets. *Icarus* **156**, 562–569.

Rampino, M. R., Self, S. & Fairbridge, R. W. (1979). Can rapid climatic change cause volcanic eruptions? *Science* **206**, 826–829.

Rampino, M. R., Self, S. & Stothers, R. B. (1988). Volcanic winters. *Annual Review of Earth and Planetary Sciences* **16**, 73–99.

Ramsay, W. R. H., Collett, G. S., Ramsay, E. G. & Kerby, G. (2021). Three scoria blocks from Māori occupation sites, New Zealand: their petrography, mineralogy, geochemistry and attribution to a source locality in eastern Polynesia. *Geoarchaeology* **36**, 455–473.

Rapp, R. P., Watson, E. B. & Miller, C. F. (1991). Partial melting of amphibolite/eclogite and the origin of Archean trondhjemites and tonalites. *Precambrian Research* **51**, 1–25.

Ratajeski, K., Sisson, T. W. & Glazner, A. F. (2005). Experimental and geochemical evidence for derivation of the El Capitan Granite, California, by partial melting of hydrous gabbroic lower crust. *Contributions to Mineralogy and Petrology* **149**, 713–734.

Rau, C. J. & Forsyth, D. W. (2011). Melt in the mantle beneath the amagmatic zone, southern Nevada. *Geology* **39**, 975–978.

Reay, A. & Harris, P. G. (1964). The partial fusion of peridotite. *Bulletin of Volcanology* **27**, 115–127.

Reches, Z. & Fink, J. (1988). The mechanism of intrusion of the Inyo Dike, Long Valley Caldera, California. *Journal of Geophysical Research* **93**, 4321–4334.

Reeder, J. W. (2016). The active Yakutat (Kula?) plate and its southcentral Alaska megathrust and intraplate earthquakes. *Journal of Earth Science and Engineering* **6**, 18–58.

Reidel, S. P., Camp, V. E., Tolan, T. L. & Martin, B. S. (2013). The Columbia River flood basalt province: stratigraphy, areal extent, volume, and physical volcanology. *Geological Society of America Special Paper* **497**, 1–43.

Renfrew, C. & Peacey, J. S. (1968). Aegean marble: a petrological study. *Annual of the British School at Athens* **63**, 45–66.

Retallack, G. J., Veevers, J. J. & Morante, R. (1996). Global coal gap between Permian–Triassic extinction and Middle Triassic recovery of peat-forming plants. *Geological Society of America Bulletin* **108**, 195–207.

Reynolds, D. L. (1946). The sequence of geochemical changes leading to granitization. *Quarterly Journal of the Geological Society* **102**, 389–446.

Richards, M. A., Alvarez, W., Self, S., Karlstrom, L., Renne, P. R., Manga, M., Sprain, C. J., Smit, J., Vanderkluysen, L. & Gibson, S. A. (2017). Triggering of the largest Deccan eruptions by the Chicxulub impact: reply. *Geological Society of America Bulletin* **129**, 256.

Richards, M. J. & Günther, J. (2019). The past, present and future values of the Polynesian stone adzes and pounders collected on the Pandora. *Bulletin of the History of Archaeology* **29**, 1–15.

Riel, N., Duarte, J. C., Almeida, J., Kaus, B. J. P., Rosas, F., Rojas-Agramonte, Y. & Popov, A. (2023). Subduction initiation triggered the Caribbean large igneous province. *Nature Communications* **14**, doi.org/10.1038/s41467-023-36419-x.

Ringwood, A. E. (1975). *Composition and Petrology of the Earth's Mantle*. McGraw-Hill, 618 p.

Rioux, M., Farmer, G. L., Bowring, S. A., Wooton, K. M., Amato, J. M., Coleman, D. S. & Verplanck, P. L. (2016). The link between volcanism and plutonism in epizonal magma systems: high-precision U–Pb zircon geochronology from the Organ Mountains caldera and batholith, New Mexico. *Contributions to Mineralogy and Petrology* **171**, 1–22.

Rivadeneyra-Vera, C., Bianchi, M., Assumpção, M., Cedraz, V., Julià, J., Rodríguez, M., Sánchez, L., Sánchez, G., Lopez-Murua, L., Fernandez, G. & Fugarazzo, R. (2019). An updated crustal thickness map of central South America based on receiver function measurements in the region of the Chaco, Pantanal, and Paraná Basins, southwestern Brazil. *Journal of Geophysical Research* **124**, 8491–8505.

Robie, R. A. & Hemingway, B. S. (1995). Thermodynamic properties of minerals and related substances at 298.15 K and 1 bar (10^5 Pascals) pressure and at higher temperatures. *US Geological Survey Bulletin* **2131**, 461 p.

Robin, P. Y. F. & Cruden, A. R. (1994). Strain and vorticity patterns in ideally ductile transpression zones. *Journal of Structural Geology* **16**, 447–466.

Robinson, D. M., DeCelles, P. G. & Copeland, P. (2006). Tectonic evolution of the Himalayan thrust belt in western Nepal: implications for channel flow models. *Geological Society of America Bulletin* **118**, 865–885.

Rocchi, S., Westerman, D. S., Dini, A., Innocenti, F. & Tonarini, S. (2002). Two-stage growth of laccoliths at Elba Island, Italy. *Geology* **30**, 9883–986.

Roeder, P. L. & Emslie, R. F. (1970). Olivine-liquid equilibrium. *Contributions to Mineralogy and Petrology* **29**, 275–289.

Rollinson, H. R. (2013). *Using Geochemical Data: Evaluation, Presentation, Interpretation*. Taylor & Francis, 384 p.

Rosas J. C. & Korenaga J. (2018). Rapid crustal growth and efficient crustal recycling in the early Earth: implications for Hadean and Archean geodynamics. *Earth and Planetary Science Letters* **494**, 42–49.

Roscoe, R. (1952). The viscosity of suspensions of rigid spheres. *British Journal of Applied Physics* **3**, 267–269.

Rosenberg, C. L., Berger, A. & Schmid, S. M. (1995). Observations from the floor of a granitoid pluton: inferences on the driving force of final emplacement. *Geology* **23**, 443–446.

Rosenberg, C. L. & Handy, M. R. (2005). Experimental deformation of partially melted granite revisited: implications for the continental crust. *Journal of Metamorphic Geology* **23**, 19–28.

Rosenfeld, J. L. & Chase, A. B. (1961). Pressure and temperature of crystallization from elastic effects around solid inclusions in minerals? *American Journal of Science* **259**, 519–541.

Rotenberg, E., Davis, D. W., Amelin, Y., Ghosh, S. & Bergquist, B. A. (2012). Determination of the decay-constant of ^{87}Rb by laboratory accumulation of ^{87}Sr. *Geochemica et Cosmochimica Acta* **85**, 41–57.

Rowland, S. K. & Walker, G. P. L. (1990). Pahoehoe and aa in Hawaii: volumetric flow rate controls the lava structure. *Bulletin of Volcanology* **52**, 615–628.

Rubatto, D. & Hermann, J. (2001). Exhumation as fast as subduction? *Geology* **29**, 3–6.

Rudnick, R. L. & Gao, S. (2014). Composition of the continental crust. In: Turekian, K. K. & Holland, H. D. (eds) *Treatise on Geochemistry: Second Edition*, Vol. 4. Elsevier, 1–51.

Russell, J. K., Sparks, R. S. J. & Kavanagh, J. L. (2019). Kimberlite volcanology: transport, ascent, and eruption. *Elements* **15**, 405–410.

Saccocia, P. J., Seewald, J. S. & Shanks, W. C. (2009). Oxygen and hydrogen isotope fractionation in serpentine-water and talc-water systems from 250 to 450 °C, 50 MPa. *Geochimica et Cosmochimica Acta* **73**, 6789–6804.

Sack, R. O., Carmichael, I. S. E., Rivers, M. & Ghiorso, M. S. (1980). Ferric-ferrous equilibria in natural silicate liquids at 1 bar. *Contributions to Mineralogy and Petrology* **75**, 369–376.

Sakuyama, M. (1979). Evidence of magma mixing: petrological study of Shirouma-Oike calc-alkaline andesite volcano, Japan. *Journal of Volcanology and Geothermal Research* **5**, 179–208.

Salmonsen, L. P. & Tegner, C. (2013). Crystallization sequence of the Upper Border Series of the Skaergaard Intrusion: revised subdivision and implications for chamber-scale magma homogeneity. *Contributions to Mineralogy and Petrology* **165**, 1155–1171.

Salters, V. J. M. & Stracke, A. (2004). Composition of the depleted mantle. *Geochemistry, Geophysics, Geosystems* **5**, doi.org/10.1029/2003GC000597.

Samperton, K. M., Schoene, B., Cottle, J. M., Keller, C. B., Crowley, J. L. & Schmitz, M. D. (2015). Magma emplacement, differentiation and cooling in the middle crust: Integrated zircon geochronological-geochemical constraints from the Bergell Intrusion, Central Alps. *Chemical Geology* **417**, 322–340.

Sarna-Wojcicki, A. M., Shipley, S., Waitt, R. B., Jr., Dzurisin, D. & Wood, S. H. (1981). Areal distribution, thickness, mass, volume, and grain size of air-fall ash from the six major eruptions of 1980. *US Geological Survey Professional Paper* **1250**, 577–600.

Satake, K., Smith, J. R. & Shinozaki, K. (2002). Three-dimensional reconstruction and tsunami model of the Nuuanu and Wailau giant landslides, Hawaii. *Geophysical Monograph Series* **128**, 333–346.

Sato, H. (2005). Viscosity measurement of subliquidus magmas: 1707 basalt of Fuji volcano. *Journal of Mineralogical and Petrological Sciences* **100**, 133–142.

Sato, M., Hickling, N. L. & McLane, J. E. (1973). Oxygen fugacity values of Apollo 12, 14, and 15 lunar samples and reduced state of lunar magmas. *Proceedings of the Lunar Science Conference* **4**, 1061.

Saunders, A. D. (2005). Large igneous provinces: origin and environmental consequences. *Elements* **1**, 259–263.

Scaillet, B. & Evans, B. W. (1999). The 15 June 1991 eruption of Mount Pinatubo. I. Phase equilibria and pre-eruption P–T–fO_2–fH_2O conditions of the dacite magma. *Journal of Petrology* **40**, 381–411.

Scaillet, B., Holtz, F. & Pichavant, M. (1998). Phase equilibrium constraints on the viscosity of silicic magmas: 1. Volcanic-plutonic comparison. *Journal of Geophysical Research* **103**, 27257–27266.

Schairer, J. F. & Bowen, N. L. (1955). The system K_2O-Al_2O_3-SiO_2. *American Journal of Science* **253**, 681–746.

Schmidt, M. W. & Poli, S. (2014). Devolatilization during subduction. In: Turekian, K. K. & Holland, H. D. (eds) *Treatise on Geochemistry: Second Edition*, Vol. 4. Elsevier, 669–701.

Schmieder, M., Schwarz, W. H., Trieloff, M., Tohver, E., Buchner, E., Hopp, J. & Osinski, G. R. (2015). New $^{40}Ar/^{39}Ar$ dating of the Clearwater Lake impact structures (Québec, Canada)–not the binary asteroid impact it seems? *Geochimica et Cosmochimica Acta* **148**, 304–324.

Schmitt, A. K., Marks, M. A. W., Nesbor, H. D. & Markl, G. (2007). The onset and origin of differentiated Rhine Graben volcanism based on U-Pb ages and oxygen isotopic composition of zircon. *European Journal of Mineralogy* **19**, 849–857.

Schoene, B. (2014). U-Th-Pb geochronology. In: Turekian, K. K. & Holland, H. D. (eds) *Treatise on Geochemistry: Second Edition*, Vol. 4. Elsevier, 341–378.

Schoene, B., Eddy, M. P., Samperton, K. M., Keller, C. B., Keller, G., Adatte, T. & Khadri, S. F. R. (2019). U-Pb constraints on pulsed eruption of the Deccan Traps across the end-Cretaceous mass extinction. *Science* **363**, 862–866.

Scoates, J. S., Wall, C. J., Friedman, R. M., Weis, D., Mathez, E. A. & VanTongeren, J. A. (2021). Dating the Bushveld Complex: timing of crystallization, duration of magmatism, and cooling of the world's largest layered intrusion and related rocks. *Journal of Petrology* **62**, 1–39.

Scoon, R. N. (2018). *Geology of National Parks of Central/Southern Kenya and Northern Tanzania: Geotourism of the Gregory Rift Valley, Active Volcanism and Regional Plateaus.* Springer, 220 p.

Scotese, C. R. (2021). An atlas of Phanerozoic paleogeographic maps: the seas come in and the seas go out. *Annual Review of Earth and Planetary Sciences* **49**, 679–728.

Scott, E. R. D. (2007). Impact origins for pallasites. *Abstracts of Papers Submitted to the Lunar and Planetary Science Conference* **38**, Abstract 2284.

Seitz, M. G. & Kushiro, I. (1974). Melting relations of the Allende meteorite. *Science* **183**, 954–957.

Self, S. (1992). Krakatau revisited: the course of events and interpretation of the 1883 eruption. *GeoJournal* **28**, 109–121.

Self, S., Keszthelyi, L. & Thordarson, T. (1998). The importance of pahoehoe. *Annual Review of Earth and Planetary Sciences* **26**, 81–110.

Self, S., Rampino, M. R. & Barbera, J. J. (1981). The possible effects of large 19th and 20th century volcanic eruptions on zonal and hemispheric surface temperatures. *Journal of Volcanology and Geothermal Research* **11**, 41–60.

Self, S., Thordarson, T. & Widdowson, M. (2005). Gas fluxes from flood basalt eruptions. *Elements* **1**, 283–287.

Selverstone, J. (1985). Petrologic constraints on imbrication, metamorphism and uplift in the SW Tauern Window, Eastern Alps. *Tectonics* **4**, 687–704.

Shahabpour, J. (1998). Liesegang blocks from sandstone beds of the Hojedk Formation, Kerman, Iran. *Geomorphology* **22**, 93–106.

Shaw, H. R. (1969). Rheology of basalt in the melting range. *Journal of Petrology* **10**, 510–535.

Shelley, M. W. (1818). *Frankenstein; or, the modern Prometheus.* Lackington, Hughes, Harding, Mavor, & Jones.

Shirley, D. N. (1987). Differentiation and compaction in the Palisades Sill, New Jersey. *Journal of Petrology* **28**, 835–865.

Shoemaker, E. M. (1983). Asteroid and comet bombardment of the Earth. *Annual Review of Earth and Planetary Sciences* **11**, 461–494.

Sibson, R. H. (1977). Fault rocks and fault mechanisms. *Journal of the Geological Society* **133**, 191–213.

Sigurdsson, H., Cashdollar, S. & Sparks, S. R. J. (1982). The eruption of Vesuvius in A. D. 79: reconstruction from historical and volcanological evidence. *American Journal of Archaeology* **86**, 39–51.

Sigurdsson, H. & Sparks, R. S. J. (1981). Petrology of rhyolitic and mixed magma ejecta from the 1875 eruption of Askja, Iceland. *Journal of Petrology* **22**, 41–84.

Simkin, T. & Siebert, L. (2000). Earth's volcanoes and eruptions: an overview. In: Sigurdsson, H. (ed) *Encyclopedia of Volcanoes.* Academic Press, 249–262.

Simoes, M., Avouac, J. P., Beyssac, O., Goffé, B., Farley, K. A. & Chen, Y. G. (2007). Mountain building in Taiwan: a thermokinematic model. *Journal of Geophysical Research* **112**, 1–25.

Singer, B. S., Andersen, N. L., Le Mével, H., Feigl, K. L., DeMets, C., Tikoff, B., Thurber, C. H., Jicha, B. R., Cardona, C., Córdova, L. & Gil, F. (2014). Dynamics of a large, restless, rhyolitic magma system at Laguna del Maule, southern Andes, Chile. *GSA Today* **24**, 4–10.

Singleton, J. S. & Mosher, S. (2012). Mylonitization in the lower plate of the Buckskin-Rawhide detachment fault, west-central Arizona: Implications for the geometric evolution of metamorphic core complexes. *Journal of Structural Geology* **39**, 180–198.

Sinton, J. M. & Detrick, R. S. (1992). Mid-ocean ridge magma chambers. *Journal of Geophysical Research* **97**, 197–216.

Sisson, T. W. & Bacon, C. R. (1999). Gas-driven filter pressing in magmas. *Geology* **27**, 613–616.

Sisson, T. W., Ratajeski, K., Hankins, W. B. & Glazner, A. F. (2005). Voluminous granitic magmas from common basaltic sources. *Contributions to Mineralogy and Petrology* **148**, 635–661.

Sisson, T. W., Vallance, J. W. & Pringle, P. T. (2001). Progress made in understanding Mount Rainier's hazards. *Eos* **82**, 5–9.

Skoglund, P., Posth, C., Sirak, K., Spriggs, M., Valentin, F., Bedford, S., Clark, G. R., Reepmeyer, C., Petchey, F., Fernandes, D. & Fu, Q. (2016). Genomic insights into the peopling of the Southwest Pacific. *Nature* **538**, 510–513.

Sleep, N. H. (2002). Ridge-crossing mantle plumes and gaps in tracks. *Geochemistry, Geophysics, Geosystems* **3**, 1–33.

Sleep, N. H. (2015). Evolution of the Earth: plate tectonics through time. In: Schubert, G. (ed) *Treatise on Geophysics*, 2nd ed., Vol. 9. Elsevier, 145–172.

Sleep, N. H., Zahnle, K. J., Kasting, J. F. & Morowitz, H. J. (1989). Annihilation of ecosystems by large asteroid impacts on the early Earth. *Nature* **342**, 139–142.

Smit, K., Shirey, S., Pearson, G., Stachel, T., Nestola, F. & Moses, T. (eds). (2022). Diamond: Genesis, mineralogy and geochemistry. *Reviews in Mineralogy and Geochemistry*, **88**, 875 p.

Smith, C. S. (1964). Some elementary principles of polycrystalline microstructure. *Metallurgical Reviews* **9**, 1–48.

Smith, D. C. (1984). Coesite in clinopyroxene in the Caledonides and its implications for geodynamics. *Nature* **310**, 641–644.

Smith, E. M., Ni, P., Shirey, S. B., Richardson, S. H., Wang, W. & Shahar, A. (2021). Heavy iron in large gem diamonds traces deep subduction of serpentinized ocean floor. *Science Advances* **7**, doi.org/10.1126/sciadv.abe9773.

Smith, E. M., Shirey, S. B., Nestola, F., Bullock, E. S., Wang, J., Richardson, S. H. & Wang, W. (2016). Large gem diamonds from metallic liquid in Earth's deep mantle. *Science* **354**, 1403–1405.

Smith, E. M., Shirey, S. B., Richardson, S. H., Nestola, F., Bullock, E. S., Wang, J. & Wang, W. (2018). Blue boron-bearing diamonds from Earth's lower mantle. *Nature* **560**, 84–87.

Smyth, J. R. & Jacobsen, S. D. (2006). Nominally anhydrous minerals and Earth's deep water cycle. *Geophysical Monograph* **168**, 1–11.

Snow, E. & Yund, R. A. (1985). Thermal history of a Bishop Tuff section as determined from the width of cryptoperthite lamellae. *Geology* **13**, 50–53.

Sobolev, N. V & Shatsky, V. S. (1990). Diamond inclusions in garnets from metamorphic rocks: a new environment for diamond formation. *Nature* **343**, 742–746.

Soderberg, E. R. & Wolff, J. A. (2023). Mantle source lithologies for the Columbia River flood basalt province. *Contributions to Mineralogy and Petrology* **178**, doi.org/10.1007/s00410-023-01993-2.

Solgadi, F. & Sawyer, E. W. (2008). Formation of igneous layering in granodiorite by gravity flow: a field, microstructure and geochemical study of the Tuolumne Intrusive Suite at Sawmill Canyon, California. *Journal of Petrology* **49**, 2009–2042.

Sparks, R. S. J., Folkes, C. B., Humphreys, M. C. S., Barfod, D. N., Clavero, J., Sunagua, M. C., McNutt, S. R. & Pritchard, M. E. (2008). Uturuncu volcano, Bolivia: volcanic unrest due to mid-crustal magma intrusion. *American Journal of Science* **308**, 727–769.

Sparks, R. S. J. & Wilson, L. (1976). A model for the formation of ignimbrite by gravitational column collapse. *Journal of the Geological Society* **132**, 441–451.

Spear, F. S. (1995). *Metamorphic Phase Equilibria and Pressure-Temperature-Time Paths.* Mineralogical Society of America, 799 p.

Spear, F. S. & Cheney, J. T. (1989). A petrogenetic grid for pelitic schists in the system SiO_2-Al_2O_3-FeO-MgO-K_2O-H_2O. *Contributions to Mineralogy and Petrology* **101**, 149–164.

Spear, F. S., Kohn, M. J., Cheney, J. T. & Florence, F. (2002). Metamorphic, thermal, and tectonic evolution of central New England. *Journal of Petrology* **43**, 2097–2120.

Spear, F. S. & Markussen, J. C. (1997). Mineral zoning, P-T-X-M phase relations, and metamorphic evolution of some Adirondack granulites, New York. *Journal of Petrology* **38**, 757–783.

Sprain, C. J., Renne, P. R., Vanderkluysen, L., Pande, K., Self, S. & Mittal, T. (2019). The eruptive tempo of Deccan volcanism in relation to the Cretaceous-Paleogene boundary. *Science* **363**, 866–870.

Stakes, D. S. & Taylor Jr, H. P. (1992). The northern Samail ophiolite: an oxygen isotope, microprobe, and field study. *Journal of Geophysical Research* **97**, 7043–7080.

Stanciu, C. A., Russo, R. M., Mocanu, V. I., Bremner, P. M., Hongsresawat, S., Torpey, M. E., VanDecar, J. C., Foster, D. A. & Hole, J. A. (2016). Crustal structure beneath the Blue Mountains terranes and cratonic North America, eastern Oregon and Idaho, from teleseismic receiver functions: *Journal of Geophysical Research* **121**, 5049–5067.

Stearns, M. A., Bartley, J. M., Bowman, J. R., Forster, C. W., Beno, C. J., Riddle, D. D., Callis, S. J. & Udy, N. D. (2020). Simultaneous magmatic and hydrothermal regimes in Alta–Little Cottonwood stocks, Utah, USA, recorded using multiphase U-Pb petrochronology. *Geosciences* **10**, doi.org/10.3390/geosciences10040129.

Steffe, J. F. (1996). *Rheological Methods in Food Process Engineering.* Freeman Press, 418 p.

Steinmann, G. (1905). Geologische Beobachtungen in den Alpen, II. Die Schardtsche Ueberfaltungstheorie und die geologische Bedeutung der Tiefseeabsätze und der ophiolithischen Massengesteine. *Berichte der Naturforschenden Gesellschaft zu Freiburg im Breisgau* **16**, 18–67.

Stephens, W. E. & Halliday, A. N. (1980). Discontinuities in the composition surface of a zoned pluton, Criffell, Scotland. *Geological Society of America Bulletin* **91**, 165–170.

Stern, C. R., Huang, W. L. & Wyllie, P. J. (1975). Basalt-andesite-rhyolite-H_2O: crystallization intervals with excess H_2O and H_2O-undersaturated liquidus surfaces to 35 kilobars, with implications for magma genesis. *Earth and Planetary Science Letters* **28**, 189–196.

Stevens, L. M., Baldwin, J. A., Cottle, J. M. & Kylander-Clark, A. R. C. (2015). Phase equilibria modelling and LASS monazite petrochronology: P-T-t constraints on the evolution of the Priest River core complex, northern Idaho. *Journal of Metamorphic Geology* **33**, 385–411.

Stevens, L. M., Baldwin, J. A., Crowley, J. L., Fisher, C. M. & Vervoort, J. D. (2016). Magmatism as a response to exhumation of the Priest River complex, northern Idaho: constraints from zircon U–Pb geochronology and Hf isotopes. *Lithos* **262**, 285–297.

Stevens, L. M., Bendick, R. & Baldwin, J. A. (2017). Synconvergent exhumation of metamorphic core complexes in the northern North American Cordillera. *Geology* **45**, 495–498.

Stevenson C. T. E., Hutton D. H. W. & Price A. R. (2008). The Trawenagh Bay Granite and a new model for the emplacement of the Donegal Batholith. *Transactions of the Royal Society of Edinburgh Earth Science* **97**, 455–477.

Stevenson, D. J. (2008). A planetary perspective on the deep Earth. *Nature* **451**, 261–265.

Stock, G. M., Anderson, R. S. & Finkel, R. C. (2004). Pace of landscape evolution in the Sierra Nevada, California, revealed by cosmogenic dating of cave sediments. *Geology* **32**, 193–196.

Stoddard, E. F. & Blake, D. E. (eds). (1994). *Geology and Field Trip Guide, Western Flank of the Raleigh Metamorphic Belt, North Carolina*. North Carolina Geological Society, 110 p.

Stracke, A. (2021). A process-oriented approach to mantle geochemistry. *Chemical Geology* **579**, doi.org/10.1016/j.chemgeo.2021.120350.

Streck, M. J., Broderick, C. A., Thornber, C. R., Clynne, M. A. & Pallister, J. S. (2008). Plagioclase populations and zoning in dacite of the 2004–2005 Mount St. Helens eruption: constraints for magma origin and dynamics. *US Geological Survey Professional Paper* **1750**, 791–808.

Streckeisen, A. (1976). To each plutonic rock its proper name. *Earth Science Reviews* **12**, 1–33.

Strohmeier, B. R., Huntington, J. C., Bunker, K. L., Sanchez, M. S., Allison, K. & Lee, R. J. (2010). What is asbestos and why is it important? Challenges of defining and characterizing asbestos. *International Geology Review* **52**, 801–872.

Sumner, A. L., Menke, E. J., Dubowski, Y., Newberg, J. T., Penner, R. M., Hemminger, J. C., Wingen, L. M., Brauers, T. & Finlayson-Pitts, B. J., (2004).The nature of water on surfaces of laboratory systems and implications for heterogeneous chemistry in the troposphere. *Physical Chemistry Chemical Physics : PCCP* **6**, 604–613.

Svensen, H., Planke, S., Chevallier, L., Malthe-Sørenssen, A., Corfu, F. & Jamtveit, B. (2007). Hydrothermal venting of greenhouse gases triggering Early Jurassic global warming. *Earth and Planetary Science Letters* **256**, 554–566.

Swamy, V., Saxena, S. K., Sundman, B. & Zhang, J. (1994). A thermodynamic assessment of silica phase diagram. *Journal of Geophysical Research* **99**, 11787–11794.

Swindles, G. T., Savov, I. P., Schmidt, A., Hooper, A., Connor, C. B. & Carrivick, J. L. (2018). Climatic control on Icelandic volcanic activity during the mid-Holocene. *Geology* **46**, 47–50.

Sylvester, A. G. (1998). Magma mixing, structure, and re-evaluation of the emplacement mechanism of Vrådal pluton, central Telemark, southern Norway. *Norsk Geologisk Tidsskrift* **78**, 259–276.

Sylvester, A. G., Oertel, G., Nelson, C. A. & Christie, J. M. (1978). Papoose Flat pluton: a granitic blister in the Inyo Mountains, California. *Geological Society of America Bulletin* **89**, 1205–1219.

Syracuse, E. M. & Abers, G. A. (2006). Global compilation of variations in slab depth beneath arc volcanoes and implications. *Geochemistry, Geophysics, Geosystems* **7**, doi.org/10.1029/2005GC001045.

Syracuse, E. M., van Keken, P. E. & Abers, G. A. (2010). The global range of subduction zone thermal models. *Physics of the Earth and Planetary Interiors* **183**, 73–90.

Taagepera, R. (1979). Size and duration of empires: growth-decline curves, 600 BC to 600 AD. *Social Science History* **3**, 115–138.

Taagepera, R. (1997). Expansion and contraction patterns of large polities: context for Russia. *International Studies Quarterly* **41**, 475–504.

Takahashi, E. & Kushiro, I. (1983). Melting of a dry peridotite at high pressures and basalt magma genesis. *American Mineralogist* **68**, 859–879.

Takahashi, E., Shimazaki, T., Tsuzaki, Y. & Yoshida, H. (1993). Melting study of a peridotite KLB-1 to 6.5 GPa, and the origin of basaltic magmas. *Philosophical Transactions of the Royal Society of London. Series A: Mathematical, Physical and Engineering Sciences* **342**, 105–120.

Tappa, M. J., Coleman, D. S., Mills, R. D. & Samperton, K. M. (2011). The plutonic record of a silicic ignimbrite from the Latir volcanic field, New Mexico. *Geochemistry, Geophysics, Geosystems* **12**, doi.org/10.1029/2011GC003700.

Tarduno, J. A., Duncan, R. A., Scholl, D. W., Cottrell, R. D., Steinberger, B., Thordarson, T., Kerr, B. C., Neal, C. R., Frey, F. A., Torii, M. & Carvallo, C. (2003). The Emperor Seamounts: southward motion of the Hawaiian hotspot plume in Earth's mantle. *Science* **301**, 1064–1069.

Taylor, B. (2006). The single largest oceanic plateau: Ontong Java-Manihiki-Hikurangi. *Earth and Planetary Science Letters* **241**, 372–380.

Taylor, C. D., Schulz, K. J., Doebrich, J. L., Orris, G. J., Denning, P. D. & Kirschbaum, M. J. (2005). Geology and nonfuel mineral deposits of Africa and the Middle East. *US Geological Survey Open-File Report* 1294-E, 247 p.

Taylor, S. R. & McLennan, S. (2008). *Planetary Crusts: Their Composition, Origin and Evolution.* Cambridge University Press, 378 p.

Tera, F., Brown, L., Morris, J., Sacks, I. S., Klein, J. & Middleton, R. (1986). Sediment incorporation in island-arc magmas: inferences from ^{10}Be. *Geochimica et Cosmochimica Acta* **50**, 535–550.

Thompson, J. B. (1957). The graphical analysis of mineral assemblages in pelitic schists. *American Mineralogist* **42**, 842–858.

Thompson, J. B. (1959). Local equilibrium in metasomatic processes. In: Abelson, P. H. (ed) *Researches in Geochemistry.* John Wiley and Sons, 427–457.

Thordarson, T. & Self, S. (2003). Atmospheric and environmental effects of the 1783–1784 Laki eruption: a review and reassessment. *Journal of Geophysical Research: Atmospheres* **108**, doi.org/10.1029/2001JD002042.

Thurston, P. (2011). Canadian Precambrian shield. In: Gargaud, M., Amils, R., Cernicharo Quintanilla, J., Cleaves, H. J., Irvine, W. M., Pini, D. L. & Viso, M. (eds) *Encyclopedia of Astrobiology.* Springer, 230–232.

Tikoff, B. & Teyssier, C. (1992). Crustal-scale, en echelon "P-shear" tensional bridges—a possible solution to the batholithic room problem. *Geology* **20**, 927–930.

Till, C. B., Grove, T. L., Carlson, R. W., Donnelly-Nolan, J. M., Fouch, M. J., Wagner, L. S. & Hart, W. K. (2013). Depths and temperatures of <10.5 Ma mantle melting and the lithosphere-asthenosphere boundary below southern Oregon and northern California. *Geochemistry, Geophysics, Geosystems* **14**, 864–879.

Tobisch, O. T., Saleeby, J. B. & Fiske, R. S. (1986). Structural history of continental volcanic arc rocks, eastern Sierra Nevada, California: a case for extensional tectonics. *Tectonics* **5**, 65–94.

Tomlinson, E. L. & Holland, T. J. B. (2021). A thermodynamic model for the subsolidus evolution and melting of peridotite. *Journal of Petrology* **62**, 1–23.

Tornos, F., Peter, J. M., Allen, R. & Conde, C. (2015). Controls on the siting and style of volcanogenic massive sulphide deposits. *Ore Geology Reviews* **68**, 142–163.

Treiman, A. H. & Essene, E. J. (1985). The Oka carbonatite complex, Quebec: geology and evidence for silicate-carbonate liquid immiscibility. *American Mineralogist* **70**, 1101–1113.

Tsuchiyama, A. (1985). Dissolution kinetics of plagioclase in the melt of the system diopside-albite-anorthite, and origin of dusty plagioclase in andesites. *Contributions to Mineralogy and Petrology* **89**, 1–16.

Turcotte, D. L. & Schubert, G. (2002). *Geodynamics.* Cambridge University Press, 456 p.

Tuttle, O. F. & Bowen, N. L. (1958). Origin of granite in the light of experimental studies in the system $NaAlSi_3O_8$-$KAlSi_3O_8$-SiO_2-H_2O. *Geological Society of America Memoir* **74**, 153 p.

Tykot, R. H. (2002). Chemical fingerprinting and source tracing of obsidian: the central Mediterranean trade in black gold. *Accounts of Chemical Research* **35**, 618–627.

Ulmer, P. & Trommsdorff, V. (1995). Serpentine stability to mantle depths and subduction-related magmatism. *Science* **268**, 858–861.

Urey, H. C. (1952). *The Planets, Their Origin and Development.* Yale University Press, 245 p.

Ussler, W. (1988). Experimental and theoretical studies of mixed magmas. PhD thesis, University of North Carolina at Chapel Hill.

Ussler, W. & Glazner, A. F. (1989). Phase equilibria along a basalt-rhyolite mixing line: implications for the origin of calc-alkaline intermediate magmas. *Contributions to Mineralogy and Petrology* **101**, 232–244.

Valentine, G. A. & Krogh, K. E. C. (2006). Emplacement of shallow dikes and sills beneath a small basaltic volcanic center—the role of pre-existing structure (Paiute Ridge, southern Nevada, USA). *Earth and Planetary Science Letters* **246**, 217–230.

van Keken, P. E., Kiefer, B. & Peacock, S. M. (2002). High-resolution models of subduction zones: implications for mineral dehydration reactions and the transport of water into the deep mantle. *Geochemistry, Geophysics, Geosystems* **3**, doi.org/10.1029/2001GC000256.

VanTongeren, J. A. & Mathez, E. A. (2012). Large-scale liquid immiscibility at the top of the Bushveld Complex, South Africa. *Geology* **40**, 491–494.

Veksler, I. V., Dorfman, A. M., Danyushevsky, L. V., Jakobsen, J. K. & Dingwell, D. B. (2006). Immiscible silicate liquid partition coefficients: implications for crystal-melt element partitioning and basalt petrogenesis. *Contributions to Mineralogy and Petrology* **152**, 685–702.

Venezky, D. Y. & Rutherford, M. J. (1997). Preeruption conditions and timing of dacite-andesite magma mixing in the 2.2 ka eruption at Mount Rainier. *Journal of Geophysical Research* **102**, 20069–20086.

Vera, E. E., Mutter, J. C., Buhl, P., Orcutt, J. A., Harding, A. J., Kappus, M. E., Detrick, R. S. & Brocher, T. M. (1990). The structure of 0- to 0.2-m.y.-old oceanic crust at 9°N on the East Pacific Rise from expanded spread profiles. *Journal of Geophysical Research* **95**, 15529–15556.

Vernon, R. H. & Collins, W. J. (2011). Structural criteria for identifying granitic cumulates. *Journal of Geology* **119**, 127–142.

Vernon, R. H. & Paterson, S. R. (2008). How late are K-feldspar megacrysts in granites? *Lithos* **104**, 327–336.

Viete, D. R., Hermann, J., Lister, G. S. & Stenhouse, I. R. (2011). The nature and origin of the Barrovian metamorphism, Scotland: diffusion length scales in garnet and inferred thermal time scales. *Journal of the Geological Society* **168**, 115–131.

Vines, J. A. & Law, R. D. (2000). Emplacement of the Santa Rita Flat pluton as a pluton-scale saddle reef. *Geology* **28**, 1115–1118.

Visscher, H., Brinkhuis, H., Dilcher, D. L., Elsik, W. C., Eshet, Y., Looy, C. V., Rampino, M. R. & Traverse, A. (1996). The terminal Paleozoic fungal event: evidence of terrestrial ecosystem destabilization and collapse. *Proceedings of the National Academy of Sciences* **93**, 2155–2158.

Visscher, H., Looy, C., Collinson, M., Brinkhuis, H., Kürschner, W. & Sephton, M. (2004). Environmental mutagenesis at the time of the end Permian ecological crisis. *Proceedings of the National Academy of Sciences* **101**, 12952–12956.

von Eckermann, H. (1948). The Alkaline District of Alnö Island (Alnö alkalina område). *Sveriges Geologiska Undersökning* **36**, 176 p.

Vorhies, S. H. & Ague, J. J. (2011). Pressure-temperature evolution and thermal regimes in the Barrovian zones, Scotland. *Journal of the Geological Society* **168**, 1147–1166.

Wadati, K. (1928). Shallow and deep earthquakes. *Geophysical Magazine* **1**, 162–202.

Wadati, K. (1935). On the activity of deep-focus earthquakes in the Japan islands and neighbourhoods. *Geophysical Magazine* **8**, 305–325.

Waff, H. S. & Bulau, J. R. (1979). Equilibrium fluid distribution in an ultramafic partial melt under hydrostatic stress conditions. *Journal of Geophysical Research* **84**, 6109–6114.

Wager, L. R. & Bailey, E. B. (1953). Basic magma chilled against acid magma. *Nature* **172**, 68–69.

Wager, L. R., Brown, G. M. & Wadsworth, W. J. (1960). Types of igneous cumulates. *Journal of Petrology* **1**, 73–85.

Wager, L. R. & Deer, W. A. (1939). Geological investigations in East Greenland. Part III. The petrology of the Skaergaard intrusion, Kangerdlugssuaq, East Greenland. *Meddelelser om Grønland* **105**, 1–352.

Wakita, H. & Schmitt, R. A. (1970). Lunar anorthosites: rare-earth and other elemental abundances. *Science* **170**, 969–974.

Waldbaum, D. R. & Thompson, J. B., Jr. (1969). Mixing properties of sanidine crystalline solutions: IV. phase diagrams from equations of state. *American Mineralogist* **54**, 1274–1298.

Walker, G. P. L. (1989). Gravitational (density) controls on volcanism, magma chambers and intrusions. *Australian Journal of Earth Sciences* **36**, 149–165.

Wall, C. J., Scoates, J. S., Weis, D., Friedman, R. M., Amini, M. & Meurer, W. P. (2018). The Stillwater complex: integrating zircon geochronological and geochemical constraints on the age, emplacement history and crystallization of a large, open-system layered intrusion. *Journal of Petrology* **59**, 153–190.

Wallace, P. J. (2005). Volatiles in subduction zone magmas: concentrations and fluxes based on melt inclusion and volcanic gas data. *Journal of Volcanology and Geothermal Research* **140**, 217–240.

Wallmann, P. C., Mahood, G. A. & Pollard, D. D. (1988). Mechanical models for correlation of ring-fracture eruptions at Pantelleria, Strait of Sicily, with glacial sea level drawdown. *Bulletin of Volcanology* **50**, 327–339.

Warren, John K. (2016). *Evaporites: A Geological Compendium*. Springer, 1813 p.

Washington, H. S. (1917). Chemical analyses of igneous rocks published from 1884 to 1913, inclusive, with a critical discussion of the character and use of analyses. *US Geological Survey Professional Paper* 99, 1201 p.

Waters, L. E. & Lange, R. A. (2017). Why aplites freeze and rhyolites erupt: controls on the accumulation and eruption of high-SiO_2 (eutectic) melts. *Geology* **45**, 1019–1022.

Watson, E. B. & Harrison, T. M. (1983). Zircon saturation revisited: temperature and composition effects in a variety of crustal magma types. *Earth and Planetary Science Letters* **64**, 295–304.

Watson, E. B., Wark, D. A. & Thomas, J. B. (2006). Crystallization thermometers for zircon and rutile. *Contributions to Mineralogy and Petrology* **151**, 413–433.

Watson, J. D. & Crick, F. H. (1953). Molecular structure of nucleic acids: a structure for deoxyribose nucleic acid. *Nature* **171**, 737–738.

Wedepohl, K. H. (1995). The composition of the continental crust. *Geochimica et Cosmochimica Acta* **59**, 1217–1232.

Wegener, A. (1912). Die Entstehung der Kontinente. *Geologische Rundschau* **3**, 276–292.

Weidendorfer, D., Schmidt, M. W. & Mattsson, H. B. (2017). A common origin of carbonatite magmas. *Geology* **45**, 507–510.

Weill, D. F., Hon, R. & Navrotsky, A. (1980). The igneous system $CaMgSi_2O_6$-$CaAl_2Si_2O_8$-$NaAlSi_3O_8$: variations on a classic theme by Bowen. In: Hargraves, R. B. (ed) *Physics of Magmatic Processes*, Princeton University Press, 49–92.

Weinberg, R. F., Sial, A. N. & Pessoa, R. R. (2001). Magma flow within the Tavares Pluton, northeastern Brazil: compositional and thermal convection. *Geological Society of America Bulletin* **113**, 508–520.

Weisler, M. I., Bolhar, R., Ma, J., Pierre, E. S., Sheppard, P., Walter, R. K., Feng, Y., Zhao, J. X. & Kirch, P. V. (2016). Cook Island artifact geochemistry demonstrates spatial and temporal extent of pre-European interarchipelago voyaging in East Polynesia. *Proceedings of the National Academy of Sciences* **113**, 8150–8155.

Wernicke, B. & Burchfiel, B. C. (1982). Modes of extensional tectonics. *Journal of Structural Geology* **4**, 105–115.

White, W. M. (2010). Oceanic island basalts and mantle plumes: the geochemical perspective. *Annual Review of Earth and Planetary Sciences* **38**, 133–160.

White, W. M. (2013). *Geochemistry*. Wiley-Blackwell, 672 p.

White, W. M. & Dupré, B. (1986). Sediment subduction and magma genesis in the Lesser Antilles: isotopic and trace element constraints. *Journal of Geophysical Research* **91**, 5927–5941.

White, W. M. & Klein, E. M. (2014). Composition of the oceanic crust. In: Turekian, K. K. & Holland, H. D. (eds) *Treatise on Geochemistry: Second Edition*, Vol. 4. Elsevier, 457–496.

Whitney, D. L. & Evans, B. W. (2010). Abbreviations for names of rock-forming minerals. *American Mineralogist* **95**, 185–187.

Wiebe, R. (1973). Relations between coexisting basaltic and granitic magmas in a composite dike. *American Journal of Science* **273**, 130–151.

Wiebe, R. A., Jellinek, A. M. & Hodge, K. F. (2017). New insights into the origin of ladder dikes: implications for punctuated growth and crystal accumulation in the Cathedral Peak granodiorite. *Lithos* **277**, 241–258.

Wignall, P. (2005). The link between large igneous province eruptions and mass extinctions. *Elements* **1**, 293–297.

Wignall, P. B. (2011). Lethal volcanism. *Nature* **477**, 285–286.

Wilcox, R. E. (1944). The rhyolite-basalt complex on Gardiner River, Yellowstone Park, Wyoming. *Geological Society of America Bulletin* **55**, 1047–1080.

Wilcox, R. E. (1954). Petrology of Parícutin region. *US Geological Survey Bulletin* **965–C**, 281–353.

Wilde, S. A., Valley, J. W., Peck, W. H. & Graham, C. M. (2001). Evidence from detrital zircons for the existence of continental crust and oceans on the Earth 4.4 Gyr ago. *Nature* **409**, 175–178.

Williams, H., Turner, F. J. & Gilbert, C. M. (1982). *Petrography: An Introduction to the Study of Rock in Thin Sections*, 2nd ed. W. H. Freeman, 626 p.

Williams, M. L. & Karlstrom, K. E. (1996). Looping PT paths and high-T, low-P middle crustal metamorphism: Proterozoic evolution of the southwestern United States. *Geology* **24**, 1119–1122.

Williams, P. F. (1990). Differentiated layering in metamorphic rocks. *Earth-Science Reviews* **29**, 267–281.

Williams, Q. & Tobisch, O. T. (1994). Microgranitic enclave shapes and magmatic strain histories: constraints from drop deformation theory. *Journal of Geophysical Research* **99**, 24359–24368.

Wilshire, H. G. (1969). Mineral layering in the Twin Lakes granodiorite, Colorado. *Geological Society of America Memoir* **115**, 235–261.

Wilshire, H. G., Meyer, C. E., Nakata, J. K., Calk, L. C., Shervais, J. W., Nielson, J. E. & Schwarzman, E. C. (1988). Mafic and ultramafic xenoliths from volcanic rocks of the western United States. *US Geological Survey Professional Paper* **1443**, 179 p.

Wilshire, H. G. & Shervais, J. W. (1975). Al-augite and Cr-diopside ultramafic xenoliths in basaltic rocks from western United States. *Physics and Chemistry of the Earth* **9**, 257–272.

Wilson, C. J. N. & Hildreth, W. H. (1997). The Bishop Tuff: new insights from eruptive stratigraphy. *Journal of Geology* **105**, 407–439.

Wilson, J. T. (1963). A possible origin of the Hawaiian Islands. *Canadian Journal of Physics* **41**, 863–870.

Wilson, J. T. (1968). Static or mobile earth: the current scientific revolution. *Proceedings of the American Philosophical Society* **112**, 309–320.

Wilson, L. (1976). Explosive volcanic eruptions—III. Plinian eruption columns. *Geophysical Journal of the Royal Astronomical Society* **45**, 543–556.

Wilson, L. (1980). Relationships between pressure, volatile content and ejecta velocity in three types of volcanic explosion. *Journal of Volcanology and Geothermal Research* **8**, 297–313.

Wilson, M. & Downes, H. (2006). Tertiary-Quaternary intra-plate magmatism in Europe and its relationship to mantle dynamics. *Geological Society Memoir* **32**, 147–166.

Winter, J. D. (2010). *An Introduction to Igneous and Metamorphic Petrology*. Prentice Hall, 702 p.

Wolery, T. J. & Sleep, N. H. (1976). Hydrothermal circulation and geochemical flux at mid-ocean ridges. *Journal of Geology* **84**, 249–275.

Wolfe, E. W. (1988). The Puu Oo eruption of Kilauea Volcano, Hawaii: Episode 1 through 20, January 3, 1983, through June 8, 1984. *US Geological Survey Professional Paper* **1463**, 251 p.

Wong, T. F. & Brace, W. F. (1979). Thermal expansion of rocks: some measurements at high pressure. *Tectonophysics* **57**, 95–117.

Wood, B. J. & Fraser, D. G. (1978). *Elementary Thermodynamics for Geologists*. Oxford University Press, 303 p.

Wotzlaw, J.-F., Bindeman, I. N., Schaltegger, U., Brooks, C. K. & Naslund, H. R. (2012). High-resolution insights into episodes of crystallization, hydrothermal alteration and remelting in the Skaergaard intrusive complex. *Earth and Planetary Science Letters* **355**, 199–212.

Wright, J. V, Smith, A. L. & Self, S. (1980). A working terminology of pyroclastic deposits. *Journal of Volcanology and Geothermal Research* **8**, 315–336.

Wyllie, P. J. & Tuttle, O. F. (1959). Synthetic carbonatite magma. *Nature* **183**, 770.

Yang, W. & Ahrens, T. J. (1998). Shock vaporization of anhydrite and global effects of the K/T bolide. *Earth and Planetary Science Letters* **156**, 125–140.

Yaxley, G. M., Anenburg, M., Tappe, S., Decree, S. & Guzmics, T. (2022). Carbonatites: classification, sources, evolution, and emplacement. *Annual Review of Earth and Planetary Sciences* **50**, 261–293.

Yaxley, G. M., Kjarsgaard, B. A. & Jaques, A. L. (2021). Evolution of carbonatite magmas in the upper mantle and crust. *Elements* **17**, 315–320.

Yoder, H. S. (1976). *Generation of Basaltic Magma*. National Academies Press, 281 p.

Yoder, H. S. & Tilley, C. E. (1962). Origin of basalt magmas: an experimental study of natural and synthetic rock systems. *Journal of Petrology* **3**, 342–529.

Yoshinobu, A. & Barnes, C. G. (2008). Is stoping a volumetrically significant pluton emplacement process?: Discussion. *Geological Society of America Bulletin* **120**, 1080–1081.

Yoshinobu, A. S., Fowler, T. K., Jr., Paterson, S. R., Llambias, E., Tickyj, H. & Sato, A. M. (2003). A view from the roof: magmatic stoping in the shallow crust, Chita Pluton, Argentina. *Journal of Structural Geology* **25**, 1037–1048.

Young, F. E. & Jones, F. T. (1949). Sucrose hydrates. The sucrose–water phase diagram. *Journal of Physical Chemistry* **53**, 1334–1350.

Yuan, H. & Romanowicz, B. (2010). Lithospheric layering in the North American craton. *Nature* **466**, 1063–1068.

Zandt, G., Gilbert, H., Owens, T. J., Ducea, M., Saleeby, J. & Jones, C. H. (2004). Active foundering of a continental arc root beneath the southern Sierra Nevada in California. *Nature* **431**, 41–46.

Zeitler, P. K., Meltzer, A. S., Brown, L., Kidd, W. S. F., Lim, C. & Enkelmann, E. (2014). Tectonics and topographic evolution of Namche Barwa and the easternmost Lhasa block, Tibet. *Geological Society of America Special Paper* **507**, 23–58.

Zen, E. A. (1986). Aluminum enrichment in silicate melts by fractional crystallization: some mineralogic and petrographic constraints. *Journal of Petrology* **27**, 1095–1117.

Zerefos, C. S., Gerogiannis, V. T., Balis, D., Zerefos, S. C. & Kazantzidis, A. (2007). Atmospheric effects of volcanic eruptions as seen by famous artists and depicted in their paintings. *Atmospheric Chemistry and Physics* **7**, 4027–4042.

Zhao, D. (2004). Global tomographic images of mantle plumes and subducting slabs: insight into deep Earth dynamics. *Physics of the Earth and Planetary Interiors* **146**, 3–34.

Zindler, A. & Hart, S. (1986). Chemical geodynamics. *Annual Review of Earth and Planetary Sciences* **14**, 493–571.

Zou, Y., Li, Q., Chu, X., Zhai, M., Mitchell, R. N., Zhao, L., Zhou, L., Wang, Y. & Liu, B. (2021). Older orogens cooled slower: new constraints on Orosirian tectonics from garnet diffusion modeling of metamorphic timescales, Jiaobei terrain, North China Craton. *Contributions to Mineralogy and Petrology* **176**, doi.org/10.1007/s00410-021-01846-w.

Index

References to figures and tables are indicated by "f" and "t" following page numbers.

T

U